T0406141

Handbook of Nonlocal Continuum Mechanics for Materials and Structures

George Z. Voyiadjis
Editor

Handbook of Nonlocal Continuum Mechanics for Materials and Structures

Volume 2

With 572 Figures and 59 Tables

Editor
George Z. Voyiadjis
Department of Civil and Environmental Engineering
Louisiana State University
Baton Rouge, LA, USA

ISBN 978-3-319-58727-1 ISBN 978-3-319-58729-5 (eBook)
ISBN 978-3-319-58728-8 (print and electronic bundle)
https://doi.org/10.1007/978-3-319-58729-5

Library of Congress Control Number: 2018960886

This Springer imprint is published by the registered company Springer Nature Switzerland AG
The registered company address is: Gewerbestrasse 11, 6330 Cham, Switzerland

Preface

This handbook discusses the integral and gradient formulations of nonlocality, computational aspects, micromechanical considerations, and comparison of approaches and emphasizes recent developments in the bridging of material length and time scales. The contributions in this handbook are on nonlocal continuum plasticity in terms of the experimental, theoretical, and numerical investigations. This handbook presents a comprehensive treatment of the most important areas of nonlocality (integral and gradient) of time-dependent inelastic deformation behavior and heat transfer responses. The following aspects of the advanced material modeling are presented: enhanced (generalized) continuum mechanics, microscopic mechanisms and micro-mechanical aspects responsible for size effect and micro-scale heat transfer, thermodynamic framework, and multiscale computational aspects with detailed nonlocal computational algorithms in the context of finite element. Measures for length scales are introduced together with their appropriate evolution relations.

The work addresses the thermal and mechanical responses of small-scale metallic compounds under fast transient processes based on small and large deformation framework. This handbook presents a comprehensive treatment of the most important areas of nonlocality (integral and gradient) of time-dependent inelastic deformation behavior and heat transfer responses.

From the experimental aspects, a wide spectrum of materials are included in these chapters. For the case of glassy polymers, their nanostructural responses and nanoindentation measurements were discussed in detail by Voyiadjis et al. For metals, the hydrogen embrittlement cracking was discussed by Yonezu and Chen, whereas its size effect of nanoindentation was elucidated by Voyiadjis et al. Composites and their nonlocal behaviors have received particular attention, including the studies of their cracking initiation (Xu et al.), interface stability (Meng et al.), buckling (Chen et al.), and dynamic properties (Tomar et al.). Indentation and its nonlocal characteristics were also emphasized through modeling (Mills et al., Liu et al.) and fatigue testing (Xu et al.).

The micromorphic approach has aroused strong interest from the materials science and computational mechanics communities because of its regularization power in the context of softening plasticity and damage. The micromorphic and Cosserat theories in gradient plasticity are introduced and analyzed in these chapters to address the instabilities in the materials and structures. The micromorphic approach for the gradient continuum plasticity/damage and crystal plasticity is elucidated by Forest et al. The micropolar theory for the crystal plasticity is also introduced (Mayeur et al.). The micromorphic and Cosserat approach are applied to localization in geomaterials (Stefanou et al.) and dispersion of waves in metamaterials (Madeo et al.) in these chapters.

The section on "Mathematical Methods in Nonlocal Continuum Mechanics" combines the original concepts for the description of nonlocal effects, including applications to nonstandard mathematical models. Both quasi-static and dynamic processes are included, together with micro- and macro-level of description, and furthermore multiphysics (in the sense of, e.g., thermomechanical interaction) is covered accordingly. In the chapter by Lazopoulos et al., the fractional calculus is applied to obtain the space-fractional nonlocal continuum mechanics formulation and then applied to the analysis of fractional Zener viscoelastic model. The contribution by Ostoja-Starzewski et al. includes modeling of fractal materials utilizing fractional integrals and application of homogenized continuum mechanics together with the framework of calculus in non-integer dimensional spaces. The work by Sumelka et al. presents implicit time nonlocal modeling of metallic materials including damage anisotropy, and furthermore stress-fractional extensions are suggested. Next, the chapter by Tarasov includes modeling of physical lattices with long-range interactions utilizing exact fractional-order difference operators. Finally, in the chapter by Voyiadjis et al., the strain gradient plasticity is presented with the appropriate flow rules of the grain interior and grain boundary areas within the thermodynamically consistent framework and applied for modeling metallic materials. One can recapitulate that the overall section contents open new fields of investigations of mathematical models for bodies exhibiting strong scale effect.

The section on "Computational Modeling for Gradient Plasticity in Both Temporal and Spatial Scales" introduces a variety of numerical examples for the gradient-enhanced plasticity. The failure mechanisms of metallic materials for high-velocity impact loading are simulated based on the nonlocal approach (Voyiadjis et al.). The gradient plasticity is combined to micro-/mesoscale crystal plasticity (Yalcinkaya et al., Ozdemir et al.) and fracture mechanics (Lancioni et al.). The transverse vibration of microbeams and axial vibration of micro-rods are studied by Civalek et al. using the strain gradient plasticity theory. In addition, in this section, the temperature-driven ductile-to-brittle transition fracture in Ferritic steels is modeled and analyzed (Deliktaş et al.).

Nonlocal peridynamic models for damage and fracture are introduced and analyzed in these chapters. The theory guaranteeing the existence of solution for continuum models of fracture evolution is essential for the development of mesh-independent discretizations. A bond is exhibited based on peridynamic models with damage and softening and shows that they are well-posed evolutions both over the space of Holder continuous functions and Sobolev functions (Jha and Lipton). This feature is used to develop numerical convergence rates in space and time for the associated finite element and finite-difference schemes (Jha and Lipton). These are the first convergence rates for numerical schemes applied to fully nonlinear and nonlocal peridynamic models. A more general state-based peridynamic model is developed for free fracture evolution. In the limit of vanishing nonlocal interaction, this model is shown to converge to the equation of elastic momentum balance away from the crack set (Said et al.). In the final chapter, a state-based and history-dependent dynamic damage model is developed. Several numerical examples are provided illustrating the theory (Said et al.).

This handbook integrates knowledge from the theoretical, numerical, and experimental areas of nonlocal continuum plasticity. This book is focused mainly for graduate students of nonlocal continuum plasticity, researchers in academia and industry who are active or intend to become active in this field, and practicing engineers and scientists who work in this topic and would like to solve problems utilizing the tools offered by nonlocal mechanics. This handbook should serve as an excellent text for a series of graduate courses in mechanical engineering, civil engineering, materials science, engineering mechanics, aerospace engineering, applied mathematics, applied physics, or applied chemistry.

This handbook is basically intended as a textbook for university courses as well as a reference for researchers in this field. It will serve as a timely addition to the literature on nonlocal mechanics and will serve as an invaluable resource to members of the international scientific and industrial communities.

It is hoped that the reader will find this handbook a useful resource as he/she progresses in their study and research in nonlocal mechanics. Each of the individual sections of this handbook could be considered as a compact self-contained mini-book right under its own title. However, these topics are presented in relation to the basic principles of nonlocal mechanics.

What is finally presented in the handbook is the work contributed by celebrated international experts for their best knowledge and practices on specific and related topics in nonlocal mechanics.

The editor would like to thank all the contributors who wrote chapters for this handbook. Finally, the editor would like to acknowledge the help and support of his family members and the editors at Springer who made this handbook possible.

Baton Rouge, USA Dr. George Z. Voyiadjis
December 2018

Contents

Volume 1

Volume 2

About the Editor

George Z. Voyiadjis is the Boyd Professor at the Louisiana State University in the Department of Civil and Environmental Engineering. This is the highest professorial rank awarded by the Louisiana State University System. He is also the holder of the Freeport-McMoRan Endowed Chair in Engineering. He joined the faculty of Louisiana State University in 1980. He is currently the Chair of the Department of Civil and Environmental Engineering. He holds this position since February of 2001. He currently also serves since 2012 as the Director of the Louisiana State University Center for GeoInformatics (LSU C4G; http://c4gnet.lsu.edu/c4g/).

Voyiadjis is a Foreign Member of both the Polish Academy of Sciences, Division IV (Technical Sciences), and the National Academy of Engineering of Korea. He is the recipient of the 2008 Nathan M. Newmark Medal of the American Society of Civil Engineers and the 2012 Khan International Medal for outstanding lifelong contribution to the field of plasticity. He was also the recipient of the Medal for his significant contribution to Continuum Damage Mechanics, presented to him during the Second International Conference on Damage Mechanics (ICDM2), Troyes, France, July, 2015. This is sponsored by the *International Journal of Damage Mechanics* and is held every 3 years.

Voyiadjis was honored in April of 2012 by the International Symposium on "Modeling Material Behavior at Multiple Scales" sponsored by Hanyang University, Seoul, Korea, chaired by T. Park and X. Chen (with a dedicated special issue in the *Journal of Engineering Materials and Technology* of the ASME). He was also honored by an International Mini-Symposium on "Multiscale and Mechanism Oriented Models:

Computations and Experiments" sponsored by the International Symposium on Plasticity and Its Current Applications, chaired by V. Tomar and X. Chen, in January 2013.

He is a Distinguished Member of the American Society of Civil Engineers; Fellow of the American Society of Mechanical Engineers, the Society of Engineering Science, the American Academy of Mechanics and the Engineering Mechanics Institute of ASCE; and Associate Fellow of the American Institute of Aeronautics and Astronautics. He was on the Board of Governors of the Engineering Mechanics Institute of the American Society of Civil Engineers, and Past President of the Board of Directors of the Society of Engineering Science. He was also the Chair of the Executive Committee of the Materials Division (MD) of the American Society of Mechanical Engineers. Dr. Voyiadjis is the Founding Chief Editor of the Journal of Nanomechanics and Micromechanics of the ASCE and is on the editorial board of numerous engineering journals. He was also selected by Korea Science and Engineering Foundation (KOSEF) as one of the only two World-Class University foreign scholars in the area of civil and architectural engineering to work on nanofusion in civil engineering. This is a multimillion research grant.

Voyiadjis' primary research interest is in plasticity and damage mechanics of metals, metal matrix composites, polymers, and ceramics with emphasis on the theoretical modeling, numerical simulation of material behavior, and experimental correlation. Research activities of particular interest encompass macro-mechanical and micro-mechanical constitutive modeling, experimental procedures for quantification of crack densities, inelastic behavior, thermal effects, interfaces, damage, failure, fracture, impact, and numerical modeling.

Dr. Voyiadjis' research has been performed on developing numerical models that aim at simulating the damage and dynamic failure response of advanced engineering materials and structures under high-speed impact loading conditions. This work will guide the development of design criteria and fabrication processes of high-performance materials and structures under severe loading conditions. Emphasis is placed on survivability area that aims to develop and field a contingency armor that is thin and lightweight, but with a very high level of an overpressure protection system that provides low penetration depths. The formation of cracks and voids in the adiabatic shear bands, which are the precursors to fracture, is mainly investigated.

He has 2 patents, over 320 refereed journal articles, and 19 books (11 as editor) to his credit. He gave over 400 presentations as plenary, keynote, and invited speaker as well as other talks. Over 62 graduate students (36 Ph.D.) completed their degrees under his direction. He has also supervised numerous postdoctoral associates. Voyiadjis has been extremely successful in securing more than $25.0 million in research funds as a principal investigator/investigator from the National Science Foundation, the Department of Defense, the Air Force Office of Scientific Research, the Department of Transportation, and major companies such as IBM and Martin Marietta.

He has been invited to give plenary presentations and keynote lectures in many countries around the world. He has also been invited as guest editor in numerous volumes of the *Journal of Computer Methods in Applied Mechanics and Engineering*, *International Journal of Plasticity*, *Journal of Engineering Mechanics of the ASCE*, and *Journal of Mechanics of Materials*. These special issues focus in the areas of damage mechanics, structures, fracture mechanics, localization, and bridging of length scales.

He has extensive international collaborations with universities in France, the Republic of Korea, and Poland.

Associate Editors

Xi Chen
Department of Earth and Environmental Engineering
Columbia Nanomechanics Research Center
Columbia University
New York, NY, USA

Samuel Forest
Centre des Materiaux
Mines ParisTech CNRS
PSL Research University
Paris, Evry Cedex, France

Wojciech Sumelka
Institute of Structural Engineering
Poznan University of Technology
Poznan, Poland

Babür Deliktaş
Faculty of Engineering-Architecture
Department of Civil Engineering
Uludag University
Bursa, Görükle, Turkey

Michael L. Parks
Center for Computing Research
Sandia National Laboratories
Albuquerque, NM, USA

Contributors

Bekir Akgöz Civil Engineering Department, Division of Mechanics, Akdeniz University, Antalya, Turkey

Burak Aksoylu Department of Mathematics, Wayne State University, Detroit, MI, USA

Florin Bobaru Mechanical and Materials Engineering, University of Nebraska–Lincoln, Lincoln, NE, USA

Pavel Bochev Center for Computing Research, Sandia National Laboratories, Albuquerque, NM, USA

Ferhun C. Caner School of Industrial Engineering, Institute of Energy Technologies, Universitat Politècnica de Catalunya, Barcelona, Spain

Department of Materials Science and Metallurgical Engineering, Universitat Politècnica de Catalunya, Barcelona, Spain

Fatih Celiker Department of Mathematics, Wayne State University, Detroit, MI, USA

Xi Chen Department of Earth and Environmental Engineering, Columbia Nanomechanics Research Center, Columbia University, New York, NY, USA

Youlong Chen International Center for Applied Mechanics, State Key Laboratory for Strength and Vibration of Mechanical Structures, School of Aerospace, Xi'an Jiaotong University, Xi'an, China

Ziguang Chen Department of Mechanics, Huazhong University of Science and Technology, Wuhan, Hubei Sheng, China

Hubei Key Laboratory of Engineering, Structural Analysis and Safety Assessment,Wuhan, China

N. Chiba National Defense Academy of Japan, Yokosuka, Japan

Ömer Civalek Civil Engineering Department, Division of Mechanics, Akdeniz University, Antalya, Turkey

Qiang Cui Department of Chemistry and Theoretical Chemistry Institute, University of Wisconsin-Madison, Madison, WI, USA

Marta D'Elia Optimization and Uncertainty Quantification Department Center for Computing Research, Sandia National Laboratories, Albuquerque, NM, USA

Valentín de Carlos Blasco School of Industrial Engineering, Institute of Energy Technologies, Universitat Politècnica de Catalunya, Barcelona, Spain

Babür Deliktaş Faculty of Engineering-Architecture, Department of Civil Engineering, Uludag University, Bursa, Görükle, Turkey

Paul N. Demmie Sandia National Laboratories, Albuquerque, NM, USA

Qiang Du Department of Applied Physics and Applied Mathematics, Columbia University, New York, NY, USA

Samuel Forest Centre des Materiaux, Mines ParisTech CNRS, PSL Research University, Paris, Evry Cedex, France

Mikil D. Foss Department of Mathematics, University of Nebraska-Lincoln, Lincoln, NE, USA

Mercè Ginjaume Egido School of Industrial Engineering, Institute of Energy Technologies, Universitat Politècnica de Catalunya, Barcelona, Spain

Max Gunzburger Department of Scientific Computing, Florida State University, Tallahassee, FL, USA

Prashant K. Jha Department of Mathematics, Louisiana State University, Baton Rouge, LA, USA

Orsan Kilicer Department of Mathematics, Wayne State University, Detroit, MI, USA

Giovanni Lancioni Dipartimento di Ingegneria Civile, Edile e Architettura, Università Politecnica delle Marche, Ancona, Italy

A. K. Lazopoulos Mathematical Sciences Department, Hellenic Army Academy, Vari, Greece

K. A. Lazopoulos National Technical University of Athens, Rafina, Greece

Jun Li Department of Mechanical Engineering, University of Massachusetts, Dartmouth, MA, USA

Robert Lipton Department of Mathematics and Center for Computation and Technology, Louisiana State University, Baton Rouge, LA, USA

David J. Littlewood Center for Computing Research, Sandia National Laboratories, Albuquerque, NM, USA

L. Liu Department of Mechanical and Aerospace Engineering, Utah State University, Logan, UT, USA

Yilun Liu International Center for Applied Mechanics, State Key Laboratory for Strength and Vibration of Mechanical Structures, School of Aerospace, Xi'an Jiaotong University, Xi'an, China

Tomasz Łodygowski Institute of Structural Engineering, Poznan University of Technology, Poznan, Poland

Angela Madeo SMS-ID, INSA-Lyon, Université de Lyon, Villeurbanne cedex, Lyon, France

Institut universitaire de France, Paris Cedex 05, Paris, France

Leila Malekmotiei Department of Civil and Environmental Engineering, Louisiana State University, Baton Rouge, LA, USA

J. R. Mayeur Theoretical Division, Los Alamos National Laboratory, Los Alamos, NM, USA

D. L. McDowell Woodruff School of Mechanical Engineering, School of Materials Science and Engineering, Georgia Institute of Technology, Atlanta, GA, USA

Wen Jin Meng Department of Mechanical and Industrial Engineering, Louisiana State University, Baton Rouge, LA, USA

James A. Mills Department of Civil Engineering and Engineering Mechanics, Columbia University, New York, NY, USA

Patrizio Neff Fakultät für Mathematik, Universität Duisburg-Essen, Essen, Germany

Sina Niazi Mechanical and Materials Engineering, University of Nebraska–Lincoln, Lincoln, NE, USA

N. Ogasawara Department of Mechanical Engineering, National Defense Academy of Japan, Yokosuka, Japan

Martin Ostoja-Starzewski Department of Mechanical Science and Engineering, Institute for Condensed Matter Theory and Beckman Institute, University of Illinois at Urbana–Champaign, Urbana, IL, USA

İzzet Özdemir Department of Civil Engineering, İzmir Institute of Technology, İzmir, Turkey

Mauro Perego Center for Computing Research, Sandia National Laboratories, Albuquerque, NM, USA

Chandra Prakash School of Aeronautics and Astronautics, Purdue University, West Lafayette, IN, USA

Petronela Radu Department of Mathematics, University of Nebraska-Lincoln, Lincoln, NE, USA

Hadrien Rattez Navier (CERMES), UMR 8205, Ecole des Ponts, IFSTTAR, CNRS, Champs-sur-Marne, France

Eyad Said Department of Mathematics, Louisiana State University, Baton Rouge, LA, USA

Aref Samadi-Dooki Computational Solid Mechanics Laboratory, Department of Civil and Environmental Engineering, Louisiana State University, Baton Rouge, LA, USA

Pablo Seleson Computer Science and Mathematics Division, Oak Ridge National Laboratory, Oak Ridge, TN, USA

Shuai Shao Department of Mechanical and Industrial Engineering, Louisiana State University, Baton Rouge, LA, USA

S. A. Silling Sandia National Laboratories, Albuquerque, NM, USA

Yooseob Song Department of Civil and Environmental Engineering, Louisiana State University, Baton Rouge, LA, USA

Ioannis Stefanou Navier (CERMES), UMR 8205, Ecole des Ponts, IFSTTAR, CNRS, Champs-sur-Marne, France

Jean Sulem Navier (CERMES), UMR 8205, Ecole des Ponts, IFSTTAR, CNRS, Champs-sur-Marne, France

Wojciech Sumelka Institute of Structural Engineering, Poznan University of Technology, Poznan, Poland

Vasily E. Tarasov Skobeltsyn Institute of Nuclear Physics, Lomonosov Moscow State University, Moscow, Russia

Vikas Tomar School of Aeronautics and Astronautics, Purdue University, West Lafayette, IN, USA

Ismail Cem Turtuk Mechanical Design Department, Meteksan Defence, Ankara, Turkey

Department of Civil Engineering, Uludag Univeristy, Bursa, Turkey

Devendra Verma School of Aeronautics and Astronautics, Purdue University, West Lafayette, IN, USA

George Z. Voyiadjis Department of Civil and Environmental Engineering, Louisiana State University, Baton Rouge, LA, USA

Hong Wang Department of Mathematics, University of South Carolina, Columbia, SC, USA

Hang Xiao School of Chemical Engineering, Northwest University, Xi'an, China

Baoxing Xu Department of Mechanical and Aerospace Engineering, University of Virginia, Charlottesville, VA, USA

Jun Xu Department of Automotive Engineering, School of Transportation Science and Engineering, Beihang University, Beijing, China

Advanced Vehicle Research Center (AVRC), Beihang University, Beijing, China

Mohammadreza Yaghoobi Department of Civil and Environmental Engineering, Louisiana State University, Baton Rouge, LA, USA

Tuncay Yalçinkaya Aerospace Engineering Program, Middle East Technical University Northern Cyprus Campus, Guzelyurt, Mersin, Turkey

Department of Aerospace Engineering, Middle East Technical University, Ankara, Turkey

Yuan Yan School of Chemical Engineering, Northwest University, Xi'an, China

Akio Yonezu Department of Precision Mechanics, Chuo University, Tokyo, Japan

Zhufeng Yue Department of Engineering Mechanics, Northwestern Polytechnical University, Xi'an, Shaanxi, China

Cheng Zhang Medtronic, Inc., Tempe, AZ, USA

Guanfeng Zhang Mechanical and Materials Engineering, University of Nebraska–Lincoln, Lincoln, NE, USA

Yanting Zheng China Automotive Technology and Research Center, Tianjin, China

Liangliang Zhu Columbia Nanomechanics Research Center, Department of Earth and Environmental Engineering, Columbia University, New York, NY, USA

International Center for Applied Mechanics, State Key Laboratory for Strength and Vibration of Mechanical Structures, School of Aerospace, Xi'an Jiaotong University, Xi'an, China

Part III
Mathematical Methods in Nonlocal Continuum Mechanics

Implicit Nonlocality in the Framework of Viscoplasticity

21

Wojciech Sumelka and Tomasz Łodygowski

Contents

Abstract

The considerations are addressed to the notion of implicit nonlocality in mechanical models. The term *implicit* means that there is no direct measure of nonlocal action in a model (like classical or fractional gradients, etc. in explicit nonlocal models), but some phenomenological material parameters can be interpreted as

W. Sumelka (✉) · T. Łodygowski
Institute of Structural Engineering, Poznan University of Technology, Poznan, Poland
e-mail: wojciech.sumelka@put.poznan.pl

G. Z. Voyiadjis (ed.), *Handbook of Nonlocal Continuum Mechanics for Materials and Structures*, https://doi.org/10.1007/978-3-319-58729-5_17

one that maps some experimentally observed phenomena responsible for the scale effects.

The overall discussion is conducted in the framework of the Perzyna Theory of Viscoplasticity where the role of the implicit length scale parameter plays the relaxation time of the mechanical disturbance. In this sense, in the viscoplastic range of the material behavior, the deformation at each material point contributes to the finite surrounding. The important consequence is that the solution of the IBVP described by Perzyna's theory is unique – the relaxation time is the regularizing parameter.

Keywords
Implicit nonlocality · Viscoplasticity · Anisotropic damage

Introduction

Fifty years after publishing the fundamental paper on the theory of thermo-viscoplasticity (TTV) by Perzyna (1963), the concept of material overstress function is still vivid nowadays in mechanics and enables new findings (cf. Sumelka 2014; Glema et al. 2014; Sumelka and Nowak 2016). One can say that the viscoplasticity concept is one of the most fruitful ideas in modeling plastic (irreversible) processes. It is obvious that TTV formulation evolved during the years to obtain the mature form in the early 1990s (Perzyna 2005) – the one strongly influenced by the works of J.E. Marsden group (cf. Marsden and Hughes 1983; Abraham et al. 1988) – where both detailed experimental observations and robust mathematical modeling have aimed to obtain the unique solution of the posed thermomechanical problem with clear physical interpretation.

The last mentioned aspect of uniqueness of initial boundary value problem (IBVP) formulated in the framework of TTV plays the fundamental role in the following discussion. This crucial aspect involves the concept of the relaxation time – the parameter which controls viscoplasticity effects in TTV and maps implicitly the physical length scale. On the other hand, the main drawback of TTV is complexity of the formulation that requires the necessity of dynamically solving what is computationally time-consuming (Łodygowski 1996). Herein it should be pointed out that TTV and its inherent properties strongly differ from the classical plasticity with application of yield strength limit in a rate form (Heeres et al. 2002) – in this sense the term 'viscoplasticity' is used misleadingly in the literature (thus the reader must carefully analyze the definition of plastic strain in a specific formulation to judge whether it is consistent with original TTV).

TTV was extensively verified and validated in the literature. Let us mention herein some crucial aspects of TTV discussed by the authors in a series of papers: (i) mathematically well-posedness was presented in Łodygowski et al. (1994), Łodygowski (1996) and Glema et al. (1997); (ii) solution existence and uniqueness of softening problem were subjects of Perzyna (2005); (iii) viscosity defined by material parameter was analyzed in Glema (2000) and Glema et al. (2003); (iv) propagation of mechanical and thermal waves was clarified in Glema and Łodygowski (2002) and Glema (2004); (v) dispersive material character was presented in Glema (2000); (vi) diverse way of energy dissipation was discussed in Glema et al. (2003); (vii) smooth and non-smooth distributions within damage and failure were concluded in Glema et al. (2009); (viii) damage anisotropy was included into TTV in Perzyna (2008) and Glema et al. (2009) and developed in Sumelka (2009); (ix) transition from ductile to brittle type of damage and role of covariance in damage mechanics was analyzed in Łodygowski and Sumelka (2015) and Sumelka (2013); (x) or finally the generalization of classical TTV by fractional calculus (Podlubny 1999; Kilbas et al. 2006) application to obtain *fractional viscoplasticity* was recently proposed in Sumelka (2014) and since then has been under continuous development, e.g., Sumelka and Nowak (2017), Sun and Shen (2017) and Xiao et al. (2017). All these results prove that TTV is a reliable tool for modeling the varied types of materials including geometerials, concrete materials, and especially metallic-like materials for a broad range of strains, strain rates, and temperatures.

It should be stated that the explicit nonlocal mechanical models like those proposed by Mindlin and Tiersten (1962), Kröner (1963), Toupin (1963, 1964), Green and Rivlin (1964), Mindlin (1964; 1965) and Mindlin and Eshel (1968), Yang et al. (2002), and Park and Gao (2008) and the nonlocal continuum mechanics initiated by Eringen and coworkers (Eringen 1972a,b, 1983), generalized continuum formulations (Polyzos and Fotiadis 2012; Tarasov 2014), or the one by de Borst and Pamin (1996), Fleck and Hutchinson (1997), Aifantis (1999), and Voyiadjis and Abu Al-Rub (2005) are more robust than implicit TTV (other implicit models can be found in, e.g., Voyiadjis and Abed 2005, 2006; Voyiadjis and Kattan 2007). This means that for the explicit nonlocal formulations, the advantages of inclusion of the length scale parameter are more pronounced and physical interpretation is more straightforward. Nonetheless, the serious drawback of explicit models is the high number of material parameters; recall that Mindlin theory contains 1764 coefficients in total (903 independent, Morán 2016) and these concepts make the numerical implementation more involved and computation time can be significantly higher due to the additional variables (e.g., higher order stresses).

The remaining part of this paper is organized as follows.

In section "Relaxation Time: Implicit Length Scale Parameter," the physical interpretation of the relaxation time of a mechanical perturbation as a length scale parameter is provided.

Section "The Thermo-viscoplasticity Model" describes the TTV accounting for anisotropic damage nucleation and growth (Sumelka 2009).

In section "Definition of Material Functions for Adiabatic Process" the material function for an adiabatic processes are identified.

Finally, in section "Numerical Examples: Modeling of Spalling Phenomena," the numerical results of spalling phenomena including the effects of anisotropic damage nucleation and growth are presented.

Section "Conclusions" concludes the document.

Relaxation Time: Implicit Length Scale Parameter

Herein the original concepts by Perzyna are recalled (2005, 2010, 2012).

The attention is focused on metallic materials. It is important that basic results come from the single crystal deformation analysis (micro scale of observation) and afterward they are generalized for the polycrystalline solids (meso-macro scale of observation), the one captured by the continuum description.

From the theory of crystal dislocations, the inelastic shear strain rate can be expressed as

$$\dot{\epsilon}^p = \alpha b \mathrm{v}, \tag{1}$$

where α denotes the mean density of mobile dislocations, b is the Burgers vector (the displacement per dislocation line), and v is the mean dislocation velocity (Asaro 1983). Next, based on the experimental observations that thermally activated and phonon damping mechanisms are most pronounced, the mean dislocation velocity can be expressed by

$$\mathrm{v} = \frac{AL^{-1}}{t_S + t_B}, \tag{2}$$

where AL^{-1} is the average distance of dislocation movement after each thermal activation, t_S is the time a dislocation has spent at the obstacle, and t_B denotes the time of traveling between the barriers. Applying Eq. (2) in Eq. (1), it can be shown that (Teodosiu and Sidoroff 1976)

$$\begin{aligned}\dot{\epsilon}^p &= \frac{1}{T_{mT}} \left\langle \exp\left\{ \frac{U[(\tau - \tau_\mu)Lb]}{k\vartheta} \right\} + \frac{BAL^{-1}v}{(\tau - \tau_B)b} \right\rangle^{-1} \\ &= \frac{1}{T} \left\langle \Phi\left[\frac{\tau}{\tau_Y} - 1 \right] \right\rangle \mathrm{sgn}\tau, \end{aligned}\tag{3}$$

where T_{mT} is the relaxation time for the thermally activated mechanism, U is the activation energy (Gibbs free energy), τ is the applied stress, τ_μ is the athermal stress, L is the distance between obstacle dislocation, k is the Boltzmann constant, B is called the dislocation drag coefficient, ϑ is the absolute temperature, v is the frequency of dislocation vibration, τ_B denotes the stress needed to overcome the forest dislocation barriers to the dislocation motion (the back stress), $< \cdot >$ denotes the Macaulay bracket, T is the relaxation time, Φ is the empirical overstress function, and τ_Y is a static yield stress. From the point of view of herein discussion, it is crucial that

$$T_{mT} = \frac{1}{\alpha b A L^{-1} v}. \tag{4}$$

Therefore, the relaxation time for the thermally activated mechanism is a function of the average distance of dislocation movement, thus including the information on the characteristic size of plastic deformation, identified with crystal (micro) length scale characteristic for specific material. For completeness it should be stated that the relaxation time for the phonon damping mechanism (T_{mD}) is

$$T_{mD} = \frac{B}{\alpha b^2 \tau_B} = T_{mT} \frac{BAL^{-1}v}{b\tau_B}. \tag{5}$$

The final postulate is that for a polycrystalline material (continuum level), the meso-macro viscoplastic strain is described by the analogous formula to Eq. (1). As will be discussed in the following section, the macroscopic relaxation time T_m is then

$$T_m = \frac{\ell}{\beta c}, \tag{6}$$

where ℓ is a macroscopic length scale, β is a proportionality factor, and c denotes the velocity of the propagation of the elastic wave. It can be shown Sluys (1992) that for 1D longitudinal wave propagation in the elasto-viscoplastic material

$$T_m = \frac{\ell E}{2\sigma_0 c}, \tag{7}$$

where E is the Young modulus and σ_0 is the yield stress; thus, $\beta = \frac{2\sigma_0}{E}$.

The last relation shows clearly that the relaxation time maps the material length scale, thus implicitly introducing length scale to the continuum model. Furthermore, the relaxation time T_m can be viewed not only as a microstructural parameter to be determined from experimental observations but also as a mathematical regularization parameter, as mentioned.

The Thermo-viscoplasticity Model

The assumed kinematics induces the form of stresses (Dłużewski 1996); hence, first the kinematics of the body is described. Next, following the balance principles and necessary constitutive axioms, the constitutive model for a thermomechanical process is obtained. It should be pointed out that the description follows the concepts presented in Sumelka (2009).

Basic Definitions

The following fundamental definitions hold in continuum mechanics (Truesdell and Noll 1965; Rymarz 1993; Perzyna 1978; Ostrowska-Maciejewska 1994).

Definition 1. The **material continuum** is a three-dimensional differentiable manifold $\mathcal{M}$:

$$\mathcal{M} = (\mathbb{R}^3, S)$$

where $\mathbb{R}^3$ is a three-dimensional continuum and S denotes the structure of manifold.

Definition 2. The material **body** $\mathcal{B}$ is a subset of material continuum $\mathcal{M}$:

$$\mathcal{B} \subset \mathcal{M}$$

and is characterized by the following structure S:

1. The set of material bodies is a material body:

$$\mathcal{B} = \mathcal{B}_1 \cup \mathcal{B}_2 \cup \ldots \cup \mathcal{B} \subset \mathcal{M},$$

2. Material body $\mathcal{B}$ or its arbitrary part can be uniquely mapped in the finite domain D of the Euclidean point space E^3:

$$\boldsymbol{\varkappa} : \mathcal{B} \rightarrow D,$$

   where $\boldsymbol{\varkappa} \subset E^3$. The map $\boldsymbol{\varkappa}$ is called the **configuration** of the body and $\boldsymbol{\varkappa} \in K$, where K is a set of a homeomorphic maps.
3. The smallest part of the material body is a **material point** X. The map $\boldsymbol{\varkappa}$ assigns to every material point, X, its geometrical point in D

$$\mathbf{X} = \boldsymbol{\varkappa}(X), \quad X = \boldsymbol{\varkappa}^{-1}(\mathbf{X}), \quad \mathbf{X} \in E^3, \quad X \in \mathcal{B}.$$

4. For every pair of configurations $\boldsymbol{\varkappa}, \boldsymbol{\gamma}$, there exist a homeomorphic map:

$$\boldsymbol{\phi}_{\varkappa\gamma} : \boldsymbol{\gamma} \circ \boldsymbol{\varkappa}^{-1}, \quad \boldsymbol{\phi}_{\varkappa\gamma} : E^3 \rightarrow E^3.$$

   The map $\boldsymbol{\phi}_{\varkappa\gamma}$ is called **Deformative configuration** and maps the domain E^3 in E^3. Those domains are occupied by the body $\mathcal{B}$ in different configurations.
5. In the E^3 space, the **additive Borelian measure** M corresponds to every configuration $\boldsymbol{\varkappa}$:

$$M : D \rightarrow \mathbb{R}^1 \quad or \quad M : \boldsymbol{\varkappa}(\mathcal{B}) \rightarrow \mathbb{R}^1.$$

6. The constitutive relations describing material of the body are fulfilled in every material point X of the body $\mathcal{B}$.

Definition 3. **Absolute time** is a time which runs at the same rate for all the observers in the universe.

Those definitions state a passage from an abstract mathematical description of the body, $\mathcal{B}$, to physical one.

Kinematics

Thus, the real material body become the abstract body, modeled as a manifold. The deformation of a real material body is treated in this 'mathematical world' as a mapping between manifolds. The observer in an abstract setting can occupy different positions; by analogy to the real observer, however, some of them are most convenient from many point of views. Two of them are most important for description of the material body motion, namely, the Lagrangean (material, referential) and the Eulerian (spatial, current) ones. These descriptions span two manifolds, as mentioned, and will be denoted by $\mathcal{B}$ and $\mathcal{S}$, respectively (Marsden and Hughes 1983).

Points in $\mathcal{B}$ are denoted by $\mathbf{X}$ while in $\mathcal{S}$ by $\mathbf{x}$. Coordinate system for $\mathcal{B}$ is denoted by $\{X^A\}$ with base $\mathbf{E}_A$ and for $\mathcal{S}$ by $\{x^a\}$ with base $\mathbf{e}_a$. Dual bases in those coordinate systems are denoted by $\mathbf{E}^A$ and $\mathbf{e}^a$, respectively. The tangent spaces in $\mathcal{B}$ and $\mathcal{S}$ are written as $T_{\mathbf{X}}\mathcal{B} = \{\mathbf{X}\} \times V^3$ and $T_{\mathbf{x}}\mathcal{S} = \{\mathbf{x}\} \times V^3$. They are understood as Euclidean vector space V^3, regarded as vectors emanating from points $\mathbf{X}$ and $\mathbf{x}$, respectively (Marsden and Hughes 1983).

For measuring purposes in the abstract space, the Riemannian space on manifolds $\mathcal{B}$ and $\mathcal{S}$ is introduced, i.e., $\{\mathcal{B}, \mathbf{G}\}$ and $\{\mathcal{S}, \mathbf{g}\}$ where metric tensors are defined, as $\mathbf{G} : T\mathcal{B} \to T^*\mathcal{B}$ and $\mathbf{g} : T\mathcal{S} \to T^*\mathcal{S}$ where $T\mathcal{B}$ and $T\mathcal{S}$ denote the tangent bundles of $\mathcal{B}$ and $\mathcal{S}$, respectively, while $T^*\mathcal{B}$ and $T^*\mathcal{S}$ denote their dual tangent bundles. Explicit definitions for metric tensors are then $G_{AB}(\mathbf{X}) = (\mathbf{E}_A, \mathbf{E}_B)_{\mathbf{X}}$ and $g_{ab}(\mathbf{x}) = (\mathbf{e}_a, \mathbf{e}_b)_{\mathbf{x}}$ where $(,)_{\mathbf{X}}$ and $(,)_{\mathbf{x}}$ denote inner product in $\mathcal{B}$ and $\mathcal{S}$, respectively.

The regular motion of the material body is treated as a series of the immersing of the abstract body $\mathcal{B}$ in the Euclidean point space E^3 (Rymarz 1993), and can be written as

$$\mathbf{x} = \phi(\mathbf{X}, t). \tag{8}$$

Thus, $\phi_t : \mathcal{B} \to \mathcal{S}$ is a C^1 current configuration of $\mathcal{B}$ in $\mathcal{S}$, at time t. For the analysis of abstract body deformation, the tangent of ϕ, which defines the two point tensor field $\mathbf{F}$, called deformation gradient, is considered. $\mathbf{F}$ describes all local deformation properties and is the primary measure of deformation (Perzyna 1978; Holzapfel 2000), thus

$$\mathbf{F}(\mathbf{X}, t) = T\phi = \frac{\partial \phi(\mathbf{X}, t)}{\partial \mathbf{X}}, \tag{9}$$

and using the notion of tangent space

$$\mathbf{F}(\mathbf{X}, t) : T_{\mathbf{X}}\mathcal{B} \to T_{\mathbf{x}=\phi(\mathbf{X},t)}\mathcal{S}, \tag{10}$$

so $\mathbf{F}$ is a linear transformation for each $\mathbf{X} \in \mathcal{B}$ and $t \in I \subset \mathbb{R}^1$.

The important properties of $\mathbf{F}$ are a consequence of the assumption that map ϕ is uniquely invertible (smooth homeomorphism) ($\mathbf{X} = \phi^{-1}(\mathbf{x}, t)$); hence, there exists the inverse of deformation gradient:

$$\mathbf{F}^{-1}(\mathbf{x}, t) = \frac{\partial \phi^{-1}(\mathbf{x}, t)}{\partial \mathbf{x}}. \tag{11}$$

Thus, the tensor field $\mathbf{F}$ is non-singular ($\det(\mathbf{F}) \neq 0$), and because of the impenetrability of matter $\det(\mathbf{F}) > 0$. Furthermore $\mathbf{F}$ can be uniquely decomposed into pure stretch and pure rotation, called polar decomposition, namely

$$\mathbf{F} = \mathbf{R}\mathbf{U} = \mathbf{v}\mathbf{R}, \tag{12}$$

where $\mathbf{R}$ is the rotation tensor (unique, proper orthogonal) which measures local orientation and $\mathbf{U}$ and $\mathbf{v}$ define unique, positive define, symmetric tensors called the right (or material) stretch tensor and the left (or spatial) stretch tensor, respectively (stretch tensors measure the local shape). Using the notion of tangent space, the result is obtained that for each $\mathbf{X} \in \mathcal{B}$, $\mathbf{U}(\mathbf{X}) : T_{\mathbf{X}}\mathcal{B} \to T_{\mathbf{X}}\mathcal{B}$ and for each $\mathbf{x} \in \mathcal{S}$, $\mathbf{v}(\mathbf{x}) : T_{\mathbf{x}}\mathcal{S} \to T_{\mathbf{x}}\mathcal{S}$.

From the general class of the Lagrangean and the Eulerian strain measures, defined through one single scale function given by (cf. Hill 1978; Xiao et al. 1998),

$$\mathbf{E} = g(\mathbf{C}) = \sum_{i=1}^{3} g(\breve{\chi}_i)\mathbf{C}_i,$$

and

$$\mathbf{e} = g(\mathbf{B}) = \sum_{i=1}^{3} g(\breve{\chi}_i)\mathbf{B}_i,$$

where the scale function $g(\cdot)$ is a smooth increasing function with the normalized property $g(1) = g'(1) - 1 = 0$, $\breve{\chi}_i$ is used to denote distinct eigenvalues of the right and left Cauchy-Green tensors $\mathbf{C}$ and $\mathbf{B}$, respectively, and $\mathbf{C}_i$ and $\mathbf{B}_i$ are the corresponding subordinate eigenprojections; the Green-Lagrange and the Euler-Almansi definitions were accepted. As will be presented, such choice is crucial for the clear meaning of the rate-type constitutive structure.

The Green-Lagrange strain tensor is defined as Perzyna (2005) ($\mathbf{E} : T_{\mathbf{X}}\mathcal{B} \to T_{\mathbf{X}}\mathcal{B}$):

$$2\mathbf{E} = \mathbf{C} - \mathbf{I}, \tag{13}$$

where $\mathbf{E}$ stands for the Green-Lagrange strain tensor, $\mathbf{I}$ denotes the identity on $T_{\mathbf{X}}\mathcal{B}$, and

$$\mathbf{C} = \mathbf{F}^T \cdot \mathbf{F} = \mathbf{U}^2 = \mathbf{B}^{-1}. \tag{14}$$

On the other side, the Euler-Almansi strain tensor definition is ($\mathbf{e} : T_{\mathbf{x}}\mathcal{S} \rightarrow T_{\mathbf{x}}\mathcal{S}$)

$$2\mathbf{e} = \mathbf{i} - \mathbf{c}, \tag{15}$$

where $\mathbf{e}$ stands for the Euler-Almansi strain tensor and $\mathbf{i}$ denotes the identity on $T_{\mathbf{x}}\mathcal{S}$. One has also

$$\mathbf{c} = \mathbf{b}^{-1} \quad \text{and} \quad \mathbf{b} = \mathbf{F} \cdot \mathbf{F}^{T} = \mathbf{v}^2, \tag{16}$$

where tensor $\mathbf{b}$ is sometimes referred to as the Finger deformation tensor.

For the purpose of the objective tensor rate definition, the push-forward and the pullback operations are introduced as

$$\phi_*((\cdot)^\flat) = \mathbf{F}^{-T}(\cdot)^\flat\mathbf{F}^{-1}, \tag{17}$$

$$\phi^*((\cdot)^\flat) = \mathbf{F}^{T}(\cdot)^\flat\mathbf{F}, \tag{18}$$

respectively, where $\flat$ indicates that a tensor has all its indices lowered (Marsden and Hughes 1983). In view of the above definitions, the following holds:

$$\mathbf{e}^\flat = \phi_*(\mathbf{E}^\flat) = \mathbf{F}^{-T}\mathbf{E}^\flat\mathbf{F}^{-1}, \tag{19}$$

$$\mathbf{E}^\flat = \phi^*(\mathbf{e}^\flat) = \mathbf{F}^{T}\mathbf{e}^\flat\mathbf{F}. \tag{20}$$

The generality of possible deformations of a real material body assumes that also in an abstract world it can be arbitrary, called commonly finite. With this respect, the finite elasto-viscoplastic deformation is assumed, manifested by the multiplicative decomposition of the total deformation gradient, namely

$$\mathbf{F}(\mathbf{X}, t) = \mathbf{F}^e(\mathbf{X}, t) \cdot \mathbf{F}^p(\mathbf{X}, t). \tag{21}$$

This decomposition is justified by the micromechanics of single crystal plasticity (Perzyna 1998) and states that the component $\mathbf{F}^e$ is a lattice contribution to $\mathbf{F}$ while $\mathbf{F}^p$ describes the deformation solely due to plastic shearing on crystallographic slip systems.

The inverse of the local elastic deformation $\mathbf{F}^{e^{-1}}$ releases from the stress state in every surrounding ($\mathcal{N}(\mathbf{x}) \subset \phi(\mathcal{B})$) in an actual configuration. The configuration obtained by the linear map $\mathbf{F}^{e^{-1}}$ from actual configuration $\mathcal{S}$ is called unstressed configuration and is denoted by $\mathcal{S}'$. Thus one can write (see Fig. 1)

$$\mathbf{F}^e : T_{\mathbf{y}}\mathcal{S}' \rightarrow T_{\mathbf{x}}\mathcal{S}, \quad \mathbf{F}^p : T_{\mathbf{X}}\mathcal{B} \rightarrow T_{\mathbf{y}}\mathcal{S}', \tag{22}$$

where the material point in the configuration $\mathcal{S}'$ is characterized by $\mathbf{y}$.

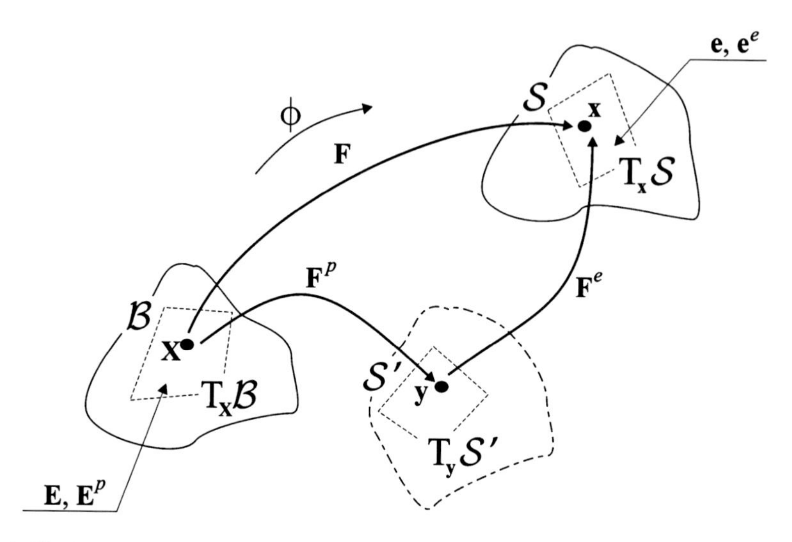

Fig. 1 The interpretation of the multiplicative decomposition of $\mathbf{F}$

Based on the decomposition Eq. (21), the viscoplastic strain tensor $\mathbf{E}^p : T_{\mathbf{X}}\mathcal{B} \to T_{\mathbf{X}}\mathcal{B}$ can be written as

$$2\mathbf{E}^p = \mathbf{C}^p - \mathbf{I}, \tag{23}$$

where

$$\mathbf{C}^p = \mathbf{F}^{pT} \cdot \mathbf{F}^p = \mathbf{U}^{p2} = \mathbf{B}^{p-1} \quad \text{and} \quad \mathbf{E}^e = \mathbf{E} - \mathbf{E}^p, \tag{24}$$

whereas the elastic strain tensor $\mathbf{e}^e : T_{\mathbf{x}}\mathcal{S} \to T_{\mathbf{x}}\mathcal{S}$ is

$$2\mathbf{e}^e = \mathbf{i} - \mathbf{c}^e, \tag{25}$$

where

$$\mathbf{c}^e = \mathbf{b}^{e-1} \quad \text{and} \quad \mathbf{b}^e = \mathbf{F}^e \cdot \mathbf{F}^{eT} = \mathbf{v}^{e2} \quad \text{and} \quad \mathbf{e}^p = \mathbf{e} - \mathbf{e}^e. \tag{26}$$

To define TTV, being the rate-type constitutive structure, the rate of deformation needs to be defined. Starting with the spatial velocity $\boldsymbol{\upsilon}$,

$$\boldsymbol{\upsilon}(\mathbf{x}, t) = \dot{\mathbf{x}} = \frac{\partial \phi}{\partial t}, \tag{27}$$

the gradient of $\boldsymbol{\upsilon}$, the tensor field (nonsymmetric, second order), called spatial velocity gradient, can be computed (Holzapfel 2000):

$$\mathbf{l}(\mathbf{x}, t) = \frac{\partial \boldsymbol{\upsilon}(\mathbf{x}, t)}{\partial \mathbf{x}}, \tag{28}$$

where $\mathbf{l}$ stands for spatial velocity gradient. Next, based on the definitions given by Eqs. (9), (21), and (28) the fundamental relation is obtained (Perzyna 2005):

$$\mathbf{l} = \dot{\mathbf{F}} \cdot \mathbf{F}^{-1} = \dot{\mathbf{F}}^e \cdot \mathbf{F}^{e-1} + \mathbf{F}^e \cdot (\dot{\mathbf{F}}^p \cdot \mathbf{F}^{p-1}) \cdot \mathbf{F}^{e-1} = \mathbf{l}^e + \mathbf{l}^p, \tag{29}$$

which introduces the elastic $\mathbf{l}^e$ and plastic $\mathbf{l}^p$ parts of spatial velocity gradient. On the other hand, the additive decomposition of spatial velocity gradient to symmetric and antisymmetric parts generates covariant tensor field $\mathbf{d}$ called rate of deformation tensor and also covariant tensor field $\mathbf{w}$ called spin tensor, with the definitions:

$$\mathbf{l} = \mathbf{d} + \mathbf{w} = \mathbf{d}^e + \mathbf{w}^e + \mathbf{d}^p + \mathbf{w}^p, \tag{30}$$

$$\mathbf{d} = \frac{1}{2}(\mathbf{l} + \mathbf{l}^T), \tag{31}$$

$$\mathbf{w} = \frac{1}{2}(\mathbf{l} - \mathbf{l}^T). \tag{32}$$

To define the objective rate, the Lie-type derivative (assuring diffeomorphisms) is accepted. Its definition for an arbitrary spatial tensor field φ is obtained using the following concept:

(i) compute the pullback operation of φ – the material field Φ is obtained,
(ii) take the material time derivative of Φ,
(iii) carry out the push-forward operation of the result field from (ii).

The scheme can be summarized as

$$\mathrm{L}_v(\varphi) = \phi_* \left(\frac{\mathrm{D}}{\mathrm{D}t} \phi^*(\varphi) \right), \tag{33}$$

where L_v stands for Lie derivative.

In view of the above definition, Lie derivative of the Euler-Almansi strain measure results in

$$\mathbf{d}^\flat = \mathrm{L}_v(\mathbf{e}^\flat). \tag{34}$$

Thus, Lie derivative states a direct relationship between the stretching $\mathbf{d}$ and the Eulerian strain $\mathbf{e}$. It should be pointed out that Eq. (34) holds for $\mathbf{e}^\flat$ only; thus, the choice of $\mathbf{e}^\flat$ to be the strain measure is the most proper from the point of view of physical interpretation of the mathematical model.

Furthermore based on Eq. (34), one can write

$$\mathbf{d}^{e\flat} = \mathrm{L}_v(\mathbf{e}^{e\flat}), \quad \mathbf{d}^{p\flat} = \mathrm{L}_v(\mathbf{e}^{p\flat}). \tag{35}$$

Constitutive Postulates

The constitutive structure must fulfill all basic experimentally proven balance principles. Nonetheless, some constitutive postulates are still needed for completeness. Thus, assuming that conservation of mass, balance of momentum, balance of moment of momentum, and balance of energy and entropy production hold, four constitutive postulates are accepted (Perzyna 1986a, 2005):

(i) Existence of the free energy function ψ being formally the following scalar function of tensorial argument:

$$\psi = \hat{\psi}(\mathbf{e}, \mathbf{F}, \vartheta; \boldsymbol{\mu}), \tag{36}$$

where $\boldsymbol{\mu}$ denotes a set of internal state variables governing the description of dissipation effects and ϑ represents temperature. It is clear that for nonempty $\boldsymbol{\mu}$ state vector, a dissipation exists in the model; otherwise, the presented model describes thermoelasticity.

(ii) Axiom of objectivity. The material model should be invariant with respect to diffeomorphism (any superposed motion) – (cf. Frewer 2009) for interesting discussion.

(iii) The axiom of the entropy production. For every regular process, the constitutive functions should satisfy the second law of thermodynamics.

(iv) The evolution equation for the internal state variables vector $\boldsymbol{\mu}$ should be of the form

$$\mathrm{L}_{v}\boldsymbol{\mu} = \hat{\mathbf{m}}(\mathbf{e}, \mathbf{F}, \vartheta, \boldsymbol{\mu}). \tag{37}$$

where evolution function $\hat{\mathbf{m}}$ has to be determined based on the experimental observations. This postulate states the hardest part of modeling; others are nowadays broadly accepted standards.

Constitutive Relations

To obtain the stress and the entropy relations, the reduced dissipation inequality is applied in the form (Marsden and Hughes 1983; Sumelka 2009)

$$\frac{1}{\rho_{\mathrm{Ref}}}\boldsymbol{\tau} : \mathbf{d} - (\eta\dot{\vartheta} + \dot{\psi}) - \frac{1}{\rho\vartheta}\mathbf{q} \cdot \mathrm{grad}\vartheta \geq 0, \tag{38}$$

where ρ denotes current and ρ_{Ref} reference densities, $\boldsymbol{\tau}$ denotes Kirchhoff stress, ψ is the free energy function, ϑ is absolute temperature, η denotes the specific (per unit mass) entropy, and $\mathbf{q}$ is the heat flux. Using postulate (i), Eq. (38) can be rewritten in the form:

$$\left(\frac{1}{\rho_{\text{Ref}}}\boldsymbol{\tau} - \frac{\partial \hat{\psi}}{\partial \mathbf{e}}\right) : \mathbf{d} - \left(\eta + \frac{\partial \hat{\psi}}{\partial \vartheta}\right)\dot{\vartheta} - \frac{\partial \hat{\psi}}{\partial \boldsymbol{\mu}}\mathrm{L}_{\upsilon}\boldsymbol{\mu} - \frac{1}{\rho\vartheta}\mathbf{q}\cdot \text{grad}\vartheta \geq 0, \tag{39}$$

so because of arbitrariness

$$\boldsymbol{\tau} = \rho_{\text{Ref}}\frac{\partial \hat{\psi}}{\partial \mathbf{e}}, \tag{40}$$

$$\eta = -\frac{\partial \hat{\psi}}{\partial \vartheta}. \tag{41}$$

Thus, Eq. (39) reduces to

$$-\frac{\partial \hat{\psi}}{\partial \boldsymbol{\mu}}\mathrm{L}_{\upsilon}\boldsymbol{\mu} - \frac{1}{\rho\vartheta}\mathbf{q}\cdot \text{grad}\vartheta \geq 0. \tag{42}$$

The specific form of the constitutive relations depends on the assumed internal state vector components. Herein we assume that $\boldsymbol{\mu}$ consists of two variables, namely, (Perzyna 2008; Glema et al. 2009; Sumelka and Łodygowski 2011)

$$\boldsymbol{\mu} = (\in^p, \boldsymbol{\xi}), \tag{43}$$

where $\in^p$ is the equivalent plastic deformation $\dot{\in^p} = \left(\frac{2}{3}\mathbf{d}^p : \mathbf{d}^p\right)^{\frac{1}{2}}$, which describes the dissipation effects generated by viscoplastic deformation and $\boldsymbol{\xi}$ being the microdamage tensor which takes into account the anisotropic microdamage effects.

The assumption Eq. (43), together with the application of Lie derivative to formula Eq. (40), with internal state vector constant (or in other words keeping the history constant – thermoelastic process), gives us the evolution equation for Kirchhoff stress tensor in the form (Duszek–Perzyna and Perzyna 1994)

$$(\mathrm{L}_{\upsilon}\boldsymbol{\tau})^e = \mathcal{L}^e : \mathbf{d}^e - \mathcal{L}^{th}\dot{\vartheta}, \tag{44}$$

where

$$\mathcal{L}^e = \rho_{\text{Ref}}\frac{\partial^2 \hat{\psi}}{\partial \mathbf{e}^2}, \tag{45}$$

$$\mathcal{L}^{th} = -\rho_{\text{Ref}}\frac{\partial^2 \hat{\psi}}{\partial \mathbf{e}\partial \vartheta}, \tag{46}$$

in above $\mathcal{L}^e$ denotes elastic constitutive tensor and $\mathcal{L}^{th}$ is thermal operator. Using the relations

$$(\mathrm{L}_v\boldsymbol{\tau})^e = \dot{\boldsymbol{\tau}} - \boldsymbol{\tau}\cdot\mathbf{d}^e - \mathbf{d}^e\cdot\boldsymbol{\tau}, \tag{47}$$

and

$$\mathbf{d} = \mathbf{d}^e + \mathbf{d}^p, \tag{48}$$

final form of the Kirhchoff stress rate is

$$\mathrm{L}_v\boldsymbol{\tau} = \mathcal{L}^e : \mathbf{d} - \mathcal{L}^{th}\dot{\vartheta} - (\mathcal{L}^e + \mathbf{g}\boldsymbol{\tau} + \boldsymbol{\tau}\mathbf{g}) : \mathbf{d}^p. \tag{49}$$

On the other side, the energy balance in the form (Perzyna 2005; Sumelka 2009)

$$\rho\vartheta\dot{\eta} = -\mathrm{div}\mathbf{q} - \rho\frac{\partial\hat{\psi}}{\partial\boldsymbol{\mu}}\cdot\mathrm{L}_v\boldsymbol{\mu}, \tag{50}$$

together with the rate of entropy, defined by Eq. (41), defines the evolution equation for temperature, namely,

$$\rho c_p\dot{\vartheta} = -\mathrm{div}\mathbf{q} + \vartheta\frac{\rho}{\rho_{\mathrm{Ref}}}\frac{\partial\boldsymbol{\tau}}{\partial\vartheta} : \mathbf{d} + \chi^*\boldsymbol{\tau} : \mathbf{d}^p + \chi^{**}\mathbf{k} : \mathrm{L}_v\boldsymbol{\xi}, \tag{51}$$

where the specific heat and the irreversibility coefficients χ^* and χ^{**} are determined by

$$c_p = -\vartheta\frac{\partial^2\hat{\psi}}{\partial\vartheta^2}, \tag{52}$$

$$\chi^* = -\rho\left(\frac{\partial\hat{\psi}}{\partial\in^p} - \vartheta\frac{\partial^2\hat{\psi}}{\partial\vartheta\partial\in^p}\right)\sqrt{\frac{2}{3}}\frac{1}{\boldsymbol{\tau}:\mathbf{p}}, \tag{53}$$

$$\chi^{**} = -\rho\left(\frac{\partial\hat{\psi}}{\partial\boldsymbol{\xi}} - \vartheta\frac{\partial^2\hat{\psi}}{\partial\vartheta\partial\boldsymbol{\xi}}\right) : \frac{1}{\mathbf{k}},$$

where $\mathbf{p}$ defines the tensor of viscoplastic flow direction.

The Microdamage Tensor

The microdamage tensor is postulated to be the state variable in Eq. (43) and maps the experimentally observed micro-voiding in the zones of the severe plastic

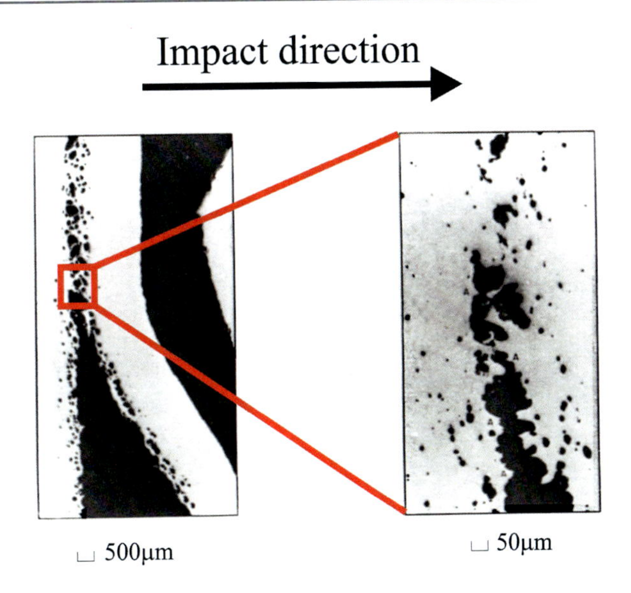

Fig. 2 Cracks anisotropy in 1145 aluminum after flat plate impact experiment (Seaman et al. 1976)

deformation. As an example in Fig. 2, the effects of a flat plate impact experiment in 1145 aluminum are considered (Seaman et al. 1976). It is observed that three stages of damage evolution are crucial: nucleation, growth, and coalescence. Another important fact is that the microdefects have directional geometry – in this experiment they have approximately an ellipsoidal shape. This damage anisotropy is important and should be included in the continuum description (at meso-macro scale), keeping simultaneously the information about level of material porosity.

The microdamage tensor (the state variable at meso-macro scale) Fig. 3 is introduced through the logic presented below, with the most important properties that components of this tensor are proportional to the damage area on the representative volume element (RVE) (its principal values), whereas its Euclidean norm defines porosity.

Considering the RVE, being the abstract point P_i in the model (cf. Fig. 3), one can introduce three ratios:

$$\frac{A_i^p}{A}, \tag{54}$$

where A_i^p is a damaged area and A denotes assumed characteristic area of the RVE. Next, three vectors are obtained based on the introduced ratios – theirs modules are equal to those ratios and are normal to the RVE's planes. From all possible spatial configuration of the RVE, the one is chosen, in which the resultant module is largest. This resultant is called the *main microdamage vector* and is denoted by $\hat{\boldsymbol{\xi}}^{(m)}$ (Sumelka and Glema 2007), i.e.,

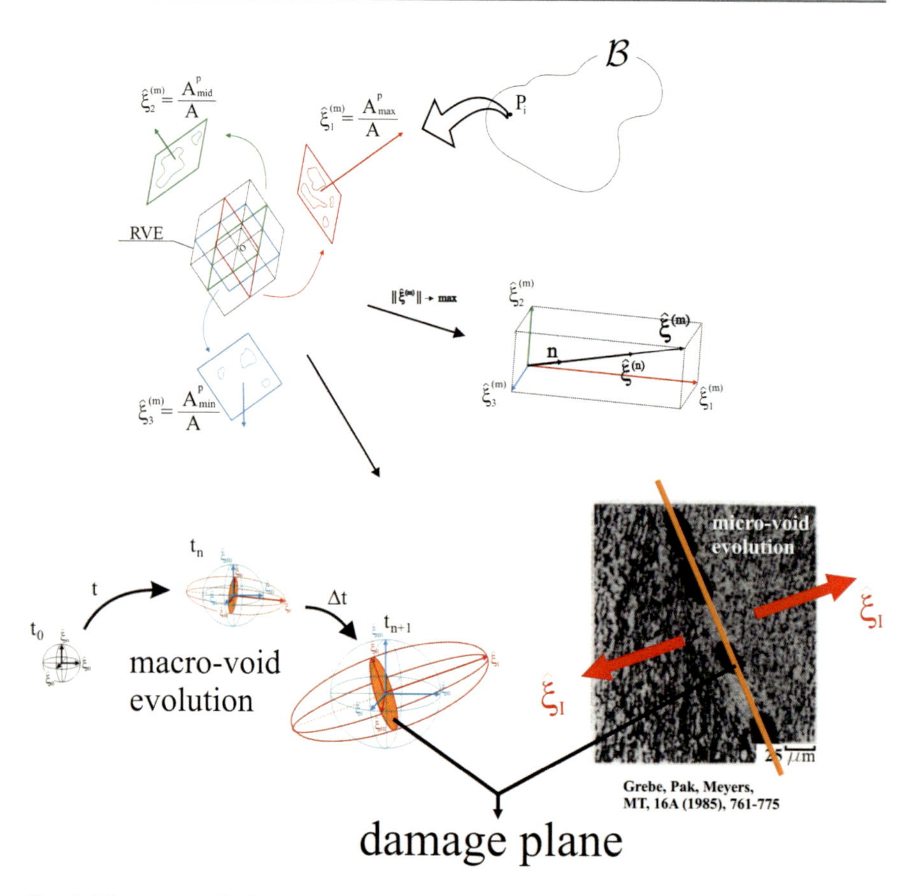

Fig. 3 The concept of microdamage tensor

$$\hat{\boldsymbol{\xi}}^{(m)} = \frac{A_1^p}{A}\hat{\mathbf{e}}_1 + \frac{A_2^p}{A}\hat{\mathbf{e}}_2 + \frac{A_3^p}{A}\hat{\mathbf{e}}_3, \tag{55}$$

where $\hat{(\cdot)}$ denotes the principal directions of microdamage with $A_1^p \geq A_2^p \geq A_3^p$.

In the following step, based on the main microdamage vector, the *microdamage vector* is built, denoted by $\hat{\boldsymbol{\xi}}^{(n)}$ (Sumelka and Glema 2007):

$$\hat{\boldsymbol{\xi}}^{(n)} = \frac{1}{\left\|\hat{\boldsymbol{\xi}}^{(m)}\right\|}\left(\left(\frac{A_1^p}{A}\right)^2\hat{\mathbf{e}}_1 + \left(\frac{A_2^p}{A}\right)^2\hat{\mathbf{e}}_2 + \left(\frac{A_3^p}{A}\right)^2\hat{\mathbf{e}}_3\right), \tag{56}$$

which states clear connection to microdamage tensor (Sumelka and Glema 2007)

$$\hat{\boldsymbol{\xi}}^{(n)} = \hat{\boldsymbol{\xi}}\mathbf{n}, \tag{57}$$

where

$$\mathbf{n} = \sqrt{3}\left|\left|\hat{\boldsymbol{\xi}}^{(m)}\right|\right|^{-1}\left(\hat{\xi}_1^{(m)}\hat{\mathbf{e}}_1 + \hat{\xi}_2^{(m)}\hat{\mathbf{e}}_2 + \hat{\xi}_3^{(m)}\hat{\mathbf{e}}_3\right), \tag{58}$$

giving in the principal directions of $\boldsymbol{\xi}$ the fundamental relation

$$\hat{\boldsymbol{\xi}} = \frac{\sqrt{3}}{3}\begin{bmatrix} \hat{\xi}_1^{(m)} & 0 & 0 \\ 0 & \hat{\xi}_2^{(m)} & 0 \\ 0 & 0 & \hat{\xi}_3^{(m)} \end{bmatrix}. \tag{59}$$

Therefore, as mentioned, the physical interpretation of the microdamage tensor components is that the diagonal components ξ_{ii} of the microdamage tensor $\boldsymbol{\xi}$, in its principal directions, are proportional to the components of the main micro-damage vector $\xi_i^{(m)}$ which defines the ratio of the damaged area to the assumed characteristic area of the RVE, on the plane perpendicular to the i direction. As a consequence, the damage plane is the one perpendicular to the maximal principal value of $\boldsymbol{\xi}$.

Moreover, the Euclidean norm from the microdamage field $\hat{\boldsymbol{\xi}}$ results in additional physical interpretation for microdamage tensor, namely, it defines the scalar quantity called the *volume fraction porosity* or simply *porosity* (Perzyna 2008):

$$\sqrt{\boldsymbol{\xi} : \boldsymbol{\xi}} = \xi = \frac{V - V_s}{V} = \frac{V_p}{V}, \tag{60}$$

where ξ denotes porosity (scalar damage parameter), V is the volume of a material element, V_s is the volume of the solid constituent of that material element, and V_p denotes void volume:

$$V_p = \frac{\sqrt{3}\,l}{3}\sqrt{\left(A_1^p\right)^2 + \left(A_2^p\right)^2 + \left(A_3^p\right)^2}. \tag{61}$$

The interpretations of the microdamage tensorial field impose the mathematical bounds for the microdamage evolution, as

$$\xi \in < 0, 1 >, \quad \text{and} \quad \hat{\boldsymbol{\xi}}_{ii} \in < 0, 1 > . \tag{62}$$

However, the physical bounds are different and are rate dependent (Cochran and Banner 1988; Meyers and Aimone 1983), e.g., under extreme loading is of the order $0.09 \div 0.35$ (Dornowski and Perzyna 2002, 2006). Furthermore, on continuum level, the initial porosity exists in metals (denoted by ξ_0) which is in the order of $\xi_0 \cong 10^{-4} \div 10^{-3}$ (Nemes and Eftis 1991).

Definition of Material Functions for Adiabatic Process

Evolution Equations for Internal State Variables

The evolution equations for the internal state variables are postulated in the following rate form:

$$\mathbf{d}^p = \Lambda \mathbf{p}, \tag{63}$$

$$\mathrm{L}_v \boldsymbol{\xi} = \Lambda^h \frac{\partial h^*}{\partial \boldsymbol{\tau}} + \Lambda^g \frac{\partial g^*}{\partial \boldsymbol{\tau}}, \tag{64}$$

where Λ, Λ^h, and Λ^g define the intensity of the viscoplastic flow, the microdamage nucleation, and the microdamage growth, respectively, while $\mathbf{p}$, $\frac{\partial h^*}{\partial \boldsymbol{\tau}}$, and $\frac{\partial g^*}{\partial \boldsymbol{\tau}}$ define viscoplastic flow direction, microdamage nucleation direction, and microdamage growth direction, respectively.

Material Functions

Thermoelastic Range

Because of existence of the initial microdamage state, the microdamage tensor also influences thermoelastic range. Mathematically it is expressed by putting the stiffness operator to be a function of the microdamage state (Sumelka and Glema 2007, 2008):

$$\mathcal{L}^e = \mathcal{L}^e(\boldsymbol{\xi}). \tag{65}$$

The explicit definition of Eq. (65) is given under the assumption that the operator $\mathcal{L}^e$ in damage directions is coaxial with principal directions of microdamage tensor $\boldsymbol{\xi}$, namely,

$$\hat{\mathcal{L}}^e_{ijkm} = (\hat{\mathcal{C}})_{ijkm}\delta_{ij}\delta_{km} + (\hat{\mathcal{D}})_{ijkm}(\delta_{ik}\delta_{jm} + \delta_{im}\delta_{jk}) + (\hat{\mathcal{E}})_{ijkm}(\delta_{ik}\delta_{jm} - \delta_{im}\delta_{jk}), \tag{66}$$

where $\mathcal{C}$, $\mathcal{D}$, and $\mathcal{E}$ are the fourth-order material parameters tensors; $(\hat{\mathcal{C}})_{ijkm}$, $(\hat{\mathcal{D}})_{ijkm}$, and $(\hat{\mathcal{E}})_{ijkm}$ are projections of those tensors onto the orthonormal basis vectors $\hat{e}_i$ (Holzapfel 2000) (alternatively we could use the notation introduced in Mura 1987, e.g., instead of $(\hat{\mathcal{C}})_{ijkm}$, we could write $\hat{\mathcal{C}}_{IJKM}$ with the annotation that uppercase indices take on the same numbers as the corresponding lowercase ones but are not summed up – see also Skolnik et al. 2008.); δ denotes the Kronecker symbol; and $\hat{(\cdot)}$ denotes damage directions. The tensors in Eq. (66) which describe the material parameters are defined in the following way:

$$\hat{\mathcal{C}}_{ijkm} = (\hat{\mathcal{G}})_{ijkm}, \tag{67}$$

$$\hat{\mathcal{D}}_{ijkm} = (\hat{\mathcal{G}})_{ijkm} \frac{1}{2(1-2(\hat{\boldsymbol{\nu}})_{ijkm})}, \tag{68}$$

$$\hat{\mathcal{E}}_{ijkm} = (\hat{\mathcal{G}})_{ijkm} \frac{1-4(\hat{\boldsymbol{\nu}})_{ijkm}}{2(1-2(\hat{\boldsymbol{\nu}})_{ijkm})}, \tag{69}$$

where

$$\hat{\mathcal{G}}_{ijkm} = G_0(1-(\hat{\boldsymbol{\Upsilon}})_{ijkm}) \left(1 - \frac{6K_0+12G_0}{9K_0+8G_0}(\hat{\boldsymbol{\Upsilon}})_{ijkm}\right), \tag{70}$$

$$\hat{K}_{ijkm} = \frac{4G_0K_0(1-(\hat{\boldsymbol{\Upsilon}})_{ijkm})}{4G_0+3K_0(\hat{\boldsymbol{\Upsilon}})_{ijkm}}, \tag{71}$$

$$\hat{\nu}_{ijkm} = \frac{1}{2}\frac{3\hat{K}_{ijkm} - 2\hat{\mathcal{G}}_{ijkm}}{3\hat{K}_{ijkm} + \hat{\mathcal{G}}_{ijkm}}. \tag{72}$$

The parameters G_0 and K_0 are the shear and bulk modulus of the material matrix, respectively, and $\hat{\boldsymbol{\Upsilon}}$ is defined as

$$\hat{\Upsilon}_{ijkm} = \sqrt{3}\sqrt{\hat{\xi}_{ij}\hat{\xi}_{km}}\hat{\mathbf{e}}_i \otimes \hat{\mathbf{e}}_j \otimes \hat{\mathbf{e}}_k \otimes \hat{\mathbf{e}}_m. \tag{73}$$

Finally, applying Voigt notation, the matrix representation of the constitutive tensor $\hat{\mathcal{L}}^e$ is

$$\hat{\mathcal{L}}^e = \begin{bmatrix} \hat{\mathcal{L}}^e\left(\hat{\Upsilon}_{1111}\right) & \hat{\mathcal{L}}^e(\hat{\Upsilon}_{1122}) & \hat{\mathcal{L}}^e(\hat{\Upsilon}_{1133}) & 0 & 0 & 0 \\ \hat{\mathcal{L}}^e(\hat{\Upsilon}_{2211}) & \hat{\mathcal{L}}^e\left(\hat{\Upsilon}_{2222}\right) & \hat{\mathcal{L}}^e(\hat{\Upsilon}_{2233}) & 0 & 0 & 0 \\ \hat{\mathcal{L}}^e(\hat{\Upsilon}_{3311}) & \hat{\mathcal{L}}^e(\hat{\Upsilon}_{3322}) & \hat{\mathcal{L}}^e\left(\hat{\Upsilon}_{3333}\right) & 0 & 0 & 0 \\ 0 & 0 & 0 & \hat{\mathcal{L}}^e\left(\hat{\Upsilon}_{1212}\right) & 0 & 0 \\ 0 & 0 & 0 & 0 & \hat{\mathcal{L}}^e\left(\hat{\Upsilon}_{2323}\right) & 0 \\ 0 & 0 & 0 & 0 & 0 & \hat{\mathcal{L}}^e\left(\hat{\Upsilon}_{3131}\right) \end{bmatrix}. \tag{74}$$

Notice that, $\hat{\Upsilon}_{1212} = \hat{\Upsilon}_{2323} = \hat{\Upsilon}_{3131} = 0$. It is important that for $\hat{\boldsymbol{\xi}} \to \hat{\mathbf{0}}$ we arrive at

$$\begin{aligned} \lim_{\hat{\boldsymbol{\xi}}\to\hat{\mathbf{0}}} \hat{\mathcal{G}}_{ijkm} &\equiv G_0, \\ \lim_{\hat{\boldsymbol{\xi}}\to\hat{\mathbf{0}}} \hat{K}_{ijkm} &\equiv K_0, \\ \lim_{\hat{\boldsymbol{\xi}}\to\hat{\mathbf{0}}} \hat{\nu}_{ijkm} &\equiv \tfrac{1}{2}\tfrac{3K_0-2G_0}{3K_0+G_0}. \end{aligned} \tag{75}$$

Thus, the tensor $\mathcal{L}^e$ becomes the well-known (isotropic) Hooke's elastic one. For $\hat{\boldsymbol{\xi}} \to \hat{\mathbf{1}}$, we obtain material annihilation:

$$\lim_{\hat{\xi}\to\hat{\mathbf{1}}} \hat{\mathcal{G}}_{ijkm} \equiv 0,$$

$$\lim_{\hat{\xi}\to\hat{\mathbf{1}}} \hat{K}_{ijkm} \equiv 0, \quad (76)$$

$$\lim_{\hat{\xi}\to\hat{\mathbf{1}}} \hat{\nu}_{ijkm} \equiv \frac{45K_0^2+16G_0^2+48K_0G_0}{117K_0^2-16G_0^2+96K_0G_0}.$$

For isotropic microdamage, when we substitute the microdamage tensor by scalar damage parameter, the Mackenzie formulation is obtained (Mackenzie 1950; Perzyna 1986a).

Nonetheless, in following discussion, it is assumed that the elastic range is isotropic and independent of the microdamage state; thus, elastic constitutive tensor $\mathcal{L}^e$ takes the form

$$\mathcal{L}^e = 2\mu\mathcal{I} + \lambda(\mathbf{g}\otimes\mathbf{g}), \quad (77)$$

where μ and λ are Lamé constants. Such assumption is justified because of small influence of elastic range in the extremely dynamic processes.

Because of above assumption, the thermal expansion effects are also isotropic; thus, thermal operator $\mathcal{L}^{th}$ is

$$\mathcal{L}^{th} = (2\mu + 3\lambda)\theta\mathbf{g}, \quad (78)$$

where θ is thermal expansion coefficient.

Viscoplastic Range

We accept common assumptions for TTV theory concerning the rate of viscoplastic strains $\mathbf{d}^p$ Perzyna (1963, 1966), namely,

$$\Lambda^{vp} = \frac{1}{T_m}\left\langle \Phi^{vp}\left(\frac{f}{\kappa}-1\right)\right\rangle = \frac{1}{T_m}\left\langle \left(\frac{f}{\kappa}-1\right)^{m_{pl}}\right\rangle, \quad (79)$$

$$f = \left\{J_2' + \left[n_1(\vartheta) + n_2(\vartheta)(\boldsymbol{\xi}:\boldsymbol{\xi})^{\frac{1}{2}}\right]J_1^2\right\}^{\frac{1}{2}}, \quad (80)$$

$$n_1(\vartheta) = 0, \qquad n_2(\vartheta) = n = \text{const.}, \quad (81)$$

$$\kappa = \{\kappa_s(\vartheta) - [\kappa_s(\vartheta) - \kappa_0(\vartheta)]\exp[-\delta(\vartheta)\,\epsilon^p]\}\left[1 - \left(\frac{(\boldsymbol{\xi}:\boldsymbol{\xi})^{\frac{1}{2}}}{\xi^F}\right)^{\beta(\vartheta)}\right], \quad (82)$$

$$\overline{\vartheta} = \frac{\vartheta - \vartheta_0}{\vartheta_0}, \quad \kappa_s(\vartheta) = \kappa_s^* - \kappa_s^{**}\overline{\vartheta}, \quad \kappa_0(\vartheta) = \kappa_0^* - \kappa_0^{**}\overline{\vartheta},$$

$$\delta(\vartheta) = \delta^* - \delta^{**}\overline{\vartheta}, \quad \beta(\vartheta) = \beta^* - \beta^{**}\overline{\vartheta}. \quad (83)$$

$$\mathbf{p} = \left.\frac{\partial f}{\partial \boldsymbol{\tau}}\right|_{\xi=\mathrm{const}} \left(\left\|\frac{\partial f}{\partial \boldsymbol{\tau}}\right\|\right)^{-1} = \frac{1}{[2J_2' + 3A^2(\mathrm{tr}\boldsymbol{\tau})^2]^{\frac{1}{2}}}[\boldsymbol{\tau}' + A\mathrm{tr}\boldsymbol{\tau}\boldsymbol{\delta}], \tag{84}$$

where yield surface f is considered in the form suggested in Shima and Oyane (1976), Perzyna (1986a,b) and Glema et al. (2009), κ is the work hardening-softening function (Perzyna 1986a; Nemes and Eftis 1993), $\boldsymbol{\tau}'$ represents the stress deviator, J_1 and J_2' are the first and the second invariants of Kirchhoff stress tensor and deviatoric part of the Kirchhoff stress tensor, respectively, and $A = 2(n_1 + n_2(\boldsymbol{\xi} : \boldsymbol{\xi})^{\frac{1}{2}})$.

Microdamage Nucleation and Growth

The material functions for the microdamage tensor evolution are Dornowski (1999) and Glema et al. (2009) as follows:

- for the anisotropic microdamage nucleation effects,

$$\frac{\partial h^*}{\partial \boldsymbol{\tau}} = \left\langle \frac{\partial \hat{h}}{\partial \boldsymbol{\tau}} \right\rangle \| \left\langle \frac{\partial \hat{h}}{\partial \boldsymbol{\tau}} \right\rangle \|^{-1}, \tag{85}$$

$$\hat{h} = \frac{1}{2}\mathbf{e} : \mathcal{N} : \mathbf{e}, \quad \mathcal{N} = \mathcal{I}^s, \qquad \mathcal{N}_{ijkl} = \frac{1}{2}\left(\delta_{ik}\delta_{jl} + \delta_{il}\delta_{jk}\right), \tag{86}$$

$$I_n = a_1 J_1 + a_2 \left(J_2'\right)^{\frac{1}{2}} + a_3 \left(J_3'\right)^{\frac{1}{3}}, \tag{87}$$

$$\tau_n = \left(1 - (\boldsymbol{\xi} : \boldsymbol{\xi})^{\frac{1}{2}}\right)\left(\tau_n^* - \tau_n^{**}(\vartheta) + \tau_n^{***}(\dot{\epsilon}^p)\right), \tag{88}$$

$$\tau_n^{**}(\vartheta) = \tau_n^{**}\frac{\vartheta - \vartheta_0}{\vartheta_0}, \quad \tau_n^{***}(\dot{\epsilon}^p) = \tau_n^{***}\log\frac{\dot{\epsilon}^p - \dot{\epsilon}_Q^p}{\dot{\epsilon}_Q^p}, \tag{89}$$

$$\Phi^n\left(\frac{I_n}{\tau_n} - 1\right) = \left(\frac{I_n}{\tau_n} - 1\right)^{m_n}. \tag{90}$$

  where $\bar{a}_i$ $(i = 1, 2, 3)$ are the material parameters, J_3' is the third invariant of deviatoric part of the Kirchhoff stress tensor, and τ_n is the void nucleation threshold stress.
- for the anisotropic microdamage growth effects,

$$\frac{\partial g^*}{\partial \boldsymbol{\tau}} = \left\langle \frac{\partial \hat{g}}{\partial \boldsymbol{\tau}} \right\rangle \left\| \left\langle \frac{\partial \hat{g}}{\partial \boldsymbol{\tau}} \right\rangle \right\|^{-1}, \tag{91}$$

$$\hat{g} = \frac{1}{2}\boldsymbol{\tau} : \mathcal{G} : \boldsymbol{\tau}, \quad \mathcal{G} = \mathcal{I}^s, \qquad \mathcal{G}_{ijkl} = \frac{1}{2}\left(\delta_{ik}\delta_{jl} + \delta_{il}\delta_{jk}\right), \tag{92}$$

$$I_g = b_1 J_1 + b_2 \left(J_2'\right)^{\frac{1}{2}} + b_3 \left(J_3'\right)^{\frac{1}{3}}, \tag{93}$$

$$\tau_{eq} = \tau_{eq}(\boldsymbol{\xi}, \vartheta, \in^p) = c(\vartheta)(1 - (\boldsymbol{\xi} : \boldsymbol{\xi})^{\frac{1}{2}}) \ln \frac{1}{(\boldsymbol{\xi} : \boldsymbol{\xi})^{\frac{1}{2}}}$$
$$\{2\kappa_s(\vartheta) - [\kappa_s(\vartheta) - \kappa_0(\vartheta)]F(\xi_0, \boldsymbol{\xi}, \vartheta)\}, \tag{94}$$

$$F = F(\xi_0, \boldsymbol{\xi}, \vartheta) = \left(\frac{\xi_0}{1-\xi_0}\frac{1-(\boldsymbol{\xi} : \boldsymbol{\xi})^{\frac{1}{2}}}{(\boldsymbol{\xi} : \boldsymbol{\xi})^{\frac{1}{2}}}\right)^{\frac{2}{3}\delta(\vartheta)} + \left(\frac{1-(\boldsymbol{\xi} : \boldsymbol{\xi})^{\frac{1}{2}}}{1-\xi_0}\right)^{\frac{2}{3}\delta(\vartheta)}, \tag{95}$$

$$\Phi^g\left(\frac{I_g}{\tau_{eq}} - 1\right) = \left(\frac{I_g}{\tau_{eq}} - 1\right)^{m_g}, \tag{96}$$

where τ_{eq} is the void growth threshold stress and $\bar{b}_i$ $(i = 1, 2, 3)$ are the material parameters.

- for thermal effects,

$$\mathrm{div}\mathbf{q} = \mathbf{0}, \tag{97}$$

$$\mathbf{k} = \boldsymbol{\tau}. \tag{98}$$

It should be noticed that for the microdamage mechanism, it is assumed that Dornowski (1999) and Glema et al. (2009):

- velocity of the microdamage nucleation is coaxial with the principal directions of strain state,
- velocity of the microdamage growth is coaxial with the principal directions of stress state,
- only positive (tension) principal strain (stress) induces the nucleation (growth) of the microdamage.

Concluding, for an adiabatic process, one needs to identify 36 material parameters:

- 5 parameters related to viscoplastic evolution,
- 17 parameters related to anisotropic microdamage evolution (nucleation 8, growth 5, fracture porosity 4),
- 7 parameters shared with viscoplastic and microdamage evolution,
- 4 parameters related to thermal evolution,
- 2 elasticity parameters, and
- reference density.

Numerical Examples: Modeling of Spalling Phenomena

Introductory Remarks

Herein the results discussed in Łodygowski and Sumelka (2015) are developed to map the damage nucleation effects, also. From the TTV properties like: (i) the description is invariant with respect to any diffeomorphism (covariant material model), (ii) the obtained evolution problem is well posed, (iii) sensitivity to the rate of deformation, (iv) finite elasto-viscoplastic deformations, (v) plastic non-normality, (vi) dissipation effects, (vii) thermomechanical couplings, and (viii) length scale sensitivity; the last one plays herein the exposed role. Namely, precise modeling of the geometry and intensity of the localized deformation zones, the initiation time of macrodamage, its direction, and the final fracture pattern are aimed to be modeled in details.

A series of 3D tension analyses were included to understand the influence of mesh density, element type, and/or the relaxation time on numerical results – cf. Figs. 4, 5, 6, and 7. Fully integrated elements are the most proper choice considering damage analysis where damaged elements are removed from mesh causing a considerable change in the numerical model. The reduced elements in which single integration point exists are more sensitive – herein damage of an element is controlled in a single point whereas in a full integrated element in eight points, damage criterion must be fulfilled to delete the element. Furthermore, the relaxation time controls truly the rate of viscoplastic deformation and at the same time the initiation and growth of damage. Nonetheless, because of extreme dynamic regime (the velocity of tension was $60\frac{\mathrm{m}}{\mathrm{s}}$ and the specimen length c.a. 20 mm), the shear band zones were localized close to loaded end.

In the remaining part of this section before analysis of the spalling phenomenon, initial boundary value problem (IBVP) is emphasized together with some comments on computer implementation of TTV in the Abaqus/Explicit finite element program.

Initial Boundary Value Problem

To solve the thermomechanical problem in terms of the presented TTV means to find $\phi, \boldsymbol{\upsilon}, \rho, \boldsymbol{\tau}, \xi$, and ϑ as functions of t and position $\mathbf{x}$ such that the following equations are satisfied Perzyna (1994), Łodygowski (1996) and Łodygowski and Perzyna (1997a,b):

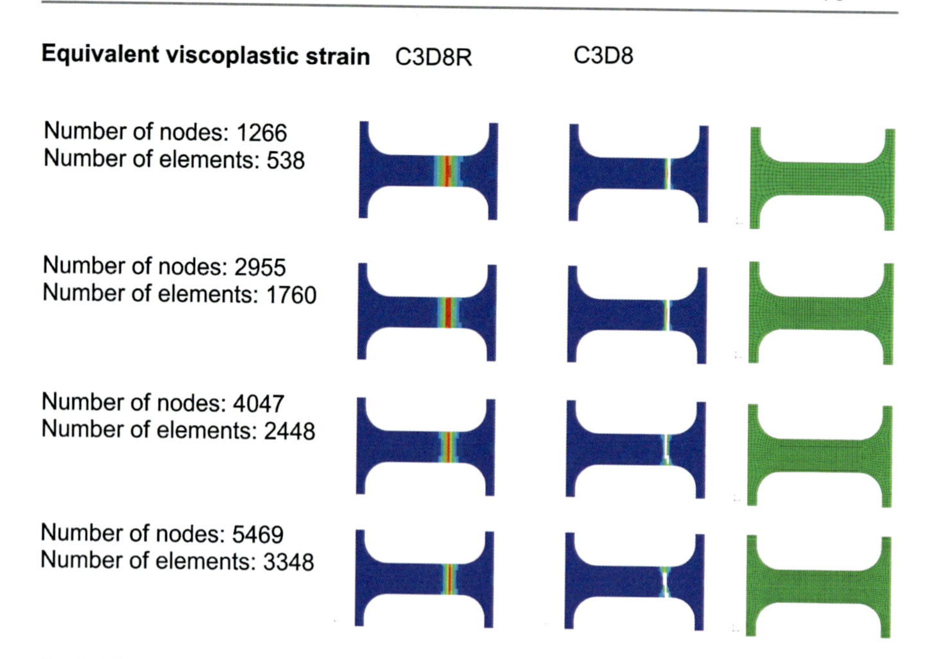

Fig. 4 The comparison of equivalent plastic strain localization zones vs. element type and mesh size

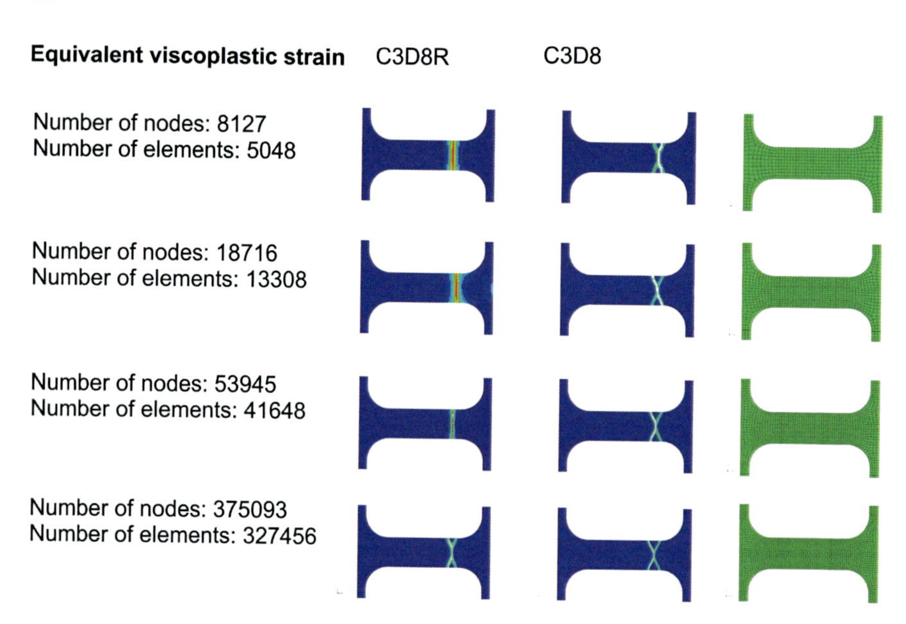

Fig. 5 The comparison of equivalent plastic strain localization zones vs. element type and mesh size

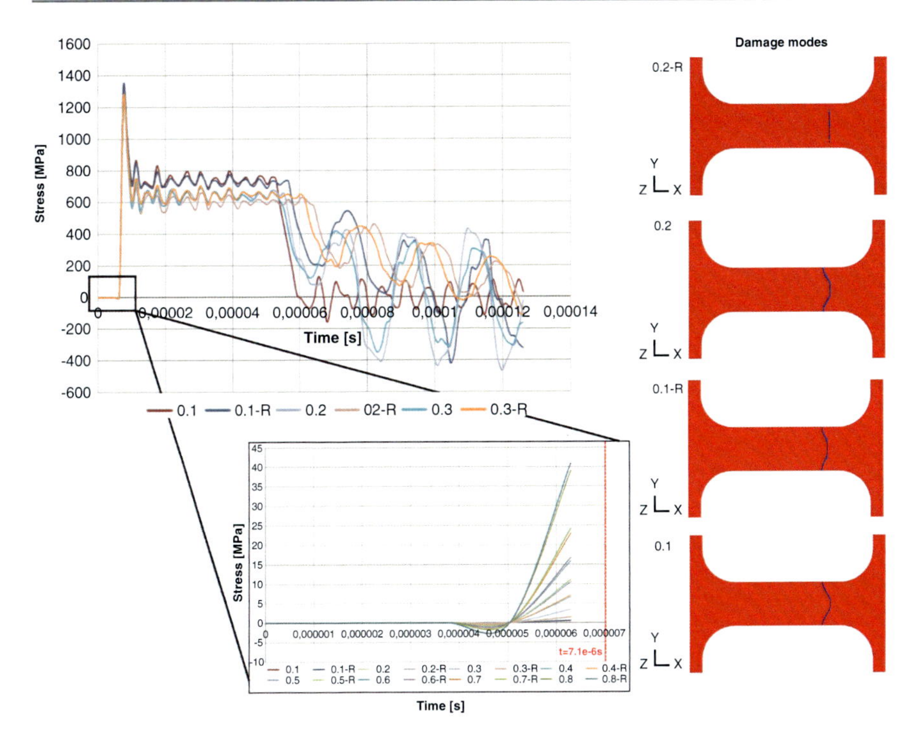

Fig. 6 The comparison of strain and time at the fixed end curves dependently on element type and mesh size

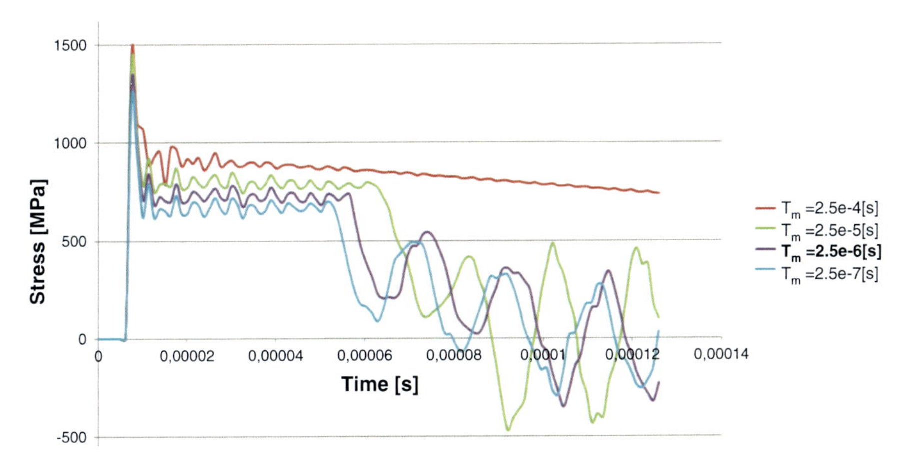

Fig. 7 The comparison of strain and time at the fixed end curves as a function of relaxation time

(i) the field equations:

$$
\begin{aligned}
\dot{\phi} &= \boldsymbol{\upsilon}, \\
\dot{\boldsymbol{\upsilon}} &= \frac{1}{\rho_{\mathrm{Ref}}}\left(\mathrm{div}\boldsymbol{\tau} + \frac{\boldsymbol{\tau}}{\rho}\cdot \mathrm{grad}\rho - \frac{\boldsymbol{\tau}}{1-(\boldsymbol{\xi}:\boldsymbol{\xi})^{\frac{1}{2}}}\mathrm{grad}(\boldsymbol{\xi}:\boldsymbol{\xi})^{\frac{1}{2}}\right), \\
\dot{\rho} &= -\rho\mathrm{div}\boldsymbol{\upsilon} + \frac{\rho}{1-(\boldsymbol{\xi}:\boldsymbol{\xi})^{\frac{1}{2}}}(\mathrm{L}_{\upsilon}\boldsymbol{\xi}:\mathrm{L}_{\upsilon}\boldsymbol{\xi})^{\frac{1}{2}}, \\
\dot{\boldsymbol{\tau}} &= \mathcal{L}^e:\mathbf{d} + 2\boldsymbol{\tau}\cdot\mathbf{d} - \mathcal{L}^{th}\dot{\vartheta} - (\mathcal{L}^e + \mathbf{g}\boldsymbol{\tau} + \boldsymbol{\tau}\mathbf{g}):\mathbf{d}^p, \\
\dot{\boldsymbol{\xi}} &= 2\boldsymbol{\xi}\cdot\mathbf{d} + \frac{\partial g^*}{\partial\boldsymbol{\tau}}\frac{1}{T_m}\left\langle\Phi^g\left[\frac{I_g}{\tau_{eq}(\boldsymbol{\xi},\vartheta,\in^p)}-1\right]\right\rangle \\
&\quad + \frac{\partial h^*}{\partial\boldsymbol{\tau}}\frac{1}{T_m}\left\langle\Phi^n\left[\frac{I_n}{\tau_n(\boldsymbol{\xi},\vartheta,\in^p)}-1\right]\right\rangle, \\
\dot{\vartheta} &= \frac{\chi^*}{\rho c_p}\boldsymbol{\tau}:\mathbf{d}^p + \frac{\chi^{**}}{\rho c_p}\mathbf{k}:\mathrm{L}_{\upsilon}\boldsymbol{\xi},
\end{aligned}
\tag{99}
$$

(ii) the boundary conditions:
   (a) displacement ϕ is prescribed on a part Γ_ϕ of $\Gamma(\mathcal{B})$ and tractions $(\boldsymbol{\tau}\cdot\mathbf{n})^a$ are prescribed on a part Γ_τ of $\Gamma(\mathcal{B})$, where $\Gamma_\phi \cap \Gamma_\tau = 0$ and $\Gamma_\phi \cup \Gamma_\tau = \Gamma(\mathcal{B})$,
   (b) heat flux $\mathbf{q}\cdot\mathbf{n} = 0$ is prescribed on $\Gamma(\mathcal{B})$,

(iii) and the initial conditions $\phi, \boldsymbol{\upsilon}, \rho, \boldsymbol{\tau}, \boldsymbol{\xi}$, and ϑ are given for each particle $\mathbf{X} \in \mathcal{B}$ at $t = 0$,

are satisfied.

In Eq. $(99)_6$, because of adiabatic regime assumption, the first two terms in temperature evolution law Eq. (51) are omitted. It should be emphasized that adiabatic condition assumption weakens the robustness of modeling due to the fact that the first term in Eq. (51) introduces in a natural way the nonlocality. Nevertheless, recall that in the viscoplasticity, the nonlocality results implicitly from the relaxation time parameter (T_m).

This evolution form is even better seen when the operator equation is involved (Łodygowski 1995). Hence, the inhomogeneous abstract Cauchy problem can be formulated in the following form:

$$
\dot{\boldsymbol{\varphi}}(t,\mathbf{x}) = \mathcal{A}(t,\mathbf{x})\cdot\boldsymbol{\varphi}(t,\mathbf{x}) + \mathbf{f}(t,\mathbf{x},\boldsymbol{\varphi}) \quad \text{for} \quad t\in(0,T] \quad \text{and} \quad \mathbf{x}\in\Omega \tag{100}
$$

with the initial condition

$$
\boldsymbol{\varphi}(0,\mathbf{x}) = \boldsymbol{\varphi}_0. \tag{101}
$$

where

$$\boldsymbol{\varphi} = \begin{bmatrix} \phi \\ \boldsymbol{\upsilon} \\ \rho \\ \boldsymbol{\tau} \\ \boldsymbol{\xi} \\ \vartheta \end{bmatrix}, \tag{102}$$

$$\mathcal{A} = \begin{bmatrix} 0 & 0 & 0 & 0 & 0 & 0 \\ 0 & 0 & \frac{\boldsymbol{\tau}\,\text{grad}}{\rho_{\text{Ref}}\rho} & \frac{\text{div}}{\rho_{\text{Ref}}} & \frac{-\boldsymbol{\tau}\,\text{grad}}{\rho_{\text{Ref}}(1-\xi)} & 0 \\ 0 & -\rho\text{div} & 0 & 0 & 0 & 0 \\ 0 & \mathcal{L}^e : \text{sym}\frac{\partial}{\partial \mathbf{x}} + 2\text{sym}(\boldsymbol{\tau}\cdot\frac{\partial}{\partial \mathbf{x}}) & 0 & 0 & 0 & 0 \\ 0 & 2\text{sym}(\boldsymbol{\xi}\cdot\frac{\partial}{\partial \mathbf{x}}) & 0 & 0 & 0 & 0 \\ 0 & 0 & 0 & 0 & 0 & 0 \end{bmatrix}, \tag{103}$$

$$\mathbf{f} = \begin{bmatrix} \boldsymbol{\upsilon} \\ \mathbf{0} \\ \frac{\rho}{1-\xi}(\Xi : \Xi)^{\frac{1}{2}} \\ -\left[\left(\frac{\chi^*}{\rho c_p}\mathcal{L}^{th}\boldsymbol{\tau} + \mathcal{L}^e + \mathbf{g}\boldsymbol{\tau} + \boldsymbol{\tau}\mathbf{g}\right) : \mathbf{p}\right]\frac{1}{T_m}\left\langle\left(\frac{f}{\kappa} - 1\right)^{m_{pl}}\right\rangle - \frac{\chi^{**}\mathcal{L}^{th}}{\rho c_p}\mathbf{k} : \Xi \\ \Xi \\ \frac{\chi^*}{\rho c_p}\boldsymbol{\tau} : \mathbf{p}\frac{1}{T_{m_{pl}}}\left\langle\left(\frac{f}{\kappa} - 1\right)^{m}\right\rangle + \frac{\chi^{**}}{\rho c_p}\mathbf{k} : \Xi \end{bmatrix}, \tag{104}$$

where $\Xi = \mathrm{L}_{\upsilon}\boldsymbol{\xi}$.

Now, it is necessary to split the discussion into two branches. One is so called the abstract Cauchy problem (ACP), which takes into consideration only the initial conditions, and the other one more significant for responsible computations initial boundary value problem (IBVP). For detailed description and discussion, please refer to the works Łodygowski et al. (1994), Łodygowski (1995, 1996), and Łodygowski and Perzyna (1997a). The well-posedness of the ACP or IBVP means that the solution for the components of vector $\dot{\boldsymbol{\varphi}}$ exists, is unique, and is stable in Hadamard or Lyapunov sense. It seems to be possible to prove the conditions for well-posedness of ACP and, using the Lax theorem, define the conditions for finite element approximation. Unfortunately, for IBVP until now it is extremely difficult in a strict mathematical way to prove the well-posedness of the system of governing equations (for 1D case it was done by Ionescu and Sofonea 1993). One cannot count for the general proof for arbitrary IBVP. Fortunately, we are not completely helpless. Using the numerical computations, one is able to discover the pathological mesh dependency (using different meshes, e.g., more dense, the typical convergence is not achieved) which is for softening behavior the implicit proof that the governing set of equations is not well posed. For this purpose, any finite element computational results have to be verified for different meshes including specific refinement in the localized areas.

Table 1 Material parameters for HSLA-65 steel

$\lambda = 121.154\,\text{GPa}$	$\mu = 80.769\,\text{GPa}$	$\chi^* = 0.8$	$\chi^{**} = 0.1$
$\kappa_s^* = 570\,\text{MPa}$	$\kappa_s^{**} = 129\,\text{MPa}$	$\kappa_0^* = 457\,\text{MPa}$	$\kappa_0^{**} = 103\,\text{MPa}$
$\beta^* = 11.0$	$\beta^{**} = 2.5$	$\delta^* = 6.0$	$\delta^{**} = 1.4$
$a_1 = 0.7$	$a_2 = 0.85$	$a_3 = 0$	$c = 0.067$
$b_1 = 0.02$	$b_2 = 0.5$	$b_3 = 0$	$m_g = 1$
$\|\|\mathrm{L}_\upsilon \boldsymbol{\xi}_c\|\| - \text{s}^{-1}$	$\xi^{F^*} = 0.36$	$\xi^{F^{**}} = -$	$m_n = 1$
$\tau_n^* = 671\,\text{MPa}$	$\tau_n^{**} = 265\,\text{MPa}$	$\tau_n^{***} = 38\,\text{MPa}$	$m_F -$
$c_p = 470\,\text{J/kgK}$	$\theta = 10^{-6}\,\text{K}^{-1}$	$\dot{\epsilon}_O^p = 10^{-5}\,\text{s}^{-1}$	$m_{pl} = 0.14$
$\rho_{\text{Ref}} = 7800\,\text{kg/m}^3$	$n_1 = 0$	$n_2 = 0.25$	$T_m = 2.5\,\mu\text{s}$

Material Parameter Identification for HSLA-65 Steel

As in Łodygowski and Sumelka (2015), the HSLA-65 steel experimentally analyzed in Nemat-Nasser and Guo (2005) is considered. Herein, precise identification of such a great number of material parameters is even more troublesome, because of damage nucleation term. Thus parameters presented in Table 1 should be thought as a compromise; hence, small fluctuations of them are possible (dependently on detailed experimental results showing competition of fundamental variables, e.g., temperature, viscoplastic strain, microdamage). For a proposition of a method to reduce the number of material parameters using soft computing methods cf. (Sumelka and Łodygowski 2013).

Computer Implementation in Abaqus/Explicit

The Abaqus/Explicit commercial finite element code has been adapted as a solver. The Abaqus/ Explicit utilizes central difference time integration rule along with the diagonal ("lumped") element mass matrices. To remove damaged elements from the mesh (elements in which for every integration point fracture porosity was reached, or equivalently load carrying capacity is zero $\kappa \to 0$) the so-called element deletion method is applied (Song et al. 2008). The model has been implemented in the software, by taking advantage of a user subroutine VUMAT, which is coupled with the Abaqus system (Abaqus 2012).

Because of Lie derivative application in TTV, special attention is paid to the stress update in VUMAT. Namely, the Green-Naghdi rate is calculated by default in Abaqus/Explicit VUMAT user subroutine; thus, the following formula was applied to enforce spatial covariance (Sumelka 2009):

$$\tilde{\boldsymbol{\tau}}\ |_{i+1} = \mathbf{R}^T\ |_{i+1}\ [\boldsymbol{\tau}\ |_i + \Delta t\ (2\boldsymbol{\tau}\ |_i \cdot \mathbf{d}\ |_i + \mathrm{L}_\upsilon \boldsymbol{\tau}\ |_i) + \boldsymbol{\Upsilon}\ |_i]\ \mathbf{R}\ |_{i+1}, \tag{105}$$

where $\boldsymbol{\Upsilon}\ |_i = -\Delta t\ (\boldsymbol{\Omega}\ |_i \cdot \boldsymbol{\tau}\ |_i - \boldsymbol{\tau}\ |_i \cdot \boldsymbol{\Omega}\ |_i)$ and $\boldsymbol{\tau}\ |_i = \mathbf{R}\ |_i\ \tilde{\boldsymbol{\tau}}\ |_i\ \mathbf{R}^T\ |_i$, where $\boldsymbol{\Omega}$ is the spin tensor. The importance of the subject of matter lays in fact that the applied objective rate can influence the results considerably – (cf. Sumelka 2013). Furthermore, some of them (e.g., Zaremba 1903) can lead to the nonphysical solutions (Dienes 1979; Lehmann 1972; Nagtegaal and de Jong 1982; Xiao et al. 1997).

Spall Fracture Phenomenon Modeling

In this section, the results presented in Łodygowski and Sumelka (2015) (as mentioned) are resolved including the additional effects which come from microdamage nucleation effects – previously in all elements initial porosity was assumed ($\xi_0 = 6 \cdot 10^{-4}$), namely,

$$\boldsymbol{\xi}_0 = \begin{bmatrix} 34.64 \cdot 10^{-5} & 0 & 0 \\ 0 & 34.64 \cdot 10^{-5} & 0 \\ 0 & 0 & 34.64 \cdot 10^{-5} \end{bmatrix}.$$

As before, spalling will be the result of flat plate impact test (Meyers and Aimone 1983; Curran 1987; Klepaczko 1990; Hanim and Klepaczko 1999; Boidin et al. 2006) (cf. Fig. 8). In this test two, plates with high velocities impact each other. As a result of the impact (caused by the flyer plate), a complete or partial separation of the material can appear in the target plate Fig. 8. This is due to tension in the target plate, induced by the interaction of two waves, incident and reflected. In spall zone, the intensive evolution of microdamage appears and changes dependently on velocity of flayer plate. Furthermore, spall test (among others, e.g., Moćko and Kowalewski 2011, 2013) is an important technique to analyze strain-stress behavior of a material.

The numerical model mapping the flat plate impact experiment is presented in Fig. 9. The dimensions of flayer plate are diameter $\phi_{\text{fla}} = 114\,\text{mm}$ and thickness $t_{\text{fla}} = 5\,\text{mm}$, while for target plate diameter is $\phi_{\text{tar}} = 114\,\text{mm}$ and thickness is $t_{\text{tar}} = 10\,\text{mm}$. Due to the microdamage anisotropy, full spatial modeling was necessary. The numerical analyses were conducted in Abaqus/Explicit code including user subroutine VUMAT, as mentioned. The C3D8R finite elements (eight-node linear brick, reduced integration element) were applied – in total c.a. $3M$ finite elements were used. The initial velocity of the flyer plate was $v_0 = 500\frac{\text{m}}{\text{s}}$, and the initial temperature 296 K was assumed.

The initial microdamage state needs detailed discussion. Because of lack of the experimental data concerning the initial microdamage distribution in the specimen, it was assumed as follows (cf. Figs. 10 and 11):

- for a random set of elements, $\xi_0 = 0$ was assumed,
- for majority of elements, $\xi_0 = 6.0e - 4$ was prescribed,
- and finally for few elements (randomly chosen), the porosity close to the fractured one was taken into account.

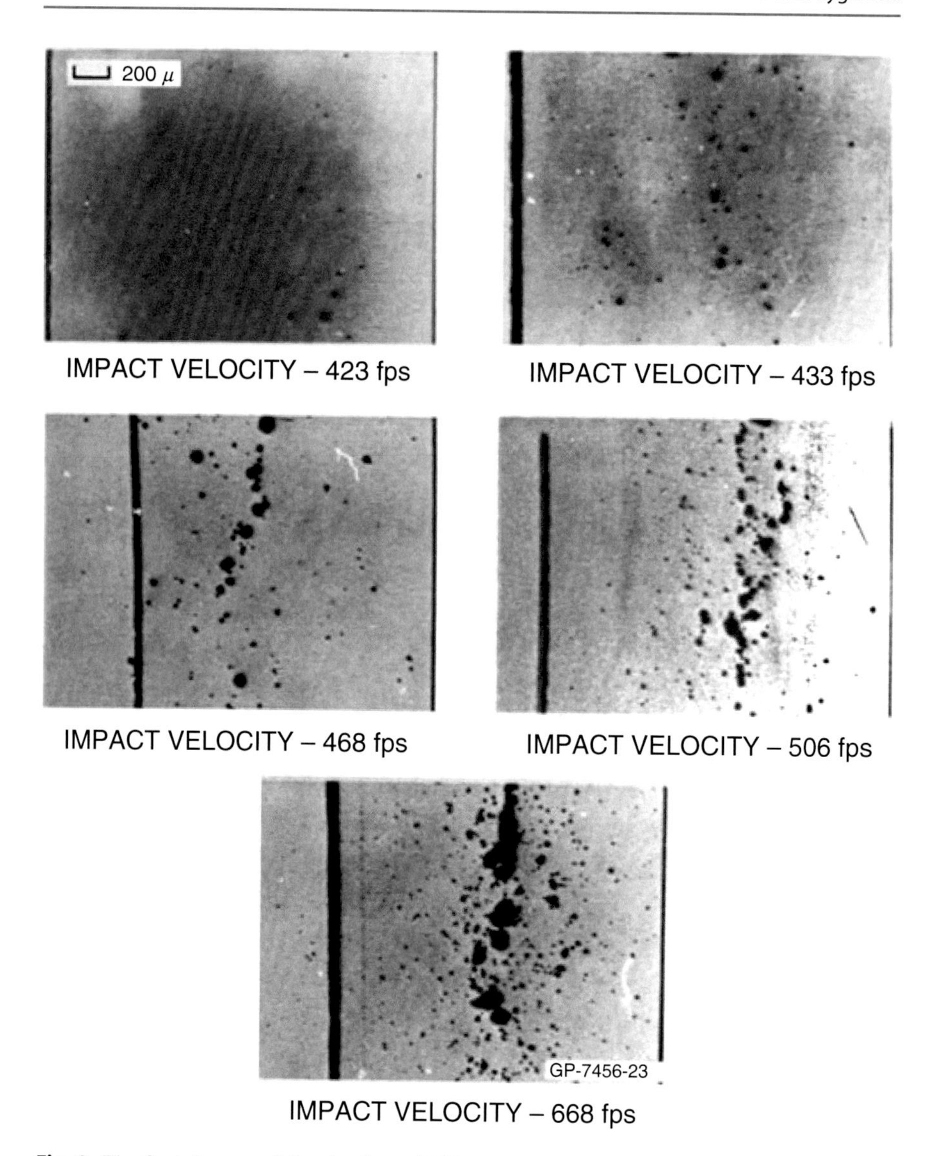

Fig. 8 The final damage of the aluminum 1145 target plate for a constant shot geometry but for different impact velocities (After Barbee et al. 1972)

It is clear that because of tensorial nature of the microdamage state variable, the components of initial microdamage tensor field were provided in every integration point. For simplicity, it was assumed that in the initial configuration the microdamage tensor is isotropic.

Material parameters for TTV model were taken as discussed in section "Material Parameter Identification for HSLA-65 Steel."

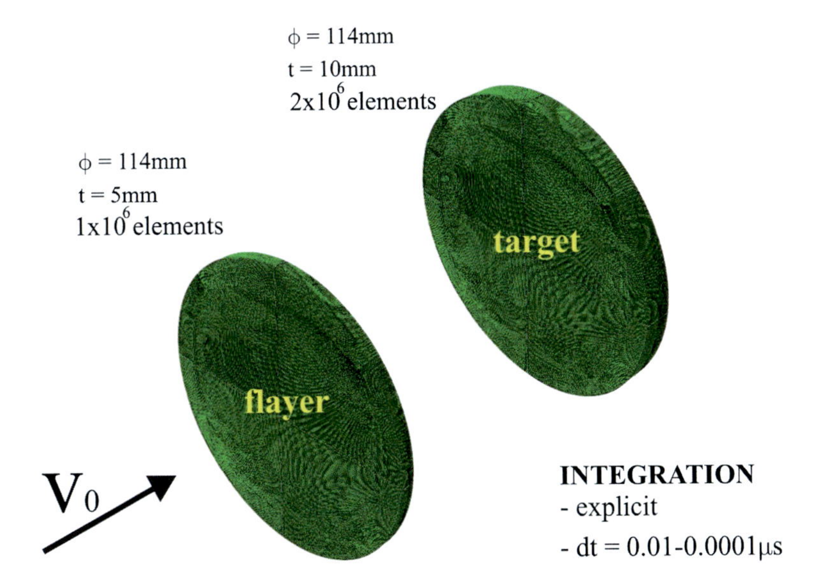

Fig. 9 Numerical model for modelling spall phenomenon

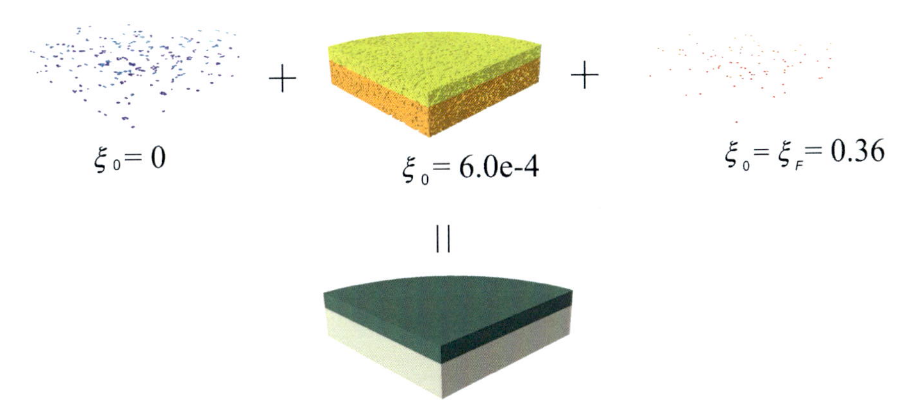

Fig. 10 Initial configuration for spalling analysis: (*left*) finite elements without initial porosity; (*middle*) finite elements with initial porosity; (*middle*) finite elements with initial porosity close to fracture porosity

In Figs. 12 and 13, the numerical results are presented. The initial velocity of flyer plate, equals $v_0 = 500\frac{\mathrm{m}}{\mathrm{s}}$, was high enough to cause spalling in the target plate (cf. Łodygowski and Sumelka 2015). The sample fields of strain rates, microdamage growth velocity, fracture porosity, and porosity during the process (the analysis of other variables like stresses, thermal stresses, strains, temperatures, etc. is not included) show almost immediate localization of a separation zone as in the experiment. In this sense, the TTV in a presented form is verified and validated.

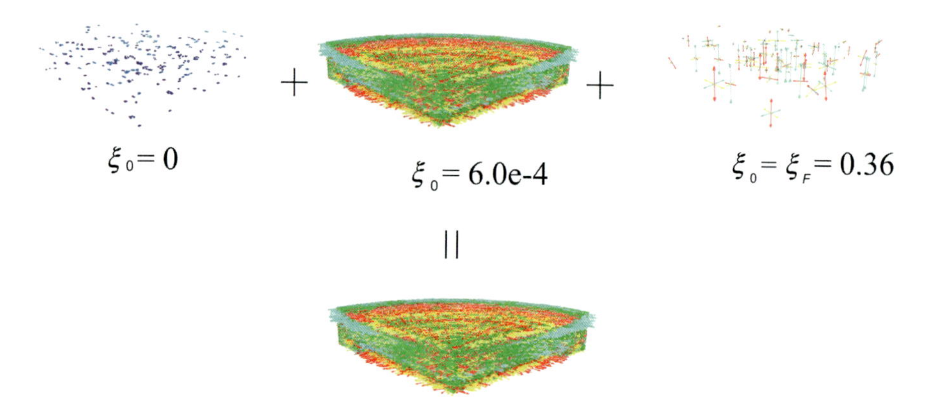

Fig. 11 Initial configuration of a microdamage directions for spalling analysis: (*left*) finite elements without initial porosity; (*middle*) finite elements with initially isotropic porosity; (*middle*) finite elements with initial porosity close to fracture porosity

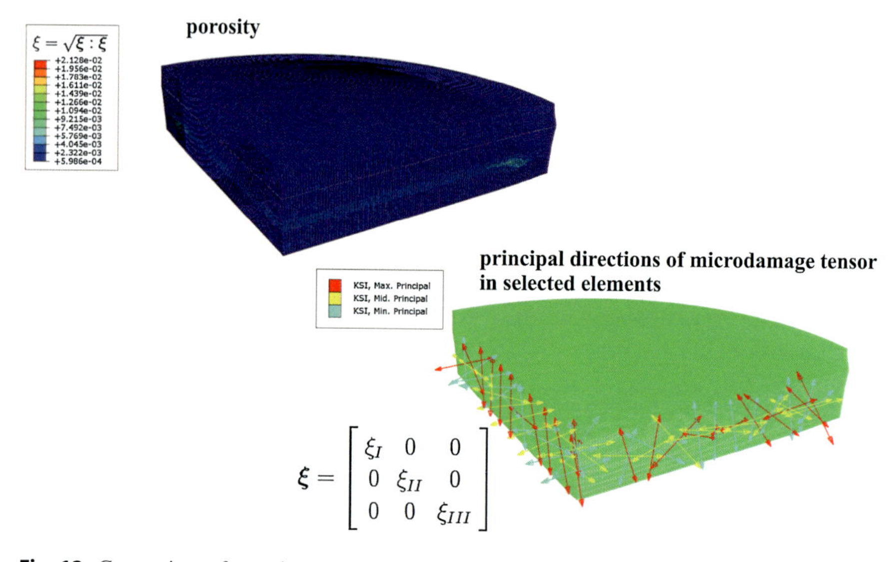

Fig. 12 Comparison of porosity (norm of the microdamage tensor) and principal directions of the microdamage tensor

Conclusions

The presented Perzyna Theory of Thermo-Viscoplasticity (TTV) is motivated by the micromechanical experimental results and ensures unique solution of the posed IBVP. On the other side, the attractiveness of TTV model is reduced due to the high number of material parameters needed for practical applications. Nonetheless, such adverse factor will always appear in the phenomenological modeling if qualitative and quantitative results are needed.

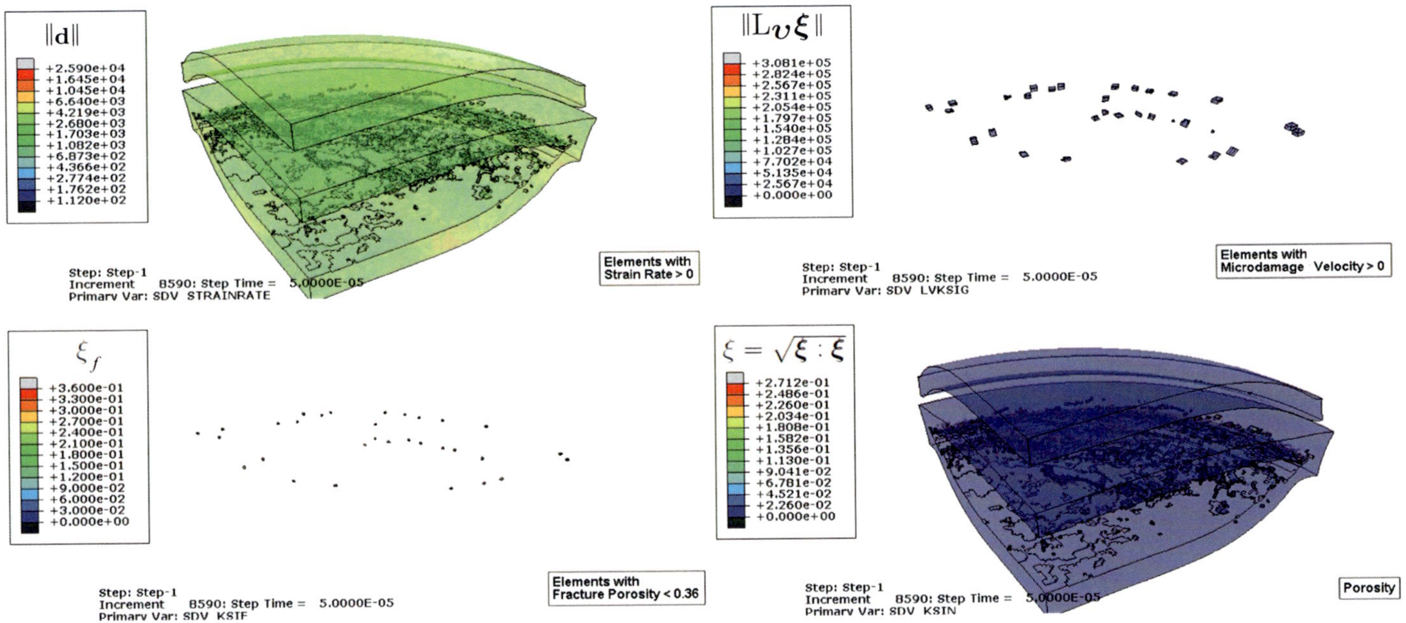

Fig. 13 The plots of strain rates, microdamage growth velocity, fracture porosity, and porosity for time process $5 \cdot 10^{-6}$ s (flayer velocity $v_0 = 550 \frac{m}{s}$)

It is commonly recognized that nonlocal models are more robust than local ones. TTV belongs to the class of implicit nonlocal models, in the sense that there is no direct measure of nonlocal action in the model but some phenomenological material parameters can be interpreted as one that control the scale effects. In TTV this parameter is the relaxation time – deduced from the analysis of the single crystal deformation. As presented, the relaxation time (T_m) can be viewed not only as a microstructural parameter (which includes the information about the characteristic size of plastic deformation) but also as a mathematical regularization parameter.

Finally, it is important to emphasize that today over 50 years after publishing the basic paper on the theory of thermo-viscoplasticity by Perzyna, its concepts are still vivid and stimulate new findings, e.g., *fractional viscoplasticity*.

References

R. Abraham, J.E. Marsden, T. Ratiu, *Manifolds, Tensor Analysis and Applications* (Springer, Berlin, 1988)

Abaqus, *Abaqus Version 6.12 Collection* (SIMULIA Worldwide Headquarters, Providence, 2012)

E.C. Aifantis, Strain gradient interpretation of size effects. Int. J. Fract. **95**, 299–314 (1999)

R.J. Asaro, Crystal plasticity. J. Appl. Mech. **50**, 921–934 (1983)

T.W. Barbee, L. Seaman, R. Crewdson, D. Curran, Dynamic fracture criteria for ductile and brittle metals. J. Mater. **7**, 393–401 (1972)

X. Boidin, P. Chevrier, J.R. Klepaczko, H. Sabar, Identification of damage mechanism and validation of a fracture model based on mesoscale approach in spalling of titanium alloy. Int. J. Solids Struct. **43**(14–15), 4029–4630 (2006)

D.R. Curran, L. Seaman, D.A. Shockey, Dynamic failure of solids. Phys. Rep. **147**(5–6), 253–388 (1987)

S. Cochran, D. Banner, Spall studies in uranium. J. Appl. Phys. **48**(7), 2729–2737 (1988)

R. de Borst, J. Pamin, Some novel developments in finite element procedures for gradient-dependent plasticity. Int. J. Numer. Methods Eng. **39**, 2477–2505 (1996)

J.K. Dienes, On the analysis of rotation and stress rate in deforming bodies. Acta Mech. **32**, 217–232 (1979)

P. Dłużewski, *Continuum Theory of Dislocations as a Theory of Constitutive Modelling of Finite Elastic-Plastic Deformations*. Volume 13 of IFTR Reports. Institute of Fundamental Technological Research – Polish Academy of Science, 1996. (D.Sc. Thesis – in Polish)

W. Dornowski, Influence of finite deformations on the growth mechanism of microvoids contained in structural metals. Arch. Mech. **51**(1), 71–86 (1999)

W. Dornowski, P. Perzyna, Analysis of the influence of various effects on cycle fatigue damage in dynamic process. Arch. Appl. Mech. **72**, 418–438 (2002)

W. Dornowski, P. Perzyna, Numerical investigation of localized fracture phenomena in inelastic solids. Found. Civil Environ. Eng. **7**, 79–116 (2006)

M.K. Duszek–Perzyna, P. Perzyna, Analysis of the influence of different effects on criteria for adiabatic shear band localization in inelastic solids, vol. 50. *Material Instabilities: Theory and Applications* (ASME, New York, 1994)

A.C. Eringen, Linear theory of nonlocal elasticity and dispersion of plane-waves. Int. J. Eng. Sci. **10**(5), 233–248 (1972a)

A.C. Eringen, Nonlocal polar elastic continua. Int. J. Eng. Sci. **10**, 1–16 (1972b)

A.C. Eringen, On differential-equations of nonlocal elasticity and solutions of screw dislocation and surface-waves. J. Appl. Phys. **54**(9), 4703–4710 (1983)

N.A. Fleck, J.W. Hutchinson, Strain gradient plasticity. Adv. Appl. Mech. **33**, 295–361 (1997)

M. Frewer, More clarity on the concept of material frame-indifference in classical continuum mechanics. Acta Mech. **202**(1–4), 213–246 (2009)

A. Glema, Analysis of Wave Nature in Plastic Strain Localization in Solids. Volume 379 of Rozprawy, Publishing House of Poznan University of Technology, 2004 (in Polish)

A. Glema, T. Łodygowski, On importance of imperfections in plastic strain localization problems in materials under impact loading. Arch. Mech. **54**(5–6), 411–423 (2002)

A. Glema, W. Kakol, T. Łodygowski, Numerical modelling of adiabatic shear band formation in a twisting test. Eng. Trans. **45**(3–4), 419–431 (1997)

A. Glema, T. Łodygowski, P. Perzyna, Interaction of deformation waves and localization phenomena in inelastic solids. Comput. Methods Appl. Mech. Eng. **183**, 123–140 (2000)

A. Glema, T. Łodygowski, P. Perzyna, Localization of plastic deformations as a result of wave interaction. Comput. Assist. Mech. Eng. Sci. **10**(1), 81–91 (2003)

A. Glema, T. Łodygowski, W. Sumelka, P. Perzyna, The numerical analysis of the intrinsic anisotropic microdamage evolution in elasto-viscoplastic solids. Int. J. Damage Mech. **18**(3), 205–231 (2009)

A. Glema, T. Lodygowski, W. Sumelka, Piotr perzyna – scientific conductor within theory of thermo-viscoplasticity. Eng. Trans. **62**(3), 193–219 (2014)

A.E. Green, R.S. Rivlin, Multipolar continuum mechanics. Arch. Ration. Mech. Anal. **17**(2), 113–147 (1964)

S. Hanim, J.R. Klepaczko, Numerical study of spalling in an aluminum alloy 7020 – T6. Int. J. Impact Eng. **22**, 649–673 (1999)

O.M. Heeres, A.S.J. Suiker, R. de Borst, A comparison between the perzyna viscoplastic model and the consistency viscoplastic model. Eur. J. Mech. A. Solids **21**(1), 1–12 (2002)

R. Hill, Aspects of invariance in solid mechanics. Adv. Appl. Mech. **18**, 1–75 (1978)

G.A. Holzapfel, *Nonlinear Solid Mechanics – A Continuum Approach for Engineering* (Chichester, England, 2000)

I.R. Ionescu, M. Sofonea, *Functional and Numerical Methods in Viscoplasticity* (Oxford University Press, Oxford/New York/Tokyo, 1993)

A.A. Kilbas, H.M. Srivastava, J.J. Trujillo, *Theory and Applications of Fractional Differential Equations* (Elsevier, Amsterdam, 2006)

J.R. Klepaczko, Dynamic crack initiation, some experimental methods and modelling, in *Crack Dynamics in Metallic Materials*, ed. by J.R. Klepaczko (Springer, Vienna, 1990), pp. 255–453

E. Kröner, On the physical reality of torque stresses in continuum mechanics. Int. J. Eng. Sci. **1**, 261–278 (1963)

Th. Lehmann, Anisotrope plastische Formänderungen. Romanian J. Tech. Sci. Appl. Mech. **17**, 1077–1086 (1972)

T. Łodygowski, On avoiding of spurious mesh sensitivity in numerical analysis of plastic strain localization. Comput. Assist. Mech. Eng. Sci. **2**, 231–248 (1995)

T. Łodygowski, Theoretical and Numerical Aspects of Plastic Strain Localization. Volume 312 of D.Sc. Thesis, Publishing House of Poznan University of Technology, 1996

T. Łodygowski, P. Perzyna, Localized fracture of inelastic polycrystalline solids under dynamic loading process. Int. J. Damage Mech. **6**, 364–407 (1997a)

T. Łodygowski, P. Perzyna, Numerical modelling of localized fracture of inelastic solids in dynamic loading process. Int. J. Numer. Methods Eng. **40**, 4137–4158 (1997b)

T. Łodygowski, W. Sumelka, Anisotropic damage for extreme dynamics, in *Handbook of Damage Mechanics Nano to Macro Scale for Materials and Structures*, ed. by G.Z. Voyiadjis (Springer, New York, 2015), pp. 1185–1220

T. Łodygowski, P. Perzyna, M. Lengnick, E. Stein, Viscoplastic numerical analysis of dynamic plastic shear localization for a ductile material. Arch. Mech. **46**(4), 541–557 (1994)

J.K. Mackenzie, The elastic constants of a solids containing spherical holes. Proc. Phys. Soc. **63B**, 2–11 (1950)

J.E. Marsden, T.J.H. Hughes, *Mathematical Foundations of Elasticity* (Prentice-Hall, New Jersey, 1983). https://www.sciencedirect.com/science/article/pii/0079642583900038

M.A. Meyers, C.T. Aimone, Dynamic fracture (Spalling) of materials, *Progress in Material Science*, **28**(1), 1–96 (1983)
R.D. Mindlin, Micro-structure in linear elasticity. Arch. Ration. Mech. Anal. **16**(1), 51–78 (1964)
R.D. Mindlin, Second gradient of strain and surface-tension in linear elasticity. Int. J. Solids Struct. **1**(4), 417–438 (1965)
R.D. Mindlin, N. Eshel, On first strain-gradient theories in linear elasticity. Int. J. Solids Struct. **4**(1), 109–124 (1968)
R.D. Mindlin, H.F. Tiersten, Effects of couple-stresses in linear elasticity. Arch. Ration. Mech. Anal. **11**(5), 415–448 (1962)
W. Moćko, Z.L. Kowalewski, Mechanical properties of a359/sicp metal matrix composites at wide range of strain rates. Appl. Mech. Mater. **82**, 166–171 (2011)
W. Moćko, Z.L. Kowalewski, Perforation test as an accuracy evaluation tool for a constitutive model of austenitic steel. Arch. Metall. Mater. **58**(4), 1105–1110 (2013)
J.V. Morán, Continuum Models for the Dynamic Behavior of 1D Nonlinear Structured Solids. Doctoral Thesis, Publishing House of the Universidad Carlos III de Madrid, 2016
T. Mura, *Micromechanics of Defects in Solids* (Kluwer Academic, Dordrecht, 1987)
J.C. Nagtegaal, J.E. de Jong, Some aspects of non-isotropic work-hardening in finite strain plasticity, in *Proceedings of the Workshop on Plasticity of Metals at Finite Strain: Theory, Experiment and Computation*, ed. by E.H. Lee, R.L. Mallet (Stanford University, 1982), pp. 65–102
S. Nemat-Nasser, W.-G. Guo, Thermomechanical response of HSLA-65 steel plates: experiments and modeling. Mech. Mater. **37**, 379–405 (2005)
J.A. Nemes, J. Eftis, Several features of a viscoplastic study of plate-impact spallation with multidimensional strain. Comput. Struct. **38**(3), 317–328 (1991)
J.A. Nemes, J. Eftis, Constitutive modelling of the dynamic fracture of smooth tensile bars. Int. J. Plast. **9**(2), 243–270 (1993)
J. Ostrowska-Maciejewska, *Mechanika ciał odkształcalnych* (PWN, Warszawa, 1994)
S.K. Park, X.-L. Gao, Variational formulation of a modified couple stress theory and its application to a simple shear problem. Zeitschrift fur angewandte Mathematik und Physik **59**, 904–917 (2008)
P. Perzyna, The constitutive equations for rate sensitive plastic materials. Q. Appl. Math. **20**, 321–332 (1963)
P. Perzyna, Fundamental problems in viscoplasticity. Adv. Appl. Mech. **9**, 243–377 (1966)
P. Perzyna, *Termodynamika materiałów niesprężystych* (PWN, Warszawa, 1978) (in Polish)
P. Perzyna, Internal state variable description of dynamic fracture of ductile solids. Int. J. Solids Struct. **22**, 797–818 (1986a)
P. Perzyna, Constitutive modelling for brittle dynamic fracture in dissipative solids. Arch. Mech. **38**, 725–738 (1986b)
P. Perzyna, Instability phenomena and adiabatic shear band localization in thermoplastic flow process. Acta Mech. **106**, 173–205 (1994)
P. Perzyna, Constitutive modelling of dissipative solids for localization and fracture, in *Localization and Fracture Phenomena in Inelastic Solids*, Chapter 3. CISM Course and Lectures, vol. 386, ed. by P. Perzyna (Springer, 1998), pp. 99–241
P. Perzyna, The thermodynamical theory of elasto-viscoplasticity. Eng. Trans. **53**, 235–316 (2005)
P. Perzyna, The thermodynamical theory of elasto-viscoplasticity accounting for microshear banding and induced anisotropy effects. Mechanics **27**(1), 25–42 (2008)
P. Perzyna, The thermodynamical theory of elasto-viscoplasticity for description of nanocrystalline metals. Eng. Trans. **58**(1–2), 15–74 (2010)
P. Perzyna, Multiscale constitutive modelling of the influence of anisotropy effects on fracture phenomena in inelastic solids. Eng. Trans. **60**(3), 225–284 (2012)
I. Podlubny, Fractional differential equations, in *Mathematics in Science and Engineering*, vol. 198 (Academin Press, USA, 1999)
D. Polyzos, D.I. Fotiadis, Derivation of Mindlin's first and second strain gradient elastic theory via simple lattice and continuum models. Int. J. Solids Struct. **49**, 470–480 (2012)

C. Rymarz, *Mechanika ośrodków* (PWN, Warszawa, 1993) (in Polish)
L. Seaman, D.R. Curran, D.A. Shockey, Computational models for ductile and brittle fracture. J. Appl. Phys. **47**(11), 4814–4826 (1976)
S. Shima, M. Oyane, Plasticity for porous solids. Int. J. Mech. Sci. **18**, 285–291 (1976)
D.A. Skolnik, H.T. Liu, H.C. Wu, L.Z. Sun, Anisotropic elastoplastic and damage behavior of sicp/al composite sheets. Int. J. Damage Mech. **17**, 247–272 (2008)
L.J. Sluys, Wave Propagation, Localization and Dispersion in Softening Solids. Doctoral Thesis, Delft University Press, Delft, 1992
J.-H. Song, H. Wang, T. Belytschko, A comparative study on finite element methods for dynamic fracture. Comput. Mech. **42**, 239–250 (2008)
W. Sumelka, The Constitutive Model of the Anisotropy Evolution for Metals with Microstructural Defects, Publishing House of Poznan University of Technology, Poznań, 2009
W. Sumelka, Role of covariance in continuum damage mechanics. ASCE J. Eng. Mech. **139**(11), 1610–1620 (2013)
W. Sumelka, Fractional viscoplasticity. Mech. Res. Commun. **56**,31–36 (2014)
W. Sumelka, A. Glema, The evolution of microvoids in elastic solids, in *17th International Conference on Computer Methods in Mechanics CMM-2007*, Łódź-Spała, 19–22 June 2007, pp. 347–348
W. Sumelka, A. Glema, Intrinsic microstructure anisotropy in elastic solids, in *GAMM 2008 79th Annual Meeting of the International Association of Applied Mathematics and Mechanics*, Bremen, 31 Mar–4 Apr 2008
W. Sumelka, T. Łodygowski, The influence of the initial microdamage anisotropy on macrodamage mode during extremely fast thermomechanical processes. Arch. Appl. Mech. **81**(12), 1973–1992 (2011)
W. Sumelka, T. Łodygowski, Reduction of the number of material parameters by ANN approximation. Comput. Mech. **52**, 287–300 (2013)
W. Sumelka, M. Nowak, Non-normality and induced plastic anisotropy under fractional plastic flow rule: a numerical study. Int. J. Numer. Anal. Methods Geomech. **40**, 651–675 (2016)
W. Sumelka, M. Nowak, On a general numerical scheme for the fractional plastic flow rule. Mech. Mater. (2017). https://doi.org/10.1016/j.mechmat.2017.02.005
Y. Sun, Y. Shen, Constitutive model of granular soils using fractional-order plastic-flow rule. Int. J. Geomech. **17**(8), 04017025 (2017)
V.E. Tarasov, General lattice model of gradient elasticity. Mod. Phys. Lett. B **28**(17), 1450054 (2014)
C. Teodosiu, F. Sidoroff, A theory of finite elastoviscoplasticity of single crystals. Int. J. Eng. Sci. **14**(2), 165–176 (1976)
R.A. Toupin, Elastic materials with couple-stresses. Arch. Ration. Mech. Anal. **11**(5), 385–414 (1963)
R.A. Toupin, Theories of elasticity with couple-stress. Arch. Ration. Mech. Anal. **17**(2), 85–112 (1964)
C. Truesdell, W. Noll, The non-linear field theories of mechanics, in *Handbuch der Physik*, vol. III/3, ed. by S. Flügge (Springer, Berlin, 1965)
G.Z. Voyiadjis, F.H. Abed, Microstructural based models for bcc and fcc metals with temperature and strain rate dependency. Mech. Mater. **37**, 355–378 (2005)
G.Z. Voyiadjis, F.H. Abed, Implicit algorithm for finite deformation hypoelastic-viscoplasticity in fcc metals. Int. J. Numer. Methods Eng. **67**, 933–959 (2006)
G.Z. Voyiadjis, R.K. Abu Al-Rub, Gradient plasticity theory with a variable length scale parameter. Int. J. Solids Struct. **42**(14), 3998–4029 (2005)
G.Z. Voyiadjis, P.I. Kattan, Evolution of fabric tensors in damage mechanics of solids with micro-cracks: part I – theory and fundamental concepts. Mech. Res. Commun. **34**, 145–154 (2007)
H. Xiao, O.T. Bruhns, A. Meyers, Hypo-elasticity model based upon the logarithmic stress rate. J. Elast. **47**, 51–68 (1997)
H. Xiao, O.T. Bruhns, A. Meyers, Strain rates and material spin. J. Elast. **52**, 1–41 (1998)

R. Xiao, H. Sun, W. Chen, A finite deformation fractional viscoplastic model for the glass transition behavior of amorphous polymers. Int. J. Non Linear Mech. **93**, 7–14 (2017)

F. Yang, A.C.M. Chong, D.C.C. Lam, P. Tong, Couple stress based strain gradient theory for elasticity. Int. J. Solids Struct. **39**, 2731–2743 (2002)

S. Zaremba, Sur une forme perfectionée de la théorie de la relaxation. Bull. Int. Acad. Sci. Cracovie 594–614 (1903)

Finite Element Analysis of Thermodynamically Consistent Strain Gradient Plasticity Theory and Applications

22

George Z. Voyiadjis and Yooseob Song

Contents

G. Z. Voyiadjis (✉) · Y. Song
Department of Civil and Environmental Engineering, Louisiana State University, Baton Rouge, LA, USA
e-mail: voyiadjis@eng.lsu.edu; ysong17@lsu.edu

G. Z. Voyiadjis (ed.), *Handbook of Nonlocal Continuum Mechanics for Materials and Structures*, https://doi.org/10.1007/978-3-319-58729-5_51

Abstract

In this chapter, a coupled thermomechanical gradient-enhanced continuum plasticity theory containing the flow rules of the grain interior and grain boundary areas is developed within the thermodynamically consistent framework. Two-dimensional finite element implementation for the proposed gradient plasticity theory is carried out to examine the micro-mechanical and thermal characteristics of small-scale metallic volumes. The proposed model is conceptually based on the dislocation interaction mechanisms and thermal activation energy. The thermodynamic conjugate microstresses are decomposed into dissipative and energetic components; correspondingly, the dissipative and energetic length scales for both the grain interior and grain boundary are incorporated in the proposed model, and an additional length scale related to the geometrically necessary dislocation-induced strengthening is also included. Not only the partial heat dissipation caused by the fast transient time but also the distribution of temperature caused by the transition from the plastic work to the heat is included into the coupled thermomechanical model by deriving a generalized heat equation. The derived constitutive framework and two-dimensional finite element model are validated through the comparison with the experimental observations conducted on microscale thin films. The proposed enhanced model is examined by solving the simple shear problem and the square plate problem to explore the thermomechanical characteristics of small-scale metallic materials. Finally, some significant conclusions are presented.

Keywords

Strain gradient plasticity · Thermomechanical coupling · Grain boundary · 2D FEM · Validation · Size effect

Introduction

The conventional continuum plasticity model is characteristically size-independent and is not capable of capturing the size effects, in particular, when the material is subjected to the nonhomogeneous (heterogeneous) plastic deformation under the fast transient time and its size ranges from a few hundreds of nanometers to a few tens of micrometers. The evidence of such a behavior is found in many micromechanical experimental observations such as nano-/micro-indentation hardness (Almasri and Voyiadjis 2010; Kim and Park 1996; Lim et al. 2017; Park et al. 1996; Voyiadjis et al. 2010; Voyiadjis and Peters 2010; Voyiadjis and Zhang 2015; Zhang and Voyiadjis 2016), nano/micro-pillars (Hwang et al. 1995), torsion of micron-dimensioned metal wires (Fleck and Hutchinson 1997; Kim et al. 1994), bending of microscale single and polycrystalline thin films (Venkatraman et al. 1994), thin beams under micron-dimensioned tension/bending (Huang et al. 2006; Parilla et al. 1993), flow strength of nanocrystalline metals (Bergeman et al. 2006), and microscale reverse extrusion of copper (Zhang et al. 2018).

It is commonly accepted that the interaction between the statistically stored dislocations (SSDs) and the geometrically necessary dislocations (GNDs) gives rise to the size effect observed in micro-/nanoscale metallic volumes. The SSDs are stored by random trapping each other and increase with the plastic strain, whereas the GNDs are stored to preserve the compatibility of diverse material components and increase with the gradient of the plastic strain. As the size of material specimen decreases, the GNDs increase the resistance to deformation by acting as the blockages to the SSDs (Fleck et al. 1994). This mechanism is called glide control because the existence of GNDs, caused by nonuniform deformation or the prescribed boundary conditions, restrains the slip of dislocation glide (Muhlhaus and Aifantis 1991; Nicola et al. 2006; Xiang and Vlassak 2006). Another mechanism for the size effect is the dislocation starvation caused by the insufficient amount of dislocations that arises from the small volumes (Bergeman et al. 2006; Giacomazzi et al. 2004; Yaghoobi and Voyiadjis 2016).

Numerous theoretical and numerical works have been carried out to explore the aforementioned phenomena based on the gradient-enhanced nonlocal plasticity theory (Fleck and Hutchinson 1997; Gudmundson 2004; Gurtin and Anand 2009; Hutchinson 2012; Ivanitsky and Kadakov 1983; Lele and Anand 2008; Song and Voyiadjis 2018; Voyiadjis and Song 2017) since the pioneering investigations of Aifantis (1984, 1987) which incorporate the gradient term in the conventional flow rule. McDowell (2010) reviewed the trends in plasticity research for metals over the 25 years prior to the publication year (2010) in terms of the multiscale kinematics, the effect of material length scale, the role of grain boundaries, and so forth.

Hutchinson (2012) classified the strain gradient version of J2 flow theories into two classes: incremental theory developed by Fleck and Hutchinson and non-incremental theory developed by Gudmundson, Gurtin, and Anand (c.f. see Fleck et al. (2014, 2015); Gudmundson (2004); Gurtin and Anand (2005, 2009); Hutchinson (2012) for details). Fleck and co-workers (Fleck et al. 2014, 2015) then pointed out that the specific phenomenon, which exhibits a significant stress jump due to infinitesimal variation in the direction of plastic strain that may occur under the nonproportional loading, arises in the non-incremental theory and also discussed its physical acceptance in their work. Fleck and co-workers (Fleck et al. 2014, 2015) have shown this phenomenon with two plane strain problems, stretch-bending problem and stretch-surface passivation problem, for nonproportional loading condition. In their work, it is noted that the dissipative higher-order microforces always cause the significant stress gap or jump under the nonproportional loading conditions. Recently, Voyiadjis and Song (2017) examined the stress jump phenomenon with the stretch-surface passivation problem using the one-dimensional finite element implementation, and an extensive parametric study is also performed in order to investigate the characteristics of the stress jump phenomenon.

Another important issue in the strain gradient plasticity (SGP) theory is the thermal effect. In the nano/micro systems, the effect of temperature gradient needs to be considered for the fast transient time. If the mean free path of phonons is approximately the medium size, heat transfer is partly ballistic rather than purely diffusive. This is caused by the small depth of the zone influenced by heat or small size of the structure and the non-equilibrium transition of thermodynamic circumstances related to reducing the response time (Tzou and Zhang 1995). Moreover, when the response time in nano-/microscale materials decreases to the thermalization time range, it results in the non-equilibrium conversion of thermodynamic states between phonons and electrons (Brorson et al. 1990; Tzou and Zhang 1995). The conventional heat equation is not capable of capturing the effect of electron-phonon interaction in this time frame; thus, the microscopic generalized heat equation has to be employed to interpret these phenomena.

Voyiadjis and co-workers (Voyiadjis and Deliktas 2009a, b; Voyiadjis and Faghihi 2012; Voyiadjis et al. 2014, 2017; Voyiadjis and Song 2017) have developed the thermodynamically consistent and coupled thermomechanical SGP models to study the characteristics of nano-/microscale metallic materials. All those works, however, have been limited to one-dimensional finite element implementation. As it is well known, there is bound to be a fundamental difference between one-dimensional finite element implementation and two-dimensional one. For example, in one-dimensional case, some special complications, e.g., the resonance between the physical scale and mesh scale, cannot be considered during the simulation. It should be noted that there is no difference between the different dimensions from the variational point of view; however, the difference exists in the description of the finite-dimensional approximation spaces. The finite element implementation in multidimension is based on the same principle of the one in the one dimension; thus the piecewise-polynomial functions of low degree with many terms are considered for accuracy. In the two-dimensional model, it is significantly more complicated since the polynomial functions have more variables and the open sets are much more varied than in the one-dimensional model. In terms of the dimensional extension, there is the simple modification from one-dimensional finite element implementation for the strain gradient plasticity model to the two-dimensional one in Voyiadjis and Song (2017). However, in that work, the effects of temperature and its gradient were not considered, but just addressed the effect of the mechanical component of the thermodynamic microforces in terms of the stress jump phenomenon. Recently, in Song and Voyiadjis (2018), the two-dimensional finite element implementation of the coupled thermomechanical strain gradient plasticity model is performed. In that work, two null boundary conditions, i.e., microscopically free and hard boundary conditions, are considered at the grain boundary to describe the dislocation movement and the plastic flow at the grain boundary areas.

It is well known that the free surface may act as the source for the defect development and its propagation toward the grain inside, whereas the grain boundaries block this dislocation movement and, consequently, give rise to the strain gradients to accommodate the geometrically necessary dislocations (Hirth and Lothe 1982). In addition, the grain boundaries can be the source of dislocation through the transmission of plastic slip to the neighboring grains (Clark et al. 1992). Besides these physical remarks, from the mathematical viewpoint, the nonstandard boundary conditions are necessary at the external boundary of a region for the well-posed governing equations in the implementation of higher-order strain gradient plasticity models. Therefore, careful modeling of the grain boundary is important in the continued development of higher-order strain gradient plasticity models.

The experimental observations on slip transmission motivate one to assume that the effect of surface/interfacial energy and the global nonlocal energy residual should be nonvanishing. Examples can be found from the in situ TEM direct observations, e.g., Lee et al. (1989), or using the geometrically necessary dislocation (GND) concept in the description of observations in bicrystallines, e.g., Sun et al. (2000), and nanoindentation tests close to the grain boundary, e.g., Soer et al. (2005). This results in a new type of boundary condition, in the context of strain gradient plasticity incorporating the interfacial energy, accounting for the surface resistance to the slip transfer due to the grain boundary misalignment (see, e.g., Aifantis and Willis (2005); Cermelli and Gurtin (2002); Fredriksson and Gudmundson (2007); Gudmundson (2004); Gurtin (2008)).

Also, in this chapter, two-dimensional numerical simulation in the context of the small deformation framework is developed incorporating the temperature- and rate-dependent flow rules for the grain interior and grain boundary, and the proposed model is validated by comparing against two sets of small-scale experiments showing the size effects. The reason for choosing these two experiments is that these experiments were also used in Voyiadjis and Song (2017) for validating the one-dimensional model, so it will be possible to compare the numerical results and validated material properties directly. The simple shear problem is solved based on the validated model in order to examine the size effect in the small-scale metallic materials. The square plate problem with the various grain boundary conditions is also solved to investigate the grain boundary effect in conjunction with the length scales.

Additionally, a generalized structure for modeling polycrystals from micro- to nano-size range is introduced Voyiadjis and Deliktas (2010). The polycrystal structure is defined in terms of the grain core, the grain boundary, and the triple junction regions with their corresponding volume fractions. Depending on the size of the crystal from micro to nano, different types of analyses are used for the respective different regions of the polycrystal. The analyses encompass local and nonlocal continuum or crystal plasticity. Depending on the physics of the region, dislocation-based inelastic deformation and/or slip/separation is used to characterize the behavior of the material. The analyses incorporate interfacial energy with grain boundary sliding and grain boundary separation. Certain state variables are appropriately decomposed into energetic and dissipative components to accurately describe the size effects.

Kinematics

In this chapter, tensors are denoted by the subscripts i, j, k, l, m, and n. The superscripts e, p, int, $ext.$, en, dis, etc. imply specific quantities such as elastic state, plastic state, internal, external, energetic, dissipative, etc., respectively. Also, the superimposed dot represents derivative with respect to time, and the indices after a comma represent the partial derivatives.

In the conventional continuum plastic theory of the isotropic solids for the small deformation assumption, the displacement gradient $u_{i,j}$ is decomposed into elastic $u^e_{i,j}$ and plastic counterparts $u^p_{i,j}$ as follows:

$$u_{i,j} = u^e_{i,j} + u^p_{i,j} \text{ where } u^p_{k,k} = 0 \tag{1}$$

where the elastic distortion $u^e_{i,j}$ and the plastic distortion $u^p_{i,j}$ indicate the recoverable stretching and rotation and the evolution of dislocations in the material structure, respectively.

For the small deformation framework in the conventional theory, the strain tensor ε_{ij} is also decomposed into the elastic and plastic elements as follows:

$$\varepsilon_{ij} = \varepsilon^e_{ij} + \varepsilon^p_{ij} = \frac{1}{2}\left(u^e_{i,j} + u^e_{j,i}\right) + \frac{1}{2}\left(u^p_{i,j} + u^p_{j,i}\right) \tag{2}$$

Here, the plastic strain is assumed to be deviatoric, $\varepsilon^p_{kk} = 0$, since the material volume is not changed by the dislocation glide-induced plastic deformation (Gurtin and Anand 2009). The plastic rotation is then given as follows:

$$W^p_{ij} = \frac{1}{2}\left(u^p_{i,j} - u^p_{j,i}\right) \tag{3}$$

In the conventional isotropic plasticity theory, since the plastic rotations (the rotation of material relative to the lattice) may be absorbed by their elastic counterparts without any influence on the field equations, they are fundamentally irrelevant to the theory. In this sense, the plasticity irrotational assumption is principally adopted in this chapter as indicated in Eq. (4) (see Gurtin et al. (2010)).

$$W^p_{ij} = 0 \tag{4}$$

Accordingly, $u^p_{i,j} = \varepsilon^p_{ij}$.

Meanwhile, the direction of plastic flow N_{ij} is given by

$$N_{ij} = \frac{\dot{\varepsilon}^p_{ij}}{\left\|\dot{\varepsilon}^p_{ij}\right\|} = \frac{\dot{\varepsilon}^p_{ij}}{\dot{e}^p} \Rightarrow \dot{\varepsilon}^p_{ij} = \dot{e}^p N_{ij} \tag{5}$$

with the accumulated plastic strain rate $\dot{e}^p$ given by

$$\dot{e}^p = \left\|\dot{\varepsilon}^p_{ij}\right\| = \sqrt{\dot{\varepsilon}^p_{ij}\dot{\varepsilon}^p_{ij}} \tag{6}$$

Correspondingly, the accumulated plastic strain e^p can be obtained by

$$e^p = \int_0^t \dot{e}^p dt \tag{7}$$

Principle of Virtual Power: Grain Interior

Recently, the thermodynamically consistent SGP theories have been developed by Voyiadjis and his co-workers to account for the characteristics of small-scale metallic material behavior based on the principle of virtual power (Voyiadjis and Song 2017; Voyiadjis et al. 2017).

The internal power $\mathcal{P}^{int}$ is presented with a combination of three energy contributions, i.e., the macro-, micro-, and thermal-energy contributions, in the arbitrary region Ω_0 as follows:

$$\mathcal{P}^{int} = \int_{\Omega_0} \left(\underbrace{\sigma_{ij}\dot{\varepsilon}^e_{ij}}_{\text{Macro}} + \underbrace{\mathcal{X}\dot{e}^p + \mathcal{Q}_i\dot{e}^p_{,i}}_{\text{Micro}} + \underbrace{\mathcal{A}\dot{\mathcal{T}} + \mathcal{B}_i\dot{\mathcal{T}}_{,i}}_{\text{Thermal}} \right) dV \tag{8}$$

where $\mathcal{X}$ and $\mathcal{Q}_i$ are the thermodynamic microforces conjugate respectively to $\dot{e}^p$ and $\dot{e}^p_{,i}$ and σ_{ij} is the Cauchy stress tensor. It is assumed in this chapter that extra contributions to the internal power exist from the temperature. $\mathcal{A}$ and $\mathcal{B}_i$ are the micromorphic scalar and vector generalized stress-like variables conjugate to the temperature rate $\dot{\mathcal{T}}$ and the first gradient of the temperature rate $\dot{\mathcal{T}}_{,i}$, respectively. These terms are introduced in a micromorphic fashion to lead the additional thermal balance equations considering the nonlocal thermal effects (Liu et al. 2017). Therefore, the purely mechanical part of the internal power is complemented by the thermal contributions that represent the thermal part of the power of work, which is the convectively performed power. Note that only the first gradient of temperature rate is considered, for the sake of simplicity. Meanwhile, one can refer to Faghihi and Voyiadjis (2014) to see the model in the absence of the temperature-related terms $\mathcal{A}$ and $\mathcal{B}_i$.

The internal power $\mathcal{P}^{int}$ for Ω_0 is equated with the external power $\mathcal{P}^{ext}$ expended by the macro- and microtractions (t_i, m) on the external surface $\partial\Omega_0$ and the body forces acting within Ω_0 as shown below:

$$\mathcal{P}^{ext} = \int_{\Omega_0} \underbrace{b_i\dot{u}_i}_{\text{Macro}} dV + \int_{\partial\Omega_0} \left(\underbrace{t_i\dot{u}_i}_{\text{Macro}} + \underbrace{m\dot{e}^p}_{\text{Micro}} + \underbrace{a\dot{\mathcal{T}}}_{\text{Thermal}} \right) dS \tag{9}$$

where b_i is the generalized external body force conjugate to the macroscopic velocity $\dot{u}_i$. Furthermore, it is assumed for the external power to have the term of a, conjugate to $\dot{\mathcal{T}}$ for the thermal effect.

By using the equation, $\mathcal{P}^{int} = \mathcal{P}^{ext}$, in conjunction with the divergence theorem and factoring out the common terms, the balance equations for the macroscopic linear momentum, nonlocal microforce, and generalized stresses $\mathcal{A}$ and $\mathcal{B}_i$ for volume Ω_0 can be obtained, respectively, as follows (see Appendix A for detailed derivations):

$$\sigma_{ij,j} + \mathscr{b}_i = 0 \tag{10}$$

$$\overline{\sigma}_{ij} = (\mathscr{x} - \mathcal{Q}_{k,k})\, N_{ij} \tag{11}$$

$$\mathcal{B}_{i,i} - \mathcal{A} = 0 \tag{12}$$

where $\overline{\sigma}_{ij}$ is the deviatoric part of σ_{ij} with the Kronecker delta δ_{ij} $\left(\overline{\sigma}_{ij} = \sigma_{ij} - \sigma_{kk}\delta_{ij}/3\right)$.

On $\partial\Omega_0$, the balance equations for the local surface traction and the nonlocal microtraction are expressed with the outward unit normal vector to $\partial\Omega_0$, n_i, respectively, as

$$\mathscr{t}_j = \sigma_{ij} n_i \tag{13}$$

$$\mathscr{m} = \mathcal{Q}_i n_i \tag{14}$$

$$\mathscr{a} = \mathcal{B}_i n_i \tag{15}$$

The first law of thermodynamics is considered here to derive the thermodynamically consistent formulation to account for the thermo-viscoplastic small-scale behavior of the metallic volumes during the fast transient time. The adiabatic viscoplastic deformation for metals is affected by the initial temperature, the loading rate, and the temperature evolution caused by the transition from plastic work to heat. The enhanced gradient theory is employed for the mechanical part of the formulation, whereas the micromorphic model is employed for the thermal part as follows (see the work of Forest and Amestoy (2008)):

$$\rho\dot{\mathcal{E}} = \sigma_{ij}\dot{\varepsilon}_{ij}^{e} + \mathscr{x}\dot{e}^{p} + \mathcal{Q}_i\dot{e}_{,i}^{p} + \mathcal{A}\dot{\mathcal{T}} + \mathcal{B}_i\dot{\mathcal{T}}_{,i} - q_{i,i} + \rho\mathcal{H}^{ext} \tag{16}$$

where ρ is the mass density, $\mathcal{E}$ is the specific internal energy, q_i is the thermal flux vector, and $\mathcal{H}^{ext}$ is the specific heat from the external source.

The second law of thermodynamics introduces a physical base to account for the GNDs' distribution in the body. The following entropy production inequality can be obtained based on the basic statement of this law that the free energy must increase at a rate less than the one at which the work is carried out, with the specific entropy $\mathscr{s}$ and the micromorphic approach by Forest (2009):

$$-\rho\dot{\mathcal{E}} + \rho\dot{\mathscr{s}}\mathcal{T} + \sigma_{ij}\dot{\varepsilon}_{ij}^{e} + \mathscr{x}\dot{e}^{p} + \mathcal{Q}_i\dot{e}_{,i}^{p} + \mathcal{A}\dot{\mathcal{T}} + \mathcal{B}_i\dot{\mathcal{T}}_{,i} - q_i\frac{\mathcal{T}_{,i}}{\mathcal{T}} \geq 0 \tag{17}$$

The entropy production vector is assumed in this chapter to be equal to the thermal flux vector divided by the temperature, as given in Coleman and Noll (1963).

Energetic and Dissipative Thermodynamic Microforces: Grain Interior

The Helmholtz free energy Ψ (per unit volume) is obtained with the entropy $\mathcal{s}$, internal energy $\mathcal{E}$, and temperature $\mathcal{T}$ describing a current state of the material as follows:

$$\Psi = \mathcal{E} - \mathcal{T}\mathcal{s} \tag{18}$$

By using Eqs. (17) and (18), the Clausius-Duhem inequality is derived as follows:

$$\sigma_{ij}\dot{\varepsilon}_{ij}^{e} + \mathcal{x}\dot{e}^{p} + \mathcal{Q}_i\dot{e}_{,i}^{p} + \mathcal{A}\dot{\mathcal{T}} + \mathcal{B}_i\dot{\mathcal{T}}_{,i} - \rho\dot{\Psi} - \rho\mathcal{s}\dot{\mathcal{T}} - q_i\frac{\mathcal{T}_{,i}}{\mathcal{T}} \geq 0 \tag{19}$$

For deriving the constitutive equations within a small-scale framework, an attempt to address the effect of the nonuniform microdefect distribution with the temperature is carried out in the present work with the functional form of the Helmholtz free energy given as:

$$\Psi = \Psi\left(\varepsilon_{ij}^{e}, e^{p}, e_{,i}^{p}, \mathcal{T}, \mathcal{T}_{,i}\right) \tag{20}$$

In the process of developing the constitutive equations, the plastic dissipation work must be nonnegative. By taking the time derivative of the Helmholtz free energy, $\dot{\Psi}$ is expressed as follows:

$$\dot{\Psi} = \frac{\partial\Psi}{\partial\varepsilon_{ij}^{e}}\dot{\varepsilon}_{ij}^{e} + \frac{\partial\Psi}{\partial e^{p}}\dot{e}^{p} + \frac{\partial\Psi}{\partial e_{,i}^{p}}\dot{e}_{,i}^{p} + \frac{\partial\Psi}{\partial\mathcal{T}}\dot{\mathcal{T}} + \frac{\partial\Psi}{\partial\mathcal{T}_{,i}}\dot{\mathcal{T}}_{,i} \tag{21}$$

Substituting Eq. (21) into Eq. (19) and factoring the common terms out give the inequality as follows:

$$\left(\sigma_{ij} - \rho\frac{\partial\Psi}{\partial\varepsilon_{ij}^{e}}\right)\dot{\varepsilon}_{ij}^{e} + \left(\mathcal{x} - \rho\frac{\partial\Psi}{\partial e^{p}}\right)\dot{e}^{p} + \left(\mathcal{Q}_i - \rho\frac{\partial\Psi}{\partial e_{,i}^{p}}\right)\dot{e}_{,i}^{p} + \left(\mathcal{A} - \rho\mathcal{s} - \rho\frac{\partial\Psi}{\partial\mathcal{T}}\right)\dot{\mathcal{T}}$$
$$+ \left(\mathcal{B}_i - \rho\frac{\partial\Psi}{\partial\mathcal{T}_{,i}}\right)\dot{\mathcal{T}}_{,i} - \frac{q_i}{\mathcal{T}}\mathcal{T}_{,i} \geq 0 \tag{22}$$

Meanwhile, the thermodynamic conjugate microforces $\mathcal{x}$, $\mathcal{Q}_i$, and $\mathcal{A}$ are assumed to be decomposed into the energetic and the dissipative elements as follows:

$$\mathcal{x} = \mathcal{x}^{en} + \mathcal{x}^{dis} \tag{23}$$

$$\mathcal{Q}_i = \mathcal{Q}_i^{en} + \mathcal{Q}_i^{dis} \tag{24}$$

$$\mathcal{A} = \mathcal{A}^{en} + \mathcal{A}^{dis} \tag{25}$$

Substituting Eqs. (23), (24), and (25) into Eq. (22) and rearranging them in accordance with the energetic and the dissipative parts result in the following expression:

$$\begin{aligned}&\left(\sigma_{ij}-\rho\frac{\partial\Psi}{\partial\varepsilon_{ij}^{e}}\right)\dot{\varepsilon}_{ij}^{e}+\left(\mathcal{X}^{en}-\rho\frac{\partial\Psi}{\partial e^{p}}\right)\dot{e}^{p}+\left(\mathcal{Q}_{i}^{en}-\rho\frac{\partial\Psi}{\partial e_{,i}^{p}}\right)\dot{e}_{,i}^{p}\\&+\left(\mathcal{A}^{en}-\rho s-\rho\frac{\partial\Psi}{\partial\mathcal{T}}\right)\dot{\mathcal{T}}+\left(\mathcal{B}_{i}-\rho\frac{\partial\Psi}{\partial\mathcal{T}_{,i}}\right)\dot{\mathcal{T}}_{,i}+\mathcal{X}^{dis}\dot{e}^{p}\\&+\mathcal{Q}_{i}^{dis}\dot{e}_{,i}^{p}+\mathcal{A}^{dis}\dot{\mathcal{T}}-\frac{q_{i}}{\mathcal{T}}\mathcal{T}_{,i}\geq0\end{aligned}\tag{26}$$

By assuming that the fifth term in Eq. (26) is strictly energetic, the energetic components of the thermodynamic microforces are defined as follows:

$$\sigma_{ij}=\rho\frac{\partial\Psi}{\partial\varepsilon_{ij}^{e}}\tag{27}$$

$$\mathcal{X}^{en}=\rho\frac{\partial\Psi}{\partial e^{p}}\tag{28}$$

$$\mathcal{Q}_{i}^{en}=\rho\frac{\partial\Psi}{\partial e_{,i}^{p}}\tag{29}$$

$$\mathcal{A}^{en}=\rho\left(s+\frac{\partial\Psi}{\partial\mathcal{T}}\right)\tag{30}$$

$$\mathcal{B}_{i}=\rho\frac{\partial\Psi}{\partial\mathcal{T}_{,i}}\tag{31}$$

The dissipation density per unit time $\mathcal{D}$ is then obtained as:

$$\mathcal{D}=\mathcal{X}^{dis}\dot{e}^{p}+\mathcal{Q}_{i}^{dis}\dot{e}_{,i}^{p}+\mathcal{A}^{dis}\dot{\mathcal{T}}-\frac{q_{i}}{\mathcal{T}}\mathcal{T}_{,i}\geq0\tag{32}$$

The dissipative counterparts of the thermodynamic microforces are obtained from the dissipation potential $\mathcal{D}\left(\dot{e}^{p},\dot{e}_{,i}^{p},\dot{\mathcal{T}},\mathcal{T}_{,i}\right)$ as follows:

$$\mathcal{X}^{dis}=\frac{\partial\mathcal{D}}{\partial\dot{e}^{p}}\tag{33}$$

$$\mathcal{Q}_{i}^{dis}=\frac{\partial\mathcal{D}}{\partial\dot{e}_{,i}^{p}}\tag{34}$$

$$\mathcal{A}^{dis}=\frac{\partial\mathcal{D}}{\partial\dot{\mathcal{T}}}\tag{35}$$

$$-\frac{q_i}{\mathcal{T}} = \frac{\partial \mathcal{D}}{\partial \mathcal{T}_{,i}} \tag{36}$$

It is necessary to define the proper formulation of Helmholtz free energy Ψ because it establishes the basis for the derivation of the constitutive relations. It has to take not only the material type such as fcc metals, bcc metals, polymer, steel alloys, concrete, etc. but also the deformation condition such as rate dependency of the material into consideration. In the current work, the Helmholtz free energy function is put forward with three main counterparts, i.e., elastic energy Ψ^e, defect energy Ψ^d, and thermal energy Ψ^{th}, as follows (Voyiadjis and Song 2017; Voyiadjis et al. 2017):

$$\Psi\left(\varepsilon_{ij}^e, e^p, e_{,i}^p, \mathcal{T}, \mathcal{T}_{,i}\right) = \Psi^e\left(\varepsilon_{ij}^e, \mathcal{T}\right) + \Psi^d\left(e^p, e_{,i}^p, \mathcal{T}\right) + \Psi^{th}\left(\mathcal{T}, \mathcal{T}_{,i}\right) \tag{37}$$

with

$$\Psi^e\left(\varepsilon_{ij}^e, \mathcal{T}\right) = \frac{1}{2\rho}\varepsilon_{ij}^e E_{ijkl}\varepsilon_{kl}^e - \frac{\alpha^{th}}{\rho}\left(\mathcal{T} - \mathcal{T}_r\right)\varepsilon_{ij}^e\delta_{ij} \tag{38}$$

$$\Psi^d\left(e^p, e_{,i}^p, \mathcal{T}\right) = \underbrace{\frac{\mathcal{H}_0}{\rho\left(r+1\right)}\left[1 - \left(\frac{\mathcal{T}}{\mathcal{T}_y}\right)^n\right]\left(e^p\right)^{r+1}}_{\Psi_1^d} + \underbrace{\frac{\sigma_0}{\rho\left(\vartheta+1\right)}\left[\ell_{en}^2\left(e_{,i}^p e_{,i}^p\right)\right]^{\frac{\vartheta+1}{2}}}_{\Psi_2^d} \tag{39}$$

$$\Psi^{th}\left(\mathcal{T}, \mathcal{T}_{,i}\right) = -\frac{1}{2}\frac{c_\varepsilon}{\mathcal{T}_r}\left(\mathcal{T} - \mathcal{T}_r\right)^2 - \frac{1}{2\rho}a\mathcal{T}_{,i}\mathcal{T}_{,i} \tag{40}$$

where α^{th} is the thermal expansion coefficient, E_{ijkl} is the elastic modulus tensor, $\mathcal{H}_0$ is the standard isotropic hardening parameter, $r\ (0 < r < 1)$ is the isotropic hardening material parameter, $\mathcal{T}_y$ and n are the thermal material parameters, $\sigma_0 > 0$ is the stress-dimensioned scaling parameter to explain the initial slip resistance, ℓ_{en} is the energetic material length scale describing the feature of the short-range interaction of the GNDs, a is the material constant for the isotropic heat conduction which accounts for the interaction of the energy carriers, ϑ is the parameter for governing the nonlinearity of the gradient-dependent defect energy, $\mathcal{T}_r > 0$ is the reference temperature, and c_ε is the specific heat capacity at the constant stress. In this chapter, $\mathcal{T}_y$ is determined by the calibration with experimental data.

The second term of the defect energy Ψ_2^d is postulated as a function of $e_{,i}^p$, and the condition $\vartheta > 0$ ensures the convexity of Ψ_2^d. (The short-range interaction of the coupling dislocations shifting on the adjacent slip planes, so-called GND core energy, is characterized by the defect energy Ψ_2^d. Ψ_2^d may be assumed to be a function of the GND density. It is assumed in this work that the plastic strain gradient is viewed as a macroscopic measure of GNDs.) In particular, by setting $\vartheta = 1$, the function Ψ_2^d turns to the quadratic formula (see Bardella (2006); Gurtin

(2004)), while the L_2-norm of $e^p_{,i}$ is used in Garroni et al. (2010) and Ohno and Okumura (2007) by setting $\vartheta = 0$. It is worth mentioning that Ψ^d_2 is independent of the temperature since it indicates the energy carried by the dislocations; thus it is energetic in nature (Lele and Anand 2008).

Since the central objective of the present work is to account for thermal variation, thermal terms are included in the free energy. The elastic and defect parts of the Helmholtz free energy functions, Ψ^e and Ψ^d, are locally convex with respect to strain-related terms at all points of the body in the equilibrium state. However, Ψ^t is a concave function of the temperature (Lubarda 2008). The second-order variation of Ψ is related to the second-order variation of $\mathcal{E}$ by using Eq. (18) and taking the virtual variations of temperature and entropy states into account such as $\partial^2\Psi/\partial\mathcal{T}^2(\delta\mathcal{T})^2 = -\partial^2\mathcal{E}/\partial\mathcal{s}^2(\delta\mathcal{s})^2$. Thus, $\mathcal{E}$ is convex with respect to entropy $\mathcal{s}$, and Ψ is concave with respect to temperature since $\partial^2\mathcal{E}/\partial\mathcal{s}^2(\delta\mathcal{s})^2 > 0 \rightarrow \partial^2\Psi/\partial\mathcal{T}^2(\delta\mathcal{T})^2 < 0$. In addition, it can be assumed that entropy increases monotonically with respect to temperature; thus $\partial\mathcal{s}/\partial\mathcal{T} = -\partial^2\Psi/\partial\mathcal{T}^2$ (Callen 1985).

One can now obtain the energetic thermodynamic forces by using the definitions in Eqs. (27), (28), (29), (30), and (31) in conjunction with Eqs. (37), (38), (39), and (40) as follows:

$$\sigma_{ij} = E_{ijkl}\varepsilon^e_{kl} - \alpha^{th}\left(\mathcal{T} - \mathcal{T}_r\right)\delta_{ij} \tag{41}$$

$$\mathcal{x}^{en} = \mathcal{H}_0\left[1 - \left(\frac{\mathcal{T}}{\mathcal{T}_y}\right)^n\right](e^p)^{\mathcal{r}} \tag{42}$$

$$\mathcal{Q}^{en}_i = \sigma_0\ell^2_{en}\left[\ell^2_{en}\left(e^p_{,k}e^p_{,k}\right)\right]^{\frac{\vartheta-1}{2}}e^p_{,i} \tag{43}$$

$$\mathcal{A}^{en} = \rho\mathcal{s} - \alpha^{th}\left(\mathcal{T} - \mathcal{T}_r\right)\varepsilon^e_{ij}\delta_{ij} - \frac{c_\varepsilon}{\mathcal{T}_r}\left(\mathcal{T} - \mathcal{T}_r\right) - \frac{\mathcal{H}_0(e^p)^{\mathcal{r}+1}}{\mathcal{r}+1}\frac{\mathcal{T}}{\mathcal{T}_y}\left(\frac{\mathcal{T}}{\mathcal{T}_y}\right)^{n-1} \tag{44}$$

$$\mathcal{B}_i = -a\mathcal{T}_{,i} \tag{45}$$

It is assumed here that the dissipation potential function is composed of two parts, the mechanical part which is dependent on the plastic strain and plastic strain gradient and the thermal counterpart which shows the purely thermal effect such as the heat conduction. In this sense, and in the context of Eq. (32), the functional form of the dissipation potential, which is dependent on $e^p_{,i}$, can be put forward as:

$$\mathcal{D} = \mathcal{D}^p\left(e^p, \varepsilon^p, \dot{e}^p, \mathcal{T}\right) + \mathcal{D}^g\left(\dot{e}^p_{,i}, \mathcal{T}\right) + \mathcal{D}^{th}\left(\dot{\mathcal{T}}, \mathcal{T}_{,i}\right) \tag{46}$$

where $\mathcal{D}^p$ and $\mathcal{D}^g$ are the mechanical parts and $\mathcal{D}^{th}$ accounts for the purely thermal effect. The functional forms of each part are given as follows:

$$\mathcal{D}^{p}\left(e^{p},\varepsilon^{p},\dot{e}^{p},\mathcal{T}\right)=\sigma_{0}\sqrt{\mathcal{H}^{2}\left(e^{p}\right)+\xi^{2}\left(\frac{\mu}{\sigma_{0}}\right)^{2}b\varepsilon^{p}}\left[1-\left(\frac{\mathcal{T}}{\mathcal{T}_{y}}\right)^{n}\right]\left(\frac{\dot{e}^{p}}{\dot{\mathcal{p}}_{1}}\right)^{m_{1}}\dot{e}^{p} \tag{47}$$

$$\mathcal{D}^{g}\left(\dot{e}^{p}_{,i},\mathcal{T}\right)=\sigma_{0}\left[1-\left(\frac{\mathcal{T}}{\mathcal{T}_{y}}\right)^{n}\right]\left(\frac{\dot{\mathcal{p}}}{\dot{\mathcal{p}}_{2}}\right)^{m_{2}}\dot{\mathcal{p}} \tag{48}$$

$$\mathcal{D}^{th}\left(\dot{\mathcal{T}},\mathcal{T}_{,i}\right)=-\frac{\varsigma}{2}\dot{\mathcal{T}}^{2}-\frac{1}{2}\frac{\mathcal{k}\left(\mathcal{T}\right)}{\mathcal{T}}\mathcal{T}_{,i}\mathcal{T}_{,i} \tag{49}$$

where ξ and b are the numerical parameter and magnitude of the Burgers vector, which are characteristically given as $0.2 \leq \xi \leq 0.5$ and $b \approx 0.3$ nm for metals, respectively. The parameter μ is the shear modulus, $\dot{\mathcal{p}}_1$ and $\dot{\mathcal{p}}_2$ are the nonnegative reference rates, m_1 and m_2 are the nonnegative strain rate sensitivity parameters, ς is the material constant characterizing the energy exchange between phonon and electron, and $\mathcal{k}\left(\mathcal{T}\right)$ is the thermal conductivity coefficient. The dimensionless function $\left(\dot{e}^{p}/\dot{\mathcal{p}}_1\right)^{m_1}$ in Eq. (47) leads to different physical effects from the term $\left(\dot{\mathcal{p}}/\dot{\mathcal{p}}_2\right)^{m_2}$ in Eq. (48), in spite of the similar forms (see Lele and Anand (2008) for more details).

The parameter $\dot{\mathcal{p}}$ is a scalar measuring the plastic strain rate gradient, which is defined by

$$\dot{\mathcal{p}} \stackrel{\text{def}}{=} \ell_{dis}\left\|\dot{e}^{p}_{,i}\right\| = \ell_{dis}\sqrt{\dot{e}^{p}_{,i}\dot{e}^{p}_{,i}} \tag{50}$$

where ℓ_{dis} is the dissipative length scale.

The dimensionless function $\mathcal{H}\left(e^{p}\right)$ is related to the strain hardening/softening behavior with $\mathcal{H}(0) = 1$. In the current work, the following form of mixed hardening function is adopted (Voce 1955):

$$\mathcal{H}\left(e^{p}\right)=1+(\chi-1)\left[1-\exp\left(-\omega e^{p}\right)\right]+\frac{\mathcal{H}_{0}}{\sigma_{0}}e^{p} \tag{51}$$

where χ and ω are the material parameters. The strain hardening, strain softening, and strain hardening/softening can be modeled based on the particular choices for these parameters.

The Nye dislocation density tensor α_{ij}, which indicates the i-component of the resultant Burgers vector related to GNDs of line vector j, is exploited here to account for the effects of plastic strain gradient (Arsenlis and Parks 1999; Fleck and Hutchinson 1997). Nonvanishing α_{ij} indicates that the GNDs exist and the net Burgers vector b_i can be obtained by using the Stokes' theorem as follows:

$$b_{i}=\oint_{\mathcal{C}}u^{p}_{i,k}dx_{k}=\int_{\mathcal{S}}\epsilon_{jkl}u^{p}_{i,lk}n_{j}dS \tag{52}$$

where ϵ_{jkl} is the permutation tensor and n_j is the unit vector normal to the surface $\mathcal{S}$ whose boundary is the curve $\mathcal{C}$. Under the assumption that the plastic flow is irrotational, the Nye dislocation density tensor α_{ij} is given by

$$\alpha_{ij} = \epsilon_{ikl} u^p_{j,lk} = \epsilon_{ikl} \varepsilon^p_{jl,k} \tag{53}$$

in which, in the work of Gurtin (2004), α_{ji} is indicated as the Burgers tensor.

With neglecting the interaction between different slip systems, the total accumulation of GNDs is calculated as follows:

$$\varepsilon^p \overset{\text{def}}{=} \left\| \alpha_{ij} \right\| = b \rho_G \tag{54}$$

where ε^p is a scalar measure of an effective plastic strain gradient and ρ_G is the total GND density. In order to present the microstructural hardening induced by GNDs, another length scale parameter, designated as the N-G length scale parameter, is defined here:

$$\ell_{N-G} \overset{\text{def}}{=} \xi^2 \left(\frac{\mu}{\sigma_0} \right)^2 b \tag{55}$$

where the N-G length scale parameter was first introduced by Nix and Gao (1998). With the definition of ℓ_{N-G}, Eq. (47) can be expressed by

$$\mathcal{D}^p \left(e^p, \varepsilon^p, \dot{e}^p, \mathcal{T} \right) = \sigma_0 \sqrt{\mathcal{H}^2 \left(e^p \right) + \ell_{N-G} \varepsilon^p} \left[1 - \left(\frac{\mathcal{T}}{\mathcal{T}_y} \right)^n \right] \left(\frac{\dot{e}^p}{\dot{p}_1} \right)^{m_1} \dot{e}^p \tag{56}$$

In the special case $\ell_{N-G} = 0$ and $\mathcal{H}\left(e^p\right) = 1$, Eq. (56) reduces to $\mathcal{D}^p = \sigma_0 \left(1 - \left(\mathcal{T}/\mathcal{T}_y\right)^n\right) \left(\dot{e}^p/\dot{p}_1\right)^{m_1} \dot{e}^p$, a form in Voyiadjis and Song (2017).

Using the dissipative potential given in Eqs. (47), (48), (49), and (56) along with Eqs. (33), (34), (35), and (36) and considering $k\left(\mathcal{T}\right)/\mathcal{T} = k_0 = constant$, the constitutive relations for the dissipative microforces are obtained as follows:

$$x^{dis} = \sigma_0 \sqrt{\mathcal{H}^2 \left(e^p \right) + \ell_{N-G} \varepsilon^p} \left[1 - \left(\frac{\mathcal{T}}{\mathcal{T}_y} \right)^n \right] \left(\frac{\dot{e}^p}{\dot{p}_1} \right)^{m_1} \tag{57}$$

$$Q_i^{dis} = \sigma_0 \ell_{dis}^2 \left(m_2 + 1 \right) \left[1 - \left(\frac{\mathcal{T}}{\mathcal{T}_y} \right)^n \right] \left(\frac{\dot{p}}{\dot{p}_2} \right)^{m_2} \frac{\dot{e}^p_{,i}}{\dot{p}} \tag{58}$$

$$\mathcal{A}^{dis} = -\varsigma \dot{\mathcal{T}} \tag{59}$$

$$\frac{q_i}{\mathcal{T}} = k_0 \mathcal{T}_{,i} \tag{60}$$

Flow Rule: Grain Interior

The flow rule in the present framework is established based on the nonlocal microforce balance, given in Eq. (11), and strengthened by thermodynamically consistent constitutive relations for both energetic and dissipative microforces. By considering the backstress, the microforce equilibrium can be expressed as follows:

$$\overline{\sigma}_{ij} - \underbrace{\left(-\mathcal{Q}^{en}_{k,k}\right) N_{ij}}_{\text{Backstress}} = \left(\varkappa - \mathcal{Q}^{dis}_{k,k}\right) N_{ij} \tag{61}$$

By substituting Eqs. (42), (43), (57), and (58) into Eq. (61), one can obtain a second-order partial differential flow rule as follows:

$$\begin{aligned} &\overline{\sigma}_{ij} - \left[-\sigma_0 \ell_{en}^2 \left\{\ell_{en}^2 \left(e^p_{,i} e^p_{,i}\right)\right\}^{\frac{\vartheta-1}{2}} e^p_{,kk}\right] N_{ij} \\ &= \left[\mathcal{H}_0 \left[1 - \left(\frac{\mathcal{T}}{\mathcal{T}_y}\right)^n\right] (e^p)^r + \sigma_0 \sqrt{\mathcal{H}^2 \left(e^p\right) + \ell_{N-G} \varepsilon^p} \right. \\ &\left[1 - \left(\frac{\mathcal{T}}{\mathcal{T}_y}\right)^n\right] \left(\frac{\dot{e}^p}{\dot{p}_1}\right)^{m_1} - \sigma_0 \ell_{dis}^2 \left(m_2 + 1\right) \\ &\left.\left[1 - \left(\frac{\mathcal{T}}{\mathcal{T}_y}\right)^n\right] \left(\frac{\dot{p}}{\dot{p}_2}\right)^{m_2} \frac{\dot{e}^p_{,kk}}{\dot{p}}\right] N_{ij} \end{aligned} \tag{62}$$

It is required to accompany the initial conditions for e^p and ε^p in the flow rule. A standard initial condition, for the behavior starting at time $t = 0$ from a virgin state, is assumed here such that

$$e^p_{t=0} = \varepsilon^p_{t=0} = 0 \tag{63}$$

Thermodynamic Derivations of the Heat Evolution Equation

Heat flow is controlled by the first law of thermodynamics, i.e., the energy conservation law, given in Eq. (16). The temperature field is governed by the heat flow generated through the inelastic dissipation and thermomechanical coupling effect. By considering the law of energy conservation given in Eq. (16) along with the dissipation potential given in Eqs. (32), (46), (47), (48), and (49) in conjunction with the equations for the energetic and dissipative components of the thermodynamic microforces given, respectively, by Eqs. (27), (28), (29), (30), and (31) and Eqs. (33), (34), (35), and (36), the relationship for the evolution of the entropy, which describes the irreversible process, can be derived as follows:

$$\rho \dot{s} \mathcal{T} = \mathcal{D} + \rho \mathcal{H}^{ext} \tag{64}$$

By using Eq. (44) for solving the rate of the entropy $\dot{s}$ and assuming the specific heat capacity at the constant volume c_0 as $c_0 = constant \cong c_\varepsilon \mathcal{T}/\mathcal{T}_r$, the temperature evolution can be obtained as follows:

$$\rho c_0 \dot{\mathcal{T}} = \underbrace{\chi^{dis}\dot{e}^p + \mathcal{Q}_i^{dis}\dot{e}_{,i}^p}_{\text{irreversible mechanical process}} \underbrace{-\alpha^{th}(\mathcal{T}-\mathcal{T}_r)\dot{\varepsilon}_{ij}^e\delta_{ij}\mathcal{T}}_{\text{elastic-thermal coupling}} - \underbrace{\dot{\mathcal{P}}\mathcal{T}}_{\text{plastic-thermal coupling}} + \underbrace{\frac{k_0}{2}\mathcal{T}_{,i}\mathcal{T}_{,i}}_{\text{heat conduction}} + \underbrace{\rho\mathcal{H}^{ext}}_{\text{heat source}} \tag{65}$$

where

$$\dot{\mathcal{P}} = \mathcal{H}_0(e^p)^r\left(\frac{\mathcal{T}}{\mathcal{T}_y}\right)^n\dot{e}^p + \frac{n\mathcal{H}_0(e^p)^{r+1}}{(r+1)\mathcal{T}_y}\left(\frac{\mathcal{T}}{\mathcal{T}_y}\right)^{n-1}\dot{\mathcal{T}} \tag{66}$$

It should be noted that the heat conduction term can be generalized to the microscale heat equation by considering the effects of the temperature gradient on the stored energy and the temperature on the energy dissipation individually in terms of the two extra material intrinsic time scale parameters (Voyiadjis and Faghihi 2012).

By substituting the constitutive equations of the dissipative microforces into Eq. (65) and assuming that the external heat source is absent, the temperature evolution is consequently obtained as:

$$\begin{aligned}
&\left[1 + \frac{n\mathcal{H}_0(e^p)^{r+1}}{\rho c_0(r+1)\mathcal{T}_y}\left(\frac{\mathcal{T}}{\mathcal{T}_y}\right)^{n-1}\right]\dot{\mathcal{T}} \\
&= \frac{1}{\rho c_0}\left[\left\{\sigma_0\sqrt{\mathcal{H}^2(e^p) + \ell_{N-G}\varepsilon^p}\left\{1-\left(\frac{\mathcal{T}}{\mathcal{T}_y}\right)^n\right\}\left(\frac{\dot{e}^p}{\dot{p}_1}\right)^{m_1}\right.\right. \\
&\left.-\mathcal{H}_0(e^p)^r\mathcal{T}\left(\frac{\mathcal{T}}{\mathcal{T}_y}\right)^n\right\}\dot{e}^p \\
&+\sigma_0\ell_{dis}^2(m_2+1)\left\{1-\left(\frac{\mathcal{T}}{\mathcal{T}_y}\right)^n\right\}\left(\frac{\dot{p}}{\dot{p}_2}\right)^{m_2}\frac{\dot{e}_{,i}^p\dot{e}_{,i}^p}{\dot{p}} \\
&\left.-\alpha^{th}(\mathcal{T}-\mathcal{T}_r)\dot{\varepsilon}_{ij}^e\delta_{ij}\mathcal{T}\right] + t_{eff}\mathcal{T}_{,ii}
\end{aligned} \tag{67}$$

where the additional term t_{eff} is defined as $t_{eff} = k_0/2\rho c_0$.

Principle of Virtual Power: Grain Boundary

One of the main goals in this study is to develop the thermodynamically consistent gradient-enhanced plasticity model for the grain boundary, which should be also consistent with the one for the grain interior. Hereafter, the superscript *GB* and the expression GB will be used to denote the specific variables at the grain boundary.

Two grains $\mathcal{G}_1$ and $\mathcal{G}_2$ separated by the grain boundary are taken into account in this chapter, and the displacement field is assumed to be continuous, i.e., $u_i^{\mathcal{G}_1} = u_i^{\mathcal{G}_2}$, across the grain boundary. The internal part of the principle of virtual power for the grain boundary is assumed to depend on the GB accumulated plastic strain rates $\dot{e}^{p^{GB\mathcal{G}_1}}$ at $S^{GB\mathcal{G}_1}$ and $\dot{e}^{p^{GB\mathcal{G}_2}}$ at $S^{GB\mathcal{G}_2}$ in the arbitrary surface S^{GB} of the grain boundary as follows:

$$\mathcal{P}^{int\,GB} = \int_{S^{GB}} \left(\mathbb{M}^{GB\mathcal{G}_1} \dot{e}^{p^{GB\mathcal{G}_1}} + \mathbb{M}^{GB\mathcal{G}_2} \dot{e}^{p^{GB\mathcal{G}_2}} \right) dS^{GB} \tag{68}$$

where the GB microscopic moment tractions $\mathbb{M}^{GB\mathcal{G}_1}$ and $\mathbb{M}^{GB\mathcal{G}_2}$ are assumed to expend power over $\dot{e}^{p^{GB\mathcal{G}_1}}$ and $\dot{e}^{p^{GB\mathcal{G}_2}}$, respectively. In addition, the GB external power $\mathcal{P}^{ext\,GB}$ is expended by the macrotractions $\sigma_{ij}^{\mathcal{G}_1}\left(-n_j^{GB}\right)$ and $\sigma_{ij}^{\mathcal{G}_2}\left(n_j^{GB}\right)$ conjugate to the macroscopic velocity $\dot{u}_i$ and the microtractions $\mathbb{Q}_k^{\mathcal{G}_1}\left(-n_k^{GB}\right)$ and $\mathbb{Q}_i^{\mathcal{G}_2}\left(n_k^{GB}\right)$ that are conjugate to $\dot{e}^{p^{GB\mathcal{G}_1}}$ and $\dot{e}^{p^{GB\mathcal{G}_2}}$, respectively, as follows:

$$\mathcal{P}^{ext\,GB} = \int_{S^{GB}} \left\{ \left(\sigma_{ij}^{\mathcal{G}_2} n_j^{GB} - \sigma_{ij}^{\mathcal{G}_1} n_j^{GB} \right) \dot{u}_i + \mathbb{Q}_k^{\mathcal{G}_2} n_k^{GB} \dot{e}^{p^{GB\mathcal{G}_2}} - \mathbb{Q}_k^{\mathcal{G}_1} n_k^{GB} \dot{e}^{p^{GB\mathcal{G}_1}} \right\} dS^{GB} \tag{69}$$

where $\mathbf{n}^{GB}$ is the unit outward normal vector of the grain boundary surface. From $\mathcal{P}^{int\,GB} = \mathcal{P}^{ext\,GB}$, the macro- and microforce balances for the grain boundary are obtained as follows:

$$\left(\sigma_{ij}^{\mathcal{G}_1} - \sigma_{ij}^{\mathcal{G}_2} \right) n_j^{GB}; \quad \mathbb{M}^{GB\mathcal{G}_1} + \mathbb{Q}_k^{\mathcal{G}_1} n_k^{GB} = 0; \quad \mathbb{M}^{GB\mathcal{G}_2} - \mathbb{Q}_k^{\mathcal{G}_2} n_k^{GB} = 0 \tag{70}$$

The first and second laws of thermodynamics are considered to construct the thermodynamically consistent gradient- and temperature-enhanced framework for the grain boundary as follows:

$$\dot{\mathcal{E}}^{GB} = \mathbb{M}^{GB} \dot{e}^{p^{GB}} + q_i^{GB} n_i^{GB} \tag{71}$$

$$\dot{\mathcal{s}}^{GB} \mathcal{T}^{GB} - q_i^{GB} n_i^{GB} \geq 0 \tag{72}$$

where $\mathcal{E}^{GB}$ is the GB surface energy density, q_i^{GB} is the GB heat flux vector, and $\mathcal{s}^{GB}$ is the surface density of the entropy of the grain boundary.

Energetic and Dissipative Thermodynamic Microforces: Grain Boundary

By using the time derivative of the equation, $\Psi^{GB} = \mathcal{E}^{GB} - \mathcal{T}^{GB}\mathcal{s}^{GB}$, and substituting it into Eqs. (71) and (72), the following Clausius-Duhem inequality for the grain boundary is obtained:

$$\mathbb{M}^{GB}\dot{e}^{p^{GB}} - \dot{\Psi}^{GB} - \mathcal{s}^{GB}\dot{\mathcal{T}}^{GB} \geq 0 \tag{73}$$

One assumes the isothermal condition for the grain boundary ($\dot{\mathcal{T}}^{GB} = 0$) and the Helmholtz free energy for the grain boundary is given by $\Psi^{GB} = \Psi^{GB}(e^{pGB})$. Substituting the time derivative of Ψ^{GB} into Eq. (73) gives the following inequality:

$$\mathbb{M}^{GB}\dot{e}^{p^{GB}} - \rho\frac{\partial\Psi^{GB}}{\partial e^{pGB}}\dot{e}^{p^{GB}} \geq 0 \tag{74}$$

The GB thermodynamic microforce quantity $\mathbb{M}^{GB}$ is further assumed to be decomposed into the energy and dissipative components such as $\mathbb{M}^{GB} = \mathbb{M}^{GB,en} + \mathbb{M}^{GB,dis}$. The components $\mathbb{M}^{GB,en}$ and $\mathbb{M}^{GB,dis}$ indicate the mechanisms for the pre- and post-slip transfer and thus involve the plastic strain at the grain boundary prior to the slip transfer $e^{pGB(pre)}$ and the one after the slip transfer $e^{pGB(post)}$, respectively, ($e^{pGB} = e^{pGB(pre)} + e^{pGB(post)}$). From Eq. (74) one obtains

$$\left(\mathbb{M}^{GB,en} - \rho\frac{\partial\Psi^{GB}}{\partial e^{pGB}}\right)\dot{e}^{p^{GB}} + \mathbb{M}^{GB,dis}\dot{e}^{p^{GB}} \geq 0 \tag{75}$$

The GB energetic microforce can be obtained as:

$$\mathbb{M}^{GB,en} = \rho\frac{\partial\Psi^{GB}}{\partial e^{pGB}} \tag{76}$$

Hence the GB dissipative microforce can then be obtained as:

$$\mathbb{M}^{GB,dis} = \frac{\partial\mathcal{D}^{GB}}{\partial\dot{e}^{p^{GB}}} \tag{77}$$

where $\mathcal{D}^{GB}$ is the nonnegative dissipation density per unit time for the grain boundary, given by $\mathcal{D}^{GB} = \mathbb{M}^{GB,dis}\dot{e}^{p^{GB}} \geq 0$. This nonnegative plastic dissipation condition can be satisfied when the GB plastic dissipation potential is a convex function of the GB accumulated plastic strain rate.

In this chapter, it is assumed by following Fredriksson and Gudmundson (2007) that the GB Helmholtz free energy per unit surface has the form of the general power law as follows:

$$\Psi^{GB}\left(e^{pGB}\right) = \frac{1}{2} G \ell_{en}^{GB} \left(e^{pGB(pre)}\right)^2 \tag{78}$$

where G is the shear modulus in the case of isotropic linear elasticity and ℓ_{en}^{GB} is the GB energetic length scale. By substituting Eq. (78) into Eq. (76), the GB energetic microforce quantity can be obtained as follows:

$$\mathbb{M}^{GB,en} = G \ell_{en}^{GB} e^{pGB(pre)} \tag{79}$$

Note that $\mathbb{M}^{GB,en}$ is independent of the plastic strain rate and temperature since this variable comes from the recoverable stored energy.

Meanwhile, two major factors might be identified affecting the energy dissipation when the dislocations move in the grain boundary area (Aifantis and Willis 2005). When dislocations encounter a grain boundary, they pile up there. Slip can transmit to the adjacent grain only when the stress field ahead of the pileup is high enough. Direct observation of the process using transmission electron microscopy (TEM) also shows that the main mechanisms for the aforementioned slip transmission are the dislocation absorption and reemission for the low-angle boundaries (Soer et al. 2005) and the dislocation nucleation in the adjacent grain for the high-angle boundaries (Ohmura et al. 2004), respectively. As soon as deformation initiates in the adjacent grain, the grain boundary begins to deform and the plastic strain on the grain boundary increases. The energy associated with the deformation of the grain boundary in this case is taken to be mainly due to the energy dissipation as dislocations move in the grain boundary region. In addition to considering the resistance force to dislocation motion being temperature and rate dependent, this energy dissipation can be taken as a linear function of GB plastic strain.

Moreover, change in the grain boundary area can also affect the energy dissipation. The macroscopic accumulated plastic strain at the grain boundary, e^{pGB}, can be related to the microscopically deformation of the grain boundary through the root-mean-square of the gradient of this deformation. In addition, the energy change after the grain boundary has yielded, i.e., onset of slip transmission, can be approximated by a quadratic function of the aforementioned displacement gradient at microscale and hence the GB plastic strain at macroscale.

Combining both aforementioned mechanisms, i.e., change in the grain boundary area and deformation of the grain boundary due to the dislocation movement, involved in the energy dissipation due to the plastic strain transfer across the grain boundary, one can postulate the following generalized expression for the GB dissipation potential:

$$\mathcal{D}^{GB} = \frac{\ell_{dis}^{GB}}{m^{GB}+1}\left(\sigma_0^{GB} + \mathcal{H}_0^{GB} e^{pGB(post)}\right)$$
$$\left(1 - \frac{\mathcal{T}^{GB}}{\mathcal{T}_y^{GB}}\right)^{n^{GB}} \left(\frac{\dot{e}^{pGB(post)}}{\dot{\wp}^{GB}}\right)^{m^{GB}} \dot{e}^{pGB(post)} \geq 0 \tag{80}$$

where ℓ_{dis}^{GB} is the GB dissipative length scale, m^{GB} and $\dot{\wp}^{GB}$ are the viscous related material parameters, σ_0^{GB} is a constant accounting for the GB yield stress, $\mathcal{H}_0^{GB}$ is the GB hardening parameter, $\mathcal{T}_y^{GB}$ is the scale-independent GB thermal parameter at the onset of yield, and n^{GB} is the GB thermal parameter. The temperature and rate dependencies of the GB energy are shown, respectively, in the terms $\left(1 - \mathcal{T}^{GB}/\mathcal{T}_y^{GB}\right)^{n^{GB}}$ and $\left(\dot{e}^{pGB(post)}/\dot{\wp}^{GB}\right)^{m^{GB}}$.

By using Eqs. (77) and (80), the GB dissipative microforce $\mathbb{M}^{GB,dis}$ can be obtained as:

$$\mathbb{M}^{GB,dis} = \ell_{dis}^{GB}\left(\sigma_0^{GB} + \mathcal{H}_0^{GB} e^{pGB(post)}\right)\left(1 - \frac{\mathcal{T}^{GB}}{\mathcal{T}_y^{GB}}\right)^{n^{GB}} \left(\frac{\dot{e}^{pGB(post)}}{\dot{\wp}^{GB}}\right)^{m^{GB}} \tag{81}$$

Therefore, the GB thermodynamic microforce $\mathbb{M}^{GB}$ can be obtained as:

$$\mathbb{M}^{GB} = G\ell_{en}^{GB} e^{pGB(pre)} + \ell_{dis}^{GB}\left(\sigma_0^{GB} + \mathcal{H}_0^{GB} e^{pGB(post)}\right)$$
$$\left(1 - \frac{\mathcal{T}^{GB}}{\mathcal{T}_y^{GB}}\right)^{n^{GB}} \left(\frac{\dot{e}^{pGB(post)}}{\dot{\wp}^{GB}}\right)^{m^{GB}} \tag{82}$$

Flow Rule: Grain Boundary

The flow rule for the grain boundary can be derived by substituting Eq. (82) into the microforce balances for the grain boundary, Eq. (70), such as:
for $S^{GB\mathcal{G}_1}$,

$$\left[\sigma_0\ell_{en}^2\left[\ell_{en}^2\left(e_{,k}^p e_{,k}^p\right)\right]^{\frac{\vartheta-1}{2}} e_{,i}^p + \sigma_0\ell_{dis}^2\left(m_2+1\right)\left[1-\left(\frac{\mathcal{T}}{\mathcal{T}_y}\right)^n\right]\left(\frac{\dot{\wp}}{\dot{\wp}_2}\right)^{m_2}\frac{\dot{e}_{,i}^p}{\dot{\wp}}\right] n_k^{GB}$$
$$+ G\ell_{en}^{GB} e^{pGB(pre)} = -\ell_{dis}^{GB}\left(\sigma_0^{GB} + \mathcal{H}_0^{GB} e^{pGB(post)}\right)$$
$$\left(1 - \frac{\mathcal{T}^{GB}}{\mathcal{T}_y^{GB}}\right)^{n^{GB}} \left(\frac{\dot{e}^{pGB(post)}}{\dot{\wp}^{GB}}\right)^{m^{GB}} \tag{83}$$

for $S^{GB\mathcal{G}_2}$,

$$\left[\sigma_0\ell_{en}^2\left[\ell_{en}^2\left(e_{,k}^pe_{,k}^p\right)\right]^{\frac{\vartheta-1}{2}}e_{,i}^p+\sigma_0\ell_{dis}^2\left(m_2+1\right)\left[1-\left(\frac{\mathcal{T}}{\mathcal{T}_y}\right)^n\right]\left(\frac{\dot{p}}{\dot{p}_2}\right)^{m_2}\frac{\dot{e}_{,i}^p}{\dot{p}}\right]n_k^{GB}$$
$$-G\ell_{en}^{GB}e^{pGB(pre)}=\ell_{dis}^{GB}\left(\sigma_0^{GB}+\mathcal{H}_0^{GB}e^{pGB(post)}\right)$$
$$\left(1-\frac{\mathcal{T}^{GB}}{\mathcal{T}_y^{GB}}\right)^{n^{GB}}\left(\frac{\dot{e}^{pGB(post)}}{\dot{p}^{GB}}\right)^{m^{GB}} \tag{84}$$

where the second term in LHS of both equations represents the backstress. Note that, in general case, the grain boundary model parameters are not identical on each side; however in this chapter, the same values are assumed to be considered for simplification.

Considering the GB flow rules as the boundary conditions of the grain interior flow rule, Eq. (62), results in a yield condition accounting for the temperature- and rate-dependent barrier effect of grain boundaries on the plastic slip and consequently the influence on the GND evolution in the grain interior.

Finite Element Formulation of the Proposed SGP Model

A two-dimensional finite element model for the proposed SGP model is developed to account for the size-dependent response for microscopic structures. The boundary value problem consists of solving the flow rules for the grain interior/boundary given in sections "Flow Rule: Grain Interior" and "Flow Rule: Grain Boundary" in conjunction with the constitutive equations given in sections "Energetic and Dissipative Thermodynamic Microforces: Grain Interior" and "Energetic and Dissipative Thermodynamic Microforces: Grain Boundary" subjected to the prescribed displacement conditions $u_t^{\dagger}$ on part of the boundary $\partial\Omega_0'$ and traction free condition on the remaining boundary part of the body. The microscopic and macroscopic force balances can then be described in the global weak form after utilizing the principle of virtual power and applying the corresponding boundary conditions, i.e., arbitrary virtual displacement fields $\delta u=0$ on $\partial\Omega_0'$ and arbitrary virtual plastic strain fields $\delta e^p=0$ on $\partial\Omega_0''$ as follows:

$$\int_{\Omega_0}\left(\sigma_{ij}\delta u_{i,j}\right)dV=0 \tag{85}$$

$$\int_{\Omega_0}\left[\left(\mathcal{X}-\overline{\sigma}\right)\delta e^p+\mathcal{Q}_i\delta e_{,i}^p\right]dV=0 \tag{86}$$

where $\overline{\sigma}$ is the resolved shear stress defined by

$$\overline{\sigma} = \sigma_{ij} N_{ij} = \overline{\sigma}_{ij} N_{ij} \tag{87}$$

The UEL subroutine in the finite element software ABAQUS/Standard (2012) is built in this chapter for numerically solving the global weak forms of macroscopic and microscopic force balances, Eqs. (85) and (86), respectively. In this finite element formulation, the plastic strain field e^p and the displacement field u_i are discretized independently, and both of the fields are taken as fundamental unknown nodal degrees of freedom. In this regard, the displacement and corresponding strain field, ε_{ij}, and the plastic strain and corresponding plastic strain gradient field $e^p_{,i}$ are obtained by using the interpolation as follows:

$$u_i = \sum_{\eta=1}^{n_u} \mathcal{U}^{\eta}_{u_i} \mathbb{N}^{\eta} \tag{88}$$

$$\varepsilon_{ij} = \frac{1}{2}\left(\frac{\partial u_i}{\partial x_j} + \frac{\partial u_j}{\partial x_i}\right) = \frac{1}{2}\sum_{\eta=1}^{n_u}\left(\mathcal{U}^{\eta}_{u_i}\frac{\partial \mathbb{N}^{\eta}}{\partial x_j} + \mathcal{U}^{\eta}_{u_j}\frac{\partial \mathbb{N}^{\eta}}{\partial x_i}\right) \tag{89}$$

$$e^p = \sum_{\eta=1}^{n_{e^p}} \mathcal{E}^{\eta}_{e^p} \mathbb{N}^{\eta} \tag{90}$$

$$e^p_{,j} = \frac{\partial e^p}{\partial x_j} = \sum_{\eta=1}^{n_{e^p}} \mathcal{E}^{\eta}_{e^p} \frac{\partial \mathbb{N}^{\eta}}{\partial x_j} \tag{91}$$

where $\mathbb{N}^{\eta}_{u}$ and $\mathbb{N}^{\eta}_{e^p}$ denote the interpolation functions and $\mathcal{U}^{\eta}_{u_i}$ and $\mathcal{E}^{\eta}_{e^p}$ denote the nodal values of the plastic strains and displacements at node η, respectively. The terms n_u and n_{e^p} represent the number of nodes per single element for displacement and plastic strain, respectively. Since a two-dimensional quadratic 9-node element is used in this chapter, n_u and n_{e^p} are set up as nine. It should be noted that n_u and n_{e^p} do not necessarily have to be the same as each other in the present finite element implementation, even though both the displacement and plastic strain fields are calculated by using the standard isoparametric interpolation functions.

The body is approximated using finite elements, $\Omega = \cup\, \Omega_{el}$. By substituting Eqs. (88), (89), (90), and (91) into Eqs. (85) and (86), the nodal residuals for the displacement $\mathbb{R}_{u_i}$ and the plastic strain $\mathbb{R}_{e^p}$ for each element Ω_{el} can be obtained as:

$$(\mathbb{R}_{u_i})_{\eta} = -\int_{\Omega_{el}} \left(\sigma_{ij}\frac{\partial \mathbb{N}^{\eta}_{u}}{\partial x_j}\right) dV \tag{92}$$

$$(\mathbb{R}_{e^p})_{\eta} = -\int_{\Omega_{el}} \left[(\mathcal{X} - \overline{\sigma})\, \mathbb{N}^{\eta}_{e^p} + \mathcal{Q}_i \frac{\partial \mathbb{N}^{\eta}_{e^p}}{\partial x_i}\right] dV \tag{93}$$

The system of linear equations, $(\mathbb{R}_{u_i})_\eta = 0$ and $(\mathbb{R}_{e^p})_\eta = 0$, are solved using ABAQUS/Standard (2012) based on the Newton-Raphson iterative method. Occasionally, the modified Newton-Raphson method, referred to as quasi-Newton-Raphson method, is employed in the case that the numerical solution suffers a divergence during the initial increment immediately after an abrupt change in loading. In the quasi-Newton-Raphson method, a specific correction factor, which is less than one, is multiplied by one portion of the stiffness matrix. By using this method, a divergence problem can be overcome; however, convergence is expected to be slow because of the expensive computational cost. The Taylor expansion of the residuals with regard to the current nodal values can be expressed by assuming the nodal displacement and the plastic strain in iteration ζ as $\mathcal{U}^\zeta_{u_i}$ and $\mathcal{E}^\zeta_{e^p}$ are, respectively, as follows:

$$\left(\mathbb{R}_{u_i}\big|_{\mathcal{U}^{\zeta+1}_{u_i},\mathcal{E}^{\zeta+1}_{e^p}}\right)_\eta = \left(\mathbb{R}_{u_i}\big|_{\mathcal{U}^{\zeta}_{u_i},\mathcal{E}^{\zeta}_{e^p}}\right)_\eta + \left(\frac{\partial \mathbb{R}_{u_i}}{\partial \mathcal{U}^\eta_{u_k}}\bigg|_{\mathcal{U}^\zeta_{u_i}}\right)\Delta\mathcal{U}^\eta_{u_i} + \left(\frac{\partial \mathbb{R}_{u_i}}{\partial \mathcal{E}^\eta_{e^p}}\bigg|_{\mathcal{E}^\zeta_{e^p}}\right)\Delta\mathcal{E}^\eta_{e^p} + \mathcal{O}\left(\left(\Delta\mathcal{U}^\eta_{u_i}\right)^2, \left(\Delta\mathcal{E}^\eta_{e^p}\right)^2\right) \tag{94}$$

$$\left(\mathbb{R}_{e^p}\big|_{\mathcal{U}^{\zeta+1}_{u_i},\mathcal{E}^{\zeta+1}_{e^p}}\right)_\eta = \left(\mathbb{R}_{e^p}\big|_{\mathcal{U}^{\zeta}_{u_i},\mathcal{E}^{\zeta}_{e^p}}\right)_\eta + \left(\frac{\partial \mathbb{R}_{e^p}}{\partial \mathcal{U}^\eta_{u_k}}\bigg|_{\mathcal{U}^\zeta_{u_i}}\right)\Delta\mathcal{U}^\eta_{u} + \left(\frac{\partial \mathbb{R}_{e^p}}{\partial \mathcal{E}^\eta_{e^p}}\bigg|_{\mathcal{E}^\zeta_{e^p}}\right)\Delta\mathcal{E}^\eta_{e^p} + \mathcal{O}\left(\left(\Delta\mathcal{U}^\eta_{u_i}\right)^2, \left(\Delta\mathcal{E}^\eta_{e^p}\right)^2\right) \tag{95}$$

where $\Delta\mathcal{U}^\eta_{u_i} = \left(\mathcal{U}^{\zeta+1}_{u_i}\right)_\eta - \left(\mathcal{U}^{\zeta}_{u_i}\right)_\eta$, $\Delta\mathcal{E}^\eta_{e^p} = \left(\mathcal{E}^{\zeta+1}_{e^p}\right)_\eta - \left(\mathcal{E}^{\zeta}_{e^p}\right)_\eta$, and $\mathcal{O}\left(\left(\Delta\mathcal{U}^\eta_{u_i}\right)^2, \left(\Delta\mathcal{E}^\eta_{e^p}\right)^2\right)$ is the big O notation to represent the terms of higher order than the second degree. These residuals are repetitively calculated at every time step, and the calculated numerical results are updated during the whole iterations. The increments in the nodal displacement and plastic strains can be obtained by computing the system of linear equations shown in Eq. (96):

$$\underbrace{\begin{bmatrix} K^{\Omega_{el}}_{u_i u_k} & K^{\Omega_{el}}_{u_i e^p} \\ K^{\Omega_{el}}_{e^p u_k} & K^{\Omega_{el}}_{e^p e^p} \end{bmatrix}}_{\mathbf{K}^{\Omega_{el}}} \begin{Bmatrix} \Delta\mathcal{U}^\eta_{u_k} \\ \Delta\mathcal{E}^\eta_{e^p} \end{Bmatrix} = \begin{Bmatrix} \left(\mathbb{R}_{u_i}\big|_{\mathcal{U}^{\zeta}_{u_i},\mathcal{E}^{\zeta}_{e^p}}\right)_\eta \\ \left(\mathbb{R}_{e^p}\big|_{\mathcal{U}^{\zeta}_{u_i},\mathcal{E}^{\zeta}_{e^p}}\right)_\eta \end{Bmatrix} \tag{96}$$

where $\mathbf{K}^{\Omega_{el}}$ is the Jacobian (stiffness) matrix.

From the functional forms of the thermodynamic microforces defined in sections "Energetic and Dissipative Thermodynamic Microforces: Grain Interior" and "Energetic and Dissipative Thermodynamic Microforces: Grain Boundary" and Eqs. (94) and (95) along with the Eqs. (88), (89), (90), and (91) at the end of each time step, each component of the Jacobian matrix can be obtained, respectively, as follows:

$$K_{u_i u_k}^{\Omega_{el}} = -\frac{\partial \mathbb{R}_{u_i}}{\partial \mathcal{U}_{u_k}^{\eta}}\bigg|_{\mathcal{U}_{u_i}^{\zeta}} = \int_{\Omega_{el}} \left(E_{ijkl} \frac{\partial \mathrm{N}_u^{\eta}}{\partial x_j} \frac{\partial \mathrm{N}_u^{\eta}}{\partial x_l} \right) dV \tag{97}$$

$$K_{u_i e^p}^{\Omega_{el}} = -\frac{\partial \mathbb{R}_{u_i}}{\partial \mathcal{E}_{e^p}^{\eta}}\bigg|_{\mathcal{E}_{e^p}^{\zeta}} = \int_{\Omega_{el}} \left(E_{ijkl} \frac{e^p}{\varepsilon_{kl}} \frac{\partial \mathrm{N}_u^{\eta}}{\partial x_j} \mathrm{N}_{e^p}^{\eta} \right) dV \tag{98}$$

$$K_{e^p u_k}^{\Omega_{el}} = -\frac{\partial \mathbb{R}_{e^p}}{\partial \mathcal{U}_{u_k}^{\eta}}\bigg|_{\mathcal{U}_{u_i}^{\zeta}} = \int_{\Omega_{el}} \left(E_{ijkl} \frac{e^p}{\varepsilon_{ij}} \mathrm{N}_u^{\eta} \frac{\partial \mathrm{N}_{e^p}^{\eta}}{\partial x_l} \right) dV \tag{99}$$

$$\begin{aligned}
K_{e^p e^p}^{\Omega_{el}} &= -\frac{\partial \mathbb{R}_{e^p}}{\partial \mathcal{E}_{e^p}^{\eta}}\bigg|_{\mathcal{E}_{e^p}^{\zeta}} \\
&= \int_{\Omega_{el}} \Bigg[\Bigg(r\mathcal{H}_0 (e^p)^r + \mathcal{H}_0 \left(\frac{\dot{e}^p}{\dot{p}_1} \right)^{m_1} + \sigma_0 \sqrt{\mathcal{H}^2 (e^p) + \ell_{N-G} \varepsilon^p} \frac{m_1 \dot{e}^{p^{m_1 - 1}}}{\dot{p}_1^{m_1} \Delta t} \Bigg) \\
&\left[1 - \left(\frac{\mathcal{T}}{\mathcal{T}_y} \right)^n \right] \mathrm{N}_{e^p}^{\eta} \mathrm{N}_{e^p}^{\eta} + \sigma_0 \ell_{en}^2 \left[\ell_{en}^2 \left(e_{,k}^p e_{,k}^p \right) \right]^{\frac{\vartheta - 1}{2}} \frac{\partial \mathrm{N}_{e^p}^{\eta}}{\partial x_j} \frac{\partial \mathrm{N}_{e^p}^{\eta}}{\partial x_j} + \sigma_0 \ell_{dis}^4 \left(m_2^2 - 1 \right) \\
&\left[1 - \left(\frac{\mathcal{T}}{\mathcal{T}_y} \right)^n \right] \frac{\dot{p}^{m_2 - 3}}{\dot{p}_2^{m_2} \Delta t} \left(\dot{e}_{,j}^p \frac{\partial \mathrm{N}_{e^p}^{\eta}}{\partial x_j} \right) \left(\dot{e}_{,j}^p \frac{\partial \mathrm{N}_{e^p}^{\eta}}{\partial x_j} \right) \\
&+ \sigma_0 \ell_{dis}^2 (m_2 + 1) \left[1 - \left(\frac{\mathcal{T}}{\mathcal{T}_y} \right)^n \right] \frac{\dot{p}^{m_2 - 1}}{\dot{p}_2^{m_2} \Delta t} \frac{\partial \mathrm{N}_{e^p}^{\eta}}{\partial x_j} \frac{\partial \mathrm{N}_{e^p}^{\eta}}{\partial x_j} \Bigg] dV \\
&- \Bigg[G \ell_{en}^{GB} + \ell_{dis}^{GB} \frac{\left(\sigma_0^{GB} + \mathcal{H}_0^{GB} e^{pGB(post)} \right)}{\left(\Delta t \dot{p}^{GB} \right)^{m^{GB}}} \\
&\left(1 - \frac{\mathcal{T}^{GB}}{\mathcal{T}_y^{GB}} \right)^{n^{GB}} \dot{e}^{p^{GB(post)} m^{GB} - 1} \Bigg] \mathrm{N}_{e^p}^{\eta} \mathrm{N}_{e^p}^{\eta}
\end{aligned} \tag{100}$$

where Δt is a time step. The grain boundary terms in Eq. (100) are only applied for nodes on the grain boundary area.

Validation of the Proposed SGP Model

In this section, the proposed SGP theory and corresponding finite element implementation are validated through the comparison against the experimental measurements from two sets of size effect tests: aluminum thin film experiments by Haque and Saif (2003) and nickel thin film experiments by Han et al. (2008). The material parameters for two different metals, aluminum and nickel, are also calibrated by using the experimental data. The parameters σ_0 and $\mathcal{H}_0$ are determined by extrapolating the experimental data, and the material length scales are determined based on the suggestion by Anand et al. (2005). In their work, an initial assumption for the material length scales is suggested by matching the numerical results from the proposed flow rule to the yield strength and backstress experimental measurements under the assumption of $\ell_{en} = 0$ and $\ell_{dis} = 0$ at each case, respectively. The rest of the material parameters come from the literature (Han et al. 2008; Haque and Saif 2003; Voyiadjis and Song 2017). The grain boundary flow rule is not considered in sections "Validation of the Proposed SGP Model," "Microfree Grain Boundary," and "Microhard Grain Boundary."

Haque and Saif (2003) developed the micro-electromechanical system (MEMS)-based testing skill for the nanoscale aluminum (Al) thin films under the uniaxial tensile loading to investigate the strain gradient effect in 100 nm, 150 nm, 200 nm, and 485 nm-thick specimens, which have the average grain sizes of 50 nm, 65 nm, 80 nm, and 212 nm, respectively. The specimens with 99.99% pure sputter-deposited freestanding Al thin films are 10 μm wide and 275 μm long. All experiments are carried out in situ in SEM, and the strain and stress resolutions for the tests are set as 0.03% and 5 MPa, respectively. The general and calibrated material parameters are shown in Table 1. Figure 1 displays the direct comparison between the proposed model and the experimental observations. As clearly shown in this figure, the size effect, "smaller is stronger," is observed on the stress-strain

Table 1 The general and calibrated material parameters used for the validation of the proposed strain gradient plasticity model. (Reprinted with permission from Song and Voyiadjis (2018a))

General		Aluminum	Nickel
$E\ (GPa)$	Elastic modulus for isotropic linear elasticity	70	115
ν	Poisson's ratio	0.30	0.31
$\mu\ (GPa)$	Shear modulus for isotropic linear elasticity	27	44
$\rho\ (g\cdot cm^{-3})$	Density	2.702	8.902
$c_{\varepsilon}\ (J/g\cdot{}^{\circ}K)$	Specific heat capacity at constant stress	0.910	0.540
$\alpha^{th}(\mu m/m\cdot{}^{\circ}K)$	Thermal expansion coefficient	24.0	13.1
$\dot{p}_1, \dot{p}_2\ (s^{-1})$	Reference plastic strain rate	0.04	0.04
r	Nonlinear hardening material constant	0.6	0.2
m_1	Nonnegative strain rate sensitivity parameter	0.05	0.05
m_2	Nonnegative strain rate sensitivity parameter	0.2	0.2
$\mathcal{T}_y\ ({}^{\circ}K)$	Thermal material parameter	933	890
n	Temperature sensitivity parameter	0.3	0.3
Calibrated		Aluminum	Nickel
$\sigma_0\ (MPa)$	Stress-dimensioned scaling constant	1,000	950
$\mathcal{H}_0\ (MPa)$	Isotropic hardening parameter	1,000	1,000
$\ell_{en}\ (\mu m)$	Energetic length scale	1.2	1.6
$\ell_{dis}\ (\mu m)$	Dissipative length scale	0.7	0.5
$\ell_{N-G}\ (\mu m)$	N-G length scale	1.0	2.0

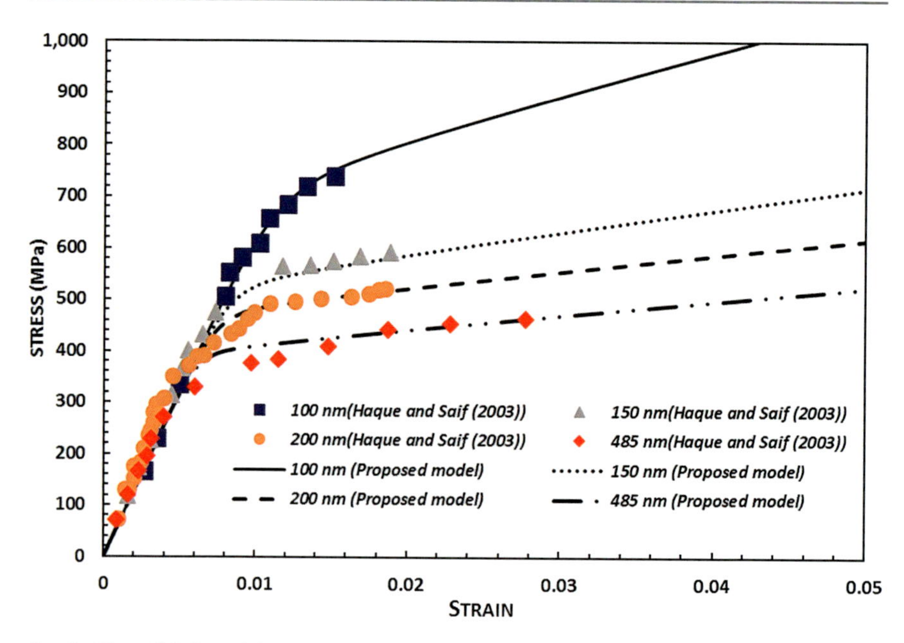

Fig. 1 The validation of the proposed SGP model by comparing the numerical results from the proposed model with the experimental measurements from Haque and Saif (2003) on the stress-strain response of the sputter-deposited Al thin films. (Reprinted with permission from Song and Voyiadjis (2018a))

Fig. 2 The schematic illustration of the dog bone specimen and its main dimensions (Han et al. 2008). (Reprinted with permission from Song and Voyiadjis (2018a))

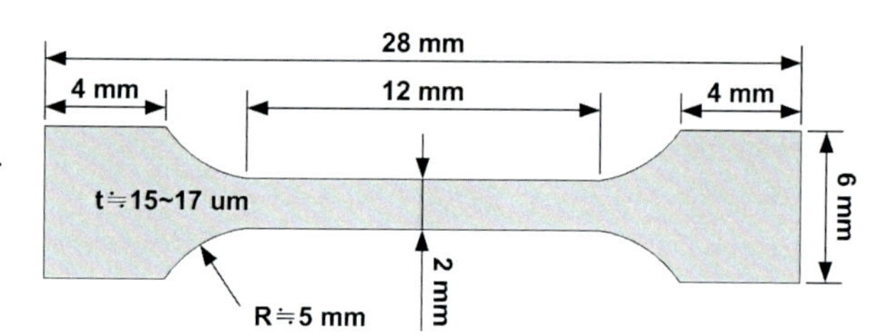

curves; furthermore, the numerical results from the proposed SGP model and the experimental data correspond closely with each other.

The microscale tensile experiment skill for evaluating the mechanical and thermal properties of the nickel (Ni) thin films at high temperatures is developed by Han et al. (2008). The dog bone-shaped specimens used in their experiments were made by micro-electromechanical system (MEMS) processes, and the primary dimensions of the specimen are given in Fig. 2. The general and calibrated material parameters for Ni are presented in Table 1. The experimental measurements at four different temperatures, i.e., 25 °C, 75 °C, 145 °C, and 218 °C, and corresponding numerical values from the proposed model are presented in Fig. 3. As shown in this figure, it is obvious from both the numerical and experimental results that the Young's modulus is not affected by variations in temperature

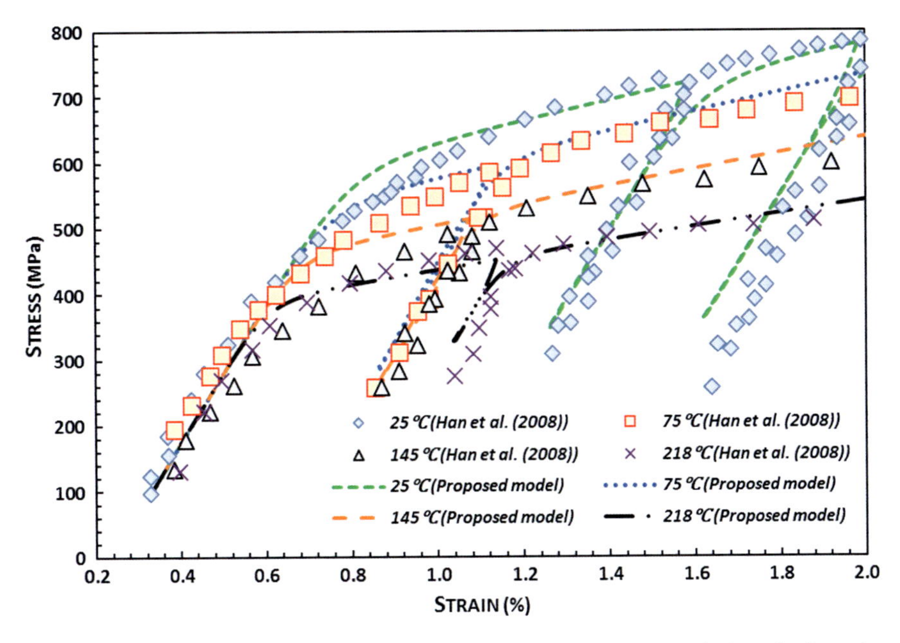

Fig. 3 The validation of the proposedSGP model by comparing the numerical results from the proposed model with the experimental measurements from Han et al. (2008) on the stress-strain response of the Ni thin films. (Reprinted with permission from Song and Voyiadjis (2018a))

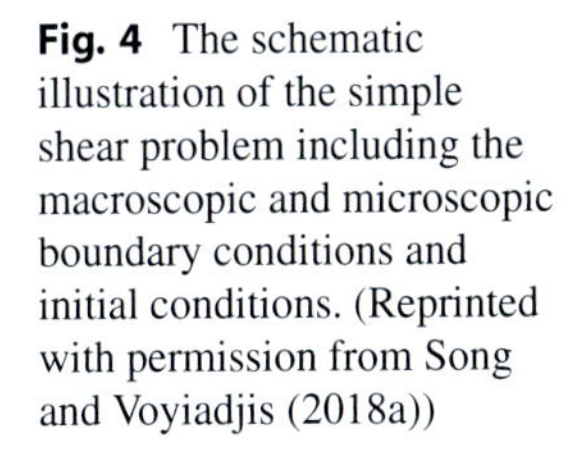

Fig. 4 The schematic illustration of the simple shear problem including the macroscopic and microscopic boundary conditions and initial conditions. (Reprinted with permission from Song and Voyiadjis (2018a))

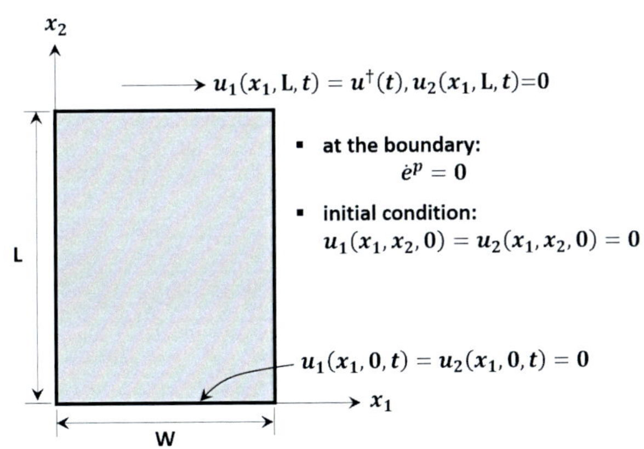

while the yield strength decreases as the specimen temperature increases; in addition, Fig. 3 clearly shows that the Bauschinger effect is not affected very much by variations in the specimen temperature. All these observations are not much different from those observed in one-dimensional finite element simulations (Voyiadjis and Song 2017).

Microfree Grain Boundary

The characteristics of the proposed SGP theory under the microfree boundary condition at the grain boundary is addressed in this section by solving the shear problem of a rectangular plate with varying material parameters.

By following Gurtin (2003), the simple class of microscopically free boundary condition on a prescribed subsurface is employed from Eq. (14) as follows:

$$m = Q_i n_i = 0 \tag{101}$$

The schematic illustration of the problem, the initial conditions, and the macroscopic and microscopic boundary conditions are shown in Fig. 4. The parameter $u^{\dagger}(t)$ represents the prescribed displacement. The stress-strain behavior and the distributions of the temperature and accumulated plastic strain across the plate in x_2 direction are investigated for the various material length scales (ℓ_{en}, $\ell_{dis,}$ and ℓ_{N-G}), the hardening parameter $\mathcal{H}_0$, and the temperature-related parameters (n and $\mathcal{T}_y$). The material parameters in Table 2 are used for this section unless stated otherwise.

Energetic Gradient Hardening

For exploring the characteristics of the energetic gradient hardening only, the dissipative gradient and GND gradient terms are assumed to vanish by imposing $\ell_{dis} = \ell_{N-G} = 0$. The stress-strain behaviors and the distributions of e^p and $\mathcal{T}$

Table 2 The material parameters used in sections "Microfree Grain Boundary" and "Microhard Grain Boundary." (Reprinted with permission from Song and Voyiadjis (2018a))

General		Values
$E\ (GPa)$	Elastic modulus for isotropic linear elasticity	110
ν	Poisson's ratio	0.33
$\mu\ (GPa)$	Shear modulus for isotropic linear elasticity	48
$\rho\ (g \cdot cm^{-3})$	Density	8.960
$c_\varepsilon\ (J/g \cdot {}^{\circ}K)$	Specific heat capacity at constant stress	0.385
$\alpha^{th}(\mu m/m \cdot {}^{\circ}K)$	Thermal expansion coefficient	24.0
$\dot{p}_1, \dot{p}_2\ (s^{-1})$	Reference plastic strain rate	0.04
r	Nonlinear hardening material constant	0.6
m_1	Nonnegative strain rate sensitivity parameter	0.05
m_2	Nonnegative strain rate sensitivity parameter	0.2
$\mathcal{T}_y({}^{\circ}K)$	Thermal material parameter	1,358
n	Temperature sensitivity parameter	0.3
$\sigma_0\ (MPa)$	Stress-dimensioned scaling constant	1,000
$\mathcal{H}_0\ (MPa)$	Isotropic hardening parameter	1,000

along the height of the plate for $\ell_{en}/L = 0$, 0.1, 0.2, 0.3, 0.5, 0.7, and 1.0 with $\mathcal{H}_0 = 0$ MPa are shown in Fig. 5. For the case of $\ell_{en}/L = 0$, the stress-strain curve does not show hardening as expected, and the distribution of e^p is uniform. As ℓ_{en}/L increases, the rates of the hardening increase as shown in Fig. 5a, and the distributions of e^p and $\mathcal{T}$ become more parabolic. When the dissipative length scale ℓ_{dis} is absent, the heat generation is governed by the amount of e^p; therefore the distribution of $\mathcal{T}$ is identical to the one of e^p as observed in Fig. 5b, c. Due to the microhard boundary condition, the temperature does not rise at the boundary, and the maximum temperature is observed at the center of the plate.

Dissipative Gradient Strengthening

It is now assumed that the energetic and GND hardening disappear by setting $\ell_{en} = \ell_{N-G} = 0$. The isotropic hardening parameter is also set as $\mathcal{H}_0 = 0\ MPa$. The stress-strain curves and the distributions of e^p and $\mathcal{T}$ across the height of the plate for $\ell_{dis}/L = 0$, 0.05, 0.1, 0.2, 0.3, 0.5, and 0.7 are shown in Fig. 6. As ℓ_{dis}/L increases, it is shown in Fig. 6a that the initial yield strength increases without strain hardening. The non-monotonic behavior of e^p with ℓ_{dis}/L is elucidated by plotting the maximal values of e^p at the center of the plate for varying ℓ_{dis}/L as shown in Fig. 6c. These numerical results are in line with the works of Bardella (2006) and Fleck and Hutchinson (2001). The drastic change shown in the accumulated plastic strain distribution according to varying ℓ_{dis}/L prominently affects the temperature distribution. According to Eq. (67), both the plastic strain and its gradient are involved in the evolution of $\mathcal{T}$. As observed from Fig. 6b, the large amounts of plastic strain gradient raise the temperature at the boundary; on the other hand, the temperature at the center of the plate is primarily influenced by plastic strain.

GND Hardening

It is now assumed that the energetic and dissipative gradient terms disappear by setting $\ell_{en} = \ell_{dis} = 0$. The stress-strain curves and the distributions of e^p and $\mathcal{T}$ across the height of the plate for $\ell_{N-G}/L = 0$, 0.1, 0.2, 0.5, 1.0, 1.5, and 2.0 with $\mathcal{H}_0 = 0$ MPa are shown in Fig. 7. As ℓ_{N-G}/L increases, the rates of the strain hardening increase more significantly, while no initial yield strength is observed. Similar to the energetic gradient hardening case, in the absence of ℓ_{dis}, the distribution of temperature is identical to the one of plastic strain.

Temperature-Related Parametric Study

The effects are now investigated of the two material parameters, n and $\mathcal{T}_y$, on the stress-strain behavior, the distributions of the temperature and accumulated plastic

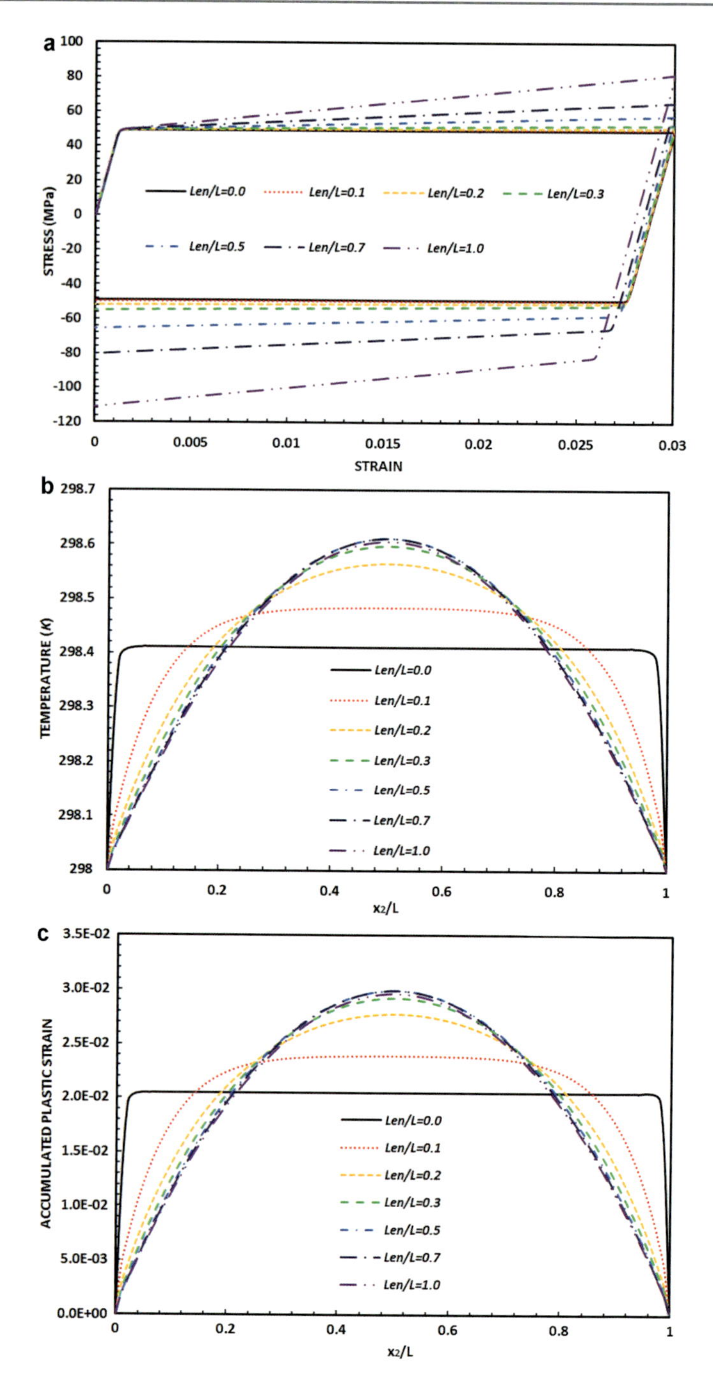

Fig. 5 The effects of the energetic gradient hardening with the energetic length scale only ($\ell_{dis} = \ell_{N-G} = 0$): (**a**) the stress-strain response, (**b**) the temperature distribution, and (**c**) the accumulated plastic strain distribution. (Reprinted with permission from Song and Voyiadjis (2018a))

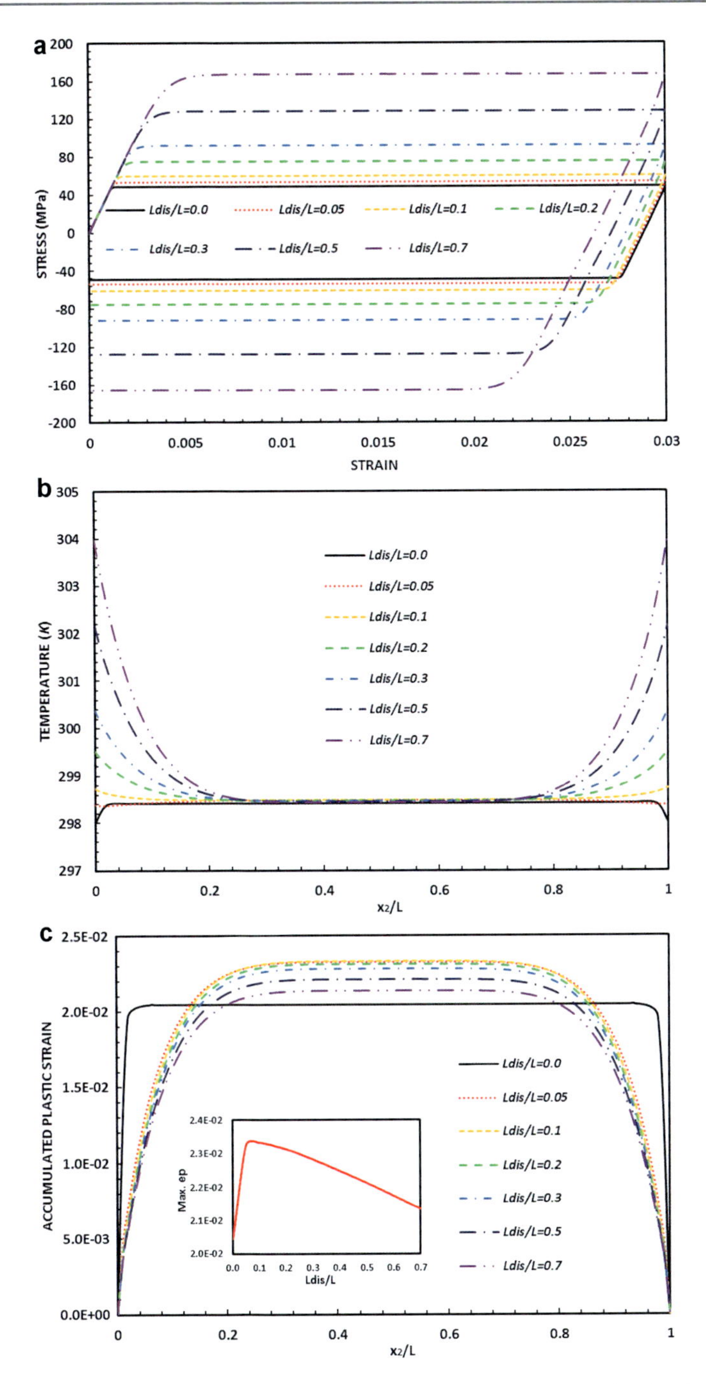

Fig. 6 The effects of the dissipative gradient strengthening with the dissipative length scale only ($\ell_{en} = \ell_{N-G} = 0$): (**a**) the stress-strain response, (**b**) the temperature distribution, and (**c**) the accumulated plastic strain distribution. (Reprinted with permission from Song and Voyiadjis (2018a))

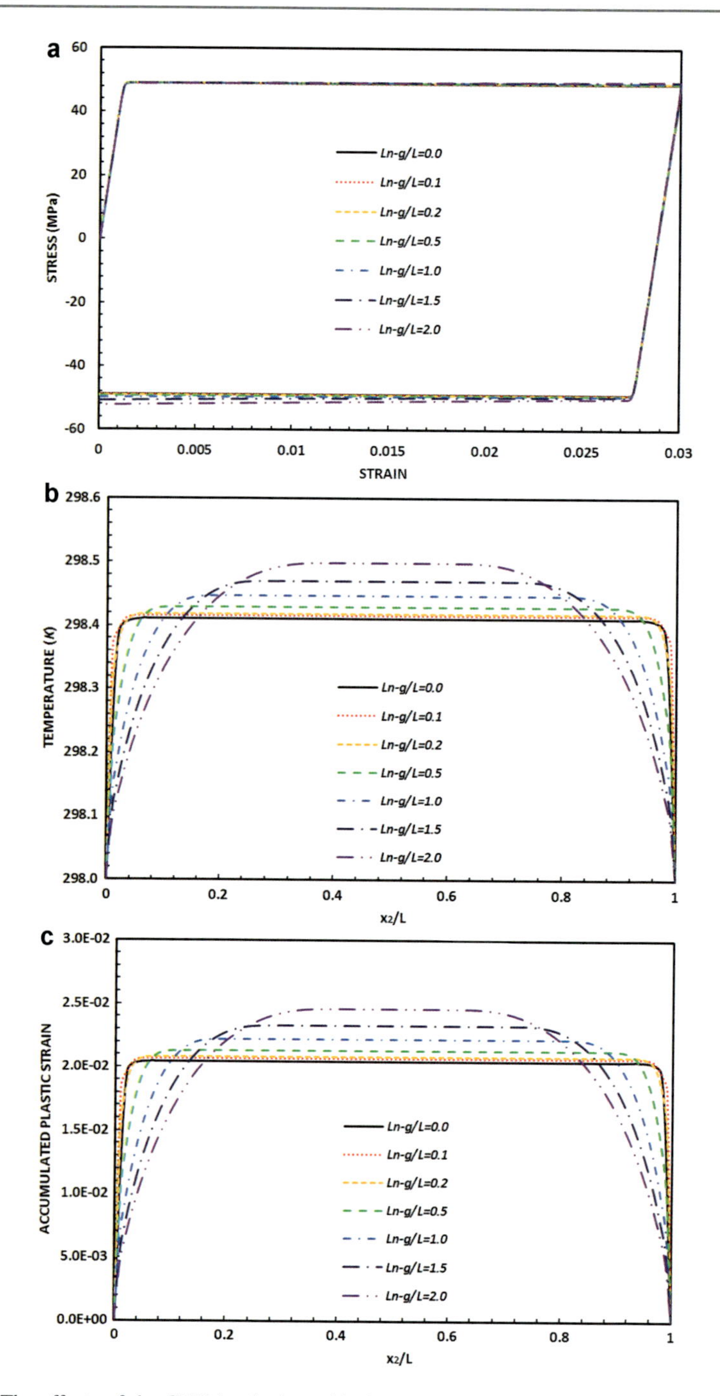

Fig. 7 The effects of the GND hardening with the N-G length scale only ($\ell_{en} = \ell_{dis} = 0$): (**a**) the stress-strain response, (**b**) the temperature distribution, and (**c**) the accumulated plastic strain distribution. (Reprinted with permission from Song and Voyiadjis (2018a))

strain, and the temperature evolution in conjunction with $\ell_{en}/L = 1.0$, $\ell_{dis}/L = 0.5$, and $\ell_{N-G}/L = 1.0$.

Figures 8 and 9 show the effect of the material parameters n and $\mathcal{T}_y$ applying the temperature-dependent behavior of the present model on the stress-strain curve, the distributions of the temperature and accumulated plastic strain distribution, and the evolution of the temperature at the midpoint of the plate. As shown in these figures, the effects of the two parameters on the thermal and mechanical material response are similar to each other due to the fact that two parameters affect the temperature-related term $\left(1 - \left(\mathcal{T}/\mathcal{T}_y\right)^n\right)$ in the flow rule, which explains the thermal activation mechanism for overcoming the local barriers to the dislocation movement, in a similar way. It is worth mentioning that, in some literatures, the parameter $\mathcal{T}_y$ is assumed to be equal to the melting temperature of the specific material in order not to bring in the material parameter additionally. In this chapter, however, the normal parameter is employed as a normalizing constant that should be calibrated by using the experimental data.

It is clearly observed in Figs. 8a and 9a that, as n and $\mathcal{T}_y$ increase, the strain hardening and the initial yield strength also increase. The effect, nevertheless, is more prominent in initial yield strength, because strain hardening mechanism is affected by temperature through the dislocation forest barriers; on the other hand, the backstress does not depend on the temperature. For both n and $\mathcal{T}_y$, it is shown in Figs. 8b, c and 9b, c that the accumulated plastic strains at the boundary ($0 < x_2/L < 0.2$ and $0.8 < x_2/L < 1$) are almost identical with respect to the different values of n and $\mathcal{T}_y$; on the other hand, the most significant differences in the temperature profiles occur at the same range for both parameters. In contrast to this, the most substantial differences in the accumulated plastic strain profiles according to the values of n and $\mathcal{T}_y$ occur in the middle ($0.2 < x_2/L < 0.8$) of the plate; on the other hand, the temperature profiles at this part do not show considerable difference.

Microhard Grain Boundary

In this section, a microhard boundary condition, which describes that the dislocation movements are blocked completely at the grain boundary, is assumed to hold on the grain boundary as follows:

$$e^p = 0 \tag{102}$$

The square plate with an edge of W is solved to investigate the effect of this grain boundary condition. The top edge of the plate is subject to a prescribed condition in terms of the displacement $\left(u_1^{top}(x_1, W, t) = u_1^*(t), u_2^{top}(x_1, W, t) = 0\right)$, while the bottom edge is fixed $\left(u_1^{bot}(x_1, 0, t) = u_2^{bot}(x_1, 0, t) = 0\right)$. The whole square is meshed using 1,600 (40×40) elements and split into several grains by the four different grain boundary areas, which are indicated by the bold lines as shown in Fig. 10. For these simulations, the stress-dimensioned scaling constant σ_0 and the

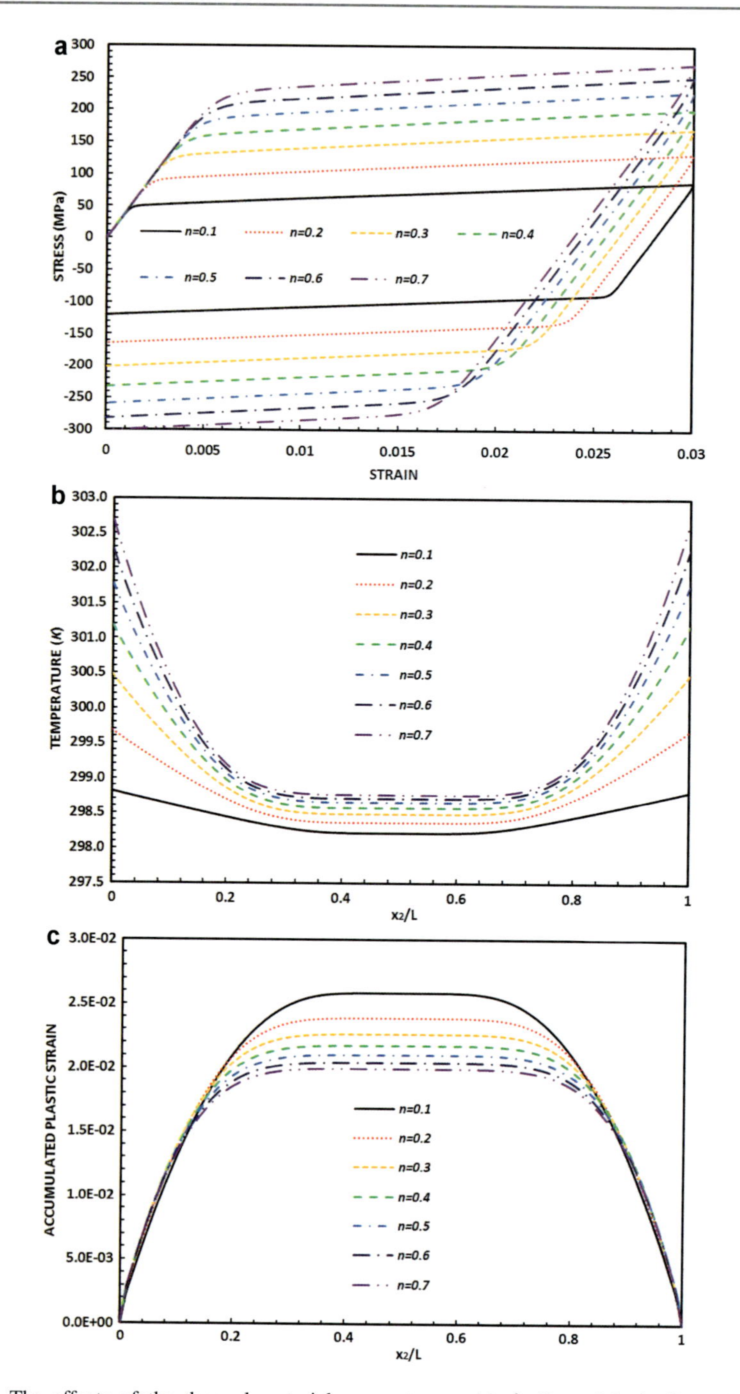

Fig. 8 The effects of the thermal material parameter n with $\ell_{en}/L = 1.0$, $\ell_{dis}/L = 0.5$, and $\ell_{N-G}/L = 1.0$: (**a**) the stress-strain response, (**b**) the temperature distribution, and (**c**) the accumulated plastic strain distribution. (Reprinted with permission from Song and Voyiadjis (2018a))

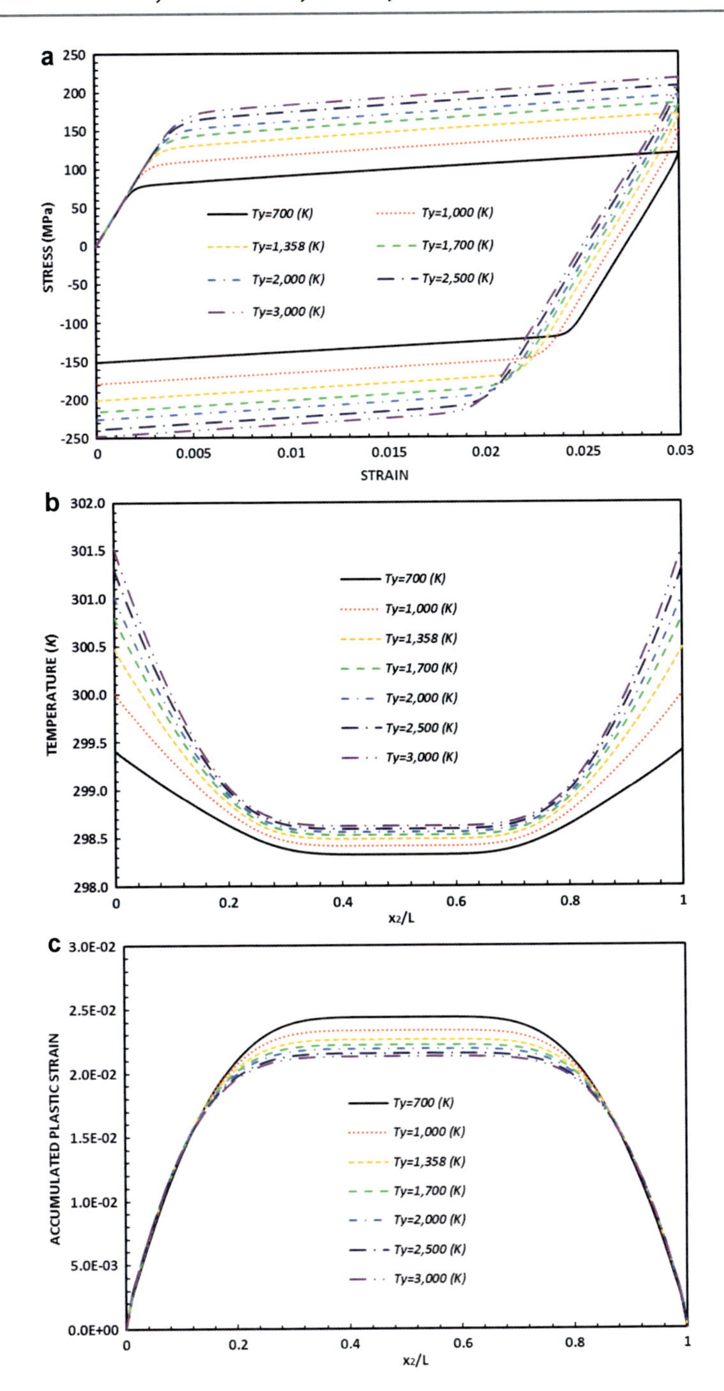

Fig. 9 The effects of the thermal material parameter $\mathcal{T}_y$ with $\ell_{en}/L = 1.0$, $\ell_{dis}/L = 0.5$, and $\ell_{N-G}/L = 1.0$: (**a**) the stress-strain response, (**b**) the temperature distribution, and (c) the accumulated plastic strain distribution. (Reprinted with permission from Song and Voyiadjis (2018a))

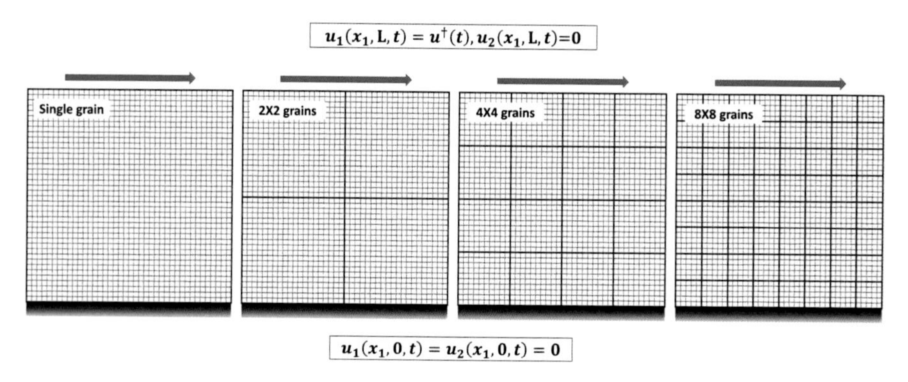

Fig. 10 The schematic illustration of the square plate problem with four different grain boundary areas. (Reprinted with permission from Song and Voyiadjis (2018a))

isotropic hardening parameter $\mathcal{H}_0$ are set as 195 *MPa* and 0 *MPa*, respectively, and the values in Table 2 are used again for the rest of the material parameters.

The comparison of the microscopic boundary condition is addressed through the single grain in Fig. 11. Figure 11a shows the distributions of the accumulated plastic strain e^p and temperature $\mathcal{T}$, respectively, with the microhard boundary condition at the grain boundaries, while Fig. 11b shows those with the microfree boundary condition. The terminologies "NT11" and "UVARM6" in Fig. 11 indicate the accumulated plastic strain and the temperature, respectively, and are used continually in the rest of this work. For this example, $\ell_{en}/L = 0.3$, $\ell_{dis}/L = 0.0$, and $\ell_{N-G}/L = 0.0$ are used. As can be seen in Fig. 11a, each edge of the plate with microhard boundary condition obstructs the dislocation movement, which results in $e^p = 0$, whereas in the case of microfree boundary condition, e^p and $\mathcal{T}$ are evenly spread across the grain.

Figure 12 shows the distributions of e^p and $\mathcal{T}$ with no gradient effect. This can be regarded as the reference case for the purpose of the comparison to other cases with the gradient effect.

The effects of the grain boundary area in conjunction with the energetic gradient hardening, dissipative gradient strengthening, GND hardening, and no gradient effect are shown in Fig. 13 through the stress-strain responses. As can be seen in Fig. 13a, no significant hardening or strengthening is observed under the conventional plasticity theory. On the other hand, by considering the gradient effect, the energetic hardening, dissipative strengthening, and GND hardening are observed, respectively, in Fig. 13b, c, d with varying grain boundary areas.

The effects of the grain boundary area in conjunction with the energetic gradient hardening, dissipative gradient strengthening, and GND hardening on the distributions of e^p and $\mathcal{T}$ are also shown, respectively, in Figs. 14, 15, and 16. For the simulations in Fig. 14, the energetic length scales vary from 0.1 to 0.3, and the dissipative and N-G length scales are set as zero. It should be noted that the distributions of e^p and $\mathcal{T}$ are not identical in all grains since the simple shear loading is applied to the top edge of the plate and not to each grain individually. It

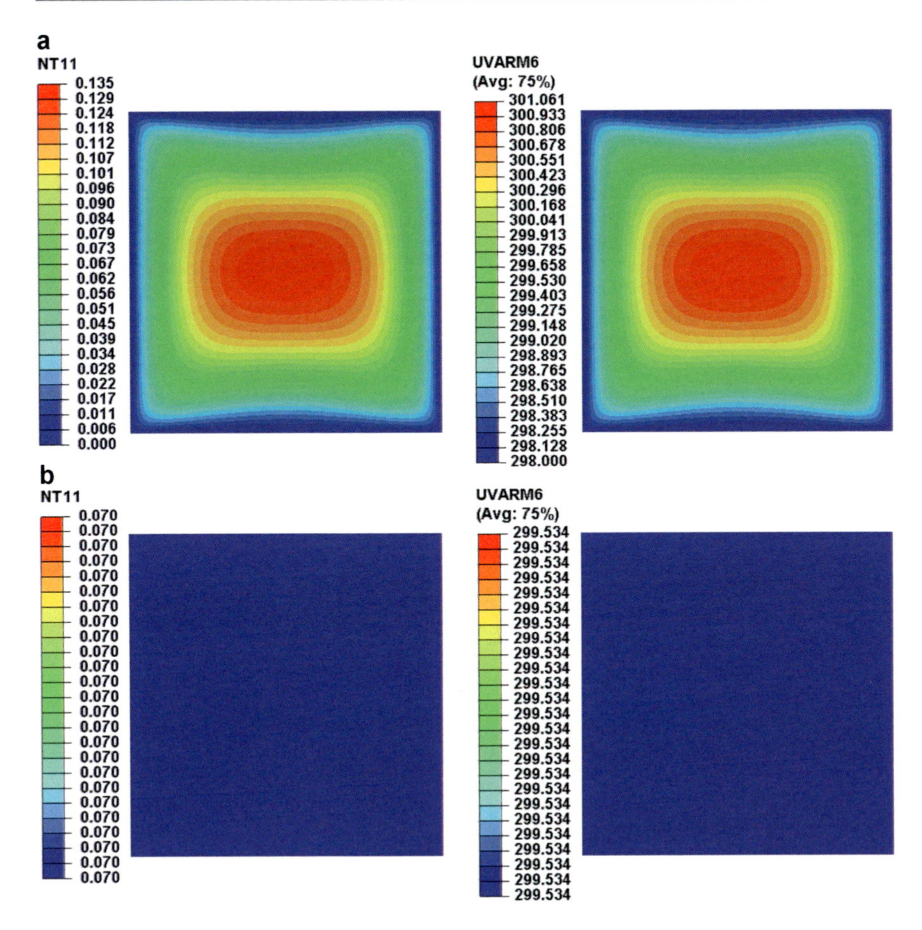

Fig. 11 The distributions of the accumulated plastic strain and the temperature with (**a**) the microhard and (**b**) microfree boundary conditions. (Reprinted with permission from Song and Voyiadjis (2018a))

is clearly observed in Fig. 14 by comparing to Fig. 12 that the distributions of e^p and $\mathcal{T}$ with the energetic hardening are significantly different from those with no gradient effect. This tendency is also observed with the dissipative strengthening effects when comparing Fig. 15 with Fig. 12 as well as with the GND hardening effects when comparing Fig. 16 with Fig. 12.

Intermediate (Deformable) Grain Boundary

The assumption of two null boundary conditions, microfree and microhard boundary conditions, is used at the grain boundary in the previous two sections. In this section, the governing differential equation is solved by imposing the proposed grain

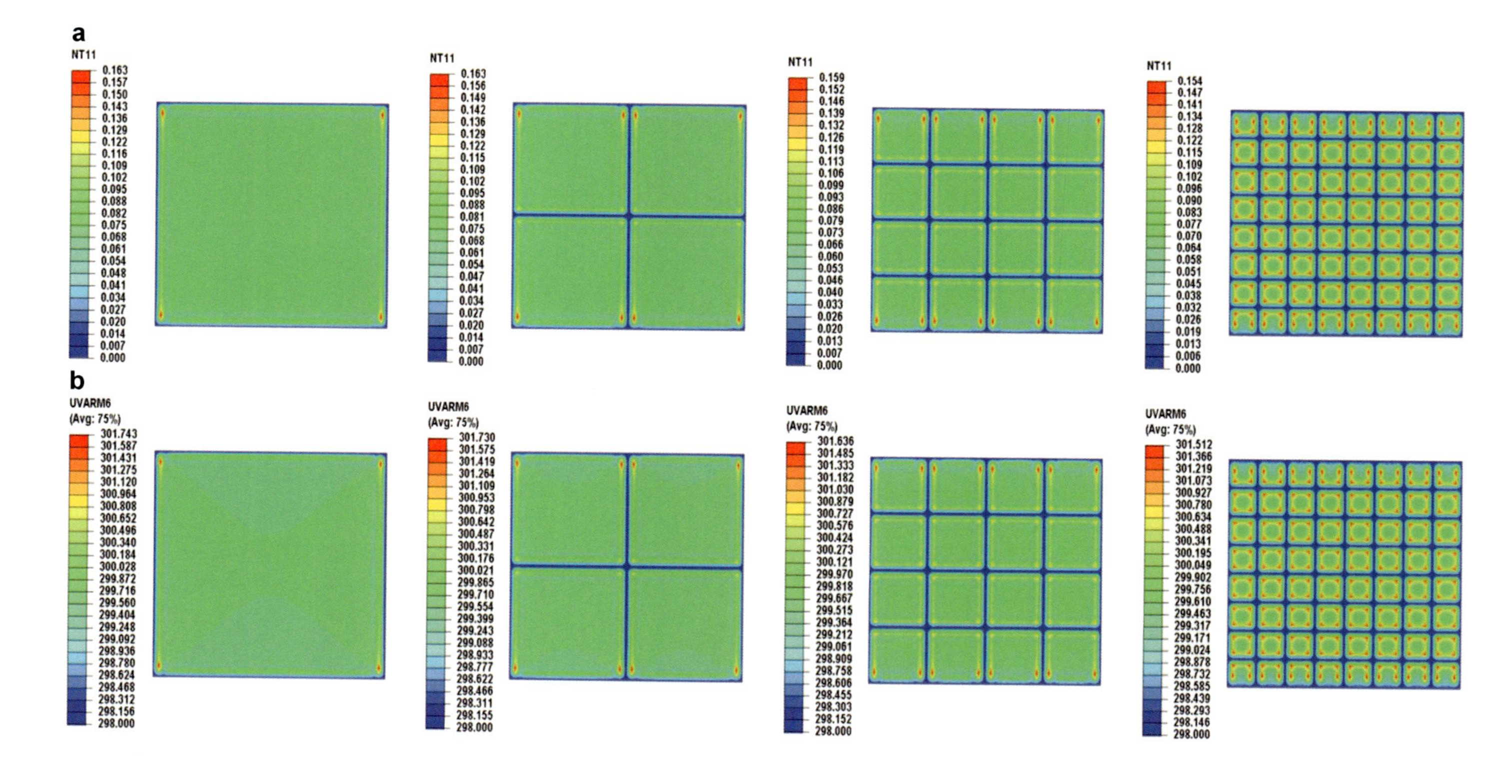

Fig. 12 The distributions of the accumulated plastic strain and the temperature in conjunction with varying grain boundary areas with $\ell_{en} = \ell_{dis} = \ell_{N-G} = 0$: (**a**) e^p and (**b**) $\mathcal{T}$. (Reprinted with permission from Song and Voyiadjis (2018a))

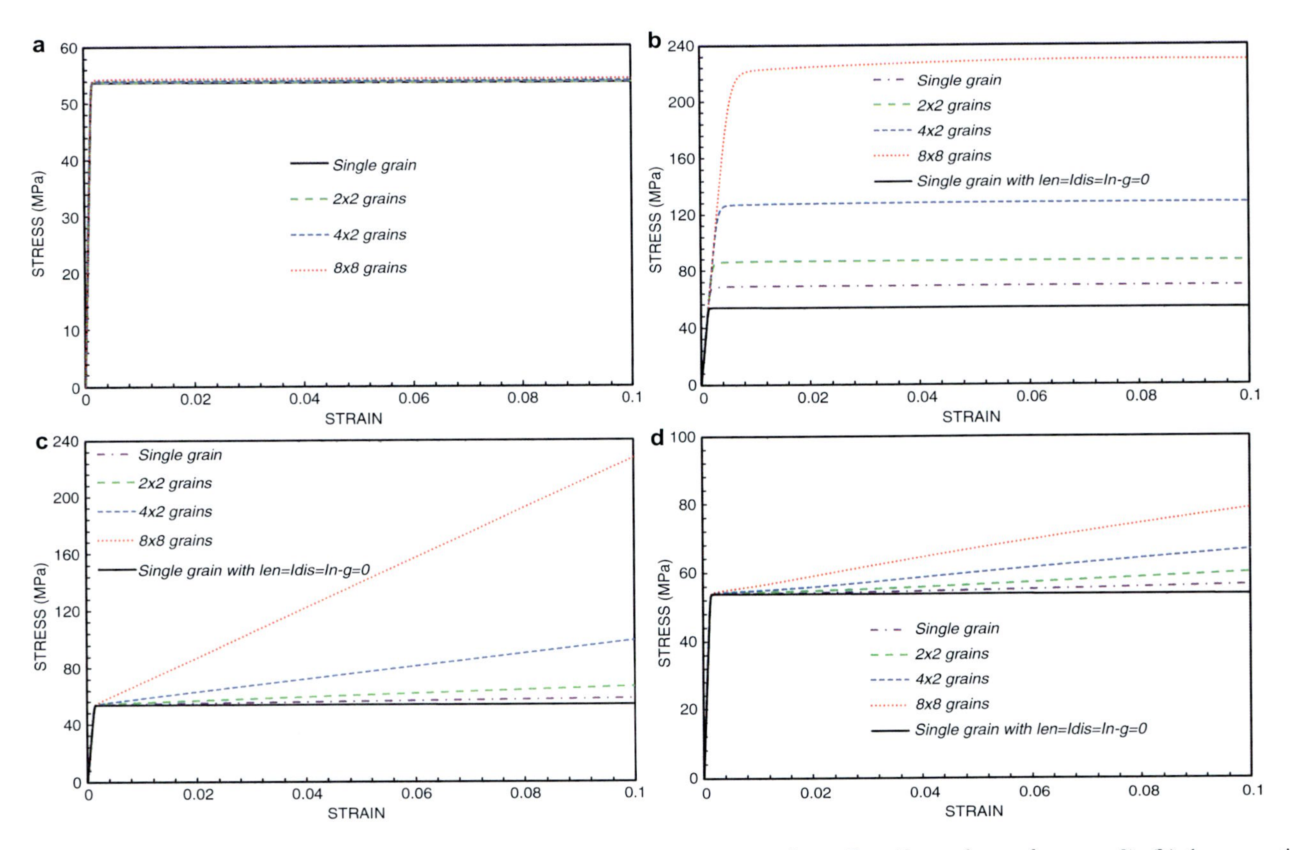

Fig. 13 The effects of the grain boundary area on the stress-strain response with (**a**) no gradient effect ($\ell_{en} = \ell_{dis} = \ell_{N-G} = 0$), (**b**) the energetic length scale only ($\ell_{en}/W = 0.1$, $\ell_{dis} = \ell_{N-G} = 0$), (**c**) the dissipative length scale only ($\ell_{dis}/W = 0.05$, $\ell_{en} = \ell_{N-G} = 0$), and (**d**) the N-G length scale only ($\ell_{N-G}/W = 0.5$, $\ell_{en} = \ell_{dis} = 0$). (Reprinted with permission from Song and Voyiadjis (2018a))

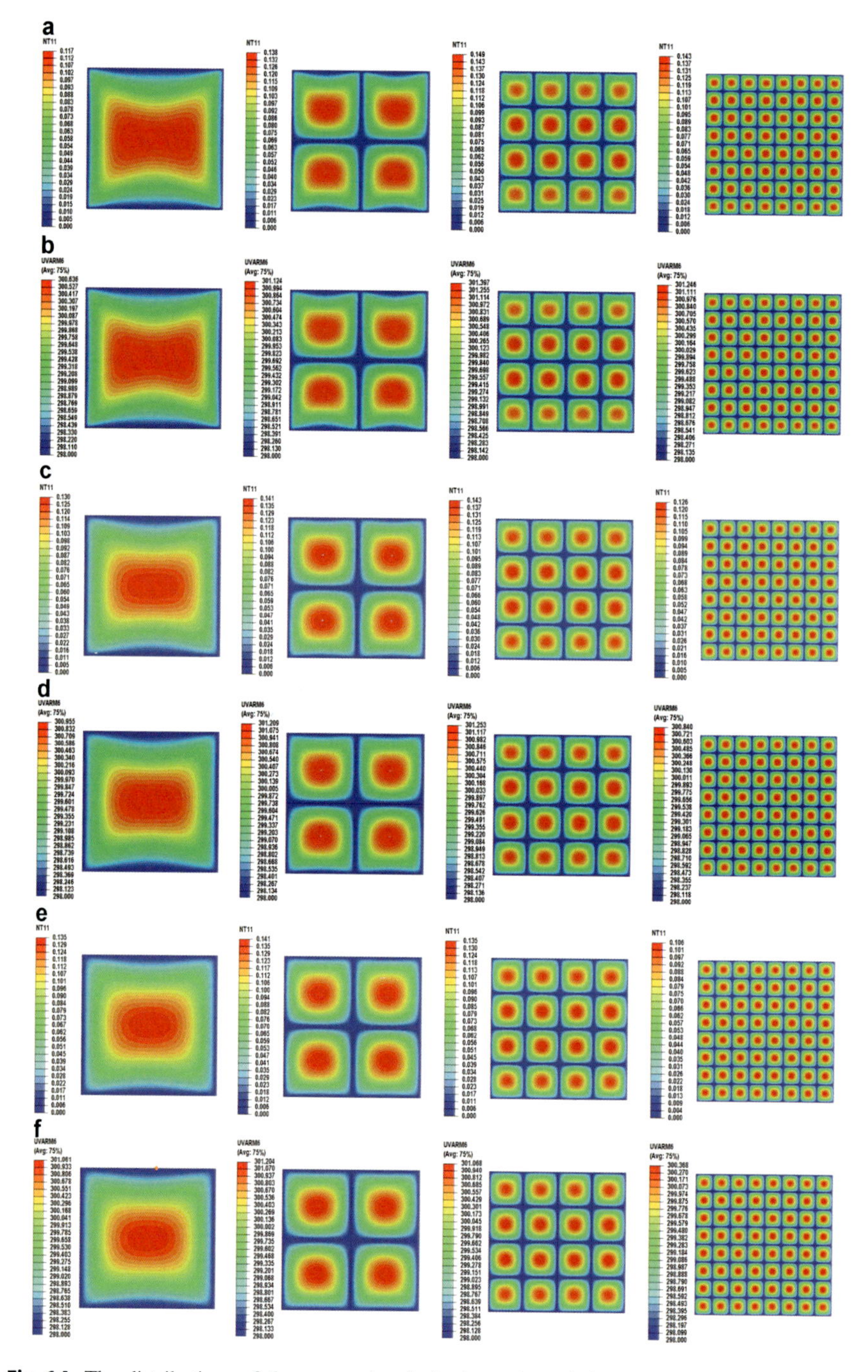

Fig. 14 The distributions of the accumulated plastic strain and the temperature in conjunction with varying grain boundary areas with the energetic length scale only ($\ell_{dis} = \ell_{N-G} = 0$): (**a**) e^p with $\ell_{en}/W = 0.1$, (**b**) $\mathcal{T}$ with $\ell_{en}/W = 0.1$, (**c**) e^p with $\ell_{en}/W = 0.2$, (**d**) $\mathcal{T}$ with $\ell_{en}/W = 0.2$, (**e**) e^p with $\ell_{en}/W = 0.3$, and (**f**) $\mathcal{T}$ with $\ell_{en}/W = 0.3$. (Reprinted with permission from Song and Voyiadjis (2018a))

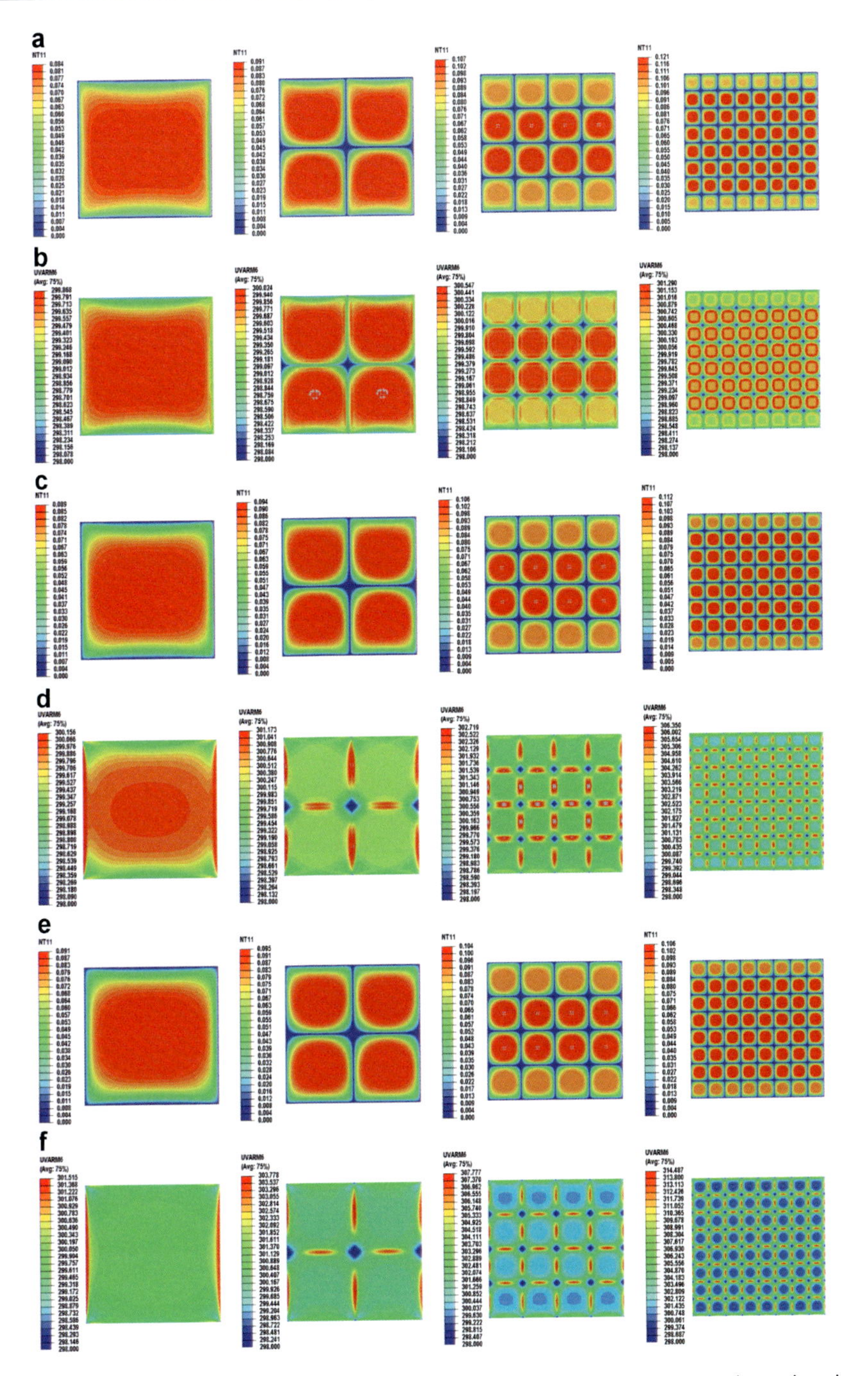

Fig. 15 The distributions of the accumulated plastic strain and the temperature in conjunction with varying grain boundary areas with the dissipative length scale only ($\ell_{en} = \ell_{N-G} = 0$): (**a**) e^p with $\ell_{dis}/W = 0.01$, (**b**) $\mathcal{T}$ with $\ell_{dis}/W = 0.01$, (**c**) e^p with $\ell_{dis}/W = 0.03$, (**d**) $\mathcal{T}$ with $\ell_{dis}/W = 0.03$, (**e**) e^p with $\ell_{dis}/W = 0.05$, and (**f**) $\mathcal{T}$ with $\ell_{dis}/W = 0.05$. (Reprinted with permission from Song and Voyiadjis (2018a))

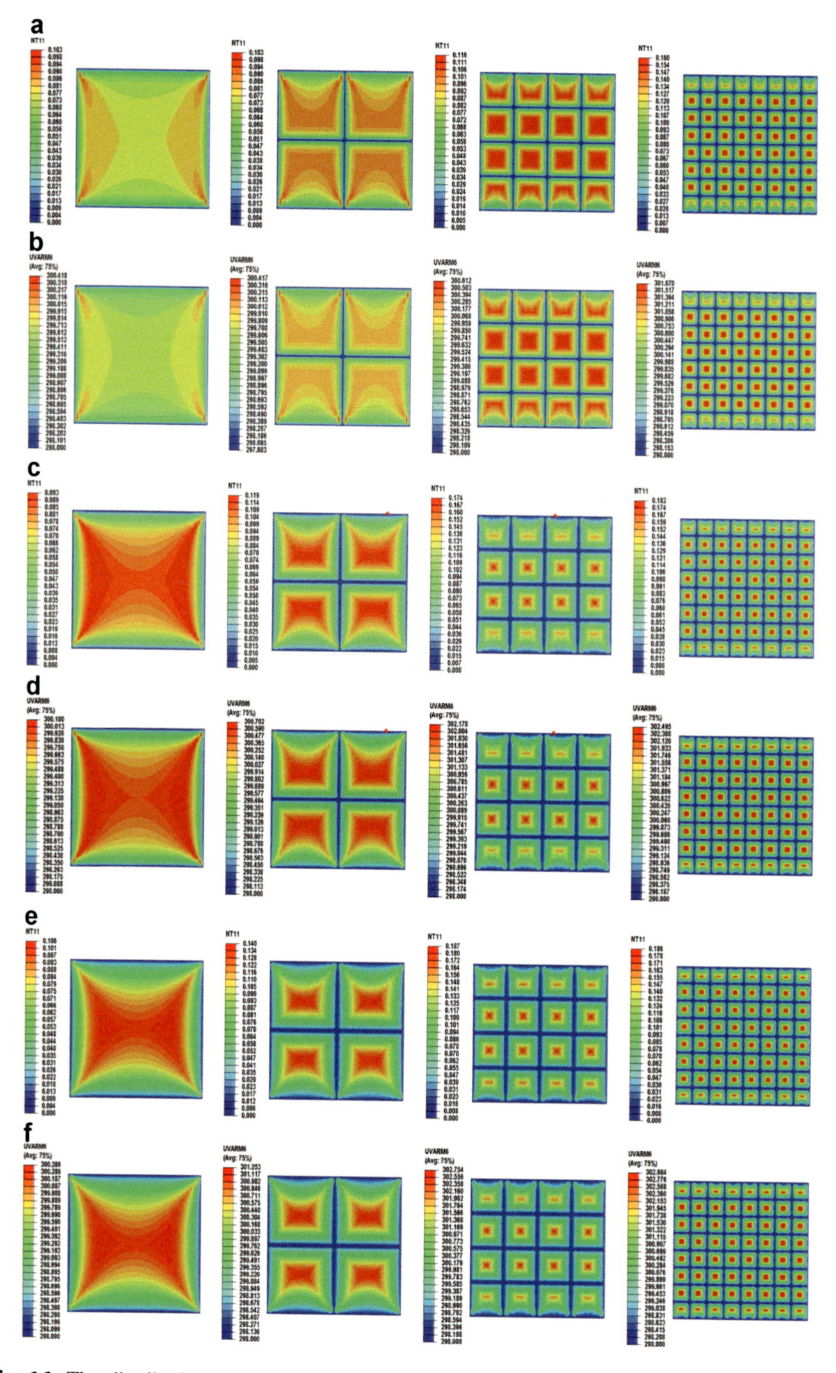

Fig. 16 The distributions of the accumulated plastic strain and the temperature in conjunction with varying grain boundary areas with the N-G length scale only ($\ell_{en} = \ell_{dis} = 0$): (**a**) e^p with $\ell_{N-G}/W = 0.1$, (**b**) $\mathcal{T}$ with $\ell_{N-G}/W = 0.1$, (**c**) e^p with $\ell_{N-G}/W = 0.3$, (**d**) $\mathcal{T}$ with $\ell_{N-G}/W = 0.3$, (**e**) e^p with $\ell_{N-G}/W = 0.5$, and (**f**) $\mathcal{T}$ with $\ell_{N-G}/W = 0$. (Reprinted with permission from Song and Voyiadjis (2018a))

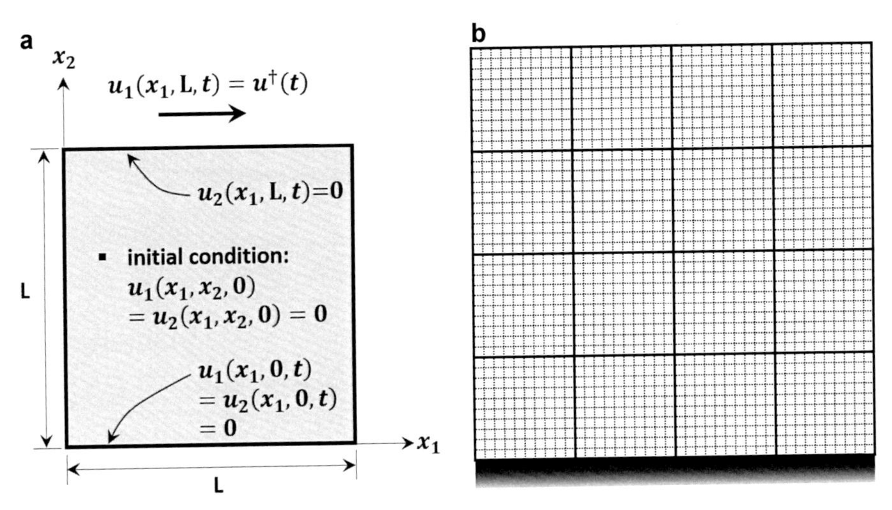

Fig. 17 The schematic illustration of the simple shear problem: (**a**) the macroscopic, microscopic boundary conditions and initial conditions (**b**) 4 × 4 grains. (Reprinted with permission from Song and Voyiadjis (2018b))

boundary flow rule to account for the deformable grain boundary. Furthermore, the characteristics of the proposed strain gradient plasticity theory incorporating the flow rules of both the grain interior and grain boundary are addressed in this section by solving the shear problem of a square plate with an edge of L. The schematic illustration of the problem, the initial conditions, and the macroscopic and microscopic boundary conditions as well as the grain boundary area are shown in Fig. 17. The parameter $u^\dagger(t)$ represents the prescribed displacement. The whole square is meshed using 1,600 (40 × 40) elements and split into the 16 (4 × 4) grains by the grain boundary area, which is indicated by the bold lines. The material parameters for the grain interior and grain boundary are presented in Table 3.

As can be seen from Eq. (82), the grain boundary may act like a free surface, i.e., microfree boundary condition, when $\ell_{en}^{GB} = \ell_{dis}^{GB} = 0$. On the other hand, the microhard boundary condition can be compelled under the conditions $\ell_{en}^{GB} \to \infty$ and $\ell_{dis}^{GB} \to \infty$. Firstly, the validity of these conditions is examined in this section. Next, the direct comparison between the classical plasticity theory ($\ell_{en}/L = \ell_{dis}/L = \ell_{N-G}/L = 0.0$) and the gradient-enhanced plasticity theory ($\ell_{en}/L = \ell_{dis}/L = \ell_{N-G}/L = 0.1$) is given in order to check the ability of the proposed flow rule on the size effect. The numerical results in terms of the accumulated plastic strain profile and the stress-strain curves are shown in Figs. 18 and 19. As can be seen in these figures, the microfree and microhard boundary conditions are well captured under the classical plasticity theory as well as the gradient-enhanced plasticity theory. In addition, in Fig. 18c, no size effect is observed in the classical plasticity theory with varying normalized material length scales as expected. In Fig. 19c, on the other hand, strain hardening and strengthening are more pronounced as the dimensions of the shear plate height are reduced

Table 3 The material parameters used in section "Intermediate (Deformable) Grain Boundary." (Reprinted with permission from Song and Voyiadjis (2018b))

Grain interior		Values
$E\ (GPa)$	Elastic modulus for isotropic linear elasticity	110
ν	Poisson's ratio	0.33
$\mu\ (GPa)$	Shear modulus for isotropic linear elasticity	48
$\rho\ (g \cdot cm^{-3})$	Density	8.960
$c_\varepsilon\ (J/g \cdot {}^\circ K)$	Specific heat capacity at constant stress	0.385
$\alpha^{th} (\mu m/m \cdot {}^\circ K)$	Thermal expansion coefficient	16.0
$\dot{p}_1, \dot{p}_2\ \left(s^{-1}\right)$	Reference plastic strain rate	0.04
r	Nonlinear hardening material constant	0.6
m_1	Nonnegative strain rate sensitivity parameter	0.05
m_2	Nonnegative strain rate sensitivity parameter	0.2
$\mathcal{T}_y\ \left({}^\circ K\right)$	Thermal material parameter	1,358
n	Temperature sensitivity parameter	0.3
$\sigma_0\ (MPa)$	Stress-dimensioned scaling constant	195
$\mathcal{H}_0\ (MPa)$	Isotropic hardening parameter	0
Grain boundary		Values
$\sigma_0^{GB}\ (MPa)$	Constant accounting for the GB yield stress	195
$\mathcal{H}_0^{GB}\ (MPa)$	GB hardening parameter	0
$\dot{p}^{GB}$	Viscous related material parameters for GB	0.04
m^{GB}	Viscous related material parameters for GB	1
$\mathcal{T}_y^{GB}$	GB thermal parameter at the onset of yield	700
n^{GB}	GB thermal parameter	0.4

$(\ell_{en}^{GB}/L \to \infty, \ell_{dis}^{GB}/L \to \infty)$. In Fig. 20, the effects of each material length scale parameter, i.e., ℓ_{en}, ℓ_{dis}, and ℓ_{N-G}, along with the microscopically hard boundary condition, are also examined through the profile of accumulated plastic strain. In addition, the contributions of each length scale parameter on the stress-strain responses are shown in Fig. 20c.

Variations in the stress-strain responses and the evolutions of maximum temperature are investigated for the various values of the normalized energetic and dissipative grain boundary material length scales as shown in Figs. 21 and 22. It is assumed by setting $\ell_{dis}^{GB}/\ell_{dis} = 0$ that all plastic work at the grain boundary is stored as surface energy which depends on the plastic strain state at the surface. In this case, ℓ_{en}^{GB}/ℓ_{en} reflects the grain boundary resistance to plastic deformation. Figures 21b and 22b show the size effects on the strain hardening and temperature evolution due to the grain boundary energetic length scale, and it is more pronounced in the more strongly constrained material, i.e., increasing ℓ_{en}^{GB}/ℓ_{en}. On the other hand, by setting $\ell_{en}^{GB}/\ell_{en} = 0$, it is assumed that the work performed at the grain boundary is dissipated in the absence of surface energy. In this case, $\ell_{dis}^{GB}/\ell_{dis}$ reflects the grain boundary resistance to slip transfer. As can be seen in Fig. 21c, the initial yield strength increases without strain hardening as $\ell_{dis}^{GB}/\ell_{dis}$ increases.

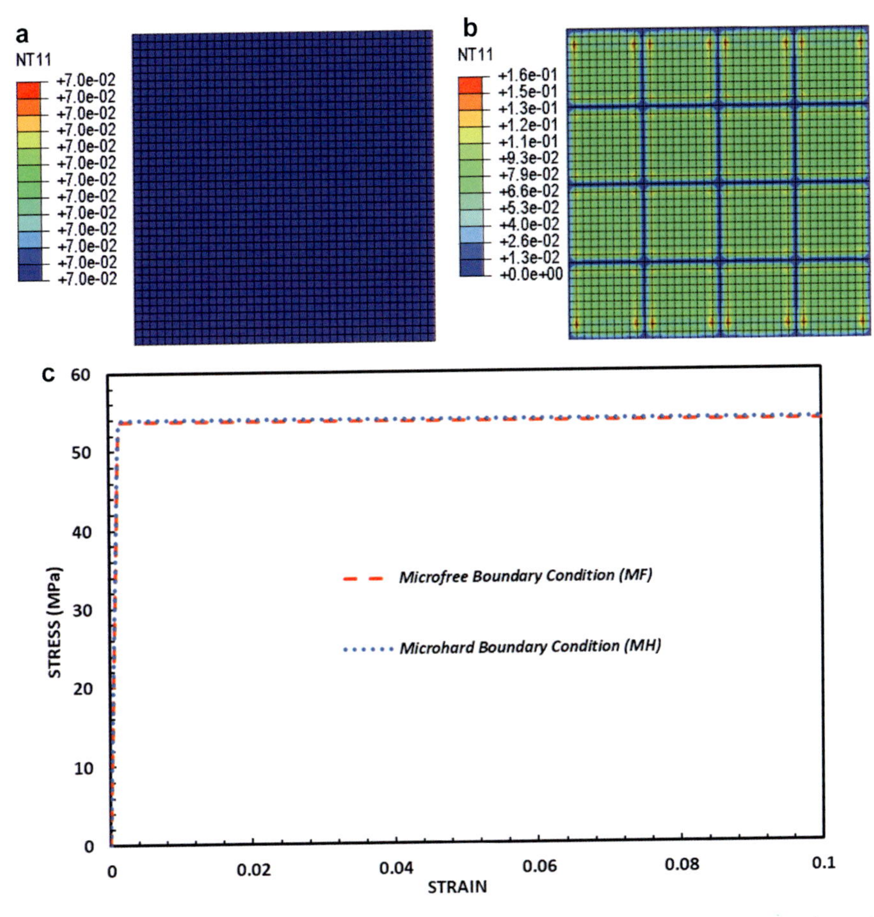

Fig. 18 Classical plasticity theory ($\ell_{en}/L = \ell_{dis}/L = \ell_{N-G}/L = 0.0$). The distributions of the accumulated plastic strain with (**a**) the microscopically free ($\ell_{en}^{GB} = \ell_{dis}^{GB} = 0$) and (**b**) microscopically hard boundary conditions ($\ell_{en}^{GB} \rightarrow \infty$, $\ell_{dis}^{GB} \rightarrow \infty$) and (**c**) the stress-strain responses. (Reprinted with permission from Song and Voyiadjis (2018b))

Generalized Structure for Modeling Polycrystals from Micro- to Nanoscale Range

In this section, modeling of strengthening in inelastic nanocrystalline materials with reference to the triple junction and grain boundaries using SGP is introduced based on the work of Voyiadjis and Deliktas (2010). Voyiadjis and Deliktas (2010) not only provide the internal interface energies but also introduce two additional internal state variables for the internal surfaces (contact surfaces). By using these internal state variables together with displacement and temperature, the constitutive model is formulated as usual by state laws utilizing free energies and complimentary laws based on the dissipation potentials. One of these new state

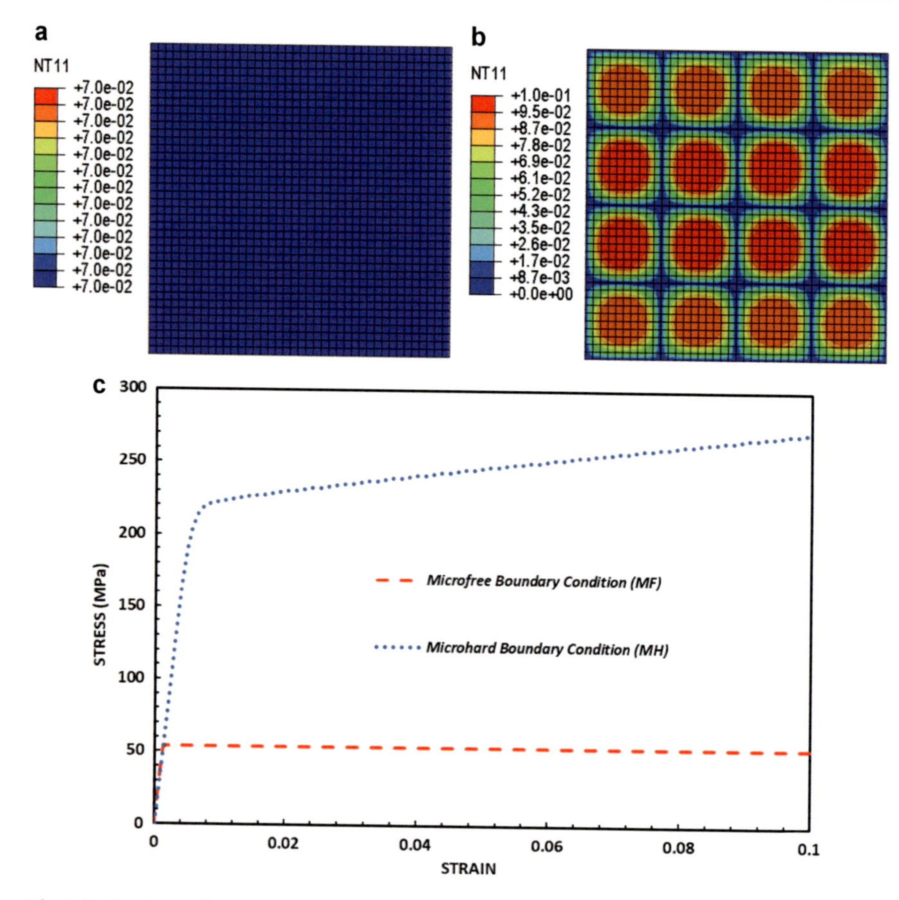

Fig. 19 Strain gradient plasticity theory ($\ell_{en}/L = \ell_{dis}/L = \ell_{N-G}/L = 0.1$). The distributions of the accumulated plastic strain with (**a**) the microscopically free ($\ell_{en}^{GB} = \ell_{dis}^{GB} = 0$) and (**b**) microscopically hard boundary conditions ($\ell_{en}^{GB} \rightarrow \infty$, $\ell_{dis}^{GB} \rightarrow \infty$) and (**c**) the stress-strain responses. (Reprinted with permission from Song and Voyiadjis (2018b))

variables measures tangential sliding between the grain boundaries and the other measures the respective separation. A homogenization technique is developed to describe the local stress and strain in the material. The material is characterized as a composite with three phases: the grain core, the grain boundaries, and the triple junctions.

Nanocrystalline materials are structurally characterized by a large volume fraction of grain boundaries which may significantly alter their physical, mechanical, and chemical properties in comparison with conventional coarse-grained polycrystalline materials which have grain size usually in the range 10–300μm (Meyers et al. 2006). The grain size of nanocrystalline materials is usually below 10 nm. The nanocrystalline structure described by Gleiter (2000) is composed of structural elements with a characteristic size of a few nanometers (see Fig. 23).

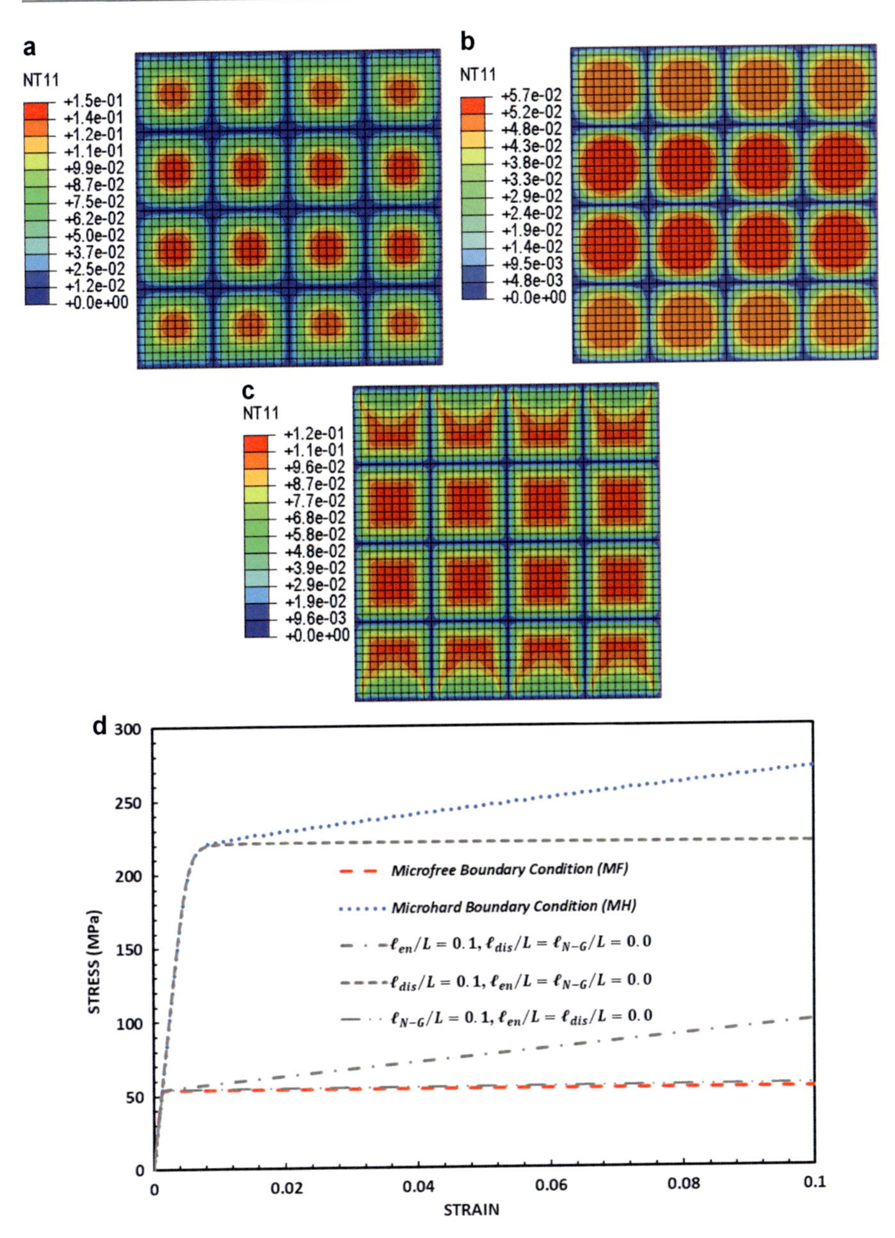

Fig. 20 The distributions of the accumulated plastic strain with the microscopically hard boundary condition ($\ell_{en}^{GB}/L \rightarrow \infty$, $\ell_{dis}^{GB}/L \rightarrow \infty$) under (**a**) the energetic length scale only ($\ell_{en}/L = 0.1$, $\ell_{dis}/L = \ell_{N-G}/L = 0.0$), (**b**) the dissipative length scale only ($\ell_{dis}/L = 0.1$, $\ell_{en}/L = \ell_{N-G}/L = 0.0$), (**c**) the N-G length scale only ($\ell_{N-G}/L = 0.1$, $\ell_{en}/L = \ell_{dis}/L = 0.0$), and (**d**) the stress-strain responses. (Reprinted with permission from Song and Voyiadjis (2018b))

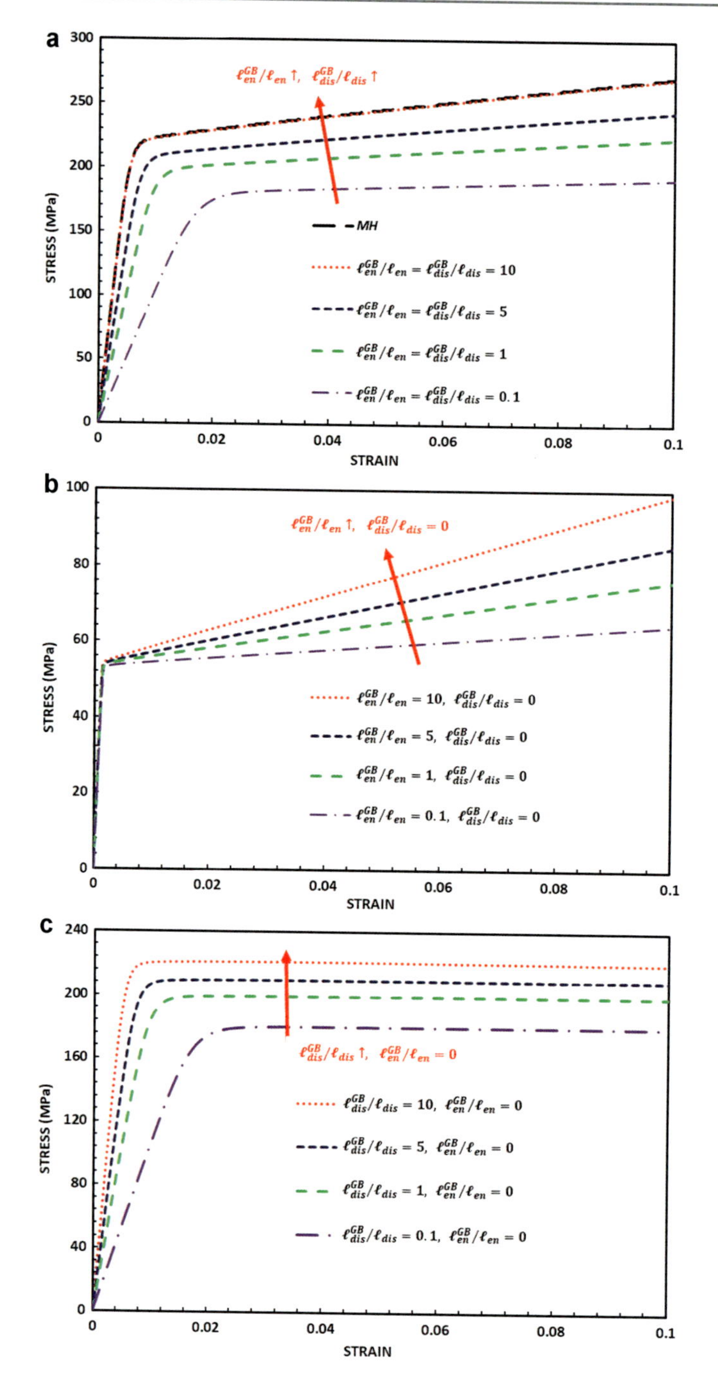

Fig. 21 The distributions of the accumulated plastic strain according to the various values of ℓ_{en}^{GB}/ℓ_{en} and $\ell_{dis}^{GB}/\ell_{dis}$: (**a**) combined ℓ_{en}^{GB} and ℓ_{dis}^{GB}, (**b**) ℓ_{en}^{GB} only, and (**c**) ℓ_{dis}^{GB} only. (Reprinted with permission from Song and Voyiadjis (2018b))

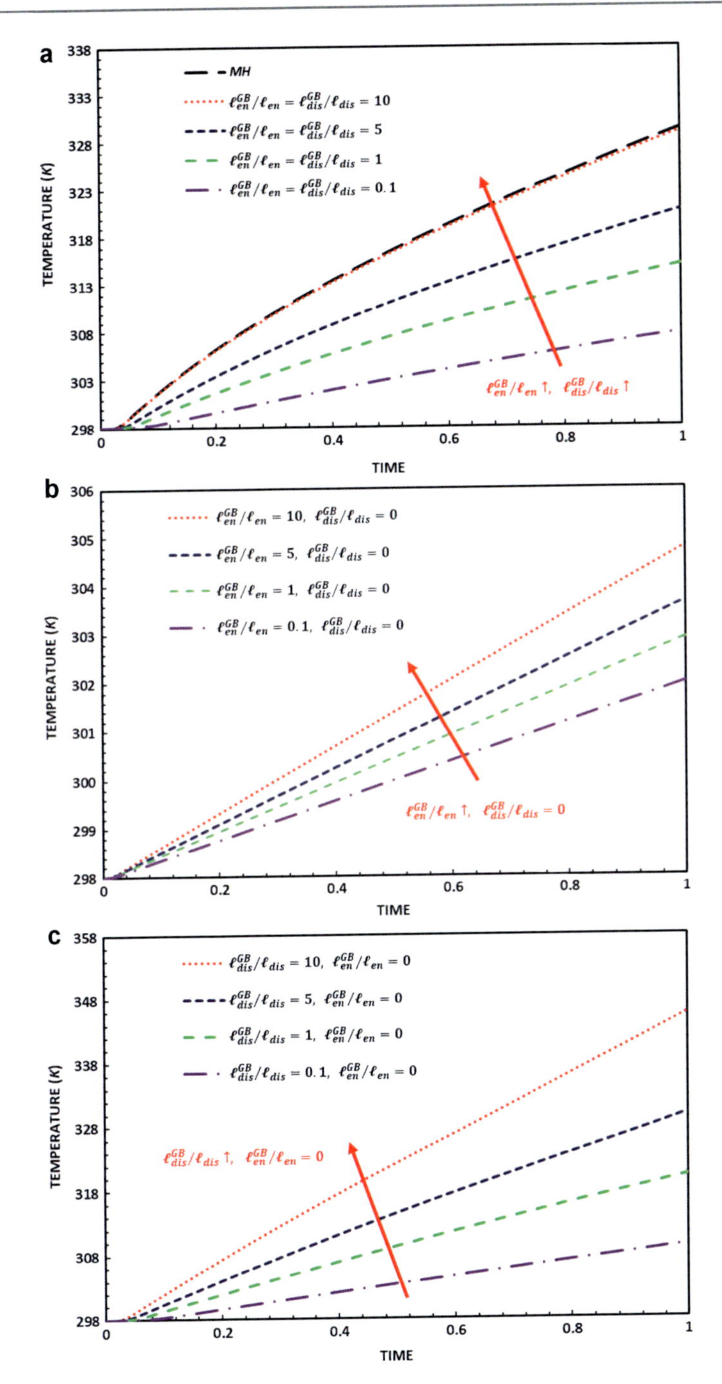

Fig. 22 The evolutions of the maximum temperature according to the various values of ℓ_{en}^{GB}/ℓ_{en} and $\ell_{dis}^{GB}/\ell_{dis}$: (**a**) combined ℓ_{en}^{GB} and ℓ_{dis}^{GB}, (**b**) ℓ_{en}^{GB} only, and (**c**) ℓ_{dis}^{GB} only. (Reprinted with permission from Song and Voyiadjis (2018b))

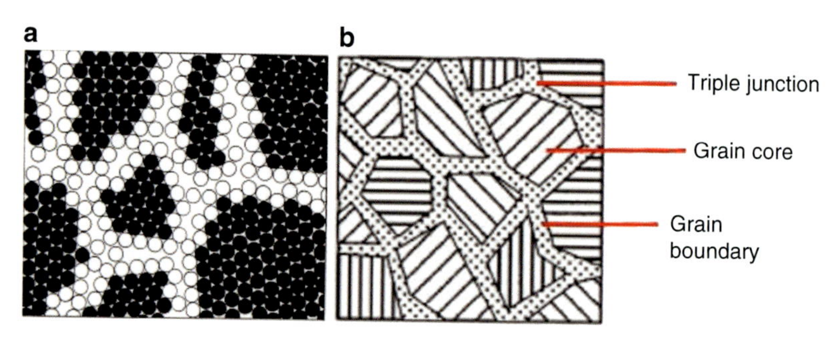

Fig. 23 Two-dimensional model of nanostructured materials (Gleiter 2000; Meyers et al. 2006). (Reprinted with permission from Voyiadjis and Deliktas (2010))

In Fig. 23a, the image represents crystal atoms with neighbor configurations corresponding to the lattice and the boundary atoms with a wide variety of interatomic spacing. The atoms in the centers of the crystals are indicated in black and the one in the boundary regions are represented as open circles. Figure 23b shows the simplified representation of the image of the nanocrystalline structure by comprising grain core, grain boundary, and triple junctions. As one notes from the image in Fig. 23a, many atoms reside in the grain boundary region, and in this case the volume fraction of the interfacial region will not be zero. The small sizes involved in this case limit the conventional operation of the dislocation sources.

The geometrical representation of the RVE proposed by different authors (Mecking and Kocks 1981; Pipard et al. 2009) can be conceptually described by three regions such as the grain core, the grain boundary, and the triple junctions with their corresponding internal interfaces, respectively. In this work the simplified nanocrystalline structure shown in Fig. 23b is represented as a 2D triangle representative volume element (RVE) of a composite material with three phases; grain core, grain boundary, and triple junction (see Fig. 24).

In Fig. 24, if one uses r as the indicator for the phase, then it can be named one of three regions such as the grain core ($r = gc$), the grain boundary ($r = gb$), or the triple junction ($r = tj$). Similarly, one may use the symbol rs to represent the interface surfaces that surround the regions presented in Fig. 24 such that ($rs = cb$) indicates the internal interface between the grain core and grain boundary and ($rs = btj$) shows the internal interface between the grain boundary and the triple junction.

The deformation of each region is attributed to different involved dislocation storage mechanisms and grain boundary sliding or slip. The region called the grain core represents the interior of the grain for which strain hardening only results from the evolution of statistically stored dislocations (SSDs), ρ_s. These are due to multiplication and annihilation processes between dislocations with net Burgers vector equal to zero (Arsenlis and Parks 1999). The SSDs are inherently random within the whole grain. The region with thickness w is termed the grain boundary region and is bounding the grain core. Triple junction is defined as a separate region.

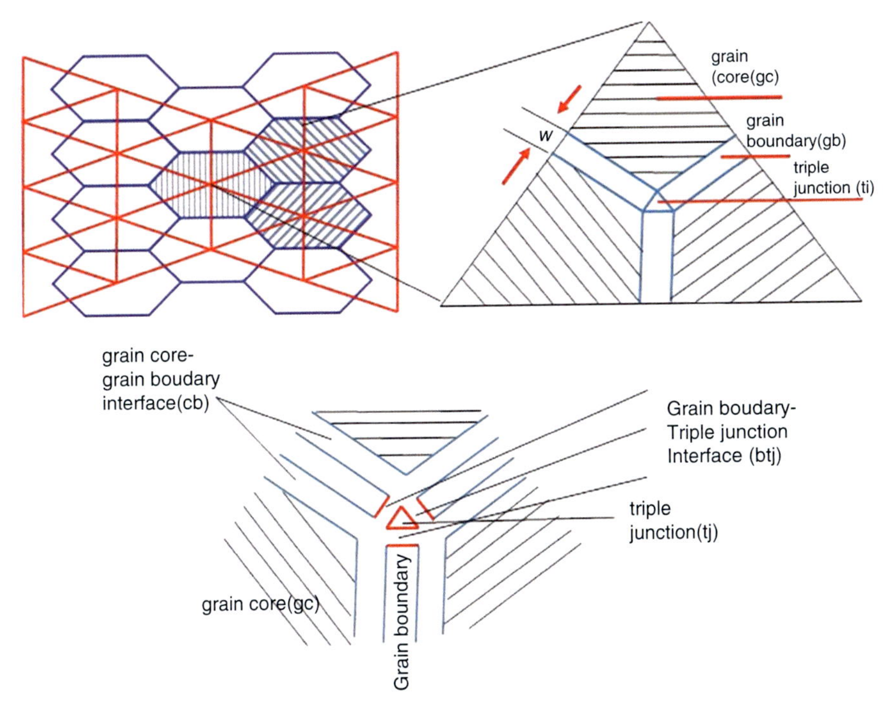

Fig. 24 Representative volume element for the description of the simplified nanocrystalline structure. (Reprinted with permission from Voyiadjis and Deliktas (2010))

The physical origin of the presence of GNDs in this region is due at the mesoscopic scale due to the concentration of slip bands at the grain boundaries such as dislocation pileups. This one has to be relaxed with the help of GNDs. The layer thickness w probably depends on the grain size (since the slip line patterns may vary with grain size) and may vary with strain essentially due to an expansion of the layer with increasing dislocation density. For the sake of simplicity, w will be here assumed to be constant. In addition to these three zones, two interfaces are defined to characterize interactions among the grain core, the grain boundary, and the triple junction. The interface between grain core and the grain boundary that is attributed to the dislocation blocking and pileups can be described as hard, soft, or intermediate interfaces where interfacial hardening and strengthening parameters are introduced based on the theory presented by Voyiadjis and Deliktas (2009b). The second interface which is defined between the grain boundary and the triple junction grain is attributed to the grain boundary sliding or slip which could be described through modeling of the behavior of the triple junction. Full details of the thermodynamical framework for modeling the inelastic behavior of the material and the homogenization scheme for material modeling of the heterogeneous solids developed with a view to address the heterogeneous microstructure and effect of volume fraction of the phases can be found in Voyiadjis and Deliktas (2010).

Multilevel homogenization scheme presented in Voyiadjis and Deliktas (2010) is used for the evaluation of the material model behavior. The local stresses in the grain core, grain boundary, and triple junction are evaluated to better understand the physical mechanisms responsible for the grain size effects on the overall inelastic behavior of the material. The inverse Hall-Petch effects are also investigated, and simulated results are compared with the reported experimental ones. The studied material is Ni-P alloy with a fixed grain boundary width, 4 nm, and varying grain sizes from 10 nm to 200 nm. The material response of each phase is predicted by the constitutive relation given in Voyiadjis and Deliktas (2010). These constitutive relations represent the generalized case of the formulation. For example, if one considers the polycrystalline structured material, the inelastic deformation is mostly dominated in the grain core by dislocation mechanisms (movement and storage), and both plastic strain and its gradient are used in the formulation characterized by the strain gradient formulation. In the case of nanocrystalline structured materials, for the grain size range (Pipard et al. 2009), 30 nm $< d \leq$ 100 nm, it is assumed that the grain core carries only statistically stored dislocations, whereas the grain boundary and the triple junction regions have geometrically necessary dislocations. Therefore, in this case, while the plastic strain gradient is dropping out from the grain core phase, it should be considered in the grain boundary and triple junction phases. In the last case where nanocrystalline grain size is less than 30 nm (Meyers et al. 2006), the dislocation-based inelastic deformation in the grain core phase is essentially shut off, and the inelastic deformation occurs primarily by slip and separation of the grain boundaries. The material constants of the Ni-P are obtained from experimental studies (Benson et al. 2001; Zhao et al. 2003) reported by Qing and Xingming (2006).

The overall stress-strain curves of the Ni-P nanocrystalline material for various grain sizes are presented in Fig. 25. Clearly materials with large grain sizes show more plastic hardening than that with smaller grain sizes. For material grain size that is bigger, the volume fraction of the interphases is negligible, and therefore, the material response is dominated by the behavior of the grain core. However, when the grain size is getting smaller, the overall response is more complex. In this case the material responses of both the grain boundary and the triple junction along with that of the grain core play essential role in the overall behavior of the nanocrystalline material.

Nanocrystalline material exhibits almost an order of magnitude higher strain rate sensitivity than their microcrystalline counterparts. This enhanced strain rate sensitivity of the nanocrystalline materials is captured qualitatively by proposed theory here (Fig. 26).

As one notes clearly from Fig. 26, the nanocrystalline material shows the strong strain rate sensitivity at the grain size of 20 nm. However, as the grain size is increased that of microcrystalline counterpart it is observed that material does not exhibit the strain rate sensitivity.

Finally, in Fig. 27, the relation between the yield strength and the grain size in nanometer Ni-P is shown. By comparison with experimental results, when the critical size of the crystalline grains of nanometer Ni-P is greater than 10 nm, the

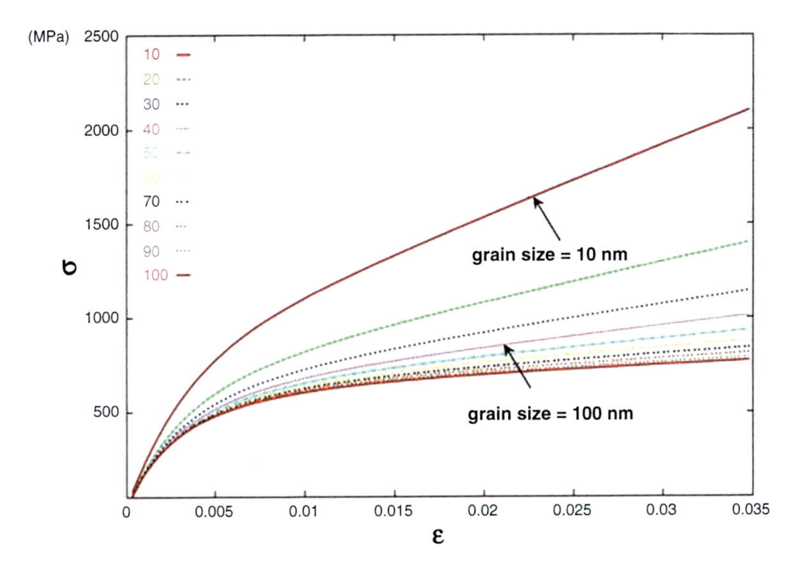

Fig. 25 Overall stress-strain response of nanocrystalline with various grain sizes. (Reprinted with permission from Voyiadjis and Deliktas (2010))

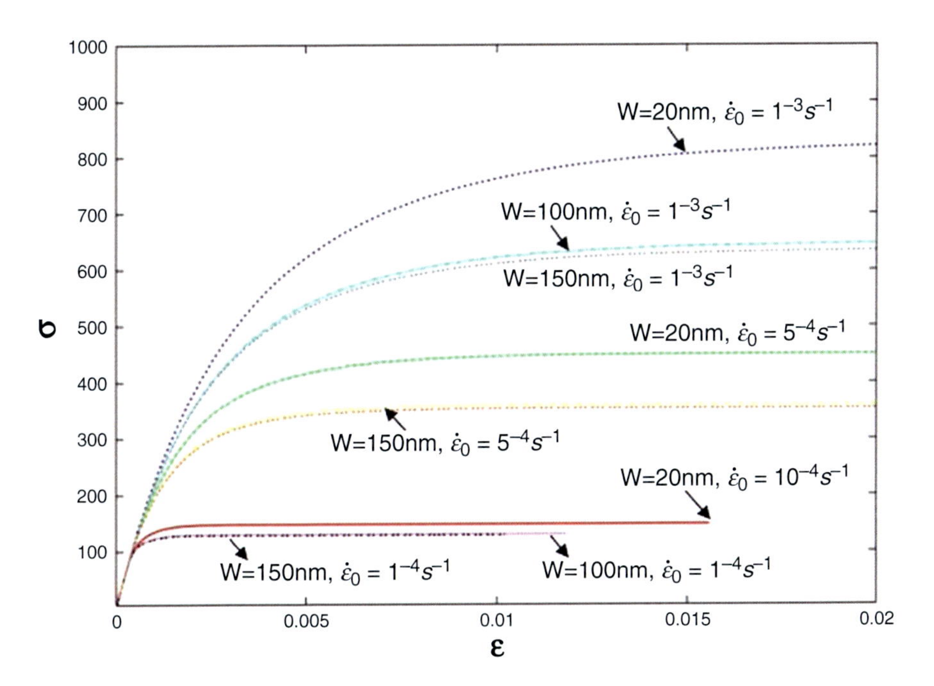

Fig. 26 Stress-strain curves for the nanocrystalline with the grain sizes 20 nm, 100 nm, and 150 nm at the different strain rates. (Reprinted with permission from Voyiadjis and Deliktas (2010))

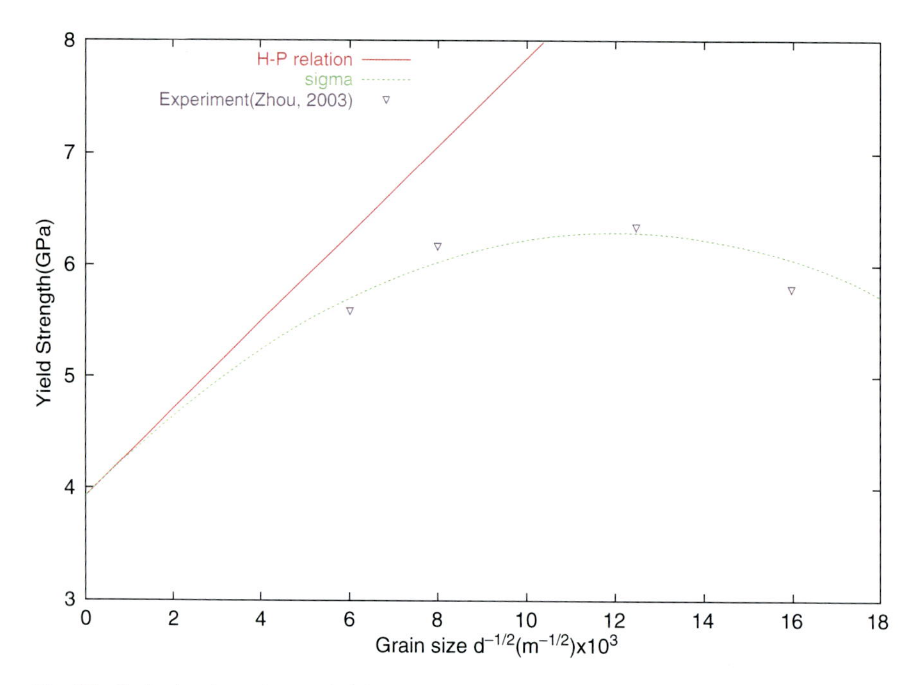

Fig. 27 Grain size dependency of yield strength for the Ni-P material. (Reprinted with permission from Voyiadjis and Deliktas (2010))

positive H-P slope can be seen. However, when the size of the crystalline grain is less than 10 nm, a negative H-P slope can be observed (Fig. 27).

Conclusions

The two-dimensional finite element analysis for the thermodynamically consistent thermomechanical coupled gradient-enhanced plasticity model is proposed within the areas of grain interior and grain boundary and validated by comparing against two sets of small-scale experiments demonstrating the size effects. The proposed formulation is developed based on the concept of dislocation interaction mechanisms and thermal activation energy. The thermodynamic microstresses are assumed to be divided in two components, i.e., the energetic and dissipative components, which in turn, both energetic and dissipative material length scale parameters are incorporated in the governing constitutive equations. These two thermodynamic microstresses can be respectively obtained in a direct way from the Helmholtz free energy and rate of dissipation potential by taking maximum entropy production into account. The concept of GND density is additionally employed in this work to interpret the microstructural strengthening mechanisms induced by the nonhomogeneous deformation. Correspondingly, the model in this work incorporates the terms related to GND-induced strengthening and the additional material length scale parameter. Not only the partial heat dissipation due to fast transient time but also the distribution of the temperature caused by the transition from plastic work to heat is included into the coupled thermomechanical model by deriving a generalized heat equation.

The proposed SGP model and the correspondingly developed finite element code are validated through the comparisons against the experimental measurements from small-scale aluminum and nickel thin film tests. The material parameters for two kinds of metals are also calibrated by using the experimental measurements. The numerical results show good agreement with the experimental measurements in terms of both tests. The simple shear problem and the square plate problem are solved based on the validated model in order to examine size effect in small-scale metallic materials and two null boundary conditions, respectively. The energetic hardening, dissipative strengthening, and GND hardening are well observed, respectively, with varying ℓ_{en}, ℓ_{dis}, and ℓ_{N-G}. The parametric study is carried out to investigate the effect of the temperature-related parameters on the stress-strain curve, the distributions of the temperature and the accumulated plastic strain, and the temperature evolution. The effect of the microhard boundary condition at the grain boundary on the stress-strain curve and the distributions of the plastic strain and the temperature are presented by solving the square plate problem. The strengthening effect due to the microhard boundary condition at the grain boundary is clearly observed with increasing grain boundary areas in this simulation. In addition, the microfree and microhard boundary conditions are well captured by using the proposed grain boundary flow rule. Lastly, the size effects on the stress-strain responses and the evolutions of maximum temperature are well observed with the cases of (a) combined ℓ_{en}^{GB} and ℓ_{dis}^{GB}, (b) ℓ_{en}^{GB} only, and (c) ℓ_{dis}^{GB} only.

A generalized structure for modeling polycrystals from micro- to nano-size range is presented in the last section. The polycrystalline structure is defined in terms of the grain core, the grain boundary, and the triple junction regions with their corresponding volume fractions. This is achieved by describing a simplified nanocrystalline structure as a 2D triangle representative volume element (RVE) of a composite material. It is shown that the developed model is able to capture qualitatively the enhanced strain rate sensitivity of the nanocrystalline materials that exhibit almost an order of magnitude higher strain rate sensitivity than their microcrystalline counterparts. It is also shown that the inverse Hall-Petch is captured from the model prediction that fits quite well to the experimental observations.

Appendix A. Deriving the Balance Equations

The total strain rate $\dot{\varepsilon}_{ij}$ is defined as follows:

$$\dot{\varepsilon}_{ij} = \frac{1}{2}\left(\dot{u}_{i,j} + \dot{u}_{j,i}\right) \tag{103}$$

with the velocity gradient $\dot{u}_{i,j} = \partial \dot{u}_i / \partial x_j$.

By substituting Eq. (2) into Eq. (8), one obtains

$$\mathcal{P}^{int} = \int_{\Omega_0} \left(\sigma_{ij}\dot{\varepsilon}_{ij} - \sigma_{ij}\dot{\varepsilon}^p_{ij} + \mathcal{X}\dot{e}^p + \mathcal{Q}_i \dot{e}^p_{,i} + \mathcal{A}\dot{\mathcal{T}} + \mathcal{B}_i \dot{\mathcal{T}}_{,i} \right) dV \tag{104}$$

From the plastic incompressibility ($\dot{\varepsilon}^p_{kk} = 0$), $\sigma_{ij}\dot{\varepsilon}^p_{ij} = \overline{\sigma}_{ij}\dot{\varepsilon}^p_{ij}$. The divergence theorem can be used here in Eq. (104) together with Eq. (103) to obtain the following expression:

$$\begin{aligned} \mathcal{P}^{int} = & \int_{\partial\Omega_0} \left(\sigma_{ij} n_j \dot{u}_i + \mathcal{Q}_i n_i \dot{e}^p + \mathcal{B}_i n_i \right) dS \\ & - \int_{\Omega_0} \left(\sigma_{ij,j}\dot{u}_i - \overline{\sigma}_{ij}\dot{\varepsilon}^p_{ij} + \mathcal{X}\dot{e}^p - \mathcal{Q}_{i,i}\dot{e}^p + \mathcal{A}\dot{\mathcal{T}} - \mathcal{B}_{i,i}\dot{\mathcal{T}} \right) dV \end{aligned} \tag{105}$$

By equating the external power given in Eq. (9) to the internal power ($\mathcal{P}^{int} = \mathcal{P}^{ext}$), the following expression is obtained:

$$\begin{aligned} & \int_{\partial\Omega_0} \left\{ \left(\sigma_{ij} n_j - t_i \right) \dot{u}_i + \left(\mathcal{Q}_i n_i - m \right) \dot{e}^p + \left(\mathcal{B}_i n_i - a \right) \right\} dS \\ & - \int_{\Omega_0} \left\{ \left(\sigma_{ij,j} + b_i \right) \dot{u}_i + \left(\overline{\sigma}_{ij} N_{ij} - \mathcal{X} + \mathcal{Q}_{i,i} \right) \dot{e}^p + \left(\mathcal{A} - \mathcal{B}_{i,i} \right) \dot{\mathcal{T}} \right\} dV \end{aligned} \tag{106}$$

Here, $\dot{u}_i$, $\dot{e}^p$, and $\dot{\mathcal{T}}$ can be designated randomly when the following conditions are satisfied:

$$\sigma_{ij,j} + b_i = 0 \tag{107}$$

$$\overline{\sigma}_{ij} = \left(\mathcal{X} - \mathcal{Q}_{k,k} \right) N_{ij} \tag{108}$$

$$\mathcal{B}_{i,i} - \mathcal{A} = 0 \tag{109}$$

$$t_j = \sigma_{ij} n_i \tag{110}$$

$$m = \mathcal{Q}_i n_i \tag{111}$$

$$a = \mathcal{B}_i n_i \tag{112}$$

References

Abaqus, *User's Manual (Version 6.12)* (Dassault Systemes Simulia Corporation, Providence, 2012)
E.C. Aifantis, J. Eng. Mater-T Asme. **106**, 4 (1984)
E.C. Aifantis, Int. J. Plast. **3**, 3 (1987)
K.E. Aifantis, J.R. Willis, J. Mech. Phys. Solids **53**, 5 (2005)
A.H. Almasri, G.Z. Voyiadjis, Acta Mech. **209**, 1–2 (2010)
L. Anand, M.E. Gurtin, S.P. Lele, C. Gething, J. Mech. Phys. Solids **53**, 8 (2005)
A. Arsenlis, D.M. Parks, Acta Mater. **47**, 5 (1999)
L. Bardella, J. Mech. Phys. Solids **54**, 1 (2006)
D.J. Benson, H.H. Fu, M.A. Meyers, Mater. Sci. Eng. A **319**, 319–321 (2001)
T. Bergeman, J. Qi, D. Wang, Y. Huang, H.K. Pechkis, E.E. Eyler, P.L. Gould, W.C. Stwalley, R.A. Cline, J.D. Miller, D.J. Heinzen, J. Phys. B-Atom. Mol. Opt. Phys. **39**, 19 (2006)
S.D. Brorson, A. Kazeroonian, J.S. Moodera, D.W. Face, T.K. Cheng, E.P. Ippen, M.S. Dresselhaus, G. Dresselhaus, Phys. Rev. Lett. **64**, 18 (1990)
H.B. Callen, *Thermodynamics and an Introduction to Thermostatistics* (Wiley, Hoboken, 1985)
P. Cermelli, M.E. Gurtin, Int. J. Solids Struct. **39**, 26 (2002)
W.A.T. Clark, R.H. Wagoner, Z.Y. Shen, T.C. Lee, I.M. Robertson, H.K. Birnbaum, Scr. Met. Mater. **26**, 2 (1992)
B.D. Coleman, W. Noll, Arch. Ration. Mech. Anal. **13**, 3 (1963)
D. Faghihi, G.Z. Voyiadjis, J. Eng. Mater-T Asme. **136**, 1 (2014)
N.A. Fleck, J.W. Hutchinson, Adv. Appl. Mech. **33**, 295 (1997)
N.A. Fleck, J.W. Hutchinson, J. Mech. Phys. Solids **49**, 10 (2001)
N.A. Fleck, G.M. Muller, M.F. Ashby, J.W. Hutchinson, Acta Metall. Mater. **42**, 2 (1994)
N.A. Fleck, J.W. Hutchinson, J.R. Willis, P. Roy. Soc. A-Math. Phy. **470**, 2170 (2014)
N.A. Fleck, J.W. Hutchinson, J.R. Willis, J. Appl. Mech-T Asme. **82**, 7 (2015)
S. Forest, J. Eng. Mech-Asce. **135**, 3 (2009)
S. Forest, M. Amestoy, Cr. Mecanique **336**, 4 (2008)
P. Fredriksson, P. Gudmundson, Model. Simul. Mater. Sci. **15**, 1 (2007)
A. Garroni, G. Leoni, M. Ponsiglione, J. Eur. Math. Soc. **12**, 5 (2010)
S. Giacomazzi, F. Leroi, C. L'Henaff, J.J. Joffraud, Lett. Appl. Microbiol. **38**, 2 (2004)
H. Gleiter, Acta Mater. **48**, 1 (2000)
P. Gudmundson, J. Mech. Phys. Solids **52**, 6 (2004)
M.E. Gurtin, Int. J. Plast. **19**, 1 (2003)
M.E. Gurtin, J. Mech. Phys. Solids **52**, 11 (2004)
M.E. Gurtin, J. Mech. Phys. Solids **56**, 2 (2008)
M.E. Gurtin, L. Anand, J. Mech. Phys. Solids **53**, 7 (2005)
M.E. Gurtin, L. Anand, J. Mech. Phys. Solids **57**, 3 (2009)
M.E. Gurtin, E. Fried, L. Anand, *The Mechanics and Thermodynamics of Continua* (Cambridge University Press, Cambridge, 2010)
S. Han, T. Kim, H. Lee, H. Lee, Electronics System-Integration Technology Conference, 2008. ESTC 2008. 2nd (2008)
M.A. Haque, M.T.A. Saif, Acta Mater. **51**, 11 (2003)
J.P. Hirth, J. Lothe, *Theory of Dislocations* (Krieger Publishing Company, 1982). ISBN: 0894646176, 9780894646171
Y. Huang, J. Qi, H.K. Pechkis, D. Wang, E.E. Eyler, P.L. Gould, W.C. Stwalley, J. Phys. B-Atom. Mol. Opt. Phys. **39**, 19 (2006)
J.W. Hutchinson, Acta. Mech. Sinica-Prc. **28**, 4 (2012)
J.S. Hwang, H.L. Park, T.W. Kim, H.J. Lee, Phys. Status Solidi a-Appl. Res. **148**, 2 (1995)
A.M. Ivanitsky, D.A. Kadakov, Izv. Vyssh. Uchebn. Zaved Radiofiz. **26**, 9 (1983)
T.W. Kim, H.L. Park, J. Cryst. Growth **159**, 1–4 (1996)

T.W. Kim, H.L. Park, J.Y. Lee, Appl. Phys. Lett. **64**, 19 (1994)
T.C. Lee, I.M. Robertson, H.K. Birnbaum, Scr. Metall. **23**, 5 (1989)
S.P. Lele, L. Anand, Philos. Mag. **88**, 30–32 (2008)
J.M. Lim, K. Cho, M. Cho, Appl. Phys. Lett. **110**, 1 (2017)
W.J. Liu, K. Saanouni, S. Forest, P. Hu, J. Non-Equil. Thermody. **42**, 4 (2017)
V.A. Lubarda, Int. J. Solids Struct. **45**, 1 (2008)
D.L. McDowell, Int. J. Plast. **26**, 9 (2010)
H. Mecking, U.F. Kocks, Acta Metall. **29**, 1865–1875 (1981)
M.A. Meyers, A. Mishra, D.J. Benson, Prog. Mater. Sci. **51**, 427 (2006)
H.B. Muhlhaus, E.C. Aifantis, Int. J. Solids Struct. **28**, 7 (1991)
L. Nicola, Y. Xiang, J.J. Vlassak, E. Van der Giessen, A. Needleman, J. Mech. Phys. Solids **54**, 10 (2006)
W.D. Nix, H.J. Gao, J. Mech. Phys. Solids **46**, 3 (1998)
T. Ohmura, A.M. Minor, E.A. Stach, J.W. Morris, J. Mater. Res. **19**, 12 (2004)
N. Ohno, D. Okumura, J. Mech. Phys. Solids **55**, 9 (2007)
P.A. Parilla, M.F. Hundley, A. Zettl, Solid State Commun. **87**, 6 (1993)
H.L. Park, S.H. Lee, T.W. Kim, Compd. Semicond. **1995**, 145 (1996)
J.M. Pipard, N. Nicaise, S. Berbenni, O. Bouaziz, M. Berveiller, Comput. Mater. Sci. **45**, 604–610 (2009)
X. Qing, G. Xingming, Int. J. Solids Struct. **43**, 25–26 (2006)
W.A. Soer, K.E. Aifantis, J.T.M. De Hosson, Acta Mater. **53**, 17 (2005)
Y. Song, G.Z. Voyiadjis, Int. J. Solids Struct. **134**, 195–215 (2018a)
Y. Song, G.Z. Voyiadjis, J. Theor. App. Mech-Pol. **56**, 2 (2018b)
S. Sun, B.L. Adams, W.E. King, Philos. Mag. A **80**, 1 (2000)
D.Y. Tzou, Y.S. Zhang, Int. J. Eng. Sci. **33**, 10 (1995)
R. Venkatraman, P.R. Besser, J.C. Bravman, S. Brennan, J. Mater. Res. **9**, 2 (1994)
E. Voce, Meta **51**, 219 (1955)
G.Z. Voyiadjis, B. Deliktas, Int. J. Plast. **25**, 10 (2009a)
G.Z. Voyiadjis, B. Deliktas, Int. J. Eng. Sci. **47**, 11–12 (2009b)
G.Z. Voyiadjis, B. Deliktas, Acta Mech. **213**, 1–2 (2010)
G.Z. Voyiadjis, D. Faghihi, Int. J. Plast. **30–31**, 218 (2012)
G.Z. Voyiadjis, R. Peters, Acta Mech. **211**, 1–2 (2010)
G.Z. Voyiadjis, Y. Song, Philos. Mag. **97**, 5 (2017)
G.Z. Voyiadjis, C. Zhang, Mat. Sci. Eng. A-Struct. **621**, 218 (2015)
G.Z. Voyiadjis, A.H. Almasri, T. Park, Mech. Res. Commun. **37**, 3 (2010)
G.Z. Voyiadjis, D. Faghihi, Y.D. Zhang, Int. J. Solids Struct. **51**, 10 (2014)
G.Z. Voyiadjis, Y. Song, T. Park, J. Eng. Mater. Technol. **139**, 2 (2017)
Y. Xiang, J.J. Vlassak, Acta Mater. **54**, 20 (2006)
M. Yaghoobi, G.Z. Voyiadjis, Acta Mater. **121**, 190 (2016)
C. Zhang, G.Z. Voyiadjis, Mat. Sci. Eng. A-Struct. **659**, 55 (2016)
B. Zhang, Y. Song, G.Z. Voyiadjis, W.J. Meng, J. Mater. Res. **361**(1–2), 160–164 (2018)
J.R. Zhao, J.C. Li, Q. Jiang, J. Alloys Compd. **361**, 160 (2003)

Fractional Nonlocal Continuum Mechanics and Microstructural Models

23

Vasily E. Tarasov

Contents

Abstract

Models of physical lattices with long-range interactions for nonlocal continuum are suggested. The lattice long-range interactions are described by exact fractional-order difference operators. Continuous limit of suggested lattice operators gives continuum fractional derivatives of non-integer orders. The proposed approach gives a new microstructural basis to formulation of theory of nonlocal materials with power-law nonlocality. Moreover these lattice models, which is based on exact fractional differences, allow us to have a unified microscopic description of fractional nonlocal and standard local continuum.

V. E. Tarasov (✉)
Skobeltsyn Institute of Nuclear Physics, Lomonosov Moscow State University, Moscow, Russia
e-mail: tarasov@theory.sinp.msu.ru

G. Z. Voyiadjis (ed.), *Handbook of Nonlocal Continuum Mechanics for Materials and Structures*, https://doi.org/10.1007/978-3-319-58729-5_15

Keywords
Non-local continuum · Lattice model · Long-range interaction · Fractional derivatives · Exact differences

Introduction

Continuum mechanics can be considered as a continuous limit of lattice models, where the length scales of a continuum element are much larger than the distances between the lattice particles. In general, continuum models, which are described by differential equations with a finite number of integer-order derivatives with respect to coordinates, cannot be considered as nonlocal models. An application of general form infinite series with integer derivatives to describe media with nonlocality is a difficult problem. This problem can be solved for power-law type of nonlocality by using fractional derivatives of non-integer orders. It is important to note that the use of the derivatives of non-integer orders is actually equivalent to using an infinite number of derivatives of integer orders (e.g., see Lemma 15.3 in Samko et al. 1993). The fractional-order derivatives allow us to describe continuum with nonlocality of power-law type. First time the fractional derivatives with respect to space coordinates have been applied to mechanics of nonlocal continuum by Gubenko (1957) and Rostovtsev (1959) in 1957. Recently, the fractional-order derivatives are actively used to describe continua with power-law type of nonlocality in Di Paola et al. (2009a,b, 2013), Di Paola and Zingales (2011), Drapaca and Sivaloganathan (2012), Challamel et al. (2013), Atanackovic et al. (2014a,b), Tarasov (2010, 2013, 2014a,b,c,d, 2015a,b,c,d,e,g, 2016b,c,e, 2017), Carpinteri et al. (2009, 2011), Sapora et al. (2013), Cottone et al. (2009a,b), Sumelka and Blaszczyk (2014), Sumelka et al. (2015), and Sumelka (2015), where the microscopic models of fractional continuum mechanics are also discussed. Fractional calculus is a powerful tool to describe processes in continuously distributed media with nonlocality of power-law type. As it was shown in Tarasov (2010, 2006a,b), the continuum equations with fractional derivatives are directly connected to lattice models with long-range interactions. As it was shown in Tarasov (2006a,b), the differential equations with fractional derivatives of non-integer orders can be derived from equation for lattice particles with long-range interactions in the continuous limit, where the distance between the lattice particles tends to zero. A direct connection between the lattice with long-range interaction and nonlocal continuum has been proved by using the special transform operation (Tarasov 2006a,b, 2014e, 2015f) (see also Tarasov and Zaslavsky 2006a,b).

In works Tarasov (2014e, 2015f,h, 2016f,g,h,i, 2017b), it has been suggested exact lattice (discrete) analogs of fractional differential operators of integer and non-integer orders. This mathematical tool allows us to formulate lattice models that are exact discrete (microstructural) analogs of continuum models. The models of lattices with long-range interactions and corresponding models of continua with power-law nonlocality have been suggested in Tarasov (2013, 2014a,b,c,d, 2015a,b,c,e,d, 2016b,c,e, 2017).

Long-Range Interactions of Lattice Particles

Let us consider three-dimensional physical lattices. These lattices are characterized by space periodicity. For unbounded lattices we can use three noncoplanar vectors $\mathbf{a}_1$, $\mathbf{a}_2$, $\mathbf{a}_2$, which are the shortest vectors by which a lattice can be displaced to coincidence with itself. Sites of this lattice can be characterized by the number vector $\mathbf{n} = (n_1, n_2, n_3)$, where n_i $(j = 1, 2, 3)$ are integer. For simplification, we consider a lattice with mutually perpendicular primitive lattice vectors $\mathbf{a}_j$, $(j = 1, 2, 3)$. This means that we use a primitive orthorhombic Bravais lattice. We choose directions of the axes of the Cartesian coordinate system coincide with the vector $\mathbf{a}_j$, such that $\mathbf{a}_j = a_j\,\mathbf{e}_j$, where $a_i = |\mathbf{a}_j| > 0$ and $\mathbf{e}_j$ are the basis vectors of the Cartesian coordinate system. Then the vector $\mathbf{n}$ can be represented as $\mathbf{n} = n_1\mathbf{e}_1 + n_2\mathbf{e}_2 + n_3\mathbf{e}_3$.

Choosing a coordinate origin at one of the lattice sites, then the positions of all other site with $\mathbf{n} = (n_1, n_2, n_3)$ is described by the vector $\mathbf{r}(\mathbf{n}) = n_1\mathbf{a}_1 + n_2\mathbf{a}_2 + n_3\mathbf{a}_3$. The lattice sites are numbered by $\mathbf{n}$. Therefore the vector $\mathbf{n}$ is called the number vector of the corresponding particle. We assume that the equilibrium positions of particles coincide with the lattice sites $\mathbf{r}(\mathbf{n})$. Coordinates $\mathbf{r}(\mathbf{n})$ of lattice sites differ from the coordinates of the corresponding particles, when particles are displaced with respect to their equilibrium positions. To define the coordinates of a particle, we define displacement of this particle from its equilibrium position by the vector field $\mathbf{u}(\mathbf{n}, t) = \sum_{i=1}^{3} u_i(\mathbf{n}, t)\,\mathbf{e}_i$, where $u_i(\mathbf{n}, t) = u_i(n_1, n_2, n_3, t)$ are components of the displacement vector for lattice particle.

To simplify our consideration, we use some assumptions for microscopic (lattice) structure. We assume that lattice n-particle interacts by pair manner with all lattice m-particles that reflects the long-range nature of the interaction in the suggested nonlocal model of material.

In general, it is possible to consider the long-range interactions that are characterized by different orders α_j in different directions $\mathbf{e}_j = \mathbf{a}_j/|\mathbf{a}_j|$. In these models we should use the difference operators of orders α_j, where $\alpha_1 \neq \alpha_2, \alpha_1 \neq \alpha_3, \alpha_2 \neq \alpha_3$.

Let us give a definition of the long-range interaction of power-law type (for details see Tarasov (2006a,b) and Sect. 8 of Tarasov 2010). An interaction of lattice particles is called the interaction of power-law type if the kernels $K(n - m)$ of this interaction satisfy the conditions

$$\lim_{k\to 0} \frac{\hat{K}_\alpha(k) - \hat{K}_\alpha(0)}{|k|^\alpha} = A_\alpha, \quad \alpha > 0, \quad 0 < |A_\alpha| < \infty, \tag{1}$$

where

$$\hat{K}_\alpha(k\Delta x) = \sum_{n=-\infty}^{+\infty} e^{-ikn\Delta x} K(n) = 2\sum_{n=1}^{\infty} K(n)\cos(kn\Delta x). \tag{2}$$

The examples of power-law type interaction are considered in Tarasov (2006a,b, 2010, 2014c,e).

The power-law type of interactions is suggested in the papers Tarasov (2006a,b) (see also Sect. 8 of Tarasov 2010, 2014e, 2015f). This type of interactions is characterized by the power-law asymptotic behavior of spatial dispersions in lattice. We assume that the power-law spatial dispersion in the lattice can be caused by non-Debye screening of electromagnetic interatomic interactions. Some aspects of the theory of this screening are described in the papers Tarasov and Trujillo (2013) and Tarasov (2016a), where fractional-order power-law spatial dispersion in electrodynamics of continuum is discussed. Some elasticity models of materials with power-law spatial dispersion are discussed in Tarasov (2013, 2014a).

Lattices Fractional Integro-Differentiation

In the lattice models, the long-range interactions can be described by using the lattice fractional integro-differential operators that has been suggested in Tarasov (2014e, 2015f, 2016f,g).

Let us give a definition of the lattice fractional integro-differentiation: *Lattice integro-differential operators* ${}^T\mathbb{D}_L\left[\begin{smallmatrix}\alpha\\ j\end{smallmatrix}\right]$ *of order* $\alpha > -1$ *in the direction* $\mathbf{e}_j = \mathbf{a}_j/|\mathbf{a}_j|$ *is an operator that is defined by the equation*

$$\left({}^T\mathbb{D}_L\begin{bmatrix}\alpha\\ j\end{bmatrix} f\right)(\mathbf{n}) = \frac{1}{a_j^{\alpha}} \sum_{m_j=-\infty}^{+\infty} K_{\alpha}(n_j - m_j)\, f(\mathbf{m}), \quad (j = 1, \ldots, N), \tag{3}$$

where $\mathbf{n}, \mathbf{m} \in \mathbb{Z}^N$, *and* $\mathbf{n} = \mathbf{m} + (n_j - m_j)\mathbf{e}_j$. *The kernel* $K_\alpha(n)$ *is a real-valued function of integer variable* $n \in \mathbb{Z}$, *that is defined by the equation*

$$K_{\alpha}(n) = \cos\left(\frac{\pi\,\alpha}{2}\right) \frac{\pi^{\alpha}}{\alpha+1}\, {}_1F_2\left(\frac{\alpha+1}{2}; \frac{1}{2}, \frac{\alpha+3}{2}; -\frac{\pi^2 n^2}{4}\right)$$

$$-\sin\left(\frac{\pi\,\alpha}{2}\right) \frac{\pi^{\alpha+1} n}{\alpha+2}\, {}_1F_2\left(\frac{\alpha+2}{2}; \frac{3}{2}, \frac{\alpha+4}{2}; -\frac{\pi^2 n^2}{4}\right), \tag{4}$$

where $\alpha > -1$. *If* $\alpha > 0$, *then the lattice operator (3) will be called the lattice fractional derivative. For* $-1 < \alpha < 0$, *then the lattice operator (3) is called the lattice fractional integral (antiderivative).*

In kernel (4) of the lattice derivatives, we use the generalized hypergeometric function ${}_1F_2$ that is defined (see Sect. 1.6 of Kilbas et al. 2006) by the equation

$${}_1F_2(a; b, c; z) := \sum_{k=0}^{\infty} \frac{(a)_k}{(b)_k\,(c)_k} \frac{z^k}{k!} = \sum_{k=0}^{\infty} \frac{\Gamma(a+k)\,\Gamma(b)\,\Gamma(c)}{\Gamma(a)\,\Gamma(b+k)\,\Gamma(c+k)} \frac{z^k}{k!}, \tag{5}$$

where $(a)_k$ is the Pochhammer symbol (rising factorial) that is defined by $(a)_0 = 1$, and $(a)_k = \Gamma(a+k)/\Gamma(a)$.

Note that using (5) and $\Gamma(z+1) = z\,\Gamma(z)$, expression (4) of the kernel $K_\alpha(n)$ can be represented in the form

$$K_\alpha(n) = \sum_{k=0}^{\infty} \frac{(-1)^k\, \pi^{2k+1/2+\alpha}\, n^{2k}}{2^{2k}\, k!\, \Gamma(k+1/2)} \left(\frac{\cos\left(\frac{\pi\,\alpha}{2}\right)}{\alpha+2k+1} - \frac{\pi\, n\, \sin\left(\frac{\pi\,\alpha}{2}\right)}{(\alpha+2k+2)\,(2k+1)} \right), \tag{6}$$

where $\alpha > -1$.

Lattice Derivatives of Integer Orders

Let us give exact expressions of lattice fractional derivatives for integer orders (Tarasov 2014e, 2015h,f, 2016f). The lattice derivatives of integer orders are defined by the equations

$$^{\mathcal{T}}\boldsymbol{\Delta}^{2s}\, f[n] := \sum_{\substack{m=-\infty \\ m\neq 0}}^{+\infty} K_{2s}(m)\, f[n-m] + K_{2s}(0)\, f[n], \quad (s \in \mathbb{N}), \tag{7}$$

$$^{\mathcal{T}}\boldsymbol{\Delta}^{2s-1}\, f[n] := \sum_{\substack{m=-\infty \\ m\neq 0}}^{+\infty} K_{2s-1}(m)\, f[n-m], \quad (s \in \mathbb{N}), \tag{8}$$

where the kernels $K_{2s}(m)$ and $K_{2s-1}(m)$ are defined by Eqs. (3) and (6).

The kernels of these lattice operators can be represented (Tarasov 2016f) in a simpler form.

$$K_{2s}(m) = \sum_{k=0}^{s-1} \frac{(-1)^{m+k+s}\,(2s)!\,\pi^{2s-2k-2}}{(2s-2k-1)!}\,\frac{1}{m^{2k+2}} \quad (m \in \mathbb{Z}, \quad m \neq 0), \tag{9}$$

$$K_{2s-1}(m) = \sum_{k=0}^{s-1} \frac{(-1)^{m+k+s+1}\,(2s-1)!\,\pi^{2s-2k-2}}{(2s-2k-1)!}\,\frac{1}{m^{2k+1}} \quad (m \in \mathbb{Z}, \quad m \neq 0). \tag{10}$$

and

$$K_{2s}(0) = \frac{(-1)^s\,\pi^{2s}}{2s+1}, \quad K_{2s-1}(0) = 0. \tag{11}$$

Note that $K_{2s}(0)$ describes a self-interaction of lattice particles. The interaction of different particles is described by $K_s(n-m)$ with $n - m \neq 0$.

The first-order lattice operators is

$$ {}^{\mathcal{T}}\mathbf{\Delta}^{1} f[n] := \sum_{\substack{m=-\infty \\ m\neq 0}}^{+\infty} \frac{(-1)^m}{m} f[n-m]. \tag{12} $$

The lattice antiderivative is defined by the equation

$$ {}^{\mathcal{T}}\mathbf{\Delta}^{-1} f[n] := \sum_{\substack{m=-\infty \\ m\neq 0}}^{+\infty} \pi^{-1}\, Si(\pi\, m)\, f[n-m], \tag{13} $$

where $Si(z)$ is the sine integral.

From Lattice Models to Continuum Models

Using the methods suggested in Tarasov (2006a,b, 2014e, 2015f, 2016f), we can define the operation that transforms a lattice field $u(\mathbf{n})$ into a field $u(\mathbf{r})$ of continuum. For this transformation, we will consider the lattice scalar field $u(\mathbf{n})$ as Fourier series coefficients of some function $\hat{u}(\mathbf{k})$ for $k_j \in [-k_{j0}/2, k_{j0}/2]$, where $j = 1, 2, 3$. At the next step, we use the continuous limit $\mathbf{k}_0 \to \infty$ to obtain $\tilde{u}(\mathbf{k})$. Finally we apply the inverse Fourier integral transformation to obtain the continuum scalar field $u(\mathbf{r})$. Let us describe these steps with details:

Step 1: The discrete Fourier series transform $u(\mathbf{n}) \to \mathcal{F}_{\Delta}\{u(\mathbf{n})\} = \hat{u}(\mathbf{k})$ of the lattice scalar field $u(\mathbf{n})$ is defined by

$$ \hat{u}(\mathbf{k}) = \mathcal{F}_{\Delta}\{u(\mathbf{n})\} = \sum_{n_1,n_2,n_3=-\infty}^{+\infty} u(\mathbf{n})\, e^{-i(\mathbf{k},\mathbf{r}(\mathbf{n}))}, \tag{14} $$

where $\mathbf{r}(\mathbf{n}) = \sum_{j=1}^{3} n_j\, \mathbf{a}_j$, and $a_j = 2\pi/k_{j0}$ are distance between lattice particles in the direction $\mathbf{a}_j$.

Step 2: The passage to the limit $\hat{u}(\mathbf{k}) \to \mathrm{Lim}\{\hat{u}(\mathbf{k})\} = \tilde{u}(\mathbf{k})$, where we use $a_j \to 0$ (or $k_{j0} \to \infty$), allows us to derive the function $\tilde{u}(\mathbf{k})$ from $\hat{u}(\mathbf{k})$. By definition $\tilde{u}(\mathbf{k})$ is the Fourier integral transform of the continuum field $u(\mathbf{r})$, and the function $\hat{u}(k)$ is the Fourier series transform of the lattice field $u(\mathbf{n})$, where

$$ u(\mathbf{n}) = \frac{(2\pi)^3}{k_{10}k_{20}k_{30}} u(\mathbf{r}(\mathbf{n})), $$

and $\mathbf{r}(\mathbf{n}) = \sum_{j=1}^{3} n_j a_j = \sum_{j=1}^{3} 2\pi n_j/k_{j0} \to \mathbf{r}$.

Step 3: The inverse Fourier integral transform $\tilde{u}(\mathbf{k}) \to \mathcal{F}^{-1}\{\tilde{u}(\mathbf{k})\} = u(\mathbf{r})$ is defined by

$$u(\mathbf{r}) = \frac{1}{(2\pi)^3} \iiint_{-\infty}^{+\infty} dk_1 dh_2 dk_3 \, e^{i \sum_{j=1}^{3} k_j x_j} \tilde{u}(\mathbf{k}) = \mathcal{F}^{-1}\{\tilde{u}(\mathbf{k})\}. \tag{15}$$

The combination $\mathcal{F}^{-1} \circ \mathrm{Lim} \circ \mathcal{F}_{\Delta}$ of the operations $\mathcal{F}^{-1}$, Lim, and $\mathcal{F}_{\Delta}$ define the lattice-continuum transform operation

$$\mathcal{T}_{L \to C} = \mathcal{F}^{-1} \circ \mathrm{Lim} \circ \mathcal{F}_{\Delta} \tag{16}$$

that maps lattice models into the continuum models (Tarasov 2006a,b).

The lattice-continuum transform operation $\mathcal{T}_{L \to C}$ as the combination of three operations $\mathcal{F}^{-1} \circ \mathrm{Limit} \circ \mathcal{F}_{\Delta}$ can be applied not only for lattice fields but also for lattice operators. The operation $\mathcal{T}_{L \to C}$ allows us to map of lattice derivatives ${}^{T}\mathbb{D}_L \left[{\alpha \atop i} \right]$ into continuum derivatives ${}^{RT}\mathbb{D}_C \left[{\alpha \atop i} \right]$, by the equation

$${}^{RT}\mathbb{D}_C \left[{\alpha \atop j} \right] f(\mathbf{r}) = \mathcal{T}_{L \to C} \left(\left({}^{T}\mathbb{D}_L \left[{\alpha \atop j} \right] f \right) (\mathbf{n}) \right) \quad (\alpha > -1). \tag{17}$$

The transform operation $\mathcal{T}_{L \to C}$, which maps the lattice operator to the continuum operators, has been defined in Tarasov (2006a,b, 2010, 2016f).

For integer order $\alpha > 0$, the suggested lattice fractional derivatives are directly related to the partial derivatives of integer orders

$${}^{RT}\mathbb{D}_C \left[{m \atop j} \right] f(\mathbf{r}) = \mathcal{T}_{L \to C} \left(\left({}^{T}\mathbb{D}_L \left[{m \atop j} \right] f \right) (\mathbf{n}) \right) = \frac{\partial^m f(\mathbf{r})}{\partial x_j^m}, \tag{18}$$

where $m \in \mathbb{N}$. This means that the lattice fractional derivatives (3) of integer positive values of $\alpha = m \in \mathbb{N}$ are standard partial derivatives of integer orders m.

Lattice Derivatives Are Exact Discretization of Continuum Derivatives

In paper Tarasov (2016f), it has been suggested an exact discretization for derivatives of integer and non-integer orders.

Let us give the definition of the exact discretization of integro-differentiation: *Let* $\mathcal{A}(\mathbb{R}^N)$ *be a function space and* ${}^{RT}\mathbb{D}_C \left[{\alpha \atop j} \right]$ *be an differential operator on a function space* $\mathcal{A}(\mathbb{R}^N)$ *such that*

$${}^{RT}\mathbb{D}_C \left[{\alpha \atop j} \right] f(\mathbf{r}) = g(\mathbf{r}) \quad (\mathbf{r} \in \mathbb{R}^N) \tag{19}$$

for all $f(\mathbf{r}) \in \mathcal{A}(\mathbb{R}^N)$, *where* $g(\mathbf{r}) \in \mathcal{A}(\mathbb{R}^N)$. *The lattice operator* ${}^{T}\mathbb{D}_L \begin{bmatrix} \alpha \\ j \end{bmatrix}$ *will be called an exact discretization of* ${}^{RT}\mathbb{D}_C \begin{bmatrix} \alpha \\ j \end{bmatrix}$ *if the equation*

$$ {}^{T}\mathbb{D}_L \begin{bmatrix} \alpha \\ j \end{bmatrix} f(\mathbf{n}) = g(\mathbf{n}) \quad (\mathbf{n} \in \mathbb{Z}^N) \tag{20} $$

is satisfied for all $f(\mathbf{n}) \in \mathcal{A}(\mathbb{Z}^N)$, *where* $g(\mathbf{n}) \in \mathcal{A}(\mathbb{Z}^N)$.

Condition (20) means that the equality

$$ \left({}^{RT}\mathbb{D}_C \begin{bmatrix} \alpha \\ j \end{bmatrix} f(\mathbf{r}) \right)_{\mathbf{r}=\mathbf{n}} = {}^{T}\mathbb{D}_L \begin{bmatrix} \alpha \\ j \end{bmatrix} f(\mathbf{n}) \tag{21} $$

holds for all $\mathbf{n} \in \mathbb{Z}^N$.

In work Tarasov (2016f), it has been proposed a general principle of algebraic correspondence for an exact discretization. For the lattice fractional calculus, this principle can be formulated in the following form.

Principle of exact correspondence between lattice and continuum theories: *The correspondence between the theories of difference (lattice) and differential (integro-differential) equations lies not so much in the limiting condition when the primitive lattice vectors* $a_j \to 0$ *as in the fact that mathematical operations on these two theories should obey in many cases the same laws.*

This principle is similar to the Dirac's principle of correspondence between the quantum and classical theories. To specify our consideration, the suggested principle of correspondence can be formulated in the following form: *The lattice integro-differential operators, which are exact discretization of continuum integro-differential operators of integer or non-integer orders, should satisfy the same algebraic characteristic relations as the continuum integro-differential operators.* For detail see Tarasov (2016f).

Let us give the necessary condition of exact discretization of derivatives: *The lattice operator (difference)* ${}^{T}\mathbb{D}_L \begin{bmatrix} \alpha_j \\ j \end{bmatrix}$ *is the exact discretization of the continuum operator* ${}^{RT}\mathbb{D}_C \begin{bmatrix} \alpha_j \\ j \end{bmatrix}$ *of order* α_j *if the following condition*

$$ \frac{1}{a_j} \mathcal{F}^{-1} \mathcal{F}_{a,\Delta} \left({}^{T}\mathbb{D}_L \begin{bmatrix} \alpha_j \\ j \end{bmatrix} \right) = {}^{RT}\mathbb{D}_C \begin{bmatrix} \alpha_j \\ j \end{bmatrix} \tag{22} $$

is satisfied for arbitrary values of $a_j > 0$. *If the condition (22) is not satisfied, but the condition*

$$ \lim_{a_j \to 0+} \frac{1}{a_j} \mathcal{F}^{-1} \mathcal{F}_{a,\Delta} \left({}^{T}\mathbb{D}_L \begin{bmatrix} \alpha_j \\ j \end{bmatrix} \right) = {}^{RT}\mathbb{D}_C \begin{bmatrix} \alpha_j \\ j \end{bmatrix} \tag{23} $$

holds, then the lattice operator ${}^{T}\mathbb{D}_L\left[\begin{smallmatrix}\alpha_j\\ j\end{smallmatrix}\right]$ *is the asymptotic (approximate) discretization of the continuum operator* ${}^{RT}\mathbb{D}_C\left[\begin{smallmatrix}\alpha_j\\ j\end{smallmatrix}\right]$. *Here* $\mathcal{F}^{-1}$ *is the inverse Fourier integral transform and* $\mathcal{F}_{a,\Delta}$ *is the Fourier series transform* . It is obvious that the lattice operator, which is an exact discretization, satisfies the condition (23) of the asymptotic discretization also.

Using the suggested lattice derivatives and antiderivative, we can obtain exact lattice analogs of the differential equations of continuum models without using approximations. It should be noted that this discretization allows to obtain lattice (difference) equations whose solutions are equal to the solutions of corresponding differential equations (Tarasov 2016f,h, 2017a). The suggested lattice fractional derivatives greatly simplify the construction of microstructural (lattice) models with long-range interactions in fractional nonlocal theories of continua, media, and fields (Tarasov 2013, 2014a,b,c,d, 2015a,b,c, 2016b,c).

References

T. Atanackovic, S. Pilipovic, B. Stankovic, D. Zorica, *Fractional Calculus with Applications in Mechanics: Vibrations and Diffusion Processes* (Wiley-ISTE, Hoboken, 2014a)

T.M. Atanackovic, S. Pilipovic, B. Stankovic, D. Zorica, *Fractional Calculus with Applications in Mechanics: Wave Propagation, Impact and Variational Principles* (Wiley-ISTE, Hoboken, 2014b)

A. Carpinteri, P. Cornetti, A. Sapora, Static-kinematic fractional operators for fractal and non-local solids. Zeitschrift für Angewandte Mathematik und Mechanik. Appl. Math. Mech. **89**(3), 207–217 (2009)

A. Carpinteri, P. Cornetti, A. Sapora, A fractional calculus approach to nonlocal elasticity. Eur. Phys. J. Spec. Top. **193**, 193–204 (2011)

N. Challamel, D. Zorica, T.M. Atanackovic, D.T. Spasic, On the fractional generalization of Eringen's nonlocal elasticity for wave propagation. C. R. Mec. **341**(3), 298–303 (2013)

G. Cottone, M. Di Paola, M. Zingales, Elastic waves propagation in 1D fractional non-local continuum. Physica E **42**(2), 95–103 (2009a)

G. Cottone, M. Di Paola, M. Zingales, Fractional mechanical model for the dynamics of non-local continuum, in *Advances in Numerical Methods*. Lecture Notes in Electrical Engineering, vol. 11 (Springer, New York, 2009b), Chapter 33. pp. 389–423

M. Di Paola, M. Zingales, Fractional differential calculus for 3D mechanically based non-local elasticity. Int. J. Multiscale Comput. Eng. **9**(5), 579–597 (2011)

M. Di Paola, F. Marino, M. Zingales, A generalized model of elastic foundation based on long-range interactions: integral and fractional model. Int. J. Solids Struct. **46**(17), 3124–3137 (2009a)

M. Di Paola, G. Failla, M. Zingales, Physically-based approach to the mechanics of strong non-local linear elasticity theory. J. Elast. **97**(2), 103–130 (2009b)

M. Di Paola, G. Failla, A. Pirrotta, A. Sofi, M. Zingales, The mechanically based non-local elasticity: an overview of main results and future challenges. Philos. Trans. R. Soc. A. **371**(1993), 20120433 (2013)

C.S. Drapaca, S. Sivaloganathan, A fractional model of continuum mechanics. J. Elast. **107**(2), 105–123 (2012)

V.S. Gubenko, Some contact problems of the theory of elasticity and fractional differentiation. J. Appl. Math. Mech. **21**(2), 279–280 (1957, in Russian)

A.A. Kilbas, H.M. Srivastava, J.J. Trujillo, *Theory and Applications of Fractional Differential Equations* (Elsevier, Amsterdam, 2006) p. 353

N.A. Rostovtsev, Remarks on the paper by V.S. Gubenko, Some contact problems of the theory of elasticity and fractional differentiation. J. Appl. Math. Mech. **23**(4), 1143–1149 (1959)

S.G. Samko, A.A. Kilbas, O.I. Marichev, *Fractional Integrals and Derivatives Theory and Applications* (Gordon and Breach, New York, 1993), p. 1006

A. Sapora, P. Cornetti, A. Carpinteri, Wave propagation in nonlocal elastic continua modelled by a fractional calculus approach. Commun. Nonlinear Sci. Numer. Simul. **18**(1), 63–74 (2013)

W. Sumelka, Non-local KirchhoffLove plates in terms of fractional calculus. Arch. Civil Mech. Eng. **15**(1), 231–242 (2015)

W. Sumelka, T. Blaszczyk, Fractional continua for linear elasticity. Arch. Mech. **66**(3), 147–172 (2014)

W. Sumelka, R. Zaera, J. Fernández-Sáez, A theoretical analysis of the free axial vibration of non-local rods with fractional continuum mechanics. Meccanica. **50**(9), 2309–2323 (2015)

V.E. Tarasov, Continuous limit of discrete systems with long-range interaction. J. Phys. A. **39**(48), 14895–14910 (2006a). arXiv:0711.0826

V.E. Tarasov, Map of discrete system into continuous. J. Math. Phys. **47**(9), 092901 (2006b). arXiv:0711.2612

V.E. Tarasov, *Fractional Dynamics: Applications of Fractional Calculus to Dynamics of Particles, Fields and Media* (Springer, New York, 2010)

V.E. Tarasov, Lattice model with power-law spatial dispersion for fractional elasticity. Centr. Eur. J. Phys. **11**(11), 1580–1588 (2013)

V.E. Tarasov, Fractional gradient elasticity from spatial dispersion law. ISRN Condens. Matter Phys. **2014**, 794097 (13 pages) (2014a)

V.E. Tarasov, Lattice model of fractional gradient and integral elasticity: long-range interaction of Grunwald-Letnikov-Riesz type. Mech. Mater. **70**(1), 106–114 (2014b). arXiv:1502.06268

V.E. Tarasov, Lattice with long-range interaction of power-law type for fractional non-local elasticity. Int. J. Solids Struct. **51**(15–16), 2900–2907 (2014c). arXiv:1502.05492

V.E. Tarasov, Fractional quantum field theory: from lattice to continuum. Adv. High Energy Phys. **2014**, 957863 (14 pages) (2014d)

V.E. Tarasov, Toward lattice fractional vector calculus. J. Phys. A. **47**(35), 355204 (51 pages) (2014e)

V.E. Tarasov, General lattice model of gradient elasticity. Mod. Phys. Lett. B. **28**(7), 1450054 (2014f). arXiv:1501.01435

V.E. Tarasov, Three-dimensional lattice approach to fractional generalization of continuum gradient elasticity. Prog. Frac. Differ. Appl. **1**(4), 243–258 (2015a)

V.E. Tarasov, Fractional-order difference equations for physical lattices and some applications. J. Math. Phys. **56**(10), 103506 (2015b)

V.E. Tarasov, Discretely and continuously distributed dynamical systems with fractional nonlocality, in *Fractional Dynamics*, ed. by C. Cattani, H.M. Srivastava, X.-J. Yang (De Gruyter Open, Berlin, 2015c), Chapter 3, pp. 31–49. https://doi.org/10.1515/9783110472097-003

V.E. Tarasov, Variational principle of stationary action for fractional nonlocal media. Pac. J. Math. Ind. **7**(1), Article 6. [11 pages] (2015d)

V.E. Tarasov, Non-linear fractional field equations: weak non-linearity at power-law non-locality. Nonlinear Dyn. **80**(4), 1665–1672 (2015e)

V.E. Tarasov, Lattice fractional calculus. Appl. Math. Comput. **257**, 12–33 (2015f)

V.E. Tarasov, Lattice model with nearest-neighbor and next-nearest-neighbor interactions for gradient elasticity. Discontinuity Nonlinearity Complex **4**(1), 11–23 (2015g). arXiv:1503.03633

V.E. Tarasov, Exact discrete analogs of derivatives of integer orders: differences as infinite series. J. Math. **2015**, Article ID 134842 (2015h)

V.E. Tarasov, Electric field in media with power-law spatial dispersion. Mod. Phys. Lett. B **30**(10), 1650132 (11 pages) (2016a). https://doi.org/10.1142/S0217984916501323

V.E. Tarasov, Discrete model of dislocations in fractional nonlocal elasticity. J. King Saud Univ. Sci. **28**(1), 33–36 (2016b)

V.E. Tarasov, Three-dimensional lattice models with long-range interactions of Grunwald-Letnikov type for fractional generalization of gradient elasticity. Meccanica. **51**(1), 125–138 (2016c)
V.E. Tarasov, Fractional mechanics of elastic solids: continuum aspects. J. Eng. Mech. **143**(5), (2017). https://doi.org/10.1061/(ASCE)EM.1943-7889.0001074
V.E. Tarasov, Partial fractional derivatives of Riesz type and nonlinear fractional differential equations. Nonlinear Dyn. **86**(3), 1745–1759 (2016e). https://doi.org/10.1007/s11071-016-2991-y
V.E. Tarasov, Exact discretization by Fourier transforms. Commun. Nonlinear Sci. Numer. Simul. **37**, 31–61 (2016f)
V.E. Tarasov, United lattice fractional integro-differentiation. Frac. Calc. Appl. Anal. **19**(3), 625–664 (2016g). https://doi.org/10.1515/fca-2016-0034
V.E. Tarasov, Exact discretization of Schrodinger equation. Phys. Lett. A. **380**(1–2), 68–75 (2016h)
V.E. Tarasov, What discrete model corresponds exactly to gradient elasticity equation?. J. Mech. Mater. Struct. **11**(4), 329–343 (2016i). https://doi.org/10.2140/jomms.2016.11.329
V.E. Tarasov, Exact solution of T-difference radial Schrodinger equation. Int. J. Appl. Comput. Math. (2017a). https://doi.org/10.1007/s40819-016-0270-8
V.E. Tarasov, Exact discretization of fractional Laplacian. Comput. Math. Appl. **73**(5), 855–863 (2017b). https://doi.org/10.1016/j.camwa.2017.01.012
V.E. Tarasov, J.J. Trujillo, Fractional power-law spatial dispersion in electrodynamics. Ann. Phys. **334**, 1–23 (2013). arXiv:1503.04349
V.E. Tarasov, G.M. Zaslavsky, Fractional dynamics of coupled oscillators with long-range interaction. Chaos. **16**(2), 023110 (2006a). arXiv:nlin.PS/0512013
V.E. Tarasov, G.M. Zaslavsky, Fractional dynamics of systems with long-range interaction. Commun. Nonlinear Sci. Numer. Simul. **11**(8), 885–898 (2006b). arXiv:1107.5436

Fractional Differential Calculus and Continuum Mechanics

24

K. A. Lazopoulos and A. K. Lazopoulos

Contents

The present essay is dedicated to Pepi Lazopoulou M.D., adorable wife and mother of the authors for her devotion in our family

K. A. Lazopoulos (✉)
National Technical University of Athens, Rafina, Greece
e-mail: kolazop@mail.ntua.gr

A. K. Lazopoulos
Mathematical Sciences Department, Hellenic Army Academy, Vari, Greece
e-mail: Orfeakos74@gmail.com

G. Z. Voyiadjis (ed.), *Handbook of Nonlocal Continuum Mechanics for Materials and Structures*, https://doi.org/10.1007/978-3-319-58729-5_16

Abstract

The present essay is an attempt to present a meaningful continuum mechanics formulation into the context of fractional calculus. The task is not easy, since people working on various fields using fractional calculus take for granted that a fractional physical problem is set up by simple substitution of the conventional derivatives to any kind of the plethora of fractional derivatives. However, that procedure is meaningless, although popular, since laws in science are derived through differentials and not through derivatives. One source of that mistake is that the fractional derivative of a variable with respect to itself is different from one. The other source of the same mistake is that the well-known derivatives are not able to form differentials. This leads to erroneous and meaningless quantities like fractional velocity and fractional strain. In reality those quantities, that nobody understands what physically represent, alter the dimensions of the physical quantities. In fact the dimension of the fractional velocity is L/T^{α}, contrary to the conventional L/T. Likewise, the dimension of the fractional strain is $L^{-\alpha}$, contrary to the conventional L^0. That fact cannot be justified. Imagine that even in relativity theory, where everything is changed, like time, lengths, velocities, momentums, etc., the dimensions remain constant. Fractional calculus is allowed up to now to change the dimensions and to accept derivatives that are not able to form differentials, according to differential topology laws. Those handicaps have been pointed out in two recent conferences dedicated to fractional calculus by the authors, (K.A. Lazopoulos, in *Fractional Vector Calculus and Fractional Continuum Mechanics, Conference "Mechanics though Mathematical Modelling", celebrating the 70th birthday of Prof. T. Atanackovic,* Novi Sad, 6–11 Sept, Abstract, p. 40, 2015; K.A. Lazopoulos, A.K. Lazopoulos, Fractional vector calculus and fractional continuum mechanics. Prog. Fract. Diff. Appl. **2**(1), 67–86, 2016a) and were accepted by the fractional calculus community. The authors in their lectures (K.A. Lazopoulos, in *Fractional Vector Calculus and Fractional Continuum Mechanics, Conference "Mechanics though Mathematical Modelling", celebrating the 70th birthday of Prof. T. Atanackovic,* Novi Sad, 6–11 Sept, Abstract, p. 40, 2015; K.A. Lazopoulos, in *Fractional Differential Geometry of Curves and Surfaces, International Conference on Fractional Differentiation and Its Applications (ICFDA 2016)*, Novi Sad, 2016b; A.K. Lazopoulos, On *Fractional Peridynamic Deformations, International Conference on Fractional Differentiation and Its Applications, Proceedings ICFDA 2016*, Novi Sad, 2016c) and in the two recently published papers concerning fractional differential geometry of curves and surfaces (K.A. Lazopoulos, A.K. Lazopoulos, On the fractional differential geometry of curves and surfaces. Prog. Fract. Diff. Appl., No **2**(3), 169–186, 2016b) and fractional continuum mechanics (K.A. Lazopoulos, A.K. Lazopoulos, Fractional vector calculus and fractional continuum mechanics. Prog. Fract. Diff. Appl. **2**(1), 67–86, 2016a) added in the plethora of fractional derivatives one more, that called Leibniz L-fractional derivative. That derivative is able to yield differential and formulate fractional differential geometry. Using that derivative the dimensions of the various quantities remain constant and are equal to the dimensions of the conventional quantities. Since the establishment of fractional differential geometry is necessary for dealing with continuum mechanics, fractional differential geometry of curves and surfaces with the fractional field theory will be discussed first. Then the quantities and principles concerning fractional continuum mechanics will be derived. Finally, fractional viscoelasticity Zener model will be presented as application of the proposed theory, since it is of first priority for the fractional calculus people. Hence the present essay will be divided into two major chapters, the chapter of fractional differential geometry, and the chapter of the fractional continuum mechanics. It is pointed out that the well-known historical events concerning the evolution of the fractional calculus will be circumvented, since the goal of the authors is the presentation of the fractional analysis with derivatives able to form differentials, formulating not only fractional differential geometry but also establishing the fractional continuum mechanics principles. For instance, following the concepts of fractional differential and Leibniz's L-fractional derivatives, proposed by the author (K.A. Lazopoulos, A.K. Lazopoulos, Fractional vector calculus and fractional continuum mechanics. Prog. Fract. Diff. Appl. **2**(1), 67–86, 2016a), the L-fractional chain rule is introduced. Furthermore, the theory of curves and surfaces is revisited, into the context of fractional calculus. The fractional tangents, normals, curvature vectors, and radii of curvature of curves are defined. Further, the Serret-Frenet equations are revisited, into the context of fractional calculus. The proposed theory is implemented into a parabola and the curve configured by the Weierstrass function as well. The fractional bending problem of an inhomogeneous beam is also presented, as implementation of the proposed theory. In addition, the theory is extended on manifolds, defining the fractional first differential (tangent) spaces, along with the revisiting first and second fundamental forms for the surfaces. Yet, revisited operators like fractional gradient, divergence, and rotation are introduced, outlining revision of the vector field theorems. Finally, the viscoelastic mechanical Zener system is modelled with the help of Leibniz fractional derivative. The compliance and relaxation behavior of the viscoelastic systems is revisited and comparison with the conventional systems and the existing fractional viscoelastic systems are presented.

Keywords
Fractional Derivative · Fractional Differential · Fractional Stress · Fractional Strain · Fractional Principles in Mechanics · Fractional Continuum Mechanics

Introduction

Fractional Calculus, originated by Leibnitz (1849), Liouville (1832), and Riemann (1876) has recently applied to modern advances in physics and engineering. Fractional derivative models account for long-range (nonlocal) dependence of phenomena, resulting in better description of their behavior. Various material models, based upon fractional time derivatives, have been presented, describing their viscoelastic interaction (Atanackovic 2002; Mainardi 2010). Lazopoulos (2006) has proposed an elastic uniaxial model, based upon fractional derivatives for lifting Noll's axiom of local-action (Carpinteri et al. 2011) have also proposed a fractional approach to nonlocal mechanics. Applications in various physical areas may also be found in various books (Kilbas et al. 2006; Samko et al. 1993; Poldubny 1999; Oldham and Spanier 1974).

Since the need for Fractional Differential Geometry has extensively been discussed in various places, researchers have presented different aspects, concerning fractional geometry of manifolds (Tarasov 2010; Calcani 2012) with applications in fields of mechanics, quantum mechanics, relativity, finance, probabilities, etc. Nevertheless, researchers are raising doubtfulness about the existence of fractional differential geometry and their argument is not easily rejected.

Basically, the classical differential $df(x) = f'(x)dx$ has been substituted by the fractional one introduced by Adda (2001, 1998) in the form:

$$d^a f = g(x)(dx)^a$$

Nevertheless that definition of the differential is valid in the case of positive increments dx, whereas in the case of negative increments, the differential $d^\alpha f(x)$

may be complex. That is exactly the reason why many researchers reasonably reject the existence of fractional differential geometry. However the variable x accepts its own fractional differential:

$$d^{\alpha}x = \sigma(x)(dx)^{\alpha}$$

with $\sigma(x) \neq 1$, differently of the conventional case when $a = 1$, where $\sigma(x)$ is always one. Relating both equations, it appears that:

$$d^{a}f = \frac{g(x)}{\sigma(x)}d^{a}x$$

In this case $d^{\alpha}x$ is always a real quantity accepting positive or negative incremental real values alike. On these bases, the development of fractional differential geometry may be established.

Further, fractal functions exhibiting self-similarity are nondifferentiable functions, but they exhibit fractional differentiability of order $0<\alpha<1$ (see Yao et al. 2005; Carpinteri et al. 2009 Goldmankhaneh et al. 2013; Liang and Su 2007). Goldmankhaneh et al. (2013) introduced the generalized fractional Riemann-Liouville and Caputo-like derivatives for functions defined on fractal sets.

Fractional Calculus in mechanics has been suggested by many researchers, Tarasov (2010, 2008), Drapaca and Sivaloganathan (2012), Sumelka (2014), and Lazopoulos and Lazopoulos (2016a), in problems of continuum mechanics with microstructure where nonlocal elasticity is necessary. Fractional continuum mechanics has been applied to various problems in hydrodynamics (Tarasov 2010; Balankin and Elizarrataz 2012). Recently fractional calculus has been introduced by the author (Lazopoulos 2016a) for the description of peridynamic theory (Silling 2003, 2010). Yet, fractional calculus has been considered as the best frame for describing viscoelastic problems (Atanackovic 2002; Beyer and Kempfle 1995). In addition fractional differential geometry affects rigid body dynamics, in holonomic and nonholonomic systems (Riewe 1996, 1997; Baleanu et al. 2013). Recent applications in quantum mechanics, physics, and relativity demand differential geometry revisited by fractional calculus (Golmankhaneh Ali et al. 2015; Baleanu et al. 2009).

In the present work, the fractional differential established in Lazopoulos and Lazopoulos (2016a) will be recalled along with the introduced Leibniz's L-fractional derivatives. Those differentials are always real and proper for establishing the fractional differential geometry. Correcting the picture of fractional differential of a function, the fractional tangent space of a manifold was defined, introducing also Leibniz's L-fractional derivative that is the only one having physical meaning. Moreover, the present work reviews description of fractional geometry of curves, describing their tangent (first differential) spaces, their normals, the curvature vectors, and the corresponding radii of curvature. In addition, the Serret-Frenet equations will be revisited into the fractional calculus context. The theory is implemented to a parabola, to the Weierstrass function and the beam bending (Lazopoulos et al. 2015), considered as applications of the curves' theory to the solid mechanics. Yet, the theory is extended on manifolds, just to describe the fractional differential geometry of surfaces. Finally outline of fractional vector field theory is included, along with the revisited fractional vector field theorems.

The mechanics researchers have been motivated by the mechanical behavior of disordered (nonhomogeneous) materials with microstructure. Porous materials (Vardoulakis et al. 1998, Ma et al. 2002), colloidal aggregates (Wyss et al. 2005), ceramics, etc., are materials with microstructure that exert strong influence in their deformation. Major factors in determining the material deformation are microcracks, voids, material phases, etc. The nonhomogeneity of the heterogeneous materials has been tackled by various homogenization theories (Bakhalov and Panasenko 1989). Nevertheless, these materials require the lifting of the basic local action axiom of continuum mechanics (Truesdell 1977; Truesdell and Noll 1965). As defined by Noll (1958, 1959) simple materials satisfy the three fundamental axioms:

(a) The principle of determinism
(b) The principle of local action
(c) The principle of material frame-indifference.

Truesdell and Noll (1965) points out in his classic continuum mechanics book: "The motion of body-points at a finite distance from a point $\boldsymbol{x}$ in some shape may be disregarded in calculating the stress at $\boldsymbol{x}$." Material microstructure, inhomogeneities, microcracks, etc., are some of the various important factors that affect the material deformation with nonlocal action. These factors are not considered in the simple materials formulation.

Various theories have been proposed just to introduce a long distance action in the deformation of the materials. One direction considers Taylor's expansion of the strain tensor in the neighborhood of a point, taking in consideration one or two most important terms. Hence gradient strain theories have been appeared in nonlinear form (Toupin 1965), and in linear deformation (Mindlin 1965). Eringen (2002) has also proposed a theory dealing with micropolar elasticity. Mindlin introduced a simpler version of linear gradient theories and an even simpler model has been presented by Aifantis (1999) with his GRADELA model. In these theories, the authors introduced intrinsic material lengths that accompany the higher order derivatives of the strain. Many problems have been solved employing those theories concerning size effects, lifting of various singularities, porous materials (Aifantis 1999, 2003, 2011; Askes and Aifantis 2011), mechanics of microbeams, microplates, and microsheets (Lazopoulos 2004; Lazopoulos et al. 2010).

However various nonlocal elastic theories have been introduced, that are more reliable in taking care of nonsmooth deformations, since integrals are friendlier than derivatives to take care of various nonsmooth phenomena.

Lazopoulos (2006) introduced fractional derivatives of the strain in the strain energy density function in an attempt to introduce nonlocality in the elastic response of materials. Fractional calculus was used by many researchers, not only in the field of Mechanics but mainly in Physics and especially in Quantum Mechanics, to develop the idea of introducing nonlocality. In fact, the history of fractional calculus is dated since seventeenth century. Particle physics, electromagnetics, mechanics of materials, hydrodynamics, fluid flow, rheology, viscoelasticity, optics, electrochemistry and corrosion, and chemical physics are some fields where fractional calculus has been introduced.

Fractional calculus in material deformations has been adopted in solving various types of problems. First we may consider the deformation problems with nonsmooth strain field. Second heterogeneous material deformations may also be studied. Furthermore, time fractional derivative is proved to be more suitable in viscoelastic deformations, since viscoelastic deformations with retarded memory materials may also be discussed. The nonlocal strain effects of deformation problems are concerned by the last type of those problems. There are many studies considering fractional elasticity theory, introducing fractional strain (Drapaca and Sivaloganathan 2012; Carpinteri et al. 2001, 2011; Di Paola et al. 2009; Atanackovic et al. 2008; Agrawal 2008). Recently Sumelka (2014) has presented applications of Fractional Calculus in the nonlocal elastic deformation of Kirchhoff-Love plates and in the rate-independent plasticity. Nevertheless a different definition of fractional strain is yielded in the present work. Jumarie (2012) has proposed modified Riemann-Liouvile derivative of fractional order with an approach to differential geometry of fractional order. In addition Meerschaert et al. (2006) have presented fractional order vector calculus for fractional advection-dispersion. Recent applications of fractional calculus have been appeared in peridynamic theory (Silling 2000; Silling et al. 2003; Evangelatos 2011; Evangelatos and Spanos 2012). Tarasov (2010) has also presented a book including fractional mathematics and its applications to various physics areas. In addition Tarasov (2008) has presented a fractional vector fields theory combining fractals (Feder 1988), and fractional calculus.

Lazopoulos and Lazopoulos (2016a, b) have clarified the geometry of the fractional differential resulting in fractional tangent spaces of the manifolds quite different from the conventional ones. Hence the fractional differential geometry has been established, indispensable for the development of fractional mechanics. It is evident that the definition of the stress and the strain is greatly affected by the tangent spaces. Hence the fractional stress tensors and the fractional strain tensors are quite different from the conventional ones. The linear strain tensors are also revisited. Those basic concepts are important for establishing fractional continuum mechanics.

In the present work, fractional vector calculus is revisited, since the fractional differential of a function is not linearly dependent upon the conventional differential of the variables. Furthermore, the fractional derivative of a variable with respect to itself is different from one. The fractional vector calculus is revisited along with the basic field theorems of Green, Stokes, and Gauss. Applications of the fractional vector calculus to continuum mechanics are presented. The revision in the right and left Cauchy-Green deformation tensors and Green (Lagrange) and Euler-Almanssi strain tensors are exhibited. The change of volume and the surface due to deformation (change of configuration) of a deformable body is also discussed. Further the revisited fractional continuum mechanics principles yielding the fractional continuity and motion equations are also derived.

Moreover, linearization of the strain tensors is performed. The change of fractional volume and fractional surface due to deformation (change of configuration) of a deformable body is also discussed. in addition, the revisited fractional continuum mechanics principles, such as mass conservation (fractional continuity equation) and motion equations (conservation of fractional linear and rotational momentum), are also discussed.

In addition, Bagley et al. (1986) introduced Fractional Calculus in viscoelasticity, and Atanackovic et al. (2002, 2002a) pursue the idea in many applications and in fractional variational problems (Atanackovic et al. 2008) with fractional derivatives. Mainardi (2010) has also discussed the application of fractional calculus in linear viscoelasticity. In Sabatier et al. (2007), there exists a section concerning viscoelastic disordered media. As an application to the present theory, the viscoelastic behavior of Zener model will be revisited using Leibniz fractional time derivatives. Comparison of the proposed model to the existing fractional ones will be discussed. Further the behavior of the proposed model concerning its compliance and relaxation is discussed and compared to the existing fractional ones and the conventional as well.

It is pointed out that the present essay will be divided into two major chapters, the chapter of fractional differential geometry and the chapter of the fractional continuum mechanics. The viscoelasticity Zener model will be discussed into the context of the proposed theory.

Basic Properties of Fractional Calculus

Fractional Calculus has recently become a branch of pure mathematics, with many applications in Physics and Engineering, (Tarasov 2008, 2010). Many definitions of fractional derivatives exist. In fact, fractional calculus originated by Leibniz, looking for the possibility of defining the derivative $\frac{d^n g}{dx^n}$ when $n = \frac{1}{2}$. The various types of the fractional derivatives exhibit some advantages over the others. Nevertheless they are almost all nonlocal, contrary to the conventional ones.

The detailed properties of fractional derivatives may be found in Kilbas et al. (2006), Podlubny (1999), and Samko et al. (1993). Starting from Cauchy formula for the n-fold integral of a primitive function $f(x)$

$$I^n f(x) = \int_{a}^{x} f(s)\,(ds)^n = \int_{a}^{x} dx_n \int_{a}^{x_n} dx_{n-1} \int_{a}^{x_{n-1}} dx_{n-2} \ldots \int_{a}^{x_2} f(x_1)\,dx_1 \quad (1)$$

expressed by:

$${}_{a}I_{x}^{n} f(x) = \frac{1}{(n-1)!} \int_{a}^{x} (x-s)^{n-1} f(s)ds, \quad x > 0, n \in N \quad (2)$$

and

$$ {}_xI_b^n f(x) = \frac{1}{(n-1)!}\int_x^b (s-x)^{n-1} f(s)ds,\ x > 0, n \in N \tag{3} $$

the left and right fractional integral of f are defined as:

$$ {}_aI_x^a f(x) = \frac{1}{\Gamma(\alpha)}\int_a^x \frac{f(s)}{(x-s)^{1-a}}ds \tag{4} $$

$$ {}_xI_b^a f(x) = \frac{1}{\Gamma(\alpha)}\int_x^b \frac{f(s)}{(s-x)^{1-a}}ds \tag{5} $$

In Eqs. 4 and 5 we assume that α is the order of fractional integrals with $0 < a \leq 1$, considering $\Gamma(x) = (x-1)!$ with $\Gamma(\alpha)$ Euler's Gamma function.

Thus the left and right Riemann-Liouville (R-L) derivatives are defined by:

$$ {}_aD_x^a f(x) = \frac{d}{dx}\left({}_aI_x^{1-a}\ f(x)\right) \tag{6} $$

and

$$ {}_xD_b^a f(x) = -\frac{d}{dx}\left({}_bI_x^{1-a}\ f(x)\right) \tag{7} $$

Pointing out that the R-L derivatives of a constant c are nonzero, Caputo's derivative has been introduced, yielding zero for any constant. Thus, it is considered as more suitable in the description of physical systems.

In fact Caputo's derivative is defined by:

$$ {}_a^cD_x^a f(x) = \frac{1}{\Gamma(1-\alpha)}\int_a^x \frac{f'(s)}{(x-s)^a}ds \tag{8} $$

$$ \text{and}\ {}_x^cD_b^a f(x) = -\frac{1}{\Gamma(1-\alpha)}\int_x^b \frac{f'(s)}{(s-x)^a}ds \tag{9} $$

Evaluating Caputo's derivatives for functions of the type: $f(x) = (x-a)^n$ or $f(x) = (b-x)^n$ we get:

$$ {}_a^cD_x^a (x-a)^{\nu} = \frac{\Gamma(\nu+1)}{\Gamma(-\alpha+\nu+1)}(x-a)^{\nu-\alpha}, \tag{10} $$

and for the corresponding right Caputo's derivative:

$${}_{x}D_{b}^{a}(b-x)^{\nu}=\frac{\Gamma(\nu+1)}{\Gamma(-\alpha+\nu+1)}(b-x)^{\nu-\alpha}$$

Likewise, Caputo's derivatives are zero for constant functions:

$$f(x)=c. \tag{11}$$

Finally, Jumarie's derivatives are defined by,

$${}_{a}^{J}D_{x}^{a}f(x)=\frac{1}{\Gamma(1-\alpha)}\frac{d}{dx}\int_{a}^{x}\frac{f(s)-f(a)}{(x-s)^{a}}ds$$

and

$${}_{x}^{J}D_{b}^{a}f(x)=-\frac{1}{\Gamma(1-\alpha)}\frac{d}{dx}\int_{x}^{b}\frac{f(s)-f(b)}{(s-x)^{a}}ds$$

Although those derivatives are accompanied by some derivation rules that are not valid, the derivatives themselves are valid and according to our opinion are better than Caputo's, since they accept functions less smooth than the ones for Caputo. Also Jumarie's derivative is zero for constant functions, basic property, advantage, of Caputo derivative. Nevertheless, Caputo's derivative will be employed in the present work, having in mind that Jumarie's derivative may serve better our purpose.

The Geometry of Fractional Differential

It is reminded, the n-fold integral of the primitive function $f(x)$, Eq. 1, is

$$I^{n}f(x)=\int_{\mathrm{a}}^{x}f(s)\,(ds)^{n} \tag{12}$$

which is real for any positive or negative increment ds. Passing to the fractional integral

$$I^{\alpha}(f(x))=\int_{\mathrm{a}}^{x}f(s)\,(ds)^{\alpha} \tag{13}$$

the integer n is simply substituted by the fractional number α. Nevertheless, that substitution is not at all straightforward. The major difference between passing from

Eqs. 11 to 12 is that although $(ds)^n$ is real for negative values of ds, $(ds)^\alpha$ is complex. Therefore, the fractional integral, Eq. 13, is not compact for any increment ds. Hence the integral of Eq. 13 is misleading. In other words, the differential, necessary for the existence of the fractional integral, Eq. 13, is wrong. Hence, a new fractional differential, real and valid for positive and negative values of the increment ds, should be established.

It is reminded that the a-Fractional differential of a function $f(x)$ is defined by, Adda (1998):

$$d^a f(x) = {}_{\mathrm{a}}^{c}D_x^a f(x)(dx)^a \tag{14}$$

It is evident that the fractional differential, defined by Eq. 14, is valid for positive incremental dx, whereas for negative ones, that differential might be complex. Hence considering for the moment that the increment dx is positive, and recalling that ${}_a^cD_x^a x \neq 1$, the α-fractional differential of the variable x is:

$$d^a x = {}_{\mathrm{a}}^{c}D_x^a x(dx)^a \tag{15}$$

Hence

$$d^a f(x) = \frac{{}_{\mathrm{a}}^{c}D_x^a f(x)}{{}_{\mathrm{a}}^{c}D_x^a x} d^a x \tag{16}$$

It is evident that $d^a f(x)$ is a nonlinear function of dx, although it is a linear function of $d^a x$. That fact suggests the consideration of the fractional tangent space that we propose. Now the definition of fractional differential, Eq. 16, is imposed either for positive or negative variable differentials $d^\alpha x$. In addition the proposed L-fractional (in honor of Leibniz) derivative ${}_a^LD_x^a f(x)$ is defined by,

$$d^a f(x) = {}_a^LD_x^a f(x) d^a x \tag{17}$$

with the Leibniz L-fractional derivative,

$${}_a^LD_x^a f(x) = \frac{{}_a^cD_x^a f(x)}{{}_a^cD_x^a x} \tag{18}$$

Hence only Leibniz's derivative has any geometrical of physical meaning.

In addition, Eq. 3, is deceiving and the correct form of Eq. 3 should be substituted by,

$$f(x) - f(\mathrm{a}) = {}_{\mathrm{a}}^{L}I_x^a \left({}_{\mathrm{a}}^{L}D_x^\alpha f(x) \right) = \frac{1}{\Gamma(\alpha)\Gamma(2-a)} \int_{\mathrm{a}}^{x} \frac{(s-\mathrm{a})^{1-\alpha}}{(x-s)^{1-a}} {}_{\mathrm{a}}^{L}D_x^\alpha f(s) ds \tag{19}$$

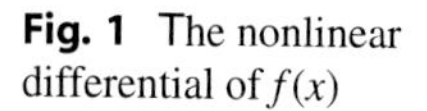
Fig. 1 The nonlinear differential of $f(x)$

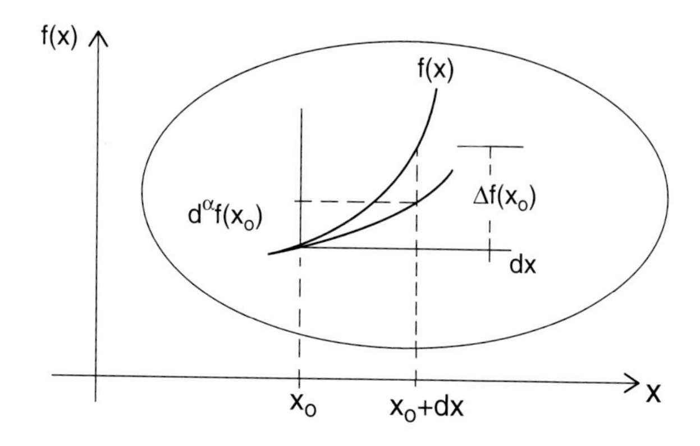

It should be pointed out that the correct forms are defined for the fractional differential by Eq. 17, the Leibniz derivative, Eq. 18 and the fractional integral by Eq. 19. All the other forms are misleading.

Configuring the fractional differential, along with the first fractional differential space (fractional tangent space), the function $y = f(x)$ has been drawn in Fig. 1, with the corresponding first differential space at a point x, according to Adda's definition, Eq. 14.

The tangent space, according to Adda's (1998) definition, Eq. 14, is configured by the nonlinear curve $d^a f(x)$ versus dx. Nevertheless, there are some questions concerning the correct picture of the configuration, (Fig. 1), concerning the fractional differential presented by Adda (1998). Indeed,

(a) The tangent space should be linear. There is not conceivable reason for the nonlinear tangent spaces.
(b) The differential should be configured for positive and negative increments dx. However, the tangent spaces, in the present case, do not exist for negative increments dx.
(c) The axis $d^a f(x)$, in Fig. 1, presents the fractional differential of the function $f(x)$, however dx denotes the conventional differential of the variable x. It is evident that both axes along x and $f(x)$ should correspond to differentials of the same order.

Therefore, the tangent space (first differential space), should be configured in the coordinate system with axes $(d^\alpha x, d^\alpha f(x))$. Hence, the fractional differential, defined by Eq. 17, is configured in the plane $(d^\alpha x, d^\alpha f(x))$ by a line, as it is shown in Fig. 2.

It is evident that the differential space is not tangent (in the conventional sense) to the function at x_0, but intersects the figure $y = f(x)$ at least at one point x_0. This space, we introduce, is the tangent space. Likewise, the normal is perpendicular to the line of the fractional tangent. Hence we are able to establish fractional differential geometry of curves and surfaces with the fractional field theory. Consequently when $\alpha = 1$, the tangent spaces, we propose, coincide with

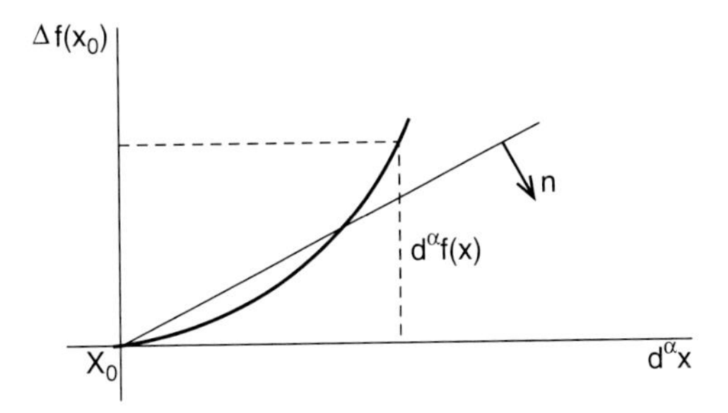

Fig. 2 The virtual tangent space of the $f(x)$ at the point $x = x_0$

the conventional tangent spaces. As a last comment concerning the proposed L-fractional derivative, the physical dimensions of the various quantities remain unaltered from the conventional to any order Fractional Calculus.

Differentiation

The following chapter is a summary from the paragraphs 2.4–2.6 of the book, Differential topology with a view to applications (Chillingworth 1976).

Let E and F be two normed linear spaces with respective norms $\|\cdot\|_E$ and $\|\cdot\|_F$, suppose $f: E \to F$ is a given continuous (not necessarily linear) map, and let x be a particular point of E.

Definition
The map f is differentiable if there exists a linear map L: E→F which approximates f at x in the sense that for all h in E we have:

$$f\,(x+h) - f\,(x) = L(h) + \|h\|_E \eta(h) \tag{20}$$

where $h(x)$ is an element of F with,

$$\|\eta(h)\|_F \to 0 \text{ as } \|h\|_E \to 0 \tag{21}$$

In this definition, E and F are two normed linear spaces with respective norms $\|\cdot\|_E$ and $\|\cdot\|_F$, and $f: E \to F$ is a given continuous (not necessarily linear) map.

Furthermore, let x be a particular point of E and U an open subset of E. The properties and uses of the derivative are:

Linear Combinations

If f, g: $U \to F$ are differentiable then so is the map $\alpha f + \beta g$ for any constants α, β and $D(\alpha f + \beta g) = \alpha Df + \beta Dg$ as maps $U \to L(E,F)$. At this point we must point out that constant maps themselves have derivative zero.

Linear Maps

If f is the restriction to U of a continuous linear map: $E \to F$, then $Df(x) = L$ for every $x \varepsilon U$, i.e. L is already its own linear approximation.

In the special case $E = F = R$ we have $f(x) = \zeta x$ for some number ζ, and $Df(x) = \zeta$ regarded as the linear map $R \to R$: $x \to \zeta x$.

Bilinear Maps

If $E = E_1 \times E_2$ and f is the restriction to U of a continuous bilinear map B: $E_1 \times E_2 \to F$ (i.e. B is linear in each factor separately) then $Df(x)h = B(x_1, h_2) + B(h_1, x_2)$ where $x = (x_1, x_2)$, $h = (h_1, h_2)$ with x_1, h_1 in E_1 and x_2, h_2 in E_2.

Cartesian Products

If f_1: $U_1 \to F_1$ and f_2: $U_2 \to F_2$ are differentiable then so is $f = f_1 \times f_2 : U_1 \times U_2 \to F_1 \times F_2$ where $(f_1 \times f_2)(x_1, x_2)$ means $(f_1\,(x_1), f_2\,(x_2))$, and we have:

$$Df\,(x_1, x_2)\,(h_1 \times h_2) = (Df_1\,(x_1)\,h_1, Df_2\,(x_2)\,h_2) \tag{22}$$

That is, derivatives operate coordinate-wise.

Compositions

Chain Rule

Suppose E, F, and G are three normed linear spaces, and U, V are open sets in E, F, respectively. Let f: $U \to F$ and g: $V \to G$ be continuous maps, and suppose the image of f lies in V so that the composition $g \bullet f$: $U \to G$ exists.

If f is differentiable at x and g is differentiable at $f(x)$ then $g \bullet f$ is differentiable

$$D\,(g \bullet f)\,(x) = Dg\,(f\,(x))\Big) \bullet Df\,(x) \tag{23}$$

In other words, the derivative of a composition is the composition of the derivatives.

If $E = R^n$, $F = R^m$, $G = R^p$, and $y = f(x)$, $z = z(y)$ the chain rule states that:

$$\frac{\partial\left(z_1, z_2, \ldots\ldots, z_p\right)}{\partial\left(x_1, x_2, \ldots\ldots, x_n\right)} = \frac{\partial\left(z_1, z_2, \ldots\ldots, z_p\right)}{\partial\left(y_1, y_2, \ldots\ldots, y_m\right)} \cdot \frac{\partial\left(y_1, y_2, \ldots\ldots, y_m\right)}{\partial\left(x_1, x_2, \ldots\ldots, x_n\right)} \tag{24}$$

Product Rule

For the same maps *f,g* holds:

$$D\,(f\cdot g)=f\cdot D(g)+g\cdot D(f) \tag{25}$$

Let's assume derivative of a product $f(t)=A(t)\cdot b(t)$ where $A(t)$ is an $m\times n$ matrix and $b(t)$ is an n vector. Then:

$$Df\,(t)=A(t)\cdot Db(t)+DA(t)\cdot b(t) \tag{26}$$

Now the question arises whether L-fractional derivative satisfies all the conditions, Eqs. 22, 23, 24, 25, and 26 required by the demand of differentiation according to differential topology rules. The answer is no. There are some rules that the L-fractional derivative satisfies, like the linearity condition, Eq. 17 that cannot be satisfied by the common fractional derivatives like Caputo etc., since the nonlinear Eq. 14 holds for them. However, there are other conditions, such as the chain rule that are not valid. Now we have to make a choice. Either to define differential that is necessary for establishing fractional differential geometry that is necessary for dealing with problems in Physics, or to forget fractional calculus. On that dilemma we make the choice to *impose* the necessary rules, like fractional chain rule, just for forming fractional differential geometry. In that case the physical problem leaves its print or trace on mathematics, that in conventional calculus is not necessary, since all differential rules are satisfied by themselves, in the conventional case.

The Fractional Arc Length

Let $y=f(x)$ be a function, which may be non-differentiable but has a fractional derivative of order α, $0<\alpha<1$. The fractional differential of $y=f(x)$ in the differential space is defined by:

$$d^a y=\frac{{}_aD_x^a f(x)}{{}_aD_x^a x}d^a x={}_a^L D_x^a f(x)d^a x \tag{27}$$

Therefore the arc length is expressed by:

$$s_1\,(x,a)={}_aI_x^a\left[(d^a y)^2+(d^a x)^2\right]^{1/2}={}_aI_x^a\left[\left(\frac{{}_aD_x^a f(x)}{{}_aD_x^a x}\right)^2+1\right]^{\frac{1}{2}}d^a x \tag{28}$$

Furthermore, for parametric curves of the type:

$$y=f(t),\ x=g(t) \tag{29}$$

The fractional α-differentials are defined by:

$$\begin{aligned} d^a x &= \frac{{}_aD_t^a g(t)}{{}_aD_t^a t} d^a t \\ d^a y &= \frac{{}_aD_t^a f(t)}{{}_aD_t^a t} d^a t \end{aligned} \tag{30}$$

and the fractional differential of the arc length is expressed by:

$$d^a s = \sqrt{(d^a y)^2 + (d^a x)^2 = \left[\left(\frac{{}_aD_t^a f(t)}{{}_aD_t^a t}\right)^2 + \left(\frac{{}_aD_t^a g(t)}{{}_aD_t^a t}\right)^2\right]^{\frac{1}{2}} d^a t} \tag{31}$$

and

$$\begin{aligned} s &= {}_a^L I_x^a d^a s = {}_a^L I_t^a \left[\left(\frac{{}_aD_t^a f(t)}{{}_aD_x^a t}\right)^2 + \left(\frac{{}_aD_t^a g(t)}{{}_aD_x^a t}\right)^2\right]^{\frac{1}{2}} d^\alpha t \\ &= {}_a^L I_t^a \left[\left({}_a^L D_t^a f(t)\right)^2 + \left({}_a^L D_t^a g(t)\right)^2\right]^{\frac{1}{2}} d^\alpha t \end{aligned} \tag{32}$$

The Fractional Tangent Space

Let $\mathbf{r} = \mathbf{r}(s)$ be a natural representation of a curve C, where s is the α-fractional length of the curve. Since the velocity of a moving material point on the curve $r(s)$ defines the tangent space, the fractional tangent space of the curve $\mathbf{r} = \mathbf{r}(s)$ is defined by the first derivative:

$$\mathbf{r}_1 = \frac{d^a \mathbf{r}}{d^a s} = \frac{{}_aD_s^a \mathbf{r}}{{}_aD_s^a s} = {}_a^L D_s^a \mathbf{r} \tag{33}$$

Recalling

$$d^a \left|\mathbf{r}\right| = d^a s \tag{34}$$

the length $|\mathbf{r}_1|$ of the fractional tangent vector is unity.

The tangent space line of the curve $\mathbf{r} = \mathbf{r}(s)$ at the point $\mathbf{r}_0 = \mathbf{r}(s_0)$ is defined by:

$$\mathbf{r} = \mathbf{r}_0 + k\mathbf{t}_0 \quad 0 < k < \infty \tag{35}$$

where $\mathbf{t}_0 = \mathbf{t}(s_0)$ is the unit tangent vector at $\mathbf{r}$.

The plane through $\mathbf{r}_0$, orthogonal to the tangent line at $\mathbf{r}_o$, is called the normal plane to the curve C at s_0.The points $\mathbf{y}$ of that orthogonal plane are defined by:

$$(\mathbf{y} - \mathbf{r}_0) \cdot \mathbf{t}\,(s_0) = (\mathbf{y} - \mathbf{r}_0) \cdot \mathbf{r}_1\,(s_0) = 0 \tag{36}$$

Fractional Curvature of Curves

Considering the fractional tangent vector:

$$\mathbf{t} = \mathbf{r}_1(s) = \frac{{}_aD_s^a \mathbf{r}}{{}_aD_s^a s} = {}_a^L D_s^a \mathbf{r} \tag{37}$$

its fractional derivative may be considered:

$$\mathbf{r}_2(s) = \frac{d^a \mathbf{t}}{d^a s} = \frac{{}_aD_s^a \mathbf{t}}{{}_aD_s^a s} = {}_a^L D_s^a \mathbf{t} = \mathbf{t}_1(s) \tag{38}$$

The vector $\mathbf{t}_1(\mathrm{s})$ is called the fractional curvature vector on C at the point $\mathbf{r}(s)$ and is denoted by $\boldsymbol{\kappa} = \boldsymbol{\kappa}(s) = \mathbf{t}_1(s)$.

Since $\mathbf{t}$ is a unit vector

$$\mathbf{t} \cdot \mathbf{t} = 1 \tag{39}$$

Restricted to fractional derivatives that yield zero for a constant function, such as Caputo's derivatives, the curvature vector $\mathbf{t}_1(s)$ on C is orthogonal to $\mathbf{t}$ and parallel to the normal plane. The magnitude of the fractional curvature vector:

$$\kappa = |\boldsymbol{\kappa}(s)| \tag{40}$$

is called the fractional curvature of C at $\mathbf{r}(\mathrm{s})$.The reciprocal of the curvature κ is the fractional radius of curvature at $\mathbf{r}(\mathrm{s})$:

$$\rho = \frac{1}{\kappa} = \frac{1}{|\boldsymbol{\kappa}(s)|} \tag{41}$$

The Fractional Radius of Curvature of a Curve

Following Porteous (1994), for the fractional curvature of a plane curve $\mathbf{r}$, we study at each point $\mathbf{r}(t)$ of the curve, how closely the curve approximates there to a parameterized circle. Now in the tangent or first differential space at a point $\mathbf{r}(t_0)$, the circle, with center c and radius ρ, consists of all $\mathbf{r}(t)$ in the differential space such that:

$$(\mathbf{r} - \mathbf{c}) \cdot (\mathbf{r} - \mathbf{c}) = \rho^2 \tag{42}$$

Further Eq. 42 yields:

$$\mathbf{c} \cdot \mathbf{r} - \frac{1}{2}\mathbf{r} \cdot \mathbf{r} = \frac{1}{2}\left(\mathbf{c} \cdot \mathbf{c} - \rho^2\right) \tag{43}$$

with the right hand side been constant.

Therefore the derivation of the function

$$\mathbf{V}(\mathbf{c}) : t \to \mathbf{c} \cdot \mathbf{r}(t) - \frac{1}{2}\mathbf{r}(t) \cdot \mathbf{r}(t) \tag{44}$$

Hence:

$$V(\mathbf{c})_1 = (\mathbf{c} - \mathbf{r}(t)) \cdot \mathbf{r}_1(t) = 0 \tag{45}$$

$$V(\mathbf{c})_2 = (\mathbf{c} - \mathbf{r}(t)) \cdot \mathbf{r}_2(t) - \mathbf{r}_1(t) \cdot \mathbf{r}_1(t) = 0 \tag{46}$$

Suppose that r is a parametric curve with $\mathbf{r}(t)$ in the virtual tangent space. Then $V(\mathbf{c})_1(t) = 0$ when the vector $\mathbf{c} - \mathbf{r}(t)$ in the tangent space is orthogonal to the tangent vector $\mathbf{r}_1(t)$. Indeed when the point $\mathbf{c}$, in the tangent space, lies on the normal to $\mathbf{r}_1(t)$ at t, the line through $\mathbf{r}(t)$ is orthogonal to the tangent line.

When $\mathbf{r}_2(t)$ is not linearly dependent upon $\mathbf{r}_1(t)$, there will be a unique point $\mathbf{c} \neq \mathbf{r}(t)$,on the normal line, such that also $V(\mathbf{c})_2(t) = 0$.

The Serret-Frenet Equations

Let $\mathbf{r}$ be a curve with unit speed, where the fractional velocity vector, (Porteous 1994),

$$\mathbf{t}(s) = \mathbf{r}_1(s) = \frac{{}_a^c D_s^a \mathbf{r}(s)}{{}_a^c D_s^a s} = {}_a^L D_s^a \mathbf{r}(s) \tag{47}$$

is of unit length.

Let $\mathbf{r}(s)$ be such a curve. The vector

$$\mathbf{t}_1(s) = \mathbf{r}_2(s) = \frac{{}_a^c D_s^a \mathbf{r}_1(s)}{{}_a^c D_s^a s} = \frac{{}_a^c D_s^a}{{}_a^c D_s^a s}\left(\frac{{}_a^c D_s^a \mathbf{r}(s)}{{}_a^c D_s^a s}\right) = {}_a^L D_s^a \left({}_a^L D_s^a \mathbf{r}(s)\right) \tag{48}$$

is normal to the curve $\mathbf{r} = \mathbf{r}(s)$ since $\mathbf{t}(s) \cdot \mathbf{r}(s) = 1$ and

$$\mathbf{t}_1(s) \cdot \mathbf{t}(s) = 0 \tag{49}$$

since for Caputo's derivative ${}_a D_s^a c = 0$ for any constant c.

Consider $\mathbf{t}_1(s)\ \kappa(s)\mathbf{n}(s)$, where $\mathbf{n}(s)$is the unit principal normal to $\mathbf{r}$ at s, provided that $\kappa(s) \neq 0$ where $\kappa(s)$ is the curvature of $\mathbf{r}$ at s.

Hence the equations for the focal line are defined by

$$\begin{aligned} (\mathbf{c} - \mathbf{r}(s)) \cdot \mathbf{r}_1(s) &= 0 \\ (\mathbf{c} - \mathbf{r}(s)) \cdot \kappa(s)\mathbf{n}(s) &= 1 \end{aligned} \tag{50}$$

Thus, the principal center of curvature $\mathbf{c}$ at s is the point $\mathbf{r}(s) + \rho(s)\mathbf{n}(s)$, where $\rho(s) = \frac{1}{\kappa(s)}$. Furthermore, the principal normal vector $\mathbf{n}(s)$ orthogonal to the tangent line is pointing towards the focal line (locus of the curvature centers). Likewise, the (unit) binormal $\mathbf{b}(s)$ is defined to be the vector $\mathbf{t}(s) \times \mathbf{n}(s)$, the triad of unit vectors $\mathbf{t}(s)$, $\mathbf{n}(s)$, $\mathbf{b}(s)$ forming a right-handed orthonormal basis for the tangent vector space to the curvature $\mathbf{r}(s)$. Each of the derivative vectors $\mathbf{t}_1(s)$, $\mathbf{n}_1(s)$, $\mathbf{b}_1(s)$ linearly depends on $\mathbf{t}(s)$, $\mathbf{n}(s)$, $\mathbf{b}(s)$. Considering the equations: $\mathbf{t_1}\,\mathbf{t} = 0$ and $\mathbf{t_1}\,\mathbf{n} = 0$ with $\mathbf{t_1}\,\mathbf{n} + \mathbf{n_1} \cdot \mathbf{t} = 0$, we get the fractional Sarret-Frenet equations:

$$\begin{aligned} \mathbf{t_1} &= \kappa\mathbf{n} \\ \mathbf{n_1} &= -\kappa\mathbf{t} + \tau\mathbf{b} \\ \mathbf{b_1} &= -\tau\mathbf{n} \end{aligned} \tag{51}$$

The coefficient τ is defined to be the torsion of the curve $\mathbf{r}$. These equations are the fractional equations for the fractional Serret-Frenet system. Considering plane curves,

$$\mathbf{r}(x) = x\mathbf{i} + y(x)\mathbf{j} \tag{52}$$

Equations 45 and 46 defining the fractional centers of curvature $\mathbf{c} = c_1\mathbf{i} + c_2\mathbf{j}$ become,

$$\begin{aligned} &(c_1 - x) + (c_2 - y(x))\, {}_a^L D_x^\alpha y(x) = 0 \\ &(c_2 - y(x))\, {}_a^L D_x^\alpha \left({}_a^L D_x^\alpha y(x)\right) - \left(1 + {}_a^L D_x^\alpha y(x)^2\right) = 0 \end{aligned} \tag{53}$$

Since the fractional radius of curvature is defined by

$$\boldsymbol{\rho}^\alpha = \rho_1^\alpha \mathbf{i} + \rho_2^\alpha \mathbf{j} = (c_1 - x)\,\mathbf{i} + (c_2 - y(x))\,\mathbf{j} \tag{54}$$

the components of the fractional curvature are given by

$$\begin{aligned} \rho_1^\alpha &= -\frac{1 + {}_a^L D_s^\alpha y(x)^2}{{}_a^L D_s^\alpha \left({}_a^L D_s^\alpha y(x)\right)}\, {}_a^L D_s^\alpha y(x) \\ \rho_2^\alpha &= \frac{1 + {}_a^L D_s^\alpha y(x)^2}{{}_a^L D_s^\alpha \left({}_a^L D_s^\alpha y(x)\right)} \end{aligned} \tag{55}$$

Further, for the case of $|y(x)| << 1$, that we consider in linear bending, Eq. 28 yields,

$$d^a s = d^a x + o(d^a x)^2 \tag{56}$$

with

$${}_a^L D_s^\alpha () = {}_a^L D_x^\alpha () \tag{57}$$

and

$$\rho^\alpha = |\mathbf{r}^\alpha| \approx \frac{1}{{}_a^L D_x^\alpha \left({}_a^L D_x^\alpha y(x) \right)} \tag{58}$$

Let us consider a fractional beam with its source point $(x,y,z) = (0,0,0)$. That means, the fractional u Caputo's derivatives of any function, concerning the beam, are defined by

$${}_0^c D_u^a f(u) = \frac{1}{\Gamma(1-\alpha)} \int_0^u \frac{f'(s)}{(u-s)^a} \mathrm{ds}$$

where, u might be one of the variables (x, y, z).

Applications

The Fractional Geometry of a Parabola

Let $\mathbf{r}$ be a parabola $t \to (t, t^2)$.Then we have

$$\mathbf{r}(t) = t\mathbf{e}_1 + t^2\mathbf{e}_2 \tag{59}$$

Hence:

$$\mathbf{r}_1(t) = \mathbf{e}_1 + \frac{{}_a^c D_t^a (t^2)}{{}_a^c D_t^a t}\mathbf{e}_2 = \mathbf{e}_1 + \frac{2t}{2-a}\mathbf{e}_2 \tag{60}$$

and

$$\mathbf{r}_2(t) = \frac{2}{2-a}\mathbf{e}_2$$

Then the centers of curvature of the parabola describe a curve:

$$\mathbf{c}(t) = c_1(t)\mathbf{e}_1 + c_2(t)\mathbf{e}_2 \tag{61}$$

Satisfying Eqs. 45 and 46 with

$$c_1 + \frac{2t}{2-a}c_2 = t + \frac{2t^3}{2-a} \tag{62}$$

$$\frac{2}{2-a}c_2 = \frac{2t^2}{2-a} + 1 + \frac{4t^2}{(2-a)^2} \tag{63}$$

Solving the system of Eqs. 62 and 63 we get:

$$c_1 = -\frac{4t^3}{(-2+a)^2} \tag{64}$$

$$c_2 = -\frac{4 + a^2 + 8t^2 - 2a\left(2+t^2\right)}{(4-2a)} \tag{65}$$

Figure 3 shows the tangent space of the parabola at the point $t = 1.5$ for various values of the fractional dimension $\alpha = (1, 0.7, 0.3)$.

It is clear that the tangent spaces for $\alpha = 0.7$ and $\alpha = 0.3$ intersect the parabola at the point $t = 1.5$, although the conventional tangent space with fractional dimension $\alpha = 1.0$ touches the parabola at $t = 1.5$.

Furthermore the centers of curvature for various values of the fractional dimension α the point $t = 1{,}5$ are (for the conventional case):

$\alpha = 1.0$	$c_1 = -13.5$	and $c_2 = 7.25$
$\alpha = 0.7$	$c_1 = -7.98$	and $c_2 = 6.36$
$\alpha = 0.3$	$c_1 = -4.67$	and $c_2 = 5.75$

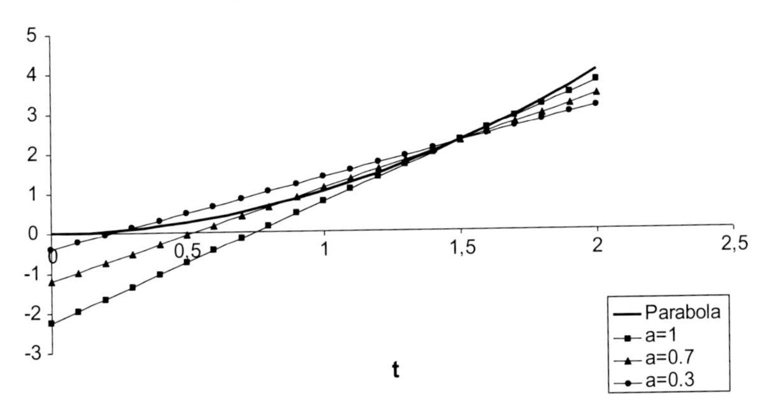

Fig. 3 The parabola with its tangent spaces at $t = 1.5$ for $\alpha = 1$, $\alpha = 0.7$, $\alpha = 0.3$

The Tangent and Curvature Center of the Weierstrass Function

Let us consider the function,

$$W(t) = \sum_{n=1}^{\infty} \lambda^{-\alpha n} \left\{ \sin\left(\frac{\lambda^n t}{2}\right) - \sin\left(2\lambda^n t\right) \right\} \tag{66}$$

the well-known Weierstrass function, continuous with discontinuous conventional derivatives at any point, (Liang and Su 2007). The parameter α has been proved to be related to the fractional dimension of the function $W(t)$. Restricting the function to $w(t)$ with

$$w(t) = \sum_{n=1}^{6} \lambda^{-\alpha n} \left\{ \sin\left(\frac{\lambda^n t}{2}\right) - \sin\left(2\lambda^n t\right) \right\} \tag{67}$$

and for $\alpha = 0.5$ and $\lambda = 2$, the fractional tangent to the curve at the point $t = 1.0$ has been drawn, (Fig. 4), with the help of the Mathematica computerized pack.

Bending of Fractional Beams

Considering the pure fractional bending problem of a beam with microcracks, microvoids, various other defects, we get the fractional strain:

$$\varepsilon_{xx}^{\alpha} = -\frac{y}{\rho^{\alpha}} \tag{68}$$

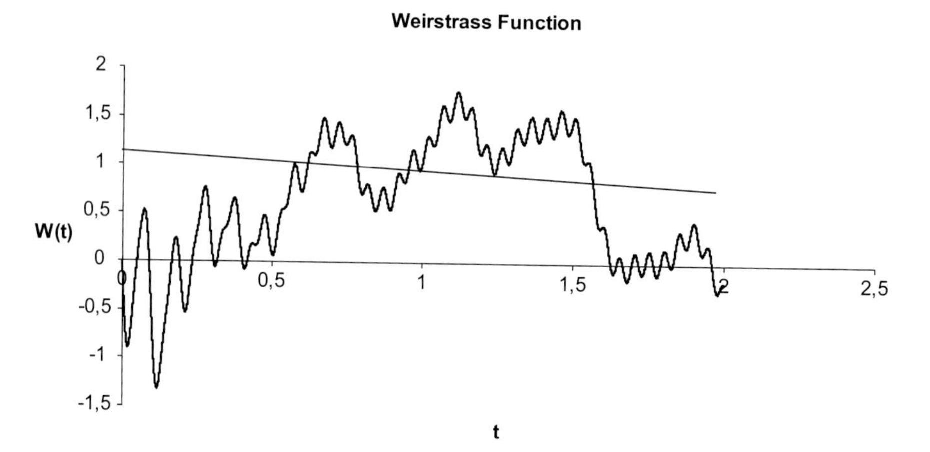

Fig. 4 The function $w(t)$ with its fractional ($\alpha = 0.5$) tangent at $t = 1.0$

where the fractional curvature is defined by,

$$\frac{1}{\rho^{\alpha}} = D_2 w(x) = \frac{1}{{}_aD_x^{\alpha} x} {}_aD_x^{\alpha} \left(\frac{{}_aD_x^{\alpha} w(x)}{{}_aD_x^{\alpha} x} \right), \tag{69}$$

with $w(x)$ denoting the elastic line of the beam.

Likewise, the fractional bending moment is expressed by:

$$M = -2 \int_0^{h/2} \sigma_{xx}^{\alpha} y \, d^{\alpha} y = \frac{EI^{\alpha}}{\rho^{\alpha}} \tag{70}$$

with the fractional stress, (see Lazopoulos et al. 2015),

$$\sigma_{xx}^{\alpha} = -\frac{M}{I^{\alpha}} y \tag{71}$$

Hence the fractional bending of beams formula is revisited and expressed by:

$$M = EI^{\alpha} D_2 w(x) = \frac{EI^{\alpha}}{{}_0^cD_x^{\alpha} x} {}_0^cD_x^{\alpha} \left(\frac{{}_0^cD_x^{\alpha} w(x)}{{}_0^cD_x^{\alpha} x} \right) \tag{72}$$

Therefore, the deflection curve $w(x)$ is defined by:

$$w(x) = \int_0^x \left(\int_0^s \frac{M(t)}{\mathrm{EI}^{\alpha}} d^{\alpha} t \right) d^{\alpha} s + c_1 x + c_2 \tag{73}$$

In conventional integration, the deflection curve is defined by:

$$w(x) = \int_0^x \frac{s^{1-\alpha}}{\Gamma(2-\alpha)} \left(\int_0^s \frac{M(t)}{\mathrm{EI}^{\alpha}} \frac{t^{1-\alpha}}{\Gamma(2-\alpha)} \frac{1}{\Gamma(\alpha)(s-t)^{1-\alpha}} \mathrm{dt} \right) \frac{\mathrm{ds}}{(x-s)^{1-\alpha}} + c_1 x + c_2 \tag{74}$$

The Fractional Tangent Plane of a Surface

Let us consider a manifold, with points $M(u,v)$, defined by the vectors

$$M(u, v) = \mathbf{x}(u, v) \tag{75}$$

with,

$$x_i = x_i(u, v), \; u_1 \le u \le u_2, \; v_1 \le v \le v_2, \; i = 1, 2, 3 \tag{76}$$

The infinitesimal distance between two points P and Q on the manifold M is defined by,

$$d^{\alpha}\mathbf{x} = \frac{{}_{a}^{c}D_{u}^{\alpha}\mathbf{x}}{{}_{a}^{c}D_{u}^{\alpha}u} d^{\alpha}u + \frac{{}_{a}^{c}D_{u}^{\alpha}\mathbf{x}}{{}_{a}^{c}D_{u}^{\alpha}v} d^{\alpha}v \tag{77}$$

In fact for the surface

$$z = u^2 v^2 \tag{78}$$

see, Fig. 2, the tangent space according to Eq. 77 is expressed by:

$$d^{\alpha}\mathbf{r} = d^{\alpha}x\,\mathbf{i} + d^{\alpha}y\,\mathbf{j} + \frac{2xy}{(2-\alpha)}\left(yd^{\alpha}x + xd^{\alpha}y\right)\mathbf{k} \tag{79}$$

Figures 5 and 6 shows the surface defined by Eq. 78 with its fractional tangent

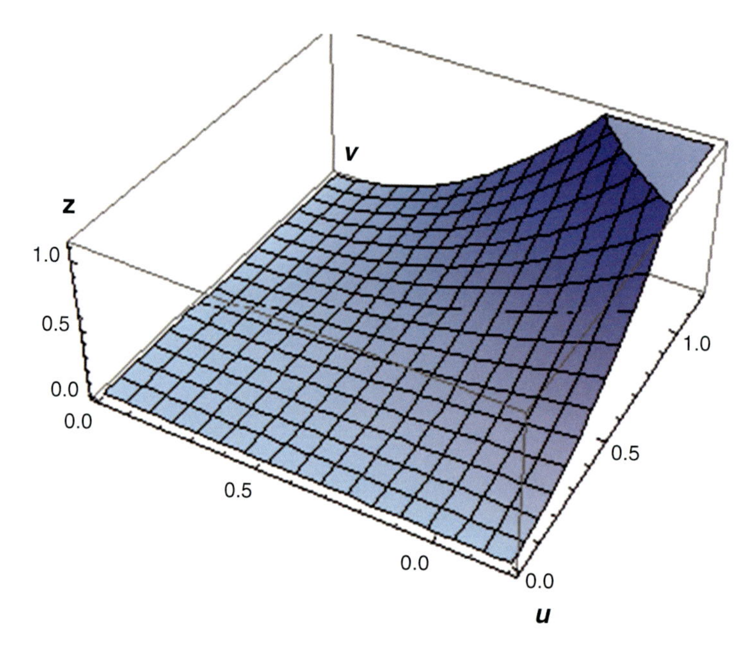

Fig. 5 The surface $z = u^2v^2$ plane (space) at the point $(u,v) = (0.5, 0.5)$ for two fractional dimensions, $\alpha = 1$ (the conventional case) and $\alpha = 0.3$. It is clear that the fractional tangent plane is different from the conventional one ($\alpha = 1$).

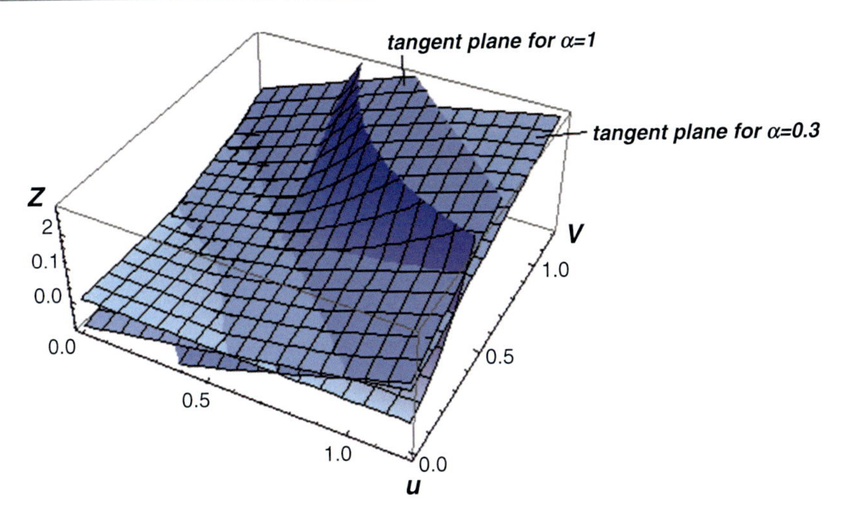

Fig. 6 The tangent planes for various values of the fractional dimension α

Fundamental Differential Forms on Fractional Differential Manifolds

The First Fractional Fundamental Form

Following formal procedure (Guggenheimer 1977), the quantity

$$
\begin{aligned}
I^{\alpha} = d^{\alpha}\mathbf{x} \cdot d^{\alpha}\mathbf{x} &= \left(\frac{{}_{a}^{c}D_{u}^{\alpha}\mathbf{x}}{{}_{a}^{c}D_{u}^{\alpha}u} d^{\alpha}u + \frac{{}_{\mathrm{a}}^{c}D_{v}^{\alpha}\mathbf{x}}{{}_{\mathrm{a}}^{c}D_{v}^{\alpha}v} d^{\alpha}v \right) \cdot \left(\frac{{}_{\mathrm{a}}^{c}D_{u}^{\alpha}\mathbf{x}}{{}_{\mathrm{a}}D_{u}^{\alpha}u} d^{\alpha}u + \frac{{}_{\mathrm{a}}^{c}D_{v}^{\alpha}\mathbf{x}}{{}_{\mathrm{a}}D_{v}^{\alpha}v} d^{\alpha}v \right) \\
&= E\ d^{\alpha}u^{2} + 2F d^{\alpha}u\ d^{\alpha}v + G\ d^{\alpha}v^{2}
\end{aligned}
\tag{80}
$$

defined upon the tangent space of the manifold, as it has been clarified earlier, the I^{α} stands for the first fractional differential form, with the dot meaning the inner product.

$$
\begin{aligned}
E &= \frac{{}_{a}^{c}D_{u}^{\alpha}\mathbf{x}}{{}_{a}^{c}D_{u}^{\alpha}u} \cdot \frac{{}_{a}^{c}D_{u}^{\alpha}\mathbf{x}}{{}_{a}^{c}D_{u}^{\alpha}u} \\
F &= \frac{{}_{a}^{c}D_{u}^{\alpha}\mathbf{x}}{{}_{a}^{c}D_{u}^{\alpha}u} \cdot \frac{{}_{a}^{c}D_{v}^{\alpha}\mathbf{x}}{{}_{a}^{c}D_{v}^{\alpha}v} \\
G &= \frac{{}_{a}^{c}D_{v}^{\alpha}\mathbf{x}}{{}_{a}^{c}D_{v}^{\alpha}v} \cdot \frac{{}_{a}^{c}D_{v}^{\alpha}\mathbf{x}}{{}_{a}^{c}D_{v}^{\alpha}v}
\end{aligned}
\tag{81}
$$

corresponding to

$$I^{\alpha} = E\ du^2 + 2F du\ dv + G\ dv^2$$

Furthermore the first fundamental form is positive definite, i.e., $0 \leq I^{\alpha}$ with $I^{\alpha} = 0$ if and only if $d^{\alpha}u$ and $d^{\alpha}v$ are equal to zero. Hence,
$EG - F^2 > 0$.

The Second Fractional Fundamental Form

Consider the manifold $M(u, v) = \mathbf{x}(u, v)$. Then, at each point of the manifold, there is a fractional unit normal N to the fractional tangent plane

$$N = \frac{\frac{{}_aD_u^{\alpha}\mathbf{x}}{{}_aD_u^{\alpha}u} \times \frac{{}_aD_v^{\alpha}\mathbf{x}}{{}_aD_v^{\alpha}v}}{\left|\frac{{}_aD_u^{\alpha}\mathbf{x}}{{}_aD_u^{\alpha}u} \times \frac{{}_aD_v^{\alpha}\mathbf{x}}{{}_aD_v^{\alpha}v}\right|} \tag{82}$$

that is a function of u and v with the fractional differential

$$d^{\alpha}\mathbf{N} = \frac{{}_aD_u^{\alpha}\mathbf{N}}{{}_aD_u^{\alpha}u} d^{\alpha}u + \frac{{}_aD_u^{\alpha}\mathbf{N}}{{}_aD_u^{\alpha}v} d^{\alpha}v \tag{83}$$

Restricting only to Caputo's fractional derivatives with the property of zero fractional derivative of any constant, and taking into consideration that $\mathbf{N}{\cdot}\mathbf{N} = 1$,we get,

$$d^{\alpha}\mathbf{N} \cdot \mathbf{N} = 0 \tag{84}$$

where the vector d^{α}N is parallel to the fractional tangent space.

The second fractional fundamental form is defined by Guggenheimer (1977),

$$\begin{aligned} \mathrm{II}^{\alpha} &= -d^{\alpha}\mathbf{x} \cdot d^{\alpha}\mathbf{N} = -\left(\frac{{}_a^cD_u^{\alpha}\mathbf{x}}{{}_aD_u^{\alpha}u} d^{\alpha}u + \frac{{}_a^cD_u^{\alpha}\mathbf{x}}{{}_aD_u^{\alpha}v} d^{\alpha}v\right) \cdot \left(\frac{{}_a^cD_u^{\alpha}\mathbf{N}}{{}_aD_u^{\alpha}u} d^{\alpha}u + \frac{{}_a^cD_u^{\alpha}\mathbf{N}}{{}_aD_u^{\alpha}v} d^{\alpha}v\right) \\ &= L\ d^{\alpha}u^2 + 2M d^{\alpha}u\ d^{\alpha}v + N d^{\alpha}v^2 \end{aligned} \tag{85}$$

with,

$$\begin{aligned} L &= -\frac{{}_a^cD_u^{\alpha}\mathbf{x}}{{}_aD_u^{\alpha}u} \cdot \frac{{}_a^cD_u^{\alpha}\mathbf{N}}{{}_aD_u^{\alpha}u} \\ M &= -\frac{1}{2}\left(\frac{{}_a^cD_u^{\alpha}\mathbf{x}}{{}_a^cD_u^{\alpha}u} \cdot \frac{{}_a^cD_v^{\alpha}\mathbf{N}}{{}_a^cD_v^{\alpha}v} + \frac{{}_a^cD_u^{\alpha}\mathbf{N}}{{}_a^cD_u^{\alpha}u} \cdot \frac{{}_a^cD_v^{\alpha}\mathbf{x}}{{}_a^cD_v^{\alpha}v}\right) \\ N &= -\frac{{}_a^cD_v^{\alpha}\mathbf{x}}{{}_a^cD_v^{\alpha}v} \cdot \frac{{}_a^cD_u^{\alpha}\mathbf{N}}{{}_a^cD_v^{\alpha}v} \end{aligned} \tag{86}$$

It is pointed again that the geometric procedures, that use quantities not defined upon the correct tangent spaces, are questionable. Even if analytically may yield the same results, geometrically are confusing.

The Fractional Normal Curvature

Let P be a point on a surface $\mathbf{x} = \mathbf{x}(u, v)$ and $\mathbf{x}\,(t) = \mathbf{x}(u(t), v(t))$ a regular curve C at P. The fractional curvature of curves has been discussed in ▶ Chap. 6, "Uniqueness of Elastoplastic Properties Measured by Instrumented Indentation" . The normal curvature $\mathbf{k}_n^\alpha$ vector of C at P is the vector projection of the curvature vector $\mathbf{k}^\alpha$ onto the normal vector $\mathbf{N}$ at P. The component of $\mathbf{k}^\alpha$ in the direction of the normal $\mathbf{N}$ is called the normal fractional curvature of C at P and is denoted by k_n^α. Therefore,

$$k_n^\alpha = \mathbf{k}^\alpha \cdot \mathbf{N} \tag{87}$$

Since the unit tangent to C at P is the vector

$$\mathbf{t} = \frac{d^\alpha \mathbf{x}}{d^\alpha s} = \frac{d^\alpha \mathbf{x}}{d^\alpha t} / \left| \frac{d^\alpha \mathbf{x}}{d^\alpha t} \right| \tag{88}$$

where s denotes the fractional arc length of the curve and $\mathbf{t}$ is the unit perpendicular to the normal $\mathbf{N}$ along the curve, we get:

$$0 = \frac{d^\alpha\,(\mathbf{t}\cdot\mathbf{N})}{d^\alpha t} = \frac{d^\alpha \mathbf{t}}{d^\alpha t}\cdot\mathbf{N} + \mathbf{t}\cdot\frac{d^\alpha \mathbf{N}}{d^\alpha t} \tag{89}$$

Therefore, the normal curvature of a curve is equal to:

$$\begin{aligned}
k_n^\alpha &= \mathbf{k}\cdot\mathbf{N} = \frac{d^\alpha \mathbf{t}}{d^\alpha t}\cdot\mathbf{N} / \left|\frac{d^\alpha \mathbf{x}}{d^\alpha t}\right| = -\mathbf{t}\cdot\frac{d^\alpha \mathbf{N}}{d^\alpha t} / \left|\frac{d^\alpha \mathbf{x}}{d^\alpha t}\right| \\
&= -\frac{d^\alpha \mathbf{x}}{d^\alpha t}\cdot\frac{d^\alpha \mathbf{N}}{d^\alpha t} / \left|\frac{d^\alpha \mathbf{x}}{d^\alpha t}\right|^2 \\
&= -\frac{\left(\frac{{}_a^c D_u^\alpha \mathbf{x}}{{}_a^c D_u^\alpha u}\frac{d^\alpha u}{d^\alpha t} + \frac{{}_a^c D_u^\alpha \mathbf{x}}{{}_a^c D_u^\alpha v}\frac{d^\alpha v}{d^\alpha t}\right)\cdot\left(\frac{{}_a^c D_u^\alpha \mathbf{N}}{{}_a^c D_u^\alpha u}\frac{d^\alpha u}{d^\alpha t} + \frac{{}_a^c D_u^\alpha \mathbf{N}}{{}_a^c D_u^\alpha v}\frac{d^\alpha v}{d^\alpha t}\right)}{\left(\frac{{}_a^c D_u^\alpha \mathbf{x}}{{}_a^c D_u^\alpha u}\frac{d^\alpha u}{d^\alpha t} + \frac{{}_a^c D_u^\alpha \mathbf{x}}{{}_a^c D_u^\alpha v}\frac{d^\alpha v}{d^\alpha t}\right)\cdot\left(\frac{{}_a^c D_u^\alpha \mathbf{x}}{{}_a^c D_u^\alpha u}\frac{d^\alpha u}{d^\alpha t} + \frac{{}_a^c D_u^\alpha \mathbf{x}}{{}_a^c D_u^\alpha v}\frac{d^\alpha v}{d^\alpha t}\right)} \\
&= \frac{L(d^\alpha u/d^\alpha t)^2 + 2M\,(d^\alpha u/d^\alpha t)\,(d^\alpha v/d^\alpha t) + N(d^\alpha v/d^\alpha t)^2}{E(d^\alpha u/d^\alpha t)^2 + 2F\,(d^\alpha u/d^\alpha t)\,(d^\alpha v/d^\alpha t) + G(d^\alpha v/d^\alpha t)^2}
\end{aligned} \tag{90}$$

Recalling Eqs. 87 and 90, the normal curvature is defined by:

$$k_n^\alpha = \frac{\mathrm{II}^\alpha}{I^\alpha} \tag{91}$$

Fractional Vector Operators

In the present section the fractional tangent spaces along with their fractional normal vectors should be reminded, as they were defined in the preceding sections "The Geometry of Fractional Differential," "The Fractional Arc Length," and "The Fractional Tangent Space."

For Cartesian coordinates, fractional generalizations of the divergence or gradient operators are defined by:

$$\nabla^{(a)} f(\mathbf{x}) = \mathrm{grad}^{(a)} f(\mathbf{x}) = \nabla_i^{(\alpha)} f(x)\mathbf{e}_i = \frac{{}_{\omega}^{c}D_i^a f(\mathbf{x})}{{}_{\omega}^{c}D_{\underline{i}}^a x_{\underline{i}}} + \mathbf{e}_i = {}_{\omega}^{L}D_i^a f(x)\mathbf{e}_i \tag{92}$$

where ${}_{\omega}^{c}D_i^a$ are Caputo's fractional derivatives of order α and the subline meaning no contraction. Further, ${}_{\omega}^{L}D_i^\alpha f(\mathbf{x})$ is Leibniz's derivative, Eq. 18. Hence, the gradient of the vector $\mathbf{x}$ is

$$\nabla^{(\alpha)}\mathbf{x} = \mathbf{I} \tag{93}$$

with $\mathbf{I}$ denoting the identity matrix. Consequently for a vector field

$$\mathbf{F}(x_1, x_2, x_3) = \mathbf{e}_1 F_1(x_1, x_2, x_3) + \mathbf{e}_2 F_2(x_1, x_2, x_3) + \mathbf{e}_3 F_3(x_1, x_2, x_3) \tag{94}$$

where $F_i(\mathrm{x}_1,\mathrm{x}_2,\mathrm{x}_3)$ are absolutely integrable, the circulation is defined by:

$$C_L^{(\alpha)}(\mathbf{F}) = \left({}_{\omega}I_L^{(a)}, \mathbf{F}\right) = \int_L (\mathrm{dL}, \mathbf{F}) = {}_{\omega}I^{(a)}{}_L(F_1 d^\alpha x_1) + {}_{\omega}I^{(a)}{}_L(F_2 d^\alpha x_2) + {}_{\omega}I^{(a)}{}_L(F_3 d^\alpha x_3) \tag{95}$$

Furthermore, the divergence of a vector $\mathbf{F(x)}$ is defined by:

$$\nabla^{(a)} \cdot \mathbf{F}(x) = \mathrm{div}^{(a)}\mathbf{F}(x) = \frac{{}_{\omega}^{c}D_k^a F_k(\mathbf{x})}{{}_{\omega}^{c}D_{\underline{k}}^a x_{\underline{k}}} = {}_{\omega}^{L}D_k^a F_k(\mathbf{x}) \tag{96}$$

where the subline denotes no contraction.

Moreover, the fractional curl $\boldsymbol{F}$ ($curl^{(a)}\mathbf{F}(\mathbf{x})$) of a vector $\mathbf{F}$ is defined by:

$$\mathrm{curl}^{(a)}\mathbf{F} = \mathbf{e}_l\ \varepsilon_{\mathrm{lmn}} \frac{{}_{\omega}{}^{c}D^{a}_{m}F_n}{{}_{\omega}{}^{c}D^{a}_{\underline{m}}x_{\underline{m}}} = \mathbf{e}_l\ \varepsilon_{\mathrm{lmn}\omega}{}^{L}D^{a}_{m}F_n \tag{97}$$

A fractional flux of the vector $\mathbf{F}$ expressed in Cartesian coordinates across surface S is a fractional surface integral of the field with:

$$\Phi^{\alpha}_{s}(\mathbf{F}) = \left({}_{\omega}I^{\alpha}_{s}, \mathbf{F}\right) = {}^{(\alpha)}\iint\limits_{\omega\ S} \left(F_1 d^{\alpha}x_2 d^{\alpha}x_3 + F_2 d^{\alpha}x_3 d^{\alpha}x_1 + F_3 d^{\alpha}x_2 d^{\alpha}x_3\right) \tag{98}$$

A fractional volume integral of a triple fractional integral of a scalar field $f = f(x_1, x_2, x_3)$ is defined by:

$${}_{\omega}V^{(a)}_{\Omega}[f] = {}_{\omega}I^{(a)}_{\Omega}[x_1, x_2, x_3]\, f(x_1, x_2, x_3) = {}^{(\alpha)}\iiint\limits_{\omega\ \Omega} f(x_1, x_2, x_3)\, d^{\alpha}x_1 d^{\alpha}x_2 d^{\alpha}x_3 \tag{99}$$

It should be pointed out that the triple fractional integral is not a volume integral, since the fractional derivative of a variable with respect to itself is different from one. So there is a clear distinction between the simple, double, or triple integrals and the line, surface, and volume integrals respectively.

Fractional Vector Field Theorems

Fractional Green's Formula

Green's theorem relates a line integral around a simple closed curve ∂B and a double integral over the plane region B with boundary ∂B. With positively oriented boundary ∂B, the conventional Greens theorem for a vector field $\mathbf{F} = F_1\mathbf{e}_1 + F_2\mathbf{e}_2$ is expressed by:

$$\int\limits_{\partial B} (F_1 dx_1 + F_2 dx_2) = \int\int\limits_{B} \left(\frac{\partial (F_1)}{\partial x_2} - \frac{\partial (F_2)}{\partial x_1}\right) dx_1 dx_2 \tag{100}$$

Recalling that:

$$d^{a}\mathbf{x} = (d^{a}x_1, d^{a}x_2) = \left({}^{c}_{\omega}D^{a}_{x_1}[x_1]\, d^{\alpha}x_1, {}^{c}_{\omega}D^{a}_{x_2}[x_2]\, d^{\alpha}x_2\right) \tag{100a}$$

and substituting into conventional Green's theorem Eq. 100 we get:

$$\int_{\omega \partial W}^{(\alpha)} (F_1 d^\alpha x_1 + F_2 d^\alpha x_2) = \int_{\omega}^{(\alpha)}\int_W \left(\frac{{}_{\omega}^{c}D_{x_2}^{a}(F_1)}{{}_{\omega}^{c}D_{x_2}^{a}(x_2)} - \frac{{}_{\omega}^{c}D_{x_1}^{a}(F_2)}{{}_{\omega}^{c}D_{x_1}^{a}(x_1)} \right) d^\alpha x_1 d^\alpha x_2 \quad (101)$$

Fractional Stoke's Formula

Restricting in the consideration of a simple surface W, if we denote its boundary by ∂W and if $\mathbf{F}$ is a vector field defined on W, then the conventional Stokes' Theorem asserts that:

$$\oint_W \mathbf{F} \cdot \mathbf{dL} = \oiint_W curl \;\; \mathbf{F} \cdot \mathbf{dS} \quad (102)$$

In Cartesian coordinates it yields:

$$\begin{aligned} \int_{\partial W} (F_1 dx_1 + F_2 dx_2 + F_3 dx_3) &= \iint_W \left(\tfrac{\partial (F_3)}{\partial x_2} - \tfrac{\partial (F_2)}{\partial x_3} \right) dx_2 dx_3 \\ &+ \left(\tfrac{\partial (F_1)}{\partial x_3} - \tfrac{\partial (F_3)}{\partial x_1} \right) dx_3 dx_1 + \left(\tfrac{\partial (F_2)}{\partial x_1} - \tfrac{\partial (F_1)}{\partial x_2} \right) dx_1 dx_2 \end{aligned} \quad (103)$$

where $\mathbf{F}(x_1, x_2, x_3) = e_1 F_1(x_1, x_2, x_3) + \mathbf{e_2} F_2(x_1, x_2, x_3) + \mathbf{e_3} F_3(x_1, x_2, x_3)$.

In this case, the fractional curl operation is defined by:

$$\begin{aligned} curl_w^\alpha(\mathbf{F}) = e_l \varepsilon_{lmn} {}_{\omega}^{\alpha}D_{x_m}^{a}(F_n) / {}_{\omega}^{c}D_{x_k}^{a}(x_i)\, \delta_{mk} &= \mathbf{e}_1 \left(\frac{{}_{\omega}^{\alpha}D_{x_2}^{a} F_3}{{}_{\omega}^{\alpha}D_{x_2}^{a} x_2} - \frac{{}_{\omega}^{\alpha}D_{x_3}^{a} F_2}{{}_{\omega}^{\alpha}D_{x_3}^{a} x_3} \right) \\ + \mathbf{e}_2 \left(\frac{{}_{\omega}^{\alpha}D_{x_3}^{a} F_1}{{}_{\omega}^{\alpha}D_{x_3}^{a} x_3} - \frac{{}_{\omega}^{\alpha}D_{x_1}^{a} F_3}{{}_{\omega}^{\alpha}D_{x_1}^{a} x_1} \right) &+ \mathbf{e}_3 \left(\frac{{}_{\omega}^{\alpha}D_{x_1}^{a} F_2}{{}_{\omega}^{\alpha}D_{x_1}^{a} x_1} - \frac{{}_{\omega}^{\alpha}D_{x_2}^{a} F_1}{{}_{\omega}^{\alpha}D_{x_2}^{a} x_2} \right) \end{aligned} \quad (104)$$

Therefore transforming the conventional Stokes' theorem into the fractional form we get:

$$\begin{aligned} {}^{(\alpha)}\int_{\omega}{}_W (F_1 d^\alpha x_1 + F_2 d^\alpha x_2 + F_3 d^\alpha x_3) = {}^{(\alpha)}\iint_{\omega}{}_W &\left\{ \left(\frac{{}_{\omega}^{\alpha}D_{x_2}^{a} F_3}{{}_{\omega}^{\alpha}D_{x_2}^{a} x_2} - \frac{{}_{\omega}^{\alpha}D_{x_3}^{a} F_2}{{}_{\omega}^{\alpha}D_{x_3}^{a} x_3} \right) d^\alpha x_2 d^\alpha x_3 \right. \\ &+ \left(\frac{{}_{\omega}^{\alpha}D_{x_3}^{a} F_1}{{}_{\omega}^{\alpha}D_{x_3}^{a} x_3} - \frac{{}_{\omega}^{\alpha}D_{x_1}^{a} F_3}{{}_{\omega}^{\alpha}D_{x_1}^{a} x_1} \right) d^\alpha x_3 d^\alpha x_1 \\ &+ \left. \left(\frac{{}_{\omega}^{\alpha}D_{x_1}^{a} F_2}{{}_{\omega}^{\alpha}D_{x_1}^{a} x_1} - \frac{{}_{\omega}^{\alpha}D_{x_2}^{a} F_1}{{}_{\omega}^{\alpha}D_{x_2}^{a} x_2} \right) d^\alpha x_1 d^\alpha x_2 \right\} \end{aligned} \quad (105)$$

Fractional Gauss' Formula

For the conventional fields theory, let $\mathbf{F} = \mathbf{e_1}F_1+\mathbf{e_2}\ F_{2+}\mathbf{e_3}\ F_{3.}$ be a continuously differentiable real-valued function in a domain W with boundary ∂W. Then the conventional divergence Gauss' theorem is expressed by:

$$\iint\limits_{\partial W} \mathbf{F}\cdot\mathbf{dS} = \iiint\limits_{W} div\mathbf{F}dV \tag{106}$$

Since

$$\mathbf{d}^{(\alpha)}\mathbf{S} = \mathbf{e}_1 d^\alpha x_2 d^\alpha x_3 + \mathbf{e}_2 d^\alpha x_3 d^\alpha x_1 + \mathbf{e}_3 d^\alpha x_1 d^\alpha x_2 \tag{107}$$

where $d^a x_i\ i = 1{,}2{,}3$ is expressed by Eq. 15,

$$d^{(\alpha)}V = d^\alpha x_1 d^\alpha x_2 d^\alpha x_3 \tag{108}$$

Furthermore, see Eq. 96,

$$div^{(a)}\mathbf{F}(x) = \frac{{}_{\omega}^{c}D_k^a F_k\,(\mathbf{x})}{{}_{\omega}^{c}D_k^a x_i}\delta_{km} \tag{109}$$

The Fractional Gauss divergence theorem becomes:

$$\overset{(\alpha)}{\underset{\omega}{\iint}}{}_{\partial W}\mathbf{F}\cdot d^{(\alpha)}\mathbf{S} = \overset{(\alpha)}{\underset{\omega}{\iiint}}{}_{W}\operatorname{div}^{(\alpha)}\mathbf{F}d^{(\alpha)}V \tag{110}$$

Remember that the differential $d^\alpha S = \mathbf{n}^\alpha d^\alpha S$, where $\mathbf{n}^\alpha$ is the unit normal of the fractional tangent space as it has been defined in section "The Geometry of Fractional Differential."

Fractional Deformation Geometry

Outlining the Fractional Deformation Geometry presented in Lazopoulos and Lazopoulos (2016a), we assume the description in the Euclidean space, considering the reference configuration B with the boundary ∂B of a body displaced to its current configuration b with the boundary ∂b (see Truesdell 1977). The points in the reference placement B, defining the material points, are described by $\mathbf{X}$, whereas the set of the displaced points $\mathbf{y}$ describe the current configuration b with the boundary ∂b. The coordinate system in B is denoted by $\{X_A\}$, while the corresponding to the current configuration b is reflected to $\{y_i\}$. In the present description, both systems have the same axial directions with base vectors $\{\mathbf{e}_A\},\{\mathbf{e}_i\}$ whether the reference concerns the current or the initial (unstressed) configuration. The motion of a reference point X is described by the function

$$\mathbf{y} = \boldsymbol{\Psi}(\mathbf{X}, t) \tag{111}$$

the conventional gradient of the deformation is defined by:

$$\mathbf{F}(\boldsymbol{\Psi}, t) = \frac{\partial \boldsymbol{\Psi}(\mathbf{X}, t)}{\partial \mathbf{X}} \text{ or } F_{iA} = \frac{\partial \Psi_i}{\partial X_A} \tag{112}$$

with

$$\mathbf{F}^{(\alpha)} = F_{ij}^{(\alpha)} = \nabla_{\mathbf{X}}^{(\alpha)} \mathbf{y}_i = {}_a^c D_{X_j}^{\alpha} \mathbf{y}_i = \frac{{}_a^c D_{X_j}^{\alpha} (\mathbf{X} + \mathbf{u})}{{}_a^c D_{X_j}^{\alpha} \mathbf{X}} = \nabla_{\mathbf{X}}^{(\alpha)} \mathbf{X} + \nabla_{\mathbf{X}}^{(\alpha)} \mathbf{u} \tag{113}$$

Furthermore, the right Cauchy-Green fractional deformation tensor,

$$\mathbf{C}^{(\alpha)} = \mathbf{F}^{(\alpha)T} \mathbf{F}^{(\alpha)} \tag{114}$$

and

$$\mathbf{B}^{(\alpha)} = \mathbf{F}^{(\alpha)} \mathbf{F}^{(\alpha)T} \tag{115}$$

is the left Cauchy-Green fractional deformation tensor.

Likewise, the fractional nonlinear fractional Green-Lagrange strain tensor $\mathbf{E}^{(\alpha)}$ may be defined by:

$$E_N^{(\alpha)} = \mathbf{N} \cdot \overset{\alpha}{\mathbf{E}} \mathbf{N} \tag{116}$$

where $\mathbf{N}$ is the unit vector of the considered fiber in the reference placement, with

$$\mathbf{E}^{(\alpha)} = \frac{1}{2} \left(\mathbf{C}^{(\alpha)} - \mathbf{I} \right) \tag{117}$$

Recalling that the current placement $\mathbf{y} = \mathbf{X} + \mathbf{u}$, where $\mathbf{u}$ denotes the displacement vector, the fractional Green-Lagrange deformation tensor becomes:

$$\mathbf{E}^{(\alpha)} = \frac{1}{2} \left[\left(\nabla_{\mathbf{X}}^{(\alpha)} \mathbf{u} \right)^T + \nabla^{(\alpha)} \mathbf{u} + \left(\nabla_{\mathbf{X}}^{(\alpha)} \mathbf{u} \right)^T \nabla^{(\alpha)} \mathbf{u} \right] \tag{118}$$

Proceeding to define the fractional strain tensor referred to the current placement, i.e., Euler-Almansi strain tensor:

$$\mathbf{A}^{(\alpha)} = \frac{1}{2} \left(\mathbf{I} - \left(\mathbf{B}^{(\alpha)} \right)^{-1} \right) \tag{119}$$

with the strain

$$\varepsilon_{\mathbf{n}} = \mathbf{n}^T \cdot \mathbf{A}^{(\alpha)}\mathbf{n} \tag{120}$$

where **n** is the unit vector along the deformed fiber, corresponding to the **N** unit vector along the reference placement fiber. It is evident that in the conventional case with $a = 1$ the fractional Euler Almansi strain tensor $\mathbf{A}^{(\alpha)}$ reduces to the conventional strain tensor **A**. It should be pointed out that the fractional stretches $\lambda^{(\alpha)}$, adopting the right Cauchy-Green strain tensor $\mathbf{C}^{(\alpha)}$, are defined by:

$$\lambda^{(\alpha)} = \frac{d^\alpha s}{d^\alpha S} \tag{121}$$

as the ratio of the measures of the final infinitesimal length over the corresponding length of the fractional differential $d^\alpha\mathbf{X}$ vector with $d^\alpha S = (d^\alpha\mathbf{X} \cdot d^\alpha\mathbf{X})^{\frac{1}{2}}$. In fact the stretches $\lambda^{(\alpha)}$ are defined by:

$$\left(\lambda^{(\alpha)}\right)^2 = \mathbf{N}^T \cdot \mathbf{C}^{(\alpha)}\mathbf{N} \tag{122}$$

where $\mathbf{N} = N_A e_A$ is the unit vector directed along the material reference fiber.

Furthermore for

$$\frac{1}{\left(\lambda^{(\alpha)}\right)^2} = \left(\frac{d^{(\alpha)}S}{d^{(\alpha)}s}\right)^2 = \mathbf{n}^T \cdot \left(\mathbf{B}^{(\alpha)}\right)^{-1}\mathbf{n} \tag{123}$$

where **n** the unit vector along the deformed fiber.

Polar Decomposition of the Deformation Gradient

It is well known that every nonsingular matrix may be decomposed into a product of an orthogonal and a symmetric positive tensor. Applying the property to the deformation gradient we get:

$$\mathbf{F}^{(\alpha)} = \mathbf{R}^{(\alpha)}\mathbf{U}^{(\alpha)} = \mathbf{V}^{(\alpha)}\mathbf{R}^{(\alpha)} \tag{124}$$

where **R** is orthogonal $\mathbf{R}=\mathbf{R}^{-\mathbf{T}}$ and **U** and **V** are symmetric positive ($\mathbf{U}=\mathbf{U}^{\mathbf{T}}$ and $\mathbf{V}=\mathbf{V}^{\mathbf{T}}$). Therefore:

$$\mathbf{C}^{(\alpha)} = \left(\mathbf{U}^{(\alpha)}\right)^2 \text{ and } \mathbf{B}^{(\alpha)} = \left(\mathbf{V}^{(\alpha)}\right)^2 \tag{125}$$

Moreover, the eigenvalues of $\mathbf{C}^{(\alpha)}$ and $\mathbf{V}^{(\alpha)}$ are the same, but the eigenvectors $\mathbf{u}^{(\alpha)}$ of $\mathbf{U}^{(\alpha)}$ and $\mathbf{v}^{(\alpha)}$ of $\mathbf{V}^{(\alpha)}$ are related by $\mathbf{v}^{(\alpha)} = \mathbf{R}^{(\alpha)}\mathbf{u}^{(\alpha)}$. In fact $\mathbf{v}^{(\alpha)}$ is directed along a principal direction (eigenvector) of the strain tensor $\mathbf{V}^{(\alpha)}$ with $\mathbf{u}^{(\alpha)}$ been the eigenvector of $\mathbf{U}^{(\alpha)}$. In other words, principal directions refer to the vectors $\mathbf{u}^{(\alpha)} = u_A^{(a)} e_A$ and $\mathbf{v}^{(\alpha)} = v_A^{(a)} e_A$.

Deformation of Volume and Surface

Consider three noncoplanar line elements $d^\alpha \mathbf{X}^{(1)}$, $d^\alpha \mathbf{X}^{(2)}$, $d^\alpha \mathbf{X}^{(3)}$ at the point $\mathbf{X}$ in B so that:

$$\mathbf{d}^\alpha \mathbf{y}^{(i)} = \mathbf{F}^{(\alpha)} d^\alpha \mathbf{X}^{(i)} \tag{126}$$

with $d^\alpha \mathbf{y}^i$ the corresponding fractional differential vectors in the current placement.

Further, the volume $d^\alpha V$ is derived by

$$d^\alpha V = d^\alpha \mathbf{X}^{(1)} \cdot \left(d^\alpha \mathbf{X}^{(2)} \wedge d^\alpha \mathbf{X}^{(3)} \right) \tag{127}$$

Alternatively

$$d^\alpha V = \det \left(d^\alpha \mathbf{X}^{(1)}, d^\alpha \mathbf{X}^{(2)}, d^\alpha \mathbf{X}^{(3)} \right) \tag{128}$$

in which $d\mathbf{X}^{(1)}$ denotes a column vector ($i = 1,2,3$). The corresponding volume $d^\alpha \boldsymbol{v}$ in the deformed configuration is

$$d^\alpha v = \det \left(d^\alpha \mathbf{y}^{(1)}, d^\alpha \mathbf{y}^{(2)}, d^\alpha \mathbf{y}^{(3)} \right) \tag{129}$$

and

$$d^\alpha v = \det \left(\mathbf{F}^{(\alpha)} \right) d^\alpha V \equiv J^{(\alpha)} d^\alpha V \tag{130}$$

where

$$J^{(\alpha)} = \det \ \mathbf{F}^{(\alpha)} \ \text{ and } \ d^\alpha V = d^\alpha \mathbf{X}^{(1)} \cdot d^\alpha \mathbf{X}^{(2)} \wedge d^\alpha \mathbf{X}^{(3)} \tag{131}$$

Consider, further, an infinitesimal vector element of material surface dS in the neighborhood of the point $\mathbf{X}$ in B with $d^\alpha \mathbf{S} = \mathbf{N}^\alpha d^\alpha S$ the fractional tangent surface vector corresponding to the normal fractional normal vector $\mathbf{N}^\alpha$. Likewise, $d^\alpha \mathbf{X}$ is an arbitrary fiber cutting the edge d^αS such that $d^\alpha \mathbf{X} \cdot d^\alpha \mathbf{S} > 0$. The volume of the cylinder with base $d^\alpha \mathbf{S}$ and generators $d^\alpha \mathbf{X}$ has volume $d^\alpha V = d^\alpha \mathbf{X} \cdot d^\alpha \mathbf{S}$. If $d^\alpha \mathbf{x}$ and $d^\alpha \mathbf{s}$ are the deformed configurations of $d^\alpha \mathbf{X}$ and $d^\alpha \mathbf{S}$ respectively, with $d^\alpha \mathbf{s} = \mathbf{n}^\alpha d^\alpha s$, where $\mathbf{n}^\alpha$ is the normal vector to the deformed surface, the volume $d^\alpha V$ in the reference placement corresponds to the volume $d^\alpha v = d^\alpha \mathbf{x} \cdot d^\alpha \mathbf{s}$ in the current configuration so that:

$$d^{\alpha}v = d^{\alpha}\mathbf{y} \cdot d^{\alpha}\mathbf{s} = J^{(\alpha)} d^{\alpha}\mathbf{X} \cdot d^{\alpha}\mathbf{S} \tag{132}$$

Since $d^{\alpha}\mathbf{y} = \mathbf{F}^{(\alpha)} d^{\alpha}\mathbf{X}$, we obtain

$$\mathbf{F}^{(\alpha)T} d^{\alpha}\mathbf{X} \cdot d^{\alpha}\mathbf{s} = J^{(\alpha)} d^{\alpha}\mathbf{X} \cdot d^{\alpha}\mathbf{S} \tag{133}$$

removing the arbitrary $d^{\alpha}\mathbf{X}$. Consequently,

$$d^{\alpha}\mathbf{s} = J^{(\alpha)} \left(\mathbf{F}^{(\alpha)}\right)^{-T} d^{\alpha}\mathbf{S} \tag{134}$$

and

$$\mathbf{n}^{\alpha} d^{\alpha}s = J^{(\alpha)} \left(\mathbf{F}^{(\alpha)}\right)^{-T} \mathbf{N}^{\alpha} d^{\alpha}S \tag{135}$$

The relation between the area elements corresponding to reference and current configurations is the well-known Nanson's formula for the fractional deformations.

Examples of Deformations

Homogeneous Deformations

The most general homogeneous deformation of the body B from its reference configuration is expressed by:

$$\mathbf{x} = \mathbf{A}\mathbf{X} \tag{136}$$

Choosing as fractional derivatives Caputo ones that are given by:

$${}_{a}^{C}D_{t}^{a}(t-a)^{v} = \frac{\Gamma(v+1)}{\Gamma(-\alpha+v+1)}(t-a)^{v-a}$$

and

$${}_{\mathrm{a}}^{C}D_{X}^{a}X = \frac{\Gamma(2)}{\Gamma(2-a)}X^{1-a} \tag{137}$$

and specializing the homogeneous deformations with the example of simple shear, we discuss the deformation,

$$\begin{aligned} x_1 &= X_1 + \gamma X_2 \\ x_2 &= X_2 \\ x_3 &= X_3 \end{aligned} \tag{138}$$

Therefore the Cauchy-Green deformation tensors $\mathbf{C}^{(\alpha)}$ and $\mathbf{B}^{(\alpha)}$ become:

$$\mathbf{C}^{(\alpha)} = \left(\mathbf{F}^{(\alpha)}\right)^T \mathbf{F}^{\alpha} = \begin{vmatrix} 1 & \gamma & 0 \\ \gamma & (\gamma^2+1) & 0 \\ 0 & 0 & 1 \end{vmatrix} \tag{139}$$

and

$$\mathbf{B} = \mathbf{F}^{(\alpha)}\left(\mathbf{F}^{(\alpha)}\right)^T = \begin{vmatrix} 1+\gamma^2 & \gamma & 0 \\ \gamma & 1 & 0 \\ 0 & 0 & 1 \end{vmatrix} \tag{140}$$

The Green (Langrange) Fractional strain tensor is:

$$\mathbf{E}^{(\alpha)} = \frac{1}{2}\left(\mathbf{C}^{(\alpha)} - \mathbf{I}\right) = \begin{bmatrix} 0 & \gamma/2 & 0 \\ \gamma/2 & \gamma^2/2 & 0 \\ 0 & 0 & 0 \end{bmatrix} \tag{141}$$

And the Euler-Almansi strain tensor is given by:

$$\mathbf{A}^{(\alpha)} = \frac{1}{2}\left(\mathbf{I} - \left(\mathbf{B}^{(\alpha)}\right)^{-1}\right) = \begin{bmatrix} 0 & \gamma/2 & 0 \\ \gamma/2 & -\gamma^2/2 & 0 \\ 0 & 0 & 0 \end{bmatrix} \tag{142}$$

It would seem strange that the results are exactly the same as the ones of the conventional elasticity. However, there is mathematical explanation for the homogeneous deformations. Just taking into consideration the definition of the fractional derivative and differential, the differential for a linear function of the form:

$$f(x) = Ax \tag{143}$$

The fractional differential is given by:

$$d^{\alpha} f(x) = A\, d^{\alpha} x \tag{144}$$

the fractional differential of the function has almost the same form as the conventional one. Nevertheless, for the nonlinear function:

$$f(x) = x^4 \tag{145}$$

the fractional differential

$$d^{\alpha} f(x) = {}_a^C D_x^{\alpha} x^4 \cdot \left({}_a^C D_x^{\alpha} x\right)^{-1} d^{\alpha} x \tag{146}$$

Is equal to

$$d^{\alpha}f(x) = \frac{\Gamma(5)\Gamma(2-\alpha)}{\Gamma(5-\alpha)}x^{3}d^{\alpha}x \tag{147}$$

with coefficient depending upon the α fractional dimension. That makes the difference in nonhomogeneous deformations discussed in the next section.

The Nonhomogeneous Deformations

The nonhomogeneous deformation is defined by the equations:

$$\begin{aligned} x_1 &= X_1 + \gamma X_2^4 \\ x_2 &= X_2 \\ x_3 &= X_3 \end{aligned} \tag{148}$$

Taking into consideration Eqs. 124 and 125, the fractional Cauchy-Green deformation tensors are expressed by:

$$\mathbf{C}^{(\alpha)} = \left(\mathbf{F}^{(\alpha)}\right)^{T}\mathbf{F}^{(\alpha)} = \begin{vmatrix} 1 & \frac{24\gamma\Gamma(2-\alpha)X_2^3}{\Gamma(5-\alpha)} & 0 \\ \frac{24\gamma\Gamma(2-\alpha)X_2^3}{\Gamma(5-\alpha)} & 1+\frac{576\gamma^2\Gamma(2-\alpha)^2X_2^6}{\Gamma(5-\alpha)^2} & 0 \\ 0 & 0 & 1 \end{vmatrix} \tag{149}$$

and

$$\mathbf{B}^{(\alpha)} = \mathbf{F}^{(\alpha)}\left(\mathbf{F}^{(\alpha)}\right)^{T} = \begin{vmatrix} 1+\frac{576\gamma^2\Gamma(2-\alpha)^2X_2^6}{\Gamma(5-\alpha)^2} & \frac{24\gamma\Gamma(2-\alpha)X_2^3}{\Gamma(5-\alpha)} & 0 \\ \frac{24\gamma\Gamma(2-\alpha)X_2^3}{\Gamma(5-\alpha)} & 1 & 0 \\ 0 & 0 & 1 \end{vmatrix} \tag{150}$$

Hence Green-Lagrange strain tensor is expressed by:

$$\mathbf{E}^{(\alpha)} = \begin{vmatrix} 0 & \frac{12\gamma\Gamma(2-\alpha)X_2^3}{\Gamma(5-\alpha)} & 0 \\ \frac{12\gamma\Gamma(2-\alpha)X_2^3}{\Gamma(5-\alpha)} & \frac{288\gamma^2\Gamma(2-\alpha)^2X_2^6}{\Gamma(5-\alpha)^2} & 0 \\ 0 & 0 & 0 \end{vmatrix} \tag{151}$$

and Euler-Almansi strain tensor is defined by:

$$\mathbf{A}^{(\alpha)} = \begin{vmatrix} 0 & \frac{12\gamma\Gamma(2-\alpha)x_2^3}{\Gamma(5-\alpha)} & 0 \\ \frac{12\gamma\Gamma(2-\alpha)x_2^3}{\Gamma(5-\alpha)} & -\frac{288\gamma^2\Gamma(2-\alpha)^2x_2^6}{\Gamma(5-\alpha)^2} & 0 \\ 0 & 0 & 0 \end{vmatrix} \tag{152}$$

It is evident that the strain tensors strongly depend upon the fractional dimension α for the present nonhomogeneous deformation. However, the present deformation also depends upon the source ω of the fractional analysis.

The Infinitesimal Deformations

Since there has been pointed out in the introduction, fractional strain tensors have been considered in the literature, mainly in infinitesimal deformations, simply by substituting the common derivatives to fractional ones. It would be wise to study whether that idea is valid or not. Unfortunately it is proven a mistake. Fractional strain with simple substitution of derivatives does not have any physical meaning.

Considering the fractional Euler-Lagrange strain tensor, Eq. 118, where $\mathbf{u}$ is the small displacement vector with $|\mathbf{u}| \ll 1$ and $\left|\nabla_{\mathbf{X}}^{(\alpha)}\mathbf{u}\right| \ll 1$, we restrict into the linear deformation analysis, and the infinitesimal (linear) fractional Euler-Lagrange strain tensor $\overset{\alpha}{\mathbf{E}}_{lin}$ becomes:

$$\mathbf{E}_{lin}^{(\alpha)} = \frac{1}{2}\left[(\nabla_{\mathbf{X}}) \overset{a}{\mathbf{u}}^{T} + \overset{a}{\nabla}_{\mathbf{X}}\mathbf{u} \right] \tag{153}$$

It is recalled from Eq. 92 that:

$$\nabla^{(a)} f(\mathbf{x}) = grad^{(a)} f(\mathbf{x}) = \nabla_i^{(\alpha)} f(x)\mathbf{e}_i = \frac{{}_{\omega}^{c}D_i^{a} f(\mathbf{x})}{{}_{\omega}^{c}D_{\underline{i}}^{a} x_{\underline{i}}}\mathbf{e}_i, \tag{154}$$

Hence the linear fractional strain is not simply the half of the sum of the fractional derivatives of the displacement vector and its transport, but the half of the fractional gradient (as it has been defined by Eq. 92) and its transport.

Fractional Stresses

Pointing out that the fractional tangent space of a surface has different orientation of the conventional one, the fractional normal vector $\mathbf{n}^{\alpha}$ does not coincide with the conventional normal vector $\mathbf{n}$. Hence we should expect the stresses and consequently the stress tensor to differ from the conventional ones not only in the values.

If $d^{\alpha}\mathbf{P}$ is a contact force acting on the deformed area $d^{\alpha}\mathbf{a} = \mathbf{n}^{\alpha} d^{\alpha}$a, lying on the fractional tangent plane where $\mathbf{n}^{\alpha}$ is the unit outer normal to the element of area $d^{\alpha}\mathbf{a}$, then the α-fractional stress vector is defined by:

$$\mathbf{t}^{\alpha} = \lim_{d^{\alpha}a \geq 0} \frac{d^{\alpha}\mathbf{P}}{d^{\alpha}a} \tag{155}$$

However, the α-fractional stress vector does not have any connection with the conventional one

$$\mathbf{t} = \lim_{\mathrm{da} \geq 0} \frac{d\mathbf{P}}{da} \tag{156}$$

since the conventional tangent plane has different orientation from the α-fractional tangent plane and the corresponding normal vectors too.

Following similar procedures as the conventional ones we may establish Cauchy's fundamental theorem (see Truesdell 1977).

If $\mathbf{t}^{\alpha}(\cdot, \mathbf{n}^{\alpha})$ is a continuous function of the transplacement vector **x**, there is an α-fractional Cauchy stress tensor field

$$\mathbf{T}^{\alpha} = \left[\sigma_{\mathrm{ij}}^{\alpha}\right] \tag{157}$$

The Balance Principles

Almost all balance principles are based upon Reynold's transport theorem. Hence the modification of that theorem, just to conform to fractional analysis is presented. The conventional Reynold's transport theorem is expressed by:

$$\frac{d}{dt}\int_{W} A dV = \int_{W} \frac{dA}{dt} dV + \int_{\partial W} A\mathbf{v_n} dS \tag{158}$$

For a vector field A applied upon region W with boundary ∂W and $\mathbf{v_n}$ is the normal velocity of the boundary ∂W.

Material Derivatives of Volume, Surface, and Line Integrals

For any scalar, vector or tensor property that may be represented by:

$$P_{ij}(t) = \int_{V} P_{ij}^{*}(x, t)\, dV \tag{159}$$

where V is the volume of the current placement in the conventional calculus. The material time derivative of $P_{\mathrm{ij}}^{*}(x, t)$ is expressed by:

$$\frac{dP_{ij}^{*}}{dt} = \frac{\partial P_{ij}^{*}}{\partial t} + v\frac{\partial P_{ij}^{*}}{\partial x_j} = \frac{\partial P_{ij}^{*}}{\partial t} + \mathbf{v} \cdot \nabla_{\mathbf{x}} P_{ij}^{*} \tag{160}$$

recalling that we consider constant material points during time derivation. Similarly in fractional calculus, the material time derivative is given, for any tensor field P_{ij} by:

$$\frac{dP_{ij}^{*}}{dt}=\frac{\partial P_{ij}^{*}}{\partial t}+\frac{\partial^{\alpha}P_{ij}^{*}}{\partial x_{k}^{a}}\frac{d^{a}x_{l}}{dx_{k}^{a}}\frac{dg\left(x_{m}\right)}{dt}\delta_{km} \tag{161}$$

Since

$$v_{k}=\frac{\partial^{a}x_{e}}{\partial x_{k}^{a}}\frac{dg\left(x_{m}\right)}{dt}\delta_{km} \tag{162}$$

The material derivative, into the context of fractional calculus, is expressed by:

$$\frac{d\left(\ \right)}{dt}=\frac{\partial\left(\ \right)}{\partial t}+\mathbf{v}\cdot\overset{a}{\nabla}_{\mathbf{x}}() \tag{163}$$

Hence, the acceleration is defined by:

$$\mathbf{a}=\frac{d\mathbf{v}}{dt}=\frac{\partial\mathbf{v}}{\partial t}+\mathbf{v}\cdot\overset{\alpha}{\nabla}_{\mathbf{x}}\mathbf{v} \tag{164}$$

Furthermore the material time derivative of $P_{ij}(t)$ is expressed in conventional analysis by:

$$\frac{d}{dt}\left[P_{ij}(t)\right]=\frac{d}{dt}\int\limits_{V}P_{ij}^{*}\left(x,t\right)dV \tag{165}$$

where b is the current placement of the region. It is well known that: $\left(d\overset{\bullet}{V}\right)=JdV$, and the Eq. 165 yields:

$$\frac{d}{dt}\int\limits_{V}P_{ij}^{*}\left(x,t\right)dV=\int\limits_{V}\left[\frac{\partial P_{ij}^{*}\left(x,t\right)}{dt}+P_{ij}^{*}\left(x,t\right)\frac{\partial v_{p}}{\partial x_{p}}\right]dV \tag{166}$$

Recalling the material derivative operator Eqs. 160 and 166 yields,

$$\frac{d}{dt}\int\limits_{V}P_{ij}^{*}\left(x,t\right)dV=\int\limits_{b}\left[\frac{\partial P_{ij}^{*}\left(x,t\right)}{\partial t}+\frac{\partial\left(u_{p}P_{ij}^{*}\left(x,t\right)\right)}{\partial x_{p}}\right]dV \tag{167}$$

yielding Reynold's Transport theorem:

$$\frac{d}{dt}\int_{v} P_{ij}^{*}(x,t)\,dV = \int_{V} \frac{\partial P_{ij}^{*}(x,t)}{\partial t}dV + \int_{S} v_p\left[P_{ij}^{*}(x,t)\right]dS_p \tag{168}$$

Expressing Reynold's Transfer Theorem in fractional form we get:

$$\frac{d}{dt}\int_{\omega V} P_{ij}^{*}(x,t)\,d^{\alpha}V = \int_{\omega V} \frac{\partial P_{ij}^{*}(x,t)}{\partial t}d^{\alpha}V + \int_{\omega S} u_p\left[P_{ij}^{*}(x,t)\right]d^{\alpha}S_p \tag{169}$$

where $d^{\alpha}V$ and $d^{\alpha}S$ the infinitessimal fractional volume and surface, respectively.

The volume integral of the material time derivative of $P_{ij}(t)$ may also be expressed by:

$$\frac{d}{dt}\int_{\omega V}^{(\alpha)} P_{ij}^{*}(\mathbf{x},t)\,d^{\alpha}V = \int_{\omega V}^{(\alpha)}\left(\frac{\partial P_{ij}^{*}(x,t)}{\partial t} + div^{\alpha}\left[\mathbf{v}P_{ij}^{*}\right]\right)d^{\alpha}V \tag{170}$$

The Balance of Mass

The conventional balance of mass, expressing the mass preservation, is expressed by:

$$\frac{d}{dt}\int_{W} \rho dV = 0 \tag{171}$$

In the fractional form it is given by:

$$\frac{d}{dt}\int_{\omega W}^{(\alpha)} \rho\, d^{\alpha}V = 0 \tag{172}$$

where $\frac{d}{dt}$ is the total time derivative.

Recalling the fractional Reynold's Transport Theorem, we get:

$$\frac{d}{dt}\int_{\omega V}^{(\alpha)} \rho d^{\alpha}V = \int_{\omega V}^{(\alpha)}\left(\frac{\partial\rho(x,t)}{\partial t} + div^{\alpha}\left[\mathbf{v}\rho\right]\right)d^{\alpha}V \tag{173}$$

Since Eq. 173 is valid for any volume V, the continuity equation is:

$$\frac{\partial \rho}{\partial t} + div^{a}\left[\mathbf{v}\rho\right] = 0 \qquad (174)$$

where div^{α} has already been defined by Eq. 96. That is the continuity equation expressed in fractional form.

Balance of Linear Momentum Principle

It is reminded that the conventional balance of linear momentum is expressed in continuum mechanics by:

$$\frac{d}{dt}\int\limits_{\Omega} \rho \mathbf{v} dV = \int\limits_{\partial\Omega} \mathbf{t}^{(n)} dS + \int\limits_{\Omega} \rho \mathbf{b} dV \qquad (175)$$

where $\mathbf{v}$ is the velocity, $\mathbf{t}^{(n)}$ is the traction on the boundary, and $\mathbf{b}$ is the body force per unit mass. Likewise that principle in fractional form is expressed by:

$$\frac{d}{dt}\int\limits_{\omega\,\Omega}^{(\alpha)} \rho \mathbf{v} d^{(\alpha)}V = \int\limits_{\omega\,\Omega}^{(\alpha)} \left[\rho \mathbf{b} + div^{a}\left(\mathbf{T}^{\alpha}\right)\right] d^{(\alpha)}V \qquad (176)$$

Hence the equation of linear motion, expressing the balance of linear momentum, is defined by:

$$div^{a}\left[\mathbf{T}^{\alpha}\right] + \rho \mathbf{b} - \rho \overset{\bullet}{\mathbf{v}} = 0 \qquad (177)$$

It should be pointed out that div^{α} has already been defined by Eq. 96 and is different from the common definition of the divergence.

Following similar steps as in the conventional case, the balance of rotational momentum yields the symmetry of Cauchy stress tensor.

Balance of Rotational Momentum Principle

Following similar procedure as in the conventional case, we may end up to the symmetry property of the fractional stress tensor, i.e.:

$$\mathbf{T}^{\alpha} = \left(\mathbf{T}^{\alpha}\right)^{T} \qquad (178)$$

Fractional Zener Viscoelastic Model

The Integer Viscoelastic Model

It was proposed by Zener consists of the elastic springs with elastic constraints E_1 and E_2 and the viscous dashpot having a viscosity constant C. Then the standard linear solid model called Zener viscoelastic model is indicated in Fig. 7.

The three-parameter Zener model (Fig. 7) yields the following constitutive equation:

$$\left[1+\frac{C}{E_1+E_2}\frac{d}{dt}\right]\sigma(t)=\frac{E_2}{E_1+E_2}\left[E_1+C\frac{d}{dt}\right]\varepsilon(t) \tag{179}$$

In this case the creep and relaxation functions take the form:

$$J(t)=\frac{E_1+E_2}{E_1E_2}\{1-\exp\left[-\left(t/\tau_\varepsilon\right)\right]\} \tag{180}$$

$$G(t)=\frac{E_1E_2}{E_1+E_2}\{1+\exp\left[-\left(t/\tau_\varepsilon\right)\right]\} \tag{181}$$

where

$$\tau_\varepsilon=\frac{C}{E_1},\tau_\sigma=\frac{C}{E_1+E_2}. \tag{182}$$

For the viscoelastic materials with $E_1 = 0.16\ 10^9$ N/m, $E_2 = 1.5\ 10^9$ N/m and $C = 16\ 10^9$ Ns/m, and the initial condition $y_o = 0$, the function of the compliance with respect to time is shown in Fig. 8.

Furthermore, the function of the relaxation modulus G(t) with respect to time t is shown in Fig. 9:

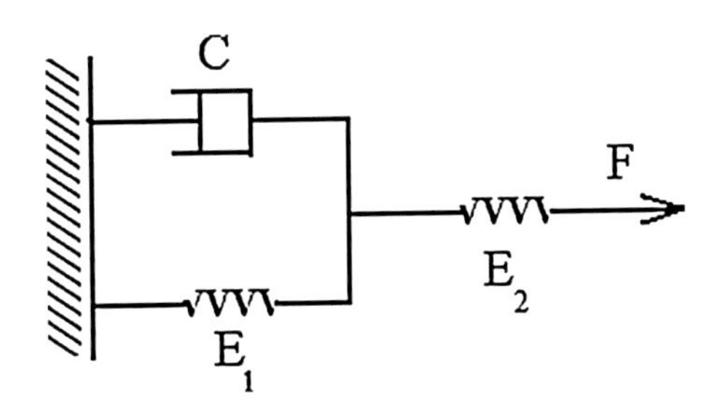

Fig. 7 The three-parameter Zener model

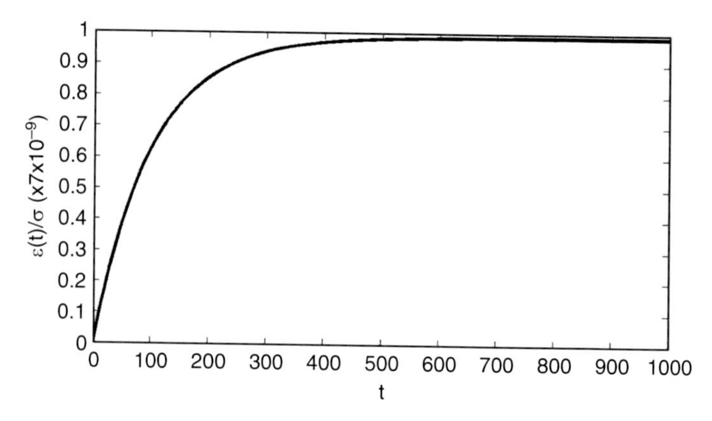

Fig. 8 The variation of the compliance with respect to time for the integer Zener viscoelastic model

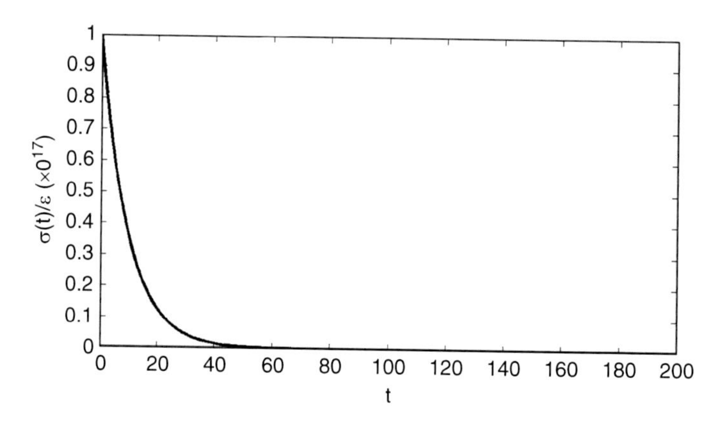

Fig. 9 Variation of the relaxation modulus with respect to time for the integer Zener viscoelastic model

The Fractional Order Derivative Viscoelastic Models

Viscoelastic models have been proposed using fractional time derivatives, just to take into account the material memory effects. The fractional Zener model introduced by Bagley and Torvik (1986) has been extensively studied in the literature (see Atanackovic 2002 and Sabatier et al. 2007). Consequently, the fractional Zener model has been proposed, using fractional time derivatives and more specifically the Caputo time derivatives. Therefore the constitutive equation for the model is:

$$\left[1+\frac{C}{E_1+E_2}\frac{d^\alpha}{dt^\alpha}\right]\sigma(t)=\frac{E_2}{E_1+E_2}\left[E_1+C\frac{d^\alpha}{dt^\alpha}\right]\varepsilon(t) \tag{183}$$

with creep compliance and relaxation modulus:

$$J(t) = \frac{1}{E_2} + \frac{1}{E_1}\{1 - E_\alpha[-(t/\tau_\varepsilon)^\alpha]\} \tag{184}$$

$$G(t) = \frac{E_2}{E_1 + E_2}\{E_0 + E_1 E_\alpha[-(t/\tau_\varepsilon)^\alpha]\} \tag{185}$$

where $\tau_\varepsilon^\alpha = \frac{C}{E_1}$, $\tau_\sigma^\alpha = \frac{C}{E_1+E_2}$.

For the viscoelastic materials with $E_1 = 0.16\ 10^9$ N/m, $E_2 = 1.5\ 10^9$ N/m, $C = 16\ 10^9$ Ns/m, and the initial condition $y_o = 0$, the function of the compliance with respect to time is shown in Fig. 10.

Furthermore, the function of the relaxation modulus $G(t)$ with respect to time t is shown in Fig. 11. It is shown that the lower the fractional order the slower the convergence to the final zero value of the relaxation.

Proposed Fractional Viscoelastic Zener Model

Since only Leibnitz L-fractional derivatives have physical sense, the fractional viscoelastic equations should be expressed in terms of L-fractional derivatives. Therefore the L-fractional Zener model should be expressed by:

$$\left[1 + \frac{C}{E_1 + E_2}\,{}_0^L D_t^\alpha\right]\sigma(t) = \frac{E_2}{E_1 + E_2}\left[E_1 + C\,{}_0^L D_t^\alpha\right]\varepsilon(t) \tag{186}$$

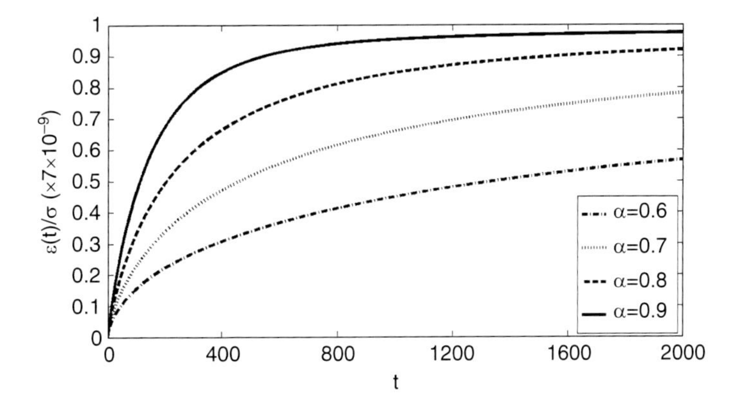

Fig. 10 The variation of the compliance with respect to time for various values of the fractional order for the existing fractional models

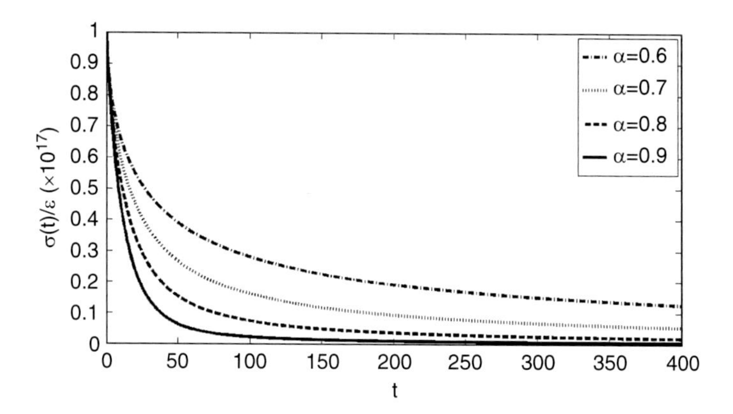

Fig. 11 The variation of the relaxation with respect to time for various values of the fractional order for the existing fractional models

Hence for constant applied stress the compliance

$$J(t) = \frac{\varepsilon(t)}{\sigma} \tag{187}$$

expressing the variation of strain with respect to time is given by the equation:

$$\left[E_1 + C{}_0^L D_t^\alpha\right] J(t) = \frac{E_1 + E_2}{E_2} \tag{188}$$

Therefore we have to solve the fractional linear equation:

$$C{}_0^L D_t^\alpha y(t) + E_1 y(t) = 1 + \frac{E_1}{E_2} \tag{189}$$

Looking now for a solution of the type:

$$y(t) = \sum_{k=0}^{\infty} y_k t^k \tag{190}$$

and substituting $y(t)$ from Eq. 190 to the governing compliance, Eq. 189 we get:

$$\sum_{k=0}^{\infty} C y_{k+1} \frac{\Gamma(2-\alpha)\,\Gamma(\kappa+2)}{\Gamma(\kappa+2-\alpha)} t^k + \sum_{k=0}^{\infty} E_1 y_k t^k = 1 + \frac{E_1}{E_2} \tag{191}$$

Since the algebraic Eq. 191 is valid for any t the various coefficients y_i are defined by:

$$y_1 = \frac{1}{C}\left(-E_1 y_0 + 1 + \frac{E_1}{E_2}\right) \tag{192}$$

$$y_{k+1} = -\frac{E_1}{C}\frac{\Gamma(\kappa + 2 - \alpha)}{\Gamma(\kappa + 2)\,\Gamma(2 - \alpha)} y_k, \forall k \geq 1 \tag{193}$$

Hence

$$y_k = \left(-\frac{E_1}{C\Gamma(2-\alpha)}\right)^{k-1} \prod_{m-1}^{k-1} \frac{\Gamma(m + 2 - a)}{\Gamma(m + 2)} y_1, \forall k \geq 2 \tag{194}$$

For the viscoelastic materials with $E_1 = 0.16\ 10^9$ N/m, $E_2 = 1.5\ 10^9$ N/m, $C = 16\ 10^9$ Ns/m, and the initial condition $y_o = 0$; the function of the compliance with respect to time is shown in Fig. 6.

It is clear from Fig. 6 that the fractional order has influence upon the time of convergence of the compliance modulus to the final value. The lower the value of the fractional order the slower the convergence of the compliance modulus to the final value.

Now, proceeding to the relaxation behavior of the Fractional Zener Viscoelastic model, we consider constant strain $\varepsilon(t) = \varepsilon$, then the governing Eq. 185 becomes:

$$C{}_0^L D_t^\alpha \left(\frac{\sigma(t)}{\varepsilon}\right) + (E_1 + E_2)\left(\frac{\sigma(t)}{\varepsilon}\right) = E_1 E_2 \tag{195}$$

For the relaxation modulus $y(t) = G(t) = \frac{\sigma(t)}{\varepsilon}$, the Eq. 197 above takes the form:

$$C{}_0^L D_t^\alpha y(t) + (E_1 + E_2)\, y(t) = E_1 E_2 \tag{196}$$

Looking for solution of the type

$$y(t) = \sum_{k=0}^{\infty} y_k t^k \tag{197}$$

and substituting in Eq. 196 we get:

$$\sum_{k=0}^{\infty} C\ \ y_{k+1} \frac{\Gamma(2-\alpha)\,\Gamma(k+2)}{\Gamma(k+2-\alpha)} t^k + \sum_{k=0}^{\infty} (E_1 + E_2)\, y_k t^k = E_1 E_2 \tag{198}$$

Since Eq. 198 is valid for any t, it is an identity. Hence:

$$y_1 = -\frac{E_1 + E_2}{C} y_0 + \frac{E_1 E_2}{C} \tag{199}$$

$$y_{k+1} = -\frac{E_1 + E_2}{C} \frac{\Gamma(m + 2 - \alpha)}{\Gamma(2 - \alpha)\Gamma(k + 2)} y_1, \forall k \geq 2 \tag{200}$$

Those relations yield:

$$y_k = \left(-\frac{E_1 + E_2}{c\,\Gamma(2 - \alpha)}\right)^{k-1} \prod_{m=1}^{k-1} \frac{\Gamma(m + 2 - \alpha)}{\Gamma(m + 2)} y_1, \forall k \geq 2. \tag{201}$$

For the viscoelastic materials with $E_1 = 0.16\ 10^9$ N/m, $E_2 = 1.5\ 10^9$ N/m, and $C = 16\ 10^9$ Ns/m, the function of the relaxation modulus with respect to time is shown in Fig. 7.

It is clear from Fig. 7 that the fractional order has influence upon the time of convergence of the relaxation modulus to the zero value. The lower the value of the fractional order the slower the convergence of the relaxation modulus to the zero value.

Comparison of the Three Viscoelastic Models

For the viscoelastic materials with $E_1 = 0.16\ 10^9$ N/m, $E_2 = 1.5\ 10^9$ N/m, and $C = 16\ 10^9$ Ns/m, and for two different values of the fractional orders $a = 0.5$, $a = 0.7$, and $a = 0.9$ the (Figs. 8, 9, and 10) show the behavior of the compliance modulus and (Figs. 11, 12, 13, 14, 15, and 16) the behavior of the relaxation modulus.

Comparing (Figs. 17, 18, and 19) the behavior of the proposed model exhibits a more mild convergence for the compliance and relaxation moduli to the final values as far as to Caputo's viscoelastic models are concerned. Further, the lower

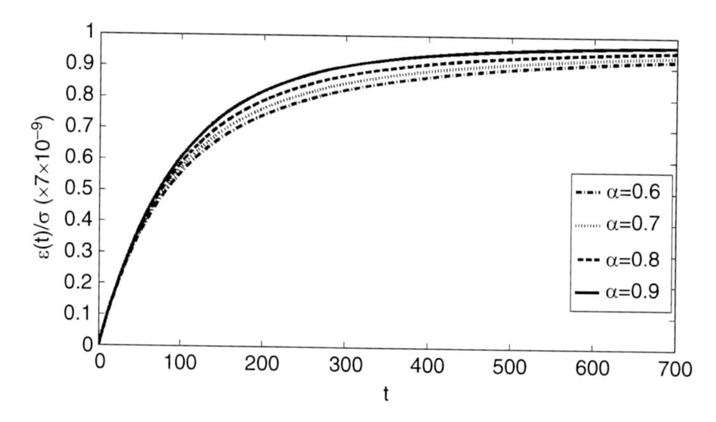

Fig. 12 The variation of the compliance with respect to time for various values of the fractional order for the existing fractional models

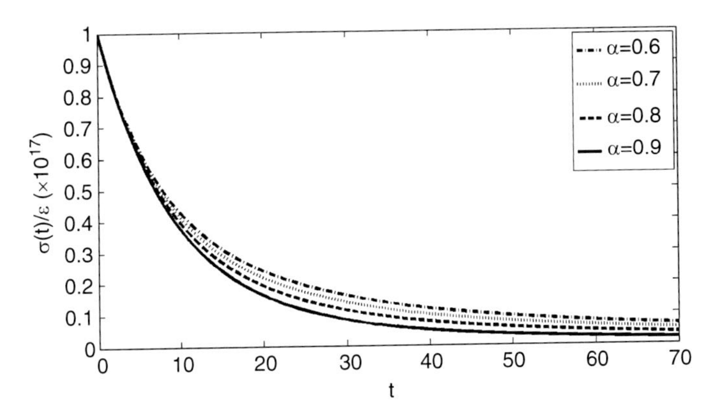

Fig. 13 Variation of the relaxation modulus with respect to time for various values of the fractional order for the existing fractional models

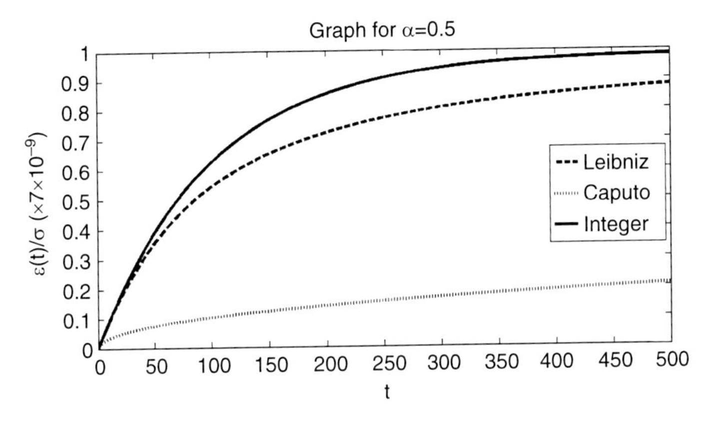

Fig. 14 The variation of the compliance with respect to time for fractional order $a = 0.5$

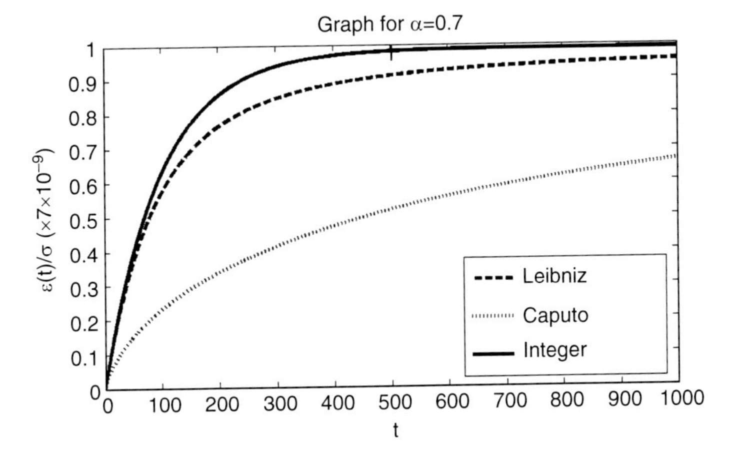

Fig. 15 The variation of the compliance with respect to time for fractional order $a = 0.7$

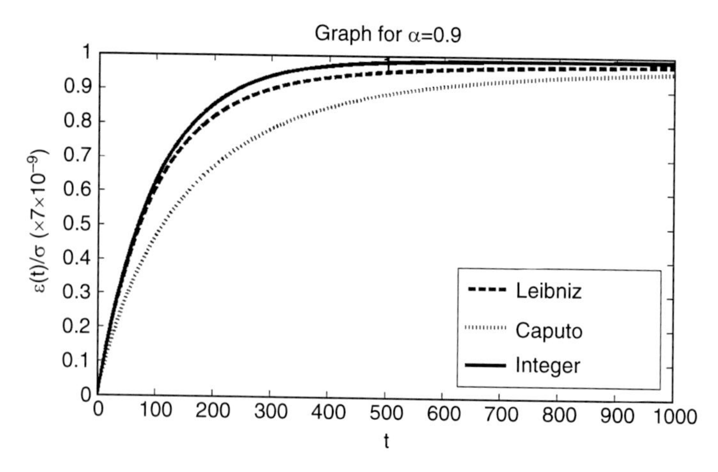

Fig. 16 The variation of the compliance with respect to time for fractional order $a = 0.9$

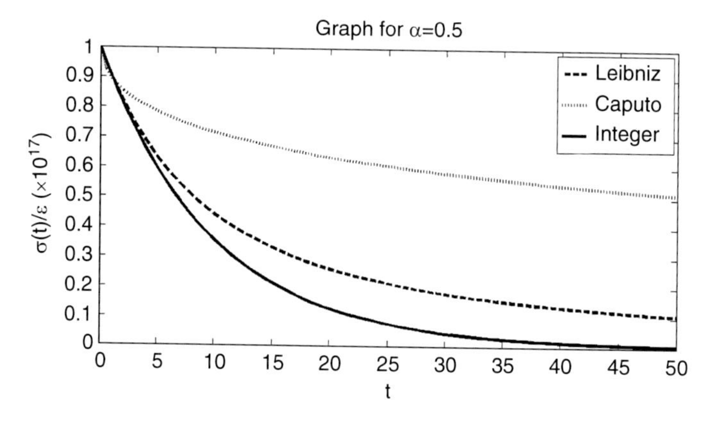

Fig. 17 Variation of the relaxation modulus with respect to time for fractional order $a = 0.5$

the fractional order of the model the slower the convergence to the final values for the same proposed model.

Conclusion: Further Research

Correcting the picture of fractional differential of a function, the fractional tangent space of a manifold was defined, introducing also Leibniz's L-fractional derivative that is the only one having physical meaning. Further, the L-fractional chain rule is imposed, that is necessary for the existence of fractional differential. After establishing the fractional differential of a function, the theory of fractional differential geometry of curves is developed. In addition, the basic forms concerning the first and second differential forms of the surfaces were defined, through the

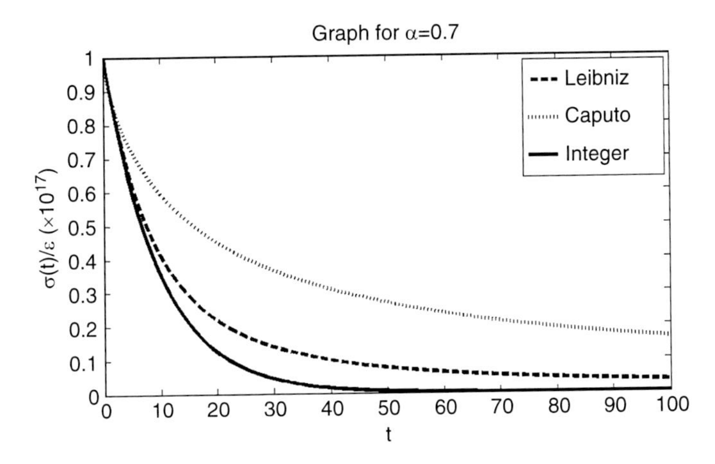

Fig. 18 Variation of the relaxation modulus with respect to time for fractional order $a = 0.7$

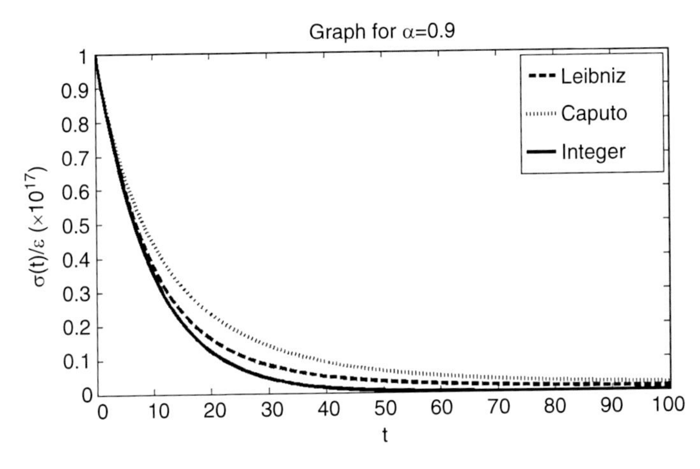

Fig. 19 Variation of the relaxation modulus with respect to time for fractional order $a = 0.9$

tangent spaces defined earlier, having mathematical meaning without any confusion, contrary to the existing procedures. Further the field theorems have been outlined in an accurate manner that may not cause confusion in their applications. The present work will help in discussing many applications concerning mechanics, quantum mechanics, and relativity, that need a clear description, based upon the fractional differential geometry. Moreover, the basic theorems of fractional vector calculus have been revised along with the basic concepts of fractional continuum mechanics. Especially the concept of fractional strain has been pointed out, because it is widely used in various places in a quite different way. The present analysis may be useful for solving updated problems in Mechanics and especially for lately proposed theories such as peridynamic theory. Finally, viscoelasticity models are studied using L-fractional derivatives. Those models exhibit milder behavior from the ones formulated through Caputo derivatives, compared to the conventional (integer) models. In the present study, only the Zener viscoelastic model was discussed and its behavior concerning the compliance and relaxation moduli was studied.

References

F.B. Adda, Interpretation geometrique de la differentiabilite et du gradient d'ordre reel. CR Acad. Sci. Paris **326**(Serie I), 931–934 (1998)

F.B. Adda, The differentiability in the fractional calculus. Nonlinear Anal. **47**, 5423–5428 (2001)

O.P. Agrawal, A general finite element formulation for fractional variational problems. J. Math. Anal. Appl. **337**, 1–12 (2008)

E.C. Aifantis, Strain gradient interpretation of size effects. Int. J. Fract. **95**, 299–314 (1999)

E.C. Aifantis, Update in a class of gradient theories. Mech. Mater. **35**, 259–280 (2003)

E.C. Aifantis, On the gradient approach – relations to Eringen's nonlocal theory. Int. J. Eng. Sci. **49**, 1367–1377 (2011)

H. Askes, E.C. Aifantis, Gradient elasticity in statics and dynamics: an overview of formulations, length scale identification procedures, finite element implementations and new results. Int. J. Solids Struct. **48**, 1962–1990 (2011)

T.M. Atanackovic, A generalized model for the uniaxial isothermal deformation of a viscoelastic body. Acta Mech. **159**, 77–86 (2002)

T.M. Atanackovic, B. Stankovic, Dynamics of a viscoelastic rod of fractional derivative type. ZAMM **82**(6), 377–386 (2002)

T.M. Atanackovic, S. Konjik, S. Philipovic, Variational problems with fractional derivatives. Euler-Lagrange equations. J. Phys. A Math. Theor. **41**, 095201 (2008)

R.L. Bagley, P.J. Torvik, Fractional calculus model of viscoelastic behavior. J. Rheol. **30**, 133–155 (1986)

N. Bakhvalov, G. Panasenko, *Homogenisation: Averaging Processes in Periodic Media* (Kluwer, London, 1989)

A. Balankin, B. Elizarrataz, Hydrodynamics of fractal continuum flow. Phys. Rev. E **85**, 025302(R) (2012)

D. Baleanu, K. Golmankhaneh Ali, K. Golmankhaneh Alir, M.C. Baleanu, Fractional electromagnetic equations using fractional forms. Int. J. Theor. Phys. **48**(11), 3114–3123 (2009)

D.D. Baleanu, H. Srivastava, V. Daftardar-Gezzi,C. Li,J.A.T. Machado, Advanced topics in fractional dynamics. Adv. Mat. Phys. Article ID 723496 (2013)

H. Beyer, S. Kempfle, Definition of physically consistent damping laws with fractional derivatives. ZAMM **75**(8), 623–635 (1995)

G. Calcani, Geometry of fractional spaces. Adv. Theor. Math. Phys. **16**, 549–644 (2012)

A. Carpinteri, B. Chiaia, P. Cornetti, Static-kinematic duality and the principle of virtual work in the mechanics of fractal media. Comput. Methods Appl. Mech. Eng. **191**, 3–19 (2001)

A. Carpinteri, P. Cornetti, A. Sapora, Static-kinematic fractional operator for fractal and non-local solids. ZAMM **89**(3), 207–217 (2009)

A. Carpinteri, P. Cornetti, A. Sapora, A fractional calculus approach to non-local elasticity. Eur. Phys. J. Spec Top **193**, 193–204 (2011)

D.R.J. Chillingworth, *Differential Topology with a View to Applications* (Pitman, London, 1976)

M. Di Paola, G. Failla, M. Zingales, Physically-based approach to the mechanics of strong non-local linear elasticity theory. J. Elast. **97**(2), 103–130 (2009)

C.S. Drapaca, S. Sivaloganathan, A fractional model of continuum mechanics. J. Elast. **107**, 107–123 (2012)

A.C. Eringen, *Nonlocal Continuum Field Theories* (Springer, New York, 2002)

G.I. Evangelatos, *Non Local Mechanics in the Time and Space Domain-Fracture Propagation via a Peridynamics Formulation: A Stochastic\Deterministic Perspective*. Thesis, Houston Texas, 2011

G.I. Evangelatos, P.D. Spanos, Estimating the 'In-Service' modulus of elasticity and length of polyester mooring lines via a non linear viscoelastic model governed by fractional derivatives. ASME 2012 Int. Mech. Eng. Congr. Expo. **8**, 687–698 (2012)

J. Feder, *Fractals* (Plenum Press, New York, 1988)

A.K. Goldmankhaneh, A.K. Goldmankhaneh, D. Baleanu, Lagrangian and Hamiltonian mechanics. Int. J. Theor. Rhys. **52**, 4210–4217 (2013)

K. Golmankhaneh Ali, K. Golmankhaneh Alir, D. Baleanu, About Schrodinger equation on fractals curves imbedding in R^3. Int. J. Theor. Phys. **54**(4), 1275–1282 (2015)

H. Guggenheimer, *Differential Geometry* (Dover, New York, 1977)

G. Jumarie, An approach to differential geometry of fractional order via modified Riemann-Liouville derivative. Acta Math. Sin. Engl. Ser. **28**(9), 1741–1768 (2012)

A.A. Kilbas, H.M. Srivastava, J.J. Trujillo, *Theory and Applications of Fractional Differential Equations* (Elsevier, Amsterdam, 2006)

K.A. Lazopoulos, On the gradient strain elasticity theory of plates. Eur. J. Mech. A/Solids **23**, 843–852 (2004)

K.A. Lazopoulos, Nonlocal continuum mechanics and fractional calculus. Mech. Res. Commun. **33**, 753–757 (2006)

K.A. Lazopoulos, in *Fractional Vector Calculus and Fractional Continuum Mechanics, Conference "Mechanics though Mathematical Modelling", celebrating the 70th birthday of Prof. T. Atanackovic,* Novi Sad, 6–11 Sept, Abstract, p. 40 (2015)

A.K. Lazopoulos, On fractional peridynamic deformations. Arch. Appl. Mech. **86**(12), 1987–1994 (2016a)

K.A. Lazopoulos, in *Fractional Differential Geometry of Curves and Surfaces, International Conference on Fractional Differentiation and Its Applications (ICFDA 2016)*, Novi Sad (2016b)

A.K. Lazopoulos, *On Fractional Peridynamic Deformations, International Conference on Fractional Differentiation and Its Applications, Proceedings ICFDA 2016*, Novi Sad (2016c)

K.A. Lazopoulos, A.K. Lazopoulos, Bending and buckling of strain gradient elastic beams. Eur. J. Mech. A/Solids **29**(5), 837–843 (2010)

K.A. Lazopoulos, A.K. Lazopoulos, On fractional bending of beams. Arch. Appl. Mech. (2015). https://doi.org/10.1007/S00419-015-1083-7

K.A. Lazopoulos, A.K. Lazopoulos, Fractional vector calculus and fractional continuum mechanics. Prog. Fract. Diff. Appl. **2**(1), 67–86 (2016a)

K.A. Lazopoulos, A.K. Lazopoulos, On the fractional differential geometry of curves and surfaces. Prog. Fract. Diff. Appl., No **2**(3), 169–186 (2016b)

G.W. Leibnitz, Letter to G. A. L'Hospital. Leibnitzen Mathematishe Schriftenr. **2**, 301–302 (1849)

Y. Liang, W. Su, Connection between the order of fractional calculus and fractional dimensions of a type of fractal functions. Anal. Theory Appl. **23**(4), 354–362 (2007)

J. Liouville, Sur le calcul des differentielles a indices quelconques. J. Ec. Polytech. **13**, 71–162 (1832)

H.-S. Ma, J.H. Prevost, G.W. Sherer, Elasticity of dlca model gels with loops. Int. J. Solids Struct. **39**, 4605–4616 (2002)

F. Mainardi, *Fractional Calculus and Waves in Linear Viscoelasticity* (Imperial College Press, London, 2010)

M. Meerschaert, J. Mortensen, S. Wheatcraft, Fractional vector calculus for fractional advection–dispersion. Physica A **367**, 181–190 (2006)

R.D. Mindlin, Second gradient of strain and surface tension in linear elasticity. Int. Jnl. Solids & Struct. **1**, 417–438 (1965)

W. Noll, A mathematical theory of the mechanical behavior of continuous media. Arch. Rational Mech. Anal. **2**, 197–226 (1958/1959)

K.B. Oldham, J. Spanier, *The Fractional Calculus* (Academic, New York, 1974)

I. Podlubny, *Fractional Differential Equations (An Introduction to Fractional Derivatives Fractional Differential Equations, Some Methods of Their Solution and Some of Their Applications)* (Academic, San Diego, 1999)

I.R. Porteous, *Geometric Differentiation* (Cambridge University Press, Cambridge, 1994)

B. Riemann, Versuch einer allgemeinen Auffassung der Integration and Differentiation, in *Gesammelte Werke*, vol. 62 (1876)

F. Riewe, Nonconservative Lagrangian and Hamiltonian mechanics. Phys. Rev. E **53**(2), 1890–1899 (1996)

F. Riewe, Mechanics with fractional derivatives. Phys. Rev. E **55**(3), 3581–3592 (1997)

J. Sabatier, O.P. Agrawal, J.A. Machado, *Advances in Fractional Calculus (Theoretical Developments and Applications in Physics and Engineering)* (Springer, The Netherlands, 2007)

S.G. Samko, A.A. Kilbas, O.I. Marichev, *Fractional Integrals and Derivatives: Theory and Applications* (Gordon and Breach, Amsterdam, 1993)

S.A. Silling, Reformulation of elasticity theory for discontinuities and long-range forces. J. Mech. Phys. Solids **48**, 175–209 (2000)

S.A. Silling, R.B. Lehoucq, Peridynamic theory of solid mechanics. Adv. Appl. Mech. **44**, 175–209 (2000)

S.A. Silling, M. Zimmermann, R. Abeyaratne, Deformation of a peridynamic bar. J. Elast. **73**, 173–190 (2003)

W. Sumelka, Non-local Kirchhoff-Love plates in terms of fractional calculus. Arch. Civil and Mech. Eng. **208** (2014). https://doi.org/10.1016/j.acme2014.03.006

V.E. Tarasov, Fractional vector calculus and fractional Maxwell's equations. Ann. Phys. **323**, 2756–2778 (2008)

V.E. Tarasov, *Fractional Dynamics: Applications of Fractional Calculus to Dynamics of Particles, Fields and Media* (Springer, Berlin, 2010)

R.A. Toupin, Theories of elasticity with couple stress. Arch. Ration. Mech. Anal. **17**, 85–112 (1965)

C. Truesdell, *A First Course in Rational Continuum Mechanics*, vol 1 (Academic, New York, 1977)

C. Truesdell, W. Noll, The non-linear field theories of mechanics, in *Handbuch der Physik*, vol. III/3, ed. by S. Fluegge (Springer, Berlin, 1965)

I. Vardoulakis, G. Exadactylos, S.K. Kourkoulis, Bending of a marble with intrinsic length scales: a gradient theory with surface energy and size effects. J. Phys. IV **8**, 399–406 (1998)

H.M. Wyss, A.M. Deliormanli, E. Tervoort, L.J. Gauckler, Influence of microstructure on the rheological behaviour of dense particle gels. AIChE J. **51**, 134–141 (2005)

K. Yao, W.Y. Su, S.P. Zhou, On the connection between the order of fractional calculus and the dimensions of a fractal function. Chaos, Solitons Fractals **23**, 621–629 (2005)

Continuum Homogenization of Fractal Media

25

Martin Ostoja-Starzewski, Jun Li, and Paul N. Demmie

Contents

Sandia National Laboratories is a multimission laboratory operated by Sandia Corporation, a Lockheed Martin Company, for the United States Department of Energy under Contract DE-AC04-94AL85000.

M. Ostoja-Starzewski (✉)
Department of Mechanical Science and Engineering, Institute for Condensed Matter Theory and Beckman Institute, University of Illinois at Urbana–Champaign, Urbana, IL, USA
e-mail: martinos@illinois.edu

J. Li
Department of Mechanical Engineering, University of Massachusetts, Dartmouth, MA, USA

P. N. Demmie
Sandia National Laboratories, Albuquerque, NM, USA

G. Z. Voyiadjis (ed.), *Handbook of Nonlocal Continuum Mechanics for Materials and Structures*, https://doi.org/10.1007/978-3-319-58729-5_18

Abstract

This chapter reviews the modeling of fractal materials by homogenized continuum mechanics using calculus in non-integer dimensional spaces. The approach relies on expressing the global balance laws in terms of fractional integrals and, then, converting them to integer-order integrals in conventional (Euclidean) space. Via localization, this allows development of local balance laws of fractal media (continuity, linear and angular momenta, energy, and second law) and, in case of elastic responses, formulation of wave equations in several settings (1D and 3D wave motions, fractal Timoshenko beam, and elastodynamics under finite strains). Next, follows an account of extremum and variational principles, and fracture mechanics. In all the cases, the derived equations for fractal media depend explicitly on fractal dimensions and reduce to conventional forms for continuous media with Euclidean geometries upon setting the dimensions to integers.

Keywords

Balance laws · Fractal · Fractional calculus · Fractal derivative · Homogenization

Introduction

It has been observed by Benoît Mandelbrot (1982) that many natural objects are statistically self-similar and "broken" in space (or time) and exhibit non-smooth or highly irregular features. Mandelbrot called such objects *fractals*. Examples include coastlines, porous media, cracks, turbulent flows, clouds, mountains, lightning bolts, snowflakes, melting ice, and even parts of living entities such as the neural structure or the surface of the human brain (Barnsley 1993; Le Méhauté 1991; Hastings and Sugihara 1993; Falconer 2003; Tarasov 2005a). There are also fractals in time – signals, processes, and musical compositions – but we shall not concern ourselves with them here.

Mathematical fractal sets are characterized by a Hausdorff dimension D, which is the scaling exponent characterizing the fractal pattern's power law. Physical fractals can be modeled only for some finite range of length scales within the lower and upper cutoffs by mathematical ones. These objects are called *pre-fractals*. While the mechanics of fractal and pre-fractal media is still in an early developing stage, the new field of *fractal mechanics* can already generate elegant models (Tarasov 2005b; Ostoja-Starzewski 2007, 2008a, 2009; Joumaa and Ostoja-Starzewski 2011; Li and Ostoja-Starzewski 2009a).

The approach we employ is based on calculus in non-integer dimensional spaces (i.e., spaces embedded in Euclidean space), thus allowing a homogenization; by a slight abuse of terminology stemming from quantum mechanics (Li and Ostoja-Starzewski 2009b), this method has originally been called "dimensional regularization." With this approach the fractional integrals over fractal sets (i.e., fractal material domains) are transformed to equivalent continuous integrals over Euclidean sets (Jumarie 2009). This transformation produces balance laws that are expressed in continuous form, thereby simplifying their mathematical manipulation both analytically and computationally. A product measure is used to achieve this transformation.

The idea goes back to V.E. Tarasov (2005a,b,c), who developed continuum-type equations for conservation of mass, linear and angular momentum, and energy of fractals and studied several fluid mechanics and wave problems (Tarasov 2005a,b,c, 2010). An advantage of this approach is that it admits upper and lower cutoffs of fractal scaling, so that one effectively deals with a physical pre-fractal rather than a purely mathematical fractal lacking any cutoffs. The original formulation of Tarasov was based on the Riesz measure, which is more appropriate for isotropic fractal media than for anisotropic fractal media. To represent more general heterogeneous media, Li and Ostoja-Starzewski introduced a model based on a product measure (Li and Ostoja-Starzewski 2010, 2011; Ostoja-Starzewski et al. 2016). Since this measure has different fractal dimensions in different directions, it grasps the anisotropy of fractal geometry better than the Tarasov formulation for a range of length scales between the lower and upper cutoffs (Li and Ostoja-Starzewski 2010; Ostoja-Starzewski et al. 2016). The great promise is that the conventional requirement of continuum mechanics, the separation of scales, can be removed with continuum-type field equations still employed. This approach was applied, among others, to thermomechanics with internal variables, extremum principles of elasticity and plasticity, turbulence in fractal porous media, dynamics of fractal beams, fracture mechanics, and thermoelasticity (Ostoja-Starzewski 2007, 2008a, 2009; Jumarie 2009; Oldham and Spanier 1974; Ostoja-Starzewski et al. 2014).

Homogenization of Fractal Media

Mass Power Law and Product Measure

The basic approach to homogenization of fractal media by continua originated with Tarasov (2005a,b, 2010). He started to work in the setting where the mass obeys a power law

$$m(R) \sim R^D, \quad D < 3, \tag{1}$$

with R being the length scale of measurement (or resolution) and D the fractal dimension of mass in the three-dimensional (3D) Euclidean space $\mathbb{E}^3$. Note that the relation (1) can be applied to a *pre-fractal*, i.e., a fractal-type, physical object with lower and upper cutoffs. More specifically, Tarasov used a fractional integral to represent mass in a region $\mathcal{W}$ embedded in $\mathbb{E}^3$. In the subsequent work, we focused on a general anisotropic, fractal medium governed, in place of (1), by a more general power law relation with respect to each coordinate (Li and Ostoja-Starzewski 2010, 2011; Ostoja-Starzewski et al. 2016) (which, in fact, had originally been recognized by Tarasov)

$$m(x_1, x_2, x_3) \sim {x_1}^{\alpha_1} {x_2}^{\alpha_2} {x_3}^{\alpha_3}. \tag{2}$$

Then, the mass is specified via a *product measure*

$$m(\mathcal{W}) = \int_{\mathcal{W}} \rho(x_1, x_2, x_3) dl_{\alpha_1}(x_1) dl_{\alpha_2}(x_2) dl_{\alpha_3}(x_3), \tag{3}$$

while the length measure along each coordinate is given through transformation coefficients $c_1^{(k)}$

$$dl_{\alpha_k}(x_k) = c_1^{(k)}(\alpha_k, x_k) dx_k, \quad k = 1, 2, 3 \quad \text{(no sum)}. \tag{4}$$

Equation (3) implies that the mass fractal dimension D equals $\alpha_1 + \alpha_2 + \alpha_3$ along the diagonals, $|x_1| = |x_2| = |x_3|$, where each α_k plays the role of a fractal dimension in the direction x_k. While it is noted that, in other directions, the anisotropic fractal body's fractal dimension is not necessarily the sum of projected fractal dimensions, an observation from an established text on mathematics of fractals is recalled here (Falconer 2003): *Many fractals encountered in practice are not actually products, but are product-like.* In what follows, we expect the equality between $D = \alpha_1 + \alpha_2 + \alpha_3$ to hold for fractals encountered in practice, whereas a rigorous proof of this property remains an open research topic.

This formulation has four advantages over the original approach (Li and Ostoja-Starzewski 2009b, 2011):

1. It does not involve a left-sided fractional derivative (Riemann-Liouville), which does not give zero when applied to a constant function, a rather unphysical property.
2. The mechanics-type derivation of wave equations (in 3D, 2D, or 1D) yields the same result as that obtained from the variational-type derivation.
3. The 3D wave equation cleanly reduces to the 1D wave equation.
4. It offers a consistent way of handling not only generally anisotropic fractals but also isotropic ones, overall developing a systematic formulation of continuum mechanics. In particular, see the "fractal derivative" (13) below.

Product Measure

The relation (4) implies that the infinitesimal fractal volume element, dV_D, is

$$dV_D = dl_{\alpha_1}(x_1)dl_{\alpha_2}(x_2)dl_{\alpha_3}(x_3) = c_1^{(1)}c_1^{(2)}c_1^{(3)}dx_1dx_2dx_3 = c_3dV_3, \quad \text{with} \quad c_3 = c_1^{(1)}c_1^{(2)}c_1^{(3)}. \tag{5}$$

In fact, it plays the role of a fractal representative volume element (RVE), which is mapped into the RVE of non-fractal (conventional) type. Note that (i) this map is different from the scaling of a random microstructure toward a deterministic volume element in non-fractal media (Ostoja-Starzewski et al. 2016), and (ii) the material spatial randomness is not explicitly introduced into the formulation discussed here.

For the surface transformation coefficient $c_2^{(k)}$, we consider a cubic volume element, $dV_3 = dx_1dx_2dx_3$, whose surface elements are specified by the normal vector along the axes i, j, or k in Fig. 1. Therefore, $c_2^{(k)}$ associated with the surface $S_d^{(k)}$ is

$$c_2^{(k)} = c_1^{(i)}c_1^{(j)} = c_3/c_1^{(k)}, \quad i \neq j, \quad i, j \neq k. \tag{6}$$

The sum $d^{(k)} = \alpha_i + \alpha_j$, $i \neq j, i, j \neq k$, is the fractal dimension of the surface $S_d^{(k)}$ along the diagonals $|x_i| = |x_j|$ in $S_d^{(k)}$. This equality is not necessarily true elsewhere, but is expected to hold for fractals encountered in practice (Falconer 2003) as discussed previously for the relationship between D and $\alpha_1 + \alpha_2 + \alpha_3$. Figure 1 illustrates the relationship among the line transformation coefficients $c_1^{(k)}$ and respective surface ($c_2^{(k)}$) and volume (c_3) transformation coefficients. We note that, when $D \to 3$, with each $\alpha_i \to 1$, the conventional concept of mass is recovered (Ostoja-Starzewski et al. 2014).

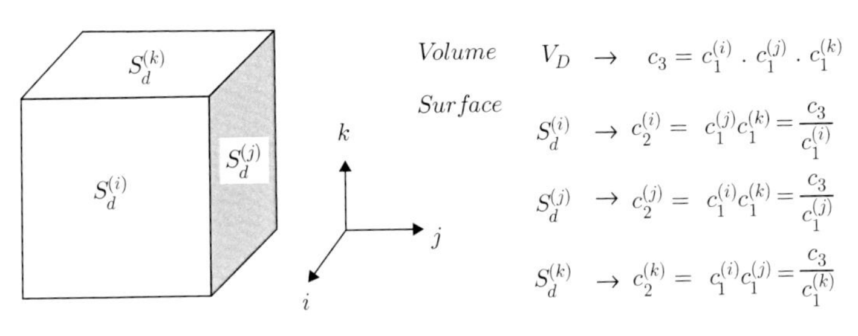

Fig. 1 Roles of the transformation coefficients $c_1^{(i)}$, $c_2^{(k)}$, and c_3 in homogenizing a fractal body of volume dV_D, surface dS_d, and lengths dl_α into a Euclidean parallelpiped of volume dV_3, surface dS_2, and lengths dx_k

We adopt the modified Riemann-Liouville fractional integral of Jumarie (2009, 2005) for $c_1^{(k)}$:

$$c_1^{(k)} = \alpha_k \left(\frac{l_k - x_k}{l_{k0}}\right)^{\alpha_k - 1}, \quad k = 1, 2, 3, \quad \text{(no sum on } k\text{)}, \tag{7}$$

where l_k is the total length (integral interval) along x_k, while l_{k0} is the characteristic length in the given direction (e.g., the mean pore size). In the product measure formulation, the resolution length scale is

$$R = \sqrt{l_k l_k}. \tag{8}$$

Examining $c_1^{(k)}$ in two special cases, we observe:

1. Uniform mass: The mass is distributed isotropically in a cubic region with a power law relation (9). Denoting the reference mass density by ρ_0 and the cubic length by l, we obtain

$$m(\mathcal{W}) = \rho_0 l^{\alpha_1} l^{\alpha_2} l^{\alpha_3} / l_0^{D-3} = \rho_0 l^{\alpha_1 + \alpha_2 + \alpha_3} / l_0^{D-3} = \rho_0 l^D / l_0^{D-3}, \tag{9}$$

which is consistent with the mass power law (1). In general, however, $D \neq \alpha_1 + \alpha_2 + \alpha_3$.
2. Point mass: The distribution of mass is concentrated at one point, so that the mass density is denoted by the Dirac function $\rho(x_1, x_2, x_3) = m_0 \delta(x_1)\delta(x_2)\delta(x_3)$. The fractional integral representing mass becomes

$$m(\mathcal{W}) = \alpha_1 \alpha_2 \alpha_3 \frac{l^{\alpha_1 - 1} l^{\alpha_2 - 1} l^{\alpha_3 - 1}}{l_0^{D-3}} m_0 = \alpha_1 \alpha_2 \alpha_3 \left(\frac{l}{l_0}\right)^{D-3} m_0, \tag{10}$$

When $D \to 3$ ($\alpha_1, \alpha_2, \alpha_3 \to 1$), $m(\mathcal{W}) \to m_0$ and the conventional concept of point mass is recovered (Ostoja-Starzewski et al. 2014). Note that using the Riesz fractional integral is not well defined except when $D = 3$ (by letting $0^0 = 1$ in $m(\mathcal{W}) = \alpha_1 \alpha_2 \alpha_3 0^{D-3} m_0$), which, on the other hand, shows a non-smooth transition of mass with respect to its fractal dimension. This fact also supports our choice of the non-Riesz-type expressions for $c_1^{(k)}$ in (14).

Note that our expression for $c_1^{(k)}$ shows that the length dimension, and hence the mass m would involve an unusual physical dimension if it were replaced by $c_1^{(k)} = \alpha_k (l_k - x_k)^{\alpha_k - 1}$. This behavior is understandable since, mathematically, a fractal curve only exhibits a finite measure with respect to a fractal dimensional length unit (Mandelbrot 1982). Of course, in practice, we prefer physical quantities to have usual dimensions, and so we work with nondimensionalized coefficients $c_1^{(k)}$.

A simple generic example of an anisotropic, product-like fractal is the so-called *Carpinteri column* (Carpinteri et al. 2004), a parallelepiped domain in $\mathbb{E}^3$, having mathematically well-defined Hausdorff dimensions in all three directions. It has been proposed as a model of concrete columns which are essentially composite structures featuring oriented fractal-type microstructures. The square cross section of the column is a *Sierpiński carpet*, which is "fractally" swept along the longitudinal direction in conjunction with a *Cantor ternary set* (see Ostoja-Starzewski et al. 2014 for a discussion).

Another example is offered by the rings of Saturn. If considered as random fields, the rings possess (i) statistical stationarity in time, (ii) statistical isotropy in space, and (iii) statistical spatial nonstationarity. The reason for (i) is an extremely slow decay of rings relative to the time scale of orbiting around Saturn. The reason for (ii) is the obviously circular, albeit radially disordered and fractal pattern of rings; fractality is present in the radial but not polar coordinate (Li and Ostoja-Starzewski 2015). The reason for (iii) is the lack of invariance with respect to arbitrary shifts in Cartesian space which, on the contrary, holds true in, say, a basic model of turbulent velocity fields.

Fractional Integral Theorems and Fractal Derivatives

In order to develop continuum mechanics of fractal media, we introduce the notion of fractal derivatives with respect to the coordinate x_k and time t. The definitions of these derivatives follow naturally from the fractional generalization of two basic integral theorems that are employed in continuum mechanics (Tarasov 2005c, 2010; Li and Ostoja-Starzewski 2009b; Demmie and Ostoja-Starzewski 2011): Gauss theorem, which relates a volume integral to the surface integral over its bounding surface, and the Reynolds transport theorem, which provides an expression for the time rate of change of any volume integral in a continuous medium.

Consider the surface integral

$$\int_{\partial\mathcal{W}} \mathbf{f}\cdot\mathbf{n}dS_d = \int_{\partial\mathcal{W}} f_k n_k dS_d, \tag{11}$$

where $\mathbf{f}(=f_k)$ is any vector field and $\mathbf{n}(=n_k)$ is the outward normal vector field to the surface $\partial\mathcal{W}$ which is the boundary surface for some volume $\mathcal{W}$, and dS_d is the surface element in fractal space. The notation $(=A_k)$ is used to indicate that the A_k are components of the vector $\mathbf{A}$.

To compute (11), we relate the integral element $\mathbf{n}dS_d$ to its conventional surface element $\mathbf{n}dS_2$ in $\mathbb{E}^3$ via the fractal surface coefficients $c_2^{(k)}$, $k=1,2,3$, as shown in Fig. 1. This figure shows that the infinitesimal element $\mathbf{n}dS_d$ can be expressed as a linear combination of the $n_k c_2^{(k)} dS_2$, $k=1,2,3$ (no sum).

By the conventional Gauss theorem, and noting that $c_2^{(k)}$ does not depend on the coordinate x_k, (11) becomes

$$\int_{\partial\mathcal{W}} \mathbf{f}\cdot\mathbf{n}dS_d = \int_{\partial\mathcal{W}} f_k n_k dS_d = \int_{\mathcal{W}} [f_k c_2^{(k)}]_{,k}\, dV_3 = \int_{\mathcal{W}} \frac{f_{k,k}}{c_1^{(k)}} dV_D. \tag{12}$$

In (12) and elsewhere, we employ the usual convention that $(\cdot)_{,k}$ is the partial derivative of $(\cdot)$ with respect to x_k. Next, based on (11) and the above, we define the *fractal derivative* (*fractal gradient*), ∇_k^D as

$$\nabla^D \phi = \mathbf{e}_k \nabla_k^D \phi \quad \text{or} \quad \nabla_k^D \phi = \frac{1}{c_1^{(k)}} \frac{\partial \phi}{\partial x_k} \quad (\text{no sum on } k), \tag{13}$$

where $\mathbf{e}_k$ is the base vector. With this definition, the *Gauss theorem for fractal media* becomes

$$\int_{\partial \mathcal{W}} \mathbf{f} \cdot \mathbf{n} dS_d = \int_{\mathcal{W}} \nabla_k^D f_k dV_D = \int_{\mathcal{W}} \left(\nabla^D \cdot \mathbf{f} \right) dV_D. \tag{14}$$

It is straightforward to show that the fractal operator, ∇_k^D, commutes with the fractional integral operator, is its inverse, and satisfies the product rule for differentiation (the Leibnitz property). Furthermore, the fractal derivative of a constant is zero. This latter property shows that a fractal derivative and a fractional derivative are not the same since the fractional derivative of a constant does not always equal to zero (Oldham and Spanier 1974).

To define the fractal material time derivative, we consider the fractional generalization of Reynolds transport theorem. Consider any quantity, P, accompanied by a moving fractal material system, $\mathcal{W}_t$, with velocity vector field $\mathbf{v}$ $(= \upsilon_k)$. The time derivative of the volume integral of P over $\mathcal{W}_t$ is

$$\frac{d}{dt} \int_{\mathcal{W}_t} P dV_D. \tag{15}$$

Using the Jacobian (J) of the transformation between the current configuration (x_k) and the reference configuration (X_k), the relationship between the corresponding volume elements $(dV_D = JdV_D^0)$, and the expression for the time derivative of J, it is straightforward to show that

$$\begin{aligned} \frac{d}{dt} \int_{\mathcal{W}_t} P dV_D &= \int_{\mathcal{W}_t} \left[\frac{\partial P}{\partial t} + (\upsilon_k P)_{,k} \right] dV_D \\ &= \int_{\mathcal{W}_t} \left[\frac{\partial P}{\partial t} + c_1^{(k)} \nabla_k^D (\upsilon_k P) \right] dV_D. \end{aligned} \tag{16}$$

The result given by the first equality is identical to the conventional representation. Hence, the fractal material time derivative and the conventional material time derivative are the same,

$$\left(\frac{d}{dt} \right)_D P = \frac{\partial P}{\partial t} + \upsilon_k P_{,k} = \frac{\partial P}{\partial t} + c_1^{(k)} \upsilon_k \nabla_k^D P. \tag{17}$$

Equation (17) is the *Reynolds theorem for fractal media*; its form is similar to the conventional one; an alternative form (Li and Ostoja-Starzewski 2010) of the fractional Reynolds transport theorem that involves surface integrals is different from the conventional one and rather complicated. This difference results from the fractal volume coefficient c_3 depending on all the coordinates, whereas in the derivation of the Gauss theorem (14), $c_2^{(k)}$ is independent of x_k.

Vector Calculus on Anisotropic Fractals

Motivated by the fractal derivative (13), we have the *fractal divergence* of a vector field (**f**)

$$\mathrm{div}\mathbf{f} = \nabla^D \cdot \mathbf{f} \quad \text{or} \quad \nabla_k^D f_k = \frac{1}{c_1^{(k)}} \frac{\partial f_k}{\partial x_k} \tag{18}$$

and the *fractal curl* operator

$$\mathrm{curl}\mathbf{f} = \nabla^D \times \mathbf{f} \quad \text{or} \quad e_{jki} \nabla_k^D f_i = e_{jki} \frac{1}{c_1^{(k)}} \frac{\partial f_i}{\partial x_k}. \tag{19}$$

The four fundamental identities of the conventional vector calculus can now be shown to carry over in terms of these new operators:

(i) The divergence of the curl of a vector field **f**:

$$\mathrm{div} \cdot \mathrm{curl}\mathbf{f} = \mathbf{e}_m \nabla_m^D \cdot \mathbf{e}_j e_{jki} \nabla_k^D f_i = \frac{1}{c_1^{(j)}} \frac{\partial f}{\partial x_j} \left[e_{jki} \frac{1}{c_1^{(k)}} \frac{\partial f_i}{\partial x_k} \right] = 0. \tag{20}$$

where e_{ijk} is the permutation tensor.

(ii) The curl of the gradient of a scalar field ϕ:

$$\mathrm{curl} \times (\mathrm{grad}\phi) = \mathbf{e}_i e_{ijk} \nabla_j^D (\nabla_k^D \phi) = \mathbf{e}_i e_{ijk} \frac{1}{c_1^{(j)}} \frac{\partial}{\partial x_j} \left[\frac{1}{c_1^{(k)}} \frac{\partial \phi}{\partial x_k} \right] = 0. \tag{21}$$

In both cases above, we can pull $1/c_1^{(k)}$ in front of the gradient because the coefficient $c_1^{(k)}$ is independent of x_j.

(iii) The divergence of the gradient of a scalar field ϕ is written in terms of the fractal gradient as

$$\mathrm{div} \cdot (\mathrm{grad}\phi) = \nabla_j^D \cdot \nabla_k^D \phi = \frac{1}{c_1^{(j)}} \frac{\partial}{\partial x_j} \left[\frac{1}{c_1^{(j)}} \frac{\partial \phi}{\partial x_j} \right] = \frac{1}{c_1^{(j)}} \left[\frac{\partial \phi_{,j}}{c_1^{(j)}} \right]_{,j}, \tag{22}$$

which gives an explicit form of the *fractal Laplacian*.

(iv) The curl of the curl operating on a vector field $\mathbf{f}$:

$$\mathrm{curl}\times(\mathrm{curl}\mathbf{f}) = \mathbf{e}_p e_{prj} \nabla_r^D (e_{jki} \nabla_k^D f_i) = \mathbf{e}_p \nabla_r^D \left(\nabla_p^D f_r\right) - \mathbf{e}_p \nabla_r^D \nabla_r^D f_p \, . \quad (23)$$

Next, the Helmholtz decomposition for fractals can be proved just as in the conventional case: a vector field $\mathbf{F}$ with known divergence and curl, none of which equal to zero and which is finite and uniform and vanishes at infinity, may be expressed as the sum of a lamellar vector U and a solenoidal vector V

$$\mathbf{F} = \mathbf{U} + \mathbf{V} \quad (24)$$

with the operations

$$\mathrm{curl}\mathbf{U} = \mathbf{0}, \quad \mathrm{div}\mathbf{V} = 0 \quad (25)$$

understood in the sense of (18) and (19), respectively.

These results have recently been used (Ostoja-Starzewski 2012) to obtain Maxwell equations modified to generally anisotropic fractal media using two independent approaches: a conceptual one (involving generalized Faraday and Ampère laws) and the one directly based on a variational principle for electromagnetic fields. In both cases the resulting equations are the same, thereby providing a self-consistent verification of our derivations. Just as Tarasov (2010), we have found that the presence of anisotropy in the fractal structure leads to a source/disturbance as a result of generally unequal fractal dimensions in various directions, although, in the case of isotropy, our modified Maxwell equations are different. For most recent developments, see Tarasov (2014, 2015a,b).

Homogenization Process for Fractal Media

The formula (3) for fractal mass expresses the mass power law using fractional integrals. From a homogenization standpoint, this relationship allows an interpretation of the fractal medium as an intrinsically discontinuous continuum with a fractal metric embedded in the equivalent homogenized continuum model as shown in Fig. 2. In this figure, dl_{α_i}, dS_d, and dV_D represent the line, surface, and volume elements in the fractal medium, while dx_i, dS_2, and dV_3, respectively, denote these elements in the homogenized continuum model. The coefficients $c_1^{(i)}$, $c_2^{(k)}$, and c_3 provide the relationship between the fractal medium and the homogenized continuum model:

$$dl_{\alpha_i} = c_1^{(i)} dx_i, \quad dS_d = c_2^{(k)} dS_2, \quad dV_D = c_3 dV_3 \quad \text{(no sum)}. \quad (26)$$

Standard image analysis techniques (such as the "box method" or the "sausage method" Stoyan and Stoyan 1994) allow a quantitative calibration of these coefficients for every direction and every cross-sectional plane. In a non-fractal medium

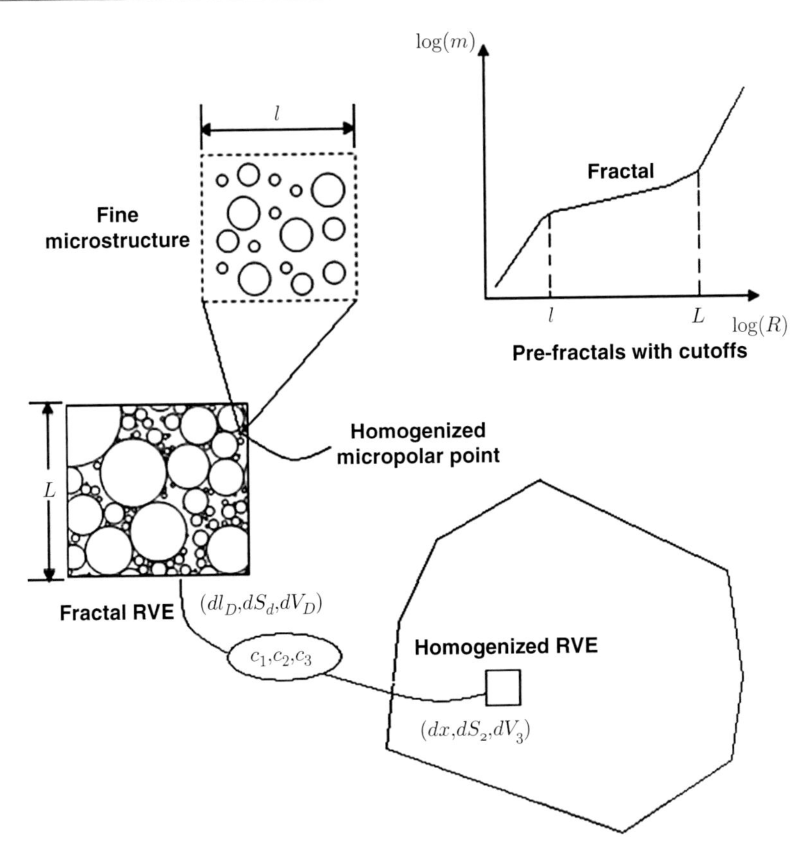

Fig. 2 Illustration of the two-level homogenization processes: fractal effects are present between the resolutions l and L in a fractal RVE with an Apollonian packing porous microstructure

where all the c coefficients in (26) are unity, one recovers conventional forms of the transport and balance equations of continuum mechanics. As discussed in section "Fractal Angular Momentum Equation," the presence of fractal geometric anisotropy ($c_1^{(j)} \neq c_1^{(k)}$, $j \neq k$, in general), as reflected by differences between the α's, leads to micropolar effects; see also Li and Ostoja-Starzewski (2010, 2011).

The above formulations provide one choice of calculus on fractals, i.e., through product integral (3), reflecting the mass scaling law (2) of fractal media. The advantage of our approach is that it is connected with conventional calculus through coefficients c_1, c_2, and c_3 and therefore well suited for development of continuum mechanics and partial differential equations on fractal media as we shall see in the next sections. Besides, the product formulation allows a decoupling of coordinate variables, which profoundly simplifies the Gauss theorem (14) and many results thereafter. Other choices of calculus on fractals have been discussed in Ostoja-Starzewski et al. (2014).

Continuum Mechanics of Fractal Media

In the preceding section, we discussed product measures and fractional integrals, generalized the Gauss and Reynolds theorems to fractal media, and introduced fractal derivatives. We now have the framework to develop continuum mechanics in a fractal setting. We proceed just like it in classical continuum mechanics but employ fractional integrals expressed in terms of fractal derivatives.

In light of the discussion in section "Homogenization Process for Fractal Media" of the difference between fractal media and classical continuum mechanics, the definitions of stress and strain are to be modified appropriately. First, with reference to the modified concept of surface elements, we have the Cauchy tetrahedron of Fig. 3. Next, we specify the relationship between surface force, $\mathbf{F}^S$ $(= F_k^S)$, and the Cauchy stress tensor $\boldsymbol{\sigma}$ $(= \sigma_{kl})$ using fractional integrals as

$$F_k^S = \int_S \sigma_{lk} n_l dS_d, \tag{27}$$

where n_l are the components of the outward normal $\mathbf{n}$ to S. On account of $(26)_2$, this force becomes

$$F_k^S = \int_S \sigma_{lk} n_l dS_d = \int_S \sigma_{lk} n_l c_2^{(l)} dS_2. \tag{28}$$

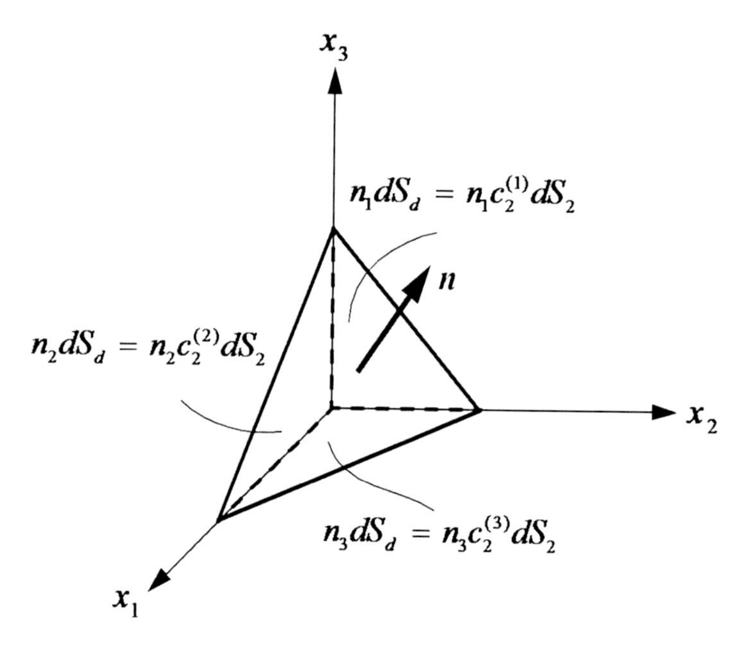

Fig. 3 Cauchy's tetrahedron of a fractal body interpreted via product measures

To specify the strain, we observe, again using $(26)_1$ and the definition of fractal derivative (13), that

$$\frac{\partial}{\partial l_{\alpha_k}} = \frac{\partial x_k}{\partial l_{\alpha_k}} \frac{\partial}{\partial x_k} = \frac{1}{c_1^{(k)}} \frac{\partial}{\partial x_k} = \nabla_k^D. \tag{29}$$

Thus, for small deformations, we define the strain, ε_{ij}, in terms of the displacement u_k as

$$\varepsilon_{ij} = \frac{1}{2}\left(\nabla_j^D u_i + \nabla_i^D u_j\right) = \frac{1}{2}\left[\frac{1}{c_1^{(j)}} u_{i,j} + \frac{1}{c_1^{(i)}} u_{j,i}\right] \quad \text{(no sum).} \tag{30}$$

As shown in Li and Ostoja-Starzewski (2010), this definition of strain results in the same equations governing wave motion in linear elastic materials when derived by a variational approach as when derived by a mechanical approach, also see section "Fractal Wave Equations."

In the following, we apply the balance laws for mass, linear momentum, energy, and entropy production to the fractal medium in order to derive the corresponding continuity equations.

Fractal Continuity Equation

Consider the equation for conservation of mass for $\mathcal{W}$

$$\frac{d}{dt} \int_{\mathcal{W}} \rho d V_D = 0, \tag{31}$$

where ρ is the density of the medium. Using the fractional Reynolds transport theorem (17), since $\mathcal{W}$ is arbitrary, we find, in terms of the fractal derivative (13),

$$\frac{d\rho}{dt} + \rho c_1^{(k)} \nabla_k^D v_k = 0. \tag{32}$$

Fractal Linear Momentum Equation

Consider the balance law of linear momentum for $\mathcal{W}$,

$$\frac{d}{dt} \int_{\mathcal{W}} \rho \mathbf{v} d V_D = \mathbf{F}^B + \mathbf{F}^S, \tag{33}$$

where $\mathbf{F}^B$ is the body force, and $\mathbf{F}^S$ is the surface force given by (28). In terms of the components of velocity, v_k, and body force density, X_k, (33) can be written as

$$\frac{d}{dt}\int_{\mathcal{W}} \rho v_k dV_D = \int_{\mathcal{W}} X_k dV_D + \int_{\partial\mathcal{W}} \sigma_{lk} n_l dS_d. \tag{34}$$

Using the Reynolds transport theorem and the continuity equation (32), the left-hand side is changed to

$$\begin{aligned} \frac{d}{dt}\int_{\mathcal{W}} \rho v_k dV_D &= \int_{\mathcal{W}} \left[\frac{\partial \rho v_k}{\partial t} + (v_k v_l \rho)_{,l}\right] dV_D \\ &= \int_{\mathcal{W}} \rho\left[\frac{\partial v_k}{\partial t} + v_l v_{k,l}\right] dV_D = \int_{\mathcal{W}} \rho \frac{dv_k}{dt} dV_D. \end{aligned} \tag{35}$$

Next, by the fractal Gauss theorem (14) and localization, we obtain the *fractal linear momentum equation*

$$\rho \dot{v}_k = X_k + \nabla_l^D \sigma_{lk}. \tag{36}$$

Fractal Angular Momentum Equation

The conservation of angular momentum in a fractal medium is stated as

$$\frac{d}{dt}\int_{\mathcal{W}} \rho e_{ijk} x_j v_k dV_D = \int_{\mathcal{W}} e_{ijk} x_j X_k dV_D + \int_{\partial\mathcal{W}} e_{ijk} x_j \sigma_{lk} n_l dS_d. \tag{37}$$

Using (37) and (14) yields

$$e_{ijk}\frac{\sigma_{jk}}{c_1^{(j)}} = 0. \tag{38}$$

It was shown in Li and Ostoja-Starzewski (2011) and Ostoja-Starzewski et al. (2016) that the presence of an anisotropic fractal structure is reflected by differences in the fractal dimensions α_i in different directions, which implies that $c_1^{(j)} \neq c_1^{(k)}$, $j \neq k$, in general. Therefore, the Cauchy stress is generally asymmetric in fractal media, indicating that the micropolar effects should be accounted for and (38) should be augmented by the presence of couple stresses. It is important to note here that a material may have anisotropic fractal structure yet be isotropic in terms of its constitutive laws (Joumaa and Ostoja-Starzewski 2011).

Focusing now on physical fractals (so-called pre-fractals), we consider a body that obeys a fractal mass power law (4) between the lower and upper cutoffs. The choice of the continuum approximation is specified by the resolution R. Choosing the upper cutoff, we arrive at the fractal representative volume element (RVE) involving a region up to the upper cutoff, which is mapped onto a homogenized continuum element in the whole body. The micropolar point homogenizes the very fine microstructures into a rigid body (with 6 degrees of freedom) at the lower cutoff. The two-level homogenization processes are illustrated in Fig. 2.

To determine the inertia tensor $\mathbf{i}$ at any micropolar point, we consider a rigid particle p having a volume $\mathcal{P}$, whose angular momentum is

$$\sigma_A = \int_{\mathcal{P}} (\mathbf{x} - \mathbf{x}_A) \times \mathbf{v}(\mathbf{x}, t) d\mu(\mathbf{x}). \tag{39}$$

Taking $\mathbf{v}(\mathbf{x}, t)$ as a helicoidal vector field (for some vector $\boldsymbol{\omega} \in \mathbb{R}^3$), $\mathbf{v}(\mathbf{x}, t) = \mathbf{v}(\mathbf{x}_A, t) + \boldsymbol{\omega} \times (\mathbf{x} - \mathbf{x}_A)$, we have found (Ostoja-Starzewski et al. 2016) all (diagonal and off-diagonal) components of i_{kl} as

$$I_{kl} = \int_{\mathcal{P}} [x_m x_m \delta_{kl} - x_k x_l] \rho(\mathbf{x}) d V_D. \tag{40}$$

Here we use I_{kl} (Eringen uses i_{kl}) in the current state so as to distinguish it from I_{KL} in the reference state, the relation between both being given by Eringen (1999) $I_{kl} = I_{KL} \chi_{kK} \chi_{lL}$, where χ_{kK} is a microdeformation tensor, or deformable director. For the entire fractal particle $\mathcal{P}$, we have

$$\frac{d}{dt} \int_{\mathcal{P}} \rho I_{KL} d V_D = 0, \tag{41}$$

which, in view of (17), results in the *fractal conservation of microinertia*

$$\frac{d}{dt} I_{kl} = I_{kr} v_{lr} + I_{lr} v_{kr}. \tag{42}$$

In micropolar continuum mechanics (Nowacki 1986; Eringen 1999), one needs a couple-stress tensor $\boldsymbol{\mu}$ and a rotation vector $\boldsymbol{\varphi}$ augmenting, respectively, the Cauchy stress tensor $\boldsymbol{\tau}$ (thus denoted so as to distinguish it from the symmetric $\boldsymbol{\sigma}$) and the deformation vector $\mathbf{u}$. The surface force and surface couple in the fractal setting can be specified by fractional integrals of $\boldsymbol{\tau}$ and $\boldsymbol{\mu}$, respectively, as

$$T_k^S = \int_{\partial\mathcal{W}} \tau_{ik} n_i d S_d, \quad M_k^S = \int_{\partial\mathcal{W}} \mu_{ik} n_i d S_d. \tag{43}$$

The above is consistent with the relation of force tractions and couple tractions to the force stresses and couple stresses on any surface element dS_d

$$t_k = \tau_{ik} n_i, \quad m_k = \mu_{ik} n_i. \tag{44}$$

Now, proceeding in a fashion similar as before, we obtain (36) and the *fractal angular momentum equation*

$$\frac{e_{ijk}}{c_1^{(j)}}\tau_{jk} + \nabla_j^D \mu_{ji} + Y_i = I_{ij}\dot{w}_j. \tag{45}$$

In the above, Y_i is the body force couple, while v_k $(= \dot{u}_k)$ and w_k $(= \dot{\varphi}_k)$ are the deformation and rotation velocities, respectively.

Fractal Energy Equation

Globally, the conservation of energy has the following form

$$\frac{d}{dt}\int_{\mathcal{W}} (e+k)dV_D = \int_{\mathcal{W}} (X_i v_i + Y_i w_i)\, dV_D + \int_{\partial\mathcal{W}} (t_i v_i + m_i w_i - q_i n_i)\, dS_d, \tag{46}$$

where $k = \frac{1}{2}\left(\rho v_i v_i + I_{ij} w_i w_j\right)$ is the kinetic energy density, e the internal energy density, and $\mathbf{q}$ $(= q_i)$ heat flux through the boundary of $\mathcal{W}$. As an aside we note that, just like in conventional (non-fractal media) continuum mechanics, the balance equations of linear momentum (36) and angular momentum (47) can be consistently derived from the invariance of energy (48) with respect to rigid body translations $(v_i \to v_i + b_i,\ w_i \to w_i)$ and rotations $(v_i \to v_i + e_{ijk}x_j\omega_k,\ w_i \to w_i + \omega_i)$, respectively.

To obtain the expression for the rate of change of internal energy, we start from

$$\begin{gathered}\int_{\mathcal{W}} (\dot{e} + \rho\, v_i \dot{v}_i + I_{ij} w_i \dot{w}_i) dV_D = \\ \int_{\mathcal{W}} \left[X_i v_i + Y_i w_i + \nabla_j^D(\tau_{ji} v_j + \mu_{ji} w_j)\right] dV_D - \int_{\mathcal{W}} \nabla_i^D q_i\, dV_D,\end{gathered} \tag{47}$$

and note (36) and (47), to find

$$\dot{e} = \tau_{ji}\left(\nabla_j^D v_i - e_{kji} w_k\right) + \mu_{ji}\nabla_j^D w_i - \nabla_i^D q_i. \tag{48}$$

Next, introducing the infinitesimal strain tensor and the curvature tensor in fractal media

$$\gamma_{ji} = \nabla_j^D u_i - e_{kji}\varphi_k, \quad \kappa_{ji} = \nabla_j^D \varphi_i, \tag{49}$$

we find that the energy balance (52) can be written as

$$\dot{e} = \tau_{ij}\dot{\gamma}_{ij} + \mu_{ij}\dot{\kappa}_{ij}. \tag{50}$$

Assuming e to be a state function of γ_{ij} and κ_{ij} only and assuming τ_{ij} and μ_{ij} not to be explicitly dependent on the temporal derivatives of γ_{ij} and κ_{ij}, we find

$$\tau_{ij} = \frac{\partial e}{\partial \gamma_{ij}} \qquad \mu_{ij} = \frac{\partial e}{\partial \kappa_{ij}}. \tag{51}$$

which shows that, just like in non-fractal continuum mechanics, also in the fractal setting, (τ_{ij}, γ_{ij}) and (μ_{ij}, κ_{ij}) are conjugate pairs.

Fractal Second Law of Thermodynamics

To derive the field equation of the second law of thermodynamics in a fractal medium $B(\omega)$, we begin with the global form of that law in the volume V_D, having a Euclidean boundary $\partial \mathcal{W}$, that is

$$\dot{S} = \dot{S}^{(r)} + \dot{S}^{(i)} \text{ with } \dot{S}^{(r)} = \frac{\dot{Q}}{T}, \; \dot{S}^{(i)} \geq 0, \tag{52}$$

where $\dot{S}$, $\dot{S}^{(r)}$ and $\dot{S}^{(i)}$ stand, respectively, for the total, reversible, and irreversible entropy production rates in V_D. Equivalently,

$$\dot{S} \geq \dot{S}^{(r)}. \tag{53}$$

Since these two rates are extensive quantities, we obtain

$$\int_W \rho \frac{d}{dt} s \, dV_D = \frac{d}{dt} \int_W \rho s \, dV_D \geq - \int_{\partial W} \frac{q_k n_k}{T} dS_d = - \int_W \nabla_k^D \left(\frac{q_k}{T} \right) dV_D, \tag{54}$$

which yields the local form of the second law

$$\rho \frac{ds}{dt} \geq -\nabla_k^D \left(\frac{q_k}{T} \right) - \frac{\nabla_k^D q_{k,k}}{T} + \frac{q_k \nabla_k^D T}{T^2}. \tag{55}$$

Just like in thermomechanics of non-fractal bodies (Ziegler 1983; Maugin 1999), we now introduce the rate of irreversible entropy production $\rho \dot{s}^{(i)}$ which, in view of (57), gives

$$0 \leq \rho \dot{s}^{(i)} = \rho \dot{s} + \frac{\nabla_k^D q_{k,k}}{T} - \frac{q_k \nabla_k^D T}{T^2} + \rho h. \tag{56}$$

Here with s we denote specific entropies (i.e., per unit mass). Next, we recall the classical relation between the free energy density ψ, the internal energy density e, the entropy s, and the absolute temperature T: $\psi = e - Ts$. This allows us to write for time rates of these quantities $\dot{\psi} = \dot{e} - s\dot{T} - T\dot{s}$. On the other hand, with ψ being a function of the strain γ_{ji} and curvature-torsion κ_{ji} tensors, the internal variables α_{ij} (strain type) and ζ_{ij} (curvature-torsion type), and temperature T, we have

$$\rho\dot{\psi} = \rho\frac{\partial\psi}{\partial\gamma_{ij}}\dot{\gamma}_{ij} + \rho\frac{\partial\psi}{\partial\alpha_{ij}}\dot{\alpha}_{ij} + \rho\frac{\partial\psi}{\partial\kappa_{ij}}\dot{\kappa}_{ij} + \rho\frac{\partial\psi}{\partial\zeta_{ij}}\dot{\zeta}_{ij} + \rho\frac{\partial\psi}{\partial T}\dot{T}. \tag{57}$$

In the above we have adopted the conventional relations giving the (external and internal) quasi-conservative Cauchy and Cosserat (couple) stresses as well as the entropy density as gradients of ψ

$$\tau_{ij}^{(q)} = \rho\frac{\partial\psi}{\partial\varepsilon_{ij}}\ \beta_{ij}^{(q)} = \rho\frac{\partial\psi}{\partial\alpha_{ij}}\ \mu_{ij}^{(q)} = \rho\frac{\partial\psi}{\partial\kappa_{ij}}\ \eta_{ij}^{(q)} = \rho\frac{\partial\psi}{\partial\zeta_{ij}}\ s = -\frac{\partial\psi}{\partial T} \tag{58}$$

This is accompanied by a split of total Cauchy and micropolar stresses into their quasi-conservative and dissipative parts

$$\tau_{ij} = \tau_{ij}^{(q)} + \tau_{ij}^{(d)}, \quad \mu_{ij} = \mu_{ij}^{(q)} + \mu_{ij}^{(d)}, \tag{59}$$

along with relations between the internal quasi-conservative and dissipative stresses

$$\beta_{ij}^{(q)} = -\beta_{ij}^{(d)}, \quad \eta_{ij}^{(q)} = -\eta_{ij}^{(d)}. \tag{60}$$

In view of (59) and (60), we obtain

$$\rho\left(\dot{\psi} + s\dot{T}\right) = \rho\left(\dot{e} - T\dot{s}\right) = \tau_{ij}^{(q)}\dot{\gamma}_{ij} + \beta_{ij}^{(q)}\alpha_{ij} + \mu_{ij}^{(q)}\dot{\kappa}_{ij} + \eta_{ij}^{(q)}\dot{\zeta}_{ij} + \rho h. \tag{61}$$

On account of the energy balance, this is equivalent to

$$T\rho\dot{s} = \tau_{ij}^{(d)}\dot{\gamma}_{ij} + \beta_{ij}^{(d)}\dot{\alpha}_{ij} + \mu_{ij}^{(d)}\dot{\kappa}_{ij} + \eta_{ij}^{(d)}\dot{\zeta}_{ij} - \nabla_k^D q_k + \rho h. \tag{62}$$

Recalling (61), we find the local form of the second law in terms of time rates of strains and internal parameters

$$0 \leq T\rho\dot{s}^{(i)} = \tau_{ij}^{(d)}\dot{\gamma}_{ij} + \beta_{ij}^{(d)}\dot{\alpha}_{ij} + \mu_{ij}^{(d)}\dot{\kappa}_{ij} + \eta_{ij}^{(d)}\dot{\zeta}_{ij} - \frac{q_k\nabla_k^D T}{T} + \rho h. \tag{63}$$

The above is a generalization of the Clausius-Duhem inequality to fractal dissipative media with internal parameters. Upon dropping the internal parameters (which may well be the case for a number of materials), the terms $\beta_{ij}^{(d)}\dot{\alpha}_{ij}$ and $\eta_{ij}^{(d)}\dot{\zeta}_{ij}$ drop out, whereas, upon neglecting the micropolar effects, the terms $\mu_{ij}^{(d)}\dot{\kappa}_{ij}$ and $\eta_{ij}^{(d)}\dot{\zeta}_{ij}$ drop out.

For non-fractal bodies without internal parameters, the stress tensor τ_{ij} reverts back to σ_{ij}, and (63) reduces to the simple well-known form Ziegler (1983)

$$0 \leq T\rho\dot{s}^{(i)} = \sigma_{ij}^{(d)}\dot{\gamma}_{ij} - \frac{T_{,k}\,q_k}{T} + \rho h. \tag{64}$$

Interestingly, the fractal gradient (13) appears only in the thermal dissipation term. In fact, this derivative arises in processes of heat transfer in a fractal rigid conductor and coupled thermoelasticity of fractal deformable media (Oldham and Spanier 1974).

Fractal Wave Equations

Just like in conventional continuum mechanics, the basic continuum equations for fractal media presented above have to be augmented by constitutive relations. At this point, we expect that the fractal geometry influences configurations of physical quantities like stress and strain, but does not affect the physical laws (like conservation principles) and constitutive relations that are inherently due to material properties. This expectation is supported in two ways:

- A study of scale effects of material strength and stress from the standpoint of fractal geometry which is confirmed by experiments involving both brittle and plastic materials (Carpinteri and Pugno 2005)
- Derivations of wave equations through mechanical and variational approaches, respectively (Demmie and Ostoja-Starzewski 2011), as shown below

Elastodynamics of a Fractal Timoshenko Beam

Analogous results, also exhibiting self-consistency, were obtained in elastodynamics of a fractally structured Timoshenko beam (Li and Ostoja-Starzewski 2009a). First recall that such a beam model has two degrees of freedom (q_1, q_2) at each point: the transverse displacement $q_1 = w$ and the rotation $q_2 = \varphi$. In the mechanical approach, the beam equation can be derived from the force and moment balance analysis. Thus, beginning with the expressions of shear force (V) and bending moment (M)

$$V = \kappa\mu A\left(\nabla_x^D w - \varphi\right), \qquad M = -EI\nabla_x^D \varphi, \tag{65}$$

we find

$$\rho A\ddot{w} = \nabla_x^D V, \qquad \rho I\ddot{\varphi} = V - \nabla_x^D M. \tag{66}$$

which lead to

$$\begin{aligned} \rho A\ddot{w} &= \nabla_x^D\left[\kappa\mu A\left(\nabla_x^D w - \varphi\right)\right], \\ \rho I\ddot{\varphi} &= \nabla_x^D\left(EI\nabla_x^D\varphi\right) + \kappa\mu A\left(\nabla_x^D w - \varphi\right). \end{aligned} \tag{67}$$

The kinetic energy is

$$T = \frac{1}{2}\rho \int_0^l \left[I \left(\dot{\varphi}\right)^2 + A \left(\dot{w}\right)^2 \right] dl_D, \tag{68}$$

while the potential energy is

$$\begin{aligned} U &= \frac{1}{2}\int_0^l \left[EI \left(\frac{\partial \varphi}{\partial l_D}\right)^2 + \kappa\mu A \left(\frac{\partial w}{\partial l_D} - \varphi\right)^2 \right] dl_D \\ &= \frac{1}{2}\int_0^l \left[EI c_1^{-2} \left(\varphi_{,x}\right)^2 + \kappa\mu A \left(c_1^{-1} w_{,x} - \varphi\right)^2 \right] c_1 dx. \end{aligned} \tag{69}$$

Now, the Euler-Lagrange equations

$$\frac{\partial}{\partial t}\left[\frac{\partial L}{\partial \dot{q}_i}\right] + \sum_{j=1}^{3} \frac{\partial}{\partial x_j}\left[\frac{\partial L}{\partial \left(q_{i,j}\right)}\right] - \frac{\partial L}{\partial q_i} = 0 \tag{70}$$

result in the same as above.

In the case of elastostatics and when the rotational degree of freedom ceases to be independent ($\varphi = \partial w / \partial l_D = \nabla_x^D w$), we find the equation of a fractal Euler-Bernoulli beam

$$\nabla_x^D \nabla_x^D \left(EI \nabla_x^D \nabla_x^D w\right) = 0, \tag{71}$$

which shows that

$$M = EI \nabla_x^D \nabla_x^D w. \tag{72}$$

The relationship between the bending moment (M) and the curvature ($\nabla_x^D \nabla_x^D w$) still holds, while c_1 enters the determination of curvature $\left(\nabla_x^D \nabla_x^D w = c_1^{-1}\left(c_1^{-1} w_{,x}\right)_{,x}\right)$.

In a nutshell, the fractional power law of mass implies a fractal dimension of scale measure, so the derivatives involving spatial scales should be modified to incorporate such effect by postulating c_1, c_2, and c_3 coefficients, according to the material body being embedded in a 1D, 2D, or 3D Euclidean space.

More work was done on waves in linear elastic fractal solids under small motions. Several cases of isotropic (Joumaa and Ostoja-Starzewski 2011) or anisotropic (with micropolar effects) (Joumaa et al. 2014; Joumaa and Ostoja-Starzewski 2016) media have been considered through analytical and computational methods. It was found on the mathematical side that fractal versions of harmonic, Bessel, and Hankel functions. The study (Joumaa and Ostoja-Starzewski 2011) provided the basis for a study of wave motion in a human head, where the brain, while protected by skull bones, is actually surrounded by a cerebrospinal fluid (so-called CSF) and has fractal geometric characteristics (Joumaa and Ostoja-Starzewski 2013).

Elastodynamics in 3D

First, the equations of motion of 3D elastodynamics can be determined for finite strain motions by extending the procedure of classical continuum mechanics. To this end, begin with the action functional of a fractal solid $\mathcal{W}$ isolated from external interactions

$$\delta \mathcal{I} = \delta \int_{t_1}^{t_2} [\mathcal{K} - \mathcal{E}] dt = 0, \tag{73}$$

where $\mathcal{K}$ and $\mathcal{E}$ are the kinetic and internal energies

$$\mathcal{K} = \frac{1}{2} \int_{\mathcal{W}} \rho v_i v_i dV_D \quad \mathcal{E} = \int_{\mathcal{W}} \rho e dV_D. \tag{74}$$

Thus, we have a functional, which can be rewritten in fractal space-time (e being the specific, per unit mass, internal energy density) as

$$0 = \delta \mathcal{I} = \delta \int_{t_1}^{t_2} \int_{\mathcal{W}} \left[\frac{1}{2} \rho v_k^2 - \rho e \right] dV_D dt = \delta \int_{t_1}^{t_2} \int_{\mathcal{W}} \left[\rho \frac{1}{2} v_k^2 - \rho e \right] dV_D dt. \tag{75}$$

Analogous to the strain of (30), the deformation gradient is

$$F_{kI} = \nabla_I^D x_k \equiv \frac{1}{c_1^{(I)}} x_{k,I}. \tag{76}$$

Assuming that the specific energy density e depends only on the deformation gradient, and that ρ has no explicit dependence on time, leads to:

(i) The boundary conditions

$$\partial \mathcal{W} \rho \frac{\partial e}{\partial F_{kI}} N_I = 0 \quad or \quad \delta x_k = 0 \quad on \quad \partial \mathcal{W} \tag{77}$$

(ii) The kinematic constraints $\delta x_k = 0$ at $t = t_1$ and $t = t_2$

The Hamilton principle for (73) implies the equation governing motion in a fractal solid under finite strains

$$\nabla_I^D \left[\rho \frac{\partial e}{\partial F_{kI}} \right] - \rho \frac{d v_k}{dt} = 0. \tag{78}$$

Next, in fractal bodies without internal dissipation, e plays the role of a potential

$$T_{kI} = \rho \frac{\partial e}{\partial F_{kI}}, \tag{79}$$

where T_{kI} is the first Piola-Kirchhoff stress tensor, and (78) becomes

$$\nabla_I^D T_{kI} - \rho \frac{dv_k}{dt} = 0 \quad \text{or} \quad \frac{1}{c_1^{(I)}} T_{kI},_I - \rho \frac{dv_k}{dt} = 0. \tag{80}$$

Restricting the motion to small deformation gradients, T_{kI} becomes the Cauchy stress tensor, and we recover the linear momentum equation (37). Generalizing this to the situation of $\mathcal{W}$ interacting with the environment (i.e., subject to body forces), $\delta(\mathcal{I} + W - \mathcal{P}) = 0$, we obtain

$$\nabla_I^D T_{kI} + \rho \left(b_k - \frac{dv_k}{dt} \right) = 0. \tag{81}$$

The above formulation was specialized in Demmie and Ostoja-Starzewski (2011) to 1D models and focused on nonlinear waves, demonstrating how the governing equations might be solved by the method of characteristics in fractal space-time. We also studied shock fronts in linear viscoelastic solids under small strains. We showed that the discontinuity in stress across a shock front in a fractal medium is identical to the classical result.

Related Topics

Extremum and Variational Principles in Fractal Bodies

Just like in preceding sections, the dimensional regularization approach can also be applied to other statements in continuum/solid mechanics involving integral relations. For example, the Maxwell-Betti reciprocity relation of linear elasticity $\int_{\partial W} t_i^* u_i dS_2 = \int_{\partial W} t_i u_i^* dS_2$ is generalized for fractal media to

$$\int_{\partial W} t_i^* u_i dS_d + \int_{\partial W} m_i^* \varphi_i dS_d = \int_{\partial W} t_i u_i^* dS_d + \int_{\partial W} m_i \varphi_i^* dS_d, \tag{82}$$

so as to read in $\mathbb{E}^3$:

$$\int_{\partial W} t_i^* u_i c_2 dS_2 + \int_{\partial W} m_i^* \varphi_i c_2 dS_2 = \int_{\partial W} t_i u_i^* c_2 dS_2 + \int_{\partial W} m_i^* \varphi_i c_2 dS_2. \tag{83}$$

The reciprocity relation (82) is proved by appealing to the Green-Gauss theorem and the Hooke law ($\sigma_{ij} = C_{ijkl}\varepsilon_{kl}$) and proceeding just like in the conventional continuum elasticity. As an application, consider the classical problem of calculation of the reduction in volume ΔV of a linear elastic isotropic body (of bulk modulus κ) due to two equal, collinear, opposite forces F, separated by a distance L. Clearly, in the classical case, one does not need the micropolar term and, as discussed in Li and Ostoja-Starzewski (2009a), finds $\Delta V = FL/3\kappa$. On the other hand, for a fractal body, given the loading $\tau_{ij} = -p\delta_{ij}$ and $\mu_{ij} = 0$, the integrals involving scalar products of couple traction with rotation vanish, and we obtain

$$p\ \Delta V\ c_3 = p\ \frac{FL}{3\kappa} c_1. \tag{84}$$

Recalling (6) and assuming that the opposite forces are applied parallel to one of the axis of the coordinate system (say, x_k) yield $\Delta V = FL/3\kappa c_2^{(k)}$. This can be evaluated numerically for a specific material according to (6).

Next, suppose we focus on situations where couple-stress effects are negligible. Then, we recall the concept of a *statically admissible field* as a tensor function $\sigma_{ij}(\mathbf{x})$, such that $\sigma_{ij} = \sigma_{ji}$ and $(F_k = \rho f_k)$

$$F_k + \nabla_l^D \sigma_{kl} = 0 \tag{85}$$

in W and the boundary conditions

$$\sigma_{kl} n_l = t_k \tag{86}$$

on ∂W_t. Similarly, we recall a *kinematically admissible displacement field* as a vector function $\mathbf{u}(\mathbf{x})$ satisfying the boundary conditions

$$u_i = f_i \tag{87}$$

on ∂W_u. We can now consider the *principle of virtual work*: "The virtual work of the internal forces equals the virtual work of the external forces." Let $\boldsymbol{\sigma}(\mathbf{x})$ be a statically admissible stress field and $\mathbf{u}(\mathbf{x})$ a kinematically admissible displacement field. Define $\varepsilon_{ij}(\mathbf{u}) = u_{(i,j)}$. Then

$$\int_W \sigma_{ij}\varepsilon_{ij} d V_D = \int_W F_i u_i d V_D + \int_{\partial W_t} t_i u_i d S_d + \int_{\partial W_u} \sigma_{ij} n_j f_i d S_d. \tag{88}$$

The proof follows by substitution from the fractional equation of static equilibrium and boundary conditions after integrating by parts, and using the Gauss theorem, all conducted over the fractal domain W. Of course, the above can be rewritten in terms of conventional integrals:

$$\begin{aligned}\int_W \sigma_{ij}\varepsilon_{ij} c_3 d V_3 = \int_W F_i u_i c_3 d V_3 + \int_{\partial W_t} c_2 t_i u_i c_2 d S_2 \\ + \int_{\partial W_t} \sigma_{ij} n_j f_i c_2 d S_2.\end{aligned} \tag{89}$$

In a similar way, we can adapt to fractal elastic bodies the principle of virtual displacement, principle of virtual stresses, principle of minimum potential energy, principle of minimum complementary energy, and related principles for elastic-plastic or rigid bodies.

Relation to other studies of complex systems: While the calculus in non-integer spaces outlined in this brief review has been employed for fractal porous media, the approach can potentially be extended to microscopically heterogeneous physical systems (made up of many different micromechanical systems, having different physical properties, and interacting under the influence of different kinds of phenomena). Several such possibilities have been discussed in Ostoja-Starzewski et al. (2014), and one paradigm is given in the next section.

Fracture in Elastic-Brittle Fractal Solids

General Considerations

According to Griffith's theory of elastic-brittle solids, the strain energy release rate G is given by Gdoutos (1993)

$$G = \frac{\partial W}{\partial A} - \frac{\partial \mathcal{E}^e}{\partial A} = 2\gamma, \tag{90}$$

where A is the crack surface area formed, W is the work performed by the applied loads, $\mathcal{E}^e$ is the elastic strain energy, and γ is the energy required to form a unit of new material surface. The material parameter γ is conventionally taken as constant, but, given the presence of a randomly microheterogeneous material structure, its random field nature is sometimes considered explicitly (Ostoja-Starzewski 2008b, 2004). Recognizing that the random material structure also affects the elastic moduli (such as E), the computation of $\mathcal{E}^e$ and G in (90) needs to be reexamined (Ostoja-Starzewski 2004); see also Balankin et al. (2011) in the context of paper mechanics. With reference to Fig. 4, we consider a 3D material body described by D and $d$and having a crack of depth a and a fractal dimension DF.

Focusing on a fractal porous material, we have

$$U^e = \int_W \rho\, u\, dV_D = \int_W \rho\, u\, c_3\, dV_3. \tag{91}$$

By revising Griffith's derivation for a fractal elastic material, we then obtain

$$U^e = \frac{\pi a^2 c_1^2 \sigma^2}{8\mu}(K + 1)c_3, \tag{92}$$

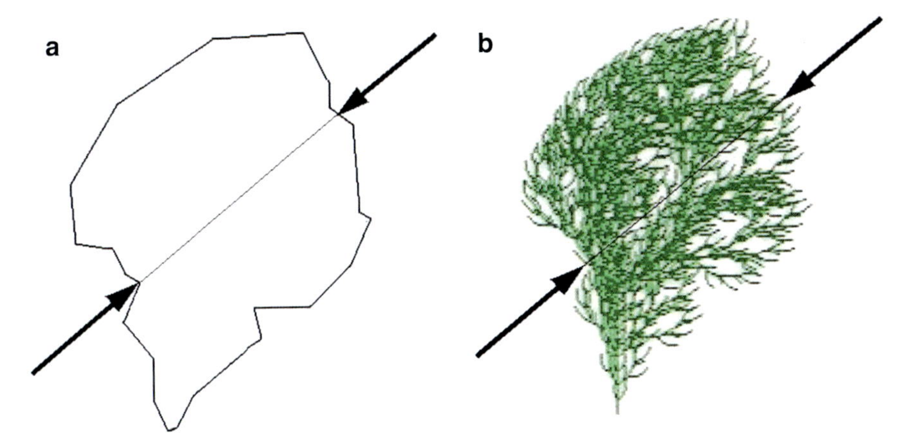

Fig. 4 (**a**) A fractal body subjected to two equal, collinear, opposite forces F and (**b**) its homogenized equivalent via dimensional regularization

with ν being the Poisson ratio, and

$$K = \begin{cases} 3 - 4\nu & \text{for plane strain} \\ \dfrac{3-\nu}{1+\nu} & \text{for plane stress} \end{cases} \tag{93}$$

the Kolosov constant.

Dead-load conditions. Equation (90) becomes

$$G = \frac{\partial U^e}{\partial A} = 2\gamma. \tag{94}$$

If $A = 2a \times 1$, this gives the critical stress

$$\sigma_c = \sqrt{\frac{2\gamma E}{(1-\nu^2)\pi a\, c_3\, c_1^2}}. \tag{95}$$

However, if the fracture surface is fractal of a dimension α, then we should use $\partial/\partial l_{DF}$ instead of $\partial/\partial a$. Now, since we have (note Fig. 5) $dl_{DF} = c_1 da$, the new partial derivative becomes

$$\frac{\partial}{\partial l_{DF}} = \frac{\partial}{c_1 \partial a}, \tag{96}$$

so that

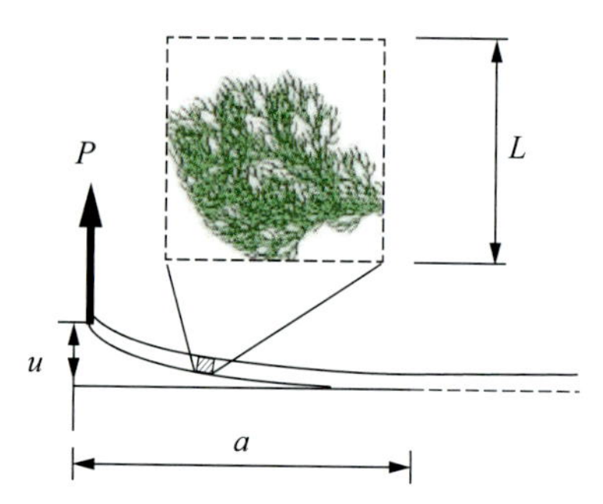

Fig. 5 Fracture and peeling of a microbeam of thickness L off a substrate. A representative volume element dV_3 imposed by the pre-fractal structure characterized by upper cutoff scale L is shown. Thus, the beam is homogeneous above the length scale L. By introducing random variability in that structure, one obtains a random beam; see (103)

$$\sigma_c = \sqrt{\frac{2\gamma E}{(1-\nu^2)\pi a\, c_3\, c_1}}. \tag{97}$$

Fixed-grip conditions. We consider the case of a crack of depth a and width B in plane strain. In this case the displacement is constant (i.e., non-random), and the load is random. Now, only the second term in (90) remains, so that

$$G = -\frac{\partial \mathcal{E}^e(a)}{B\partial l_{DF}} = -\frac{\partial \mathcal{E}^e(a)}{Bc_1\partial a}. \tag{98}$$

Peeling a Layer Off a Substrate

Dead-load conditions. As a specific case, we take an Euler-Bernoulli beam, so that the strain energy is

$$\mathcal{E}(a) = \int_0^a \frac{M^2}{2IE} dx, \tag{99}$$

where a is crack length, M is bending moment, and I is beam's moment of inertia. Henceforth, we simply work with $a = A/B$, where B is the constant beam (and crack) width. In view of Clapeyron's theorem, the strain energy release rate may be written as

$$G = \frac{\partial \mathcal{E}}{B\partial a}. \tag{100}$$

For a layer modeled as a fractal Euler-Bernoulli beam (section "Peeling a Layer Off a Substrate)," we have

$$\mathcal{E}(a) = \int_0^a \frac{M^2}{2IE} dl_D = \int_0^a \frac{M^2}{2IE} c_1 dx, \tag{101}$$

so that

$$G = \frac{\partial \mathcal{E}}{c_1 B \partial a}. \tag{102}$$

Generalization to a Statistical Ensemble

Now, if the beam's material is random, E is a random field parametrized by x, which we can write as a sum of a constant mean $\langle E \rangle$ and a zero-mean fluctuation $E'(x)$

$$E(\omega, x) = \langle E \rangle + E'(\omega, x) \quad \omega \in \Omega, \tag{103}$$

where ω is an elementary event in Ω, a sample space. Clearly, $\mathcal{E}$ is a random integral, such that, for each and every realization $\omega \in \Omega$, we should consider

$$\mathcal{E}(a, E(\omega)) = \int_0^a \frac{M^2 c_1 dx}{2IE(\omega, x)}. \tag{104}$$

Upon ensemble averaging, this leads to an average energy

$$\langle \mathcal{E}(a, E) \rangle = \left\langle \int_0^a \frac{M^2 c_1 dx}{2I \left[\langle E \rangle + E'(\omega, x) \right]} \right\rangle. \tag{105}$$

In the conventional formulation of deterministic fracture mechanics, random microscale heterogeneities $E'(x, \omega)$ are disregarded, and (104) is evaluated by simply replacing the denominator by $\langle E \rangle$, so that

$$\mathcal{E}(a, \langle E \rangle) = \int_0^a \frac{M^2 c_1 dx}{2I \langle E \rangle}. \tag{106}$$

Clearly, this amounts to postulating that the response of an idealized homogeneous material is equal to that of a random one on average. To make a statement about $\langle \mathcal{E}(a, E) \rangle$ versus $\mathcal{E}(a, \langle E \rangle)$, and about $\langle G(E) \rangle$ versus $G(\langle E \rangle)$, first, note the random field E is positive valued almost surely. Then, Jensen's inequality yields a relation between harmonic and arithmetic averages of the random variable $E(\omega)$

$$\frac{1}{\langle E \rangle} \leq \left\langle \frac{1}{E} \right\rangle. \tag{107}$$

whereby the x-dependence is immaterial in view of the assumed wide-sense stationary of field E. With (105) and (106), and assuming that the conditions required by Fubini's theorem are met, this implies that

$$\begin{aligned}\mathcal{E}(a,\langle E\rangle) &= \int_0^a \frac{M^2 c_1 dx}{2I\,\langle E\rangle} \leq \int_0^a \frac{M^2 c_1}{2I}\left\langle \frac{1}{E}\right\rangle dx \\ &= \left\langle \int_0^a \frac{M^2 c_1 dx}{2IE(\omega,x)}\right\rangle = \langle \mathcal{E}(a,E)\rangle\,,\end{aligned} \tag{108}$$

Now, defining the strain energy release rate $G(a,\langle E\rangle)$ in a reference material specified by $\langle E\rangle$, and the strain energy release rate $\langle G(a,E)\rangle$ properly ensemble averaged in the random material $\{E(\omega,x);\omega\in\Omega,x\in[0,a]\}$

$$G(a,\langle E\rangle) = \frac{\partial\mathcal{E}(a,\langle E\rangle)}{Bc_1\partial a} \qquad \langle G(a,E)\rangle = \frac{\partial\,\langle\mathcal{E}(a,E)\rangle}{Bc_1\partial a}, \tag{109}$$

and noting that the side condition is the same in both cases

$$\mathcal{E}(a,\langle E\rangle)\mid_{a=0}=0 \qquad \langle\mathcal{E}(a,E)\rangle\mid_{a=0}=0, \tag{110}$$

we find

$$G(a,\langle E\rangle) \leq \langle G(a,E)\rangle\,. \tag{111}$$

This provides a formula for the ensemble average G under dead-load conditions using deterministic fracture mechanics for Euler-Bernoulli beams made of fractal random materials.

Just like in the case of non-fractal materials (Ostoja-Starzewski 2004), the inequality (111) shows that G computed under the assumption that the random material is directly replaced by a homogeneous material ($E(x,\omega)=\langle E\rangle$) is lower than G computed with E taken explicitly as a spatially varying material property.

Fixed-grip conditions. On account of (4), assuming that there is loading by a force P at the tip, we obtain

$$G = -\frac{u}{2Bc_1}\frac{\partial P}{\partial a}. \tag{112}$$

Take now a cantilever beam problem implying $P=3uEI/(c_1a)^3$. Then, we find

$$\langle G\rangle = -\frac{u}{2Bc_1}\left\langle\frac{\partial P}{\partial a}\right\rangle = -\frac{u}{2Bc_1}\frac{\partial\,\langle P\rangle}{\partial a} = \frac{9u^2I\,\langle E\rangle}{2B(c_1a)^4}. \tag{113}$$

Since the load – be it a force and/or a moment – is always proportional to E, this indicates that G can be computed by direct ensemble averaging of E under fixed-grip loading, and, indeed, the same conclusion carries over to Timoshenko beams. This analysis may be extended to mixed-loading conditions and stochastic crack stability by generalizing the study of non-fractal, random beams carried out in Li and Ostoja-Starzewski (2009a).

Closure

While fractals are abundant in nature, the mechanics of fractal media is still in its infancy. In this brief review, we showed how fundamental balance laws for fractal porous media can be developed using a homogenization method using calculus in non-integer dimensional spaces, a special version of dimensional regularization. The developments are confined to materials whose mass is specified by a product measure, where the overall mass fractal dimension D equals $\alpha_1 + \alpha_2 + \alpha_3$ along the diagonals, $|x_1| = |x_2| = |x_3|$, where each α_k plays the role of a fractal dimension in the direction x_k. The strategy is to express the balance laws for fractal media in terms of fractional integrals and, then, to convert them to integer-order integrals in conventional (Euclidean) space. This leads to balance laws expressed in continuous forms albeit modified by the presence of coefficients responsible for dependencies on α_1's; this makes mathematical treatments more tractable. It is shown how to develop balance laws of fractal media (continuity, linear and angular momenta, first and second law of thermodynamics) as well as the elastodynamics equations of a fractal Timoshenko beam and 3D solids under finite strains. In general, if the fractal geometry is described by a product measure, the angular momentum balance cannot be satisfied unless the stress tensor is asymmetric. Therefore, a generally anisotropic fractal medium necessitates adoption of a nonclassical continuum, the Cosserat model, for the balance laws to be satisfied (Maugin 2016). We then discuss extremum and variational principles and fracture mechanics in fractal media. In all the cases, the derived equations for fractal media depend explicitly on fractal dimensions and reduce to conventional forms for continuous media with Euclidean geometries upon setting the dimensions to integers. The fractal model discussed is useful for solving complex mechanics problems involving fractal materials composed of microstructures of inherent length scale, generalizing the universally applied classical theory of elastodynamics for continuous bodies. This modeling approach offers a foundation upon which one can study (thermo)mechanical phenomena involving fractals analytically and computationally while determining fractal dimensions by image analyses.

Acknowledgements This work was made possible by the support from NSF (grant CMMI-1462749).

References

A.S. Balankin, O. Susarrey, C.A. Mora Santos, J. Patíno, A. Yogues, E.I. García, Stress concentration and size effect in fracture of notched heterogeneous material. Phys. Rev. E **83**, 015101(R) (2011)

M.F. Barnsley, *Fractals Everywhere* (Morgan Kaufmann, San Francisco, 1993)
A. Carpinteri, B. Chiaia, P.A. Cornetti, A disordered microstructure material model based on fractal geometry and fractional calculus. ZAMP **84**, 128–135 (2004)
A. Carpinteri, N. Pugno, Are scaling laws on strength of solids related to mechanics or to geometry? Nat. Mater. **4**, 421–23 (2005)
P.N. Demmie, Ostoja-Starzewski, Waves in fractal media. J. Elast. **104**, 187–204 (2011)
A.C. Eringen, *Microcontinuum Field Theories I* (Springer, New York, 1999)
K. Falconer, *Fractal Geometry: Mathematical Foundations and Applications* (Wiley, Chichester, 2003).
E.E. Gdoutos, *Fracture Mechanics: An Introduction* (Kluwer Academic Publishers, Dordrecht, 1993)
H.M. Hastings, G. Sugihara, *Fractals: A User's Guide for the Natural Sciences* (Oxford Science Publications, Oxford, 1993)
H. Joumaa, M. Ostoja-Starzewski, On the wave propagation in isotropic fractal media. ZAMP **62**, 1117–1129 (2011)
H. Joumaa, M. Ostoja-Starzewski, Acoustic-elastodynamic interaction in isotropic fractal media. Eur. Phys. J. Spec. Top. **222**, 1949–1958 (2013)
H. Joumaa, M. Ostoja-Starzewski, P.N. Demmie, Elastodynamics in micropolar fractal solids. Math. Mech. Solids **19**(2), 117–134 (2014)
H. Joumaa, M. Ostoja-Starzewski, On the dilatational wave motion in anisotropic fractal solids. Math. Comput. Simul. **127**, 114–130 (2016)
G. Jumarie, On the representation of fractional Brownian motion as an integral with respect to $(dt)^a$. Appl. Math. Lett. **18**, 739–748 (2005)
G. Jumarie, Table of some basic fractional calculus formulae derived from a modified Riemann-Liouville derivative for non-differentiable functions. Appl. Math. Lett. **22**(3), 378–385 (2009)
A. Le Méhauté, *Fractal Geometry: Theory and Applications* (CRC Press, Boca Raton, 1991)
J. Li, M. Ostoja-Starzewski, Fractal materials, beams and fracture mechanics. ZAMP **60**, 1–12 (2009a)
J. Li, M. Ostoja-Starzewski, Fractal solids, product measures and fractional wave equations. Proc. R. Soc. A **465**, 2521–2536 (2009b); Errata (2010)
J. Li, M. Ostoja-Starzewski, Fractal solids, product measures and continuum mechanics, chapter 33, in *Mechanics of Generalized Continua: One Hundred Years After the Cosserats*, ed. by G.A. Maugin, A.V. Metrikine (Springer, New York, 2010), pp. 315–323
J. Li, M. Ostoja-Starzewski, Micropolar continuum mechanics of fractal media. Int. J. Eng. Sci. (A.C. Eringen Spec. Issue) **49**, 1302–1310 (2011)
J. Li, M. Ostoja-Starzewski, Edges of Saturn's rings are fractal. SpringerPlus **4**, 158 (2015). arXiv:1207.0155 (2012)
B.B. Mandelbrot, *The Fractal Geometry of Nature* (W.H. Freeman & Co, NewYork, 1982)
G.A. Maugin, *The Thermomechanics of Nonlinear Irreversible Behaviours* (World Scientific Pub. Co., Singapore, 1999)
G.A. Maugin, *Non-classical Continuum Mechanics: A Dictionary* (Springer, Singapore, 2016)
W. Nowacki, *Theory of Asymmetric Elasticity* (Pergamon Press/PWN – Polish Sci. Publ., Oxford/Warszawa, 1986)
K.B. Oldham, J. Spanier, *The Fractional Calculus* (Academic Press, San Diego, 1974)
M. Ostoja-Starzewski, Fracture of brittle micro-beams. ASME J. Appl. Mech. **71**, 424–427 (2004)
M. Ostoja-Starzewski, Towards thermomechanics of fractal media. ZAMP **58**(6), 1085–1096 (2007)
M. Ostoja-Starzewski, On turbulence in fractal porous media. ZAMP **59**(6), 1111–1117 (2008a)
M. Ostoja-Starzewski, *Microstructural Randomness and Scaling in Mechanics of Materials* (CRC Press, Boca Raton, 2008b)
M. Ostoja-Starzewski, Extremum and variational principles for elastic and inelastic media with fractal geometries. Acta Mech. **205**, 161–170 (2009)
M. Ostoja-Starzewski, Electromagnetism on anisotropic fractal media. ZAMP **64**(2), 381–390 (2013)

M. Ostoja-Starzewski, J. Li, H. Joumaa, P.N. Demmie, From fractal media to continuum mechanics. ZAMM **94**(5), 373–401 (2014)
M. Ostoja-Starzewski, S. Kale, P. Karimi, A. Malyarenko, B. Raghavan, S.I. Ranganathan, J. Zhang, Scaling to RVE in random media. Adv. Appl. Mech. **49**, 111–211 (2016)
D. Stoyan, H. Stoyan, *Fractals, Random Shapes and Point Fields* (Wiley, Chichester, 1994)
V.E. Tarasov, Fractional hydrodynamic equations for fractal media. Ann. Phys. **318**(2), 286–307 (2005a)
V.E. Tarasov, Wave equation for fractal solid string. Mod. Phys. Lett. B **19**(15), 721–728 (2005b)
V.E. Tarasov, Continuous medium model for fractal media. Phys. Lett. A **336**, 167–174 (2005c)
V.E. Tarasov, *Fractional Dynamics: Applications of Fractional Calculus to Dynamics of Particles, Fields and Media* (Springer, Berlin, 2010)
V.E. Tarasov, Anisotropic fractal media by vector calculus in non-integer dimensional space. J. Math. Phys. **55**, 083510-1-20 (2014)
V.E. Tarasov, Electromagnetic waves in non-integer dimensional spaces and fractals. Chaos, Solitons Fractals **81**, 38–42 (2015a)
V.E. Tarasov, Vector calculus in non-integer dimensional space and its applications to fractal media. Commun. Nonlinear Sci. Numer. Simul. **20**, 360–374 (2015b)
H. Ziegler, *An Introduction to Thermomechanics* (North-Holland, Amsterdam, 1983)

Part IV
Computational Modeling for Gradient Plasticity in Both Temporal and Spatial Scales

Modeling High-Speed Impact Failure of Metallic Materials: Nonlocal Approaches

26

George Z. Voyiadjis and Babür Deliktaş

Contents

Abstract

Development and application of advanced, computationally intensive multiscale (macro-, meso-, and micro-mechanically) physically based models to describe physical phenomena associated with friction and wear in heterogeneous solids, particularly under high velocity impact loading conditions. Emphasis will be placed on the development of fundamental, thermodynamically consistent theories to describe *high-velocity material wear failure* processes in combinations of ductile and brittle materials for wear damage-related problems. The wear failure

G. Z. Voyiadjis (✉)
Department of Civil and Environmental Engineering, Louisiana State University, Baton Rouge, LA, USA
e-mail: voyiadjis@eng.lsu.edu

B. Deliktaş
Faculty of Engineering-Architecture, Department of Civil Engineering, Uludag University, Bursa, Görükle, Turkey
e-mail: bdeliktas@uludag.edu.tr

G. Z. Voyiadjis (ed.), *Handbook of Nonlocal Continuum Mechanics for Materials and Structures*, https://doi.org/10.1007/978-3-319-58729-5_5

criterion will be based on dissipated energies due to plastic strains at elevated temperatures. Frictional coefficients will be identified for the contact surfaces based on temperature, strain rates, and roughness of the surfaces. In addition failure models for microstructural effects, such as shear bands and localized deformations, will be studied.

The computations will be carried with Abaqus Explicit as a dynamic temperature-displacement analysis. The contact between sliding against each other's surfaces is specified as surface-to-surface contact on the master-slave basis. The tangential behavior is defined as kinematic contact with finite sliding. The validation of computations utilizing the novel approach presented in this work is going to be conducted on the continuum level while comparing the obtained numerical results with the experimental results obtained in the laboratories in Metz, France. Reaction forces due to friction between the two specimens and temperature resulting from the dissipated energy during the friction experiment are going to be compared and discussed in detail. Additionally the indentation response at the macroscale, for decreasing the size of the indenter, will be used to critically assess and evaluate the length scale parameters.

Keywords
Failure in metals · Frictional coefficient · High speed impact · High velocity material wear · Micro structural effects · Multiscale modeling

Introduction

The need to improve the reliability and life of tribological components that are prone to severe contact stress during some of the engineering applications such as cutting tool for metals, gun tubes, engine exhaust valves, engine turbocharger components, rail gun environment, jet engine gear box splines and gears, and slippers on high-speed test track sleds. Severe contact stresses of such applications generate high temperature and create thermomechanical gouging and wear due to high-velocity sliding between contacting materials (Ireman et al. 2003; Ireman and Nguyen 2004; Arakawa 2014, 2017; Hernandez et al. 2015; Bayart et al. 2016). High-velocity sliding between dissimilar materials can result in thermomechanical gouging of the materials. This problem can be considered in the field of nonlocal modeling of heterogeneous media that assesses a strong coupling between rate-dependent plasticity (viscoplasticity) and rate-dependent damage (viscodamage or creep damage) under high-velocity gouging (Lodygowski et al. 2011; Voyiadjis et al. 2010).

The field of research dealing with wear modeling is still very elusive, yet there has been progress made in the understanding of many aspects of the wear mechanisms. The definition of the wear is stated in the work of Johansson and Klarbring (1993) as "the loss of material from a surface, transfer of material from one surface to another or movement of material within a single surface." This definition can be simplified as "damage to a solid surface generally involving progressive loss of materials due to relative motion between that surface and a coating surface or substance." (Johnson et al. 1947; Klarbring 1986, 1990; Johansson and Klarbring 2000; Varga et al. 2013). A close look at the activity leading to wear is at the dimensional level of 10 nm where material structure can be seen as nanocrystalline. In addition to this small-scale phenomenon, local high spots (asperities) produce high stresses over short time scales of order of microseconds, whereas strain rate are of the order of 10^4–10^7 s^{-1}.

Another important aspect of wear problem is the time dependency and high temperature produced by energy dissipation through the contact interaction. Due to both high strain rate and temperature, it is possible to observe in the wear of metals the formation of adiabatic shear bands at the microscale. Many of the features described here make it obvious that the highly complex nature of the wear problem cannot be solely treated by using macroscale phenomenological models. Therefore, a clear need exists for the development of a realistic and reliable physically based material model within the framework of multiscale modeling that can be utilized in the sever contact stress applications.

The major consideration of this research is to develop an experimental/theoretical model for the material in order to better characterize and predict the internal failure surrounding the gouging and wear events. This research is to be carried out by first investigating the phenomenon of the wear, and later it will be extended to incorporate gauging problems.

The theoretical model is based on a nonlocal theory of crystal plasticity that incorporates macroscale interstate variables and their higher-order gradients in order to describe the change in the internal structure and investigate the size effect of statistical inhomogeneity of the evolution-related plasticity and damage-hardening variables. The gradients are introduced here as the hardening internal state variables and are considered independent of their local counterparts. It also incorporates the thermomechanical coupling effects as well as the internal dissipative effects through the rate-type covariance constitutive structure with a finite set of internal state variables.

Crystal plasticity models (Asaro 1983) that account for the crystalline microstructure by distinguishing between different crystallographic orientations have become successful in modeling the anisotropic plastic deformation of single crystals. The discrete slip systems that depend on the crystal structure are incorporated into these models, but the plastic part of deformation is still modeled in a continuum sense. The characteristic length of the deformed microstructure becomes significant in the analysis of the material at a scale where the microstructure characteristic length is no longer negligible with respect to the material size. This triggers an important question, whether and how macroscopic overall mechanical properties as strength, hardness, etc. depend upon a natural internal length scale related to the characteristic size of the microstructures in the material. A nonlocal crystal plasticity (NCP) model is developed in this work using thermodynamically consistent higher-order strain gradient theory. Multiple length scales are introduced at the NCP to account for the microstructure. These length scales are used to characterize size effects, strengthening, hardening, values of dislocation line energies, back-stress magnitudes, and the thickness of the plastic boundary layer.

The theoretical model is formulated for the crystalline system with multiple symmetric double slip systems. The boundary value problem analyzed here is the simple shear of a constrained crystalline strip in order to investigate the effects of the dissipative and energetic parameters on the size effect response of the constrained crystalline strip systems. The developed theory described quantitatively the thickness of the boundary layer, hardening, and strengthening response of the system.

The classical crystal plasticity, which inherently includes no material length scales, cannot predict size effects. Strain gradient plasticity (SGP) theories extend the classical plasticity models by including an intrinsic material length scale and are therefore appropriate for problems involving small dimensions (Aifantis and Willis 2006; Anand et al. 2005; Fleck and Willis 2008; Fredriksson and Gudmundson 2007). These material length scales are necessary both for dimensional consistency when strain gradients are used in the formulation and in crystal plasticity based on a continuum description of the dislocation behavior. The physical basis of the SGP for metals has been founded on theoretical developments concerning geometrically necessary dislocations (GNDs) (Nye 1953). Standard micromechanical modeling of metals for the inelastic material behavior of single crystals and polycrystals is commonly based on the premise that resistance to glide is due mainly to the random trapping of mobile dislocations during locally homogeneous deformations. Such trapped dislocations are commonly referred to as statistically stored dislocations (SSDs) and act as obstacles to stop further dislocation motion, resulting in hardening. An additional contribution to the density of immobile dislocations and to hardening can arise when the continuum length scale approaches that of the dominant microstructural features.

One of the open issues of the SGP is the ongoing discussion of the energetic and dissipative nature of the dislocation network that accounts for many plasticity phenomena. Gurtin (2003) argued that the density of geometrically necessary dislocations is quantified by Nye's tensor that leads to increase in the free energy. However, Fleck and Willis (2008) discussed this issue by questioning whether the additional strengthening is mainly energetic or dissipative. They assumed that the core energy of dislocations stored during plastic deformation is much smaller than the plastic work dissipated in dislocation motion. Based on this observation, they concluded that both statistically stored and geometrically necessary dislocations contribute more to plastic dissipation than to a change in energy. Bardella (2006, 2007) also pointed out that modeling involving only energetic material length scales may not be sufficient to describe the size effects exhibited in metals. He reasoned the fact that energetic length scales, defined through a function of Nye's dislocation density tensor, allow the description of the increase in strain hardening accompanied with diminishing size, but they do not help in capturing the related strengthening. He proposed that at least one dissipative length scale is required in modeling in order to capture the strengthening effect.

In this work both energetic and dissipative material length scales are introduced, and their physically based relations as a function of the accumulated plastic deformation, strain rate, temperature, and corresponding microstructural features are proposed. It was observed that plasticity and damage phenomena at small-scale levels dictate the necessity of more than one length scale parameters in the gradient description. Energetic material length scales quantify the size effects due to GNDs. The need of more energetic material length scales in order to have a better description of the material behavior should be first evaluated in light of the experimental results (or discrete dislocation simulations). Two main energy storage mechanisms are used here. The first is that energy stored in dislocation network uses dislocation density that depends on the inelastic strain path. Dislocation density increases with inelastic deformation and eventually reaches a saturation point. The second storage mechanism is due to the presence of the second hard phase.

Nonlocal Strain Gradient Crystal Plasticity Formulation

In this section the nonlocal strain gradient crystal plasticity theory is presented. This theory is developed based on the thermodynamically consistent higher-order strain gradient theory developed by Voyiadjis and coworkers (Voyiadjis and Deliktas 2009a, b; Voyiadjis and Abu Al-Rub 2007). In order to adapt this theory to a crystalline system, they introduced in the virtual work the following fields $X^{(\beta)}$ and $\mathbb{X}^{(\beta)}$ that are work-conjugates to the slip, $\gamma^{(\beta)}$, and the slip gradient, $\gamma_{,i}^{(\beta)}$, respectively. The principle of virtual work can be expressed in terms of these fields as follows:

$$\int_{V}\left(\sigma_{ij}\delta\varepsilon_{ij}^{e}+\sum_{\beta}X^{(\beta)}\delta\gamma^{(\beta)}+\sum_{\beta}\mathbb{X}_{,i}^{(\beta)}\delta\gamma_{,i}^{(\beta)}\right)dV=\int_{S}t_{i}\delta u_{i}dS+\int_{S}\mathbb{t}^{(\beta)}\delta\gamma^{(\beta)}dS \tag{1}$$

where σ_{ij} is the standard stress tensor with $\sigma_{ij} = \sigma_{ji}$., t_i is the standard traction stress vector, while $\mathbb{t}^{(\beta)}$ is the nonstandard traction force work-conjugate to the slip, $\gamma^{(\beta)}$. In order to derive balance laws, integration by parts is used on some terms of the integral in Eq. (1) along with the divergence theorem ($\int_V grad\ (F)dV = \int_V \nabla\ \mathbf{F}\ dV = \int_S F\ \mathbf{n}\ dS$). The gradient of the displacement is given by $\varepsilon_{ij} = 1/2(u_{i,j} + u_{j,i})$, and the resolved shear stress is $\tau^{(\beta)} = m_i^{(\beta)}\sigma_{ij}s_j^{(\beta)}$ where $m_i^{(\beta)}$ and $s_j^{(\beta)}$ are unit vectors characterizing the normal to the slip plane β and its slip direction, respectively. Equation (1) is now expressed as follows:

$$\int_{V}\left(\sigma_{ij,j}\delta u_i+\sum_{\beta}\left(X^{(\beta)}-\tau^{(\beta)}-\mathbb{X}_{i,i}^{(\beta)}\right)\delta\gamma^{(\beta)}\right)dV+\int_{S}\left(t_i-\sigma_{ij}n_j\right)\delta u_i dS$$
$$+\int_{S}\left(\mathfrak{t}^{(\beta)}-\mathbb{X}_i^{(\beta)}n_i\right)\delta\gamma^{(\beta)}dS \tag{2}$$

Equation (2) must hold for any variations in δu_i and $\delta\gamma^{(\beta)}$, and therefore one obtains the classical equilibrium equation

$$\sigma_{ij,j}=0 \tag{3}$$

and the micro-force balance equation

$$X^{(\beta)}-\tau^{(\beta)}-\mathbb{X}_{i,i}^{(\beta)}=0 \tag{4}$$

Standard boundary conditions are now obtained that can be defined by prescribing either traction force or displacement, respectively, at each point on the boundary

$$t_i=\sigma_{ij}n_j \quad \text{or} \quad \overline{u}_i \tag{5}$$

Nonstandard boundary conditions can be given by prescribing either micro-traction force or slip at each point of the boundary

$$\mathfrak{t}^{(\beta)}=\mathbb{X}_i^{(\beta)}n_i \quad \text{or} \quad \overline{\gamma}^{(\beta)} \tag{6}$$

A boundary condition of $\overline{\gamma}^{(\beta)}=0$ form is referred to as a micro-clamped boundary condition, and a boundary condition of $\mathfrak{t}^{(\beta)}=0$ is termed a micro-free boundary condition.

In fact, the plastic strain, ε_{ij}^{p}, in a single crystal can be expressed as the symmetric part of the plastic distortion, γ, by the following expression:

$$\gamma_{ij}=\sum_{\beta=1}^{N}\gamma^{(\beta)}m_i^{(\beta)}s_j^{(\beta)} \tag{7}$$

Therefore, the plastic deformation can be expressed in terms of the plastic distortion as

$$\varepsilon_{ij}^{p}=\sum_{\beta=1}^{N}\gamma^{(\beta)}\mu_{ij}^{(\beta)} \tag{8}$$

where

$$\mu_{ij}^{(\beta)} = \frac{1}{2}\left(m_i^{(\beta)} s_j^{(\beta)} + m_j^{(\beta)} s_i^{(\beta)}\right) \tag{9}$$

and $\mu_{ij}^{(\beta)}$ is the symmetric Schmid tensor of the slip system β.

A motivation for the nonlocal formulation proposed by Gurtin (2003) stems from the fact that plastic deformation in crystalline solids arises from the motion of dislocations, which are line defects characterized by the Burgers vector b_i. This observation allows one to characterize the net Burgers vector B_i with respect to a closed loop C in the crystal through the following integral

$$B_i = \oint_C u_{ij}^p dx_j = \oint_S \alpha_{ij} v_j dS \tag{10}$$

in which S is any surface with boundary C and v_j is the unit normal field for S suitably oriented with respect to C. The tensor α_{ij}, which is often referred to as Nye's dislocation density tensor (Nye 1953) or as the density of geometrically necessary dislocations, is given by the explicit relation

$$\alpha_{ij} = \sum_{\beta=1}^{N} e_{jkl} \gamma_{,k}^{(\beta)} m_i^{(\beta)} s_l^{(\beta)} \tag{11}$$

where e_{jkl} is the alternating tensor.

Helmholtz free energy that accounts only for stored energy allows one to obtain energetic components of the thermodynamic forces σ_{ij}, $X^{(\beta)}$, and $\mathbb{X}^{(\beta)}$ work-conjugates to the elastic strain, ε_{ij}^e, slip, $\gamma^{(\beta)}$, and slip gradient, $\gamma_{,i}^{(\beta)}$, respectively. The functional form of the Helmholtz free energy is assumed to be

$$\Psi = \frac{1}{2}\varepsilon_{ij}^e E_{ijkl} \varepsilon_{kl}^e + \sum_{\beta=1}^{N} \frac{1}{2} a_1^{(\beta)} \left(\gamma^{(\beta)}\right)^2 + \frac{1}{2} a_2 \alpha_{ij} \alpha_{ij} \tag{12}$$

In this formulation one does not assume that Nye's tensor α_{ij} is an independent state variable of the Helmholtz free energy but it is dependent on the slip gradient, $\gamma_{,i}$, of the free energy. The energetic components of the thermodynamic forces can be obtained as follows:

$$\begin{aligned} \sigma_{ij} &= \frac{\partial \Psi}{\partial \varepsilon_{kl}^e} = E_{ijkl} \varepsilon_{kl}^e \\ {}^{en}X^{(\beta)} &= \frac{\partial \Psi}{\partial \gamma^{(\beta)}} = a_1^{(\beta)} \gamma^{(\beta)} \\ {}^{en}\mathbb{X}_i^{(\beta)} &= \frac{\partial \Psi}{\partial \gamma_{,i}^{(\beta)}} = \frac{\partial \Psi}{\partial \alpha_{kl}} \frac{\partial \alpha_{kl}}{\partial \gamma_{,i}^{(\beta)}} = a_2 \alpha_{kl} N_{kil}^{(\beta)} \end{aligned} \tag{13}$$

where $N_{kjl} = e_{ijl} m_i^{(\beta)} s_k^{(\beta)}, a_1^{(\beta)} = h_o$ and $a_2 = G\ell_{en}^2$. The thermodynamic force given by $(13)_3$ is similar to the expression for $\xi_i^{(\beta)}$ that is defined by Bittencourt et al. (2003). ℓ_{en} is the energetic material length scale that quantifies the size effect due to GND (related to the Burgers vector length). This characteristic length scale from a microstructural point of view can be attributed to dislocation network such as dislocation spacing, dislocation cell size, and misorientation of the cells in the metallic materials.

In order to satisfy the second law of thermodynamics, one requires the dissipation to be positive. This implies that $\sum_\beta X^{(\beta)}\dot{\gamma}^{(\beta)} + \mathbb{X}_{,i}^{(\beta)}\dot{\gamma}_{,i}^{(\beta)} \geq 0$. This inequality leads to the existence of the dissipation potential that should be a convex functional form in terms of its variables. One possible choice of such potential can be given by the following power law (Bardella 2007)

$$\mathcal{D} = \frac{h_o \gamma_o}{N+1}\left(\frac{\gamma_{eff}^{(\beta)}}{\gamma_o}\right)^{N+1} \tag{14}$$

where $\gamma_{eff}^{(\beta)} = \sqrt{\left|\gamma^{(\beta)}\right|^2 + \ell_{dis}^2 \gamma_{,i}^{(\beta)} \gamma_{,i}^{(\beta)}}$. ℓ_{dis} is the dissipative material length scale that quantifies the strengthening due to the size effect. This characteristic length scale from a microstructural point of view can be attributed to the movement of the mobile dislocations.

The dissipative components of each thermodynamic force can be obtained from Eq. (14) by taking partial derivatives of this potential with respect to its flux respective variables

$$\begin{aligned} {}^{dis}X^{(\beta)} &= h_o\left(\frac{\gamma_{eff}}{\gamma_o}\right)^{N-1}\frac{\gamma^{(\beta)}}{\gamma_o} \\ {}^{dis}\mathbb{X}_i^{(\beta)} &= h_o \ell_{dis}^2\left(\frac{\gamma_{\text{eff}}}{\gamma_o}\right)^{N-1}\frac{\gamma_{,k}^{(\beta)}}{\gamma_o} \end{aligned} \tag{15}$$

Making use of Eqs. $(13)_2$, $(13)_3$, and (15) into (4) and (6), the following nonlocal differential equation for the flow stress and the nonstandard boundary condition can be obtained, respectively,

$$\begin{aligned} \tau^{(\beta)} = a_1^{(\beta)}\gamma^{(\beta)} &+ h_o\left(\frac{\gamma_{eff}}{\gamma_o}\right)^{N-1}\frac{\gamma^{(\beta)}}{\gamma_o} \\ &- \ell_{en}^2 G\left[\alpha_{ij} N_{ikj}^{(\beta)}\right]_{,k} - h_o \ell_{dis}^2\left[\left(\frac{\gamma_{eff}}{\gamma_o}\right)^{N-1}\frac{\gamma_{,k}^{(\beta)}}{\gamma_o}\right]_{,k} \end{aligned} \tag{16}$$

and for the corresponding nonstandard boundary condition

$$\mathfrak{t}^{(\beta)} = \ell_{en}^2 G \alpha_{ij} N_{ikj}^{(\beta)} + h_o \ell_{dis}^2\left[\left(\frac{\gamma_{eff}}{\gamma_o}\right)^{N-1}\frac{\gamma_{,k}^{(\beta)}}{\gamma_o}\right] n_i \tag{17}$$

The elasticity is taken to be isotropic, characterized by Young's modulus E and Poisson's ratio v, with the shear modulus given by $2G = E/(1 + v)$. Thus there are seven positive material parameters that characterize the mechanical behavior: E, v, h_o, Γ_o, N and material length scales, ℓ_{dis} and ℓ_{en}, respectively. In addition, for each crystal the number of slip systems and the respective orientation of each system specified by $m_i^{(\beta)}, s_i^{(\beta)}$ $\left(m_i^{(\beta)} s_i^{(\beta)} = 0\right)$ need to be identified.

Simple Shear of a Constrained Crystalline Strip with Double Slips

The boundary value problem analyzed here is the simple shear of a constrained crystalline strip, of height H in the x_2 direction and unbounded in the x_1 and x_3 directions. The crystal is characterized by incompressible isotropic linear elastic material with plain strain and quasi-static loading conditions. It is also assumed that the crystalline strip has multiple double β slip systems that consist of two possible glides symmetrically oriented with respect to any plane constant x_2 by angle θ_β. The strip is sheared by applying to the plane $x_2 = H$ a uniform displacement equal to $\Upsilon(t)H$ in the x_1 direction where $\Upsilon(t)$ is the prescribed shear strain. The schematic description of this problem is illustrated in Fig. 1.

The standard *macroscopic boundary conditions* are

$$\begin{aligned} u_1 &= 0, \ \text{along} \ x_2 = 0 \\ u_1 &= U(t) = H\Upsilon(t), \ u_2 = 0 \ \text{along} \ x_2 = H \end{aligned} \tag{18}$$

In the constrained layer problem, one restricts to monotonic loading, so that the prescribed shear rate satisfies $\dot{\Upsilon} > 0$.

The nonstandard boundary conditions due to the nonlocality are defined as *micro-free boundary conditions on the sides* of the strip

$$\mathbb{t}^{(\beta)} = \mathbb{X}_i^{(\beta)} n_i = 0 \ \text{along} \ x_1 = \pm w \tag{19}$$

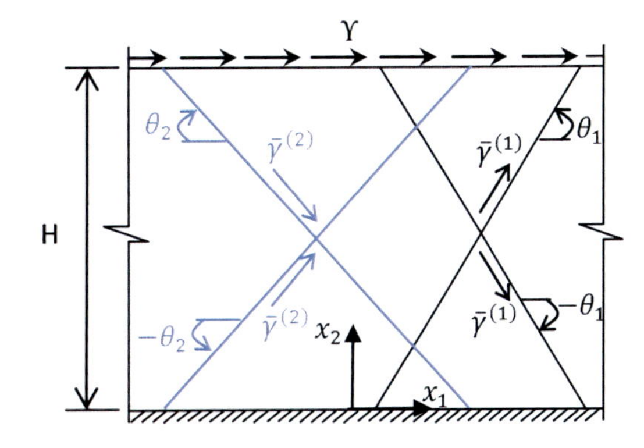

Fig. 1 A constrained crystalline strip with multiple symmetric double slip systems subjected to the simple shear

and *micro-clamped boundary conditions* on the bottom and top of the strip

$$\gamma^{(\beta)} = 0, \text{ along } x_2 = 0, H \tag{20}$$

One now considers the solution of this boundary value problem with field quantities that are independent of x_1. Macroscopic equilibrium requires that σ_{12} be spatially uniform. One seeks the solution for which σ_{11} and σ_{22} vanish. The elasticity solution in simple shear has a spatially uniform stress. Thus the entire strip reaches yield at the same time. The orientation of the first slip of the double system is given by

$$m_i^1 = \cos\theta_\beta \mathbf{e_1} + \sin\theta_\beta \mathbf{e_2}, \;\; s_i^1 = -\sin\theta_\beta \; \mathbf{e_1} + \cos\theta_\beta \mathbf{e_2} \tag{21}$$

The orientation of the second slip of the double system can be obtained by replacing θ_β by $-\theta_\beta$ and recalling that the following equalities hold due to symmetry$\left(\gamma_1^{(\beta)} = \gamma_2^{(\beta)} = \overline{\gamma}^{(\beta)}\right)$ and $\left(\tau_1^{(\beta)} = \tau_2^{(\beta)} = \overline{\tau}^{(\beta)}\right)$. The sole nonvanishing component of the strain tensor is given by

$$\varepsilon_{12}^p = \sum_\beta \overline{\gamma}^{(\beta)} \cos 2\theta_\beta \tag{22}$$

Since all the fields must be independent of x_1 and x_3 and due to symmetry, therefore the nonvanishing component of the Nye's tensor is in the x_2 direction due to the edge dislocation lying along x_3 and is given by

$$\alpha_{23} = \sum_\beta \overline{\gamma}_{,2}^{(\beta)} \sin^2\theta_\beta \tag{23}$$

Hence the only nonvanishing components of N_{ijk} tensor can be obtained as

$$N_{223} = \frac{\partial\alpha_{23}}{\partial\gamma_{,2}^{(\beta)}} = \sin^2\theta_\beta \tag{24}$$

One first considers the linear case by setting $N = 1$ in Eqs. (16) and (17) and making use of Eqs. (21–24); the simplified form of the Eqs. (16) and (17) can be obtained as follows:

$$\overline{\tau}^{(\beta)} = h_o \overline{\gamma}^{(\beta)} + h_o \frac{\overline{\gamma}^{(\beta)}}{\gamma_o} - \ell_{en}^2 G \sin^2\theta_\beta \sum_\zeta \overline{\gamma}_{,22}^{\zeta} \sin^2\theta_\zeta - h_o \ell_{dis}^2 \frac{\overline{\gamma}_{,22}^{(\beta)}}{\gamma_o} \tag{25}$$

and for the corresponding nonstandard boundary condition

$$\mathfrak{t}^{(\beta)} = \left[\ell_{en}^2 G \sin^2\theta_\beta \sum_\zeta \overline{\gamma}_{,2}^{\zeta} \sin^2\theta_\zeta + h_o \ell_{dis}^2 \frac{\overline{\gamma}_{,2}^{(\beta)}}{\gamma_o} \right] n_2 \tag{26}$$

Equation (25) can be written in the matrix form as

$$\mathcal{M}^{(\beta\zeta)}\gamma_{,22}^{(\zeta)} - \gamma^{(\beta)} = -\mathcal{F}^{(\beta)} \tag{27}$$

where $\mathcal{M}^{(\beta\zeta)}$ is defined as

$$\mathcal{M}^{(\beta\zeta)} = \begin{bmatrix} a_1^2 + b_1^2 & b_1 b_2 & \cdots\cdots & b_1 b_\zeta \\ b_2 b_1 & a_2^2 + b_2^2 & & \\ \vdots & & \ddots & \\ \vdots & & & \ddots \\ b_\beta b_1 & & & a_\beta^2 + b_\beta^2 \end{bmatrix} \tag{28}$$

and its components are defined as

$$\begin{aligned} a_\beta = \tfrac{\ell_{dis}}{\sqrt{1+\gamma_o}}, \, b_\beta &= \sqrt{\tfrac{G\gamma_o}{(1+\gamma_o)h_o}}\ell_{en}\sin^2\theta_\beta \\ \mathcal{F}^\beta &= \tfrac{\tau^{(\beta)}\gamma_o}{(1+\gamma_o)h_o} \end{aligned} \tag{29}$$

For the sake of simplicity, the strip system is assumed to have only one double system, and the dissipative material scale is set to zero. Equation (27) reduces to the following one dimensional form

$$\overline{\gamma}_{,22} - \lambda^2\overline{\gamma} = -\lambda^2 \tag{30}$$

where the coefficients λ^2 and F are defined as

$$\lambda^2 = \frac{(1+\gamma_o)\,h_o}{(\ell)^2 G\gamma_o \sin^4\theta}, \quad F = \frac{\overline{\tau}}{(\ell)^2 G \sin^4\theta} \tag{31}$$

Case 1 No dissipative hardening ($h_o = 0$, such that $\lambda^2 = 0$) is considered here. In this case the solution of Eq. (30) with the boundary conditions given by Eq. (20) yields the following quadratic relation for the plastic slip as a function of x_2:

$$\overline{\gamma} = \frac{F}{2}\left(x_2 H - x_2^2\right) \tag{32}$$

The remaining field quantities ε_{12} and σ_{12} can be computed using the following steps. The nonvanishing components of the plastic strain given by Eq. (22) can be rewritten as follows:

$$\varepsilon_{12}^p = \gamma\cos 2\theta \tag{33}$$

and the resolved shear stress is given by $\overline{\tau} = \cos 2\theta\sigma_{12}$ where shear stress is defined as $\sigma_{12} = 2G\left(\varepsilon_{12} - \overline{\gamma}\cos 2\theta\right)$. Averaging shear stress with respect to x_2 over the interval $[0, H]$ and knowing that σ_{12} is spatially constant and the average of ε_{12} over the interval $[0, H]$ is Y/2, then one obtains

$$\sigma_{12} = G\left(\Upsilon - 2\cos 2\theta \gamma_{ave}\right) \tag{34}$$

The solution of the differential Eq. (30) that is expressed by Eq. (32) is a function of $\overline{\tau}$. By eliminating σ_{12} from Eq. (34), one can solve the resulting linear equation for $\overline{\tau}$ such as

$$\overline{\tau}\,(\mathrm{r}) = \frac{6G\Upsilon\cos 2\theta \sin^4 2\theta}{\left(6\sin^4 2\theta + \cos^2 2\theta \gamma_o r^2\right)} \tag{35}$$

where $r = H/\ell$. Substituting Eq. (35) into (32) and (34) along with setting $(x_2 = zH)$ yields the following relation:

$$\overline{\gamma}\,(\mathrm{r}) = \frac{\overline{\tau}\,(\mathrm{r})\, r^2}{2G\sin^4 2\theta}\left(z - z^2\right) \tag{36}$$

This solution shows that the slip distribution is in quadratic form in the normalized thickness. The shear stress component, σ_{12}, can be obtained in terms of

$$\sigma_{12}\,(\mathrm{r}) = \frac{6G\Upsilon\sin^4 2\theta}{\left(6\sin^4 2\theta + \cos^2 2\theta \gamma_o r^2\right)} \tag{37}$$

These two equations are plotted in Fig. 2a, b, respectively.

The size of the thickness is defined by $(r = H/\ell)$. Since ℓ is the material property and has a constant value, the value of r represents the size of the thickness. It is clear from both figures that the material indicates strengthening with decrease in size. Figure 2a, b illustrate this fact by showing the decrease in thickness results in an increase in shear stress which in turn provides less plastic deformation. Another observation from these analyses is the formation of the boundary layer with decreasing thickness (region of gradients). It is obvious from Fig. 2a that the boundary layer starts forming after r reduces to 20. It is clearer when r is 15.

Case 2 In the case of the dissipative hardening where ($h_o > 0$), the solution of the differential equation in Eq. (30) along with the boundary conditions given in Eq. (20) results to the following expression:

$$\overline{\gamma} = \frac{F}{\lambda^2}\left[1 - \cosh\left[\lambda x_2\right] + \sinh\left[\lambda x_2\right]\tanh\left[H\lambda/2\right]\right] \tag{38}$$

By averaging the Eq. (38) over the interval $[0, H]$ and substituting the resulting relation into Eq. (34), one can solve the resulting equality for $\overline{\tau}$

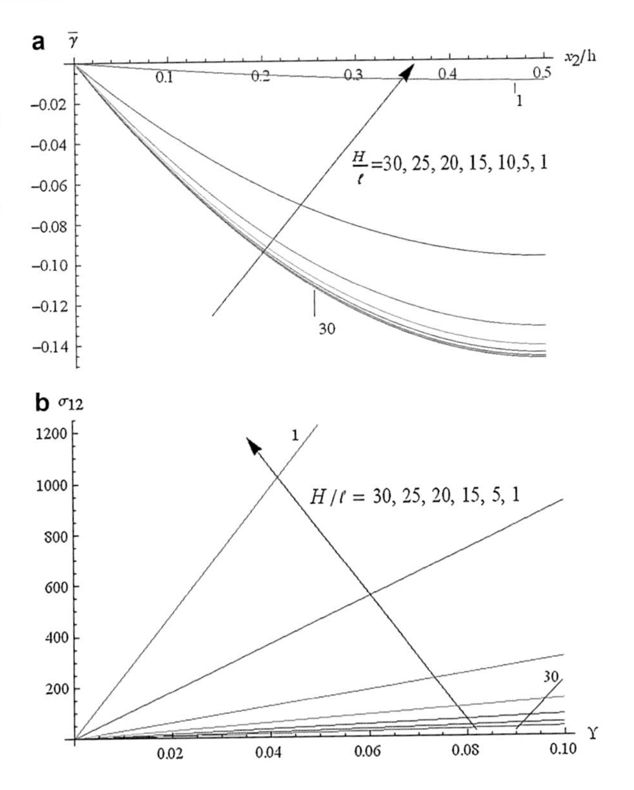

Fig. 2 Ideal plastic case where ($h_o = 0$) at various length scales ($H/\ell = 1, 5, 10, 15, 20, 25, 30$). Crystal strip has two slip systems with ($\theta^1 = 60, \theta^2 = -60$), $G = 26300$ MPa, $\gamma_o = 0.01$, $\ell_{dis} = 0$, and $N = 1$. (**a**) Shear slip distribution vs. normalized thickness. (**b**) Shear stress distribution vs. imposed shear strain

$$\overline{\tau} = \frac{G\Upsilon}{\left(\frac{1}{\cos 2\theta} + \frac{2G\gamma_o}{h_o(\gamma_o+1)}\cos 2\theta\left\{1 - \frac{\tanh[H\lambda/2]}{H\lambda/2}\right\}\right)} \tag{39}$$

Substituting Eq. (39) into (32) and (34) yields the following relations. Shear stress component, σ_{12}, can be obtained as a function of $r = H/\ell$ by replacing the term $H\lambda/2$ with Ar where $A = \sqrt{h_o\left(\gamma_o + 1\right)/G}/2\sin^2\theta$

$$\sigma_{12}(r) = \frac{G\Upsilon}{\left[1 + \frac{2G\gamma_o}{h_o(\gamma_o+1)}\cos^2 2\theta\left\{1 - \frac{\tanh[\mathrm{Ar}]}{\mathrm{Ar}}\right\}\right]} \tag{40}$$

In the case when $r \to 0$, the response of the shear stress is $\sigma_{12}(r) = G\Upsilon$ which is the linear elastic solution. For the case when $r \to \infty$, the response of the shear stress is obtained as $\sigma_{12}(r) = G\Upsilon/(1 + 2G\gamma_o \cos 2\theta/h_o(\gamma_o + 1))$ which shows size-dependent behavior. Figure 3 shows the curves for the plastic strain versus normalized thickness for various length scales.

Although the predictions of the models show similar trends, there is variation in their hardening behaviors. This is because the proposed model introduces decomposition for each thermodynamics force. However, in the works of Bardella

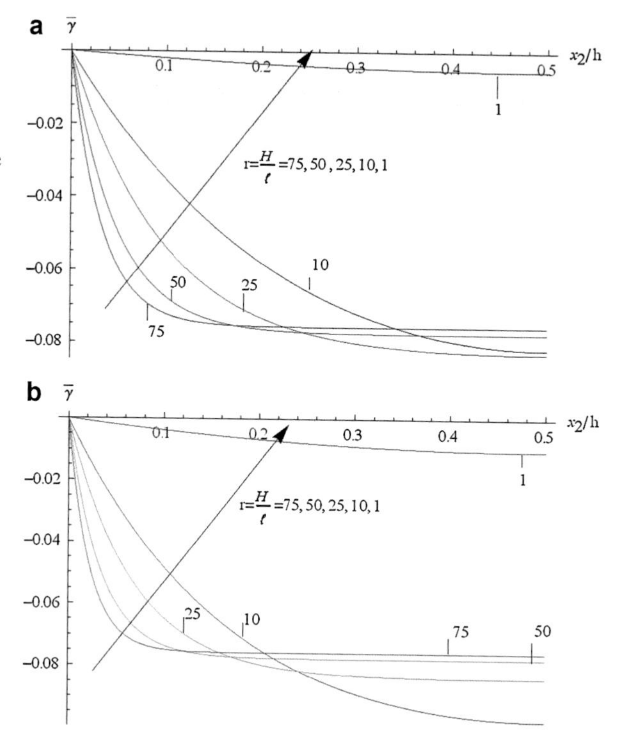

Fig. 3 Comparison of two models of the plastic slip distribution vs. normalized thickness at various length scales ($H/\ell = 1, 10, 25, 50, 75$) that are obtained from the (**a**) proposed theory and (**b**) formulation by Bardella (2006). Crystal strip has two slip system with ($\theta^1 = 60$, $\theta^2 = -60$), $G = 26300$ MPa, $h_o = 50$ MPa, $\gamma_o = 0.01$, $\mathrm{Y} = 0.1$, $\ell_{dis} = 0$, and $N = 1$

(2007) and Anand et al. (2005), they decomposed the thermodynamic forces that are related to the gradient. The current model indicates that the hardening becomes effective when the length scale parameter is reduced to 25, whereas the prediction of the other model (Bardella 2007) shows hardening at a value of 10. This explains that the current model predicts the thickness of the boundary layer higher than the prediction of that by Bardella (2007). Figure 4 shows the predictions of both models on the response of the shear stress versus imposed shear strain.

If one looks at the marked dashed lines in both Fig. 4a, b, it is clear that both models agree in their predictions of strengthening due to size effects.

A parametric study is performed by plotting the shear stress versus the slip orientation. This result is presented in Fig. 5.

As observed from Fig. 5, the influence of the orientation of the slip planes with variation in θ indicates clearly the length scale effects. In the case when the orientation of the slips is $\theta \cong 0$, the strip becomes less stiff, and the most complaint response is observed. However, when the orientation of the slips is around $\theta \cong 45$, one obtains a stiffer response, and the maximum shear stress is obtained. In both cases since the material response is linear elastic, no size effect is observed.

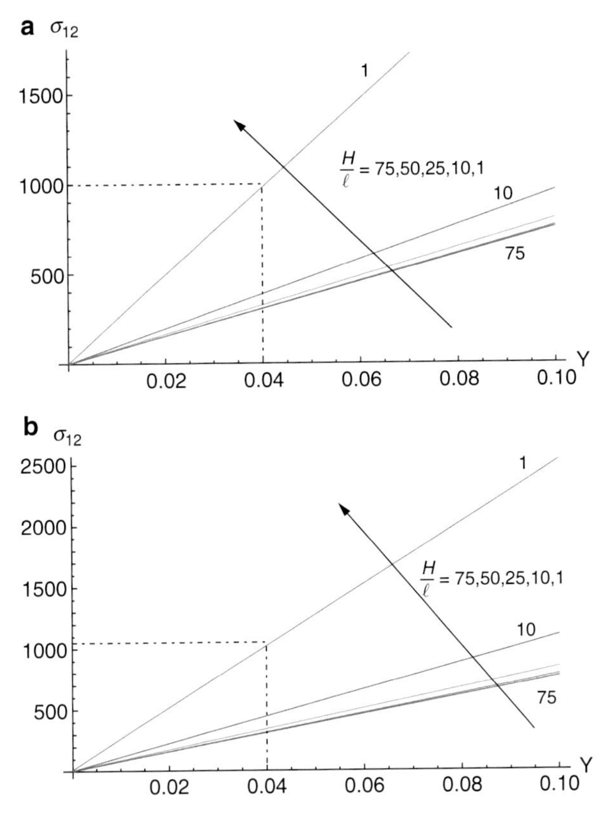

Fig. 4 Comparison between two models for the shear stress vs. imposed shear strain at various length scales ($H/\ell = 1, 10, 25, 50, 75$) that are obtained from the (**a**) proposed theory and (**b**) formulation by Bardella (2006). Crystal strip has two slip systems with ($\theta^1 = 60, \theta^2 = -60$), $G = 26300$ MPa, $h_o = 50$ MPa, $\gamma_o = 0.01$, $\ell_{dis} = 0$, and $N = 1$

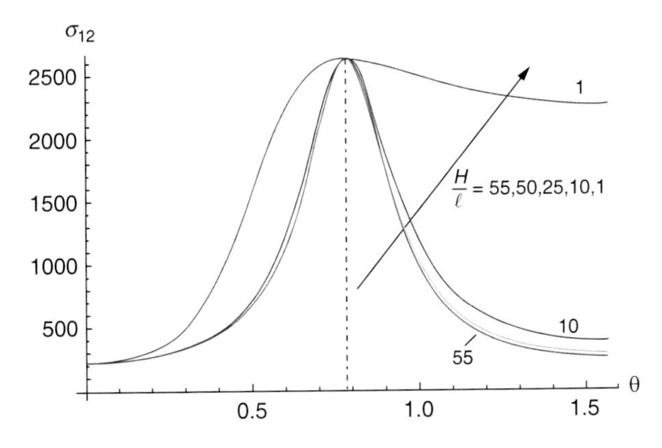

Fig. 5 Shear stress distribution versus orientation of the two symmetric slip systems at various length scales ($H/\ell = 1, 10, 25, 50, 55$). Other parameters used are $G = 26300$ MPa, $h_o = 50$ MPa, $\gamma_o = 0.01$, $\Upsilon = 0.1$, $\ell_{dis} = 0$, and $N = 1$. Angles are measured in radians

Table 1 Specimen designations and measured actual cross-sectional dimensions

Specimen designation	Specimen material	Section dimensions (in × in)	Applied loading
1080_04_1	AISI 1080	0.039 × 0.4	Monotonic increasing
1080_04_2	AISI 1080	0.039 × 0.4	Cyclic
4130N_05_1	AISI 4130	0.056 × 0.315	Monotonic increasing
4130N_05_2	AISI 4130	0.056 × 0.315	Cyclic
4130N_05_3	AISI 4130	0.056 × 0.316	Cyclic
VAX300_06_1	VascoMax C-300	0.064 × 0.313	Monotonic increasing
VAX300_06_2	VascoMax C-300	0.065 × 0.316	Cyclic
VAX300_06_3	VascoMax C-300	0.064 × 0.314	Cyclic

Experimental Procedure for Material Evaluation

The experimental procedure consists of three sets of tests conducted on AISI 1080, AISI 4130, and VascoMax C-300 steels. The first set of experiments involve monotonic tensile loading of the specimens up to failure, while they were unloaded at certain levels and reloaded in the second and the third sets of experiments.

An MTS 810 Hydraulic Materials Testing System was used to apply uniaxial stresses on the specimens. The strain data were obtained using a 1 in. gauge length MTS extensometer (Model No. 634.11E–24) and 1 mm gauge length Omega general-purpose pre-wired 1-axis strain gauges. The actual cross-sectional dimensions of the specimens were measured prior to tests using a digital caliper, and the measured dimensions are presented in Table 1 with the corresponding designations. The three sets of experiments and their results are presented next.

Result of the First Set of Experiments

In the first phase of the experimental program, one specimen from each material type was tested under monotonic increasing tensile loads in order to obtain mechanical properties (yield strength and ultimate strength) of the materials. During the tests, only the longitudinal strain values were recorded in the midsection of the specimens with a 1 in. gauge length MTS extensometer (Fig. 6).

The rate of loading of the MTS testing machine was set to be 0.01 in./min for specimen 4130N_05_1; however, due to the large size of the output file, it was decreased to 0.02 in./min for specimens 1080_04_1 and Vax300_06_01. Determinations of these loading rate values were made based on ASTM E8, Standard Test Methods for Tension Testing of Metallic Materials. Even though the recommended rates of loading values in ASTM E8 are higher than the ones used in this study, slower loading rates were used in this study because the cross sections of

Fig. 6 Measurement of longitudinal strains

the specimens are considerably smaller than the standard tension testing specimens addressed in the standard.

Using the output data file obtained from the software controlling the MTS testing machine, the engineering stress values were obtained by dividing the actual stress output data by the measured initial cross-sectional areas, and the stress-strain curves for all three tested materials are presented in Figs. 7, 8, and 9.

Result of the Second Set of Experiments

For the second set of experiments, transverse strain values were also recorded in addition to the longitudinal strains. For this purpose, a pair of pre-wired Omega strain gauges was attached on opposite surfaces of one specimen from each material type (Fig. 10). Due to the small width of the specimens, 1 mm gauge length (total length of 4 mm) strain gauges were used. The strain gauges were placed at the midsections of the specimens using commercial strain-gauge glue. Through a data logger, the strain data from both of the strain gauges are stored in micro-strains.

The second phase of the experimental program differs from the first one not only for measuring transverse strains but also in terms of the loading procedure. The specimens, in the second experimental phase, were loaded up to certain stress/strain levels, and then they were unloaded until the applied force is equal to zero.

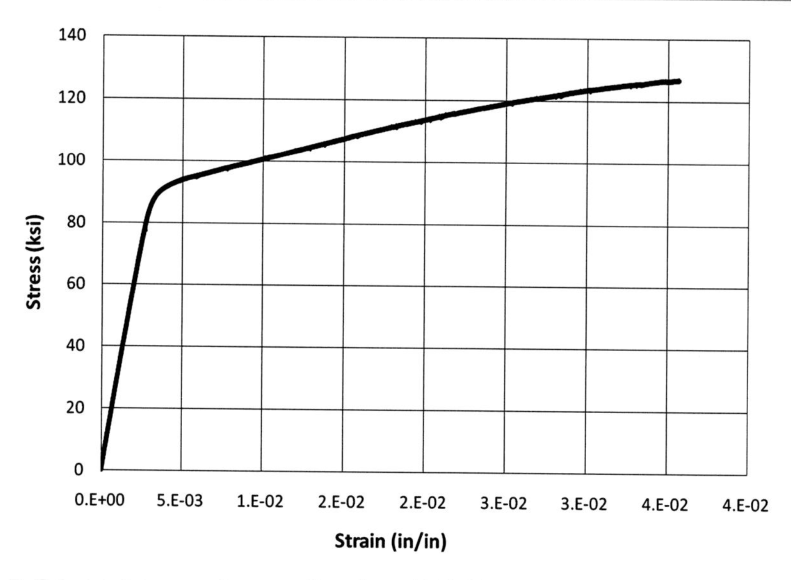

Fig. 7 Calculated stress-strain curve of specimen 1080_04_1 under monotonic loading

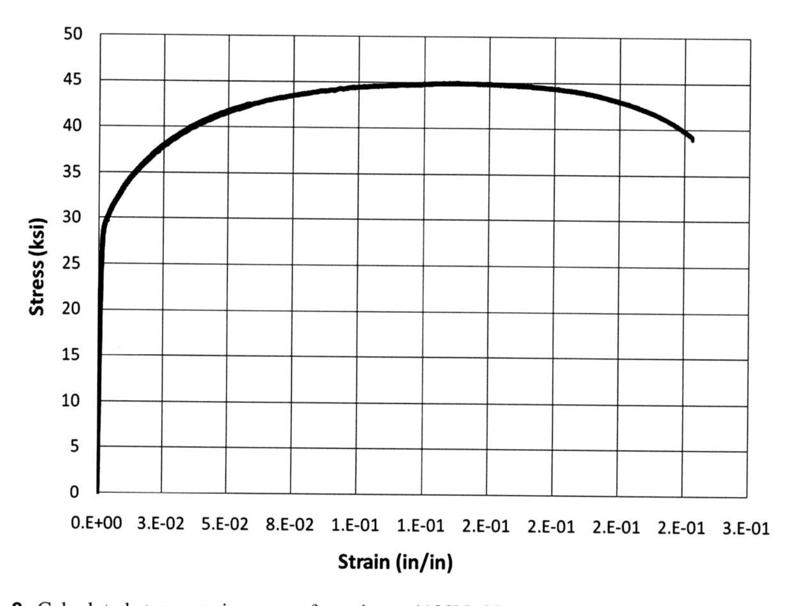

Fig. 8 Calculated stress-strain curve of specimen 4130N_05_1 under monotonic loading

Several cycles were decided to be applied in such a way that the peak strain value of each cycle exceeds the one that belongs to the previous cycle. For this purpose, special test procedures were defined using the MTS controller software, and the time, machine displacement, applied force, and strain values were stored.

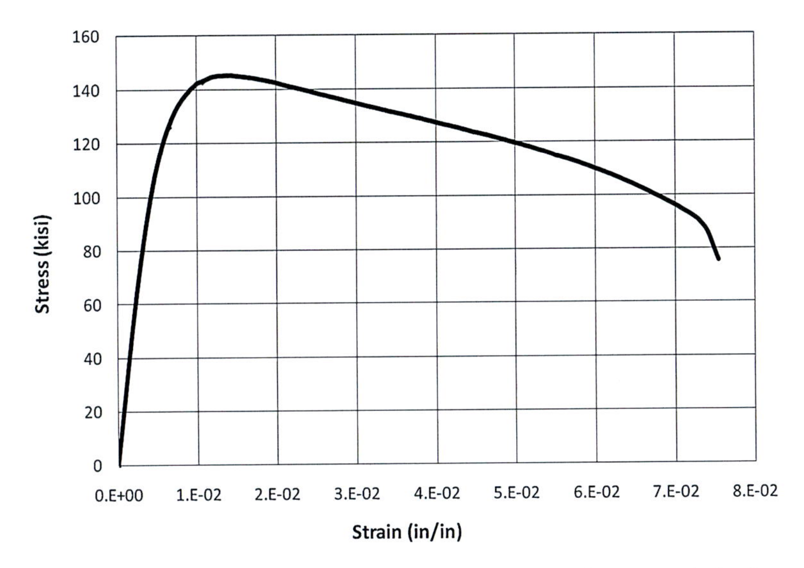

Fig. 9 Calculated stress-strain curve of specimen VAX300_06_1 under monotonic loading

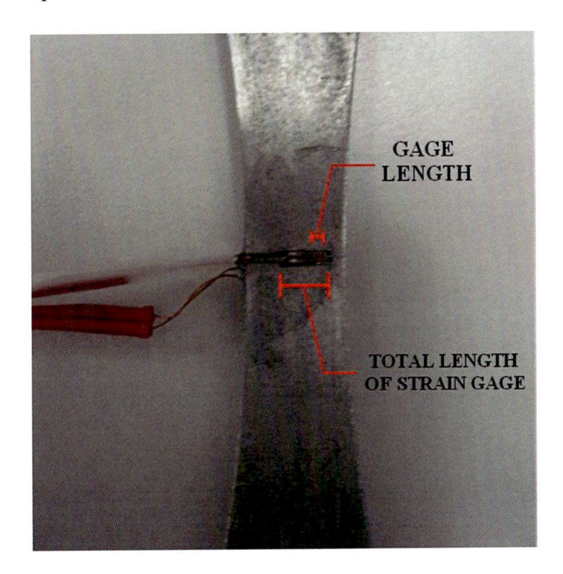

Fig. 10 A pair of pre-wired strain gauges glued on opposite sides of a specimen

In order to calculate the applied stresses, the average of the transverse strain obtained from the two strain gauges was used. The transverse strain along the thickness of the specimen was assumed to be equal to the transverse strain along the width. Figure 11 illustrates how the extensometer and the strain gauges were used to measure longitudinal and transverse strains.

The transverse strain data obtained from the attached strain gauges were averaged to be used for calculating the change in the cross section of the specimens, thus the applied stresses. The strain data collected from the strain gauges and their average

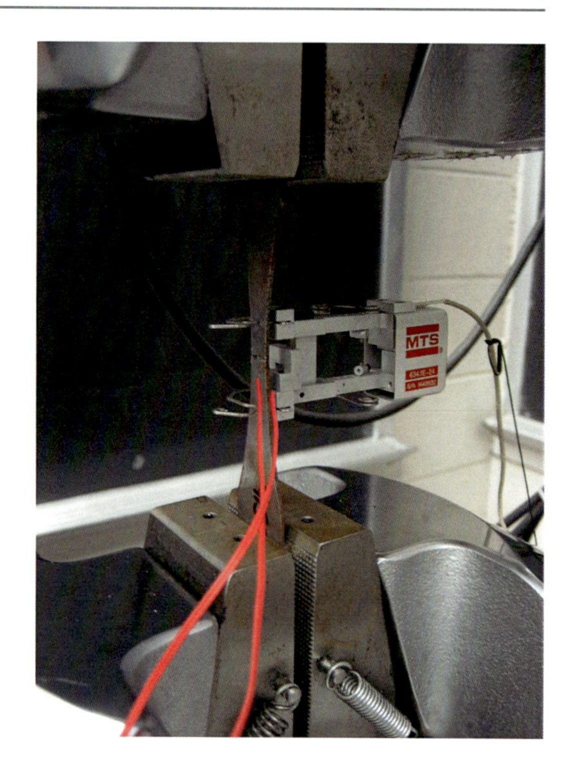

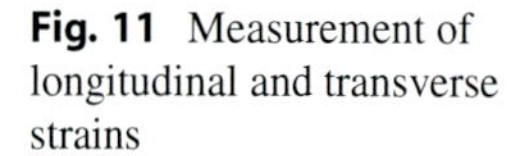
Fig. 11 Measurement of longitudinal and transverse strains

values are plotted and shown in Figs. 8 and 12 for specimens 4130N_05_2 and VAX300_06_2. During these two tests, visual observations were made, and it was noticed that the strain gauges were detached after the first few load cycles because of the small size of the glued area. Furthermore, the cyclic loading has also an effect on how quickly the gauges get detached and collect meaningless data.

It can be seen in Figs. 12 and 13 that the strain gauges attached to specimen 4130N_05_2 start collecting useless data after the first loading cycle, while the strain gauges of specimen VAX300_06_2 are detached approximately at the end of the fourth cycle. The strain data before the strain gauges were detached were used to calculate the actual applied stresses. It should be noted that the strains after that were considered to be constant and equal to the last meaningful values. The stress-strain relationships for the two specimens tested under cyclic tensile forces are shown in Figs. 14 and 15. It should be noted that the small spikes in Fig. 15 are caused when the applied loads (and selected target strain levels for the cycles) were being adjusted by the MTS testing machine.

Result of the Third Set of Experiments

Similar to the second phase, transverse strain values were also recorded in addition to the longitudinal strains in the third set of experiments. However, different

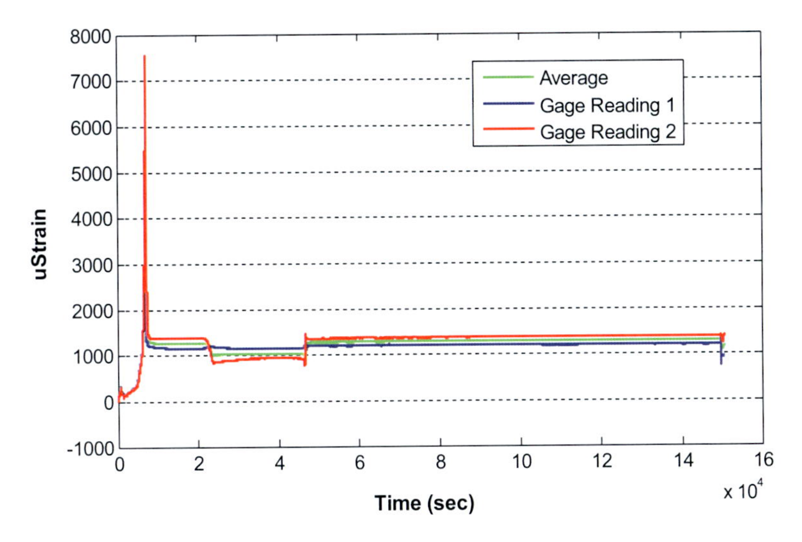

Fig. 12 Transverse strain-gauge readings of specimen 4130N_05_2

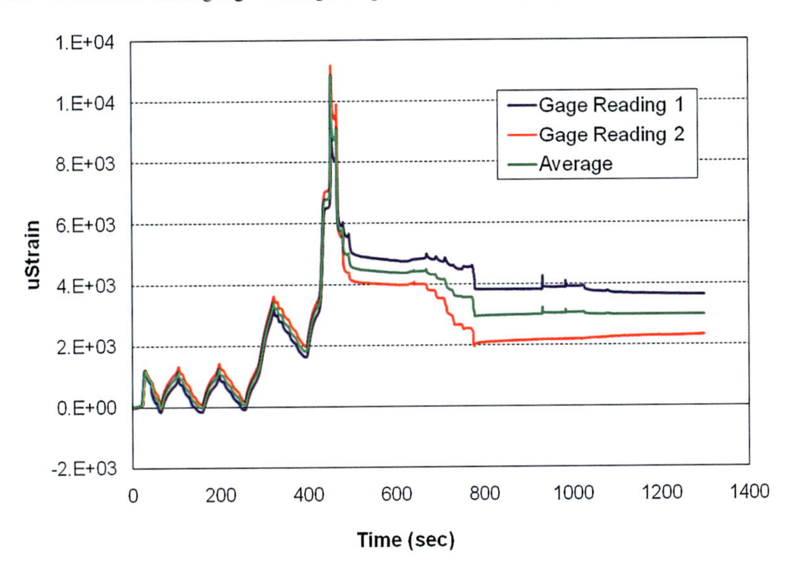

Fig. 13 Transverse strain-gauge readings of specimen VAX300_06_2

than the second set, a single strain gauge was attached to the midsection of one specimen from each material type (Fig. 16). The transverse strain along the thickness of the specimen was assumed to be equal to the transverse strain along the width.

The specimens (1080_04_2, 4130N_05_3 and VAX300_06_3) were, again, loaded up to certain stress/strain levels, and then they were unloaded until the

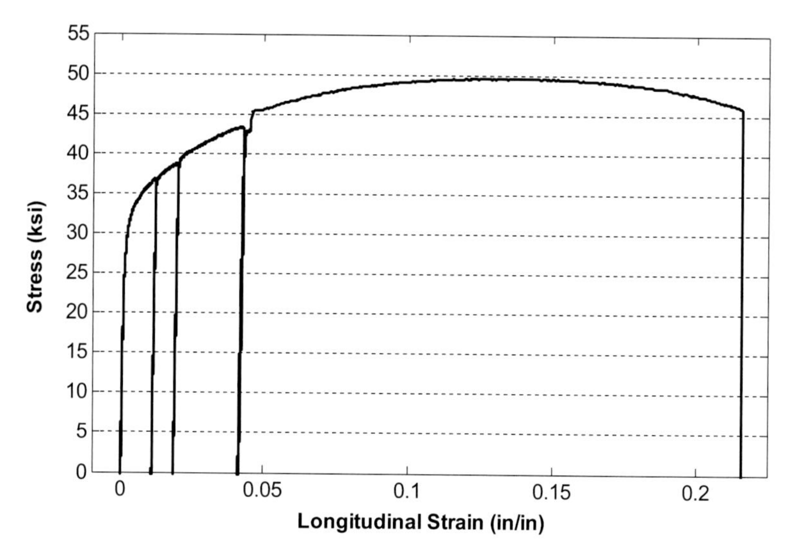

Fig. 14 Calculated stress-strain curve of specimen 4130N_05_2

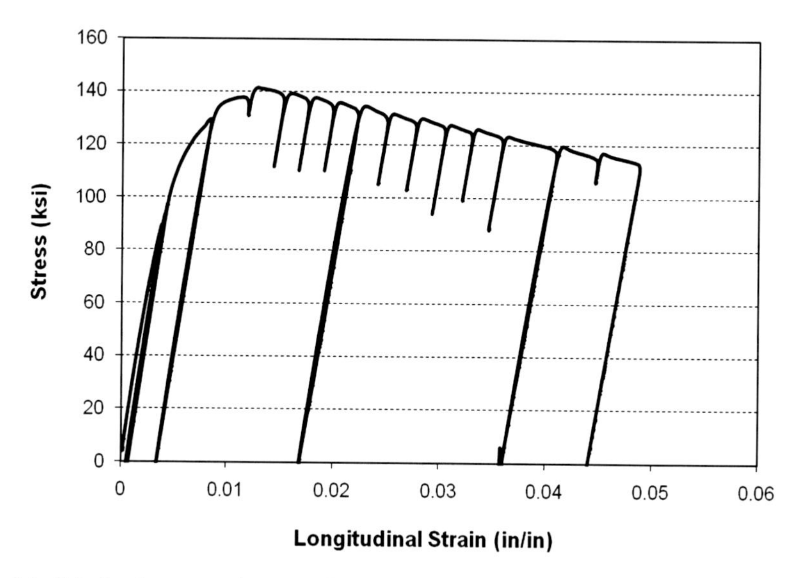

Fig. 15 Calculated stress-strain curve of specimen VAX300_06_2

applied force is equal to zero before the following loading. In order to calculate the applied stresses, the transverse strain data obtained from the single strain gauge were used. The strain recordings for all three tested specimens are presented in Figs. 17, 18, and 19. It can be seen from Fig. 17 that the strain gauge on the specimen 1080_04_2 kept recording valuable data until the end of the test while the ones

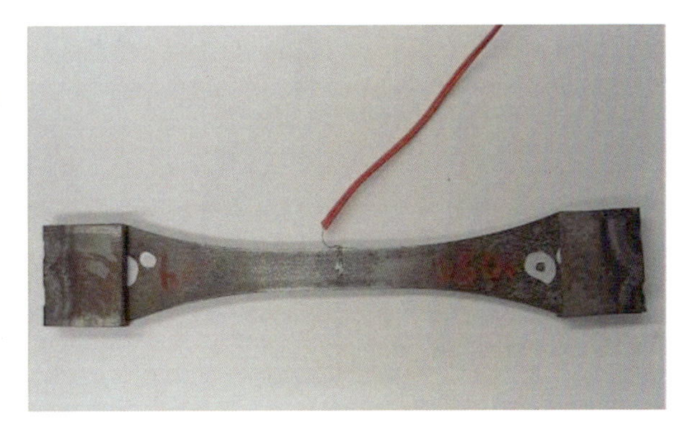

Fig. 16 A single strain gauge attached to the midsection of a test specimen

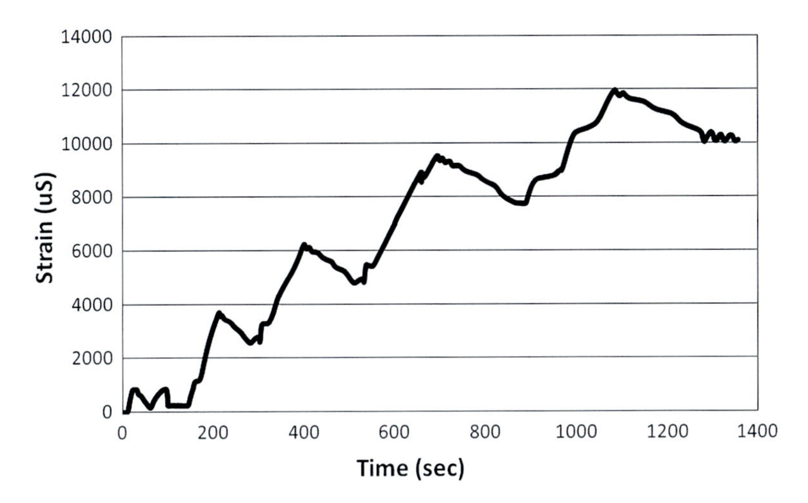

Fig. 17 Transverse strain-gauge readings of specimen 1080_04_2

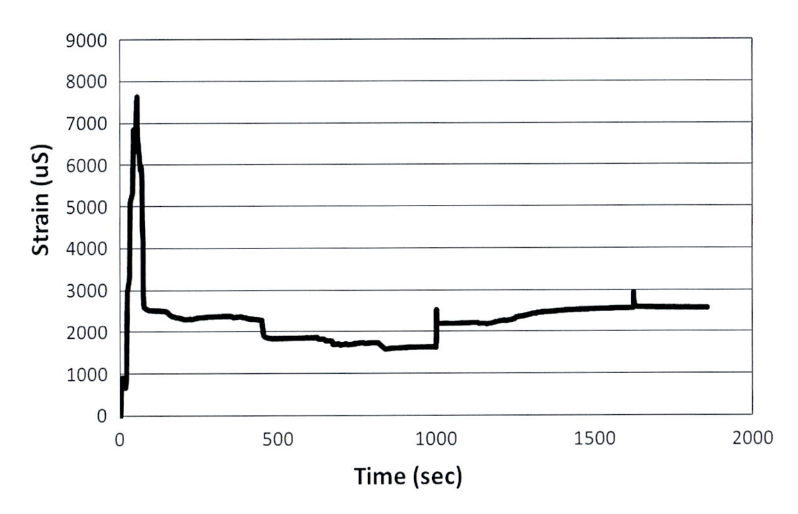

Fig. 18 Transverse strain-gauge readings of specimen 4130N_05_3

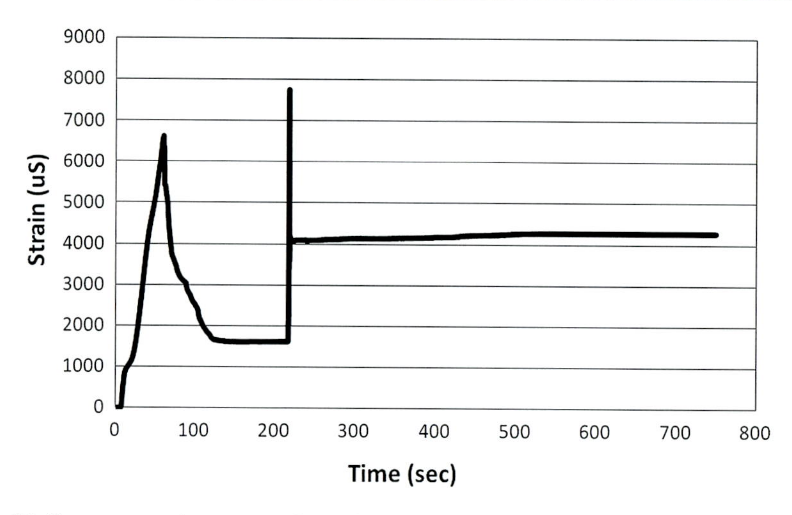

Fig. 19 Transverse strain-gauge readings of specimen VAX300_06_3

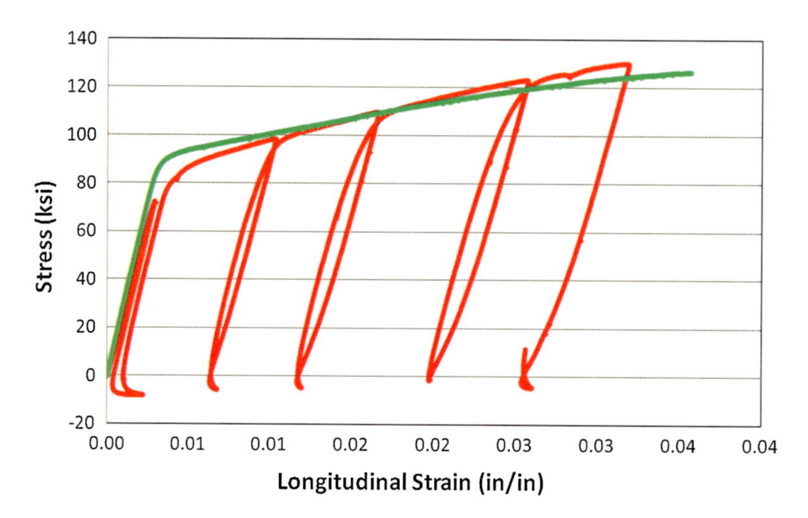

Fig. 20 Calculated stress-strain curve of specimen 1080_04_02 (red line) with the monotonic stress-strain curve of specimen 1080_04_1 (green line)

on 4130N and VascoMax C300 specimens detached by the end of the first loading cycle.

The meaningful strain data were used to calculate the actual applied stresses, and the stress-strain relationships for the three specimens tested under cyclic tensile forces are shown in Figs. 21 and 22 with red lines. The stress-strain relationships for the monotonic loading are also included in these figures with green lines (Fig. 20).

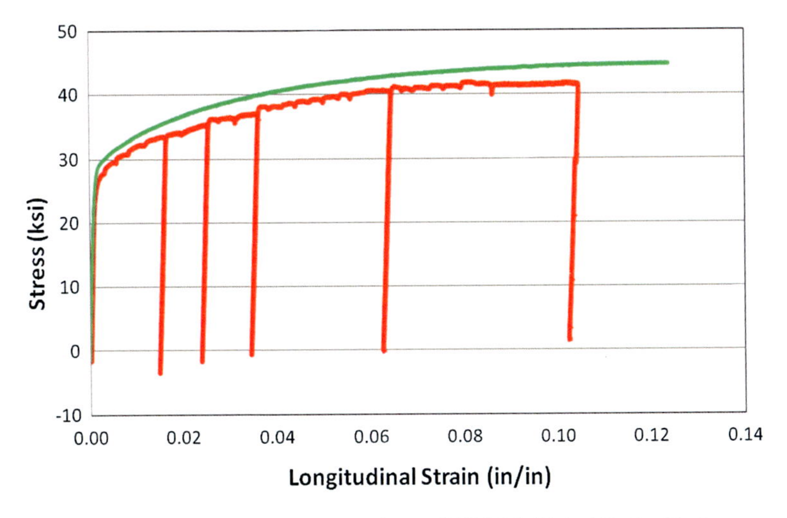

Fig. 21 Calculated stress-strain curve of specimen 4130N_05_03 (red line) with the monotonic stress-strain curve of specimen 4130N_05_1 (green line)

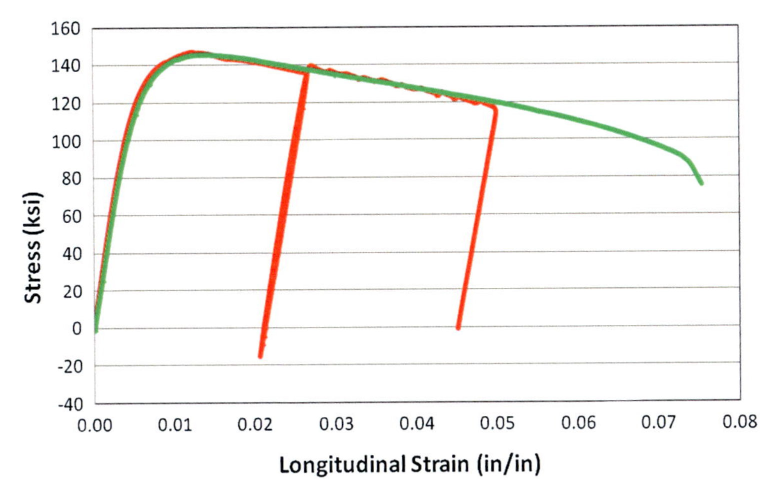

Fig. 22 Calculated stress-strain curve of specimen VAX300_06_03 (red line) with the monotonic stress-strain curve of specimen VAX300_06_1 (green line)

Numerical Simulations and Validating the Experimental Approach

The experimental work on the study of friction between metallic surfaces will be conducted by Voyiadjis and coworkers (Voyiadjis and Abu Al-Rub 2007; Voyiadjis et al. 2010; Voyiadjis and Deliktas 2009a, b; Lodygowski et al. 2011) with its counterpart at Ecole de' Nationale Institut der Mechanic (ENIM), at Metz, France. A modified Hopkinson's bar experiment shown schematically in Fig. 23a, developed by Phillipon at ENIM, simulates high-velocity contact between two surfaces. In this experiment, ultrahigh-speed projectiles of the 1020 steel (for sled shoe surface) will be fired onto a stationary VASOMAX steel surface (for the sled track). The contact force in this experiment is measured by a dynamometer ring shown in Fig. 23b. The contact force exerted on the dynamometer (in Fig. 23c) can be predicted by measuring the change in length of the friction device using strain. The results from the experiment will include (a) the measured contact force (both normal and frictional) response from the dynamometer ring and (b) the final contact surface between the two specimens. The latter (along with the initial surface) will be

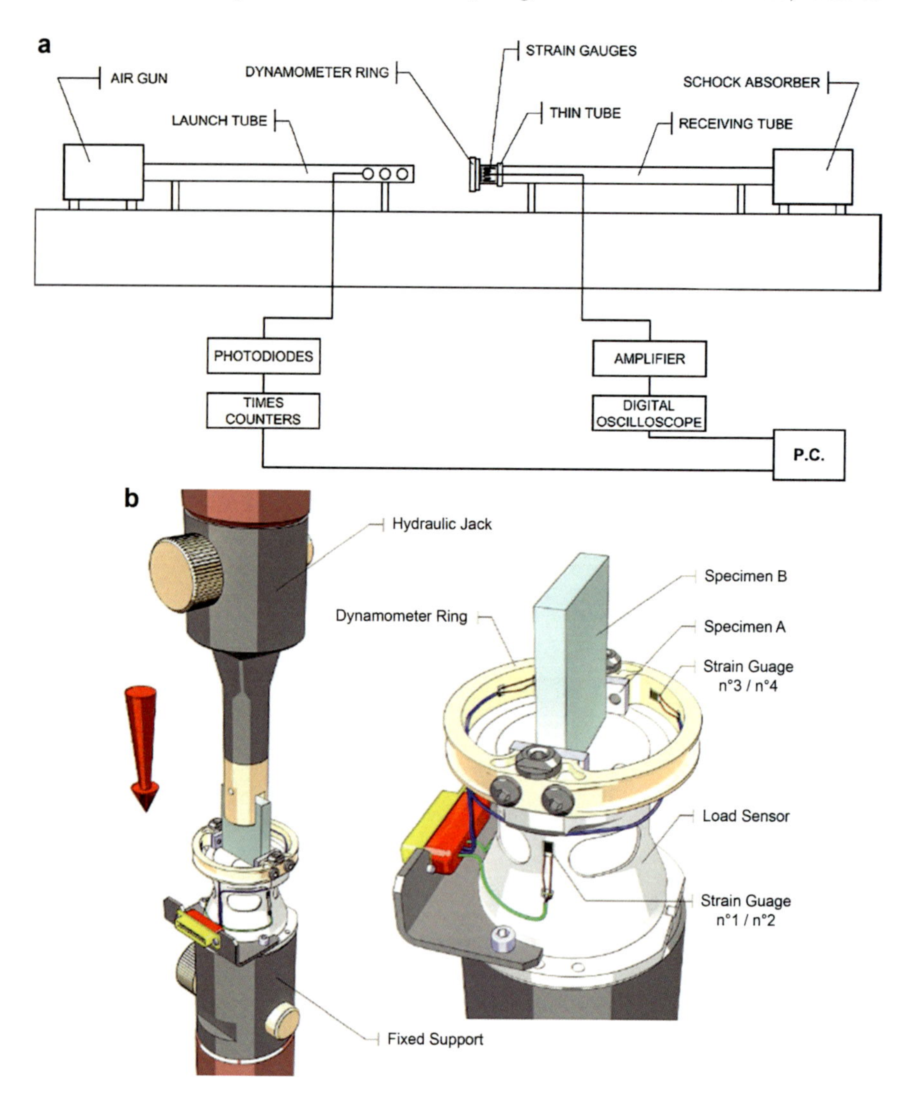

Fig. 23 (**a**) Schematic of modified Hopkinson experiment, developed by Phillipon et al. at (**b**) zoom-in view of friction device used to experimentally measure contact forces during high-velocity contact and (**c**) measured contact force on dynamometer

processed using SEM to characterize the topography of the surfaces before/after and consequently the size of contacting asperities. In this experiment, roughness measured through the SEM is characterized by software package (Vision32) provided by Veeco Metrology Group. The presliding characterization of the sled shoe and track surfaces indicates a presliding surface roughness of 4–8 mm, and it is postulated that damage phenomena are generated from dislocation generation at between crystallographically misoriented single crystals (at the microscale). In order to further characterize the sliding interfaces, experiments were conducted.

ENIM, used to measure contact forces during high-velocity contact between polycrystalline surfaces.

The simulations were performed with the FE commercial software ABAQUS (REFERENCE) according to the theory developed and coded as a VUMAT subroutine. The set of simulation was performed for different values of normal force and sliding velocity capturing the development of surface temperature as well as plastic deformation. The different mesh sizes were also considered to present the mesh independency thanks to the implemented theory. In the following figures, a couple of main results are presented.

In Fig. 24 the following stages of the simulation are presented using the coarse and fine mesh approach. The initial stage (Figs. 24a and 25a) presents the

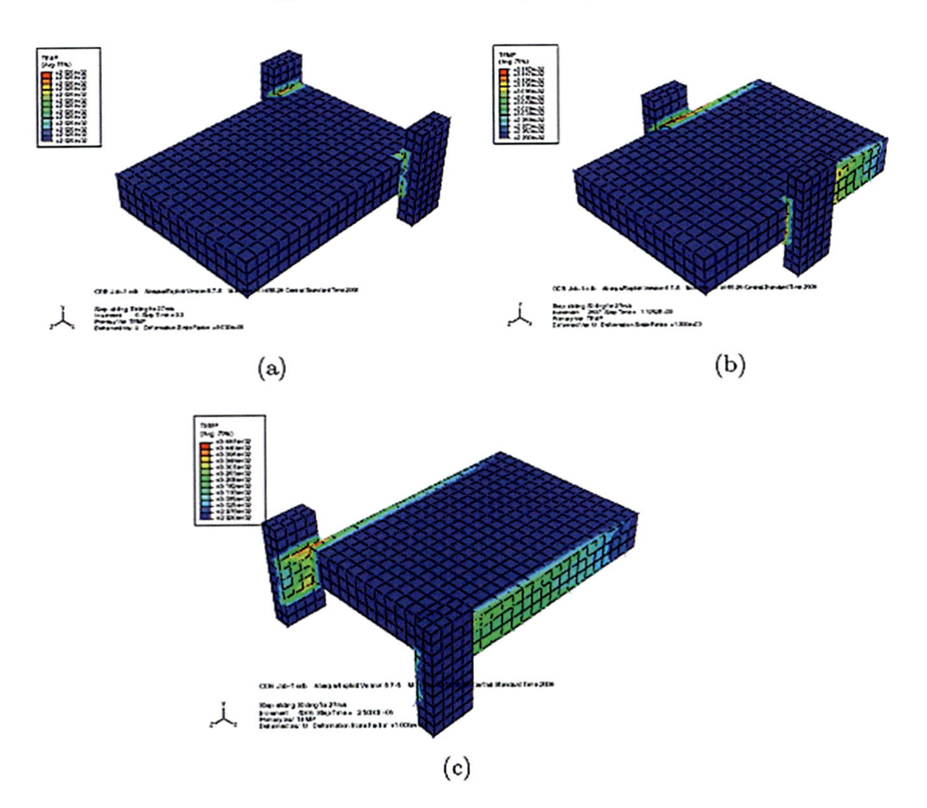

Fig. 24 FEM with coarse mesh and fine mesh for high-speed friction experiment with (**a**) boundary condition, (**b**) development temperature, (**c**) development of the moment

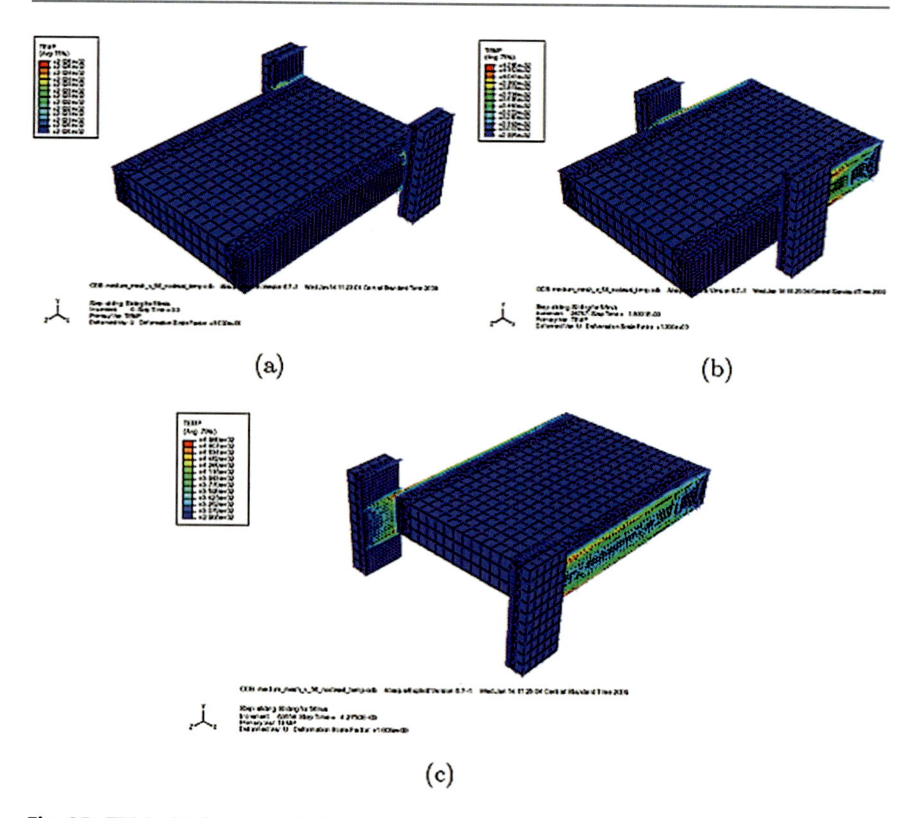

Fig. 25 FEM with fine mesh for high-speed friction experiment with (**a**) boundary condition, (**b**) development temperature, (**c**) development of the moment

application of all boundary conditions just before applying the sliding velocity to the specimen B. Figures 24b and 25b present the development of the temperature as one of the mainly considered variables. Lastly Figs. 24c and 25c show the moment when the contact between the specimens is lost. Figure 24 with the coarse mesh presents the results for the sliding velocity of 27 m=s, whereas Fig. 25 presents the fine mesh for the sliding velocity 56 m=s. Figure 24a presents the initial configuration of the simulation just before the sliding begins. Figure 24b presents the middle stage of the simulation process and the surface temperature development. Figure 24c shows the final stage of the simulation.

Discussions and Conclusions

Metal-to-metal friction problem is addressed in this work and revisited, specifically in investigating the coefficient of the dry friction for steel in the high-velocity range. The following contributions are made by this study:

- A physical model based on the nonlocal crystal plasticity model is proposed to investigate the coefficient of the dry friction. Results are presented from the simulations of the *simple shear of a constrained crystalline strip with double slips* and *high-speed friction experiment.*
- It is shown that the material responses obtained from the simulation of the physically based constitutive model agree with the real behavior of the metals.
- This work investigates also the effects of two thermodynamic processes, energetic and dissipative, involved in strain gradient crystal plasticity. The improved theory by Voyiadjis and Deliktas (2009b) is enhanced by decomposition for each thermodynamic force. This decomposition, in turn, introduces two different types of material length scales which provide better description of the dislocation network. The theoretical model is formulated for the crystalline system with multiple symmetric double slip systems. The boundary value problem analyzed here is the case of simple shear of a constrained crystalline strip in order to investigate the effects of the dissipative and energetic parameters on the size effect response of the constrained crystalline strip systems. The developed theory describes quantitatively the thickness of the boundary layer, hardening, and strengthening response of the system.
- However, there are still open issues of great interest in the field of SGP such as the following: (i) *dissipative and energetic nature of the dislocation network* (Anand et al. 2005; Bardella 2007; Bittencourt et al. 2003; Gurtin 2008; Voyiadjis and Deliktas 2009a) (the important question is how energy depends on the characteristics of the dislocation distribution), (ii) *evolution of the microstructure* which is attributed to many plasticity phenomena (Berdichevsky 2006; Gurtin 2008; Molinari and Ravichandran 2005; Nijs et al. 2008), and (iii) *physical bases of the material length scale* (Abu Al-Rub and Voyiadjis 2004; Anand et al. 2005; Mughrabi 2004).

References

R.K. Abu Al-Rub, G.Z. Voyiadjis, Analytical and experimental determination of the material intrinsic length scale of strain gradient plasticity theory from micro- and nano-indentation experiments. Int. J. Plast. **20**, 1139–1182 (2004)

K.E. Aifantis, J.R. Willis, Scale effects induced by strain gradient plasticity and interfacial resistance in periodic and randomly heterogeneous media. Mech. Mater. **38**, 702–716 (2006)

L. Anand, M.E. Gurtin, S.P. Lele, C. Gething, A one-dimensional theory of strain-gradient plasticity: formulation, analysis, numerical results. J. Mech. Phys. Solids **53**, 1789–1826 (2005)

K. Arakawa, Effect of time derivative of contact area on dynamic friction. Appl. Phys. Lett. **104**, 241603 (2014)

K. Arakawa, An analytical model of dynamic sliding friction during impact. Sci. Rep. **7**, 40102 (2017)

R.J. Asaro, Crystal plasticity. J. Appl. Mech. Trans. Asme. **50**, 921–934 (1983)

L. Bardella, A deformation theory of strain gradient crystal plasticity that accounts for geometrically necessary dislocations. J. Mech. Phys. Solids **54**, 128–160 (2006)

L. Bardella, Some remarks on the strain gradient crystal plasticity, with particular reference to the material length scales involved. Int. J. Plast. **23**, 296–322 (2007)

E. Bayart, I. Svetlizky, J. Fineberg, Fracture mechanics determine the lengths of interface ruptures that mediate frictional motion. Nat. Phys. **12**, 166–170 (2016)

V.L. Berdichevsky, On thermodynamics of crystal plasticity. Scr. Mater. **54**, 711–716 (2006)

E. Bittencourt, A. Needleman, M.E. Gurtin, E. Van der Giessen, A comparison of nonlocal continuum and discrete dislocation plasticity predictions. J. Mech. Phys. Solids **51**, 281–310 (2003)

R.J. Clifton, J. Duffy, K.A. Hartley, T.G. Shawki, On critical conditions for shear band formation at high strain rates. Scr. Etall. **18**, 443–448 (1984)

N.A. Fleck, J.R. Willis, A mathematical basis for strain gradient plasticity theory. Part I: scalar plastic multiplier. J. Mech. Phys. Solids (2008). https://doi.org/10.1016/j.jmps.2008.09.010

P. Fredriksson, P. Gudmundson, Competition between interface and bulk dominated plastic deformation in strain gradient plasticity. Model. Simul. Mat. Sci. Eng. **15**, S61–S69 (2007)

M.E. Gurtin, On a framework for small-deformation viscoplasticity: free energy, microforces, strain gradient. Int. J. Plast. **19**, 47–90 (2003)

M.E. Gurtin, A theory of grain boundaries that accounts automatically for grain misorientation and grain-boundary orientation. J. Mech. Phys. Solids **56**, 640–662 (2008)

S. Hernandez, J. Hardell, H. Winkelmann, M.R. Ripoll, B. Prakash, Influence of temperature on abrasive wear of boron steel and hot forming tool steels. Wear **338–339**, 27–35 (2015)

P. Ireman, Q.S. Nguyen, Using the gradients of temperature and internal parameters in continuum thermodynamics. C. R. Mecanique **332**, 249–255 (2004)

P. Ireman, A. Klarbring, N. Stromberg, A model of damage coupled to wear. Int. J. Solids Struct. **40**, 2957–2974 (2003)

L. Johansson, A. Klarbring, Thermoelastic frictional contact problems: modelling, FE-approximation and numerical realization. Comput. Methods Appl. Mech. Eng. **105**, 181–210 (1993)

L. Johansson, A. Klarbring, Study of frictional impact using a nonsmooth equations solver. J. Appl. Mech. Trans. ASME **67**, 267–273 (2000)

R.L. Johnson, M.A. Swikert, E.E. Bisson, Friction at high sliding velocities; naca-tn-1442 (1947)

A. Klarbring, A mathematical-programming approach to 3-dimensional contact problems with friction. Comput. Methods Appl. Mech. Eng. **58**, 175–200 (1986)

A. Klarbring, Examples of nonuniqueness and nonexistence of solutions to quasi-static contact problems with friction. Ingenieur Arch. **60**, 529–541 (1990)

A. Lodygowski, G.Z. Voyiadjis, B. Deliktas, A. Palazotto, Non-local and numerical formulations for dry sliding friction and wear at high velocities. Int. J. Plast. **27**, 1004–1024 (2011)

A. Molinari, G. Ravichandran, Constitutive modeling of high-strain-rate deformation in metals based on the evolution of an effective microstructural length. Mech. Mater. **37**, 737–752 (2005)

H. Mughrabi, On the current understanding of strain gradient plasticity. Mat. Sci. Eng. A Struct. Mat. Prop. Microstruct. Process. **387–89**, 209–213 (2004)

O. Nijs, B. Holmedal, J. Friis, E. Nes, Sub-structure strengthening and work hardening of an ultra-fine grained aluminum-magnesium alloy. Mat. Sci. Eng. A Struct. Mat. Prop. Microstruct. Process. **483**, 51–53 (2008)

J.F. Nye, Some geometrical relations in dislocated crystals. Acta Metall. **1**, 153–162 (1953)

M. Varga, M. Rojacz, H. Winkelman, H. Mayer, E. Badisch, Wear reducing effects and temperature dependence of tribolayers formation in harsh environment. Tribol. Int. **65**, 190–199 (2013)

G.Z. Voyiadjis, R.K. Abu Al-Rub, Nonlocal gradient-dependent thermodynamics for modeling scale-dependent plasticity. Int. J. Multiscale Comput. Eng. **5**, 295–323 (2007)

G.Z. Voyiadjis, B. Deliktas, Formulation of strain gradient plasticity with interface energy in a consistent thermodynamic framework. Int. J. Plast. **25**(10), 1997–2024 (2009a)

G.Z. Voyiadjis, B. Deliktas, Mechanics of strain gradient plasticity with particular reference to decomposition of the state variables into energetic and dissipative components. Int. J. Eng. Sci. **47**(11–12), 1405–1423 (2009b)
G.Z. Voyiadjis, B. Deliktas, D. Faghihi, A. Lodygowski, Friction coefficient evaluation using physically based viscoplasticity model at the contact region during high velocity sliding. Acta Mech. **213**, 39–52 (2010)

Strain Gradient Plasticity: Deformation Patterning, Localization, and Fracture

27

Giovanni Lancioni and Tuncay Yalçinkaya

Contents

Abstract

In this chapter, two different strain gradient plasticity models based on non-convex plastic energies are presented and compared through analytical estimates

G. Lancioni (✉)
Dipartimento di Ingegneria Civile, Edile e Architettura, Università Politecnica delle Marche, Ancona, Italy
e-mail: g.lancioni@univpm.it

T. Yalçinkaya
Aerospace Engineering Program, Middle East Technical University Northern Cyprus Campus, Guzelyurt, Mersin, Turkey

Department of Aerospace Engineering, Middle East Technical University, Ankara, Turkey
e-mail: yalcinka@metu.edu.tr

G. Z. Voyiadjis (ed.), *Handbook of Nonlocal Continuum Mechanics for Materials and Structures*, https://doi.org/10.1007/978-3-319-58729-5_43

and numerical experiments. The models are formulated in the simple one-dimensional setting, and their ability to reproduce heterogeneous plastic strain processes is analyzed, focusing on strain localization phenomena observed in metallic materials at different length scales. In a geometrically linear context, both models are based on the additive decomposition of the strain into elastic and plastic parts. Moreover, they share the same non-convex plastic energy, and they are both characterized by the same nonlocal plastic energy as well, i.e., a quadratic form of the plastic strain gradient. In the first model, proposed in Yalçinkaya et al. (Int J Solids Struct 49:2625–2636, 2012) and Yalcinkaya (Microstructure evolution in crystal plasticity: strain path effects and dislocation slip patterning. Ph.D. thesis, Eindhoven University of Technology, 2011), the plastic energy is assumed to be conservative, and plastic dissipation is introduced through a viscous term, which makes the formulation rate-dependent. In the second model, developed in Del Piero et al. (J Mech Mater Struct 8(2–4):109–151, 2013), the plastic term is supposed to be totally dissipative. As a result, plastic deformations are not recoverable, and the resulting framework is rate-independent, contrary to the first model. First, the evolution problems resulting from the two theories are analytically solved in a special simplified case, and correlations between the shape of the plastic potential and the modeling predictions are established. Then, the models are numerically implemented by finite elements, and numerical solutions of two different one-dimensional problems, associated with different plastic energies, are determined. In the first problem, a double-well plastic energy is considered, and the evolution of plastic slip patterning observed in crystals at the mesoscale is reproduced. In the second problem, a convex-concave plastic energy is used to simulate the macroscopic response of a tensile steel bar, which experiences the so-called necking process, with plastic strains localization and final coalescing into fracture. Numerical results provided by the two models are analyzed and compared.

Keywords
Strain gradient plasticity · Size effect · Localization · Deformation patterning · Damage · Fracture

Introduction

In recent years, developments in constitutive modeling have illustrated the ability of strain gradient plasticity theories to capture different strain localization processes by incorporating non-convex plastic energies into the model in a thermodynamically consistent way (see, e.g., Yalçinkaya et al. 2011, 2012; Klusemann and Yalçinkaya 2013; Klusemann et al. 2013 for the modeling of intragranular microstructure evolution resulting in plastic anisotropy and Del Piero et al. 2013; Lancioni 2015 for strain localization leading to failure). Since processes like strain localization and microstructures evolution are stress-softening processes, the corresponding evolution boundary value problems undergo convergence troubles due to loss of ellipticity. In order to remedy these problems, several methods have been proposed including variational regularization methods, nonlocal methods, viscous regularization techniques, and Cosserat theories. However, nowadays, the modeling of softening mechanisms is still an open issue, and it is subject to intense research activities. In order to contribute to this, in this chapter two strain gradient plasticity models incorporating the phase-field idea for the evolution of plastic slip are proposed and analyzed. The numerical comparison study of rate-dependent and rate-independent models has been addressed recently in Yalçinkaya and Lancioni (2014) and Lancioni et al. (2015a). In the current chapter, we focus on the similarities and differences of both models in a variational format and present analytical and numerical solutions for that purpose. Models are formulated in the simple one-dimensional setting, which preserves simplicity and allows to focus on the key physical and mathematical aspects, avoiding the complicated technicalities of multidimensional frameworks, and the hypothesis of linear elastoplasticity is assumed, which allows to decompose the total strain into the sum of elastic and plastic contributions. Extensions to multidimensional setting, which would result in (strain gradient) crystal plasticity models (see, e.g., Yalçinkaya et al. 2012; Lancioni et al. 2015b) and/or finite elasticity frameworks, are not included.

The first model, proposed in Yalçinkaya et al. (2011, 2012), is deduced by following a classical thermodynamical approach (see Gurtin et al. 2010). The principle of virtual work is used to get balance equations at macro and micro level, while the dissipation inequality is exploited to define micro-stresses. In order to satisfy the dissipation inequality, a viscous plastic stress is introduced, which makes the model rate-dependent (RD). The free energy is assumed to be sum of a non-convex plastic term (which does not exist in Gurtin et al. 2010) and two quadratic terms with respect to the elastic deformation and the plastic deformation gradient. The inclusion of a non-convex plastic term in free energy that drives the localization and patterning is arguable, yet it is physical for the illustration of deformation patterning under monotonic loading, which is the main scope of the current work.

In the second model, developed in Del Piero et al. (2013), discussed in Del Piero (2013), and extended in Lancioni (2015), the plastic evolution is determined by incremental minimization of an energy functional, which is composed by three contributions equal to those considered in the abovementioned free energy of the RD model. However, in this case, the plastic term is supposed to be totally dissipative, and, therefore, plastic deformations are not recoverable. The resulting framework is rate-independent (RD), contrary to the previous model.

The different dissipative properties assigned to the plastic energy in two formulations reflect on different descriptions of the plastic strain evolution. In the RD model, plastic strains are partially recoverable, because of the plastic free energy term, and partially dissipate, since the viscous stress contributes to dissipate plastic power. On the contrary, in the RI model, plastic strains are totally irreversible, being the corresponding energy dissipative. In both the two models, stability issues related to stress-softening regimes are solved by incorporation of the plastic strain gradient term into the expression of the free energy. This nonlocal term not only allows to regularize the evolution problem but also plays the fundamental role of strain localization limiter against abrupt fractures (see Bazant and Jirásek 2002). Moreover, it also gives the opportunity to work at different length scales due to the internal length scale parameter included intrinsically in its formulations.

The aim of this study is to analyze the influence of the form of the plastic energy function on the description of heterogeneous plastic processes. In this respect, analytical solutions are found in the simplified case of evolution from homogeneous strain configurations, providing precise indications on the shape to give to the plastic energy in order to reproduce specific evolution processes. Then, finite element codes are developed by implementing the RD and RI models and used to solve numerically two different problems of strain localization occurring at different length scales, as described in the following:

(i) The first problem is the evolution of heterogeneous plastic shear strains in metallic materials. When metals are subject to considerable loadings (e.g., during forming processes), mechanisms of dislocation self-organization activate that lead to the formation of patterning observable at the mesoscale, where regions of high dislocation density (dislocation walls) envelop areas of low dislocation density (dislocation cell interiors), which can also be regarded as domains of high plastic slip and low plastic slip activity, as shown in the picture of Fig. 1a. To reproduce this patterning process, the plastic energy is assumed to be a Landau-Devonshire type of double-well functional, where the second well is shifted up, similar to the form used in Yalçinkaya et al. (2011). Using such a type of plastic potential results in a Ginzburg-Landau phase-field-like relation for the evolution of plastic slip, where the different phases are identified as regions with high plastic and low plastic strain.

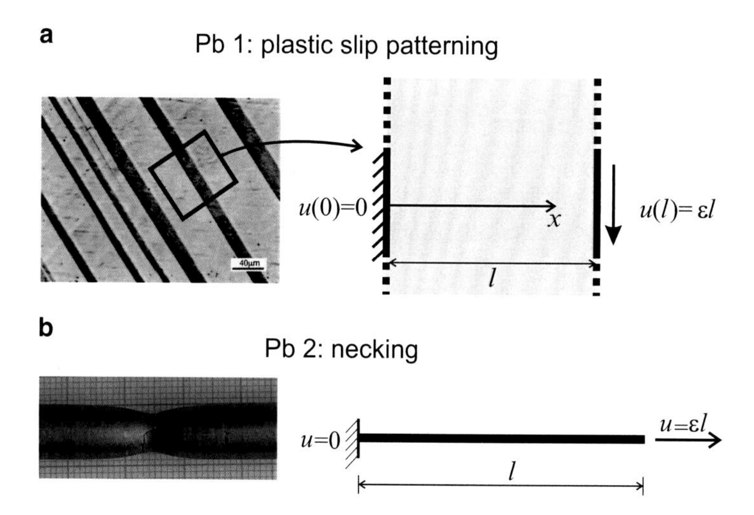

Fig. 1 Plastic slip patterning: (**a**) straight slip bands on a single crystal of 3.25% silicon iron (Hull 1963); (**b**) one-dimensional geometrical scheme of the shear problem. Necking: (**c**) neck in a tensile steel bar; (**d**) one-dimensional scheme of a tensile bar

(ii) The second problem refers to the process of strain localization that leads to final fracture, observed in tensile metals, the so-called necking process (see Fig. 1c). Necking occurs at a length scale which is almost at the macroscopic engineering level, thus completely different from that of the previous problem. For this problem, a convex-concave plastic functional has been chosen. Localization initiates and evolves when the amount of plastic deformation reaches values in the concave part of the energy, and it is accompanied by substantial stress-softening. Localization continues on smaller and smaller regions, until final fracture, which corresponds to the solution loss of stability. The response of steel bars to tensile loadings has been studied in Lancioni (2015), by using an enriched version of the RI model. In the present study, it is solved by the RD model as well, and comparisons are established.

The chapter is organized as follows. In section "Nonlocal Rate-Dependent and Rate-Independent Models," the two models are formulated within a common unified framework, which allows to easily point out analogies and differences. In section "Analytical Solutions," analytical solution is found, and in section "Numerical Results" numerical results are presented and commented. Concluding remarks are summarized in section "Conclusions."

Nonlocal Rate-Dependent and Rate-Independent Models

Problem Statement

Consider a one-dimensional domain $(0, l)$ of length l. The displacement of a point $x \in (0, l)$ at the time instant t is denoted by $u = u_t(x)$. Here and in the following, the dependence on time is indicated by a subscript, and the following notation for derivatives is used: for any function $v = v_t(w)$, which depends on a certain space-dependent variable $w = w(x)$ and on time t, a prime indicates derivative with respect to w, $v' = dv/dw$, and a dot means time derivative, $\dot{v} = dv/dt$.

The boundary conditions assigned at the domain endpoints are

$$u_t(0) = 0, \quad u_t(l) = l\varepsilon_t, \tag{1}$$

where ε_t is an imposed deformation, function of time. We suppose that the deformation is decomposed additively into an elastic part ε^e and a plastic part ε^p, i.e.,

$$u' = \varepsilon^e + \varepsilon^p. \tag{2}$$

Since positive deformations are applied ($\varepsilon_t \geq 0$), the plastic deformation coincides with the cumulative plastic strain, here denoted by γ. Thus we can write

$$u' = \varepsilon^e + \gamma, \tag{3}$$

In addition to (1), two possible boundary conditions on γ can be assigned, which are

$$\gamma_t(0) = \gamma_t(l) = 0 \ \ (\text{hard b.c.}), \ \ \text{or} \ \ \gamma_t'(0) = \gamma_t'(l) = 0 \ \ (\text{soft b.c.}). \tag{4}$$

As shown in the numerical simulations, hard boundary conditions force the plastic strain γ to evolve in the central part of the body, and a boundary layer develops at the both ends of the bar. On the other hand, soft boundary conditions lead to evolution of a homogeneous γ field. An interesting discussion is proposed in Jirásek and Rolshoven (2009) on possible assignment of boundary conditions for the plastic strain in gradient plasticity models.

Model Equations: Dissipation, Equilibrium, and Evolution

Both theories are formulated within a common unified framework, which allows an easy comparison from a theoretical viewpoint. In here, the general modeling ingredients are introduced that are needed to deduce balance and evolution equations in a thermodynamically consistent way. In the next sections "Rate-Dependent Model" and "Rate-Independent Model," these relations are specialized for the models, according the specific constitutive assumptions of each theory.

Given the triplet (u, γ, γ') of independent internal variable, the material is assumed to be endowed with the free energy

$$\psi(\varepsilon^e, \gamma, \gamma') = \psi_e(\varepsilon^e) + \psi_\gamma(\gamma) + \psi_{\gamma'}(\gamma'), \tag{5}$$

which is sum of elastic, plastic, and stored contributions, analogously to the free energy considered in Gurtin and Anand (2009). Here, we assume that the elastic and the nonlocal terms are quadratic functions

$$\psi_e(\varepsilon^e) = \frac{1}{2}E\varepsilon^{e2}, \qquad \psi_{\gamma'}(\gamma') = \frac{1}{2}A\gamma'^2, \tag{6}$$

but, in general, different expressions can be considered. The plastic potential ψ_γ is a monotonic increasing function of γ, which can assume any expression. The macroscopic stress and the microscopic stress and hyperstress power-conjugated to $\dot{\varepsilon}^e$, $\dot{\gamma}$ and $\dot{\gamma}'$, respectively, are defined as

$$\sigma = \psi_e'(\varepsilon^e) = E\varepsilon^e, \qquad \pi = \psi_\gamma'(\gamma), \qquad \xi = \psi_{\gamma'}'(\gamma') = A\gamma'. \tag{7}$$

These stresses are energetic. The next stage concerns the definition of the local dissipated power.

Dissipation. If we suppose that dissipation is only due to plastic strains, the dissipated power has the form

$$D(\gamma, \dot{\gamma}) = \sigma^d(\gamma, \dot{\gamma})\dot{\gamma}, \tag{8}$$

where $\sigma^d = \sigma^d(\gamma, \dot{\gamma})$ is a dissipated micro-stress power-conjugated to $\dot{\gamma}$. The dependence of σ^d on the rate $\dot{\gamma}$ allows to introduce viscous plastic micro-stresses within the model, as done in the RD model of section "Rate-Dependent Model." Since the dissipation inequality for isothermal processes reads

$$D = P - \dot{\psi} \geq 0, \tag{9}$$

with P the local internal power, σ^d must be such that the dissipation D in (8) is non-negative. A further way to introduce dissipative behavior is through a dissipation potential, according to the theory of generalized standard materials (Mielke 2006). If $\theta(\gamma)$ is a dissipative plastic potential, the dissipated micro-stress is $\theta'(\gamma)$, and the dissipation inequality (9) becomes

$$D(\gamma, \dot{\gamma}) = \theta'(\gamma)\dot{\gamma} \geq 0. \tag{10}$$

Thus, summing up, the total micro-stress conjugated to γ is

$$\sigma_\gamma = \pi + \sigma^d + \theta', \tag{11}$$

where the first term is energetic, and the second and third terms are dissipated. If we assume that volume forces are null, the body total energy is

$$E(u, \gamma) = \int_0^l \left(\psi\left(u' - \gamma, \gamma, \gamma'\right) + \theta(\gamma)\right) dx, \tag{12}$$

which is used in the following to deduce the equilibrium and the evolution equations in a variational format.

Equilibrium. A configuration (u, γ) is equilibrated if

$$\delta E(u, \gamma; \delta u, \delta\gamma) + W_{NC}(\gamma, \dot{\gamma}; \delta\gamma) \geq 0, \tag{13}$$

for any admissible perturbation $(\delta u, \delta\gamma)$, where δE is the first variation of E, and W_{NC} is the infinitesimal virtual work of σ^d. Using (7), the two terms in the above inequality are

$$\begin{aligned} \delta E &= \int_0^l \left(\sigma\delta u' + (\pi + \theta' - \sigma)\delta\gamma + \xi\delta\gamma'\right) dx, \\ W_{NC} &= \int_0^l \sigma^d \delta\gamma \, dx. \end{aligned} \tag{14}$$

When the unknown pair (u, γ) is not constrained through inequality conditions, the equilibrium condition (13) reduces to an equality. Indeed, in the case of unconstrained (u, γ), for any admissible perturbation $(\delta u, \delta\gamma)$, also the opposite perturbation $(-\delta u, -\delta\gamma)$ is admissible, and thus (13) must be satisfied as an equality, in order to be satisfied for any pair of opposite perturbations. The resulting equation is the equation of virtual power (see Gurtin et al. 2010). This situation is found in the RD model. Differently, in the RI model, an inequality constrain on γ leads to an equilibrium inequality.

Evolution. The evolution problem is solved by following a variational incremental procedure, which allows to determine the solution $(u_{t+\tau}, \gamma_{t+\tau})$ at the time instant $t+\tau$, with τ a given time increment, when the solution (u_t, γ_t) at the previous instant t is known.

The solution at $t+\tau$ is approximated by the first-order Taylor expansion

$$u_{t+\tau} = u_t + \tau\dot{u}_t, \qquad \gamma_{t+\tau} = \gamma_t + \tau\dot{\gamma}_t, \tag{15}$$

where the velocity pair $(\dot{u}_t, \dot{\gamma}_t)$ represents the unknown to be determined. The total energy (12) is approximated by the second-order development

$$E_{t+\tau}(\dot{u}_t, \dot{\gamma}_t) = E_t + \tau\dot{E}_t(\dot{u}_t, \dot{\gamma}_t) + \frac{1}{2}\tau^2\ddot{E}_t(\dot{u}_t, \dot{\gamma}_t), \tag{16}$$

and

$$\Delta E_t(\dot{u}_t, \dot{\gamma}_t) = E_{t+\tau} - E_t = \tau\dot{E}_t(\dot{u}_t, \dot{\gamma}_t) + \frac{1}{2}\tau^2\ddot{E}_t(\dot{u}_t, \dot{\gamma}_t) \tag{17}$$

is the approximated energy increment within a time step $t \to t+\tau$. The pair $(\dot{u}_t, \dot{\gamma}_t)$ solves the variational inequality

$$\delta\Delta E_t(\dot{u}_t, \dot{\gamma}_t; \delta\dot{u}, \delta\dot{\gamma}) + \Delta W_{NC}(\dot{\gamma}_t; \delta\dot{\gamma}) \geq 0, \tag{18}$$

for any admissible perturbations pair $(\delta\dot{u}, \delta\dot{\gamma})$, where $\Delta W_{NC}(\dot{\gamma}_t; \delta\dot{\gamma})$ is the virtual power expended by σ^d within the time interval $(t, t+\tau)$, i.e.,

$$\Delta W_{NC} = \int_0^l (\sigma^d_{t+\tau} - \sigma^d_t)\delta\dot{\gamma}\, dx. \tag{19}$$

As above commented for the equilibrium relation (13) and (18) reduces to an equality, when the unknowns $(\dot{u}_t, \dot{\gamma}_t)$ are not constrained by inequality conditions.

The formulation proposed in this section allows to obtain equilibrium and evolution equations, provided that certain constitutive assumptions are made. In particular, the quantities to be assigned are the plastic free energy $\psi_\gamma(\gamma)$, the dissipative micro-stress σ^d, and the dissipation potential $\theta(\gamma)$, which must satisfy the dissipation inequality (10). We recall that the elastic and the nonlocal free energies are assigned in (6). In the next sections “Rate-Dependent Model” and “Rate-Independent Model,” specific constitutive assumptions are made, and both models are formulated.

Rate-Dependent Model

In this section, we formulate the RD model, which was first proposed in Yalçinkaya et al. (2011) by using the principle of virtual power.
Constitutive assumptions. The basic constitutive assumptions are:

1. The plastic free energy $\psi_\gamma(\gamma)$ satisfies the conditions $\psi_\gamma(0) = 0$, $\psi_\gamma'(0) \geq 0$, and $\psi_\gamma'(\gamma) > 0$ for any $\gamma > 0$. Two different non-convex expressions of $\psi_\gamma(\gamma)$ will be given for the two problems numerically solved in sections “Plastic Slip Patterning” and “Necking in Tensile Steel Bars.”
2. A viscous dissipative micro-stress of the form

$$\sigma^d = c\dot{\gamma} \tag{20}$$

   is assumed, where c is a viscous coefficient. Notice that, with this stress, the dissipation inequality (10) is automatically satisfied. The reader is referred to Yalçinkaya et al. (2011) for a discussion on more complex expressions of σ^d.
3. The dissipative potential $\theta(\gamma)$ is neglected.

With these assumptions, the micro-stress power-conjugated to $\dot{\gamma}$ is

$$\sigma_\gamma = \psi_\gamma' + c\dot{\gamma}, \tag{21}$$

where the first term is energetic, and the second is dissipative. The viscous contribution constitutes the source of rate dependency of the model.
Equilibrium. Under these assumptions, the equilibrium condition (13) rewrites

$$\int_0^l \left(\sigma(\delta u' - \delta\gamma) + \psi_\gamma'\delta\gamma + \xi\delta\gamma' + c\dot{\gamma}\delta\gamma\right) dx = 0, \tag{22}$$

for any admissible perturbation pair $(\delta u, \delta\gamma)$, from which the macroscopic and microscopic force balance equations are obtained by assuming $\delta\gamma = 0$ and $\delta u' = 0$ (i.e., $\delta\gamma = -\delta\varepsilon^e$), respectively. They are

$$\sigma' = 0, \qquad \sigma - \psi_\gamma' + \xi' - c\dot{\gamma} = 0. \tag{23}$$

Remark 1. If we assume $\psi_\gamma'(0) > 0$, solution of the above equilibrium equations at the initial instant $t = 0$ is

$$\varepsilon_0^e = \frac{\psi_\gamma'(0)}{E}, \quad \gamma_0 = 0. \tag{24}$$

This initial strain corresponds to the prestress $\sigma_0 = \psi_\gamma'(0)$ associated with the pre-imposed deformation $\varepsilon_0 = \psi_\gamma'(0)/E$. From such an initial state, the elastoplastic deformation evolves according to the evolution equation

$$E\varepsilon^e - \left(\psi_\gamma' - \psi_\gamma'(0)\right) + A\gamma'' - c\dot{\gamma} = 0, \tag{25}$$

which differs from $(23)_2$ for the second term within the brackets, which is null at the beginning of the evolution process. The constitutive assumption $\psi_\gamma'(0) > 0$ is made to reproduce initial purely elastic behaviors.

Evolution. The evolution equation is obtained from (18), with

$$\Delta W_{NC} = \int_0^l (c\dot{\gamma} - \sigma_t^d)\delta\dot{\gamma}\, dx. \tag{26}$$

The macroscopic and microscopic evolution equations are determined by assuming $\delta\dot{\gamma} = 0$ and $\delta\dot{u}' = 0$, respectively, and they are

$$(\dot{u}_t' - \dot{\gamma}_t)' = 0, \quad E(\dot{u}_t' - \dot{\gamma}_t) - \left(\psi_\gamma''(\gamma_t) + \frac{c}{\tau}\right)\dot{\gamma}_t + A\dot{\gamma}_t'' = -\frac{\sigma_t^d}{\tau}, \tag{27}$$

with the boundary conditions

$$\begin{aligned} &\dot{u}_t(0) = 0, \qquad \dot{u}_t(l) = l\dot{\varepsilon}_t, \\ &\dot{\gamma}_t(0) = \dot{\gamma}_t(l) = 0 \ \ (\text{hard b.c.}), \ \ \text{or} \ \ \dot{\gamma}_t'(0) = \dot{\gamma}_t'(l) = 0 \ \ (\text{soft b.c.}). \end{aligned} \tag{28}$$

Equations (27) are obtained in Yalçinkaya et al. (2011) by linearizing equations (23). Equation $(27)_1$ states that $(\dot{u}_t' - \dot{\gamma}_t)$ is constant, and thus it can be written in the form

$$\dot{u}_t' - \dot{\gamma}_t = \frac{1}{l}\int_0^l (\dot{u}_t' - \dot{\gamma}_t)\, dx = \dot{\varepsilon}_t - \bar{\dot{\gamma}}_t, \tag{29}$$

where

$$\bar{\dot{\gamma}} = \frac{1}{l}\int_0^l \dot{\gamma}_t\, dx \tag{30}$$

is the mean value of $\dot{\gamma}_t$ in $(0, l)$. Substituting (29) in $(27)_2$, we get the following set of equations, alternative to (27),

$$A\dot{\gamma}_t'' - \left(\psi_\gamma''(\gamma_t) + \frac{c}{\tau}\right)\dot{\gamma}_t - E\dot{\bar{\gamma}}_t = -E\dot{\varepsilon}_t - \frac{\sigma_t^d}{\tau},$$
$$\dot{u}_t' = \dot{\varepsilon}_t + \dot{\gamma}_t - \dot{\bar{\gamma}}_t, \tag{31}$$

where the first equation depends only on $\dot{\gamma}_t$. Equations (31) are equivalent to (27), and they are solved by determining $\dot{\gamma}_t$ from the first equation and, then, $\dot{u}_t$ from the second equation.

Rate-Independent Model

In this section, we deduce the governing equations of the RI model first proposed in Del Piero et al. (2013).
Constitutive assumptions. We make the following constitutive assumptions:

1. The plastic free energy $\psi_\gamma(\gamma)$ is neglected.
2. The dissipative micro-stress σ^d is neglected.
3. A dissipative potential $\theta(\gamma)$ is considered, which is assumed to be equal to the plastic free energy of the RD model, defined at point (1) of section "Rate-Dependent Model." Thus the dissipative potential must satisfy the conditions $\theta(0) = 0$, $\theta'(0) \geq 0$, and $\theta'(\gamma) > 0$, for any $\gamma > 0$.

With these hypotheses, the plastic micro-stress is

$$\sigma_\gamma = \theta'(\gamma), \tag{32}$$

which is totally dissipative, in agreement with Aifantis theory (Aifantis 1984) and the reformulations proposed in Gudmundson (2004) and Gurtin and Anand (2009) in a thermodynamically consistent format, where the local term of the flow rule is dissipative and the nonlocal term is energetic.

Since $\theta'(\gamma) \geq 0$ for any γ, according to the assumptions given in the above point 3, the dissipation inequality (10) is satisfied if

$$\dot{\gamma} \geq 0, \tag{33}$$

which represents a constraint for the plastic strain and imposes that plastic strain can never decrease in an evolution process.
Equilibrium. From (33) an admissible perturbation is such that $\delta\gamma \geq 0$. Since $W_{NC} = 0$, inequality (13) rewrites

$$\int_0^l \left(\sigma(\delta u' - \delta\gamma) + \xi\delta\gamma' + \theta'\delta\gamma\right) dx \geq 0, \tag{34}$$

for any admissible plastic strain perturbation such that $\delta\gamma \geq 0$. Assuming $\delta\gamma = 0$, we obtain the balance equation $(23)_1$, while, assuming $\delta u' = 0$, we found the plastic yield condition

$$\sigma \leq \theta' - \xi', \tag{35}$$

which states that the stress σ cannot be greater than the yield limit $\theta' - \xi'$.
Evolution. Evolution is governed by inequality (18), where $\Delta W_{NC} = 0$. Assuming $\delta\dot{\gamma} = 0$, we get the macroscopic evolution equation $(27)_1$, and, setting $\delta\dot{u}' = 0$, we obtain

$$\delta\Delta E_t(\dot{u}_t, \dot{\gamma}_t, 0, \delta\dot{\gamma}) = \int_0^l \left(f_t + \tau\dot{f}_t\right)\delta\dot{\gamma}\,dx + [(\xi_t + \tau\dot{\xi}_t)\delta\dot{\gamma}]_0^l \geq 0, \tag{36}$$

where

$$f_t = \theta_t' - \sigma_t - \xi_t', \tag{37}$$

and $\delta\dot{\gamma}$ is an arbitrary perturbation that satisfies the condition

$$\dot{\gamma}_t + \delta\dot{\gamma} \geq 0, \tag{38}$$

imposed by the irreversibility relation (33). The boundary terms are null, if the boundary conditions (28) are assigned. For non-negativeness of the integral term, the two possible cases $\dot{\gamma}_t = 0$ and $\dot{\gamma}_t > 0$ are separately considered. (i) if $\dot{\gamma}_t = 0$, then $\delta\dot{\gamma} \geq 0$ for (38), and the integral term in (36) is nonnegative if $f_t + \tau\dot{f}_t \geq 0$; (ii) if $\dot{\gamma}_t > 0$, $\delta\dot{\gamma}$ can have any sign, and the integral term in (36) is nonnegative if $f_t + \tau\dot{f}_t = 0$. Thus, summing up, evolution is governed by the Kuhn-Tucker conditions

$$\dot{\gamma}_t \geq 0, \quad f_t + \tau\dot{f}_t \geq 0, \quad (f_t + \tau\dot{f}_t)\dot{\gamma}_t = 0, \tag{39}$$

which represent the flow rule of plasticity. It states that the stress maintains equal to the yield stress when γ evolves.

Equations $(27)_1$, (39) and the boundary conditions (28) are necessary conditions for a minimum of the total energy $E_{t+\tau}$, i.e., for $(\dot{u}_t, \dot{\gamma}_t)$ to be solution of the following constrained quadratic programming problem

$$(\dot{u}_t, \dot{\gamma}_t) = \text{argmin}\{\Delta E_t(\dot{u}_t, \dot{\gamma}_t),\ \dot{\gamma}_t \geq 0,\ \text{and b.c. on } \dot{u}_t \text{ and } \dot{\gamma}_t\}. \tag{40}$$

In the numerical code, problem (40) is solved by implementing a constrained quadratic programming algorithm.

Analytical Solutions

In this section, the evolution problems described in the previous sections are analytically solved in the simplified special case of evolution from homogeneous configurations. The analytical solutions found in Del Piero et al. (2013) and in Lancioni (2015) for the RI model and in Lancioni et al. (2015a) for the RD model are recalled and commented, since they provide useful criteria for the choice of the plastic energy functionals, which are the free energy ψ_γ in the RD model and the dissipative potential θ in the RI model, whose shape is still unspecified.

At the instant t, we assume that γ_t is homogenous and that, for the RI model, the yield condition (35) is satisfied as an equality, i.e., $\sigma_t = \theta_t'$, since $\xi_t' = 0$ for constant γ. The boundary conditions are

$$\dot{u}_t(0) = 0, \quad \dot{u}_t(l) = l\dot{\varepsilon}_t, \quad \dot{\gamma}_t(0) = \dot{\gamma}_t(l) = 0, \tag{41}$$

where the most interesting case of hard boundary conditions for $\dot{\gamma}_t$ is considered.

RD Model

Under the above simplified assumptions, the evolution equation $(31)_1$ reduces to a differential equation with constant coefficients, which can be written in the form

$$A\dot{\gamma}_t'' - \left(\psi_\gamma'' + \frac{c}{\delta\varepsilon_t}\dot{\varepsilon}_t\right)\dot{\gamma}_t = -\left(\dot{\sigma}_t + \frac{\sigma^d}{\delta\varepsilon_t}\dot{\varepsilon}_t\right), \tag{42}$$

where $\psi_\gamma'' = \psi_\gamma''(\gamma_t)$ is constant. In order to make explicit the dependence on the deformation rate $\dot{\varepsilon}$, we have used the relation $\tau = \delta\varepsilon/\dot{\varepsilon}$, with $\delta\varepsilon$ the deformation increment in the time step τ. Equation (42) is solved in two steps. In the first step, the solution $\dot{\gamma}_t$ is found as a function of $\dot{\sigma}_t$, and, in the second step, $\dot{\sigma}_t$ is explicitly determined by using the relation

$$\dot{\sigma}_t = E(\dot{\varepsilon}_t - \bar{\dot{\gamma}}_t), \tag{43}$$

which is obtained by differentiating $(7)_1$ and using (29). Omitting subscript t for brevity, the solution of the boundary value problem (41) and (42) is

$$\dot{\gamma} = \frac{\dot{\sigma} + \frac{\sigma^d}{\delta\varepsilon}\dot{\varepsilon}}{\psi_\gamma'' + \frac{c\dot{\varepsilon}}{\delta\varepsilon}}\left(1 - \frac{\cosh k_{\dot{\varepsilon}}(l/2 - x)}{\cosh k_{\dot{\varepsilon}}l/2}\right),$$

$$\dot{\sigma} = E\frac{\psi_\gamma'' + \frac{c\dot{\varepsilon} - \sigma^d}{\delta\varepsilon}}{\psi_\gamma'' + \frac{c\dot{\varepsilon}}{\delta\varepsilon} + E\varphi_1(k_{\dot{\varepsilon}}l)}\dot{\varepsilon}, \text{ if } \psi_\gamma'' > -\frac{c\dot{\varepsilon}}{\delta\varepsilon},$$

$$\dot{\gamma}=\frac{\dot{\sigma}+\frac{\sigma^d}{\delta\varepsilon}\dot{\varepsilon}}{2A}x(l-x),\ \dot{\sigma}=E\frac{12A-\frac{\sigma^d l^2}{\delta\varepsilon}}{12A+El^2}\ \text{if}\ \psi_\gamma''=-\frac{c\dot{\varepsilon}}{\delta\varepsilon},$$

$$\dot{\gamma}=\frac{\dot{\sigma}+\frac{\sigma^d}{\delta\varepsilon}\dot{\varepsilon}}{\psi_\gamma''+\frac{c\dot{\varepsilon}}{\delta\varepsilon}}\left(1-\frac{\cos k_{\dot{\varepsilon}}(l/2-x)}{\cos k_{\dot{\varepsilon}}l/2}\right),$$

$$\dot{\sigma}=E\frac{\psi_\gamma''+\frac{c\dot{\varepsilon}-\sigma^d}{\delta\varepsilon}}{\psi_\gamma''+\frac{c\dot{\varepsilon}}{\delta\varepsilon}+E\varphi_2(k_{\dot{\varepsilon}}l)}\dot{\varepsilon},\ \text{if}\ \psi_\gamma''<-\frac{c\dot{\varepsilon}}{\delta\varepsilon}, \tag{44}$$

with

$$k_{\dot{\varepsilon}}=\sqrt{\frac{1}{A}\left|\psi_\gamma''+\frac{c\dot{\varepsilon}}{\delta\varepsilon}\right|}, \tag{45}$$

and

$$\varphi_1(y)=1-\frac{2}{y}\tanh\frac{y}{2},\quad \varphi_2(y)=1-\frac{2}{y}\tan\frac{y}{2}. \tag{46}$$

Functions φ_1 and φ_2 are graphed in Fig. 2.

Formulas (44) refer to three different types of solutions, associated with three different intervals of values of ψ_γ'', as shown by the graphs of Fig. 3a. Hyperbolic solutions characterize the semiaxis $\psi_\gamma'' > -\frac{c\dot{\varepsilon}}{\delta\varepsilon}$. In the semiaxis $\psi_\gamma'' < -\frac{c\dot{\varepsilon}}{\delta\varepsilon}$, two types of trigonometric solutions distinguish: for $-\frac{4\pi^2 A}{l^2}-\frac{c\dot{\varepsilon}}{\delta\varepsilon}\le\psi_\gamma''<-\frac{c\dot{\varepsilon}}{\delta\varepsilon}$, the plastic strain rate is positive in any points of the domain, while, for $\psi_\gamma''<-\frac{4\pi^2 A}{l^2}-\frac{c\dot{\varepsilon}}{\delta\varepsilon}$, the plastic strain rate is positive in a central part of the domain and negative in portions close to the endpoints. This latter situation describes a process of strain localization in the center of the domain and plastic strain recovery on the sides. Plastic strain unloading is possible since the plastic energy is assumed to be stored.

Notice that the sign of $\dot{\sigma}$ depends on the deformation rate $\dot{\varepsilon}$. If we suppose that σ^d is so small that it can be neglected, then $\dot{\sigma}$ is positive for $\psi_\gamma''>-(\pi^2 A/l^2-c\dot{\varepsilon}/\delta\varepsilon)$, and it is negative for $\psi_\gamma''<-(\pi^2 A/l^2-c\dot{\varepsilon}/\delta\varepsilon)$. In the former case, the evolution regime is stress-hardening, while, in the latter case, it is stress-softening.

RI Model

Since we have assumed that $\sigma_t=\psi_\gamma'$, then $f_t=0$, and the evolution relations (39) simplify as follows:

$$\dot{\gamma}_t\ge 0,\quad \dot{\sigma}_t\le\theta_t''\dot{\gamma}_t-\alpha\dot{\gamma}_t'',\quad (\theta_t''\dot{\gamma}_t-\dot{\sigma}_t-\alpha\dot{\gamma}_t'')\dot{\gamma}_t=0. \tag{47}$$

We introduce the coefficient

$$k_0 = \sqrt{\frac{|\theta''|}{\alpha}}, \tag{48}$$

which is equal to the coefficient (45), with $\dot{\varepsilon} = 0$ and $\psi_\gamma(\gamma) = \theta(\gamma)$. First, we suppose $\theta'' \geq 0$. Integrating $(47)_3$ over $(0, l)$, we obtain

$$\dot{\sigma} \int_0^l \dot{\gamma}\, dx = \int_0^l \left(\theta'' \dot{\gamma}^2 + A\dot{\gamma}'^2\right) dx > 0, \tag{49}$$

from which $\dot{\sigma} > 0$. If we suppose $\dot{\gamma} = 0$ in some intervals of $(0, l)$, from $(47)_2$, $\dot{\sigma} \leq 0$, in contradiction with the above result. Thus $\dot{\gamma}$ must be positive almost everywhere in $(0, l)$, and the Khun-Tucker conditions (47) are satisfied if

$$\dot{\gamma}'' - k_0^2 \dot{\gamma} = -\frac{\dot{\sigma}}{A}, \quad \text{for any } x \in (0, l), \tag{50}$$

where index t is omitted. Solutions of (50) satisfying the boundary conditions (41) are found by following the procedure of section "RD Model": first the solution $\dot{\gamma}_t$ is found in terms of $\dot{\sigma}_t$, and, then, $\dot{\sigma}_t$ is explicitly determined from (43). We obtain

$$\dot{\gamma} = \frac{\dot{\sigma}}{\theta''}\left(1 - \frac{\cosh k_0\left(\frac{l}{2} - x\right)}{\cosh k_0 \frac{l}{2}}\right), \quad \dot{\sigma} = E\frac{\theta''}{\theta'' + E\varphi_1(k_0 l)}\dot{\varepsilon} > 0, \tag{51}$$

with φ_1 equal to $(46)_1$. Notice that $\dot{\gamma} > 0$ for any $x \in (0, l)$. Also $\dot{\sigma} > 0$, since numerator and denominator in $(51)_2$ are positive. It follows that, if $\theta'' > 0$, plastic strain evolves in the whole domain (*full-size* solution) under a regime of stress-hardening. If $\theta'' = 0$, Eq. (50) reduces to $A\dot{\gamma}'' = \dot{\sigma}$, whose solution is

$$\dot{\gamma} = \frac{\dot{\sigma}}{2A}x(l - x), \quad \dot{\sigma} = E\frac{12A}{12A + El^2}\dot{\varepsilon} > 0. \tag{52}$$

Even in this case, $\dot{\gamma}$ is full-size, and the evolution process is stress-hardening. For $\theta'' < 0$, Eq. $(47)_3$ becomes elliptic

$$\dot{\gamma}'' + k_0^2 \dot{\gamma} = -\frac{\dot{\sigma}}{A}, \tag{53}$$

and it admits the solution

$$\dot{\gamma} = \frac{\dot{\sigma}}{\theta''}\left(1 - \frac{\cos k_0(\frac{l}{2} - x)}{\cos k_0 \frac{l}{2}}\right), \quad \dot{\sigma} = E\frac{\theta''}{\theta'' + E\varphi_2(k_0 l)}\dot{\varepsilon}, \tag{54}$$

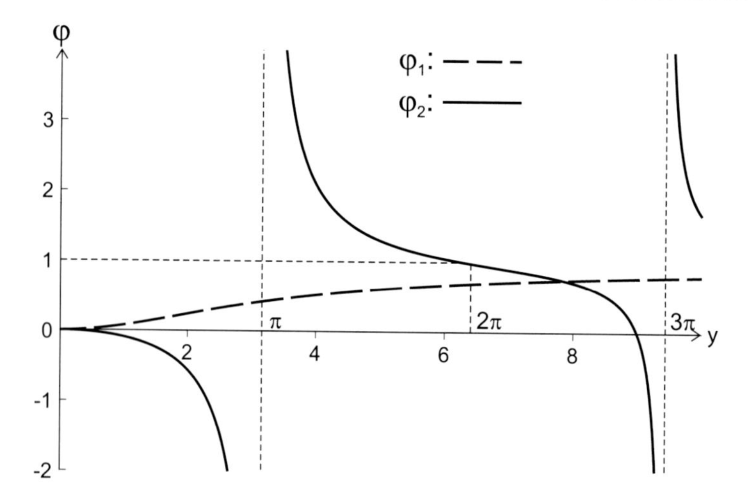

Fig. 2 Graphs of functions $\varphi_1 = \varphi_1(y)$ and $\varphi_2 = \varphi_2(y)$

with φ_2 as in $(46)_2$. Solution (54) is positive for any $x \in (0, l)$, if $-4\pi^2 A/l^2 \leq \theta'' < 0$. For $\theta'' < -4\pi^2 A/l^2$, the above solution is negative in portions of $(0, l)$ close to the endpoints, in contradiction with the irreversibility condition (33), and, thus, localized solutions must be looked for. Before determining localized solutions, we discuss the sign of $\dot{\sigma}$ in $(54)_2$. If $-\pi^2 A/l^2 < \theta'' < 0$ (i.e., $0 < k_0 l < \pi$), then $\dot{\sigma} > 0$. Indeed, being $\varphi_2 < 0$ (see Fig. 2), both numerator and denominator in $(54)_2$ are negative, and the evolution regime is stress-hardening. For $-4\pi^2 A/l^2 \leq \theta'' < -\pi^2 A/l^2$ ($\pi < k_0 l \leq 2\pi$), φ_2 is positive, and $\dot{\sigma} \leq 0$, i.e., stress-softening regime, if

$$\theta'' + E\varphi_2(k_0 l) > 0. \tag{55}$$

It was proven in Del Piero et al. (2013) that inequality (55) is a sufficient condition for stability of solution (54). Since $\varphi_2 \geq 1$, if $-4\pi^2 A/l^2 \leq \theta'' < -\pi^2 A/l^2$, the above inequality is satisfied if $E > 4\pi^2 A/l^2$, which is a condition usually met by real applications.

Now, we consider the case $\theta'' < -4\pi^2 A/l^2$, and we look for localized solutions. We assume that $\dot{\gamma}$ localizes in a portion $(0, \hat{l})$, adjacent to the bar left endpoint, and the supplementary continuity conditions $\dot{\gamma}(\hat{l}) = \dot{\gamma}'(\hat{l}) = 0$ are assigned. Equation (53) is solved within the localization zone $(0, \hat{l})$, and the inequality $\dot{\sigma} \leq 0$ is satisfied outside it, where $\dot{\gamma} = 0$. The resulting solution is

$$\begin{cases} \dot{\gamma} = \dfrac{\dot{\sigma}}{\theta''}\left(1 - \cos k_0 x\right), & \text{for } 0 < x \leq \hat{l}, \text{ with } \hat{l} = \dfrac{2\pi}{k_0}, \\ \dot{\gamma} = 0, & \text{for } \hat{l} < x < l, \end{cases} \tag{56}$$

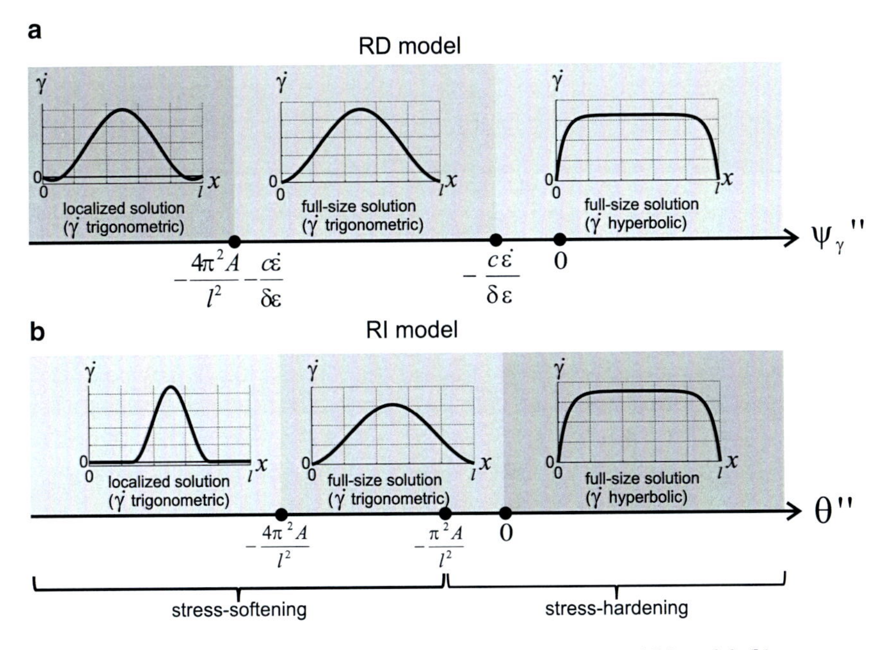

Fig. 3 Scheme of analytical solutions in the case of RD model (**a**), and RI model (**b**)

where the length $\hat{l}$ is determined from the supplementary condition $\dot{\gamma}'(\hat{l}) = 0$, and

$$\dot{\sigma} = E \frac{\theta''}{\theta'' + 2\pi \frac{E}{k_0 l}} \dot{\varepsilon}. \tag{57}$$

Notice that any translation of the localized solution (56) is also a solution. The stress rate $\dot{\sigma}$ is negative (stress-softening regime), if

$$\theta'' + 2\pi \frac{E}{k_0 l} > 0, \tag{58}$$

which, as proved in Del Piero et al. (2013), is a sufficient condition for stability of (56), analogously to the stability relation (55). When θ'' approaches $-2\pi E/(k_0 l)$ from above, then $\dot{\sigma} \to -\infty$, and $\dot{\gamma} \to \infty$ in a small region of the domain. The total energy associated with this configuration drops to minus infinity. This situation describes the occurrence of fracture. Detailed analyses of fractured configurations are proposed in Del Piero et al. (2013), Del Piero (2013), and Lancioni (2015).

Figure 3b schematizes the different solutions described above, which depend on the value of θ''.

Numerical Results

The RD and RI evolution problems, governed by Eqs. (28) and (31) and the constrained quadratic programming problem (40), respectively, are implemented numerically by finite elements. In each one-dimensional finite element, the plastic strain γ is approximated by a linear shape function, depending on two nodal unknowns defined at the element endpoints, while the displacement u is approximated by a quadratic shape function, where three nodal variables are defined at the element endpoints and midpoint. The solution of the evolution problems is refined at each time step by means of a Newton-Raphson iterative scheme in the RD formulation and by implementing a sequential quadratic programming algorithm in the RI model.

In the following, the numerical tests presented in Lancioni et al. (2015a) are discussed, where two distinct non-convex plastic energies are assigned in order to reproduce two different processes: plastic shear slip patterning in metals (section "Plastic Slip Patterning") and necking in tensile steel bars ("section "Necking in Tensile Steel Bars").

Plastic Slip Patterning

The problem of formation and evolution of microstructures in metallic materials is addressed in the simplified one-dimensional geometrical scheme proposed in Fig. 1b. A semi-infinite layer with thickness l is subjected to a shear deformation ε. The evolution of the plastic shear strain γ through the thickness is analyzed in both RD and RI frameworks, by first assuming the soft boundary conditions $(4)_1$ for γ and then the hard boundary conditions $(4)_2$. The layer thickness is $l = 1$ mm. The Young modulus E used in the above theoretical sections is replaced by the shear modulus $G = 78.947$ GPa, which corresponds to the Young modulus $E = 210$ GPa and Poisson's ratio $\nu = 0.33$ of steel. As in Yalçinkaya et al. (2011), we fix $c = 7$ MPa/s and $A = 147.29$ N. Coefficient A is related to the internal length scale by the formula $A = ER^2/(16(1-\nu^2))$ proposed in Bayley et al (2006), where the length scale R represents the radius of the dislocation domain contributing to the internal stress field. In this example $R = 0.1$ mm. A Landau-Devonshire type of potential $f_1 = f_1(\gamma)$ is assigned to the plastic energy densities ψ_γ (RD model) and θ (RI model), whose expression is

$$f_1(\gamma) = 1.525 \times 10^8\gamma^4 - 5.2 \times 10^6\gamma^3 + 5.25 \times 10^4\gamma^2 \text{ MPa}. \tag{59}$$

The graphs of f_1 and of its first derivative are represented in Fig. 4. Values of binodal and spinodal points are indicated in the graphs, as well the stress corresponding to the Maxwell line.

When soft boundary conditions are applied, the response of the layer to the shear deformation ε is described by the stress versus strain curves of Fig. 5. The RI model

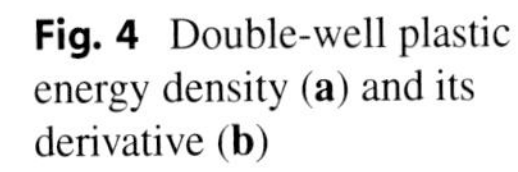

Fig. 4 Double-well plastic energy density (**a**) and its derivative (**b**)

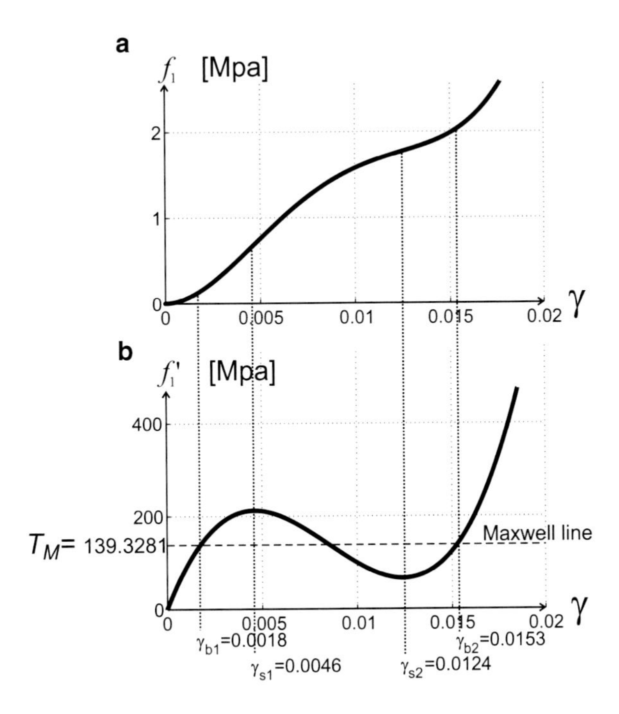

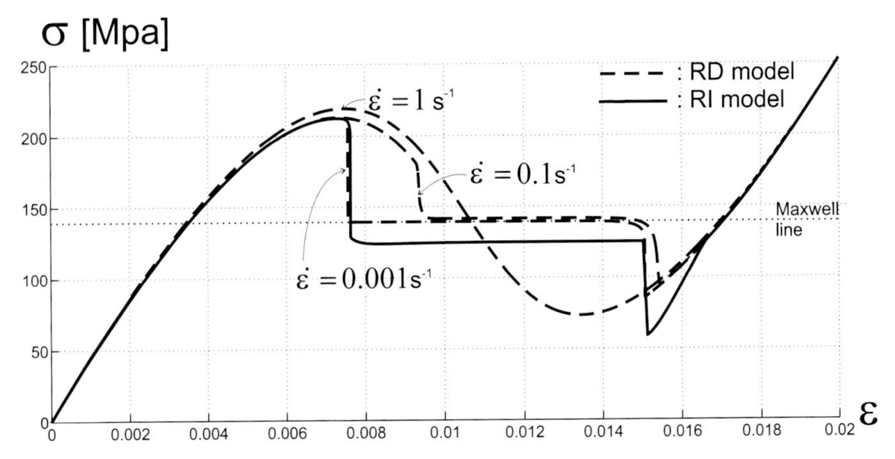

Fig. 5 Response curves in case of soft boundary conditions. Deformation rates $\dot{\varepsilon}$ = 0.001, 0.1, 1 s^{-1} are considered in the RD problem

provides the solid line curve, while the dashed line curves are obtained from the RD model, by assigning three different deformation rates, i.e., $\dot{\varepsilon} = 0.001, 0.1,$ and 1 s^{-1}. Notice that the response curves of the RD model get closer to the curve of the RI model, as $\dot{\varepsilon}$ decreases, because the effect of the viscous stress reduces. Three branches characterizes the response curves: (i) an initial hardening branch;

(ii) an intermediate softening phase, in the RI case and in RD cases with low rates, exhibiting sharp initial and final drops, and central plateaus; and (iii) a final hardening curve. These three steps of the curves are associated with different plastic strain regimes, as described in Fig. 6, which show different profiles of γ at different values of the imposed deformation.

In the initial hardening phase, the plastic strain grows homogeneously in the whole domain, which is common to all the four simulations. This phase ends when the plastic strain (plastic slip in this one-dimensional case) γ_{1s}, corresponding to the first spinodal point, is reached. From then on, different evolutions are observed. The RI simulation and the RD simulations with sufficiently low deformations rates ($\dot{\varepsilon} = 0.001, 0.1\ \mathrm{s}^{-1}$) exhibit strain localization on the right side of the domain, corresponding to the stress dropping down in the response curves. This stress drop occurs and is postponed as the deformation rate increases. The viscous stress contribution taken into account in the RD model produces a delay of the localization process. This can be clearly noticed by comparing the profiles of γ in Fig. 6b, c corresponding to the localization stage. Afterward, the plastic band moves from the right to the left. The plastic wave propagation corresponds to the stress plateau of the $\sigma - \varepsilon$ curves. The stress value of the plateau lowers as $\dot{\varepsilon}$ decreases, approaching to the Maxwell line. Differently, in the RI case, the plateau is below the Maxwell line.

The evolutions predicted by the two models in this intermediate phase of heterogeneous strain evolution exhibit a further crucial difference pointed out in the following. In the RD model, plastic strain localizes till the saturation value γ_{b2}, corresponding to the second binodal point, is reached in the right part of the body, while at the left side γ rapidly reduces to the value γ_{b1} of the first binodal point. Although the plastic wave front for the RI model has a similar width, the wave peak has a value slightly lower than γ_{b2}, and the strain downstream remains constant at the value γ_{s1} of the first spinodal point, the value at which localization is initiated. This is explained by the fact that the plastic energy is totally dissipated in the RI model, and thus γ can only grow. On the other hand, the energy ψ_γ of the RD model is stored, and therefore γ can be partially recovered, which justifies the reduction of γ outside the localization zone. Plastic energy recovery is also evident at the end of the slip patterning evolution, when the deformation becomes homogeneous. The evolution step from $\varepsilon = 0.0150$ to $\varepsilon = 0.0151$ results in a significant reduction in γ over the entire domain. The change to homogeneous strain is accompanied by a further stress drop, as shown by the curves of Fig. 5. This strain patterning is totally lost by the RD model when sufficiently large deformation rates are considered. For $\dot{\varepsilon} = 1\ \mathrm{s}^{-1}$, evolution is always homogeneous, as shown in Fig. 6d, even if the stress-strain curve exhibits a softening branch.

Constitutive response curves in the case of hard boundary conditions are plotted in Fig. 7. Their shapes are similar in type to those of the previous case of soft boundary conditions, that is, they present two hardening branches and an interposed softening stage. In this case they present a smoother shape. However, also the curves of Fig. 5 are always continuous, and the drops, which apparently look discontinuous, are smooth quickly decreasing curves.

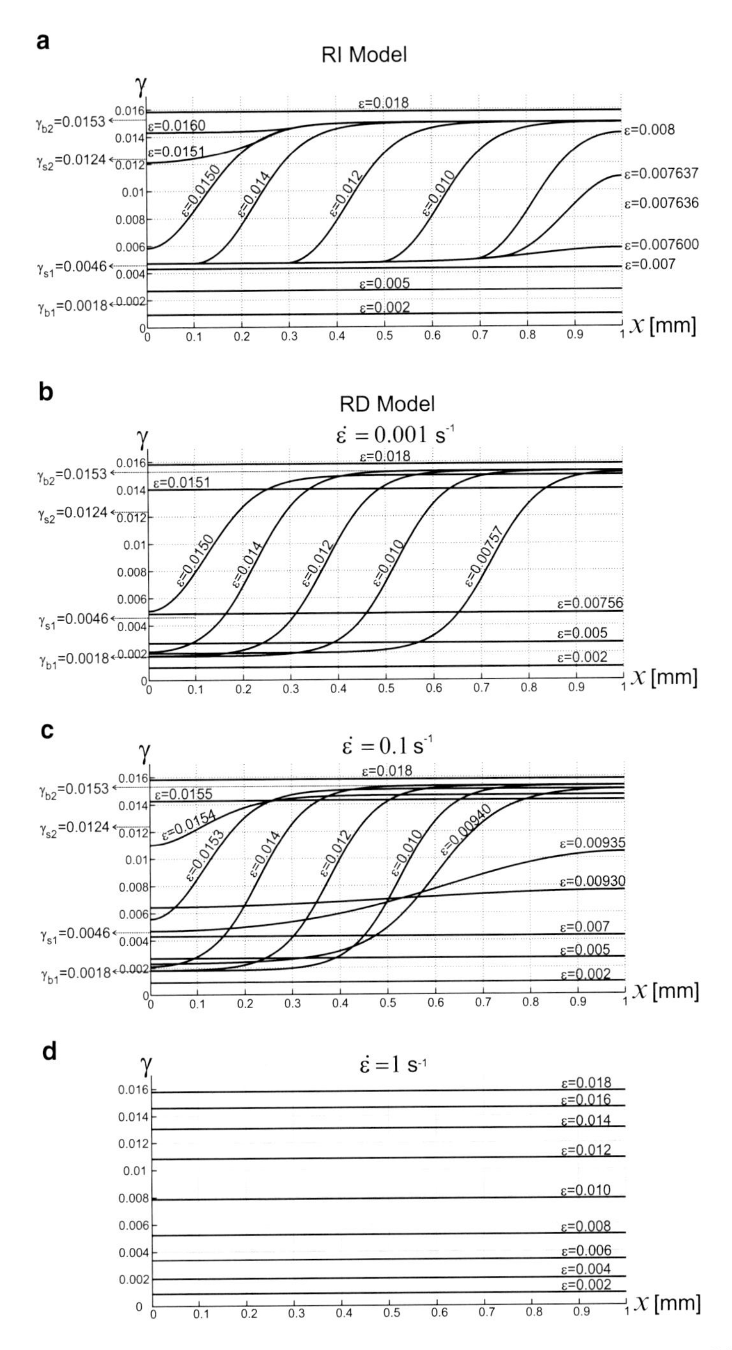

Fig. 6 Profiles of γ at different values of $\dot{\varepsilon}$ in case of soft boundary conditions. (**a**) RI model, (**b**) RD model with $\dot{\varepsilon} = 0.001\ \mathrm{s}^{-1}$, (**c**) RD model with $\dot{\varepsilon} = 0.1\ \mathrm{s}^{-1}$, (**d**) RD model with $\dot{\varepsilon} = 1\ \mathrm{s}^{-1}$

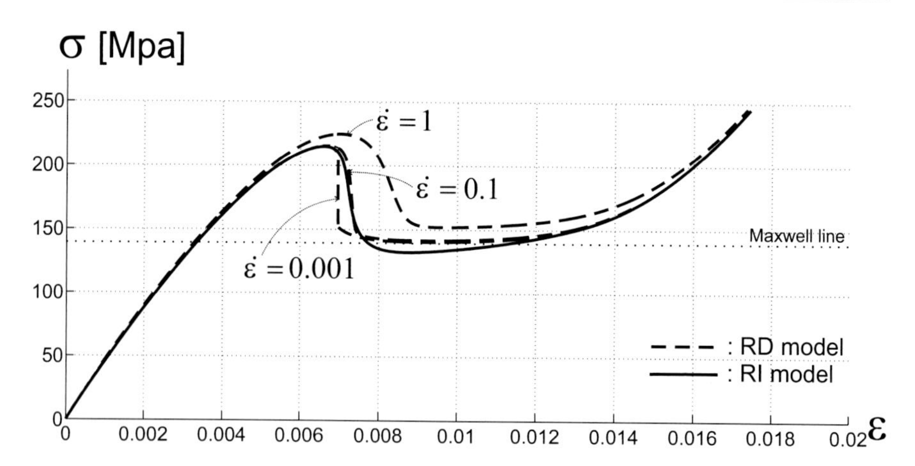

Fig. 7 Response curves in case of hard boundary conditions. Deformation rates $\dot{\varepsilon} = 0.001, 0.1, 1\ \mathrm{s}^{-1}$ are considered in the RD problem

In Fig. 8, snapshots of γ are taken at deformation increment $\Delta\varepsilon = 10^{-3}$. The evolution process that they describe is clearly composed by four phases: (1) an initial hardening phase in which deformation is homogeneous, except in the boundary layers developed at each end; (2) a softening process of strain localization in the middle of the domain; (3) a subsequent stress plateau characterized by plastic slip spreading toward the boundaries; and (4) a final hardening branch, where the deformation field evolves in a similar way to the initial hardening phase. If the evolution profiles of Fig. 8b, c are compared, it can be noticed that the increase of the deformation rate produces a delay of the localization process, which only partially takes place. The size of the zone where γ initially localizes becomes larger and larger as $\dot{\varepsilon}$ increases.

Necking in Tensile Steel Bars

Now, we consider an homogeneous bar of length $l = 140\,\mathrm{mm}$, clamped at the left endpoint and subjected to the tensile displacement εl at the right endpoint (see Fig. 1b). The bar is made of steel, with Young's Modulus: $E = 210\,\mathrm{GPa}$. For the plastic energy (ψ_γ in the RD model and θ in the RI model), we consider the piecewise cubic function considered in Lancioni (2015). This function, that we indicate with $f_2 = f_2(\gamma)$, is graphed in Fig. 9, with its first and second derivatives, and its analytical expression is

$$f_2(\gamma) = \begin{cases} c_1\gamma + \dfrac{1}{2}c_2\gamma^2\left(1 - \dfrac{\gamma}{3\gamma_1}\right), & \text{if } 0 < \gamma \leq \gamma_1, \\ c_3\gamma + c_4(\gamma - \gamma_1) + c_5(\gamma - \gamma_1)^3, & \text{if } \gamma_1 < \gamma \leq \gamma_2, \end{cases} \tag{60}$$

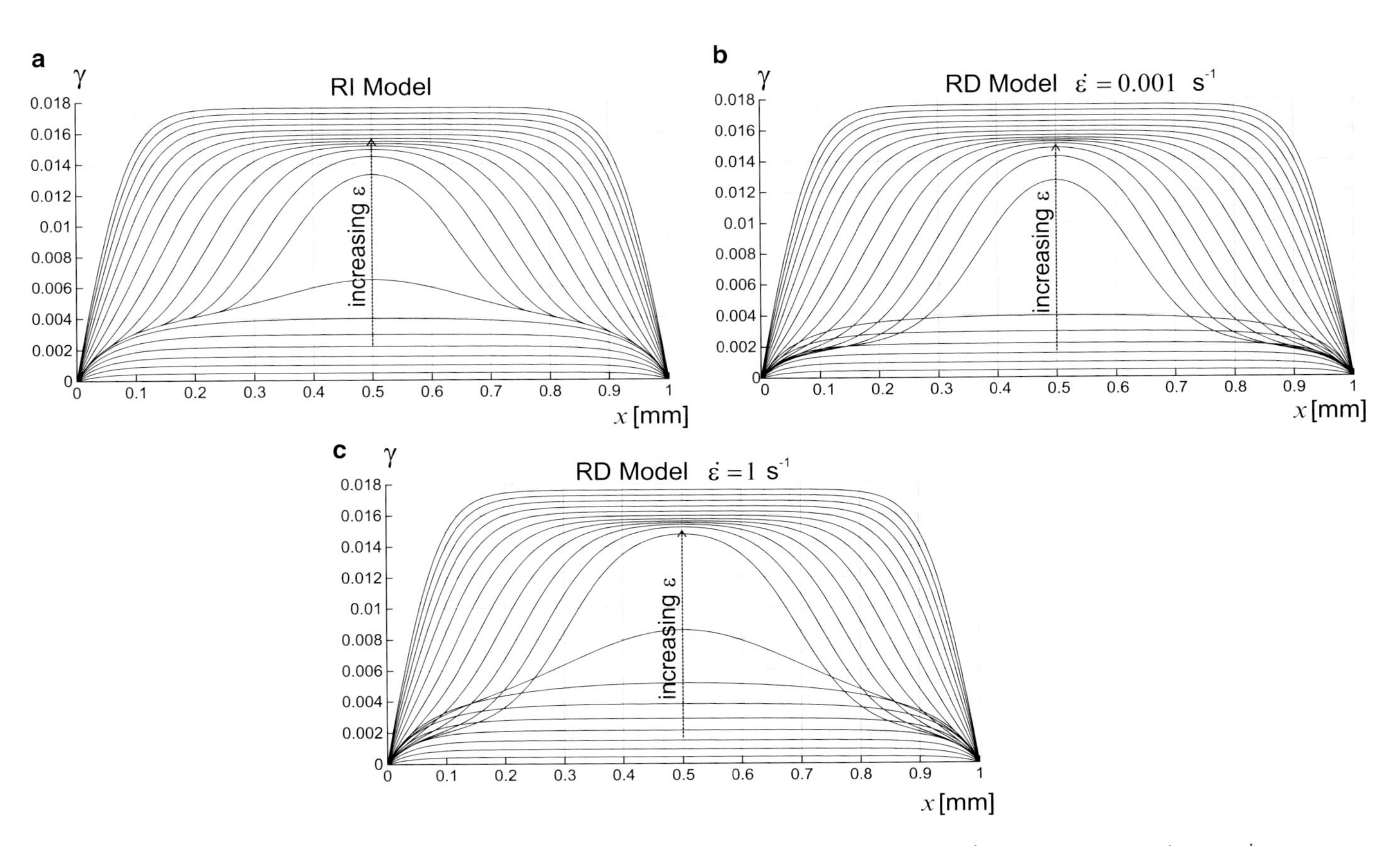

Fig. 8 Snapshots of γ taken at increment of ε equal to 10^{-3}. (**a**) RI model, (**b**) RD model with $\dot{\varepsilon} = 0.001\ \text{s}^{-1}$, (**c**) RD model with $\dot{\varepsilon} = 1\ \text{s}^{-1}$

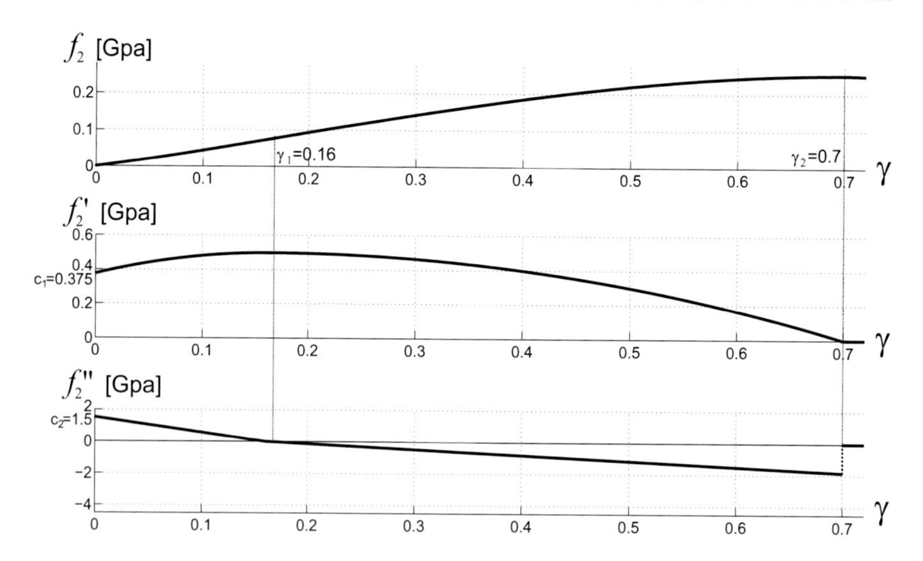

Fig. 9 Plastic energy density of the steel bar and its first and second derivatives

where $c_3 = c_1\gamma_1 + c_2\gamma_1^2/6$, $c_4 = c_1 + c_2\gamma_1/2$, and $c_5 = -(2c_1 + c_2\gamma_1)/(6(\gamma_2 - \gamma_1)^2)$ to guarantee the continuity of θ and θ' at γ_1 and the condition $f_2' \geq 0$. Coefficients c_1, c_2,γ_1, and γ_2 are related to easily measurable experimental quantities through the relations

$$c_1 = \sigma^c, \quad c_2 = \frac{2}{\gamma_1}(\sigma^m - \sigma^c), \quad \gamma_1 = \varepsilon^m - \frac{\sigma^m}{E}, \quad \gamma_2 = \frac{2(d_0 - d_{break})}{d_0}, \tag{61}$$

where σ^c is the yield stress, σ^m is the peak stress, ε^m is the corresponding strain, d_0 are the initial diameters of the bar cross section, and d_{break} is the diameter of the broken cross section. We assign the following values obtained from tensile tests

$$\sigma^c = 0.376 \text{ GPa}, \ \sigma^m = 0.496 \text{ GPa}, \ \varepsilon^m = 0.15, \ d_0 = 10 \text{ mm}, \ d_{brak} = 6.5 \text{ mm}. \tag{62}$$

Deduction and detailed description of formulas (61) are given in Lancioni (2015). The energy functional f_2 is convex in the interval $0 \leq \gamma < \gamma_1$, concave for $\gamma_1 \leq \gamma < \gamma_2$, and constant for $\gamma > \gamma_2$. For $\gamma > \gamma_2$, the bar deforms without spending plastic energy, and this corresponds to complete breaking. Thus, the simulations presented in the following are interrupted when γ reaches the breaking value $\gamma_2 = 0.7$. For the nonlocal parameter, we assume $A = 2$ kN, as in Lancioni (2015), and, for the viscous coefficient of the RD model, we fix $c = 0.015$ GPa s^{-1} as in Lancioni et al. (2015a).

Macroscopic constitutive response ($\sigma - \varepsilon$) curves are compared in Fig. 10. For the RD model, the four different deformation rates $\dot{\varepsilon} = 10^{-3}, \ 10^{-1}, \ 1$, and 10 s^{-1} are considered. In the enlargement on the left side of Fig. 10, the experimental curve is

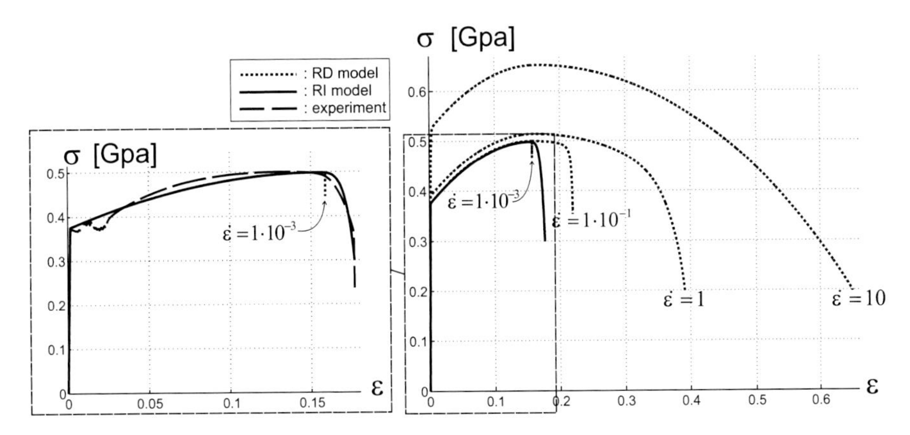

Fig. 10 Response curves of a tensile steel bar

also plotted (dashed line). The simulations predict three phases: an initial perfectly elastic phase, which interrupts when σ reaches the yield value σ^c, a hardening phase, and a final softening phase. In Lancioni (2015), a more sophisticated plastic energy was considered to reproduce the yielding plateau observed at the onset of the inelastic regime (see the experimental curve in the left Fig. 10).

For sufficiently low deformation rates ($\dot{\varepsilon} = 10^{-3}$ and 10^{-1} s^{-1}), the RI and RD models give practically the same hardening branches, with equal yield and peak stresses, but they differ in predicting the softening curves. The branch of the RI model is very close to the experimental curve, while those of the RD model strongly depends on $\dot{\varepsilon}$, getting longer as the deformation rate increases. Since $\dot{\varepsilon} = 1 \cdot 10^{-3}$ is a very small rate, the corresponding curve is a good approximation of the response curve in the limit static case $\dot{\varepsilon} = 0$. The discrepancies between this curve and that of the RI model clearly prove that the RI model is not the limit case of the RD model for $\dot{\varepsilon} \rightarrow 0$, although the RD model gives results close to those of the RI model when small deformation rates are considered. For large values of the deformation rate ($\dot{\varepsilon} = 1$ and 10 s^{-1}), the response curves of the RD model largely deviates from the curve of the RI model. As $\dot{\varepsilon}$ grows, the yield and peak stresses increases, and the softening branches extends. Also the plastic strain evolution is completely different than that of the RI case, as described in the following.

The evolution of γ is described in Fig. 11, where profiles of γ at different values of ε are plotted for the RI case and for the RD one as well with $\dot{\varepsilon} = 10^{-3},\ 1,\ 10$ s^{-1}. Results of the RI model and RD model with $\dot{\varepsilon} = 10^{-3}$ s^{-1} are comparable. In both the cases, the hardening phase is associated with the evolution of homogeneous plastic strains (with the exception of small boundary layers where γ decreases to zero), and the softening phase is characterized by a progressive localization of γ in smaller and smaller portions in the center of the bar, up to the final fracture, occurring when γ reaches the breaking value $\gamma_2 = 0.7$ in the bar midpoint. Both models describe fracture as the ending stage of a strain localization process; however, they differ in predicting the evolution of γ outside the localization zone:

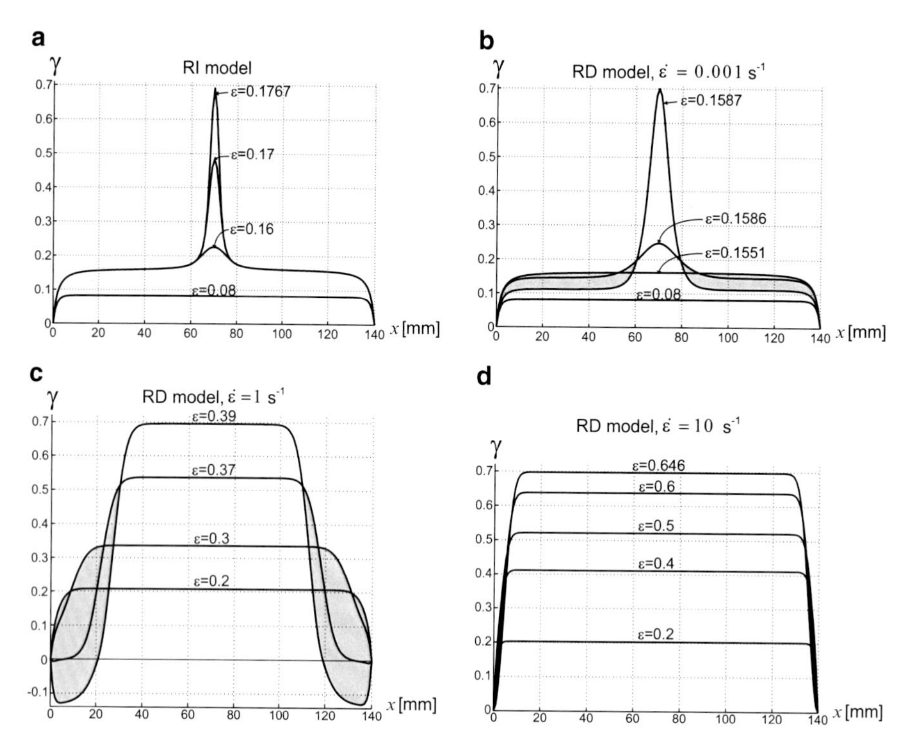

Fig. 11 Profiles of γ at different values of $\dot{\varepsilon}$. (**a**) RI model, (**b**) RD model with $\dot{\varepsilon} = 0.001\ \mathrm{s}^{-1}$, (**c**) RD model with $\dot{\varepsilon} = 1\ \mathrm{s}^{-1}$, (**d**) RD model with $\dot{\varepsilon} = 10\ \mathrm{s}^{-1}$

γ maintains constant in the RI case, and only the elastic deformation ε reduces, since $\delta\varepsilon = \delta\sigma/E < 0$ (elastic unloading), while both γ and ε^e reduce in the RD process (elastoplastic unloading). Gray areas in Fig. 11b highlight the zones where γ is recovered. The partial strain recovery outside the localization portion makes the softening branch of the response curve much steeper than that of the RI curve. The evolution of γ is very different if large $\dot{\varepsilon}$ are applied, as illustrated in Fig. 11c, d. Since the plastic strain has no time to localize, the limit value $\gamma = 0.7$ that leads to failure is reached in large zones in the center of the bar. On the contrary, plastic strain is recovered in small zones close to the endpoints. For $\dot{\varepsilon} = 1\ \mathrm{s}^{-1}$, plastic recovery leads to local compressive strain states in the final stages of the evolution (see the gray regions in Fig. 11c).

We notice that the above results agree with the analytical solutions found in sections "RD Model" and "RI Model." The analytical results basically state that solutions are full-size if the plastic energy is convex and localized if it is concave. Accordingly, numerical simulations predict an initial full-size evolution of γ, spreading in the whole bar. The initial hardening phase terminates when γ reach the value $\gamma_1 = 0.16$, at which the plastic energy changes from concave to convex. From this point on, the bar experiences localization. The fact that γ localizes on

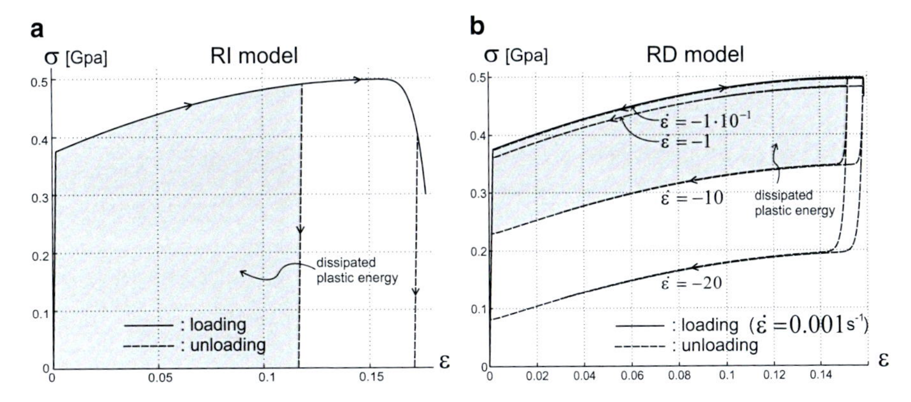

Fig. 12 Loading (solid line) and unloading (dashed line) response curves. (**a**) RI model, (**b**) RD model with different unloading plastic strains

smaller and smaller portions is also expected by the analytical results, according to which the size of the localization zone is inversely proportional to $|\psi_{\gamma}''|$ (see solution $(44)_3$) or to $|-\theta''|$ (solution (56)). In the plastic energy (60) implemented in the simulations, $|f_2''|$ is increasing with γ.

The bar response at unloading is analyzed in Fig. 12. For the RI model, unloading is always elastic, since γ does not reduce. The dissipated plastic energy is represented by the area below the response curve (gray area in Fig. 12a). For the RD model, unloading depends on the deformation rate. For small deformation rates, the unloading curves practically coincide with the loading curve, with no plastic dissipation, but, for larger deformation rates, hysteretic loops are obtained, and the area of the resulting closed curves represents the dissipated viscous energy. These areas are larger and larger as $|\dot{\varepsilon}|$ increases. The unloading process given by the RD model exhibits a first elastic unloading, followed by a second elastoplastic unloading, which are clearly recognizable by the different slope of the corresponding branches. The different responses at unloading described by the two models constitute a further distinguishing feature of the two proposed formulations.

Conclusions

In this chapter, two nonlocal plasticity models based on non-convex plastic potentials are addressed, by following a variational procedure which is thermodynamically consistent. Their ability in capturing the evolution of heterogeneous plastic patterning is analyzed in a simplified one-dimensional mathematical setting. The first model is rate-dependent, and it accounts for partially recoverable plastic strains. The second one is rate-independent, and it assumes that plastic strains are totally dissipative. Both plasticity models are enhanced by a gradient energy contribution, which introduces a length parameter into the models allowing studies at different length scales. Non-convex plastic potentials are assumed that lead to heterogeneous strain distributions and to processes of plastic localization. The instability due to the presence of a non-convex plastic energy is stabilized by the gradient energy term.

The RD and RI problems governing the strain evolution are solved analytically, for a simplified case. Analytical solutions give useful information on the shape that should be assigned to the plastic energy in order to predict specific behaviors, such as stress-hardening with a diffused plastic strain evolution, or stress-softening with strain localization. Both formulations are incorporated into finite element solution procedure and two different problems are solved, each one characterized by a specific non-convex plastic energy and by a specific length scale.

Numerical results have pointed out clearly the similarities and differences of the models, in some cases confirming the analytical evidences. At low deformation rates, the results of the RD model approach to those of the RI formulation, because of the effect of the viscous stress contribution. Differently, at high deformation rates, viscosity produces a delay of the strain localization, which, for particularly large deformation rates, is even missed. Consequently, results largely deviate from those of the RI model.

Differences that are independent on the deformation rate have been also observed, which are related to the different assumptions on the dissipative nature of the plastic energy made within each formulation. Two main differences deserve to be mentioned. (1) The different strain recovery processes that activate outside the strain localization zone: in the RI case, only the elastic strain reduces, while, in the RD case, both elastic and plastic strains are recovered. (2) The different responses at unloading: purely elastic unloading is reproduced by the RI model, with plastic strains cumulated in the body, and elastoplastic unloading is described by the RD case. In this latter case, hysteretic loops are observed and no residual strains cumulate in the body.

References

E.C. Aifantis, On the microstructural origin of certain inelastic models. J. Eng. Mater. Technol. **106**, 326–330 (1984)

C.J. Bayley, W.A.M. Brekelmans, M.G.D. Geers, A comparison of dislocation induced back stress formulations in strain gradient crystal plasticity. Int. J. Solids Struct. **43**, 7268–7286 (2006)

Z.P. Bazant, M. Jirásek, Nonlocal integral formulations of plasticity and damage: survey of progress. J. Eng. Mech. **128**, 1119–1149 (2002)

G. Del Piero, A variational approach to fracture and other inelastic phenomena. J. Elast. **112**(1), 3–77 (2013)

G. Del Piero, G. Lancioni, R. March, A diffuse cohesive energy approach to fracture and plasticity: the one-dimensional case. J. Mech. Mater. Struct. **8**(2–4), 109–151 (2013)

P. Gudmundson, A unified treatment of strain gradient plasticity. J. Mech. Phys. Solids **52**, 1379–1406 (2004)

M. Gurtin, L. Anand, Thermodynamics applied to gradient theories involving the accumulated plastic strain: the theories of Aifantis and Fleck & Hutchinson and their generalization. J. Mech. Phys. Solids **57**, 405–421 (2009)

M.E. Gurtin, E. Fried, L. Anand, *The Mechanics and Thermodinamics of Continua* (Cambridge University Press, New York, 2010)
D. Hull, Orientation and temperature dependence of plastic deformation processes in 3·25 percent silicon iron. Proc. R. Soc. A **274**, 5–24 (1963)
M. Jirásek, S. Rolshoven, Localization properties of strain-softening gradient plasticity models. Part II. Theories with gradients of internal variables. Int. J. Solids Struct. **46**, 2239–2254 (2009)
B. Klusemann, T. Yalçinkaya, Plastic deformation induced microstructure evolution through gradient enhanced crystal plasticity based on a non-convex Helmholtz energy. Int. J. Plast. **48**, 168–188 (2013)
B. Klusemann, T. Yalçinkaya, M.G.D. Geers, B. Svendsen, Application of non-convex rate dependent gradient plasticity to the modeling and simulation of inelastic microstructure development and inhomogeneous material behavior. Comput. Mater. Sci. **80**, 51–60 (2013)
G. Lancioni, Modeling the response of tensile steel bars by means of incremental energy minimization. J. Elast. **121**, 25–54 (2015)
G. Lancioni, T. Yalcinkaya, A. Cocks, Energy based non-local plasticity models for deformation patterning, localization and fracture. Proc. R. Soc. A **471**, 20150275:1–20150275:23 (2015a)
G. Lancioni, G. Zitti, T. Yalcinkaya, Rate-independent deformation patterning in crystal plasticity. Key Eng. Mater. **651–653**, 944–949 (2015b)
A. Mielke, A mathematical framework for generalized standard materials in the rate-independent case, in *Multifield Problems in Solid and Fluid Mechanics*, ed. by R. Helmig, A. Mielke, B. Wohlmuth. Lecture Notes in Applied and Computational Mechanics, vol. 28 (Springer, Berlin/Heidelberg, 2006), pp. 399–428
T. Yalçinkaya, Microstructure evolution in crystal plasticity: strain path effects and dislocation slip patterning. Ph.D. thesis, Eindhoven University of Technology (2011)
T. Yalçinkaya, G. Lancioni, Energy-based modeling of localization and necking in plasticity. Procedia Mater. Sci. **3**, 1618–1625 (2014)
T. Yalçinkaya, W.A.M. Brekelmans, M.G.D. Geers, Deformation patterning driven by rate dependent nonconvex strain gradient plasticity. J. Mech. Phys. Solids **59**, 1–17 (2011)
T. Yalçinkaya, W.A.M. Brekelmans, M.G.D. Geers, Non-convex rate dependent strain gradient crystal plasticity and deformation patterning. Int. J. Solids Struct. **49**, 2625–2636 (2012)

Strain Gradient Crystal Plasticity: Thermodynamics and Implementation

28

Tuncay Yalçinkaya

Contents

Abstract

This chapter studies the thermodynamical consistency and the finite element implementation aspects of a rate-dependent nonlocal (strain gradient) crystal plasticity model, which is used to address the modeling of the size-dependent behavior of polycrystalline metallic materials. The possibilities and required updates for the simulation of dislocation microstructure evolution, grain boundary-dislocation interaction mechanisms, and localization leading to necking and fracture phenomena are shortly discussed as well. The development of the model is conducted in terms of the displacement and the plastic slip, where the coupled fields are updated incrementally through finite element method. Numerical examples illustrate the size effect predictions in polycrystalline materials through Voronoi tessellation.

T. Yalçinkaya (✉)
Aerospace Engineering Program, Middle East Technical University Northern Cyprus Campus, Guzelyurt, Mersin, Turkey

Department of Aerospace Engineering, Middle East Technical University, Ankara, Turkey
e-mail: yalcinka@metu.edu.tr

G. Z. Voyiadjis (ed.), *Handbook of Nonlocal Continuum Mechanics for Materials and Structures*, https://doi.org/10.1007/978-3-319-58729-5_2

Keywords
Size effect · Strain gradient · Crystal plasticity · Finite element · Polycrystalline plasticity · Thermodynamics

Introduction

Strain gradient frameworks have been the most popular plasticity models for the last three decades, since the early work Aifantis (1984) due to their intrinsic capability to capture the size-dependent behavior of metallic materials through the incorporated length scale parameter which is absent in the classical models that are local and has no reference to the microstructural characteristic lengths. There is a large body of literature illustrating the size effect in the different materials under different loading conditions such as torsion (see, e.g., Fleck et al. 1994; Aifantis 1999), bending (see, e.g., Aifantis 1999; Stölken and Evans 1998; Haque and Saif 2003), indentation (see, e.g., Nix and Gao 1998; Swadenera et al. 2002), and compression experiments such as Wang et al. (2006) and Volkert and Lilleodden (2006).

There have been developments in both phenomenological-type isotropic strain gradient plasticity models and physics-based strain gradient crystal plasticity frameworks in order to model the size-dependent response of materials and the evolution of the inhomogeneous strain distribution at the interfaces and boundaries. All of them incorporate a length scale, to capture the size effect, however there is no unified structure of such nonlocal frameworks. Depending on the chosen internal state variables, the global degrees of freedom, the way the strain gradients are incorporated, and the thermodynamical work conjugate nature of the models, there have been various classifications in the literature.

Considering the phenomenological models, Hutchinson (2012) mentions that it has not been a simple matter to obtain a sound extension of the classical J2 flow theory of plasticity that incorporates a dependence on plastic strain gradients. Two classes of J2-type extensions have been proposed: one with increments in higher-order stresses related to increments of strain gradients and the other characterized by the higher-order stresses themselves expressed in terms of increments of strain gradients. The theory in Mühlhaus and Aifantis (1991) and Fleck and Hutchinson (2001) are in the first class, and these do not always satisfy the thermodynamic consistency. The other class includes Gudmundson (2004) and Gurtin and Anand (2009) which are thermodynamically consistent; however they have the physical deficiency that the higher-order stress quantities can change discontinuously for bodies subject to arbitrarily small load changes. Hutchinson (2012) presents a sound phenomenological extension of the rate-independent theory of the first class with a modification of the Fleck-Hutchinson model ensuring its thermodynamic integrity. The problems including nonproportional loading have recently been solved using the updated thermodynamically consistent model in Fleck et al. (2014). Reddy (2011a) has recently studied the flow rules of rate-independent gradient plasticity models and analyzed the uniqueness and the existence of solutions. In addition to these higher-order theories, lower-order theories have been developed based on the plastic strain gradient-dependent work-hardening rules such as Chen and Wang (2000) and Huang et al. (2004). The recent work by Panteghini and Bardella (2016) presents a detailed finite element implementation of isotropic gradient plasticity frameworks.

Initiated by Taylor (1938) crystal plasticity frameworks have been the main physically based approach in plasticity to address the intrinsic anisotropy and the localization phenomena in metals. It offers a kinematically representative description of first-order crystallographic phenomena (slip planes and directions, elastic anisotropy) and based on the statistical representation of the kinetics of groups of dislocations via flow and hardening rules at the individual slip system level. Even though the main application of such models is the small-scale engineering, the conventional ones do not include a length scale to address the realistic size-dependent behavior. Moreover, slip band formation, plastic slip microstructure evolution, and the effect of the grain boundaries are the limiting features of these models. The extension of crystal plasticity models to capture such effects will be discussed in the current and upcoming chapters.

Since the crystal plasticity extension of Fleck and Hutchinson (1997) to illustrate the grain size effect on strength (Shu and Fleck 1999), various strain gradient crystal plasticity models have been proposed. One of the simplest ways to incorporate the length scale into the constitutive frameworks has been the enhancement of the work-hardening laws with strain gradients such as lattice incompatibility-based models (see, e.g., Acharya and Bassani 2000; Bassani 2001), mechanism-based strain gradient crystal plasticity models (see, e.g., Han et al. 2005a,b), and others such as Ohashi (2005), Dunne et al. (2007), and Liang and Dunne (2009). In such models plastic strain gradients correspond to the geometrically necessary dislocation (GND) densities (Ashby 1970), and they contribute to the hardening behavior in addition to the statistically stored dislocation (SSD) densities. This type of lower-order crystal plasticity enhancement does not bring any additional boundary condition to the conventional boundary value problem formulation.

The other approach is the higher-order extension of crystal plasticity frameworks (see Gurtin 2000, 2002; Yefimov et al. 2004; Evers et al. 2004; Arsenlis et al. 2004; Bayley et al. 2006; Ma et al. 2006; Borg 2007; Geers et al. 2007; Levkovitch and Svendsen 2006; Kuroda and Tvergaard 2008; Reddy 2011b; Yalcinkaya 2011; Yalcinkaya et al. 2012; Klusemann and Yalçinkaya 2013; Klusemann et al. 2013) that would allow the treatment of plastic slip/dislocation density boundary conditions, and it is the main focus of the present paper. The higher-order theories can further be subclassified into work-conjugate and non-work-conjugate types (see Kuroda and Tvergaard 2008). In the former one, the plastic slip gradients are accompanied by the thermodynamically conjugate higher-order stress terms in the virtual work expression, and there exists an external contribution of work arising from a corresponding higher-order surface traction (see, e.g., Gurtin 2002; Borg 2007; Yalcinkaya et al. 2012). This unconventional virtual work expression leads to slip or higher-order traction at the boundaries which would allow us to solve problems where strong barriers of a material interface are impenetrable for dislocations. In the non-work-conjugate type of works (such as Yefimov et al. 2004;

Evers et al. 2004; Arsenlis et al. 2004; Bayley et al. 2006; Ma et al. 2006), a physically based slip evolution is influenced by higher-order stress contribution which is written in terms of gradients of plastic slip, where the arguments of virtual work have not been applied. These extended crystal plasticity models could further be classified according to chosen global degrees of freedom for the finite element solution procedure. In this context, there have been two different approaches where authors take plastic slip (see, e.g., Borg 2007; Yalcinkaya 2011; Yalcinkaya et al. 2012; Klusemann and Yalçinkaya 2013; Klusemann et al. 2013) or the dislocation density (see, e.g., Evers et al. 2004; Bayley et al. 2006; Geers et al. 2007) as degrees of freedom together with displacement field. For the sake of simplicity, the current study considers the most elementary representation of a strain gradient crystal plasticity theory in terms of plastic slip and displacement which are taken as coupled global degrees of freedom.

In addition to their success in predicting size effects, strain gradient crystal plasticity models have been improved further to simulate the intragranular microstructure evolution, intergranular grain boundary behavior, and macroscopic strain localization leading to necking in metallic materials (see, e.g., also Yalcinkaya et al. 2011, 2012; Özdemir and Yalçinkaya 2014; Yalcinkaya and Lancioni 2014; Lancioni et al. 2015a). A thermodynamically consistent incorporation of a proper plastic potentials results in different responses in terms of both strain distribution and the global constitutive response. This chapter addresses the thermodynamically consistent development and extension of higher-order, work-conjugate, plastic slip-based strain gradient crystal plasticity frameworks to capture the size effect and inhomogeneous strain and stress distribution in polycrystalline materials, in a simple and illustrative setting. It also studies different approaches to develop these models, the boundary conditions, and the solution algorithms and addresses some future challenges.

The chapter is organized as follows. First, in section "Plastic Slip-Based Strain Gradient Crystal Plasticity," the classical thermodynamical derivation of the rate-dependent strain gradient crystal plasticity framework is discussed. Then in section "Rate Variational Formulation of Strain Gradient Crystal Plasticity," the rate variational formulation of the same model is addressed. In section "Finite Element Solution Procedure of Strain Gradient Crystal Plasticity Framework," the finite element solution procedure of the framework is studied in detail. Then in section "Simulation of Polycrystalline Behavior," numerical examples are solved to illustrate the size effect in polycrystalline metals. Last, in section "Conclusion and Outlook" the concluding remarks are summarized.

Plastic Slip-Based Strain Gradient Crystal Plasticity

In this section the complete formulation and thermodynamical consistency of the higher-order, work-conjugate, plastic slip-based rate-dependent strain gradient crystal plasticity model are studied, and the weak form of equations is derived for the finite element implementation. Depending on the plastic slip potential for the energetic hardening, the models are distinguished as convex and non-convex in nature. The convex type of model, which is used in this chapter, is essentially used for the size effect predictions, while the non-convex type of models could be employed for the simulation of intrinsic microstructure evolution or macroscopic localization and necking (see, e.g., Yalcinkaya et al. 2011, 2012; Yalçinkaya 2013; Yalcinkaya and Lancioni 2014). The purpose of this section is to study the thermodynamics and the derivation of the convex-type strain gradient crystal plasticity model and to address the ways to incorporate different physical phenomena into the developed framework.

The theoretical framework is developed in a geometrically linear context, with small displacements, strains, and rotations. The time-dependent displacement field is denoted by $\boldsymbol{u} = \boldsymbol{u}(\boldsymbol{x}, t)$, where the vector $\boldsymbol{x}$ indicates the position of a material point. The strain tensor $\boldsymbol{\varepsilon}$ is defined as $\boldsymbol{\varepsilon} = \frac{1}{2}(\nabla \boldsymbol{u} + (\nabla \boldsymbol{u})^T)$, and the velocity vector is represented as $\boldsymbol{v} = \dot{\boldsymbol{u}}$. The strain is decomposed additively as

$$\boldsymbol{\varepsilon} = \boldsymbol{\varepsilon}^e + \boldsymbol{\varepsilon}^p \tag{1}$$

into an elastic part $\boldsymbol{\varepsilon}^e$ and a plastic part $\boldsymbol{\varepsilon}^p$. The plastic strain rate can be written as a summation of plastic slip rates on the individual slip systems, $\dot{\boldsymbol{\varepsilon}}^p = \sum_\alpha \dot{\gamma}^\alpha \boldsymbol{P}^\alpha$ with $\boldsymbol{P}^\alpha = \frac{1}{2}(\boldsymbol{s}^\alpha \otimes \boldsymbol{n}^\alpha + \boldsymbol{n}^\alpha \otimes \boldsymbol{s}^\alpha)$ the symmetrized Schmid tensor, where $\boldsymbol{s}^\alpha$ and $\boldsymbol{n}^\alpha$ are the unit slip direction vector and unit normal vector on slip system α, respectively.

Next, the selection of the internal state variables is discussed. The choice of the state variables controls the formulation of the free energy and the governing equations which would be needed to obtain the constitutive response and the evolution of the deformation field. The state variables are chosen to be given by the set

$$state = \{\boldsymbol{\varepsilon}^e, \gamma^\alpha, \nabla\gamma^\alpha\} \tag{2}$$

where γ^α contains the plastic slips on the different slip systems α and $\nabla\gamma^\alpha$ represents the gradient of the slips on these slip systems. The chosen variables describing the state are to be regarded as mesoscale internal state variables. At that scale, the glide plane slip and their gradients are natural candidates. They both describe the physical state on a slip plane in the mean field sense. The slip characterizes the average plastic deformation accumulated on a glide plane, whereas their gradients characterize the amount of geometrically necessary dislocations that accompany that process, also an important characteristic of the mean dislocation configuration. The use of these state variables naturally entails the other quantities in the constitutive description, like the gradient of the dislocation density, where the divergence of the microstress directly involves the gradient of GND.

There are a number of choices possible for the state variables. At the mesoscale, the choice made is quite appropriate in representing a spatiotemporal ensemble of microscopic states. The mesoscale state variables are measures of the current dislocation state in the material relative to which energy storage and hardening in the material are modeled. Microscopic processes like phase transitions or dislocation interaction may survive coarse-graining (from the microscale to the mesoscale), resulting in non-convex contributions to the free energy as in phase field models and a transition from spatial or material homogeneity or inhomogeneity. The use of slip-like quantities as mesoscale state variables is quite common in crystal plasticity. Several examples can be found in the literature, e.g., Rice (1971), using a continuum slip model, which characterizes the state of the crystal in terms of the shear strains on each slip system. Later on Perzyna (1988) considered plastic slips together with the slip resistance as internal state variables. For the numerical problems with monotonic loading histories, the adopted mesoscopic state variables are well capable of bridging the microscopic and mesoscopic states of the material.

Following the arguments of Gurtin (e.g., Gurtin 2000, 2002), the power expended by each independent rate-like kinematical descriptor is expressible in terms of an associated force consistent with its own balance. However, the basic kinematical fields of rate variables, namely, $\dot{\boldsymbol{\varepsilon}}^e$, $\dot{\boldsymbol{u}}$, and $\dot{\gamma}^\alpha$ are not spatially independent. It is therefore not immediately clear how the associated force balances are to be formulated, and, for that reason, these balances are established using the principle of virtual power.

Assuming that at a fixed time the fields $\boldsymbol{u}$, $\boldsymbol{\varepsilon}^e$, and γ^α are known, we consider $\delta\dot{\boldsymbol{u}}$, $\delta\dot{\boldsymbol{\varepsilon}}^e$, and $\delta\dot{\gamma}^\alpha$ as virtual rates, which are collected in the generalized virtual velocity $\mathcal{V} = \{\delta\dot{\boldsymbol{u}}, \delta\dot{\boldsymbol{\varepsilon}}^e, \delta\dot{\gamma}^\alpha\}$. $\mathscr{P}_{ext}$ is the power expended on the domain Ω and $\mathscr{P}_{int}$ a concomitant expenditure of power within Ω, given by

$$\begin{aligned} \mathscr{P}_{ext}(\Omega, \mathcal{V}) &= \int_S \boldsymbol{t}(\boldsymbol{n}) \cdot \delta\dot{\boldsymbol{u}}\, dS + \int_S \textstyle\sum_\alpha (\chi^\alpha(\boldsymbol{n})\, \delta\dot{\gamma}^\alpha)\, dS \\ \mathscr{P}_{int}(\Omega, \mathcal{V}) &= \int_\Omega \boldsymbol{\sigma} : \delta\dot{\boldsymbol{\varepsilon}}^e\, d\Omega + \int_\Omega \textstyle\sum_\alpha (\pi^\alpha\, \delta\dot{\gamma}^\alpha) d\Omega + \int_\Omega \sum_\alpha (\boldsymbol{\xi}^\alpha \cdot \nabla\delta\dot{\gamma}^\alpha)\, d\Omega \end{aligned} \tag{3}$$

where the stress tensor $\boldsymbol{\sigma}$, the scalar internal forces π^α, and the microstress vectors $\boldsymbol{\xi}^\alpha$ are the thermodynamical forces conjugate to the internal state variables $\boldsymbol{\varepsilon}^e$, γ^α, and $\nabla\gamma^\alpha$, respectively. In $\mathscr{P}_{ext}$, $\boldsymbol{t}(\boldsymbol{n})$ is the macroscopic surface traction, while $\chi^\alpha(\boldsymbol{n})$ represents the microscopic surface traction conjugate to γ^α at the boundary S with $\boldsymbol{n}$ indicating the boundary normal. The principle of virtual power states that for any generalized virtual velocity $\mathcal{V}$, the corresponding internal and external power are balanced, i.e.,

$$\mathscr{P}_{ext}(\Omega, \mathcal{V}) = \mathscr{P}_{int}(\Omega, \mathcal{V}) \qquad \forall\, \mathcal{V} \tag{4}$$

Now the consequences are derived following the restriction $\nabla\delta\dot{\boldsymbol{u}} = \delta\dot{\boldsymbol{\varepsilon}}^e + \delta\dot{\boldsymbol{\omega}}^e + \sum_\alpha \delta\dot{\gamma}^\alpha \boldsymbol{s}^\alpha \otimes \boldsymbol{m}^\alpha$. First a generalized virtual velocity without slip rate is considered, namely, $\delta\dot{\gamma}^\alpha = 0$ which means $\nabla\delta\dot{\boldsymbol{u}} = \delta\dot{\boldsymbol{\varepsilon}}^e + \delta\dot{\boldsymbol{\omega}}^e$. Considering that Cauchy stress is symmetric, the power balance becomes

$$\int_{\Omega} \boldsymbol{\sigma} : \nabla \delta \dot{\boldsymbol{u}} d\Omega = \int_{S} \boldsymbol{t}(\boldsymbol{n}) \cdot \delta \dot{\boldsymbol{u}} dS \tag{5}$$

After applying divergence theorem, the conditions derived from equation (5) are the traction condition

$$\boldsymbol{t}(\boldsymbol{n}) = \boldsymbol{\sigma} \boldsymbol{n} \tag{6}$$

and the classical linear momentum balance

$$\nabla \cdot \boldsymbol{\sigma} = \boldsymbol{0}. \tag{7}$$

The microscopic counterparts of the conditions are obtained through a consideration of a generalized virtual velocity with $\delta \dot{\boldsymbol{u}} = \boldsymbol{0}$ with arbitrary $\delta \dot{\gamma}^{\alpha}$ field which results in

$$\sum_{\alpha} \delta \dot{\gamma}^{\alpha} (\boldsymbol{s}^{\alpha} \otimes \boldsymbol{m}^{\alpha}) = -\delta \dot{\boldsymbol{\varepsilon}}^{e} - \delta \dot{\boldsymbol{\omega}}^{e} \tag{8}$$

And the term $\boldsymbol{\sigma} : \delta \dot{\boldsymbol{\varepsilon}}^{e}$ becomes

$$\boldsymbol{\sigma} : \delta \dot{\boldsymbol{\varepsilon}}^{e} = -\boldsymbol{\sigma} : \sum_{\alpha} \delta \dot{\gamma}^{\alpha} (\boldsymbol{s}^{\alpha} \otimes \boldsymbol{m}^{\alpha}) - \boldsymbol{\sigma} : \delta \dot{\boldsymbol{\omega}}^{e} \tag{9}$$

Using the symmetry of the Cauchy stress, the following relation is obtained $\boldsymbol{\sigma} : \delta \dot{\boldsymbol{\varepsilon}}^{e} = -\sum_{\alpha} \delta \dot{\gamma}^{\alpha} \tau^{\alpha}$ where τ^{α} is the Schmid resolved stress. For this case the power balance (4) utilized again

$$\begin{aligned} \int_S \textstyle\sum_{\alpha} (\chi^{\alpha}(\boldsymbol{n}) \delta \dot{\gamma}^{\alpha}) dS = &- \textstyle\sum_{\alpha} \int_{\Omega} \tau^{\alpha} \delta \dot{\gamma}^{\alpha} d\Omega + \sum_{\alpha} \int_{\Omega} \pi^{\alpha} \delta \dot{\gamma}^{\alpha} d\Omega \\ &+ \textstyle\sum_{\alpha} \int_{\Omega} \boldsymbol{\xi}^{\alpha} \cdot \nabla \delta \dot{\gamma}^{\alpha} d\Omega \end{aligned} \tag{10}$$

or equivalently

$$\textstyle\sum_{\alpha} \int_{\Omega} (-\tau^{\alpha} + \pi^{\alpha} - \nabla \cdot \boldsymbol{\xi}^{\alpha}) \delta \dot{\gamma}^{\alpha} d\Omega + \sum_{\alpha} \int_S (\boldsymbol{\xi}^{\alpha} \cdot \boldsymbol{n} - \chi^{\alpha}(\boldsymbol{n})) \delta \dot{\gamma}^{\alpha} dS = 0 \tag{11}$$

which should be satisfied for all $\delta \dot{\gamma}^{\alpha}$. This argument yields the microscopic traction condition on the outer boundary of the bulk material

$$\chi^{\alpha}(\boldsymbol{n}) = \boldsymbol{\xi}^{\alpha} \cdot \boldsymbol{n} \tag{12}$$

and the microscopic force balance inside the bulk material on each slip system α

$$\tau^{\alpha} - \pi^{\alpha} + \nabla \cdot \boldsymbol{\xi}^{\alpha} = 0 \tag{13}$$

on each slip system α.

Next, the thermodynamical consistent derivation of global system of equations is addressed, starting with the local internal power expression

$$P_i = \boldsymbol{\sigma} : \dot{\boldsymbol{\varepsilon}}^e + \sum_{\alpha} (\pi^\alpha \dot{\gamma}^\alpha + \boldsymbol{\xi}^\alpha \cdot \nabla \dot{\gamma}^\alpha) \tag{14}$$

The local dissipation inequality results in

$$D = P_i - \dot{\psi} = \boldsymbol{\sigma} : \dot{\boldsymbol{\varepsilon}}^e + \sum_{\alpha} (\pi^\alpha \dot{\gamma}^\alpha + \boldsymbol{\xi}^\alpha \cdot \nabla \dot{\gamma}^\alpha) - \dot{\psi} \geq 0 \tag{15}$$

For a single crystal material without grain boundaries, the material is assumed to be endowed with a free energy with different contributions according to

$$\psi = \psi_e + \psi_\gamma + \psi_{\nabla\gamma} \tag{16}$$

The time derivative of the free energy is expanded and equation (15) is elaborated to

$$\begin{aligned} D &= \boldsymbol{\sigma} : \dot{\boldsymbol{\varepsilon}}^e + \textstyle\sum_\alpha (\pi^\alpha \dot{\gamma}^\alpha + \boldsymbol{\xi}^\alpha \cdot \nabla \dot{\gamma}^\alpha - \dfrac{\partial \psi}{\partial \boldsymbol{\varepsilon}^e} : \dot{\boldsymbol{\varepsilon}}^e - \dfrac{\partial \psi}{\partial \gamma^\alpha} \dot{\gamma}^\alpha - \dfrac{\partial \psi}{\partial \nabla \gamma^\alpha} \cdot \nabla \dot{\gamma}^\alpha) \\ &= \underbrace{(\boldsymbol{\sigma} - \frac{d\psi_e}{d\boldsymbol{\varepsilon}^e})}_{0} : \dot{\boldsymbol{\varepsilon}}^e + \textstyle\sum_\alpha (\pi^\alpha - \dfrac{\partial \psi_\gamma}{\partial \gamma^\alpha}) \dot{\gamma}^\alpha + \textstyle\sum_\alpha \underbrace{(\boldsymbol{\xi}^\alpha - \frac{\partial \psi_{\nabla\gamma}}{\partial \nabla \gamma^\alpha})}_{0} \cdot \nabla \dot{\gamma}^\alpha \geq 0 \end{aligned} \tag{17}$$

The stress $\boldsymbol{\sigma}$ and the microstress vectors $\boldsymbol{\xi}^\alpha$ are regarded as energetic quantities having no contribution to the dissipation

$$\begin{aligned} \boldsymbol{\sigma} &= \frac{d\psi_e}{d\boldsymbol{\varepsilon}^e} \\ \boldsymbol{\xi}^\alpha &= \frac{\partial \psi_{\nabla\gamma}}{\partial \nabla \gamma^\alpha} \end{aligned} \tag{18}$$

whereas π^α does have a dissipative contribution. Note that considering the microstress vectors, $\boldsymbol{\xi}^\alpha$ as non-dissipative is an assumption. It is also possible to take dissipative and non-dissipative parts of the microstresses due to the gradient of the plastic deformation as done in some of the recent works (see, e.g., Gurtin 2002, 2008). In that case the dissipative parts would appear in the following reduced dissipation inequality, while in the current framework it does not exist

$$D = \sum_{\alpha} (\pi^\alpha - \frac{\partial \psi_\gamma}{\partial \gamma^\alpha}) \dot{\gamma}^\alpha \geq 0. \tag{19}$$

The multipliers of the plastic slip rates are identified as the set of dissipative stresses $\boldsymbol{\sigma}^{\alpha}_{\text{dis}}$

$$\boldsymbol{\sigma}^{\alpha}_{\text{dis}} = \pi^{\alpha} - \frac{\partial \psi_{\gamma}}{\partial \gamma^{\alpha}} \tag{20}$$

In order to satisfy the reduced dissipation inequality at the slip system level, the following constitutive equation is proposed

$$\sigma^{\alpha}_{\text{dis}} = \varphi^{\alpha} \operatorname{sign}(\dot{\gamma}^{\alpha}) \tag{21}$$

where φ^{α} represents the mobilized slip resistance of the slip system under consideration

$$\varphi^{\alpha} = \frac{s^{\alpha}}{\dot{\gamma}^{\alpha}_{0}} |\dot{\gamma}^{\alpha}| \tag{22}$$

where s^{α} is the resistance to dislocation slip which is assumed to be constant and $\dot{\gamma}_0$ is the reference slip rate. Substituting (22) into (21) gives

$$\dot{\gamma}^{\alpha} = \frac{\dot{\gamma}^{\alpha}_{0}}{s^{\alpha}} \sigma^{\alpha}_{\text{dis}} \tag{23}$$

Substitution of $\sigma^{\alpha}_{\text{dis}}$ according to (20) into (23) reveals

$$\dot{\gamma}^{\alpha} = \frac{\dot{\gamma}^{\alpha}_{0}}{s^{\alpha}} \left(\pi^{\alpha} - \frac{\partial \psi_{\gamma}}{\partial \gamma^{\alpha}} \right) \tag{24}$$

Using the microforce balance (13) results in the plastic slip equation

$$\dot{\gamma}^{\alpha} = \frac{\dot{\gamma}^{\alpha}_{0}}{s^{\alpha}} \left(\tau^{\alpha} + \nabla \cdot \boldsymbol{\xi}^{\alpha} - \frac{\partial \psi_{\gamma}}{\partial \gamma^{\alpha}} \right). \tag{25}$$

Note that the derivation of the thermodynamically consistent constitutive equations (plastic slip evolution) are based on an assumption which results in an expression identical to classical power-law relation of crystal plasticity frameworks with an exponent $m = 1$. This choice was of course made for simplicity only, yet with a large similarity to discrete dislocation studies using linear drag relations. The general form of equations would follow by including the rate sensitivity exponent in (22) as $\varphi^{\alpha} = s^{\alpha} (\dot{\gamma}^{\alpha}_{0} / |\dot{\gamma}^{\alpha}|)^{m}$ which would yield the following general form

$$\dot{\gamma}^{\alpha} = \dot{\gamma}^{\alpha}_{0} \left(|\tau^{\alpha} + \nabla \cdot \boldsymbol{\xi}^{\alpha} - \frac{\partial \psi_{\gamma}}{\partial \gamma^{\alpha}}| / s^{\alpha} \right)^{\frac{1}{m}} \operatorname{sign}(\pi^{\alpha} - \frac{d \psi_{\gamma}}{d \gamma^{\alpha}}) \tag{26}$$

The identification of the model as crystal plasticity, strain gradient crystal plasticity, or non-convex strain gradient crystal plasticity depends on the terms entering the equations (25) and (26). Comparing equations (23) and (25) resolves $\sigma^{\alpha}_{\text{dis}} = \tau^{\alpha} + \boldsymbol{\nabla} \cdot \boldsymbol{\xi}^{\alpha} - \partial\psi_{\gamma}/\partial\gamma^{\alpha}$. In the recently developed non-convex strain gradient crystal plasticity frameworks (see, e.g., Yalcinkaya et al. 2011, 2012), the driving force for the dislocation slip evolution is $\sigma^{\alpha}_{\text{dis}}$ which physically means that, in addition to the resolved shear stress τ^{α}, the back stress due to the gradients of the geometrically necessary dislocation densities $\boldsymbol{\nabla} \cdot \boldsymbol{\xi}^{\alpha}$, and the internal force leading to the accumulation of plastic slip $\partial\psi_{\gamma}/\partial\gamma^{\alpha}$, is affecting the plastic flow. In classical crystal plasticity frameworks, it is only the resolved shear stress τ^{α} which determines the plastic flow, while in strain gradient type of models it is the effective resolved shear stress $\tau^{\alpha}_{eff} = \tau^{\alpha} + \boldsymbol{\nabla} \cdot \boldsymbol{\xi}^{\alpha}$. In addition to the explicit contribution of ψ_{γ}, other contributions of the free energies defined in (16) enter the slip equation via (18) with $\tau^{\alpha} = d\psi_e/d\boldsymbol{\varepsilon}^e : \boldsymbol{P}^{\alpha}$ and $\boldsymbol{\xi}^{\alpha} = \partial\psi_{\nabla\gamma}/\partial\boldsymbol{\nabla}\gamma^{\alpha}$.

It is necessary to comment on the particular choices for energy potentials $(\psi_e, \psi_{\gamma}, \psi_{\nabla\gamma})$ which would result in the expressions for the related stresses with $\tau^{\alpha} = d\psi_e/d\boldsymbol{\varepsilon}^e : \boldsymbol{P}^{\alpha}$, $\boldsymbol{\xi}^{\alpha} = \partial\psi_{\nabla\gamma}/\partial\boldsymbol{\nabla}\gamma^{\alpha}$ and $\partial\psi_{\gamma}/\partial\gamma^{\alpha}$. In the numerical examples of this chapter, quadratic forms are used for the elastic free energy ψ_e and the plastic slip gradients free energy contribution $\psi_{\nabla\gamma}$, i.e.,

$$\psi_e = \frac{1}{2}\boldsymbol{\varepsilon}^e : {}^4\boldsymbol{C} : \boldsymbol{\varepsilon}^e \quad \text{and} \quad \psi_{\nabla\gamma} = \sum_{\alpha} \frac{1}{2} A \boldsymbol{\nabla}\gamma^{\alpha} \cdot \boldsymbol{\nabla}\gamma^{\alpha} \tag{27}$$

where A is a scalar quantity, which includes an internal length scale parameter, governing the effect of the plastic slip gradients on the internal stress field, and it could be introduced in different ways. One choice is $A = ER^2/(16(1-\nu^2))$ as, e.g., used in Bayley et al. (2006) and Geers et al. (2007), where R is a typical length scale for dislocation interactions and it physically represents the radius of the dislocation domain contributing to the internal stress field. If the dislocation interaction is limited to nearest neighbor interactions only, then R equals the dislocation spacing. Moreover, ν is Poisson's ratio and E is Young's modulus. Depending on the problem, different relations for A could be introduced or it could even be used as a parameter itself in a more phenomenological way. The definitions in (27) result in the following stress expressions

$$\begin{aligned} \boldsymbol{\sigma} &= \frac{d\psi_e}{d\boldsymbol{\varepsilon}^e} = {}^4\boldsymbol{C} : \boldsymbol{\varepsilon}^e \\ \boldsymbol{\xi}^{\alpha} &= \frac{\partial\psi_{\nabla\gamma}}{\partial\boldsymbol{\nabla}\gamma^{\alpha}} = A\boldsymbol{\nabla}\gamma^{\alpha} \end{aligned} \tag{28}$$

The contributions from the elastic and gradient potentials constitute the *convex strain gradient crystal plasticity framework*. The crucial part is the determination of the plastic potential ψ_{γ} which governs the energetic hardening behavior of the model, the spatial distribution, and the localization of the strain and deformation.

While a convex ψ_γ influences solely the hardening behavior of the model, a non-convex contribution incudes deformation localization and macroscopic softening. Due to the incorporation of a non-convex potential, we refer the resulting framework the *non-convex strain gradient crystal plasticity* having the capability to model heterogeneous distribution of deformation (microstructure evolution). The effect of incorporation of both phenomenological and physically based non-convex plastic potentials on the spatial distribution of deformation and the macroscopic constitutive response within the scope of plasticity, damage, and fracture has been illustrated recently in, e.g., Yalcinkaya et al. (2011, 2012), Klusemann and Yalçinkaya (2013), Klusemann et al. (2013), Lancioni et al. (2015a,b).

Another thermodynamically consistent approach to reach the identical constitutive model and plastic slip evolution relation is to work in the framework of continuum thermodynamics and rate variational methods (see, e.g., Svendsen 2004; Reddy 2011a,b) for history-dependent material behavior, which is addressed in the next section. In this case the evolution equations are derived from a dissipation potential. Same results could be obtained through variational formulation and incremental minimization procedure as well (see Lancioni et al. 2015a).

In order to solve initial boundary value problems using the rate-dependent strain gradient crystal plasticity framework, a fully coupled finite element solution algorithm is employed where both the displacement $\boldsymbol{u}$ and the plastic slips γ^α are considered as primary variables. These fields are determined in the solution domain by solving simultaneously the linear momentum balance (7) and the slip evolution equation (26) for $m = 1$ and constant slip resistance, which constitute the local strong form of the balance equations

$$\begin{aligned} \nabla\cdot\boldsymbol{\sigma} &= \boldsymbol{0} \\ \dot{\gamma}^\alpha - \frac{\dot{\gamma}_0^\alpha}{s^\alpha}\tau^\alpha - \frac{\dot{\gamma}_0^\alpha}{s^\alpha}\nabla\cdot\boldsymbol{\xi}^\alpha + \frac{\dot{\gamma}_0^\alpha}{s^\alpha}\frac{\partial\psi_\gamma}{\partial\gamma^\alpha} &= 0 \end{aligned} \tag{29}$$

In order to obtain variational expressions representing the weak forms of the governing equations given above, these equations are multiplied by weighting functions $\delta_{\boldsymbol{u}}$ and δ_γ^α and integrated over the domain Ω. Using the Gauss theorem (S is the boundary of Ω) results in

$$\begin{aligned} G_u &= \int_\Omega \nabla\delta_{\boldsymbol{u}} : \boldsymbol{\sigma}\, d\Omega - \int_S \delta_{\boldsymbol{u}}\cdot\boldsymbol{t}\, dS \\ G_\gamma^\alpha &= \int_\Omega \delta_\gamma^\alpha\,\dot{\gamma}^\alpha\, d\Omega - \int_\Omega \delta_\gamma^\alpha\frac{\dot{\gamma}_0^\alpha}{s^\alpha}\tau^\alpha\, d\Omega + \int_\Omega \nabla\delta_\gamma^\alpha\cdot\frac{\dot{\gamma}_0^\alpha}{s^\alpha}\boldsymbol{\xi}^\alpha d\Omega \\ &+ \int_\Omega \delta_\gamma^\alpha\frac{\dot{\gamma}_0^\alpha}{s^\alpha}\frac{\partial\psi_\gamma}{\partial\gamma^\alpha}\, d\Omega - \int_S \delta_\gamma^\alpha\frac{\dot{\gamma}_0^\alpha}{s^\alpha}\chi^\alpha\, dS \end{aligned} \tag{30}$$

where $\boldsymbol{t}$ is the external traction vector on the boundary S, and $\chi^\alpha = \boldsymbol{\xi}\cdot\boldsymbol{n}$. The domain Ω is subdivided into finite elements, where the unknown fields of the displacement and slips and the associated weighting functions within each element are approximated by their nodal values multiplied with the interpolation shape functions stored in the $\underline{N}^u$ and $\underline{N}^\gamma$ matrices,

$$\begin{aligned} \delta_u &= \underline{N}^u \underline{\delta}_u & \qquad u &= \underline{N}^u \underline{u} \\ \delta_\gamma^\alpha &= \underline{N}^\gamma \underline{\delta}_\gamma^\alpha & \qquad \gamma^\alpha &= \underline{N}^\gamma \underline{\gamma}^\alpha \end{aligned} \tag{31}$$

where $\underline{u}$, $\underline{\delta}_u$, $\underline{\gamma}^\alpha$, and $\underline{\delta}_\gamma^\alpha$ are columns containing the nodal variables. Bilinear interpolation functions for the slip field and quadratic interpolation functions for the displacement field are used. An implicit backward Euler time integration scheme is used for $\dot{\gamma}^\alpha$ in a typical time increment $[t_n, t_{n+1}]$ which gives $\dot{\gamma}^\alpha = [\gamma_{n+1}^\alpha - \gamma_n^\alpha]/\Delta t$. The discretized element weak forms read

$$\begin{aligned} G_u^e &= \underline{\delta}_u^T \left[\int_{\Omega^e} \underline{B}^u \underline{\sigma}\, d\Omega^e - \int_{S^e} \underline{N}^u \underline{t}\, dS^e \right] \\ G_\gamma^{\alpha e} &= \underline{\delta}_\gamma^\alpha \left[\int_{\Omega^e} \underline{N}^{\gamma T} \underline{N}^\gamma \left[\frac{\underline{\gamma}_{n+1}^\alpha - \underline{\gamma}_n^\alpha}{\Delta t} \right] d\Omega^e - \int_{\Omega^e} \frac{\dot{\gamma}_0^\alpha}{s^\alpha} \underline{N}^{\gamma T}\, \underline{\tau}^\alpha\, d\Omega^e \right] \\ &+ \underline{\delta}_\gamma^\alpha \frac{\dot{\gamma}_0^\alpha}{s^\alpha} \left[\int_{\Omega^e} \underline{B}^\gamma\, \underline{\xi}^\alpha\, d\Omega^e + \int_{\Omega^e} \underline{N}^{\gamma T} \frac{\partial \psi_{\underline{\gamma}^\alpha}}{\partial \gamma^\alpha}\, d\Omega^e - \int_{S^e} \underline{N}^{\gamma T}\, \underline{\chi}^\alpha\, dS^e \right] \end{aligned} \tag{32}$$

The weak forms of the balance equations (32) are linearized with respect to the variations of the primary variables $\underline{u}$ and $\underline{\gamma}^\alpha$ and solved by means of a Newton-Raphson solution scheme for the increments of the displacement field $\Delta \underline{u}$ and the plastic slips $\Delta \gamma^\alpha$. The procedure results in a system of linear equations which can be written in the following matrix format

$$\begin{bmatrix} \underline{K}^{uu} & \underline{K}^{u\gamma} \\ \underline{K}^{\gamma u} & \underline{K}^{\gamma\gamma} \end{bmatrix} \begin{bmatrix} \Delta \underline{u} \\ \Delta \underline{\gamma}^\alpha \end{bmatrix} = \begin{bmatrix} -\underline{R}^u + \underline{R}_u^{ext} \\ -\underline{R}^\gamma + \underline{R}_\gamma^{ext} \end{bmatrix} \tag{33}$$

where $\underline{K}^{uu}$, $\underline{K}^{u\gamma}$, $\underline{K}^{\gamma u}$, and $\underline{K}^{\gamma\gamma}$ represent the global tangent matrices, while $\underline{R}^u$ and $\underline{R}^\gamma$ are the global residual columns. The contributions $\underline{R}_u^{ext}$ and $\underline{R}_\gamma^{ext}$ originate from the boundary terms.

The global degrees of freedom in this framework are the displacement and the plastic slips, in terms of which the boundary conditions are defined. Without an explicit grain boundary model, in the current setting, there are two types of conditions that could be used at grain boundaries during polycrystal simulations. The first one is the soft boundary condition for the plastic slip which does not restrict the transfer of dislocations to the neighboring grain, and the other one is the hard boundary condition which blocks the transmission of the dislocations and results in the boundary layer in terms of plastic slip. The reality is in between, and in order to model the proper behavior of transmission, emission, and dissociation of dislocations within the grain boundary, an explicit grain boundary model should be included in the framework (see very recent examples Özdemir and Yalçinkaya 2014; van Beers et al. 2015a,b; Gottschalk et al. 2016; Bayerschen et al. 2016). For a more detailed discussion on the GB modelling through strain gradient crystal plasticity, please refer to the upcoming chapter by Yalcinkaya and Ozdemir.

Rate Variational Formulation of Strain Gradient Crystal Plasticity

In this section the derivation of the constitutive equations of the rate-dependent strain gradient crystal plasticity framework is conducted via rate variational formulation. The same set of coupled equations are eventually obtained to be solved through finite element method.

In here, the formulation is carried out in the framework of continuum thermodynamics and rate variational methods (see, e.g., Svendsen 2004; Svendsen and Bargmann 2010) for history-dependent behavior. The attention is confined to quasi-static, infinitesimal deformation processes. Let B represent sample domain with boundary ∂B. Besides the displacement field u, the principle global unknowns are taken as the plastic slips γ^α on each glide system α. The model behavior is identified with energetic and dissipative processes. The energetic processes are represented by the free energy density ψ. In the case of general non-convex gradient crystal plasticity (Yalcinkaya et al. 2011, 2012), this consists of an elastic ψ_e, non-convex ψ_γ, and the gradient $\psi_{\nabla\gamma}$ parts (see equation 16).

The dissipative/kinetic processes are represented by a simple rate-dependent power-law form of dissipation potential φ, which would result in the same constitutive relation for the evolution of plastic slip field

$$\varphi = \sum_{\alpha} \frac{1}{m+1} \, s \, \dot{\gamma}_0 \left| \frac{\dot{\gamma}^\alpha}{\dot{\gamma}_0} \right|^{m+1} \tag{34}$$

The dissipation potential φ is nonnegative and convex in $\dot{\gamma}$; therefore it satisfies the dissipation principle Silhavy (1997) sufficiently. This form tacitly assumes zero activation energy or stress for initiation of inelastic deformation. Since the current work is concerned with purely qualitative effects, $m = 1$ is chosen for simplicity, analogous to discrete dislocation studies based on linear drag relations. When γ is modeled as a global variable, the current framework results in a Ginzburg-Landau-/phase-field-like relation for γ when ψ_γ is introduced as non-convex. Other choices for m, including those used in classical crystal plasticity studies, would result in a non-Ginzburg-Landau form. Other values of m influence the strength, but not the qualitative effect of the rate-dependent material behavior and microstructure development (Yalcinkaya et al. 2011).

Next, the finite element solution procedure of the continuum thermodynamics variational formulation of the evolution field relations for the initial-boundary value problem is addressed briefly. The formulation begins with the following rate functional

$$R = \int_B r_v \mathrm{d}V + \int_{\partial B} r_s \mathrm{d}S \tag{35}$$

which is based on the corresponding volumetric and boundary rate potentials

$$\begin{aligned} r_v &:= \zeta_v + \varphi_v \\ r_s &:= \zeta_s + \varphi_s \end{aligned} \tag{36}$$

The energetic and the dissipative terms are represented by ζ and φ, respectively. The energy storage density ζ could be determined by

$$\zeta_v = \frac{\partial \psi}{\partial \nabla \boldsymbol{u}} \cdot \nabla \dot{\boldsymbol{u}} + \sum\nolimits_\alpha \frac{\partial \psi}{\partial \gamma^\alpha} \dot{\gamma}^\alpha + \sum\nolimits_\alpha \frac{\partial \psi}{\partial \nabla \gamma^\alpha} \cdot \nabla \dot{\gamma}^\alpha \tag{37}$$

Using the definition of the stresses due to elastic and gradient free energy in (18)

$$\zeta_v = \boldsymbol{\sigma} \cdot \nabla \dot{\boldsymbol{u}} + \sum\nolimits_\alpha \left(\frac{\partial \psi_\gamma}{\partial \gamma^\alpha} - \tau^\alpha\right) \dot{\gamma}^\alpha + \sum\nolimits_\alpha \boldsymbol{\xi}^\alpha \cdot \nabla \dot{\gamma}^\alpha \tag{38}$$

The surface rate potential r_s consists of energetic ζ_s and kinetic or dissipative φ_s parts which are linear and nonlinear, respectively, in the rates $\dot{\boldsymbol{u}}$ and $\dot{\gamma}^\alpha$ and determines the flux boundary conditions

$$-\boldsymbol{t} = \frac{\partial r_s}{\partial \dot{\boldsymbol{u}}} \quad \text{and} \quad -\chi^\alpha = \frac{\partial r_s}{\partial \dot{\gamma}^\alpha} \tag{39}$$

associated with $\dot{\boldsymbol{u}}$ and $\dot{\gamma}_\alpha$, respectively. Considering flux free boundary conditions, the rate potentials are

$$\begin{aligned} r_v &= \boldsymbol{\sigma} \cdot \nabla \dot{\boldsymbol{u}} + \sum\nolimits_\alpha \left(\left(\frac{\partial \psi_\gamma}{\partial \gamma^\alpha} - \tau^\alpha\right) \dot{\gamma}^\alpha + \boldsymbol{\xi}^\alpha \cdot \nabla \dot{\gamma}^\alpha + \tfrac{1}{2} s \dot{\gamma}_0 \left| \tfrac{\dot{\gamma}^\alpha}{\dot{\gamma}_0} \right|^2 \right) \\ r_s &= \zeta_s + \varphi_s = 0 \quad (\text{flux} - \text{free}). \end{aligned} \tag{40}$$

Now we get the first variation of R in $\dot{\boldsymbol{u}}$ and $\dot{\gamma}_\alpha$ by using integration by parts, the divergence theorem, and the variational derivative $\dfrac{\delta f}{\delta x} = \dfrac{\partial f}{\partial x} - \nabla \cdot \dfrac{\partial f}{\partial \nabla x}$

$$\delta R = \int_B \frac{\delta r_{\mathrm{v}}}{\delta \dot{\boldsymbol{u}}} \delta \dot{\boldsymbol{u}} \mathrm{d}V + \int_{\partial B} (\frac{\partial r_{\mathrm{v}}}{\partial \nabla \dot{\boldsymbol{u}}} \boldsymbol{n} + \frac{\partial r_s}{\partial \dot{\boldsymbol{u}}}) \cdot \delta \dot{\boldsymbol{u}} \mathrm{d}S$$
$$+ \sum_\alpha \left\{ \int_B \frac{\delta r_{\mathrm{v}}}{\delta \dot{\gamma}^\alpha} \delta \dot{\gamma}^\alpha \mathrm{d}V + \int_{\partial B} (\frac{\partial r_{\mathrm{v}}}{\partial \nabla \dot{\gamma}^\alpha} \cdot \boldsymbol{n} + \frac{\partial r_s}{\partial \dot{\gamma}^\alpha}) \delta \dot{\gamma}^\alpha \mathrm{d}S \right\} \tag{41}$$

Note that this form is independent of the gradients of $\delta \dot{\boldsymbol{u}}$ and $\delta \dot{\gamma}_\alpha$.

R is stationary with respect to all admissible variations of $\dot{\boldsymbol{u}}$ when

$$\frac{\delta r_{\mathrm{v}}}{\delta \dot{\boldsymbol{u}}} = \mathbf{0} \quad \text{in } B$$
$$\frac{\partial r_{\mathrm{v}}}{\partial \nabla \dot{\boldsymbol{u}}} \boldsymbol{n} + \frac{\partial r_s}{\partial \dot{\boldsymbol{u}}} = \mathbf{0} \quad \text{on } \partial B_t \tag{42}$$

hold in the bulk and at the flux part of the boundary ∂B_t. These relations represent the momentum balance in B and on ∂B_t in (rate) variational form.

The stationarity of R with respect to admissible variations of $\dot{\gamma}_\alpha$ is given when

$$\frac{\delta r_{\mathrm{v}}}{\delta \dot{\gamma}^\alpha} = 0 \quad \text{in } B$$
$$\frac{\partial r_{\mathrm{v}}}{\partial \nabla \dot{\gamma}^\alpha} \cdot \boldsymbol{n} + \frac{\partial r_s}{\partial \dot{\gamma}^\alpha} = 0 \quad \text{on } \partial B_\varphi \tag{43}$$

hold. The physical interpretation of these relations would simply be the generalized flow rule on each glide system.

Finally the field relations for the deformation

$$\nabla \cdot \boldsymbol{\sigma} = \mathbf{0} \quad \text{in } B$$
$$\boldsymbol{\sigma} \boldsymbol{n} = \boldsymbol{t} \quad \text{on } \partial B_t \tag{44}$$

and for each glide system

$$\frac{s^\alpha}{\dot{\gamma}_0^\alpha} \dot{\gamma}^\alpha - \tau^\alpha - \nabla \cdot \boldsymbol{\xi}^\alpha + \frac{\partial \psi_\gamma}{\partial \gamma^\alpha} = \mathbf{0} \quad \text{in } B$$
$$\boldsymbol{\xi}^\alpha \cdot \boldsymbol{n} = \chi^\alpha \quad \text{on } \partial B_\varphi \tag{45}$$

are obtained. Equations (44) and (45) form the same set of strong form of equations of the previous section (29) to be solved by the finite element method. We follow the same strategy to obtain the weak form of the equations and to solve numerically for the macroscopic response and the evolution of the deformation field. Both approaches result in a thermodynamically consistent strain gradient crystal plasticity model. For the sake of mechanistic understanding, the constitutive relation/the dissipation potential is chosen to be the most simple one; however other choices could easily be implemented in the framework as well following the steps above. After studying the thermodynamics and formulation of single crystal strain gradient plasticity, we focus on the details of the coupled finite element implementation of the model in the next section.

Finite Element Solution Procedure of Strain Gradient Crystal Plasticity Framework

In this section the detailed finite element implementation of the coupled slip-based rate-dependent (convex) strain gradient crystal plasticity model with three slip systems is presented. Considering constant slip resistance, without slip interactions, and the linear case with $m = 1$, the plastic slip-dependent part of the free energy ψ_γ drops, and slip equation becomes

$$\dot{\underline{\gamma}} - \frac{\dot{\gamma}_0}{s}\underline{\tau} - \frac{\dot{\gamma}_0}{s}\nabla \cdot \boldsymbol{\xi} = 0 \tag{46}$$

Substituting $\boldsymbol{\xi}$ and collecting constants in matrix form results in

$$\dot{\underline{\gamma}} - \underline{C}\,\underline{\tau} - \nabla \cdot (\underline{D}\nabla\underline{\gamma}) = 0 \tag{47}$$

where the plastic slips, resolved Schmid stress, and the gradient of the slip systems are included as

$$\underline{\gamma} = \begin{bmatrix} \gamma_1 \\ \gamma_2 \\ \gamma_3 \end{bmatrix} \quad \underline{\tau} = \begin{bmatrix} \tau_1 \\ \tau_2 \\ \tau_3 \end{bmatrix} \quad \nabla\underline{\gamma} = \begin{bmatrix} \frac{\partial \gamma_1}{\partial x} & \frac{\partial \gamma_1}{\partial y} \\ \frac{\partial \gamma_2}{\partial x} & \frac{\partial \gamma_2}{\partial y} \\ \frac{\partial \gamma_3}{\partial x} & \frac{\partial \gamma_3}{\partial y} \end{bmatrix} \tag{48}$$

and the parameters read

$$\underline{C} = \begin{bmatrix} \frac{\dot{\gamma}_0}{s} & 0 & 0 \\ 0 & \frac{\dot{\gamma}_0}{s} & 0 \\ 0 & 0 & \frac{\dot{\gamma}_0}{s} \end{bmatrix} \quad \underline{D} = \begin{bmatrix} \frac{\dot{\gamma}_0}{s}A & 0 & 0 \\ 0 & \frac{\dot{\gamma}_0}{s}A & 0 \\ 0 & 0 & \frac{\dot{\gamma}_0}{s}A \end{bmatrix} \tag{49}$$

Together with the slip equation in (47), the linear momentum balance

$$\nabla \cdot \sigma = \mathbf{0} \tag{50}$$

enters the finite element solution procedure as strong form equations. The weak forms of the equations are obtained in a standard manner, using a Galerkin procedure. Firstly, the balance of linear momentum is tested with a field of virtual displacements $\delta_{\boldsymbol{u}}$ and integrated over the domain Ω, which results in $G_{\boldsymbol{u}}$

$$G_{\boldsymbol{u}} = \int_{\Omega} \delta_{\boldsymbol{u}} \cdot (\nabla \cdot \boldsymbol{\sigma}) d\Omega \tag{51}$$

which could be written as

$$G_{\boldsymbol{u}} = \int_{\Omega} \nabla \delta_{\boldsymbol{u}} : \boldsymbol{\sigma} \, d\Omega - \int_{\Gamma} \boldsymbol{t} \cdot \delta_{\boldsymbol{u}} d\Gamma \tag{52}$$

where $\boldsymbol{t}$ is the traction vector $\boldsymbol{t} = \boldsymbol{\sigma} \cdot \boldsymbol{n}$. Then, $G_{\underline{\gamma}}$ is obtained in a similar way by testing the slip equation with virtual plastic slip and integrating over the domain

$$G_{\gamma} = \int_{\Omega} \underline{\dot{\gamma}} \cdot \underline{\delta}_{\gamma} \, d\Omega - \int_{\Omega} \underline{C} \, \underline{\tau} \cdot \underline{\delta}_{\gamma} \, d\Omega - \int_{\Omega} [\nabla \cdot (\underline{D} \nabla \underline{\gamma})] \cdot \underline{\delta}_{\gamma} d\Omega \tag{53}$$

which could be written as

$$G_{\gamma} = \int_{\Omega} \underline{\dot{\gamma}} \cdot \underline{\delta}_{\gamma} \, d\Omega - \int_{\Omega} \underline{C} \, \underline{\tau} \cdot \underline{\delta}_{\gamma} \, d\Omega + \int_{\Omega} \nabla \underline{\delta}_{\gamma} : \underline{D} \nabla \underline{\gamma} d\Omega - \int_{\Gamma} \underline{t} \cdot \underline{\delta}_{\gamma} d\Gamma \tag{54}$$

with

$$\underline{t} = \underline{D} \nabla \underline{\boldsymbol{\gamma}} \cdot \boldsymbol{n} \tag{55}$$

Next, the equations are linearized through Newton-Raphson iterations using the linearization operator Δ which is basically defined as $\Delta_x F = \dfrac{\partial F}{\partial x} \Delta x$

$$\begin{aligned} \text{Lin} G_{\boldsymbol{u}} &= \Delta_{\boldsymbol{u}} G_{\boldsymbol{u}} + \Delta_{\gamma} G_{\boldsymbol{u}} + G^*_{\boldsymbol{u}} = 0 \\ \text{Lin} G_{\underline{\gamma}} &= \Delta_{\underline{\gamma}} G_{\underline{\gamma}} + \Delta_{\boldsymbol{u}} G_{\underline{\gamma}} + G^*_{\underline{\gamma}} = 0 \end{aligned} \tag{56}$$

Now each term is linearized

$$\Delta_{\boldsymbol{u}} G_{\boldsymbol{u}} = \int_{\Omega} \frac{\partial \boldsymbol{\sigma}}{\partial \boldsymbol{\varepsilon}} : \Delta_{\boldsymbol{u}} \boldsymbol{\varepsilon} : \nabla \delta_{\boldsymbol{u}} \, d\Omega \tag{57}$$

with $\boldsymbol{\sigma} = {}^4\boldsymbol{C} : (\boldsymbol{\varepsilon} - \boldsymbol{\varepsilon}^p)$

$$\frac{\partial \boldsymbol{\sigma}}{\partial \boldsymbol{\varepsilon}} = {}^4\boldsymbol{C}$$
$$\Delta_u \boldsymbol{\varepsilon} = \frac{1}{2}(\nabla \Delta \boldsymbol{u} + (\nabla \Delta \boldsymbol{u})^T) \tag{58}$$

and the first term of the linearized displacement Galerkin reads

$$\boxed{\Delta_u G_u = \int_\Omega \nabla \delta_{\boldsymbol{u}} : {}^4\boldsymbol{C} : \frac{1}{2}(\nabla \Delta \boldsymbol{u} + (\nabla \Delta \boldsymbol{u})^T)\, d\Omega.} \tag{59}$$

Then the second term of the linearized displacement Galerkin is derived as follows

$$\Delta_{\underline{\gamma}} G_u = \int_\Omega \frac{\partial \boldsymbol{\sigma}}{\partial \underline{\gamma}} \Delta \underline{\gamma} : \nabla \delta_{\boldsymbol{u}} d\Omega \tag{60}$$

In order to calculate $\dfrac{\partial \boldsymbol{\sigma}}{\partial \underline{\gamma}}$ the rate of plastic slip is written as

$$\dot{\boldsymbol{\varepsilon}}^p = \sum_\alpha \dot{\gamma}^\alpha \boldsymbol{P}^\alpha \quad \text{with} \quad \boldsymbol{P}^\alpha = \frac{1}{2}(\boldsymbol{s}^\alpha \otimes \boldsymbol{n}^\alpha + \boldsymbol{n}^\alpha \otimes \boldsymbol{s}^\alpha) \tag{61}$$

and the integration of the plastic slip rate gives

$$\boldsymbol{\varepsilon}^p_{n+1} = \boldsymbol{\varepsilon}^p_n + \sum_\alpha (\gamma^\alpha_{n+1} - \gamma^\alpha_n) \boldsymbol{P}^\alpha \tag{62}$$

then the regarding derivatives read as

$$\frac{\partial \boldsymbol{\varepsilon}^p}{\partial \gamma^\beta} = \boldsymbol{P}^\beta$$
$$\frac{\partial \boldsymbol{\sigma}}{\partial \gamma^\beta} = \frac{\partial \boldsymbol{\sigma}}{\partial \boldsymbol{\varepsilon}^p} : \frac{\partial \boldsymbol{\varepsilon}^p}{\partial \gamma^\beta} = -{}^4\boldsymbol{C} : \boldsymbol{P}^\beta \tag{63}$$

Finally the second term of the linearized displacement Galerkin expression is expressed as follows

$$\boxed{\Delta_{\underline{\gamma}} G_u = -\int_\Omega \nabla \delta_{\boldsymbol{u}} : {}^4\boldsymbol{C} : \boldsymbol{P}^\beta \, \Delta \gamma^\beta \, d\Omega.} \tag{64}$$

Then the terms related to the linearization of $G_{\underline{\gamma}}$ are handled. The operator with respect to the displacement is given by

$$\Delta_{u} G_{\underline{\gamma}} = - \int_{\Omega} \underline{C} \frac{\partial \underline{\tau}}{\partial \boldsymbol{\varepsilon}} : \Delta_{u} \boldsymbol{\varepsilon}\, \underline{\delta}_{\gamma}\, d\Omega \tag{65}$$

By using the derivative $\dfrac{\partial \tau^{\alpha}}{\partial \boldsymbol{\varepsilon}} = {}^{4}\boldsymbol{C} : \boldsymbol{P}^{\alpha}$ it could be written as

$$\boxed{\Delta_{u} G_{\underline{\gamma}} = - \int_{\Omega} \underline{C}\, \frac{1}{2} (\nabla \Delta \boldsymbol{u} + (\nabla \Delta \boldsymbol{u})^{T}) : {}^{4}\boldsymbol{C} : \boldsymbol{P}^{\alpha} \underline{\delta}_{\gamma} d\Omega.} \tag{66}$$

The last term is

$$\Delta_{\underline{\gamma}} G_{\underline{\gamma}} = \int_{\Omega} \Delta_{\underline{\gamma}}\, \underline{\dot{\gamma}}\, \underline{\delta}_{\gamma} d\Omega - \int_{\Omega} \underline{C}\, \Delta_{\underline{\gamma}} \underline{\tau}\, \underline{\delta}_{\gamma} d\Omega + \int_{\Omega} \underline{D} \Delta_{\underline{\gamma}} (\nabla \underline{\gamma}) \cdot \nabla \underline{\delta}_{\gamma} d\Omega \tag{67}$$

The following two terms are necessary

$$\begin{aligned} \Delta_{\underline{\gamma}} \underline{\tau} &= \frac{\partial \tau^{\alpha}}{\partial \gamma^{\beta}} \Delta \underline{\gamma} = \frac{\partial}{\partial \gamma^{\beta}} (\boldsymbol{\sigma} : \boldsymbol{P}^{\alpha}) \Delta \underline{\gamma} = -\boldsymbol{P}^{\alpha} : {}^{4}\boldsymbol{C} : \boldsymbol{P}^{\beta} \Delta \underline{\gamma} \\ \Delta_{\underline{\gamma}} (\nabla \underline{\gamma}) &= \nabla \Delta \underline{\gamma} \end{aligned} \tag{68}$$

Finally, equation (67) is written as

$$\boxed{\Delta_{\underline{\gamma}} G_{\underline{\gamma}} = \int_{\Omega} \frac{1}{\Delta t}\, \underline{\delta}_{\gamma}\, \Delta \underline{\gamma}\, d\Omega + \int_{\Omega} \underline{C} \boldsymbol{P}^{\alpha} : {}^{4}\boldsymbol{C} : \boldsymbol{P}^{\beta} \Delta \underline{\gamma}\, \underline{\delta}_{\gamma}\, d\Omega + \int_{\Omega} \underline{D} \nabla (\Delta \underline{\gamma}) \cdot \nabla \underline{\delta}_{\gamma} d\Omega} \tag{69}$$

After the linearization procedure, the above system of equations is discretized. The problem is solved in 2D under plane strain assumption. Quadratic interpolation is used for the displacement field, and linear interpolation is used for the plastic slip field. The degree of freedoms in an element is explicitly illustrated in Fig. 1. The interpolation of the fields is conducted through the matrix of shape functions $\underline{N}$

$$\begin{aligned} \delta_{u} &= \underline{N}^{u}\, \underline{\delta}_{u} & \boldsymbol{u} &= \underline{N}^{u}\, \underline{\boldsymbol{u}} \\ \underline{\delta}_{\gamma} &= \underline{N}^{\gamma} \underline{\underline{\delta}}_{\gamma} & \underline{\gamma} &= \underline{N}^{\gamma} \underline{\underline{\gamma}} \end{aligned} \tag{70}$$

and the interpolations of the gradients of the fields are done through the matrix of shape functions' derivatives

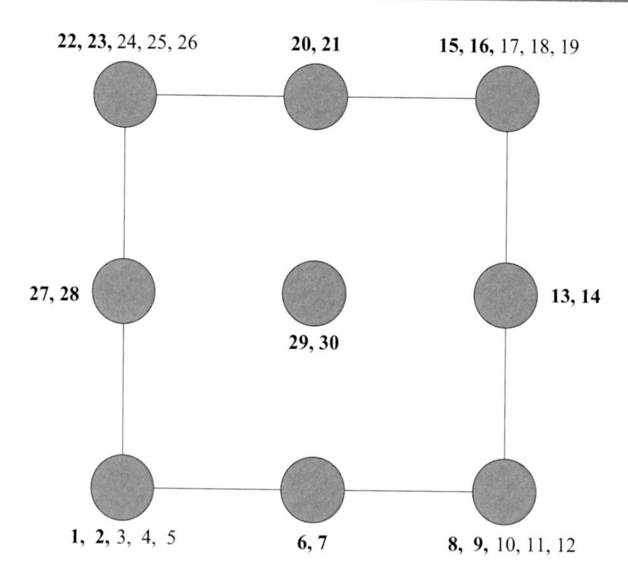

Fig. 1 The displacement degree of freedom numbers is represented with bold, while the other numbers (placed only at the corner nodes) illustrate the plastic dof in an element. This representation corresponds to the quadratic interpolation in $\boldsymbol{u}$ and linear interpolation in γ

$$\begin{aligned} \nabla \delta_{\boldsymbol{u}} &= \underline{B}^u \, \underline{\delta}_{\boldsymbol{u}} \qquad & \nabla \boldsymbol{u} &= \underline{B}^u \, \underline{\boldsymbol{u}} \\ \nabla \underline{\delta}_{\gamma} &= \underline{B}^{\gamma} \, \underline{\underline{\delta}}_{\gamma} \qquad & \nabla \underline{\gamma} &= \underline{B}^{\gamma} \, \underline{\underline{\gamma}} \end{aligned} \tag{71}$$

The second-order tensors are stored in Voigt notation as follows: $\boldsymbol{A} = [A_{11} \;\; A_{12} \;\; A_{21} \;\; A_{22}]'$. The operations in equations (70) and (71) are explicitly illustrated in the matrix format as follows

$$\begin{bmatrix} u_x \\ u_y \end{bmatrix} = \underbrace{\begin{bmatrix} N_1^u & 0 & N_2^u & 0 & \dots & N_9^u & 0 \\ 0 & N_1^u & 0 & N_2^u & \dots & 0 & N_9^u \end{bmatrix}}_{\underline{N}^u} \begin{bmatrix} u_{1x} \\ u_{1y} \\ u_{2x} \\ u_{2y} \\ \cdot \\ \cdot \\ \cdot \\ u_{9x} \\ u_{9y} \end{bmatrix} \tag{72}$$

$$\begin{bmatrix} \gamma^1 \\ \gamma^2 \\ \gamma^3 \end{bmatrix} = \underbrace{\begin{bmatrix} N_1^\gamma & N_2^\gamma & N_3^\gamma & N_4^\gamma & 0 & 0 & 0 & 0 & 0 & 0 & 0 & 0 \\ 0 & 0 & 0 & 0 & N_1^\gamma & N_2^\gamma & N_3^\gamma & N_4^\gamma & 0 & 0 & 0 & 0 \\ 0 & 0 & 0 & 0 & 0 & 0 & 0 & 0 & N_1^\gamma & N_2^\gamma & N_3^\gamma & N_4^\gamma \end{bmatrix}}_{\underline{N}^\gamma} \begin{bmatrix} \gamma_1^1 \\ \gamma_2^1 \\ \gamma_3^1 \\ \gamma_4^1 \\ \gamma_1^2 \\ \gamma_2^2 \\ . \\ . \\ . \\ \gamma_3^3 \\ \gamma_4^3 \end{bmatrix} \tag{73}$$

$$\begin{bmatrix} u_{x,x} \\ u_{x,y} \\ u_{y,x} \\ u_{y,y} \end{bmatrix} = \underbrace{\begin{bmatrix} N_{1,x}^u & 0 & N_{2,x}^u & 0 & \dots & N_{9,x}^u & 0 \\ N_{1,y}^u & 0 & N_{2,y}^u & 0 & \dots & N_{9,y}^u & 0 \\ 0 & N_{1,x}^u & 0 & N_{2,x}^u & \dots & 0 & N_{9,x}^u \\ 0 & N_{1,y}^u & 0 & N_{2,y}^u & \dots & 0 & N_{9,y}^u \end{bmatrix}}_{\underline{B}^u} \begin{bmatrix} u_{1x} \\ u_{1y} \\ u_{2x} \\ u_{2y} \\ . \\ . \\ . \\ u_{9x} \\ u_{9y} \end{bmatrix}, \begin{bmatrix} u_{x,x} \\ u_{x,y} \\ u_{y,x} \\ u_{y,y} \end{bmatrix}$$

$$= \underbrace{\begin{bmatrix} N_{1,x}^u & 0 & N_{2,x}^u & 0 & \dots & N_{9,x}^u & 0 \\ 0 & N_{1,x}^u & 0 & N_{2,x}^u & \dots & 0 & N_{9,x}^u \\ N_{1,y}^u & 0 & N_{2,y}^u & 0 & \dots & N_{9,y}^u & 0 \\ 0 & N_{1,y}^u & 0 & N_{2,y}^u & \dots & 0 & N_{9,y}^u \end{bmatrix}}_{\underline{B}_*^u} \begin{bmatrix} u_{1x} \\ u_{1y} \\ u_{2x} \\ u_{2y} \\ . \\ . \\ . \\ u_{9x} \\ u_{9y} \end{bmatrix} \tag{74}$$

$$\begin{bmatrix} \gamma^1_{,1} \\ \gamma^1_{,2} \\ \gamma^2_{,1} \\ \gamma^2_{,2} \\ \gamma^3_{,1} \\ \gamma^3_{,2} \end{bmatrix} = \underbrace{\begin{bmatrix} N^\gamma_{1,x} & N^\gamma_{2,x} & N^\gamma_{3,x} & N^\gamma_{4,x} & 0 & 0 & 0 & 0 & 0 & 0 & 0 & 0 \\ N^\gamma_{1,y} & N^\gamma_{2,y} & N^\gamma_{3,y} & N^\gamma_{4,y} & 0 & 0 & 0 & 0 & 0 & 0 & 0 & 0 \\ 0 & 0 & 0 & 0 & N^\gamma_{1,x} & N^\gamma_{2,x} & N^\gamma_{3,x} & N^\gamma_{4,x} & 0 & 0 & 0 & 0 \\ 0 & 0 & 0 & 0 & N^\gamma_{1,y} & N^\gamma_{2,y} & N^\gamma_{3,y} & N^\gamma_{4,y} & 0 & 0 & 0 & 0 \\ 0 & 0 & 0 & 0 & 0 & 0 & 0 & 0 & N^\gamma_{1,x} & N^\gamma_{2,x} & N^\gamma_{3,x} & N^\gamma_{4,x} \\ 0 & 0 & 0 & 0 & 0 & 0 & 0 & 0 & N^\gamma_{1,y} & N^\gamma_{2,y} & N^\gamma_{3,y} & N^\gamma_{4,y} \end{bmatrix}}_{\underline{B}^\gamma} \begin{bmatrix} \gamma^1_1 \\ \gamma^1_2 \\ \gamma^1_3 \\ \gamma^1_4 \\ \gamma^2_1 \\ \gamma^2_2 \\ . \\ . \\ . \\ \gamma^3_3 \\ \gamma^3_4 \end{bmatrix} \tag{75}$$

The following matrix will be needed as well

$$\underline{B}^u_S = \frac{1}{2}(\underline{B}^u + \underline{B}^u_*) \tag{76}$$

A commonly used operation regarding the fourth- and second-order tensors is $\boldsymbol{D} : {}^4\boldsymbol{C} : \boldsymbol{B}$ which can be written in index notation as $A = D_{ij}\, C_{ijkl}\, B_{kl}$. The following matrix multiplication is used for this operation

$$[A] = \underbrace{\begin{bmatrix} D_{11}\ D_{12}\ D_{21}\ D_{22} \end{bmatrix}}_{\underline{\boldsymbol{D}}^T} \underbrace{\begin{bmatrix} C_{1111} & C_{1112} & C_{1121} & C_{1122} \\ C_{1211} & C_{1212} & C_{1221} & C_{1222} \\ C_{2111} & C_{2112} & C_{2121} & C_{2122} \\ C_{2211} & C_{2212} & C_{2221} & C_{2222} \end{bmatrix}}_{\underline{\boldsymbol{C}}} \underbrace{\begin{bmatrix} B_{11} \\ B_{12} \\ B_{21} \\ B_{22} \end{bmatrix}}_{\underline{\boldsymbol{B}}} \tag{77}$$

After the discretization the linearization operator of the displacement Galerkin could be written as

$$\begin{aligned} \Delta_{\boldsymbol{u}} G_{\boldsymbol{u}} &= \int_{\Omega^e} \underline{\delta}^T_{\boldsymbol{u}}\ \underline{B}^{uT}\ \underline{\boldsymbol{C}}\ \underline{B}^u_S\ \Delta\boldsymbol{u}\, d\,\Omega^e \\ \Delta_{\underline{\gamma}} G_{\boldsymbol{u}} &= -\int_{\Omega^e} \underline{\delta}^T_{\boldsymbol{u}}\ \underline{B}^{uT}\ \underline{E}\ \underline{N}^\gamma \Delta\underline{\gamma}\, d\,\Omega^e \end{aligned} \tag{78}$$

where

$$\begin{aligned} \underline{E} &= [\underline{E1} \quad \underline{E2} \quad \underline{E3}] \\ \underline{E1} &= \underline{\boldsymbol{C}}\ \underline{P1}^T \\ \underline{E2} &= \underline{\boldsymbol{C}}\ \underline{P2}^T \\ \underline{E3} &= \underline{\boldsymbol{C}}\ \underline{P3}^T \end{aligned} \tag{79}$$

where $\underline{P1}^T$, $\underline{P2}^T$, and $\underline{P3}^T$ represent the column version of the $\boldsymbol{P}^1$, $\boldsymbol{P}^2$, and $\boldsymbol{P}^3$ tensors. $\underline{\boldsymbol{C}}$ is the matrix version of the ${}^4\boldsymbol{C}$ as defined above. The displacement linearization operator of the discretized plastic slip Galerkin reads

$$\Delta_{\boldsymbol{u}} G_\gamma = - \int_{\Omega^e} \delta_\gamma \underline{N^\gamma}^T \underline{C_l}\ \underline{C_p}\ \Delta \boldsymbol{u}\, d\Omega^e \tag{80}$$

with

$$\underline{C_l} = \begin{bmatrix} \frac{\dot{\gamma}_0}{s} & 0 & 0 \\ 0 & \frac{\dot{\gamma}_0}{s} & 0 \\ 0 & 0 & \frac{\dot{\gamma}_0}{s} \end{bmatrix} \quad \text{and} \quad \underline{C_p} = \begin{bmatrix} \underline{C_p^1} \\ \underline{C_p^2} \\ \underline{C_p^3} \end{bmatrix} \quad \text{and} \quad \begin{aligned} \underline{C_p^1} &= \underline{P1}\ \underline{\boldsymbol{C}}\ \underline{B_S^u} \\ \underline{C_p^2} &= \underline{P2}\ \underline{\boldsymbol{C}}\ \underline{B_S^u} \\ \underline{C_p^3} &= \underline{P3}\ \underline{\boldsymbol{C}}\ \underline{B_S^u} \end{aligned} \tag{81}$$

where $\underline{B_S^u}$ is 4×18; $\underline{\boldsymbol{C}}$ is 4×4; $\underline{P1}$, $\underline{P2}$, and $\underline{P3}$ are 1×4; $\underline{C_p^1}$, $\underline{C_p^2}$, and $\underline{C_p^3}$ are 1×18, $\underline{C_p}$ is 3×18; and $\underline{C_l}$ is 3×3 matrices. The slip linearization operator of the discretized plastic slip Galerkin is

$$\begin{aligned} \Delta_\gamma G_\gamma = & \int_{\Omega^e} \frac{1}{\Delta t} \underline{\delta_\gamma} \underline{N^\gamma}^T \underline{N^\gamma} \underline{\Delta\gamma} d\Omega^e + \int_{\Omega^e} \underline{\delta_\gamma}\ \underline{N^\gamma}^T \underline{P^p}\ \underline{N^\gamma} \underline{\Delta\gamma} d\Omega^e \\ & + \int_{\Omega^e} \underline{\delta_\gamma} \underline{B^\gamma}^T \underline{D^d}\ \underline{B^\gamma} \underline{\Delta\gamma} d\Omega^e \end{aligned} \tag{82}$$

where

$$\underline{P^p} = \begin{bmatrix} \frac{\dot{\gamma}_0}{s} \underline{P1}\ \underline{\boldsymbol{C}}\ \underline{P1}^T & \frac{\dot{\gamma}_0}{s} \underline{P1}\ \underline{\boldsymbol{C}}\ \underline{P2}^T & \frac{\dot{\gamma}_0}{s} \underline{P1}\ \underline{\boldsymbol{C}}\ \underline{P3}^T \\ \frac{\dot{\gamma}_0}{s} \underline{P2}\ \underline{\boldsymbol{C}}\ \underline{P1}^T & \frac{\dot{\gamma}_0}{s} \underline{P2}\ \underline{\boldsymbol{C}}\ \underline{P2}^T & \frac{\dot{\gamma}_0}{s} \underline{P2}\ \underline{\boldsymbol{C}}\ \underline{P3}^T \\ \frac{\dot{\gamma}_0}{s} \underline{P3}\ \underline{\boldsymbol{C}}\ \underline{P1}^T & \frac{\dot{\gamma}_0}{s} \underline{P3}\ \underline{\boldsymbol{C}}\ \underline{P2}^T & \frac{\dot{\gamma}_0}{s} \underline{P3}\ \underline{\boldsymbol{C}}\ \underline{P3}^T \end{bmatrix} \quad \text{and} \quad \underline{D^d}$$

$$
= \begin{bmatrix} \frac{\dot{\gamma}_0}{s} A & 0 & 0 & 0 & 0 & 0 \\ 0 & \frac{\dot{\gamma}_0}{s} A & 0 & 0 & 0 & 0 \\ 0 & 0 & \frac{\dot{\gamma}_0}{s} A & 0 & 0 & 0 \\ 0 & 0 & 0 & \frac{\dot{\gamma}_0}{s} A & 0 & 0 \\ 0 & 0 & 0 & 0 & \frac{\dot{\gamma}_0}{s} A & 0 \\ 0 & 0 & 0 & 0 & 0 & \frac{\dot{\gamma}_0}{s} A \end{bmatrix} \tag{83}
$$

And the discretized Galerkins of previous step are written as

$$
\begin{aligned}
G^*_{\boldsymbol{u}} &= \int_{\Omega^e} \underline{\delta}^T_{\boldsymbol{u}} \, \underline{B}^u_S \, \underline{\boldsymbol{\sigma}} \, d\Omega^e \\
G^*_{\underline{\gamma}} &= \int_{\Omega^e} \underline{\underline{\delta}}_{\gamma} \underline{N}^{\gamma T} \underline{N}^{\gamma} \left(\frac{\underline{\gamma}_{n+1} - \underline{\gamma}_n}{\Delta t} \right) d\Omega^e - \int_{\Omega^e} \underline{\underline{\delta}}_{\gamma} \underline{N}^{\gamma T} \, \underline{C} \, \underline{\tau} \, d\Omega^e \\
&+ \int_{\Omega^e} \underline{\underline{\delta}}_{\gamma} \, \underline{B}^{\gamma T} \, \underline{D}^d \, \underline{B}^{\gamma} \underline{\underline{\gamma}} d\Omega^e.
\end{aligned} \tag{84}
$$

Eventually, the element stiffness tangents would be

$$
\begin{aligned}
k_{\boldsymbol{uu}} &= \int_{\Omega^e} \underline{B}^{uT} \, \underline{\boldsymbol{C}} \, \underline{B}^u_S \, d\Omega^e \\
k_{\boldsymbol{u}\underline{\gamma}} &= - \int_{\Omega^e} \underline{B}^{uT} \, \underline{E} \, \underline{N}^{\gamma} d\Omega^e \\
k_{\underline{\gamma}\boldsymbol{u}} &= - \int_{\Omega^e} \underline{N}^{\gamma T} \underline{C}_l \, \underline{C}_p \, d\Omega^e \\
k_{\underline{\gamma}\underline{\gamma}} &= \int_{\Omega^e} \tfrac{1}{\Delta t} \, \underline{N}^{\gamma T} \underline{N}^{\gamma} \Omega^e + \int_{\Omega^e} \underline{N}^{\gamma T} \underline{P}^p \, \underline{N}^{\gamma} d\Omega^e + \int_{\Omega^e} \underline{B}^{\gamma T} \underline{D}^d \, \underline{B}^{\gamma} d\Omega^e.
\end{aligned} \tag{85}
$$

Element residuals are calculated by using the values from the previous estimate

$$
\begin{aligned}
r_{\boldsymbol{u}} &= \int_{\Omega^e} \underline{B}^u_S \, \underline{\boldsymbol{\sigma}} \, d\Omega^e \\
r_{\underline{\gamma}} &= \int_{\Omega^e} \underline{N}^{\gamma T} \underline{N}^{\gamma} \left(\frac{\underline{\underline{\gamma}}_{n+1} - \underline{\underline{\gamma}}_n}{\Delta t} \right) d\Omega^e - \int_{\Omega^e} \underline{N}^{\gamma T} \, \underline{C}_l \, \underline{\tau} \, d\Omega^e \\
&+ \int_{\Omega^e} \underline{B}^{\gamma T} \, \underline{D}^d \, \underline{B}^{\gamma} \underline{\underline{\gamma}} d\Omega^e.
\end{aligned} \tag{86}
$$

The assembly operation gives the global tangent and the residual

$$\begin{aligned} K^{uu} &= \mathbb{A}\{k_{uu}\} & K^{u\underline{\gamma}} &= \mathbb{A}\{k_{u\underline{\gamma}}\} & R^{u} &= \mathbb{A}\{r_{u}\} \\ K^{\underline{\gamma}u} &= \mathbb{A}\{k_{\underline{\gamma}u}\} & K^{\underline{\gamma}\underline{\gamma}} &= \mathbb{A}\{k_{\underline{\gamma}\underline{\gamma}}\} & R^{\underline{\gamma}} &= \mathbb{A}\{r_{\underline{\gamma}}\} \end{aligned} \tag{87}$$

and the system of equations to be solved reads

$$\begin{bmatrix} K^{uu} & K^{u\underline{\gamma}} \\ K^{\underline{\gamma}u} & K^{\underline{\gamma}\underline{\gamma}} \end{bmatrix} \begin{bmatrix} \Delta \boldsymbol{u} \\ \Delta \underline{\gamma} \end{bmatrix} = \begin{bmatrix} -R^{u} + R^{ext}_{\boldsymbol{u}} \\ -R^{\underline{\gamma}} + R^{ext}_{\underline{\gamma}} \end{bmatrix} \tag{88}$$

where $R^{ext}_{\boldsymbol{u}}$ and $R^{ext}_{\underline{\gamma}}$ originate from the boundary terms in the equilibrium. Above procedure is implemented in Matlab and Abaqus software, and numerical problems illustrating the behavior of polycrystalline materials are solved and presented in the next section.

Simulation of Polycrystalline Behavior

In this section the performance of the strain gradient crystal plasticity model is tested through size effect simulations using Voronoi tessellation (see, e.g., Aurenhammer 1991), which is used to obtain basic geometries of polycrystalline aggregates containing 14, 110 or 212 grains, presented in Fig. 2. For each aggregate three different average grain sizes are used: $D_{avg} = 50$, 100 and 150 μm. This compares well with the average grain sizes of AISI 304 stainless steels (1 to 47 μm, Di Schino and Kenny 2003) and AISI 316L (13 to 107 μm Feaugas and Haddou 2003).

Displacement load is applied to the left and right edge in the global -X direction (left edge) and the global +X direction (right edge), resulting in macroscopic $<\epsilon_{11}> = 10\%$. The symbol $\langle\,\rangle$ represents the Macaulay bracket, indicating a macroscopically averaged value. Rigid body movement is prevented by fixing the bottom left and bottom right nodes of the model in global Y direction. If one imagines that the aggregate is embedded in a larger piece of material, the applied boundary condition lies between the two bordering cases of constraints, imposed to the aggregate by the surrounding material: (a) stresses or (b) displacements imposed on the aggregate boundaries. The latter condition results in straight edges. Plastic slip at the boundaries of the model is not constrained.

Material parameters are taken from Yalcinkaya et al. (2012) and are used to demonstrate the strain gradient effects in the polycrystalline aggregates. They are not directly related to any engineering materials. Table 1 lists the used strain gradient crystal plasticity material properties. Crystallographic orientations of grains are randomly distributed (0–360°) using uniform probability distribution. Three slip systems are considered in each grain.

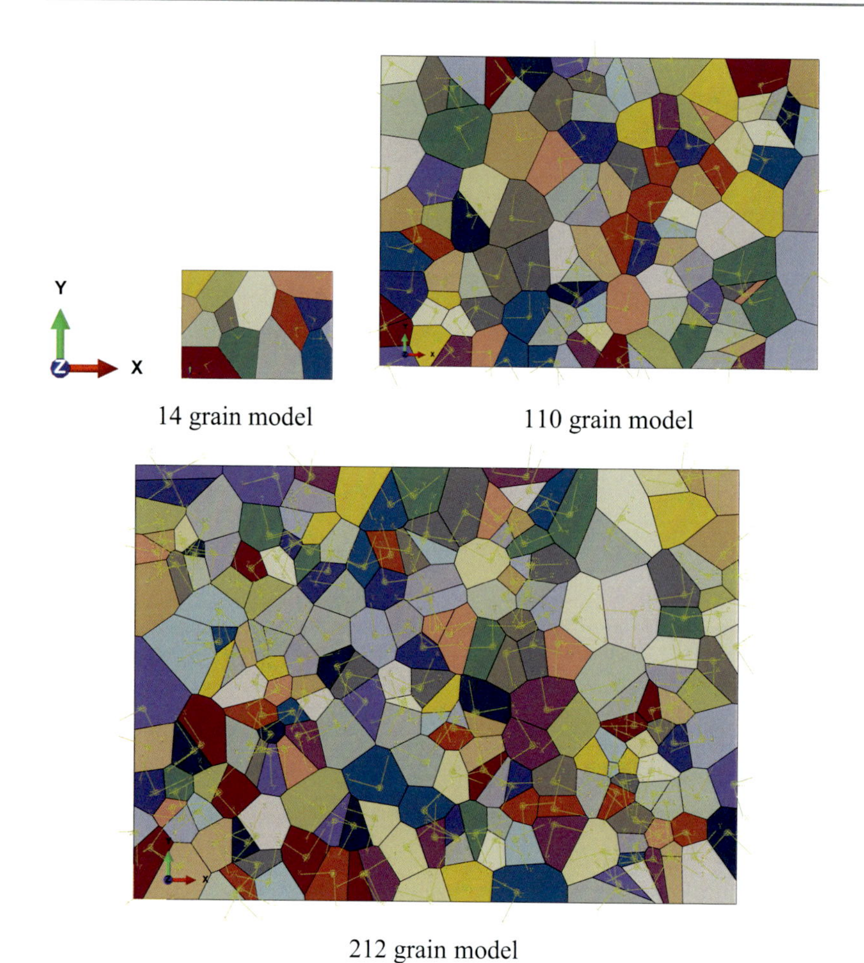

Fig. 2 14, 110, and 212 grain models. Colors represent individual grains; grain orientations are given by local coordinate systems

Table 1 Material properties of the strain gradient single crystal plasticity model

Young modulus E [MPa]	Poisson ratio ν [/]	Reference slip rate $\dot{\gamma}_0$ [s^{-1}]	Slip resistance s [MPa]	Orientations [°]	Material length scale R [μm]
210000.0	0.33	0.15	20.0	120, 60, 45	0.0, 1, 5, 7.5, 10

Material length scale parameter R or the average grain size has substantial effect on the plastic behavior of the material. R is related to different microstructural features such as dislocation spacing (see, e.g., Nix and Gao 1998), dislocation source distance (see, e.g., Aifantis et al. 2009), or grain size (see, e.g., Voyiadjis and Abu Al-Rub 2005). Here we relate R to a certain percentage of the average grain size, D_{avg}. The ratio R/D_{avg} has a significant effect on the macroscopic stress-strain response. Larger R/D_{avg} values lead to stiffer responses; see Fig. 3.

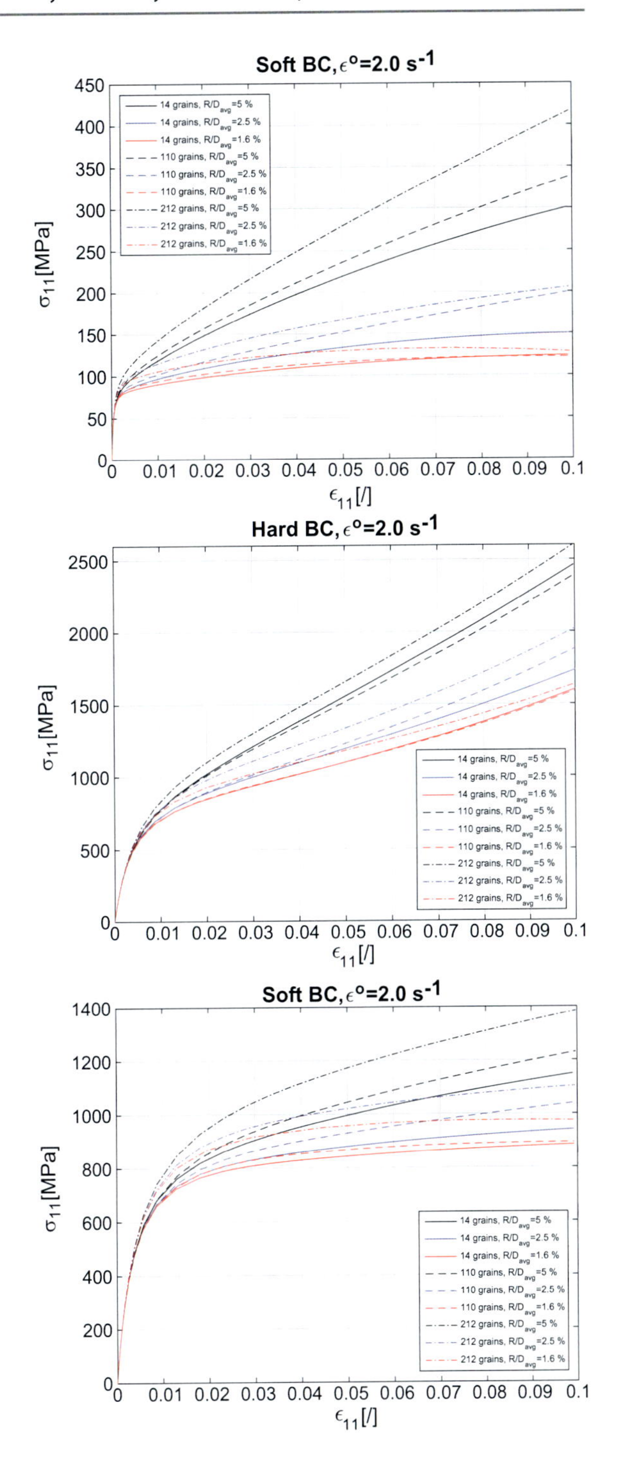

Fig. 3 The effect of the material length scale parameter R for strain rates of $\dot{\epsilon} = 0.2$ and $2\,\mathrm{s}^{-1}$

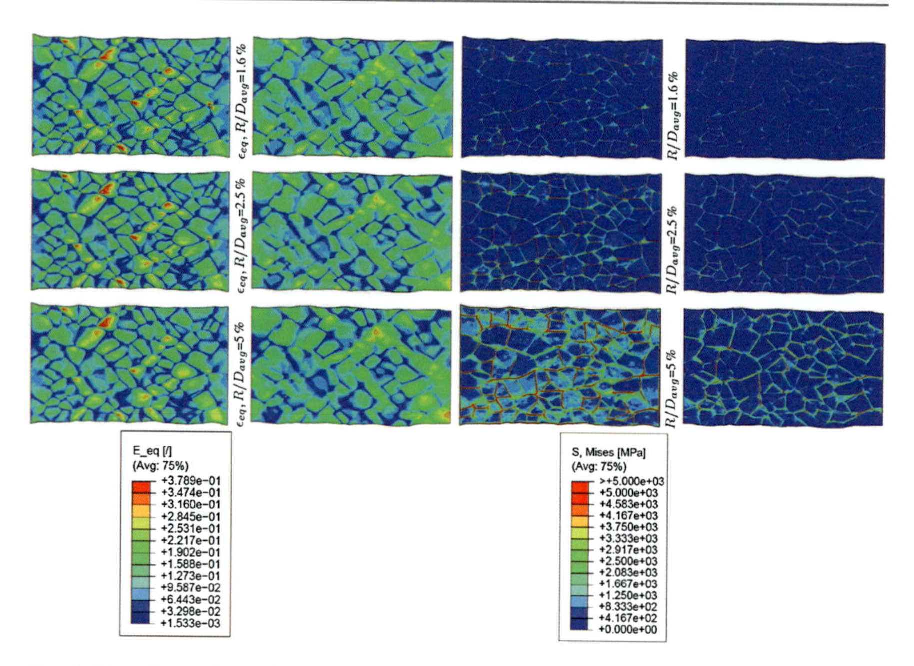

Fig. 4 The effect of hard (*left*) and soft (*right*) BC on the strain and stress fields in a 110 grain aggregate with $R/D_{avg} = 1.6\,\%$, $2.5\,\%$, and $5\,\%$ and $\dot{\epsilon} = 0.2\,\mathrm{s}^{-1}$

The same trend is observed for the two strain rates $\dot{\epsilon} = 0.2,\ 2\,\mathrm{s}^{-1}$. Stiffer responses are obtained at higher strain rates which is in line with expectation. For soft BC higher R/D_{avg} ratios also lead to larger differences in the response between aggregates of different number of grains. So grain number influence is larger. For hard BC, the grain number influence on the macroscopic response is significantly smaller, while at the same time the macroscopic responses are stiffer compared to the soft BC responses. For a given R/D_{avg} ratio, both local responses of a polycrystalline aggregate are the same even though the grain sizes are different, resulting in the same macroscopic responses as well.

Hard and soft BCs have strong effect on the local stress/strain distribution as well. With hard BC the plastic slips at the grain boundaries are imposed to be zero, which constrains the deformations at the grain boundaries significantly. Consequently, the grain boundary stresses are higher; see Figs. 4 and 5 for two different strain rates where Mises equivalent stress and strains are plotted. Increasing the R/D_{avg} ratios increases the grain boundary stresses and widens the areas of higher stresses at the grain boundaries. This is especially evident for the hard BC but can also be clearly observed for the soft BC. In general, the grain boundary stress concentrations can be observed already at relative small R/D_{avg} ratio of 1.6%.

Both higher strain rates and R/D_{avg} ratios result in stiffer macroscopic responses, while R/D_{avg} significantly impacts the grain boundary stresses, where the influence of the rate on the grain boundary stress is considerably smaller.

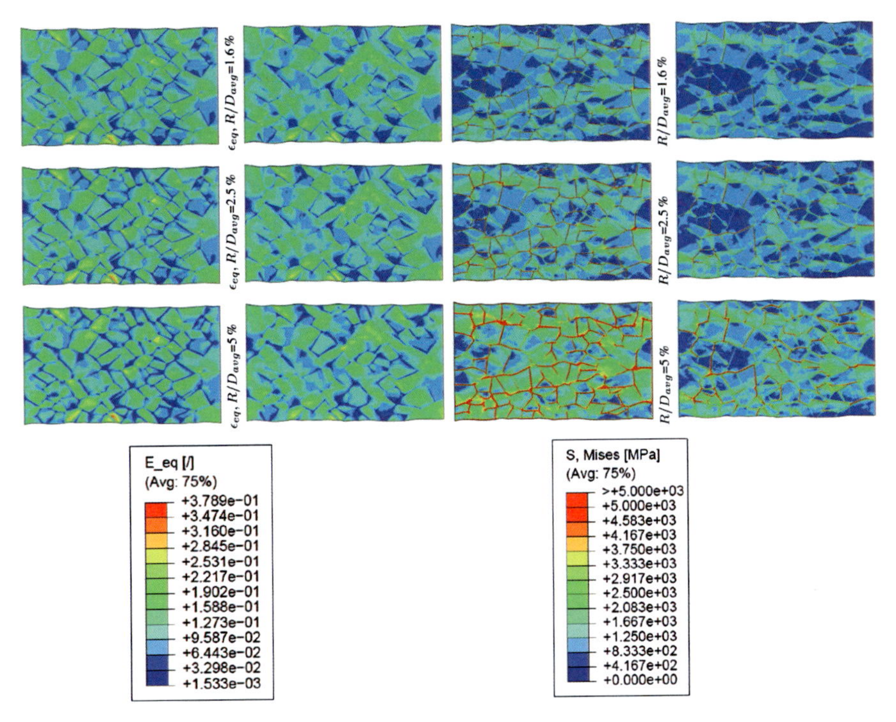

Fig. 5 The effect of hard (*left*) and soft (*right*) BC on the strain and stress fields in a 110 grain aggregate with $R/D_{avg} = 1.6\,\%$, $2.5\,\%$, and $5\,\%$ and $\dot{\epsilon} = 2\,\mathrm{s}^{-1}$

As for the material parameters, the rates used here are not related to a specific experiment, and they are quite high compared to 0.5 mm/mm/min ($0.0083\,\mathrm{s}^{-1}$) – 0.05 mm/mm/m ($0.00083\,\mathrm{s}^{-1}$) which ASTM E 8/E 8M-08 specifies for the strain rate during standard tensile testing of metallic materials ASTM 2009. Both material parameters and loading conditions should be identified for more realistic simulations.

Conclusion and Outlook

In this chapter, the thermodynamically consistent derivation of a higher-order, work-conjugate, plastic slip-based rate-dependent strain gradient crystal plasticity model is presented through both classical thermodynamics and rate variational formulation. After obtaining the coupled strong form of equations, the weak forms are derived through Galerkin method, and the complete finite element implementation is discussed in detail. The model presented here does not include any energetic hardening term. For the sake of simplicity and clarity, a linear dependence for the slip evolution equation is employed. Any extension of the model to include nonlinear terms could easily be done by following the presented steps above. The performance of the model is illustrated through polycrystalline examples using

Voronoi tessellation, and the influence of the boundary conditions, the loading rate, and the number and the size of the grains are studied. The current study considers only the hard and soft boundary conditions which are two extreme cases restricting the plastic slip completely or allowing the dislocation slip transfer without any resistance. Even though it is possible to capture the size effect due to the plasticity activity at the grain boundaries with the current nonlocal model, the boundary conditions are not completely physical, and the real behavior is somewhere in between these two situations. A grain boundary model is needed in order to incorporate the complex dislocation-grain boundary interaction mechanisms into the plasticity model. Therefore, the next chapter studies a particular grain boundary model and illustrates explicitly the effect of the misorientation between the grains and the orientation of the grain boundary. Due to its convex nature, the current model does not predict the intragranular plastic/dislocation microstructure formation and evolution. This requires the incorporation of an additional non-convex energy term.

Acknowledgements Tuncay Yalçinkaya gratefully acknowledges the support by the Scientific and Technological Research Council of Turkey (TÜBİTAK) under the 3001 Programme (Grant No. 215M381).

References

A. Acharya, J.L. Bassani, Lattice incompatibility and a gradient theory of crystal plasticity. J. Mech. Phys. Solids **48**, 1565–1595 (2000)

E.C. Aifantis, On the microstructural origin of certain inelastic models. J. Eng. Mater. Technol. **106**, 326–330 (1984)

E.C. Aifantis, Strain gradient interpretation of size effects. Int. J. Fract. **95**, 299–314 (1999)

K. Aifantis, J. Senger, D. Weygand, M. Zaiser, Discrete dislocation dynamics simulation and continuum modeling of plastic boundary layers in tricrystal micropillars. IOP Conf. Ser. Mater. Sci. Eng. **3**, 012025 (2009)

A. Arsenlis, D.M. Parks, R. Becker, V.V. Bulatov, On the evolution of crystallographic dislocation density in non-homogeneously deforming crystals. J. Mech. Phys. Solids **52**, 1213–1246 (2004)

M.F. Ashby, The deformation of plastically non-homogeneous materials. Philos. Mag. **21**, 399–424 (1970)

ASTM, *Annual Book of ASTM Standards* (ASTM International, West Conshohocken, 2009)

F. Aurenhammer, Voronoi diagrams – a survey of a fundamental geometric data structure. ACM Comput. Surv. **23**(3), 345–405 (1991)

J.L. Bassani, Incompatibility and a simple gradient theory. J. Mech. Phys. Solids **49**, 1983–1996 (2001)

E. Bayerschen, A.T. McBride, B.D. Reddy, T. Böhlke, Review on slip transmission criteria in experiments and crystal plasticity models. J. Mater. Sci. **51**(5), 2243–2258 (2016)

C.J. Bayley, W.A.M. Brekelmans, M.G.D. Geers, A comparison of dislocation induced back stress formulations in strain gradient crystal plasticity. Int. J. Solids Struct. **43**, 7268–7286 (2006)

P. van Beers, V. Kouznetsova, M. Geers, Defect redistribution within a continuum grain boundary plasticity model. J. Mech. Phys. Solids **83**, 243–262 (2015a)

P. van Beers, V. Kouznetsova, M. Geers, Grain boundary interfacial plasticity with incorporation of internal structure and energy. Mech. Mater. **90**, 69–82 (2015b). Proceedings of the IUTAM Symposium on Micromechanics of Defects in Solids

U. Borg, A strain gradient crystal plasticity analysis of grain size effects in polycrystals. Eur. J. Mech. A-Solid. **26**, 313–324 (2007)

S.H. Chen, T.C. Wang, A new hardening law for strain gradient plasticity. Acta Mater. **48**, 3997–4005 (2000)
A. Di Schino, J. Kenny, Grain size dependence of the fatigue behaviour of a ultrafine-grained AISI 304 stainless steel. Mater. Lett. **57**(21), 3182–3185 (2003)
F.P.E. Dunne, D. Rugg, A. Walker, Lengthscale-dependent, elastically anisotropic, physically-based HCP crystal plasticity: application to cold-dwell fatigue in Ti alloys. Int. J. Plast. **23**, 1061–1083 (2007)
L.P. Evers, W.A.M. Brekelmans, M.G.D. Geers, Non-local crystal plasticity model with intrinsic SSD and GND effects. J. Mech. Phys. Solids **52**, 2379–2401 (2004)
X. Feaugas, H. Haddou, Grain-size effects on tensile behavior of nickel and AISI 316l stainless steel. Metall. Mater. Trans. A **34A**, 2329–2340 (2003)
N.A. Fleck, J.W. Hutchinson, Strain gradient plasticity. Adv. Appl. Mech. **33**, 184–251 (1997)
N.A. Fleck, J.W. Hutchinson, A reformulation of strain gradient plasticity. J. Mech. Phys. Solids **49**, 2245–2271 (2001)
N.A. Fleck, G.M. Muller, M.F. Ashby, J.W. Hutchinson, Strain gradient plasticity: theory and experiment. Acta Metall. Mater. **42**, 475–487 (1994)
N.A. Fleck, J.W. Hutchinson, J.R. Willis, Strain gradient plasticity under non-proportional loading. Proc. R. Soc. A **470**, 20140267 (2014)
M.G.D. Geers, W.A.M. Brekelmans, C.J. Bayley, Second-order crystal plasticity: internal stress effects and cyclic loading. Modell. Simul. Mater. Sci. Eng. **15**, 133–145 (2007)
D. Gottschalk, A. McBride, B. Reddy, A. Javili, P. Wriggers, C. Hirschberger, Computational and theoretical aspects of a grain-boundary model that accounts for grain misorientation and grain-boundary orientation. Comput. Mater. Sci. **111**, 443–459 (2016)
P. Gudmundson, A unified treatment of strain gradient plasticity. J. Mech. Phys. Solids **52**, 1379–1406 (2004)
M.E. Gurtin, On the plasticity of single crystals: free energy, microforces, plastic-strain gradients. J. Mech. Phys. Solids **48**, 989–1036 (2000)
M.E. Gurtin, A gradient theory of single-crystal viscoplasticity that accounts for geometrically necessary dislocations. J. Mech. Phys. Solids **50**, 5–32 (2002)
M.E. Gurtin, A finite-deformation, gradient theory of single-crystal plasticity with free energy dependent on densities of geometrically necessary dislocations. Int. J. Plast. **24**, 702–725 (2008)
M.E. Gurtin, L. Anand, Thermodynamics applied to gradient theories involving the accumulated plastic strain: the theories of Aifantis and Fleck and Hutchinson and their generalization. J. Mech. Phys. Solids **57**, 405–421 (2009)
C.S. Han, H. Gao, Y. Huang, W.D. Nix, Mechanism-based strain gradient crystal plasticity – I. Theory. J. Mech. Phys. Solids **53**, 1188–1203 (2005a)
C.S. Han, H. Gao, Y. Huang, W.D. Nix, Mechanism-based strain gradient crystal plasticity – II. Analysis. J. Mech. Phys. Solids **53**, 1204–1222 (2005b)
M.A. Haque, M.T.A. Saif, Strain gradient effect in nanoscale thin films. Acta Mater. **51**, 3053–3061 (2003)
Y. Huang, S. Qu, K.C. Hwang, M. Li, H. Gao, A conventional theory of mechanism-based strain gradient plasticity. Int. J. Plast. **20**, 753–782 (2004)
J.W. Hutchinson, Generalizing j2 flow theory: fundamental issues in strain gradient plasticity. Acta Mech. Sinica **28**, 1078–1086 (2012)
B. Klusemann, T. Yalçinkaya, Plastic deformation induced microstructure evolution through gradient enhanced crystal plasticity based on a non-convex helmholtz energy. Int. J. Plast. **48**, 168–188 (2013)
B. Klusemann, T. Yalçinkaya, M.G.D. Geers, B. Svendsen, Application of non-convex rate dependent gradient plasticity to the modeling and simulation of inelastic microstructure development and inhomogeneous material behavior. Comput. Mater. Sci. **80**, 51–60 (2013)
M. Kuroda, V. Tvergaard, On the formulations of higher-order strain gradient crystal plasticity models. J. Mech. Phys. Solids **56**, 1591–1608 (2008)
G. Lancioni, T. Yalçinkaya, A. Cocks, Energy-based non-local plasticity models for deformation patterning, localization and fracture. Proc. R. Soc. A **471**: 20150275 (2015a)

G. Lancioni, G. Zitti, T. Yalcinkaya, Rate-independent deformation patterning in crystal plasticity. Key Eng. Mater. **651–653**, 944–949 (2015b)

V. Levkovitch, B. Svendsen, On the large-deformation- and continuum-based formulation of models for extended crystal plasticity. Int. J. Solids Struct. **43**, 7246–7267 (2006)

L. Liang, F.P.E. Dunne, GND accumulation in bi-crystal deformation: crystal plasticity analysis and comparison with experiments. Int. J. Mech. Sci. **51**, 326–333 (2009)

A. Ma, F. Roters, D. Raabe, A dislocation density based constitutive model for crystal plasticity FEM including geometrically necessary dislocations. Acta Mater. **54**, 2169–2179 (2006)

H.B. Mühlhaus, E.C. Aifantis, A variational principle for gradient plasticity. Int. J. Solids Struct. **28**, 845–857 (1991)

W. Nix, H. Gao, Indentation size effects in crystalline materials: a law for strain gradient plasticity. J. Mech. Phys. Solids **46**(3), 411–425 (1998)

T. Ohashi, Crystal plasticity analysis of dislocation emission from micro voids. Int. J. Plast. **21**, 2071–2088 (2005)

I. Özdemir, T. Yalçinkaya, Modeling of dislocation-grain boundary interactions in a strain gradient crystal plasticity framework. Comput. Mech. **54**, 255–268 (2014)

A. Panteghini, L. Bardella, On the finite element implementation of higher-order gradient plasticity, with focus on theories based on plastic distortion incompatibility. Comput. Methods Appl. Mech. Eng. (2016). https://doi.org/10.1016/j.cma.2016.07.045

P. Perzyna, Temperature and rate dependent theory of plasticity of crystalline solids. Revue Phys. Appl. **23**, 445–459 (1988)

B.D. Reddy, The role of dissipation and defect energy in variational formulations of problems in strain-gradient plasticity. Part 1: polycrystalline plasticity. Contin. Mech. Thermodyn. **23**, 527–549 (2011a)

B.D. Reddy, The role of dissipation and defect energy in variational formulations of problems in strain-gradient plasticity. Part 2: single-crystal plasticity. Contin. Mech. Thermodyn. **23**, 551–572 (2011b)

J.R. Rice, Inelastic constitutive relations for solids: an internal variable theory and its application to metal plasticity. J. Mech. Phys. Solids **19**, 433–455 (1971)

J.Y. Shu, N.A. Fleck, Strain gradient crystal plasticity: size-dependent deformation of bicrystals. J. Mech. Phys. Solids **47**, 297–324 (1999)

M. Silhavy, *The Mechanics and Thermodynamics of Continuous Media*, 1st edn. (Springer, Berlin, 1997)

J.S. Stölken, A.G. Evans, A microbend test method for measuring the plasticity length scale. Acta Mater. **46**, 5109–5115 (1998)

B. Svendsen, On thermodynamic- and variational-based formulations of models for inelastic continua with internal length scales. Comput. Methods Appl. Mech. Eng. **193**, 5429–5452 (2004)

B. Svendsen, S. Bargmann, On the continuum thermodynamic rate variational formulation of models for extended crystal plasticity at large deformation. J. Mech. Phys. Solids **58**, 1253–1271 (2010)

J.G. Swadenera, E.P. Georgea, G.M. Pharra, The correlation of the indentation size effect measured with indenters of various shapes. J. Mech. Phys. Solids **50**, 681–694 (2002)

G.I. Taylor, Plastic strain in metals. J. Inst. Met. **62**, 307–325 (1938)

C.A. Volkert, E.T. Lilleodden, Size effects in the deformation of sub-micron au columns. Philos. Mag. **86**, 5567–5579 (2006)

G. Voyiadjis, R. Abu Al-Rub, Gradient plasticity theory with a variable length scale parameter. Int. J. Solids Struct. **42**(14), 3998–4029 (2005)

J. Wang, J. Lian, J.R. Greer, W.D. Nix, K.S. Kim, Size effect in contact compression of nano- and microscale pyramid structures. Acta Mater. **54**, 3973–3982 (2006)

T. Yalcinkaya, Microstructure evolution in crystal plasticity : strain path effects and dislocation slip patterning. PhD Thesis, Eindhoven University of Technology, 2011

T. Yalçinkaya, Multi-scale modeling of microstructure evolution induced anisotropy in metals. Key Eng. Mater. **554–557**, 2388–2399 (2013)

T. Yalcinkaya, G. Lancioni, Energy-based modeling of localization and necking in plasticity. Proc. Mater. Sci. **3**, 1618–1625 (2014)
T. Yalcinkaya, W.A.M. Brekelmans, M.G.D. Geers, Deformation patterning driven by rate dependent nonconvex strain gradient plasticity. J. Mech. Phys. Solids **59**, 1–17 (2011)
T. Yalcinkaya, W.A.M. Brekelmans, M.G.D. Geers, Non-convex rate dependent strain gradient crystal plasticity and deformation patterning. Int. J. Solids Struct. **49**, 2625–2636 (2012)
S. Yefimov, I. Groma, E. van der Giessena, A comparison of a statistical-mechanics based plasticity model with discrete dislocation plasticity calculations. J. Mech. Phys. Solids **52**, 279–300 (2004)

Strain Gradient Crystal Plasticity: Intergranular Microstructure Formation

29

İzzet Özdemir and Tuncay Yalçinkaya

Contents

Abstract

This chapter addresses the formation and evolution of inhomogeneous plastic deformation field between grains in polycrystalline metals by focusing on continuum scale modeling of dislocation-grain boundary interactions within a strain gradient crystal plasticity (SGCP) framework. Thermodynamically consistent extension of a particular strain gradient plasticity model, addressed previously

İ. Özdemir (✉)
Department of Civil Engineering, İzmir Institute of Technology, İzmir, Turkey
e-mail: izzetozdemir@iyte.edu.tr

T. Yalçinkaya
Aerospace Engineering Program, Middle East Technical University Northern Cyprus Campus, Guzelyurt, Mersin, Turkey

Department of Aerospace Engineering, Middle East Technical University, Ankara, Turkey
e-mail: yalcinka@metu.edu.tr

G. Z. Voyiadjis (ed.), *Handbook of Nonlocal Continuum Mechanics for Materials and Structures*, https://doi.org/10.1007/978-3-319-58729-5_4

(see also, e.g., Yalcinkaya et al, J Mech Phys Solids 59:1–17, 2011), is presented which incorporates the effect of grain boundaries on plastic slip evolution explicitly. Among various choices, a potential-type non-dissipative grain boundary description in terms of grain boundary Burgers tensor (see, e.g., Gurtin, J Mech Phys Solids 56:640–662, 2008) is preferred since this is the essential descriptor to capture both the misorientation and grain boundary orientation effects. A mixed finite element formulation is used to discretize the problem in which both displacements and plastic slips are considered as primary variables. For the treatment of grain boundaries within the solution algorithm, an interface element is formulated. The capabilities of the framework is demonstrated through 3D bi-crystal and polycrystal examples, and potential extensions and currently pursued multi-scale modeling efforts are briefly discussed in the closure.

Keywords
Strain gradient plasticity · Grain boundary · Grain boundary-dislocation interaction · Misorientation · Grain boundary Burgers tensor

Introduction

At the grain scale, polycrystalline materials tend to develop heterogeneous plastic deformation fields due to variation of grain orientations, geometries, and defects. Grain boundaries are natural locations for plastic slip accumulation and geometrically necessary dislocations accommodating the gradients of inhomogeneous plastic strain. Grain boundaries have been subject to extensive research for decades due to their major influence on the mechanical properties of polycrystalline metals. Investigations on grain boundaries within polycrystalline materials date back to the observations of Hall and Petch. According to the so-called Hall-Petch effect, the yield strength of a polycrystalline metal specimen scales linearly with the inverse square root of the grain size. This is explained by the fact that grain boundaries behave as barriers against dislocation mobility, and they limit the mean free path of the dislocations thereby increasing strain hardening. Obviously, the grain size represents an intrinsic length scale of the polycrystalline system which inevitably influences the material response significantly. Conventional plasticity models do not incorporate a material length scale, and predictions of such theories involve only the lengths associated with the geometry of the whole solid body. In order to capture the grain size-dependent response, an internal length scale must be incorporated in the constitutive law. In the last two decades, the number of phenomenological and physically based strain gradient-type plasticity models has increased significantly. The intention of these models was to address the issue of internal length scale properly. Some of these models indeed consider the grain boundaries in a direct or smeared way; however understanding of dislocation-grain boundary interaction is far from being complete and constitutive models, and their numerical implementations are still subject of ongoing discussions. As stated by McDowell (2008), capturing the role of grain boundaries accurately within continuum scale plasticity models is a challenging task. To this end, researchers have utilized different models and tools at different scales. In the context of grain boundary-dislocation interaction modeling, most of the existing efforts fall under one of the four categories identified as: (i) atomistic level studies, (ii) dislocation dynamics simulations, (iii) macroscale phenomenological modeling, and (iv) (multi-scale) crystal plasticity modeling.

When resolved at atomic scale, grain boundary is a thin transition layer from one grain (orientation) to the other, and it has its own structure. This structure endows the grain boundary with an initial energy (equilibrium energy) which is the focus of a considerable fraction of the literature on atomistic level grain boundary models; see, for example, Tschopp and McDowell (2007) and McDowell (2008). Although these studies mostly focus on specific grain boundaries (e.g., symmetric tilt boundaries) and a limited number of materials, the findings might be used effectively in multi-scale strain gradient plasticity type models. Developing a quantitative understanding through experiments on dislocation nucleation and slip transfer reactions at grain boundaries is extremely difficult. Therefore, atomistic models are the ideal tools to develop an understanding on unit mechanisms involved in dislocation absorption, transmission, and reflection within/from a grain boundary as exemplified by de Koning et al. (2002, 2003) and Spearot and McDowell (2009). However, significant challenges exist such as extending the number of slip transfer events to realistically large numbers, modeling nonequilibrium grain boundaries, and understanding dislocation activity in polycrystalline metals with heterogeneous compositions, e.g., impurities at grain boundaries or metallic alloys with hetero-phase interfaces.

Along with atomistic models, there have been a number of studies which have tried to embed grain boundaries within discrete dislocation dynamics framework (DDD, Van der Giessen and Needleman 1995) in a consistent way. At first, the focus was primarily on the treatment of slip transmission through grain boundaries within DDD. In this context, dislocation transmission rules based on transmission electron microscope (TEM) observations (see, e.g., Shen et al. 1986; Lee et al. 1989) were introduced into DDD codes (Li et al. 2009). As an alternative, the findings of atomistic level studies such as the one by de Koning et al. (2002) are adopted to introduce slip transmission criteria in DDD; see Kumar et al. (2010). However, the transmission of dislocations is not the only interaction mechanism to be considered. The absorption and emission of dislocations are also possible, but the rules for the incorporation of these phenomena are not clear yet.

The last two groups of studies mainly consist of strain gradient-type plasticity models taking into account the effect of inhomogeneous plastic strain distribution due to grain boundaries, i.e., high slip gradients in the vicinity of grain boundaries and related back-stress evolution, with either grain boundary conditions or with a decomposition of the grain into a core and a grain boundary effected zone. The majority of the strain gradient models could only treat the grain boundaries in two limiting cases as either impenetrable surfaces with no slip (hard boundaries) or slip without any resistance (soft boundaries). However, the proper interpretation of grain boundary effects could only be done by incorporating a grain boundary description into strain gradient plasticity models which could cover other scenarios of dislocation-grain boundary interactions.

It was first recognized by Gudmundson (2004) that jump conditions across interfaces in strain gradient theories can be introduced and an additional contribution to the internal virtual work has to be considered along the interface between two plastically deforming phases. The interface conditions and traction expressions are derived, yet a polycrystal plasticity simulation was not performed. Later on Fredriksson and Gudmundson (2005) used this strain gradient framework with the interface model to capture dislocation built up at elastic/plastic interfaces. In a following work, Aifantis and Willis (2005) presented a similar approach which incorporates the physical properties of interfaces by means of interface potential appended to a phenomenological strain gradient plasticity model. To explore the capabilities of the model, Aifantis et al. (2006) calibrated the proposed interface potential through indentation tests, where the grain boundary "yield" reveals itself in the form of strain/displacement bursts observed in the load vs. tip displacement diagrams. In a parallel work, Abu Al-Rub (2008) extended a strain gradient plasticity framework by including an interface potential to investigate thin film mechanics. In the work of Massart and Pardoen (2010), interface elements were used to control/prescribe higher order boundary conditions stemming from the use of Fleck-Hutchinson gradient plasticity model (Fleck and Hutchinson 2001). The rate of plastic slip is initially set to zero at these interfaces and evolves with deformation mimicking the grain boundary relaxation mechanisms. Apart from these phenomenological gradient plasticity theories, there have been a number of attempts which have consolidated grain boundary effects within SGCP frameworks. One of the earliest ones was proposed by Ma et al. (2006) which incorporated the mechanical interaction between mobile dislocations and grain boundaries in a dislocation-based crystal plasticity model. The influence of grain boundaries is taken into account in terms of an additional energy affecting the dislocation velocity (plastic slip) in the bulk as opposed to the interfacial nature of grain boundaries. Nevertheless, the study reveals that the misorientation of the neighboring grains alone is not sufficient to describe the influence of grain boundaries. Borrowing the ideas of Aifantis and Willis (2005), Borg (2007) and Borg and Fleck (2007) incorporated an energy potential that penalizes crystallographic slip at grain boundaries into a SGCP theory for finite deformations. This approach does not take into account the misorientation of the neighboring grains or the grain boundary orientation, reflecting a rather phenomenological description. In a more recent study by Ekh et al. (2011), interface conditions sensitive to misorientation between grains were introduced in an ad hoc manner, where the slip resistance at grain boundaries decreases as the level of slip system alignment increases. However Gurtin (2008), which is also the point of departure for this chapter and has influenced many other recent works by van Beers et al. (2013), Özdemir and Yalçinkaya (2014), and Gottschalk et al. (2016), intrinsically includes the effect of both the misorientation between neighboring grains and the orientation of the grain boundary through grain boundary Burgers tensor.

Departing from the structure outlined by Gurtin (2008), in this chapter an extension of a rate-dependent strain gradient crystal plasticity model (convex counterpart of Yalcinkaya et al. 2011 and Yalçinkaya et al. 2012) is presented in which grain boundary response is embedded in a thermodynamically consistent manner. Furthermore a mixed finite element-based solution algorithm is elaborated where both the plastic slips and displacement fields are taken as degrees of freedom.

The chapter is organized as follows. First, in section "Strain Gradient Crystal Plasticity Framework and Grain Boundary Model," the SGCP framework is recapitulated, and its consistent extension is presented. The particular form of the grain boundary potential is described in detail as well. In the following section, the finite element-based solution algorithm and the treatment of grain boundaries in a discrete setting are worked out. In section "Numerical Examples," the proposed formulation is assessed on the basis of three-dimensional examples. The chapter is closed by conclusion and outlook section, commenting on current efforts of extensions within a multi-scale modeling perspective.

Strain Gradient Crystal Plasticity Framework and Grain Boundary Model

In this section, within the context of continuum thermodynamics, crystal plasticity model is presented where the bulk behavior and the interface are treated separately. First, the force balances are derived via the principle of virtual power followed by the consideration of dissipation inequality for the bulk material and the interface successively.

Principle of Virtual Power: Macroscopic and Microscopic Energy Balances

In a geometrically linear kinematics setting, the time-dependent displacement field of a body with a grain boundary, as shown in Fig. 1, is denoted by $\boldsymbol{u} = \boldsymbol{u}(\boldsymbol{x}, t)$, where $\boldsymbol{x}$ indicates the position of a material point. It has to be noted that the displacement field is continuous across the grain boundary. Furthermore, although the discussions are based on a body composed of two grains as depicted in Fig. 1, the following derivations are valid for bodies composed of multiple grains.

The strain tensor $\boldsymbol{\varepsilon}$ is defined as $\boldsymbol{\varepsilon} = \frac{1}{2}(\nabla \boldsymbol{u} + (\nabla \boldsymbol{u})^T)$, and the velocity vector is represented as $\boldsymbol{v} = \dot{\boldsymbol{u}}$. The strain is decomposed additively as

$$\boldsymbol{\varepsilon} = \boldsymbol{\varepsilon}^e + \boldsymbol{\varepsilon}^p \tag{1}$$

into an elastic part $\boldsymbol{\varepsilon}^e$ and a plastic part $\boldsymbol{\varepsilon}^p$.

The plastic strain rate can be written as the summation of plastic slip rates on the individual slip systems, $\dot{\boldsymbol{\varepsilon}}^p = \sum_\alpha \dot{\gamma}^\alpha \boldsymbol{P}^\alpha$ with $\boldsymbol{P}^\alpha = \frac{1}{2}(\boldsymbol{s}^\alpha \otimes \boldsymbol{m}^\alpha + \boldsymbol{m}^\alpha \otimes \boldsymbol{s}^\alpha)$ the symmetrized Schmid tensor, where $\boldsymbol{s}^\alpha$ and $\boldsymbol{m}^\alpha$ are the unit slip direction vector and unit normal vector on slip system α, respectively. The set

$$state = \boldsymbol{\varepsilon}^e, \gamma^\alpha, \nabla\gamma^\alpha \tag{2}$$

is chosen to be the state variables where γ^α and $\nabla\gamma^\alpha$ are the plastic slip and gradient of plastic slip on system α. Following the arguments of Gurtin (e.g., Gurtin 2000, 2002), the power expended by each independent rate-like kinematical descriptor is expressible in terms of an associated force consistent with its own balance. However, the basic kinematical fields of rate variables, namely, $\dot{\varepsilon}^e$, $\dot{u}$, and $\dot{\gamma}^\alpha$ are not independent. It is therefore not immediately clear how the associated force balances are to be formulated, and, for that reason, these balances are established using the principle of virtual power.

The classical macroscopic system is defined by a traction $\boldsymbol{t}(\bar{\boldsymbol{n}})$ that expends power over the velocity $\dot{\boldsymbol{u}}$ and stress $\boldsymbol{\sigma}$ that expends power over the elastic strain rate $\dot{\boldsymbol{\varepsilon}}^e$. There are no body forces acting on the system. The microscopic system is composed of a scalar microscopic stress π^α for each slip system that expends power over the slip rate $\dot{\gamma}^\alpha$, a vector of microscopic stress $\boldsymbol{\xi}^\alpha$ that expends power over the slip-rate gradient $\nabla\dot{\gamma}^\alpha$, a scalar microscopic traction $\chi^\alpha(\bar{\boldsymbol{n}})$ that expends power over $\dot{\gamma}^\alpha$, and a scalar interfacial microscopic grain boundary stress λ^α that expends power over the grain boundary slip rates $\dot{\gamma}^\alpha_A$ and $\dot{\gamma}^\alpha_B$ where approaching the grain boundary from different grains is designated by subscripts A and B. The virtual rates are collected in the generalized virtual velocity $\mathscr{V} = (\delta\dot{\boldsymbol{u}}, \delta\dot{\boldsymbol{\varepsilon}}^e, \delta\dot{\gamma}^\alpha, \delta\dot{\gamma}^\alpha_A, \delta\dot{\gamma}^\alpha_B)$. The force systems are characterized through their work-conjugated nature with respect to the state variables. $\mathscr{P}_{ext}$ is the power expended on the domain Ω, $\mathscr{P}_{int}$ a concomitant expenditure of power within Ω stated as

$$\mathscr{P}_{ext}(\Omega,\mathscr{V}) = \int_S \boldsymbol{t}(\bar{\boldsymbol{n}})\cdot\delta\dot{\boldsymbol{u}}\,dS + \sum_\alpha \int_S \chi^\alpha(\bar{\boldsymbol{n}})\delta\dot{\gamma}^\alpha dS + \sum_\alpha \int_{S_{GB}} \left(\lambda^\alpha_A\,\delta\dot{\gamma}^\alpha_A + \lambda^\alpha_B\,\delta\dot{\gamma}^\alpha_B\right) dS \quad (3)$$

$$\mathscr{P}_{int}(\Omega,\mathscr{V}) = \int_\Omega \boldsymbol{\sigma} : \delta\dot{\boldsymbol{\varepsilon}}^e d\Omega + \sum_\alpha \int_\Omega \pi^\alpha\delta\dot{\gamma}^\alpha d\Omega + \sum_\alpha \int_\Omega \boldsymbol{\xi}^\alpha\cdot\nabla\delta\dot{\gamma}^\alpha d\Omega$$

where λ_A and λ_B are the microscopic grain boundary stresses associated with the grain boundary of grain A and grain B, respectively. The boundary S is the union of outer boundaries of grains A and B; and S_{GB} is the grain boundary (the interface) as shown in Fig. 1.

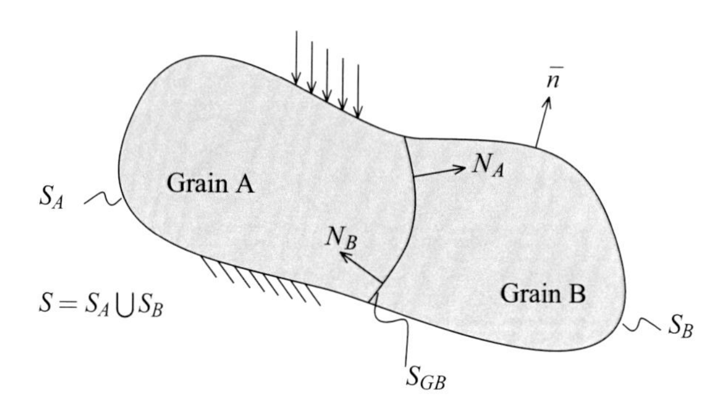

Fig. 1 A body with a grain boundary

Postulation of principle of virtual power states that given any set of virtual fields, the corresponding internal and external powers are balanced

$$\mathscr{P}_{ext}(\Omega, \mathscr{V}) = \mathscr{P}_{int}(\Omega, \mathscr{V}) \qquad \forall \mathscr{V} \tag{4}$$

Now the consequences are derived following the restriction $\nabla \delta \dot{\boldsymbol{u}} = \delta \dot{\boldsymbol{\varepsilon}}^e + \delta \dot{\boldsymbol{\omega}}^e + \sum_\alpha \delta \dot{\gamma}^\alpha \boldsymbol{s}^\alpha \otimes \boldsymbol{m}^\alpha$. First a generalized virtual velocity without slip is considered, namely, $\delta \dot{\gamma}^\alpha = 0$ which means $\nabla \delta \dot{\boldsymbol{u}} = \delta \dot{\boldsymbol{\varepsilon}}^e + \delta \dot{\boldsymbol{\omega}}^e$. Considering that Cauchy stress is symmetric, the power balance becomes

$$\int_\Omega \boldsymbol{\sigma} : \nabla \delta \dot{\boldsymbol{u}} d\Omega = \int_S \boldsymbol{t}(\bar{\boldsymbol{n}}) \cdot \delta \dot{\boldsymbol{u}} dS \tag{5}$$

The conditions derived from equation (5) are well-known traction conditions

$$\boldsymbol{t}(\bar{\boldsymbol{n}}) = \boldsymbol{\sigma} \bar{\boldsymbol{n}} \tag{6}$$

and the classical linear momentum balance

$$\nabla \cdot \boldsymbol{\sigma} = \boldsymbol{0} \tag{7}$$

The microscopic counterparts of the conditions are obtained through a consideration of a generalized virtual velocity with $\delta \dot{\boldsymbol{u}} = \boldsymbol{0}$ with arbitrary $\delta \dot{\gamma}^\alpha$ field which results in

$$\sum_\alpha \delta \dot{\gamma}^\alpha (\boldsymbol{s}^\alpha \otimes \boldsymbol{m}^\alpha) = -\delta \dot{\boldsymbol{\varepsilon}}^e - \delta \dot{\boldsymbol{\omega}}^e \tag{8}$$

and the term $\boldsymbol{\sigma} : \delta \dot{\boldsymbol{\varepsilon}}^e$ becomes $\boldsymbol{\sigma} : \delta \dot{\boldsymbol{\varepsilon}}^e = -\boldsymbol{\sigma} : \sum_\alpha \delta \dot{\gamma}^\alpha (\boldsymbol{s}^\alpha \otimes \boldsymbol{m}^\alpha) - \boldsymbol{\sigma} : \delta \dot{\boldsymbol{\omega}}^e$. Using the symmetry of the Cauchy stress and the definition of the Schmid resolved stress τ^α

$$\boldsymbol{\sigma} : \delta \dot{\boldsymbol{\varepsilon}}^e = -\sum_\alpha \delta \dot{\gamma}^\alpha \tau^\alpha \tag{9}$$

For this case the power balance (4) is utilized again and the following form is obtained

$$\begin{aligned}&\textstyle\sum_\alpha \int_S \chi^\alpha(\bar{\boldsymbol{n}}) \delta \dot{\gamma}^\alpha dS + \sum_\alpha \int_{S_{GB}} \left(\lambda_A^\alpha \delta \dot{\gamma}_A^\alpha + \lambda_B^\alpha \delta \dot{\gamma}_B^\alpha\right) dS = \\ &\textstyle -\sum_\alpha \int_\Omega \tau^\alpha \delta \dot{\gamma}^\alpha d\Omega + \sum_\alpha \int_\Omega \pi^\alpha \delta \dot{\gamma}^\alpha d\Omega + \sum_\alpha \int_\Omega \boldsymbol{\xi}^\alpha \cdot \nabla \delta \dot{\gamma}^\alpha d\Omega\end{aligned} \tag{10}$$

which could be written as

$$\sum_\alpha \int_\Omega (-\tau^\alpha + \pi^\alpha - \nabla \cdot \boldsymbol{\xi}^\alpha)\delta\dot{\gamma}^\alpha d\Omega + \sum_\alpha \int_S (\boldsymbol{\xi}^\alpha \cdot \bar{\boldsymbol{n}} - \chi^\alpha(\bar{\boldsymbol{n}}))\delta\dot{\gamma}^\alpha dS$$
$$+ \sum_\alpha \int_{S_{GB}} \left(\boldsymbol{\xi}_A^\alpha \cdot \boldsymbol{N}_A - \lambda_A^\alpha\right) \delta\dot{\gamma}_A^\alpha dS + \sum_\alpha \int_{S_{GB}} \left(\boldsymbol{\xi}_B^\alpha \cdot \boldsymbol{N}_B - \lambda_B^\alpha\right) \delta\dot{\gamma}_B^\alpha dS = 0 \tag{11}$$

and has to be satisfied for all $\delta\dot{\gamma}^\alpha$. This argument yields the microscopic traction condition on the outer boundary of the bulk material

$$\chi^\alpha(\bar{\boldsymbol{n}}) = \boldsymbol{\xi}^\alpha \cdot \bar{\boldsymbol{n}} \tag{12}$$

and the microscopic force balance inside the bulk material on each slip system α

$$\tau^\alpha - \pi^\alpha + \nabla \cdot \boldsymbol{\xi}^\alpha = 0 \tag{13}$$

and microscopic grain boundary interface conditions on both side of the grain boundary

$$\begin{aligned} \boldsymbol{\xi}_A^\alpha \cdot \boldsymbol{N}^A &= \lambda_A^\alpha \\ \boldsymbol{\xi}_B^\alpha \cdot \boldsymbol{N}^B &= \lambda_B^\alpha \end{aligned} \tag{14}$$

Since $\boldsymbol{N}^A = -\boldsymbol{N}^B$, the interface conditions given by equation (14) can be expressed as

$$\begin{aligned} \lambda_A^\alpha &= \boldsymbol{\xi}_A^\alpha \cdot \boldsymbol{N}^A \\ \lambda_B^\alpha &= -\boldsymbol{\xi}_B^\alpha \cdot \boldsymbol{N}^A \end{aligned} \tag{15}$$

The details of the previous derivations for the case of single crystals (i.e., bulk material without GB terms) could be followed explicitly from Gurtin (2000, 2002) and Gurtin et al. (2007) and including the grain boundary terms presented in detail in Gurtin (2008).

Free Energy Imbalance: Bulk Material

The local internal power expression in the bulk material can be written as

$$P_i = \boldsymbol{\sigma} : \dot{\boldsymbol{\varepsilon}}^e + \sum_\alpha (\pi^\alpha \dot{\gamma}^\alpha + \boldsymbol{\xi}^\alpha \cdot \nabla \dot{\gamma}^\alpha) \tag{16}$$

and the local dissipation inequality results in

$$D = P_i - \dot{\psi} = \boldsymbol{\sigma} : \dot{\boldsymbol{\varepsilon}}^e + \sum_\alpha (\pi^\alpha \dot{\gamma}^\alpha + \boldsymbol{\xi}^\alpha \cdot \nabla \dot{\gamma}^\alpha) - \dot{\psi} \geq 0 \tag{17}$$

The material is assumed to be endowed with a free energy with different contributions according to

$$\psi(\boldsymbol{\varepsilon}^e, \nabla\gamma^\alpha) = \psi_e(\boldsymbol{\varepsilon}^e) + \psi_{\nabla\gamma}(\nabla\gamma^\alpha) \tag{18}$$

The free energy is assumed to depend only on the elastic strain and the dislocation density. In comparison to recent works of the authors (see, e.g., Yalcinkaya et al. 2011, Yalçinkaya et al. 2012, Yalçinkaya 2013, Yalcinkaya and Lancioni 2014, Lancioni et al. 2015a,b) where the framework is developed and used for modeling the inhomogeneous deformation field (microstructure) formation and evolution, the current chapter does not include the energetic hardening term ψ_γ in the bulk material, which could be convex or non-convex in nature. The details of such modeling approaches are discussed in one of the chapters here.

The time derivative of the free energy is expanded and equation (17) is elaborated to

$$\begin{aligned} D &= \boldsymbol{\sigma} : \dot{\boldsymbol{\varepsilon}}^e + \textstyle\sum_\alpha (\pi^\alpha \dot{\gamma}^\alpha + \boldsymbol{\xi}^\alpha \cdot \nabla\dot{\gamma}^\alpha - \dfrac{\partial\psi}{\partial\boldsymbol{\varepsilon}^e} : \dot{\boldsymbol{\varepsilon}}^e - \dfrac{\partial\psi}{\partial\gamma^\alpha}\dot{\gamma}^\alpha - \dfrac{\partial\psi}{\partial\nabla\gamma^\alpha} \cdot \nabla\dot{\gamma}^\alpha) \\ &= \underbrace{(\boldsymbol{\sigma} - \frac{d\psi_e}{d\boldsymbol{\varepsilon}^e})}_{0} : \dot{\boldsymbol{\varepsilon}}^e + \textstyle\sum_\alpha (\pi^\alpha)\dot{\gamma^\alpha} + \textstyle\sum_\alpha \underbrace{(\boldsymbol{\xi}^\alpha - \frac{\partial\psi_{\nabla\gamma}}{\partial\nabla\gamma^\alpha})}_{0} \cdot \nabla\dot{\gamma^\alpha} \geq 0 \end{aligned} \tag{19}$$

The stress $\boldsymbol{\sigma}$ and the microstress vectors $\boldsymbol{\xi}^\alpha$ are regarded as energetic quantities

$$\begin{aligned} \boldsymbol{\sigma} &= \frac{d\psi_e}{d\boldsymbol{\varepsilon}^e} \\ \boldsymbol{\xi}^\alpha &= \frac{\partial\psi_{\nabla\gamma}}{\partial\nabla\gamma^\alpha} \end{aligned} \tag{20}$$

whereas π^α does have a dissipative contribution

$$D = \sum_\alpha (\pi^\alpha)\dot{\gamma^\alpha} \geq 0 \tag{21}$$

The multipliers of the plastic slip rates are identified as the set of dissipative stresses defined as $\sigma^\alpha_{\mathrm{dis}}$

$$\sigma^\alpha_{\mathrm{dis}} = \pi^\alpha \tag{22}$$

In order to satisfy the reduced dissipation inequality at the slip system level, the following constitutive equation is proposed

$$\sigma^\alpha_{\mathrm{dis}} = \varphi^\alpha \operatorname{sign}(\dot{\gamma}^\alpha) \tag{23}$$

where φ^α represents the mobilized slip resistance of the slip system under consideration

$$\varphi^\alpha = \frac{s^\alpha}{\dot{\gamma}_0^\alpha}|\dot{\gamma}^\alpha| \tag{24}$$

with s^α the resistance to dislocation slip which is assumed to be constant and with $\dot{\gamma}_0$ a reference slip rate. Note that this relation assumes a linear relationship between strain rate and stress. In more quantitative simulations, small values of strain rate sensitivity exponent would be required. This choice was of course made for simplicity only, yet with a large similarity to discrete dislocation studies using linear drag relations. Simulations with other strain rate exponent do not change the qualitative nature of the examples, yet they will affect the rate-dependent (time-dependent) behavior. Substituting (24) into (23) gives

$$\dot{\gamma}^\alpha = \frac{\dot{\gamma}_0^\alpha}{s^\alpha}\sigma_{\text{dis}}^\alpha \tag{25}$$

which, upon substitution of $\sigma_{\text{dis}}^\alpha$ according to (22), takes the following form

$$\dot{\gamma}^\alpha = \frac{\dot{\gamma}_0^\alpha}{s^\alpha}(\pi^\alpha) \tag{26}$$

Finally, using the microforce balance (13) results in the plastic slip equation

$$\dot{\gamma}^\alpha = \frac{\dot{\gamma}_0^\alpha}{s^\alpha}(\tau^\alpha + \nabla \cdot \boldsymbol{\xi}^\alpha) \tag{27}$$

Contributions of the free energies defined in (18) enter the slip equation via (20) with $\tau^\alpha = d\psi_e/d\boldsymbol{\varepsilon}^e : \boldsymbol{P}^\alpha$ and $\boldsymbol{\xi}^\alpha = \partial\psi_{\nabla\gamma}/\partial\nabla\gamma^\alpha$. Quadratic forms are used for the elastic free energy ψ_e and the plastic slip gradients free energy contribution $\psi_{\nabla\gamma}$, i.e.,

$$\begin{aligned} \psi_e &= \frac{1}{2}\boldsymbol{\varepsilon}^e : {}^4\boldsymbol{C} : \boldsymbol{\varepsilon}^e \\ \psi_{\nabla\gamma} &= \textstyle\sum_\alpha \frac{1}{2}A\nabla\gamma^\alpha \cdot \nabla\gamma^\alpha \end{aligned} \tag{28}$$

where ${}^4\boldsymbol{C}$ is the fourth-order elasticity tensor and A is a scalar quantity, which includes an internal length scale parameter, governing the effect of the plastic slip gradients on the internal stress field. It may be, e.g., expressed as $A = ER^2/(16(1-\nu^2))$ as used in Bayley et al. (2006) and Geers et al. (2007), where R physically represents the radius of the dislocation domain contributing to the internal stress field, ν is Poisson's ratio, and E is Young's modulus.

Free Energy Imbalance: Interface

Similarly, the dissipation associated with the grain boundary can be written as,

$$D_{GB} = P_{GB} - \dot{\psi}_{GB} \geq 0 \tag{29}$$

where

$$P_{GB} = \sum_{\alpha} \int_{S_{GB}} (\lambda_A^\alpha \dot{\gamma}_A^\alpha + \lambda_B^\alpha \dot{\gamma}_B^\alpha)\, dS \tag{30}$$

and ψ_{GB} is the free energy of the grain boundary which is essentially the coarse-grained representation of complex grain boundary-dislocation slip interaction mechanism such as dislocation transmission, emission, and dissociation of dislocations within the grain boundary. Transmission electron microscopy (TEM) studies and dislocation level models reveal that energetic and dissipative character of grain boundaries is dictated by the local geometrical features such as misorientation of grains and the grain boundary orientation. Therefore, an energy potential which is sensitive to noncoherent slip systems of neighboring grains and grain boundary orientation would reflect the underlying physics sufficiently well to a certain extent.

In order to account for essential geometrical features, the kinematic characterization presented in Gurtin (2008) is adopted in this work. In Gurtin (2008), the slip incompatibility of the neighboring grains is described in terms of the grain boundary Burgers tensor defined as

$$\boldsymbol{G} = \sum_{\alpha} [\gamma_B^\alpha \boldsymbol{s}_B^\alpha \otimes \boldsymbol{n}_B^\alpha - \gamma_A^\alpha \boldsymbol{s}_A^\alpha \otimes \boldsymbol{n}_A^\alpha](\boldsymbol{N}\times) \tag{31}$$

where for any vector $\boldsymbol{N}$, $\boldsymbol{N}\times$ is the tensor with components $(\boldsymbol{N}\times)_{ij} = \varepsilon_{ikj} N_k$. In equation (31), the relative misorientation of grains is reflected by the difference term, and the grain boundary orientation is accounted for by the tensor $\boldsymbol{N}\times$. Furthermore, using (31), the magnitude of grain boundary Burgers tensor can be expressed in the following form

$$|\boldsymbol{G}|^2 = \sum_{\alpha} \sum_{\beta} \left(C_{AA}^{\alpha\beta} \gamma_A^\alpha \gamma_A^\beta + C_{BB}^{\alpha\beta} \gamma_B^\alpha \gamma_B^\beta + -2 C_{AB}^{\alpha\beta} \gamma_A^\alpha \gamma_B^\beta \right) \tag{32}$$

in which the slip interaction moduli are introduced such that

$$\begin{aligned} C_{AA}^{\alpha\beta} &= \left(\boldsymbol{s}_A^\alpha \cdot \boldsymbol{s}_A^\beta\right) \left(\boldsymbol{n}_A^\alpha \times \boldsymbol{N}\right) \cdot \left(\boldsymbol{n}_A^\beta \times \boldsymbol{N}\right) \\ C_{AB}^{\alpha\beta} &= \left(\boldsymbol{s}_A^\alpha \cdot \boldsymbol{s}_B^\beta\right) \left(\boldsymbol{n}_A^\alpha \times \boldsymbol{N}\right) \cdot \left(\boldsymbol{n}_B^\beta \times \boldsymbol{N}\right) \\ C_{BB}^{\alpha\beta} &= \left(\boldsymbol{s}_B^\alpha \cdot \boldsymbol{s}_B^\beta\right) \left(\boldsymbol{n}_B^\alpha \times \boldsymbol{N}\right) \cdot \left(\boldsymbol{n}_B^\beta \times \boldsymbol{N}\right) \end{aligned} \tag{33}$$

Since $C_{AA}^{\alpha\beta}$ and $C_{BB}^{\alpha\beta}$ represent interactions between slip systems within grain A and grain B, respectively, they are called intra-grain interaction moduli, whereas $C_{AB}^{\alpha\beta}$ represent the interaction between slip systems of the two grains and called inter-grain interaction moduli.

Furthermore, by decomposing slip system normal vector $(\boldsymbol{n}_A^{\alpha}, \boldsymbol{n}_A^{\beta}, \boldsymbol{n}_B^{\alpha}, \boldsymbol{n}_B^{\beta})$ into normal and tangential components with respect to grain boundary plane as

$$\boldsymbol{n}_A^{\alpha} = \left(\boldsymbol{n}_A^{\alpha} \cdot \boldsymbol{N}\right) \boldsymbol{N} + \left(\boldsymbol{I} - \boldsymbol{n}_A^{\alpha} \cdot \boldsymbol{N}\right) \boldsymbol{N} = \boldsymbol{n}_A^{\alpha,nor} + \boldsymbol{n}_A^{\alpha,tan} \tag{34}$$

and using $\varepsilon - \delta$ identity ($\varepsilon_{ijk}\varepsilon_{ipq} = \delta_{jp}\delta_{kq} - \delta_{jq}\delta_{kp}$), it can be shown that

$$\left(\boldsymbol{n}_A^{\alpha} \times \boldsymbol{N}\right) \cdot \left(\boldsymbol{n}_A^{\beta} \times \boldsymbol{N}\right) = \boldsymbol{n}_A^{\alpha,tan} \cdot \boldsymbol{n}_A^{\beta,tan} \tag{35}$$

Therefore, the interaction moduli can be expressed in the following alternative form

$$\begin{aligned} C_{AA}^{\alpha\beta} &= \left(\boldsymbol{s}_A^{\alpha} \cdot \boldsymbol{s}_A^{\beta}\right) \left(\boldsymbol{n}_A^{\alpha,tan} \cdot \boldsymbol{n}_A^{\beta,tan}\right) \\ C_{AB}^{\alpha\beta} &= \left(\boldsymbol{s}_A^{\alpha} \cdot \boldsymbol{s}_B^{\beta}\right) \left(\boldsymbol{n}_A^{\alpha,tan} \cdot \boldsymbol{n}_B^{\beta,tan}\right) \\ C_{BB}^{\alpha\beta} &= \left(\boldsymbol{s}_B^{\alpha} \cdot \boldsymbol{s}_B^{\beta}\right) \left(\boldsymbol{n}_B^{\alpha,tan} \cdot \boldsymbol{n}_B^{\beta,tan}\right) \end{aligned} \tag{36}$$

which allows a clear geometric interpretation as illustrated in Fig. 2. For the sake of clarity, a two-dimensional bi-crystal with slip system α located in grain A and slip system β located in grain B is considered, and different geometric configurations are depicted in Fig. 2.

For the intra-grain moduli ($C_{AA}^{\alpha\alpha}$) and ($C_{BB}^{\beta\beta}$), the first inner product yields unity, and the second inner product term measures the contribution of this slip system to the slip incompatibility depending on the orientation of slip system with respect to the grain boundary orientation. If the slip system orientation coincides with the orientation of the grain boundary, in other words if the slip system is parallel to the grain boundary, then there is no interaction between the slip system and the grain boundary; therefore the slip system does not contribute to slip incompatibility.

In the inter-grain interaction moduli ($C_{AA}^{\alpha\alpha}$) expression, the first inner product term is the measure of geometric coherency of the slip systems, taking the maximum value of unity when the systems (slip system α in grain A and slip system β in grain B) are parallel. Similar to the intra-grain moduli, the second inner product term is the influence of the orientation of slip systems α and β with respect to the grain boundary orientation.

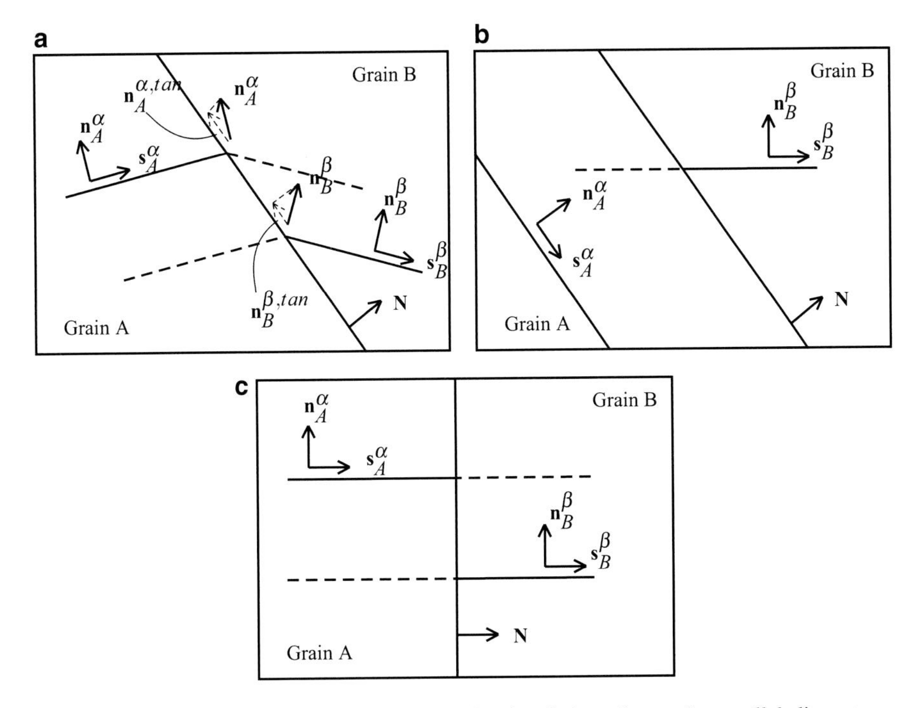

Fig. 2 (**a**) An arbitrary configuration, (**b**) an inclined grain boundary and a parallel slip system, and (**c**) coherent slip systems with a vertical grain boundary

At this stage ignoring the dissipative effects, a simple free energy potential of the form

$$\psi_{GB} = \frac{1}{2}\kappa|\boldsymbol{G}|^2 \tag{37}$$

is proposed where κ is a positive constant modulus. With this particular form in hand (32), rate of free energy can be expressed as

$$\dot{\psi}_{GB} = \kappa \sum_{\alpha}\sum_{\beta}\left[\left(\gamma_B^\beta C_{BB}^{\alpha\beta} - \gamma_A^\beta C_{BA}^{\alpha\beta}\right)\dot{\gamma}_B^\alpha - \left(\gamma_B^\beta C_{AB}^{\alpha\beta} - \gamma_A^\beta C_{AA}^{\alpha\beta}\right)\dot{\gamma}_A^\alpha\right] \tag{38}$$

Then the dissipation inequality reads as

$$\left[\lambda_A^\alpha - \kappa\sum_{\beta}\left(\gamma_A^\beta C_{AA}^{\alpha\beta} - \gamma_B^\beta C_{AB}^{\alpha\beta}\right)\right]\dot{\gamma}_A^\alpha + \left[\lambda_B^\alpha - \kappa\sum_{\beta}\left(\gamma_B^\beta C_{BB}^{\alpha\beta} - \gamma_A^\beta C_{BA}^{\alpha\beta}\right)\right]\dot{\gamma}_B^\alpha \geq 0 \tag{39}$$

Since a non-dissipative grain boundary response is assumed, equation (39) leads to following identifications

$$\begin{aligned} \lambda_A^\alpha &= \kappa \sum_\beta \left(\gamma_A^\beta C_{AA}^{\alpha\beta} - \gamma_B^\beta C_{AB}^{\alpha\beta} \right) \\ \lambda_B^\alpha &= \kappa \sum_\beta \left(\gamma_B^\beta C_{BB}^{\alpha\beta} - \gamma_A^\beta C_{BA}^{\alpha\beta} \right) \end{aligned} \tag{40}$$

In case of single slip system (α) in grain A and single slip system (β) in grain B, for an arbitrary orientation other than the orientation where one of the slip systems (α or β) is parallel to the grain boundary, the interaction moduli read as

$$\begin{aligned} C_{AA}^{\alpha\alpha} &= \boldsymbol{n}_A^{\alpha,tan} \cdot \boldsymbol{n}_A^{\alpha,tan} \\ C_{BB}^{\beta\beta} &= \boldsymbol{n}_B^{\beta,tan} \cdot \boldsymbol{n}_B^{\beta,tan} \\ C_{AB}^{\alpha\beta} &= \left(\boldsymbol{s}_A^\alpha \cdot \boldsymbol{s}_B^\beta \right) \left(\boldsymbol{n}_A^{\alpha,tan} \cdot \boldsymbol{n}_A^{\beta,tan} \right) \end{aligned} \tag{41}$$

In case of coherent slip systems, the inter-grain interaction moduli reduce to

$$C_{AB}^{\alpha\beta} = \boldsymbol{n}_A^{\alpha,tan} \cdot \boldsymbol{n}_B^{\beta,tan} \tag{42}$$

which implies

$$\begin{aligned} \lambda_A^\alpha &= \bar{\kappa} \left(\gamma_A^\alpha - \gamma_B^\beta \right) \\ \lambda_B^\beta &= \bar{\kappa} \left(\gamma_B^\beta - \gamma_A^\alpha \right) \end{aligned} \tag{43}$$

where $\bar{\kappa} = \left(\boldsymbol{n}_A^{\alpha,tan} \cdot \boldsymbol{n}_B^{\beta,tan} \right) \kappa$. If the slip systems are coherent and the grain boundary is perpendicular to these parallel slip systems, then the moduli take the following values

$$C_{AA}^{\alpha\alpha} = 1.0,\ C_{BB}^{\alpha\alpha} = 1.0,\ C_{AB}^{\alpha\alpha} = 1.0 \tag{44}$$

implying that

$$\begin{aligned} \lambda_A^\alpha &= \kappa \left(\gamma_A^\alpha - \gamma_B^\alpha \right) \\ \lambda_B^\alpha &= \kappa \left(\gamma_B^\alpha - \gamma_A^\alpha \right) \end{aligned} \tag{45}$$

The difference between the slips is being penalized by the grain boundary strength κ. When the slip system α of grain A is parallel to the grain boundary, the interaction moduli simplify to

$$C_{AA}^{\alpha\alpha} = 0, C_{AB}^{\alpha\beta} = 0, C_{BB}^{\beta\beta} = 1.0 \tag{46}$$

yielding

$$\lambda_A = 0, \lambda_B = \kappa\, \gamma_B^{\beta} \tag{47}$$

Since, in this configuration, grain A is not interacting with the grain boundary and grain B, the corresponding interaction moduli turn out to be zero. Depending on the value of κ ("the strength of the interface"), the grain boundary may act as a soft ($\kappa = 0$) or hard ($\kappa = \infty$) boundary for grain B as reflected by equation (47).

Finite Element Implementation

In order to solve the initial boundary value problem for the grain boundary enhanced strain gradient crystal plasticity framework, a mixed finite element formulation is used. The displacement $\boldsymbol{u}$ and plastic slips γ^α are taken as primary variables, and these fields are determined within the problem domain by solving simultaneously the linear momentum balance and the microscopic force balance. Therefore the strong form of these equations reads as

$$\begin{aligned} \nabla \cdot \boldsymbol{\sigma} &= \boldsymbol{0} \\ \dot{\gamma}^\alpha - \frac{\dot{\gamma}_0^\alpha}{s^\alpha}\, \tau^\alpha - \frac{\dot{\gamma}_0^\alpha}{s^\alpha}\, \nabla \cdot \boldsymbol{\xi}^\alpha &= 0 \end{aligned} \tag{48}$$

complemented with the boundary conditions (6), (12) on the associated outer boundaries and the interface conditions (14) on the grain boundary, respectively. It has to be noted that the slip evolution equation is in fact the microforce balance (equation (13)) reexpressed in an alternative form with the aid of equations (22), (23), and (24). The weak forms of the balance equations are obtained through multiplication by weighting functions $\delta_{\boldsymbol{u}}$ and $\delta\gamma^\alpha$ and integration over the domain Ω, yielding

$$\begin{aligned} G_u &= \int_\Omega \nabla\delta\boldsymbol{u} : \boldsymbol{\sigma}\, d\Omega - \int_S \delta\boldsymbol{u} \cdot \boldsymbol{t}\, dS \\ G_\gamma^\alpha &= \int_\Omega \delta\gamma^\alpha\, \dot{\gamma}^\alpha\, d\Omega - \int_\Omega \frac{\dot{\gamma}_0^\alpha}{s^\alpha}\, \delta\gamma^\alpha\, \tau^\alpha\, d\Omega + \int_\Omega \frac{\dot{\gamma}_0^\alpha}{s^\alpha}\, \nabla\delta\gamma^\alpha \cdot A\, \nabla\gamma^\alpha d\Omega \\ &\quad - \int_S \frac{\dot{\gamma}_0^\alpha}{s^\alpha}\, \delta\gamma^\alpha\, \chi^\alpha\, dS \quad - \int_{S_{GB}} \frac{\dot{\gamma}_0^\alpha}{s^\alpha}\, \delta\gamma_A^\alpha \lambda_A^\alpha\, dS - \int_{S_{GB}} \frac{\dot{\gamma}_0^\alpha}{s^\alpha}\, \delta\gamma_B^\alpha \lambda_B^\alpha\, dS \end{aligned} \tag{49}$$

where the grain boundary contributions appear as additional terms in the weak form of microstress balance. Furthermore, $\boldsymbol{t}$ in equation (49) is the external traction vector on the boundary S, and $\chi^\alpha = A\nabla\gamma^\alpha \cdot \bar{\boldsymbol{n}}$. Using a standard Galerkin approach, the unknown fields of displacements, slips, and associated weighting functions are interpolated via

$$\begin{aligned} \delta \boldsymbol{u} &= \underline{N}^{u} \underline{\delta \tilde{u}} \qquad & \boldsymbol{u} &= \underline{N}^{u} \underline{\tilde{u}} \\ \delta \gamma^{\alpha} &= \underline{N}^{\gamma} \underline{\delta \tilde{\gamma}}^{\alpha} \qquad & \gamma^{\alpha} &= \underline{N}^{\gamma} \underline{\tilde{\gamma}}^{\alpha} \end{aligned} \tag{50}$$

where $\underline{\tilde{u}}$, $\underline{\delta \tilde{u}}$, $\underline{\tilde{\gamma}}^{\alpha}$, and $\underline{\delta \tilde{\gamma}}^{\alpha}$ are the columns containing the nodal variables of a particular element. Adopting a discretization by ten-noded tetrahedra elements, quadratic interpolation for the displacement field and linear interpolation for the slips are used. Referring to Fig. 3, for a ten-noded tetrahedra, only the corner nodes have the slip degrees of freedom, whereas all nodes have displacement degrees of freedom.

To facilitate the integration of the grain boundary contributions, 12-noded zero thickness interface elements are used which are inserted along the grain boundaries; please see Fig. 3. By means of these elements, one has the access to the slip values along the grain boundary as approached from grain A and grain B. However, it is important to note that the interface elements do not possess any kind of mechanical cohesive behavior and do not cause discontinuity in displacement field. In the solution phase, the displacement continuity across the grain boundary is fulfilled by means of equality constraints (rigid ties) enforcing the same dis-

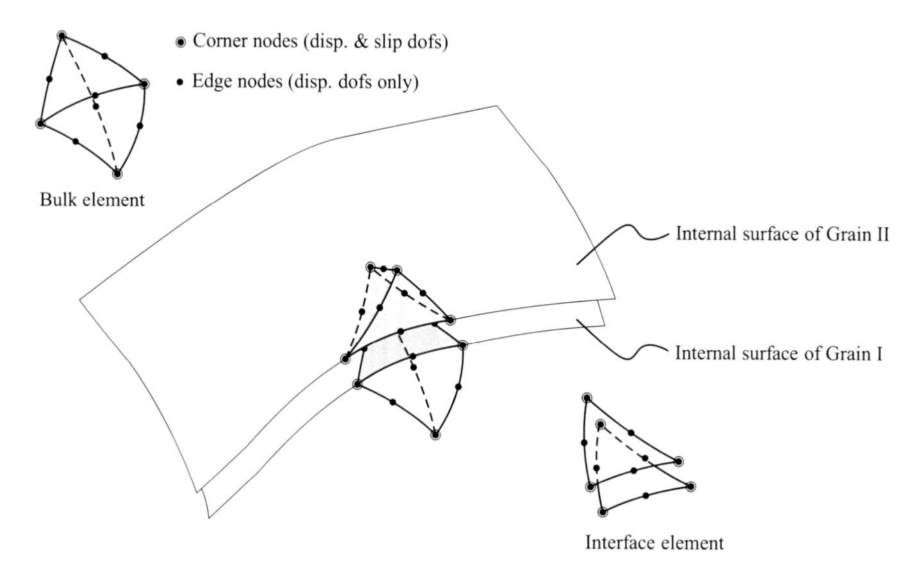

Fig. 3 Element types used for discretization. For ease of illustration, grain boundary and interface element are shown with a gap

placement field for the corresponding nodes of on the two sides of an interface element.

With a 12-noded interface element, $\delta\gamma_A^\alpha$ and $\delta\gamma_B^\alpha$ can be expressed as

$$\begin{bmatrix} \delta\gamma_A^\alpha \\ \delta\gamma_B^\alpha \end{bmatrix} = \begin{bmatrix} N_1 & N_2 & N_3 & 0 & 0 & 0 \\ 0 & 0 & 0 & N_1 & N_2 & N_3 \end{bmatrix} \begin{bmatrix} \delta\gamma_1^\alpha \\ \delta\gamma_2^\alpha \\ \delta\gamma_3^\alpha \\ \delta\gamma_4^\alpha \\ \delta\gamma_5^\alpha \\ \delta\gamma_6^\alpha \end{bmatrix} = \underline{N}_{GB}\,\underline{\delta\gamma}^\alpha_{GB} \tag{51}$$

in terms of nodal virtual slip variables $\delta\gamma_i$ and the standard shape functions (N_1, N_2, N_3) for a linear triangle. In the columns of nodal quantities such as $\underline{\delta\gamma}^\alpha_{GB}$, subscripts designate the node number, and the superscripts are reserved for slip system identity. Introducing $\underline{\lambda} = \left[\lambda_A^\alpha \quad \lambda_B^\alpha\right]^T$, the grain boundary integral for the particular slip system α can be expressed as

$$\int_{S_{GB}} \frac{\dot{\gamma}_0^\alpha}{s^\alpha} \left(\lambda_A^\alpha \delta\gamma_A^\alpha + \lambda_B^\alpha \delta\gamma_B^\alpha\right) dS = \underline{\delta\gamma}_\alpha^T \int_{S_{GB}} \frac{\dot{\gamma}_0^\alpha}{s^\alpha}\, \underline{N}_{GB}^T\, \underline{\lambda}\, dS \tag{52}$$

Introducing the $2n \times 6n$ $\underline{T}$ matrix (where n is the number of slip systems)

$$\underline{T} = \begin{bmatrix} N_1 & N_2 & N_3 & 0 & 0 & 0 & 0\,0\,0 \ldots & 0 & 0 & 0 & 0 & 0 & 0 \\ 0 & 0 & 0 & N_1 & N_2 & N_3 & 0\,0\,0 \ldots & 0 & 0 & 0 & 0 & 0 & 0 \\ \vdots & \vdots & \vdots & \vdots & \vdots & \vdots & \vdots\,\vdots\,\vdots\;\vdots & \vdots & \vdots & \vdots & \vdots & \vdots & \vdots \\ 0 & 0 & 0 & 0 & 0 & 0 & 0\,0\,0 \ldots & N_1 & N_2 & N_3 & 0 & 0 & 0 \\ 0 & 0 & 0 & 0 & 0 & 0 & 0\,0\,0 \ldots & 0 & 0 & 0 & N_1 & N_2 & N_3 \end{bmatrix} \tag{53}$$

and the rows

$$\begin{aligned} \underline{C}_A &= \left[C_{AA}^{\alpha 1} \; -C_{AB}^{\alpha 1} \; C_{AA}^{\alpha 2} \; -C_{AB}^{\alpha 2} \; \ldots \; C_{AA}^{\alpha n} \; -C_{AB}^{\alpha n}\right] \\ \underline{C}_B &= \left[-C_{BA}^{\alpha 1} \; C_{BB}^{\alpha 1} \; -C_{BA}^{\alpha 2} \; C_{BB}^{\alpha 2} \; \ldots \; -C_{BA}^{\alpha n} \; C_{BB}^{\alpha n}\right] \\ \underline{\gamma}^T &= \left[\gamma_1^1 \; \gamma_2^1 \; \gamma_3^1 \; \gamma_4^1 \; \gamma_5^1 \; \gamma_6^1 \ldots \gamma_1^n \; \gamma_2^n \; \gamma_3^n \; \gamma_4^n \; \gamma_5^n \; \gamma_6^n\right] \end{aligned} \tag{54}$$

for slip system α, the interface element contribution reads as

$$G_{GB,\gamma}^{\alpha,e} = -\underline{\delta\gamma}_\alpha^T \int_{S_{GB}^e} \underline{N}_{GB}^T\, \underline{D}\, \underline{\gamma}\, dS^e \tag{55}$$

where

$$\underline{D} = \kappa \, \frac{\dot{\gamma}_0^\alpha}{s^\alpha} \begin{bmatrix} \underline{C}_A \underline{T} \\ \underline{C}_B \underline{T} \end{bmatrix} \tag{56}$$

By employing backward Euler time integration for $\dot{\gamma}^\alpha$, one can write $\dot{\gamma}^\alpha = [\gamma_{n+1}^\alpha - \gamma_n^\alpha]/\Delta t$ with $\Delta t = t_{n+1} - t_n$, and fully discrete weak forms for element e read

$$\begin{aligned}
G_{\boldsymbol{u}}^e &= \underline{\delta \boldsymbol{u}}^T \left[\int_{\Omega^e} \underline{B}^u \, \underline{\boldsymbol{\sigma}} \, d\Omega^e - \int_{S^e} \underline{N}^u \, \underline{\boldsymbol{t}} \, dS^e \right] \\
G_\gamma^{\alpha,e} &= \underline{\delta \tilde{\gamma}}^\alpha \left[\int_{\Omega^e} \underline{N}^{\gamma T} \underline{N}^\gamma \left[\frac{\underline{\gamma}_{n+1}^\alpha - \underline{\gamma}_n^\alpha}{\Delta t} \right] d\Omega^e - \int_{\Omega^e} \frac{\dot{\gamma}_0^\alpha}{s^\alpha} \underline{N}^{\gamma T} \, \underline{\tau}^\alpha \, d\Omega^e \right] \\
&\quad + \underline{\delta \tilde{\gamma}}^\alpha \left[\int_{\Omega^e} \frac{\dot{\gamma}_0^\alpha}{s^\alpha} A \, \underline{B}^{\gamma T} \, \underline{B}^\gamma \, \underline{\gamma}^\alpha \, d\Omega^e - \int_{S^e} \frac{\dot{\gamma}_0^\alpha}{s^\alpha} \underline{N}^{\gamma T} \, \underline{\chi}^\alpha \, dS^e \right]
\end{aligned} \tag{57}$$

By adding the integrals for the interface and bulk discretizations ((56) and (57), respectively) and enforcing $\sum_{i=1}^{n_{el}} G_{GB,\gamma}^{\alpha,e} + \sum_{i=1}^{n_{el}} G_\gamma^{\alpha,e} = 0$, the weak form for the whole domain is obtained. The weak forms of the balance equations are linearized with respect to the increments of the primary variables $\boldsymbol{u}$ and γ^α and solved by means of a Newton-Raphson solution procedure for the corrective terms $\Delta \boldsymbol{u}$ and $\Delta \gamma^\alpha$. The resulting system of linear equations can be written in the following compact form

$$\begin{bmatrix} \underline{K}^{uu} & \underline{K}^{u\gamma} \\ \underline{K}^{\gamma u} & \underline{K}^{\gamma\gamma} + \underline{K}_{GB}^{\gamma\gamma} \end{bmatrix} \begin{bmatrix} \Delta \underline{\boldsymbol{u}} \\ \Delta \underline{\gamma}^\alpha \end{bmatrix} = \begin{bmatrix} -\underline{R}^u + \underline{R}_u^{ext} \\ -\underline{R}^\gamma + \underline{R}_\gamma^{ext} \end{bmatrix} \tag{58}$$

where $\underline{K}^{uu}$, $\underline{K}^{u\gamma}$, $\underline{K}^{\gamma u}$, and $\underline{K}^{\gamma\gamma}$ represent the global tangent matrices of the bulk. $\underline{K}_{GB}^{\gamma\gamma}$ is the contribution of the interface elements, while $\underline{R}^u$ and $\underline{R}^\gamma$ are the global residual columns. The contributions $\underline{R}_u^{ext}$ and $\underline{R}_\gamma^{ext}$ originate from the boundary terms.

Numerical Examples

The presented framework is implemented both for 2D and 3D problems in Abaqus V 6.12-1 through user element capabilities. Since the 2D implementation and corresponding examples were documented in Özdemir and Yalçinkaya (2014), in this work the focus is on 3D formulation. To this end, a ten-noded tetrahedron with quadratic displacement and linear slip interpolations is developed and used for the discretization of the grain interiors. The slip degrees of freedom are defined on the corner nodes, whereas the edge nodes have only displacement degrees of freedom. For the treatment of grain boundaries, 12-noded zero thickness interface element is implemented by following the formulation presented in section "Finite Element Implementation." Although the interface element has 12 nodes, only the corner nodes and the associated slip degrees of freedom are used for interpolation (see Fig. 3). As mentioned before, the displacement continuity is achieved by using equality constraints (rigid links) enforcing the same displacements on the two sides of a grain boundary.

Bi-crystal Specimen with Single Slip System

To assess the basic characteristics of the proposed formulation, a relatively simple geometry of a bi-crystal specimen is considered. Mechanical and slip boundary conditions are schematically shown in Fig. 4, and the material parameters are given in Table 1. The front face is displaced with a constant rate in positive z-direction by 10 μm in 0.5 s. For the sake of clarity, a single slip system with $\vec{m} = (0, \sqrt{2}/2, \sqrt{2}/2)$ and $\vec{s} = (0, -\sqrt{2}/2, \sqrt{2}/2)$ is considered. In fact, this is a slip system solely defined in YZ plane and making an angle of 45° with the positive Z-axis. As far as grain boundary (GB) orientation is concerned, two different cases, namely, a vertical GB and a tilted one, are treated separately; please see Fig. 4. In what follows, slip profiles along the specimen are plotted with reference to the longitudinal path (from the front face to the back face of the specimen) shown in Fig. 4.

To start with, the case where the GB is vertical and slip systems of the two grains are perfectly coherent is considered. In other words, the misorientation between the

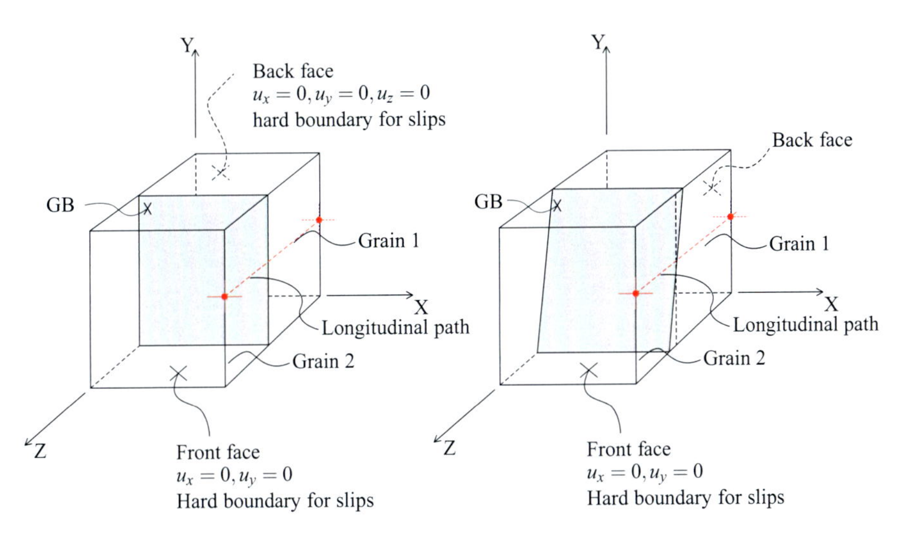

Fig. 4 Bi-crystal example with a vertical (*left*) and a tilted (*right*) grain boundary. Each grain is cubic and 50 by 50 by 50 μm in dimensions

Table 1 Geometry and material parameters for example 4.1 and example 4.2

	Example 4.1	Example 4.2
E (MPa)	70,000.0	70,000.0
ν (−)	0.33	0.33
$\dot{\gamma}_0$ (s^{-1})	0.0115	0.0115
s (MPa)	20	20
r (mm)	0.005	0.005
$\vec{m}_1$	$(0,\sqrt{2}/2,\sqrt{2}/2)$	$(1/\sqrt{3},1/\sqrt{3},1/\sqrt{3})$
$\vec{s}_1$	$(0,-\sqrt{2}/2,\sqrt{2}/2)$	$(-\sqrt{2}/2,\sqrt{2}/2,0)$
$\vec{m}_2$	–	$(1/\sqrt{3},1/\sqrt{3},1/\sqrt{3})$
$\vec{s}_2$	–	$(0,\sqrt{2}/2,-\sqrt{2}/2)$
$\vec{m}_3$	–	$(1/\sqrt{3},1/\sqrt{3},1/\sqrt{3})$
$\vec{s}_3$	–	$(\sqrt{2}/2,0,-\sqrt{2}/2)$

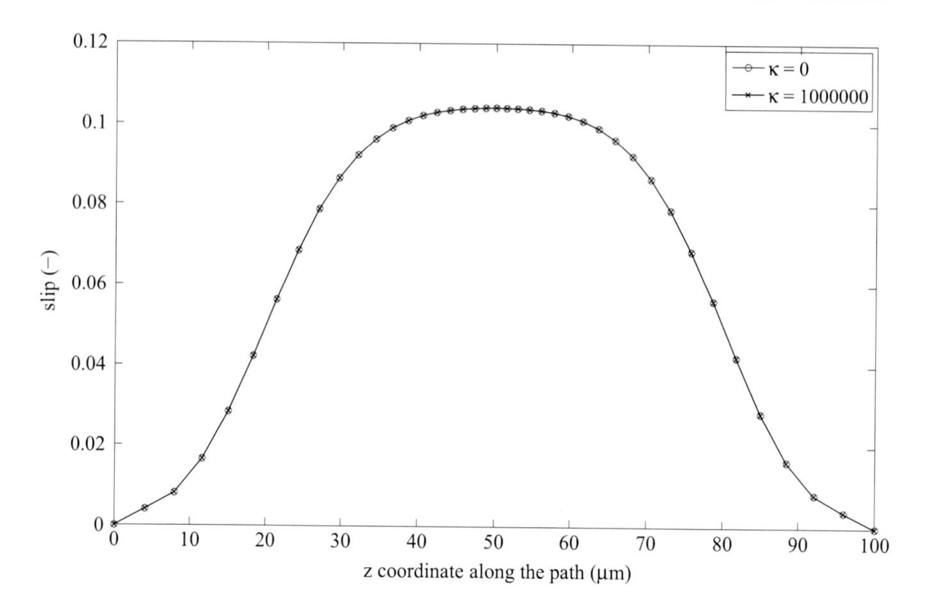

Fig. 5 Slip distribution along the longitudinal path for two different κ values

two grains is zero. For this case, two analyses with different GB strength (κ) are conducted. The slip distributions along the longitudinal path are shown in Fig. 5.

$\kappa = 0$ corresponds to a bi-crystal without a grain boundary, and $\kappa = 1{,}000{,}000$ is very close to a hard boundary as is going to be shown shortly. In fact the two solutions are identical which are consistent with the finding of section "Finite Element Implementation" stating that when the misorientation is zero, the grains do not "feel" the existence of a grain boundary. According to the analysis conducted in section "Finite Element Implementation," the grains should not "feel" even an inclined grain boundary as long as the slip systems of the two grains are aligned. In order to test this, a set of analysis with crystallographically aligned bi-crystal with a tilted GB (by 30°, see Fig. 4) is conducted. In Fig. 6, the slip contours within the bi-crystal specimen cut by a vertical GB and a tilted GB are depicted. For both cases, $\kappa = 1{,}000{,}000$ and from Fig. 5, it is known that for a vertical GB with aligned grains, the response is insensitive to κ. Identical contour seen in Fig. 6 verifies that an inclined grain boundary does not have any influence on the slip distribution provided that there is no misorientation.

In the next stage, the problem is slightly modified by rotating grain 2 around x axis by 15°. In this configuration, there is a misalignment between the slip systems of the two grains, and the slip distribution along the longitudinal path for different κ values are shown in Fig. 7. The influence of the grain boundary is clearly visible, and as the interface strength is increased, the slip distribution in the vicinity of the

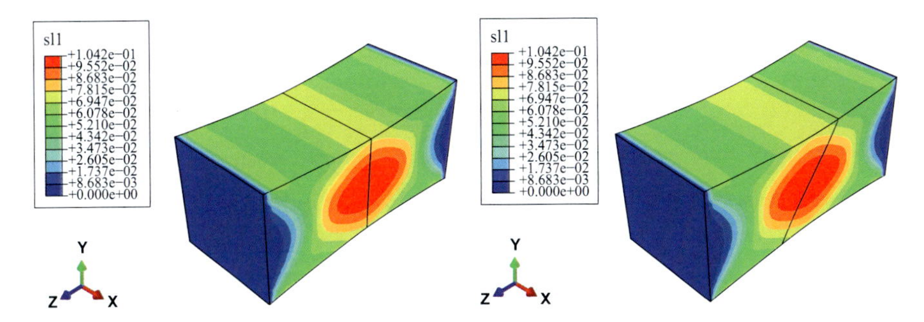

Fig. 6 Slip contours for a flat GB (top) and for a tilted GB. For both cases the misorientation angle between the grains is zero and $\kappa = 1{,}000{,}000$

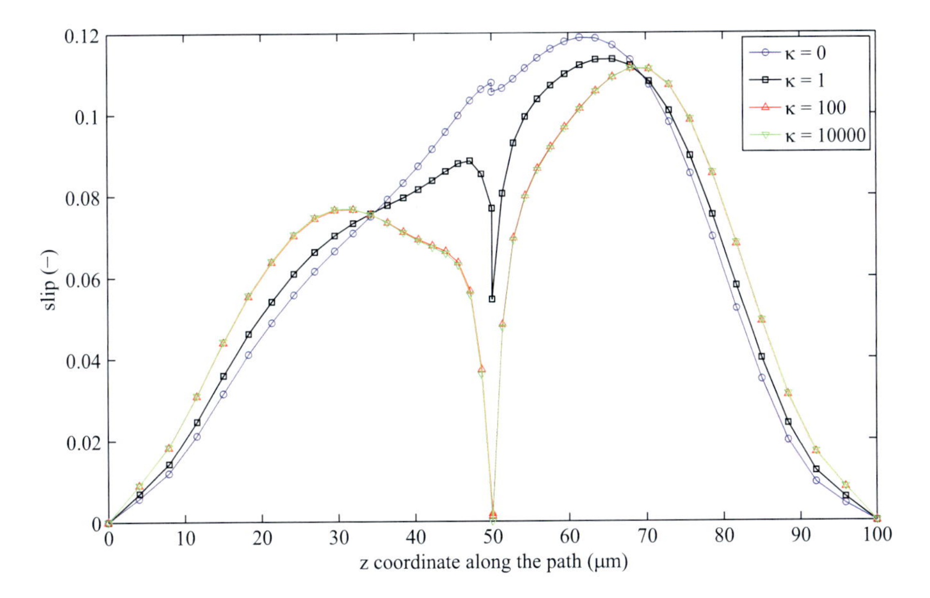

Fig. 7 Slip distribution along the longitudinal path for different κ values

grain boundary resembles to that observed around a hard boundary. The contour plot comparison shown in Fig. 8 indicates that the hard boundary like behavior grows in the vicinity of the whole grain boundary.

To investigate the influence of the "GB strength" on the mechanical response of the system, in Fig. 9, the total reaction force in z-direction is drawn for different κ values. The increase in reaction force and the slope of the force-displacement curves clearly demonstrates the hardening effect of the GB. Due to significant changes in slip distribution, there is a considerable increase in the reaction force. It is also noteworthy to mention that over a certain threshold, the increase in the reaction is

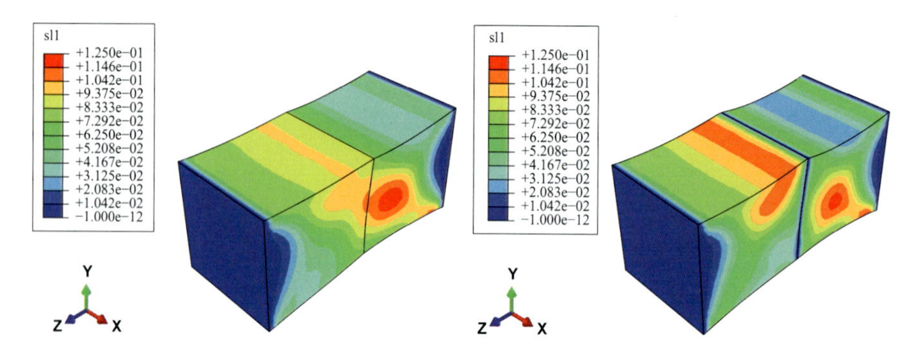

Fig. 8 Slip contours within a crystallographically misaligned bi-crystal for $\kappa = 0$ (*left*) and $\kappa = 10{,}000$ (*right*)

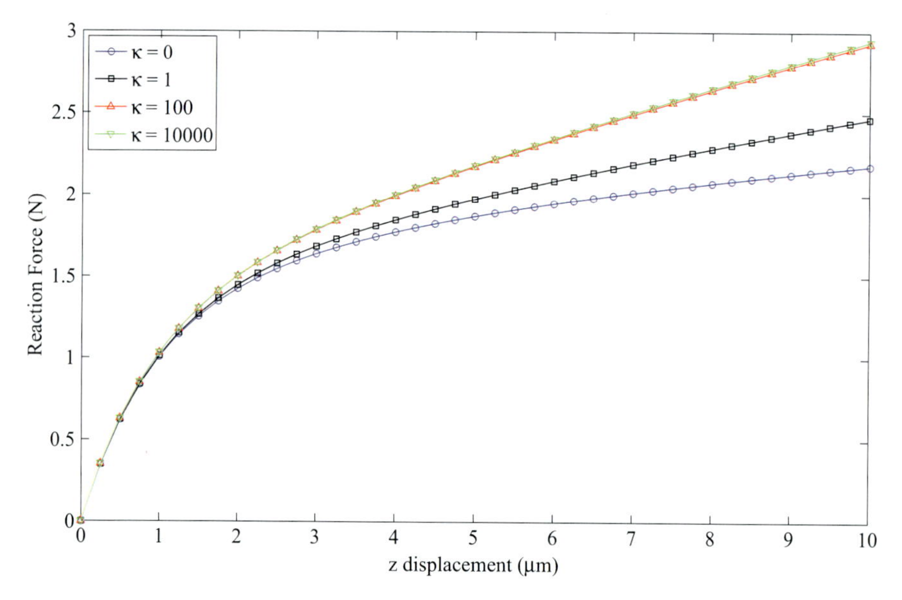

Fig. 9 Total force versus displacement for different κ values when there is a misalignment between the two grains

becoming insignificant which indicates that the grain boundary is sufficiently close to a hard boundary.

Thereafter, the tilted grain boundary with misaligned grains is considered, and the same set of analysis is reiterated. An important and interesting case can be constructed if one of the grains is rotated such that the slip system within that grain becomes parallel to the grain boundary. To set up such a configuration, grain 1 is rotated by 15° counterclockwise. In this case, the slip system in grain 1 is not going to interact with the grain boundary since $\vec{s}$ vector of the slip plane becomes parallel to the GB. Contour plots and the slip profile along the longitudinal path presented in Figs. 10 and 11 are obtained by setting $\kappa = 0$ and $\kappa = 1{,}000{,}000$, respectively.

Comparison of the slip distribution for grain 1 for $\kappa = 0$ and $\kappa = 1{,}000{,}000$ indicates that the slip distributions are very close in these two cases. In other words, the slip profile in grain 1 that would be obtained without the grain boundary ($\kappa = 0$) is almost identical to the profile obtained when there is almost a hard boundary ($\kappa = 1{,}000{,}000$). Such a situation can arise if and only if the crystallographic system of the grain does not interact with the GB. This is in fact the case since the slip system of grain 1 is parallel to the GB. The slight disagreement between the slip profiles of grain 1 for $\kappa = 0$ and $\kappa = 1{,}000{,}000$ stems from the fact that the slip distribution in grain 2 is severely modified (for $\kappa = 1{,}000{,}000$) which has some consequences on elastic strains and stresses in both grains due to coupled nature of the problem. Due to changes in elastic strains within grain 2, the slip distribution within grain 1 for $\kappa = 0$ and $\kappa = 1{,}000{,}000$ is slightly altered although the slip system of this grain and the GB are parallel.

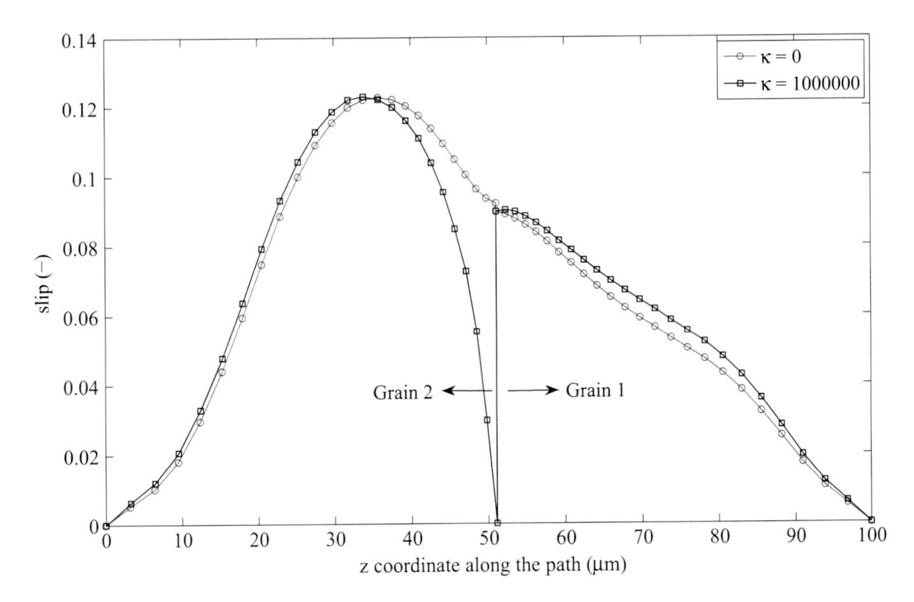

Fig. 10 Slip distribution along the longitudinal axis for $\kappa = 0$ and $\kappa = 1{,}000{,}000$ values. Slip distributions within grain 1 are very close to each other for two different κ values

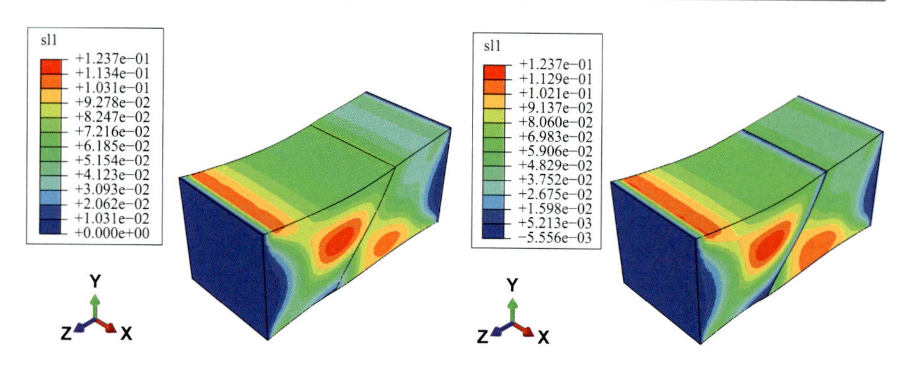

Fig. 11 Slip contours for $\kappa = 0$ (top) and $\kappa = 1{,}000{,}000$ (bottom) when the slip system in grain 1 is parallel to the GB

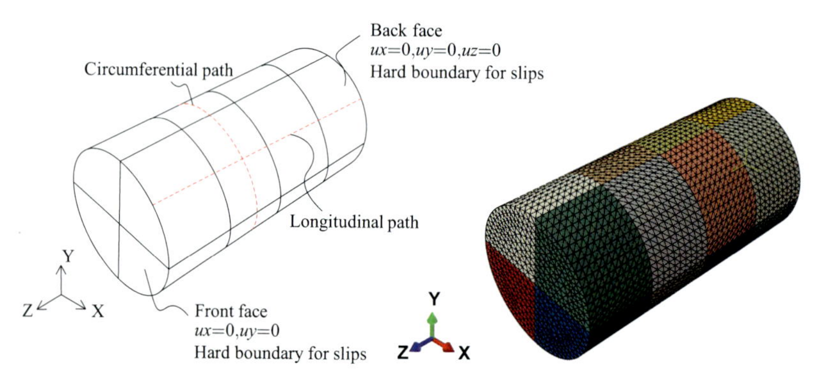

Fig. 12 A cylindrical specimen with 16 grains. The specimen is discretized by 69235 bulk and 74719 interface elements

Cylindrical Specimen with Three Slip System

Although bi-crystal specimens are ideal for investigating grain boundary mechanics, even the smallest engineering components typically consist of multiple grains. Furthermore, a single slip system is far from being realistic. To assess the potential consequences of embedding a grain boundary model within a polycrystalline specimen with multiple active slip systems, a cylindrical specimen consisting of 16 regular grains is considered in this example. The specimen is 12.5 μm in radius and 50 μm in length and stretched in positive Z-direction by 5 μm in 1 s. Please see Fig. 12 for displacement and slip boundary conditions. The parameters of the problem are given in Table 1, and it is assumed that three slip systems (octahedral plane slip systems) are active at any material point as tabulated in the same table. The orientations of the grains are described by Euler angles, and they are generated randomly using the Bunge convention. Therefore the slip systems of the neighboring grains are misaligned in a random manner.

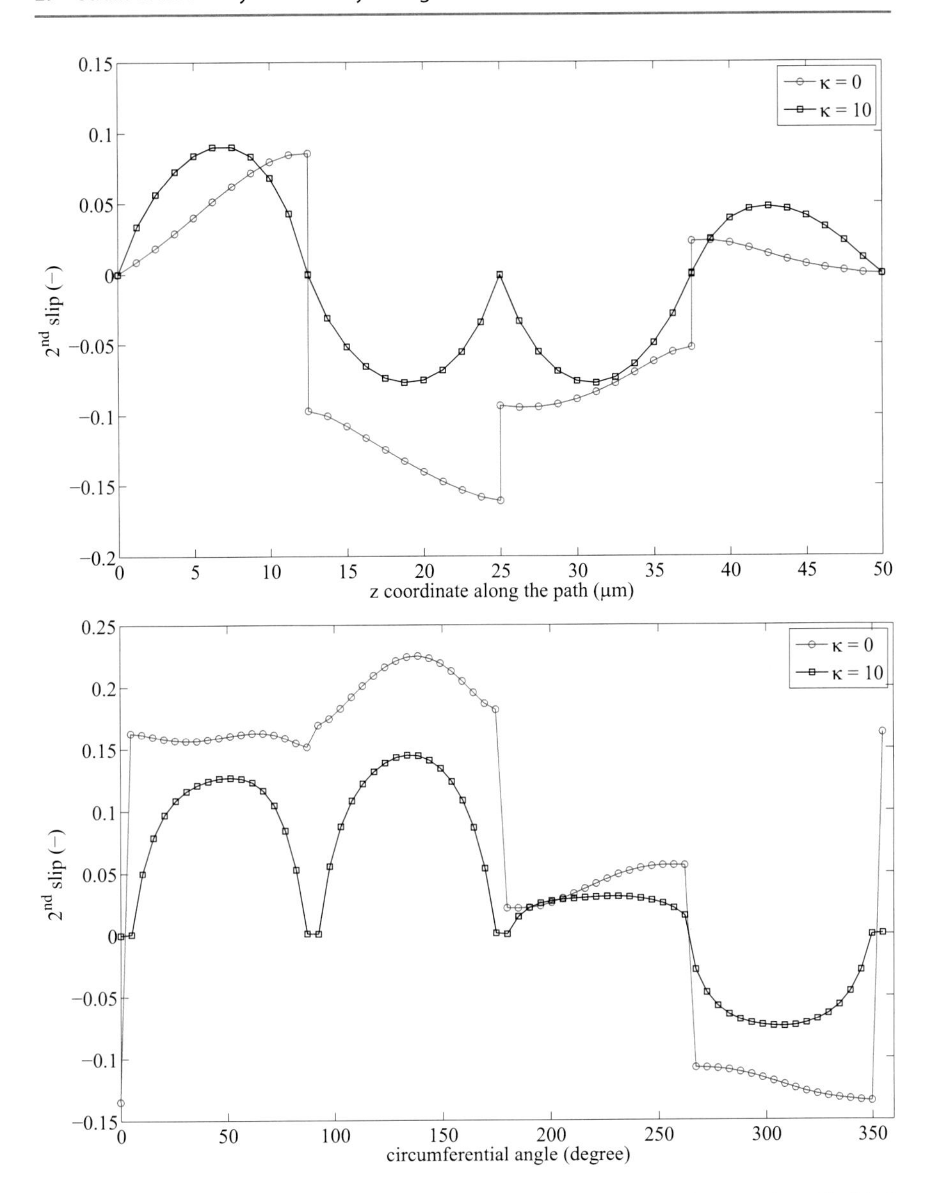

Fig. 13 Slip profile along longitudinal (*top*) and circumferential path (*bottom*) for cylindrical specimen

Two analyses with $\kappa = 0$ and $\kappa = 10$ are conducted, and the corresponding slip distributions (second slip system) along a representative longitudinal and a circumferential path (please see Fig. 12 for the chosen paths) are presented in Fig. 13. Although the same set of Euler angles are used in the two analyses, a stronger grain boundary yields a significant change in slip distribution. The discontinuities

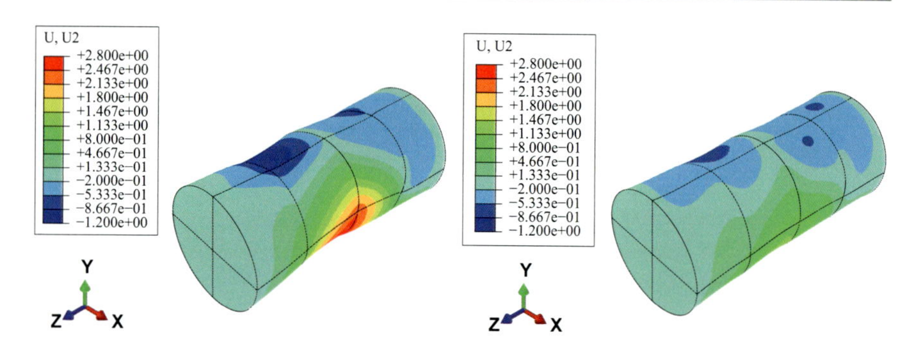

Fig. 14 Y-displacement as obtained by $\kappa = 0$ (*left*) and $\kappa = 10$ (*right*). The orientations of the grains in both analyses are kept the same

disappear for both longitudinal and circumferential paths, and a sort of pattern induced by grain boundaries is visible for strong GBs. Most of the grain boundaries behave very similar to hard boundaries although the GB strength is not that high (as compared to the bi-crystal example). Therefore, it seems that not only the strength of the GB but also the misorientation angle between the grains is influential on the slip profiles. Although the second slip system is presented, similar patterns are observed also for slip systems 1 and 3.

So far the discussion has focused on the slip response and the way it is altered as a function of GB strength and misorientations. However, due to coupled nature of the problem, it is natural to expect that GBs would also influence the strain and displacement fields of the specimen. In order to underscore this fact, in Fig. 14, contours of Y-displacement as obtained from the two analyses are presented. The significance of the difference is clearly visible, and in fact this comparison suggests that the GBs might have a strong influence on localization bands in polycrystalline specimens. Although the results of the current analysis are quite limited, the framework has the potential to investigate the influence of GBs on overall behavior to a certain extent through a statistical perspective. Furthermore, by switching on and off the GBs in combination with some specific textures (in terms of grain orientation distributions), the current formulation might be useful to quantify the relative contribution of GBs on mechanical response indicators such as load capacity corresponding to a "macroscopic" strain and/or "macroscopic" strain necessary to reach a certain "microscopic" strain within the polycrystalline solid body.

Conclusion and Outlook

In this chapter, a grain boundary model, which was rigorously derived in Gurtin (2008), is incorporated into a strain gradient crystal plasticity framework in a thermodynamically consistent manner. The description of the grain boundary model is based on the grain boundary Burgers tensor which takes into account both the effect of the mismatch between the grains and the orientation of the grain boundary.

Macroscopically, the geometric structure of a grain boundary is described by the grain misorientation and the grain boundary normal vector, Wolf and Yip (1992). Therefore the current formulation fully captures the geometric structure of a grain boundary. However, as long as intrinsic energy and energy stemming from interaction of lattice dislocations and grain boundary are concerned, the quadratic energy form used here is not really based on physical arguments. Nonetheless, this form can still be used to investigate the basic properties and limiting cases in a systematic manner. At this stage, it has to be noted that one can construct situations in which grain boundary Burgers tensor is zero although there exist nonzero slips within the grains. A physical explanation of such a situation might correspond to annihilation. However, in that case the model cannot reproduce the hard boundary conditions regardless of the magnitude of κ; see Gottschalk et al. (2016).

In order to capture the role of grain boundaries more accurately within continuum scale plasticity models, reflecting complicated physics of grain boundaries through proper upscaling strategies is essential. A particular approach in this context makes use of distributed disclination model to describe the initial energy of the symmetric tilt boundaries, Fressengeas et al. (2014). As compared to the results of atomistic models, the predictions of such defect mechanics based models are satisfactory. However, evolution of grain boundary intrinsic energy and incorporation of the effect of lattice dislocation interactions with grain boundaries and the associated energy still need to be addressed. Atomistic studies on grain boundary structures in various material systems are quite numerous, and their results can be exploited to construct computationally feasible multi-scale models. With regard to this, a systematic approach is presented in a series of papers by van Beers et al. (2015a,b) where a logarithmic relation for the initial intrinsic energy with two fitting parameters identified by atomistic simulations is embedded into a strain gradient crystal plasticity model. Furthermore, the interaction with lattice dislocations and grain boundary is quantified in terms of a net defect density measure (consistent with the grain boundary Burgers tensor), and defect redistribution within the grain boundary is addressed through net defect balance equation. Although the model has some phenomenological aspects, some qualitative agreement with small-scale experiments is achieved. Other mechanisms such as grain boundary sliding and opening resulting in the change of the structure of the grain boundary are intimately coupled with defect redistribution. Such changes involving relative sliding and opening call for a proper description of the mechanical response of the grain boundary. The treatment of GBs under finite strains (see, e.g., McBride et al. 2016) and coupling defect redistribution with mechanical cohesive behavior in a thermodynamically consistent manner (see, e.g., Mosler and Scheider 2011) are some open problems to be addressed. Here defect redistribution is primarily concerned with the evolution of the initial structure of the GBs and does not deal with the slip transmission events. It seems that continuum level description of such transmission events requires some special attention on different aspects ranging from the determination of the essential interactions (e.g., dislocation interactions across a GB, see Stricker et al. 2016) to proper set of transmission criteria; see Bayerschen et al. (2016). Apart from all these potential improvements, it has to be noted that quantitative validation of GB models at continuum scale is a challenge due to intrinsic coupling between defect transport including redistribution and the mechanical integrity of GBs.

References

R.K. Abu Al-Rub, Interfacial gradient plasticity governs scale-dependent yield strength and strain hardening rates in micro/nano structured metals. Int. J. Plast. **24**, 1277–1306 (2008)

K.E. Aifantis, J.R. Willis, The role of interfaces in enhancing the yield strength of composites and polycrystals. J. Mech. Phys. Solids **53**, 1047–1070 (2005)

K.E. Aifantis, W.A. Soer, J.T. de Hosson, J. Willis, Interfaces within strain gradient plasticity: theory and experiments. Acta Mater. **54**, 5077–5085 (2006)

E. Bayerschen, A.T. McBride, B.D. Reddy, T. Böhlke, Review of slip transmission criteria in experiments and crystal plasticity models. J. Mater. Sci. **51**, 2243–2258 (2016)

C.J. Bayley, W.A.M. Brekelmans, M.G.D. Geers, A comparison of dislocation induced back stress formulations in strain gradient crystal plasticity. Int. J. Solids Struct. **43**, 7268–7286 (2006)

U. Borg, A strain gradient crystal plasticity analysis of grain size effects in polycrystals. Eur. J. Mech. A. Solids **26**, 313–324 (2007)

T. Borg, N.A. Fleck, Strain gradient effects in surface roughening. Model. Simul. Mater. Sci. Eng. **15**, S1–S12 (2007)

M. de Koning, R. Miller, V.V. Bulatov, F.F. Abraham, Modelling grain boundary resistence in intergranular dislocation slip transmission. Philos. Mag. A **82**, 2511–2527 (2002)

M. de Koning, R.J. Kurtz, V.V. Bulatov, C.H. Henager, R.G. Hoagland, W. Cai, M. Nomura, Modeling of dislocation-grain boundary interactions. J. Nucl. Mater. **323**, 281–289 (2003)

M. Ekh, S. Bargmann, M. Grymer, Influence of grain boundary conditions on modeling of size-dependence in polycrystals. Acta Mech. **218**, 103–113 (2011)

N.A. Fleck, J.W. Hutchinson, A formulation of strain gradient plasticity. J. Mech. Phys. Solids **49**, 2245–2271 (2001)

P. Fredriksson, P. Gudmundson, Size-dependent yield strength of thin films. Int. J. Plast. **21**, 1834–1854 (2005)

C. Fressengeas, V. Taupin, L. Capolunga, Continuous modeling of the structure f symmetric tilt boundaries. Int. J. Solids Struct. **51**, 1434–1441 (2014)

M.G.D. Geers, W.A.M. Brekelmans, C.J. Bayley, Second-order crystal plasticity: internal stress effects and cyclic loading. Model. Simul. Mater. Sci. Eng. **15**, 133–145 (2007)

D. Gottschalk, A. McBride, B.D. Reddy, A. Javili, P. Wriggers, C.B. Hirschberger, Computational and theoretical aspects of a grain-boundary model that accounts for grain misorientation and grain-boundary orientation. Comput. Mater. Sci. **111**, 443–459 (2016)

P. Gudmundson, A unified treatment of strain gradient plasticity. J. Mech. Phys. Solids **52**, 1379–1406 (2004)

M.E. Gurtin, On the plasticity of single crystals: free energy, microforces, plastic-strain gradients. J. Mech. Phys. Solids **48**, 989–1036 (2000)

M.E. Gurtin, A gradient theory of single-crystal viscoplasticity that accounts for geometrically necessary dislocations. J. Mech. Phys. Solids **50**, 5–32 (2002)

M.E. Gurtin, A theory of grain boundaries that accounts automatically for grain misorientation and grain-boundary orientation. J. Mech. Phys. Solids **56**, 640–662 (2008)

M.E. Gurtin, L. Anand, S.P. Lele, Gradient single-crystal plasticity with free energy dependent on dislocation densities. J. Mech. Phys. Solids **55**, 1853–1878 (2007)

R. Kumar, F. Szekely, E. Van der Giessen, Modelling dislocation transmission across tilt grain boundaries in 2D. Comput. Mater. Sci. **49**, 46–54 (2010)

G. Lancioni, T. Yalçinkaya, A. Cocks, Energy-based non-local plasticity models for deformation patterning, localization and fracture. Proc. R. Soc. Lond. A Math. Phys. Eng. Sci. **471**(2180) (2015a)

G. Lancioni, G. Zitti, T. Yalcinkaya, Rate-independent deformation patterning in crystal plasticity. Key Eng. Mater. **651–653**, 944–949 (2015b)

T.C. Lee, I.M. Robertson, H.K. Birnbaum, Prediction of slip transfer mechanisms across grain boundaries. Scr. Metall. **23**(5), 799–803 (1989)

Z. Li, C. Hou, M. Huang, C. Ouyang, Strengthening mechanism in micro-polycrystals with penetrable grain boundaries by discrete dislocation dynamics simulation and Hall-Petch effect. Comput. Mater. Sci. **46**, 1124–1134 (2009)

A. Ma, F. Roters, D. Raabe, On the consideration of interactions between dislocations and grain boundaries in crystal plasticity finite element modeling – theory, experiments, and simulations. Acta Mater. **54**, 2181–2194 (2006)

T.J. Massart, T. Pardoen, Strain gradient plasticity analysis of the grain-size-dependent strength and ductility of polycrystals with evolving grain boundary confinement. Acta Mater. **58**, 5768–5781 (2010)

A.T. McBride, D. Gottschalk, B.D. Reddy, P. Wriggers, A. Javili, Computational and theoretical aspects of a grain-boundary model at finite deformations. Tech. Mech. **36**, 102–119 (2016)

D.L. McDowell, Viscoplasticity of heterogeneous metallic materials. Mater. Sci. Eng. R **62**, 67–123 (2008)

J. Mosler, I. Scheider, A thermodynamically and variationally consistent class of damage-type cohesive models. J. Mech. Phys. Solids **59**(8), 1647–1668 (2011)

I. Özdemir, T. Yalçinkaya, Modeling of dislocation-grain boundary interactions in a strain gradient crystal plasticity framework. Comput. Mech. **54**, 255–268 (2014)

Z. Shen, R.H. Wagoner, W.A.T. Clark, Dislocation pile-up and grain boundary interactions in 304 stainless steel. Scr. Metall. **20**(6), 921–926 (1986)

D.E. Spearot, D.L. McDowell, Atomistic modeling of grain boundaries and dislocation processes in metallic polycrystalline materials. J. Eng. Mater. Tech. **131**, 041,204 (2009)

M. Stricker, J. Gagel, S. Schmitt, K. Schulz, D. Weygand, P.B. Gumbsch, On slip transmission and grain boundary yielding. Meccanica **51**, 271–278 (2016)

M.A. Tschopp, D.L. McDowell, Asymmetric tilt grain boundary structure and energy in copper and aluminium. Philos. Mag. **87**, 3871–3892 (2007)

P.R.M. van Beers, G.J. McShane, V.G. Kouznetsova, M.G.D. Geers, Grain boundary interface mechanics in strain gradient crystal plasticity. J. Mech. Phys. Solids **61**, 2659–2679 (2013)

P.R.M. van Beers, V.G. Kouznetsova, M.G.D. Geers, Defect redistribution within continuum grain boundary plasticity model. J. Mech. Phys. Solids **83**, 243–262 (2015a)

P.R.M. van Beers, V.G. Kouznetsova, M.G.D. Geers, M.A. Tschopp, D.L. McDowell, A multiscale model to grain boundary structure and energy: from atomistics to a continuum description. Acta Mater. **82**, 513–529 (2015b)

E. Van der Giessen, A. Needleman, Discrete dislocation plasticity: a simple planar model. Model. Simul. Mater. Sci. Eng. **3**, 689–735 (1995)

D. Wolf, S Yip, *Material Interfaces: Atomic-Level Structure and Properties* (Chapman and Hall, London, 1992)

T. Yalçinkaya, Multi-scale modeling of microstructure evolution induced anisotropy in metals. Key Eng. Mater. **554–557**, 2388–2399 (2013)

T. Yalçinkaya, W.A.M. Brekelmans, M.G.D. Geers, Non-convex rate dependent strain gradient crystal plasticity and deformation patterning. Int. J. Solids Struct. **49**, 2625–2636 (2012)

T. Yalcinkaya, G. Lancioni, Energy-based modeling of localization and necking in plasticity. Procedia Mater. Sci. **3**, 1618–1625 (2014)

T. Yalcinkaya, W.A.M. Brekelmans, M.G.D. Geers, Deformation patterning driven by rate dependent non-convex strain gradient plasticity. J. Mech. Phys. Solids **59**, 1–17 (2011)

analysis as in classical plasticity formulations in which the formulation is developed in terms of tensor invariants and their combinations. However, to model the general three-dimensional constitutive behavior of the so-called geomaterials at arbitrary nonproportional load paths that frequently arise in dynamic loadings, such direct approaches do not yield models with desired accuracy. Instead, microplane approach prescribes the constitutive behavior on planes of various orientations of the material microstructure independently, and the second-order stress tensor is obtained by imposing the equilibrium of second-order stress tensor with the microplane stress vectors. In this work, particular attention is devoted to the milestone microplane models for plain concrete, namely, the model M4 and the model M7. Furthermore, a novel autocalibrating version of the model M7 called the model M7Auto is presented as an alternative to both differential and integral type nonlocal formulations since the model M7Auto does not suffer from the shortcomings of these classical nonlocal approaches. Examples of the performance of the models M7 and M7Auto are shown by simulating well-known benchmark test data like three-point bending size effect test data of plain concrete beams using finite element meshes of the same element size and Nooru-Mohamed test data obtained at different load paths using finite element meshes having different element sizes, respectively.

Keywords
Constitutive model · Microplane model · Crack band model · Concrete · Auto-calibrating Microplane model · Three dimensional finite element analysis

Introduction

The classical approach to the constitutive modeling of engineering materials is to relate the second-order strain tensor to the second-order stress tensor meanwhile satisfying tensorial invariance requirements. Often this approach requires either the use of tensor polynomial or scalar loading functions and inelastic potential functions that vary only with the invariants of the stress or strain tensors. In addition, in the case of pressure-sensitive dilatant quasi-brittle materials such as concretes, rocks, and stiff soils, the principal directions of second-order stress and strain tensors almost never coincide when a general dynamic loading causing significant wave propagation is considered. Thus, the classical approach to constitutive modeling becomes extremely difficult if not impossible in the case of such materials. A multitude of models exist that follow this classical approach, but almost all of them fall short of a complete description of the behavior of quasi-brittle materials at general nonproportional load paths.

In microplane approach, however, the material behavior is prescribed on planes of many different orientations independently, and the response is assembled to yield the second-order stress or strain tensor by arguing equilibrium of stresses that act at different scales in the microstructure of the material. This indirect approach is computationally much more demanding than the classical approach, but at the same time, it offers many more possibilities for modeling complex material behavior. For example, the so-called kinematic constraint employed for quasi-brittle materials, which implies that the microplane strains on different microplanes are projections of the second-order strain tensor, combined with equivalence of the virtual work of second-order stress and strain tensor over the volume of a unit hemisphere with that of microplane stress and strains over the surface of the same unit hemisphere (called the "stress equilibrium") automatically yields a microplane model in which the principal directions of the second-order stress and strain tensors almost never coincide. Moreover this microplane model captures the so-called vertex effect automatically (Caner et al. 2002). Similarly, using the so-called static constraint employed for polycrystalline metals, which implies that the microplane stresses on different microplanes are projections of the second-order stress tensor, leads to a multisurface plasticity model after imposing the stress equilibrium. Furthermore, in such a multisurface plasticity model with as many yield surfaces as the number of microplanes, there is no need to invent new, invariant rules for the activation and deactivation of the many yield surfaces for a multitude of possible load paths under dynamic loads; that is automatically determined by the magnitude of the deviatoric part of the projected microplane stress vectors.

For many materials, the microplane can physically be related to the material microstructure. For example, in polycrystalline plasticity, the microplanes represent compact planes over which dislocation motion takes place more easily than on other planes. Although there is a finite number of such planes for many metals and alloys and thus the material response appears to be likely anisotropic, many different grains oriented in many different directions actually lead to isotropic behavior in the structural scale with dimensions on the order of 0.01 mm or larger. In the case of concretes, there are no dislocations; however, as shown in Fig. 1b, the contact surfaces between coarse and fine aggregates and cement paste within the material mesostructure are where the inelastic phenomena take place, and thus microplanes represent a collection of all such contact planes oriented in all possible directions creating again initially isotropic response in the structural scale with dimensions on the order of 5 mm or larger. In biological soft tissue, for example, the material is made of a matrix of elastin fibers forming a network reinforced with collagen fibers with a mean direction oriented close to the maximum stress direction in the tissue. The distribution of collagen fiber directions can easily be projected on to the microplanes oriented in all possible directions which results in anisotropic behavior. The elastin fiber network is almost completely isotropic and can readily be represented by these microplanes.

The focus of this study is microplane models for plain concrete, arguably one of the most challenging engineering materials to model mathematically. Only the most important milestones in the development of microplane models for concrete are dealt with. In particular, the model M4 (Bažant et al. 2000; Caner and Bažant 2000), the model M7 (Caner and Bažant 2013a,b), and a novel nonlocal extension of the model M7 called the model M7Auto are studied in this work. The models M4 and M7 are continuum models developed to predict fracture, damage, and plasticity in plain concrete using the finite element method in which the damage

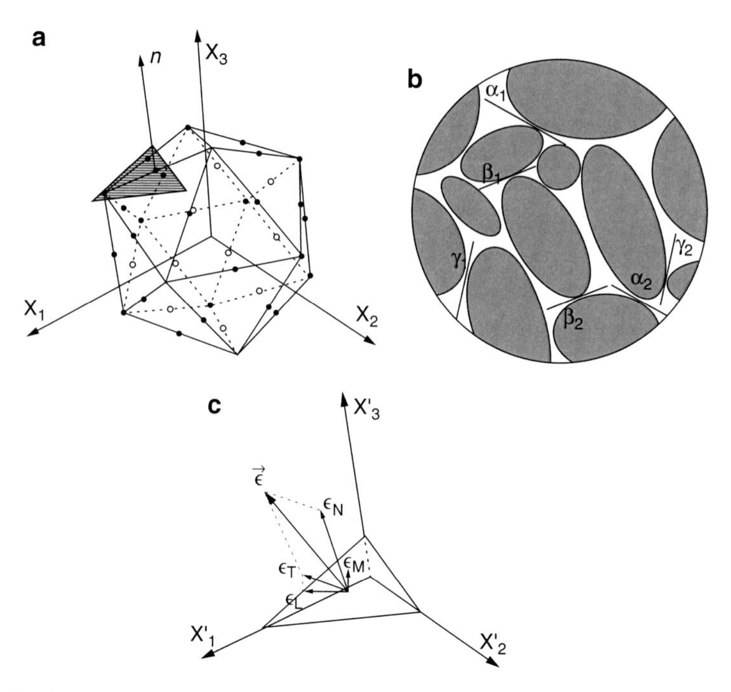

Fig. 1 (**a**) The Gaussian quadrature points representing the microplanes over a unit sphere, (**b**) the weak planes shown in the mesostructure of concrete, (**c**) the microplane strain vectors shown on a microplane

and fracture localize to a one element wide band. Such models are also called crack band models. In these models the response depends on the size and shape of the elements in the finite element mesh employed. In crack band models, the calibration of the model is carried out for a fixed element size, and in the analyses a finite element mesh of the same size elements with an aspect ratio of approximately 1 must be employed to obtain accurate predictions of the concrete response to an arbitrary three-dimensional stress state. By contrast, the classical approach to remove the mesh size dependence of crack band models is in the form of either (1) an integral type approach or (2) a differential type approach in both of which a characteristic material size acts as a localization limiter. Both of these approaches suffer from significant issues such as unobjective treatment of the Gauss points near the boundary of the structure, prediction of cracks initiating ahead of the crack tip, errors introduced due to shielding of Gauss points by a partially open cohesive crack, physically difficult to interpret higher-order stresses, and associated boundary conditions. In this study, the model M7Auto is described to overcome all these difficulties simply by taking into account the element size, shape, and type during the calibration process to extract the dependence of the microplane stress-strain boundaries on the finite element size so as to dissipate a constant fracture energy at quasistatic loading rates. It is shown that to successfully employ the model

M7Auto in the finite element analysis of concrete structures, it is necessary that (1) the size of the elements in the mesh does not exceed the range of autocalibration of the model, (2) the elements have an aspect ratio close to 1 for the simulation of general three-dimensional crack propagation, and (3) the element type should be the same in the mesh. The predictive capabilities of the models M7 and M7Auto are demonstrated by simulating some well-known benchmark test data.

Other than microplane models for cementitious materials, there are a number of microplane models formulated for polymer foams (Brocca et al. 2001), polycrystalline metals (Brocca and Bažant 2000), fiber-reinforced polymer composites (Kirane et al. 2016; Caner et al. 2011; Cusatis et al. 2008), rocks (Bažant and Zi 2003; Chen and Bažant 2014), soils (Prat and Bažant 1991; Chang and Sture 2006), shape memory alloys (Brocca et al. 2002), and biological soft tissue (Caner et al. 2007). Details of these formulations are excluded from the present study.

The Principal Microplane Formulations

Essentially the formulation for the constitutive response of a material involves determination of the stress tensor given the strain tensor and, in the inelastic range, the history variables that describe the earlier state of the material. In general, this relation must be a direct relation as in the case of elastic behavior of materials. Often an incremental approach must be employed in the inelastic range due to several mathematical and physical reasons. A typical example is the classical J_2 plasticity formulation that separates the plastic part from the strain tensor and determines the current stress tensor that corresponds to the current strain tensor. The microplane approach is fundamentally different in that the response quantities are first determined on the microplanes and integrated over all possible microplanes oriented in different directions to yield the familiar second-order tensor form. Due to the relationship between the material microstructure and the mechanical response, the application of the microplane approach to the constitutive modeling of materials with very different mechanical behaviors is of interest. Consequently, several formulations of microplane models proposed for several different engineering materials exist; among them are the so-called geomaterials that include cement-based materials, rocks, and soils. These materials exhibit pressure sensitivity, dilatancy, and softening due to fracturing. To capture this strain-softening behavior, the so-called kinematic constraint must be applied instead of the so-called static constraint. In the kinematically constrained microplane formulations, both the strain tensor and its increment are projected over all possible microplanes as shown in Fig. 1c, and the stress vector acting on each of these microplanes is evaluated using the constitutive law prescribed on the microplane level. Finally, the stress tensor is calculated by integrating the stress vector over all possible microplanes as shown in Fig. 1a. By contrast, in the statically constrained microplane formulations used to describe polycrystalline plasticity of metals and their alloys, both the stress tensor and its increment are projected over all possible microplanes, and the corresponding strain vectors are calculated using the constitutive law prescribed at the microplane level. Finally, the second-order tensor form of the strain tensor is obtained by integrating the strain vector over all possible microplanes. Thus, the microplane approach is an indirect one that employs simpler constitutive relations between stress and strain vectors at the microplane level to model complex material behavior. The kinematically constrained formulations can further be divided into the formulations with the so-called volumetric-deviatoric split and those without the volumetric-deviatoric split. The split in these formulations refers to the separation of the strain tensor into volumetric and deviatoric parts.

In the remainder of this section, the most commonly used formulations, namely, the kinematic no-split, the kinematic split, and the static split formulations, are discussed.

The Kinematic No-Split Formulation

Given the strain tensor, the first step is to project it onto the microplanes defined by discretizing a unit hemisphere at special locations for optimal integration (e.g., using Gaussian quadrature) over the surface of the hemisphere:

$$\epsilon_N = N_{ij}\epsilon_{ij} \tag{1}$$

where $N_{ij} = n_i n_j$ and the normal vector $\mathbf{n}$ is the normal to the microplane defined for each of the microplanes involved in the model.

To treat the microplane shear strain vector, it is convenient to define the directions $\mathbf{m}$ and $\mathbf{l}$ so that the base vectors in the directions $\mathbf{n}$, $\mathbf{m}$, and $\mathbf{l}$ form an orthogonal rectangular coordinate system on the microplane. These directions should be defined randomly so that no directional bias is created across the microplanes that form the unit hemisphere. Using these vectors, we can define the second-order tensors:

$$\begin{aligned} M_{ij} &= \frac{1}{2}\left(m_i n_j + m_j n_i\right) \\ L_{ij} &= \frac{1}{2}\left(l_i n_j + l_j n_i\right) \end{aligned} \tag{2}$$

Using the tensors $\mathbf{M}$ and $\mathbf{L}$, the components of shear strain vector along the directions $\mathbf{m}$ and $\mathbf{l}$ can be evaluated as

$$\begin{aligned} \epsilon_M &= M_{ij}\epsilon_{ij} \\ \epsilon_L &= L_{ij}\epsilon_{ij} \end{aligned} \tag{3}$$

The stress quantities that correspond to these microplane strain quantities are evaluated using the microplane constitutive laws:

$$\begin{aligned} \sigma_N &= \mathcal{F}_N\left(\epsilon_N, \ldots\right) \\ \tau_M &= \mathcal{F}_\tau\left(\epsilon_N, \epsilon_M, \ldots\right) \\ \tau_L &= \mathcal{F}_\tau\left(\epsilon_N, \epsilon_L, \ldots\right) \end{aligned} \tag{4}$$

where the functions $\mathcal{F}_N$ and $\mathcal{F}_\tau$ denote the microplane normal and shear constitutive laws.

Finally to calculate the stress tensor, the following stress equilibrium equation may be employed:

$$\int_V \sigma_{ij}\delta\epsilon_{ij}\, dV = \int_S \left[\sigma_N \delta\epsilon_N + \tau_M \delta\epsilon_M + \tau_L \delta\epsilon_L\right] dS \tag{5}$$

$$\Rightarrow \sigma_{ij}\delta\epsilon_{ij}\int_V dV = \int_S \left[\sigma_N N_{ij}\delta\epsilon_{ij} + \tau_M M_{ij}\delta\epsilon_{ij} + \tau_L L_{ij}\delta\epsilon_{ij}\right] dS$$

$$\Rightarrow \left[\sigma_{ij}\frac{2\pi}{3} - \int_S \left[\sigma_N N_{ij} + \tau_M M_{ij} + \tau_L L_{ij}\right] dS\right]\delta\epsilon_{ij} = 0 \;\forall\; \delta\epsilon_{ij} \neq 0$$

$$\Rightarrow \sigma_{ij} = \frac{3}{2\pi}\int_S \left[\sigma_N N_{ij} + \tau_M M_{ij} + \tau_L L_{ij}\right] dS \tag{6}$$

Eq. (5) means that the virtual work of stress and strain tensor over the volume of the unit hemisphere is equal to the virtual work of the microplane stresses and strains over the surface of the unit hemisphere.

As a useful exercise, the small strain linearly elastic behavior may be obtained using this formulation as follows. First, the microplane stress and strains are assumed to be related as

$$\begin{aligned} \sigma_N &= E_N \epsilon_N \\ \tau_M &= E_\tau \epsilon_M \\ \tau_L &= E_\tau \epsilon_L \end{aligned} \tag{7}$$

Substituting Eqs. (7), (1) and (3) into Eq. (6) and factoring out the strain tensor $\boldsymbol{\epsilon}$ yield the stiffness tensor as obtained by the microplane no-split formulation. This tensor may be compared to the classical isotropic elastic stiffness tensor, and for the two tensors to be identical, the microplane elastic moduli turn out to be

$$\begin{aligned} E_N &= \frac{E}{1-2\nu} \\ E_\tau &= \frac{E}{1-2\nu}\frac{1-4\nu}{1+\nu} \end{aligned} \tag{8}$$

According to the above relations, the thermodynamically admissable range of the coefficient of the Poisson is limited to $-1 < \nu < 1/4$. Thus for a material for which the coefficient of Poisson lies in the range $[1/4, 1/2)$, the model cannot be used to predict the behavior of that material. However, for many quasi-brittle materials such as concretes, fiber-reinforced concretes, and many other ceramics, the coefficient of Poisson lies in the range $[-1, 1/4]$, and this formulation can be used. For example, recently this formulation has been successfully extended to model the complete range of the mechanical behavior of normal strength and high strength concretes as well as fiber-reinforced concretes.

The Kinematic Split Formulation

In this formulation, the microplane normal strain is decomposed into volumetric and deviatoric parts:

$$\begin{aligned} \epsilon_N &= \epsilon_V + \epsilon_D \\ \sigma_N &= \sigma_V + \sigma_D \end{aligned} \tag{9}$$

where

$$\epsilon_V = \epsilon_{kk}/3 \tag{10}$$

and thus

$$\epsilon_D = \left(N_{ij} - \frac{\delta_{ij}}{3}\right)\epsilon_{ij} \tag{11}$$

The shear strains remain as given in Eq. (3). However, in this formulation Eqs. (4) should be rewritten as

$$\begin{aligned} \sigma_D &= \mathcal{F}_D\left(\epsilon_D, \ldots\right) \\ \sigma_V &= \mathcal{F}_V\left(\epsilon_V, \ldots\right) \\ \tau_M &= \mathcal{F}_\tau\left(\epsilon_D, \epsilon_V, \epsilon_M, \ldots\right) \\ \tau_L &= \mathcal{F}_\tau\left(\epsilon_D, \epsilon_V, \epsilon_L, \ldots\right) \end{aligned} \tag{12}$$

Furthermore, Eq. (6) must also be rewritten as

$$\begin{aligned} &\int_V \sigma_{ij}\delta\epsilon_{ij}\,dV = \int_S \left[\sigma_V\delta\epsilon_V + \sigma_D\delta\epsilon_D + \tau_M\delta\epsilon_M + \tau_L\delta\epsilon_L\right] dS \\ &\Rightarrow \sigma_{ij}\delta\epsilon_{ij}\int_V dV = \int_S \left[\sigma_V\frac{\delta_{ij}}{3}\delta\epsilon_{ij} + \sigma_D\left(N_{ij} - \frac{\delta_{ij}}{3}\right)\delta\epsilon_{ij} + \tau_M M_{ij}\delta\epsilon_{ij} + \tau_L L_{ij}\delta\epsilon_{ij}\right] dS \\ &\Rightarrow \left[\sigma_{ij}\frac{2\pi}{3} - \sigma_V\delta_{ij}\frac{2\pi}{3} - \int_S \left[\sigma_D\left(N_{ij} - \frac{\delta_{ij}}{3}\right) + \tau_M M_{ij} + \tau_L L_{ij}\right] dS\right]\delta\epsilon_{ij} = 0\ \forall\ \delta\epsilon_{ij} \neq 0 \\ &\Rightarrow \sigma_{ij} = \sigma_V\delta_{ij} + \frac{3}{2\pi}\int_S \left[\sigma_D\left(N_{ij} - \frac{\delta_{ij}}{3}\right) + \tau_M M_{ij} + \tau_L L_{ij}\right] dS \end{aligned} \tag{13}$$

because $\int_S N_{ij} dS = \delta_{ij} 2\pi/3$ and $\int_S dS = 2\pi$. Instead of Eq. (13) the microplane split formulations prior to the microplane model M4 used simply

$$\sigma_{ij} = \sigma_V \delta_{ij} + \frac{3}{2\pi} \int_S \left[\sigma_D N_{ij} + \tau_M M_{ij} + \tau_L L_{ij}\right] dS \tag{14}$$

which violates the work conjugacy of volumetric stress. This can be conveniently shown by simply setting $i = j = k$ in Eq. (14) which yields

$$\sigma_{kk} = 3\sigma_V + \frac{3}{2\pi} \int_S \sigma_D dS \tag{15}$$

in which generally $\int_S \sigma_D dS \neq 0$ except in the elastic range. Thus, the pressure depends on deviatoric stress when Eq. (14) is employed in the calculation of the second-order stress tensor.

Using this formulation the isotropic linearly elastic behavior may be simulated as follows: Let the relationships between the microplane stress and strains be given by

$$\begin{aligned} \sigma_V &= E_V \epsilon_V \\ \sigma_D &= E_\tau \epsilon_D \\ \tau_M &= E_\tau \epsilon_M \\ \tau_L &= E_\tau \epsilon_L \end{aligned} \tag{16}$$

Substituting Eqs. (16), (3), (10), and (11) into Eq. (13) and again factoring out the strain tensor ϵ result in the stiffness tensor as obtained by the microplane split formulation. This tensor may again be compared to the classical elastic stiffness tensor for an isotropic material, and for the two tensors to be identical, the microplane elastic moduli turn out to be

$$\begin{aligned} E_V &= \frac{E}{1 - 2\nu} \\ E_D &= E_\tau = \frac{E}{1 + \nu} \end{aligned} \tag{17}$$

The above equations clearly show that in this formulation, the full thermodynamically admissable range $[-1, 1/2]$ of the coefficient of Poisson can be represented. Several microplane models for concrete and fiber-reinforced concrete have been developed by different authors using this formulation. Due to the volumetric-deviatoric split of the normal strain both in the elastic and in the inelastic ranges, this formulation results in similar material behavior both in tension and in compression. On the other hand, for quasi-brittle materials like concrete, such similarity in the predicted response is not helpful because the behavior in tension is completely different than behavior in compression for such materials.

The Static Split Formulation

The static split formulation is obtained by projecting the stress tensor over the microplanes:

$$\begin{aligned} \sigma_N &= N_{ij}\sigma_{ij} \\ \tau_M &= M_{ij}\sigma_{ij} \\ \tau_L &= L_{ij}\sigma_{ij} \end{aligned} \tag{18}$$

In this formulation, typically the stress tensor being projected is the trial stress tensor as in the classical plasticity theory. Furthermore, the projected normal stress is divided into volumetric and deviatoric parts:

$$\sigma_N = \sigma_V + \sigma_D \tag{19}$$

where $\sigma_V = \delta_{ij}\sigma_{ij}/3$ and $\sigma_D = (N_{ij} - \delta_{ij}/3)\sigma_{ij}$. The projected and split microplane stresses are employed to determine the microplane inelastic strains using the constitutive laws $\mathcal{H}_D, \mathcal{H}_V$, and $\mathcal{H}_\tau$ defined at the microplane level:

$$\begin{aligned} \epsilon_D^{in} &= \mathcal{H}_D\left(\sigma_D, \sigma_V, \ldots\right) \\ \epsilon_V^{in} &= \mathcal{H}_V\left(\sigma_V, \ldots\right) \\ \epsilon_M^{in} &= \mathcal{H}_\tau\left(\sigma_D, \sigma_V, \sigma_M, \ldots\right) \\ \epsilon_L^{in} &= \mathcal{H}_\tau\left(\sigma_D, \sigma_V, \sigma_L, \ldots\right) \end{aligned} \tag{20}$$

Finally to calculate the inelastic strain tensor, the following stress equilibrium equation may be employed:

$$\int_V \epsilon_{ij}^{in}\delta\sigma_{ij}\,dV = \int_S \left[\epsilon_V^{in}\delta\sigma_V + \epsilon_D^{in}\delta\sigma_D + \epsilon_M^{in}\delta\tau_M + \epsilon_L^{in}\delta\tau_L\right] dS \tag{21}$$

$$\Rightarrow \epsilon_{ij}^{in}\delta\sigma_{ij}\int_V dV = \int_S \left[\epsilon_V^{in}\frac{\delta_{ij}}{3}\delta\sigma_{ij} + \epsilon_D^{in}\left(N_{ij} - \frac{\delta_{ij}}{3}\right)\delta\sigma_{ij} + \epsilon_M^{in}M_{ij}\delta\sigma_{ij} + \epsilon_L^{in}L_{ij}\delta\sigma_{ij}\right] dS$$

$$\Rightarrow \left[\epsilon_{ij}^{in}\frac{2\pi}{3} - \epsilon_V^{in}\delta_{ij}\frac{2\pi}{3} - \int_S \left[\epsilon_D^{in}\left(N_{ij} - \frac{\delta_{ij}}{3}\right) + \epsilon_M^{in}M_{ij} + \epsilon_L^{in}L_{ij}\right] dS\right]\delta\sigma_{ij} = 0\ \forall\ \delta\sigma_{ij}\neq 0$$

$$\Rightarrow \epsilon_{ij}^{in} = \epsilon_V^{in}\delta_{ij} + \frac{3}{2\pi}\int_S \left[\epsilon_D^{in}\left(N_{ij} - \frac{\delta_{ij}}{3}\right) + \epsilon_M^{in}M_{ij} + \epsilon_L^{in}L_{ij}\right] dS \tag{22}$$

Eq. (22) determines the inelastic part of the total strain tensor which readily facilitates the calculation of the elastic part of the strain tensor which in turn leads to the actual stress tensor. For example, assuming a series coupling of elastic and plastic zones in an elastoplastic material, once the inelastic strain tensor is calculated using Eq. (22), the elastic part of the total strain tensor can easily be determined which readily facilitates the calculation of the actual stress tensor.

The microplane isotropic linearly elastic flexibility tensor can be obtained by treating the inelastic strains in Eq. (22) as elastic strains and substituting Eq. (18) and

$$\begin{aligned} \epsilon_V &= \frac{1}{E_V}\sigma_V \\ \epsilon_D &= \frac{1}{E_\tau}\sigma_D \\ \epsilon_M &= \frac{1}{E_\tau}\sigma_M \\ \epsilon_L &= \frac{1}{E_\tau}\sigma_L \end{aligned} \tag{23}$$

into Eq. (22) and factoring out the stress tensor. By comparing this microplane flexibility tensor with the standard isotropic linearly elastic flexibility tensor, the same relations given in Eq. (17) are obtained.

Microplane Models for Cementitious Materials

Several authors have proposed microplane models for constitutive behavior of cement-based materials since the microplane approach with a formulation applicable to tensile fracturing of cement-based materials was first outlined in 1984 (Bažant and Gambarova 1984; Bažant and Oh 1985). In the remainder of this section, the most important formulations of microplane approach for cementitious materials are discussed.

Earlier Microplane Models with Limited Predictive Capabilities

The first microplane no-split-type formulation presented in Bažant and Gambarova (1984) and Bažant and Oh (1985) is based on the hypotheses that the microplane normal strains are the projections of the second-order strain tensor as shown in Eq. (1) and that the microplane normal stress on any given microplane is a function of the microplane normal strain on that microplane as shown in the first of Eq. (4). The shear resistance on the microplanes is neglected. The microplane system then is coupled to an elastic solid in series so as to yield the desired Poisson's ratio for concrete. To numerically calculate the integral in Eq. (6) (where τ_M and τ_L are taken as zero in this formulation) as efficiently as possible, the optimum Gaussian integration over the surface of a sphere is employed. Using this model, the experimentally observed dilatancy in concrete when cracks are sheared is successfully simulated. In addition, this formulation is shown to be capable of simulating the tensile fracture of concrete. On the other hand, the compression failure of concrete cannot be correctly predicted using this formulation.

In the microplane split-type formulation presented in Bažant and Prat (1988a) and verified in Bažant and Prat (1988b), the microplane shear resistance is taken into account (as in Eqs. (3) with (12)), and the invariants of the stress and strain tensors are introduced in the formulation so as to be able to model the general three-dimensional pressure-sensitive dilatant behavior of concrete in compression. The volumetric-deviatoric split of the microplane normal strain and stress (as in Eqs. (9) and (19)) is introduced leading to the simulation of the complete thermodynamically admissable range of Poisson's ratio. With this formulation, the microplane formulation is able to predict a considerably wide range of inelastic behavior of concrete, including tensile softening, dilatancy under shear, dilatancy and friction in compression, compression softening, ductile-brittle transition in compression, and hydrostatic compression behaviors. However, this formulation cannot accurately predict the concrete response to load cycles in compression and in tension. Furthermore, this formulation suffers from excessively long softening tails in the model response.

The microplane split-type formulation presented in Ožbolt and Bažant (1992) refines and extends the model presented in Bažant and Prat (1988a) to cyclic loading and the strain rate effect in concrete. The rate effect is modeled as in the Maxwell series coupling model of a nonlinear spring and a dashpot which ignores the real physical mechanisms behind this phenomenon. Although some success is achieved in simulating the load cycles in compression, the load cycles in tension could not be correctly simulated. In particular, this formulation fails to predict damage in tension manifested by unloading to origin with little observed plasticity. In this formulation the microplane damage variables are treated as nonlocal variables as described in Bažant and Ožbolt (1990) and in Bažant and Pijaudiercabot (1988).

The studies Bažant et al. (1996a,b) present and verify yet another split-type formulation that introduces the so-called stress-strain boundaries. These boundaries should be considered as yield limits that vary with microplane strains and other history variables defined on the microplane. The basic idea is to describe the monotonic loading in inelastic range using these boundaries and prescribe the loading, unloading, and reloading responses between the boundaries as elastic. It is verified against different test data from different researchers as well as several test data obtained using specimens from the same concrete at different proportional and nonproportional load paths by the same researchers. Remarkably all these data are simulated relatively well using this formulation. However, in this model the prediction of concrete response to load cycles in tension still remains unsolved. Furthermore, the stiffness degradation in tensile softening cannot be simulated. Also, in tensile softening predictions, the lateral strains are observed to be excessively large which cannot be physically justified.

The Microplane Model M4

The microplane model M4 deserves more attention as it is currently the most widely used microplane model for concrete even though it is *not* the most capable microplane model currently available for concrete. It has been implemented in many commercial and free finite element analysis codes including ATENA, OOFEM, MARS, DIANA, SBETA, and EPIC with an ever-growing user community because it satisfies the needs of many engineers and researchers by providing consistently outstanding prediction of concrete behavior at most commonly used load paths.

One important feature of the model M4 is, according to the model M4 formulation presented in Bažant et al. (2000) and verified in Caner and Bažant (2000), that the work conjugacy of volumetric stress is satisfied leading to the stress equilibrium equation given by Eq. (13). All split-type formulations prior to the model M4 suffer from a lack of work conjugacy of volumetric stress, the correction of which is the main motivation behind the model M4. When Eq. (14) is used in the calculation of the second-order stress tensor instead of Eq. (13), a volumetric-deviatoric coupling occurs automatically in the inelastic range. This, on the other hand, is *not* an error per se because the experimentally observed dilatant behavior of concrete under shear in fact implies the existence of such a coupling. However, this coupling must be introduced explicitly in the boundaries of the model so as to control the model behavior to fit the test data more accurately. Consequently, the model M4 formulation employs Eq. (13) that satisfies the work conjugacy of the volumetric stress.

Another feature of the model M4 is that the microplane boundaries that control softening are enhanced with plateaus so that the sharp peaks observed in the softening predictions by the earlier formulations become rounded in the softening response of the model. In the earlier formulations, the inelastic strains localize to a few microplanes which create a sharp peak which is observed in the experiments as a round one. In the formulation of the model M4, the plateaus incorporated into the microplane boundaries that control the softening response allow more microplanes to reach the yield limit, and thus a rounded peak results.

Yet another major feature embedded in the model M4 is the introduction of a completely new nonlinear friction boundary. Earlier formulations use a linear dependence of the frictional stress on the normal stress which spuriously strengthens the concrete at large normal stresses. The newly introduced boundary starts as nearly linear at small normal stresses but quickly reaches a completely horizontal asymptote at large normal stresses. Thus, at large confining pressures, a maximum frictional resistance at the microplanes is reached. This feature allows the prediction of both the entry and exit crater shapes in the dynamic projectile penetration simulations of concrete walls of different thicknesses which involve only moderately large microplane normal stresses (Bažant et al. 2000). With the microplane formulations prior to the model M4, these crater shapes are not predicted correctly at all.

Furthermore, the unloading and reloading laws which remain elastic in the formulation presented in Bažant et al. (1996b) are redefined to fit the experimental data better in compression load cycles as well as in unloading at high confining pressures.

In tension, a crack-closing boundary is proposed which allows the response to return to origin when unloading occurs. However, the resulting load cycle loops are too large compared to the experimentally observed loops. Furthermore, the stiffness degradation in tension is not captured at all in this formulation. Yet another problem that plagues the model M4 as well as the earlier split formulations is that in the tensile softening response, excessive lateral contractions which cannot be physically justified are predicted.

The model M4 is extended to strain rate effects and finite strain as well, but the important feature of this extension given in Bažant et al. (2000) is that the fracturing strain rate effects are based on the theory of Arrhenius-type fracture process with an activation energy. Although this theory is empirical, it can describe an incredibly wide range of physical phenomena including elastic deformations as well as diffusion phenomena in the material microstructure. The finite strain theory presented and applied to the microplane model M4 in Bažant et al. (2000) proposes as the finite strain tensor the Green's Lagrangian strain tensor and as the stress tensor the back-rotated Cauchy stress tensor which are *not* work conjugate. However, the error that results from the lack of work conjugacy is predicted to be small.

The Microplane Model M7

The microplane model M7 presented in Caner and Bažant (2013a) and verified in Caner and Bažant (2013b) is currently the most capable of the microplane models for concrete. Although it has a kinematic no-split-type formulation in the elastic range, in the inelastic range, the compressive stress-strain boundaries employ the volumetric-deviatoric split of microplane stress and strains, while the tensile boundary still employs the total microplane normal strain, and thus its formulation is of mixed type. The formulation of this model resolves the three aforementioned issues that have been persistent for several decades, namely, (1) the excessive lateral contraction during softening in tension is prevented (now the lateral strains return to origin as the tensile softening progresses), (2) the stiffness degradation during tensile softening is correctly captured, and (3) the tension-compression load cycles are correctly represented. In the rest of the load paths, the response predicted by the model M7 is either slightly better or about the same as that of the model M4. Thus the model M7 appears to be a significant improvement over its predecessors. The price to pay to achieve it is that the Poisson's ratio must stay in the interval $(-1, 1/4)$, which typically is the case for all kinematic no-split microplane formulations that do not ignore the shear resistance on the microplanes. This situation, however, could be remedied by coupling an elastic solid in series to the microplane solid as shown in Caner and Bažant (2013a) which allows the representation of the full thermodynamically admissable range of Poisson's ratio by this series coupling model as a whole.

The stress equilibrium equation employed in this formulation is the same as in Eq. (6). In the elastic range, Eqs. (7) and (8) hold true. In the inelastic range, the microplane stress and strains are separated into their volumetric and deviatoric parts as in Eqs. (9) and used in the microplane constitutive laws given by

$$\mathcal{F}_D(\epsilon_D, \epsilon_V) = -\frac{E k_1 \beta_3}{1 + (\langle -\epsilon_D \rangle / (k_1 \beta_2))^2}$$

$$\mathcal{F}_V(\epsilon_V, \epsilon_I, \epsilon_{III}) = -E k_1 k_3 \exp\left(-\frac{\epsilon_V}{k_1 \alpha(\epsilon_I, \epsilon_{III})} \right)$$

$$\mathcal{F}_N^-(\epsilon_D, \epsilon_V, \epsilon_I, \epsilon_{III}) = \mathcal{F}_D(\epsilon_D, \epsilon_V) + \mathcal{F}_V(\epsilon_V, \epsilon_I, \epsilon_{III}) \tag{24}$$

$$\mathcal{F}_N^+(\epsilon_N, \sigma_V) = E k_1 \beta_1 \exp\left(-\langle \epsilon_N - \beta_1 c_2 k_1 \rangle / \left(c_4 \epsilon_e^0 + k_1 c_3 \right) \right)$$

$$\mathcal{F}_\tau(\sigma_N, \epsilon_V) = \frac{c_{10} \langle \hat{\sigma}_N^0(\epsilon_V) - \langle -\sigma_N \rangle \rangle}{1 + c_{10} \langle \hat{\sigma}_N^0(\epsilon_V) - \langle -\sigma_N \rangle \rangle / (E_\tau k_1 k_2)}$$

where $\epsilon_e^0 = \sigma_V^0 / E_{N0}$, $\beta_1 = -c_1 + c_{17} \exp\left[-\left(c_{19} \langle -\sigma_V^0 - c_{18} \rangle \right) / E_{N0} \right]$, $\beta_2 = c_5 \exp\left(f'_{c0}/E_0 - f'_c/E \right) \times \tanh\left(c_9 \langle -\epsilon_V \rangle / k_1 \right) + c_7$, $\beta_3 = c_6 \exp\left(f'_{c0}/E_0 - f'_c/E \right) \tanh\left(c_9 \langle -\epsilon_V \rangle / k_1 \right) + c_8$, $\alpha(\epsilon_I, \epsilon_{III}) = \left[k_5 / \left(1 + \min(\langle -\sigma_V^0 \rangle, c_{21}) / E_{N0} \right) \right] \times \left[\left(\epsilon_I^0 - \epsilon_{III}^0 \right) / k_1 \right]^{c_{20}} + k_4$, and $\hat{\sigma}_N^0(\epsilon_V) = E_\tau \langle k_1 c_{11} - c_{12} \langle \epsilon_V \rangle \rangle$. In these equations, $k_1, k_2, \ldots, k_5$ are the free parameters that need to be calibrated for any given concrete, while $c_1, c_2, \ldots, c_{20}$ are fixed constants for which the values are given in Table 1 of Caner and Bažant (2013b). Moreover, in tension the microplane normal modulus of elasticity degrades according to

$$E_N = E_{N0} \exp\left(-c_{13} \epsilon_N^{0+} \right) f(\xi) \tag{25}$$

where ϵ_N^{0+} is the maximum tensile fracturing strain ever reached on a microplane, a history variable; $f(\xi) = 1/(1 + 0.1\xi)$ where $\xi = \int \langle d\epsilon_V \rangle$ is a correction function to take into account the fatigue of concrete under tensile load cycles (Kirane and Bažant 2015). In compression, the microplane normal elastic modulus degradation occurs according to

$$E_N = E_{N0} \left[\exp\left(-c_{14} |\epsilon_N^{0-}| / \left[1 + c_{15} \epsilon_e^0 \right] \right) + c_{16} \epsilon_e^0 \right] \tag{26}$$

where ϵ_N^{0-} is the maximum compressive fracturing strain ever reached on a microplane, another history variable. In the foregoing equations; E_{N0} is the undamaged microplane normal elastic modulus, E_N is the (possibly damaged) microplane normal elastic modulus.

The elastic microplane stresses are calculated incrementally by

$$\begin{aligned} \sigma_N^e &= \sigma_N^0 + E_N \Delta\epsilon_N \\ \tau_L^e &= \tau_L^0 + E_\tau \Delta\epsilon_L \\ \tau_M^e &= \tau_M^0 + E_\tau \Delta\epsilon_M \\ \tau^e &= \sqrt{\tau_L^{e\,2} + \tau_M^{e\,2}} \end{aligned} \tag{27}$$

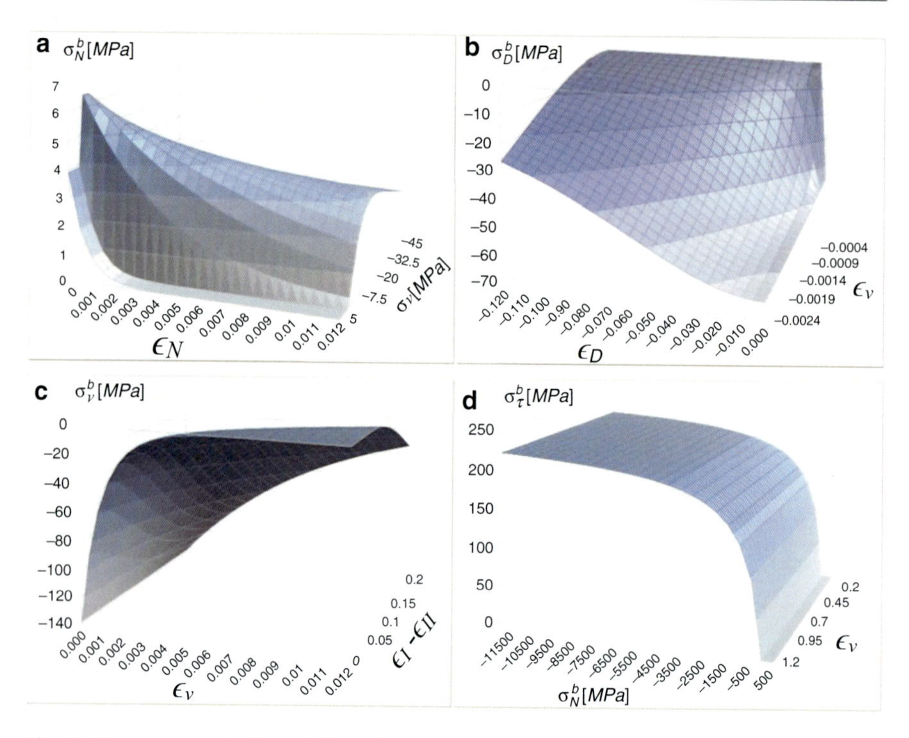

Fig. 2 The stress-strain boundaries of the model M7: (**a**) normal boundary, (**b**) deviatoric boundary, (**c**) volumetric boundary, and (**d**) friction boundary

where the microplane stresses from the previous load step σ_N^0, τ_L^0, and τ_M^0 are also history variables, leading to a total of five history variables per microplane. But the stress response on a microplane cannot exceed the corresponding stress-strain boundary:

$$\begin{aligned}\sigma_N &= \max\left(\min\left(\sigma_N^e, \mathcal{F}_N^+\right), \mathcal{F}_N^-\right)\\ \tau_L &= \frac{\tau_L^e}{\tau^e}\min\left(\tau^e, \mathcal{F}_\tau\right)\\ \tau_M &= \frac{\tau_M^e}{\tau^e}\min\left(\tau^e, \mathcal{F}_\tau\right)\end{aligned} \tag{28}$$

Finally the second-order stress tensor is obtained by evaluating the integral in Eq. (6) using Gaussian quadrature. The strain-dependent yield limits of the model M7 given by the aforementioned equations are depicted in Fig. 2.

The experimental data first reported in Figs. 12 through 15 of Bažant et al. (1996a) features many different load paths in compression for the same concrete. To simulate these test data, once the model is calibrated, the model parameters cannot be changed from one load path prediction to the other, and thus these tests

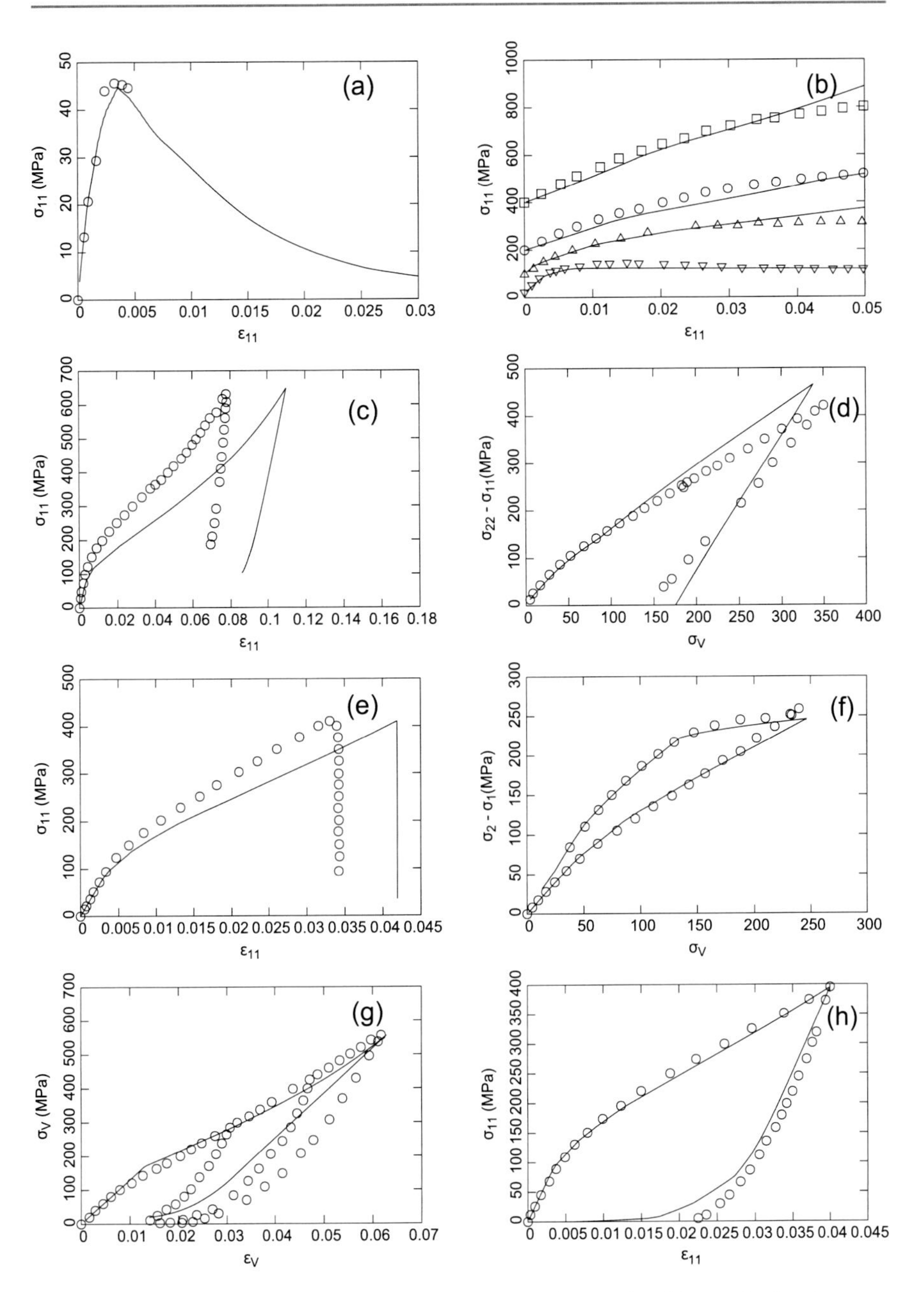

Fig. 3 The test data obtained at Waterways Experiment Station for different load paths using the same concrete (*the symbols*) and predictions by the model M7 (*the curves*): (**a**) uniaxial compression, (**b**) triaxial compression, (**c, d**) load path with lateral to axial strain ratio of -0.2, (**e, f**) axial loading and lateral unloading, (**g**) hydrostatic compression, and (**h**) confined compression

are a major challenge for any concrete model. The model M7 successfully predicts response of this plain concrete at nearly all load paths as shown in Fig. 3 except for Fig. 3c. The test data for this load path for which the predicted response of the model M7 seems to be off could never be predicted correctly neither by the earlier microplane formulations nor by mesoscale models for plain concrete (see Cusatis et al. 2011) suggesting a likely problem with the test data. The performance of the model M7 in predicting the test data shown in these figures is about the same as that of the model M4. However, the model M7 calibration in this case also satisfies a most likely uniaxial tension behavior of this type of concrete, whereas the model M4 calibration did not take into account the tensile behavior of this concrete. Unfortunately a uniaxial tension test data for this concrete is not available, and thus the predicted response shown in this figure cannot be compared to the test data.

In Fig. 3 the parameter set of the model M7 employed for all simulations is given by $E = 25,000$ MPa, $\nu = 0.18, k_1 = 160.10^6, k_2 = 100, k_3 = 15, k_4 = 72, k_5 = 10.10^6$, $f'_{c0} = 15.08$ MPa, $E_0 = 20,000$MPa, $c_1 = 0,089, c_2 = 0.4, c_3 = 0.4, c_4 = 50, c_5 = 3500, c_6 = 40, c_7 = 30, c_8 = 8, c_9 = 0.012, c_{10} = 0.4, c_{11} = 1.9, c_{12} = 0.18, c_{13} = 2500, c_{14} = 500, c_{15} = 7000, c_{16} = 100, c_{17} = 1, c_{18} = 1.6 \cdot 10^3$MPa, $c_{19} = 1000, c_{20} = 1.8, c_{21} = 250$ MPa. Elements are deleted when the maximum microplane normal strain in tension exceeds 90% in all simulations.

A further extensive verification of the model M7 is given in Fig. 4. In these figures, the experimental data from three-point bending size effect tests originally published in Hoover et al. (2013) are simulated using the model M7. Once the model is calibrated using the load-displacement curve of the beam with 9.3 cm depth and a relative notch length of 30%, the model parameters remain fixed for the rest of the simulations. Thus, these size effect tests are a major challenge for any constitutive model for plain concrete. As shown in these figures, the performance in tension of the model M7 is outstanding. In addition, the model M7 can predict compression-tension load cycles unlike the earlier formulations of microplane models as shown in Fig.4 of Caner and Bažant (2013b) and in Kirane and Bažant (2015).

Furthermore, in Fig. 5 the compression test data published in Wendner et al. (2015) are predicted using the model M7. The test data given in Fig. 5a is for a cylindrical specimen aged for 31 days, and the data given in Fig. 5b is for a cylindrical specimen aged for 400 days. Other than this difference, the two concretes come from the same batch as the specimens of Fig. 4. The test data of Fig. 5c, d are for cubic specimens aged for 470 days. The compressive response parameters of the model M7 are calibrated in Fig. 5c, and for each case, the reported value of elastic modulus is employed. In addition, in all compression simulations, the contact conditions between test platens and the specimens are simulated using a band of cohesive elements. These contact conditions turn out to be critical to obtain the result shown in Fig. 5d. The tensile response of the model remained the same as in Fig. 4. As can be inferred from the fits in these figures, the model M7 can reproduce the compressive response of concrete to a large extent simultaneously with the tensile response using the same set of parameters.

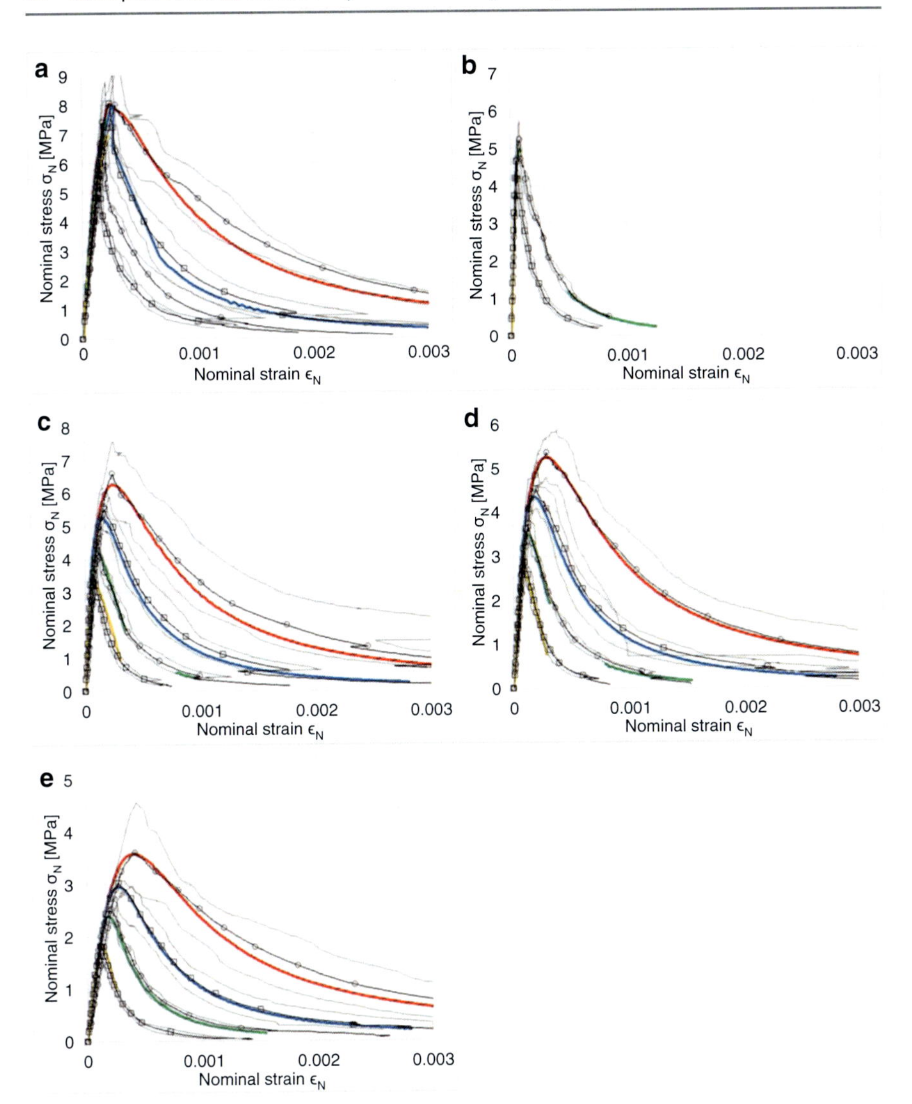

Fig. 4 *Curves with symbols* show the size effect test data from three-point bending tests of plain concrete specimens with various notch lengths relative to beam depth: (**a**) 0% notch depth (unnotched), (**b**) 2.5% notch depth, (**c**) 7.5% notch depth, (**d**) 15% notch depth, and (**e**) 30% notch depth. *Light colored thin lines* show the range of test data. The predictions by the model M7 are shown by *thick curves in color*

The specimens with predicted cracks are shown in Fig. 6. The cylindrical specimens exhibit a diagonal band of axial splitting cracks at failure. The 4 cm cubic specimen shown in Fig. 6c develops extensive distributed splitting cracks at failure. In contrast, the major mode of failure in the 15 cm cubic specimen

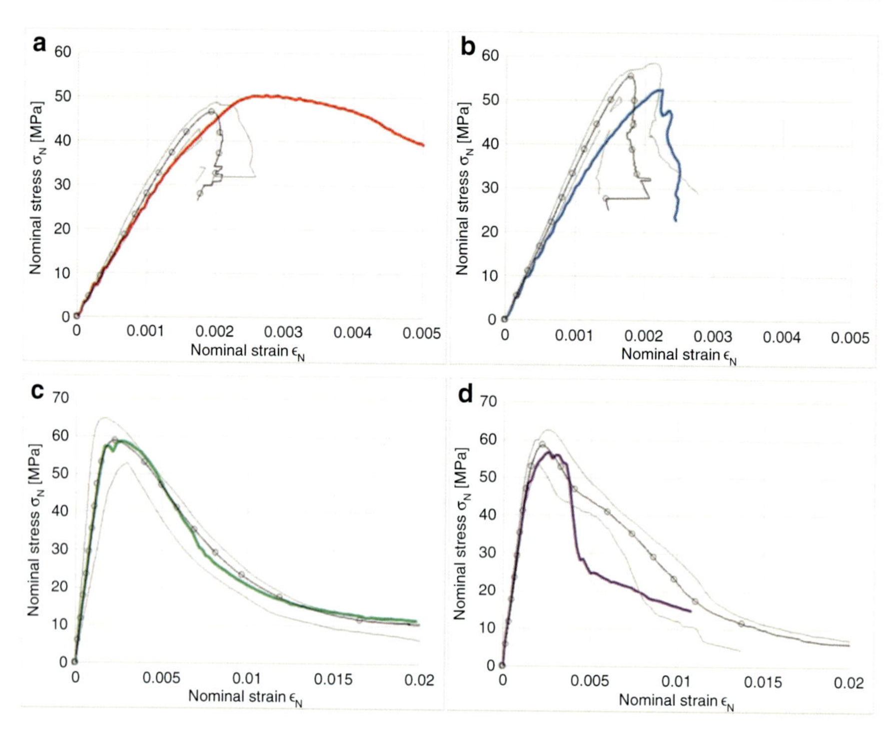

Fig. 5 Compression test data obtained from the specimens from an overcured version of the same concrete as the three-point bending plain concrete specimens of Fig. 4: (**a**) Cylindrical specimen of concrete aged 31 days with dimensions 7.5 by 15 cm, (**b**) cylindrical specimen of concrete aged 400 days with the same dimensions, (**c**) cubic specimen with a dimension of 4 cm, and finally (**d**) cubic specimen with a dimension of 15 cm. The *curve with symbols* is the test data; *light colored thin lines* show the range of test data. The predictions by the model M7 are shown by *thick curves in color*

shown in Fig. 6d is a large splitting crack in the middle. The hexahedral mesh used in all these calculations of compression failure has an average size of 5 mm.

In Figs. 4 through 6 the parameter set of the model M7 employed for all simulations is given by $\nu = 0.18, k_1 = 125{\cdot}10^6, k_2 = 15, k_3 = 15, k_4 = 20, k_5 = 16{\cdot}10^6$, $f'_{c0} = 15.08$ MPa, $E_0 = 20,000$ MPa, $c_1 = 0,089, c_2 = -1.318, c_3 = 8.4, c_4 = 70, c_5 = 3500, c_6 = 20, c_7 = 80, c_8 = 15, c_9 = 0.012, c_{10} = 1, c_{11} = 0.5, c_{12} = 2.36, c_{13} = 4500, c_{14} = 300, c_{15} = 4000, c_{16} = 60, c_{17} = 1.4, c_{18} = 17.5\,\text{MPa}, c_{19} = 14,000, c_{20} = 1.8, c_{21} = 250$ MPa. Elements are deleted when the maximum microplane normal strain in tension exceeds 90%. The Young modulus employed to calculate the curves of Fig. 4 is given by $E = 41,290$ MPa. In Figs. 5a and 6a $E = 27,735$ MPa, in Figs. 5b and 6b $E = 31,382$ MPa, in Figs. 5c and 6c $E = 40,509$ MPa and in Figs. 5d and 6d $E = 38,063$ MPa are used.

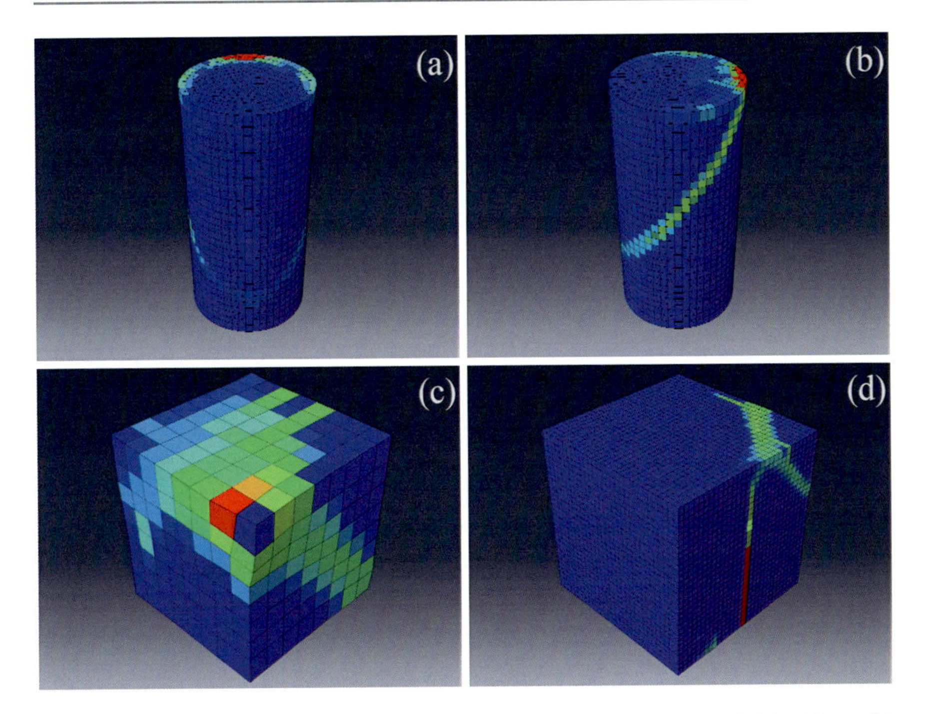

Fig. 6 (**a**) Cylindrical specimen of concrete aged 31 days with dimensions 7.5 by 15 cm, (**b**) cylindrical specimen of concrete aged 400 days with the same dimensions, (**c**) cubic specimen with a dimension of 4 cm, and finally (**d**) cubic specimen with a dimension of 15 cm

Nonlocal Microplane Models

In the analysis of structures by the finite element method, the microplane model is meant to be used to calculate the second-order stress tensor that corresponds to a given second-order strain tensor at the Gauss points of a finite element mesh made of continuum elements. The fracture predicted by the microplane model at the Gauss points is distributed across the width of an element creating a crack band. This approach to modeling fracture has a number of advantages including automatic tracking of crack propagation direction and automatic detection of crack nucleation anywhere in the mesh. It also allows the use of the well-developed continuum mechanics theories in the simulation of fracture and damage. However, one disadvantage to crack band modeling is that the calculated response depends on the width of the crack band. The smaller the width, the smaller is the energy dissipated by fracture; thus in the limit of vanishing element width, fracture energy also vanishes. Obviously this is unrealistic. There seems to be three choices to get the correct energy dissipation as the cracks propagate in the material: (1) the element width may be considered as a material parameter in which case, the finite element mesh should consist of elements of one single size and one single aspect ratio close to that of a cube for a general 3-D fracture propagation and the microplane model should be calibrated for that element size, (2) a localization limiter may be introduced by which the average of the variables of the model that govern the fracturing of the material or equivalently of the fracturing part of the total strains over a finite neighborhood of every Gauss point may be calculated, and finally (3) using principle of virtual work, the strongly nonlocal partial differential stress equations of motion with a characteristic length which also yields the corresponding boundary conditions may be developed.

The nonlocal integral generalization of the microplane split-type formulation given in Bažant and Prat (1988a) is presented in Bažant and Ožbolt (1990) as the first ever nonlocal microplane model. In this formulation, the localization of the fracture process zone into zero width as the mesh is refined is prevented by introducing the volume averages of only damage variables, equivalent to averaging the fracturing part of the total strains. Thus, the elastic part of the response remains local. The calculation of the volume average when part of the volume protrudes off the boundary, the integration weights are rescaled so as to produce a true average. This kind of treatment of the boundary zone is one of the reasons for the stiffness matrix to become nonsymmetric. The weight function employed is a bell-shaped one, similar to the Gaussian probability distribution function, with a cutoff distance beyond which the value of the function becomes zero. The fact that many other weight functions can produce satisfactory fits of experimental data means a unique weight function does not exist. This formulation leads to a mesh size-independent model (excluding the discretization error that results when coarse meshes are used). In Bažant and Ožbolt (1992), the nonlocal formulation presented in Bažant and Ožbolt (1990) is employed to simulate the compression fracture of concrete which is more complex than the tensile fracture of concrete.

The nonlocal formulation presented in Bažant and Di Luzio (2004) extends the microplane model M4 given in Bažant et al. (2000) to nonlocality. The over-nonlocal formulation in which the strains in the stress-strain boundaries of the model M4 are replaced by the total microplane strains averaged over a finite neighborhood of the Gauss point, multiplied by an empirical coefficient m and added to total microplane strain multiplied by $(1-m)$, is presented in di Luzio (2007) and Di Luzio and Bažant (2005). This formulation has the advantage that it does not require a variable that governs the fracture nor the separation of the total microplane strains into fracturing and elastic parts. However, it also suffers from the typical problems of the classical integral type nonlocal formulations.

A Mesh Size-Independent Microplane Model: The Model M7Auto

Although the nonlocal integral type crack band models achieve the mesh size independence of the calculated response, they have a number of disadvantages and problems. For example, the nonlocal crack band models cannot capture the closely spaced distributed cracking due to either the reinforcement in a deforming reinforced concrete beam or the drying shrinkage of concrete surface, whereas the mesh size-dependent crack band models can indeed capture those cracks accurately provided that they are calibrated for the element size employed in the mesh. Another important problem in the nonlocal crack band modeling is the treatment of the boundary. When the averaging volume protrudes the boundary of the structure, the averaging volume has to be reduced, which means that at the boundary, the averaging becomes unobjective. The least problematic treatment of the boundary in the sense of nonlocal crack band modeling is probably keeping a sufficiently thick boundary layer where the model is kept mesh size *dependent* with a calibration for that mesh size employed in the boundary layer (Bažant et al. 2010). Yet another problem in such nonlocal crack band models is that cracks initiate ahead of the crack tip which is not physically realistic. Furthermore, in dynamic applications, the separation of the particles smaller than the averaging size cannot be captured. Moreover, there is no practical way to detect the shielding effect of a partially formed cohesive crack between two Gauss points within the same averaging volume. Yet another difficulty is that a varying size of the averaging volume is needed for correctly capturing crack initiation and crack growth. A computational difficulty in Abaqus (Simulia Corporation 2014) is that for large meshes, it is not possible to correctly calculate the volume averages of the relevant fracturing variable of a crack band model because the user subroutine is evaluated in separate blocks of Gauss points.

However, if the continuum model can calibrate itself for different element size, aspect ratio, and type, not only the model response becomes independent of mesh size, but also all the aforementioned difficulties associated with classical nonlocal integral type approach could be avoided. Thus, in what follows, an autocalibrating version of the model M7 called the model M7Auto is described, and its performance is illustrated using several examples.

In the model M7, the softening behavior in tension and in compression is governed by the parameters c_4, c_{11}, c_{12}, and c_{13}. Thus, with varying element size, these four fixed parameters must vary so as to dissipate the same fracture energy. The resulting element size-dependent stress-strain boundaries are obtained by substituting

$$
\begin{aligned}
\mathcal{F}_N^{+}(\epsilon_N, \sigma_V) &= E k_1 \beta_1 \exp\left(-\langle \epsilon_N - \beta_1 c_2 k_1 \rangle / \left(\left(c_4 + \frac{s_9 - c_4}{\left(1 + \left(\frac{l_{ch}}{s_{10}}\right)^{s_{11}}\right)^{s_{12}}}\right) \epsilon_e^0 + k_1 c_3\right)\right) \\
\hat{\sigma}_N^0(\epsilon_V) &= E_\tau \langle k_1 \left(c_{11} + \frac{s_1 - c_{11}}{\left(1 + \left(\frac{l_{ch}}{s_2}\right)^{s_3}\right)^{s_4}}\right) - \left(c_{12} + \frac{s_5 - c_{12}}{\left(1 + \left(\frac{l_{ch}}{s_6}\right)^{s_7}\right)^{s_8}}\right) \langle \epsilon_V \rangle \rangle \qquad (29) \\
E_N &= E_{N0} \exp\left(-(112 l_{ch} + 1780) \epsilon_N^{0+}\right) f(\xi)
\end{aligned}
$$

in Eqs. (24) and (25). In Figs. 7 through 10 the parameter set of the model M7 employed for all simulations is given by $E = 30,500$ MPa, $\nu = 0.18, k_1 =$

$189 \cdot 10^6, k_2 = 110, k_3 = 24, k_4 = 6, k_5 = 100 \cdot 10^6, f'_{c0} = 15.08$ MPa, $E = 20,000$MPa, $c_1 = 0,089, c_2 = -1.9, c_3 = 3.7538, c_4 = 25.85335, c_5 = 3500, c_6 = 20, c_7 = 1, c_8 = 8, c_9 = 0.012, c_{10} = 0.33, c_{11} = 184,434.5, c_{12} = 264,009.5, c_{13} = 112\, l_{ch} + 1780, c_{14} = 300, c_{15} = 4000, c_{16} = 60, c_{17} = 1.4, c_{18} = 1.6 \cdot 10^{-3}$MPa, $c_{19} = 1000, c_{20} = 1.8, c_{21} = 250$ MPa. The newly introduced "s" parameters are given by $s_1 = 0.981856, s_2 = 21.67212, s_3 = 4.291001, s_4 = 1.51 \cdot 10^{-6}, s_5 = 0.020228, s_6 = 16.85614, s_7 = 7.696828, s_8 = 1.8 \cdot 10^{-6}, s_9 = 124.9338, s_{10} = 7,625,177, s_{11} = 1.15299$ and $s_{12} = 3,048,460$. Elements are deleted when the maximum microplane normal strain in tension exceeds 90% in all simulations. The parameter l_{ch} is the characteristic length of the element that corresponds to the variable "charLength" in Abaqus user subroutine. The reference fracture energy can be obtained by calibrating the model using a mesh with a fixed element size, type, and aspect ratio close to 1. During this initial calibration phase, the free parameters of the model k_1 through k_5 also need to be varied. Next, a new mesh is generated with the desired element size keeping the element type the same and the aspect ratio as close to 1 as possible. Now the free parameters are kept fixed, and the four fixed parameters are varied until the same fracture energy is dissipated both in tension and in uniaxial and triaxial compression. This procedure is repeated for different element sizes, and the resulting trends in those four fixed parameters are approximated using appropriate functions. These variations of the fixed parameters as functions of element size are programmed into the model M7 in the form of a user subroutine for Abaqus. This allows the model to recalibrate itself for different element sizes within a range of 5 to 20 mm. The upper limit of this range can further be increased using the approach given in Červenka et al. (2005) but that requires a new user element to be defined in Abaqus.

In concrete the characteristic size is considered to be roughly the half of the size of the fracture process zone. In crack band models, the width, w, of the crack band corresponds to the size of the fracture process zone and thus is a measure of the characteristic size of the material. In the finite element calculations, it seems reasonable to use elements of size about three to four times the maximum aggregate size of the concrete of interest, and it seems unlikely that smaller elements can be used to simulate the softening behavior. However, as shown in this study, much smaller crack band widths can be employed in the finite element calculations to obtain correct material response provided that the model is properly calibrated for these widths. Thus, for a different element size w, the predicted energy per unit volume dissipated by fracture, γ_f, should vary in order to satisfy

$$G_f = w\gamma_f \tag{30}$$

where G_f is constant in quasi-static loading conditions (see Fig. 1 in Bažant and Caner 2005). Although it is found out in this study that for much smaller values of w than three to four times the maximum aggregate size the model can be successfully recalibrated in order to satisfy Eq. (30), when the width of the element (or the crack band) is much larger, satisfying this equation becomes very difficult and a different strategy as in Červenka et al. (2005) may be needed.

In Abaqus documentation (Simulia Corporation 2014), the precise calculation of the characteristic length of an element in three-dimensional space is not given. In the finite element calculations, however, it is found out that for both bilinear hexahedral and linear tetrahedral elements, the characteristic length made available to the user subroutine as the variable "charLength" coincides with the cube root of the element volume:

$$\text{charLength} = \sqrt[3]{V} \tag{31}$$

For a hexahedral element with an aspect ratio of 1, the characteristic length is the same as its side length. This is convenient for the model calibration in problems in which the cracks propagate along the mesh lines in a hexahedral mesh using Abaqus. For quadratic tetrahedral elements, the characteristic length is given by

$$\text{charLength} = \sqrt[3]{V/4} \tag{32}$$

The tetrahedral elements not only allow meshing of more complex geometries compared to hexahedral elements, but also in crack band models, they pose less resistance to a possible change in the direction of cracks, and thus potentially they are more useful than hexahedral elements. However in this study both hexahedral structured meshes and tetrahedral random meshes have been employed in the calibration of the model. For the case of a priori knowledge of crack paths, the hexahedral elements are preferred because the characteristic size is more closely related to element dimensions in these elements than in the tetrahedral elements.

In the following finite element analyses, the aspect ratio of the elements is kept as close to 1 as possible. For an arbitrary geometry, even for a geometry as simple as a cylindrical one, it is not trivial to obtain a mesh made of elements with an aspect ratio very close to unity. However, the aspect ratio of the elements must be as close to unity as possible to be able capture the correct energy dissipation as well as the correct crack paths.

The Nooru-Mohamed-mixed mode fracture test data set presented in Nooru-Mohamed (1992) for different load paths is a well-known challenge for constitutive models for plain concrete. The finite element model required to fit these test data involves a correct representation of the test setup including the rigidities of the three frames used in the tests in addition to the other restrictions related to finite element mesh used in the sense of crack band model to discretize the concrete specimens. First, the model M7Auto is calibrated by optimally fitting the experimental load-displacement data with a mesh oriented along the experimentally observed crack paths (Fig. 7d). Next, the performance of the model M7Auto is tested by employing finite element random meshes with different element sizes and types. The load-displacement predictions are compared to the test data in Fig. 7f. In this figure, the dark-blue curve is associated with the mesh in Fig. 7a, the thick gray curve

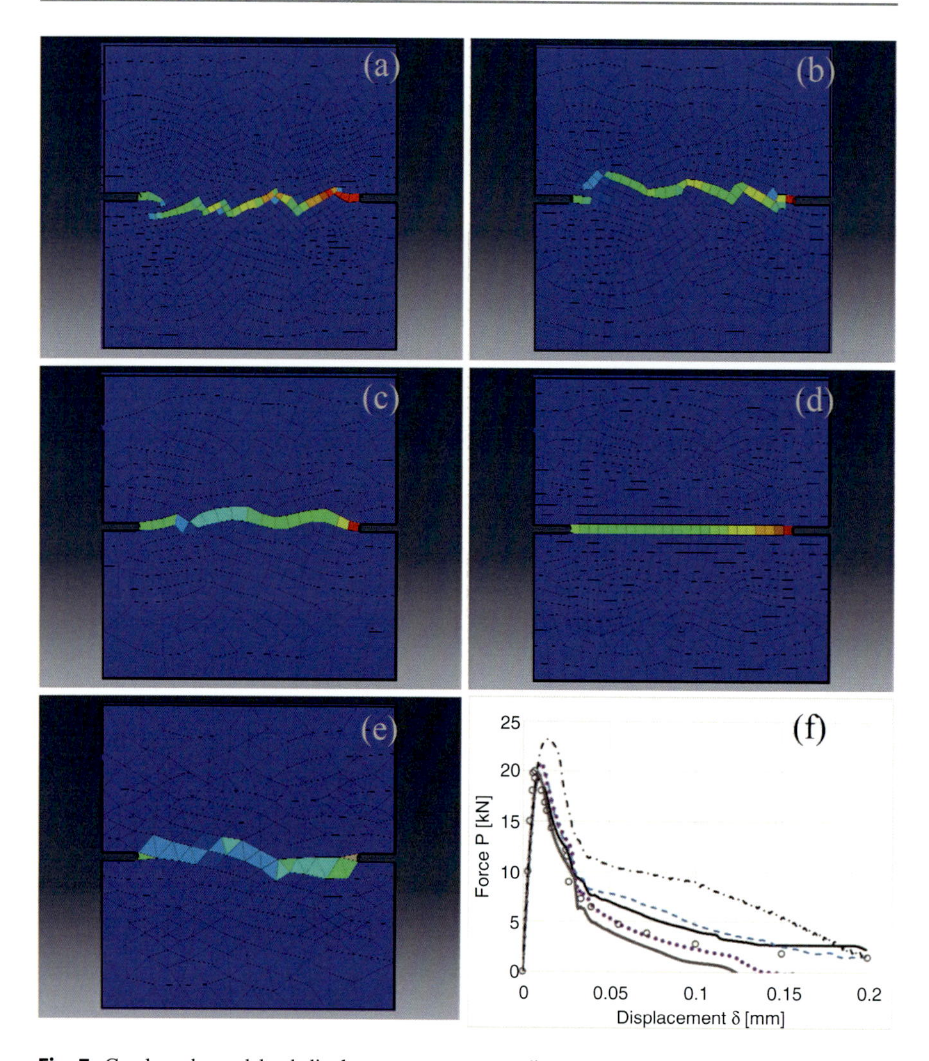

Fig. 7 Crack paths and load displacement curves predicted by the model M7Auto in uniaxial tension of the Nooru-Mohamed specimens using different mesh sizes: (**a**) 5 mm hexahedral random mesh, (**b**) 6.25 mm hexahedral random mesh, (**c**) 10 mm hexahedral random mesh, (**d**) 6.25 mm hexahedral mesh oriented along the experimentally observed crack paths, (**e**) 5 mm random tetrahedral mesh, (**f**) the load-displacement experimental data shown as *symbols* and the predictions by the model M7Auto

with the mesh of Fig. 7d, the dotted purple curve with the mesh of Fig. 7c, the dashed blue curve with the mesh in Fig. 7b, and finally the blue dashed dotted curve with the mesh in Fig. 7e. The crack paths predicted in this simple load path are in general reasonable, and the load-displacement predictions agree very well with the experimental data with the exception of the case of Fig. 7e in which relatively large tetrahedral elements are employed.

In Fig. 8a–e, the experimentally observed crack patterns that correspond to the load path 4a and the crack paths predicted by the already calibrated model M7Auto are shown. The average mesh sizes used in the calculations are approximately 5, 6.25, and 10 mm as shown in the figures. The meshes are random in Fig. 8a, b, c, e, whereas in Fig. 8d the mesh is oriented along the experimentally observed crack paths. Clearly the crack paths in Fig. 8d are captured almost exactly by the model as the cracks do not have to propagate skew to the mesh in a zigzag pattern. In Fig. 8a, b, c, e the crack paths are still captured reasonably well, but especially more accurately when the element size is small enough. In Fig. 8e a random mesh of linear tetrahedral elements are used, and comparing this figure to Fig. 8a in which bilinear hexahedral elements are used, it is observed that tetrahedral elements prevent the zigzag pattern of crack propagation skew to the mesh. The predicted load-displacement response corresponding to Fig. 8a–e and test data are given in Fig. 8f. In this figure, the dark-blue curve is associated with the mesh in Fig. 8a, the thick gray curve with the mesh of Fig. 8d, the dotted purple curve with the mesh of Fig. 8c, the dashed blue curve with the mesh in Fig. 8b, and finally the blue dashed dotted curve with the mesh in Fig. 8e. The peak load is generally overestimated, but the response predicted by the model M7Auto in general agrees well with the test data with the exception of the case of Fig. 8e.

Figure 9a–e show the crack patterns that result from the load path 4b and the crack paths predicted by the model M7Auto. Similar to Fig. 8, the average mesh sizes are approximately 5, 6.25, and 10 mm in these figures. The meshes are random in Fig. 9a, b, c, e, whereas in Fig. 9d the mesh is oriented along the experimentally observed crack paths. Clearly the crack paths in Fig. 9d are simulated very well by the model as the cracks do not have to propagate skew to the mesh. In Fig. 9a, b, c, e the crack paths are still simulated reasonably well, but the accuracy is higher when the element size is small enough. In Fig. 9e a random mesh of linear tetrahedral elements are used. Comparing this figure to Fig. 9a in which bilinear hexahedral elements are used, it is observed that tetrahedral elements facilitate the crack propagation skew to the mesh. The predicted load-displacement response corresponding to Fig. 9a–e and test data are shown in Fig. 9f. In this figure, the dark-blue curve corresponds to the mesh in Fig. 9a, the thick gray curve to the mesh of Fig. 9d, the dotted purple curve to the mesh of Fig. 9c, the dashed blue curve to the mesh in Fig. 9b, and finally the blue dashed dotted curve to the mesh in Fig. 9e. The peak load is again mostly overestimated, but the response predicted by the model M7Auto in general is in a reasonable agreement with the test data although the case of Fig. 9e shows a significantly large deviation from the test data.

Similarly, Fig. 10a–e show the experimentally observed crack patterns that correspond to the load path 4c and the crack paths predicted by the already calibrated model M7Auto. Similar to Fig. 8, the average mesh sizes shown are approximately 5, 6.25, and 10 mm in these figures. The meshes are random in Fig. 10a, b, c, e, whereas in Fig. 10d the mesh is oriented along the experimentally observed crack paths. Clearly the crack paths in Fig. 10d are captured almost exactly by the model as the cracks do not have to propagate skew to the mesh in a zigzag pattern. In Fig. 10a, b, c, e the crack paths are still captured

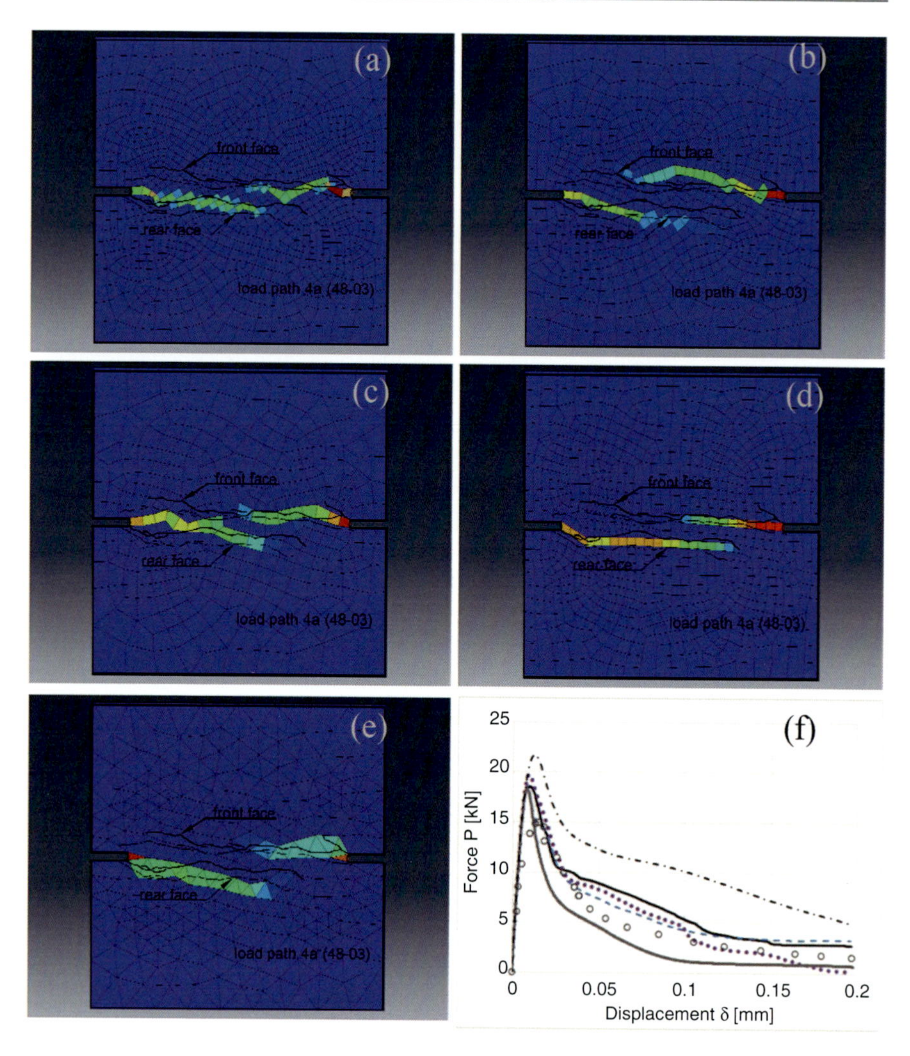

Fig. 8 Crack paths and load displacement curves predicted by the model M7Auto as the response of the Nooru-Mohamed specimens to the load path 4a using different mesh sizes: (**a**) 5 mm hexahedral random mesh, (**b**) 6.25 mm hexahedral random mesh, (**c**) 10 mm hexahedral random mesh, (**d**) 6.25 mm hexahedral mesh oriented along the experimentally observed crack paths, (**e**) 5 mm random tetrahedral mesh, (**f**) the load-displacement experimental data shown as *symbols* and the predictions by the model M7Auto

reasonably well, but especially more accurately when the element size is small enough. In Fig. 10e a random mesh of linear tetrahedral elements are used, and comparing this figure to Fig. 10a in which bilinear hexahedral elements are used, it is observed that tetrahedral elements prevent the zigzag pattern of crack propagation skew to the mesh. The predicted load-displacement response corresponding to Fig. 10a–e and test data are given in Fig. 10f. In this figure,

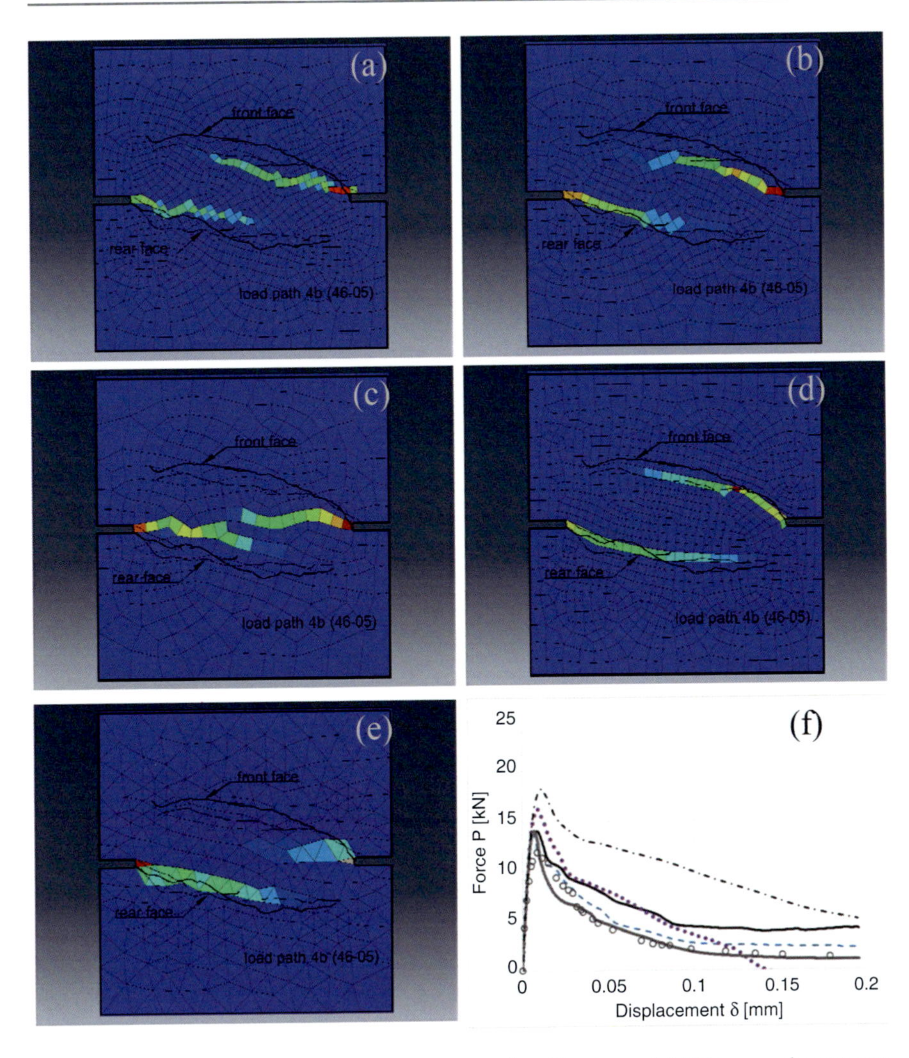

Fig. 9 Crack paths and load displacement curves predicted by the model M7Auto as the response of the Nooru-Mohamed specimens to the load path 4b using different mesh sizes: (**a**) 5 mm hexahedral random mesh, (**b**) 6.25 mm hexahedral random mesh, (**c**) 10 mm hexahedral random mesh, (**d**) 6.25 mm hexahedral mesh oriented along the experimentally observed crack paths, (**e**) 5 mm random tetrahedral mesh, (**f**) the load-displacement experimental data shown as *symbols* and the predictions by the model M7Auto

the dark-blue curve is associated with the mesh in Fig. 10a, the thick gray curve with the mesh of Fig. 10d, the dotted purple curve with the mesh of Fig. 10c, the dashed blue curve with the mesh in Fig. 10b, and finally the blue dashed dotted curve with the mesh in Fig. 10e. The peak load is generally overestimated, but the response predicted by the model M7Auto in general agrees well with the test data with the exception of the case of Fig. 10e.

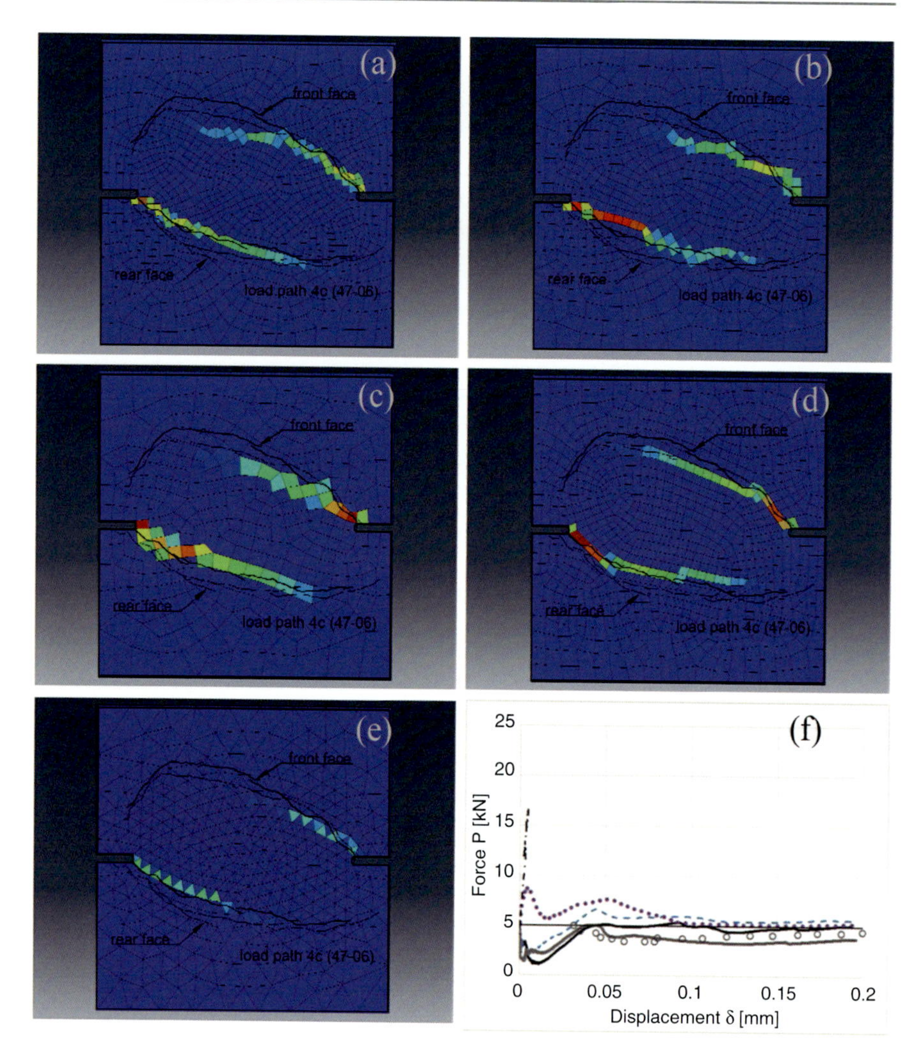

Fig. 10 Crack paths and load displacement curves predicted by the model M7Auto as the response of the Nooru-Mohamed specimens to the load path 4c using different mesh sizes: (**a**) 5 mm hexahedral random mesh, (**b**) 6.25 mm hexahedral random mesh, (**c**) 10 mm hexahedral random mesh, (**d**) 6.25 mm hexahedral mesh oriented along the experimentally observed crack paths, (**e**) 5 mm random tetrahedral mesh, (**f**) the load-displacement experimental data shown as symbols and the predictions by the model M7Auto

Although the predictions corresponding to the cases given in Figs. 7e, 8e, 9e, and 10e seem to be too inaccurate, in fact the coarse random mesh can cause this kind of deviations in response due to discretization error; see Figs. 5 and 6 in Červenka et al. (2005). These calculation results reinforce the idea that if the crack paths are not known in advance, a two-step calculation procedure should be followed: in the trial step, the calculations should be carried out using a random mesh, and in the final step, the calculations should be done using a mesh oriented along the crack paths observed in the trial step.

Conclusions

In this study three types of common microplane formulations that can be applied to a wide range of material constitutive behavior are discussed. The focus is on the microplane models for plain concrete, a challenging engineering material to model mathematically. Only the major milestones in the development of microplane models for concrete are explicitly treated. Thus, the model M4, the model M7, and a novel nonlocal extension of the model M7 called the model M7Auto are discussed in this study. The predictive capabilities of the model M7 are demonstrated by simulating the size effect test data obtained by three-point bending tests of geometrically similar beams as well as the compression test data obtained using specimens from an overcured version of the same concrete at different proportional and nonproportional load paths. Thus, it is shown that the model M7 fits all these experimental data very accurately.

Furthermore, it is shown that the model M7Auto overcomes the difficulties associated with classical approaches to nonlocal modeling simply by recalibrating itself according to the element size, shape, and type. The requirements for a successful use of the model M7Auto are that (1) the size of the elements in the mesh should not exceed the range of autocalibration of the model, (2) the elements must have an aspect ratio close to 1 for the simulation of general three-dimensional crack propagation, (3) the element type should be the same in the mesh. The predictive capabilities of the model M7Auto are demonstrated by simulating some well-known load paths in the benchmark test data of Nooru-Mohamed. It is shown that the model M7Auto can handle a significant range of element sizes from 5 to 20 mm and still predict the concrete response to complex load paths quite accurately.

References

Z. Bažant, M. Adley, I. Carol, M. Jirasek, S. Akers, B. Rohani, J. Cargile, F. Caner, Large-strain generalization of microplane model for concrete and application. J. Eng. Mech. ASCE **126**(9), 971–980 (2000)

Z. Bažant, F. Caner, Microplane model M5 with kinematic and static constraints for concrete fracture and anelasticity. II: computation. J. Eng. Mech. ASCE **131**(1), 41–47 (2005)

Z. Bažant, F. Caner, M. Adley, S. Akers, Fracturing rate effect and creep in microplane model for dynamics. J. Eng. Mech. ASCE **126**(9), 962–970 (2000)

Z. Bažant, F. Caner, I. Carol, M. Adley, S. Akers, Microplane model M4 for concrete. I: formulation with work-conjugate deviatoric stress. J. Eng. Mech. ASCE **126**(9), 944–953 (2000)

Z. Bažant, G. Di Luzio, Nonlocal microplane model with strain-softening yield limits. Int. J. Solids Struct. **41**(24–25), 7209–7240 (2004)

Z. Bažant, P. Gambarova, Crack shear in concrete – crack band microplane model. J. Eng. Mech. ASCE **110**(9), 2015–2035 (1984)

Z. Bažant, J.-L. Le, C. Hoover, Nonlocal boundary layer (NBL) model: overcoming boundary condition problems in strength statistics and fracture analysis of quasibrittle materials, in *Proceedings of FraMCoS-7*, 2010, pp. 135–143

Z. Bažant, B. Oh, Microplane model for progressive fracture of concrete and rock. J. Eng. Mech. ASCE **111**(4), 559–582 (1985)

Z. Bažant, J. Ožbolt, Nonlocal microplane model for fracture, damage, and size effect in structures. J. Eng. Mech. ASCE **116**(11), 2485–2505 (1990)

Z. Bažant, J. Ožbolt, Compression failure of quasibrittle material – nonlocal microplane model. J. Eng. Mech. ASCE **118**(3), 540–556 (1992)

Z. Bažant, G. Pijaudiercabot, Nonlocal continuum damage, localization instability and convergence. J. Appl. Mech. Trans. ASME **55**(2), 287–293 (1988)

Z. Bažant, P. Prat, Microplane model for brittle-plastic material. 1. Theory. J. Eng. Mech. ASCE **114**(10), 1672–1687 (1988a)

Z. Bažant, P. Prat, Microplane model for brittle-plastic material. 2. Verification. J. Eng. Mech. ASCE **114**(10), 1689–1702 (1988b)

Z. Bažant, Y. Xiang, M. Adley, P. Prat, S. Akers, Microplane model for concrete. 2. Data delocalization and verification. J. Eng. Mech. ASCE **122**(3), 255–262 (1996a)

Z. Bažant, Y. Xiang, P. Prat, Microplane model for concrete. 1. Stress-strain boundaries and finite strain. J. Eng. Mech. ASCE **122**(3), 245–254 (1996b)

Z. Bažant, G. Zi, Microplane constitutive model for porous isotropic rocks. Int. J Numer Anal. methods Geomech. **27**(1), 25–47 (2003)

M. Brocca, Z. Bažant, Microplane constitutive model and metal plasticity. Appl. Mech. Rev. ASME **53**(10), 265–281 (2000)

M. Brocca, Z. Bažant, I. Daniel, Microplane model for stiff foams and finite element analysis of sandwich failure by core indentation. Int. J. Solids Struct. **38**(44-45), 8111–8132 (2001)

M. Brocca, L. Brinson, Z. Bažant, Three-dimensional constitutive model for shape memory alloys based on microplane model. J. Mech. Phys. solids **50**(5), 1051–1077 (2002)

F. Caner, Z. Bažant, Microplane model M4 for concrete. II: algorithm and calibration. J. Eng. Mech. ASCE **126**(9), 954–961 (2000)

F. Caner, Z. Bažant, J. Cervenka, Vertex effect in strain-softening concrete at rotating principal axes. J. Eng. Mech. ASCE **128**(1), 24–33 (2002)

F.C. Caner, Z.P. Bažant, Microplane model M7 for plain concrete. I: formulation. J. Eng. Mech. **139**(12), 1714–1723 (2013a)

F.C. Caner, Z.P. Bažant, Microplane model M7 for plain concrete. II: calibration and verification. J. Eng. Mech. **139**(12), 1724–1735 (2013b)

F.C. Caner, Z.P. Bažant, C.G. Hoover, A.M. Waas, K.W. Shahwan, Microplane model for fracturing damage of triaxially braided fiber-polymer composites. J. Eng. Mater. Tech. Trans. ASME **133**(2) 021024-1–021024-12 (2011)

F.C. Caner, Z. Guo, B. Moran, Z.P. Bažant, I. Carol, Hyperelastic anisotropic microplane constitutive model for annulus fibrosus. J. Biomech. Eng. Trans. ASME **129**(5), 632–641 (2007)

J. Červenka, Z. Bažant, M. Wierer, Equivalent localization element for crack band approach to mesh-sensitivity in microplane model. Int. J. Numer. Methods Eng. **62**(5), 700–726 (2005)

K.-T. Chang, S. Sture, Microplane modeling of sand behavior under non-proportional loading. Comput. Geotech. **33**(3), 177–187 (2006)

X. Chen, Z.P. Bažant, Microplane damage model for jointed rock masses. Int. J. Numer. Anal. Methods Geomech. **38**(14), 1431–1452 (2014)

G. Cusatis, A. Beghini, Z.P. Bažant, Spectral stiffness microplane model for quasibrittle composite laminates – part I: theory. J. Appl. Mech. Trans. ASME **75**(2) 021009-1–021009-9 (2008)

G.Cusatis, A. Mencarelli, D. Pelessone, J. Baylot, Lattice discrete particle model (LDPM) for failure behavior of concrete. II: calibration and validation. Cem. Concr. Compos. **33**(9), 891–905 (2011)

G. di Luzio, A symmetric over-nonlocal microplane model {M4} for fracture in concrete. Int. J. Solids Struct. **44**(13), 4418–4441 (2007)

G. Di Luzio, Z. Bažant, Spectral analysis of localization in nonlocal and over-nonlocal materials with softening plasticity or damage. Int. J. Solids Struct. **42**(23), 6071–6100 (2005)
C.G. Hoover, Z.P. Bazant, J. Vorel, R. Wendner, M.H. Hubler, Comprehensive concrete fracture tests: description and results. Eng. Fract. Mech. **114**, 92–103 (2013)
K. Kirane, Z. Bažant, Microplane damage model for fatigue of quasibrittle materials: sub-critical crack growth, lifetime and residual strength. Int. J. Fatigue **70**, 93–105 (2015)
K. Kirane, M. Salviato, Z.P. Bažant, Microplane-triad model for elastic and fracturing behavior of woven composites. J. Appl. Mech. Trans. ASME**83**(4) 041006-1–041006-14 (2016)
M.B. Nooru-Mohamed, Mixed-mode fracture of concrete: an experimental approach. Ph.D. thesis, Civil Engineering and Geosciences, Universiteitsdrukkerij, Delft University of Technology, Delft, The Netherlands. (1992, 5)
J. Ožbolt, Z. Bažant, Microplane model for cyclic triaxial behavior of concrete. J. Eng. Mech. ASCE **118**(7), 1365–1386 (1992)
P. Prat, Z. Bažant, Microplane model for triaxial deformation of saturated cohesive soils. J. Geotech. Eng. ASCE **117**(6), 891–912 (1991)
Simulia Corporation, Abaqus User Subroutines Reference Guide. Dassault Systèmes (2014)
R. Wendner, J. Vorel, J. Smith, C.G. Hoover, Z.P. Bazant, G. Cusatis, Characterization of concrete failure behavior: a comprehensive experimental database for the calibration and validation of concrete models. Mater. Struct. **48**(11), 3603–3626 (2015)

Modeling Temperature-Driven Ductile-to-Brittle Transition Fracture in Ferritic Steels

31

Babür Deliktaş, Ismail Cem Turtuk, and George Z. Voyiadjis

Contents

Abstract

The most catastrophic brittle failure in ferritic steels is observed as their tendency of losing almost all of their toughness when the temperature drops below their ductile-to-brittle transition (DBT) temperature. There have been put large efforts

B. Deliktaş (✉)
Faculty of Engineering-Architecture, Department of Civil Engineering, Uludag University, Bursa, Görükle, Turkey
e-mail: bdeliktas@uludag.edu.tr

I. Cem Turtuk
Mechanical Design Department, Meteksan Defence, Ankara, Turkey

Department of Civil Engineering, Uludag Univeristy, Bursa, Turkey
e-mail: ismail.turtuk@gmail.com

G. Z. Voyiadjis
Department of Civil and Environmental Engineering, Louisiana State University, Baton Rouge, LA, USA
e-mail: cgzv1@lsu.edu

G. Z. Voyiadjis (ed.), *Handbook of Nonlocal Continuum Mechanics for Materials and Structures*, https://doi.org/10.1007/978-3-319-58729-5_6

in experimental and theoretical studies to clarify the controlling mechanism of this transition; however, it still remains unclear how to model accurately the coupled ductile/brittle fracture behavior of ferritic steels in the region of ductile-to-brittle transition.

Therefore, in this study, an important attempt is made to model coupled ductile/brittle fracture by means of blended micro-void and micro-cracks. To this end, a thermomechanical finite strain-coupled plasticity and continuum damage mechanics models which incorporate the blended effects of micro-heterogeneities in the form of micro-cracks and micro-voids are proposed.

In order to determine the proposed model material constant, a set of finite element model, where the proposed unified framework, which characterizes ductile-to-brittle fracture behavior of ferritic steels, is implemented as a VUMAT, is performed by modeling the benchmark experiment given in the experimental research published by Turba et al., then, using these models as a departure point, the fracture response of the small punch fracture testing is investigated numerically at 22°C and −196°C and at which the fracture is characterized as ductile and brittle, respectively.

Keywords

Ductile-Brittle transition · Porous plasticity · Damage mechanics · Small punch test · Ferritic steels

Introduction

Large-scale ferritic steel structures are widely used in advanced engineering applications such as pressure vessels, fusion reactor structures, construction of nuclear reactors, line pipes used for gas and oil transportation, and welded steel ships. It is well known that during the low-temperature climate operation, the catastrophic breakdown of structure and machine elements often occurred. As noted by Sutar et al. (2014) between 1942 and 1952, around 250 large welded steel ships were lost due to catastrophic brittle failure. Although the most of steel structure is normally capable of sustaining great loads and capable of ductility above certain degree of temperature, they become so brittle and have tendency to lose almost all of their toughness when the temperature drops below their ductile-to-brittle transition temperature (Baloso et al. 2017). Depending on loading rate, ambient temperature, as well as triaxiality, the mode of fracture in metals switches from ductile to brittle or vice versa, which is called as ductile-to-brittle transition phenomenon. The ductile-to-brittle transition temperature (DBTT) is a phenomenon that is widely observed in metals especially in severe steel structures in severe weather (Renevey et al. 1996; Tanguy et al. 2007; Hütter 2013b).

In metals, the fracture mode in the upper-temperature range (toward 22°C) is ductile with micro-voids and dimples. At the lowest-temperature range (toward −196°C), the fracture is almost completely brittle, while plasticity governed ductile failure is observed at temperatures as low as −158°C. In that case, the plasticity properties fall, and the character of fracture changes from ductile to brittle (Hütter 2013b). In the transition region, both ductile and brittle fracture, consisting of transgranular cleavage marks, exist. The amount of ductile crack growth decreases gradually with temperature decrease. At lower temperature, brittle failure is linked with the inter- or intragranular cleavage with nucleation, growth, and coalescence of micro-cracks and is identified as dominant unstable failure mechanisms, while at the higher temperature ductile failure is characterized by the nucleation, growth, and coalescence of micro-voids leading to rupture (Anderson 2004; Chakraborty and Biner 2013).

The controlling mechanism of this transition still remains unclear despite large efforts made in experimental and theoretical investigation. Thus understanding the triggering mechanism of such kind of transition is of great importance. Two common failure mechanisms, namely, ductile and brittle fracture in metallic materials, should be characterized well in the constitutive model for better understanding the triggering mechanism of such kind of transition.

There have been significant efforts put forward to develop unified models that take into account the different failure mechanisms that are active at different temperature regime as well as at transition zone (see, e.g., Shterenlikht 2003; Hutter et al. 2014; Needleman and Tvergaard 2000; Batra and Lear 2004; Xia and Fong Shih 1996; Chakraborty and Biner 2013; McAuliffe and Waisman 2015; Turtuk and Deliktas 2016; Soyarslan et al. 2016).

In those approaches, ductile failure mechanism is modeled by the strip line model of Freund and Lee (1990) and Rousselier (1987)'s model, or mostly by Gurson-Tvergaard-Needleman model (GTN-model Needleman and Tvergaard 2000), whereas cleavage failure in metallic materials has been described primarily by two models. The deterministic model by Ritchie-Knott-Rice (Ritchie et al. 1973) relies on a critical stress over the critical distance principle. In other words, brittle fracture occurs once the principal stresses averaged over a region within a characteristic length exceed a temperature- and rate-independent threshold, while it may or may not be accompanied by plastic flow. Drawback of these models is that the experimentally observed scatter caused by cleavage in fracture toughness cannot be captured. Curry and Knott (1979) modified this theory by incorporating a critical volume ahead of the crack tip to describe the scatter in fracture toughness due to cleavage failure.

The statistical model proposed by Beremin Research Group (see, e.g., Beremin 1983 and Mudry 1987), on the other hand, incorporates Wiebull distribution (Weibull 1953) to describe fracture stress in terms of failure probability at lower and transition temperatures. Statistical nature of this model helps one to capture the scatter in fracture toughness; however, it fails to model accurately stable crack growth near transition regime. Turtuk and Deliktas (2016) proposed a new approach specifically to model temperature-driven ductile brittle transition regime to introduce internal softening associated with micro-crack initiation and propagation due to cleavage mechanisms.

Many researchers proposed approaches that use a constitutive model such as the GTN model which accounts for the damage-induced nonlocality (see, e.g., Bergheau et al. 2014; Zhang et al. 2000; Rivalin et al. 2001a, b; Pardoen and Hutchinson 2000). Linse et al. (2012) has made nonlocal modification on the GTN model by employing an implicit gradient-enriched formulation where the volumetric plastic strain is introduced as nonlocal, damage-driving state variable. Ramaswamy and Aravas (1998) has been discussed and implemented the gradient-based extension of the Gurson model.

There are only few studies in the literature where the particular cleavage softening is taken into account. Samal et al. (2008) modified Rousselier model by considering a nonlocal Rousselier model (Samal et al. 2008) in conjunction with Beremin model (Beremin 1983). Chakraborty and Biner (2013) proposed a unified model that incorporates both ductile damage and cleavage failure mechanisms through temperature- and failure probability-dependent parameters. The flow strength of the bulk material is varied to obtain the temperature-dependent bulk material behavior. It is assumed that without cleavage, the cohesive law follows a traction-separation behavior of ductile damage as described in Scheider and Brocks (2003). Modeling cleavage failure by a cohesive zone is a heuristic approach which combines two relevant features. Firstly, the softening initiates under pure mode-I when the maximum principal stress in the ligament reaches a critical level, the so-called cohesive strength, which is a common assumption. In addition, the work of cohesive separation can be interpreted as the work required to drive the micro-crack which initiated at a broken second-phase particle into the neighboring grain and through the next grain boundary (Kabir et al. 2007; Hardenacke et al. 2012)

In the work of Hutter (2013a), a unified model is proposed by means of a cohesive zone in addition to the ductile material degradation, where the crack propagation is simulated by incorporating the softening associated with cleavage initiation in the ductile-brittle transition region. He introduced two length scales to investigate the material behavior. On the macroscopic scale, the ductile mechanism is modeled with a nonlocal modification of the Gurson-Tvergaard-Needleman model (GTN model), whereas the discrete voids are resolved in the microscopic model. Soyarslan et al. (2016) proposed a model for the ductile-brittle transition at the macroscale where two length scales are incorporated into the constitutive relations. The ductile fracture length scale is based on the average inclusion distance and associated with the nonlocal evolution equation for the porosity. The brittle fracture length scale is based on the average grain size and associated with the material region at which the maximum principal stress is averaged out.

However, these approaches suffer from the mentioned inherent weaknesses pertaining to brittle fracture models. An important attempt to model ductile fracture by means of blended micro-void and micro-cracks is given in literature by Chaboche et al. (2006). In their work, micro-void- and micro-crack-driven separate damage variables have been defined in order to model ductile fracture at isothermal conditions, meaning that no thermal coupling exists in their work. Recently Turtuk and Deliktas (2016) published a work that presents present thermomechanical finite strain-coupled plasticity and continuum damage mechanics models which incorporate the blended effects of micro-heterogeneities in the form of micro-cracks and micro-voids. They applied presented models in small punch fracture testing in order to simulate temperature-driven ductile-to-brittle transition fracture of P91 steel. While void growth-driven damage is represented by Gurson plasticity, micro-crack-driven damage is formulated with continuum damage mechanics framework. The presented work also takes up blended modeling approach and extends further, incorporating thermal coupling, Gurson's porous plasticity, brittle fracture modeling, and application to ductile-to-brittle transition fracture. The developed unified framework, which characterizes ductile, brittle, and ductile-to-brittle transition fractures, was implemented as a VUMAT in ABAQUS to describe fracture response. Taking the experimental research of notched specimens published by Turba et.al, the finite element analyses have been performed to simulate fracture response of the small punch fracture specimens made of P91 steel at 22°C and −196°C and at which the fracture is characterized as ductile and brittle, respectively.

Theoretical Background

Theoretical framework presented here based on the thermomechanical finite strain-coupled plasticity and continuum damage mechanics models which incorporate the blended effects of micro-heterogeneities in the form of micro-cracks and micro-voids. While void growth-driven damage is represented by Gurson plasticity, micro-crack-driven damage is formulated with continuum damage mechanics framework. The presented work also takes up blended modeling approach and extends further, incorporating thermal coupling, Gurson's porous plasticity, brittle fracture modeling, and application to ductile-to-brittle transition fracture. The developed unified framework, which characterizes ductile, brittle and ductile to brittle transition

Ductile Fracture: Gurson Porous Plasticity

The modeling of fracture in ductile metals due to damage is often based on the micromechanical model of Gurson (1977) for the growth of a single void in an ideal elastoplastic matrix. This model devises a hydrostatic stress-dependent yield potential derived using homogenization over void rigid plastic matrix and limit analysis. This potential is later modified by Tvergaard and Needleman, by the introduction of void shape effects as well as acceleration in the void growth during void coalescence, to be named as Gurson-Tvergaard-Needleman porous plasticity model (Tvergaard and Needleman 1984), along with other contributors, e.g., Tvergaard (1981, 1982a,b), Needleman and Tvergaard (1998), Nahshon and Hutchinson (2008), Nahshon and Xue (2009), Wen et al. (2005), and Malcher et al. (2012), also with extensions to nonlocal formulation at hyper-elastic setting as in Hakansson et al. (2006).

$$\phi^p = \left[\frac{\bar{\bar{\sigma}}_{eq}}{\sigma_y}\right]^2 + 2q_1 f^* \cosh\left(\frac{q_2}{2\sigma_y}\mathrm{tr}\left[\bar{\bar{\sigma}}\right]\right) - \left[1 + \left[q_1 f^*\right]^2\right] = 0. \tag{1}$$

$$\bar{\bar{\sigma}}_{eq} = \sqrt{\frac{3}{2}\left[\mathrm{dev}\bar{\bar{\sigma}} : \mathrm{dev}\bar{\bar{\sigma}}\right]}$$

where f is the void volume fraction that evolves with plastic strains, $\boldsymbol{\sigma}_y$ is the current yield stress of material, and $\bar{\tilde{\boldsymbol{\sigma}}}$ is the so-called *effective stress*, which is defined within continuum damage mechanics formalism and introduced in detail in the following section. Note that the proposed coupling with continuum damage mechanics formalism introduces dev $\left[\bar{\tilde{\boldsymbol{\sigma}}}\right]$ as the *deviatoric part of the effective stress tensor at rotationally neutralized configuration* in the yield potential $\boldsymbol{\phi}^p$, onto which both void growth and micro-crack-driven damage mechanisms are reflected in order to account for softening effects of ductile and brittle failure phenomena. Since the aim is to model combined effect of micro-voids and micro-cracks on fracture, the effective stress-based plastic potential can be considered as a Gurson's plastic matrix with built-in micro-cracks. A similar usage of effective stress in the plastic dissipation potential has been introduced by Chaboche et al. (2006), by using two damage state variables (one for pressure-dependent plastic flow response and the other for generalized damage situations) and focusing only on ductile fracture with isotropic damage. Further details on effective stress concept and micro-crack-driven damage are given in the following section.

Evolution of the plastic flow is assumed to follow normality rule and is calculated from:

$$\dot{\bar{\boldsymbol{\varepsilon}}}^p = \lambda \partial_{\tilde{\sigma}} \chi = \lambda \partial_{\tilde{\sigma}} \boldsymbol{\phi}^p \tag{2}$$

where λ is the plastic multiplier, χ is the dissipation potential function which is defined as $\chi = \boldsymbol{\phi}^p(\tilde{\boldsymbol{\sigma}}, \mathbf{D}) + \boldsymbol{\phi}^d(\mathbf{Y}, \mathbf{D})$, with $\boldsymbol{\phi}^d$ being the damage dissipation potential representing micro-crack-driven damage contribution linked to brittle fracture. Note that the micro-crack-driven damage is coupled to plastic flow upon λ, considering the plastic flow existence in ductile-to-brittle transition state. Further elaboration of damage potential and its parameters are given in subsection on continuum damage mechanics. Tvergaard and Needleman's criteria for void growth and coalescence have been defined by:

$$f^*(f) = \begin{cases} f & , f \leq f_c \\ f_c + \left[f_u^* - f_c\right] \dfrac{[f - f_c]}{[f_F - f_c]} & , f > f_c \end{cases} \tag{3}$$

where $f_u = 1/q_1$, f_c is the threshold value of void volume ratio at which void growth and coalescence start to accelerate and f_F is the critical value of void volume ratio at which the load-carrying capacity of the material is completely lost. The evolution equation for void volume ratio f is calculated using principle of conservation of mass and incompressibility of plastic flow. It is composed of two separate phases, namely, nucleation and growth under fully developed plastic flow, meaning that evolution of void volume fraction can mathematically be defined by:

$$\dot{f} = \dot{f}^{\text{nuc}} + \dot{f}^{gr} \tag{4}$$

The void evolution due to nucleation is statistically dependent and is defined as in Chu and Needleman (1980):

$$\dot{f}^{\text{nuc}} = A_N \dot{\widetilde{\varepsilon}}^p_{\text{eq}}; \quad A_N = A_N(\widetilde{\varepsilon}^p_{\text{eq}}) = \frac{f_N}{S_N\sqrt{2\pi}} \exp\left(-\frac{\left[\widetilde{\varepsilon}^p_{\text{eq}} - \widetilde{\varepsilon}^p_N\right]^2}{2[S_N]^2}\right) \tag{5}$$

where $\widetilde{\varepsilon}^p_N$ and S_N denote the mean equivalent plastic strain at the nucleation and its standard deviation. Employing the principle of plastic work equivalence, the equivalent effective plastic strain rate reads as follows.

$$\dot{\widetilde{\varepsilon}}^p_{\text{eq}} = \frac{\bar{\bar{\boldsymbol{\sigma}}} : \dot{\boldsymbol{\varepsilon}}^p}{[1 - f]\boldsymbol{\sigma}_y} \tag{6}$$

After Nahshon and Hutchinson (2008), $\dot{f}^{\text{gr}}$ has further been decomposed into void growth under shear and under normal stresses such that $\dot{f}^{\text{gr}}_{\text{normal}}$ controls void growth under hydrostatic stresses and $\dot{f}^{\text{gr}}_{\text{shear}}$ enhances behavior of softening due to void growth when material is under shear stresses.

$$\begin{aligned} \dot{f}^{\text{gr}} &= \dot{f}^{\text{gr}}_{\text{normal}} + \dot{f}^{\text{gr}}_{\text{shear}} \\ \dot{f}^{\text{gr}}_{\text{normal}} &= [1 - f]\text{tr}\left[\dot{\boldsymbol{\varepsilon}}^p\right] \\ \dot{f}^{\text{gr}}_{\text{shear}} &= k_w f \frac{w}{\bar{\bar{\boldsymbol{\sigma}}}_{\text{eq}}} \text{dev}\left[\bar{\bar{\boldsymbol{\sigma}}}\right] : \dot{\boldsymbol{\varepsilon}}^p \end{aligned} \tag{7}$$

where k_w is a material parameter governing shear-related growth and is suggested in Nahshon and Hutchinson (2008) to be within $0 \leq k_w \leq 3$ and w is a function of deviatoric stress within $0 \leq w \leq 1$ that is responsible for distinguishing the axisymmetric stress states from generalized plane strain and is calculated as:

$$w = \hat{w}\left(\text{dev}\left[\bar{\bar{\boldsymbol{\sigma}}}\right]\right) = -\left[\frac{27\bar{J}_3}{2\bar{\bar{\boldsymbol{\sigma}}}^3_{eq}}\right]^2 \tag{8}$$

in which $\bar{J}_3$ is the third deviatoric invariant of the stress tensor. Note that, although a generalized shear-driven evolution is included within the formalism of continuum damage mechanics, within the formulation of the present work, Nahshon and Hutchinson shear correction is required to describe damage for ductile fracture, while CDM formulation is taken to model brittle damage evolution. Finally, the balance of energy is used to compute the time rate of change of temperature using the plastic work equation:

$$\rho c_p \dot{\theta} = \varsigma \bar{\bar{\boldsymbol{\sigma}}} : \dot{\boldsymbol{\varepsilon}}^p \tag{9}$$

where ς denotes the so-called Taylor-Quinney coefficient (Taylor and Quinney 1934) that controls the amount of plastic work converted to heat during plastic flow, in which ρ and c_p are the density and the heat capacity of the material, respectively. Extension of the plastic hardening model for strain rate and temperature dependency has been performed considering Johnson-Cook-type multiplicative decomposition of the material yield stress on strain hardening, strain rate dependency, and thermal softening. However, the original Johnson-Cook hardening law has been replaced with Hockett-Sherby hardening equation (Hockett and Sherby 1975), for better representation of the post-yielding and pre-peak response in small punch testing simulations of P91. That is:

$$\begin{aligned} \boldsymbol{\sigma}_y(\widetilde{\varepsilon}^p_{\text{eq}}, \dot{\widetilde{\varepsilon}}^p_{\text{eq}}, \theta) &= h_y(\widetilde{\varepsilon}^p_{\text{eq}}) r_y(\dot{\widetilde{\varepsilon}}^p_{\text{eq}}) t_y(\theta) \\ h_y(\widetilde{\varepsilon}^p_{\text{eq}}) &= h\widetilde{\varepsilon}^p_{\text{eq}} + b - [b - \boldsymbol{\sigma}_{y0}] \exp(-m \left[\widetilde{\varepsilon}^p_{\text{eq}}\right]^n) \\ r_y(\dot{\widetilde{\varepsilon}}^p_{\text{eq}}) &= 1 + C \log\left(\frac{\dot{\widetilde{\varepsilon}}^p_{\text{eq}}}{\dot{\varepsilon}^p_0}\right) \\ t_y(\theta) &= 1 - \Omega^r \end{aligned} \tag{10}$$

where $\boldsymbol{\sigma}_{y0}, b, C, n, m$ and r are material parameters. The reference strain rate has been denoted with $\dot{\varepsilon}^p_0$, and Ω is defined as $\Omega := [\theta - \theta_0]/[\theta_M - \theta_0]$ where θ_0 and θ_M represent the reference and melting temperatures, respectively. Here, only dependency of the yield stress on strain rate and temperature is considered within Johnson-Cook formalism, i.e., employed formulation is not viscoplastic. Note that theoretical basis presented here and numerical implementation performed cover also rate effects. However, it is excluded within the scope of this work, and only temperature-driven effects are investigated.

Brittle Fracture Model: Continuum Damage Mechanics

Concept of damaged material has initially been introduced by Kachanov (1958) and Rabotnov (1969), with an attempt to describe creep behavior of materials. Following these initial proposals, the concept of damage mechanics has been developed through works of Chaboche (1977), Lemaitre (1985), Germain et al. (1983), Murakami and Ohno (1980), Chow and Wang (1987a,b), and Krajcinovic and Fonseka (1981), within consistency of continuum thermodynamics. An exhaustive literature cite is not aimed here; however further literature is given in Besson et al. (2010) and Skrzypek et al. (2008).

For a general case, damage is physically defined as the ratio of the surface area of micro-cracks or cavities situated on the plane passing through the cross section of a representative volume element (RVE) to the total surface area of the RVE. Damage is assumed to act over an RVE such that $D = 0$ describes undamaged material, while $D = 1$ represents fully damaged material, at which the load-carrying capacity of RVE has totally been lost. In phenomenological approach to damage, also called as continuum damage mechanics, a damage state variable affects the material stress state, with the introduction of the so-called *effective* stress tensor. Definition of the effective stress tensor is either based on the strain equivalence principle introduced by Chaboche (1977) or by the energy equivalence principle developed by Cordebois and Sidoroff (1982). In strain equivalence principle, effective stress is defined as the stress tensor, when applied to a fictitious undamaged material generates the same strain tensor obtained when net (or homogenized) stress tensor is applied to a damaged material. In energy equivalence principle, the equivalence is formulated over elastic strain energies of damaged and undamaged materials. For the case of isotropic damage, effective stress is defined as $\bar{\tilde{\boldsymbol{\sigma}}} = \tilde{\boldsymbol{\sigma}}/(1-d)$, where d is a scalar damage variable and $\boldsymbol{\sigma}$ is the net stress tensor. For the tensorial damage variable, also employed in the present work, several definitions are given in literature (Skrzypek et al. 2008).

The effective stress tensor employed in the present work is based on the volumetric-deviatoric split and symmetrization of effective stress tensor given by Cordebois and Sidoroff (1982) and Chow and Wang (1987a,b) and as:

$$\bar{\tilde{\boldsymbol{\sigma}}} = \frac{\mathrm{tr}\,[\tilde{\boldsymbol{\sigma}}]}{3-\mathrm{tr}\,[\mathbf{D}]}\mathbf{1} + \left[[\mathbf{1}-\mathbf{D}]^{-\frac{1}{2}}\mathrm{dev}\,[\tilde{\boldsymbol{\sigma}}]\,[\mathbf{1}-\mathbf{D}]^{-\frac{1}{2}}\right] : \mathbb{P} \tag{11}$$

$$\bar{\tilde{\boldsymbol{\sigma}}} = \frac{\mathrm{tr}\,[\tilde{\boldsymbol{\sigma}}]}{3-\mathrm{tr}\,[\mathbf{D}]}\mathbf{1} + \left[[\mathbf{1}-\mathbf{D}]^{-\frac{1}{2}}\mathrm{dev}\,[\tilde{\boldsymbol{\sigma}}]\,[\mathbf{1}-\mathbf{D}]^{-\frac{1}{2}}\right] : \mathbb{P} \tag{12}$$

in which “tr” and “dev” denote trace and deviator of the tensorial quantity and $\mathbb{P}$ is the 4th order deviatoric projection tensor and $\mathbf{D}$ is the 2nd order tensorial damage variable representing the micro-crack-driven damage evolution linked to brittle fracture. The coupling and dependency between micro-crack-driven brittle damage evolution and micro-void-driven ductile damage is achieved by assuming the normality rule for damage evolution, using plastic multiplier λ and potential $\chi = \boldsymbol{\phi}^p(\tilde{\boldsymbol{\sigma}}, \mathbf{D}) + \boldsymbol{\phi}^d(\mathbf{Y}, \mathbf{D})$ introduced previously, with a specific choice of damage potential $\boldsymbol{\phi}^d$ given in Hayakawa and Murakami (1997) and Abu Al-Rub and Voyiadjis (2003):

$$\begin{aligned} \dot{\mathbf{D}} &= \lambda \partial_{\mathbf{Y}} \chi = \lambda \partial_{\mathbf{Y}} \boldsymbol{\phi}^d \\ \boldsymbol{\phi}^d &= \frac{1}{S^m(m+1)} \sqrt{\left[[\mathbf{1}-\mathbf{D}]^{-1} : [\mathbf{1}-\mathbf{D}]^{-1}\right]} \sqrt{[\mathbf{Y}:\mathbf{Y}]^{m+1}} \end{aligned} \tag{13}$$

where $\mathbf{Y}$ is stress-like variable, controlling brittle damage initiation and evolution, and is conjugate to $\mathbf{D}$. Note that, the specific choice for $\boldsymbol{\phi}^d$ is in terms of the invariants of stress-like damage conjugate variable $\mathbf{Y}$ so that the brittle damage formulation is analogous to ductile damage part with $\boldsymbol{\phi}^p$ and f. In a thermodynamically consistent framework, one can obtain the expression for damage conjugate variable named as strain energy release rate $\mathbf{Y}$, by taking its derivative with respect to the free energy function defined, i.e., $\mathbf{Y} = -\partial_{\mathbf{D}}\psi$. Instead, for the hypo-elastic plasticity framework upon which this work is based, we propose two different laws to calculate $\mathbf{Y}$ and $\dot{\mathbf{D}}$. Looking from brittle fracture point of view and based on experimental evidence that brittle fracture is linked to principal stresses, principal stress-driven Leckie-Hayhurst-type and plasticity-driven Lemaitre-type expressions for $\mathbf{Y}$ and $\dot{\mathbf{D}}$ have been proposed first time specifically to model temperature-driven ductile-to-brittle transition. These proposed models are introduced and discussed next.

Modified Leckie-Hayhurst Form

Initially developed as a creep-rupture criterion (Hayhurst and Leckie 1973; Leckie and Hayhurst 1977) in terms of stress invariants and within the context of isotropy, the so-called Leckie-Hayhurst form is generalized as a function of $\bar{\bar{\boldsymbol{\sigma}}}$ as:

$$\varphi(\bar{\bar{\boldsymbol{\sigma}}}) = a\bar{\bar{\sigma}}_i + bJ_1(\bar{\bar{\boldsymbol{\sigma}}}) + [1 - a - b]J_2(\bar{\bar{\boldsymbol{\sigma}}}) \tag{14}$$

where $\bar{\bar{\sigma}}_i$ is the maximum principal component of effective stress tensor at rotationally neutralized configuration. Proposing such a form for strain energy release rate has two advantages: (a) there is direct dependency to stresses and therefore direct link to brittle fracture modeling over principal stresses and (b) straightforward numerical implementation. Furthermore, the proposed form is analogous to quantitative expression of Ritchie-Knott-Rice cleavage criterion, which requires maximum principal stress exceed a specific temperature-independent limit over a specific distance. Thus, Leckie-Hayhurst-type form has been modified to capture principal stress effects such that $Y_i = \bar{\bar{\sigma}}_i$. Since the form in general has been derived considering isotropy condition (over a scalar damage variable), the extension considering second-order damage tensor $\mathbf{D}$ has been proposed in terms of spectral representation of $\mathbf{Y}$:

$$\begin{aligned} \mathbf{Y} &= \sum_{i=1}^{3} \langle Y_i \rangle \mathbf{n}_i \otimes \mathbf{n}_i \\ \langle Y_i \rangle &= \max(0, Y_i - Y_i^{\,0}) \end{aligned} \tag{15}$$

where $Y_i^{\,0}$ is the threshold value controlling the micro-crack damage initiation and evolution and $\mathbf{n}_i$ the principal directions of effective stress tensor $\bar{\bar{\boldsymbol{\sigma}}}$. This modified form can also be considered as 2nd-order anisotropic Ritchie-Knott-Rice brittle fracture criterion which is able to model internal softening associated with micro-crack initiation, propagation, and thus cleavage. However, note that this proposed form for $\mathbf{Y}$ has no more the meaning of strain energy release rate, as it is in a thermodynamically consistent framework.

For the calibration of the material parameter Y_i^0, a mixed experimental-numerical study needs to be performed. Here, we propose a similar test setup given in Ritchie et al. (1973) and Curry and Knott (1976). In Ritchie et al. (1973) and Curry and Knott (1976), a four-point bending test has been set up, and fracture toughness of the pre-notched specimens have been measured. A finite element study of the stress field around the notch then revealed the critical fracture stress of the specimen. Since the principal stress dependency of the proposed Leckie-Hayhurst form is similar to that of Ritchie-Knott-Rice fracture law, a parameter identification setup similar to Ritchie et al. (1973) and Curry and Knott (1976), which measures fracture response in terms of fracture toughness or force-displacement curves then correlates it to principal stresses numerically, is sufficient for the determination of Y_i^0. Note that, the experimental portion of the correlations performed within the present work is based on force-displacement response of the small punch fracture experiments given in Turba et al. (2011).

Plasticity-Driven Lemaitre Form

Driven by motivation that micro-void-based dilatational damage f and micro-crack-driven brittle damage coexist in a ductile-to-brittle transition state, the kinetic damage evolution law proposed by Lemaitre et al. (2000) is adopted to describe damage evolution effects to model cleavage. The form proposed by Lemaitre et al. in (2000) is:

$$\dot{\mathbf{D}} = \left[\frac{\bar{Y}}{S}\right]^m |\dot{\boldsymbol{\varepsilon}}^p| \quad \text{where} \quad \bar{Y} = \frac{1}{2}\tilde{\boldsymbol{\varepsilon}}^e : \mathbb{C} : \tilde{\boldsymbol{\varepsilon}}^e \tag{16}$$

where $\bar{Y}$ is the effective elastic energy density, while $|\dot{\boldsymbol{\varepsilon}}^p|$ defined plastic strain rate tensor in terms of its principal components, S and m are material parameters to be correlated. $|\dot{\boldsymbol{\varepsilon}}^p|$ is formed by rewriting plastic strain rate tensor in spectral representation with its eigenvectors serving a basis and performing absolute value operation. It's to pay attention that principal directions of damage rate tensor coincide with the principal directions of plastic strain rate tensor. The evolution of damage is positive only if a certain threshold of accumulated plastic strain is reached, i.e.,

$$\dot{\mathbf{D}} = \mathbf{0} \quad \text{if} \quad \widetilde{\varepsilon}^p_{\text{eq}} \leq \widetilde{\varepsilon}^{p,\text{crit}}_{\text{eq}} \tag{17}$$

The critical value of accumulated equivalent plastic strains acts as a threshold not only for damage evolution but as a limiter to pre-, post, and in-transition zone fracture response of the material. Therefore, it should be identified along with other damage parameters. Such identification is performed numerically within the scope of the work and is given in preceding sections. A drawback of this form is this: a nonphysical damage evolution is recovered under biaxial tensile stress states. To overcome this issue, Eq. 16 can be modified to take positive part of $\dot{\boldsymbol{\varepsilon}}^p$, instead of its absolute value. However, for axisymmetric small punch fracture modeling and implementation, the original form with absolute value of principal components has been used. The experimental characterization of material parameters S,m, and $\varepsilon_{\mathrm{eq}}^{p,\mathrm{crit}}$ that define this damage evolution law is basically uniaxial test with unloadings and low-cycle fatigue tests. Details of experimental characterization are given in Lemaitre et al. (2000) and Lemaitre and Desmorat (2005).

Application to Small Punch Testing

Small punch testing is a recently developed mechanical testing method that allows the mechanical, fracture, and creep characterization of tiny specimens. It gained widespread importance and popularity since it enables the testing using specimens taken from already existing components that have been exposed to service loads and temperatures. As it is also considered a nondestructive test method, in situ characterization of fracture, creep, mechanical properties, as well as lifetime is possible.

Small Punch Experimental Setup

Small punch testing technique uses miniaturized samples, and its setup has been carried out according to the guidelines of the *Code of Practice for Small Punch Testing for Tensile and Fracture Behavior* (CEN 2006). The testing temperature range was between 22°C and −196°C. In order to achieve the lowest test temperature, the testing equipment has been placed in a vessel filled with liquid nitrogen inside its environmental chamber. The thermocouples with an accuracy of ∓2°C were used. The deformation velocity of the samples was kept at constant value of 0.005 mm/s. Specimens were disks of 8 mm diameter and 1 mm thickness, with a predefined circumferential notch. The V-notches were produced by electro-discharge machining, resulting in a notch of diameter 2.5 mm, notch tip radius of 5 microns, and notch depth of 0.5 mm, as proposed in Turba et al. (2011). The schematic of the SP creep testing equipment is presented in Fig. 1.

In this test method, a rigid spherical ended puncher penetrates through a disk specimen at a constant displacement rate. During the experiment, the puncher force F and the midspan central displacement δ are recorded. The importance of small punch testing lies in its versatility: small punch force-displacement curves can be used to estimate mechanical properties such as yield stress, maximum strength, elasticity modulus, fracture toughness, ductile-to-brittle transition temperature, high temperature creep properties, etc. Furthermore, it requires small-sized specimens so that tiny cuts from existing structures could be utilized to investigate long-term in situ mechanical or thermal response of the material. As a result of experiments performed within a wide temperature range, it has been observed that the fracture mode in the upper temperature range (toward 22°C) is ductile with micro-voids and dimples. In the transition region, both ductile and brittle fracture, consisting of transgranular cleavage marks, exist. The amount of ductile crack growth decreases

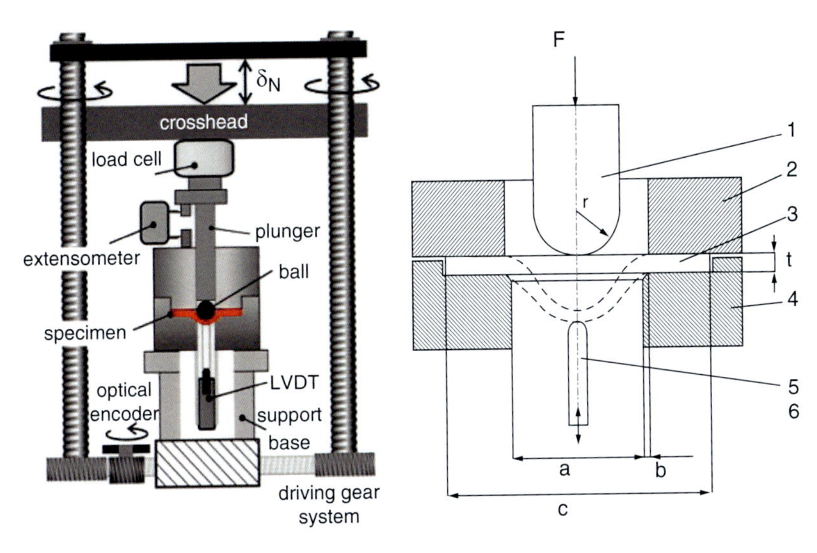

Fig. 1 Small punch test holder (Basbus et al. 2014)

gradually with reduced temperature. At the lower temperature range (toward −196°C), the fracture is almost completely brittle.

The radial symmetry of the notch enables to achieve a plane strain condition for evaluation purposes. Due to significantly small radius of the notch tip, high stress concentrations are achieved, and the crack was found to be initiating from the notch tip, for all the disks tested at different temperatures.

For the ductile case where void initiation and growth are controlled by plastic strain, the crack propagation was found to follow a path in the direction of the notch where the equivalent plastic strain is maximum. The micro-voids nucleating at second phase particles in the vicinity of the notch tip grow and coalescence as the strain level is increased, and finally a continuous crack is formed, leading to the complete fracture of the disk.

At −196°C, the crack propagates in perpendicular direction to maximum principal stress, in accordance with the maximum tensile stress theory by (Erdogan and Sih 1963).

As for the transition, the crack was found to follow an angle in-between ductile and brittle cases, while the angle approaches the one in the brittle case. Ductile, brittle, and transition fractures could be visualized in Figs. 2 and 3.

Numerical Simulations of SPT

The proposed theoretical model has been implemented in ABAQUS as a VUMAT. Numerical implementation has been performed for axisymmetric elements, due to the presence of the rotationally symmetric notch. Implemented model covers

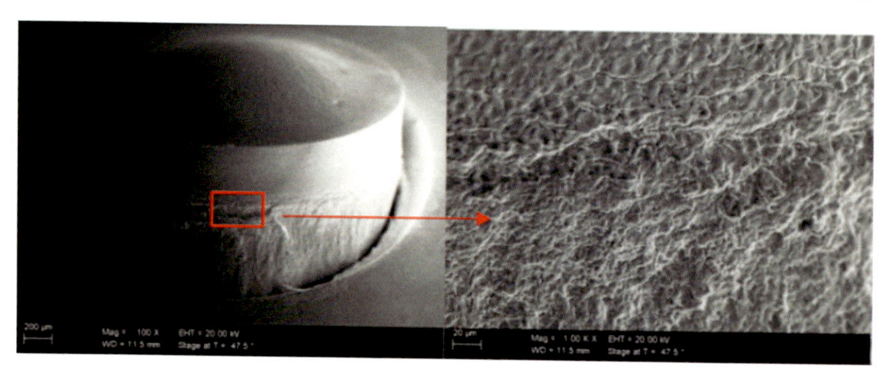

Fig. 2 After-experiment investigation of a notched specimen tested at 22°C

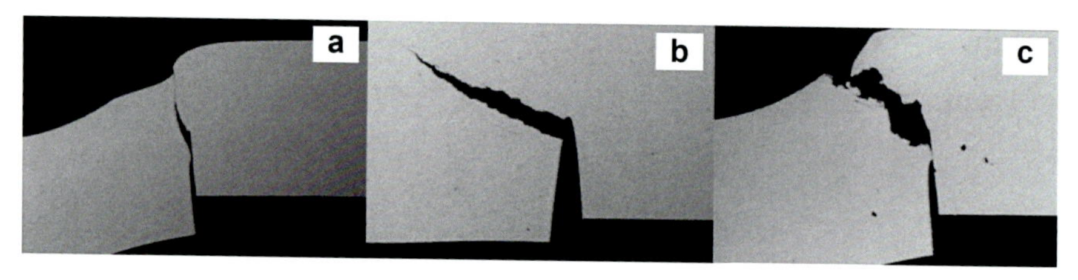

Fig. 3 Three different crack patterns observed at different temperatures (**a**): ductile fracture at 22°C, (**b**) brittle fracture at −196°C, (**c**) temperature-dependent transition fracture pattern

temperature effects on transition fracture as well as Johnson-Cook strain rate dependency and stress triaxiality effects on fracture. Axisymmetric finite elements with reduced integration and enhanced hourglassing controls (CAX4R) have been used. Clamps and puncher have been modeled as rigid instances. Artificial strain energy effects are kept at minimum thanks to sufficiently small element sizes. Small punch experiment has been modeled for 22°C and −196°C. ABAQUS setup can be visualized in Fig. 4.

The set of material parameters used to predict P91 behavior have been listed in Tables 1, 2, 3, and 4.

Simulations with the proposed material model and with stated material parameters have yielded three distinct crack patterns, which are in good correlation both with respect to theoric predictions and experimental results. If Fig. 5 is inspected, a precise correlation with simulation and numerics, in terms of crack patterns, can easily be observed. Depending on the temperature level θ and the threshold value of the principal strain energy release rate Y_i^0, direction and pattern of the transition fracture can precisely be predicted. As also inspected from experimental results, transition fracture patterns are mainly dominated by brittle fracture, while plastic strain amount and direction govern how much the crack deviates from principal stress directions. Such experimental and numerical patterns could be visualized in Fig. 5.

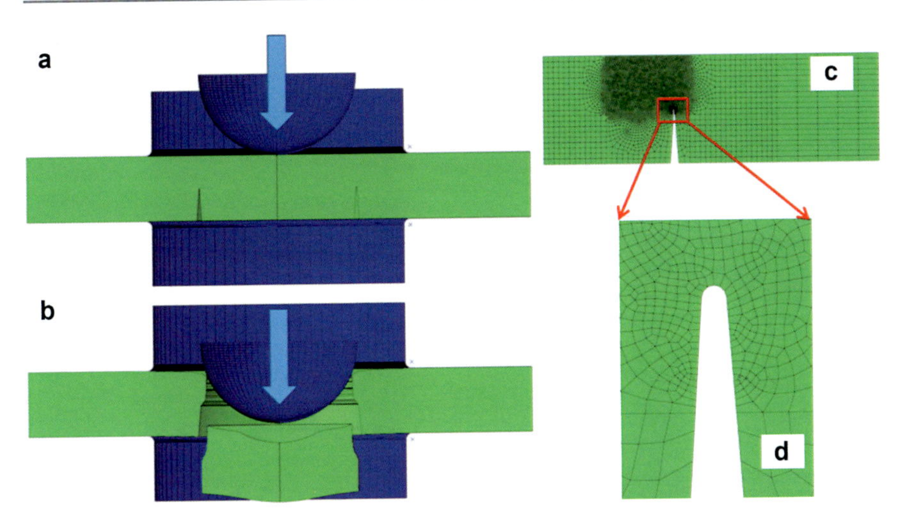

Fig. 4 (**a**) Initial configuration at the beginning of the simulation. (**b**) Final configuration at end of the simulation with ductile fracture. (**c**) Axisymmetric FE mesh of a specimen. (**d**) Close-up view of the notch tip, illustrating mesh size

Table 1 Thermoelastic properties of P91

E(GPa)	ν	c_p (m^2/[Ks^2])	ς	ε_0	θ_M (K)
210	0.3	622	0.9	1.3e-5	1717

Table 2 Flow curve data of P91 at 22°C, Johnson-Cook hardening

h (MPa)	σ_{y0} (MPa)	b (MPa)	m	n
500	346	650	25.73	1.01

Table 3 Gurson-Tvergaard-Needleman model parameters for P91

f_N	s_N	ε_N	q_1	q_2	$q_3 = 1/f_U$	f_0	k_w
0.005	0.1	0.03	1.5	1.0	2.25	0.0055	3.0

Table 4 Ductile and brittle failure criteria

f_c	f_F	$D_{i,\text{crit}}$	Y_i^0 (MPa)
0.05	0.207	0.30	1500

The force-displacement responses of the implemented models and comparison to experimental results are given in Fig. 6. As it is seen, the numerical response of the proposed models compared to experimental results is in good correlation, as a result of a change in temperature. The peak responses are accurately predicted at all three cases. The maximum pre-failure load for modified Leckie-Hayhurst model is slightly higher compared to that of Lemaitre model. The reason for such a response is natural, since micro-crack-driven damage in Lemaitre-type model is driven by equivalent plastic strain amount, leading to additional softening in load response at higher plastic strains. Still, such a deviation is within the acceptable range. Note that the brittle damage response of modified Leckie-Hayhurst model is linked only to maximum principal strain evolution.

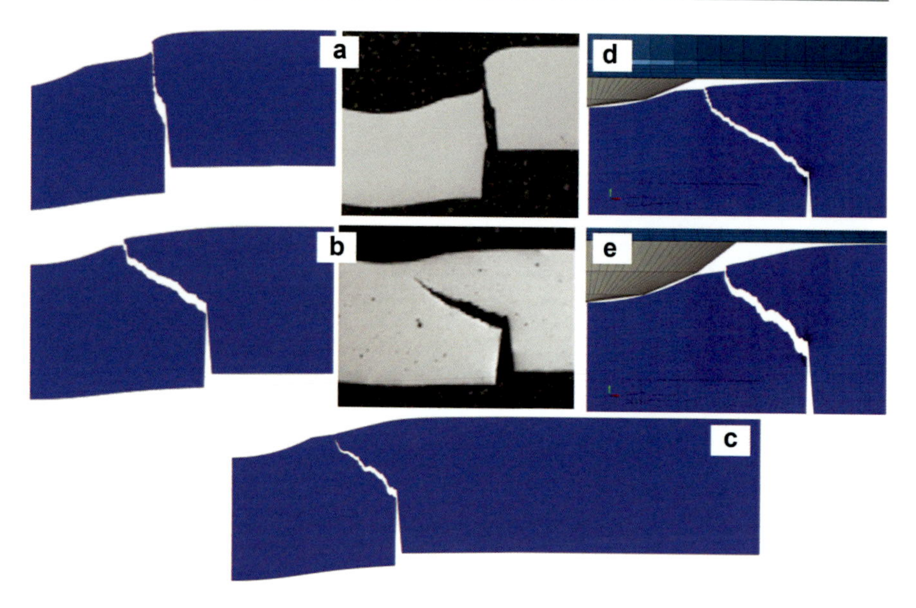

Fig. 5 (**a**) Numerical prediction and experimental investigation of ductile tearing at 22°C. (**b**) Numerical prediction and experimental investigation of brittle cleavage at −196°C. (**c**) A typical mixed-mode fracture pattern, mainly effected by brittle cleavage, while plastic strain direction deviates crack direction away from principal stress directions. (**d**) Close-up view of pure brittle crack propagation. (**e**) Close-up view of mixed-mode crack propagation. Note the difference in propagation direction compared to pure brittle fracture

Effects of Specimen Geometry on Fracture Response of SPT

The geometrical parameters of the small punch experiment such as puncher radius and notch depth (Fig. 7) have been varied, and the effects of such geometrical changes on fracture response of the specimen are investigated numerically.

A design of experiments study is performed to find the correlation between geometrical parameters.

The Effects of the Puncher Radius

The puncher radius of the loading units has been varied with the values of 2.25, 2.5, and 2.75 mm, respectively (Fig. 8), to investigate its effects on the fracture response of the notched specimen.

Usually, the radius of the puncher on the loading unit is 2.5 mm. In the case of reducing the radius of the puncher, the contact surface between the puncher and the specimen became smaller. If the puncher radius is reduced to too small value, more stress concentration occur nearby the underneath of the puncher region, and also failure of the specimen is dominated by shear due to punching effect. (See Fig. 9.)

By comparing the experimental results, it is concluded that the puncher radius should be kept between the values of 1.5 and 3.5 mm as its lower and upper limits.

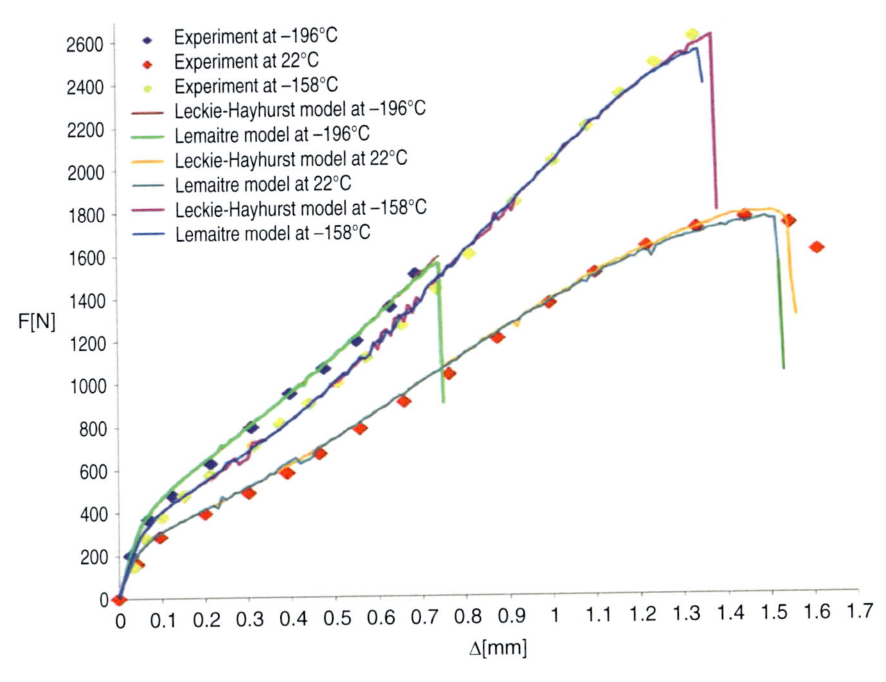

Fig. 6 Numerical simulations with two implemented material models and comparison with experimental results

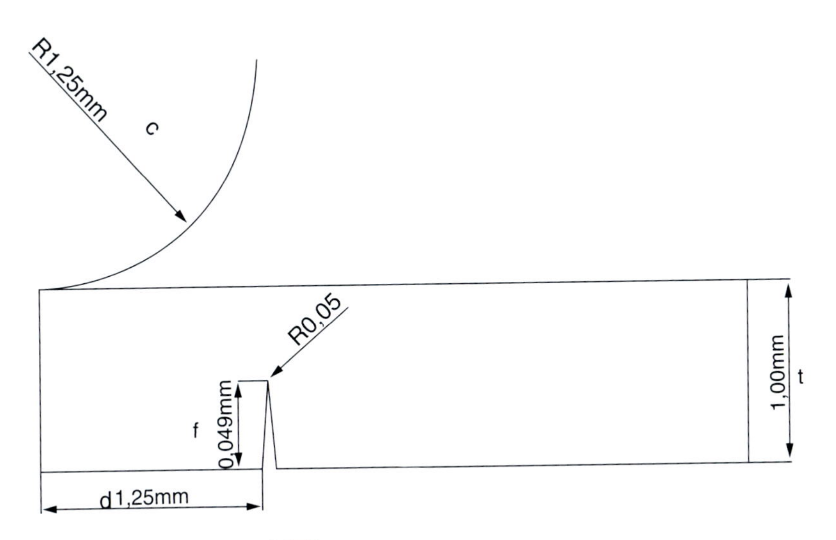

Fig. 7 Geometric parameters of SPT

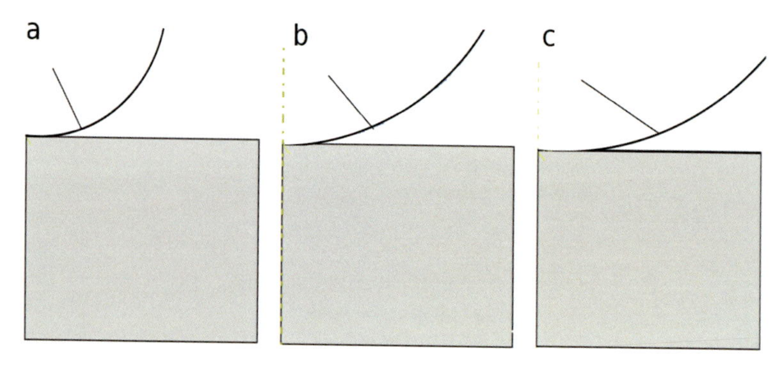

Fig. 8 Various puncher radius

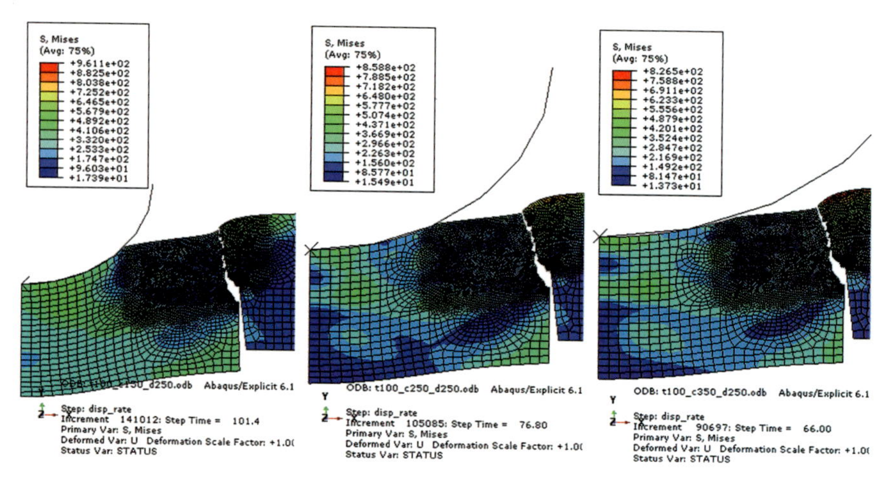

Fig. 9 (**a**) Numerical prediction of effects of the puncher radius with 2.25, 2.5, and 2.275 mm on the notched specimen

Force-displacement curves are presented in Fig. 10 in order to indicate the puncher radius effect on the fracture response of the specimen.

As it can be seen from the curves in Fig. 10, there is relation between the peak load and the radius of the puncher. As the radius of the puncher increases to a certain critical value, the region of the contacting surfaces between the puncher and the specimen becomes wider and the load reaches its peak value faster than that in the specimen loaded with the puncher having smaller radius. Therefore, this indicates that fracturing of the specimen loaded with puncher having bigger radius occurs much earlier than that of the specimen loaded with puncher having smaller radius (Fig. 9).

The Effects of Crack Depth

Depth of the crack is one of the most important fracture parameters that causes the damage and fracture in the materials. In order to investigate the effect of crack depth

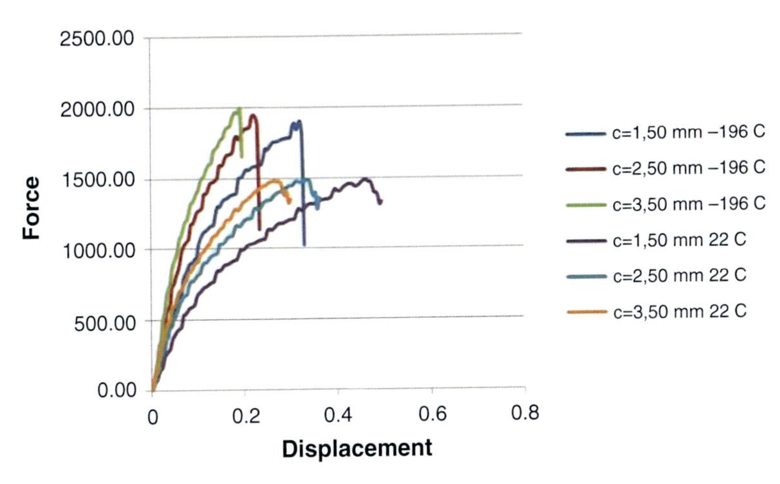

Fig. 10 Prediction of the force-displacement response of the notched specimen for various radius of the puncher

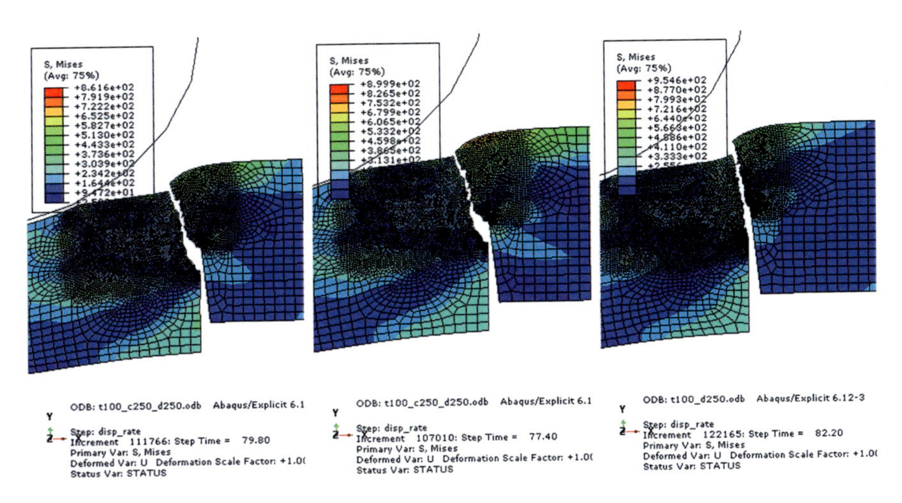

Fig. 11 Stress distribution at various crack depth

on the fracture response of the small punch test, three different crack depths, with length of 0.44, 0.49, and 0.54 mm, are defined in numerical simulation. Figure 11 mimics von Mises stress distributions as the results of the finite element simulation of SPTs at various crack depths.

As the crack depth increases along the direction of the thickness of the specimen, material becomes weaker and is much likely prone to damage and fracture. At the lower temperature −196°C for all crack lengths, the material behaves completely brittle. However, as seen in Fig. 11 for the larger crack length, material shows more ductile behavior than that of the smaller crack length.

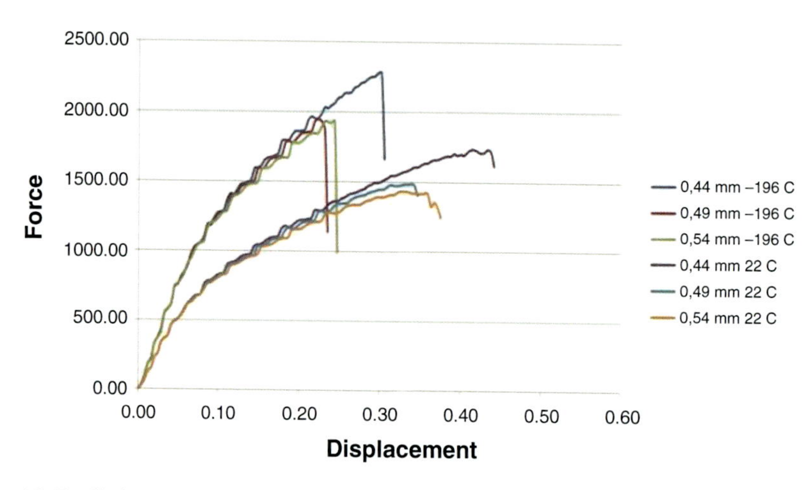

Fig. 12 Prediction of the force-displacement response of the notched specimen for various crack depth

Figure 12 shows the force-displacement curves. It is seen that the peak load of the force-displacement curves exhibit a lowering with increase in the specimen crack length at both temperatures. This implies that for decreased initial crack lengths, the specimen exhibits higher fracture resistance (Fig. 12).

Discussion of Results and Further Research

In the enclosed work, a coupled continuum damage mechanics and porous plasticity model on local scale have been presented, in order to model temperature-dependent ductile-to-brittle transition fracture in metals. Classical effective stress and damage expressions of CDM have been employed to model brittle fracture, while Gurson's porous metal plasticity has been utilized to model ductile fracture. Leckie-Hayhurst criterion has been modified to account for brittle fracture and extended to cover anisotropic material behavior. Constitutive modeling and small punch experiments have been performed for P91 steel. It has been found out that the proposed model precisely and conveniently predicts crack patterns within a relatively wide temperature range from 22°C to −196°C.

The effects of the various geometrical parameters such as puncher radius on fracture response of the SPT specimen is investigated numerically. Depth of the crack is a crucial effect parameter that causes the damage and fracture in the materials. As the crack depth increases along the direction of the thickness of the specimen, material becomes weaker and shows more ductile behavior. On the other hand the puncher radius dominates shear type failure in the specimen due to punching effects.

For the sake of completeness, the proposed model will be extended to nonlocal framework by making use of integral type nonlocal approach, complemented with associated length scales for both brittle and ductile fractures. Incorporation of the statistical effects for the brittle fracture by the stochastic distribution of material properties as well as the sensitivity study for the threshold value of the strain energy release rate Y_i^0 will also be studied.

References

CEN, Cen workshop agreement: small punch test method for metallic materials. Technical Report CWA 15627 (2006)

R. Abu Al-Rub, G. Voyiadjis, On the coupling of anisotropic damage and plasticity models for ductile materials. Int. J. Solids Struct. **40**, 2611–2643 (2003)

T. Anderson, *Fracture Mechanics: Fundamentals and Applications*, Taylor & Francis Group, Boca Raton (CRC Press, 2004)

P. Baloso, R. Madanrao, M. Mohankumar, Determination of the ductile to brittle transition temperature of various metals. Int. J. Innov. Eng. Res. Technol. **4**, 1–27 (2017)

F. Basbus, M. Moreno, A. Caneiro, L. Mogni, Effect of pr-doping on structural, electrical, thermodynamic,and mechanical properties of BaCeo3 proton conductor. J. Electrochem. Soc. **161**(10), 969–976 (2014)

R. Batra, M. Lear, Simulation of brittle and ductile fracture in an impact loaded prenotched plate. Int. J. Fract. **126**, 179–203 (2004)

F.M. Beremin, A local criterion for cleavage fracture of a nuclear pressure vessel steel. Met. Trans. A **14A**, 2277–2287 (1983)

J.-M. Bergheau, J.-B. Leblond, G. Perrin, A new numerical implementation of a second-gradient model for plastic porous solids, with an application to the simulation of ductile rupture tests. Comput. Method. Appl. M **268**, 105–125 (2014)

J. Besson, G. Cailletaud, J. Chaboche, S. Forest, *Non-linear Mechanics of Materials* (Springer, Berlin, 2010)

J. Chaboche, Sur l'utilisation des variables d'ètat interne pour la description de la viscoplasticitè cyclique avec endommagement, in *Symposium Franco-Polonais de Rhèologie et Mècanique*, 1977, pp 137–159

J. Chaboche, M. Boudifa, K. Saanouni, A CDM approach of ductile damage with plastic compressibility. Int. J. Fract. **137**, 51–75 (2006)

P. Chakraborty, B. Biner, Modeling the ductile brittle fracture transition in reactor pressure vessel steels using a cohesive zone model based approach. Int. Work. Struct. Mater. Innov. Nucl. Syst. (72):1–10 (2013)

C. Chow, J. Wang, An anisotropic theory of continuum damage mechanics for ductile fracture. Eng. Fract. Mech. **27**, 547–558 (1987a)

C. Chow, J. Wang, An anisotropic theory of elasticity for continuum damage mechanics. Int. J. Fract. **33**, 3–18 (1987b)

C. Chu, A. Needleman, Void nucleation effects in biaxially stretched sheets. J. Eng. Mater. Technol. **102**, 249–256 (1980)

J.P. Cordebois, F. Sidoroff, Damage induced elastic anisotropy. In: JP. Boehler, (eds), Mech. Behav. Anisotropic Solids / Comportment Mech. des Sol. Anisotropic. Springer, Dordrecht https://doi.org/10.1007/978-94-009-6827-1_44 761–774 (1982)

D. Curry, F. Knott, The relationship between fracture toughness and micro-structure in the cleavage fracture of mild steel. Mater. Sci. **10**, 1–6 (1976)

D. Curry, J.F. Knott, Effect of microstructure on cleavage fracture toughness in mild steel. Metal Sci. **13**, 341–345 (1979)

F. Erdogan, G.C. Sih, On the crack extension in plates under plane loading and transverse shear. J. Bas. Eng. **85**, 519–529 (1963)

L.B. Freund, Y.J. Lee, Observations on high strain rate crack growth based on a strip yield model. Int. J. Fract. **42**, 261–276 (1990)

P. Germain, Q. Nguyen, P. Suquet, Continuum thermodynamics. J. Appl. Mech. **50**, 1010–1020 (1983)

A.L. Gurson, Continuum theory of ductile rupture by void nucleation and growth: Part I- Yield Criteria and Flow Rules for Porous Ductile Media, J. Eng. Mater. Technol **99**, 2–15 (1977)

P. Hakansson, M. Wallin, M. Ristinmaa, Thermomechanical response of non-local porous material. Int. J. Plasticity **22**, 2066–2090 (2006)

V. Hardenacke, J. Hohe, V. Friedmann, D. Siegele, Enhancement of local approach models for assessment of cleavage fracture considering micromechanical aspects, in *Proceedings of the 19th European Conference on Fracture*, Kazan (72), pp. 49–72 (2012)

K. Hayakawa, S. Murakami, Thermodynamical modeling of elastic-plastic damage and experimental validation of damage potential. Int. J. Damage Mech. **6**, 333–362 (1997)

D. Hayhurst, F. Leckie, The effect of creep constitutive and damage relationships upon rupture time of solid and circular torsion bar. J. Mech. Phys. Solids **21**, 431–446 (1973)

J. Hockett, O. Sherby, Large strain deformation of polycrystalline metals at low homologous temperatures. J. Mech. Phys. Solids **23**(2), 87–98 (1975)

G. Hutter, Multi-scale simulation of crack propagation in the Ductile-Brittle transition region. PhD. Thesis, Faculty of Mechanical, Process and Energy Engineering, Technische Universitat Bergakademie Freiberg (2013a)

G. Hütter (2013b) Multi-scale simulation of crack propagation in the ductile-brittle transition region. PhD. Thesis, Faculty of Mechanical, Process and Energy Engineering, Technische Universität Bergakademie Freiberg

G. Hutter, T. Linse, S. Roth, U. Muhlich, M. Kuna, A modeling approach for the complete Ductile-to-Brittle transition region: cohesive zone in combination with a nonlocal gurson-model. Int. J. Fract. **185**, 129–153 (2014)

R. Kabir, A. Cornec, W. Brocks, Simulation of quasi-brittle fracture of lamellar tial using the cohesive model and a stochastic approach. Comput. Mater. Sci. **39**, 75–84 (2007)

L. Kachanov, Time of the rupture process under creep conditions. IsvAkadNaukSSR **8**, 26–31 (1958)

D. Krajcinovic, G. Fonseka, The continuous damage theory of brittle materials, parts 1 and 2. J. Appl. Mech. **48**, 809–824 (1981)

F. Leckie, D. Hayhurst, Constitutive equations for creep rupture. Acta Metall. **25**, 1059–1070 (1977)

J. Lemaitre, A continuous damage mechanics model for ductile fracture. J. Eng. Mater. Technol. **107**, 83–89 (1985)

J. Lemaitre, R. Desmorat, *Engineering Damage Mechanics* (Springer, Berlin/New York, 2005)

J. Lemaitre, R. Desmorat, M. Sauzay, Anisotropic damage law of evolution. Eur. J. Mech. Solids **19**, 187–208 (2000)

T. Linse, G. Hütter, M. Kuna, Simulation of crack propagation using a gradient-enriched ductile damage model based on dilatational strain. Eng. Fract. Mech. **95**, 13–28 (2012)

L. Malcher, F. Andrade Pires, J. Cesar de Sa, An assessment of isotropic constitutive models for ductile fracture under high and low stress triaxiality. Int. J. Plasticity **30–31**, 81–115 (2012)

C. McAuliffe, H. Waisman, A unified model for metal failure capturing shear banding and fracture. Int. J. Plasticity **65**, 131–151 (2015)

F. Mudry, A local approach to cleavage fracture. Nucl. Eng. Des. **105**, 65–76 (1987)

S. Murakami, N. Ohno, A continuum theory of creep and creep damage, in *3rd Creep in Structures Symposium*, Leicester, pp. 422–443 (1980)

K. Nahshon, J. Hutchinson, Modification of the gurson model to shear failure. Eur. J. Mech. A/Solids **27**, 1–17 (2008)

K. Nahshon, Z. Xue, A modified gurson model and its applications to punch-out experiments. Eng. Fract. Mech. **76**, 997–1009 (2009)

A. Needleman, V. Tvergaard, Dynamic crack growth in a nonlocal progressively cavitating solid. Eur. J. Mech. A/Solids **17**(3), 421–438 (1998)

A. Needleman, V. Tvergaard, Numerical modeling of the ductile-brittle transition. Int. J. Fract. **101**, 73–97 (2000)

T. Pardoen, J.W. Hutchinson, An extended model for void growth and coalescence, J. Mech. Phys. Solids **48**(12), 2467–2512 (2000)

Y. Rabotnov, *Creep Problems in Structural Members* (North-Holland, Amsterdam, 1969)

S. Ramaswamy, N. Aravas, Finite element implementation of gradient plasticity models. Part I: gradient-dependent yield functions. Comput. Methods Appl. Mech. Eng. **163**, 33–53 (1998)

S. Renevey, S. Carassou, B. Marini, C. Eripret, A. Pineau, Ductile – brittle transition of ferritic steels modelled by the local approach to fracture. JOURNAL DE PHYSIQUE IV Colloque C6, supplBment au Journal de Physique III **6**, 343–352 (1996)

R. Ritchie, J. Knott, J. Rice, On the relationship between critical tensile stress and fracture toughness in mild steel. J. Mech. Phys. Solids **21**, 395–410 (1973)

F. Rivalin, A. Pineau, M. Di Fant, J. Besson, Ductile tearing of Pipeline steel wide plates-I: Dynamics and quasi static experiments. Engng Fract. Mech **68**(3), 329–345 (2001a)

F. Rivalin, J. Besson, M. Di Fant, A. Pineau, Ductile tearing of Pipeline steel wide plates-II: Modeling of in-plane crack propagation. Engng Fract. Mech **68**(3), 347–364 (2001b)

G. Rousselier, Ductile fracture models and their potential in local approach of fracture. Nuc. Eng. Des. **105**, 97–111 (1987)

M. Samal, M. Seidenfuss, E. Roos, B.K. Dutta, H.S. Kushwaha, A mesh independent GTN damage model and its application in simulation of ductile fracture behaviour. ASME 2008 Pressure vessels and piping conference volume 3: Design and Analysi, Chicago, Illinois, USA, July 27–31 pp. 187–193 (2008)

I. Scheider, W. Brocks, Simulation of cup-cone fracture using the cohesive model. Eng. Fract. Mech. **70**, 1943–1961 (2003)

A. Shterenlikht, 3D CAFE modeling of transitional ductile-brittle fracture in steels. PhD. Thesis, Department of Mechanical Engineering, University of Sheffield (2003)

J. Skrzypek, A. Ganczarski, F. Rustichelli, H. Egner, *Advanced Materials and Structures for Extreme Operating Conditions* (Springer, Berlin, 2008)

C. Soyarslan, I. Turtuk, B. Deliktas, S. Bargmann, A thermomechanically consistent constitutive theory for modeling micro-void and/or micro-crack driven failure in metals at finite strains. Int. J. Appl. Mech. **8**, 1–20 (2016)

S. Sutar, G. Kale, S. Merad, Analysis of ductile-to-brittle transition temperature of mild steel. Int. J. Innov. Eng. Res. Technol. **1**, 1–10 (2014)

B. Tanguy, J. Besson, R. Piques, A. Pineau, Ductile to brittle transition of an a508 steel characterized by charpy impact test, Part I. Experimental results. Eng. Fract. Mech. **72**, 49–72 (2007)

G. Taylor, H. Quinney, The latent energy remaining in a metal after cold working. Proc. R. Soc. Lond. **A143**, 307–326 (1934)

K. Turba, B. Gulcimen, Y. Li, D. Blagoeva, P. Hähner, R. Hurst, Introduction of a new notched specimen geometry to determine fracture properties by small punch testing. Eng. Fract. Mech. **78**(16), 2826–2833 (2011)

I. Turtuk, B. Deliktas, Coupled porous plasticity – continuum damage mechanics approaches in modelling temperature driven ductile-to-brittle transition fracture in ferritic steels. Int. J. Plasticity **77**, 246–261 (2016)

V. Tvergaard, Influence of voids on shear band instabilities under plane strain conditions. Int. J. Fract. **17**, 389–407 (1981)

V. Tvergaard, On localization in ductile materials containing spherical voids. Int. J. Fract. **18**, 237–252 (1982a)

V. Tvergaard, Influence of void nucleation on ductile shear fracture at a free surface. J. Mech. Phys. Solids **30**, 399–425 (1982b)

V. Tvergaard, A. Needleman, Analysis of the cup-cone fracture in a round tensile bar. Acta Metall. **32**, 157–169 (1984)

W. Weibull, A statistical distribution function of wide applicability. Jour. App. Mech **18**, 293–297 (1953)

J. Wen, Y. Huang, K. Hwang, C. Liu, M. Li, The modified gurson model accounting for the void size effect. Int. J. Plasticity **21**, 381–395 (2005)

L. Xia, C. Fong Shih, A fracture model applied to the ductile/brittle regime. Journal de Physique IV **6**, 363–372 (1996)

Z.L. Zhang, C. Thaulow, J. Ødegard, A complete gurson model approach for ductile fracture, Eng. Fract. Mech **67**(2), 155–168 (2000)

Size-Dependent Transverse Vibration of Microbeams

32

Ömer Civalek and Bekir Akgöz

Contents

Abstract

In this chapter, a new microstructure-dependent higher-order shear deformation beam model is introduced to investigate the vibrational characteristics of microbeams. This model captures both the size and shear deformation effects without the need for any shear correction factors. The governing differential equations and related boundary conditions are derived by implementing Hamilton's principle on the basis of modified strain gradient theory in conjunction with trigonometric shear deformation beam theory. The free vibration problem for simply supported microbeams is analytically solved by employing the Navier solution procedure. Moreover, a new modified shear correction factor is firstly proposed for Timoshenko (first-order shear deformation) microbeam model. Several comparative results are presented to indicate the effects of material length-scale parameter ratio, slenderness ratio, and shear correction factor on the natural frequencies of microbeams. It is observed that effect of shear deformation becomes more considerable for both smaller slenderness ratios and higher modes.

Ö. Civalek (✉) · B. Akgöz
Civil Engineering Department, Division of Mechanics, Akdeniz University, Antalya, Turkey
e-mail: civalek@yahoo.com; bekirakgoz@akdeniz.edu.tr

G. Z. Voyiadjis (ed.), *Handbook of Nonlocal Continuum Mechanics for Materials and Structures*, https://doi.org/10.1007/978-3-319-58729-5_8

Keywords
Microbeam · Size dependency · Vibration · Small-scale effect · Modified strain gradient theory · Higher-order beam theory · Shear deformation effect · Modified shear correction factor · Length-scale parameter · Trigonometric beam model

Introduction

The miniaturized (small-sized) structures have a wide range of applications in nano- and micro-electromechanical systems (NEMS andMEMS) due to the rapid improvements in technology (Younis et al. 2003; Li and Fang 2010; Wu et al. 2010). Microbeamis one of the essential structures frequently used in MEMS/NEMS such as micro-resonators (Zook et al. 1992), atomic force microscopes (Torii et al. 1994), micro-actuators (Hung and Senturia 1999), and microswitches (Xie et al. 2003). Because of the characteristics dimensions of the microbeams (thickness, width, and length) are on the order of microns and submicrons, size effects should be taken into consideration on the determination of the mechanical characteristics of such structures. However, it has been experimentally observed for several materials that microstructural effects appear and have considerable effect on mechanical properties and deformation behavior for smaller sizes (Poole et al. 1996; Lam et al. 2003; McFarland and Colton 2005). Unfortunately, the well-known classical continuum theories, which are independent of scale of the structure's size, fail to estimate and explain of size dependency in micro- and nanoscale structures. Subsequently, various nonclassical continuum theories, which include at least one additional material length-scale parameter, have been developed like couple stress theory (Mindlin and Tiersten 1962; Koiter 1964; Toupin 1964), micropolar theory (Eringen 1967), nonlocal elasticity theory (Eringen 1972, 1983), and strain gradient theory (Fleck and Hutchinson 1993; Vardoulakis and Sulem 1995; Altan et al. 1996).

One of the higher-order continuum theories, named as strain gradient theory, developed by Fleck and Hutchinson (1993, 2001), can be viewed as extended form of the Mindlin's simplified theory (Mindlin 1965). This theory requires five additional material length-scale parameters related to second-order deformation gradients. Subsequently, Lam et al. (2003) proposed a more useful form of the strain gradient theory which is named as modified strain gradient theory (MSGT) and includes three additional material length-scale parameters for linear elastic isotropic materials.

This theory has been employed by many researchers to analyze size-dependent microbeams. For instance, Bernoulli-Euler and Timoshenko models were introduced for static bending, free vibration, and buckling behaviors of microbeams by Kong et al. (2009), Wang et al. (2010), and Akgöz and Civalek (2012, 2013a). Furthermore, Kahrobaiyan et al. (2012) and Ansari et al. (2011) introduced Bernoulli-Euler and Timoshenko beam models for functionally graded microbeams, respectively. Artan and Batra (2012) employed the method of initial values for the free vibration of Bernoulli-Euler strain gradient beams with four different boundary conditions as simply supported-simply supported, clamped-free, clamped-clamped, and clamped-simply supported. Approximate solutions for static and dynamic analyses of microbeams were also carried out by finite element method based on Bernoulli-Euler and Timoshenko beam theories, respectively (Kahrobaiyan et al. 2013; Zhang et al. 2014a).

Presently, various beam theories have been proposed and used to investigate the mechanical behaviors of beams. Influences of shear deformation can be neglected for slender beams with a large aspect ratio. However, effects of shear deformation and rotary inertia become more prominent and cannot be ignored for moderately thick beams and vibration responses on higher modes. In this manner, several shear deformation beam theories have been developed to account for the effects of transverse shear. One of the earlier shear deformation beam theories is the first-order shear deformation beam theory (commonly named as Timoshenko beam theory (TBT)) (Timoshenko 1921). This theory assumes that shear stress and strain are constant along the height of the beam. In fact, the distributions of these are not uniform, and also there are no transverse shear stress and strain at the top and bottom surfaces of the beam. For this reason, a shear correction factor is needed, as a disadvantage of the theory. After that, some higher-order shear deformation beam theories, which satisfy the condition of no shear stress and strain without any shear correction factors, have been presented such as parabolic (third-order) beam theory (Levinson 1981; Reddy 1984), trigonometric (sinusoidal) beam theory (Touratier 1991), hyperbolic beam theory (Soldatos 1992), exponential beam theory (Karama et al. 2003), and general exponential beam theory (Aydogdu 2009a). These theories have been used less than Euler-Bernoulli beam theory (EBT) and TBT on prediction of the mechanical responses of microstructures on the basis of the nonclassical continuum theories (Aydogdu 2009b; Salamat-talab et al. 2012; Şimşek and Reddy 2013a, b; Thai and Vo 2012, 2013; Akgöz and Civalek 2013b, 2014a, b, c, 2015; Zhang et al. 2014b).

In the present study, a new size-dependent trigonometric (sinusoidal) shear deformation beam model in conjunction with modified strain gradient theory is developed. This model captures both the microstructural and shear deformation effects without the need for any shear correction factors. The governing differential equations and related boundary conditions are derived by using Hamilton's principle. The free vibration response of simply supported microbeams is investigated. Analytical solutions for the first three natural frequencies are presented. In order to indicate the accuracy and validity of the present model, the results are comparatively presented with the results of other beam theories. A detailed parametric study is carried out to indicate the influences of material length-scale parameter, slenderness ratio, and shear correction factors on the natural frequencies of microbeams.

Modified Strain Gradient Theory

The modified strain gradient elasticity theory was proposed by Lam et al. (2003) in which contains not only classical strain tensor but also second-order deformation gradients (first-order strain gradients) such as dilatation gradient vector and deviatoric stretch gradient and symmetric rotation gradient tensors. The strain energy U

on the basis of the modified strain gradient elasticity theory can be written by (Lam et al. 2003; Kong et al. 2009):

$$U = \frac{1}{2}\int_{0}^{L}\int_{A}\left(\sigma_{ij}\varepsilon_{ij} + p_i\gamma_i + \tau_{ijk}^{(1)}\eta_{ijk}^{(1)} + m_{ij}^{s}\chi_{ij}^{s}\right) dA\, dx \tag{1}$$

$$\varepsilon_{ij} = \frac{1}{2}\left(u_{i,j} + u_{j,i}\right) \tag{2}$$

$$\gamma_i = \varepsilon_{mm,i} \tag{3}$$

$$\begin{aligned}\eta_{ijk}^{(1)} = \tfrac{1}{3}\left(\varepsilon_{jk,i} + \varepsilon_{ki,j} + \varepsilon_{ij,k}\right) - \tfrac{1}{15}\big[\delta_{ij}\left(\varepsilon_{mm,k} + 2\varepsilon_{mk,m}\right) \\ + \delta_{jk}\left(\varepsilon_{mm,i} + 2\varepsilon_{mi,m}\right) + \delta_{ki}\left(\varepsilon_{mm,j} + 2\varepsilon_{mj,m}\right)\big]\end{aligned} \tag{4}$$

$$\chi_{ij}^{s} = \frac{1}{2}\left(\theta_{i,j} + \theta_{j,i}\right) \tag{5}$$

$$\theta_i = \frac{1}{2}e_{ijk}u_{k,j} \tag{6}$$

where u_i, θ_i, ε_{ij}, γ_i, $\eta_{ijk}^{(1)}$ and χ_{ij}^{s} denote the components of the displacement vector $\mathbf{u}$, the rotation vector $\boldsymbol{\theta}$, the strain tensor $\boldsymbol{\varepsilon}$, the dilatation gradient vector $\boldsymbol{\gamma}$, the deviatoric stretch gradient tensor $\boldsymbol{\eta}^{(1)}$, and the symmetric rotation gradient tensor $\boldsymbol{\chi}^{s}$, respectively. Also, δ is the symbol of Kronecker delta and e_{ijk} is the permutation symbol.

Furthermore, the components of the classical stress tensor $\boldsymbol{\sigma}$ and the higher-order stress tensors $\mathbf{p}$, $\boldsymbol{\tau}^{(1)}$, and $\mathbf{m}^{s}$ defined as (Lam et al. 2003).

$$\sigma_{ij} = \lambda\varepsilon_{mm}\delta_{ij} + 2\mu\varepsilon_{ij} \tag{7}$$

$$p_i = 2\mu l_0^2\gamma_i \tag{8}$$

$$\tau_{ijk}^{(1)} = 2\mu l_1^2\eta_{ijk}^{(1)} \tag{9}$$

$$m_{ij}^{s} = 2\mu l_2^2\chi_{ij}^{s} \tag{10}$$

where l_0,l_1,l_2 are additional material length-scale parameters related to dilatation gradients, deviatoric stretch gradients, and rotation gradients, respectively. Furthermore, λ and μ are the Lamé constants defined as

$$\lambda = \frac{Ev}{(1+v)(1-2v)}, \quad \mu = \frac{E}{2(1+v)} \tag{11}$$

where E is Young's modulus and v is Poisson's ratio.

Trigonometric Shear Deformation Microbeam Model

The displacement components of an initially straight beam on the basis of trigonometric shear deformation beam theory (see Fig. 1) can be written as (Touratier 1991).

$$u_1(x,z,t) = u(x,t) - z\frac{\partial w(x,t)}{\partial x} + R(z)\phi(x,t)$$
$$u_2(x,z,t) = 0 \qquad (12)$$
$$u_3(x,z,t) = w(x,t)$$

in which

$$\phi(x,t) = \frac{\partial w(x,t)}{\partial x} - \varphi(x,t) \qquad (13)$$

where u_1, u_2 and u_3 are the x^-, y^- and z^- components of the displacement vector, and also u and w are the axial and transverse displacements, φ is the angle of rotation of the cross section about y^- axis of any point on the midplane of the beam, respectively. $R(z)$ is a function which depends on z and plays a role in determination of the transverse shear strain and stress distribution throughout the height of the beam. In order to satisfy no shear stress and strain condition at the upper ($z = -h/2$) and lower ($z = h/2$) surfaces of the beam, $R(z)$ is selected as following without need for any shear correction factors:

$$R(z) = \frac{h}{\pi}\sin\left(\frac{\pi z}{h}\right) \qquad (14)$$

It can be noted that the displacement components for EBT and TBT will be obtained by setting $R(z)$ in Eq. 12 equal to (0) and (z), respectively. With the use of Eqs. 12, 13, and 14 into Eq. 2, the nonzero strain components are obtained as

$$\varepsilon_{11} = \frac{\partial u}{\partial x} - z\frac{\partial^2 w}{\partial x^2} + R\frac{\partial \phi}{\partial x},\ \varepsilon_{13} = \frac{1}{2}S\phi \qquad (15)$$

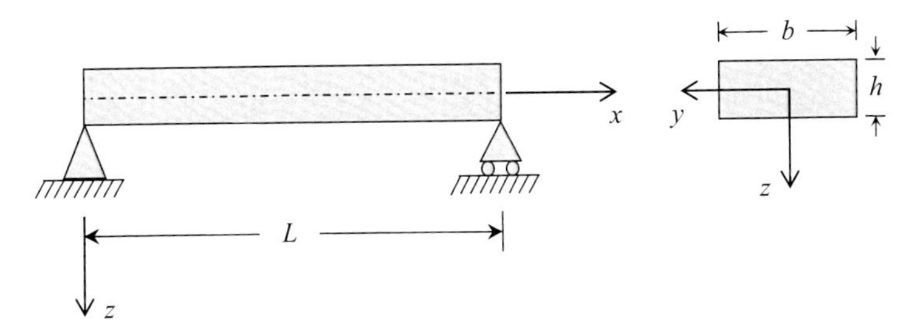

Fig. 1 Geometry, coordinate system, and cross section of a simply supported microbeam

where L is length of the microbeam, A is the area of cross section, I is the second moment of area:

$$\begin{aligned} k_1 &= l_0^2 + \tfrac{2}{5}l_1^2, k_2 = \mu A\left(\tfrac{1}{2} + \tfrac{\pi^2}{h^2}\left(\tfrac{4}{15}l_1^2 + \tfrac{1}{8}l_2^2\right)\right), \\ k_3 &= \tfrac{6}{\pi^2}EI + \mu A\left(l_0^2 + \tfrac{2}{3}l_1^2 + \tfrac{1}{8}l_2^2\right), \\ k_4 &= \tfrac{1}{\pi}\left(\tfrac{24}{\pi^2}EI + \mu A\left(4l_0^2 + \tfrac{4}{3}l_1^2 + l_2^2\right)\right), \\ k_5 &= EI + \mu A\left(2l_0^2 + \tfrac{8}{15}l_1^2 + l_2^2\right), \\ k_6 &= \tfrac{1}{5}\mu A l_1^2, k_7 = \tfrac{4}{5\pi}\mu A l_1^2 \end{aligned} \tag{27}$$

The kinetic energy of the microbeam is given by

$$T = \int_0^L \int_A \frac{1}{2}\rho\left[\left(\frac{\partial u_1}{\partial t}\right)^2 + \left(\frac{\partial u_2}{\partial t}\right)^2 + \left(\frac{\partial u_3}{\partial t}\right)^2\right] dA\, dx \tag{28}$$

where ρ is the mass density. From Eqs. 12 and 28, the first variation of the kinetic energy can be expressed as

$$\begin{aligned} \delta T = \textstyle\int_0^L \Bigg\{ & m_0\left[\frac{\partial u}{\partial t}\frac{\partial \delta u}{\partial t} + \frac{\partial w}{\partial t}\frac{\partial \delta w}{\partial t}\right] \\ & + m_2\left[\frac{\partial^2 w}{\partial x \partial t}\frac{\partial^2 \delta w}{\partial x \partial t} - \frac{24}{\pi^3}\left(\frac{\partial \phi}{\partial t}\frac{\partial^2 \delta w}{\partial x \partial t} + \frac{\partial^2 w}{\partial x \partial t}\frac{\partial \delta \phi}{\partial t}\right) + \frac{6}{\pi^2}\frac{\partial \phi}{\partial t}\frac{\partial \delta \phi}{\partial t}\right]\Bigg\} dx \end{aligned} \tag{29}$$

where (m_0, m_2) are the mass inertias as.

$$(m_0, m_2) = \rho \int_A \left(1, z^2\right) dA \tag{30}$$

The first variation of the work done by external forces can be written as

$$\begin{aligned} \delta W = \textstyle\int_0^L (f\delta u + q\delta w)\, dx + \Big[& \widehat{Q}_1\delta u + \widehat{Q}_2\delta\left(\tfrac{\partial u}{\partial x}\right) + \widehat{Q}_3\delta w \\ & + \widehat{Q}_4\delta\left(\tfrac{\partial w}{\partial x}\right) + \widehat{Q}_5\delta\left(\tfrac{\partial^2 w}{\partial x^2}\right) + \widehat{Q}_6\delta\phi + \widehat{Q}_7\delta\left(\tfrac{\partial \phi}{\partial x}\right)\Big]_0^L \end{aligned} \tag{31}$$

where $f(x, t)$ and $q(x, t)$ are the axial and transverse distributed loads, respectively. In addition, $\widehat{Q}_i$ $(i = 1, 2, \ldots, 7)$ are the specified forces or moment of forces at the end of the microbeam. After that, with the aid of Hamilton's principle as

$$0 = \int_0^T (\delta T - \delta U + \delta W)\, dt \tag{32}$$

and by substituting Eqs. 26, 29, and 31 into Eq. 32, integrating by parts, and setting the coefficients δu, δw, and $\delta\phi$ equal to zero, the governing equations of motion of the microbeam based on SBT can be obtained as (Akgöz and Civalek 2013b).

$$\delta u : -m_0\frac{\partial^2 u}{\partial t^2} + A\left(E\frac{\partial^2 u}{\partial x^2} - 2\mu k_1\frac{\partial^4 u}{\partial x^4}\right) + f = 0 \tag{33}$$

$$\begin{aligned}\delta w : &- m_0\frac{\partial^2 w}{\partial t^2} + m_2\left(\frac{\partial^4 w}{\partial x^2\partial t^2} - \frac{24}{\pi^3}\frac{\partial^3\phi}{\partial x\partial t^2}\right) - k_5\frac{\partial^4 w}{\partial x^4}\\ &+ k_4\frac{\partial^3\phi}{\partial x^3} + 2\mu I k_1\left(\frac{\partial^6 w}{\partial x^6} - \frac{24}{\pi^3}\frac{\partial^5\phi}{\partial x^5}\right) + q = 0\end{aligned} \tag{34}$$

$$\begin{aligned}\delta\phi : &\frac{24}{\pi^3}m_2\frac{\partial^3 w}{\partial x\partial t^2} - \frac{6}{\pi^2}m_2\frac{\partial^2\phi}{\partial t^2} - k_2\phi + k_3\frac{\partial^2\phi}{\partial x^2}\\ &- k_4\frac{\partial^3 w}{\partial x^3} - \frac{12}{\pi^2}\mu I k_1\left(\frac{\partial^4\phi}{\partial x^4} - \frac{4}{\pi}\frac{\partial^5 \mathrm{w}}{\partial x^5}\right) = 0\end{aligned} \tag{35}$$

and boundary conditions at $x = 0$ and $x = L$

$$\text{either } A\left(E\frac{\partial u}{\partial x} - 2\mu k_1\frac{\partial^3 u}{\partial x^3}\right) = \widehat{Q}_1 \text{ or } u = 0 \tag{36}$$

$$\text{either } 2\mu A k_1\frac{\partial^2 u}{\partial x^2} = \widehat{Q}_2 \text{ or } \frac{\partial u}{\partial x} = 0 \tag{37}$$

$$\begin{aligned}\text{either} &- k_5\frac{\partial^3 w}{\partial x^3} + k_4\frac{\partial^2\phi}{\partial x^2} + 2\mu I k_1\left(\frac{\partial^5 w}{\partial x^5} - \frac{24}{\pi^3}\frac{\partial^4\phi}{\partial x^4}\right)\\ &- m_2\left(\frac{24}{\pi^3}\frac{\partial^2\phi}{\partial t^2} - \frac{\partial^3 w}{\partial x\partial t^2}\right) = \widehat{Q}_3 \text{or } w = 0\end{aligned} \tag{38}$$

$$\text{either } k_5\frac{\partial^2 w}{\partial x^2} - k_4\frac{\partial\phi}{\partial x} - 2\mu I k_1\left(\frac{\partial^4 w}{\partial x^4} - \frac{24}{\pi^3}\frac{\partial^3\phi}{\partial x^3}\right) = \widehat{Q}_4 \text{ or } \frac{\partial w}{\partial x} = 0 \tag{39}$$

$$\text{either} - k_7\phi + 2\mu I k_1\left(\frac{\partial^3 w}{\partial x^3} - \frac{24}{\pi^3}\frac{\partial^2\phi}{\partial x^2}\right) = \widehat{Q}_5 \text{ or } \frac{\partial^2 w}{\partial x^2} = 0 \tag{40}$$

$$\text{either } (k_3 + k_6)\frac{\partial\phi}{\partial x} - (k_4 + k_7)\frac{\partial^2 w}{\partial x^2} - \frac{12}{\pi^2}\mu I k_1\left(\frac{\partial^3\phi}{\partial x^3} - \frac{4}{\pi}\frac{\partial^4 w}{\partial x^4}\right) = \widehat{Q}_6 \text{ or } \phi = 0 \tag{41}$$

$$\text{either } k_6\phi + \frac{12}{\pi^2}\mu I k_1\left(\frac{\partial^2\phi}{\partial x^2} - \frac{4}{\pi}\frac{\partial^3 w}{\partial x^3}\right) = \widehat{Q}_7 \text{ or } \frac{\partial\phi}{\partial x} = 0 \tag{42}$$

Analytical Solutions for Free Vibration Problem of Simply Supported Microbeams

Here, in order to solve free vibration problem of simply supported microbeams, the Navier solution procedure is used. The well-known geometric boundary conditions for a simply supported end can be defined as zero deflection and nonzero slope and/or rotation of the cross section as

$$w = 0, \quad \frac{\partial w}{\partial x} \neq 0, \ \phi \neq 0 \tag{43}$$

In view of Eq. 43, the left sides of Eqs. 39 and 41 must vanish. Hence, the following relations can be written by Eqs. 36, 37, 38, 39, 40, 41, 42, and 43 as

$$\frac{\partial u}{\partial x} = 0, \ \frac{\partial^2 w}{\partial x^2} = 0, \ \frac{\partial \phi}{\partial x} = 0, \ \widehat{Q}_1 = 0, \ \widehat{Q}_4 = 0, \ \widehat{Q}_6 = 0 \tag{44}$$

The following expansions of generalized displacements which include undetermined Fourier coefficients and certain trigonometric functions can be successfully employed as

$$w\left(x,t\right) = \sum_{n=1}^{\infty} W_n \sin \alpha x \ e^{i\omega_n t} \tag{45}$$

$$\phi\left(x,\mathrm{t}\right) = \sum_{n=1}^{\infty} H_n \cos \alpha x \ e^{i\omega_n t} \tag{46}$$

where W_n and H_n are the undetermined Fourier coefficients, ω_n is natural frequency, and $\alpha = \frac{n\pi}{L}$. This means that Eqs. 45 and 46 must satisfy the corresponding boundary conditions. Substituting Eqs. 45 and 46 into Eqs. 35 and 36 as the governing equations for free vibration, the following equation is obtained as

$$\left(\begin{bmatrix} K_{11} & K_{12} \\ K_{21} & K_{22} \end{bmatrix} - \omega^2 \begin{bmatrix} M_{11} & M_{12} \\ M_{21} & M_{22} \end{bmatrix} \right) \begin{Bmatrix} H_n \\ W_n \end{Bmatrix} = \begin{Bmatrix} 0 \\ 0 \end{Bmatrix} \tag{47}$$

where.

$$\begin{aligned} &K_{11} = k_2 + \alpha^2 k_3 + \alpha^4 \frac{12}{\pi^2} \mu I k_1, \quad K_{12} = K_{21} = -\alpha^3 \left(k_4 + \alpha^2 \frac{48}{\pi^3} \mu I k_1 \right), \\ &K_{22} = \alpha^4 \left(k_5 + \alpha^2 2 \mu I k_1 \right) \\ &M_{11} = \frac{6}{\pi^2} m_2, M_{12} = M_{21} = -\alpha \frac{24}{\pi^3} m_2, M_{22} = \left(m_0 + \alpha^2 m_2 \right) \end{aligned} \tag{48}$$

For a nontrivial solution, the determinant of coefficient matrix must be vanished and the characteristic equation can be reached by providing this condition. The eigenvalues are obtained by solving the characteristic equation. It can be noted that the smallest root of the characteristic equation gives the first natural (fundamental) frequency.

Numerical Results and Discussion

In this section, free vibration problem of a simply supported microbeam is analytically solved with the Navier-type solution based on trigonometric shear deformation beam theory in conjunction with modified strain gradient theory. For illustration purpose, the microbeam is taken to be made of epoxy with the following material properties: Young's modulus $E = 1.44$ GPa, Poisson's ratio $v = 0.38$, the mass density $\rho = 1{,}220\ \mathrm{kg/m^3}$ and the material length-scale parameter $l = 11.01\ \mu m$ (Kahrobaiyan et al. 2013). The microbeam has a rectangular cross section, and the width-to-thickness ratio is taken to be constant as $b/h = 2$, while the length-to-thickness ratio is taken several values as $L/h = 5{\sim}80$. All material length-scale parameters are considered to be equal to each other as $l_0 = l_1 = l_2 = l$.

As stated before, Timoshenko beam theory (TBT) needs a shear correction factor to take into consideration the nonuniformity of transverse shear strain and stress throughout the beam thickness. For rectangular cross-section beams, the most commonly used shear correction factors can be defined as $k_s = 5/6$ (used here) and $k_s = (5 + 5v)/(6 + 5v)$. The classical results evaluated by TBT and other shear deformation beam theories such as third-order (parabolic), trigonometric (sinusoidal), hyperbolic, and exponential shear deformation beam theories are in good agreement. However, this agreement may decrease for the results of higher-order continuum theories, and this situation can be seen from the previous works (Akgöz and Civalek 2013b; Şimşek and Reddy 2013a, b). Consequently, a new modified shear correction factor (k_s^*) is used for Timoshenko microbeam model (TBT*)-based MSGT as follows (Akgöz and Civalek 2014a):

$$k_s^* = k_s k_{ac}^{MSGT} \tag{49}$$

where

$$k_{ac}^{MSGT} = 15\left(\frac{l_0+l_1+l_2}{3}\right)^a \Big/ h^a \qquad a = 3\left(h \Big/ \left(\frac{l_0+l_1+l_2}{3}\right)\right)^{0.08} - 0.45 \tag{50}$$

It can be noted that k_s^* will be equal to k_s by setting material length-scale parameters equal to zero in Eq. 50. In order to demonstrate the accuracy and validity of the present analysis, some illustrative examples are comparatively given with other beam theories.

Table 1 Dimensionless fundamental frequencies ($\overline{\omega}_1 = \omega_1 L^2 \sqrt{m_0/EI}$)

l/h	Beam theory	$L = 8\ h$		$L = 40\ h$		$L = 80\ h$	
		CT	MSGT	CT	MSGT	CT	MSGT
0	EBT	9.8696	9.8696	9.8696	9.8696	9.8696	9.8696
	TBT	9.6094	9.6094	9.8587	9.8587	9.8669	9.8669
	TBT*	9.6094	9.6094	9.8587	9.8587	9.8669	9.8669
	SBT	9.6098	9.6098	9.8587	9.8587	9.8669	9.8669
0.5	EBT	9.8696	21.8020	9.8696	21.7179	9.8696	21.7153
	TBT	9.6094	19.7861	9.8587	21.6223	9.8669	21.6913
	TBT*	9.6094	21.2380	9.8587	21.6944	9.8669	21.7094
	SBT	9.6098	21.2186	9.8587	21.6933	9.8669	21.7091
1	EBT	9.8696	40.1133	9.8696	39.9305	9.8696	39.9248
	TBT	9.6094	31.3909	9.8587	39.3696	9.8669	39.7817
	TBT*	9.6094	38.9701	9.8587	39.8826	9.8669	39.9128
	SBT	9.6098	39.0201	9.8587	39.8843	9.8669	39.9132

*Timoshenko beam model with the new shear correction factor

Table 2 Dimensionless second natural frequencies ($\overline{\omega}_2 = \omega_2 L^2 \sqrt{m_0/EI}$)

l/h	Beam theory	$L = 8\ h$		$L = 40\ h$		$L = 80\ h$	
		CT	MSGT	CT	MSGT	CT	MSGT
0	EBT	39.4784	39.4784	39.4784	39.4784	39.4784	39.4784
	TBT	35.8237	35.8237	39.3048	39.3048	39.4348	39.4348
	TBT*	35.8237	35.8237	39.3048	39.3048	39.4348	39.4348
	SBT	35.8329	35.8329	39.3050	39.3050	39.4348	39.4348
0.5	EBT	39.4784	88.2502	39.4784	86.9138	39.4784	86.8718
	TBT	35.8237	66.0744	39.3048	85.4193	39.4348	86.4892
	TBT*	35.8237	80.2054	39.3048	86.5394	39.4348	86.7776
	SBT	35.8329	80.1416	39.3050	86.5229	39.4348	86.7733
1	EBT	39.4784	162.7167	39.4784	159.8137	39.4784	159.7222
	TBT	35.8237	95.8638	39.3048	151.4933	39.4348	157.4783
	TBT*	35.8237	146.5687	39.3048	159.052	39.4348	159.5306
	SBT	35.8329	147.5984	39.3050	159.0795	39.4348	159.5373

*Timoshenko beam model with the new shear correction factor

Dimensionless first three natural frequencies for various values of l/h and slenderness ratios corresponding to different beam theories are tabulated in Tables 1, 2, and 3, respectively. It can be clearly observed from the tables that the dimensionless natural frequencies predicted by both CT and TBT are lower than the other ones, while those obtained by both MSGT and EBT are larger than the other ones. Also, an increase in l/h leads to an increment in the difference between dimensionless natural frequencies corresponding to classical and nonclassical models, and also this difference becomes more prominent for higher modes. On the other hand,

Table 3 Dimensionless third natural frequencies $(\overline{\omega}_3 = \omega_3 L^2 \sqrt{m_0/EI})$

l/h	Beam theory	L = 8 h		L = 40 h		L = 80 h	
		CT	MSGT	CT	MSGT	CT	MSGT
0	EBT	88.8264	88.8264	88.8264	88.8264	88.8264	88.8264
	TBT	73.2989	73.2989	87.9565	87.9565	88.6060	88.6060
	TBT*	73.2989	73.2989	87.9565	87.9565	88.6060	88.6060
	SBT	73.3581	73.3581	87.9576	87.9576	88.6062	88.6062
0.5	EBT	88.8264	202.4110	88.8264	195.7139	88.8264	195.5009
	TBT	73.2989	126.2815	87.9565	188.4270	88.6060	193.5834
	TBT*	73.2989	167.4196	87.9565	193.8353	88.6060	195.0255
	SBT	73.3581	168.1073	87.9576	193.7568	88.6062	195.0041
1	EBT	88.8264	374.4469	88.8264	359.9240	88.8264	359.4607
	TBT	73.2989	186.4415	87.9565	322.3023	88.6060	348.4635
	TBT*	73.2989	304.9236	87.9565	356.1056	88.6060	358.4933
	SBT	73.3581	310.7247	87.9576	356.2506	88.6062	358.5275

*Timoshenko beam model with the new shear correction factor

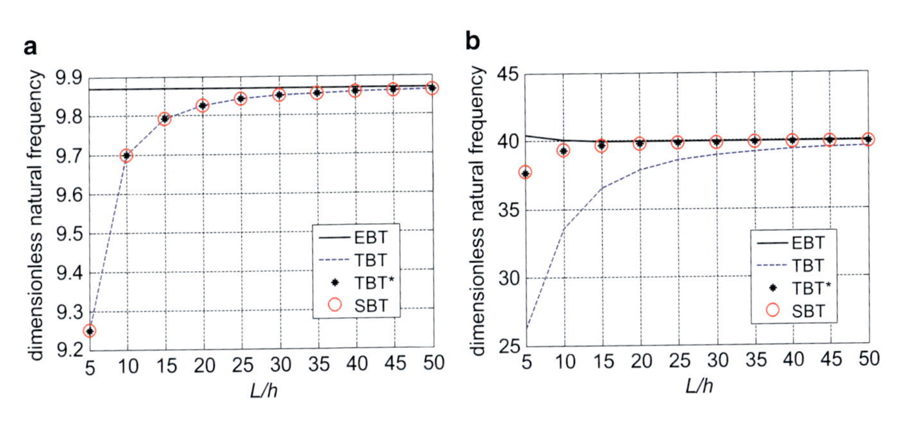

Fig. 2 Variations of the dimensionless natural frequency versus slenderness ratio (first mode). (**a**) CT (**b**) MSGT

difference between the results corresponding to EBT and shear deformation beam theories (TBT, TBT*, and SBT) is more significant for short beams. This situation can be interpreted as the effect of shear deformation is minor for slender beams with a large slenderness ratio. In addition, it can be clearly seen from the tables that the natural frequencies predicted by SBT and TBT* are in good agreement, while the divergence between the natural frequencies of SBT and TBT is considerable especially for bigger values of *l*/*h*.

Variations of the dimensionless first three natural frequencies of the simply supported microbeam with respect to the slenderness ratio corresponding to different beam models are depicted in Figs. 2, 3, and 4, respectively. It is observed that an

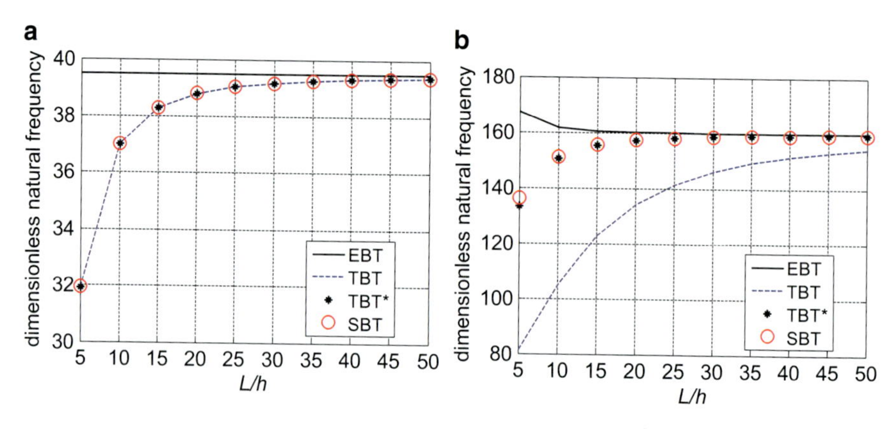

Fig. 3 Variations of the dimensionless natural frequency versus slenderness ratio (second mode). (**a**) CT (**b**) MSGT

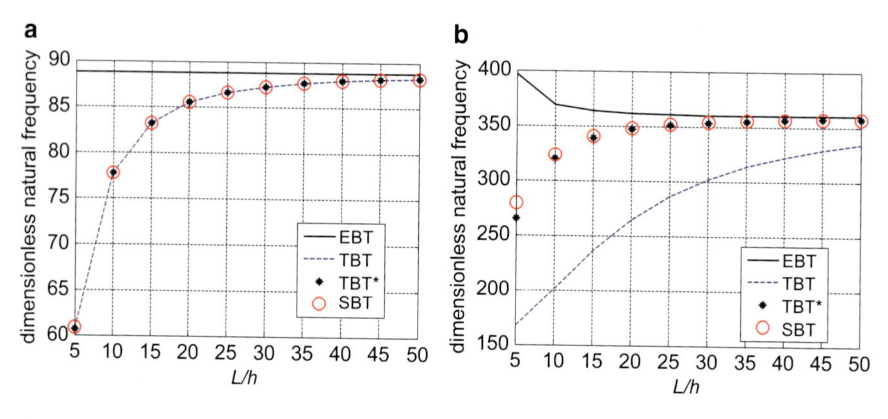

Fig. 4 Variations of the dimensionless natural frequency versus slenderness ratio (third mode). (**a**) CT (**b**) MSGT

increase in slenderness ratio leads to a decrement on effects of shear deformation, and differences between the dimensionless natural frequencies based on EBT, TBT, TBT*, and SBT are diminishing for $L/h \geq 50$. Moreover, it can be concluded that the dimensionless natural frequencies evaluated by TBT, TBT*, and SBT are nearly equal to each other for CT, but the difference between TBT and SBT is more considerable in the higher-order models for lower slenderness ratios and higher modes.

Influences of h/l ratio on the first three dimensionless natural frequencies for $L = 7h$ are illustrated in Figs. 5, 6, and 7, respectively. These figures reveal that natural frequencies based on MSGT are always bigger than CT. Also, it is found that the effects of shear deformation and small size are more considerable for smaller values of h/l and higher modes.

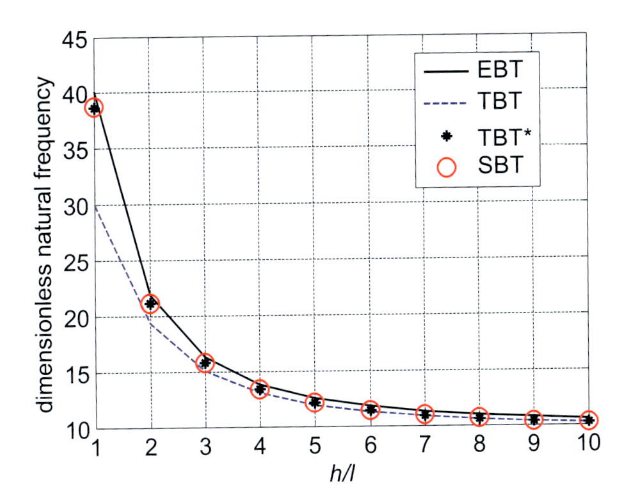

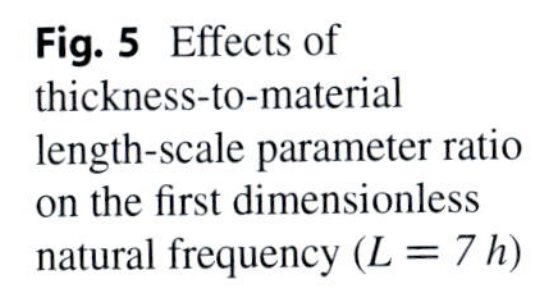
Fig. 5 Effects of thickness-to-material length-scale parameter ratio on the first dimensionless natural frequency ($L = 7\,h$)

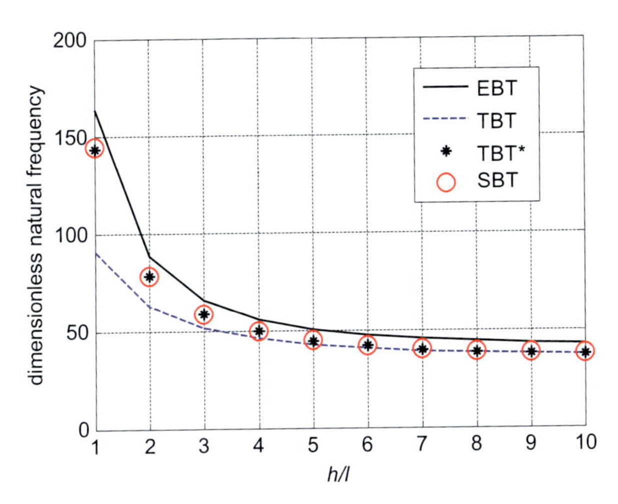

Fig. 6 Effects of thickness-to-material length-scale parameter ratio on the second dimensionless natural frequency ($L = 7\,h$)

Conclusion

In this study, a size-dependent sinusoidal shear deformation beam model in conjunction with modified strain gradient elasticity theory (MSGT) is developed. The model captures both the microstructural and shear deformation effects without any shear correction factors. The governing differential equations and corresponding boundary conditions are derived by using Hamilton's principle. The free vibration behavior of simply supported microbeams is investigated. Analytical solutions for the first three natural frequencies are presented by the Navier solution technique. The results are compared with other beam theories for the validation of the

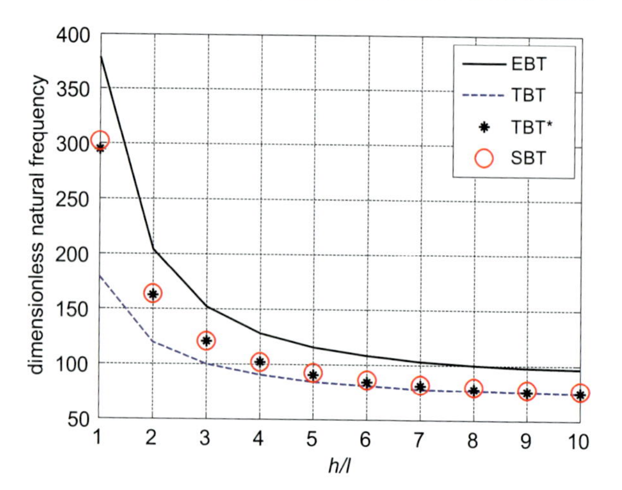

Fig. 7 Effects of thickness-to-material length-scale parameter ratio on the third dimensionless natural frequency ($L = 7\,h$)

present model. A detailed parametric study is carried out to show the influences of thickness-to-material length-scale parameter ratio, slenderness ratio, and shear deformation on the free vibration response of simply supported microbeams. The obtained results can be summarized as:

- Microbeams based on MSGT are stiffer than based on the classical theory.
- The natural frequencies obtained by both MSGT and EBT are always greater than those predicted by the other considered beam models and theories.
- The difference between the natural frequencies decreases as the thickness-to-material length-scale parameter ratio increases.
- Effect of shear deformation becomes more considerable for both smaller slenderness ratios and higher modes.
- Use of modified shear correction factors is more suitable for Timoshenko microbeam models based on higher-order continuum theories.

Acknowledgments This study has been supported by The Scientific and Technological Research Council of Turkey (TÜBİTAK) with Project No: 112M879. This support is gratefully acknowledged.

References

B. Akgöz, Ö. Civalek, Arch. Appl. Mech. **82**, 423 (2012)
B. Akgöz, Ö. Civalek, Acta Mech. **224**, 2185 (2013a)
B. Akgöz, Ö. Civalek, Int. J. Eng. Sci. **70**, 1 (2013b)
B. Akgöz, Ö. Civalek, Compos. Struct. **112**, 214 (2014a)
B. Akgöz, Ö. Civalek, Int. J. Mech. Sci. **81**, 88 (2014b)
B. Akgöz, Ö. Civalek, Int. J. Eng. Sci. **85**, 90 (2014c)
B. Akgöz, Ö. Civalek, Int. J. Mech. Sci. **99**, 10 (2015)
B.S. Altan, H. Evensen, E.C. Aifantis, Mech. Res. Commun. **23**, 35 (1996)

R. Ansari, R. Gholami, S. Sahmani, Compos. Struct. **94**, 221 (2011)
R. Artan, R.C. Batra, Acta Mech. **223**, 2393 (2012)
M. Aydogdu, Compos. Struct. **89**, 94 (2009a)
M. Aydogdu, Phys. E. **41**, 1651 (2009b)
A.C. Eringen, Z. Angew. Math. Phys. **18**, 12 (1967)
A.C. Eringen, Int. J. Eng. Sci. **10**, 1 (1972)
A.C. Eringen, J. Appl. Phys. **54**, 4703 (1983)
N.A. Fleck, J.W. Hutchinson, J. Mech. Phys. Solids **41**, 1825 (1993)
N.A. Fleck, J.W. Hutchinson, J. Mech. Phys. Solids **49**, 2245 (2001)
E.S. Hung, S.D. Senturia, J. Microelectromech. Syst. **8**, 497 (1999)
M.H. Kahrobaiyan, M. Asghari, M.T. Ahmadian, Finite Elem. Anal. Des. **68**, 63 (2013)
M.H. Kahrobaiyan, M. Rahaeifard, S.A. Tajalli, M.T. Ahmadian, Int. J. Eng. Sci. **52**, 65 (2012)
M. Karama, K.S. Afaq, S. Mistou, Int. J. Solids Struct. **40**, 1525 (2003)
W.T. Koiter, Proc. K. Ned. Akad. Wet. B **67**, 17 (1964)
S. Kong, S. Zhou, Z. Nie, K. Wang, Int. J. Eng. Sci. **47**, 487 (2009)
D.C.C. Lam, F. Yang, A.C.M. Chong, J. Wang, P. Tong, J. Mech. Phys. Solids **51**, 1477 (2003)
M. Levinson, A new rectangular beam theory. J. Sound Vib. **74**, 81 (1981)
P. Li, Y. Fang, J. Micromech. Microeng. **20**, 035005 (2010)
A.W. McFarland, J.S. Colton, J. Micromech. Microeng. **15**, 1060 (2005)
R.D. Mindlin, Int. J. Solids Struct. **1**, 417 (1965)
R.D. Mindlin, H.F. Tiersten, Arch. Ration. Mech. Anal. **11**, 415 (1962)
W.J. Poole, M.F. Ashby, N.A. Fleck, Scr. Mater. **34**, 559 (1996)
J.N. Reddy, J. Appl. Mech. **51**, 745 (1984)
J.N. Reddy, J. Mech. Phys. Solids **59**, 2382 (2011)
M. Salamat-talab, A. Nateghi, J. Torabi, Int. J. Mech. Sci. **57**, 63 (2012)
M. Şimşek, J.N. Reddy, Int. J. Eng. Sci. **64**, 37 (2013a)
M. Şimşek, J.N. Reddy, Compos. Struct. **101, 47** (2013b)
K.P. Soldatos, Acta Mech. **94**, 195 (1992)
H.T. Thai, T.P. Vo, Int. J. Eng. Sci. **54**, 58 (2012)
H.T. Thai, T.P. Vo, Compos. Struct. **96**, 376 (2013)
S.P. Timoshenko, Philos. Mag. **41**, 744 (1921)
A. Torii, M. Sasaki, K. Hane, S. Okuma, Sensors Actuators A Phys. **44**, 153 (1994)
R.A. Toupin, Arch. Ration. Mech. Anal. **17**, 85 (1964)
M. Touratier, Int. J. Eng. Sci. **29**, 901 (1991)
I. Vardoulakis, J. Sulem, *Bifurcation Analysis in Geomechanics* (Blackie/Chapman and Hall, London, 1995)
B. Wang, J. Zhao, S. Zhou, Eur. J. Mech. A/Solids **29**, 591 (2010)
Z.Y. Wu, H. Yang, X.X. Li, Y.L. Wang, J. Micromech. Microeng. **20**, 115014 (2010)
W.C. Xie, H.P. Lee, S.P. Lim, Nonlinear Dyn. **31**, 243 (2003)
M.I. Younis, E.M. Abdel-Rahman, A.H. Nayfeh, J. Microelectromech. Syst. **12**, 672 (2003)
B. Zhang, Y. He, D. Liu, Z. Gan, L. Shen, Finite Elem. Anal. Des. **79**, 22 (2014a)
B. Zhang, Y. He, D. Liu, Z. Gan, L. Shen, Eur J Mech A/Solids **47**, 211 (2014b)
J.D. Zook, D.W. Burns, H. Guckel, J.J. Sniegowski, R.L. Engelstad, Z. Feng, Sensors Actuators A Phys. **35**, 51 (1992)

Axial Vibration of Strain Gradient Micro-rods

33

Ömer Civalek, Bekir Akgöz, and Babür Deliktaş

Contents

Abstract

In this chapter, size-dependent axial vibration response of micro-sized rods is investigated on the basis of modified strain gradient elasticity theory. On the contrary to the classical rod model, the developed nonclassical micro-rod model includes additional material length scale parameters and can capture the size effect. If the additional material length scale parameters are equal to zero, the current model reduces to the classical one. The equation of motion together with initial conditions, classical and nonclassical corresponding boundary conditions, for micro-rods is derived by implementing Hamilton's principle. The resulting higher-order equation is analytically solved for clamped-free and clamped-clamped boundary conditions. Finally, some illustrative examples are presented

Ö. Civalek (✉) · B. Akgöz
Civil Engineering Department, Division of Mechanics, Akdeniz University, Antalya, Turkey
e-mail: civalek@yahoo.com; bekirakgoz@akdeniz.edu.tr

B. Deliktaş
Faculty of Engineering-Architecture, Department of Civil Engineering, Uludag University, Bursa, Görükle, Turkey
e-mail: bdeliktas@uludag.edu.tr

G. Z. Voyiadjis (ed.), *Handbook of Nonlocal Continuum Mechanics for Materials and Structures*, https://doi.org/10.1007/978-3-319-58729-5_7

to indicate the influences of the additional material length scale parameters, size dependency, boundary conditions, and mode numbers on the natural frequencies. It is found that size effect is more significant when the micro-rod diameter is closer to the additional material length scale parameter. In addition, it is observed that the difference between natural frequencies evaluated by the present and classical models becomes more considerable for both lower values of slenderness ratio and higher modes.

Keywords
Micro-rod · Size dependency · Axial vibration · Small-scale effect · Modified strain gradient theory · Length scale parameter · Higher-order rod model · Natural frequency

Introduction

Nowadays, due to the rapid advances in technologies, micro- and nano-sized mechanical systems like microbeams, microbars, biosensors, nanowires, atomic force microscope, nanotubes, micro actuators, nano probes, micro- and nano-electromechanical systems (MEMS and NEMS), and ultra-thin films have been widely used in modern applications such as mechanical, biomedical, chemical, and biological applications (Fu et al. 2003; Li et al. 2003; Najar et al. 2005; Faris and Nayfeh 2007; Moser and Gijs 2007; Kahrobaiyan et al. 2011a). The insight of the mechanical behavior characteristics of micro- and nanostructures is very important for the optimum design of such structures. Bending, buckling, and vibration responses of these structures can be investigated by experimental studies and computer simulation techniques at atomistic levels. The effects of size dependency on the deformation behaviors of the aforementioned structures have been experimentally observed (Fleck et al. 1994; Chong and Lam 1999; Senturia 2001; Haque and Saif 2003; Lam et al. 2003; Lou et al. 2006).

Due to the difficulty and computationally expensiveness of experimentation and simulation techniques at atomistic levels (e.g., molecular dynamic simulation), many scientists and researchers tended the continuum mechanics modeling as an alternative. However, the classical continuum mechanics approaches do not the ability for interpretation and explanation of the microstructural dependency of small-sized structures due to the lack of any additional (intrinsic) material length scale parameters. Then, higher-order (nonclassical) continuum theories, which include at least one additional material length scale parameter in addition to classical ones, have been proposed to predict the microstructure-dependent behavior of these small-scale structures.

Higher-order continuum theories include Cosserat elasticity by Cosserat and Cosserat (1909), strain gradient elasticity of Mindlin (1964,1965), micropolar theory (Eringen and Suhubi 1964), nonlocal elasticity (Eringen 1983), couple stress theory by Mindlin and Tiersten (1962), Toupin (1962), and Koiter (1964), strain gradient theory (Fleck and Hutchinson 1993), simple gradient elasticity with surface energy (Vardoulakis and Sulem 1995; Altan et al. 1996; Altan and Aifantis 1997), modified couple stress (Yang et al. 2002), and modified strain gradient theories (Lam et al. 2003). Some earlier studies based on these theories available in the literature have been briefly given here.

Peddieson et al. (2003) formulated a nonlocal Bernoulli-Euler beam model with nonlocal elasticity theory. Also, Reddy (2007a) investigated bending, buckling, and free vibration analysis of nonlocal beams for different beam theories. Wang et al. (2008) studied bending problem of micro- and nano-sized beams based on nonlocal Timoshenko beam theory. Aydogdu (2009) investigated the small-scale effect on longitudinal vibration of a nanorod on the basis of Eringen's nonlocal elasticity theory.

The classical couple stress theory has been used to investigate the bending analysis of a circular cylinder by Anthoine (2000). Tsepoura et al. (2002) investigated static and dynamic analysis of bars based on simple gradient elasticity theory with surface energy. Papargyri-Beskou et al. (2003a) and Lazopoulos (2012) observed dynamic analysis of gradient elastic beams. Bending and buckling analysis of gradient elastic beams is studied on the basis of Bernoulli-Euler beam model by Papargyri-Beskou et al. (2003b) and Lazopoulos and Lazopoulos (2010).

Modified couple stress theory is a higher-order continuum theory that has been elaborated by Yang et al. (2002) which contains the symmetric rotation gradient tensor and one additional material length scale parameter in addition to the conventional (classical) strain tensor. Park and Gao (2006) and Ma et al. (2008) developed new size-dependent Bernoulli-Euler and Timoshenko beam models, respectively. Kong et al. (2008) investigated free vibration analysis of the Bernoulli-Euler microbeam model based on this theory.

The modified strain gradient elasticity theory is one of the popular higher-order continuum theories, which was proposed by Lam et al. (2003), that includes dilatation and deviatoric stretch gradient tensors besides the symmetric rotation gradient and classical strain tensor and also three additional material length scale parameters for linear elastic isotropic materials. Kong et al. (2009), Akgöz and Civalek (2011), and Wang et al. (2010) used modified strain gradient elasticity for static and dynamic analyses of microbeams on the basis of Bernoulli-Euler and Timoshenko beam models, respectively. Furthermore, static torsion and torsional free vibration analyses of microbars based on this theory were presented by Kahrobaiyan et al. (2011b) and Narendar et al. (2012). Recently, longitudinal vibration responses of microbars were investigated by Akgöz and Civalek (2013, 2014), Kahrobaiyan et al. (2013), and Güven (2014).

In this chapter, size-dependent axial vibration response of micro-sized rods is investigated on the basis of modified strain gradient elasticity theory. The equation of motion together with initial conditions, classical and nonclassical corresponding boundary conditions, for micro-rods is derived with the aid of Hamilton's principle. The resulting higher-order equation is solved for two different boundary conditions as clamped-free and clamped-clamped. Influences of micro-rod characteristic lengths, slenderness ratio, additional material length scale parameters, and mode number on the vibrational response of the size-dependent micro-rod are investigated.

Formulation for Modified Strain Gradient Theory

The strain energy U in a linear elastic isotropic material occupying volume V based on the modified strain gradient elasticity theory can be written by Lam et al. (2003) and Kong et al. (2009).

$$U = \frac{1}{2}\int_V \left(\sigma_{ij}\varepsilon_{ij} + p_i\gamma_i + \tau_{ijk}^{(1)}\eta_{ijk}^{(1)} + m_{ij}^s\chi_{ij}^s\right) dv \tag{1}$$

$$\varepsilon_{ij} = \frac{1}{2}\left(u_{i,j} + u_{j,i}\right) \tag{2}$$

$$\gamma_i = \varepsilon_{mm,i} \tag{3}$$

$$\begin{aligned}\eta_{ijk}^{(1)} =& \frac{1}{3}\left(\varepsilon_{jk,i} + \varepsilon_{ki,j} + \varepsilon_{ij,k}\right) - \frac{1}{15}\left[\delta_{ij}\left(\varepsilon_{mm,k} + 2\varepsilon_{mk,m}\right)\right.\\ &\left. + \delta_{jk}\left(\varepsilon_{mm,i} + 2\varepsilon_{mi,m}\right) + \delta_{ki}\left(\varepsilon_{mm,j} + 2\varepsilon_{mj,m}\right)\right]\end{aligned} \tag{4}$$

$$\chi_{ij}^s = \frac{1}{2}\left(\theta_{i,j} + \theta_{j,i}\right) \tag{5}$$

$$\theta_i = \frac{1}{2}e_{ijk}u_{k,j} \tag{6}$$

where u_i, ε_{ij}, γ_i, $\eta_{ijk}^{(1)}$, χ_{ij}^s, and θ_i are the components of the displacement vector, the strain tensor, the dilatation gradient vector, the deviatoric stretch gradient tensor, the symmetric rotation gradient tensor, and the rotation vector, respectively. Also δ_{ij} is the Kronecker delta, and e_{ijk} is the permutation symbol. The stress measures σ_{ij}, p_i, $\tau_{ijk}^{(1)}$, and m_{ij}^s are the components of classical and higher-order stresses defined as Lam et al. (2003).

$$\sigma_{ij} = \lambda\delta_{ij}\varepsilon_{mm} + 2\mu\varepsilon_{ij} \tag{7}$$

$$p_i = 2\mu l_0^2\gamma_i \tag{8}$$

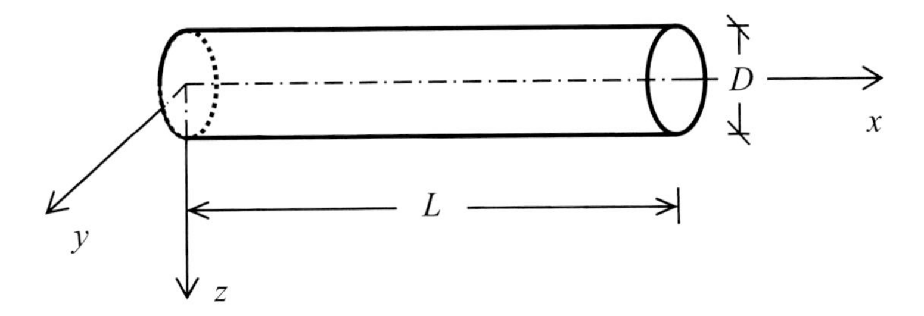

Fig. 1 Geometry and coordinate system of a straight micro-rod

$$\tau_{ijk}^{(1)} = 2\mu l_1^2 \eta_{ijk}^{(1)} \tag{9}$$

$$m_{ij}^s = 2\mu l_2^2 \chi_{ij}^s \tag{10}$$

where λ and μ are the well-known Lamé constants and l_0, l_1, l_2 are additional material length scale parameters which represent the size dependency and related to dilatation gradients, deviatoric stretch gradients, and rotation gradients, respectively.

Microstructure-Dependent Rod Model

It is considered that the case of axial vibration of a straight thin micro-rod (see Fig. 1). Due to axial vibrations take place in $x-$ direction, the deformation of the cross section in $y-$ and $z-$ directions is assumed to be negligible by a simple theory for axial vibration of thin rods. The components of displacement vector can be expressed as Rao (2007).

$$u_1 = u(x,t), u_2 = 0, u_3 = 0 \tag{11}$$

where u_1, u_2, and u_3 are the components of displacement vector in $x-$, $y-$, and $z-$ directions, respectively.

In view of Eqs. (2) and (11), the non-zero strain component of the micro-rod is

$$\varepsilon_{xx} = \frac{\partial u}{\partial x} \tag{12}$$

Use of Eqs. (12) into (3), the non-zero component of dilatation gradient vector is obtained as

$$\gamma_x = \frac{\partial^2 u}{\partial x^2} \tag{13}$$

From Eqs. (2) and (4), non-zero components of deviatoric stretch gradient tensor are achieved as

$$\eta_{xxx}^{(1)} = \frac{2}{5}\frac{\partial^2 u}{\partial x^2}, \eta_{xyy}^{(1)} = \eta_{xzz}^{(1)} = \eta_{yxy}^{(1)} = \eta_{yyx}^{(1)} = \eta_{zxz}^{(1)} = \eta_{zzx}^{(1)} = -\frac{1}{5}\frac{\partial^2 u}{\partial x^2} \tag{14}$$

Furthermore, all components of rotation vector and so symmetric rotation gradient tensor are equal to zero as

$$\theta_i = 0, (i = x, y, z) \tag{15}$$

$$\chi_{ij}^s = 0, (i, j = x, y, z) \tag{16}$$

The non-zero stress σ_{ij} can be obtained by neglecting Poisson's effect in Eq. (7) as

$$\sigma_{xx} = E\frac{\partial u}{\partial x} \tag{17}$$

where E is the elastic modulus. By inserting Eqs. (13) into (8) and Eqs. (14) into (9), the non-zero components of higher-order stresses p_i and $\tau_{ijk}^{(1)}$ can be expressed as

$$p_x = 2\mu l_0^2\frac{\partial^2 u}{\partial x^2} \tag{18}$$

$$\tau_{xxx}^{(1)} = \frac{4}{5}\mu l_1^2\frac{\partial^2 u}{\partial x^2}, \tau_{xyy}^{(1)} = \tau_{xzz}^{(1)} = \tau_{yxy}^{(1)} = \tau_{yyx}^{(1)} = \tau_{zxz}^{(1)} = \tau_{zzx}^{(1)} = -\frac{2}{5}\mu l_1^2\frac{\partial^2 u}{\partial x^2}, \tag{19}$$

Substituting above equations into Eq. (1), the strain energy U can be rewritten as

$$U = \frac{1}{2}\int_0^L \left\{ EA(u')^2 + \left(2\mu A l_0^2 + \frac{4}{5}\mu A l_1^2\right)(u'')^2 \right\} dx \tag{20}$$

where A is the cross-sectional area of the micro-rod and

$$u' = \frac{\partial u}{\partial x}, u'' = \frac{\partial^2 u}{\partial x^2} \tag{21}$$

The first variation of strain energy U in Eq. (20) on the time interval $[t_0, t_1]$ can be calculated as following expression

$$\delta U = \int_{t_0}^{t_1}\int_0^L \{Bu^{(4)} - EAu''\}\,\delta u dx dt + \int_{t_0}^{t_1}\Big\{\big[(EAu' - Bu''')\,\delta u\big]_0^L + \big[Bu''\delta u'\big]_0^L\Big\}\,dt \tag{22}$$

where

$$u''' = \frac{\partial^3 u}{\partial x^3}, u^{(4)} = \frac{\partial^4 u}{\partial x^4}, B = 2\mu A\left(l_0^2 + \frac{2}{5}l_1^2\right) \tag{23}$$

On the other hand, the first variation of the work done by external force q , axial force N, and higher-order axial force N^h on the time interval $[t_0, t_1]$ takes the following form

$$\delta W = \int_{t_0}^{t_1}\int_0^L q\delta u dx dt + \int_{t_0}^{t_1}\Big\{[N\delta u]_0^L + \big[N^h\delta u'\big]_0^L\Big\}\,dt \tag{24}$$

Also, the first variation of kinetic energy K of the micro-rod on the time interval $[t_0, t_1]$ reads as

$$\delta K = -\int_{t_0}^{t_1}\int_0^L m\ddot{u}\delta u dx dt + \int_0^L [m\dot{u}\delta u]_{t_0}^{t_1} dx \tag{25}$$

where m is the mass per unit length and

$$\dot{u} = \frac{\partial u}{\partial t}, \ddot{u} = \frac{\partial^2 u}{\partial t^2} \tag{26}$$

The following relation is written by employing Hamilton's principle with Eqs. (22, 24, and 25)

$$\delta\left\{\int_{t_0}^{t_1}[K - (U - W)]\,dt\right\} = 0 \tag{27}$$

$$\int_{t_0}^{t_1}\int_0^L \big[EAu'' - Bu^{(4)} - m\ddot{u} + q\big]\,\delta u dx dt + \int_{t_0}^{t_1}\big[\{N - (EAu' - Bu''')\}\,\delta u\big]_0^L dt + \int_{t_0}^{t_1}\big[\{N^h - Bu''\}\,\delta u'\big]_0^L dt + \int_0^L [m\dot{u}\delta u]_{t_0}^{t_1} dx = 0 \tag{28}$$

According to the fundamental lemma of calculus of variation (Reddy 2007b), the equation of motion for the micro-rod reads as

$$EAu'' - Bu^{(4)} + q = m\ddot{u} \tag{29}$$

Also, initial conditions and boundary conditions satisfy the following equations, respectively, as

$$\dot{u}(x,t_1)\,\delta u(x,t_1) - \dot{u}(x,t_0)\,\delta u(x,t_0) = 0 \tag{30}$$

$$\begin{aligned} &\left[N(L,t) - \left(EAu'(L,t) - Bu'''\left(L,t\right)\right)\right]\delta u(L,t) \\ &\quad - \left[N(0,t) - \left(EAu'(0,t) - Bu'''\left(0,t\right)\right)\right]\delta u(0,t) = 0 \end{aligned} \tag{31}$$

$$\left[N^h(L,t) - Bu''\left(L,t\right)\right]\delta u'(L,t) - \left[N^h(0,t) - Bu''\left(0,t\right)\right]\delta u'(0,t) = 0 \tag{32}$$

Solution of Axial Vibration Problem

u can be expressed as the following form by employing separation of variables method

$$u(x,t) = U(x)e^{i\omega t} \tag{33}$$

By substituting above equation into Eq. (29) in the absence of q yields

$$B\frac{d^4U}{dx^4} - EA\frac{d^2U}{dx^2} - \omega^2 mU = 0 \tag{34}$$

Analytical solution of Eq. (34) can be obtained as follows

$$U(x) = D_1 \sin\alpha x + D_2 \cos\alpha x + D_3 \sinh\beta x + D_4 \cosh\beta x \tag{35}$$

where

$$\alpha = \left(\frac{-EA + \sqrt{(EA)^2 + 4Bm\omega^2}}{2B}\right)^{1/2}, \beta = \left(\frac{EA + \sqrt{(EA)^2 + 4Bm\omega^2}}{2B}\right)^{1/2} \tag{36}$$

and D_i ($i = 1, 2, 3, 4$) are constants which can be determined by corresponding boundary conditions. For a micro-rod that both ends are clamped, classical and nonclassical boundary conditions are

$$U(0) = 0 \text{ and } U(L) = 0 \tag{37}$$

$$BU''(0) = 0 \text{ and } BU''(L) = 0 \tag{38}$$

By using of above boundary conditions in Eq. (35), solution can be written in a matrix form as

$$\begin{bmatrix} 0 & 1 & 0 & 1 \\ \sin\alpha L & \cos\alpha L & \sinh\beta L & \cosh\beta L \\ 0 & -\alpha^2 & 0 & \beta^2 \\ -\alpha^2\sin\alpha L & -\alpha^2\cos\alpha L & \beta^2\sinh\beta L & \beta^2\cosh\beta L \end{bmatrix} \begin{bmatrix} D_1 \\ D_2 \\ D_3 \\ D_4 \end{bmatrix} = 0 \tag{39}$$

For a nontrivial solution, the determinant of coefficient matrix of above equation must be vanished. This leads to the following condition as

$$\sin\alpha L = 0 \text{ namely } \alpha = \frac{n\pi}{L}, (n = 1, 2, \ldots) \tag{40}$$

By inserting Eqs. (40) in (36), the natural longitudinal frequencies of a clamped-clamped micro-rod are obtained as

$$\omega_n = \frac{n\pi}{L}\sqrt{\frac{1}{m}\left(EA + B\frac{n^2\pi^2}{L^2}\right)} \tag{41}$$

For a clamped-free micro-rod, classical and nonclassical boundary conditions are

$$U(0) = 0 \text{ and } EAU'(L) - BU'''(L) = 0 \tag{42}$$

$$BU''(0) = 0 \text{ and } U'(L) = 0 \tag{43}$$

By using of above boundary conditions in Eq. (35), solution can be given in a matrix form as

Table 1 Comparison of dimensionless natural frequencies $\left(\varpi = \omega L\sqrt{\frac{\rho}{E}}\right)$ of clamped-free micro-rod for the first three modes with various values of l/D

l/D	Mode 1			Mode 2			Mode 3	
	CT	MSGT		CT	MSGT		CT	MSGT
0	1.5708	1.5708		4.7124	4.7124		7.8540	7.8540
0.3	1.5708	1.5712		4.7124	4.7243		7.8540	7.9091
0.6	1.5708	1.5726		4.7124	4.7599		7.8540	8.0721
0.9	1.5708	1.5748		4.7124	4.8187		7.8540	8.3368
1.2	1.5708	1.5779		4.7124	4.8998		7.8540	8.6938
1.5	1.5708	1.5818		4.7124	5.0021		7.8540	9.1323

Table 2 Comparison of dimensionless natural frequencies $\left(\varpi = \omega L\sqrt{\frac{\rho}{E}}\right)$ of clamped-clamped micro-rod for the first three modes with various values of *l*/*D*

l/*D*	Mode 1		Mode 2		Mode 3	
	CT	MSGT	CT	MSGT	CT	MSGT
0	3.1416	3.1416	6.2832	6.2832	9.4248	9.4248
0.3	3.1416	3.1451	6.2832	6.3114	9.4248	9.5198
0.6	3.1416	3.1557	6.2832	6.3954	9.4248	9.7995
0.9	3.1416	3.1733	6.2832	6.5330	9.4248	10.2487
1.2	3.1416	3.1977	6.2832	6.7209	9.4248	10.8463
1.5	3.1416	3.2288	6.2832	6.9550	9.4248	11.5694

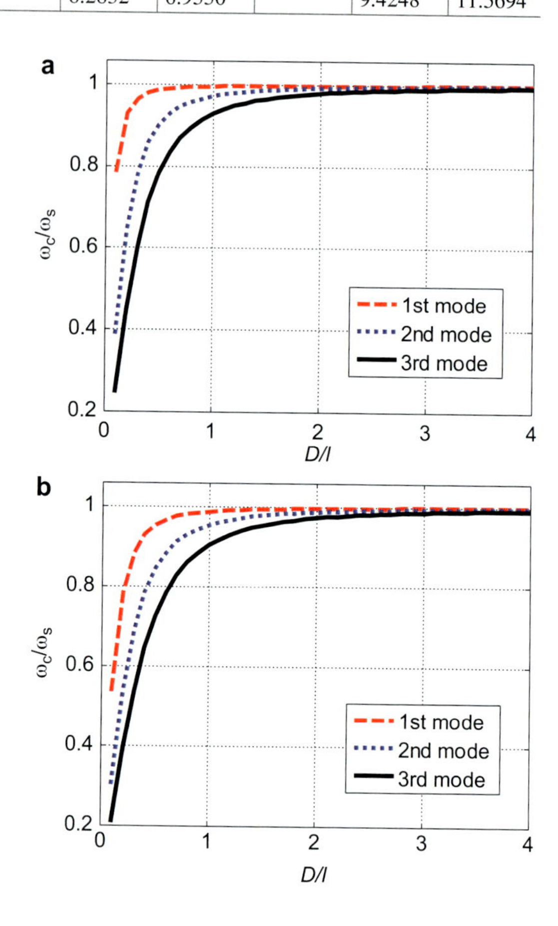

Fig. 2 Variations of the frequency ratio with respect to *D*/*l* for the first three modes. (**a**) Clamped-free (**b**) clamped-clamped

$$\begin{bmatrix} 0 & 1 & 0 & 1 \\ \alpha\lambda_1 \cos\alpha L & -\alpha\lambda_1 \sin\alpha L & \beta\lambda_2 \cosh\beta L & \beta\lambda_2 \sinh\beta L \\ 0 & -\alpha^2 & 0 & \beta^2 \\ \alpha \cos\alpha L & -\alpha \sin\alpha L & \beta \cosh\beta L & \beta \sinh\beta L \end{bmatrix} \begin{bmatrix} D_1 \\ D_2 \\ D_3 \\ D_4 \end{bmatrix} = 0 \qquad (44)$$

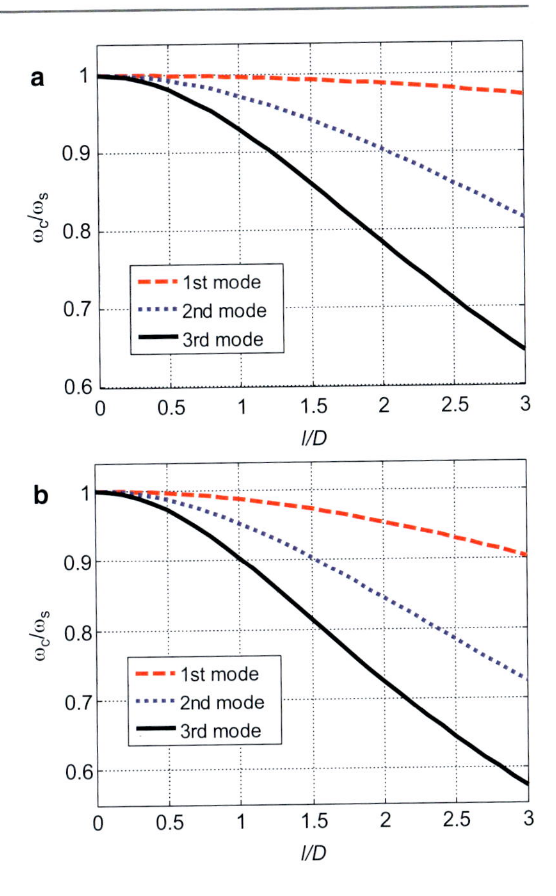

Fig. 3 Variations of the first three frequency ratios versus *l/D*. (**a**) Clamped-free (**b**) clamped-clamped

where

$$\lambda_1 = \left(EA + \alpha^2 B\right), \lambda_2 = \left(EA - \beta^2 B\right) \tag{45}$$

Similarly, the determinant of coefficient matrix of Eq. (44) must be vanished for a nontrivial solution. This leads to the following condition as

$$\cos \alpha L = 0 \text{ namely } \alpha = \frac{(2n-1)\,\pi}{2L}, (n = 1, 2, \dots) \tag{46}$$

By inserting Eqs. (46) in (36), the natural longitudinal frequencies of a clamped-free micro-rod are achieved as

$$\omega_n = \frac{(2n-1)\,\pi}{2L}\sqrt{\frac{1}{m}\left(EA + B\frac{(2n-1)^2\pi^2}{4L^2}\right)} \tag{47}$$

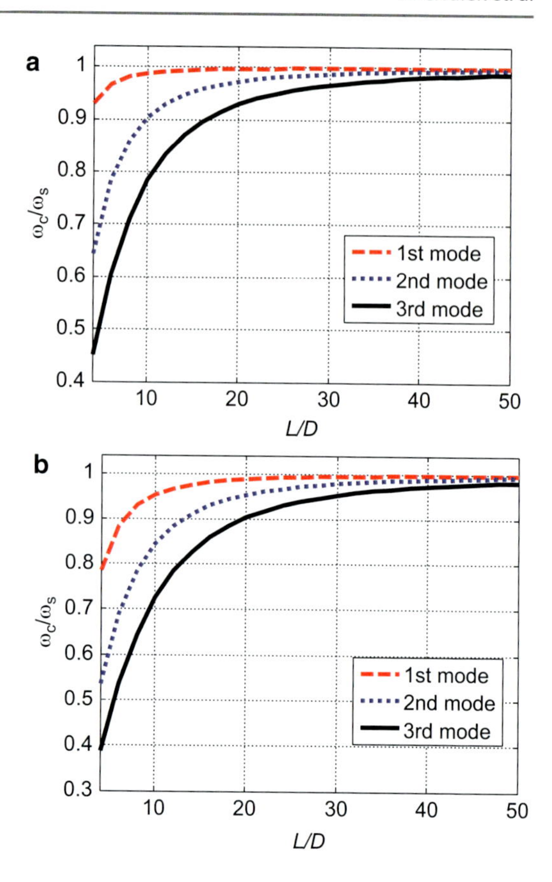

Fig. 4 Effect of slenderness ratio of the micro-rod on the frequency ratios for the first three modes ($l = D$). (**a**) Clamped-free (**b**) clamped-clamped

It is evident that if the additional material length scale parameters l_0 and l_1 are equal to zero, the natural longitudinal frequencies ω_n in Eqs. (41) and (47) will be transformed those in classical theory.

Numerical Results and Discussion

In this section, some illustrative examples for clamped-free and clamped-clamped micro-sized rods are presented. In the figures, the natural longitudinal frequencies obtained by MSGT and CT represented by ω_s and ω_c, respectively, and unless otherwise stated, $L = 20D$, $l_0 = l_1 = l$ are considered, and Poisson's ratio is chosen as 0.38.

A comparison of nondimensional first three natural frequencies of the micro-rod corresponding to various values of the additional material length scale parameters-to-diameter ratio is tabulated in Tables 1 and 2 for clamped-free and clamped-clamped boundary conditions, respectively. It is seen from the tables that the results of classical and newly developed model are identical for $l/D = 0$. It can be said that the nondimensional natural frequencies obtained by the new model increase gradually for bigger values of *l/D*, while those obtained by a classical model are not affected by the variation in *l/D*. It is also notable that differences between the results of the classical model and the current model are more prominent for larger values of *l/D* and higher modes.

Variations of the frequency ratios (ω_c / ω_s) with respect to *D/l* for the first three modes are illustrated in the Fig. 2 for clamped-free and clamped-clamped boundary conditions, respectively. It is evident that an increase in the values of *D/l* leads to an increment in the frequency ratios, and the frequency ratios are nearly equal to one for $D/l \geq 4$. Also, higher values of ω_c / ω_s are obtained for lower modes. It can be concluded that the divergence between classical and size-dependent frequencies becomes more significant for higher modes.

Variations of the first three frequency ratios are plotted versus *l/D* in Fig. 3 for clamped-free and clamped-clamped micro-rods, respectively. When the values of *l/D* increases, the frequency ratios for first, second, and third modes decrease. Also, it is noted that the frequency ratios are equal one for $l/D = 0$.

Influences of slenderness ratio on the frequency ratios are depicted for the first three modes in Fig. 4. It can be interpreted that the difference between natural frequencies predicted by the newly developed and classical models becomes more prominent for both lower slenderness ratios and higher modes. In addition, it can be said that the size dependency of the micro-rod diminishes due to an increase in the slenderness ratio.

It can be seen clearly from the present results that additional material length scale parameters are more important both clamped-free and clamped-clamped cases for smaller sizes and higher modes. Also, the values of the frequency ratios for clamped-clamped boundary condition are smaller than those of the other case.

Conclusion

A higher-order continuum theory is used for modeling of longitudinal vibration problem of micro-sized rod. Some parametric results have been presented in order to show the effect of additional length scale parameters. The results of modified strain gradient theory (MSGT) compared with those obtained by classical theory (CT). It has been seen that the frequency ratios decrease when *l/D* increases. It is also shown that the length scale parameters have some notable influences on axial vibration of the micro-sized rod. It is also possible to say that the effect of the length scale parameters is more significant for slender rods. An increase in slenderness ratio of the micro-rod leads to a decrease in the difference between natural frequencies predicting by the newly developed and classical models. It is also observed that additional material length scale parameters play an important role for smaller size of the micro-rod and higher modes. It is also notable that when additional material length scale parameters are zero, the present model directly becomes the classical model.

Acknowledgments This study has been supported by The Scientific and Technological Research Council of Turkey (TÜBİTAK) with Project No: 112 M879. This support is gratefully acknowledged.

References

B. Akgöz, Ö. Civalek, Int. J. Eng. Sci. **49**, 1268 (2011)
B. Akgöz, Ö. Civalek, Compos. Part B **55**, 263 (2013)
B. Akgöz, Ö. Civalek, J. Vib. Control. **20**, 606 (2014)
B.S. Altan, E.C. Aifantis, J. Mech. Behav. Mater. **8**, 231 (1997)
B.S. Altan, H. Evensen, E.C. Aifantis, Mech. Res. Commun. **23**, 35 (1996)
A. Anthoine, Int. J. Solids Struct. **37**, 1003 (2000)
M. Aydogdu, Phys. E. **41**, 861 (2009)
A.C.M. Chong, D.C.C. Lam, J. Mater. Res. **14**, 4103 (1999)
E. Cosserat, F. Cosserat, *Theory of Deformable Bodies* (Trans. by D.H. Delphenich) (Scientific Library, A. Hermann and Sons, Paris, 1909)
A.C. Eringen, J. Appl. Phys. **54**, 4703 (1983)
A.C. Eringen, E.S. Suhubi, Int. J. Eng. Sci. **2**, 189 (1964)
W. Faris, A.H. Nayfeh, Commun. Nonlinear Sci. Numer. Simul. **12**, 776 (2007)
N.A. Fleck, J.W. Hutchinson, J. Mech. Phys. Solids **41**, 1825 (1993)
N.A. Fleck, G.M. Muller, M.F. Ashby, J.W. Hutchinson, Acta Metall. Mater. **42**, 475 (1994)
Y.Q. Fu, H.J. Du, S. Zhang, Mater. Lett. **57**, 2995 (2003)
U. Güven, Comptes Rendus Mécanique **342**, 8 (2014)
M.A. Haque, M.T.A. Saif, Acta Mater. **51**, 3053 (2003)
M.H. Kahrobaiyan, M. Asghari, M.T. Ahmadian, Int. J. Eng. Sci. **44**, 66–67 (2013)
M.H. Kahrobaiyan, M. Rahaeifard, M.T. Ahmadian, Appl. Math. Model. **35**, 5903 (2011a)
M.H. Kahrobaiyan, S.A. Tajalli, M.R. Movahhedy, J. Akbari, M.T. Ahmadian, Int. J. Eng. Sci. **49**, 856 (2011b)
W.T. Koiter, Proc. K. Ned. Akad. Wet. B **67**, 17 (1964)
S. Kong, S. Zhou, Z. Nie, K. Wang, Int. J. Eng. Sci. **46**, 427 (2008)
S. Kong, S. Zhou, Z. Nie, K. Wang, Int. J. Eng. Sci. **47**, 487 (2009)
D.C.C. Lam, F. Yang, A.C.M. Chong, J. Wang, P. Tong, J. Mech. Phys. Solids **51**, 1477 (2003)
A.K. Lazopoulos, Int. J. Mech. Sci. **58**, 27 (2012)
K.A. Lazopoulos, A.K. Lazopoulos, Eur. J. Mech. A/Solids **29**, 837 (2010)
X. Li, B. Bhushan, K. Takashima, C.W. Baek, Y.K. Kim, Ultramicroscopy **97**, 481 (2003)
J. Lou, P. Shrotriya, S. Allameh, T. Buchheit, W.O. Soboyejo, Mater. Sci. Eng. A **441**, 299 (2006)
H.M. Ma, X.L. Gao, J.N. Reddy, J. Mech. Phys. Solids **56**, 3379 (2008)
R.D. Mindlin, Arch. Ration. Mech. Anal. **16**, 51 (1964)
R.D. Mindlin, Int. J. Solids Struct. **1**, 417 (1965)
R.D. Mindlin, H.F. Tiersten, Arch. Ration. Mech. Anal. **11**, 415 (1962)
Y. Moser, M.A.M. Gijs, Miniaturized flexible temperature sensor. J. Microelectromech. Syst. **16**, 1349 (2007)
F. Najar, S. Choura, S. El-Borgi, E.M. Abdel-Rahman, A.H. Nayfeh, J. Micromech. Microeng. **15**, 419 (2005)
S. Narendar, S. Ravinder, S. Gopalakrishnan, Int. J. Nano Dimens. **3**, 1 (2012)
S. Papargyri-Beskou, D. Polyzos, D.E. Beskos, Struct. Eng. Mech. **15**, 705 (2003a)
S. Papargyri-Beskou, K.G. Tsepoura, D. Polyzos, D.E. Beskos, Int. J. Solids Struct. **40**, 385 (2003b)
S.K. Park, X.L. Gao, J. Micromech. Microeng. **16**, 2355 (2006)
J. Peddieson, G.R. Buchanan, R.P. McNitt, Int. J. Eng. Sci. **41**, 305 (2003)
S.S. Rao, *Vibration of Continuous Systems* (Wiley Inc, Hoboken, 2007)
J.N. Reddy, Int. J. Eng. Sci. **45**, 288–307 (2007a)

J.N. Reddy, *Theory and Analysis of Elastic Plates and Shells*, 2nd edn. (Taylor & Francis, Philadelphia, 2007b)
S.D. Senturia, *Microsystem Design* (Kluwer Academic Publishers, Boston, 2001)
R.A. Toupin, Arch. Ration. Mech. Anal. **11**, 385 (1962)
K.G. Tsepoura, S. Papargyri-Beskou, D. Polyzos, D.E. Beskos, Arch. Appl. Mech. **72**, 483 (2002)
I. Vardoulakis, J. Sulem, *Bifurcation Analysis in Geomechanics, Blackie* (Chapman and Hall, London, 1995)
C.M. Wang, S. Kitipornchai, C.W. Lim, M. Eisenberger, J. Eng. Mech. **134**, 475 (2008)
B. Wang, J. Zhao, S. Zhou, Eur. J. Mech. A/Solids **29**, 591 (2010)
F. Yang, A.C.M. Chong, D.C.C. Lam, P. Tong, Int. J. Solids Struct. **39**, 2731 (2002)

Part V
Peridynamics

Peridynamics: Introduction

34

S. A. Silling

Contents

S. A. Silling (✉)
Sandia National Laboratories, Albuquerque, NM, USA
e-mail: sasilli@sandia.gov

G. Z. Voyiadjis (ed.), *Handbook of Nonlocal Continuum Mechanics for Materials and Structures*, https://doi.org/10.1007/978-3-319-58729-5_29

Abstract

The peridynamic theory is a nonlocal extension of continuum mechanics that is compatible with the physical nature of cracks as discontinuities. It avoids the need to evaluate the partial derivatives of the deformation with respect to the spatial coordinates, instead using an integro-differential equation for the linear momentum balance. This chapter summarizes the peridynamic theory, emphasizing the continuum mechanical and thermodynamic aspects. Formulation of material models is discussed, including details on the statement of models using mathematical objects called *peridynamic states* that are nonlocal and nonlinear generalizations of second-order tensors. Damage evolution is treated within a nonlocal thermodynamic framework making use of the dependence of free energy on damage. Continuous, stable growth of damage can suddenly become unstable, leading to dynamic fracture. Peridynamics treats fracture and long-range forces on the same mathematical basis as continuous deformation and contact forces, extending the applicability of continuum mechanics to new classes of problems.

Keywords

Peridynamic · Nonlocal · Damage · Elasticity · Plasticity · Eulerian

Purpose of the Peridynamic Theory

In spite of its many successes, the local theory of continuum mechanics has some limitations that have hindered its applicability to many important problems:

- Its equations cannot be applied directly on a growing discontinuity in the deformation, making it impossible to model fracture using these equations alone.
- It does not include long-range forces such as electrostatic and van der Waals forces that are important in many technologies.
- It cannot be applied to the mechanics of discrete particles, creating a fundamental divide between molecular dynamics and continuum mechanics.

The peridynamic theory addresses these limitations in the local theory. Its field equations are integro-differential equations that do not require a smooth deformation, allowing fracture to be modeled on the same basis as continuous deformation. It treats all internal forces as long range, allowing interactions such as electrostatic forces to be included in a material description in a natural way. Discrete particles can be treated as a type of peridynamic material, allowing continuous media and systems of particles to be included within the same model, following the same basic equations.

The general idea is that the peridynamic theory replaces the local equilibrium equation with a nonlocal expression as follows:

$$\nabla \cdot \boldsymbol{\sigma} + \mathbf{b} = 0 \quad \rightarrow \quad \int_{\mathcal{H}_x} \mathbf{f}(\mathbf{q}, \mathbf{x})\, dV_{\mathbf{q}} + \mathbf{b} = 0, \tag{1}$$

where $\boldsymbol{\sigma}$ is the stress field, $\mathbf{b}$ is the body force density field, and $\mathbf{f}$ is a vector field representing the force density (per unit volume squared) that material point $\mathbf{q}$ exerts on $\mathbf{x}$. $\mathcal{H}_x$ is a neighborhood of $\mathbf{x}$ to be described below.

The main purposes of this chapter are as follows:

- To explain the origin of the second of (Eq. 1)
- To show how the deformation determines $\mathbf{f}$ through the material model
- To describe material models within the framework of nonlocal thermodynamics
- To demonstrate that damage and fracture fit naturally into this framework
- To describe how some aspects of the local theory can be obtained as a limiting case of the peridynamic equations

Basic Concepts

In a peridynamic body $\mathcal{B}$, each material point $\mathbf{x}$ interacts directly with its neighbors $\mathbf{q} \in \mathcal{B}$ located within a cutoff distance δ of $\mathbf{x}$ in the reference configuration. Let $\mathbf{q}$ denote such a neighbor and define the *bond* $\boldsymbol{\xi} = \mathbf{q} - \mathbf{x}$. The *family* of $\mathbf{x}$, denoted by $\mathcal{H}_x$, consists of all the bonds with length no greater than δ:

$$\mathcal{H}_x = \{\mathbf{q} - \mathbf{x} : 0 < |\mathbf{q} - \mathbf{x}| \leq \delta, \mathbf{q} \in \mathcal{B}\}.$$

The cutoff distance δ is called the *horizon* and is assumed for purposes of this discussion to be independent of $\mathbf{x}$ (Fig. 1).

Let $\mathbf{y}(\mathbf{x})$ denote the deformation. Suppose there is a strain-energy density field $W(\mathbf{x})$, and that its value at $\mathbf{x}$ depends on the *collective deformation* of $\mathcal{H}_x$. This means that $W(\mathbf{x})$ depends not on $\partial \mathbf{y}/\partial \mathbf{x}$ (which may not exist if a crack is present) but on $\mathbf{y}(\mathbf{q})$ for all the material points within the horizon of $\mathbf{x}$.

To express the dependence of W on the collective deformation of $\mathcal{H}_x$, it is convenient to use mathematical objects called *states*. A state is simply a mapping from $\mathcal{H}_x$ to some other quantity, which can be a scalar, a vector, or a tensor. By convention, the bond that the state operates on is written in angle brackets, $\underline{\mathbf{A}}\langle\boldsymbol{\xi}\rangle$. State-valued fields depend on position and possibly time, denoted by

$$\underline{\mathbf{A}}[\mathbf{x}, t]\langle\boldsymbol{\xi}\rangle.$$

The fundamental kinematical quantity for purposes of material modeling is the *deformation state* $\underline{\mathbf{Y}}$, defined by:

$$\underline{\mathbf{Y}}[\mathbf{x}]\langle\mathbf{q} - \mathbf{x}\rangle = \mathbf{y}(\mathbf{q}) - \mathbf{y}(\mathbf{x}) \quad \forall \mathbf{q} \in \mathcal{H}_x, \quad \forall x \in \mathcal{B}$$

Geometrically, the deformation state maps each bond to its image under the deformation.

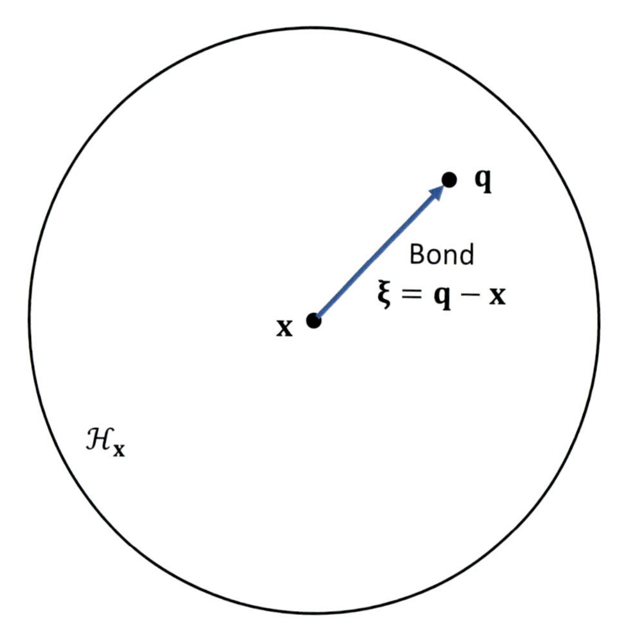

Fig. 1 Bond and family

Returning to our strain-energy density field $W(\mathbf{x})$, we can now represent its dependence on the collective deformation of $\mathcal{H}_\mathrm{x}$. We write this function in the form:

$$W(\mathbf{x}) = \widehat{W}(\underline{\mathbf{Y}}[x]) \quad \forall \mathrm{x} \in \mathcal{B}$$

where $\hat{W}(\underline{\mathbf{Y}})$ is the *strain-energy density function* for the material. The dependence of $\hat{W}$ on $\underline{\mathbf{Y}}$ contains all the material-dependent characteristics of the model. Examples of such a material model are given by:

- Isotropic *bond-based* material:

$$\widehat{W}(\underline{\mathbf{Y}}) = \frac{1}{2}\int_{\mathcal{H}_\mathrm{x}} C(|\boldsymbol{\xi}|)(|\underline{\mathbf{Y}}\langle\boldsymbol{\xi}\rangle| - |\boldsymbol{\xi}|)^2 \mathrm{d}V_{\boldsymbol{\xi}},$$

where C is a scalar-valued function of bond length.

- A possible model for a fluid (one of many, see section "Effectively Eulerian Material Models"):

$$\widehat{W}(\underline{\mathbf{Y}}) = \frac{1}{2}\left(\int_{\mathcal{H}_\mathrm{x}} C(|\boldsymbol{\xi}|)(|\underline{\mathbf{Y}}\langle\boldsymbol{\xi}\rangle| - |\boldsymbol{\xi}|)\,\mathrm{d}V_{\boldsymbol{\xi}}\right)^2.$$

Properties of States

Vector-valued states are similar to second-order tensors in that they both map vectors onto vectors. However, states can be nonlinear and even discontinuous mappings. It is useful to define the *dot product* of two vector-valued states $\underline{\mathbf{A}}$ and $\underline{\mathbf{B}}$ by:

$$\underline{\mathbf{A}} \bullet \underline{\mathbf{B}} = \int_{\mathcal{H}_{\mathbf{x}}} \underline{\mathrm{A}} \langle \xi \rangle \cdot \underline{\mathbf{B}} \langle \xi \rangle \, \mathrm{d}V_{\xi} \tag{2}$$

and of two scalar-valued states $\underline{a}$ and $\underline{b}$ by:

$$\underline{\mathrm{a}} \bullet \underline{b} = \int_{\mathcal{H}_{\mathbf{x}}} \underline{\mathrm{a}} \langle \xi \rangle \, \underline{b} \langle \xi \rangle \, \mathrm{d}V_{\xi}.$$

The *unit state* $\underline{1}$ and *zero state* $\underline{0}$ are defined by:

$$\underline{1} \langle \xi \rangle = 1, \quad \underline{0} \langle \xi \rangle = 0 \quad \forall \xi.$$

The *identity state* leaves bonds unchanged:

$$\underline{\mathrm{X}} \langle \xi \rangle = \xi \quad \forall \xi.$$

The *norm* of a state $\underline{\mathbf{A}}$ is defined by:

$$\|\underline{\mathbf{A}}\| = \sqrt{\underline{\mathbf{A}} \bullet \underline{\mathbf{A}}} \quad \text{or} \quad \|\underline{a}\| = \sqrt{\underline{a} \bullet \underline{a}}.$$

The composition of two states $\underline{\mathbf{A}}$ and $\underline{\mathbf{B}}$ is a state defined by:

$$(\underline{\mathbf{A}} \circ \underline{\mathbf{B}}) \langle \xi \rangle = \underline{\mathbf{A}} \langle \underline{\mathbf{B}} \langle \xi \rangle \rangle ;$$

The "$\circ$" symbol distinguishes this operation from the point product, to be defined below. If $\mathbf{R}$ is a second-order tensor, it may similarly be combined with a state $\underline{\mathbf{A}}$; in this case, a special symbol is not necessary:

$$(\underline{\mathbf{A}}\mathbf{R}) \langle \xi \rangle = \underline{\mathbf{A}} \langle \mathbf{R}\xi \rangle , \quad (\mathbf{R}\underline{\mathbf{A}}) \langle \xi \rangle = \mathbf{R} (\underline{\mathbf{A}} \langle \xi \rangle) .$$

The *point product* of two states, at least one of which is scalar-valued, is defined by:

$$(\underline{a\mathbf{A}}) \langle \xi \rangle = (\underline{\mathbf{A}a}) \langle \xi \rangle = \underline{a} \langle \xi \rangle \, \underline{\mathbf{A}} \langle \xi \rangle ,$$
$$(\underline{ab}) \langle \xi \rangle = (\underline{ba}) \langle \xi \rangle = \underline{a} \langle \xi \rangle \, \underline{b} \langle \xi \rangle .$$

Double states operate on two bonds rather than just one and are denoted by $\underline{\mathbb{K}}\langle\boldsymbol{\xi},\eta\rangle$, where $\boldsymbol{\xi}$ and $\boldsymbol{\eta}$ are bonds. Double states are usually tensor-valued. An example is the dyadic product double state defined by:

$$(\underline{\mathbf{A}}\star\underline{\mathbf{B}})\langle\xi,\eta\rangle=\underline{\mathbf{A}}\langle\xi\rangle\otimes\underline{\mathbf{B}}\langle\eta\rangle.$$

The dot product of a tensor-valued double state $\underline{\mathbb{K}}$ with a vector-valued state $\underline{\mathbf{A}}$ is the vector-valued state given by:

$$(\underline{\mathbb{K}}\bullet\underline{\mathbf{A}})\langle\xi\rangle=\int_{\mathcal{H}_{\mathbf{x}}}\underline{\mathbb{K}}\langle\boldsymbol{\xi},\boldsymbol{\eta}\rangle\underline{\mathbf{A}}\langle\boldsymbol{\eta}\rangle\ \mathrm{d}V_{\eta},\qquad(\underline{\mathbf{A}}\bullet\underline{\mathbb{K}})\langle\xi\rangle=\int_{\mathcal{H}_{\mathbf{x}}}\underline{\mathbf{A}}\langle\eta\rangle\underline{\mathbb{K}}\langle\boldsymbol{\eta},\boldsymbol{\xi}\rangle\ \mathrm{d}V_{\boldsymbol{\eta}}.$$

It is necessary to define a notion of differentiation of functions of states. Let $\Psi(\underline{\mathbf{A}})$ be a scalar-valued function of a vector-valued state. Consider a small increment $\Delta\underline{\mathbf{A}}$. Suppose there is a vector-valued state denoted $\Psi_{\underline{\mathbf{A}}}(\underline{\mathbf{A}})$ such that for any $\Delta\underline{\mathbf{A}}$,

$$\Psi(\underline{\mathbf{A}}+\Delta\underline{\mathbf{A}})-\Psi(\underline{\mathbf{A}})=\Psi_{\underline{\mathbf{A}}}(\underline{\mathbf{A}})\bullet\Delta\underline{\mathbf{A}}+o\left(\|\Delta\underline{\mathbf{A}}\|\right).\tag{3}$$

Then $\Psi_{\underline{\mathbf{A}}}(\underline{\mathbf{A}})$ is called the *Fréchet derivative* of Ψ at $\underline{\mathbf{A}}$. The Fréchet derivative of a function whose value is a vector-valued state is a double state.

Fréchet derivatives obey a form of the chain rule. For example, if $\underline{\mathbf{B}}$ is a state-valued function of the state $\underline{\mathbf{A}}$,

$$\Psi_{\underline{\mathbf{A}}}(\underline{\mathbf{B}}(\underline{\mathbf{A}}))=\Psi_{\underline{\mathbf{B}}}\bullet\underline{\mathbf{B}}_{\underline{\mathbf{A}}},$$

or

$$\Psi_{\underline{\mathbf{A}}}\langle\xi\rangle=\int_{\mathcal{H}_{\mathbf{x}}}\Psi_{\underline{\mathbf{B}}}\langle\eta\rangle\cdot\underline{\mathbf{B}}_{\underline{\mathbf{A}}}\langle\boldsymbol{\eta},\boldsymbol{\xi}\rangle\,\mathrm{d}V_{\boldsymbol{\eta}}.$$

Note that $\underline{\mathbf{B}}_{\underline{\mathbf{A}}}$ is a tensor-valued double state. More details about the use of states can be found in Silling et al. (2007). The symmetries of double states play an interesting role in the linearized peridynamic theory, which is beyond the scope of the present chapter (Silling 2010).

Balance of Momentum

Using the nonlocal concept of strain energy described in section "Basic Concepts," expressions for balance of linear momentum will now be derived. Assume that $\mathcal{B}$ is bounded, and consider a time-independent deformation $\mathbf{y}(\mathbf{x})$ under external body force density field $\mathbf{b}(\mathbf{x})$. Define the total potential energy of the body by:

$$\Phi_{\mathbf{y}}=\int_{\mathcal{B}}\left(W(\mathbf{x})-\mathbf{b}(\mathbf{x})\cdot\mathbf{y}(\mathbf{x})\right)\ \mathrm{d}V_{\mathbf{x}}.\tag{4}$$

Let $\Delta\mathbf{y}$ be a small increment in the deformation and assume that the Fréchet derivative $\widehat{W}_{\underline{\mathbf{Y}}}(\underline{\mathbf{Y}})$ exists for all $\underline{\mathbf{Y}}$. Define

$$\underline{\mathbf{T}}[\mathbf{x}] = \widehat{W}_{\underline{\mathbf{Y}}}(\underline{\mathbf{Y}}[\mathbf{x}]).$$

In the following, it is convenient to adopt the convention that states produce the value 0 for bonds longer than δ. This permits us to replace $\mathcal{H}_{\mathbf{x}}$ with $\mathcal{B}$ as the region of integration in some of the expressions below. From Eqs. 2, 3, and 4,

$$\begin{aligned}
\Delta\Phi_{\mathbf{y}} &= \int_{\mathcal{B}} \left(\underline{\mathbf{T}}[\mathbf{x}] \bullet \Delta\underline{\mathbf{Y}}[\mathbf{x}] - \mathbf{b}(\mathbf{x})\cdot\Delta\mathbf{y}(\mathbf{x})\right)\, \mathrm{d}V_{\mathbf{x}} \\
&= \int_{\mathcal{B}} \left(\int_{\mathcal{B}} \{\underline{\mathbf{T}}[\mathbf{x}]\langle\mathbf{q}-\mathbf{x}\rangle\cdot(\Delta\mathbf{y}(\mathbf{q}) - \Delta\mathbf{y}(\mathbf{x}))\}\, \mathrm{d}V_{\mathbf{q}} - \mathbf{b}(\mathbf{x})\cdot\Delta\mathbf{y}(\mathbf{x})\right) \mathrm{d}V_{\mathbf{x}} \\
&= \int_{\mathcal{B}} \left(\int_{\mathcal{B}} \{\underline{\mathbf{T}}[\mathbf{q}]\langle\mathbf{x}-\mathbf{q}\rangle\cdot\Delta\mathbf{y}(\mathbf{x}) - \underline{\mathbf{T}}[\mathbf{x}]\langle\mathbf{q}-\mathbf{x}\rangle\cdot\Delta\mathbf{y}(\mathbf{x})\}\, \mathrm{d}V_{\mathbf{q}} - \mathbf{b}(\mathbf{x})\cdot\mathbf{y}(\mathbf{x})\right) \mathrm{d}V_{\mathbf{x}} \\
&= \int_{\mathcal{B}} \left(\int_{\mathcal{B}} \left\{\underline{\mathbf{T}}[\mathbf{q}]\langle\mathbf{x}-\mathbf{q}\rangle - \underline{\mathbf{T}}[\mathbf{x}]\left\langle\mathbf{q}-\mathbf{x}\right\rangle\right\}\mathrm{d}V_{\mathbf{q}} - \mathbf{b}(\mathbf{x})\right)\cdot\Delta\mathbf{y}(\mathbf{x})\,\mathrm{d}V_{\mathbf{x}},
\end{aligned} \tag{5}$$

where an interchange of dummy variables of integration $\mathbf{x} \leftrightarrow \mathbf{q}$ is used in the third line. Stationary potential energy requires $\Delta\Phi\mathbf{y} = 0$ for all $\Delta\mathbf{y}$ and hence Eq. 5 can be localized to yield:

$$\int_{\mathcal{H}_{\mathbf{x}}} \{\underline{\mathbf{T}}[\mathbf{x}]\langle\mathbf{q}-\mathbf{x}\rangle - \underline{\mathbf{T}}[\mathbf{q}]\langle\mathbf{x}-\mathbf{q}\rangle\}\,\mathrm{d}V_{\mathbf{q}} + \mathbf{b}(\mathbf{x}) = 0 \quad \forall\mathbf{x}\in\mathcal{B}. \tag{6}$$

Equation 6 is the equilibrium equation in peridynamics. It is frequently written in the more suggestive form:

$$\int_{\mathcal{H}_{\mathbf{x}}} \mathbf{f}(\mathbf{q},\mathbf{x})\, \mathrm{d}V_{\mathbf{q}} + \mathbf{b}(\mathbf{x}) = 0 \quad \forall\mathbf{x}\in\mathcal{B} \tag{7}$$

where $\mathbf{f}(\mathbf{q},\mathbf{x})$ is the *pairwise bond force density* field, given by:

$$\mathbf{f}(\mathbf{q},\mathbf{x}) = \underline{\mathbf{T}}[\mathbf{x}]\langle\mathbf{q}-\mathbf{x}\rangle - \underline{\mathbf{T}}[\mathbf{q}]\langle\mathbf{x}-\mathbf{q}\rangle \quad \forall\mathbf{x},\mathbf{q}\in\mathcal{B}.$$

Observe that $\mathbf{f}(\mathbf{q},\mathbf{x})$ is comprised of two terms arising from the material models at $\mathbf{x}$ and at $\mathbf{q}$.

The pairwise bond force density field obeys the antisymmetry condition:

$$\mathbf{f}(\mathbf{x},\mathbf{q}) = -\mathbf{f}(\mathbf{q},\mathbf{x}) \quad \forall\mathbf{x},\mathbf{q}\in\mathcal{B}.$$

Mechanically, the vector $\mathbf{f}(\mathbf{q},\mathbf{x})$ can be thought of the force density (force per unit volume squared) that $\mathbf{q}$ exerts on $\mathbf{x}$. However, it is best not to take this intuitive picture too literally, because there is not necessarily a direct physical interaction between the two points.

By d'Alembert's rule, the time-dependent equation of motion is found from Eq. 7 to be:

$$\rho(\mathbf{x})\,\ddot{\mathbf{y}}(\mathbf{x},t) = \int_{\mathcal{H}_{\mathbf{x}}} \mathbf{f}(\mathbf{q},\mathbf{x},t)\;\mathrm{d}V_{\mathbf{q}} + \mathbf{b}(\mathbf{x},t) \qquad \forall \mathbf{x} \in \mathcal{B}, t \geq 0 \tag{8}$$

where ρ is the mass density field.

A more axiomatic and general derivation of the linear momentum balance in peridynamics, which does not assume the existence of a strain-energy density field, can be found in Silling and Lehoucq (2010). Also, a derivation from statistical mechanics is available in Lehoucq and Sears (2011).

Energy Balance and Thermodynamics

The laws of thermodynamics can be written in a form compatible with the nonlocal nature of the peridynamic theory. The first law of thermodynamics takes the following form:

$$\dot{\varepsilon}(\mathbf{x},t) = \underline{\mathbf{T}}[\mathbf{x},t] \bullet \dot{\underline{\mathbf{Y}}}[\mathbf{x},t] + h(\mathbf{x},t) + s(\mathbf{x},t) \qquad \forall \mathbf{x} \in \mathcal{B}, t \geq 0,$$

where ε is the internal energy density (per unit volume), h is the rate of heat transport to $\mathbf{x}$, and s is the energy source rate. More simply,

$$\dot{\varepsilon} = \underline{\mathbf{T}} \bullet \dot{\underline{\mathbf{Y}}} + h + s. \tag{9}$$

To provide an intuitive picture of this expression, recall that by Eq. 2,

$$\underline{\mathbf{T}} \bullet \dot{\underline{\mathbf{Y}}} = \int_{\mathcal{H}_{\mathbf{x}}} \underline{\mathbf{T}}\langle\boldsymbol{\xi}\rangle \cdot \dot{\underline{\mathbf{Y}}}\langle\boldsymbol{\xi}\rangle\;\mathrm{d}V_{\boldsymbol{\xi}}.$$

The energy balance (9) therefore sums up the rate of work done by the bond forces acting against the rate of extension of the individual bonds. The key point is that only the force state at $\mathbf{x}$, that is, $\underline{\mathbf{T}}[\mathbf{x},t]$, contributes to the energy change at $\mathbf{x}$. The force state $\underline{\mathbf{T}}[\mathbf{q},t]$ does not, even though it appears in the momentum balance (6). This partitioning of the pairwise bond force into energy contributions at $\mathbf{x}$ and $\mathbf{q}$ is the unique feature of the peridynamic version of thermodynamics that causes the internal energy density to be *additive*, provided that h is conserved. This fact apparently resolves a long-standing question about whether it is even possible to define a nonlocal internal energy density that is additive (Gurtin and Williams 1971); see (Silling and Lehoucq 2010) for a more complete discussion.

The energy balance (9) applies regardless of how the heat transport h is specified. This could be supplied by the Fourier heat conduction expression or by a nonlocal diffusion law such as:

$$h(\mathbf{x}, t) = \int_{\mathcal{H}_\mathbf{x}} \left[\underline{H}[\mathbf{x}, t] \langle \mathbf{q} - \mathbf{x} \rangle - \underline{H}[\mathbf{q}, t] \langle \mathbf{x} - \mathbf{q} \rangle \right] \mathrm{d}V_\mathbf{q}$$

where $\underline{H}$ is a scalar-valued state. The most intuitive form of such an $\underline{H}$ models heat diffusion along bonds as though they are conducting wires that are insulated from each other (Bobaru and Duangpanya 2010):

$$\underline{H}[\mathbf{x}, t] \langle \mathbf{q} - \mathbf{x} \rangle = \underline{K}[\mathbf{x}] \langle \mathbf{q} - \mathbf{x} \rangle (\theta(\mathbf{q}, t) - \theta(\mathbf{x}, t)) \tag{10}$$

where θ is the temperature field and K is the bond conductivity. The heat flow expression (10) is useful in modeling systems with inherently nonlocal diffusion mechanisms such as radiative heat transport, as well as in treating discontinuities and singularities in the temperature field. A model for heat transport that is more general than (Eq. 10) could have the form:

$$\underline{H}[\mathbf{x}, t] = \underline{\widehat{H}}(\underline{\Theta}[\mathbf{x}, t]), \quad \underline{\Theta}[\mathbf{x}, t] \langle \mathbf{q} - \mathbf{x} \rangle = \theta(\mathbf{q}, t) - \theta(\mathbf{x}, t). \tag{11}$$

This representation could model systems in which the heat flow in each bond can depend not only on the temperature difference between its own endpoints but also on the temperature difference in other bonds in the family.

The local form of second law of thermodynamics as implemented in peridynamics has the following form:

$$\theta \dot{\eta} \geq h + s \tag{12}$$

where η is the entropy density field (per unit volume) in the reference configuration. A restriction on the admissible forms of $\underline{\widehat{H}}$ arises from the second law, as discussed in section "Restriction on the Heat Transport Model."

Material Models

A peridynamic material model $\underline{\widehat{\mathbf{T}}}$ determines the force state at every material point $\mathbf{x}$ and time t. For most materials, the force state depends on the deformation state $\underline{\mathbf{Y}}[\mathbf{x}, t]$ and possibly other variables as well. If $\underline{\mathbf{Y}}$ is the only quantity that the material model depends on, we write:

$$\underline{\mathbf{T}}[\mathbf{x}, t] = \underline{\widehat{\mathbf{T}}}(\underline{\mathbf{Y}}[\mathbf{x}, t]).$$

If the body is heterogeneous, that is, if the material model depends explicitly on position, we write:

$$\underline{\mathbf{T}}[\mathbf{x}, t] = \underline{\widehat{\mathbf{T}}}(\underline{\mathbf{Y}}[\mathbf{x}, t], \mathbf{x}).$$

The material model can also depend on the rate of deformation, $\dot{\underline{\mathbf{Y}}}$, and other physically relevant quantities such as temperature (section

"Thermodynamic Form of A Material Model"), damage (section "Damage as a Thermodynamic Variable"), or plastic deformation (section "Plasticity").

Material models are required to satisfy certain general rules. The requirement of *nonpolarity* is written as:

$$\int_{\mathcal{H}_{\mathbf{x}}} \underline{\widehat{\mathbf{T}}}\left(\underline{\mathbf{Y}}\right)\langle\xi\rangle \times \underline{\mathbf{Y}}\langle\xi\rangle \, \mathrm{d}V_{\xi} = 0 \quad \forall \underline{\mathbf{Y}}. \tag{13}$$

Nonpolarity guarantees that global balance of angular momentum holds, that is, the material model does not nonphysically create angular momentum. It is similar to the required symmetry of the Cauchy stress tensor in the local theory. Nonpolarity is not a requirement in the micropolar versions of peridynamics that have been proposed (Gerstle et al. 2007) or in a peridynamic shell theory that includes rotational degrees of freedom (Chowdhury et al. 2016).

In the absence of external fields that provide a special physical direction that affects material response, a peridynamic material model is required to satisfy *objectivity*. This is a requirement that if the family is deformed and then rigidly rotated, then the force state undergoes the same rigid rotation. This condition is written as follows:

$$\underline{\widehat{\mathbf{T}}}\left(\mathbf{Q}\underline{\mathbf{Y}}\right) = \mathbf{Q}\underline{\widehat{\mathbf{T}}}\left(\underline{\mathbf{Y}}\right) \quad \forall \mathbf{Q}, \quad \forall \underline{\mathbf{Y}}$$

where $\mathbf{Q}$ is any proper orthogonal tensor. Unlike objectivity, *isotropy* is not a general requirement but is appropriate for modeling materials that have no internal special direction (such as embedded unidirectional reinforcement fibers). Isotropy means that if the body is first rigidly rotated, then deformed, the force state is the same as if there were no rotation:

$$\underline{\widehat{\mathbf{T}}}\left(\underline{\mathbf{Y}}\mathbf{Q}\right) = \underline{\widehat{\mathbf{T}}}\left(\underline{\mathbf{Y}}\right) \quad \forall \mathbf{Q}, \quad \forall \underline{\mathbf{Y}}$$

where $\mathbf{Q}$ is any proper orthogonal tensor.

If there is a strain-energy function $\widehat{W}_{\underline{\mathbf{Y}}}\left(\underline{\mathbf{Y}}\right)$ such that:

$$\underline{\widehat{\mathbf{T}}}\left(\underline{\mathbf{Y}}\right) = \widehat{W}_{\underline{\mathbf{Y}}}\left(\underline{\mathbf{Y}}\right) \quad \forall \underline{\mathbf{Y}},$$

then the material is *elastic*. An elastic material is objective if and only if $\widehat{W}\left(\mathbf{Q}\underline{\mathbf{Y}}\right) = \widehat{W}\left(\underline{\mathbf{Y}}\right) \quad \forall \mathbf{Q}, \quad \forall \underline{\mathbf{Y}}$. An elastic material is isotropic if and only if:

$$\widehat{W}\left(\underline{\mathbf{Y}}\mathbf{Q}\right) = \widehat{W}\left(\underline{\mathbf{Y}}\right) \quad \forall \mathbf{Q}, \quad \forall \underline{\mathbf{Y}}.$$

A material that is elastic and objective is necessarily nonpolar (Silling 2010). This property is a convenience to developers of material models, because it is often much easier to prove objectivity of $\widehat{W}$ than to prove nonpolarity directly from Eq. 13.

Peridynamic material models fall into the following categories:

- *Bond-based:* Each bond has a force response that depends *only* on its own deformation and damage, independent of all other bonds in the family. For a bond-based material model,

$$\underline{\mathbf{T}}[\mathbf{x}]\langle\mathbf{q}-\mathbf{x}\rangle = \hat{\mathbf{t}}(\underline{\mathbf{Y}}[\mathbf{x}]\langle\mathbf{q}-\mathbf{x}\rangle, \mathbf{q}, \mathbf{x}).$$

Note that the function $\hat{\mathbf{t}}$ and its arguments are all vectors rather than vector-valued states. For a homogeneous body, the equilibrium equation for a bond-based material is often written as:

$$\int_{\mathcal{H}_{\mathbf{x}}} \mathbf{f}(\mathbf{u}(\mathbf{q})-\mathbf{u}(\mathbf{x}), \mathbf{q}-\mathbf{x})\, \mathrm{d}V_{\mathbf{q}} + \mathbf{b}(\mathbf{x}) = 0,$$

with $\mathbf{f}$ rather than $\hat{\underline{\mathbf{T}}}$ as the material model. An example is given by:

$$\mathbf{f}(\eta, \xi) = \underline{C}\langle\xi\rangle(|\eta+\xi| - |\xi|)\underline{\mathbf{M}}\langle\xi\rangle, \quad \underline{\mathbf{M}}\langle\xi\rangle = \frac{\underline{\mathbf{Y}}\langle\xi\rangle}{|\underline{\mathbf{Y}}\langle\xi\rangle|}$$

where $\underline{C}$ is a scalar-valued state. In this material model, each bond acts like a linear spring.

- *State based:* Each bond has a force response that can depend on the deformation and damage in *all* the bonds in the family. Among state-based material models, there are two classes: *ordinary,* in which the bond forces are always parallel to the deformed bond direction $\underline{\mathbf{M}}\langle\xi\rangle$, and all others, which are called *non-ordinary.* Ordinary materials have the advantage of automatically being nonpolar.

Thermodynamic Form of a Material Model

This section describes a thermodynamic formulation of peridynamic material models. Material response that can be defined in terms of static variables (without rates, gradients, and loading history) can often be written in terms of a free-energy function. The advantage of this way of characterizing material response is that the resulting values for the force state, temperature, entropy, damage, coefficient of thermal expansion, and heat capacity are always consistent. It also permits us to incorporate restrictions on the material model derived from the second law of thermodynamics. Additional terms such as rate dependence can be added to the force state obtained from the thermodynamic form.

Define the free-energy density at a material point $\mathbf{x}$ and time t by:

$$\psi = \varepsilon - \theta\eta \tag{14}$$

where, as before, ε, θ, and η are the internal energy density, temperature, and entropy density, respectively. Taking the time derivative of Eq. 14 and applying the second law expression, Eq. 12 leads to:

$$\dot{\varepsilon} - \dot{\theta}\eta - \dot{\psi} \geq 0.$$

Combining this with the first law expression (9) yields:

$$\underline{\mathbf{T}} \bullet \underline{\dot{\mathbf{Y}}} + h + s - \dot{\theta}\eta - \dot{\psi} \geq 0. \tag{15}$$

Suppose the material model is expressed in terms of the dependence of free-energy density on both the deformation state and temperature, thus $\psi\ (\underline{\mathbf{Y}}, \theta)$. By the chain rule,

$$\dot{\psi} = \psi_\theta \dot{\theta} + \psi_{\underline{\mathbf{Y}}} \bullet \underline{\dot{\mathbf{Y}}}. \tag{16}$$

In this expression, ψ_θ is a partial derivative and $\psi_{\underline{\mathbf{Y}}}$ is a Fréchet derivative, and therefore, its value is a state. Combining Eq. 15 with Eq. 16 and regrouping terms yields:

$$\left(\underline{\mathbf{T}} - \psi_{\underline{\mathbf{Y}}}\right) \bullet \underline{\dot{\mathbf{Y}}} + h + \mathbf{s} - (\eta + \psi_\theta)\,\dot{\theta} \geq 0. \tag{17}$$

Following the reasoning of Coleman and Noll (1963; Gurtin et al. 2010), in principle, we can contrive an experiment in which $h = s = 0$ and the quantities $\underline{\dot{\mathbf{Y}}}$ and $\dot{\theta}$ are prescribed independently of each other.

Suppose that in such a thought experiment $\dot{\theta} = 0$ but $\underline{\dot{\mathbf{Y}}}$ is varied, and vice versa. Enforcing the inequality (17) then leads to the conclusions

$$\underline{\mathbf{T}} = \psi_{\underline{\mathbf{Y}}}, \quad \eta = -\psi_\theta. \tag{18}$$

The first of these provides the force state in the thermodynamic form of a rate-independent material model. Rate dependence can be including by assuming the dependence $\psi\left(\underline{\mathbf{Y}}, \underline{\dot{\mathbf{Y}}}, \theta, \dot{\theta}\right)$, leading to an additional term in Eq. 17:

$$\left(\underline{\mathbf{T}} - \psi_{\underline{\mathbf{Y}}}\right) \bullet \underline{\dot{\mathbf{Y}}} + h + s - (\eta + \psi_\theta)\,\dot{\theta} + \psi_{\dot{\theta}}\ddot{\theta} - \psi_{\underline{\dot{\mathbf{Y}}}} \bullet \underline{\ddot{\mathbf{Y}}} \geq 0.$$

From this, using the same reasoning as before, one concludes that $\psi_{\dot{\theta}} = \psi_{\underline{\dot{\mathbf{Y}}}} = 0$, that is, the free energy density cannot depend explicitly on $\dot{\theta}$ or $\underline{\dot{\mathbf{Y}}}$. However, without loss of generality, rate effects can be incorporated by partitioning the force state response into equilibrium and rate-dependent parts:

$$\widehat{\underline{\mathbf{T}}}\left(\underline{\mathbf{Y}}, \dot{\underline{\mathbf{Y}}}, \theta\right) = \underline{\mathbf{T}}^e\left(\underline{\mathbf{Y}}, \theta\right) + \underline{\mathbf{T}}^d\left(\underline{\mathbf{Y}}, \dot{\underline{\mathbf{Y}}}, \theta\right), \quad \underline{\mathbf{T}}^d\left(\underline{\mathbf{Y}}, \underline{0}, \theta\right) = \underline{0}.$$

From this assumption, it follows by a similar Coleman-Noll type of argument (Fried 2010) that

$$\underline{\mathbf{T}}^e = \psi_{\underline{\mathbf{Y}}}, \quad \underline{\mathbf{T}}^d \bullet \dot{\underline{\mathbf{Y}}} \geq 0. \tag{19}$$

The second of Eq. 19 is a *dissipation inequality*. It implies that

$$\underline{\mathbf{T}}^d\langle\boldsymbol{\xi}\rangle \cdot \dot{\underline{\mathbf{Y}}}\langle\boldsymbol{\xi}\rangle \geq 0, \quad \forall \boldsymbol{\xi} \in \mathcal{H}_{\mathbf{x}}.$$

Work done by both $\underline{\mathbf{T}}^e$ and $\underline{\mathbf{T}}^d$ contribute to the internal energy density according to the first law (9), in general changing the temperature. These temperature changes affect ψ indirectly, even though $\psi_{\dot{\underline{\mathbf{Y}}}} = \underline{0}$.

Thermal expansion arises from the coupling between $\underline{\mathbf{Y}}$ and θ in the form of ψ. For example, for an elastic material with strain-energy density function $\widehat{W}$, a model with thermal expansion can be defined by:

$$\psi\left(\underline{\mathbf{Y}}, \theta\right) = \frac{\widehat{W}\left((1 - \alpha\Delta\theta)\,\underline{\mathbf{Y}}\right)}{1 - \alpha\Delta\theta}, \quad \Delta\theta := \theta - \theta_0$$

where α is a linear coefficient of thermal expansion. Then by the chain rule,

$$\underline{\mathbf{T}} = \psi_{\underline{\mathbf{Y}}} = \widehat{W}_{\underline{\mathbf{Y}}}\left((1 - \alpha\Delta\theta)\,\underline{\mathbf{Y}}\right) \approx \widehat{W}_{\underline{\mathbf{Y}}}\left(\underline{\mathbf{Y}}\right) - \alpha\Delta\theta\underline{\mathbb{K}} \bullet \underline{\mathbf{X}}$$

where the approximation holds if $|\ \alpha\Delta\theta\ | \ll 1$ and $|\ \underline{\mathbf{Y}}\langle\boldsymbol{\xi}\rangle - \boldsymbol{\xi}\ | \ll |\ \boldsymbol{\xi}\ |$ for all $\boldsymbol{\xi}$. $\underline{\mathbb{K}}$ is the *micromodulus double state*, defined by:

$$\underline{\mathbb{K}} = \widehat{W}_{\underline{\mathbf{YY}}}.$$

(See Silling (2010) for details on the properties of $\underline{\mathbb{K}}$ and its role in the linearized theory.)

Restriction on the Heat Transport Model

Recall the local form of the second law of thermodynamics,

$$\theta\dot{\eta} \geq h + s. \tag{20}$$

All of us are familiar with the restriction in the local theory that heat cannot flow from cold to hot, which is a consequence of the second law. It is interesting to investigate the analogous restriction in peridynamics.

To do this, consider a bounded, nondeforming, thermodynamically isolated body. Assume that the material model is as described in the section "Thermodynamic Form of a Material Model," in which the free-energy density has the dependence $\psi(\underline{\mathbf{Y}}, \theta)$. As a consequence of the second of Eq. 18, since $\dot{\underline{\mathbf{Y}}} = \mathbf{0}$, it follows that η can depend only on θ. This implies that equality holds in Eq. 20, that is,

$$\theta \dot{\eta} = h. \tag{21}$$

(This is not a result of reversibility, which is not assumed here.) Now compute the total entropy change in the body. From Eqs. 11 and 21 (dropping t from the notation),

$$\begin{aligned}
\int_{\mathcal{B}} \dot{\eta}(\mathbf{x})\, dV_{\mathbf{x}} &= \int_{\mathcal{B}} \frac{h(\mathbf{x})}{\theta(\mathbf{x})} dV_{\mathbf{x}} \\
&= \int_{\mathcal{B}} \int_{\mathcal{B}} \frac{\underline{H}[\mathbf{x}]\langle \mathbf{q}-\mathbf{x}\rangle - \underline{H}[\mathbf{q}]\langle \mathbf{x}-\mathbf{q}\rangle}{\theta(\mathbf{x})} dV_{\mathbf{q}} dV_{\mathbf{x}} \\
&= \int_{\mathcal{B}} \int_{\mathcal{B}} \left(\frac{1}{\theta(\mathbf{x})} - \frac{1}{\theta(\mathbf{q})} \right) \underline{H}[\mathbf{x}]\langle \mathbf{q}-\mathbf{x}\rangle\, dV_{\mathbf{q}} dV_{\mathbf{x}} \\
&= -\int_{\mathcal{B}} \underline{\beta}[\mathbf{x}] \bullet \underline{H}[\mathbf{x}]\, dV_{\mathbf{x}}
\end{aligned} \tag{22}$$

where $\underline{\beta}$ is the scalar-valued state defined by:

$$\underline{\beta}[\mathbf{x}]\langle \mathbf{q}-\mathbf{x}\rangle = \frac{1}{\theta(\mathbf{q})} - \frac{1}{\theta(\mathbf{x})}.$$

Suppose that the heat transport model $\underline{\widehat{H}}$ obeys

$$\underline{\beta} \bullet \underline{\widehat{H}}(\underline{\Theta}) \leq 0 \qquad \forall \underline{\Theta}, \tag{23}$$

or equivalently

$$\int_{\mathcal{H}_{\mathbf{x}}} \underline{\beta}\langle \boldsymbol{\xi}\rangle\, \underline{\widehat{H}}\{\boldsymbol{\xi}\rangle\, dV_{\mathbf{x}} \leq 0 \qquad \forall \underline{\Theta}.$$

Working backwards through the steps in Eq. 22, evidently Eq. 23 implies that

$$\int_{\mathcal{B}} \dot{\eta}(\mathbf{x})\, dV_{\mathbf{x}} \geq 0,$$

which is the global form of the second law for an isolated body. Note that the stronger restriction

$$\underline{\beta}\langle \boldsymbol{\xi}\rangle\, \underline{\widehat{H}}\langle \boldsymbol{\xi}\rangle \leq 0 \qquad \forall \boldsymbol{\xi}$$

is sufficient but not necessary for Eq. 23 to hold.

The inequality (23) is a restriction on the constitutive model for heat transport $\underline{\widehat{H}}$; it is the peridynamic form of the rule that "heat cannot flow from cold to hot." Details on peridynamic modeling of heat transport can be found in Bobaru and Duangpanya (2010), Bobaru and Duangpanya (2012), and Oterkus et al. (2014b). A fully coupled thermomechanical treatment is discussed in Oterkus et al. (2014a). The relation of the restriction (Eq. 23) to the Clausius-Duhem inequality is discussed in section "Convergence of Peridynamics to the Local Theory" below.

Damage as a Thermodynamic Variable

In the previous section, it was assumed that the material model is stated in terms of the free-energy density function given by:

$$\psi\left(\underline{\mathbf{Y}}, \theta\right).$$

An important characteristic of the response of real materials is that they fracture and fail. To help model this aspect of material response, it is assumed that there is a scalar-valued state field called the *damage state*, denoted by $\underline{\phi}$, that has the distinguishing feature of monotonicity over time:

$$\dot{\underline{\phi}}\langle\boldsymbol{\xi}\rangle \geq 0 \qquad \forall\boldsymbol{\xi} \in \mathcal{H}_{\mathbf{x}}. \tag{24}$$

By convention, it is usually assumed that $0\leq \underline{\phi}\langle\boldsymbol{\xi}\rangle \leq 1$ for all bonds $\boldsymbol{\xi}$, with $\underline{\phi}\langle\boldsymbol{\xi}\rangle = 0$ representing an undamaged bond. By assuming a material model of the form:

$$\psi\left(\underline{\mathbf{Y}}, \theta, \underline{\phi}\right) \tag{25}$$

and working through the free-energy inequality discussed previously, one concludes (Silling and Lehoucq 2010) that the following dissipation inequality holds:

$$\psi_{\underline{\phi}} \bullet \dot{\underline{\phi}} \leq 0, \tag{26}$$

which also implies:

$$\psi_{\underline{\phi}}\langle\boldsymbol{\xi}\rangle \leq 0 \qquad \forall\boldsymbol{\xi} \in \mathcal{H}_{\mathbf{x}}. \tag{27}$$

Damage evolves according to a prescribed material-dependent *damage growth law*,

$$\dot{\underline{\phi}} = \underline{D}\left(\underline{\mathbf{Y}}, \dot{\underline{\mathbf{Y}}}, \underline{\phi}\right)$$

where $\underline{D}$ is a scalar state-valued function (section "Damage Evolution").

Examples of Material Models

The following examples of material models illustrate the connection between free-energy density and mechanical forces. They also demonstrate how to evaluate Fréchet derivatives in practice.

Bond-Based Linear Material with Damage

Consider the free-energy density function defined by:

$$\psi\left(\underline{\mathbf{Y}}, \theta, \underline{\phi}\right) = \frac{1}{2}\left(\underline{C}\left(\underline{1} - \underline{\phi}\right)\underline{e}\right) \bullet \underline{e} + c\theta\left(1 - \log\left(\theta/\theta_0\right)\right) \tag{28}$$

or equivalently

$$\psi\left(\underline{\mathbf{Y}}, \theta, \underline{\phi}\right) = \frac{1}{2}\int_{\mathcal{H}_{\mathbf{x}}} \underline{C}\left\langle\boldsymbol{\xi}\right\rangle\left(1 - \underline{\phi}\left\langle\boldsymbol{\xi}\right\rangle\right)\underline{e}^2\left\langle\boldsymbol{\xi}\right\rangle \mathrm{d}V_{\boldsymbol{\xi}} + c\theta\left(1 - \log\left(\theta/\theta_0\right)\right) \tag{29}$$

where $\underline{C}$ is the prescribed scalar-valued *micromodulus state* and c is the heat capacity (at constant $\underline{\mathbf{Y}}$). The scalar-valued state $\underline{e}$ is the *extension state*, defined by:

$$\underline{e}\left\langle\boldsymbol{\xi}\right\rangle = \left|\,\underline{\mathbf{Y}}\left\langle\boldsymbol{\xi}\right\rangle\,\right| - \left|\,\boldsymbol{\xi}\,\right|. \tag{30}$$

To explain how to evaluate the force state $\underline{\mathbf{T}} = \psi_{\underline{\mathbf{Y}}}$, we go through in detail the process of obtaining the Fréchet derivative. Recalling (Eq. 3), we seek to express a first-order approximation for incremental changes in ψ in response to any small $\Delta\underline{\mathbf{Y}}$ in the form:

$$\Delta\psi = \int_{\mathcal{H}_{\mathbf{x}}} [\text{something}] \cdot \Delta\underline{\mathbf{Y}}\left\langle\boldsymbol{\xi}\right\rangle \;\mathrm{d}V_{\boldsymbol{\xi}}.$$

The process of finding the Fréchet derivative consists of finding an expression of this form. The [something], which is a function of the dummy variable of integration $\boldsymbol{\xi}$ (and therefore is a state), is the Fréchet derivative. For the example material model (29),

$$\Delta\psi = \underline{\mathbf{T}} \bullet \Delta\underline{\mathbf{Y}} = \int_{\mathcal{H}_{\mathbf{x}}} \underline{C}\left\langle\boldsymbol{\xi}\right\rangle\left(1 - \underline{\phi}\left\langle\boldsymbol{\xi}\right\rangle\right)\underline{e}\left\langle\boldsymbol{\xi}\right\rangle \Delta\underline{e}\left\langle\boldsymbol{\xi}\right\rangle \;\mathrm{d}V_{\boldsymbol{\xi}}.$$

Since

$$\Delta\underline{e}\left\langle\boldsymbol{\xi}\right\rangle = \underline{\mathbf{M}}\left\langle\boldsymbol{\xi}\right\rangle \cdot \Delta\underline{\mathbf{Y}}\left\langle\boldsymbol{\xi}\right\rangle, \qquad \underline{\mathbf{M}}\left\langle\boldsymbol{\xi}\right\rangle = \frac{\underline{\mathbf{Y}}\left\langle\boldsymbol{\xi}\right\rangle}{\left|\,\underline{\mathbf{Y}}\left\langle\boldsymbol{\xi}\right\rangle\,\right|},$$

it follows that

$$\underline{\mathbf{T}} \bullet \Delta \underline{\mathbf{Y}} = \int_{\mathcal{H}_{\mathbf{x}}} \left[\underline{C} \langle \boldsymbol{\xi} \rangle \left(1 - \underline{\phi} \langle \boldsymbol{\xi} \rangle \right) \underline{e} \langle \boldsymbol{\xi} \rangle \underline{\mathbf{M}} \langle \boldsymbol{\xi} \rangle \right] \cdot \Delta \underline{\mathbf{Y}} \langle \boldsymbol{\xi} \rangle \, \mathrm{d}V_{\boldsymbol{\xi}}.$$

Comparing this with Eq. 3 leads to the conclusion that

$$\widehat{\underline{\mathbf{T}}} (\underline{\mathbf{Y}}) = \underline{C} \left(\underline{1} - \underline{\phi} \right) \underline{e} \underline{\mathbf{M}}. \tag{31}$$

or

$$\underline{\mathbf{T}} \langle \boldsymbol{\xi} \rangle = \underline{C} \langle \boldsymbol{\xi} \rangle \left(1 - \underline{\phi} \langle \boldsymbol{\xi} \rangle \right) \underline{e} \langle \boldsymbol{\xi} \rangle \underline{\mathbf{M}} \langle \boldsymbol{\xi} \rangle \qquad \forall \boldsymbol{\xi} \in \mathcal{H}_{\mathbf{x}}.$$

Some features of the material model (31) are as follows:

- The bond force density vector in each bond is parallel to the deformed bond.
- Holding damage fixed, the magnitude of the bond force density varies linearly with bond extension.
- The model is *bond-based*: each bond responds independently of all the others.
- The model is geometrically nonlinear: it allows for large deformation.
- Increasing the damage in each bond decreases the magnitude of each bond force density in that bond.

From Eq. 28 and the second of Eq. 18, one finds that for this material model,

$$\eta = -\psi_\theta = c \log (\theta / \theta_0) \tag{32}$$

From Eq. 14, the second of Eq. 18, and the time derivative of Eq. 32, it follows that if $\underline{\mathbf{Y}}$ and $\underline{\phi}$ are held constant,

$$\begin{aligned} \dot{\varepsilon} &= \psi_\theta \dot{\theta} + \dot{\theta} \eta + \theta \dot{\eta} \\ &= \theta \dot{\eta} \\ &= c \dot{\theta}. \end{aligned}$$

Thus, unlike the free energy, the internal energy varies linearly with temperature under these conditions, with the heat capacity as the constant of proportionality.

Ordinary State-Based Linear Material with Damage

A modification of Eq. 31 that includes a volume change may be written as:

$$\widehat{\underline{\mathbf{T}}} (\underline{\mathbf{Y}}) = \left(\underline{C} \left(\underline{1} - \underline{\phi} \right) \underline{e} + A \vartheta \underline{1} \right) \underline{\mathbf{M}}, \qquad \vartheta = \frac{\underline{1} \bullet \underline{e}}{\underline{1} \bullet \underline{1}}$$

where A is a constant and ϑ is the dilatation. (See Eq. 40 below for a more general definition of dilatation.) Note that the dilatation depends on all the bonds in the family.

Non-ordinary State-Based Material

The following material model tends to resist bending:

$$\hat{\underline{\mathbf{T}}}(\underline{\mathbf{Y}})\langle\boldsymbol{\xi}\rangle = \begin{cases} A\left(\underline{e}\langle\boldsymbol{\xi}\rangle + \underline{e}\langle-\boldsymbol{\xi}\rangle\right)\left(\underline{\mathbf{Y}}\langle\boldsymbol{\xi}\rangle - \underline{\mathbf{Y}}\langle-\boldsymbol{\xi}\rangle\right) & \text{if } \boldsymbol{\xi} \in \mathcal{H}_{\mathbf{x}} \text{ and } -\boldsymbol{\xi} \in \mathcal{H}_{\mathbf{x}}, \\ 0 & \text{otherwise.} \end{cases}$$

where A is a constant. Another way to write this is to define the *reversal state* by

$$\underline{\mathbf{R}}\langle\boldsymbol{\xi}\rangle = -\boldsymbol{\xi} \quad \forall \boldsymbol{\xi} \in \mathcal{H}_{\mathbf{x}},$$

then

$$\hat{\underline{\mathbf{T}}}(\underline{\mathbf{Y}}) = A\left(\underline{e} + \underline{e} \circ \underline{\mathbf{R}}\right)\left(\underline{\mathbf{Y}} - \underline{\mathbf{Y}} \circ \underline{\mathbf{R}}\right).$$

This material is nonpolar even though the bond forces are not necessarily parallel to the deformed bonds. It is interesting that a state-based peridynamic material model *without changing the equilibrium equation* can resist bending; see Diyaroglu et al. (2015) and Grady and Foster (2014). for details and specific material models. This is in contrast to the standard theory of beams and plates, in which a special fourth-order PDE replaces the fundamental second-order PDEs of local continuum mechanics.

Bond-Based Viscoelastic Material

The material model (31) can be modified to include a rate-dependent damping term:

$$\hat{\underline{\mathbf{T}}}\left(\underline{\mathbf{Y}}, \dot{\underline{\mathbf{Y}}}\right) = \left(\underline{C}\left(\underline{1} - \underline{\phi}\right)\underline{e} + A\dot{\underline{e}}\right)\underline{\mathbf{M}}$$

where A is a nonnegative constant. More information on viscoelastic peridynamic models can be found in Mitchell (2011a) and Weckner and Mohamed (2013).

Isotropic Bond-Based Material

In the material model (31), different bonds can have different stiffness, because $\underline{C}$ is a state (that is, its value depends on the bond). (See Hu et al. 2012a; Oterkus and

Madenci 2012). for examples of anisotropic material models.) To model isotropic materials, set

$$\underline{C}\langle\boldsymbol{\xi}\rangle = C^i\left(|\boldsymbol{\xi}|\right)$$

where C^i is a function of bond length only, thus

$$\underline{\mathbf{T}} = \left(C^i\left(\underline{1}-\underline{\phi}\right)\underline{e}\right)\underline{\mathbf{M}}.$$

The most general form for C^i is discussed in Silling (2000).

Nonconvex Bond-Based Material

If the bond force density is not a monotonic function of extension, and if it is elastic, then its strain-energy density function is called *nonconvex*. An example is:

$$\underline{\mathbf{T}}\langle\boldsymbol{\xi}\rangle = A\underline{e}\langle\boldsymbol{\xi}\rangle\exp\left(-\underline{e}^2\langle\boldsymbol{\xi}\rangle\right)\underline{\mathbf{M}}\langle\boldsymbol{\xi}\rangle \text{ or } \widehat{\underline{\mathbf{T}}}(\underline{\mathbf{Y}}) = A\underline{e}\exp\left(-\underline{e}^2\right)\underline{\mathbf{M}}.$$

The properties of nonconvex bond-based materials, including their stability and relation to brittle fracture, are discussed by Lipton (2014, 2016). More general concepts of convexity that apply to state-based materials are, to the best of the author's knowledge, a totally unexplored area.

Discrete Particles as Peridynamic Materials

Consider a set of N particles with equal mass m that interact through a multibody potential such that the potential energy of particle i is given by:

$$\Phi_i = U_i\left(\mathbf{r}_{1i}, \mathbf{r}_{2i}, \ldots, \mathbf{r}_{Ni}\right) - \mathbf{b}_i \cdot \mathbf{y}_i, \quad \mathbf{r}_{ji} := \mathbf{y}_j - \mathbf{y}_i$$

where $\mathbf{y}_i$ and $\mathbf{b}_i$ are the position of and external force on particle i. The potential energy of the entire set of particles is found from:

$$\Phi = \sum_{i=1}^{N}\Phi_i.$$

The acceleration of each particle i is obtained from Newton's second law in the form:

$$\begin{aligned}\mathbf{F}_i &= m\ddot{\mathbf{y}}_i = -\frac{\partial\Phi}{\partial\mathbf{y}_i}\\ &= \sum_{j=1}^{N}\left(\frac{\partial U_i}{\partial\mathbf{r}_{ji}} - \frac{\partial U_j}{\partial\mathbf{r}_{ij}}\right) + \mathbf{b}_i.\end{aligned} \tag{33}$$

Now consider the peridynamic body with mass-density field and body-force density field defined by:

$$\rho(\mathbf{x}) = \sum_{i=1}^{N} m\Delta(\mathbf{x}-\mathbf{x}_i), \quad \mathrm{b}(\mathbf{x}) = \sum_{i=1}^{N} \mathbf{b}_i \Delta(\mathbf{x}-\mathbf{x}_i) \tag{34}$$

where $\Delta(\cdot)$ denotes the 3D Dirac delta function. Here, the reference positions of the particles $\mathbf{x}_i$ are arbitrary and merely serve to identify the particles for purposes of the mathematics. For this peridynamic body, let the material model be elastic with strain-energy density function given by:

$$\widehat{W}(\underline{\mathbf{Y}}[\mathbf{x}], \mathbf{x}) = \sum_{i=1}^{N} U_i(\mathbf{r}_{1i}, \mathbf{r}_{2i}, \ldots, \mathbf{r}_{Ni})\Delta(\mathbf{x}-\mathbf{x}_i), \quad \mathbf{r}_{ji} = \underline{\mathbf{Y}}[\mathbf{x}_i]\langle \mathbf{x}_j - \mathbf{x}_i\rangle. \tag{35}$$

After evaluating the Fréchet derivative of this $\widehat{W}$, the force state field is found to be:

$$\underline{\mathbf{T}}[\mathbf{x}]\langle \mathbf{q}-\mathbf{x}\rangle = \sum_{i=1}^{N}\sum_{j=1}^{N} \frac{\partial U_i}{\partial \mathrm{r}_{ji}} \Delta(\mathbf{x}-\mathbf{x}_i)\Delta(\mathbf{q}-\mathbf{x}_j). \tag{36}$$

Evaluating the acceleration field using Eqs. 8, 34, and 36, the terms involving $\Delta(\mathbf{q}-\mathbf{x}_j)$ integrate to 1 and hence

$$\rho(\mathbf{x})\ddot{\mathbf{y}}(\mathbf{x}) = \int_{\mathcal{B}} \left\{\underline{\mathbf{T}}[\mathbf{x}]\langle \mathbf{q}-\mathbf{x}\rangle - \underline{\mathbf{T}}[\mathbf{q}]\langle \mathbf{x}-\mathbf{q}\rangle\right\} \mathrm{d}V_{\mathbf{q}} + \mathbf{b}(\mathbf{x})$$

becomes

$$\begin{aligned}\sum_{i=1}^{N} \left(m\ddot{\mathbf{y}}(\mathbf{x}) - \mathbf{b}(\mathbf{x})\right)\Delta(\mathbf{x}-\mathbf{x}_i) &= \sum_{i=1}^{N}\sum_{j=1}^{N} \frac{\partial U_i}{\partial \mathbf{r}_{ji}} \left(\Delta(\mathbf{x}-\mathbf{x}_i) - \Delta(\mathbf{x}-\mathbf{x}_j)\right) \\ &= \sum_{i=1}^{N}\sum_{j=1}^{N} \left(\frac{\partial U_i}{\partial \mathbf{r}_{ji}} - \frac{\partial U_j}{\partial \mathbf{r}_{ij}}\right)\Delta(\mathbf{x}-\mathbf{x}_i)\end{aligned}$$

which implies

$$m\ddot{\mathbf{y}}(\mathbf{x}_i) = \sum_{j=1}^{N} \left(\frac{\partial U_i}{\partial \mathbf{r}_{ji}} - \frac{\partial U_j}{\partial \mathbf{r}_{ij}}\right) + \mathbf{b}_i, \quad i = 1, 2, \ldots, N.$$

So, the peridynamic equation of motion for the body specified in Eqs. 34 and 35 reduces to Newton's second law, (Eq. 33).

The applicability of peridynamics to both continuous and discrete systems can be useful in modeling the interaction of particles with continuous bodies. For example, certain aspects of the mechanics of a suspension of particles in a liquid can be treated simply by adding the responses of the two media, one discrete and one continuous:

$$\widehat{W}\left(\underline{\mathbf{Y}},\mathbf{x}\right)=\sum_{i=1}^{N}U_i\left(\underline{\mathbf{Y}},\mathbf{x}\right)\Delta\left(\mathbf{x}-\mathbf{x}_i\right)+\widehat{W}_c\left(\underline{\mathbf{Y}}\right)$$

where $\widehat{W}_c$ is the strain-energy density function for the continuum. Since the machinery of peridynamics can be applied to this unconventional medium, this model could potentially be used to study interesting phenomena such as wave dispersion, attenuation, scattering in suspensions of interacting particles.

Effectively Eulerian Material Models

The material models described up to now in this chapter have been Lagrangian; they refer explicitly to a reference configuration, and the bond forces arise from movement of the bonds from their reference positions. For modeling fluids under large deformations, the Lagrangian approach becomes impractical because of the gross distortion of the families. In these cases, an Eulerian approach to material modeling may be preferable.

An effectively Eulerian material model for a fluid can be derived by letting the horizon be infinite but limiting the response to bonds that *currently* have length less than a prescribed distance δ in the deformed configuration. For example, such a model for a fluid could be specified by defining a *nonlocal density* as follows:

$$\overline{\rho}\left(\mathbf{x}\right)=\rho_0\int_{\mathcal{B}}\omega\left(\left|\underline{\mathbf{Y}}\left[\mathbf{x}\right]\langle \mathrm{q}-\mathbf{x}\rangle\right|\right)\mathbf{d}V_{\boldsymbol{\xi}} \tag{37}$$

where ρ_0 is the reference density and ω is a differentiable weighting function on $[0, \infty)$ such that:

$$\begin{aligned}&\textstyle\int_{\mathcal{B}}\omega\left(\left|\boldsymbol{\xi}\right|\right)\mathrm{d}V_{\boldsymbol{\xi}}=1,\\&\omega=0\text{ on }\left[\delta,\infty\right),\quad \omega'\leq 0\text{ on }\left[0,\delta\right].\end{aligned}$$

Even though the region of integration in Eq. 37 is $\mathcal{B}$ in the reference configuration, in effect only a neighborhood of radius δ in the deformed configuration needs to be computed. The nonlocal density Eq. 37 can be used in any conventional equation of state. The energy balance (9) continues to apply without change, since its form is independent of the material model. The pressure from the equation of state determines the bond forces through the usual Fréchet derivative:

$$\underline{\mathbf{T}}\langle\boldsymbol{\xi}\rangle = \widehat{\underline{\mathbf{T}}}(\underline{\mathbf{Y}})\langle\boldsymbol{\xi}\rangle = \psi_{\underline{\mathbf{Y}}} = \frac{\partial\psi}{\partial v}\frac{\partial v}{\partial\overline{\rho}}\overline{\rho}_{\underline{\mathbf{Y}}}$$
$$= (-p)\left(\frac{-\rho_0}{\overline{\rho}^2}\right)\left(\rho_0\omega'(|\underline{\mathbf{Y}}\langle\boldsymbol{\xi}\rangle|)\underline{\mathbf{M}}\langle\boldsymbol{\xi}\rangle\right)$$
$$= pv^2\omega'(|\underline{\mathbf{Y}}\langle\boldsymbol{\xi}\rangle|)\underline{\mathbf{M}}\langle\boldsymbol{\xi}\rangle \quad \forall\boldsymbol{\xi}\in\mathcal{B}$$

where p is the pressure and $v = \rho_0/\overline{\rho}$ is the nonlocal relative volume. This approach to modeling fluids has been successfully applied to very large deformations and strong shock waves (Silling et al. 2017).

Some Lagrangian material models can be converted to effectively Eulerian models. To do this, we must eliminate any explicit dependence of $\underline{\mathbf{T}}\langle\boldsymbol{\xi}\rangle$ on $\boldsymbol{\xi}$ except as an identifier for bonds. For example, recall the example bond-based material model (30), (31) with $\underline{\phi} = \underline{0}$:

$$\underline{\mathbf{T}}\langle\boldsymbol{\xi}\rangle = \widehat{\underline{\mathbf{T}}}(\underline{\mathbf{Y}})\langle\boldsymbol{\xi}\rangle = \underline{C}\langle\boldsymbol{\xi}\rangle\underline{e}\langle\boldsymbol{\xi}\rangle\underline{\mathbf{M}}\langle\boldsymbol{\xi}\rangle, \quad \underline{e}\langle\boldsymbol{\xi}\rangle = |\underline{\mathbf{Y}}\langle\boldsymbol{\xi}\rangle| - |\boldsymbol{\xi}|.$$

This model contains $\boldsymbol{\xi}$ explicitly through the $|\boldsymbol{\xi}|$ term, so it is Lagrangian. But consider this alternative model:

$$\widehat{\underline{\mathbf{T}}}\left(\underline{\mathbf{Y}}, \dot{\underline{\mathbf{Y}}}\right)\langle\boldsymbol{\xi}\rangle := \underline{C}\langle\underline{\mathbf{Y}}\langle\boldsymbol{\xi}\rangle\rangle\underline{E}\langle\boldsymbol{\xi}\rangle\underline{\mathbf{M}}\langle\boldsymbol{\xi}\rangle \tag{38}$$

where $\underline{E}$ is a scalar-valued state, that is a function of time, defined by:

$$\underline{E}[0] = \underline{0}, \quad \dot{\underline{E}}[t] = \begin{cases} \dot{\underline{\mathbf{Y}}}\langle\boldsymbol{\xi}\rangle\cdot\underline{\mathbf{M}}\langle\boldsymbol{\xi}\rangle & \text{if } |\underline{\mathbf{Y}}\langle\boldsymbol{\xi}\rangle| \le \delta, \\ 0 & \text{otherwise,} \end{cases}$$

or, more succinctly,

$$\widehat{\underline{\mathbf{T}}}\left(\underline{\mathbf{Y}}, \dot{\underline{\mathbf{Y}}}\right) = (\underline{C}\circ\underline{\mathbf{Y}})\underline{E}\underline{\mathbf{M}}, \quad \dot{\underline{E}} = \dot{\underline{\mathbf{Y}}}\cdot\underline{\mathbf{M}}. \tag{39}$$

Since $\boldsymbol{\xi}$ does not appear explicitly in Eq. 38 except as an identifier, this alternative model is effectively Eulerian. In Eq. 39, it is assumed that $\underline{C}\langle\mathbf{p}\rangle = 0$ whenever $|\mathbf{p}| > \delta$. Interactions can occur in bonds that start out with length greater than δ but get shorter over time. Similarly, bonds that are initially short will have zero bond force density if they elongate over time to length greater than δ. The material model (39) is effectively Eulerian but is not elastic.

Plasticity

For small deformations, it is conventional to express volume changes in the form of the *dilatation*, denoted ϑ. By linearization of Eq. 37, this is found to be:

$$v - 1 \approx \vartheta(\underline{\mathbf{Y}}) := \frac{3(\underline{\Omega}\underline{x})\bullet\underline{e}}{(\underline{\Omega}\underline{x})\bullet\underline{x}}, \quad \underline{\Omega}\langle\boldsymbol{\xi}\rangle = -|\boldsymbol{\xi}|\,\omega'(|\boldsymbol{\xi}|), \quad \underline{x}\langle\boldsymbol{\xi}\rangle = |\boldsymbol{\xi}|. \tag{40}$$

where $\underline{e}$ is the extension state defined in Eq. 30. In terms of rates, since $\dot{\underline{e}} = \underline{\mathbf{M}} \cdot \dot{\underline{\mathbf{Y}}}$, Eq. 40 can be written as

$$\dot{\vartheta}\left(\dot{\underline{\mathbf{Y}}}\right) := \frac{3\left(\underline{\Omega}\underline{\mathbf{X}}\right) \bullet \dot{\underline{\mathbf{Y}}}}{\left(\underline{\Omega}\underline{\mathbf{X}}\right) \bullet \underline{\mathbf{X}}}. \tag{41}$$

We will not make further use of the connection between $\underline{\Omega}$ and ω given in the second of Eq. 40, so $\underline{\Omega}$ can be regarded as essentially arbitrary, except that it must be nonnegative and depend only on $|\boldsymbol{\xi}|$. Similarly, for small deformations, the pressure is given by:

$$p\left(\underline{\mathbf{T}}\right) = -\frac{1}{3}\underline{\mathbf{X}} \bullet \underline{\mathbf{T}}. \tag{42}$$

Plastic deformation can be incorporated into an elastic material model by introducing a new vector-valued state called the *permanent deformation state*, denoted $\underline{\mathbf{P}}$. Given a free-energy density function $\psi^0\left(\underline{\mathbf{Y}}, \theta\right)$, define a new free-energy density function by

$$\psi\left(\underline{\mathbf{Y}}, \underline{\mathbf{P}}, \theta\right) = \psi^0\left(\underline{\mathbf{Y}} - \underline{\mathbf{P}}, \theta\right).$$

Since $\underline{\mathbf{T}} = \psi_{\underline{\mathbf{Y}}}$, it follows that

$$\widehat{\underline{\mathbf{T}}}\left(\underline{\mathbf{Y}}, \underline{\mathbf{P}}, \theta\right) = \widehat{\underline{\mathbf{T}}}^0\left(\underline{\mathbf{Y}} - \underline{\mathbf{P}}, \theta\right)$$

and similarly

$$\psi_{\underline{\mathbf{P}}} = -\underline{\mathbf{T}}. \tag{43}$$

By repeating the steps leading up to Eq. 17, one finds that

$$\left(\underline{\mathbf{T}} - \psi_{\underline{\mathbf{Y}}}\right) \bullet \dot{\underline{\mathbf{Y}}} - \psi_{\underline{\mathbf{P}}} \bullet \dot{\underline{\mathbf{P}}} + h + s - \left(\eta + \psi_\theta\right)\dot{\theta} \geq 0. \tag{44}$$

From Eqs. 43 and 44, it follows that

$$\dot{\psi}^p := \underline{\mathbf{T}} \bullet \dot{\underline{\mathbf{P}}} \geq 0 \tag{45}$$

where $\dot{\psi}^p$ is the rate of *plastic work*. Equation 45 is the dissipation inequality for plastic materials.

Plastic flow can occur when the force state is on or outside of a *yield surface* defined by:

$$\mathcal{P}\left(\underline{\mathbf{T}}\right) = 0$$

where $\mathcal{P}$ is a scalar-valued function. A possible evolution law for $\underline{\mathbf{P}}$ is given by:

$$\dot{\underline{\mathbf{P}}} = \lambda \mathcal{P}_{\underline{\mathbf{T}}}, \tag{46}$$

where $\lambda > 0$. Equation 46 can be thought of as an *associated* flow rule. The dissipation inequality (Eq. 45), which is a consequence of the second law of thermodynamics, places a restriction on $\mathcal{P}$:

$$\mathcal{P}_{\underline{\mathbf{T}}} \bullet \underline{\mathbf{T}} \geq 0 \quad \forall \underline{\mathbf{T}},$$

which is a type of convexity condition on the yield surface.

Many materials, especially metals under moderate stress, have yield surfaces that are nearly independent of the pressure. To account for this, the peridynamic yield surface can be defined to be a function of the *deviatoric force state*, which is obtained by subtracting off the hydrostatic part of the force state:

$$\underline{\mathbf{T}}^d = \underline{\mathbf{T}} - \frac{\underline{\mathbf{X}} \bullet \underline{\mathbf{T}}}{(\underline{\Omega}\underline{\mathbf{X}}) \bullet \underline{\mathbf{X}}} \underline{\Omega}\underline{\mathbf{X}}. \tag{47}$$

From Eqs. 42 and 47,

$$p\left(\underline{\mathbf{T}}^d\right) = 0,$$

that is, the deviatoric force state has zero pressure. For a material model in which the yield surface depends only on $\underline{\mathbf{T}}^d$, the associated flow rule (Eq. 46) can be evaluated using the chain rule for Fréchet derivatives, with the result:

$$\begin{aligned} \dot{\underline{\mathbf{P}}} &= \lambda \mathcal{P}_{\underline{\mathbf{T}}} \\ &= \lambda \mathcal{P}_{\underline{\mathbf{T}}^d} \bullet \underline{\mathbf{T}}^d_{\underline{\mathbf{T}}} \\ &= \lambda \left(\mathcal{P}_{\underline{\mathbf{T}}^d} - \frac{(\underline{\Omega}\underline{\mathbf{X}}) \bullet \mathcal{P}_{\underline{\mathbf{T}}^d}}{(\underline{\Omega}\underline{\mathbf{X}}) \bullet \underline{\mathbf{X}}} \underline{\mathbf{X}} \right). \end{aligned} \tag{48}$$

Comparing the structure of Eqs. 47 and 48, it is subtle but significant that in the latter, the $\underline{\Omega}$ is shifted from outside the fraction to inside the numerator. Using this fact, Eqs. 41 and 48 imply that:

$$\dot{\vartheta}\left(\dot{\underline{\mathbf{P}}}\right) = 0,$$

that is, the associated flow rule applied to a yield surface that depends only on the deviatoric force state results in zero volume change. This echoes the familiar result in the plasticity of metals that plastic strain has zero dilatation. Additional details on modeling plasticity within peridynamics can be found in Foster et al. (2010),

Madenci and Oterkus (2016), Mitchell (2011b), Sun and Sundararaghavan (2014), and Warren et al. (2009).

Damage Evolution

The monotonicity condition (24) is the only general requirement on the evolution of damage:

$$\underline{\dot{\phi}} \langle \boldsymbol{\xi} \rangle = \underline{D} \langle \boldsymbol{\xi} \rangle \geq 0 \quad \forall \boldsymbol{\xi} \in \mathcal{H}_{\mathbf{x}}.$$

Otherwise, we are free to dream up damage growth laws. The second law restriction (Eq. 26) is really a condition on the material model, not the damage growth law.

One approach to specifying how damage grows is to define a failure surface in state space:

$$\mathcal{S}\left(\underline{\mathbf{T}}, \underline{\mathbf{Y}}, \underline{\phi}\right) = 0$$

such that damage does not increase if $\mathcal{S}$ is in the interior of the surface, that is, if $\mathcal{S} < 0$.

An example of a plausible damage growth law is given by:

$$\underline{D} \langle \boldsymbol{\xi} \rangle = \begin{cases} 0 & \text{if } \mathcal{S} < 0, \\ \lambda \underline{F} \langle \boldsymbol{\xi} \rangle & \text{otherwise.} \end{cases} \tag{49}$$

where $\lambda(t)$ is a nonnegative scalar-valued function and $\underline{F}$ is the *thermodynamic force state* defined by:

$$\underline{F} = -\psi_{\underline{\phi}}. \tag{50}$$

The damage growth law (49) satisfies the monotonicity condition (24) because of the result (27), which is a consequence of the dissipation inequality for damage:

$$\underline{\dot{\phi}} \langle \boldsymbol{\xi} \rangle = \underline{D} \langle \boldsymbol{\xi} \rangle = \lambda \underline{F} \langle \boldsymbol{\xi} \rangle = -\lambda \psi_{\underline{\phi}} \langle \boldsymbol{\xi} \rangle \geq 0 \quad \forall \boldsymbol{\xi} \in \mathcal{H}_{\mathbf{x}}.$$

The remaining question is how to determine λ.

In dynamics, we can reasonably assume a dependence of the form:

$$\lambda = a\mathcal{S} \tag{51}$$

where a is a nonnegative constant. This relation allows the value of $\mathcal{S}$ to be outside the failure surface, that is, $\mathcal{S} > 0$, while damage is evolving. It is interesting to investigate the stability of the resulting damage growth. Consider a uniformly deformed body and hold the deformation fixed. Allow damage to evolve and analyze whether damage growth speeds up or slows down over time. To study this, take the total time derivative of $\mathcal{S}\left(\underline{\mathbf{T}}, \underline{\mathbf{Y}}, \underline{\phi}\right)$ holding $\dot{\underline{\mathbf{Y}}} = \underline{0}$:

$$\dot{\mathcal{S}} = \left(\mathcal{S}_{\underline{\mathbf{T}}} \bullet \underline{\mathbf{T}}_{\underline{\phi}} + \mathcal{S}_{\underline{\phi}}\right) \bullet \dot{\underline{\phi}}.$$

Using Eqs. 49 and 50 yields:

$$\dot{\lambda} = a\dot{\mathcal{S}} = a\lambda\left(\mathcal{S}_{\underline{\mathbf{T}}} \bullet \underline{\mathbf{T}}_{\underline{\phi}} + \mathcal{S}_{\underline{\phi}}\right) \bullet \underline{F}.$$

The solution to this ODE for λ is given by:

$$\lambda(t) = \lambda(0)e^{art}, \qquad r = \left(\mathcal{S}_{\underline{\mathbf{T}}} \bullet \underline{\mathbf{T}}_{\underline{\phi}} + \mathcal{S}_{\underline{\phi}}\right) \bullet \underline{F}. \tag{52}$$

Thus, the damage growth is stable if $r \leq 0$ and unstable otherwise. Mechanically, this criterion says that if $\mathcal{S}$ is being driven toward the failure surface by the $\mathcal{S}_{\underline{\mathbf{T}}}$ term faster than it is being pushed away from it by the $\mathcal{S}_{\underline{\phi}}$ term (if this term is positive), then it is unstable.

In the case of quasi-static deformation, it can be assumed that, instead of Eq. 51, the condition that determines the growth of damage is that the system always remains on the failure surface as $\underline{\mathbf{Y}}$ changes, thus:

$$\dot{\mathcal{S}} = 0.$$

Under this assumption, writing out the time derivative of $\mathcal{S}$ using Eq. 49 yields:

$$\begin{aligned} 0 &= \dot{\mathcal{S}}\left(\underline{\mathbf{T}}, \underline{\mathbf{Y}}, \underline{\phi}\right) \\ &= \mathcal{S}_{\underline{\mathbf{T}}} \bullet \left(\underline{\mathbf{T}}_{\underline{\mathbf{Y}}} \bullet \dot{\underline{\mathbf{Y}}} + \underline{\mathbf{T}}_{\underline{\phi}} \bullet \dot{\underline{\phi}}\right) + \mathcal{S}_{\underline{\mathbf{Y}}} \bullet \dot{\underline{\mathbf{Y}}} + \mathcal{S}_{\underline{\phi}} \bullet \dot{\underline{\phi}} \\ &= \mathcal{S}_{\underline{\mathbf{T}}} \bullet \left(\underline{\mathbf{T}}_{\underline{\mathbf{Y}}} \bullet \dot{\underline{\mathbf{Y}}} + \lambda\underline{\mathbf{T}}_{\underline{\phi}} \bullet \underline{F}\right) + \mathcal{S}_{\underline{\mathbf{Y}}} \bullet \dot{\underline{\mathbf{Y}}} + \lambda\mathcal{S}_{\underline{\phi}} \bullet \underline{F}. \end{aligned}$$

Solving this for λ and applying the monotonicity requirement for damage leads to:

$$\lambda = \max\left\{0, -\frac{\left(\mathcal{S}_{\underline{\mathbf{T}}} \bullet \underline{\mathbf{T}}_{\underline{\mathbf{Y}}} + \mathcal{S}_{\underline{\mathbf{Y}}}\right) \bullet \dot{\underline{\mathbf{Y}}}}{r}\right\} \quad \forall \dot{\underline{\mathbf{Y}}} \tag{53}$$

where r is given by the second of Eq. 52. Observe that Eq. 53 blows up as $r \to 0$, indicating the onset of unstable damage growth, as discussed previously. The relations (49) and (53) allow us to explicitly determine the rate of damage growth for every bond at every point in the body, provided the deformation is quasi-static and $\mathcal{S} = 0$:

$$\dot{\underline{\phi}} = \max\left\{0, -\frac{\left(\mathcal{S}_{\underline{\mathbf{T}}} \bullet \underline{\mathbf{T}}_{\underline{\mathbf{Y}}} + \mathcal{S}_{\underline{\mathbf{Y}}}\right) \bullet \dot{\underline{\mathbf{Y}}}}{\left(\mathcal{S}_{\underline{\mathbf{T}}} \bullet \underline{\mathbf{T}}_{\underline{\phi}} + \mathcal{S}_{\phi}\right) \bullet \underline{F}}\right\} \underline{F} \quad \forall \dot{\underline{\mathbf{Y}}}. \tag{54}$$

As an example of a failure surface, consider the material model (28) with a slight modification that introduces a binary variable to indicate intact bonds or *broken* bonds:

$$\underline{b}\langle \boldsymbol{\xi} \rangle = \begin{cases} 1 \text{ if } \underline{\phi}\langle \boldsymbol{\xi} \rangle < 1, \\ 0 \text{ otherwise.} \end{cases}$$

The modified free-energy expression is given by:

$$\psi\left(\underline{\mathbf{Y}}, \theta, \underline{\phi}\right) = \frac{1}{2}\left(\underline{C}\left(\underline{1} - \underline{\phi}\right)\underline{be}\right) \bullet \underline{e} + c\theta\left(1 - \log\left(\theta/\theta_0\right)\right).$$

An example of a failure surface is given by:

$$\mathcal{S}\left(\underline{\mathbf{T}}, \underline{\phi}\right) = (\underline{k}\underline{\mathbf{T}}) \bullet \underline{\mathbf{M}} + \frac{s_1}{2}\left\|\underline{\phi}\right\|^2 - s_0 \tag{55}$$

where s_0 and s_1 are constants, $s_0 \geq 0$, and $\underline{k}$ is a nonnegative, constant, scalar-valued state. The bond force density is given by:

$$\underline{\mathbf{T}} = \psi_{\underline{\mathbf{Y}}} = \underline{C}\left(\underline{1} - \underline{\phi}\right)\underline{be\mathbf{M}}.$$

Evaluating the required Fréchet derivatives in Eq. 54 leads to:

$$\underline{F} = -\psi_{\underline{\phi}} = \tfrac{1}{2}\underline{Cbe}^2, \quad \underline{\mathbf{T}}_{\underline{\mathbf{Y}}} = \underline{C}\left(\underline{1} - \underline{\phi}\right)\underline{b\mathbf{M}} \star \underline{\mathbf{M}}, \quad \underline{\mathbf{T}}_{\underline{\phi}} = -\underline{Cbe\mathbf{M}},$$
$$\mathcal{S}_{\underline{\mathbf{Y}}} = \underline{0}, \quad \mathcal{S}_{\underline{\mathbf{T}}} = \underline{k\mathbf{M}}, \quad \mathcal{S}_{\underline{\phi}} = s_1\left\|\underline{\phi}\right\|\underline{\phi}.$$

Suppose that a specimen has zero deformation and damage at time 0 and then is deformed homogeneously and quasi-statically. By Eqs. 49 and 55, damage first starts growing when the condition $\mathcal{S} = 0$ occurs, hence:

$$(\underline{k}\underline{\mathbf{T}}) \bullet \underline{\mathbf{M}} = s_0.$$

Under continued quasi-static deformation with stable damage growth, Eq. 54 then leads to:

$$\dot{\underline{\phi}} = \underline{D} = \frac{\left(\underline{kC}\left(\underline{1} - \underline{\phi}\right)\underline{b}\right) \bullet \dot{\underline{e}}}{\left(\underline{kCbe} - s_1\left\|\underline{\phi}\right\|\underline{\phi}\right) \bullet \left(\underline{Cbe}^2\right)}\underline{Cbe}^2. \tag{56}$$

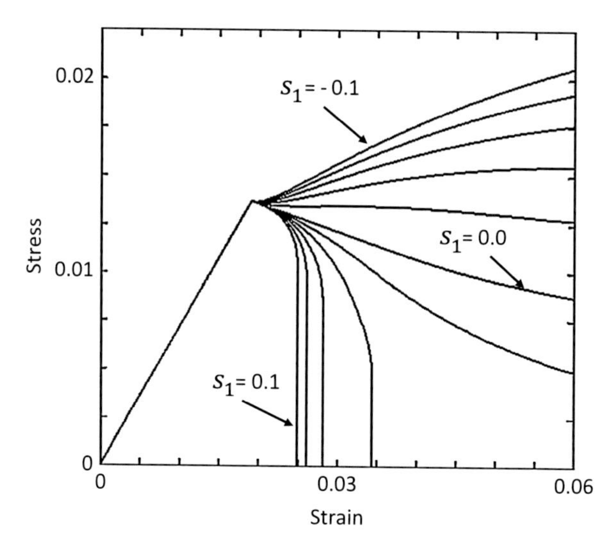

Fig. 2 Stress-strain curves for a bar under tension

Damage growth is stable until the denominator in Eq. 56 becomes nonpositive; then it becomes unstable. Figure 2 illustrates the behavior of this example material and damage growth model for a 1D body with:

$$\underline{C} = \underline{k} = \underline{1}, \quad \delta = 1, \quad s_0 = 0.02.$$

The different stress-strain curves are for different values of s_1, indicating the transition from stable to unstable damage growth as s_1 is increased. The stress is computed using Eq. 58, to be discussed below.

In 2D, similar material and damage models (for a material with a bulk modulus of 10MPa) can simulate the stable accumulation of diffuse damage near a stress singularity, as shown in Fig. 3. The specimen contains a semicircular notch and is under combined normal and transverse loading corresponding to strain rates of $\dot{\epsilon}_{22} = 2.0\text{s}^{-1}$ and $\dot{\epsilon}_{12} = 1.0\text{s}^{-1}$. Until a strain of about $\epsilon_{22} = 0.12$, there is a stable growth of damage near the notch. Then there is a sudden transition to dynamic fracture. The crack rapidly propagates to the opposite free edge of the specimen. This transition from stable to unstable can be seen in the stress-strain curve shown in Fig. 4. This curve represents the total normal load in the vertical direction divided by the cross-sectional area of the specimen (on a cross-sectional plane that does not include the notch).

These examples demonstrate the potential usefulness of peridynamic damage mechanics in modeling materials that either fracture immediately or after a period of accumulated continuous damage. The compatibility of the peridynamic field equations with both continuous and discontinuous deformations is helpful in modeling the spontaneous nucleation and growth of fractures within a damaged material.

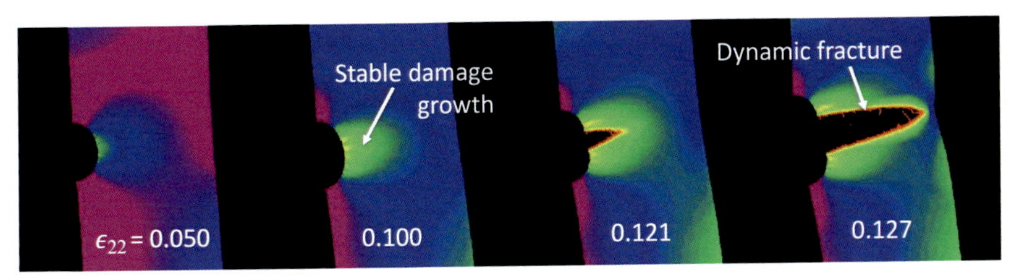

Fig. 3 Transition of stable damage growth at a stress concentration to dynamic fracture

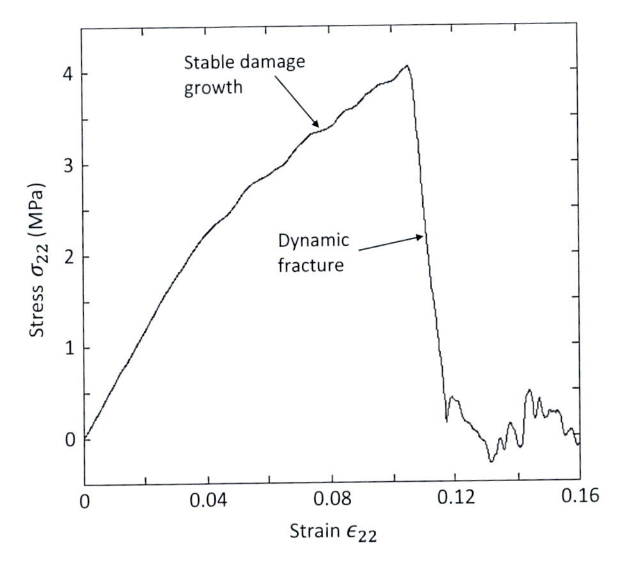

Fig. 4 Remote normal stress as a function of global strain for damage growth and fracture near a semicircular notch

Bond Breakage

A simpler approach to damage modeling in peridynamics has $\underline{\phi}\langle\boldsymbol{\xi}\rangle$ jump discontinuously from 0 to 1 according to some criterion, which can be quite general. This approach, which is called *bond breakage*, is used in the vast majority of peridynamic codes because of its simplicity and reduced memory requirements. It has the disadvantage compared with continuously varying $\underline{\phi}$ that instantaneous bond breakage can excite unwanted oscillations in a numerical grid. However, these can be suppressed in practice by applying damping forces to nodes after their bonds break.

A bond-breakage criterion can be as simple as a critical value of bond strain (Silling and Askari 2005). In this case, the critical bond strain for bond breakage can be calibrated to match a given critical energy release rate for the material. This bond-breakage strain depends on the horizon as well as the critical energy release rate. Critical bond strain as a failure criterion is a natural way to model brittle fracture, particularly in mode I. For other modes, it becomes trickier to specify the critical strain, which may depend on the conditions in other bonds in the family. Nucleation of cracks is predicted by bond breakage damage growth laws, but with a simple bond-strain criterion, it is difficult to prescribe both the critical energy release rate and the critical conditions for crack nucleation simultaneously. Postfailure response of the bonds depends on the material model; for example, it may or may not be appropriate in a given material to allow bonds to sustain compressive force after they break.

In spite of the many successes of the bond breakage approach to peridynamic damage modeling, the above considerations help motivate the development of the more general approach using the thermodynamic force and failure surfaces as described in section "Damage Evolution."

Connections with the Local Theory

This section summarizes the mathematical and conceptual connections between the nonlocal peridynamic theory and the local theory. The connections discussed here include the relations between the force state and the stress tensor, between the deformation state and the deformation gradient tensor, between nonlocal heat transport and the Clausius-Duhem inequality, the scaling and convergence of material models, and local damage mechanics. Not discussed below but also important is the peridynamic version of the Eshelby-Rice J -integral (Hu et al. 2012b; Silling and Lehoucq 2010).

Local Kinematics and Kinetics

Suppose a First Piola stress tensor $\boldsymbol{\sigma}$ is given. Let $\underline{\omega}$ be a positive scalar-valued state called the *influence function*. Consider the force state defined by:

$$\overline{\underline{\mathbf{T}}}(\sigma) = \sigma \underline{\omega} \mathbf{K}^{-1} \underline{\mathbf{X}} \quad \text{or} \quad \overline{\underline{\mathbf{T}}}(\sigma) \langle \xi \rangle = \sigma \underline{\omega} \langle \xi \rangle \mathbf{K}^{-1} \xi \quad \forall \xi \in \mathcal{H}_{\mathbf{x}} \tag{57}$$

where $\mathbf{K}$ is the *shape tensor* defined by:

$$\mathbf{K} = \int_{\mathcal{H}_{\mathbf{x}}} \underline{\omega} \langle \xi \rangle \, \xi \otimes \xi \, \mathrm{d}V_{\xi}.$$

The force state $\overline{\underline{\mathbf{T}}}(\sigma)$ defined by Eq. 57 has the property that in a uniform deformation of a homogeneous body, the force per unit area across any plane transferred by all the bonds that cross this plane is equal to:

$$\tau = \sigma \mathbf{n}$$

where $\mathbf{n}$ is a unit normal to the plane. Conversely, for a given force state $\underline{\mathbf{T}}$, the stress tensor defined by:

$$\overline{\boldsymbol{\sigma}}\left(\underline{\mathbf{T}}\right) = \int_{\mathcal{H}_{\mathbf{x}}} \underline{\mathbf{T}}\left\langle \boldsymbol{\xi} \right\rangle \otimes \boldsymbol{\xi} \mathrm{d}V_{\boldsymbol{\xi}} \tag{58}$$

has the same property. $\overline{\boldsymbol{\sigma}}$ is called the *partial stress tensor.* For any $\boldsymbol{\sigma}$,

$$\overline{\boldsymbol{\sigma}}\left(\overline{\underline{\mathbf{T}}}\left(\boldsymbol{\sigma}\right)\right) = \boldsymbol{\sigma}.$$

Analogous expressions can be derived for deformation states and deformation gradient tensors:

$$\overline{\underline{\mathbf{Y}}}\left(\mathbf{F}\right)\left\langle \boldsymbol{\xi} \right\rangle = \mathbf{F}\boldsymbol{\xi}, \qquad \overline{\mathbf{F}}\left(\underline{\mathbf{Y}}\right) = \left(\int_{\mathcal{H}_{\mathbf{x}}} \underline{\omega}\left\langle \boldsymbol{\xi} \right\rangle \underline{\mathbf{Y}}\left\langle \boldsymbol{\xi} \right\rangle \otimes \boldsymbol{\xi} \, \mathrm{d}V_{\boldsymbol{\xi}}\right) \mathbf{K}^{-1} \qquad \forall \mathbf{F}, \underline{\mathbf{Y}}. \tag{59}$$

It is easily confirmed that for any tensor $\mathbf{F}$,

$$\overline{\mathbf{F}}\left(\overline{\underline{\mathbf{Y}}}\left(\mathbf{F}\right)\right) = \mathbf{F}.$$

The relations (57) and (59) provide a way to adapt a local material model $\boldsymbol{\sigma}\left(\mathbf{F}\right)$ to peridynamics by setting

$$\underline{\mathbf{T}}\left(\underline{\mathbf{Y}}\right) = \overline{\underline{\mathbf{T}}}\left(\boldsymbol{\sigma}\left(\overline{\mathbf{F}}\left(\underline{\mathbf{Y}}\right)\right)\right) \tag{60}$$

for any $\underline{\mathbf{Y}}$. A peridynamic material model of the form (60) is called a *correspondence* model. These models have the properties that they are elastic, isotropic, and objective whenever the underlying local material model $\boldsymbol{\sigma}\left(\mathbf{F}\right)$ has these properties, in the sense of the local theory. Correspondence models generally exhibit zero-energy modes of deformation (Tupek and Radovitzky 2014) due to the noninvertibility of $\overline{\mathbf{F}}$, that is,

$$\overline{\mathbf{F}}\left(\underline{\mathbf{Y}}\right) = \overline{\mathbf{F}}\left(\underline{\mathbf{Y}}'\right)) \not\Longrightarrow \underline{\mathbf{Y}} = \underline{\mathbf{Y}}'.$$

Several practical ways of reducing this type of instability in numerical models have been proposed. One such method penalizes the departures of the deformation state from a uniform deformation within the family. In this method, the material model is modified by including an additional term as follows:

$$\underline{\mathbf{T}}\left(\underline{\mathbf{Y}}\right) = \overline{\underline{\mathbf{T}}}\left(\boldsymbol{\sigma}\left(\overline{\mathbf{F}}\left(\underline{\mathbf{Y}}\right)\right) + A\left\|\underline{\mathbf{Y}} - \overline{\mathbf{F}}\left(\underline{\mathbf{Y}}\right)\underline{\mathbf{X}}\right\|_{\underline{\mathbf{Y}}}\right.$$

where the subscript denotes the Fréchet derivative and A is a constant (Silling 2017).

For any deformation (not necessarily uniform) of any peridynamic body (not necessarily homogeneous), the *peridynamic stress tensor* is defined by:

$$\nu(\mathbf{x}) = \int_{\mathcal{U}} \int_0^\infty \int_0^\infty (y+z)^2 \left(\underline{\mathbf{T}}[\mathbf{x} - z\mathbf{m}] \langle (y+z)\mathbf{m} \rangle\right) \otimes \mathbf{m} \, \mathrm{d}y \, \mathrm{d}z \, \mathrm{d}\Omega_{\mathbf{m}}$$

where $\mathcal{U}$ is the unit sphere and $\mathrm{d}\Omega_{\mathbf{m}}$ is a differential spherical angle in the direction of the unit vector $\mathbf{m}$. ν has the surprising property (Lehoucq and Silling 2008; Lehoucq and von Lilienfeld 2010; Noll 1955) that

$$\nabla \cdot \nu = \int_{\mathcal{H}_{\mathbf{x}}} \{\underline{\mathbf{T}}[\mathbf{x}] \langle \mathbf{q} - \mathbf{x} \rangle - \underline{\mathbf{T}}[\mathbf{q}] \langle \mathbf{x} - \mathbf{q} \rangle\} \, \mathrm{d}V_{\mathbf{q}}.$$

This means that the peridynamic equilibrium equation can be written as:

$$\nabla \cdot \nu + \mathbf{b} = 0,$$

which is formally the same as in the local theory.

The partial stress tensor $\overline{\boldsymbol{\sigma}}$ is equal to $\boldsymbol{\nu}$ in the special case of a uniform deformation of a homogeneous body. Otherwise, the sense in which it approximates $\boldsymbol{\nu}$ is discussed in Silling et al. (2015).

Convergence of Peridynamics to the Local Theory

It seems reasonable to require that a proper nonlocal theory should converge, in some sense, to the local theory in the limit of "zero nonlocality." To investigate how this convergence works in the peridynamic equations, we first consider how material models scale as the horizon decreases.

As the horizon is changed, holding the bulk properties of the material fixed, peridynamic material models satisfy certain scaling relations. Let $\underline{\mathbf{T}}_\delta$ and $\underline{\mathbf{Y}}_\delta$ denote the force state and the deformation state, respectively, for any $\delta > 0$. Let δ_1 and δ_2 be two values of the horizon and $\mathcal{H}_1$ and $\mathcal{H}_2$ the corresponding families. Suppose a material model $\widehat{\underline{\mathbf{T}}}_1$ is given for the horizon δ_1. For any $\underline{\mathbf{Y}}_2$ on $\mathcal{H}_2$, define the state $\underline{\mathbf{Y}}_1$ by:

$$\underline{\mathbf{Y}}_1 \langle \boldsymbol{\xi}_1 \rangle = \frac{\delta_1}{\delta_2} \underline{\mathbf{Y}}_2 \left\langle \frac{\delta_2}{\delta_1} \boldsymbol{\xi}_1 \right\rangle \quad \forall \boldsymbol{\xi}_1 \in \mathcal{H}_1.$$

Consider the material model defined by:

$$\widehat{\underline{\mathbf{T}}}_2 (\underline{\mathbf{Y}}_2) \langle \boldsymbol{\xi}_2 \rangle = \left(\frac{\delta_1}{\delta_2}\right)^4 \widehat{\underline{\mathbf{T}}}_1 (\underline{\mathbf{Y}}_1) \left\langle \frac{\delta_1}{\delta_2} \boldsymbol{\xi}_2 \right\rangle \quad \forall \boldsymbol{\xi}_2 \in \mathcal{H}_2. \tag{61}$$

It is easily confirmed that for a uniform deformation of a homogeneous body, $\overline{\boldsymbol{\sigma}}$ defined in Eq. 58 is invariant with respect to this change in horizon:

$$\overline{\boldsymbol{\sigma}}\left(\widehat{\underline{\mathbf{T}}}_1\left(\underline{\mathbf{Y}}_1\right)\right)=\overline{\boldsymbol{\sigma}}\left(\widehat{\underline{\mathbf{T}}}_2\left(\underline{\mathbf{Y}}_2\right)\right).$$

In general, it can be shown that *if a deformation is twice continuously differentiable*, and if $\widehat{\underline{\mathbf{T}}}_\delta$ scales according to Eq. 61, then the limit:

$$\boldsymbol{\sigma}_0:=\lim_{\delta\to 0}\overline{\boldsymbol{\sigma}}\left(\widehat{\underline{\mathbf{T}}}_\delta\left(\underline{\mathbf{Y}}_\delta\right)\right)$$

exists and that:

$$\lim_{\delta\to 0}\int_{\mathcal{H}_\delta}\left\{\underline{\mathbf{T}}_\delta\left[\mathbf{x}\right]\langle\mathbf{q}-\mathbf{x}\rangle-\underline{\mathbf{T}}_\delta\left[\mathbf{q}\right]\langle\mathbf{x}-\mathbf{q}\rangle\right\}\mathrm{d}V_{\mathbf{q}}=\nabla\cdot\boldsymbol{\sigma}_0. \tag{62}$$

In summary, we now have a stress tensor field $\boldsymbol{\sigma}_0$ such that, in the limit of zero horizon, the peridynamic accelerations equal the accelerations in the local theory computed from the divergence of $\boldsymbol{\sigma}_0$; see Silling and Lehoucq (2008) for details. More rigorous results concerning convergence of peridynamics to local elasticity have been established (Emmrich et al. 2007).

Remarkably, Lipton has extended these results to discontinuous deformations; the limiting case of a peridynamic body containing a growing crack approaches a smooth solution in the local theory augmented by a Griffith crack that consumes energy at a definite rate as it grows. This result requires a peridynamic material model with a nonconvex strain-energy density function (Lipton 2014, 2016).

For heat transport, the statement analogous to Eq. 62 is as follows:

$$\lim_{\delta\to 0}\int_{\mathcal{H}_\delta}\left\{\underline{H}_\delta\left[\mathbf{x}\right]\langle\mathbf{q}-\mathbf{x}\rangle-\underline{H}_\delta\left[\mathbf{q}\right]\langle\mathbf{x}-\mathbf{q}\rangle\right\}\mathrm{d}V_{\mathbf{q}}=\nabla\cdot\mathbf{Q}_0 \tag{63}$$

where $\mathbf{Q}_0$ is the limiting heat flux vector field given by:

$$\mathbf{Q}_0:=\lim_{\delta\to 0}\overline{\mathbf{Q}}_\delta,\quad \overline{\mathbf{Q}}_\delta:=\overline{\mathbf{Q}}\left(\underline{H}_\delta\right)=-\int_{\mathcal{H}_{\mathbf{x}}}\underline{H}_\delta\langle\boldsymbol{\xi}\rangle\,\boldsymbol{\xi}\,\mathrm{d}V_{\boldsymbol{\xi}}. \tag{64}$$

The minus sign appears by convention in the second of Eq. 64 so that the heat flux $\overline{\mathbf{Q}}\cdot\mathbf{n}$ through a plane normal to a unit vector $\mathbf{n}$ will be positive if energy is flowing parallel to $\mathbf{n}$, rather than opposite to it. Recall the inequality (23) derived from the second law,

$$\underline{\beta}\bullet\underline{H}_\delta\le 0,\quad \underline{\beta}\left[\mathbf{x}\right]\langle\mathbf{q}-\mathbf{x}\rangle=\frac{1}{\theta\left(\mathbf{q}\right)}-\frac{1}{\theta\left(\mathbf{x}\right)}. \tag{65}$$

For a smooth temperature field, as $\delta \to 0$, we can use the first term of a Taylor series to write down the first-order approximation:

$$\underline{\beta}\langle\boldsymbol{\xi}\rangle \approx \boldsymbol{\xi}\cdot\nabla\beta(\mathbf{x}) \qquad \forall\boldsymbol{\xi}\in\mathcal{H}_{\mathbf{x}},$$

hence Eq. 65 can be approximated by:

$$0 \ge \underline{\beta}\bullet\underline{H}_{\delta} \approx \int_{\mathcal{H}_{\mathbf{x}}} \boldsymbol{\xi}\cdot(\nabla\beta)\,\underline{H}_{\delta}\langle\boldsymbol{\xi}\rangle = -\nabla\beta\cdot\overline{\mathbf{Q}}_{\delta}. \tag{66}$$

Similarly, for small δ, Eq. 63 is approximated by:

$$h_{\delta} := \int_{\mathcal{H}_{\delta}} \left\{\underline{H}_{\delta}[\mathbf{x}]\langle\mathbf{q}-\mathbf{x}\rangle - \underline{H}_{\delta}[\mathbf{q}]\langle\mathbf{x}-\mathbf{q}\rangle\right\}\mathrm{d}V_{\mathbf{q}} \approx \int_{\mathcal{H}_{\mathbf{x}}} \boldsymbol{\xi}\cdot\nabla\underline{H}_{\delta}\langle\boldsymbol{\xi}\rangle = -\nabla\cdot\overline{\mathbf{Q}}_{\delta}. \tag{67}$$

From Eqs. 20, 66, and 67,

$$\begin{aligned}\dot{\eta} &\ge \beta h_{\delta} + \beta s \\ &\approx -\beta\nabla\cdot\overline{\mathbf{Q}}_{\delta} + \beta s \\ &\ge -\beta\nabla\cdot\overline{\mathbf{Q}}_{\delta} - \nabla\beta\cdot\overline{\mathbf{Q}}_{\delta} + \beta s \\ &= -\nabla\left(\beta\overline{\mathbf{Q}}_{\delta}\right) + \beta s.\end{aligned}$$

Omitting some of the details of taking the limit, the result for $\delta \to 0$ is:

$$\dot{\eta} \ge \frac{s}{\theta} - \nabla\cdot\left(\frac{\mathbf{Q}_0}{\theta}\right).$$

This is a form of the Clausius-Duhem inequality of local continuum thermodynamics. It is interesting that this local form of the second law with heat transport can be derived from the peridynamic version of the second law without specifying any particular form of the constitutive model $\underline{\widehat{H}}(\underline{\Theta})$ or assuming a particular physical mechanism for heat transport (conduction, convection, radiation, etc.).

Local Continuum Damage Mechanics

The failure characteristics of engineering materials are frequently expressed in terms of failure surfaces, with or without some specification of postfailure behavior. These expressions can sometimes be converted to peridynamic failure surfaces using the partial stress and approximate deformation gradient defined in Eq. 58 and 59 above. For example, if a failure surface in terms of the stress tensor and deformation gradient tensor is given in the form $\mathcal{S}^0(\boldsymbol{\sigma}, \mathbf{F}, \varphi)$, where φ is a scalar damage variable, we can define a peridynamic failure surface by:

$$\mathcal{S}\left(\underline{\mathbf{T}}, \underline{\mathbf{Y}}, \underline{\phi}\right) = \mathcal{S}^0\left(\overline{\boldsymbol{\sigma}}\left(\underline{\mathbf{T}}\right), \overline{\mathbf{F}}\left(\underline{\mathbf{Y}}\right), \overline{\varphi}\left(\underline{\phi}\right)\right), \quad \overline{\varphi}\left(\underline{\phi}\right) = \frac{\underline{\omega} \bullet \underline{\phi}}{\underline{\omega} \bullet \underline{1}}.$$

Evaluating the required Fréchet derivatives in Eq. 54 then leads to:

$$\mathcal{S}_{\underline{\mathbf{T}}} = \frac{\partial \mathcal{S}^0}{\partial \boldsymbol{\sigma}} \underline{\mathbf{X}}, \quad \mathcal{S}_{\underline{\mathbf{Y}}} = \frac{\partial \mathcal{S}^0}{\partial \mathbf{F}} \underline{\omega} \langle \boldsymbol{\xi} \rangle \mathbf{K}^{-1} \underline{\mathbf{X}}, \quad \mathcal{S}_{\underline{\phi}} = \frac{\partial \mathcal{S}^0}{\partial \varphi} \frac{\underline{\omega}}{\underline{\omega} \bullet \underline{1}}.$$

The remaining derivatives that appear in Eq. 54 are obtained from the peridynamic material model, as before.

Discussion

Why and how does a crack grow? How does a continuous body become discontinuous? Why does nature seem to favor these discontinuities as energy minimizers, yet equip real materials with energy barriers that resist their formation and growth? How can real materials be designed or optimized to resist cracking? These and other fundamental questions possibly can be studied within the peridynamic theory.

Although peridynamics is often used to model brittle fracture and fragmentation (for example, (Hu et al. 2013)), diverse new applications are continually being discovered. These have recently included, for example:

- Electromigration in integrated circuits (Gerstle et al. 2008; Oterkus et al. 2013)
- Biological cell mechanics and tumor growth (Lejeune and Linder 2017a, b; Taylor et al. 2016)
- Damage in materials due to high voltage breakdown (Wildman and Gazonas 2015)
- Effects of residual thermal stress on fracture in glass (Jeon et al. 2015; Kilic and Madenci 2009)
- Failure of reinforced concrete (Gerstle et al. 2010)
- Mechanics of nanocomposites (Prakash and Seidel 2016)
- Fluid transport and hydraulic fracture in rocks (Katiyar et al. 2014; Nadimi 2015; Ouchi et al. 2015; Van Der Merwe 2014)
- Solitons (Silling 2016)
- Corrosion (Chen and Bobaru 2015)

and many others.

Acknowledgment Sandia National Laboratories is a multimission laboratory managed and operated by National Technology and Engineering Solutions of Sandia LLC, a wholly owned subsidiary of Honeywell International Inc. for the US Department of Energy's National Nuclear Security Administration under contract DE-NA0003525.

References

F. Bobaru, M. Duangpanya, The peridynamic formulation for transient heat conduction. Int. J. Heat Mass Transf. **53**, 4047–4059 (2010)

F. Bobaru, M. Duangpanya, A peridynamic formulation for transient heat conduction in bodies with evolving discontinuities. J. Comput. Phys. **231**, 2764–2785 (2012)

Z. Chen, F. Bobaru, Peridynamic modeling of pitting corrosion damage. J. Mech. Phys. Solids **78**, 352–381 (2015)

S.R. Chowdhury, P. Roy, D. Roy, J. Reddy, A peridynamic theory for linear elastic shells. Int. J. Solids Struct. **84**, 110–132 (2016)

B.D. Coleman, W. Noll, The thermodynamics of elastic materials with heat conduction and viscosity. Arch. Ration. Mech. Anal. **13**, 167–178 (1963)

C. Diyaroglu, E. Oterkus, S. Oterkus, E. Madenci, Peridynamics for bending of beams and plates with transverse shear deformation. Int. J. Solids Struct. **69**, 152–168 (2015)

E. Emmrich, O. Weckner, et al., On the well-posedness of the linear peridynamic model and its convergence towards the navier equation of linear elasticity. Commun. Math. Sci. **5**, 851–864 (2007)

J.T. Foster, S.A. Silling, W.W. Chen, Viscoplasticity using peridynamics. Int. J. Numer. Methods Eng. **81**, 1242–1258 (2010)

E. Fried, New insights into the classical mechanics of particle systems. Discrete Contin. Dyn. Syst. **28**, 1469–1504 (2010)

W. Gerstle, N. Sau, S.A. Silling, Peridynamic modeling of concrete structures. Nucl. Eng. Des. **237**, 1250–1258 (2007)

W. Gerstle, S. Silling, D. Read, V. Tewary, R. Lehoucq, Peridynamic simulation of electromigration. Comput. Mater. Continua **8**, 75–92 (2008)

W. Gerstle, N. Sakhavand, S. Chapman, Peridynamic and continuum models of reinforced concrete lap splice compared, in *Fracture Mechanics of Concrete and Concrete Structures, Recent Advances in Fracture Mechanics of Concrete*, ed. by B.H. Oh, et al. (2010), pp. 306–312

J. O'Grady, J. Foster, Peridynamic plates and flat shells: a non-ordinary, state-based model. Int. J. Solids Struct. **51**, 4572–4579 (2014)

M.E. Gurtin, W.O. Williams, On the first law of thermodynamics. Arch. Ration. Mech. Anal. **42**, 77–92 (1971)

M.E. Gurtin, E. Fried, L. Anand, *The mechanics and thermodynamics of continua* (Cambridge University Press, Cambridge, 2010), pp. 232–233

W. Hu, Y.D. Ha, F. Bobaru, Peridynamic model for dynamic fracture in unidirectional fiber-reinforced composites. Comput. Methods Appl. Mech. Eng. **217**, 247–261 (2012a)

W. Hu, Y.D. Ha, F. Bobaru, S.A. Silling, The formulation and computation of the nonlocal J-integral in bond-based peridynamics. Int. J. Fract. **176**, 195–206 (2012b)

W. Hu, Y. Wang, J. Yu, C.-F. Yen, F. Bobaru, Impact damage on a thin glass plate with a thin polycarbonate backing. Int. J. Impact Eng. **62**, 152–165 (2013)

B. Jeon, R.J. Stewart, I.Z. Ahmed, Peridynamic simulations of brittle structures with thermal residual deformation: strengthening and structural reactivity of glasses under impacts. Proc. R. Soc. A **471**, 20150231. (2015)

A. Katiyar, J.T. Foster, H. Ouchi, M.M. Sharma, A peridynamic formulation of pressure driven convective fluid transport in porous media. J. Comput. Phys. **261**, 209–229 (2014)

B. Kilic, E. Madenci, Prediction of crack paths in a quenched glass plate by using peridynamic theory. Int. J. Fract. **156**, 165–177 (2009)

R.B. Lehoucq, M.P. Sears, Statistical mechanical foundation of the peridynamic nonlocal continuum theory: energy and momentum conservation laws. Phys. Rev. E **84**, 031112 (2011)

R.B. Lehoucq, S.A. Silling, Force flux and the peridynamic stress tensor. J. Mech. Phys. Solids **56**, 1566–1577 (2008)

R.B. Lehoucq, O.A. von Lilienfeld, Translation of Walter Noll's derivation of the fundamental equations of continuum thermodynamics from statistical mechanics. J. Elast. **100**, 5–24 (2010)

E. Lejeune, C. Linder, Modeling tumor growth with peridynamics. Biomech. Model. Mechanobiol., 1–17 (2017a)
E. Lejeune, C. Linder, Quantifying the relationship between cell division angle and morphogenesis through computational modeling. J. Theor. Biol. **418**, 1–7 (2017b)
R. Lipton, Dynamic brittle fracture as a small horizon limit of peridynamics. J. Elast. **117**, 21–50 (2014)
R. Lipton, Cohesive dynamics and brittle fracture. J. Elast. **142**, 1–49 (2016)
E. Madenci, S. Oterkus, Ordinary state-based peridynamics for plastic deformation according to von Mises yield criteria with isotropic hardening. J. Mech. Phys. Solids **86**, 192–219 (2016)
J.A. Mitchell, A non-local, ordinary-state-based viscoelasticity model for peridynamics. Technical report SAND2011-8064, Sandia National Laboratories, Albuquerque/Livermore, October 2011a
J.A. Mitchell, A nonlocal, ordinary, state-based plasticity model for peridynamics. Technical report SAND2011-3166, Sandia National Laboratories, Albuquerque/Livermore, October 2011b
S. Nadimi, State-based peridynamics simulation of hydraulic fracture phenomenon in geological media. Master's thesis, The University of Utah, 2015
W. Noll, Die Herleitung der Grundgleichungen der Thermomechanik der Kontinua aus der statistischen Mechanik. J. Ration. Mech. Anal. **4**, 627–646 (1955.) In German, English translation available
E. Oterkus, E. Madenci, Peridynamic analysis of fiber-reinforced composite materials. J. Mech. Mater. Struct. **7**, 45–84 (2012)
S. Oterkus, J. Fox, E. Madenci, Simulation of electro-migration through peridynamics, in *2013 IEEE 63rd Electronic Components and Technology Conference* (IEEE, 2013), pp. 1488–1493
S. Oterkus, E. Madenci, A. Agwai, Fully coupled peridynamic thermomechanics. J. Mech. Phys. Solids **64**, 1–23 (2014a)
S. Oterkus, E. Madenci, A. Agwai, Peridynamic thermal diffusion. J. Comput. Phys. **265**, 71–96 (2014b)
H. Ouchi, A. Katiyar, J. Foster, M.M. Sharma, et al., A peridynamics model for the propagation of hydraulic fractures in heterogeneous, naturally fractured reservoirs. in *SPE Hydraulic Fracturing Technology Conference* (Society of Petroleum Engineers, 2015)
N. Prakash, G.D. Seidel, A coupled electromechanical peridynamics framework for modeling carbon nanotube reinforced polymer composites, in *57th AIAA/ASCE/AHS/ASC Structures, Structural Dynamics, and Materials Conference*, p. 0936, (2016)
S.A. Silling, Reformulation of elasticity theory for discontinuities and long-range forces. J. Mech. Phys. Solids **48**, 175–209 (2000)
S.A. Silling, Linearized theory of peridynamic states. J. Elast. **99**, 85–111 (2010)
S.A. Silling, Solitary waves in a peridynamic elastic solid. J. Mech. Phys. Solids **96**, 121–132 (2016)
S.A. Silling, Stability of peridynamic correspondence material models and their particle discretizations. Comput. Methods Appl. Mech. Eng. **322**, 42–57 (2017)
S.A. Silling, E. Askari, A meshfree method based on the peridynamic model of solid mechanics. Comput. Struct. **83**, 1526–1535 (2005)
S.A. Silling, R.B. Lehoucq, Convergence of peridynamics to classical elasticity theory. J. Elast. **93**, 13–37 (2008)
S.A. Silling, R.B. Lehoucq, The peridynamic theory of solid mechanics. Adv. Appl. Mech. **44**, 73–166 (2010)
S.A. Silling, M. Epton, O. Weckner, J. Xu, E. Askari, Peridynamic states and constitutive modeling. J. Elast. **88**, 151–184 (2007)
S.A. Silling, D. Littlewood, P. Seleson, Variable horizon in a peridynamic medium. J. Mech. Mater. Struct. **10**, 591–612 (2015)
S.A. Silling, M.L. Parks, J.R. Kamm, O. Weckner, M. Rassaian, Modeling shockwaves and impact phenomena with Eulerian peridynamics. Int. J. Impact Eng. **107**, 47–57 (2017)
S. Sun, V. Sundararaghavan, A peridynamic implementation of crystal plasticity. Int. J. Solids Struct. **51**, 3350–3360 (2014)

M. Taylor, I. Gözen, S. Patel, A. Jesorka, K. Bertoldi, Peridynamic modeling of ruptures in biomembranes. PLoS One **11**, e0165947 (2016)

M. Tupek, R. Radovitzky, An extended constitutive correspondence formulation of peridynamics based on nonlinear bond-strain measures. J. Mech. Phys. Solids **65**, 82–92 (2014)

C.W. Van Der Merwe, A peridynamic model for sleeved hydraulic fracture. Master's thesis, Stellenbosch University, Stellenbosch, (2014)

T.L. Warren, S.A. Silling, A. Askari, O. Weckner, M.A. Epton, J. Xu, A nonordinary state-based peridynamic method to model solid material deformation and fracture. Int. J. Solids Struct. **46**, 1186–1195 (2009)

O. Weckner, N.A.N. Mohamed, Viscoelastic material models in peridynamics. Appl. Math. Comput. **219**, 6039–6043 (2013)

R. Wildman, G. Gazonas, A dynamic electro-thermo-mechanical model of dielectric breakdown in solids using peridynamics. J. Mech. Mater. Struct. **10**, 613–630 (2015)

Recent Progress in Mathematical and Computational Aspects of Peridynamics

35

Marta D'Elia, Qiang Du, and Max Gunzburger

Contents

M. D'Elia (✉)
Optimization and Uncertainty Quantification Department Center for Computing Research, Sandia National Laboratories, Albuquerque, NM, USA
e-mail: mdelia@sandia.gov

Q. Du
Department of Applied Physics and Applied Mathematics, Columbia University, New York, NY, USA
e-mail: qd2125@columbia.edu

M. Gunzburger
Department of Scientific Computing, Florida State University, Tallahassee, FL, USA
e-mail: mgunzburger@fsu.edu

G. Z. Voyiadjis (ed.), *Handbook of Nonlocal Continuum Mechanics for Materials and Structures*, https://doi.org/10.1007/978-3-319-58729-5_30

Abstract

Recent developments in the mathematical and computational aspects of the nonlocal peridynamic model for material mechanics are provided. Based on a recently developed vector calculus for nonlocal operators, a mathematical framework is constructed that has proved useful for the mathematical analyses of peridynamic models and for the development of finite element discretizations of those models. A specific class of discretization algorithms referred to as asymptotically compatible schemes is discussed; this class consists of methods that converge to the proper limits as grid sizes and nonlocal effects tend to zero. Then, the multiscale nature of peridynamics is discussed including how, as a single model, it can account for phenomena occurring over a wide range of scales. The use of this feature of the model is shown to result in efficient finite element implementations. In addition, the mathematical and computational frameworks developed for peridynamic simulation problems are shown to extend to control, coefficient identification, and obstacle problems.

Keywords

Peridynamics · Nonlocal vector calculus · Variational forms · Multiscale methods · Finite element method · Optimal control · Obstacle problems

Introduction

The peridynamic (PD) model for solid mechanics was introduced in Silling (2000), followed by a more generally applicable version in Silling et al. (2007); see also the review (Silling and Lehoucq 2010). The main features of the model are that it is *a continuum model that is free from spatial derivatives, allows for nonlocal interactions, and allows for solutions that contain jump discontinuities across lower-dimensional manifolds.* As such, it is especially well suited for simulations of material failure phenomena such as fracture. Indeed, despite its relatively recent development, the effectiveness of PD has already been demonstrated in several sophisticated applications, including the fracture and failure of composites, crack instability, fracture of polycrystals, and nanofiber networks; see Askari et al. (2008) for a review and also Bobaru and Silling (2004), Bobaru et al. (2005), Gerstle and Sau (2004), Gerstle et al. (2005), Littlewood (2010), Silling (2003), Silling and Askari (2004), Silling and Bobaru (2005), Silling et al. (2003), Weckner and Abeyaratne (2005), and Weckner and Emmrich (2005). The successful application of PD for multiscale engineering analyses has been enabled by the development of software packages, see, e.g., Parks et al. (2008, 2012) and Silling and Askari (2005). There has also been extensive studies of peridynamic models appearing in the engineering, material science, and mathematical literatures.

An additional feature of PD is that, all on its own, it is a *multiscale* model. A *valid* material model is one that provides a faithful description of the physical phenomena. A *tractable* material model is one for which useful information can be extracted, e.g., through discretization, in an efficient manner. These two requirements allow us to define a *multiscale* material model as one that is valid and tractable over a wide range of spatial and temporal scales. In most cases, multiscale models are defined by coupling two different models that operate at different scales, e.g., molecular dynamics and classical elasticity; peridynamics provides the opportunity to do away with the need for such troublesome couplings. Thus, we refer to peridynamics as a multiscale *mono*-model in contrast to multiscale *multi*-models for which two different models, e.g., classical elasticity and molecular dynamics, are coupled to produce a multiscale model. This feature, which is discussed below, results from the introduction of a length scale $\varepsilon > 0$, often referred to as the horizon, such that points separated by a distance greater than ε do not, for all practical purposes, interact.

The rest of this section is devoted to a very brief description of peridynamic models. Then, in section "Mathematical Framework" we present a mathematical framework, based on a recently developed vector calculus for nonlocal operators, that has proved useful for the mathematical analyses of peridynamic models and for the development of finite element discretizations of those models. In particular, the recently developed concept of asymptotically compatible schemes is discussed. In section "Peridynamics as a Multiscale Mono-model," the multiscale nature of peridynamics as a mono-model for material mechanics is discussed as is the implications that feature of peridynamics has on efficient finite element implementations. In section "Control, Identification, and Obstacle Problems," we show how the mathematical and computational frameworks developed for peridynamic simulation problems can be extended to control, coefficient identification, and obstacle problems. The material presented in this paper is drawn from, among other sources, Du et al. (2012a, 2013a,b) and Du (2016a,b) for section "Mathematical Framework," Xu et al. (2016b) for section "Peridynamics as a Multiscale Mono–model," and D'Elia and Gunzburger (2014, 2016) and Guan and Gunzburger (2017) for section "Control, Identification, and Obstacle Problems."

Peridynamic Models and Their Linearized Form

The general state-based peridynamic equation of motion (Silling et al. 2007) takes the form

$$\rho\partial_{tt}\mathbf{u}(\mathbf{x},t) = \int_{\mathbb{R}^n} \{\underline{\mathrm{T}}[\mathbf{x}]\langle\mathbf{y}-\mathbf{x}\rangle - \underline{\mathrm{T}}[\mathbf{y}]\langle\mathbf{x}-\mathbf{y}\rangle\} d\mathbf{y} + \mathbf{b}(\mathbf{x}) \tag{1}$$

with $\underline{\mathrm{T}}[\mathbf{x}]\langle\mathbf{y}-\mathbf{x}\rangle$ and $\underline{\mathrm{T}}[\mathbf{y}]\langle\mathbf{x}-\mathbf{y}\rangle$ denoting the peridynamic force states, $\mathbf{u}$ the displacement vector, ρ the material density, and $\mathbf{b}$ a given body force density. Under the assumption of small deformation, (1) can be approximated by the linear integrodifferential equation

$$\rho\partial_{tt}\mathbf{u}(\mathbf{x},t) = \int_{\mathbb{R}^n} \mathrm{C}_\varepsilon(\mathbf{y},\mathbf{x})\big(\mathbf{u}(\mathbf{y},t) - \mathbf{u}(\mathbf{x},t)\big) d\mathbf{y} + \mathbf{b}(\mathbf{x}). \tag{2}$$

As an example, we have the linear bond-based model for which C_ε is a rank-one tensor of the form

$$\mathrm{C}_\varepsilon(\mathbf{y},\mathbf{x}) = \omega_\varepsilon(|\mathbf{y}-\mathbf{x}|)\big[\boldsymbol{\alpha}(\mathbf{y},\mathbf{x}) \otimes \boldsymbol{\alpha}(\mathbf{y},\ \mathbf{x})\big], \quad \text{where} \quad \boldsymbol{\alpha}(\mathbf{y},\mathbf{x}) = \frac{\mathbf{y}-\mathbf{x}}{|\mathbf{y}-\mathbf{x}|^2} \tag{3}$$

denotes a scaled vector along the bond between $\mathbf{x}$ and $\mathbf{y}$. We assume that nonlocal interactions occur only when the distance between $\mathbf{x}$ and $\mathbf{y}$ is smaller than a specified distance ε, referred to as the *horizon*. Thus, the nonlocal interaction kernel $\omega_\varepsilon(|\mathbf{y}-\mathbf{x}|)\mathrm{L}$ is supported in a spherical neighborhood of radius ε. However, for the state-based peridynamic model, the range of interactions becomes broader due to the indirect interactions (Silling 2010). Nevertheless, we still use ε as a characteristic measure of the range of nonlocal interactions.

Local limit of peridynamic models. In peridynamic models, if $\varepsilon \to 0$, then nonlocal interactions become localized. We expect that in several settings, such as linear peridynamic models, the nonlocal models reduce to classical partial differential equation (PDE) models under proper assumptions on the interaction kernels. Intuitively, these assumptions imply that $\underline{\omega}_\varepsilon(|\mathbf{z}|)$ approaches a constant multiple of the Dirac delta function at $\mathbf{z} = \mathbf{0}$. For the linear bond-based peridynamics, such studies have been carried out in Du et al. (2013b), Du and Zhou (2011), and Mengesha and Du (2013).

Mathematical Framework

Understanding and developing mathematical treatments of nonlocal models call for a new mathematical framework. In this section, we give a brief description of that framework and its use in the nonlocal mechanic setting.

A Nonlocal Vector Calculus

Because the quantities of interest in continuum mechanics are mostly vector and tensor-valued function, the nonlocal vector calculus presented in Du et al. (2013a) provides a systematic means for formulating nonlocal mechanical models such as peridynamics. Nonlocal operators are defined as the basic building blocks of the nonlocal vector calculus. For example, for any two points $\mathbf{y}$ and $\mathbf{x}$ in $\mathbb{R}^n$, we choose a two-point vector $\boldsymbol{\alpha}(\mathbf{x},\mathbf{y})\colon \mathbb{R}^n \times \mathbb{R}^n \to \mathbb{R}^m$ that is antisymmetric, i.e., we have $\boldsymbol{\alpha}(\mathbf{y},\ \mathbf{x}) = -\boldsymbol{\alpha}(\mathbf{x},\mathbf{y})$, and then, respectively, define a nonlocal divergence operator $\mathscr{D}$ acting on tensors and its adjoint operator $\mathfrak{D}^*$ acting on vectors by

$$(\mathfrak{D}\underline{\boldsymbol{\psi}})(\mathbf{x}) = \int_{\mathbb{R}^n} \big(\underline{\boldsymbol{\psi}}(\mathbf{x},\ \mathbf{y}) + \underline{\boldsymbol{\psi}}(\mathbf{y},\mathbf{x})\big)\boldsymbol{\alpha}(\mathbf{x},\mathbf{y})d\mathbf{y} \qquad \forall\, \underline{\boldsymbol{\psi}}(\mathbf{x},\mathbf{y})\colon \mathbb{R}^n \times \mathbb{R}^n \to \mathbb{R}^{n\times m}$$

$$(\mathfrak{D}^*\mathbf{v})(\mathbf{x},\mathbf{y}) = \big(\mathbf{v}(\mathbf{y}) - \mathbf{v}(\mathbf{x})\big) \otimes \boldsymbol{\alpha}(\mathbf{y},\ \mathbf{x}) \qquad \forall\, \mathbf{v}(\mathbf{y})\colon \mathbb{R}^n \to \mathbb{R}^m.$$

$\mathfrak{D}$ and $\mathfrak{D}^*$ are adjoint operators in the sense that

$$\int_{\mathbb{R}^n} \mathbf{v}(\mathbf{x}) \cdot (\mathfrak{D}\underline{\boldsymbol{\psi}})(\mathbf{x})d\mathbf{x} = \int_{\mathbb{R}^n}\int_{\mathbb{R}^n} (\mathfrak{D}^*\mathbf{v})(\mathbf{x},\mathbf{y})\underline{\boldsymbol{\psi}}(\mathbf{x},\mathbf{y})d\mathbf{y}d\mathbf{x}.$$

Note that $\mathfrak{D}^*\mathbf{v}$ is a symmetric two-point tensor-valued function, i.e., $(\mathfrak{D}^*\mathbf{v})(\mathbf{x},\mathbf{y}) = (\mathfrak{D}^*\mathbf{v})(\mathbf{y},\mathbf{x})$. On the other hand, for any two-point tensor function $\underline{\boldsymbol{\psi}}(\mathbf{x},\mathbf{y})$, $\mathfrak{D}\underline{\boldsymbol{\psi}}$ depends only on the symmetric part of $\underline{\boldsymbol{\psi}}$ with respect to its two variables $\mathbf{x}$ and $\mathbf{y}$. With respect to the dimension parameters m and n, the cases most relevant to peridynamics are $m = 1$ and $m = n$.

For notational convenience, as in Du et al. (2013a), we introduce the nonlocal divergence operator $\mathscr{D}$ acting on vectors and its adjoint operator $\mathscr{D}^*$ acting on scalars by

$$(\mathscr{D}\boldsymbol{\varphi})(\mathbf{x}) = \int_{\mathbb{R}^n} \big(\boldsymbol{\varphi}(\mathbf{x},\mathbf{y}) + \boldsymbol{\varphi}(\mathbf{y},\mathbf{x})\big)\cdot\boldsymbol{\alpha}(\mathbf{x},\mathbf{y})d\mathbf{y} \qquad \forall\ \varphi\colon \mathbb{R}^n\times\mathbb{R}^n \to \mathbb{R}^n$$

$$(\mathscr{D}^* v)(\mathbf{x},\mathbf{y}) = \big(v(\mathbf{y}) - v(\mathbf{x})\big)\boldsymbol{\alpha}(\mathbf{y},\mathbf{x}) \qquad \forall\ v\colon \mathbb{R}^n \to \mathbb{R}.$$

To account for the indirect interactions between two material points present in the state-based peridynamic models, the weighted nonlocal operators $\mathfrak{D}_{\omega_\varepsilon}$ and $\mathfrak{D}^*_{\omega_\varepsilon}$ are introduced in Du et al. (2013a). For example, letting $\omega_\varepsilon(\mathbf{x},\mathbf{y})\colon \mathbb{R}^n\times\mathbb{R}^n \to \mathbb{R}$ denote a nonnegative scalar-valued two-point function and given the function $\mathbf{u}(\mathbf{x})\colon \mathbb{R}^n \to \mathbb{R}^m$, we have the weighted divergence operator given by

$$\mathfrak{D}_{\omega_\varepsilon}(\mathbf{u})(\mathbf{x}) = \mathfrak{D}\big(\omega_\varepsilon(\mathbf{x},\mathbf{y})\mathbf{u}(\mathbf{x})\big) \qquad \text{for } \mathbf{x}\in\mathbb{R}^n.$$

These operators are more closely aligned with conventional first-order differential operators defined on vector fields compared to the operators $\mathscr{D}$, $\mathscr{D}^*$, $\mathfrak{D}$, and $\mathfrak{D}^*$. More rigorous definitions and careful modifications in bounded domains are given in Mengesha and Du (2016). Additional formal discussions about $\mathscr{D}$, $\mathscr{D}^*$, $\mathscr{D}_\omega$, and $\mathscr{D}^*_{\omega_\varepsilon}$ can be found in Du et al. (2013a) and rigorous derivations in Mengesha and Du (2016) along with nonlocal Green's identities such as

$$\int_{\mathbb{R}^n} \mathbf{u}\left(\mathfrak{D}\big(\omega(\mathfrak{D}^*\mathbf{v})\big)\right)d\mathbf{x} - \int_{\mathbb{R}^n} \left(\mathfrak{D}\big(\omega(\mathfrak{D}^*\mathbf{u})\big)\right)\mathbf{v}d\mathbf{x} = 0.$$

The nonlocal Green's identities are nonlocal analogs of the classical counterparts and are crucial in defining variational formulations of nonlocal mechanic problems.

Formulating Nonlocal Models via Nonlocal Operators

The nonlocal vector calculus allows us to reformulate linear peridynamic models in ways similar to the way local linear PDE models can be expressed in terms of basic divergence, gradient, and curl operators. For example, for a given kernel $\omega_\varepsilon =$

$\omega_\varepsilon(|\mathbf{y}-\mathbf{x}|)$ and any function $\mathbf{u} = \mathbf{u}(\mathbf{x})$ defined in $\mathbb{R}^n$, using the matrix and vector product identities

$$(\boldsymbol{\beta} \otimes \boldsymbol{\alpha})\boldsymbol{\alpha} = \boldsymbol{\beta}\boldsymbol{\alpha}^{\mathrm{T}}\boldsymbol{\alpha} = |\boldsymbol{\alpha}|^2\boldsymbol{\beta} \quad \text{and} \quad (\boldsymbol{\alpha} \otimes \boldsymbol{\alpha})\boldsymbol{\beta} = \boldsymbol{\alpha}^{\mathrm{T}}\boldsymbol{\beta}\boldsymbol{\alpha} \quad \forall\, \boldsymbol{\alpha}, \boldsymbol{\beta} \in \mathbb{R}^m \tag{4}$$

and the notation $\underline{\omega}_\varepsilon(|\mathbf{y}-\mathbf{x}|) = \omega_\varepsilon(|\mathbf{y}-\mathbf{x}|)|\boldsymbol{\alpha}(\mathbf{x}, \mathbf{y})|^2$, a composition of the operators $\mathscr{D}$ and $\mathscr{D}^*$ yields

$$\mathfrak{L}_\varepsilon(\mathbf{u})(\mathbf{x}) = -\mathfrak{D}(\omega_\varepsilon \mathfrak{D}^*\mathbf{u})(\mathbf{x}) = \int_{\mathbb{R}^n} \underline{\omega}_\varepsilon(|\mathbf{y}-\mathbf{x}|)\, ((\mathbf{u}(\mathbf{y}) - \mathbf{u}(\mathbf{x})) \otimes \boldsymbol{\alpha}(\mathbf{y}, \mathbf{x}))\, \boldsymbol{\alpha}(\mathbf{y}, \mathbf{x}) d\mathbf{y} \tag{5}$$

which corresponds to the linear Navier operator for bond-based peridynamics. On the other hand, for a scalar function u, we have

$$\mathcal{L}_\varepsilon(u)(\mathbf{x}) = -\mathscr{D}(\omega_\varepsilon \mathscr{D}^* u)(\mathbf{x}) = \int_{\mathbb{R}^n} \underline{\omega}_\varepsilon(|\mathbf{y}-\mathbf{x}|)(u(\mathbf{y}) - u(\mathbf{x})) d\mathbf{y} \tag{6}$$

which is often referred to a *nonlocal Laplacian* or a nonlocal diffusion operator.

A Linear Steady-State Problem as an Illustration

We consider a linear steady-state problem related to the nonlocal bond-based peridynamic model for which the nonlocal force is a linear Hookean spring force aligned with the bond direction. We have

$$-\mathfrak{L}_\varepsilon \mathbf{u}(\mathbf{x}) = \mathbf{b}(\mathbf{x}) \quad \forall\, \mathbf{x} \in \Omega, \tag{7}$$

where Ω denotes a bounded domain in $\mathbb{R}^n$ and the nonlocal bond-based peridynamic operator $\mathfrak{L}_\varepsilon$ is defined by

$$\begin{aligned} \mathfrak{L}_\varepsilon \mathbf{u}(\mathbf{x}) &= -\mathfrak{D}\big(\omega_\varepsilon \mathfrak{D}^*(\mathbf{u})\big)(\mathbf{x}) \\ &= \int_{\Omega\cup\Omega_I} \omega_\varepsilon(|\mathbf{y}-\mathbf{x}|)\boldsymbol{\alpha}(\mathbf{y}, \mathbf{x}) \otimes \boldsymbol{\alpha}(\mathbf{y}, \mathbf{x})\big(\mathbf{u}(\mathbf{y}) - \mathbf{u}(\mathbf{x})\big) d\mathbf{y} \\ &= \int_{\Omega\cup\Omega_I} \omega_\varepsilon(|\mathbf{y}-\mathbf{x}|) \frac{\mathbf{y}-\mathbf{x}}{|\mathbf{y}-\mathbf{x}|^2} \otimes \frac{\mathbf{y}-\mathbf{x}}{|\mathbf{y}-\mathbf{x}|^2}\big(\mathbf{u}(\mathbf{y}) - \mathbf{u}(\mathbf{x})\big) d\mathbf{y}. \end{aligned} \tag{8}$$

Here, Ω_I denotes the *interaction domain* which is defined as containing the points in $\mathbb{R}^n \backslash \Omega$ (i.e., the points outside of Ω) that interact with the points inside of Ω. Below in (9), we define the specific interaction domain that is used in peridynamic modeling. From (7) or (8), we see that points in $\mathbf{y} \in \Omega_I$ affect the nonlocal operator at points $\mathbf{x} \in \Omega$, thus the nomenclature "interaction domain." Note that Ω_I is also the region on which the solution may be constrained.

For nonlocal models such as peridynamics, an important issue is the proper understanding of the nonlocal analog of boundary conditions specified for PDEs. In Du et al. (2012a), the notion of volume-constrained problems is discussed because in general, unlike the case of local PDEs, interactions occur at a distance. Of course, this does not exclude the case of classical local boundary conditions in special cases. For example, in recent studies, linear peridynamic models with a variable horizon $\varepsilon = \varepsilon(\mathbf{x})$ for $\mathbf{x} \in \Omega$ have also been studied Silling et al. (2014) and Tian and Du (2016). In such a case, a variable horizon $\varepsilon(\mathbf{x})$ is allowed to vanish as material points approach the boundary Γ of Ω, e.g., $\varepsilon(\mathbf{x}) \propto \text{dist}(\mathbf{x}, \Gamma)$ for points $\mathbf{x}$ near the boundary Γ. In this case, displacement boundary conditions can be directly imposed on Γ based on the new trace theorem proved in Tian and Du (2016).

Concerning nonlocal problems defined on a bounded domain, we follow the discussion in Du et al. (2012a). First of all, one may define a notion of essential constraints and natural or variational conditions for nonlocal problems by drawing analogies to classical variational problems for PDEs that are defined with respect to suitable boundary conditions. Indeed, with a constant horizon, the most widely studied case is that of Dirichlet-type essential constraints, i.e., displacement constraints on the solution that are imposed on a constraint set Ω_I of nonzero measure. For example, we may impose the condition $\mathbf{u} = 0$ on Ω_I; this is an essential condition that must be built explicitly into the variational principle. Typically, we have the values of $\mathbf{u}$ specified on

$$\Omega_I = \{\mathbf{x} \in \mathbb{R}^n \setminus \Omega \ : \ \text{dist}(\mathbf{x}, \partial\Omega) < \varepsilon\} \tag{9}$$

which is a ε-collar surrounding Ω, where $\partial\Omega$ denotes the boundary of Ω and also the inner boundary of Ω_I.

The case of Neumann-type conditions that, for peridynamics, are the analog of the natural traction condition for the PDE case can take different forms. If Ω_I is an empty set, then $\mathfrak{L}_\varepsilon$ can be well-defined over all of $\Omega \cup \Omega_I = \Omega$ without the need to impose additional conditions on $\mathbf{u}$. In this case, the forcing function $\mathbf{b}$ may represent either a soft or hard loading with the distinction only showing up through the dependence on the horizon parameter. Another interpretation that is more symbolically similar to the PDE case is to have a nonempty Ω_I in the definitions of the nonlocal operators as that given by (8) and imposing (7) on both Ω that represents the nonlocal equation and Ω_I that represents the domain of nonlocal constraints or nonlocal traction conditions. Again, viewed in terms of $\Omega \cup \Omega_I$, such a separation of the equation domain and the constraint domain becomes mathematically superfluous, although $\mathbf{b}$ can be subject to different physical interpretations, i.e., in Ω it represents a body force, whereas in Ω_I it represents a nonlocal traction force.

Variational Formulation

Related to Eq. (7), we consider the nonlocal constrained value problem

$$\begin{cases} -\mathfrak{L}_\varepsilon \mathbf{u}(\mathbf{x}) = \mathbf{b}(\mathbf{x}) \;\forall\, \mathbf{x} \in \Omega \\ \quad\;\; \mathbf{u}(\mathbf{x}) = \mathbf{0} \qquad \forall\, \mathbf{x} \in \Omega_I \end{cases} \tag{10}$$

which may be derived by minimizing the corresponding energy functional

$$E(\mathbf{u}) = \int_{\Omega\cup\Omega_I}\int_{\Omega\cup\Omega_I} \frac{1}{2}\omega_\varepsilon(|\mathbf{y}-\mathbf{x}|)\,(\mathfrak{D}^*\mathbf{u}(\mathbf{x},\mathbf{y}))^2 d\mathbf{y}d\mathbf{x} - \int_{\Omega\cup\Omega_I} \mathbf{b}(\mathbf{x})\mathbf{u}(\mathbf{x})d\mathbf{x}. \tag{11}$$

It is helpful to state clearly what type of kernels are amenable to variational frameworks and treatment using the nonlocal calculus. For example, ω_ε is often assumed to satisfy

1. $\omega_\varepsilon(r) \geq 0 \quad \forall\, r \in (0, \varepsilon)$ (12)
2. $\mathrm{supp}(\omega_\varepsilon) \subset B_\varepsilon(0)$, i.e., $\underline{\omega}_\varepsilon(r) = 0 \;\; r \geq \varepsilon$, (13)
3. $\displaystyle\int_{\mathbb{R}^n} \omega_\varepsilon(|\mathbf{x}|)d\mathbf{x} = m < \infty$, (14)

where the normalization condition (14) on ω_ε implies that it has finite second-order moments, a condition that is necessary for well-defined elastic moduli (Silling 2000). It is also equivalent to the requirement that the energy is finite for a linear displacement field. A consequence is that for any square integrable vector-valued function that has square integrable first-order partial derivatives, the energy remains finite. We note that the nonnegativity assumption $\omega_\varepsilon(r) \geq 0$ can be relaxed (Mengesha and Du 2013).

The corresponding energy space V contains square integrable functions, i.e., functions in $L^2(\Omega \cup \Omega_I)$ having finite energy. This is an inner product space corresponding to the norm

$$\|\mathbf{u}\|_v^2 = |\mathbf{u}|_v^2 + \|\mathbf{u}\|_{L^2(\Omega\cup\Omega_I)}^2,$$

where the semi-norm $|\mathbf{u}|_v$ on V is defined as

$$|\mathbf{u}|_v^2 = \int_{\Omega\cup\Omega_I}\int_{\Omega\cup\Omega_I} \omega_\varepsilon(|\mathbf{x}-\mathbf{y}|)\,(\mathfrak{D}^*\mathbf{u})^2 d\mathbf{y}d\mathbf{x}.$$

The solution space $V_{c,\varepsilon}$ is the subspace of V consisting of functions satisfying the constraint $\mathbf{u}(\mathbf{x}) = \mathbf{0}$ on Ω_I. Various properties of the function spaces V and $V_{c,\varepsilon}$

can be studied in either L^2 spaces or fractional Sobolev spaces for specific kernels (Du et al. 2013a) and general abstract solution spaces (Mengesha and Du 2014a).

A variational formulation of (7) is given by the minimization problem

$$\min_{\mathbf{u}\in V_{c,\varepsilon}} E(\mathbf{u}).$$

The well-posedness of the variational problem is studied in Du et al. (2013b), Du and Zhou (2011), Mengesha and Du (2013, 2014a,b) in which additional details may be found about suitable interpretations of the operators of the nonlocal vector calculus, especially for cases involving non-integrable interaction kernels.

Note that if we have an integrable nonlocal interaction kernel, the space V and its dual V^* are both equivalent to the standard Hilbert space $L^2(\Omega \cup \Omega_I)$. In general, the nonlocal spaces may be larger than the $H^1(\Omega \cup \Omega_I)$ space so that they may be used to produce solutions having less regularity (Du et al. 2012a; Du and Zhou 2011) than that of solution of the analogous PDEs; indeed, this is a motivation for considering nonlocal mechanical models like peridynamics. As mentioned earlier, when $\varepsilon \to 0$, the nonlocal problems are closely related to classical local models. Techniques for investigating the local limit of nonlocal peridynamic models include Taylor expansions (Emmrich and Weckner 2007; Silling and Lehoucq 2008), Fourier transforms (Du et al. 2013a; Du and Zhou 2011), consistency of weak forms (Du et al. 2014), and the nonlocal calculus of variations (Mengesha and Du 2013, 2014a,b). For the linear bond-based peridynamic model, the local limit recovers the linear Navier elasticity model with a special Poisson ratio.

Discretization Schemes

There have been several approaches taken for the discretization of nonlocal models, including peridynamics. These include meshfree methods, finite difference methods, finite element methods, and numerical quadrature and collocation approaches. Mathematically rigorous studies of convergence properties of such discretizations include, for example, Tian and Du (2013), in which a number of finite difference approximations are given for one-dimensional scalar nonlocal peridynamic equations with their convergence proved; discrete maximum principles are also shown along with error estimates. For collocation schemes, rigorous studies include (Zhang et al. 2016a). For theoretical convergence results for finite element approximations, see Tian and Du (2013) and Zhou and Du (2010) for one-dimensional or special two-dimensional models; the same analysis if given in Tian and Du (2014) for general multidimensional linear peridynamic bond-based and state-based models using conforming finite element spaces defined on arbitrary grids. Nonconforming discontinuous Galerkin approximations have been discussed in Tian and Du (2015).

Quadrature rules for finite element approximations are considered in Zhang et al. (2016b).

Here, due to space limitations, we briefly review recent results about asymptotically compatible schemes.

Asymptotically Compatible Schemes

It is known in practice that to obtain consistency between nonlocal models and corresponding local PDE models, the mesh or quadrature point spacing may have to be reduced at a faster pace than the reduction of the horizon parameter (Bobaru et al. 2009; Bobaru and Duangpanya 2010). Otherwise, there could potentially be complications and inconsistent limiting solutions when the horizon parameter is coupled proportionally to the discretization parameter (Seleson et al. 2016; Tian and Du 2013, 2014). Asymptotically compatible (AC) schemes, formally introduced in Tian and Du (2014), are numerical discretization of nonlocal models that converge to nonlocal continuum models for fixed horizon parameter and to the local models as the horizon vanishes for both discrete schemes with a fixed numerical resolution and for continuum models with increasing numerical resolution.

Let $h > 0$ denote the meshing parameter (or particle spacing) and ε the horizon parameter or even a more generic model parameter. The implications of the AC property are illustrated in Fig. 1. There, u_ε denotes the solution of the continuous nonlocal problem with $\varepsilon > 0$, u_0 the solution of the corresponding continuous local problem, $u_{\varepsilon,h}$ the solution of the discretized nonlocal problem, and $u_{0,h}$ the solution of the discretized local problem.

As seen from detailed studies given in Tian and Du (2013), some popular discretization schemes of nonlocal peridynamics fail to be AC. In particular, if ε is taken to be proportional to h, then as $h \to 0$, piecewise constant conforming finite element solutions actually converge to the incorrect limit, similarly to those based on simple Riemann sum quadrature approximations to nonlocal operators. In Tian and

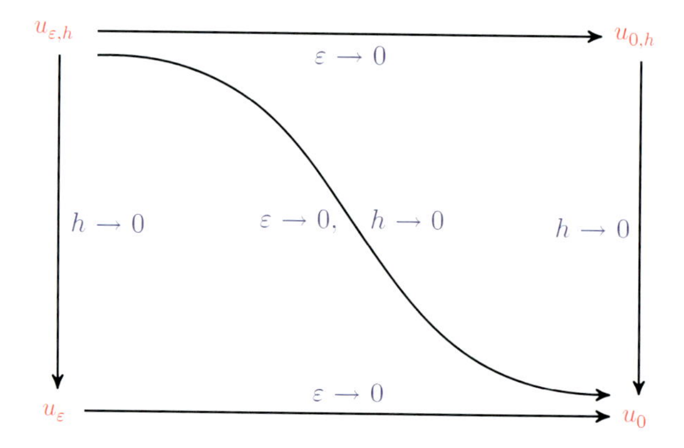

Fig. 1 A diagram for asymptomatically compatible schemes Tian and Du (2014)

Du (2014) (see also Chen and Gunzburger 2011 for a computational illustration), it is shown that as long as the condition $h = o(\varepsilon)$ is met as $\varepsilon \to 0$, then we are able to obtain the correct local limit even for discontinuous piecewise constant finite element approximations when they are of the conforming type. Practically speaking, this implies that a mild growth of bandwidth is needed as the mesh is refined to recover the correct local limit for Riemann sum like quadratures or piecewise constant finite element schemes.

On the other hand, it is shown in Tian and Du (2014) (again, see also Chen and Gunzburger 2011 for computational illustrations) that all conforming Galerkin approximations of the nonlocal models containing continuous piecewise linear functions are automatically AC. This means that they can recover the correct local limit as long as both ε and h are diminishing, even if the nonlocality measure ε is reduced at a much faster pace than the mesh spacing h. Whereas the analysis is highly technical, an intuitive explanation is that even with h larger than ε, the nonlocal features are still encoded in the stiffness matrices to ensure the correct local limit due to the incorporation of higher-order (than constant) basis functions.

Naturally, schemes using higher-order basis functions tend to provide higher-order accuracy to the nonlocal problems as well. At the moment, the theory on AC schemes does not offer any estimate of the order of convergence with respect to the different coupling of h and ε. Preliminary numerical experiments in Tian and Du (2014) offer some insight about the balance of modeling and discretization errors, but additional theoretical analyses remain to be carried out.

Asymptotically compatible (AC) schemes, being either conforming Galerkin-type approximations of weak forms Chen and Gunzburger (2011), Tian and Du (2014), and Xu et al. (2016a,b), nonconforming DG approximations (Tian and Du 2015), or collocation- and quadrature-based approximations of strong forms (Du and Tian 2014; Seleson et al. 2016; Zhang et al. 2016a,b), offer the potential to solve the desirable models with different choices of parameters to gain efficiency and to avoid the pitfall of reaching inconsistent limits.

The framework of AC schemes is very general. For example, it has also been used to prove the convergence of numerical solutions of nonlocal diffusion equations to that of fractional diffusion equations Tian et al. (2016); see also D'Elia and Gunzburger (2013). The theory also guided the development of nonconforming DG approximations (Tian and Du 2015) for peridynamic models with non-integrable interaction kernels. The framework can also be applicable to numerical studies of nonlocal models of convection and diffusion (Tian et al. 2015, 2017) that were motivated by the work on nonlocal convection (Du et al. 2014) and nonlinear nonlocal hyperbolic conservation laws (Du and Huang 2017; Du et al. 2017a) that improved the model studied in Du et al. (2012b). Furthermore, such a concept can also be used in the study of nonlocal gradient recovery for the numerical solution of nonlocal models (Du et al. 2016).

Conditioning

A good theoretical understanding of the conditioning of stiffness matrices for nonlocal problems is alluded to in the Fourier analysis of the point spectrum for the nonlocal operators as given in Du and Zhou (2016) and Zhou and Du (2010); see Aksoylu and Mengesha (2010), Aksoylu and Parks (2011), Aksoylu and Unlu (2013), Du et al. (2012a), and Seleson et al. (2009) for additional discussions. In comparison with a typical local diffusion model that yields a condition number of $O(h^{-2})$ for a discretization with meshing parameter h, the corresponding nonlocal models have their condition numbers dependent on both ε and h in general. For example, Du and Zhou (2016) provided sharp lower and upper bounds for finite element discretization of a nonlocal diffusion operator based on a quasi-uniform regular triangulation. For effective algebraic solvers of the resulting linear system, we refer to studies on the use of Toeplitz (Wang and Tian 2012) and multigrid (Du and Zhou 2016) solvers.

Adaptivity

Due to, e.g., reduced sparsity, nonlocal models such as peridynamics generally incur greater computational costs than do their local PDE-based counterparts. Thus, designing effective adaptive methodologies is important. The paper Du et al. (2013c) contains an a posteriori error analysis of conforming finite element methods for solving linear nonlocal diffusion and peridynamic mechanic models. A general abstract framework is developed for a posteriori error analysis of nonlocal volume-constrained problems for scalar equations. The reliability and efficiency of the estimators are proved, and relations between nonlocal and classical local a posteriori error estimation are also studied. In Du et al. (2013d), a convergent adaptive finite element algorithm for the numerical solution of scalar nonlocal models is developed. For problems involving certain non-integrable kernel functions, the convergence of the adaptive algorithm is rigorously derived with the help of several basic ingredients, such as an upper bound on the estimator, the estimator reduction, and the orthogonality property. How these estimators and methods work in the local limit and for general time-dependent and nonlinear peridynamic models is under current investigation.

Peridynamics as a Multiscale Mono-model

Classical PDE models for mechanics such as the Navier equations for elasticity do not feature a length scale other than those related to material properties. On the other hand, peridynamics features the horizon ε that determines the extent of nonlocal interactions, even in the case of homogeneous, isotropic materials. The appearance of a length scale implies that the nature of the peridynamic model changes as the "viewing window" through which one looks at the model changes in size relative to ε. Here, we give a brief discussion of this observation.

Let $\widehat{\Omega} \subset \Omega$. Because nonlocal interactions only occur between points within a distance ε from each other, a point $\mathbf{x} \in \widehat{\Omega}$ only interacts with points $\mathbf{y} \in \widehat{\Omega} \cup \widehat{\Omega}_\varepsilon$, where $\widehat{\Omega}_\varepsilon = \{\, \mathbf{y} \in \Omega \cup \Omega_I \setminus \widehat{\Omega} \mid |\mathbf{y} - \mathbf{x}| \le \varepsilon\}$. Thus, $\widehat{\Omega}_\varepsilon$ is the interaction domain

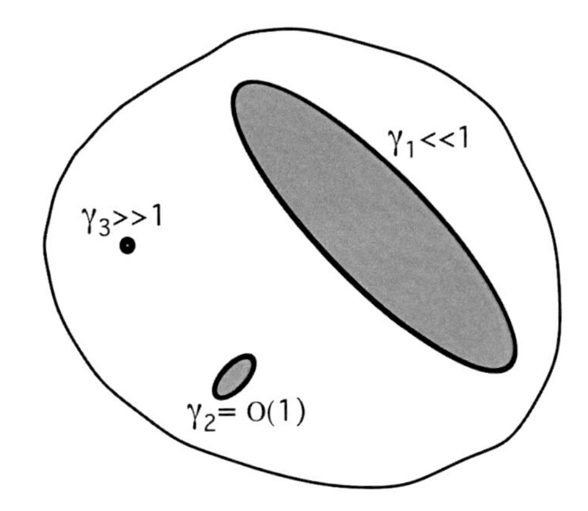

Fig. 2 The gray areas illustrate three regions Ω_i having different length scales. The black regions surrounding the gray regions are the interaction regions $\Omega_{i,\varepsilon}$, all of which have the same thickness ε

corresponding to $\widehat{\Omega}$. Note that $\widehat{\Omega}_\varepsilon$ is a collar surrounding $\widehat{\Omega}$. We denote by $|\widehat{\Omega}|$ and $|\widehat{\Omega}_\varepsilon|$ the volumes of Ω and Ω_ε, respectively.

Let $d = \text{diam}(\widehat{\Omega})$ and assume that $\widehat{\Omega}$ is "nicely shaped" so that, in $\mathbb{R}^3$, $|\widehat{\Omega}| = O(d^3)$. Then, $|\widehat{\Omega}_\varepsilon| = O(d^2\varepsilon + d\varepsilon^2 + \varepsilon^3)$. We have that the *volume ratio* $\gamma = |\widehat{\Omega}_\varepsilon|/|\widehat{\Omega}| = O(\varepsilon/d + \varepsilon^2/d^2 + \varepsilon^3/d^3)$ is an indicator of the relative nonlocality for the subdomain $\widehat{\Omega}$. If ε is a fixed material parameter, γ decreases (resp. increases) as $|\widehat{\Omega}|$ increases (resp. decreases). Because there are no self-interactions (a consequence of Newton's third law), the net internal force on $\widehat{\Omega}$ due to the action of points in $\widehat{\Omega}$ is zero so that the internal force on $\widehat{\Omega}$ is solely due to interactions with points in $\widehat{\Omega}_\varepsilon$.

We introduce three subdomains $\widehat{\Omega}_i$ with diameters d_i, interaction domains $\widehat{\Omega}_{i,\varepsilon}$, and volume ratios γ_i, $i = 1, 2, 3$; see Fig. 2. We consider three cases:

- $d_1 \gg \varepsilon$ so that $\gamma_1 = O(\varepsilon/d_1) \ll 1$ so that the internal force acting on $\widehat{\Omega}_1$ is *local*, emanating from the very thin (relative to the size of $\widehat{\Omega}_1$) layer $\widehat{\Omega}_{1,\varepsilon}$ of material points surrounding $\widehat{\Omega}$;
- $d_2 = O(\varepsilon)$ so that $\gamma_2 = O(1)$ and the internal force acting on $\widehat{\Omega}_2$ is *nonlocal*, emanating from the layer $\widehat{\Omega}_{2,\varepsilon}$ of material points having roughly the same size as $\widehat{\Omega}_2$;
- $d_3 \ll \varepsilon$ so that $\gamma_3 = O(\varepsilon^3/d_3^3) \gg 1$ so that the internal force acting on $\widehat{\Omega}_3$ is *nonlocal*, emanating from the very thick (relative to the size of $\widehat{\Omega}_3$) layer $\widehat{\Omega}_{3,\varepsilon}$ of material points surrounding $\widehat{\Omega}_3$.

Thus, depending on the scale one is operating in, peridynamics can behave as either a local or nonlocal material model so that, clearly, *peridynamics is by itself a multiscale model for material mechanics*, hence our characterization of peridynamics as a *multiscale mono-model* for mechanics. This observation has important implications for computations which we next explore.

A Multiscale Finite Element Implementation of Peridynamics

The multiscale nature of peridynamic models resulting from the horizon ε motivates a multiscale finite element implementation of peridynamics. Suppose h is a typical grid size in a discretization of a PDE model for solid mechanics in case the solution, i.e., the displacement, is known to be smooth; for example, if one is interested in achieving an L^2 error of $O(\delta)$ and one uses continuous piecewise linear finite element approximations, then h would be chosen to be of $O(\delta^{1/2})$. Even in cases where the solution has discontinuities, one can safely use the PDE model in regions away from the discontinuity, i.e., in regions where the displacement is smooth. However, near the location of discontinuities in the displacement, the PDE model breaks down, and one instead uses a peridynamic model. However, because the locations of those discontinuities are, in general, not known beforehand, one has to be able to detect where they occur. One also has to devise a strategy for coupling the peridynamic and PDE models. The overall goal of the implementation is to achieve, e.g., for piecewise linear approximations $O(h^2)$ accuracy even in the present of discontinuities.

Based on the above discussions, a multiscale implementation of the peridynamic model in one dimension is presented in Chen and Gunzburger (2011) and Xu et al. (2016a) and in two dimensions in Xu et al. (2016b). One starts with a choice for a horizon parameter ε and for a bulk grid size h, that is, what one would use should one be solving a PDE model for problems with smooth solutions. Then, the multiscale implementation includes the following components:

1. Detection of elements that contain a discontinuity in the displacement. A simple means for doing this is to examine the size of the difference in the displacements at the nodes of a triangle. More sophisticated means for this step are also considered in Xu et al. (2016b).
2. Refinement of the grid as necessary near the discontinuities. Meeting the goal of refinement requires the use of very small elements so that refinement should not result in isotropic elements containing the discontinuity because this would result in a huge number of such elements. Instead, refinement should result in highly anisotropic elements with long sides aligned with a discontinuity and very short sides crossing a discontinuity.
3. Use of discontinuous Galerkin methods (DGMs) for peridynamics in regions containing the discontinuity. The kernel functions used in peridynamics are such that such methods are conforming. In particular, unlike the PDE case, one need not worry about enforcing continuity of any kind across element edges; nonlocality takes care of a weak enforcement of such properties.
4. Use of continuous Galerkin methods (CGMs) for peridynamics in regions neighboring but not containing the discontinuity. The use of CGMs wherever possible results in a reduction in the number of degrees of freedom compared to using DGMs.

5. Use of CGMs for PDEs if sufficiently far away from the discontinuity. In regions where the displacement is smooth, PDEs provide adequate modeling and CGMs provide conforming discretizations.
6. Use of quadrature rules that can be applied for any combination of h and ε. Such quadrature rules on triangles and regions defined as the intersection of triangles with the interaction disc of radius ε are developed in Xu et al. (2016b) that are applicable to the integrable kernels appearing in two-dimensional peridynamic models.

Generically, one would choose $h \gg \varepsilon$ because then the discretized peridynamic model essentially reduces to a local model whenever the local grid size is of $O(h)$. Specifically, if one uses the same asymptotically compatible continuous finite element space and the same basis to discretize both the nonlocal peridynamic model and the corresponding local PDE model, it can be shown that the difference between the entries of the peridynamic stiffness and those of PDE stiffness matrix is of $O(\varepsilon)$. This effect greatly facilitates the coupling of a peridynamic model to a PDE model; in fact, the coupling is seamless. On the other hand, we would also choose $\varepsilon > \underline{h}$, where $\underline{h}$ denotes the grid size one would use for elements in which solution discontinuities occur. With $\varepsilon > \underline{h}$, the peridynamic model is nonlocal. To meet the goal of achieving $O(h^2)$ accuracy, one should choose $\underline{h} = O(h^4)$ because in any element containing a solution discontinuity whose position is not known exactly, the best accuracy one can achieve is $O(\underline{h}^{1/2})$ regardless of what degree polynomial one uses. It is shown in Chen and Gunzburger (2011) and Xu et al. (2016a,b) that the fine-to-coarse grid transition can, at least for kernels that allow for discontinuous displacements and for DGMs, be abrupt, i.e., there is no need for transition layer within which the element size gradually grows from small to fine. DGMs need only be used in elements that contain the discontinuity.

We illustrate the multiscale implementation strategy for an exact solution having a jump discontinuity across a circle of radius 0.5 centered within the square $[0, 1]^2$; the fixed horizon $\varepsilon = 0.01$. Starting with a uniform mesh of size h, the target grid size for regions where the solution is smooth, an adaptive grid refinement strategy that first detects elements containing the discontinuity and which ultimately constructs a grid consisting of $O(h)$ size elements except for highly anisotropic elements (the elements connecting the red nodes in Fig. 3 which is for $h = 1/16$) along the discontinuity that are of size $O(h)$ parallel to the discontinuity and of size $O(h^4)$ across the discontinuity, the actual thickness of those elements is of $O(10^{-5})$ so that they are not visible. In Fig. 3, DGM and peridynamics are used in elements with a red node, CGM and peridynamics are used in elements having both red and blue nodes, and CGM with PDE are used in the remaining elements. Note that the highly anisotropic straight-sided elements cannot, in general, contain the discontinuity throughout the element. The construction process merely insures that the discontinuity passes through the shortest side of those elements. We have empirically found that doing this is enough to preserve optimal accuracy.

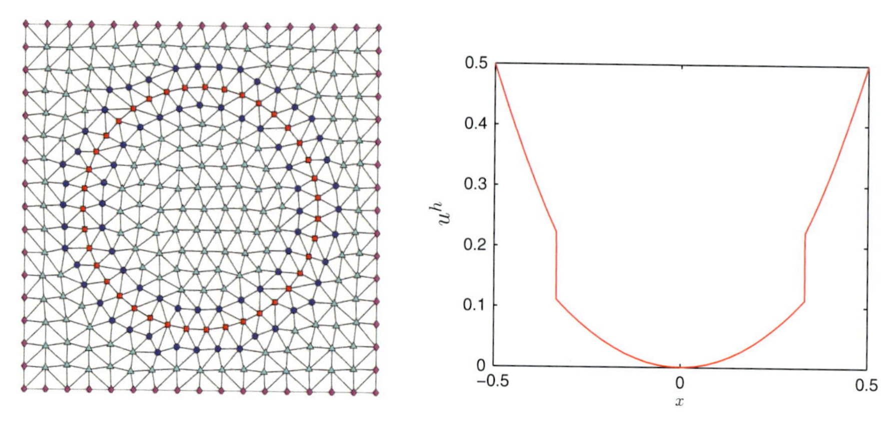

Fig. 3 For a problem with a discontinuous solution across a circle, the resulting adaptively refined mesh (left) and a cross-section of approximate solution (right). Note that all elements are of size h excepting the thin elements connecting the red nodes which are not visible because they have one side of length $O(h^4)$

As a result of these implementation choices, the number of degrees of freedom needed is only slightly greater than that for a finite element method on an $O(h)$ size grid for the PDE. The overhead incurred through the use of a nonlocal model to treat discontinuous behaviors is due to the use of such models in a layer of highly anisotropic elements surrounding the discontinuity and the use of DGMs in those elements. Because, in two dimensions, the number of such elements grows linearly, whereas the total number of elements grows quadratically, the percentage overhead actually reduces as h decreases.

The implementation of Chen and Gunzburger (2011) and Xu et al. (2016a,b) clearly illustrates the multiscale mono-model nature of peridynamics but is still a far cry from being a useful tool; much work needs to be done to achieve the latter, with the most daunting tasks remaining being extensions to three dimensions and the treatment of problems with propagating, i.e., moving, discontinuities.

Control, Identification, and Obstacle Problems

In this section, we very briefly touch upon some other recent developments for nonlocal models that go beyond standard simulation settings.

Nonlocal Optimal Control Problems

The ingredients of an optimal control problem are state and control variables, a state equation that relates the state and control variables, and a cost or objective functional which depends on the state and control. Then, the goal of the optimal control problem is to determine states and controls that minimize the cost functional, subject to the state equation being satisfied.

We consider state equations of the form

$$\begin{cases} -\mathcal{L}u(\mathbf{x}) := 2\displaystyle\int_{B_\varepsilon(\mathbf{x})} \big(u(\mathbf{y}) - u(\mathbf{x})\big)\,\beta(\mathbf{x},\mathbf{y})d\mathbf{y} = f & \text{for } \mathbf{x} \in \Omega \\ \qquad\qquad\qquad\qquad\qquad\qquad u = g & \text{for } \mathbf{x} \in \Omega_I, \end{cases} \tag{15}$$

where $u(\mathbf{x})$ denotes the state, $f(\mathbf{x})$ the control, $g(\mathbf{x})$ a given function, and $\beta(\mathbf{x},\mathbf{y})$ a symmetric kernel, i.e., $\beta(\mathbf{x},\mathbf{y}) = \beta(\mathbf{y},\mathbf{x})$. We consider the matching functional

$$J(u,f) := \frac{1}{2}\int_\Omega \big(u(\mathbf{x}) - \widehat{u}(\mathbf{x})\big)^2 d\mathbf{x} + \frac{\sigma}{2}\int_\Omega \big(f(\mathbf{x})\big)^2 d\mathbf{x}, \tag{16}$$

where $\widehat{u}(\mathbf{x})$ is a given target function and $\sigma > 0$ a given parameter. The first term in (16) is the goal of control, i.e., to match (in an L^2 sense for the functional $J(u,f)$) the solution $u(\mathbf{x})$ of (16) as well as possible to the given target function $\widehat{u}(\mathbf{x})$. Then, the optimal control problem is to

> seek an optimal state $u^*(\mathbf{x})$ and an optimal control $f^*(\mathbf{x})$ such that the functional (16) is minimized subject to (15) being satisfied.

Of course, a precise statement of this problem requires the choice of appropriate function spaces for u and f; these are defined in D'Elia and Gunzburger (2014). There, the existence and uniqueness of the optimal control and state in appropriate function spaces are also proved for several kernels relevant to peridynamics or anomalous diffusion.

The optimal state $u^*(\mathbf{x})$ is determined from the optimality system

$$\begin{cases} -\mathcal{L}u^* = -\dfrac{1}{\sigma}w^* & \text{for } \mathbf{x} \in \Omega \\ \qquad u^* = g & \text{for } \mathbf{x} \in \Omega_I \end{cases}$$

and

$$\begin{cases} -\mathcal{L}w^* = (u^* - \widehat{u}) & \text{for } \mathbf{x} \in \Omega \\ \qquad w^* = 0 & \text{for } \mathbf{x} \in \Omega_I. \end{cases} \tag{17}$$

The adjoint function $w^*(\mathbf{x})$ is a Lagrange multiplier function that is introduced to enforce the constraint (15) imposed when minimizing the functional (16).

After the optimal state $u^*(\mathbf{x})$ and optimal adjoint state $w^*(\mathbf{x})$ are obtained, the optimal control $f^*(\mathbf{x})$ is given by

$$f^* = -\frac{1}{\sigma}w^* \qquad \text{for } \mathbf{x} \in \Omega.$$

Also proved in D'Elia and Gunzburger (2014) is the convergence, as $\varepsilon \to 0$, of solutions of the nonlocal control problem to solutions of the local control problem for which the state equation (15) is replaced by the PDE problem

$$\begin{cases} -\Delta u_\ell = f_\ell & \text{for } \mathbf{x} \in \Omega \\ u_\ell = g_\ell & \text{for } \mathbf{x} \in \partial\Omega, \end{cases} \tag{18}$$

where $u_\ell(\mathbf{x})$ and $f_\ell(\mathbf{x})$ denote the local PDE solution and control, respectively, and $\partial\Omega$ denotes the boundary of Ω. The given Dirichlet data $g_\ell(\mathbf{x})$ is determined by evaluating the nonlocal data $g(\mathbf{x})$ on $\partial\Omega$; this, of course, requires $g(\mathbf{x})$ for $\mathbf{x} \in \Omega_I$ to be more regular than that needed for the nonlocal problem to be well posed. In addition, in D'Elia and Gunzburger (2014), the convergence of solutions of finite-dimensional discretizations of the nonlocal optimality system, including finite element discretizations, is proved, and optimal error estimates are derived, again for several kernels relevant to peridynamics and anomalous diffusion.

As an example, we consider a piecewise linear target function having a jump discontinuity; see the light blue curve in Fig. 4. In that figure, also plotted are the target functional $\widehat{u}$ and, for two values of ε, the finite element approximations u_N^* of the optimal state and f_N^* of the optimal control. We observe that, away from the point $x = 0.5$ at which the target function is discontinuous, the local and nonlocal approximations show a very good match to the target function. However, we observe that because the local optimal state is differentiable, its approximation cannot match the target function near $x = 0.5$. Also, because the nonlocal optimal state converges to the local optimal state as $\varepsilon \to 0$, we see that for the smaller value of ε, the nonlocal and local optimal states are in close agreement so that, in this case, the nonlocal optimal state also fails to match the target function near the discontinuity. This is to be expected because we have already stated that the solution of the nonlocal problem converges to that of the local problem as $\varepsilon \to 0$. However, for the larger value of ε for which nonlocal effects are more pronounced, we observe a difference between the approximation of the local optimal state (which does not match the target well) and the approximation of the nonlocal optimal state (which do match well), being more flexible with respect to the discontinuous behavior of the target function.

Nonlocal Parameter Identification Problems

We next consider the problem of identifying the diffusivity coefficient function $\vartheta(\mathbf{x}, \mathbf{y})$ in the nonlocal equation

$$\begin{cases} 2\displaystyle\int_{B_\varepsilon(\mathbf{x})} \big(u(\mathbf{y}) - u(\mathbf{x})\big)\, \vartheta(\mathbf{x}, \mathbf{y})\, \beta(\mathbf{x}, \mathbf{y}) d\mathbf{y} = f(\mathbf{x}) & \text{for } \mathbf{x} \in \Omega \\ u(\mathbf{x}) = g(\mathbf{x}) & \text{for } \mathbf{x} \in \Omega_I, \end{cases} \tag{19}$$

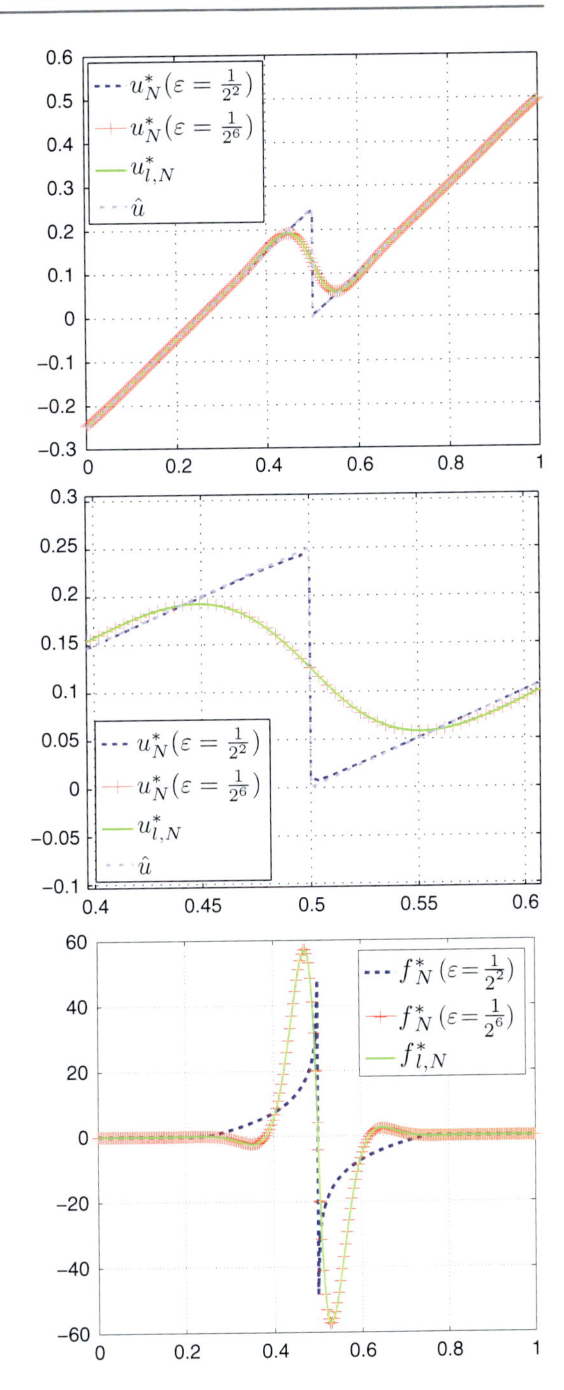

Fig. 4 The target function and finite element approximations of the optimal state (left) and control (right) for the nonlocal and local control problems. The middle plot is a zoom-in of the left plot in the vicinity of the discontinuity in the target function

where the kernel $\beta(\mathbf{x},\mathbf{y})\colon \Omega \times \Omega \to \mathbb{R}$ is a nonnegative symmetric mapping and $f(\mathbf{x})$ and $g(\mathbf{x})$ are given functions. We assume that the diffusivity function satisfies

$$0 < \vartheta_0 \leq \vartheta(\mathbf{x},\mathbf{y}) \leq \vartheta_1 < \infty \tag{20}$$

for some constants ϑ_0 and ϑ_1. Additional technical assumptions on $\vartheta(\mathbf{x},\mathbf{y})$ are needed for the analysis; see D'Elia and Gunzburger (2016).

Our approach is to

seek an optimal state $u^*(\mathbf{x})$ and an optimal diffusivity $\vartheta^*(\mathbf{x},\mathbf{y})$ such that the functional $K(u,\vartheta)$ is minimized subject to (19) and (20) being satisfied,

where, for a given target function $\widehat{u}(\mathbf{x})$, we have the functional

$$K(u,\vartheta) := \frac{1}{2}\int_\Omega \big(u(\mathbf{x}) - \widehat{u}(\mathbf{x})\big)^2 d\mathbf{x}. \tag{21}$$

Note the absence of a regularization term in (21) compared to the functional (16); that term is not needed in the analysis because of (20) and the other technical assumptions explicitly enforced on candidate solutions.

In D'Elia and Gunzburger (2016), the existence of a solution of this parameter identification problem is proved. In general, the solution is not unique. This is not due to the nonlocality; lack of uniqueness also holds for the corresponding local PDE parameter identification problem. In that paper, an optimality system whose solution provides the optimal state and optimal diffusivity coefficient is also derived. For the sake of brevity, we do not discuss this here other than to say that in addition to the state equation (19) and an adjoint equation similar to (17), complementary conditions are present that enforce the constraints in (20). Also in D'Elia and Gunzburger (2016), finite element discretizations of the optimality system are defined, and convergence proofs for the discrete solution are provided, including a priori error estimates for the approximate states and coefficients.

As an illustration, we assume the form $\vartheta(x,y) = \vartheta(\frac{x+y}{2})$ for the coefficient and then choose the particular piecewise constant coefficient

$$\vartheta_C(z) = \begin{cases} 1 & z \in (0, 0.2) \\ 0.1 & z \in (0.2, 0.6) \\ 1 & z \in (0.6, 1). \end{cases} \tag{22}$$

Piecewise constant coefficients often arise in practice; for example, in subsurface flows, disjoint portions of the subsurface may consist of rock, clay, or sand. Using that coefficient, we solve the state equation (19) with $f(x) = 5$ on $\Omega = (0,1)$, $u(x) = 0$ on $\Omega_I = (-\varepsilon, 0) \cup (1, 1+\varepsilon)$, and a peridynamic-like kernel to produce the manufactured target function $\widehat{u}(x)$ for the functional (21); in actuality, we solve for a very fine grid finite element approximation of $\widehat{u}(x)$. Having done this, we pretend we do not know either the state or coefficient but try to identify them by solving the optimality system. Note that the diffusivity coefficient we try to identify

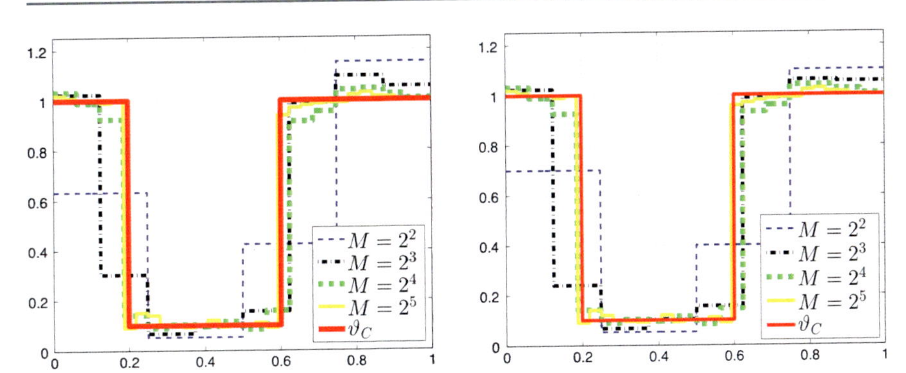

Fig. 5 For different values of M and for $N = 2^{12}$, the approximation ϑ_M^* of the optimal coefficient for $\varepsilon = 2^{-9}$ (left) and 2^{-4} (right). Also plotted is the target coefficient ϑ_C

is discontinuous so that a local PDE-based parameter identification problem would do a poor job of matching that coefficient.

Computational results are shown in Fig. 5. In that figure, M denotes the number of degrees of freedom used in a piecewise constant approximation of ϑ. Note that M is different from the number of degrees N used for a continuous piecewise linear approximation of u. In Fig. 5, we use a large value for N so as to minimize the effects of inaccuracies present in the approximate solution of the constraint equation (19). From Fig. 5, one observes that as M increases, the approximation ϑ^* resulting from the optimization process converges to the presumably unknown exact coefficient ϑ_C. We again note that nonlocality allows for much better match than is possible with PDE-based models.

Nonlocal Obstacle Problems

We consider the nonlocal obstacle problem given by the nonlocal equation (15) with $g = 0$ for which the solution $u(\mathbf{x})$ is subject to the constraint

$$u(\mathbf{x}) \geq \psi(\mathbf{x}) \qquad \text{for } \mathbf{x} \in \Omega, \tag{23}$$

where $\psi(\mathbf{x})$ is a given function. This problem is equivalent to the variational problem

$$\min_u I[u] \quad \text{subject to} \quad u \geq \psi \ \text{ in } \Omega \quad \text{and} \quad u = 0 \ \text{ on } \Omega_I \tag{24}$$

for the functional

$$I[u] := \frac{1}{2} \int_{\Omega \cup \Omega_I} \int_{\Omega \cup \Omega_I} \beta(\mathbf{x}, \mathbf{y})(u(\mathbf{x}) - u(\mathbf{y}))^2 d\mathbf{y} d\mathbf{x} - \int_{\Omega} u(\mathbf{x}) f(\mathbf{x}) d\mathbf{x}.$$

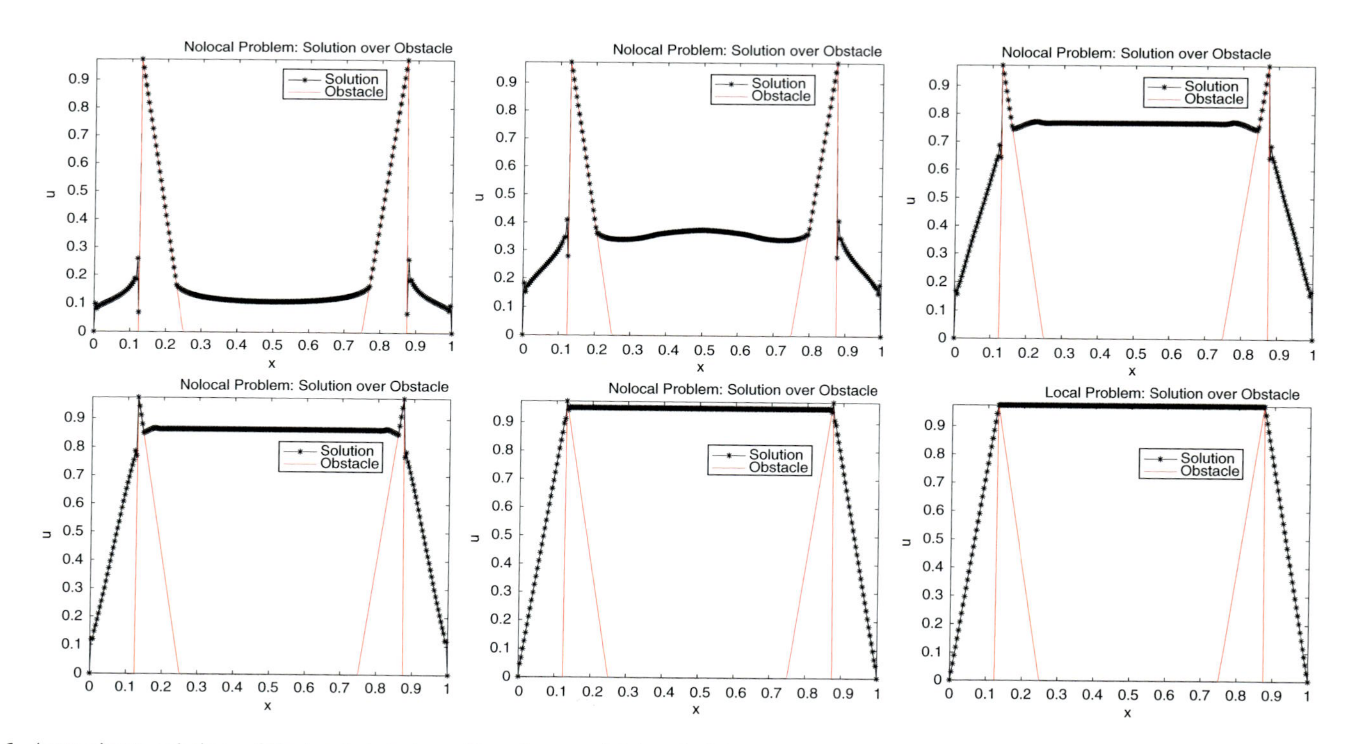

Fig. 6 Approximate solutions of the local obstacle problem (bottom-right) and of the nonlocal obstacle problem for $\varepsilon = 2, 0.5, 0.1, 0.05$, and 0.01 (left to right and top to bottom)

In Guan and Gunzburger (2017), the existence and uniqueness of solutions of the nonlocal obstacle problem (24) is proved for integrable and certain classes of non-integrable kernels. There, it is also proved that finite element discretizations are well posed as is its convergence to the solution of the continuous problem. In addition, the convergence as $\varepsilon \to 0$ of solutions of the nonlocal obstacle problem to that of the local obstacle problem is also proved under certain assumptions.

The corresponding local obstacle problem is given by Eq. (18) subject to the constraint (23). Well-posedness proofs for this problem require more smoothness for the obstacle than is required for the nonlocal obstacle problem. In particular, for the local case, obstacles with jump discontinuities are not allowable, whereas for the nonlocal case, for integrable and certain classes of non-integrable kernels with relatively weak singularities, such obstacles fall within the purview of the well-posedness and finite element analyses of Guan and Gunzburger (2017).

There is also a very significant difference in the solutions of the local and nonlocal obstacle problems, especially for non-smooth obstacles. An illustration is provided by Fig. 6 which features a piecewise linear obstacle having two jump discontinuities. The solution of the obstacle problem is given for five values of ε as is the solution of the local obstacle problem. We see that, as expected, for small values of ε, the nonlocal solution is close to the local ones, but for larger values, there is a significant difference, with the nonlocal solution being able to conform much better to the shape of the obstacle, even on its vertical edges. The local solution can only stretch over the peaks of the obstacle and completely ignores the shape of the obstacle below those peaks.

Acknowledgements MD: supported by the Collaboratory on Mathematics for Mesoscopic Modeling of Materials (CM4). Sandia National Laboratories is a multimission laboratory managed and operated by National Technology and Engineering Solutions of Sandia, LLC., a wholly owned subsidiary of Honeywell International, Inc., for the US Department of Energy's National Nuclear Security Administration under contract DE-NA-0003525.

QD: supported in part by the US NSF grant DMS-1558744, the AFOSR MURI Center for Material Failure Prediction Through Peridynamics, and the OSD/ARO/MURI W911NF-15-1-0562 on Fractional PDEs for Conservation Laws and Beyond: Theory, Numerics and Applications.

MG: supported by the US NSF grant DMS-1315259, US Department of Energy Office of Science grant DE-SC0009324, US Air Force Office of Scientific Research grant FA9550-15-1-0001, and DARPA Equips program through the Oak Ridge National Laboratory.

References

B. Aksoylu, T. Mengesha, Results on nonlocal boundary value problems. Numer. Func. Anal. Optim. **31**, 1301–1317 (2010)

B. Aksoylu, M. Parks, Variational theory and domain decomposition for nonlocal problems. Appl. Math. Comp. **217**, 6498–6515 (2011)

B. Aksoylu, Z. Unlu, Conditioning analysis of nonlocal integral operators in fractional Sobolev spaces. SIAM J. Numer. Anal. **52**(2), 653–677 (2014). https://doi.org/10.1137/13092407X

E. Askari, F. Bobaru, R. Lehoucq, M. Parks, S. Silling, O. Weckner, Peridynamics for multiscale materials modeling. J. Phys. Conf. Ser. **125**, 012078 (2008)

F. Bobaru, M. Duangpanya, The peridynamic formulation for transient heat conduction. Int. J. Heat Mass Transf. **53**, 4047–4059 (2010)

F. Bobaru, S. Silling, Peridynamic 3D problems of nanofiber networks and carbon nanotube-reinforced composites, in *Materials and Design: Proceeding of International Conference on Numerical Methods in Industrial Forming Processes* (American Institute of Physics, 2004), pp. 1565–1570

F. Bobaru, S. Silling, H. Jiang, Peridynamic fracture and damage modeling of membranes and nanofiber networks, in *Proceeding of XI International Conference on Fracture*, Turin, 2005, pp. 1–6

F. Bobaru, M. Yang, L. Alves, S. Silling, E. Askari, J. Xu, Convergence, adaptive refinement, and scaling in 1d peridynamics. Inter. J. Numer. Meth. Engrg. **77**, 852–877 (2009)

X. Chen, M. Gunzburger, Continuous and discontinuous finite element methods for a peridynamics model of mechanics. Comput. Meth. Appl. Mech. Engrg. **200**, 1237–1250 (2011) with X. Chen.

M. D'Elia, M. Gunzburger, The fractional Laplacian operator on bounded domains as a special case of the nonlocal diffusion operator. Comput. Math. Appl. **66**, 1245–1260 (2013)

M. D'Elia, M. Gunzburger, Optimal distributed control of nonlocal steady diffusion problems. SIAM J. Cont. Optim. **52**, 243–273 (2014)

M. D'Elia, M. Gunzburger, Identification of the diffusion parameter in nonlocal steady diffusion problems. Appl. Math. Optim. **73**, 227–249 (2016)

Q. Du, Nonlocal calculus of variations and well-posedness of peridynamics, in *Handbook of Peridynamic Modeling*, chapter 3 (CRC Press, Boca Raton, 2016a), pp. 61–86

Q. Du, Local limits and asymptotically compatible discretizations, in *Handbook of Peridynamic Modeling*, chapter 4 (CRC Press, Boca Raton, 2016b), pp. 87–108

Q. Du, Z. Huang, Numerical solution of a scalar one-dimensional monotonicity-preserving nonlocal nonlinear conservation law. J. Math. Res. Appl. **37**, 1–18 (2017)

Q. Du, X. Tian, Asymptotically compatible schemes for peridynamics based on numerical quadratures, in *Proceedings of the ASME 2014 International Mechanical Engineering Congress and Exposition IMECE 2014–39620*, 2014

Q. Du, K. Zhou, Mathematical analysis for the peridynamic nonlocal continuum theory. Math. Model. Numer. Anal. **45**, 217–234 (2011)

Q. Du, Z. Zhou, Multigrid finite element method for nonlocal diffusion equations with a fractional kernel (2016, preprint)

Q. Du, M. Gunzburger, R. Lehoucq, K. Zhou, Analysis and approximation of nonlocal diffusion problems with volume constraints. SIAM Rev. **56**, 676–696 (2012a)

Q. Du J. Kamm, R. Lehoucq, M. Parks, A new approach to nonlocal nonlinear coservation laws. *SIAM J. Appl. Math.* **72**, 464–487 (2012b)

Q. Du, M. Gunzburger, R. Lehoucq, K. Zhou, A nonlocal vector calculus, nonlocal volume-constrained problems, and nonlocal balance laws. Math. Mod. Meth. Appl. Sci. **23**, 493–540 (2013a)

Q. Du, M. Gunzburger, R. Lehoucq, K. Zhou, Analysis of the volume-constrained peridynamic Navier equation of linear elasticity. J. Elast. **113**, 193–217, (2013b)

Q. Du, L. Ju, L. Tian, K. Zhou, A posteriori error analysis of finite element method for linear nonlocal diffusion and peridynamic models. Math. Comput. **82**, 1889–1922 (2013c)

Q. Du, L. Tian, X. Zhao, A convergent adaptive finite element algorithm for nonlocal diffusion and peridynamic models. SIAM J. Numer. Anal. **51**, 1211–1234 (2013d)

Q. Du, Z. Huang, R. Lehoucq, Nonlocal convection-diffusion volume-constrained problems and jump processes. Disc. Cont. Dyn. Sys. B **19**, 373–389 (2014)

Q. Du, Y. Tao, X. Tian, J. Yang, Robust a posteriori stress analysis for approximations of nonlocal models via nonlocal gradient. Comput. Meth. Appl. Mech. Eng. **310**, 605–627 (2016)

Q. Du, Z. Huang, P. Lefloch, Nonlocal conservation laws. I. A new class of monotonicity-preserving models. SIAM J. Numer. Anal. **55**(5), 2465–2489 (2017)

E. Emmrich, O. Weckner, On the well-posedness of the linear peridynamic model and its convergence towards the Navier equation of linear elasticity. Commun. Math. Sci. **5**, 851–864 (2007)

W. Gerstle, N. Sau, Peridynamic modeling of concrete structures, in *Proceeding of 5th International Conference on Fracture Mechanics of Concrete Structures, Ia-FRAMCOS*, vol. 2, 2004, pp. 949–956
W. Gerstle, N. Sau, S. Silling, Peridynamic modeling of plain and reinforced concrete structures, in *SMiRT18: 18th International Conference on Structural Mechanics in Reactor Technology*, Beijing, 2005
Q. Guan, M. Gunzburger, Analysis and approximation of a nonlocal obstacle problem. J. Comput. Appl. Math. **313**, 102–118 (2017)
D. Littlewood, Simulation of dynamic fracture using peridynamics, finite element analysis, and contact, in *Proceeding of ASME 2010 International Mechanical Engineering Congress and Exposition*, Vancouver, 2010
T. Mengesha, Q. Du, Analysis of a scalar nonlocal peridynamic model with a sign changing kernel. Disc. Cont. Dyn. Syst. B **18**, 1415–1437 (2013)
T. Mengesha, Q. Du, The bond-based peridynamic system with Dirichlet type volume constraint. Proc. R. Soc. Edinb. A **144**, 161–186 (2014a)
T. Mengesha, Q. Du, Nonlocal constrained value problems for a linear peridynamic Navier equation. J. Elast. **116**, 27–51 (2014b)
T. Mengesha, Q. Du, Characterization of function spaces of vector fields via nonlocal derivatives and an application in peridynamics. Nonlinear Anal. A Theory Meth. Appl. **140**, 82–111 (2016)
M. Parks, R. Lehoucq, S. Plimpton, S. Silling, Implementing peridynamics within a molecular dynamics code. Comput. Phys. Commun. **179**, 777–783 (2008)
M. Parks, D. Littlewood, J. Mitchell, S. Silling, *Peridigm Users' Guide*, Sandia report 2012–7800, Albuquerque, 2012
P. Seleson, M. Parks, M. Gunzburger, R. Lehoucq, Peridynamics as an upscaling of molecular dynamics. Mult. Model. Simul. **8**, 204–227 (2009)
P. Seleson, Q. Du, M. Parks, On the consistency between nearest-neighbor peridynamic discretizations and discretized classical elasticity models. Comput. Meth. Appl. Mech. Engrg. **311**, 698–722 (2016)
S. Silling, Reformulation of elasticity theory for discontinuities and long-range forces. J. Mech. Phys. Solids **48**, 175–209 (2000)
S. Silling, Dynamic fracture modeling with a meshfree peridynamic code, in *Computational Fluid and Solid Mechanics* (Elsevier, Amsterdam, 2003), pp. 641–644
S. Silling, Linearized theory of peridynamic states. J. Elast. **99**, 85–111 (2010)
S. Silling, E. Askari, Peridynamic modeling of impact damage, in *PVP-Vol. 489* (ASME, New York, 2004), pp. 197–205
S. Silling, E. Askari, A meshfree method based on the peridynamic model of solid mechanics. Comput. Struct. **8**, 1526–1535 (2005)
S. Silling, F. Bobaru, Peridynamic modeling of membranes and fibers. Int. J. Nonlinear Mech. **40**, 395–409 (2005)
S. Silling, R.B. Lehoucq, Convergence of peridynamics to classical elasticity theory. J. Elast. **93**, 13–37 (2008)
S. Silling, R. Lehoucq, Peridynamic theory of solid mechanics. Adv. Appl. Mech. **44**, 73–168 (2010)
S. Silling, M. Zimmermann, R. Abeyaratne, Deformation of a peridynamic bar. J. Elast. **73**, 173–190 (2003)
S. Silling, M. Epton, O. Weckner, J. Xu, E. Askari, Peridynamic states and constitutive modeling. J. Elast. **88**, 151–184 (2007)
S. Silling, D. Littlewood, P. Seleson, Variable horizon in a peridynamic medium, Technical report No. SAND2014-19088 (Sandia National Laboratories, Albuquerque, 2014)
X. Tian, Q. Du, Analysis and comparison of different approximations to nonlocal diffusion and linear peridynamic equations. SIAM J. Numer. Anal. **51**, 3458–3482 (2013)
X. Tian, Q. Du, Asymptotically compatible schemes and applications to robust discretization of nonlocal models. SIAM J. Numer. Anal. **52**, 1641–1665 (2014)
X. Tian, Q. Du, Nonconforming discontinuous Galerkin methods for nonlocal variational problems. SIAM J. Numer. Anal. **53**(2), 762–781 (2015)

X. Tian, Q. Du, Trace theorems for some nonlocal energy spaces with heterogeneous localization. SIAM J. Math. Anal. **49**(2), 1621–1644 (2017)

H. Tian, L. Ju, Q. Du, Nonlocal convection-diffusion problems and finite element approximations. Comput. Meth. Appl. Mech. Engrg. **289**, 60–78 (2015)

X. Tian, Q. Du, M. Gunzburger, Asymptotically compatible schemes for the approximation of fractional Laplacian and related nonlocal diffusion problems on bounded domains. Adv. Comput. Math. **42**, 1363–1380 (2016)

H. Tian, L. Ju, Q. Du, A conservative nonlocal convection-diffusion model and asymptotically compatible finite difference discretization. Comput. Methods Appl. Mech. Eng. **320**, 46–67 (2017)

H. Wang, H. Tian, A fast Galerkin method with efficient matrix assembly and storage for a peridynamic model. J. Comput. Phys. **240**, 49–57 (2012)

O. Weckner, R. Abeyaratne, The effect of long-range forces on the dynamics of a bar. J. Mech. Phys. Solids **53**, 705–728 (2005)

O. Weckner, E. Emmrich, Numerical simulation of the dynamics of a nonlocal, inhomogeneous, infinite bar. J. Comput. Appl. Mech. **6**, 311–319 (2005)

F. Xu, M. Gunzburger, J. Burkardt, Q. Du, A multiscale implementation based on adaptive mesh refinement for the nonlocal peridynamics model in one dimension. Multiscale Model. Simul. **14**, 398–429 (2016a)

F. Xu, M. Gunzburger, J. Burkardt, A multiscale method for nonlocal mechanics and diffusion and for the approximation of discontinuous functions, Comput. Meth. Appl. Mech. Engrg. **307**, 117–143 (2016b)

X. Zhang, M. Gunzburger, L. Ju, Nodal-type collocation methods for hypersingular integral equations and nonlocal diffusion problems. Comput. Meth. Appl. Mech. Engrg. **299**, 401–420 (2016a)

X. Zhang, M. Gunzburger, L. Ju, Quadrature rules for finite element approximations of 1D nonlocal problems. J. Comput. Phys. **310**, 213–236 (2016b)

K. Zhou, Q. Du, Mathematical and numerical analysis of linear peridynamic models with nonlocal boundary conditions. SIAM J. Numer. Anal. **48**, 1759–1780 (2010)

Optimization-Based Coupling of Local and Nonlocal Models: Applications to Peridynamics

36

Marta D'Elia, Pavel Bochev, David J. Littlewood, and Mauro Perego

Contents

Abstract

Nonlocal continuum theories for mechanics can capture strong nonlocal effects due to long-range forces in their governing equations. When these effects cannot be neglected, nonlocal models are more accurate than partial differential

M. D'Elia (✉)
Optimization and Uncertainty Quantification Department Center for Computing Research, Sandia National Laboratories, Albuquerque, NM, USA
e-mail: mdelia@sandia.gov

P. Bochev · D. J. Littlewood · M. Perego
Center for Computing Research, Sandia National Laboratories, Albuquerque, NM, USA
e-mail: pbboche@sandia.gov; djlittl@sandia.gov; mperego@sandia.gov

G. Z. Voyiadjis (ed.), *Handbook of Nonlocal Continuum Mechanics for Materials and Structures*, https://doi.org/10.1007/978-3-319-58729-5_31

equations (PDEs); however, the accuracy comes at the price of a prohibitive computational cost, making local-to-nonlocal (LtN) coupling strategies mandatory.

In this chapter, we review the state of the art of LtN methods where the efficiency of PDEs is combined with the accuracy of nonlocal models. Then, we focus on optimization-based coupling strategies that couch the coupling of the models into a control problem where the states are the solutions of the nonlocal and local equations, the objective is to minimize their mismatch on the overlap of the local and nonlocal problem domains, and the virtual controls are the nonlocal volume constraint and the local boundary condition. The strategy is described in the context of nonlocal and local elasticity and illustrated by numerical tests on three-dimensional realistic geometries. Additional numerical tests also prove the consistency of the method via patch tests.

Keywords

Optimization-based coupling methods · Local-nonlocal coupling · Nonlocal elasticity · Classical elasticity · Peridynamics · Domain decomposition · Finite element method · Particle methods

Introduction

Nonlocal continuum theories such as peridynamics (Silling and Lehoucq 2010) and physics-based nonlocal elasticity (Di Paola et al. 2009) can capture strong nonlocal effects due to long-range forces at the mesoscale or microscale. For problems where these effects cannot be neglected, nonlocal models are more accurate than classical partial differential equations (PDEs) that only consider interactions due to contact. However, the improved accuracy of nonlocal models comes at the price of a computational cost that is significantly higher than that of PDEs.

The goal of local-to-nonlocal (LtN) coupling methods is to combine the computational efficiency of PDEs with the accuracy of nonlocal models. LtN couplings are imperative when the size of the computational domain or the extent of the nonlocal interactions is such that the nonlocal solution becomes prohibitively expensive to compute, yet the nonlocal model is required to accurately resolve small-scale features (such as crack tips or dislocations that can affect the global material behavior). In this context, the main challenge of a coupling method is the stable and accurate merging of two fundamentally different mathematical descriptions of the same physical phenomena into a physically consistent coupled formulation.

Structure of the Chapter

This chapter is organized as follows. In section "Principles of Optimization-Based Couplings" we present an abstract framework of optimization-based coupling (OBC) methods. In section "The State Models and Their Properties" we introduce the static peridynamics and the local elasticity state models and describe their properties. In section "Optimization-Based LtN Formulation of Linearized Linear

Peridynamic Solid and Classical Elasticity" we specialize the OBC approach to the state models, and in section "Discretization of the LtN Formulation" we describe its fully discrete formulation; here we also review the discretization scheme for static peridynamics. Finally, in section "Numerical Tests" we demonstrate the consistency and efficiency of the coupling method through several numerical tests using Sandia's agile software components toolkit.

Local-to-Nonlocal Coupling Methods for Continuum Mechanics

The promise of improved physical fidelity at a lower computational cost has attracted significant attention to the coupling of nonlocal and local material models in continuum mechanics. The bulk of the existing methods though is based on some form of *blending* of the two material models. This blending can involve the energies of the two models, their force balance equations, or even their material properties. We describe three examples that are representative of these types of couplings.

The extension of the Arlequin method (Ben Dhia and Rateau 2005) to LtN couplings of continuum mechanics models by Han and Lubineau (2012) is an example of an *energy-blending* approach.

Their method splits the domain into a nonlocal subdomain, where the nonlocal effects are pronounced, and an overlapping local subdomain, where such effects are negligible. The intersection of these domains forms a "gluing area" where the energy of the system is defined as a weighted average of the local and nonlocal energies. At the local and nonlocal complements of the gluing area, the energy is defined according to the models operating in these regions. A Lagrange multiplier enforces compatibility of the kinematics of both models.

The formulation in Seleson et al. (2013) provides an example of a *force-blending* coupling approach. This method couples peridynamics and classical elasticity by using a weighted average of the local and nonlocal force balance equations in the overlap or *bridging* domain. Similar to Han and Lubineau (2012), the method uses the "pure" local and nonlocal force equations in the complements of this domain. The resulting hybrid model satisfies Newton's third law and is consistent for linear fields with no external forces (i.e., the method passes a linear patch test).

Finally, the morphing approach in Lubineau et al. (2012) is an example of an LtN coupling scheme based on blending, or morphing, the *material properties* of the two models. The method consists in the definition of a single model over the entire domain with an equilibrium equation that accounts for both local and nonlocal interactions through a gradual change in the material properties characterizing the two models in a "morphing" region. In this region local and nonlocal properties are suitably weighted under the constraint of energy equivalence in the overlap of the two domains. In Azdoud et al. (2013) the same authors extend this method to anisotropic continua.

These coupling methods share two common features. First, by blending energies, forces, or material models, they effectively introduce a *hybrid* material description combining the properties of both the local and nonlocal models in the overlap regions. Second, they treat the kinematic compatibility between the models, e.g., the equality of their displacements over a suitable interface, as a constraint in a way that is reminiscent of classical domain decomposition methods. In the next section, we describe a general OBC strategy that differs fundamentally from the blending approaches discussed above and offers some distinct computational and theoretical advantages.

Principles of Optimization-Based Couplings

In contrast to the blending methods described earlier, an OBC strategy treats the coupling condition as an optimization objective, which is minimized subject to the model equations acting independently in their respective subdomains. In so doing OBC reverses the roles of the coupling conditions and the governing equations and keeps the latter separate.

In particular, the coupling of local and nonlocal models is effected by couching the LtN coupling into an optimization problem. The objective is to minimize the mismatch of the local and nonlocal solutions on the overlap of their subdomains, the constraints are the associated governing equations, and the controls are the virtual nonlocal volume constraint and the local boundary condition. This approach is inspired by nonstandard optimization-based domain decomposition methods for PDEs (Discacciati et al. 2013; Du 2001; Du and Gunzburger 2000; Gervasio et al. 2001; Gunzburger et al. 1999, 2000; Gunzburger and Lee 2000; Kuberry and Lee 2013). It has also been applied to the coupling of discrete atomistic and continuum models in Olson et al. (2014a,b). This strategy brings about valuable theoretical and computational advantages. For instance, the coupled problem passes a patch test by construction, its well-posedness typically follows from the well-posedness of the constraint equations, and its numerical solution only requires the implementation of the optimization strategy as the local and nonlocal solvers for the state equations can be used as black boxes. For this reason, we refer to OBC methods as *nonintrusive* as opposed to the coupling methods described in section "Local-to-Nonlocal Coupling Methods for Continuum Mechanics," which are *intrusive* in the sense that their implementation requires modification of the basic governing equations for the local and nonlocal models in the overlap region. In what follows, we present an abstract formulation of OBCs.

Let $\mathcal{L}_n : \mathcal{V}_n \to \mathbb{R}$ be a nonlocal operator that accurately describes the behavior of the material in a bounded body and let $\mathcal{L}_l : \mathcal{V}_l \to \mathbb{R}$ be a local operator that describes the material well enough where the nonlocal effects are negligible. We recall that the numerical solution of the accurate nonlocal model is computationally expensive, whereas the one of the local model is, in general, affordable. As in the coupling methods described in section "Local-to-Nonlocal Coupling Methods for Continuum Mechanics," we solve the nonlocal model where the nonlocality affects the global material behavior and the local problem everywhere else; the challenge is to couple those models at the interfaces or overlaps of their domains. As explained above, we tackle this by solving an optimization problem where we minimize the difference between the local and nonlocal solutions at the interfaces tuning their values on the virtual boundaries and volumes induced by the domain decomposition; see Fig. 1a.

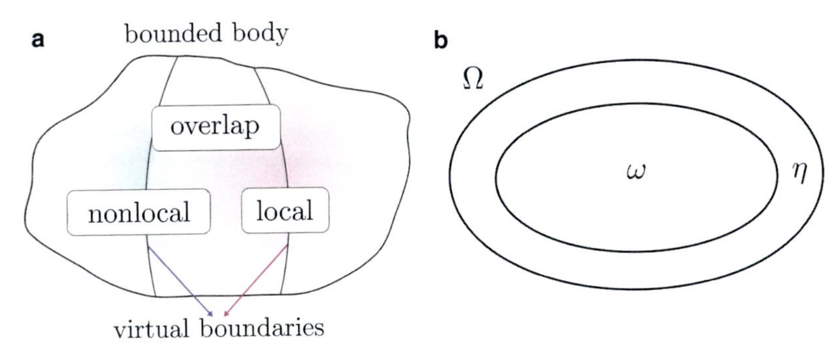

Fig. 1 Illustration of LtN OBC domain configuration for a bonded body Ω and its decomposition into ω and η. (**a**) An abstract domain configuration. (**b**) A simplified domain configuration

Formally, we state the LtN OBC as follows:

$$\min_{\mathbf{u}_n,\mathbf{u}_l,\boldsymbol{\nu}_n,\boldsymbol{\nu}_l} \mathcal{J}(\mathbf{u}_n,\mathbf{u}_l) = \frac{1}{2}\|\mathbf{u}_n - \mathbf{u}_l\|^2_{*,\text{overlap}}$$

$$\text{s.t.} \begin{cases} -\mathcal{L}_n\mathbf{u}_n = \mathbf{b}(\boldsymbol{x}) & \text{nonlocal domain} \\ \mathbf{u}_n(\boldsymbol{x}) = \mathbf{g}(\boldsymbol{x}) & \text{physical n-boundary} \\ \mathbf{u}_n(\boldsymbol{x}) = \boldsymbol{\nu}_n(\boldsymbol{x}) & \text{virtual n-boundary} \end{cases} \quad \begin{cases} -\mathcal{L}_l\mathbf{u}_l = \mathbf{b}(\boldsymbol{x}) & \text{local domain} \\ \mathbf{u}_l(\boldsymbol{x}) = \mathbf{g}(\boldsymbol{x}) & \text{physical boundary} \\ \mathbf{u}_l(\boldsymbol{x}) = \boldsymbol{\nu}_l(\boldsymbol{x}) & \text{virtual boundary,} \end{cases} \tag{1}$$

where $\mathbf{b}$ is a body force density, $\|\cdot\|_{*,\text{domain}}$ is a suitable norm on a domain, and $(\boldsymbol{\nu}_n, \boldsymbol{\nu}_l) \in \mathcal{C}$ (the control space) are the control variables. "N-boundary" stands for nonlocal boundary, usually called interaction volume (rigorously defined in section "The State Models and Their Properties"), that consists of all points outside of the domain that interact with points inside the domain. Thus, the goal of OBC is to find optimal values of the virtual controls $\boldsymbol{\nu}_n$ and $\boldsymbol{\nu}_l$ such that $\mathbf{u}_n$ and $\mathbf{u}_l$ are as close as possible on the overlap and still satisfy the model equations, which play the role of optimization constraints.

Note that this approach is very general and flexible and can be applied to any nonlocal model for continuum mechanics when a suitable local approximation is available. In this chapter, we use the OBC technique to combine nonlocal elasticity, described by a static peridynamics model, and classical linear elasticity. Our strategy is based on the recently introduced approaches (D'Elia and Bochev 2014, 2016; D'Elia et al. 2016) for local and nonlocal diffusion (Du et al. 2012).

Well-Posedness

We present a strategy for proving the well-posedness of (1) for linear operators $\mathcal{L}_l$ and $\mathcal{L}_n$. Here, without loss of generality, we consider $\mathbf{g} = \mathbf{0}$. We assume that for any pair of controls, the constraints in (1) have unique solutions $\mathbf{u}_n(\boldsymbol{\nu}_n)$ and $\mathbf{u}_l(\boldsymbol{\nu}_l)$. We introduce the reduced form of the optimization problem by eliminating the states from (1) and obtaining an optimization problem in terms of $\boldsymbol{\nu}_n$ and $\boldsymbol{\nu}_l$ only:

$$\min_{\nu_n,\nu_l} \mathcal{J}(\nu_n, \nu_l) = \frac{1}{2}\|\mathbf{u}_n(\nu_n) - \mathbf{u}_l(\nu_l)\|^2_{*,o}, \tag{2}$$

where "o" stands for "overlap." Following the approach used in Abdulle et al. (2015), D'Elia and Bochev (2016), Gervasio et al. (2001) and Olson et al. (2014b), one can show the well-posedness of (2) by splitting the solution of the state equations into the "harmonic" components $(\mathbf{v}_n(\nu_n), \mathbf{v}_l(\nu_l))$ and the homogeneous components $(\mathbf{u}_n^0, \mathbf{u}_l^0)$ such that they respectively satisfy

$$\begin{cases} -\mathcal{L}_n \mathbf{v}_n = \mathbf{0} & \text{nonlocal domain} \\ \mathbf{v}_n = \mathbf{0} & \text{physical n-boundary} \\ \mathbf{v}_n = \nu_n & \text{virtual n-boundary} \end{cases} \qquad \begin{cases} -\mathcal{L}_l \mathbf{v}_l = \mathbf{0} & \text{local domain} \\ \mathbf{v}_l = \mathbf{0} & \text{physical boundary} \\ \mathbf{v}_l = \nu_l & \text{virtual boundary,} \end{cases} \tag{3}$$

and

$$\begin{cases} -\mathcal{L}_n \mathbf{u}_n^0 = \mathbf{b} & \text{nonlocal domain} \\ \mathbf{u}_n^0 = \mathbf{0} & \text{n-boundary} \end{cases} \qquad \begin{cases} -\mathcal{L}_l \mathbf{u}_l^0 = \mathbf{b} & \text{local domain} \\ \mathbf{u}_l(x) = \mathbf{0} & \text{boundary.} \end{cases} \tag{4}$$

In terms of the components $\mathbf{u}_n = \mathbf{v}_n + \mathbf{u}_n^0$ and $\mathbf{u}_l = \mathbf{v}_l + \mathbf{u}_l^0$, the objective function and the Euler-Lagrange equations are given by

$$\begin{aligned} \mathcal{J}(\nu_n, \nu_l) = &\frac{1}{2}\|\mathbf{v}_n(\nu_n) - \mathbf{v}_l(\nu_l)\|^2_{*,o} + \frac{1}{2}\|\mathbf{u}_n^0 - \mathbf{u}_l^0\|^2_{*,o} \\ &+ (\mathbf{v}_n(\nu_n) - \mathbf{v}_l(\nu_l), \mathbf{u}_n^0 - \mathbf{u}_l^0)_{*,o}, \end{aligned}$$

and

$$Q(\boldsymbol{\sigma}_n, \boldsymbol{\sigma}_l; \boldsymbol{\beta}_n, \boldsymbol{\beta}_l) = F(\boldsymbol{\beta}_n, \boldsymbol{\beta}_l) \quad \forall\, (\boldsymbol{\beta}_n, \boldsymbol{\beta}_l) \in \mathcal{C}, \tag{5}$$

where

$$\begin{aligned} Q(\boldsymbol{\sigma}_n, \boldsymbol{\sigma}_l; \boldsymbol{\beta}_n, \boldsymbol{\beta}_l) &= (\mathbf{v}_n(\boldsymbol{\sigma}_n) - \mathbf{v}_l(\boldsymbol{\sigma}_l), \mathbf{v}_n(\boldsymbol{\beta}_n) - \mathbf{v}_l(\boldsymbol{\beta}_l))_{*,o}, \\ F(\boldsymbol{\beta}_n, \boldsymbol{\beta}_l) &= -(\mathbf{u}_n^0 - \mathbf{u}_l^0, \mathbf{v}_n(\boldsymbol{\beta}_n) - \mathbf{v}_l(\boldsymbol{\beta}_l))_{*,o}. \end{aligned}$$

The well-posedness of (2) is a consequence of the following important assumption.

Assumption .1 (Strong Cauchy-Schwarz (CS) inequality) *There exists a positive constant $\kappa < 1$ such that for all $(\boldsymbol{\sigma}_n, \boldsymbol{\sigma}_l) \in \mathcal{C}$*

$$|(\mathbf{v}_n(\boldsymbol{\sigma}_n), \mathbf{v}_l(\boldsymbol{\sigma}_l))_{*,o}| < \kappa\, \|\mathbf{v}_n(\boldsymbol{\sigma}_n)\|_{*,o}\|\mathbf{v}_l(\boldsymbol{\sigma}_l)\|_{*,o}. \tag{6}$$

This assumption, though strong, is reasonable in the context of multiscale modeling; in fact, it holds for problems such as nonlocal diffusion models (D'Elia and Bochev

2016; Olson et al. 2014b) and multiscale elliptic problems with highly oscillatory coefficients (Abdulle et al. 2015).

The following lemma establishes a fundamental property of Q.

Lemma 1. *If Assumption.1 holds, the form $Q(\cdot,\cdot)$ defines an inner product on $\mathcal{C}$.*

Proof. The bilinear form $Q(\cdot,\cdot)$ is symmetric and positive semi-definite. We show that it defines an inner product, by showing that $Q(\boldsymbol{\sigma}_n,\boldsymbol{\sigma}_l;\boldsymbol{\sigma}_n,\boldsymbol{\sigma}_l)=0$ if and only if $(\boldsymbol{\sigma}_n,\boldsymbol{\sigma}_l)=(\mathbf{0},\mathbf{0})$. Clearly, if $(\boldsymbol{\sigma}_n,\boldsymbol{\sigma}_l)=(\mathbf{0},\mathbf{0})$, then $\mathbf{v}_n(\boldsymbol{\sigma}_n)=\mathbf{0}$ and $\mathbf{v}_l(\boldsymbol{\sigma}_l)=\mathbf{0}$, implying $Q(\boldsymbol{\sigma}_n,\boldsymbol{\sigma}_l;\boldsymbol{\sigma}_n,\boldsymbol{\sigma}_l)=0$. On the other hand, if $Q(\boldsymbol{\sigma}_n,\boldsymbol{\sigma}_l;\boldsymbol{\sigma}_n,\boldsymbol{\sigma}_l)=0$,

$$
\begin{aligned}
0 = Q(\boldsymbol{\sigma}_n,\boldsymbol{\sigma}_l;\boldsymbol{\sigma}_n,\boldsymbol{\sigma}_l) &= \|\mathbf{v}_n(\boldsymbol{\sigma}_n)-\mathbf{v}_l(\boldsymbol{\sigma}_l)\|^2_{*,o} \\
&= \|\mathbf{v}_n(\boldsymbol{\sigma}_n)\|^2_{*,o}+\|\mathbf{v}_l(\boldsymbol{\sigma}_l)\|^2_{*,o}-2(\mathbf{v}_n(\boldsymbol{\sigma}_n),\mathbf{v}_l(\boldsymbol{\sigma}_l))_{*,o} \\
&\geq (1-\kappa)\big(\|\mathbf{v}_n(\boldsymbol{\sigma}_n)\|^2_{*,o}+\|\mathbf{v}_l(\boldsymbol{\sigma}_l)\|^2_{*,o}\big),
\end{aligned}
$$

where the last step is a consequence of the strong CS inequality (6) and the Young's inequality. Since $\kappa<1$, we have

$$
\big(\|\mathbf{v}_n(\boldsymbol{\sigma}_n)\|^2_{*,o}+\|\mathbf{v}_l(\boldsymbol{\sigma}_l)\|^2_{*,o}\big)\leq 0.
$$

Thus, we have that $\mathbf{v}_n(\boldsymbol{\sigma}_n)=\mathbf{0}$ and $\mathbf{v}_l(\boldsymbol{\sigma}_l)=\mathbf{0}$, which implies $(\boldsymbol{\sigma}_n,\boldsymbol{\sigma}_l)=(\mathbf{0},\mathbf{0})$. □

Note that to establish the well-posedness of problem (2), we need the completeness of $\mathcal{C}$ with respect to the norm induced by Q. However, this may not be the case; thus, as done in Abdulle et al. (2015), we may consider the completion of $\mathcal{C}$ and solve the optimization problem in the completed space, which we denote by $\mathcal{C}_c$. Then, we use the Hahn-Banach theorem to extend Q and F in $\mathcal{C}_c$ in a continuous and unique way and we denote the extensions by Q_c and F_c. The latter are such that Q_c is continuous and coercive and F_c is continuous in $\mathcal{C}_c$. The following theorem is a consequence of the considerations above.

Theorem 1. *If Assumption.1 holds, the optimization problem (2) has a unique solution $(\boldsymbol{\nu}_n^*,\boldsymbol{\nu}_l^*)\in\mathcal{C}_c$ satisfying the extended Euler-Lagrange equation*

$$
Q_c(\boldsymbol{\nu}_n^*,\boldsymbol{\nu}_l^*;\boldsymbol{\beta}_n,\boldsymbol{\beta}_l)=F_c(\boldsymbol{\beta}_n,\boldsymbol{\beta}_l)\quad\forall\,(\boldsymbol{\beta}_n,\boldsymbol{\beta}_l)\in\mathcal{C}_c.
$$

The State Models and Their Properties

Let $\Omega\subset\mathbb{R}^3$ be a bounded body with boundary $\partial\Omega=\Gamma$, the peridynamic equation of the displacement of a material point $\boldsymbol{x}\in\Omega$ at time $t\geq 0$ is given by

$$
\rho(\boldsymbol{x},t)\,\frac{\partial^2\mathbf{u}}{\partial t^2}(\boldsymbol{x},t)=\int_\Omega\{\mathbf{T}[\boldsymbol{x},t]\langle\boldsymbol{x}'-\boldsymbol{x}\rangle-\mathbf{T}[\boldsymbol{x}',t]\langle\boldsymbol{x}-\boldsymbol{x}'\rangle\}\,dV_{\boldsymbol{x}'}+\mathbf{b}(\boldsymbol{x},t),
$$

where $\rho \colon \Omega \times \mathbb{R}^+ \to \mathbb{R}^+$ is the mass density, $\mathbf{u} \colon \Omega \times \mathbb{R}^+ \to \mathbb{R}^3$ is the displacement field, $\mathbf{b} \colon \Omega \times \mathbb{R}^+ \to \mathbb{R}^3$ is a given body force density and $\mathbf{T} \colon \Omega \times \mathbb{R}^+ \to \mathbb{R}^{(3,3)}$ is the force state field, i.e, the force state at $(\boldsymbol{x}, t)$ mapping the bond $\langle \boldsymbol{x}' - \boldsymbol{x} \rangle$ to force per unit volume squared. In this work we consider the peridynamic equilibrium equation for a static problem:

$$-\mathcal{L}_n[\mathbf{u}](\boldsymbol{x}) := -\int_\Omega \left\{ \mathbf{T}[\boldsymbol{x}]\langle \boldsymbol{x}' - \boldsymbol{x} \rangle - \mathbf{T}[\boldsymbol{x}']\langle \boldsymbol{x} - \boldsymbol{x}' \rangle \right\} d V_{\boldsymbol{x}'} = \mathbf{b}(\boldsymbol{x}). \qquad (7)$$

According to the nonlocal theory, we make the assumption that a material point $\boldsymbol{x}$ interacts only with a neighborhood of points, more specifically, with material points in a ball of radius δ centered in $\boldsymbol{x}$, i.e.,

$$B_\delta(\boldsymbol{x}) = \{ \boldsymbol{x}' \in \Omega : |\boldsymbol{x} - \boldsymbol{x}'| \le \delta \},$$

where δ is a length scale referred to as the *horizon*. This implies that

$$\mathbf{T}[\boldsymbol{x}]\langle \boldsymbol{x}' - \boldsymbol{x} \rangle = 0, \quad \forall\, \boldsymbol{x}' \notin B_\delta(\boldsymbol{x}).$$

We solve (7) in $\omega \in \Omega$ and we prescribe Dirichlet volume constraints in a volumetric layer η surrounding ω so that the entire problem domain is $\Omega = \omega \cup \eta$; see Fig. 1, right. The definition of η depends on the properties of $\mathbf{T}$, and its thickness has to be large enough to guarantee the well-posedness of the problem; we provide more details below. In this work, for simplicity, we consider the linearized linear peridynamic solid (LPS) model (Silling et al. 2007) characterized by the force state field

$$\mathbf{T}[\boldsymbol{x}]\langle \boldsymbol{\xi} \rangle = \frac{w(|\boldsymbol{\xi}|)}{m} \left\{ (3K - 5G)\,\theta(\boldsymbol{x})\boldsymbol{\xi} + 15G \frac{\boldsymbol{\xi} \otimes \boldsymbol{\xi}}{|\boldsymbol{\xi}|^2} (\mathbf{u}(\boldsymbol{x} + \boldsymbol{\xi}) - \mathbf{u}(\boldsymbol{x})) \right\}, \forall\, \boldsymbol{x} \in \Omega, \qquad (8)$$

where $\boldsymbol{\xi} = \boldsymbol{x}' - \boldsymbol{x}$. Here K is the bulk modulus and G is the shear modulus. The linearized nonlocal dilatation, $\theta : \Omega \to \mathbb{R}$, is defined as

$$\theta(\boldsymbol{x}) = \frac{3}{m} \int_{B_\delta(\mathbf{0})} w(|\boldsymbol{\zeta}|)\, \boldsymbol{\zeta} \cdot (\mathbf{u}(\boldsymbol{x} + \boldsymbol{\zeta}) - \mathbf{u}(\boldsymbol{x}))\, d V_{\boldsymbol{\zeta}}, \quad \text{with } m = \int_{B_\delta(\mathbf{0})} w(|\boldsymbol{\zeta}|)\, |\boldsymbol{\zeta}|^2\, d V_{\boldsymbol{\zeta}}.$$

Here, the spherical influence function w is a scalar valued function used to determine the support of force states and to modulate the bond strength (Seleson and Parks 2011; Silling et al. 2007). Using the linearized LPS force state field in (8), we formulate the three-dimensional peridynamic problem as follows. Find $\mathbf{u} \in [L^2(\Omega)]^3$ such that

$$\begin{cases} -\mathcal{L}_{\mathrm{LPS}}[\mathbf{u}](\boldsymbol{x}) = \mathbf{b}(\boldsymbol{x}) & \boldsymbol{x} \in \omega \\ \mathbf{u}(\boldsymbol{x}) = \mathbf{g}(\boldsymbol{x}) & \boldsymbol{x} \in \eta, \end{cases} \qquad (9)$$

where $\mathbf{g} \in [L^2(\eta)]^3$ is a given displacement function and $\mathcal{L}_{\mathrm{LPS}}$ is obtained by substituting (8) into $\mathcal{L}_n$, i.e.,

$$\mathcal{L}_{\rm LPS}[\mathbf{u}](\boldsymbol{x}) := \int_{B_\delta(\mathbf{0})} \frac{w(|\boldsymbol{\xi}|)}{m} \Bigg\{ (3K - 5G)(\theta(\boldsymbol{x}) + \theta(\boldsymbol{x}+\boldsymbol{\xi}))\boldsymbol{\xi} + 30G \, \frac{\boldsymbol{\xi} \otimes \boldsymbol{\xi}}{|\boldsymbol{\xi}|^2} (\mathbf{u}(\boldsymbol{x}+\boldsymbol{\xi}) - \mathbf{u}(\boldsymbol{x})) \Bigg\} \, dV_{\boldsymbol{\xi}}. \tag{10}$$

We define the layer η as

$$\eta = \{\boldsymbol{x}' \in \Omega : \; |\boldsymbol{x}' - \boldsymbol{x}| < 2\delta\} \quad \forall \, \boldsymbol{x} \in \Gamma. \tag{11}$$

Note that the thickness is double the size of the horizon; this happens because in order to evaluate the peridynamic operator on a boundary point $\boldsymbol{x} \in \partial\omega$, we need to evaluate a double integral over $B_\delta(\mathbf{0}) \times B_\delta(\mathbf{0})$, i.e., we need values of the displacement in $B_{2\delta}(\boldsymbol{x})$.

The model (10) has two important features. First, its local limit (i.e., the limit for $\delta \to 0$ that corresponds to vanishing nonlocal interactions) is the classical Navier-Cauchy equation (NCE) of static elasticity (Seleson and Littlewood 2016):

$$-\mathcal{L}_{\rm NC}[\mathbf{u}](\boldsymbol{x}) := -\left[\left(K + \frac{1}{3} G \right) \nabla(\nabla \cdot \mathbf{u})(\boldsymbol{x}) + G \, \nabla^2 \mathbf{u}(\boldsymbol{x}) \right] = \mathbf{b}(\boldsymbol{x}), \tag{12}$$

where K, G, and $\mathbf{b}$ are defined as in (8). The latter is equivalent to the linear elasticity equation in terms of the Lamé constants (λ, μ):

$$\begin{aligned} &-\nabla \cdot \boldsymbol{\sigma}[\mathbf{u}](\boldsymbol{x}) = \mathbf{b}(\boldsymbol{x}), \quad \text{where} \\ &\boldsymbol{\sigma}[\mathbf{u}] = \lambda(\nabla \cdot \mathbf{u})\mathbf{I} + \mu(\nabla \mathbf{u} + \nabla \mathbf{u}^T) \\ &(\lambda, \mu) = \left(K - \tfrac{2G}{3}, \, G \right), \end{aligned} \tag{13}$$

where $\mathbf{I}$ is the identity tensor. This property suggests that the NC model can approximate fairly well the nonlocal model for sufficiently regular solutions; for this reason, it is the local model of choice in our coupling strategy.

Second, for a quadratic displacement field, the linearized LPS reduces to the classical NCE (see Proposition 1 in Seleson and Littlewood 2016). This property allows us to perform a quadratic patch test; see section "Patch Tests."

Optimization-Based LtN Formulation of Linearized Linear Peridynamic Solid and Classical Elasticity

Given a domain Ω representing a bounded body, we introduce a partition into a nonlocal subdomain Ω_n and a local subdomain Ω_l, with boundary Γ_l, such that $\Omega_n = \omega_n \cup \eta_n$ and $\Omega_n \cap \Omega_l = \Omega_o \neq \emptyset$; see Fig. 2 for a two-dimensional illustration.

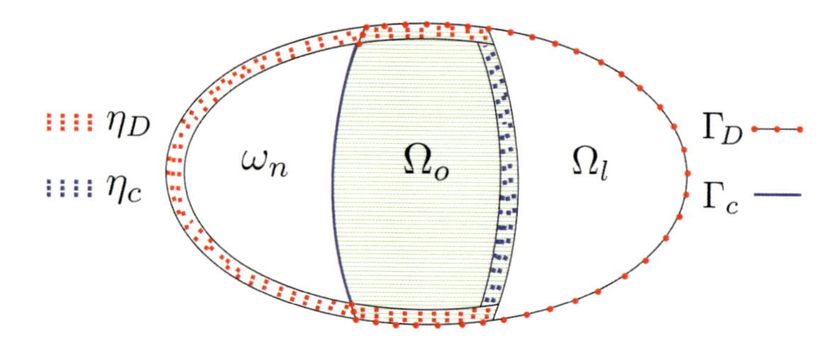

Fig. 2 An example LtN domain configuration in two dimensions

We assume that the nonlocal model (10) accurately describes the material behavior in Ω_n, while the local NC model gives a fairly reasonable representation for the rest of the domain. We formulate the coupling as an optimization problem where we minimize the difference between the nonlocal and the local solutions on the overlap Ω_o adjusting their values on the virtual interaction volume η_c and the virtual boundary Γ_c determined by the partition. Let $\eta_D = \eta \cap \eta_n$ and $\Gamma_D = \Gamma \cap \Gamma_l$ be the physical interaction volume and boundary where we prescribe the given Dirichlet data; we define the virtual control volume and boundary as $\eta_c = \eta_n \setminus \eta_D$ and $\Gamma_c = \Gamma_l \setminus \Gamma_D$. By posing the peridynamic problem on ω_n and the NC problem on Ω_l, we obtain

$$\begin{cases} -\mathcal{L}_{\text{LPS}}[\mathbf{u}_n](\boldsymbol{x}) = \mathbf{b}(\boldsymbol{x}) & \boldsymbol{x} \in \omega_n \\ \mathbf{u}_n(\boldsymbol{x}) = \mathbf{g}(\boldsymbol{x}) & \boldsymbol{x} \in \eta_D \\ \mathbf{u}_n(\boldsymbol{x}) = \boldsymbol{\nu}_n(\boldsymbol{x}) & \boldsymbol{x} \in \eta_c \end{cases} \qquad \begin{cases} -\mathcal{L}_{\text{NC}}[\mathbf{u}_l](\boldsymbol{x}) = \mathbf{b}(\boldsymbol{x}) & \boldsymbol{x} \in \Omega_l \\ \mathbf{u}_l(\boldsymbol{x}) = \mathbf{g}(\boldsymbol{x}) & \boldsymbol{x} \in \Gamma_D \\ \mathbf{u}_l(\boldsymbol{x}) = \boldsymbol{\nu}_l(\boldsymbol{x}) & \boldsymbol{x} \in \Gamma_c, \end{cases} \tag{14}$$

where $\boldsymbol{\nu}_n \in [L^2(\eta_c)]^3$ and $\boldsymbol{\nu}_l \in [H^{1/2}(\Gamma_c)]^3$ are undetermined volume constraints and boundary conditions. In our formulation (14) serve as constraints and $(\boldsymbol{\nu}_n, \boldsymbol{\nu}_l)$ as control variables of the optimization problem

$$\min_{\mathbf{u}_n, \mathbf{u}_l, \boldsymbol{\nu}_n, \boldsymbol{\nu}_l} \mathcal{J}(\mathbf{u}_n, \mathbf{u}_l) = \frac{1}{2} \int_{\Omega_o} |\mathbf{u}_n - \mathbf{u}_l|^2 \, d\boldsymbol{x} \quad \text{subject to (14).} \tag{15}$$

Given the optimal controls $\boldsymbol{\nu}_n^*$ and $\boldsymbol{\nu}_l^*$, we define the coupled solution as

$$\mathbf{u}^* = \begin{cases} \mathbf{u}_n^* & \boldsymbol{x} \in \Omega_n \\ \mathbf{u}_l^* & \boldsymbol{x} \in \Omega_l \setminus \Omega_o, \end{cases} \tag{16}$$

where $\mathbf{u}_n^* = \mathbf{u}_n(\boldsymbol{\nu}_n^*)$ and $\mathbf{u}_l^* = \mathbf{u}_l(\boldsymbol{\nu}_l^*)$.

Discretization of the LtN Formulation

For the discretization of the NC model in (12), we consider the standard finite element (FE) method. We denote the vector of values of the local discrete solution at the FE degrees of freedom by $\vec{\boldsymbol{u}}_l = [\vec{u}_l^1, \vec{u}_l^2, \vec{u}_l^3]$, with $\vec{u}_l^k \in \mathbb{R}^{N_l}$ where N_l is the number of degrees of freedom of each spatial component over the FE computational mesh.

For the peridynamic model introduced in section "The State Models and Their Properties," we utilize a meshfree discretization. For every point $\boldsymbol{x}_i$ discretizing the body Ω, we approximate the integral operator as follows

$$L[\boldsymbol{x}_i] := \sum_{j \in \mathcal{F}_i} \{\mathbf{T}[\boldsymbol{x}_i]\langle \boldsymbol{x}_j - \boldsymbol{x}_i \rangle - \mathbf{T}[\boldsymbol{x}_j]\langle \boldsymbol{x}_i - \boldsymbol{x}_j \rangle\} \; V_j^{(i)}, \tag{17}$$

where $\boldsymbol{x}_i$ and $V_j^{(i)}$ are quadrature points and weights and $\mathcal{F}_i$ represents the set of all points in Ω interacting with the ith material point. Note that the quadrature point $\boldsymbol{x}_j$ is chosen to coincide with the reference position of the jth node; the quadrature weight $V_j^{(i)}$ is the volume of the intersection between the neighborhood of $\boldsymbol{x}_j$ and the neighborhood of $\boldsymbol{x}_i$, i.e., $|B_\delta(\boldsymbol{x}_j) \cap B_\delta(\boldsymbol{x}_i)|$. For $\boldsymbol{x}_j$ near the boundary of $B_\delta(\boldsymbol{x}_i)$, $V_j^{(i)}$ represents a partial volume. Details regarding the computation of $V_j^{(i)}$ can be found in Seleson and Littlewood (2016). We denote the vector of values of the discrete nonlocal solution at the material points by $\vec{\boldsymbol{u}}_n = [\vec{u}_n^1, \vec{u}_n^2, \vec{u}_n^3]$, with $\vec{u}_n^k \in \mathbb{R}^{N_n}$, where N_n is the number of material points.

We let $S_n \in \mathbb{R}^{N_o, N_n}$ be the matrix that selects the components of $\vec{u}_n^k$ in Ω_o and $S_l \in \mathbb{R}^{N_o, N_l}$ be the operator that evaluates $\vec{u}_l^k$ at the material points in Ω_o; we define them as

$$(S_n)_{ij} := \delta_{ij} \;\text{ and }\; (S_l)_{ij} := \phi_j(\boldsymbol{x}_i), \quad \forall\, \boldsymbol{x}_i \in \Omega_o,$$

where ϕ_j is the jth FE basis function. We define the discrete functional as

$$J_d(\vec{\boldsymbol{u}}_n, \vec{\boldsymbol{u}}_l) = \frac{1}{2} \sum_{i=1}^{N_o} \sum_{k=1}^{3} |(S_n \vec{u}_n^k)_i - (S_l \vec{u}_l^k)_i|^2 \, \widetilde{V}_i, \tag{18}$$

where $\widetilde{V}_i$ is the volume associated with the ith material point, properly scaled.

Software

The example simulations are carried out using the *Albany* Salinger et al. (2013) (available at the public git repository https://github.com/gahansen/Albany) and *Peridigm* Parks et al. (2012) (available at the public git repository https://github.com/peridigm/peridigm) codes, developed in the Center for Computing Research at Sandia National Laboratories. *Albany* is an FE code for simulating a variety of physical processes governed by PDEs. It is applied for the majority of the computation, including FE assembly for the Navier-Cauchy equation, calculation of the functional and its derivative, and solution of the state and adjoint systems. *Peridigm* is a peridynamics code for solid mechanics. A software interface was developed to facilitate the linking of *Peridigm* routines with *Albany*; both *Albany* and *Peridigm* rely on several *Trilinos* packages (available at https://trilinos.org/packages), for example, *Epetra* for the management of parallel data structures, *Intrepid2* for FE assembly, and *Ifpack* and *AztecOO* for the preconditioning and solution of linear systems. We apply the LBFGS optimization algorithm, as implemented in the *Trilinos* package *ROL* (available at https://trilinos.org/packages/rol).

Numerical Tests

In this section, we demonstrate the effectiveness and consistency of our strategy through several numerical examples. We first show that the OBC method passes linear and quadratic patch tests. In these cases, the analytic solutions are available and are in agreement for the nonlocal and local models. We then apply the OBC approach to test cases in which a discontinuity is present in the nonlocal domain. For these simulations, while the nonlocal and local models behave similarly, differences in their solutions are expected in the overlap domain. We model a rectangular bar containing a crack, followed by a tensile test specimen containing a crack. The latter case represents a realistic engineering geometry that fully exercises the OBC approach in three dimensions.

Patch Tests

The patch test simulations demonstrate the effectiveness of the OBC approach on benchmark problems for which the analytic solutions are available. As mentioned previously, it was shown in Seleson and Littlewood (2016) that Eqs. (9) and (12) are equivalent for linear and quadratic displacements. As a result, for this class of problems, it is expected that numerical results obtained using the OBC approach should exhibit an excellent match between the local and nonlocal models in the overlap region, with discretization error being the only source of discrepancy.

We consider a rectangular bar in three dimensions:

$$\begin{aligned}
\Omega &= [0.0\,\text{mm},\ 100.0\,\text{mm}] \times [-12.5\,\text{mm},\ 12.5\,\text{mm}] \times [-12.5\,\text{mm},\ 12.5\,\text{mm}],\\
\omega_n \cup \eta_D \cup \eta_c &= [0.0\,\text{mm}, 62.5\,\text{mm}] \times [-12.5\,\text{mm}, 12.5\,\text{mm}] \times [-12.5\,\text{mm}, 12.5\,\text{mm}],\\
\Omega_l &= [37.5\,\text{mm},\ 100.0\,\text{mm}] \times [-12.5\,\text{mm},\ 12.5\,\text{mm}] \times [-12.5\,\text{mm},\ 12.5\,\text{mm}].
\end{aligned}$$

Following the configuration illustrated in Fig. 2, the nonlocal domains are constructed such that ω_n is fully encapsulated by $\eta_D \cup \eta_c$. The external layer provided by the domain $\eta_D \cup \eta_c$, in which volume constraints are prescribed, has a thickness equal to twice the horizon (see section "The State Models and Their Properties").

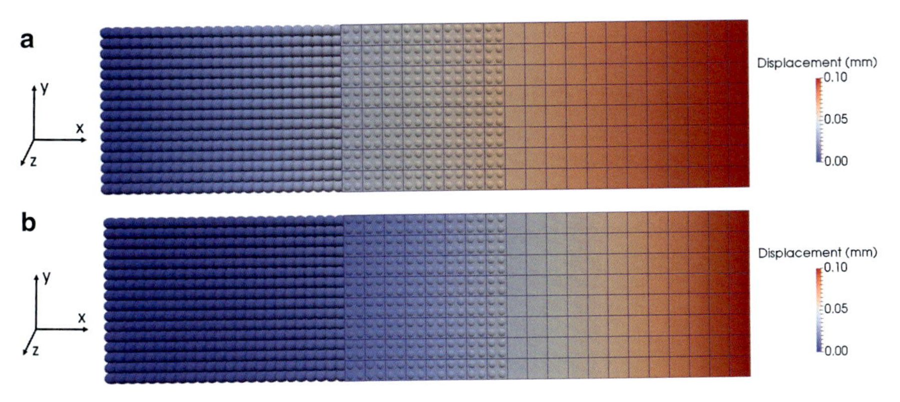

Fig. 3 Solutions for displacement in the x direction for the linear and quadratic patch tests. (**a**) Solution for the linear patch test. (**b**) Solution for the quadratic patch test

In Fig. 3, nodal volumes on the left and the FE mesh on the right represent the discretizations of $\omega_n \cup \eta_D \cup \eta_c$ and Ω_l, respectively. Further, we define

$$\textit{Linear: } \mathbf{u}(\boldsymbol{x}) = 10^{-3}(x, 0, 0),\ \mathbf{b}(\boldsymbol{x}) = \mathbf{0},\ \mathbf{g}(\boldsymbol{x}) = \mathbf{u}(\boldsymbol{x}),$$
$$\textit{Quadratic: } \mathbf{u}(\boldsymbol{x}) = 10^{-5}(x^2, 0, 0),\ \mathbf{b}(\boldsymbol{x}) = \mathbf{b}_q,\ \mathbf{g}(\boldsymbol{x}) = \mathbf{u}(\boldsymbol{x}).$$

We assign to the bulk modulus, K, a value of 150.0 GPa, and we assign to the shear modulus, G, a value of 81.496 GPa, which are representative of stainless steel. The peridynamic horizon in the nonlocal domain is assigned a value of 4.270 mm. Following Seleson and Littlewood (2016), the body force density, $\mathbf{b}_q$, producing equilibrium under the given quadratic displacement field, is given by

$$\mathbf{b}_q = 10^{-5}\left(\frac{8G}{3} + 2K\right) = 5.173\,\mathrm{N\,mm^{-3}}.$$

Simulation results for the linear and quadratic patch tests are presented in Figs. 3 and 4. Displacement solutions in the x (horizontal) direction are given in Fig. 3. In Fig. 4, we report the same variable along a horizontal line passing through the center of the bar. The patch test results are in good agreement with the expected linear and quadratic solutions, respectively, for both the nonlocal and local models.

Rectangular Bar with a Crack

We next consider a rectangular bar containing a discontinuity (crack) at its center. As illustrated in Fig. 5, OBC is utilized to connect a nonlocal domain covering the center portion of the bar with two local domains located at the ends of the bar. Under this configuration, the discontinuity is contained within the nonlocal domain, and the regions over which (non-control) Dirichlet boundary conditions are applied are restricted to the local domain. This is advantageous because, in

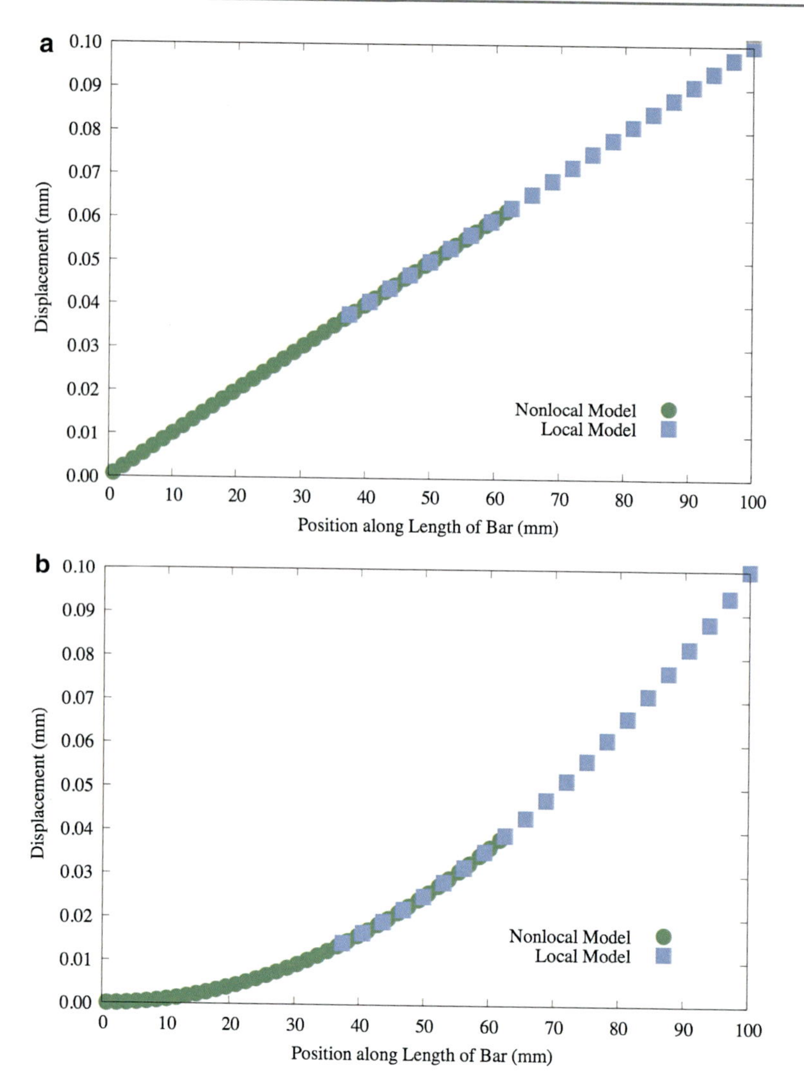

Fig. 4 Solutions for the x component of displacement along a horizontal line passing through the center of the bar for the linear and quadratic patch tests. (**a**) Solution for the linear patch test. (**b**) Solution for the quadratic patch test

practice, the determination and application of nonlocal volume constraints can be problematic (Littlewood 2015). We define the bounded body as

$$\Omega := [-50.0\,\text{mm},\ 50.0\,\text{mm}] \times [-12.5\,\text{mm},\ 12.5\,\text{mm}] \times [-12.5\,\text{mm},\ 12.5\,\text{mm}].$$

The discontinuity is inserted into the geometry via a rectangular plane defined by $x = 0.0\,\text{mm}$, $5.0\,\text{mm} \le y \le 12.5\,\text{mm}$, and $-12.5\,\text{mm} \le z \le 12.5\,\text{mm}$. The nonlocal and local domains are defined as

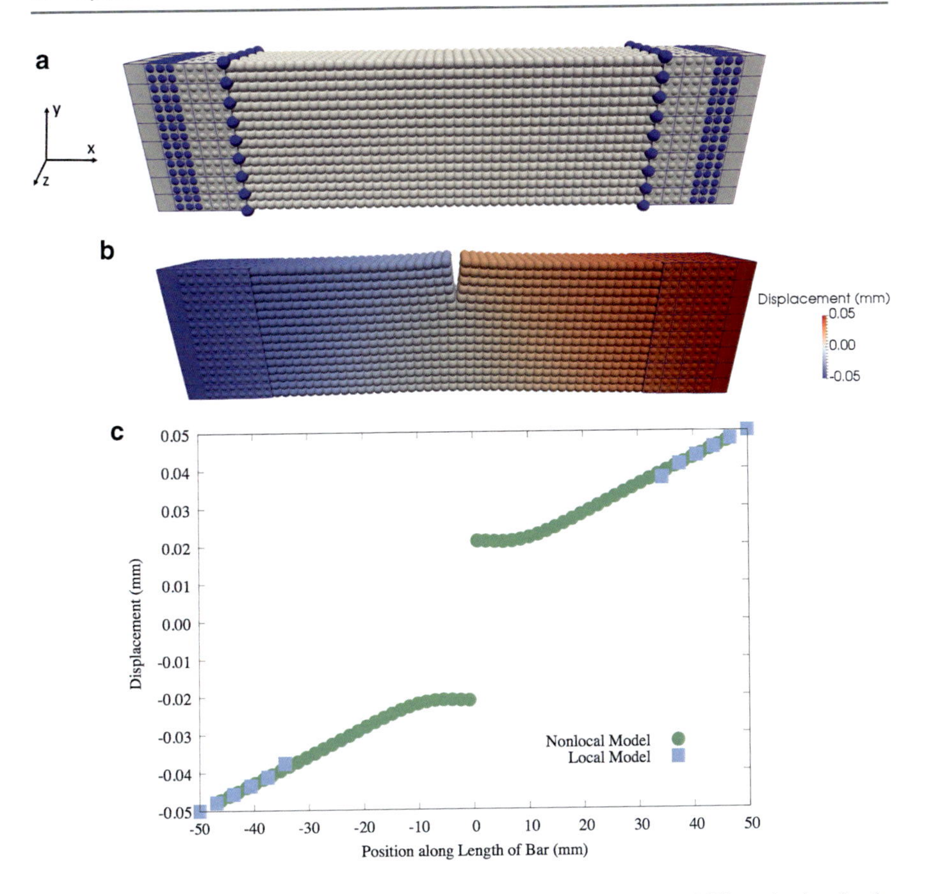

Fig. 5 Discretization and solution for the rectangular bar with a crack. (**a**) Discretization for the rectangular bar with a crack. Control nodes are highlighted in blue. (**b**) The x component of the displacement solution. Deformation is magnified by a factor of 20 to clearly illustrate the discontinuity. (**c**) The x component of the displacement solution along a horizontal line on the top edge of the bar, passing through the discontinuity

$$\begin{aligned}\omega_n \cup \eta_D \cup \eta_c &:= [-46.875\,\text{mm}, 46.875\,\text{mm}] \times [-12.5\,\text{mm}, 12.5\,\text{mm}] \times [-12.5\,\text{mm}, 12.5\,\text{mm}],\\ \Omega_{l1} &:= [-50.0\,\text{mm}, -34.375\,\text{mm}] \times [-12.5\,\text{mm}, 12.5\,\text{mm}] \times [-12.5\,\text{mm}, 12.5\,\text{mm}],\\ \Omega_{l2} &:= [34.375\,\text{mm}, 50.0\,\text{mm}] \times [-12.5\,\text{mm}, 12.5\,\text{mm}] \times [-12.5\,\text{mm}, 12.5\,\text{mm}].\end{aligned}$$

The domain η_c, over which control Dirichlet conditions for the nonlocal domain are applied, is defined by $-46.875\,\text{mm} \le x \le -42.1875\,\text{mm}$ and $42.1875\,\text{mm} \le x \le 46.875\,\text{mm}$. The control Dirichlet conditions for the local model are applied to Γ_c, defined by the planes $x = -34.375\,\text{mm}$ and $x = 34.375\,\text{mm}$. The locations of the control nodes in the discretized model are highlighted in Fig. 5a. As in the patch tests, the bulk modulus, K, is assigned a value of 150.000 GPa and the shear modulus, G, a value of 81.496 GPa. The peridynamic horizon in the nonlocal domain is assigned a value of 2.707 mm. Tensile loading is applied to the bar by prescribing displacements of −0.05 mm and 0.05 mm in the x (longitudinal) direction on the faces located at the ends of the bar defined by $x = -50.0$ mm and $x = 50.0$ mm, respectively. To eliminate rigid body modes, additional zero displacement boundary conditions are applied in the y direction along the edges defined by $x = -50.0$ mm, $y = -12.5$ mm and $x = 50.0$ mm, $y = -12.5$ mm and in the z direction along the edges defined by $x = -50.0$ mm, $z = -12.5$ mm and $x = 50.0$ mm, $z = -12.5$ mm.

Simulation results are presented in Fig. 5. The three-dimensional image in Fig. 5b shows the opening of the crack that results from tensile loading. Figure 5c gives displacement results along a horizontal line located on the top face of the bar.

Tensile Test Specimen with a Crack

The simulation of a tensile bar with a crack at its midpoint demonstrates OBC for the modeling of a common engineering geometry. As shown in Fig. 6, we restrict the use of the nonlocal model to a small subdomain in the direct vicinity of the crack. The overall height of the tensile bar specimen is 100.0 mm and the width of the bar at its midpoint is 6.25 mm. The nonlocal region, located at the midpoint of the bar and offset to the side of the bar containing the crack, has a height of 8.68 mm and width of 4.985 mm. The nodes comprising the nonlocal model control domain, η_c,

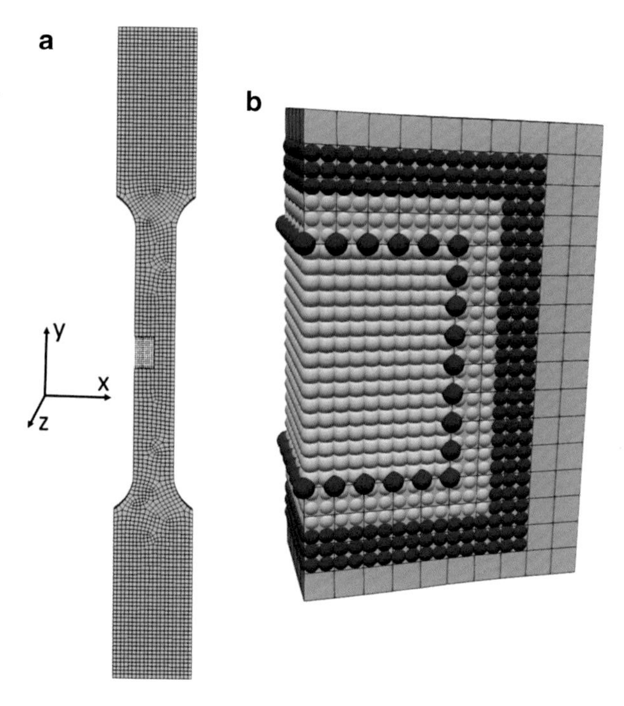

Fig. 6 Discretization of the tensile bar specimen. The nonlocal domain is restricted to a small subregion near the center of the bar. (**a**) Discretization of tensile bar specimen. (**b**) Control nodes in the overlap region

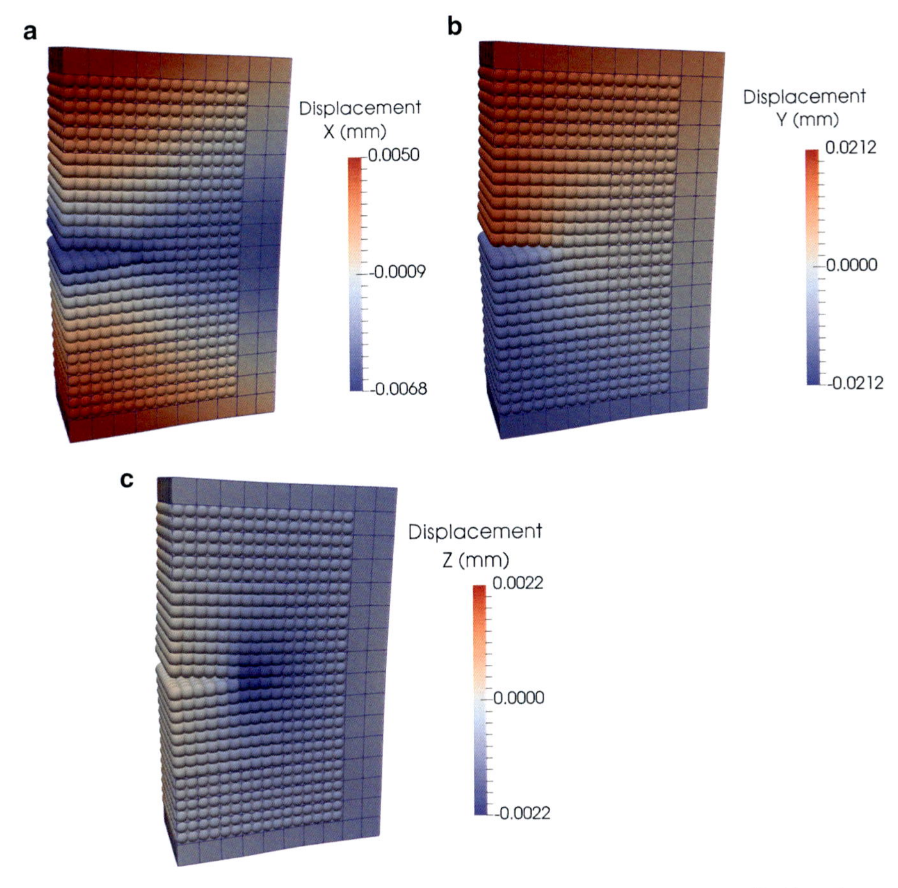

Fig. 7 Displacement solutions for the tensile test specimen. Deformation is magnified by a factor of 10 to clearly illustrate the discontinuity. (**a**) Displacement in x direction. (**b**) Displacement in y direction. (**c**) Displacement in z direction

and the local model control domain, Γ_c, are highlighted in blue in Fig. 6b. The discontinuity is inserted via a rectangular plane at the midpoint of the bar extending from the left side of the bar a distance of 1.86 mm into the bar. We employ material model parameters of 160.0 GPa for the bulk modulus and 64.0 GPa for the shear modulus. For the nonlocal model, the peridynamic horizon is assigned a value of 0.537 mm. Tensile loading is simulated via Dirichlet (displacement) boundary condition applied to the faces at the top and bottom of the bar that produce an overall engineering strain of 0.1% in the y direction. Following the strategy described in section "Rectangular Bar with a Crack," additional zero displacement boundary conditions are applied along edges on the top and bottom faces in the x and z directions to eliminate rigid body modes.

Results for the tensile bar simulation are given in Fig. 7. The influence of the crack on the displacement solution is restricted predominantly to the nonlocal region, and solutions corresponding to the nonlocal and local models are in good general agreement in the overlap domain.

Acknowledgements This material is based upon work supported by the US DOE's Laboratory Directed Research and Development (LDRD) program at Sandia National Laboratories and the US Department of Energy, Office of Science, Office of Advanced Scientific Computing Research. Part of this research was carried under the auspices of the Collaboratory on Mathematics for Mesoscopic Modeling of Materials (CM4). Sandia National Laboratories is a multimission laboratory managed and operated by National Technology and Engineering Solutions of Sandia, LLC., a wholly owned subsidiary of Honeywell International, Inc., for the US Department of Energy's National Nuclear Security Administration under contract DE-NA-0003525. SAND2017-3003 B.

References

A. Abdulle, O. Jecker, A. Shapeev, An optimization based coupling method for multiscale problems. Technical Report 36.2015, EPFL, Mathematics Institute of Computational Science and Engineering, Lausanne, Dec 2015

Y. Azdoud, F. Han, G. Lubineau, A morphing framework to couple non-local and local anisotropic continua. Int. J. Solids Struct. **50**(9), 1332–1341 (2013)

M. D'Elia, P. Bochev, Optimization-based coupling of nonlocal and local diffusion models, in *Proceedings of the Fall 2014 Materials Research Society Meeting*, ed. by R. Lipton. MRS Symposium Proceedings (Cambridge University Press, Boston, 2014)

M. D'Elia, P. Bochev, Formulation, analysis and computation of an optimization-based local-to-nonlocal coupling method. Technical Report SAND2017–1029J, Sandia National Laboratories, 2016

M. D'Elia, M. Perego, P. Bochev, D. Littlewood, A coupling strategy for nonlocal and local diffusion models with mixed volume constraints and boundary conditions. Comput. Math. Appl. **71**(11), 2218–2230 (2016)

H.B. Dhia, G. Rateau, The Arlequin method as a flexible engineering design tool. Int. J. Numer. Methods Eng. **62**(11), 1442–1462 (2005)

M. Di Paola, G. Failla, M. Zingales, Physically-based approach to the mechanics of strong non-local linear elasticity theory. J. Elast. **97**(2), 103–130 (2009)

M. Discacciati, P. Gervasio, A. Quarteroni, The interface control domain decomposition (ICDD) method for elliptic problems. SIAM J. Control. Optim. **51**(5), 3434–3458 (2013)

Q. Du, Optimization based nonoverlapping domain decomposition algorithms and their convergence. SIAM J. Numer. Anal. **39**(3), 1056–1077 (2001)

Q. Du, M.D. Gunzburger, A gradient method approach to optimization-based multidisciplinary simulations and nonoverlapping domain decomposition algorithms. SIAM J. Numer. Anal. **37**(5), 1513–1541 (2000)

Q. Du, M. Gunzburger, R. Lehoucq, K. Zhou, Analysis and approximation of nonlocal diffusion problems with volume constraints. SIAM Rev. **54**(4), 667–696 (2012)

P. Gervasio, J.-L. Lions, A. Quarteroni, Heterogeneous coupling by virtual control methods. Numerische Mathematik **90**, 241–264 (2001). https://doi.org/10.1007/s002110100303

M.D. Gunzburger, H.K. Lee, An optimization-based domain decomposition method for the Navier-Stokes equations. SIAM J. Numer. Anal. **37**(5), 1455–1480 (2000)

M.D. Gunzburger, J.S. Peterson, H. Kwon, An optimization based domain decomposition method for partial differential equations. Comput. Math. Appl. **37**(10), 77–93 (1999)

M.D. Gunzburger, M. Heinkenschloss, H.K. Lee, Solution of elliptic partial differential equations by an optimization-based domain decomposition method. Appl. Math. Comput. **113**(2–3), 111–139 (2000)

F. Han, G. Lubineau, Coupling of nonlocal and local continuum models by the Arlequin approach. Int. J. Numer. Methods Eng. **89**(6), 671–685 (2012)

P. Kuberry, H. Lee, A decoupling algorithm for fluid-structure interaction problems based on optimization. Comput. Methods Appl. Mech. Eng. **267**, 594–605 (2013)

D.J. Littlewood, Roadmap for peridynamic software implementation. Report SAND2015-9013, Sandia National Laboratories, Albuquerque, 2015

G. Lubineau, Y. Azdoud, F. Han, C. Rey, A. Askari, A morphing strategy to couple non-local to local continuum mechanics. J. Mech. Phys. Solids **60**(6), 1088–1102 (2012)

D. Olson, P. Bochev, M. Luskin, A. Shapeev, Development of an optimization-based atomistic-to-continuum coupling method, in *Proceedings of LSSC 2013*, ed. by I. Lirkov, S. Margenov, J. Wasniewski. Lecture Notes in Computer Science (Springer, Berlin/Heidelberg, 2014a)

D. Olson, P. Bochev, M. Luskin, A. Shapeev, An optimization-based atomistic-to-continuum coupling method. SIAM J. Numer. Anal. **52**(4), 2183–2204 (2014b)

M.L. Parks, D.J. Littlewood, J.A. Mitchell, S.A. Silling, Peridigm Users' Guide v1.0.0. SAND Report 2012-7800, Sandia National Laboratories, Albuquerque, 2012

A.G. Salinger, R.A. Bartlett, Q. Chen, X. Gao, G.A. Hansen, I. Kalashnikova, A. Mota, R.P. Muller, E. Nielsen, J.T. Ostien, R.P. Pawlowski, E.T. Phipps, W. Sun, Albany: a component–based partial differential equation code built on Trilinos. SAND Report 2013-8430J, Sandia National Laboratories, Albuquerque, 2013

P. Seleson, D.J. Littlewood, Convergence studies in meshfree peridynamic simulations. Comput. Math. Appl. **71**(11), 2432–2448 (2016)

P. Seleson, M.L. Parks, On the role of the influence function in the peridynamic theory. Int. J. Multiscale Comput. Eng. **9**, 689–706 (2011)

P. Seleson, S. Beneddine, S. Prudhomme, A force-based coupling scheme for peridynamics and classical elasticity. Comput. Mater. Sci. **66**, 34–49 (2013)

S.A. Silling, R.B. Lehoucq, Peridynamic theory of solid mechanics, in *Advances in Applied Mechanics*, vol. 44 (Elsevier, San Diego, 2010), pp. 73–168

S.A. Silling, M. Epton, O. Weckner, J. Xu, E. Askari, Peridynamic states and constitutive modeling. J. Elast. **88**, 151–184 (2007)

Bridging Local and Nonlocal Models: Convergence and Regularity

37

Mikil D. Foss and Petronela Radu

Contents

Abstract

As nonlocal models become more widespread in applications, we focus on their connections with their classical counterparts and also on some theoretical aspects which impact their implementation. In this context we survey recent developments by the authors and prove some new results on regularity of solutions to nonlinear systems in the nonlocal framework. In particular, we focus on semilinear problems and also on higher-order problems with applications in the theory of plate deformations.

The second author acknowledges support from the Simons Foundation.

M. D. Foss (✉) · P. Radu
Department of Mathematics, University of Nebraska-Lincoln, Lincoln, NE, USA
e-mail: mfoss@math.unl.edu; pradu@math.unl.edu

G. Z. Voyiadjis (ed.), *Handbook of Nonlocal Continuum Mechanics for Materials and Structures*, https://doi.org/10.1007/978-3-319-58729-5_32

Keywords
Nonlocal operators · Classical differentiability · Higher integrability · Weakly integrable kernels · Peridynamics

Introduction

The need for improved models in continuum mechanics has motivated rapid developments in partial differential equations (PDEs). Over the last two decades, the success of nonlocal models has given a fresh impetus to investigate partial integrodifferential equations (PIDEs), for which efforts are under way toward formulating a counterpart to the local theory in PDEs. Recent results in the nonlocal framework demonstrate the importance of robust implementations for models that depend on a length scale parameter (Tian and Du 2014). Other situations have shown the critical need for a deep understanding of theoretical aspects, such as well-posedness and regularity of solutions, and also continuity with respect to initial data which guarantee that numerical simulations produce physically relevant solutions. Toward this aim, we present below some existing and new results for the nonlocal theory, which give a correspondence between the nonlocal and the classical elliptic theory for the second- and fourth-order levels.

Recently nonlocal theories have successfully modeled singular phenomena (such as fracture), as well as phenomena with nonlocal features (aggregation models in biology (Sun et al. 2012; Mogilner and Edelstein-Keshet 1999), thermal diffusion (Oterkus et al. 2014), image processing (Gilboa and Osher 2008), sandpile formation, and more; see the monograph (Andreu-Vaillo et al. 2010) and the references therein). The theory of peridynamics (Silling 2000) introduced a model for elastic deformations in which an integral operator collects a cumulative response of pairwise interactions within a neighborhood of each material point. This formulation is applicable to both continuous deformations and those with dynamic fracture. In the absence of fracture, for a domain $\Omega \subset \mathbb{R}^n$ with smooth boundary and collar Γ (a closed set surrounding Ω of positive measure, which will be defined more precisely below), the deformation $u : \Omega \subset \mathbb{R}^n \to \mathbb{R}$ satisfies the steady-state system

$$\begin{cases} \mathcal{L}_\mu u(x) = f(x), & x \in \Omega \\ u(x) = 0, & x \in \Gamma. \end{cases} \tag{1}$$

The above operator, $\mathcal{L}_\mu$, is a nonlocal Laplacian, defined as

$$\mathcal{L}_\mu u(x) := \int_{\Omega \cup \Gamma} [u(y) - u(x)]\mu(x, y)\, dy, \tag{2}$$

with $\mu : (\Omega \cup \Gamma) \times (\Omega \cup \Gamma) \to \mathbb{R}$ a measurable kernel. The source term satisfies $f \in L^2(\Omega)$, so the equalities in (1) and in the sequel are in the almost everywhere sense.

Recent investigations have focused on the case when μ is integrable and has a finite support radius δ, usually referred to as the horizon of interaction. (If μ is not integrable, then the operator $\mathcal{L}_\mu$ needs to be defined by using the principal value). An integrable kernel (also called a weakly singular kernel) is often required in numerical implementations in order to avoid quadrature errors, especially when the function u does not have any a priori smoothness. Indeed, in the case of a highly singular kernel μ, a non-smooth or discontinuous function u will produce an unbounded $\mathcal{L}_\mu u$ with unrectifiable errors in numerical computations. Kernels with non-integrable singularity have been considered before (Caffarelli et al. 2014; Foss and Geisbauer 2012); however, their analysis is very different as it can still appeal to compactness and smoothness results that are not available when $\mu \in L^1$. Here we aim to provide a general and versatile roadmap for the study of regularity of solutions to nonlocal systems and establish connections with the local theory by formulating equivalent nonlocal theorems to existing local results. The bridge between the two theories also involves an analysis of convergence of low-regularity nonlocal solutions to the smooth solutions of the classical solutions as the support of the kernel shrinks to zero. We also prove some new regularity results which provide the groundwork for a more comprehensive regularity theory of nonlocal systems. More precisely, we establish regularity of solutions for nonlocal semilinear systems, as well as for higher-order problems.

Nonlocal models provide an *alternative* description for physical situations modeled by functions that lack any weak differentiability. Thus a nonlocal elliptic problem admits L^2- regular solutions whenever the forcing is in L^2, whereas the classical Laplacian endows the solutions with H^2- regularity (basically, existence of two weak derivatives). This fact is seen as a strong advantage to employing nonlocal models when the physics of the problems show that irregular solutions could arise. An immediate implication in the mathematical analysis shows that all estimates could be derived at the L^2-level, so no additional approximations or smoothing techniques need to be employed. The techniques for investigations of low-regularity solutions are still under development, but sufficiently many existing results indicate that the nonlocal theory mirrors in many aspects the classical PDE framework. The results that connect the nonlocal theory with the local theory are at two levels:

- Nonlocal counterparts to existing theorems in the local theory are derived (e.g., mean value theorems, regularity of "elliptic" solutions, Poincaré inequality, etc.)
- Convergence results that show that nonlocal operators and/or nonlocal solutions converge to their classical counterparts under appropriate assumptions as the horizon of interaction (the support of the kernel μ) shrinks to zero.

The new paradigm requires that one studies nonlocal solutions u_δ (dependent on the kernel's horizon, δ) with low-regularity tools. One can then investigate the convergence of u_δ to u as the horizon $\delta \to 0^+$ and identify a connection to classical results. The table below contrasts and compares the two approaches: in the local setting, smooth approximations of the solutions to the PDE will converge in H^2 to a weak solution $u \in H^2$, whereas in the nonlocal setting, the rough solutions u_δ of the counterpart PIDE will converge in L^2 to the same H^2 regular solution.

Framework	Convergence
Local	Regularized approximations $\xrightarrow{H^2}$ weak H^2 solution
Nonlocal	L^2 approximations $u_\delta \xrightarrow{L^2}$ weak H^2 solution u

The choice of the model (PDE vs. PIDE) is dictated by the physics and whether it is appropriate to require weakly differentiable solutions that satisfy a PDE or to allow solutions with stronger discontinuities as is possible with a PIDE.

Higher-Order Nonlocal Models

In Radu et al. (2017) the authors introduced and analyzed a system involving a nonlocal biharmonic operator with boundary conditions which are counterparts of the classical boundary (hinged and clamped) conditions. This nonlocal model provides an alternative to classical plate theory, which though well-developed for nice domains (convex or of class C^2) and H^2 solutions, has significant shortcomings regarding irregular solutions and domains. The diagram below summarizes the well-posedness results of Radu et al. (2017) for fourth-order nonlocal problems on arbitrary domains, even in the absence of Lipschitz regularity for the boundary of the domain. When the domains have more regularity, these results also establish a clear connection, as the horizon goes to zero, to the classical theory through L^2 strong convergence.

Regularity of domain	Results for the nonlocal biharmonic system
None	Nonlocal problem well-posed in $L^2(\Omega)$
Class C^1 (relaxation may be possible)	$L^2(\Omega)$-convergence to distributional solution
Class C^2 or convex C^1	Convergence to *weak* solution of elliptic hinged or clamped problem
Class C^4	Convergence to *regular* solution of elliptic hinged or clamped problem

For all nonlocal theories, of second- or higher-order, our goal remains to propose and investigate models that are robust in the presence of singularities. The ensuing nonlinear analysis will carry to applications such as suspension bridges where many new models and interesting advances have recently appeared (Berchio et al. 2016; Gazzola 2015; Radu et al. 2014). By introducing higher-order systems in the nonlocal setting, we are also considering applications in other disciplines where such systems play an essential role, such as nonlocal diffusion (see Andreu-Vaillo et al. 2010 and the references therein). In phase transitions, the classical Cahn-Hilliard equations are fourth-order systems, which have long been recognized to be approximations to phase separations with a binary mixture, yet the model considers it equipped with continuity or even smoothness. Different nonlocal models (Bates and Han 2005a,b) have already been considered as well, yet their convergence to the classical counterpart has not been established yet.

At a theoretical level, we state in this manuscript the first (to our knowledge) regularity results for solutions to nonlinear second-order nonlocal systems and, also, for higher-order nonlocal systems. The classical theory of these systems is very well-developed (e.g., for recent developments, see Mayboroda and Maz'ya 2014). There are several facts that motivate carrying this study in the nonlocal setting. First, as we will see below, the nonlocal elliptic theory contains many parallel results and properties comparable to the classical theory: energy minimization, comparison principles, and mean value properties (Foss and Radu 2016; Hinds and Radu 2012). Also, as discussed above, nonlocal solutions, seen as "rough" approximations of classical solutions, converge to their classical (smooth) counterparts even for higher-order systems. These convergence results have the potential to solve open classical problems where one could study first its nonlocal counterpart and then investigate the limit of the nonlocal solutions as the horizon shrinks. Finally, a major advantage of studying nonlocal problems is the fact that they obviate the need for smoothness assumptions for solutions of systems of *any* order while still capturing the physical phenomenon.

The rest of this paper is organized as follows. In the next section, we present some results regarding nonlocal ellipticity which provide a foundation for carrying out theoretical investigations of nonlocal problems. In section "Definitions and Setup" we introduce notation and some general assumptions, after which we present a method for obtaining the scaling needed for obtaining pointwise error estimates between nonlocal and local Laplacians applied to a smooth function; the section concludes with a nonlocal version of Poincaré's inequality. In section "Regularity of Nonlocal Solutions for Nonlinear Systems" we review some existing regularity results from the linear case, after which we prove some new results for nonlinear problems for second-order nonlocal problems. In the last section, we use an iteration argument to study regularity of solutions for higher-order systems studied in connection with plate deformations.

Definitions and Setup

For the remainder of the paper, the kernel μ will be a nonnegative, rotationally symmetric, and integrable kernel (Throughout the paper, with an abuse of notation, we may write $\mu(x, y) = \mu(|x - y|)$ to indicate that μ depends only on the distance between x and y). At various points in the sequel, we will refer to the following:

Assumptions and definitions for the kernels:

(A1) $\mu : [0, \infty) \to [0, \infty)$ is measurable, and there exist $\beta \in [0, n)$ and $c, \delta_0 > 0$ such that

$$
\begin{aligned}
0 < \mu(r) \leq c r^{-\beta}, \qquad & \text{for all } r \in (0, \delta_0) \\
\mu(r) = 0, \qquad & \text{for all } r \geq \delta_0.
\end{aligned}
$$

(A2) There is a $c_0 > 0$ such that $\mu(r) > c_0$, for all $r \in (0, \delta_0]$.
(A3) For every $\delta \in (0, \delta_0]$, define $\mu_\delta : [0, \infty) \to [0, \infty)$ by

$$
\mu_\delta(r) := \begin{cases} \mu(r), & 0 < r \leq \delta \\ 0, & r > \delta. \end{cases}
$$

The δ parameter is called the horizon of the kernel. The assumption (A1) ensures the integrability of the kernels μ and μ_δ. The nonlocal Poincaré inequality (Thereom 1) requires (A2).

Denote by ω_n the volume of the unit ball in n dimensions, so $n\omega_n$ will be its surface area. The n dimensional ball of radius $\varepsilon > 0$ centered at x will be denoted by $B_\varepsilon(x)$.

Let $\Omega \subset \mathbb{R}^n$ be an open and bounded domain. With δ_0 provided in (A1), the collars of Ω are the sets

$$
\Gamma := \overline{\bigcup_{x \in \Omega} B_{\delta_0}(x) \setminus \Omega}
$$

and for each $\delta \in (0, \delta_0]$

$$
\Gamma_\delta := \overline{\bigcup_{x \in \Omega} B_\delta(x) \setminus \Omega},
$$

where $\delta_0 > 0$ is specified in assumption (A1). For each $\varepsilon > 0$, set

$$
\Omega_\varepsilon := \{x \in \Omega : B_\varepsilon(x) \subseteq \Omega\}.
$$

We use ∇u for the spatial gradient of a function $u : \mathbb{R}^n \to \mathbb{R}$, provided it exists. For the mean value of u over a set $E \subset \mathbb{R}^n$ with positive and finite measure $|E|$, we will use the standard notation

$$
u_E := \fint_E u(x)\, dx = \frac{1}{|E|} \int_E u(x)\, dx. \tag{3}
$$

Given $\phi \in C^\infty(\mathbb{R}^n)$ and $u \in L^2(\Omega \cup \Gamma)$, define the convolution product as

$$
(\phi * u)(x) := \int_{\mathbb{R}^n} \phi(y - x)\bar{u}(y)\, dy
$$

where

$$\bar{u}(y) := \begin{cases} u(y), & y \in \Omega \cup \Gamma \\ 0, & y \notin \Omega \cup \Gamma. \end{cases}$$

Nonlocal and Local Laplacians

In this section, we present some results that provide a connection between the local and nonlocal Laplacian operators.

Elliptic Properties for the Nonlocal Laplacian

The classical elliptic theory has developed to a level where systems with complex nonlinearities and space-dependent coefficients are very well understood, even for domains with fairly rough boundaries. Nonlocal operators have been introduced as natural generalizations of classical differential operators, and their successful implementation in applications continues to promote their theoretical analysis to deeper levels. For second-order problems, several results have been proved that show the intimate connection between nonlocal and local frameworks. More specifically, the following elliptic-type properties hold in the nonlocal framework (Du et al. 2012; Foss and Radu 2016; Hinds and Radu 2012):

Proposition 1. *Suppose that μ satisfies assumption (A1). The operator $\mathcal{L}_\mu$ admits the following properties:*

(a) *If $u \equiv$ constant then $\mathcal{L}_\mu u = 0$ (trivially). Conversely, if $\mathcal{L}_\mu = 0$ and $u = 0$ on Γ, then $u \equiv$ constant.*
(b) *Let $x \in \Omega \cup \Gamma$. For any maximal point x_0 that satisfies $u(x_0) \geq u(x)$, we have $-\mathcal{L}_\mu u(x_0) \geq 0$. Similarly, if x_1 is a minimal point such that $u(x_1) \leq u(x)$, then $-\mathcal{L}_\mu u(x_1) \leq 0$.*
(c) *$-\mathcal{L}_\mu u$ is a positive semidefinite operator, i.e., $\langle -\mathcal{L}_\mu u, u\rangle \geq 0$, where $\langle \cdot, \cdot \rangle$ denotes the $L^2(\Omega \cup \Gamma)$ inner product.*
(d) $\displaystyle\int_{\Omega\cup\Gamma} \mathcal{L}_\mu u(x)dx = 0.$
(e) *If the kernel μ is given by a combination of derivatives of the Dirac mass, then for this particular choice of kernel, we have $\mathcal{L}_\mu = \Delta u$ in the sense of distributions (see Du et al. 2012);*
(f) *As the radius of the horizon δ goes to zero, the nonlocal solutions converge to their classical counterparts. This fact has been proven for the Laplacian (Mengesha and Du 2014), as well as for the biharmonic operator (Radu et al. 2017);*

(g) *Dirichlet's principle and variational arguments apply for nonlocal systems as proven in Hinds and Radu (2012);*

(h) *A weighted mean value theorem is available; indeed, one can easily see that a nonlocal harmonic function u, i.e., a function u for which* $\mathcal{L}_\mu u(x) = 0$ *for* $x \in \Omega$, *must also satisfy*

$$u(x) = \frac{\int_{\Omega\cup\Gamma} \mu(|x-y|)u(y)\,\mathrm{d}y}{\int_{\Omega\cup\Gamma} \mu(|x-y|)\,\mathrm{d}y}, \qquad x \in \Omega.$$

Using this result, one can prove

(i) $\Delta u = 0$ *if and only if* $\mathcal{L}_\mu u(x) = 0$ *for every kernel* $\mu(x,y) = \mu(|x-y|) \in L^1$.

We conclude this subsection by stating the following nonlocal Poincaré-type inequality, versions of which can be found in Aksoylu and Parks (2011) or Hinds and Radu (2012):

Theorem 1 (Nonlocal Poincaré inequality). *Let* Ω *be an open, bounded domain and* $\mu \in L^1(\mathbb{R}^n)$ *a nonnegative kernel that satisfies (A2) and (A3). If* $u \in L^2(\Omega\cup\Gamma)$, *then there is a constant* $\lambda_p(\delta, n, \Omega) > 0$ *s.t.*

$$\lambda_p \int_{\Omega} |u(x)|^2\,\mathrm{d}x \le \int_{\Omega}\int_{\Omega\cup\Gamma_\delta} \mu_\delta(|x-y|)|u(y)-u(x)|^2\,\mathrm{d}y\,\mathrm{d}x + \int_{\Gamma_\delta} |u(x)|^2\,\mathrm{d}x.$$

Scaling of the Nonlocal Laplacian and Pointwise Estimates

In this subsection, we will assume that (A1)–(A3) hold for the kernels μ and μ_δ. We will derive a scaling that normalizes the nonlocal Laplacian. To facilitate identification of the scaling, a modified definition for $\mathcal{L}_{\mu_\delta}$ will be used and, for convenience, denoted by $\mathcal{L}_\delta$

$$\mathcal{L}_\delta u(x) = \sigma(\delta) \int_{\Omega\cup\Gamma_\delta} [u(y)-u(x)]\mu_\delta(|x-y|)\,\mathrm{d}y. \tag{4}$$

Here $\sigma(\delta)$ is the scaling factor in terms of the support of μ_δ. This horizon-dependent scaling σ will be determined later. Moreover, we will show that the scaled $\mathcal{L}_\delta u$ converges to the local Laplacian Δu at the rate δ^2, i.e.,

$$|\mathcal{L}_\delta u - \Delta u| \le C(u)\delta^2,$$

where $C(u)$ depends on bounds for the fourth-order derivatives of u. This error bound between the local and nonlocal Laplacian has been derived before (see, e.g.,

Du et al. 2013). Below we outline the argument that was first presented in Foss and Radu (2016) which has already been utilized in deriving error estimates between nonlocal solutions and their local equivalents for systems that involve a nonlocal biharmonic operator (Radu et al. 2017) and for the newly introduced state Laplacian (Radu and Wells 2017). Additionally, the argument accommodates very general kernels (that are integrable) and shows the optimality of the δ^2 convergence that is observed for C^4 functions.

For each $\delta \in (0, \delta_0]$, we define the function $\pi_\delta : (0, \infty) \to [0, \infty)$

$$\pi_\delta(r) := \int_\delta^r \rho\, \mu_\delta(\rho)\, d\rho. \tag{5}$$

Thus, for example, if $\mu(r) = r^{-\beta}$ on $(0, \delta_0]$, then we find that

$$\pi_\delta(r) = \begin{cases} \dfrac{1}{2-\beta}[r^{2-\beta} - \delta^{2-\beta}], & 0 < r \le \delta, \beta \ne 2 \\ \ln r - \ln \delta, & 0 < r \le \delta,\ \beta = 2 \\ 0, & r > \delta. \end{cases}$$

Assume that $u \in C^4(\Omega \cup \Gamma)$. By using the fundamental theorem of calculus, we have that

$$\begin{aligned} \mathcal{L}_\delta u(x) &= \sigma(\delta) \int_{B_\delta(x)} [u(y) - u(x)]\mu_\delta(|x-y|)\, dy \\ &= \sigma(\delta) \int_{B_\delta(x)} \int_0^1 \nabla u(x + s(y-x)) \cdot (y-x)\mu_\delta(|y-x|)\, ds\, dy. \end{aligned}$$

After changing the order of integration in the double integral and changing variables $z = y - x$, we obtain that

$$\mathcal{L}_\delta u(x) = \sigma(\delta) \int_0^1 \int_{B_\delta(0)} \nabla u(x + sz) \cdot [z\mu_\delta(|z|)]\, dz\, ds.$$

With π_δ given by (5), we have that

$$\nabla_z \pi_\delta(|z|) = \pi_\delta'(z)\frac{z}{|z|} = z\mu_\delta(|z|),$$

so

$$\mathcal{L}_\delta u(x) = \sigma(\delta) \int_0^1 \int_{B_\delta(0)} \nabla u(x+sz) \cdot \nabla_z \pi_\delta(|z|)\, \mathrm{d}z\, \mathrm{d}s.$$

After an integration by parts and by using the fact that $\pi_\delta(|z|) = \pi_\delta(\delta) = 0$ for $z \in \partial B_\delta(0)$, we obtain

$$\mathcal{L}_\delta u(x) = -\sigma(\delta) \int_0^1 \int_{B_\delta(0)} s \Delta u(x+sz) \pi_\delta(|z|)\, \mathrm{d}z\, \mathrm{d}s$$

which is equivalent to

$$\begin{aligned} \mathcal{L}_\delta u = -\sigma(\delta) \int_0^1 \int_{B_\delta(0)} s[\Delta u(x+sz) - \Delta u(x)] \pi_\delta(|z|)\, \mathrm{d}z\, \mathrm{d}s \\ - \Delta u(x) \sigma(\delta) \int_0^1 \int_0^\delta \int_{\partial B_\rho(0)} s \pi_\delta(|\rho|) d\omega(z)\, \mathrm{d}\rho\, \mathrm{d}s. \end{aligned} \tag{6}$$

We compute

$$\int_0^1 \int_0^\delta \int_{\partial B_\rho(0)} s \pi_\delta(\rho)\, \mathrm{d}\omega(z)\, \mathrm{d}\rho\, \mathrm{d}s = \frac{1}{2} n \omega_n \int_0^\delta \pi_\delta(\rho) \rho^{n-1}\, \mathrm{d}\rho.$$

This results in the scaling

$$\sigma(\delta) = \frac{-2}{n\omega_n \int_0^\delta \pi_\delta(\rho)\rho^{n-1}\, \mathrm{d}\rho} \tag{7}$$

will make the coefficient of the Laplacian to be 1 on the RHS of (6) above.

For example, if $\mu(r) = \begin{cases} r^{-\beta}, & 0 < r \le \delta \\ 0, & \delta < r \end{cases}$ then for $\beta \neq 2$ we have

$$\sigma(\delta) = \frac{2(2-\beta+n)}{\omega_n \delta^{2-\beta+n}},$$

where for $n = 1$ we take $\omega_0 = 2$. With the choice of scaling given in (7), we write

$$\mathcal{L}_\delta u(x) - \Delta u(x) = -\sigma(\delta) \int_{B_\delta(0)} \int_0^1 s[\Delta u(x+sz) - \Delta u(x)]\pi_\delta(|z|)\, \mathrm{d}s\, \mathrm{d}z. \quad (8)$$

By employing the fundamental theorem of calculus and simplifying, we obtain

$$\begin{aligned}\mathcal{L}_\delta u(x) - \Delta u(x) &= -\sigma(\delta) \int_{B_\delta(0)} \int_0^1 \frac{(1-s^2)}{2}[\Delta \nabla u(x+sz)] \cdot z\pi_\delta(|z|)\, \mathrm{d}s\, \mathrm{d}z \\ &= -\sigma(\delta) \int_{B_\delta(0)} \int_0^1 \frac{(1-s^2)}{2}[\Delta \nabla u(x+sz) - \Delta \nabla u(x)] \cdot z\pi_\delta(|z|)\, \mathrm{d}s\, \mathrm{d}z,\end{aligned}$$

where we used the fact that

$$\int_{B_\delta(0)} z\pi_\delta(|z|)\, \mathrm{d}z = 0.$$

Further we have

$$\mathcal{L}_\delta u(x) - \Delta u(x) = -\sigma(\delta) \int_{B_\delta(0)} \int_0^1 \left(\frac{1}{3} - \frac{s}{2} + \frac{s^3}{6}\right) \left[\Delta \nabla^2 u(x+sz)z\right] \cdot z\pi_\delta(|z|)\, \mathrm{d}s\, \mathrm{d}z.$$

With

$$M_4 = \sup_{x \in \Omega \cup \Gamma} |D^4 u(x)|$$

we estimate

$$\begin{aligned}|\mathcal{L}_\delta u(x) - \Delta u(x)| &\leq \sigma(\delta) M_4 \int_{B_\delta(0)} \int_0^1 \left(\frac{1}{3} - \frac{s}{2} + \frac{s^3}{6}\right) |z|^2 |\pi_\delta(|z|)|\, \mathrm{d}s\, \mathrm{d}z \\ &= \frac{\sigma(\delta) M_4}{8} n\omega_n \int_0^\delta \rho^{n+1} |\pi_\delta(\rho)|\, \mathrm{d}\rho.\end{aligned}$$

Finally, we use the fact that $\rho^{n+1} \leq \rho^{n-1}\delta^2$ to obtain

$$|\mathcal{L}_\delta u(x) - \Delta u(x)| \leq \frac{M_4}{4}\delta^2. \quad (9)$$

This estimate shows that the nonlocal Laplacian approaches the classical Laplacian, as the horizon δ goes to zero, at a rate comparable to δ^2, independently of the dimension. The convergence also demonstrates that the scaling selected in (7) is the correct one.

Moreover, it was shown that besides the above convergence of the operators (applied to sufficiently smooth functions), one has the convergence of the nonlocal solutions to their classical counterparts. More precisely, it was shown in Mengesha and Du (2014) that the solutions u_δ to the problem

$$\begin{cases} \mathcal{L}_\delta u_\delta(x) = f(x), & x \in \Omega \\ u_\delta(x) = 0, & x \in \Gamma_\delta, \end{cases} \tag{10}$$

obtained for each $\delta \in (0, \delta_0]$, converge strongly in L^2 to u, the solution of the classical Laplace equation

$$\begin{cases} \Delta u(x) = f, & x \in \Omega \\ u(x) = 0, & x \in \partial\Omega. \end{cases}$$

Regularity of Nonlocal Solutions for Nonlinear Systems

In this section we provide some integrability and differentiability properties for solutions of

$$\begin{cases} \mathcal{L}_\mu u(x) = f(x, u(x)), & x \in \Omega \\ u(x) = 0, & x \in \Gamma. \end{cases} \tag{11}$$

The linear problem obtained for $x \mapsto f(x, u(x)) = f(x)$ was investigated in Foss and Radu (2016), where it was shown that a solution's integrability properties are the same as possessed by the function f. This result addressed one of the deficiencies of Poincaré's inequality, mainly, the fact that the integrability of a function cannot be improved based on bounds on its nonlocal gradient, as in the classical framework. In the same setting, we also showed in Foss and Radu (2016) that for a nonlocal Laplacian with integrable kernel (Note that if the kernel is highly singular then there are compact embedding results and Poincaré-type inequalities that do increase the integrability by using estimates on the gradient.) and $f \in C^1(\bar{\Omega})$, solutions to (1) satisfy $u \in W^{1,2}(\Omega)$. Below we will offer an extension to this result by showing that a forcing term $f \in W^{1,2}(\Omega) \cap L^\infty(\Omega)$ will yield a solution u with regularity at the same level.

Higher Integrability

We state first a result that guarantees the higher integrability of solutions to the linear problem (we refer to Foss and Radu (2016) for its proof) and then present an extension to this result in the semilinear setting.

Theorem 2 (Higher integrability of solutions to the linear problem). *Assume $\mu \in L^1(\mathbb{R}^n)$ satisfies (A1). If u is a solution of*

$$\begin{cases} \mathcal{L}_\mu u(x) = f(x), & x \in \Omega \\ u(x) = 0, & x \in \Gamma. \end{cases}$$

with a priori regularity $u \in L^p(\Omega)$, $p > 1$ and we assume that $f \in L^r(\Omega)$, $r > p$ then $u \in L^r(\Omega)$.
The proof for this result is similar, but simpler, than its generalization to a nonlinear setting, which we present below.

Theorem 3 (Higher integrability of solutions to semilinear problems). *Assume $\mu \in L^1(\mathbb{R}^n)$ satisfies (A1) and that given $f : \Omega \times \mathbb{R} \to \mathbb{R}$, the mapping*

$$u \mapsto g_x(u) = u + f(x,u)$$

is invertible on $\mathbb{R}$ with the inverse g_x^{-1} uniformly Lipschitz continuous with respect to $x \in \mathbb{R}^n$. Then any solution $u \in L^p(\Omega)$, $p > 1$ of (11) satisfies $u \in L^\infty(\Omega)$.

Proof. Without loss of generality, we assume that $\|\mu\|_{L^1} = 1$ and note that the uniform Lipschitz continuity of g_x^{-1} implies there is a constant $M < \infty$ such that

$$|g_x^{-1}(u)| \leq M|u|, \qquad \text{for all } u \in \mathbb{R},\ x \in \Omega.$$

First, extend any solution u of the above equation by zero outside Ω, and with an abuse of notation, denote the extended function by u as well. In light of our assumptions, we write the integrodifferential equation in (1) as

$$\int_{\mathbb{R}^n} u(y)\mu(|y-x|)\,dy - u(x)\int_{\mathbb{R}^n} \mu(|y-x|)\,dy = (u*\mu)(x) - u(x) = f(x,u(x)). \tag{12}$$

From the last equality of (12), we obtain that

$$u(x) = g_x^{-1}((u*\mu)(x)) \tag{13}$$

to which we apply Young's inequality for convolutions to obtain

$$\|u*\mu\|_r \leq \|u\|_p\|\mu\|_q,$$

where $\frac{1}{r} = \frac{1}{p} + \frac{1}{q} - 1$, where q is chosen appropriately.

Depending on the power β that controls the growth of the kernel μ, as given by assumption (A1), we have the following cases:

Case 1. For $\beta < n\frac{p-1}{p}$ take $q = \frac{p}{p-1}$ to obtain $r = \infty$. By (13) and using the bound on the growth of g_x^{-1} above, we have that $u \in L^\infty(\Omega)$.

Case 2. If $\beta \geq n\frac{p-1}{p}$, then let $q = \frac{n-\varepsilon}{\beta}$ for $\varepsilon > 0$ arbitrarily small. Then, the degree of integrability of $u * \mu$ is up to (but not including) $r_0 = \frac{pn}{n + p(\beta - n)}$; note that $r_0 > 1$ since $p > 1$. Again, by (13) and the bound on g_x^{-1} stated above, this integrability is transferred to u so that $u \in L^r$ for all $r < r_0$. We apply again Young's inequality

$$\|u * \mu\|_{r_1} \leq \|u\|_{r_0} \|\mu\|_{q_1},$$

with $\frac{1}{r_1} = \frac{1}{r_0} + \frac{1}{q_1} - 1$. Take $q_1 = \frac{p(n-\varepsilon)}{(n-\varepsilon)(2p-1) - p\beta}$ to obtain $r_1 = \infty$. Using one more time (13) and the growth of g_x^{-1}, we obtain, as in the previous case, that $u \in L^\infty(\Omega)$.

□

Remark 1. The above theorem is not only dimension independent, it is also applicable for unbounded, as well as bounded, domains.

The integrability result in Theorem 2 is used to establish some differentiability properties of the solution u, which are described next.

Higher Differentiability

The first result shows differentiability for solutions of the linear system (1) when the kernel is differentiable. Our final goal is to establish differentiability of solutions when μ is only assumed to be integrable.

Proposition 2. *If $\mu \in C^1(\mathbb{R}^n)$, $f \in C^1(\Omega)$ and $u \in L^1(\Omega \cup \Gamma)$ is a solution to $\mathcal{L}_\mu u = f$; i.e.,*

$$\int_{\Omega\cup\Gamma} (u(y) - u(x))\mu(|x-y|)dy = f(x), \quad \textit{for all } x \in \Omega,$$

then $u \in C^1(\Omega)$.

Proof. The theorem follows easily from the equality

$$u(x) = (u * \mu)(x) - f(x)$$

and the smoothing properties of the convolution operator. □

In the absence of differentiability assumptions for μ, we will need the following lemmas for the proof of the main result.

Lemma 1. *If $\mu \in L^1(\mathbb{R}^n)$ and $u \in C^1(\Omega \cup \Gamma)$ satisfies*

$$\mathcal{L}_\mu u(x) = f(x), \quad \textit{for all } x \in \Omega$$

for $f \in C^1(\Omega)$ then $\mathcal{L}_\mu(\nabla u)(x) = \nabla f(x)$ for all $x \in \Omega$.

Proof. The linearity of the $\mathcal{L}_\mu$ operator trivially gives the result. □

Lemma 2. *If $\mu \in L^1(\mathbb{R}^n)$, $f \in L^1(\Omega)$ and $u \in L^1(\Omega \cup \Gamma)$ satisfies*

$$\mathcal{L}_\mu u(x) = f(x), \quad \textit{for all } x \in \Omega,$$

then

$$\mathcal{L}_\mu[\phi * u](x) = (\phi * f)(x), \quad \textit{for all } x \in \Omega_\varepsilon.$$

Here $\varepsilon > 0$ and $\phi \in C^\infty(\mathbb{R}^n)$ with supp$(\phi) \subset B_\varepsilon(0)$

Proof. Again, employ the linearity of $\mathcal{L}_\mu$ and of the convolution product. Note that for $x \in \Omega \setminus \Omega_\varepsilon$, the convolution $\phi * u$ may incorporate values of u outside of Ω, so there is no assurance that the equation is satisfied outside of Ω_ε. □

We are now ready to present the regularity result for linear problems, improving the result given by Theorem 4.3 in Foss and Radu (2016).

Theorem 4 (Regularity for solutions of nonlocal equations). *Let $f \in L^\infty(\Omega) \cap W^{1,2}(\Omega)$. Suppose that $\mu \in L^1(\mathbb{R}^n)$ satisfies assumptions (A1), (A2), and (A3). Then, the unique solution u of the system* (10)

$$\begin{cases} \mathcal{L}_\mu u(x) = f(x), & x \in \Omega \\ u(x) = 0, & x \in \Gamma, \end{cases} \tag{14}$$

satisfies

$$u \in L^\infty(\Omega) \cap W^{1,2}(\Omega).$$

Remark 2. The solutions u to (14) clearly depend on μ, but we suppressed this dependence for the clarity of the argument.

Proof. The existence and uniqueness of $u \in L^2(\Omega \cup \Gamma)$ when $f \in L^2(\Omega)$ were proven before (see Hinds and Radu (2012) for a variational argument). Below we will establish uniform estimates in the gradient which will then yield the desired regularity.

Extend u and f to all of $\mathbb{R}^n$ by 0. Let $\{\phi_\varepsilon\}_{\varepsilon>0}$ be a family of mollifiers defined by $\phi_\varepsilon := \frac{1}{\varepsilon^n}\phi\left(\frac{x}{\varepsilon}\right)$, with $\phi \in C^\infty(\mathbb{R}^n)$ and $\text{supp}(\phi) \subset B_1(0)$. Fix $\varepsilon \in \left(0, \frac{\delta_0}{2}\right)$. By Lemma 2,

$$\mathcal{L}_\mu[\phi_\varepsilon * u](x) = (\phi_\varepsilon * f)(x)$$

for each $x \in \Omega_\varepsilon$. Put $u_\varepsilon := \phi_\varepsilon * u$ and $f_\varepsilon := \phi_\varepsilon * f$. Since $u_\varepsilon \in C^\infty(\Omega \cup \Gamma)$, Lemma 1 implies

$$\mathcal{L}_\mu[\nabla u_\varepsilon](x) = \nabla f_\varepsilon(x) \tag{15}$$

for each $x \in \Omega_\varepsilon$. Since $u = 0$ on Γ, we have that $u_\varepsilon = 0$ on

$$\Omega^\varepsilon := \{x \in \mathbb{R}^n | \text{dist}(x, \Omega) < \varepsilon\},$$

hence $\nabla u_\varepsilon = 0$ on $\Omega \cup \Gamma \setminus \Omega^\varepsilon$. We then have from (15) above that for all $x \in \Omega_\varepsilon$

$$\mathcal{L}_\mu[\nabla u_\varepsilon](x) = \int_{\Omega^\varepsilon} [\nabla u_\varepsilon(y) - \nabla u_\varepsilon(x)]\mu(|x-y|)dy = \nabla f_\varepsilon(x).$$

We multiply the equation by ∇u_ε and integrate on Ω_ε to obtain

$$\begin{aligned}
&\int_{\Omega_\varepsilon}\int_{\Omega^\varepsilon} [(\nabla u_\varepsilon(y) - \nabla u_\varepsilon(x))\mu(|y-x|)] \cdot \nabla u_\varepsilon(x)\, \mathrm{d}y\, \mathrm{d}x \\
&\quad = \int_{\Omega_\varepsilon} \nabla f_\varepsilon(x) \cdot \nabla u_\varepsilon(x)\, \mathrm{d}x.
\end{aligned} \tag{16}$$

We estimate the LHS above as follows:

$$\begin{aligned}
&\int_{\Omega_\varepsilon}\int_{\Omega_\varepsilon} [(\nabla u_\varepsilon(y) - \nabla u_\varepsilon(x))\mu(|y-x|)] \cdot \nabla u_\varepsilon(x)\, \mathrm{d}y\, \mathrm{d}x \\
&\quad + \int_{\Omega_\varepsilon}\int_{\Omega^\varepsilon \setminus \Omega_\varepsilon} [(\nabla u_\varepsilon(y) - \nabla u_\varepsilon(x))\mu(|y-x|)] \cdot \nabla u_\varepsilon(x)\, \mathrm{d}y\, \mathrm{d}x \\
&= \frac{1}{2}\int_{\Omega_\varepsilon}\int_{\Omega_\varepsilon} [(\nabla u_\varepsilon(y) - \nabla u_\varepsilon(x))^2 \mu(|y-x|)]\, \mathrm{d}y\, \mathrm{d}x
\end{aligned}$$

$$+\int_{\Omega_\varepsilon}\int_{\Omega^\varepsilon\setminus\Omega_\varepsilon}[(\nabla u_\varepsilon(y)-\nabla u_\varepsilon(x))\mu(|y-x|)]\cdot\nabla u_\varepsilon(x)\,\mathrm{d}y\,\mathrm{d}x$$

From (16) and above equality, we obtain

$$\int_{\Omega_\varepsilon}\int_{\Omega_\varepsilon}[(\nabla u_\varepsilon(y)-\nabla u_\varepsilon(x))^2\mu(|y-x|)]\,\mathrm{d}y\,\mathrm{d}x = 2\int_{\Omega_\varepsilon}\nabla f_\varepsilon(x)\cdot\nabla u_\varepsilon(x)\,\mathrm{d}x \qquad (17)$$

$$-2\int_{\Omega_\varepsilon}\int_{\Omega^\varepsilon\setminus\Omega_\varepsilon}[(\nabla u_\varepsilon(y)-\nabla u_\varepsilon(x))\mu(|y-x|)]\cdot\nabla u_\varepsilon(x)\,\mathrm{d}y\,\mathrm{d}x$$

By Poincaré's inequality as given by Theorem 1 applied on the domain Ω_ε with collar $\Omega^\varepsilon\setminus\Omega_\varepsilon$, we have

$$\|\nabla u_\varepsilon\|^2_{L^2(\Omega_\varepsilon)} \le C\int_{\Omega_\varepsilon}\int_{\Omega_\varepsilon}[(\nabla u_\varepsilon(y)-\nabla u_\varepsilon(x))^2\mu(|y-x|)]\,\mathrm{d}y\,\mathrm{d}x \qquad (18)$$

$$+C\int_{\Omega_\varepsilon}\int_{\Omega^\varepsilon\setminus\Omega_\varepsilon}[(\nabla u_\varepsilon(y)-\nabla u_\varepsilon(x))^2\mu(|y-x|)]\,\mathrm{d}y\,\mathrm{d}x+\|\nabla u_\varepsilon\|^2_{L^2(\Omega^\varepsilon\setminus\Omega_\varepsilon)}.$$

By using Hölder's inequality in (17) and combining it with (18), we obtain

$$\|\nabla u_\varepsilon\|^2_{L^2(\Omega_\varepsilon)} \le C\Bigg\{\|\nabla f_\varepsilon\|^2_{L^2(\Omega_\varepsilon)}+\|\nabla u_\varepsilon\|^2_{L^2(\Omega^\varepsilon\setminus\Omega_\varepsilon)} \qquad (19)$$

$$+\int_{\Omega_\varepsilon}\int_{\Omega^\varepsilon\setminus\Omega_\varepsilon}[(\nabla u_\varepsilon(y)-\nabla u_\varepsilon(x))\mu(|y-x|)]\cdot\nabla u_\varepsilon(x)\,\mathrm{d}y\,\mathrm{d}x\Bigg\}.$$

Since $\nabla u_\varepsilon(x)=(u*\nabla\phi_\varepsilon)(x)$ with $\nabla\phi_\varepsilon(x)=\frac{1}{\varepsilon^{n+1}}\nabla\phi\left(\frac{x}{\varepsilon}\right)$, we have by Young's inequality for convolutions that

$$\|\nabla u_\varepsilon\|^2_{L^2(\Omega^\varepsilon\setminus\Omega_\varepsilon)} \le \|u\|^2_{L^1(\Omega^\varepsilon\setminus\Omega_\varepsilon)}\|\nabla\phi_\varepsilon\|^2_{L^2(\Omega^\varepsilon\setminus\Omega_\varepsilon)}.$$

For $f\in L^\infty(\Omega)$, we have $u\in L^\infty(\Omega)$ by Theorem 3, and since $|\Omega^\varepsilon\setminus\Omega_\varepsilon|<C\varepsilon$ (since the boundary of Ω is smooth), the above estimate becomes

$$\|\nabla u_\varepsilon\|^2_{L^2(\Omega^\varepsilon\setminus\Omega_\varepsilon)} \le C\varepsilon^2\|u\|^2_{L^\infty(\Omega)}\frac{\|\nabla\phi\|^2_{L^1(\mathbb{R}^n)}}{\varepsilon^2}<C.$$

Finally, we estimate the last term on the RHS of (19) by using Young's inequality for convolutions.

$$\left|\int_{\Omega_\varepsilon}\int_{\Omega^\varepsilon\setminus\Omega_\varepsilon}[(\nabla u_\varepsilon(y)-\nabla u_\varepsilon(x))\mu(|y-x|)]\cdot\nabla u_\varepsilon(x)\,\mathrm{d}y\,\mathrm{d}x\right|$$

$$\leq\left|\int_{\Omega_\varepsilon}\int_{\Omega^\varepsilon\setminus\Omega_\varepsilon}\nabla u_\varepsilon(y)\cdot\nabla u_\varepsilon(x)\mu(|y-x|)\,\mathrm{d}y\,\mathrm{d}x\right|+\int_{\Omega_\varepsilon}\int_{\Omega^\varepsilon\setminus\Omega_\varepsilon}|\nabla u_\varepsilon(x)|^2\mu(|y-x|)\,\mathrm{d}y\,\mathrm{d}x$$

$$\leq\int_{\Omega_\varepsilon}|[(\chi_{\Omega^\varepsilon\setminus\Omega_\varepsilon}(y)\nabla u_\varepsilon)*\mu](x)|dx++\int_{\Omega_\varepsilon}\int_{\Omega^\varepsilon\setminus\Omega_\varepsilon}|\nabla u_\varepsilon(x)|^2\mu(|y-x|)\,\mathrm{d}y\,\mathrm{d}x$$

$$\leq\eta\|\nabla u_\varepsilon\|_{L^2(\Omega^\varepsilon)}+C(\eta)\|\mu\|^2_{L^1((\Omega_\varepsilon))}\|\nabla u_\varepsilon\|^2_{L^2(\Omega^\varepsilon\setminus\Omega_\varepsilon)}$$

$$+\int_{\Omega_\varepsilon}\int_{\Omega^\varepsilon\setminus\Omega_\varepsilon}|\nabla u_\varepsilon(x)|^2\mu(|y-x|)\,\mathrm{d}y\,\mathrm{d}x$$

for $\eta>0$, and where χ_A denotes the characteristic function for the set A. Since $\mu\in L^1(\mathbb{R}^n)$ and $|\Omega^\varepsilon\setminus\Omega_\varepsilon|<C\varepsilon$, we have that

$$J:=\int_{\Omega_\varepsilon}\int_{\Omega^\varepsilon\setminus\Omega_\varepsilon}|\nabla u_\varepsilon(x)|^2\mu(|y-x|)\,\mathrm{d}y\,\mathrm{d}x<C\varepsilon\|\nabla u_\varepsilon\|^2_{L^2(\Omega_\varepsilon)}$$

uniformly for every $x\in\mathbb{R}^n$. Thus, we can choose ε sufficiently small to absorb the term J defined above in the LHS of (19). Similarly, by choosing η sufficiently small, we obtain

$$\|\nabla u_\varepsilon\|^2_{L^2(\Omega_\varepsilon)}\leq C, \tag{20}$$

where C is independent of ε. The uniform estimate from (20) shows that given an open set $\Omega'\subset\subset\Omega$, we can extract a subsequence ∇u_{ε_k} that converges weakly to some $h\in L^2(\Omega')$. (Here $\Omega'\subset\subset\Omega$ indicates the closure of Ω' is a compact subset of Ω.) We finally argue that h is the distributional gradient of u. Let $\psi\in C_c(\Omega')$, we have

$$(h,\psi)_{L^2(\Omega')}=\lim_{k\to\infty}(\nabla u_{\varepsilon_k},\psi)_{L^2(\Omega')}=-\lim_{k\to\infty}(u_{\varepsilon_k},\nabla\psi)_{L^2(\Omega')}=-(u,\nabla\psi)_{L^2(\Omega')},$$

which verifies that $\nabla u=h$. Hence $u\in W^{1,2}(\Omega)$ since C is independent of ε. Since we already proved that $u\in L^\infty(\Omega)$ the conclusion of the theorem follows. □

Finally, we conclude this section with a theorem that establishes *infinite* regularity of solutions for semilinear problems.

Theorem 5 (Regularity of solutions to semilinear problems). *Suppose that $\mu \in L^1(\mathbb{R}^n)$ satisfies (A1) and that $\|\mu\|_{L^1(\mathbb{R}^n)} = 1$. In addition, assume that $\mu \in W^{s,2}(\mathbb{R}^n)$. Moreover, assume that the mapping*

$$u \mapsto g_x(u) = u + f(x, u)$$

is invertible on $\mathbb{R}$ with the inverse $u \mapsto g_x^{-1}(u) \in C^\infty(\mathbb{R})$ for every $x \in \Omega$. Under these assumptions, if a solution $u \in L^p(\Omega)$, $p > 1$ of (11) exists, then it must satisfy

$$u \in C^\infty(\Omega).$$

Proof. Since $\mu \in W^{s,2}(\mathbb{R}^n)$ and $u \in L^2(\Omega \cup \mathcal{B})$, we have that $u * \mu \in W^{s,2}(\Omega)$. From the equality $u(x) = g_x^{-1}(u * \mu)(x)$ and the smoothness of g_x^{-1}, we have that $u \in W^{s,2}(\Omega)$. We iterate this argument, and using the increased smoothness on u, we will obtain that $u \in W^{2s,2}(\Omega)$, and so on, to finally obtain $u \in W^{\infty,2}(\Omega)$; hence $u \in C^\infty(\Omega)$. □

Remark 3. The strength of the convolution argument that we used above is explored in full generality in the forthcoming paper (Foss et al. 2017).

Regularity for Higher-Order Nonlocal Problems

In Radu et al. (2017) we introduced the nonlocal biharmonic operator

$$\mathcal{B}_\mu[u] := \mathcal{L}_\mu[\mathcal{L}_\mu u]$$

and nonlocal versions of hinged and clamped boundary conditions and showed well-posedness of the coupled nonlocal systems. Moreover, to strengthen the connection between the local and nonlocal theories, we proved that the nonlocal solutions of the hinged and clamped plate systems converge to their respective classical counterparts.

We will focus here on the nonlocal hinged boundary value problem

$$\begin{cases} \mathcal{B}_\mu[u] = f, \; x \in \Omega_{2\delta} \\ u = 0, \qquad x \in \Omega \setminus \Omega_\delta \\ \mathcal{L}_\mu[u] = 0, \; x \in \Omega_\delta \setminus \Omega_{2\delta} \end{cases} \tag{21}$$

which we showed is well-posed. Note that the parameter $\delta > 0$ determines the thickness of the regions in Ω in which the hinged boundary conditions are imposed; in particular, it is not related to the horizon of μ. We have the following convergence result:

Theorem 6. *Let* $\Omega \subset \mathbb{R}^n$, $n \geq 2$, *be a bounded domain either of class* C^2 *or convex of class* C^1, *and let* μ *be a nonnegative and integrable kernel that satisfies assumptions (A2) and (A3). Suppose the sequence of positive scalars* $\{\delta_n\}_{n=1}^{\infty} \subset (0, \infty)$ *converges to* 0 *as* $n \to \infty$. *For* $f \in L^2(\Omega)$, *the solutions* u_{δ_n} *of the nonlocal hinged problems* (21) *converge in* $L^2(\Omega)$ *to the weak (variational) solution* $u \in W^{2,2}(\Omega) \cap W_0^{1,2}(\Omega)$ *of*

$$\begin{cases} \Delta^2 u = f, & x \in \Omega \\ u = \Delta u = 0, & x \in \partial\Omega \end{cases} \tag{22}$$

as $n \to \infty$. *Furthermore, if* Ω *is smooth, e.g., of class* C^4, *then* u *is also in* $W^{4,2}(\Omega)$.

Remark 4. The assumptions of μ are more general in Radu et al. (2017); here, we have imposed additional restrictions so we are in position to use our earlier regularity results.

By iterating the regularity result given by Theorem 4, we obtain the following regularity result.

Theorem 7 (Regularity of solutions to the linear nonlocal hinged system). *Let* $f \in W^{1,2}(\Omega)$, μ *a nonnegative and integrable kernel that satisfies assumptions (A2) and (A3). Then there exists a unique solution* u *to the system* (21) *such that* $u \in W^{1,2}(\Omega) \cap L^\infty(\Omega)$.

Proof. We decompose the system (22) into two second-order nonlocal boundary value problems:

$$\begin{cases} \mathcal{L}_\delta[u] = v, & x \in \Omega_\delta \\ u = 0, & x \in \Omega \setminus \Omega_\delta \end{cases} \tag{23}$$

$$\begin{cases} \mathcal{L}_\delta[v] = f, & x \in \Omega_{2\delta} \\ v = 0, & x \in \Omega_\delta \setminus \Omega_{2\delta} \end{cases} \tag{24}$$

and use Theorem 4 first for (24) to obtain that $v \in W^{1,2}(\Omega) \cap L^\infty(\Omega)$; then with this regularity for v, we apply the theorem to (23) to show that $u \in W^{1,2}(\Omega) \cap L^\infty(\Omega)$. □

References

B. Aksoylu, M.L. Parks, Variational theory and domain decomposition for nonlocal problems. Appl. Math. Comput. **217**(14), 6498–6515 (2011)

F. Andreu-Vaillo, J.M. Mazón, J.D. Rossi, J.J. Toledo-Melero, *Nonlocal Diffusion Problems*. Volume 165 of Mathematical Surveys and Monographs (American Mathematical Society, Providence/Real Sociedad Matemática Española, Madrid, 2010)

P.W. Bates, J. Han, The Dirichlet boundary problem for a nonlocal Cahn-Hilliard equation. J. Math. Anal. Appl. **311**(1), 289–312 (2005a)
P.W. Bates, J. Han, The Neumann boundary problem for a nonlocal Cahn-Hilliard equation. J. Differ. Equ. **212**(2), 235–277 (2005b)
E. Berchio, A. Ferrero, F. Gazzola, Structural instability of nonlinear plates modelling suspension bridges: mathematical answers to some long-standing questions. Nonlinear Anal. Real World Appl. **28**, 91–125 (2016)
L.A. Caffarelli, R. Leitão, J.M. Urbano, Regularity for anisotropic fully nonlinear integro-differential equations. Math. Ann. **360**(3–4), 681–714 (2014)
Q. Du, M. Gunzburger, R.B. Lehoucq, K. Zhou, Analysis and approximation of nonlocal diffusion problems with volume constraints. SIAM Rev. **54**(4), 667–696 (2012)
Q. Du, M. Gunzburger, R.B. Lehoucq, K. Zhou, A nonlocal vector calculus, nonlocal volume-constrained problems, and nonlocal balance laws. Math. Models Methods Appl. Sci. **23**(03), 493–540 (2013)
M. Foss, J. Geisbauer, Partial regularity for subquadratic parabolic systems with continuous coefficients. Manuscripta Math. **139**(1–2), 1–47 (2012)
M. Foss, P. Radu, Differentiability and integrability properties for solutions to nonlocal equations, in *New Trends in Differential Equations, Control Theory and Optimization: Proceedings of the 8th Congress of Romanian Mathematicians* (World Scientific, 2016), pp. 105–119
M. Foss, P. Radu, C. Wright, Regularity and existence of minimizers for nonlocal energy functionals. Differ. Integr. Equ. (2017, to appear)
F. Gazzola, *Mathematical Models for Suspension Bridges: Nonlinear Structural Instability*. Volume 15 of MS&A. Modeling, Simulation and Applications (Springer, Cham, 2015)
G. Gilboa, S. Osher, Nonlocal operators with applications to image processing. Multiscale Model. Simul. **7**(3), 1005–1028 (2008)
B. Hinds, P. Radu, Dirichlet's principle and wellposedness of solutions for a nonlocal p-Laplacian system. Appl. Math. Comput. **219**(4), 1411–1419 (2012)
S. Mayboroda, V. Maz'ya, Regularity of solutions to the polyharmonic equation in general domains. Invent. Math. **196**(1), 1–68 (2014)
T. Mengesha, Q. Du, The bond-based peridynamic system with Dirichlet-type volume constraint. Proc. R. Soc. Edinb. Sect. A **144**(1), 161–186 (2014)
A. Mogilner, L. Edelstein-Keshet, A non-local model for a swarm. J. Math. Biol. **38**(6), 534–570 (1999)
S. Oterkus, E. Madenci, A. Agwai, Peridynamic thermal diffusion. J. Comput. Phys. **265**, 71–96 (2014)
P. Radu, D. Toundykov, J. Trageser, Finite time blow-up in nonlinear suspension bridge models. J. Differ. Equ. **257**(11), 4030–4063 (2014)
P. Radu, D. Toundykov, J. Trageser, A nonlocal biharmonic operator and its connection with the classical analogue. Arch. Ration. Mech. Anal. **223**(2), 845–880 (2017)
P. Radu, K. Wells, A state-based Laplacian: properties and convergence to its local and nonlocal counterparts (2017, Preprint)
S. Silling, Reformulation of elasticity theory for discontinuities and long-range forces. J. Mech. Phys. Solids **48**, 175–209 (2000)
H. Sun, D. Uminsky, A.L. Bertozzi, Stability and clustering of self-similar solutions of aggregation equations. J. Math. Phys. **53**(11), 115610, 18 (2012)
X. Tian, Q. Du, Asymptotically compatible schemes and applications to robust discretization of nonlocal models. SIAM J. Numer. Anal. **52**(4), 1641–1665 (2014)

Dynamic Brittle Fracture from Nonlocal Double-Well Potentials: A State-Based Model

38

Robert Lipton, Eyad Said, and Prashant K. Jha

Contents

Abstract

We introduce a regularized model for free fracture propagation based on nonlocal potentials. We work within the small deformation setting, and the model is developed within a state-based peridynamic formulation. At each instant of the evolution, we identify the softening zone where strains lie above the strength of the material. We show that deformation discontinuities associated with flaws larger than the length scale of nonlocality δ can become unstable and grow. An explicit inequality is found that shows that the volume of the softening zone

R. Lipton (✉)
Department of Mathematics and Center for Computation and Technology, Louisiana State University, Baton Rouge, LA, USA
e-mail: lipton@lsu.edu

E. Said · P. K. Jha
Department of Mathematics, Louisiana State University, Baton Rouge, LA, USA
e-mail: esaid1@lsu.edu; prashant.j16o@gmail.com

G. Z. Voyiadjis (ed.), *Handbook of Nonlocal Continuum Mechanics for Materials and Structures*, https://doi.org/10.1007/978-3-319-58729-5_33

goes to zero linearly with the length scale of nonlocal interaction. This scaling is consistent with the notion that a softening zone of width proportional to δ converges to a sharp fracture set as the length scale of nonlocal interaction goes to zero. Here the softening zone is interpreted as a regularization of the crack network. Inside quiescent regions with no cracks or softening, the nonlocal operator converges to the local elastic operator at a rate proportional to the radius of nonlocal interaction. This model is designed to be calibrated to measured values of critical energy release rate, shear modulus, and bulk modulus of material samples. For this model one is not restricted to Poisson ratios of $1/4$ and can choose the potentials so that small strain behavior is specified by the isotropic elasticity tensor for any material with prescribed shear and Lamé moduli.

Keywords

Free fracture model · Nonlocal interactions · Double-well potentials · State-based peridynamics

Introduction

We address the problem of free crack propagation in homogeneous materials. The crack path is not known a priori and is found as part of the problem solution. Our approach is to use a nonlocal formulation based on double-well potentials. We will work within the small deformation setting, and the model is developed within a state-based peridynamic formulation. Peridynamics (Silling 2000; Silling et al. 2007) is a nonlocal formulation of continuum mechanics expressed in terms of displacement differences as opposed to spatial derivatives of the displacement field. These features provide the ability to simultaneously simulate both smooth displacements and defect evolution. Computational methods based on peridynamic modeling exhibit formation and evolution of sharp features associated with phase transformation (see Dayal and Bhattacharya 2006) and fracture (see Silling and Lehoucq 2008; Silling et al. 2010; Foster et al. 2011; Agwai et al. 2011; Du et al. 2013; Lipton et al. 2016; Bobaru and Hu 2012; Ha and Bobaru 2010; Silling and Bobaru 2005; Weckner and Abeyaratne 2005; Gerstle et al. 2007; Silling and Askari 2005). A recent review of the state of the art can be found in Bobaru et al. (2016).

In this work we are motivated by the recent models proposed and studied in Lipton (2014, 2016), and Lipton et al. (2016). Calibration has been investigated in Diehl et al. (2016). These models are defined by double-well two-point strain potentials. Here one potential well is centered at the origin and associated with elastic response, while the other well is at infinity and associated with surface energy. The rationale for studying these models is that they are shown to be well posed over the class of square-integrable non-smooth displacements, and in the limit of vanishing nonlocality, the dynamics localize and recover features of sharp fracture propagation (see Lipton 2014, 2016). In this work we extend this modeling approach to the state-based formulation. Our work is further motivated by the recent numerical-experimental study carried out in Diehl et al. (2016) demonstrating that the bond-based model is unable to capture the Poisson ratio for a sample of PMMA at room temperature. Here we develop a double-well state-based potential for which the Poisson ratio is no longer constrained to be 1/4. We show that for this model we can choose the potentials so that the small strain behavior is specified by the isotropic elasticity tensor for any material with prescribed shear and Lamé moduli.

Nonlocal Dynamics

We formulate the nonlocal dynamics. Here we will assume displacements u are small (infinitesimal) relative to the size of the three-dimensional body D. The tensile strain is written as $S = S(y, x, t; u)$ and given by

$$S(y, x, t; u) = \frac{u(t, y) - u(t, x)}{|y - x|} \cdot e_{y-x}, \qquad e_{y-x} = \frac{y - x}{|y - x|}, \tag{1}$$

where e_{y-x} is a unit direction vector and $\cdot$ is the dot product. It is evident that $S(y, x, t; u)$ is the tensile strain along the direction e_{y-x}. We introduce the influence function $\omega^{\delta}(|y - x|)$ such that ω^{δ} is nonzero for $|y - x| < \delta$, zero outside. Here we will take $\omega^{\delta}(|y - x|) = \omega(|y - x|/\delta)$ with $\omega(r) = 0$ for $r > 1$ nonnegative for $r < 1$ and ω is bounded.

The spherical or hydrostatic strain at x is given by

$$\theta(x, t; u) = \frac{1}{V_{\delta}} \int_{D \cap B_{\delta}(x)} \omega^{\delta}(|y - x|) S(y, x, t; u) |y - x| \, dy, \tag{2}$$

where V_{δ} is the volume of the ball $B_{\delta}(x)$ of radius δ centered at x. Here we have employed the normalization $|y - x|/\delta$ so that this factor takes values in the interval from 0 to 1.

Motivated by potentials of Lennard-Jones type, we define the force potential for tensile strain given by

$$\mathcal{W}^{\delta}(S(y, x, t; u)) = \alpha \omega^{\delta}(|y - x|) \frac{1}{\delta |y - x|} f(\sqrt{|y - x|} S(y, x, t; u)) \tag{3}$$

and the potential for hydrostatic strain

$$\mathcal{V}^{\delta}(\theta(x, t; u)) = \frac{\beta g(\theta(x, t; u))}{\delta^2} \tag{4}$$

where $\mathcal{W}^{\delta}(S(y, x, t; u))$ is the pairwise force potential per unit length between two points x and y and $\mathcal{V}^{\delta}(\theta(x, t; u))$ is the hydrostatic force potential density at x. They are described in terms of their potential functions f and g (see Fig. 1). These two potentials are double-well potentials that are chosen so that the associated forces acting between material points x and y are initially elastic and then soften and

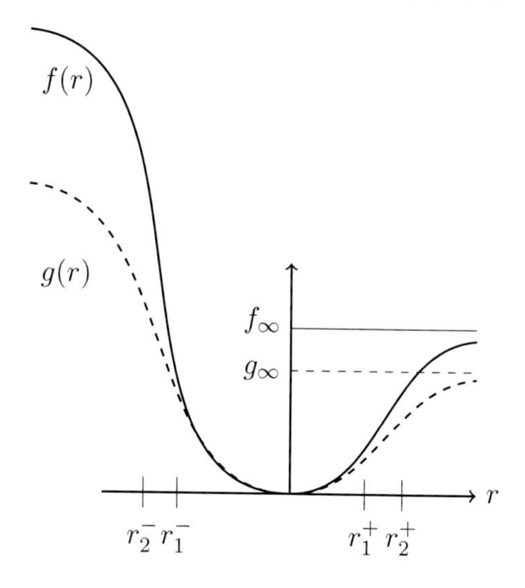

Fig. 1 Potential function $f(r)$ for tensile force and potential function $g(r)$ for hydrostatic force

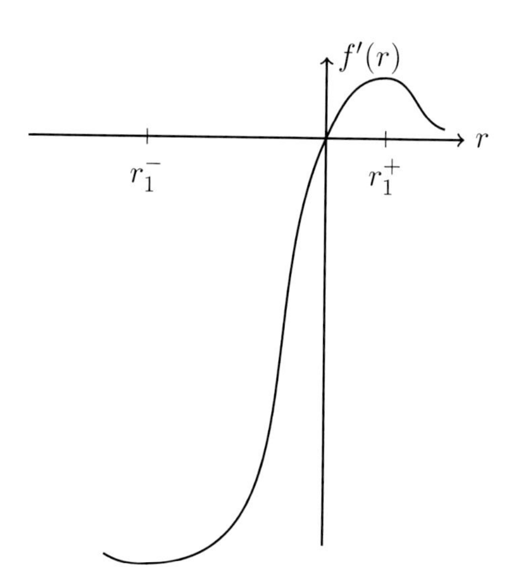

Fig. 2 Cohesive tensile force

decay to zero as the strain between points increases (see Fig. 2 for the tensile force). This force is negative for compression, and a similar force hydrostatic strain law follows from the potential for hydrostatic strain. The first well for $\mathcal{W}^\delta(S(y, x, t; u))$ and $\mathcal{V}^\delta(\theta(x, t; u))$ is at zero tensile and hydrostatic strain, respectively. With this in mind, we make the choice

$$f(0) = f'(0) = g(0) = g'(0). \tag{5}$$

The second well is at infinite tensile and hydrostatic strain and is characterized by the horizontal asymptotes $\lim_{S\to\infty} f(S) = f_\infty$ and $\lim_{\theta\to\infty} g(\theta) = g_\infty$, respectively (see Fig. 1).

The critical tensile strain $S_c^+ > 0$ for which the force begins to soften is given by the inflection point $r_1^+ > 0$ of f and is

$$S_c^+ = \frac{r_1^+}{\sqrt{|y-x|}}. \tag{6}$$

The critical negative tensile strain is chosen much larger in magnitude than S_c^+ and is

$$S_c^- = \frac{r_1^-}{\sqrt{|y-x|}}, \tag{7}$$

with $r_1^- < 0$ and $r_1^+ << |r_1^-|$. The critical value $0 < \theta_c^+$ where the force begins to soften under positive hydrostatic strain for $\theta(x,t;u) > \theta_c^+$ is given by the inflection point r_2^+ of g and is

$$\theta_c^+ = r_2^+. \tag{8}$$

The critical compressive hydrostatic strain where the force begins to soften for negative hydrostatic strain is chosen much larger in magnitude than θ_c^+ and is

$$\theta_c^- = r_2^-, \tag{9}$$

with $r_2^- < 0$ and $r_2^+ < |r_2^-|$. For this model we suppose the inflection points for g and f satisfy the ordering

$$r_2^- < r_1^- < 0 < r_1^+ < r_2^+. \tag{10}$$

This ordering is chosen to illustrate ideas for a material that is weaker in shear strain than hydrostatic strain. With this choice and the appropriate influence function ω^δ, if the hydrostatic stress is positive at x and is above the critical value θ_c^+, then there are points y in the peridynamic neighborhood for which the tensile stress between x and y is above S_c^+. This aspect of the model is established and addressed in section "Control of the Softening Zone."

The potential energy is given by

$$\begin{aligned} PD^\delta(u) =& \frac{1}{V_\delta}\int_D \int_{D\cap B_\delta(x)} |y-x|\mathcal{W}^\delta(S(y,x,t;u))\,dydx \\ &+ \int_D \mathcal{V}^\delta(\theta(x,t;u))\,dx. \end{aligned} \tag{11}$$

The material is assumed homogeneous, and the density is given by ρ, and the applied body force is denoted by $b(x, t)$. We define the Lagrangian

$$\mathrm{L}(u, \partial_t u, t) = \frac{\rho}{2}||\dot{u}||^2_{L^2(D;\mathbb{R}^3)} - PD^\delta(u) + \int_D b \cdot u dx,$$

here $\dot{u}$ is the velocity given by the time derivative of u, and $\|\dot{u}\|_{L^2(D;\mathbb{R}^3)}$ denotes the L^2 norm of the vector field $\dot{u} : D \to \mathbb{R}^3$. Applying the principle of least action together with a straightforward calculation gives the nonlocal dynamics

$$\rho\ddot{u}(x, t) = \mathcal{L}^T(u)(x, t) + \mathcal{L}^D(u)(x, t) + b(x, t), \text{ for } x \in D, \tag{12}$$

where

$$\mathcal{L}^T(u)(x, t) = \frac{2\alpha}{V_\delta} \int_{D \cap B_\delta(x)} \frac{\omega^\delta(|y - x|)}{\delta|y - x|} \partial_S f(\sqrt{|y - x|}S(y, x, t; u))e_{y-x}\, dy, \tag{13}$$

and

$$\mathcal{L}^D(u)(x, t) = \frac{\beta}{V_\delta} \int_{D \cap B_\delta(x)} \frac{\omega^\delta(|y - x|)}{\delta^2} [\partial_\theta g(\theta(y, t; u)) + \partial_\theta g(\theta(x, t; u))]\, e_{y-x}\, dy. \tag{14}$$

The dynamics is complemented with the initial data

$$u(x, 0) = u_0(x), \qquad \partial_t u(x, 0) = v_0(x). \tag{15}$$

It is readily verified that this is an ordinary state-based peridynamic model. The forces are defined by the derivatives of the potential functions, and the derivative associated with the tensile strain potential is sketched in Fig. 2. We show in the next section that this initial value problem is well posed.

Existence of Solutions

The regularity and existence of the solution depends on the regularity of the initial data and body force. In this work we choose a general class of body forces and initial conditions. The initial displacement u_0 and velocity v_0 are chosen to be integrable and belonging to $L^\infty(D; \mathbb{R}^3)$. The body force $b(x, t)$ is chosen such that for every $t \in [0, T_0]$, b takes values in $L^\infty(D, \mathbb{R}^3)$ and is continuous in time. The associated norm is defined to be $\|b\|_{C([0,T_0];L^\infty(D,\mathbb{R}^3))} = \max_{t\in[0,T_0]} \|b(x, t)\|_{L^\infty(D,\mathbb{R}^3)}$. The space of continuous functions in time taking values in $L^\infty(D; \mathbb{R}^3)$ for which this norm is finite is denoted by $C([0, T_0]; L^\infty(D, \mathbb{R}^3))$. The space of functions twice differentiable in time taking values in $L^\infty(D, \mathbb{R}^3)$ such that both derivatives belong to $C([0, T_0]; L^\infty(D, \mathbb{R}^3))$ is written as $C^2([0, T_0]; L^\infty(D, \mathbb{R}^3))$.

We will establish existence and uniqueness for the evolution by writing the second-order ODE as an equivalent first-order system. The nonlocal dynamics (12) can be written as a first-order system. Set $y = (y_1, y_2)$ where $y_1 = u$ and $y_2 = u_t$. Now, set $F^\delta(y,t) = \Big(F_1(y,t), F_2(y,t)\Big)^T$ where:

$$\begin{aligned} F_1(y,t) &= y_2 \\ F_2(y,t) &= \mathcal{L}^T(y_1)(t) + \mathcal{L}^D(y_1)(t) + b(t) \end{aligned} \tag{16}$$

And the initial value problem is given by the equivalent first-order system

$$\begin{aligned} \frac{d}{dt} y^\delta &= F^\delta(y^\delta, t) \\ y(0) &= (y_1(0), y_2(0)) = (u_0, v_0) \end{aligned} \tag{17}$$

The existence of a unique solution to the initial value problem is asserted in the following theorem.

Theorem 1. *For a body force $b(t,x)$ in $C^1\Big([0,T]; L^\infty(D, \mathbb{R}^3)\Big)$ and initial data $y_1(0)$ and $y_2(0)$ in $L_0^\infty(D;\mathbb{R}^3) \times L_0^\infty(D;\mathbb{R}^3)$, there exists a unique solution $y(t)$ such that $y_1 = u$ is in $C^2\Big([0,T]; L^\infty(D, \mathbb{R}^3)\Big)$ for the dynamics described by* (17) *with initial data in $L^\infty((D;\mathbb{R}^3) \times L^\infty(D;\mathbb{R}^3)$ and body force $b(t,x)$ in $C^1\Big([0,T]; L^\infty(D, \mathbb{R}^3)\Big)$.*

Proof. We will show that the model is *Lipschitz continuous* and then apply the theory of ODE in Banach spaces, e.g., Driver (2003), to guarantee the existence of a unique solution. It is sufficient to show that

$$||\mathcal{L}^T(u)(x,t) + \mathcal{L}^D(u)(x,t) - (\mathcal{L}^T(v)(x,t) + \mathcal{L}^D(v)(x,t))||_{L^\infty(D)} \leq C||u - v||_{L^\infty(D)} \tag{18}$$

For ease of notation, we introduce the following vectors

$$\vec{U} = u(y) - u(x),$$

$$\vec{V} = v(y) - v(x).$$

We write

$$\mathcal{L}^T(u)(x,t) + \mathcal{L}^D(u)(x,t) - (\mathcal{L}^T(v)(x,t) + \mathcal{L}^D(v)(x,t)) = \mathcal{I}_1 + \mathcal{I}_2. \tag{19}$$

Here

$$
\begin{aligned}
\mathcal{I}_1 =& \frac{2\alpha}{\delta V_\delta} \int_{D \cap B_\delta(x)} \frac{\omega^\delta(|y-x|)}{\sqrt{|y-x|}} \Big\{ f'(\sqrt{|y-x|} S(y,x,t;u)) \\
&- f'(\sqrt{|y-x|} S(y,x,t;v)) \Big\} e_{(y-x)} dy \\
\mathcal{I}_2 =& \frac{\beta}{\delta^2 V_\delta} \int_{D \cap B_\delta(x)} \omega^\delta(|y-x|) \Big(g'(\theta(y,t;u)) + g'(\theta(x,t;u)) \\
&- (g'(\theta(y,t;v)) + g'(\theta(x,t;v))) \Big) e_{(y-x)} dy
\end{aligned} \tag{20}
$$

Since f'' is bounded a straightforward calculation gives:

$$
\begin{aligned}
&|f'(\sqrt{|y-x|} S(y,x,t;u)) - f'(\sqrt{|y-x|} S(y,x,t;v)) \\
&\leq \sqrt{|y-x|} \sup_{s \in \mathbb{R}} \{|f''(s)|\} |S(y,x,t;u) - S(y,x,t;v)|,
\end{aligned}
$$

and $|e_{y-x}| = 1$, so we can bound $\mathcal{I}_1$ by

$$
|\mathcal{I}_1| \leq \frac{2\alpha}{\delta V_\delta} \int_{D \cap B_\delta(x)} \omega^\delta(|y-x|) \sup_{x \in D} \{|f''(x)|\} |S(y,x,t;u) - S(y,x,t;v)| \, dy. \tag{21}
$$

In what follows $C_1 = \sup_{s \in \mathbb{R}} \{|f''(s)|\} < \infty$ and we make the change of variable

$$
\begin{aligned}
&y = x + \delta\xi \\
&|y-x| = \sigma|\xi| \\
&dy = \delta^3 d\xi,
\end{aligned}
$$

and a straightforward calculation shows

$$
\mathcal{I}_1 \leq \frac{2\alpha C_1}{\delta^2} \int_{H_1(0) \cap \{x+\delta\xi \in D\}} |\omega(\xi)| \frac{|u(x+\delta\xi) - u(x) - (v(x+\delta\xi) - v(x))|}{|\xi|} d\xi \tag{22}
$$

Which leads to the inequality

$$
||\mathcal{I}_1||_{L^\infty(D;\mathbb{R}^3)} \leq \frac{4\alpha C_1 C_2}{\delta^2} ||u-v||_{L^\infty(D;\mathbb{R}^3)}, \tag{23}
$$

with $C_2 = \int_{H_1(0)} |\xi|^{-1} \omega(|\xi|) \, d\xi$. Now we can work on the second part, where we follow a similar approach. Noting that g'' is bounded, we let $C_3 = \sup_{\theta \in \mathbb{R}} \{|g''(\theta)|\} < \infty$ and $C_4 = \int_{H_1(0)} |\xi| \omega(|\xi|) d\xi$, to find that

$$|g'(\theta(y,t;u)) - g'(\theta(y,t;v))| \leq C_3|\theta(y,t;u) - \theta(y,t;v)|$$
$$\leq \frac{2C_3C_4}{\delta^2}\|u - v\|_{L^\infty(D;\mathbb{R}^3)},$$

and

$$|g'(\theta(x,t;u)) - g'(\theta(x,t;v))| \leq C_3|\theta(x,t;u) - \theta(x,t;v)|$$
$$\leq \frac{2C_3C_4}{\delta^2}\|u - v\|_{L^\infty(D;\mathbb{R}^3)},$$

so

$$\|\mathcal{I}_2\|_{L^\infty(D;\mathbb{R}^3)} \leq \frac{4\beta C_3C_4}{\delta^2}\|u - v\|_{L^\infty(D;\mathbb{R}^3)}. \tag{24}$$

Adding (23) and (24) gives the desired result

$$||\mathcal{L}^T(u)(x,t) + \mathcal{L}^D(u)(x,t) - (\mathcal{L}^T(v)(x,t) + \mathcal{L}^D(v)(x,t))||_{L^\infty(D;\mathbb{R}^3)}$$
$$\leq \frac{4(\alpha C_1C_2 + \beta C_3C_4)}{\delta^2}\|u - v\|_{L^\infty(D;\mathbb{R}^3)}. \tag{25}$$

□

Stability Analysis

In this section we identify a source for crack nucleation as a material defect represented by a jump discontinuity in the displacement field. To illustrate the ideas, we assume the defect is in the interior of the body and at least δ away from the boundary. This jump discontinuity can become unstable and grow in time. We proceed with a perturbation analysis and consider a time-independent body force density b and a smooth equilibrium solution u. Now assume that the defect perturbs u in the neighborhood of a point x by a piecewise constant vector field s that represents a jump in displacement across a planar surface with normal vector ν. We assume that this jump occurs along a defect of length 2δ on the planar surface.

The smooth equilibrium solution $u(x,t)$ is a solution of

$$0 = \mathcal{L}^T(u)(x,t) + \mathcal{L}^D(u)(x,t) + b(x) \tag{26}$$

Now consider a perturbed solution $u^P(x,t)$ that differs from equilibrium solution $u(x,t)$ by the jump across the planar surface which is specified by unit normal vector ν. We suppose the surface passes through x and extends across the peridynamic neighborhood centered at x. Points y for which $(y-x)\cdot\nu < 0$ are denoted by $\mathcal{E}_\nu^-$ and points for which $(y-x)\cdot\nu \geq 0$ are denoted by $\mathcal{E}_\nu^+$, see Fig. 3.

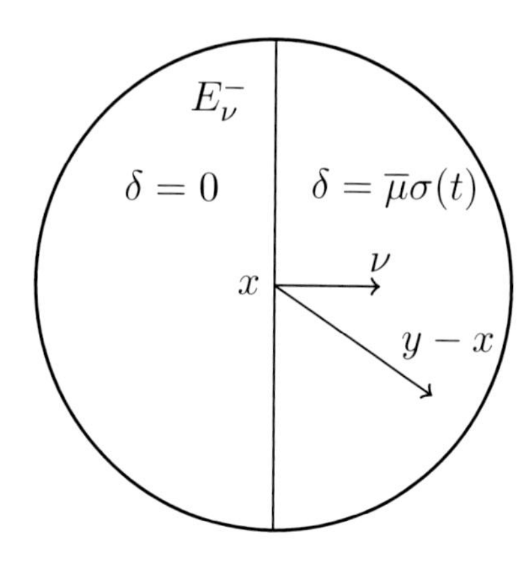

Fig. 3 Jump discontinuity

The perturbed solution u^P satisfies

$$\rho \ddot{u^P} = \mathcal{L}^T(u^P)(x,t) + \mathcal{L}^D(u^P)(x,t) + b(x) \tag{27}$$

Here the perturbed solution $u^P(x,t)$ is given by the equilibrium solution plus a piecewise constant perturbation and is written

$$u^P(y,t) = u(y,t) + s(y,t) \tag{28}$$

Where

$$s(y,t) = \begin{cases} 0 & y \in \mathcal{E}_\nu^- \\ \bar{\mu}\sigma(t) & y \in \mathcal{E}_\nu^+ \end{cases} \tag{29}$$

Subtracting (26) from (27) gives

$$\rho \ddot{u^P} = \mathcal{L}^T(u^P)(x,t) + \mathcal{L}^D(u^P)(x,t) - \mathcal{L}^T(u)(x,t) + \mathcal{L}^D(u)(x,t) \tag{30}$$

Here the second term in $\mathcal{L}^D(u)$ vanishes as we are away from the boundary, and the integrand is odd in the y variable with respect to the domain $B_\delta(x)$. Since $u^P = u + s$ and s is small, we expand $f'(\sqrt{|y-x|}(S(y,x,t;u+s)))$ in Taylor series in s. Noting that $\theta(x,t;u+s) = \theta(x,t;u) + \theta(x,t;s)$ and $\theta(x,t;s)$ is initially infinitesimal, we also expand $g'(\theta(x,t;u+s))$ in a Taylor series in $\theta(x,t;s)$. Applying the expansions to (30) shows that to leading order

$$\begin{aligned} \rho \ddot{u^P} = \rho\ddot{\sigma}\overline{\mu} = \frac{2\alpha}{\delta V_\delta} \int_{B_\delta(x)} \frac{\omega^\delta(|y-x|)}{\sqrt{|y-x|}} f''(\sqrt{|y-x|}S)(s(y,t) - s(x,t)) \cdot \\ e_{(y-x)} e_{(y-x)} dy + \frac{\beta}{V_\delta \delta^2} \int_{B_\delta(x)} \omega^\delta(|y-x|) g''(\theta(y,t;u)) \frac{1}{V_\delta} \\ \int_{B_\delta(y)} \omega^\delta(|z-y|)(s(z,t) - s(y,t)) \cdot e_{z-y}\, dz e_{y-x}\, dy = I_1 + I_2, \end{aligned} \tag{31}$$

where I_1 and I_2 are the first, second terms on the right-hand side of (31). A straightforward calculation using (29) shows that

$$I_1 = -\frac{2\alpha}{\delta V_\delta}\int_{B_\delta(x)\cap\mathcal{E}_\nu^-}\frac{J^\delta(|y-x|)}{|y-x|}f''(\sqrt{|y-x|}S)e_{(y-x)}\cdot\bar{\mu}\sigma(t)e_{(y-x)}dy \quad (32)$$

We next calculate I_2. A straightforward but delicate calculation gives

$$\frac{1}{V_\delta}\int_{B_\delta(y)}\frac{\omega^\delta(|z-y|)}{\delta}(s(z,t)-s(y,t))\cdot e_{z-y}\,dz = b(y)\cdot\overline{\mu}\sigma(t) \quad (33)$$

where

$$b(y) = \frac{1}{V_\delta}\int_0^{2\pi}\int_a^\delta\int_0^{\overline{\phi}}\omega(|z-y|)e(\theta,\phi)|z-y|^2\sin\phi\,d\phi\,d\theta\,d|z-y| \quad (34)$$

and the limits of the iterated integral are

$$a = |(y-x)\cdot\nu| \quad \overline{\phi} = \arccos\left(\frac{|(y-x)\cdot\nu|}{|z-y|}\right), \quad (35)$$

and $e(\theta,\phi)$ is the vector on the unit sphere with direction specified by the angles θ and ϕ. Calculation now gives

$$\begin{aligned} I_2 =& \frac{\beta}{V_\delta\delta^2}\Bigg(\int_{B_\delta(x)\cap\mathcal{E}_\nu^-}\omega^\delta(|y-x|)g''(\theta(y,t;u))b(y)\cdot\overline{\mu}\sigma(t)e_{y-x}dy \\ & - \int_{B_\delta(x)\cap\mathcal{E}_\nu^+}\omega^\delta(|y-x|)g''(\theta(y,t;u))b(y)\cdot\overline{\mu}\sigma(t)e_{y-x}dy\Bigg), \end{aligned} \quad (36)$$

where

$$\begin{aligned} &\int_{B_\delta(x)\cap\mathcal{E}_\nu^-}\frac{\omega^\delta(|y-x|)}{\delta}b(y)\cdot\overline{\mu}s(t)e_{y-x}dy \\ &= \int_{B_\delta(x)\cap\mathcal{E}_\nu^+}\frac{\omega^\delta(|y-x|)}{\delta}b(y)\cdot\overline{\mu}s(t)e_{y-x}dy. \end{aligned} \quad (37)$$

We now take the dot product of both sides of (31) with $\overline{\mu}$ to get

$$\rho\ddot{\sigma} = \frac{(\mathsf{A}+\mathsf{B}^{\mathrm{sym}})\overline{\mu}\cdot\overline{\mu}}{|\overline{\mu}|^2}\sigma(t), \quad (38)$$

where

$$\mathsf{A} = -\frac{2\alpha}{\delta V_\delta} \int_{B_\delta(x)\cap\mathcal{E}_v^-} \frac{J^\delta(|y-x|)}{|y-x|} f''(\sqrt{|y-x|}S)e_{(y-x)} \otimes e_{(y-x)}dy \tag{39}$$

and $\mathsf{B}^{\mathrm{sym}} = (\mathsf{B} + \mathsf{B}^T)/2$ with

$$\begin{aligned}\mathsf{B} =& \frac{\beta}{V_\delta \delta^2} \left(\int_{B_\delta(x)\cap\mathcal{E}_v^-} \omega^\delta(|y-x|) g''(\theta(y,t;u)) b(y) \otimes e_{y-x} dy \right. \\ &\left. - \int_{B_\delta(x)\cap\mathcal{E}_v^+} \omega^\delta(|y-x|) g''(\theta(y,t;u)) b(y) \otimes e_{y-x} dy \right).\end{aligned} \tag{40}$$

Inspection shows that

$$f''(\sqrt{|y-x|}S) < 0, \text{ when } S > S_c^+. \tag{41}$$

Thus the eigenvalues of A can be nonnegative whenever the tensile strain is positive and greater than S_c^+ so that the force is in the softening regime for a preponderance of points y inside $B_\delta(x)$. In general the defect will be stable if all eigenvalues of the stability matrix $\mathsf{A} + \mathsf{B}^{\mathrm{sym}}$ are negative. On the other hand, the defect will be unstable if at least one eigenvalue of the stability matrix is positive.

We collect results in the following proposition.

Proposition 1 (Fracture nucleation condition about a defect). *A condition for crack nucleation at a defect passing through a point x is that the associated stability matrix $\mathsf{A} + \mathsf{B}^{\mathrm{sym}}$ has at least one positive eigenvalue.*

If the equilibrium solution is constant, then $\theta(y,t;u) = \text{constant}$ and $\mathsf{B}^{\mathrm{sym}} = 0$. For this case the fracture nucleation condition simplifies and depends only on the eigenvalues of the matrix A. In the next section, we analyze the size of the set where the tensile strain is greater than S_c^+ so that the tensile force is in the softening regime for points y inside $B_\delta(x)$.

Control of the Softening Zone

We define the softening zone in terms of the collection of centers of peridynamic neighborhoods with tensile strain exceeding S_c^+. In what follows we probe the dynamics to obtain mathematically rigorous and explicit estimates on the size of the softening zone in terms of the radius of the peridynamic horizon. In this section we assume $\omega^\delta = 1$, $\delta < 1$, and from the definition of the hydrostatic strain $\theta(x,t;u)$, we have the following lemma.

Lemma 1 (Hydrostatic softening implies tensile softening). *If $\theta_c^+ < \theta(x,t;u)$, then $S_c^+ < S(y,x,t;u)$ for some subset of points y inside the peridynamic neighborhood centered at x and*

$$\{x \in D : \theta(x) > \theta_c^+\} \subset \{x \in D : S(x,y,t;u) > S_c^+, \textit{ for some } y \textit{ in } B_\delta(x)\}. \tag{42}$$

Proof. Suppose $\theta_c^+ < \theta(x,t;u)$, then there are points y in $B_\delta(x)$ for which

$$\theta_c^+ < |y-x|S(y,x,t;u) < \sqrt{|y-x|}S(y,x,t;u), \tag{43}$$

so

$$S_c^+ < \frac{\theta_c^+}{\sqrt{|y-x|}} < S(y,x,t;u), \tag{44}$$

since $r_1^+ < r_2^+ = \theta_c^+$. This directly implies

$$\{x \in D : \theta(x) > \theta_c^+\} \subset \{x \in D : S(x,y,t;u) > S_c^+, \text{ for some } y \text{ in } B_\delta(x)\}, \tag{45}$$

and the lemma is proved. □

This inequality shows that the collection of neighborhoods where softening is due to the hydrostatic force is also subset of the neighborhoods where there is softening due to tensile force. Motivated by this observation, we focus on peridynamic neighborhoods where the tensile strain is above critical. We start by defining the *softening zone*. The set of points y in $B_\delta(x)$ with tensile strain larger than critical can be written as

$$A_\delta^+(x) = \{y \in B_\delta(x) : S(y,x,t;u) > S_c^+\}.$$

From the monotonicity of the force potential f, we can also express this set as

$$A_\delta^+(x) = \{y \in B_\delta(x); f(\sqrt{|y-x|}S(y,x,t;u)) \geq f(r_1^+)\}.$$

We define the weighted volume of the set A_δ^+ in terms of its characteristic function $\chi_{A_\delta^+}(y)$ taking the value one for $y \in A_\delta^+$ and zero outside. The weighted volume of A_δ^+ is given by $\int_{B_\delta(x)} \chi_{A_\delta^+}(y)|y-x|\,dy$, and the weighted volume of $B_\delta(x)$ is $m = \int_{B_\delta(x)} |y-x|\,dy$. The weighted volume fraction $P_\delta(x)$ of $y \in B_\delta(x)$ with tensile strain larger than critical is given by the ratio

$$P_\delta(x) = \frac{\int_{B_\delta(x)} \chi_{A_\delta^+}(y)|y-x|\,dy}{m}.$$

Definition 1 (Softening zone). Fix any volume fraction $0 < \gamma \leq 1$, and with each time t in the interval $0 \leq t \leq T$, define the softening zone $SZ^\delta(\gamma, t)$ to be the collection of centers of peridynamic neighborhoods for which the weighted volume fraction of points y with strain $S(y, x, t; u)$ exceeding the threshold $\mathcal{S}_c$ is greater than γ, i.e.,

$$SZ^\delta(\gamma, t) = \{x \in D; P_\delta(x) > \gamma\}. \tag{46}$$

We now show that the volume of $SZ^\delta(\gamma, t)$ goes to zero linearly with the horizon δ for properly chosen initial data and body force. This scaling is consistent with the notion that a softening zone of width proportional to δ converges to a sharp fracture as the length scale δ of nonlocal interaction goes to zero. We define the sum of kinetic and potential energy as

$$W(t) = \frac{\rho}{2}||\dot{u}||^2_{L^2(D,\mathbb{R}^d)} + PD^\delta(u(t)) \tag{47}$$

and set

$$C(t) = \Big(\frac{1}{\sqrt{\rho}}\int_0^t ||b||_{L^2(D,\mathbb{R}^d)} d\tau + \sqrt{W(0)}\Big)^2. \tag{48}$$

Here $C(t)$ is a measure of the total energy delivered to the body from initial conditions and body force up to time t. The tensile toughness is defined to be the energy of tensile tension between x and y per unit length necessary for softening and is given by $f(r_1^+)/\delta$. We now state the geometric dependence of the softening zone on horizon.

Theorem 2. *The volume of the softening zone SZ^δ is controlled by the horizon δ according to the following relation expressed in terms of the total energy delivered to the system, the tensile toughness, and the weighted volume fraction of points y where the tensile strain exceeds S_c^+,*

$$\text{Volume}(SZ^\delta(\gamma, t)) \leq \frac{\delta C(t)}{\gamma m f(r_1^+)}. \tag{49}$$

Remark 1. It is clear that for zero initial data such that $u(0, x) = 0$ that $C(t)$ depends only on the body force $b(t, x)$ and initial velocity. For this choice we see that the softening zone goes to zero linearly with the horizon δ.

We now establish the theorem using Gronwall's inequality and Tchebychev's inequality. The peridynamic energy density at a point x is

$$E^\delta(x) = \frac{1}{V_\delta}\int_{D\cap B_\delta(x)} |y - x|\mathcal{W}^\delta(S(y, x, t; u))\, dy + \mathcal{V}^\delta(\theta(x, t; u)) \tag{50}$$

Which can also be rewritten with the following change of variable $y - x = \delta\xi$

$$E^{\delta}(x) = \frac{\alpha}{\delta V_1}\int_{D\cap B_1(0)} \omega(|\xi|) f(\sqrt{\delta|\xi|}S(x+\delta\xi, x, t; u))d\xi + \frac{\beta g(\theta(x,t;u))}{\delta^2} \quad (51)$$

Recall from the monotonicity of $f(r)$ that $r_1^+ < r$ implies $f(r_1^+) < f(r)$. Now define the set where the strain exceeds the threshold S_c^+

$$S^{+,\delta} = \left\{(\xi, x) \in B_1(0) \times D;\ x + \delta\xi \in D \text{ and } f(r_1^+) < f(\sqrt{\delta|\xi|}S(x+\delta\xi, x, t; u))\right\} \quad (52)$$

A straightforward calculation with $\omega(|\xi|) = 1$ shows that

$$\begin{aligned} \frac{f(r_1^+)}{\delta}\int_{S^{+,\delta}} |\xi| d\xi dx &\leq \int_{S^{+,\delta}} \frac{1}{\delta} f(\sqrt{\delta|\xi|}S(x+\delta\xi, x, t; u)d\xi dx \\ &\leq \int_D E^{\delta}(x)\,dx = PD^{\delta}(u(t)) \end{aligned} \quad (53)$$

We define the weighted volume of the set $S^{+,\delta}$ to be

$$\tilde{V}(S^{+,\delta}) = \int_{S^{+,\delta}} |\xi| d\xi dx \quad (54)$$

and inequality (53) becomes

$$\frac{f(r_1^+)}{\delta}\tilde{V}(S^{+,\delta}) \leq \int_{S^{+,\delta}} \frac{1}{\delta} f(\sqrt{\delta|\xi|}S(x+\delta\xi, x, t, u))d\xi dx \leq PD^{\delta}(u(t)) \quad (55)$$

Next we use Gronwall's inequality to prove the following theorem that shows that the kinetic and peridynamic energies of the solution $u(x,t)$ are bounded by the energy put into the system.

Theorem 3.

$$C(t) \geq \frac{\rho}{2}\|\dot{u}\|^2_{L^2(D;\mathbb{R}^2)} + PD^{\delta}(u(t)). \quad (56)$$

Proof. We start by multiplying both sides of (12) by $\dot{u}$ to get

$$\rho\ddot{u}(t)\cdot\dot{u}(t) = \left(\mathcal{L}^T(u)(x,t) + \mathcal{L}^D(u)(x,t)\right)\cdot\dot{u}(t) + b(t)\cdot\dot{u}(t).$$

Applying the product rule in the first term and integration by parts in the second term gives

$$\frac{1}{2}\frac{d}{dt}\Big[\rho||\dot{u}||^2_{L^2(D,\mathbb{R}^d)} + 2PD^\delta(u(t))\Big] = \int_D b(t)\cdot\dot{u}(t)\,dx.$$

Application of Cauchy's inequality to the right-hand side gives

$$\frac{1}{2}\frac{d}{dt}\Big[\rho||\dot{u}||^2_{L^2(D,\mathbb{R}^d)} + 2PD^\delta(u(t))\Big] = \int_D b(t)\cdot\dot{u}(t)dx$$
$$\leq ||b(t)||_{L^2(D,\mathbb{R}^d)}||\dot{u}(t)||_{L^2(D,\mathbb{R}^d)}. \tag{57}$$

Now set $\tilde{W}(t) = \rho||\dot{u}||^2_{L^2(D,\mathbb{R}^d)} + 2PD^\delta(u(t)) + \zeta$ where ζ is a positive number and can be taken arbitrarily small and (57) becomes,

$$\frac{1}{2}\tilde{W}'(t) =\leq ||b(t)||_{L^2(D,\mathbb{R}^d)}||\dot{u}(t)||_{L^2(D,\mathbb{R}^d)}$$
$$\leq ||b(t)||_{L^2(D,\mathbb{R}^d)}\frac{\sqrt{\tilde{W}(t)}}{\sqrt{\rho}}$$

Now we can write

$$\frac{1}{2}\int_0^t \frac{\tilde{W}'(\tau)}{\sqrt{\tilde{W}(\tau)}}d\tau \leq \frac{1}{\sqrt{\rho}}\int_0^t ||b||_{L^2(D,\mathbb{R}^d)}d\tau$$

Which simplifies to

$$\sqrt{\tilde{W}(t)} - \sqrt{\tilde{W}(0)} \leq \frac{1}{\sqrt{\rho}}\int_0^t ||b||_{L^2(D,\mathbb{R}^d)}d\tau. \tag{58}$$

Since ζ can be made arbitrarily small, we find that

$$\sqrt{W(t)} - \sqrt{W(0)} \leq \frac{1}{\sqrt{\rho}}\int_0^t ||b||_{L^2(D,\mathbb{R}^d)}d\tau, \tag{59}$$

and (56) follows. □

We apply inequality (55) and Theorem 3 to get the fundamental inequality.

$$\tilde{V}(S^{+,\delta}) \leq \frac{C(t)\delta}{f(\bar{r})}. \tag{60}$$

The fundamental inequality above is defined on $B_1(0)\times D$, and we now use it to bound the volume of the softening zone on D. Introducing the characteristic function $\chi^{S^{+,\delta}}(\xi,x)$ and taking the value 1 when $(\xi,x)\in S^{+,\delta}$ and 0 otherwise, we immediately have

$$mP_\delta(x)=\int_{B_1(0)}\chi^{S^{+,\delta}}(\xi,x)|\xi|d\xi.$$

So we can rewrite equation (54) as

$$\begin{aligned}\tilde{V}(S^{+,\delta})&=\int_D\int_{B_1(0)}\chi^{S^{+,\delta}}(\xi,x)|\xi|d\xi dx\\&=m\int_D P_\delta(x)\,dx.\end{aligned}\tag{61}$$

Now applying Tchebychev's inequality to (61) with $SZ^\delta(\gamma,t)$ defined by (46) gives the desired result

$$\text{Volume}(SZ^\delta(\gamma,t))\leq\frac{1}{\gamma}\int_D P_\delta(x)\,dx=\frac{\tilde{V}(S^{+,\delta})}{m\gamma}\leq\frac{C(t)\delta}{m\gamma f(r_1^+)}.\tag{62}$$

Calibration of the Model

In this section we show how to calibrate this model using the known elastic properties and energy release rate of fracture associated with a given material.

Calibrating the Peridynamic Energy to Elastic Properties

We start by considering a body D for which the strain S is small. Here *small* means for a fixed $|y-x|$ we have $|S|<<|S_c^{\pm}|$, $|\theta|<<|\theta_c^{\pm}|$. Now we proceed to calculate the peridynamic energy density inside the material due to the presence of a small deformation $u(x)$. Suppose that the *strain* at the length scale of a neighborhood of *horizon* δ is a linear function, i.e.,

$$\begin{aligned}S(u,y,x)&=\frac{u(y)-u(x)}{|y-x|}\cdot\frac{y-x}{|y-x|}\\&=F\frac{y-x}{|y-x|}\cdot\frac{y-x}{|y-x|}=Fe\cdot e,\end{aligned}\tag{63}$$

here F is a 3 by 3 matrix. We expand the first potential with respect to S and the second in θ keeping in mind that

$$f(0)=f'(0)=g(0)=g'(0)=0$$

to get

$$
\begin{aligned}
f\left(\sqrt{|y-x|}S\right) &= \frac{|y-x|}{2} f''(0)S^2 + O(S^3) \\
g\left(\theta(x,t;S)\right) &= \frac{1}{2} g''(0)\theta^2 + O(\theta^3)
\end{aligned}
\tag{64}
$$

So we write the *energy density* which was defined in Eq. (50) for points x of distance δ away from the boundary ∂D to leading order

$$
\begin{aligned}
E^\delta =& \frac{1}{V_\delta} \frac{\alpha f''(0)}{2\delta} \int_{H_\delta(x)} \omega^\delta(|y-x|)|y-x|(Fe\cdot e)^2 dy \\
&+ \frac{\beta g''(0)}{2\delta^2} \left(\frac{1}{V_\delta} \int_{H_\delta(x)} \omega^\delta(|y-x|)|y-x| Fe\cdot e\, dy \right)^2
\end{aligned}
\tag{65}
$$

The change of variable $\delta\xi = y - x$ gives to leading order

$$
\begin{aligned}
E^\delta =& \frac{\alpha f''(0)}{2V_1} \int_{H_1(0)} \omega(|\xi|)|\xi|(Fe\cdot e)^2 d\xi \\
&+ \frac{\beta g''(0)}{2V_1^2} \left(\int_{H_1(0)} \omega(|\xi|)|\xi| Fe\cdot e\, d\xi \right)^2
\end{aligned}
\tag{66}
$$

Observe that $(Fe\cdot e)^2 = \sum_{ijkl} F_{ij} F_{kl} e_i e_j e_k e_l$ and the first term in (66) is given by

$$
\sum_{ijkl} \mathbb{M}_{ijkl} F_{ij} F_{kl} \tag{67}
$$

where

$$
\mathbb{M}_{ijkl} = \frac{\alpha f''(0)}{2V_1} \int_{\mathcal{H}_1(0)} |\xi|\omega(|\xi|)\, e_i e_j e_k e_l \, d\xi = \frac{\alpha f''(0)}{2V_1} \int_0^1 |\xi|^3 \omega(|\xi|)\, d|\xi|
$$

$$
\int_{S^2} e_i e_j e_k e_l \, de. \tag{68}
$$

where de is an element of surface measure on the unit sphere. Next observe $Fe\cdot e = \sum_{kj} F_{kj} e_k e_j$ and the second term in (66) is given by

$$
\frac{\beta g''(0)}{2V_1^2} \left(\sum_{ij} \Lambda_{ij} F_{ij} \right)^2 = \frac{\beta g''(0)}{2V_1^2} \sum_{ijkl} \Lambda_{ij} \Lambda_{kl} F_{ij} F_{kl}. \tag{69}
$$

where

$$\Lambda_{jk} = \int_{\mathcal{H}_1(0)} |\xi|\omega(|\xi|)\, e_j e_k \, d\xi = \int_0^1 |\xi|^3\omega(|\xi|) d|\xi| \int_{S^2} e_j e_k \, de. \tag{70}$$

Focusing on the first term, we show that

$$\mathbb{M}_{ijkl} = 2\mu \left(\frac{\delta_{ik}\delta_{jl} + \delta_{il}\delta_{jk}}{2} \right) + \lambda \delta_{ij}\delta_{kl} \tag{71}$$

where μ and λ are given by

$$\lambda = \mu = \frac{\alpha f''(0)}{10} \int_0^1 |\xi|^3 \omega(|\xi|)\, d|\xi|. \tag{72}$$

To see this we write

$$\Gamma_{ijkl}(e) = e_i e_j e_k e_l, \tag{73}$$

to observe that $\Gamma(e)$ is a totally symmetric tensor valued function defined for $e \in S^2$ with the property

$$\Gamma_{ijkl}(Qe) = Q_{im}e_m Q_{jn}e_n Q_{ko}e_o Q_{lp}e_p = Q_{im}Q_{jn}Q_{ko}Q_{lp}\Gamma_{\mathrm{mnop}}(e) \tag{74}$$

for every rotation Q in SO^3. Here repeated indices indicate summation. We write

$$\int_{\mathcal{H}_1(0)} |\xi|^3\omega(|\xi|)\, e_i e_j e_k e_l \, d\xi = \int_0^1 |\xi|^3\omega(|\xi|) d|\xi| \int_{S^2} \Gamma_{ijkl}(e)\, de \tag{75}$$

to see that for every Q in SO^3

$$Q_{im}Q_{jn}Q_{ko}Q_{lp} \int_{S^2} \Gamma_{ijkl}(e)\, de = \int_{S^2} \Gamma_{\mathrm{mnop}}(Qe)\, de = \int_{S^2} \Gamma_{\mathrm{mnop}}(e)\, de. \tag{76}$$

Therefore we conclude that $\int_{S^2} \Gamma_{ijkl}(e)\, de$ is invariant under SO^3 and is therefore an isotropic symmetric fourth-order tensor and necessarily of the form

$$\int_{S^2} \Gamma_{ijkl}(e)\, de = a \left(\delta_{ik}\delta_{jl} + \delta_{il}\delta_{jk} \right) + b\delta_{ij}\delta_{kl}. \tag{77}$$

So $\mathbb{M}$ can be written in the form

$$\mathbb{M}_{ijkl} = 2\mu \left(\frac{\delta_{ik}\delta_{jl} + \delta_{il}\delta_{jk}}{2} \right) + \lambda \delta_{ij}\delta_{kl}, \tag{78}$$

with suitable choices of μ and λ. To evaluate μ and λ, we note the following relations between μ and λ for isotropic fourth-order tensors of the form above and their contractions

$$\mathbb{M}_{iijj} = 3(2\mu + 3\lambda), \tag{79}$$

$$\mathbb{M}_{ijij} = 3(4\mu + \lambda). \tag{80}$$

These relations can be readily verified by direct calculation.

On the other hand from the definition of $\mathbb{M}$ given by (68), we have

$$\mathbb{M}_{iijj} = \frac{\alpha f''(0)}{2V_1} \int_0^1 |\xi|^3 \omega(|\xi|)\, d|\xi| \int_{S^2} e_i^2 e_j^2\, de = \frac{4\pi\alpha f''(0)}{2V_1} \int_0^1 |\xi|^3 \omega(|\xi|)\, d|\xi|, \tag{81}$$

$$\mathbb{M}_{ijij} = \frac{\alpha f''(0)}{2V_1} \int_0^1 |\xi|^3 \omega(|\xi|)\, d|\xi| \int_{S^2} e_i^2 e_j^2\, de = \frac{4\pi\alpha f''(0)}{2V_1} \int_0^1 |\xi|^3 \omega(|\xi|)\, d|\xi|, \tag{82}$$

since $e_i^2 = \sum_i e_i^2 = 1$. Equation (72) now follows on recalling that $V_1 = \frac{4}{3}\pi$ and solving the system given by (79) and (80).

Focusing on the second term of (66) given by (69), we show that

$$\Lambda_{ij} = \frac{4\pi}{3} \int_0^1 |\xi|^3 \omega(|\xi|)\, d|\xi| \delta_{ij} \tag{83}$$

To see this we write

$$\Lambda_{ij}(e) = e_i e_j, \tag{84}$$

to observe that $\Lambda(e)$ is a totally symmetric tensor valued function defined for $e \in S^2$ with the property

$$\Lambda_{ij}(Qe) = Q_{im} e_m Q_{jn} e_n = Q_{im} Q_{jn} \Lambda_{mn}(e) \tag{85}$$

for every rotation Q in SO^3. As before repeated indices indicate summation. We consider

$$\int_{S^2} \Lambda_{ij}(e)\, de \tag{86}$$

to see that for every Q in SO^3

$$Q_{im} Q_{jn} \int_{S^2} \Lambda_{ij}(e)\, de = \int_{S^2} \Lambda_{mn}(Qe)\, de = \int_{S^2} \Lambda_{mn}(e)\, de. \tag{87}$$

Therefore we conclude that $\int_{S^2} \Lambda_{ij}(e)\,de$ is an isotropic symmetric second-order tensor and of the form

$$\int_{S^2} \Lambda_{ij}(e)\,de = a\delta_{ij}, \tag{88}$$

i.e., a multiple of the identity. So from (70) Λ is of the form

$$\Lambda_{ij} = \gamma\delta_{ij}. \tag{89}$$

To evaluate γ we take the trace of (70) and (83) as follows.

Now the second term is given by

$$\begin{aligned}\frac{\beta g''(0)}{2V_1^2}\left(\frac{4\pi}{3}\right)^2\left(\int_0^1 |\xi|^3\omega(|\xi|)d|\xi|\right)^2 \sum_{ijkl}\delta_{ij}\delta_{kl}F_{ij}F_{kl} = \\ = \beta g''(0)\frac{1}{2}\left(\int_0^1 |\xi|^3\omega(|\xi|)d|\xi|\right)^2 \sum_{ijkl}\delta_{ij}\delta_{kl}F_{ij}F_{kl} = \mathbb{K}_{ijkl}F_{ij}F_{kl}\end{aligned} \tag{90}$$

Collecting results we see that the leading order of the energy is given by

$$E^\delta = \sum_{ijkl}(\mathbb{M}_{ijkl} + \mathbb{K}_{ijkl})F_{ij}F_{kl} = \sum_{ijkl}\left(2\overline{\mu}\frac{\delta_{ik}\delta_{jl} + \delta_{il}\delta_{jk}}{2} + \overline{\lambda}\delta_{ij}\delta_{kl}\right)F_{ij}F_{kl} \tag{91}$$

where the shear modulus is given by

$$\overline{\mu} = \frac{\alpha f''(0)}{10}\int_0^1 |\xi|^3\omega(|\xi|)\,d|\xi|, \tag{92}$$

and the Lame constant is given by

$$\overline{\lambda} = \frac{\alpha f''(0)}{10}\int_0^1 |\xi|^3\omega(|\xi|)\,d|\xi| + \frac{\beta g''(0)}{2}\left(\int_0^1 |\xi|^3\omega(|\xi|)d|\xi|\right)^2. \tag{93}$$

One is free to choose α and β provided that the resulting elastic tensor satisfies the constraints of ellipticity. Here one is no longer restricted to Poisson ratios of $1/4$ as in the bond-based formulation.

An identical calculation shows that for two-dimensional problems the elastic constants are given by

$$\overline{\mu} = \frac{\alpha f''(0)}{8}\int_0^1 |\xi|^2\omega(|\xi|)\,d|\xi|, \tag{94}$$

and

$$\overline{\lambda} = \frac{\alpha f''(0)}{8} \int_0^1 |\xi|^2 \omega(|\xi|)\, d|\xi| + \frac{\beta g''(0)}{2} \left(\int_0^1 |\xi|^2 \omega(|\xi|) d|\xi| \right)^2, \tag{95}$$

and one is no longer restricted to Poisson ratio $1/3$ materials.

We note here that the two-dimensional moduli $\bar{\mu}$ and $\bar{\lambda}$ are directly related to the well-known moduli appearing in the plane strain or plane stress solutions for isotropic materials. This relationship is now well known and can be found in Jasiuk et al. (1994) and also Milton (2002).

Calibrating Energy Release Rate

In regions of large strain, the same force potentials (3) and (4) are used to calculate the amount of energy consumed by a crack per unit area of growth, i.e., the energy release rate. The energy release rate equals the work necessary to eliminate force interaction on either side of a fracture surface per unit fracture area. In this model the energy release rate has two components: one associated with the force potential for tensile strain (3) and the other associated with the force potential for hydrostatic strain (4). The critical energy release rate $\mathcal{G}s$ associated with fracture under tensile forces is found to be the same for all choices of horizon δ. However the critical energy release rate for hydrostatic fracture $\mathcal{G}_h$ increases with decreasing horizon and becomes infinite as $\delta \to 0$ at the rate $1/\delta$.

For tensile forces we use (3) and calculate the work required to eliminate interaction between two points x and y; this is given by $\mathcal{W}^\delta(\infty) = \lim_{S\to\infty} \mathcal{W}^\delta(S)$ where $\mathcal{W}^\delta(\infty) = \omega^\delta(|y-x|) f_\infty/\delta$. We suppose x gives the center of the peridynamic neighborhood located a distance z away from the planar interface separating upper and lower half spaces. We suppose x lies in the lower half space, and the points y lie in the upper half space inside the peridynamic neighborhood of x (see Fig. 4). The critical energy release rate $\mathcal{G}_s$ associated with tensile forces equals the work necessary to eliminate force interaction on either side of a fracture surface per unit fracture area. It is given in three dimensions by integration of $\mathcal{W}^\delta(\infty)$ over

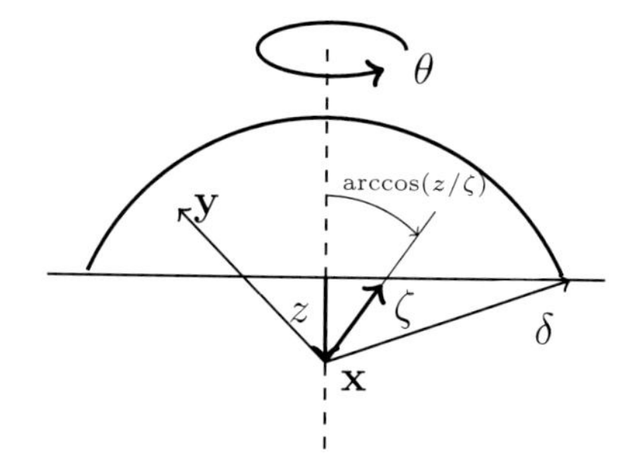

Fig. 4 Evaluation of energy release rate $\mathcal{G}_s$. For each point x along the dashed line, $0 \leq z \leq \delta$, the work required to break the interaction between x and y in the spherical cap is summed up in (96) using spherical coordinates centered at x

the intersection of the neighborhood of x and the upper half space given by the spherical cap (see Fig. 4),

$$\mathcal{G}_s = \frac{4\pi}{V_\delta} \int_0^\delta \int_z^\delta \int_0^{\cos^{-1}(z/\zeta)} \mathcal{W}^\delta(\infty, \zeta)\zeta^2 \sin\phi \, d\phi \, d\zeta \, dz \tag{96}$$

where $\zeta = |y - x|$. This integral is calculated and for d dimensions $d = 1, 2, 3$, the result is

$$\mathcal{G}_s = M \frac{2\omega_{d-1}}{\omega_d} f_\infty, \tag{97}$$

where $M = \int_0^1 r^d \omega(r) dr$ and ω_d is the volume of the d dimensional unit ball, $\omega_1 = 2, \omega_2 = \pi, \omega_3 = 4\pi/3$. We see from this calculation that the critical energy release rate is independent of δ.

For hydrostatic forces we use (4) and calculate the work required to eliminate interaction between x and the upper half plane. As before we suppose x gives the center of the peridynamic neighborhood located a distance z away from the planar interface separating upper and lower half spaces. We suppose x lies in the lower half space, and the peridynamic neighborhood of x intersects the upper half space (see Fig. 5).

The critical energy release rate $\mathcal{G}_h$ associated with hydrostatic forces equals the work necessary to eliminate force interaction on either side of a fracture surface per unit fracture area. The work per unit volume needed to eliminate hydrostatic interaction between a point x and its neighbors is

$$\mathcal{V}^\delta(\infty)(x) = \lim_{\theta \to \infty} \frac{\beta g(\theta)}{\delta^2} = \frac{\beta g_\infty}{\delta^2}. \tag{98}$$

For points $x = (0, 0, z)$, with $0 < |z| < \delta$ above and below the $z = 0$ plane, the work per unit area to eliminate hydrostatic interaction between the lower half space $z < 0$ and upper half space $z > 0$ is

$$\mathcal{G}_h = 2 \int_0^\delta \frac{\beta g_\infty}{\delta^2} dz = \frac{2\beta g_\infty}{\delta}. \tag{99}$$

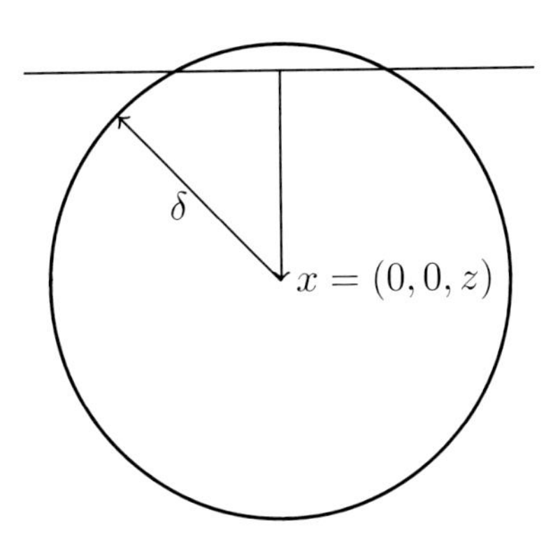

Fig. 5 Hydrostatic energy release rate $\mathcal{G}_h$

For d dimensions $d = 1, 2, 3$, the result is the same and

$$\mathcal{G}_s = \frac{2\beta g_\infty}{\delta}. \tag{100}$$

We see from this calculation that the energy release rate for hydrostatic fracture is increasing at the rate $1/\delta$.

Linear Elastic Operators in the Limit of Vanishing Horizon

In this section we consider smooth evolutions u in space and show that away from fracture set the operators $\mathcal{L}^T + \mathcal{L}^D$ acting on u converge to the operator of linear elasticity in the limit of vanishing nonlocality. We denote the fracture set by $\tilde{D}$ and consider any open un-fractured set D' interior to D with its boundary a finite distance away from the boundary of D and the fracture set $\tilde{D}$. In what follows we suppose that the nonlocal horizon δ is smaller than the distance separating the boundary of D' from the boundaries of D and $\tilde{D}$.

Theorem 4. *Convergence to linear elastic operators. Suppose that* $u(x,t) \in C^2([0,T_0], C^3(D,\mathbb{R}^3))$ *and for every* $x \in D' \subset D \setminus \tilde{D}$*, then there is a constant* $C > 0$ *independent of nonlocal horizon* δ *such that, for every* (x,t) *in* $D' \times [0,T_0]$*, one has*

$$|\mathcal{L}^T(u(t)) + \mathcal{L}^D(u(t)) - \nabla \cdot \mathbb{C}\,\mathcal{E}(u(t))| < C\delta, \tag{101}$$

where the elastic strain is $\mathcal{E}(u) = (\nabla u + (\nabla u)^T)/2$ *and the elastic tensor is isotropic and given by*

$$\mathbb{C}_{ijkl} = 2\bar{\mu}\left(\frac{\delta_{ik}\delta_{jl} + \delta_{il}\delta_{jk}}{2}\right) + \bar{\lambda}\delta_{ij}\delta_{kl}, \tag{102}$$

with shear modulus $\bar{\mu}$ *and Lamé coefficient* $\bar{\lambda}$ *given by* (92) *and* (93). *The numbers* α *and* β *can be chosen independently and can be any pair of real numbers such that* $\mathbb{C}$ *is positive definite.*

Proof. We start by showing

$$|\mathcal{L}^T(u(t)) - \frac{f''(0)}{2\omega_3}\int_{B_1(0)} e|\xi|J(|\xi|)e_i e_j e_k \, d\xi \, \partial^2_{jk} u_i(x)| < C\delta, \tag{103}$$

where $\omega_3 = 4\pi/3$ and $e = e_{y-x}$ are unit vectors on the sphere; here repeated indices indicate summation. To see this recall the formula for $\mathcal{L}^T(u)$ and write $\partial_S f(\sqrt{|y-x|}S) = f'(\sqrt{|y-x|}S)\sqrt{|y-x|}$. Now Taylor expand

$f'(\sqrt{|y-x|}S)$ in $\sqrt{|y-x|}S$ and Taylor expand $u(y)$ about x, denoting e_{y-x} by e to find that all odd terms in e integrate to zero and

$$|\mathcal{L}^T(u(t))_l - \frac{2}{V_\delta}\int_{B_\delta(x)} \frac{J^\delta(|y-x|)}{\delta|y-x|}\frac{f''(0)}{4}|y-x|^2\partial^2_{jk}u_i(x)e_ie_je_ke_l, dy| \\ < C\delta, l=1,2,3. \tag{104}$$

On changing variables $\xi = (y-x)/\delta$, we recover (103). Now we show

$$|\mathcal{L}^D(u(t))_k - \frac{1}{\omega_3}\int_{B_1(0)} |\xi|\omega(|\xi|)e_ie_j\, d\xi \frac{\beta g''(0)}{2\omega_3}\int_{B_1(0)} |\xi|\omega(|\xi|)e_ke_l\, d\xi \partial^2_{lj}u_i(x)| \\ < C\delta, k=1,2,3. \tag{105}$$

We note for $x \in D'$ that $D \cap B_\delta(x) = B_\delta(x)$ and the integrand in the second term of (14) is odd and the integral vanishes. For the first term in (14), we Taylor expand $\partial_\theta g(\theta)$ about $\theta = 0$ and Taylor expand $u(z)$ about y inside $\theta(y,t)$ noting that terms odd in $e = e_{z-y}$ integrate to zero to get

$$|\partial_\theta g(\theta(y,t)) - g''(0)\frac{1}{V_\delta}\int_{B_\delta(y)} \omega^\delta(|z-y|)|z-y|\partial_j u_i(y)e_ie_j\, dz| < C\delta^3. \tag{106}$$

Now substitution for the approximation to $\partial_\theta g(\theta(y,t))$ in the definition of $\mathcal{L}^D$ gives

$$\Big|\mathcal{L}^D(u)\frac{1}{V_\delta}\int_{B_\delta(x)} \frac{\omega^\delta(|y-x|)}{\delta^2}e_{y-x}\frac{1}{2V_\delta}\int_{B_\delta(y)} \omega^\delta(|z-y|)|z-y| \\ \beta g''(0)\partial_j u_i(y)e_ie_j\, dz\, dy\Big| < C\delta. \tag{107}$$

We Taylor expand $\partial_j u_i(y)$ about x; note that odd terms involving tensor products of e_{y-x} vanish when integrated with respect to y in $B_\delta(x)$, and we obtain (105).

We now calculate as in (Lipton 2016 equation (64) or in section "Calibrating the Peridynamic Energy to Elastic Properties" to find that

$$\frac{f''(0)}{2\omega_3}\int_{B_1(0)} |\xi|J(|\xi|)e_ie_je_ke_l\, d\xi\partial^2_{jk}u_i(x) \\ = \left(2\mu_1\left(\frac{\delta_{ik}\delta_{jl}+\delta_{il}\delta_{jk}}{2}\right) + \lambda_1\delta_{ij}\delta_{kl}\right)\partial^2_{jk}u_i(x), \tag{108}$$

where

$$\mu_1 = \lambda_1 = \frac{f''(0)}{10}\int_0^1 r^3\omega(r)\, dr. \tag{109}$$

Next observe that a straightforward calculation gives

$$\frac{1}{\omega_3}\int_{B_1(0)}|\xi|\omega(|\xi|)e_i e_j\,d\xi=\delta_{ij}\int_0^1 r^3\omega(r)\,dr, \tag{110}$$

and we deduce that

$$\begin{aligned}&\frac{1}{\omega_3}\int_{B_1(0)}|\xi|\omega(|\xi|)e_i e_j\,d\xi\frac{\beta g''(0)}{2\omega_3}\int_{B_1(0)}|\xi|\omega(|\xi|)e_k e_l\,d\xi\partial^2_{lj}u_i(x)\\&=\frac{\beta g''(0)}{2}\left(\int_0^1 r^3\omega(r)\,dr\right)^2\delta_{ij}\delta_{kl}\partial^2_{lj}u_i(x).\end{aligned} \tag{111}$$

Theorem 4 follows on adding (108) and (111) □

Conclusions

We have introduced a regularized model for free fracture propagation based on nonlocal potentials. At each instant of the evolution, we identify the softening zone where strains lie above the strength of the material. We have shown that discontinuities associated with flaws larger than the length scale of nonlocality δ can become unstable and grow. An explicit inequality is found that shows that the volume of the softening zone goes to zero linearly with the length scale of nonlocal interaction. This scaling is consistent with the notion that a softening zone of width proportional to δ converges to a sharp fracture as the length scale of nonlocal interaction goes to zero. Inside quiescent regions with no cracks, the nonlocal operator converges to the local elastic operator at a rate proportional to the radius of nonlocal interaction. We show that the model can be calibrated to measured values of critical energy release rate, shear modulus, and bulk modulus of material samples. The double-well state-based potential developed here no longer has Poisson ratio constrained to be $1/4$. For this model we can choose the potentials so that the small strain behavior is specified by the isotropic elasticity tensor for any material with prescribed shear and Lamé moduli.

The energy release rate necessary for tensile forces to create fractures is constant in δ, whereas the forces necessary to create a fracture using hydrostatic forces grows as $1/\delta$. Thus creation of fracture surfaces by hydrostatic forces will not be seen when

$$\delta<\frac{2\beta g_\infty}{G_s}. \tag{112}$$

On the other hand, the elastic properties for small strains can be made to correspond to any positive definite isotropic elastic tensor.

Acknowledgements This material is based upon the work supported by the US Army Research Laboratory and the US Army Research Office under contract/grant number W911NF1610456.

References

A. Agwai, I. Guven, E. Madenci, Predicting crack propagation with peridynamics: a comparative study. Int. J. Fract. **171**, 65–78 (2011)

F. Bobaru, W. Hu, The meaning, selection, and use of the peridynamic horizon and its relation to crack branching in brittle materials. Int. J. Fract. **176**, 215–222 (2012)

F. Bobaru, J.T. Foster, P.H. Geubelle, S.A. Silling, *Handbook of Peridynamic Modeling* (CRC Press, Boca Raton, 2016)

K. Dayal, K. Bhattacharya, Kinetics of phase transformations in the peridynamic formulation of continuum mechanics. J. Mech. Phys. Solids **54**, 1811–1842 (2006)

P. Diehl, R. Lipton, M.A. Schweitzer, Numerical verification of a bond-based softening peridynamic model for small displacements: deducing material parameters from classical linear theory. Institut für Numerische Simulation Preprint No. 1630 (2016)

B.K. Driver, *Analysis Tools with Applications*. E-book (Springer, Berlin, 2003)

Q. Du, M. Gunzburger, R. Lehoucq, K. Zhou, Analysis of the volume-constrained peridynamic Navier equation of linear elasticity. J. Elast. **113**, 193–217 (2013)

J.T. Foster, S.A. Silling, W. Chen, An energy based failure criterion for use with peridynamic states. Int. J. Multiscale Comput. Eng. **9**, 675–688 (2011)

W. Gerstle, N. Sau, S. Silling, Peridynamic modeling of concrete structures. Nucl. Eng. Des. **237**, 1250–1258 (2007)

Y.D. Ha, F. Bobaru, Studies of dynamic crack propagation and crack branching with peridynamics. Int. J. Fract. **162**, 229–244 (2010)

I. Jasiuk, J. Chen, M.F. Thorpe, Elastic moduli of two dimensional materials with polygonal and elliptical holes. Appl. Mech. Rev. **47**, S18–S28 (1994)

R. Lipton, Dynamic brittle fracture as a small horizon limit of peridynamics. J. Elast. **117**, 21–50 (2014)

R. Lipton, Cohesive dynamics and brittle fracture. J. Elast. **124**(2), 143–191 (2016)

R. Lipton, S. Silling, R. Lehoucq, Complex fracture nucleation and evolution with nonlocal elastodynamics. arXiv preprint arXiv:1602.00247 (2016)

G.W. Milton, *The Theory of Composites* (Cambridge University Press, Cambridge, 2002)

S.A. Silling, Reformulation of elasticity theory for discontinuities and long-range forces. J. Mech. Phys. Solids **48**, 175–209 (2000)

S.A. Silling, E. Askari, A meshfree method based on the peridynamic model of solid mechanics. Comput. Struct **83**, 1526–1535 (2005)

S.A. Silling, F. Bobaru, Peridynamic modeling of membranes and fibers. Int. J. Nonlinear Mech. **40**, 395–409 (2005)

S.A. Silling, R.B. Lehoucq, Convergence of peridynamics to classical elasticity theory. J. Elast. **93**, 13–37 (2008)

S. A. Silling, M. Epton, O. Weckner, J. Xu, E. Askari, Peridynamic states and constitutive modeling. J. Elast. **88**, 151–184 (2007)

S. Silling, O. Weckner, E. Askari, F. Bobaru, Crack nucleation in a peridynamic solid. Int. J. Fract. **162**, 219–227 (2010)

O. Weckner, R. Abeyaratne, The effect of long-range forces on the dynamics of a bar. J. Mech. Phys. Solids **53**, 705–728 (2005)

Nonlocal Operators with Local Boundary Conditions: An Overview

39

Burak Aksoylu, Fatih Celiker, and Orsan Kilicer

Contents

Abstract

We present novel governing operators in arbitrary dimension for nonlocal diffusion in homogeneous media. The operators are inspired by the theory of peridynamics (PD). They agree with the original PD operator in the bulk of the domain and simultaneously enforce local boundary conditions (BC). The main ingredients are periodic, antiperiodic, and mixed extensions of kernel functions together with even and odd parts of bivariate functions. We present different types of BC in 2D which include pure and mixed combinations of Neumann and Dirichlet BC. Our construction is systematic and easy to follow. We provide numerical experiments that validate our theoretical findings. When our novel operators are extended to vector-valued functions, they will allow the

---

Mathematics Subject Classification (2000): 35L05, 74B99, 47G10.

B. Aksoylu (✉) · F. Celiker · O. Kilicer
Department of Mathematics, Wayne State University, Detroit, MI, USA
e-mail: burak@wayne.edu; celiker@wayne.edu; okilicer@math.tamu.edu

G. Z. Voyiadjis (ed.), *Handbook of Nonlocal Continuum Mechanics for Materials and Structures*, https://doi.org/10.1007/978-3-319-58729-5_34

extension of PD to applications that require local BC. Furthermore, we hope that the ability to enforce local BC provides a remedy for surface effects seen in PD.

We recently proved that the nonlocal diffusion operator is a function of the classical operator. This observation opened a gateway to incorporate local BC to nonlocal problems on bounded domains. The main tool we use to define the novel governing operators is functional calculus, in which we replace the classical governing operator by a suitable function of it. We present how to apply functional calculus to general nonlocal problems in a methodical way.

Keywords
Nonlocal wave equation · Nonlocal operator · Peridynamics · Boundary condition · Integral operator

Introduction

We construct novel governing operators for nonlocal diffusion (Andreu-Vaillo et al. 2010; Du et al. 2012) in arbitrary dimension. The operators are inspired by the theory of peridynamics (PD), a nonlocal extension of continuum mechanics developed by Silling (2000). By suppressing the time variable t, we take the following operator as the original governing operator, and, in 1D, it corresponds to the original bond-based PD operator for homogeneous media. We choose the 2D domain as $\Omega := [-1,1] \times [-1,1]$ and for $(x,y) \in \Omega$,

$$\begin{aligned}\mathcal{L}_{\text{orig}}u(x,y) := &\iint_{\Omega} \widehat{C}(x'-x, y'-y)u(x,y)dx'dy' \\ &- \iint_{\Omega} \widehat{C}(x'-x, y'-y)u(x',y')dx'dy'. \end{aligned} \tag{1}$$

We define the operator that is closely related to $\mathcal{L}_{\text{orig}}$ as

$$\mathcal{L}u(x,y) := cu(x,y) - \iint_{\Omega} \widehat{C}(x'-x, y'-y)u(x',y')dx'dy', \quad (x,y) \in \Omega. \tag{2}$$

where $C := \widehat{C}|_{\Omega}$ and $c := \iint_{\Omega} C(x',y')dx'dy'$. Since $\widehat{C}$ enters into the formulation only as a function of $(x'-x, y'-y)$, the operator $\mathcal{L}$ also assumes a homogeneous medium. We will show that the two operators agree in the bulk. As the main contribution, we prove that the novel governing operators we construct agree with

$\mathcal{L}_{\text{orig}}$ in the bulk of Ω and, at the same time, enforce local pure and mixed Neumann, Dirichlet, periodic, and antiperiodic BC.

When PD is considered, the dimension of u must be equal to that of x. In that case, the governing operator in (1) restricted to 1D corresponds to the bond-based linearized PD; see Silling et al. (2003, Eq. 23) and Weckner and Abeyaratne (2005, Eq. 3). For the discussion of PD, it is implied that $u, x \in \mathbb{R}$. The case of $u \in \mathbb{R}$ and $x \in \mathbb{R}^d$ corresponds to nonlocal diffusion (Du et al. 2012; Seleson et al. 2013).

Our approach to nonlocal problems is fundamentally different because we exclusively want to use local BC. In Beyer et al. (2016), one of our major results was the finding that the governing operator of PD equation in $\mathbb{R}$ and nonlocal diffusion in $\mathbb{R}^d$ are functions of the Laplace operator. This result opened the path to the introduction of local boundary conditions into PD theory. Since PD is a nonlocal theory, one might expect only the appearance of nonlocal BC while employing $\mathcal{L}_{\text{orig}}$ as the governing operator. In the original PD formulation, the concept of local BC does not apply to PD. Instead, external forces must be supplied through the loading force density (Silling 2000, p.201). On the other hand, we demonstrate that the anticipation that local BC are incompatible with nonlocal operators is not quite correct. Our novel operators present an alternative to nonlocal BC, and we hope that the ability to enforce local BC will provide a remedy for surface effects seen in PD; see Madenci and Oterkus (2014, Chaps. 4, 5, 7, and 12) and Kilic (2008), Mitchell et al. (2015). Furthermore, our approach will provide us the capability to solve important elasticity problems that require local BC such as contact, shear, and traction.

We studied various aspects of local BC in nonlocal problems (Aksoylu et al. 2017a,b, 2016, 2017, Submitted; Beyer et al. 2016). Building on Beyer et al. (2016), we generalized the results in $\mathbb{R}$ to bounded domains (Aksoylu et al. 2017a,b), a critical feature for all practical applications. In Aksoylu et al. (2017b), we laid the theoretical foundations, and in Aksoylu et al. (2017a), we applied the foundations to prominent BC such as Dirichlet and Neumann, as well as presented numerical implementation of the corresponding wave propagation. In Aksoylu et al. (2017), we constructed the first 1D operators that agree with the original bond-based PD operator in the bulk of the domain and simultaneously enforce local Neumann and Dirichlet BC which we denote by $\mathcal{M}_{\text{N}}$ and $\mathcal{M}_{\text{D}}$, respectively. We carried out numerical experiments by utilizing $\mathcal{M}_{\text{N}}$ and $\mathcal{M}_{\text{D}}$ as governing operators in Aksoylu et al. (2017a). In Aksoylu et al. (2016), we studied other related governing operators. In Aksoylu and Kaya (2018), we study the condition numbers of the novel governing operators. Therein, we prove that the modifications made to the operator $\mathcal{L}_{\text{orig}}$ to obtain the novel operators are minor as far as the condition numbers are concerned.

Our approach is not limited to PD; the abstractness of the theoretical methods used allows generalization to other nonlocal theories. Our approach presents a unique way of combining the powers of abstract operator theory with numerical computing (Aksoylu et al. 2017a). Similar classes of operators are used in numerous applications such as nonlocal diffusion (Andreu-Vaillo et al. 2010; Du et al. 2012; Seleson et al. 2013), image processing (Gilboa and Osher 2008), population models, particle systems, phase transition, and coagulation. See the review and news articles

Du et al. (2012, 2014), and Silling and Lehoucq (2010) for a comprehensive discussion and the book (Madenci and Oterkus 2014). In addition, see the studies dedicated to conditioning analysis, domain decomposition and variational theory (Aksoylu and Kaya 2018; Aksoylu and Mengesha 2010; Aksoylu and Parks 2011; Aksoylu and Unlu 2014), discretization (Aksoylu and Unlu 2014; Emmrich and Weckner 2007; Tian and Du 2013), and kernel functions (Mengesha and Du 2013; Seleson and Parks 2011).

The rest of the paper is structured as follows. In section "The Novel Operators in 2D," first we prove that the operators $\mathcal{L}_{\text{orig}}$ and $\mathcal{L}$ agree in the bulk in 2D. We define the novel operators using orthogonal projections on bivariate functions for which we utilize the periodic, antiperiodic, and mixed extensions of the kernel function $C(x)$. We give the main theorem in 2D. In section "Operators in 1D", first we prove that the novel operators are self-adjoint. In 1D, we give the main theorem which states they all agree with $\mathcal{L}_{\text{orig}}$ in the bulk and simultaneously enforce the corresponding local BC. In section "The Construction of 2D Operators," we exploit the properties of the operators in 1D to construct the novel operators in 2D. We transfer the agreement in the bulk property established for univariate functions to bivariate ones and eventually prove that the novel operators agree with $\mathcal{L}_{\text{orig}}$ in the bulk in 2D. In section "Verifying the Boundary Conditions," we make use of the Leibniz rule, the Fubini theorem, and the Lebesgue dominated convergence theorem to prove that the novel operators enforce the local BC stated in the main theorem. In section "Operators in Higher Dimensions," we present the operators in 3D which can be easily extended to arbitrary dimension. In section "Numerical Experiments," we report the numerical experiments. In section "The Treatment of General Nonlocal Problems Using Functional Calculus," we present the treatment of general nonlocal problems using functional calculus. We conclude in section "Conclusion."

The Novel Operators in 2D

For $(x, y), (x', y') \in \Omega$[1], it follows that $(x' - x, y' - y) \in [-2, 2] \times [-2, 2]$. Hence, in (1), the domain of $\widehat{C}(x' - x, y' - y)$ is $\widehat{\Omega} := [-2, 2] \times [-2, 2]$. Furthermore, the kernel function $\widehat{C}(x, y)$ is assumed to be even. Namely,

$$\widehat{C}(-x, -y) = \widehat{C}(x, y).$$

The important choice of $\widehat{C}(x, y)$ is the *canonical* kernel function $\widehat{\chi}_\delta(x, y)$ whose only role is the representation of the nonlocal neighborhood, called the *horizon*, by a characteristic function. More precisely, for $(x, y) \in \widehat{\Omega}$,

[1]We do not explicitly denote the dimension of the domain Ω. The dimension is implied by the number of iterated integrals present in the operator. The domain Ω is equal to $[-1, 1]$, $[-1, 1] \times [-1, 1]$, and $[-1, 1] \times [-1, 1] \times [-1, 1]$ in 1D, 2D, and 3D, respectively.

$$\widehat{\chi}_\delta(x,y) := \begin{cases} 1, & (x,y) \in (-\delta,\delta)\times(-\delta,\delta) \\ 0, & \text{otherwise.} \end{cases} \tag{3}$$

The size of nonlocality is determined by δ and we assume $\delta < 1$. Since the horizon is constructed by $\widehat{\chi}_\delta(x,y)$, a kernel function used in practice is in the form

$$\widehat{C}(x,y) = \widehat{\chi}_\delta(x,y)\widehat{\mu}(x,y), \tag{4}$$

where $\widehat{\mu}(x,y) \in L^2(\widehat{\Omega})$ is even.

Throughout the paper, we assume that

$$u(x,y) \in L^2(\Omega) \cap C^1(\partial\Omega). \tag{5}$$

Inspired by the projections that give the even and odd parts of a univariate function, we define the following operators that act on a bivariate function:

$$P_{e,x'}, P_{o,x'}, P_{e,y'}, P_{o,y'} : L^2(\Omega) \to L^2(\Omega),$$

whose definitions are

$$P_{e,x'}u(x',y') := \frac{u(x',y')+u(-x',y')}{2}, \; P_{o,x'}u(x',y') := \frac{u(x',y')-u(-x',y')}{2}, \tag{6}$$

$$P_{e,y'}u(x',y') := \frac{u(x',y')+u(x',-y')}{2}, \; P_{o,y'}u(x',y') := \frac{u(x',y')-u(x',-y')}{2}. \tag{7}$$

Each operator is an orthogonal projection and possesses the following decomposition property:

$$P_{e,x'} + P_{o,x'} = I_{x'}, \quad P_{e,y'} + P_{o,y'} = I_{y'}. \tag{8}$$

One can easily check that all four orthogonal projections in (6) and (7) commute with each other. We define the following new operators obtained from the products of these projections:

$$P_{e,x'}P_{e,y'}u(x',y') := \frac{1}{4}\{[u(x',y') + u(x',-y')] + [u(-x',y') + u(-x',-y')]\},$$

$$P_{e,x'}P_{o,y'}u(x',y') := \frac{1}{4}\{[u(x',y') - u(x',-y')] + [u(-x',y') - u(-x',-y')]\},$$

$$P_{o,x'}P_{o,y'}u(x',y') := \frac{1}{4}\{[u(x',y') - u(x',-y')] - [u(-x',y') - u(-x',-y')]\},$$

$$P_{o,x'}P_{e,y'}u(x',y') := \frac{1}{4}\{[u(x',y') + u(x',-y')] - [u(-x',y') + u(-x',-y')]\}.$$

Due to the aforementioned commutativity property, these are also orthogonal projections and satisfy the following decomposition property:

$$P_{e,x'}P_{e,y'} + P_{e,x'}P_{o,y'} + P_{o,x'}P_{e,y'} + P_{o,x'}P_{o,y'} = I_{x',y'}. \tag{9}$$

These will be used in the definition of the novel operators in 2D.

Reflecting on the square support of the restricted kernel function $\chi_\delta(x, y)$, we define the bulk of the domain as follows:

$$\text{bulk} = \{(x, y) \in \Omega : (x, y) \in (-1 + \delta, 1 - \delta) \times (-1 + \delta, 1 - \delta)\}.$$

We first prove that the operators $\mathcal{L}$ and $\mathcal{L}_{\text{orig}}$ agree in the bulk. Throughout the paper, we denote the restriction of a function $\widehat{Z} : \widehat{\Omega} \to \mathbb{R}$ to Ω as Z, i.e., $Z := \widehat{Z}|_\Omega$.

Lemma 1.

$$\mathcal{L}u(x, y) = \mathcal{L}_{orig}u(x, y), \quad (x, y) \in bulk.$$

Proof. For $(x, y) \in$ bulk, we have

$$(x - \delta, x + \delta) \times (y - \delta, y + \delta) \cap \Omega = (x - \delta, x + \delta) \times (y - \delta, y + \delta).$$

Hence,

$$\iint_\Omega \widehat{C}(x' - x, y' - y)dx'dy' = \iint_\Omega \widehat{\chi}_\delta(x' - x, y' - y)\widehat{\mu}(x' - x, y - y')dx'dy'$$

$$= \int_{x-\delta}^{x+\delta}\int_{y-\delta}^{y+\delta} \widehat{\mu}(x' - x, y' - y)dx'dy' = \int_{-\delta}^{\delta}\int_{-\delta}^{\delta} \mu(x', y')dx'dy'$$

$$= \iint_\Omega \chi_\delta(x', y')\mu(x', y')dx'dy' = \iint_\Omega C(x', y')dx'dy'.$$

The result follows. □

In the construction of the novel operators, a crucial ingredient is first restricting $\widehat{C}$ to Ω and then suitably extending it back to $\widehat{\Omega}$. To this end, we define the periodic, antiperiodic, and mixed extensions of $C(x)$ from $[-1, 1]$ to $[-2, 2]$, respectively, as follows:

$$\widehat{C}_{\mathrm{p}}(x) := \begin{cases} C(x+2), & x \in [-2,-1), \\ C(x), & x \in [-1,1], \\ C(x-2), & x \in (1,2], \end{cases} \qquad \widehat{C}_{\mathrm{a}}(x) := \begin{cases} -C(x+2), & x \in [-2,-1), \\ C(x), & x \in [-1,1], \\ -C(x-2), & x \in (1,2]. \end{cases} \tag{10}$$

We also utilize the following mixed extensions of $C(x)$:

$$\widehat{C}_{\mathrm{pa}}(x) := \begin{cases} C(x+2), & x \in [-2,-1), \\ C(x), & x \in [-1,1], \\ -C(x-2), & x \in (1,2], \end{cases} \qquad \widehat{C}_{\mathrm{ap}}(x) := \begin{cases} -C(x+2), & x \in [-2,-1), \\ C(x), & x \in [-1,1], \\ C(x-2), & x \in (1,2]. \end{cases}$$

Building on our 1D construction in Aksoylu et al. (2017), in higher dimensions, we discovered the operators that enforce local pure and mixed Neumann and Dirichlet BC. We present the main theorem in 2D with the following 4 types of BC.

Theorem 1 (Main Theorem in 2D). *Let $\Omega := [-1,1] \times [-1,1]$ and the restricted kernel function be separable in the form*

$$C(x,y) = X(x)Y(y), \tag{11}$$

where X and Y are even functions. Then, the operators $\mathcal{M}_{\mathrm{N}}$, $\mathcal{M}_{\mathrm{D}}$, $\mathcal{M}_{\mathrm{ND,ND}}$, and $\mathcal{M}_{\mathrm{N,DN}}$ defined by

$$(\mathcal{M}_{\mathrm{N}} - c)u(x,y) := -\iint_{\Omega} \left[\widehat{X}_{\mathrm{p}}(x'-x)P_{e,x'} + \widehat{X}_{\mathrm{a}}(x'-x)P_{o,x'}\right] \left[\widehat{Y}_{\mathrm{p}}(y'-y)P_{e,y'} + \widehat{Y}_{\mathrm{a}}(y'-y)P_{o,y'}\right]u(x',y')dx'\,dy',$$

$$(\mathcal{M}_{\mathrm{D}} - c)u(x,y) := -\iint_{\Omega} \left[\widehat{X}_{\mathrm{a}}(x'-x)P_{e,x'} + \widehat{X}_{\mathrm{p}}(x'-x)P_{o,x'}\right] \left[\widehat{Y}_{\mathrm{a}}(y'-y)P_{e,y'} + \widehat{Y}_{\mathrm{p}}(y'-y)P_{o,y'}\right]u(x',y')dx'\,dy',$$

$$(\mathcal{M}_{\mathrm{ND,ND}} - c)u(x,y) := -\iint_{\Omega} \left[\widehat{X}_{\mathrm{ap}}(x'-x)P_{e,x'} + \widehat{X}_{\mathrm{pa}}(x'-x)P_{o,x'}\right] \left[\widehat{Y}_{\mathrm{ap}}(y'-y)P_{e,y'} + \widehat{Y}_{\mathrm{pa}}(y'-y)P_{o,y'}\right]u(x',y')dx'\,dy',$$

$$(\mathcal{M}_{\mathrm{N,DN}} - c)u(x,y) := -\iint_{\Omega} \left[\widehat{X}_{\mathrm{p}}(x'-x)P_{e,x'} + \widehat{X}_{\mathrm{a}}(x'-x)P_{o,x'}\right] \left[\widehat{Y}_{\mathrm{pa}}(y'-y)P_{e,y'} + \widehat{Y}_{\mathrm{ap}}(y'-y)P_{o,y'}\right]u(x',y')dx'\,dy',$$

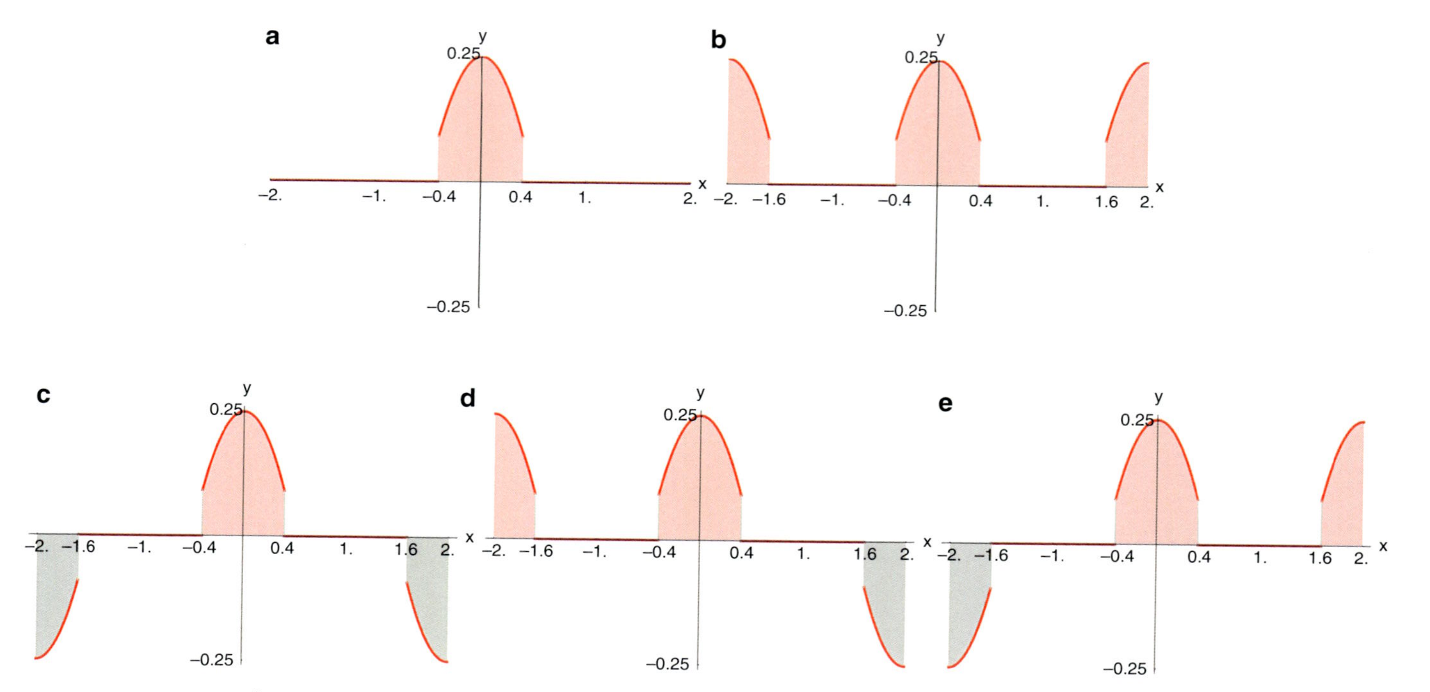

Fig. 1 The kernel function $\widehat{C}(x) = \widehat{\chi}_\delta(x)\widehat{\mu}(x)$ with $\delta = 0.4$ and $\widehat{\mu}(x) = 0.25 - x^2$. The periodic, antiperiodic, and mixed extensions of $C = \widehat{C}|_\Omega$ are denoted by $\widehat{C}_{\mathrm{p}}$, $\widehat{C}_{\mathrm{a}}$, $\widehat{C}_{\mathrm{pa}}$, and $\widehat{C}_{\mathrm{ap}}$, respectively. Plots of **(a)** $\widehat{C}(x)$, **(b)** $\widehat{C}_{\mathrm{p}}(x)$, **(c)** $\widehat{C}_{\mathrm{a}}(x)$, **(d)** $\widehat{C}_{\mathrm{pa}}(x)$, **(e)** $\widehat{C}_{\mathrm{ap}}(x)$

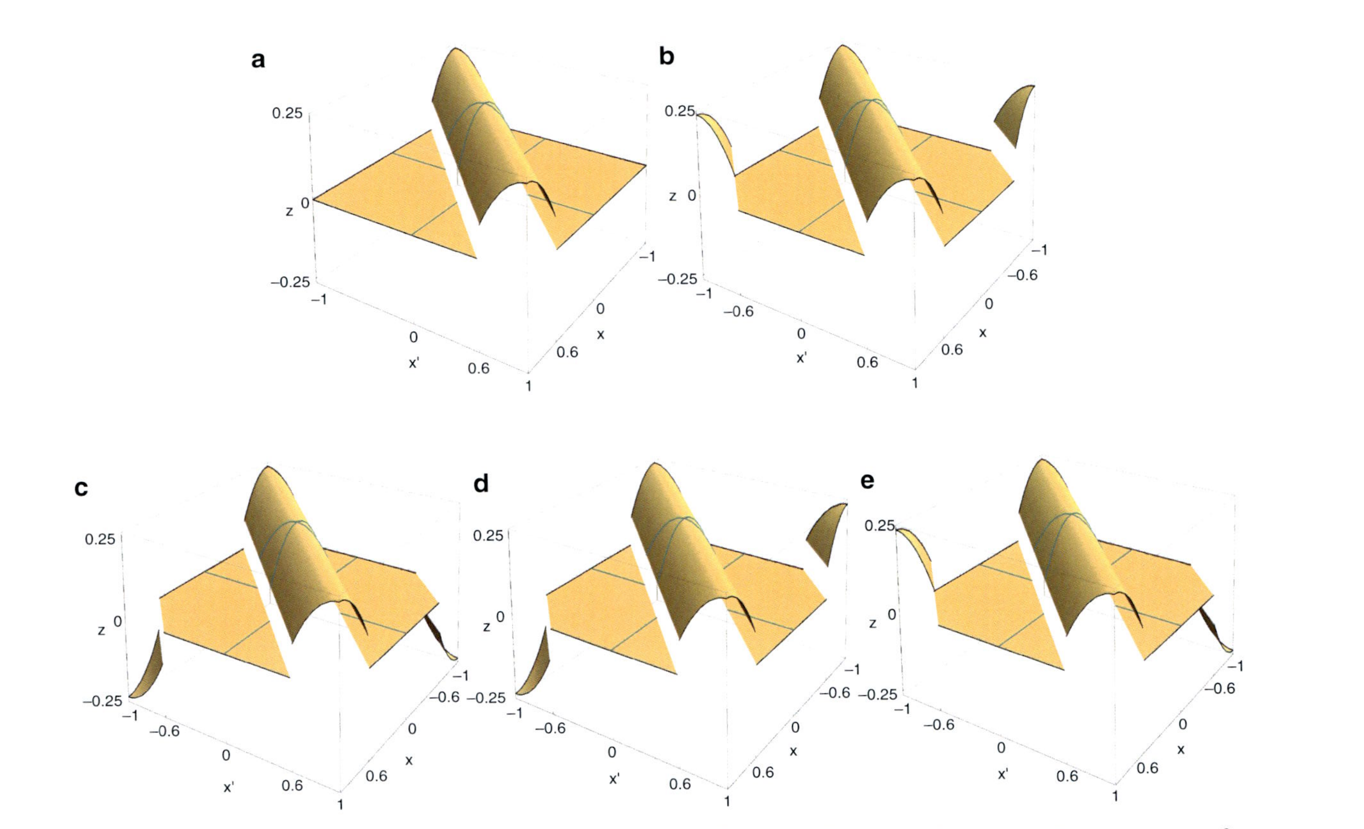

Fig. 2 For the same kernel function $\widehat{C}(x)$ in Fig. 1, we employ bivariate versions of $\widehat{C}(x'-x)$, $\widehat{C}_{\mathrm{p}}(x'-x)$, $\widehat{C}_{\mathrm{a}}(x'-x)$, $\widehat{C}_{\mathrm{pa}}(x'-x)$, and $\widehat{C}_{\mathrm{ap}}(x'-x)$ with the definition $\widehat{C}_{\mathrm{BC}}(x,x') := \widehat{C}_{\mathrm{BC}}(x'-x)$. Plots of **(a)** $\widehat{C}(x,x')$, **(b)** $\widehat{C}_{\mathrm{p}}(x,x')$, **(c)** $\widehat{C}_{\mathrm{a}}(x,x')$, **(d)** $\widehat{C}_{\mathrm{pa}}(x,x')$, **(e)** $\widehat{C}_{\mathrm{ap}}(x,x')$

agree with $\mathcal{L}u(x,y)$ *in the bulk, i.e., for* $(x,y) \in (-1+\delta, 1-\delta) \times (-1+\delta, 1-\delta)$. *Furthermore, the operators* $\mathcal{M}_{\mathrm{N}}$ *and* $\mathcal{M}_{\mathrm{D}}$ *enforce pure Neumann and Dirichlet BC, respectively:*

$$\frac{\partial}{\partial n}\big[(\mathcal{M}_{\mathrm{N}} - c)u\big](x, \pm 1) = \frac{\partial}{\partial n}\big[(\mathcal{M}_{\mathrm{N}} - c)u\big](\pm 1, y) = 0,$$

$$(\mathcal{M}_{\mathrm{D}} - c)u(x, \pm 1) = (\mathcal{M}_{\mathrm{D}} - c)u(\pm 1, y) = 0.$$

The operators $\mathcal{M}_{\mathrm{ND,ND}}$ *and* $\mathcal{M}_{\mathrm{N,DN}}$ *enforce mixed Neumann and Dirichlet BC, respectively, in the following way:*

$$\frac{\partial}{\partial n}\big[(\mathcal{M}_{\mathrm{ND,ND}} - c)u\big](-1, y) = \frac{\partial}{\partial n}\big[(\mathcal{M}_{\mathrm{ND,ND}} - c)u\big](x, -1) = 0,$$

$$(\mathcal{M}_{\mathrm{ND,ND}} - c)u(+1, y) = (\mathcal{M}_{\mathrm{ND,ND}} - c)u(x, +1) = 0,$$

and

$$\frac{\partial}{\partial n}\big[(\mathcal{M}_{\mathrm{N,DN}} - c)u\big](\pm 1, y) = \frac{\partial}{\partial n}\big[(\mathcal{M}_{\mathrm{N,DN}} - c)u\big](x, +1) = 0,$$
$$(\mathcal{M}_{\mathrm{N,DN}} - c)u(x, -1) = 0.$$

Proof. The proofs of agreement in the bulk and the verification of BC are given in sections "The Construction of 2D Operators" and "Verifying the Boundary Conditions," respectively. □

Remark 1. Although we assume a separable kernel function as in (11), note that we do not impose a separability assumption on the solution $u(x, y)$.

Operators in 1D

The construction in higher dimensions is inspired by the one in 1D. Hence, it is more instructive to provide the construction in 1D. For the convolution present in the governing operators, we use a shorthand notation and define

$$\mathcal{C}u(x) := \int_{\Omega} \widehat{C}(x' - x)u(x')dx'.$$

Furthermore, for each extension type utilized, we define the following operators which will be useful in the exposition. Following the construction in Aksoylu et al. (2017a), we assume that $u, C \in L^2(\Omega)$ and define

$$\mathcal{C}_{\mathrm{p}}u(x) := \int_{\Omega} \widehat{C}_{\mathrm{p}}(x'-x)u(x')dx', \quad \mathcal{C}_{\mathrm{a}}u(x) := \int_{\Omega} \widehat{C}_{\mathrm{a}}(x'-x)u(x')dx', \tag{12}$$

$$\mathcal{C}_{\mathrm{pa}}u(x) := \int_{\Omega} \widehat{C}_{\mathrm{pa}}(x'-x)u(x')dx', \quad \mathcal{C}_{\mathrm{ap}}u(x) := \int_{\Omega} \widehat{C}_{\mathrm{ap}}(x'-x)u(x')dx'. \tag{13}$$

The only difference in the operators $\mathcal{C}_{\mathrm{p}}, \mathcal{C}_{\mathrm{a}}, \mathcal{C}_{\mathrm{pa}}$, and $\mathcal{C}_{\mathrm{ap}}$ is the extension utilized for the kernel functions. We prove that the operators agree in the bulk by investigating how the corresponding kernel functions behave in the bulk.

Lemma 2. *Let the kernel function $\widehat{C}(x)$ be in the form*

$$\widehat{C}(x) = \widehat{\chi}_{\delta}(x)\widehat{\mu}(x),$$

where $\widehat{\mu}(x) \in L^2(\widehat{\Omega})$ is even. Let $\widehat{C}_{\mathrm{p}}(x)$, $\widehat{C}_{\mathrm{a}}(x)$, $\widehat{C}_{\mathrm{pa}}(x)$, and $\widehat{C}_{\mathrm{ap}}(x)$ denote the periodic, antiperiodic, and mixed extensions of $C(x)$ to $\widehat{\Omega}$, respectively. Then,

$$\widehat{C}(x) = \widehat{C}_{\mathrm{p}}(x) = \widehat{C}_{\mathrm{a}}(x) = \widehat{C}_{\mathrm{pa}}(x) = \widehat{C}_{\mathrm{ap}}(x), \quad x \in (-2+\delta, 2-\delta).$$

Furthermore, we have the following agreement in the bulk. Namely, for $x \in (-1+\delta, 1-\delta)$,

$$\widehat{C}(x'-x) = \widehat{C}_{\mathrm{p}}(x'-x) = \widehat{C}_{\mathrm{a}}(x'-x) = \widehat{C}_{\mathrm{pa}}(x'-x) = \widehat{C}_{\mathrm{ap}}(x'-x), \quad x' \in [-1,1]. \tag{14}$$

Proof. Let us study the definition of $\widehat{C}_{\mathrm{p}}(x)$ given in (10) by explicitly writing the expression of practical kernel (4) as follows:

$$\widehat{C}_{\mathrm{p}}(x) = \begin{cases} \chi_{\delta}(x+2)\mu(x+2), & x \in [-2,-1), \\ \chi_{\delta}(x)\mu(x), & x \in [-1,1], \\ \chi_{\delta}(x-2)\mu(x-2), & x \in (1,2]. \end{cases}$$

Let us closely look at the first expression in the above definition of $\widehat{C}_{\mathrm{p}}(x)$:

$$\widehat{C}_{\mathrm{p}}(x)|_{x\in[-2,-1)} = \chi_{\delta}(x+2)\mu(x+2). \tag{15}$$

The expression in (15) is equivalent to

$$\widehat{C}_{\mathrm{p}}(x)|_{x\in[-2,-1)} = \begin{cases} \mu(x+2), & x+2 \in (-\delta,\delta) \text{ and } x \in [-2,-1), \\ 0, & x+2 \notin (-\delta,\delta) \text{ and } x \in [-2,-1). \end{cases} \tag{16}$$

Table 1 The value of each extension of the function C

Interval	$\widehat{C}_{\mathrm{p}}(x)$	$\widehat{C}_{\mathrm{a}}(x)$	$\widehat{C}_{\mathrm{pa}}(x)$	$\widehat{C}_{\mathrm{ap}}(x)$
$x \in [-2, -2+\delta)$	$\mu(x+2)$	$-\mu(x+2)$	$\mu(x+2)$	$-\mu(x+2)$
$x \in [-2+\delta, -\delta]$	0	0	0	0
$x \in (-\delta, \delta)$	$\mu(x)$	$\mu(x)$	$\mu(x)$	$\mu(x)$
$x \in [\delta, 2-\delta]$	0	0	0	0
$x \in (2-\delta, 2]$	$\mu(x-2)$	$-\mu(x-2)$	$-\mu(x-2)$	$\mu(x-2)$

Due to the following set equivalence

$$\begin{aligned}&\{x : x+2 \in (-\delta, \delta) \text{ and } x \in [-2, -1)\}\\ &\quad = \{x : x \in (-2-\delta, -2+\delta) \cap [-2, -1) = [-2, -2+\delta)\},\end{aligned}$$

the expression (16) reduces to

$$\widehat{C}_{\mathrm{p}}(x)|_{x\in[-2,-1)} = \begin{cases} \mu(x+2), & x \in [-2, -2+\delta), \\ 0, & x \in [-2+\delta, -1). \end{cases} \tag{17}$$

Similar to (17), for $x \in (1, 2]$, we have

$$\widehat{C}_{\mathrm{p}}(x)|_{x\in(1,2]} = \begin{cases} 0, & x \in (1, 2-\delta], \\ \mu(x-2), & x \in (2-\delta, 2]. \end{cases} \tag{18}$$

Combining (17) and (18), for $x \in [-2, 2]$, we obtain the expression for $\widehat{C}_{\mathrm{p}}(x)$. Similarly, we obtain the expressions for the antiperiodic and the mixed extensions. We collect all the expressions in Table 1.

Clearly, all of the extensions agree for $x \in [-2+\delta, 2-\delta]$. Also, see Figs. 1 and 2. □

Using (14), we immediately obtain the following equivalence of operators in the bulk:

$$\mathcal{C}u(x) = \mathcal{C}_{\mathrm{p}}u(x) = \mathcal{C}_{\mathrm{a}}u(x) = \mathcal{C}_{\mathrm{pa}}u(x) = \mathcal{C}_{\mathrm{ap}}u(x), \quad x \in (-1+\delta, 1-\delta). \tag{19}$$

Even and odd parts of a univariate function $u(x)$ are used in the governing operators $\mathcal{M}_{\mathrm{N}}$, $\mathcal{M}_{\mathrm{D}}$, $\mathcal{M}_{\mathrm{ND}}$, and $\mathcal{M}_{\mathrm{DN}}$. We define the orthogonal projections that give the even and odd parts, respectively, of a univariate function by $P_e, P_o : L^2(\Omega) \to L^2(\Omega)$, whose definitions are

$$P_e u(x) := \frac{u(x) + u(-x)}{2}, \quad P_o u(x) := \frac{u(x) - u(-x)}{2}.$$

We present a commutativity property that allows us to identify the kernel functions associated with the operators $\mathcal{M}_{\rm N}$ and $\mathcal{M}_{\rm D}$.

Lemma 3.

$$\mathcal{C}_{\rm p} P_e = P_e \mathcal{C}_{\rm p}, \quad \mathcal{C}_{\rm p} P_o = P_o \mathcal{C}_{\rm p}, \quad \mathcal{C}_{\rm a} P_e = P_e \mathcal{C}_{\rm a}, \quad \mathcal{C}_{\rm a} P_o = P_o \mathcal{C}_{\rm a}. \tag{20}$$

Proof. We present the proof for $\mathcal{C}_{\rm p} P_e = P_e \mathcal{C}_{\rm p}$. The other results easily follow. We recall the definition of $\mathcal{C}_{\rm p} u(x)$ in (12). We explicitly write $P_e \mathcal{C}_{\rm p} u(x)$, and the result follows by utilizing the evenness of $\widehat{C}_{\rm p}$ and a change of variable:

$$\begin{aligned}
P_e \mathcal{C}_{\rm p} u(x) &= \frac{1}{2}\left(\int_\Omega \widehat{C}_{\rm p}(x'-x)u(x')dx' + \int_\Omega \widehat{C}_{\rm p}(x'+x)u(x')dx'\right) \\
&= \frac{1}{2}\left(\int_\Omega \widehat{C}_{\rm p}(x'-x)u(x')dx' + \int_\Omega \widehat{C}_{\rm p}(x'-x)u(-x')dx'\right) \\
&= \int_\Omega \widehat{C}_{\rm p}(x'-x) P_e u(x')dx' \\
&= \mathcal{C}_{\rm p} P_e u(x).
\end{aligned}$$

□

Remark 2. The above commutativity property plays an important role in determining the spectrum of the operators $\mathcal{M}_{\rm N}$ and $\mathcal{M}_{\rm D}$; see Aksoylu and Kaya (2018). It also helps in identifying the associated kernel functions; see (22) and (23). Note that the above commutativity property does not hold for the operators $\mathcal{C}_{\rm pa}$ and $\mathcal{C}_{\rm ap}$. Identification of the associated kernel functions can be done by direct manipulation.

Theorem 2 (Main Theorem in 1D). *Let $c = \int_\Omega C(x')dx'$. The following operators $\mathcal{M}_{\rm N}$, $\mathcal{M}_{\rm D}$, $\mathcal{M}_{\rm ND}$, and $\mathcal{M}_{\rm DN}$ defined by*

$$\begin{aligned}
(\mathcal{M}_{\rm D} - c)u(x) &:= -\int_\Omega \left[\widehat{C}_{\rm a}(x'-x)P_e u(x') + \widehat{C}_{\rm p}(x'-x)P_o u(x')\right]dx', \\
(\mathcal{M}_{\rm N} - c)u(x) &:= -\int_\Omega \left[\widehat{C}_{\rm p}(x'-x)P_e u(x') + \widehat{C}_{\rm a}(x'-x)P_o u(x')\right]dx', \\
(\mathcal{M}_{\rm ND} - c)u(x) &:= -\int_\Omega \left[\widehat{C}_{\rm ap}(x'-x)P_e u(x') + \widehat{C}_{\rm pa}(x'-x)P_o u(x')\right]dx', \\
(\mathcal{M}_{\rm DN} - c)u(x) &:= -\int_\Omega \left[\widehat{C}_{\rm pa}(x'-x)P_e u(x') + \widehat{C}_{\rm ap}(x'-x)P_o u(x')\right]dx'
\end{aligned}$$

agree with $\mathcal{L}u(x)$ in the bulk, i.e., for $x \in (-1+\delta, 1-\delta)$. Furthermore, the operators $\mathcal{M}_{\mathrm{N}}$ and $\mathcal{M}_{\mathrm{D}}$ enforce pure Neumann and Dirichlet BC, respectively:

$$\frac{d}{dx}\big[(\mathcal{M}_{\mathrm{N}} - c)u\big](\pm 1) = 0,$$
$$(\mathcal{M}_{\mathrm{D}} - c)u(\pm 1) = 0.$$

The operators $\mathcal{M}_{\mathrm{ND}}$ and $\mathcal{M}_{\mathrm{DN}}$ enforce mixed Neumann and Dirichlet BC, respectively:

$$(\mathcal{M}_{\mathrm{ND}} - c)u(+1) = \frac{d}{dx}\big[(\mathcal{M}_{\mathrm{ND}} - c)u\big](-1) = 0,$$
$$(\mathcal{M}_{\mathrm{DN}} - c)u(-1) = \frac{d}{dx}\big[(\mathcal{M}_{\mathrm{DN}} - c)u\big](+1) = 0.$$

We define the operators $\mathcal{M}_{\mathrm{N}}, \mathcal{M}_{\mathrm{D}}, \mathcal{M}_{\mathrm{ND}}$, and $\mathcal{M}_{\mathrm{DN}}$ as bounded, linear operators. More precisely, $\mathcal{M}_{\mathrm{D}}, \mathcal{M}_{\mathrm{N}}, \mathcal{M}_{\mathrm{ND}}, \mathcal{M}_{\mathrm{DN}} \in L(X, X)$ where $X = L^2(\Omega) \cap C^1(\partial\Omega)$. For $\mathcal{M}_{\mathrm{D}}$, the choice of X can be relaxed as $L^2(\Omega) \cap C^0(\partial\Omega)$. This choice is implied when we study $\mathcal{M}_{\mathrm{D}}$. The assumptions for the operators $\mathcal{M}_{\mathrm{ND}}$ and $\mathcal{M}_{\mathrm{DN}}$ are also implied in a similar way.

Imposing Neumann (also periodic and antiperiodic) BC requires differentiation. For technical details regarding differentiation under the integral sign, see the discussion on the Leibniz rule in Aksoylu et al. (2017) whose proof relies on the Lebesgue dominated convergence theorem. In addition, the limit in the definition of the Dirichlet BC can be interchanged with the integral sign, again by the Lebesgue dominated convergence theorem.

Remark 3. When we assume homogeneous Neumann and Dirichlet BC on u, then the operators $\mathcal{M}_{\mathrm{N}}$ and $\mathcal{M}_{\mathrm{D}}$ enforce homogeneous Dirichlet and Neumann BC, respectively. More precisely, for $u(\pm 1) = 0$ and $u'(\pm 1) = 0$, we obtain $\frac{d}{dx}\mathcal{M}_{\mathrm{N}}u(\pm 1) = 0$ and $\mathcal{M}_{\mathrm{D}}u(\pm 1) = 0$, respectively. The same line of argument applies to the operators $\mathcal{M}_{\mathrm{ND}}$ and $\mathcal{M}_{\mathrm{DN}}$.

Using the operators $\mathcal{C}_{\mathrm{p}}, \mathcal{C}_{\mathrm{a}}, \mathcal{C}_{\mathrm{pa}}$, and $\mathcal{C}_{\mathrm{ap}}$ given in (12) and (13), we can express the operators $\mathcal{M}_{\mathrm{N}}, \mathcal{M}_{\mathrm{D}}, \mathcal{M}_{\mathrm{ND}}$, and $\mathcal{M}_{\mathrm{DN}}$ in the following way:

$$(\mathcal{M}_{\mathrm{N}} - c) = -(\mathcal{C}_{\mathrm{p}} P_e + \mathcal{C}_{\mathrm{a}} P_o), \quad (\mathcal{M}_{\mathrm{D}} - c) = -(\mathcal{C}_{\mathrm{a}} P_e + \mathcal{C}_p P_o),$$
$$(\mathcal{M}_{\mathrm{ND}} - c) = -(\mathcal{C}_{\mathrm{ap}} P_e + \mathcal{C}_{\mathrm{pa}} P_o), \quad (\mathcal{M}_{\mathrm{DN}} - c) = -(\mathcal{C}_{\mathrm{pa}} P_e + \mathcal{C}_{\mathrm{ap}} P_o).$$

Using the commutativity property (20), we arrive at the following representation:

$$(\mathcal{M}_{\mathrm{N}} - c) = -(P_e\mathcal{C}_{\mathrm{p}} + P_o\mathcal{C}_{\mathrm{a}}), \quad (\mathcal{M}_{\mathrm{D}} - c) = -(P_e\mathcal{C}_{\mathrm{a}} + P_o\mathcal{C}_{\mathrm{p}}).$$

Now, we can identify the kernel functions associated with operators $\mathcal{M}_{\mathrm{N}}$ and $\mathcal{M}_{\mathrm{D}}$:

$$\begin{aligned}(\mathcal{M}_{\mathrm{N}}-c)u(x) &= -\int_{\Omega} K_{\mathrm{N}}(x,x')u(x')dx',\\ (\mathcal{M}_{\mathrm{D}}-c)u(x) &= -\int_{\Omega} K_{\mathrm{D}}(x,x')u(x')dx',\end{aligned} \tag{21}$$

where

$$K_{\mathrm{N}}(x,x') := \frac{1}{2}\{[\widehat{C}_{\mathrm{p}}(x'-x)+\widehat{C}_{\mathrm{p}}(x'+x)] + [\widehat{C}_{\mathrm{a}}(x'-x)-\widehat{C}_{\mathrm{a}}(x'+x)]\}, \tag{22}$$

$$K_{\mathrm{D}}(x,x') := \frac{1}{2}\{[\widehat{C}_{\mathrm{a}}(x'-x)+\widehat{C}_{\mathrm{a}}(x'+x)] + [\widehat{C}_{\mathrm{p}}(x'-x)-\widehat{C}_{\mathrm{p}}(x'+x)]\}. \tag{23}$$

We also want to identify the integrands associated with the operators $\mathcal{M}_{\mathrm{ND}}$ and $\mathcal{M}_{\mathrm{DN}}$. We proceed by direct manipulation. By writing P_e and P_o explicitly and utilizing a simple change of variable, we arrive at the following expressions:

$$\begin{aligned}(\mathcal{M}_{\mathrm{ND}}-c)u(x) &= -\int_{\Omega} K_{\mathrm{ND}}(x,x')u(x')dx',\\ (\mathcal{M}_{\mathrm{DN}}-c)u(x) &= -\int_{\Omega} K_{\mathrm{DN}}(x,x')u(x')dx',\end{aligned}$$

where

$$\begin{aligned}K_{\mathrm{ND}}(x,x') &:= \frac{1}{2}\{[\widehat{C}_{\mathrm{pa}}(x'-x)+\widehat{C}_{\mathrm{pa}}(x'+x)] + [\widehat{C}_{\mathrm{ap}}(x'-x)-\widehat{C}_{\mathrm{ap}}(x'+x)]\},\\ K_{\mathrm{DN}}(x,x') &:= \frac{1}{2}\{[\widehat{C}_{\mathrm{ap}}(x'-x)+\widehat{C}_{\mathrm{ap}}(x'+x)] + [\widehat{C}_{\mathrm{pa}}(x'-x)-\widehat{C}_{\mathrm{pa}}(x'+x)]\}.\end{aligned}$$

In order to align with the construction given in Aksoylu et al. (2017a), we assume that

$$\widehat{C}(x),\ \widehat{C}_{\mathrm{a}}(x),\ \widehat{C}_{\mathrm{p}}(x),\ \widehat{C}_{\mathrm{pa}}(x),\ \text{and}\ \widehat{C}_{\mathrm{ap}}(x) \in L^2(\widehat{\Omega}). \tag{24}$$

Remark 4. The boundedness of $\mathcal{M}_{\mathrm{N}}$, $\mathcal{M}_{\mathrm{D}}$, $\mathcal{M}_{\mathrm{ND}}$, and $\mathcal{M}_{\mathrm{DN}}$ follow from the choices of (5) and (24). In addition, all of them fall into the class of integral operators; hence, their self-adjointness follows from the fact that the corresponding kernels are symmetric (due to evenness of C), i.e., $K_{\mathrm{BC}}(x,x') = K_{\mathrm{BC}}(x',x)$ and $\mathrm{BC} \in \{\mathrm{N},\mathrm{D},\mathrm{ND},\mathrm{DN}\}$. The cases of $\mathrm{BC} \in \{\mathrm{ND},\mathrm{DN}\}$ are more involved than the rest. One useful identity is $\widehat{C}_{\mathrm{ap}}(x'-x) = \widehat{C}_{\mathrm{pa}}(-x'+x)$.

In the upcoming proofs, we want to report a minor caveat. We use $\widehat{C}_a(x'+1) = -\widehat{C}_a(x'-1)$ which holds for $x' \neq 0$. For $x' = 0$, i.e., $\widehat{C}_a(x'+1) = C(1) \neq -C(-1) = -\widehat{C}_a(x'-1)$. Since $x' = 0$ is only a point, it does not change the value of the integral. We choose not to point it out each time we run into this case.

Proof (Proof of Theorem 2). The key observation that leads to the agreement of the operators $\mathcal{M}_N$, $\mathcal{M}_D$, $\mathcal{M}_{ND}$, and $\mathcal{M}_{DN}$ with the operator $\mathcal{L}$ is the agreement of kernel functions in (14). The property (14) leads to the equivalence (19). Hence, we arrive at the following equivalence for $x \in (-1+\delta, 1-\delta)$:

$$\begin{aligned}
(\mathcal{L} - c) &= -\mathcal{C} \\
&= -\mathcal{C}(P_e + P_o) \qquad \text{(using } u = P_e u + P_o u\text{)} \\
&= -(\mathcal{C}P_e + \mathcal{C}P_o) \\
&= -(\mathcal{C}_p P_e + \mathcal{C}_a P_o) \quad \text{(using (19))} \\
&=: (\mathcal{M}_N - c).
\end{aligned} \tag{25}$$

Similar to (25), we can show that the other operators agree in the bulk as well:

$$\begin{aligned}
(\mathcal{L} - c) &= -\mathcal{C} \\
&= -\left(\mathcal{C}_a P_e + \mathcal{C}_p P_o\right) \quad =: (\mathcal{M}_D - c) && (26) \\
&= -\left(\mathcal{C}_{ap} P_e + \mathcal{C}_{pa} P_o\right) =: (\mathcal{M}_{ND} - c) && (27) \\
&= -\left(\mathcal{C}_{pa} P_e + \mathcal{C}_{ap} P_o\right) =: (\mathcal{M}_{DN} - c) && (28)
\end{aligned}$$

First, we prove that the operators $\mathcal{M}_N$ and $\mathcal{M}_D$ enforce pure Neumann and Dirichlet BC, respectively. Next, we will prove that the operators $\mathcal{M}_{ND}$ and $\mathcal{M}_{DN}$ enforce mixed Neumann and Dirichlet BC, respectively.

- **The operator** $\mathcal{M}_N$: First we remove the points at which the partial derivative of $K_N(x, x')$ does not exist from the set of integration. Note that such points form a set of measure zero and, hence, do not affect the value of the integral. We differentiate both sides of (21). In Aksoylu et al. (2017), we had proved that the differentiation in the definition of the Neumann BC can interchange with the integral. We can differentiate the integrand $K_N(x, x')$ piecewise and obtain

$$\frac{d}{dx}\left[(\mathcal{M}_N - c)u\right](x) = -\int_\Omega \frac{\partial K_N}{\partial x}(x, x')u(x')dx', \tag{29}$$

where

$$\frac{\partial K_N}{\partial x}(x, x') = \frac{1}{2}\{[-\widehat{C}'_p(x'-x) + \widehat{C}'_p(x'+x)] + [-\widehat{C}'_a(x'-x) - \widehat{C}'_a(x'+x)]\}.$$

We check the boundary values by plugging $x = \pm 1$ in (29).

$$\frac{d}{dx}[(\mathcal{M}_{\mathrm{N}} - c)u](\pm 1) = -\int_{\Omega} \frac{\partial K_{\mathrm{N}}}{\partial x}(\pm 1, x')u(x')dx'. \tag{30}$$

The functions $\widehat{C}'_{\mathrm{p}}$ and $\widehat{C}'_{\mathrm{a}}$ are 2-periodic and 2-antiperiodic because they are the derivatives of 2-periodic and 2-antiperiodic functions, respectively. Hence,

$$\widehat{C}'_{\mathrm{p}}(x' \mp 1) = \widehat{C}'_{\mathrm{p}}(x' \pm 1) \quad \text{and} \quad \widehat{C}'_{\mathrm{a}}(x' \mp 1) = -\widehat{C}'_{\mathrm{a}}(x' \pm 1).$$

Hence, the integrand in (30) vanishes, i.e.,

$$\frac{\partial K_{\mathrm{N}}}{\partial x}(\pm 1, x') = 0.$$

Therefore, we arrive at

$$\frac{d}{dx}\mathcal{M}_{\mathrm{N}}u(\pm 1) = cu'(\pm 1).$$

When we assume that u satisfies homogeneous Neumann BC, i.e., $u'(\pm 1) = 0$, we conclude that the operator $\mathcal{M}_{\mathrm{N}}$ enforces homogeneous Neumann BC as well.

- **The operator** $\mathcal{M}_{\mathrm{D}}$: By the Lebesgue dominated convergence theorem, the limit in the definition of the Dirichlet BC can be interchanged with the integral. Now, we check the boundary values by plugging $x = \pm 1$ in (23).

$$(\mathcal{M}_{\mathrm{D}} - c)u(\pm 1) = -\int_{\Omega} K_{\mathrm{D}}(\pm 1, x')u(x')dx'. \tag{31}$$

Since $\widehat{C}_{\mathrm{p}}$ and $\widehat{C}_{\mathrm{a}}$ are 2-periodic and 2-antiperiodic, respectively, we have

$$\widehat{C}_{\mathrm{p}}(x' \mp 1) = \widehat{C}_{\mathrm{p}}(x' \pm 1) \quad \text{and} \quad \widehat{C}_{\mathrm{a}}(x' \mp 1) = -\widehat{C}_{\mathrm{a}}(x' \pm 1).$$

Hence, the integrand in (31) vanishes, i.e., $K_{\mathrm{D}}(\pm 1, x') = 0$. Therefore, we arrive at

$$\mathcal{M}_{\mathrm{D}}u(\pm 1) = cu(\pm 1).$$

When we assume that u satisfies homogeneous Dirichlet BC, i.e., $u(\pm 1) = 0$, we conclude that the operator $\mathcal{M}_{\mathrm{D}}$ enforces homogeneous Dirichlet BC as well.

- **The operator** $\mathcal{M}_{\mathrm{ND}}$: First we prove that $\mathcal{C}_{\mathrm{ap}} P_e u(+1) = 0$. We use a change of variable in the second piece.

$$\mathcal{C}_{\mathrm{ap}} P_e u(+1) = \frac{1}{2}\left(\int_{\Omega} \widehat{C}_{\mathrm{ap}}(x' - 1) P_e u(x')dx' + \int_{\Omega} \widehat{C}_{\mathrm{ap}}(-x' - 1) P_e u(x')dx'\right).$$

Then, we split the integrals into two parts as follows:

$$\begin{aligned}\mathcal{C}_{\mathrm{ap}} P_e u(+1) &= \frac{1}{2}\int_{-1}^{0} \big[\widehat{C}_{\mathrm{ap}}(x'-1) + \widehat{C}_{\mathrm{ap}}(-x'-1)\big] P_e u(x') dx' \\ &+ \frac{1}{2}\int_{0}^{1} \big[\widehat{C}_{\mathrm{ap}}(x'-1) + \widehat{C}_{\mathrm{ap}}(-x'-1)\big] P_e u(x') dx'.\end{aligned} \tag{32}$$

For $x' \in [-1, 0]$, we have $x' - 1 \in [-2, -1]$. By using the definition of $\widehat{C}_{\mathrm{ap}}$ and the evenness of C, we obtain

$$\widehat{C}_{\mathrm{ap}}(x'-1) = -\widehat{C}_{\mathrm{ap}}(x'+1) = -C(x'+1) = -C(-x'-1) = -\widehat{C}_{\mathrm{ap}}(-x'-1). \tag{33}$$

For $x' \in [0, 1]$, we have $x' - 1 \in [-1, 0]$. By using the definition of $\widehat{C}_{\mathrm{ap}}$ and the evenness of C, we obtain

$$\widehat{C}_{\mathrm{ap}}(x'-1) = C(x'-1) = C(-x'+1) = -\widehat{C}_{\mathrm{ap}}(-x'-1). \tag{34}$$

Combining (33) and (34) with (32), we conclude that $\mathcal{C}_{\mathrm{ap}} P_e u(+1) = 0$. Similarly, we can conclude that $\mathcal{C}_{\mathrm{pa}} P_o u(+1) = 0$. Consequently, we arrive at

$$\mathcal{C}_{\mathrm{ND}} u(+1) = 0.$$

We prove that $\frac{d}{dx}\mathcal{C}_{\mathrm{pa}} P_o u(-1) = 0$. We use a change of variable in the second piece.

$$\begin{aligned}\frac{d}{dx}\mathcal{C}_{\mathrm{pa}} P_o u(-1) = -\frac{1}{2}\Big(&\int_{\Omega} \widehat{C}'_{\mathrm{pa}}(x'+1) P_o u(x') dx' \\ &- \int_{\Omega} \widehat{C}'_{\mathrm{pa}}(-x'+1) P_o u(x') dx'\Big).\end{aligned}$$

Then, we split the integrals into two parts as follows:

$$\begin{aligned}\frac{d}{dx}\mathcal{C}_{\mathrm{pa}} P_o u(-1) &= -\frac{1}{2}\int_{-1}^{0} \big[\widehat{C}'_{\mathrm{pa}}(x'+1) - \widehat{C}'_{\mathrm{pa}}(-x'+1)\big] P_o u(x') dx' \\ &- \frac{1}{2}\int_{0}^{1} \big[\widehat{C}'_{\mathrm{pa}}(x'+1) - \widehat{C}'_{\mathrm{pa}}(-x'+1)\big] P_o u(x') dx'.\end{aligned} \tag{35}$$

For $x' \in [-1, 0]$, we have $x' + 1 \in [0, 1]$. By using the definition of $\widehat{C}_{\mathrm{pa}}$ and the oddness of C', we obtain

$$\widehat{C}'_{\mathrm{pa}}(x'+1) = C'(x'+1) = -C'(-x'-1) = -\widehat{C}'_{\mathrm{pa}}(-x'-1) = \widehat{C}'_{\mathrm{pa}}(-x'+1). \tag{36}$$

For $x' \in [0,1]$, we have $x'+1 \in [1,2]$. By using the definition of $\widehat{C}_{\mathtt{pa}}$ and the oddness of C', we obtain

$$\widehat{C}'_{\mathtt{pa}}(x'+1) = -\widehat{C}'_{\mathtt{pa}}(x'-1) = -C'(x'-1) = C'(-x'+1) = \widehat{C}'_{\mathtt{pa}}(-x'+1). \tag{37}$$

Combining (36) and (37) with (35), we conclude that $\frac{d}{dx}\mathcal{C}_{\mathtt{pa}} P_o u(-1) = 0$. Similarly, we can conclude that $\frac{d}{dx}\mathcal{C}_{\mathtt{ap}} P_e u(-1) = 0$. Consequently, we arrive at

$$\frac{d}{dx}\mathcal{C}_{\mathrm{ND}} u(-1) = 0.$$

- **The operator $\mathcal{M}_{\mathrm{DN}}$**: The proof is similar to the case of $\mathcal{M}_{\mathrm{ND}}$.

□

Remark 5. As we prepare to construct the operators in 2D, it is useful to explicitly denote the variable x' on which P_e and P_o act in the following way.

$$\mathcal{C}_{\mathrm{N}} u(x) := \big(\mathcal{C}_{\mathrm{p}} P_{e,x'} + \mathcal{C}_{\mathrm{a}} P_{o,x'}\big) u(x)$$

$$\mathcal{C}_{\mathrm{D}} u(x) := \big(\mathcal{C}_{\mathrm{a}} P_{e,x'} + \mathcal{C}_{\mathrm{p}} P_{o,x'}\big) u(x)$$

$$\mathcal{C}_{\mathrm{ND}} u(x) := \big(\mathcal{C}_{\mathtt{ap}} P_{e,x'} + \mathcal{C}_{\mathtt{pa}} P_{o,x'}\big) u(x)$$

$$\mathcal{C}_{\mathrm{DN}} u(x) := \big(\mathcal{C}_{\mathtt{pa}} P_{e,x'} + \mathcal{C}_{\mathtt{ap}} P_{o,x'}\big) u(x).$$

Consequently, checking if the operators enforce the BC reduces to obtaining

$$\begin{aligned}
\frac{d}{dx}\mathcal{C}_{\mathrm{N}} u(\pm 1) = 0, &\qquad \mathcal{C}_{\mathrm{D}} u(\pm 1) &= 0 \\
\frac{d}{dx}\mathcal{C}_{\mathrm{ND}} u(-1) = 0, &\qquad \mathcal{C}_{\mathrm{ND}} u(+1) &= 0 \\
\frac{d}{dx}\mathcal{C}_{\mathrm{DN}} u(+1) = 0, &\qquad \mathcal{C}_{\mathrm{DN}} u(-1) &= 0.
\end{aligned} \tag{38}$$

The Construction of 2D Operators

For the convolution present in the governing operators, we use a shorthand notation and define the operator

$$\mathcal{C}u(x,y) := \iint_{\Omega} \widehat{C}(x'-x, y'-y) u(x',y') dx' dy'. \tag{39}$$

We also define the following auxiliary operators that act on a bivariate function.

$$\mathcal{X}_{\mathrm{E}}u(x,y) := \int_{\Omega} \widehat{X}_{\mathrm{E}}(x'-x)u(x',y)dx', \; \mathcal{Y}_{\mathrm{E}}u(x,y) := \int_{\Omega} \widehat{Y}_{\mathrm{E}}(y'-y)u(x,y')dy',$$

where $\mathrm{E} \in \{\mathrm{p}, \mathrm{a}, \mathrm{pa}, \mathrm{ap}\}$.

Using the separability assumption (11) on the kernel function, we have the following:

$$\widehat{C}(x,y) = \widehat{X}(x)\widehat{Y}(y). \tag{40}$$

The separability of the kernel function leads to the following important property. Using (40) and the Fubini theorem, we rewrite the operator $\mathcal{C}$ in (39).

$$\begin{aligned}
\mathcal{C}u(x,y) &= \iint\limits_{\Omega} \widehat{X}(x'-x)\widehat{Y}(y'-y)u(x',y')dx'dy' \\
&= \int_{\Omega} \widehat{X}(x'-x)\Big(\int_{\Omega} \widehat{Y}(y'-y)u(x',y')dy'\Big)dx' \\
&= \int_{\Omega} \widehat{X}(x'-x)\Big(\mathcal{Y}u(x',y)\Big)dx' \\
&= \mathcal{X}\big(\mathcal{Y}u\big)(x,y)
\end{aligned} \tag{41}$$

In other words, we proved that $\mathcal{C}$ can be decomposed into a product of two 1D operators where the action of $\mathcal{X}$ and $\mathcal{Y}$ is on the variables x and y, respectively. Furthermore, a change in the order of integration leads to

$$\mathcal{C}u(x,y) = \mathcal{Y}\mathcal{X}u(x,y). \tag{42}$$

Similar to (19), we also obtain the following equivalence of operators in the bulk. For fixed y_0, we have

$$\begin{aligned}
\mathcal{X}u(x,y_0) &= \mathcal{X}_{\mathrm{p}}u(x,y_0) = \mathcal{X}_{\mathrm{a}}u(x,y_0) \\
&= \mathcal{X}_{\mathrm{pa}}u(x,y_0) = \mathcal{X}_{\mathrm{ap}}u(x,y_0), x \in (-1+\delta, 1-\delta).
\end{aligned}$$

Also, for fixed x_0, we have

$$\begin{aligned}
\mathcal{Y}u(x_0,y) &= \mathcal{Y}_{\mathrm{p}}u(x_0,y) = \mathcal{Y}_{\mathrm{a}}u(x_0,y) = \mathcal{Y}_{\mathrm{pa}}u(x_0,y) \\
&= \mathcal{Y}_{\mathrm{ap}}u(x_0,y) \quad y \in (-1+\delta, 1-\delta).
\end{aligned}$$

The choice made in (25) leads to the construction of the operator that enforces pure Neumann BC in the x- and y-variable as follows:

$$\mathcal{X}_{\mathrm{N}} := \mathcal{X}_{\mathrm{p}} P_{e,x'} + \mathcal{X}_{\mathrm{a}} P_{o,x'} \quad \text{(in the } x\text{-variable)} \tag{43}$$

$$\mathcal{Y}_{\mathrm{N}} := \mathcal{Y}_{\mathrm{p}} P_{e,y'} + \mathcal{Y}_{\mathrm{a}} P_{o,y'} \quad \text{(in the } y\text{-variable).} \tag{44}$$

Similarly, the choice made in (26) leads to the construction of the operator that enforces pure Dirichlet BC in the x- and y-variable as follows:

$$\mathcal{X}_{\mathrm{D}} := \mathcal{X}_{\mathrm{a}} P_{e,x'} + \mathcal{X}_{\mathrm{p}} P_{o,x'} \quad \text{(in the } x\text{-variable)} \tag{45}$$

$$\mathcal{Y}_{\mathrm{D}} := \mathcal{Y}_{\mathrm{a}} P_{e,y'} + \mathcal{Y}_{\mathrm{p}} P_{o,y'} \quad \text{(in the } y\text{-variable).} \tag{46}$$

Similarly, the choices made in (27) and (28) lead to the construction of the operators that enforce mixed Neumann-Dirichlet and Dirichlet-Neumann BC in the x- and y-variable as follows:

$$\mathcal{X}_{\mathrm{ND}} := \mathcal{X}_{\mathrm{ap}} P_{e,x'} + \mathcal{X}_{\mathrm{pa}} P_{o,x'} \quad \text{(in the } x\text{-variable)} \tag{47}$$

$$\mathcal{Y}_{\mathrm{ND}} := \mathcal{Y}_{\mathrm{ap}} P_{e,y'} + \mathcal{Y}_{\mathrm{pa}} P_{o,y'} \quad \text{(in the } y\text{-variable)} \tag{48}$$

$$\mathcal{X}_{\mathrm{DN}} := \mathcal{X}_{\mathrm{pa}} P_{e,x'} + \mathcal{X}_{\mathrm{ap}} P_{o,x'} \quad \text{(in the } x\text{-variable)} \tag{49}$$

$$\mathcal{Y}_{\mathrm{DN}} := \mathcal{Y}_{\mathrm{pa}} P_{e,y'} + \mathcal{Y}_{\mathrm{ap}} P_{o,y'} \quad \text{(in the } y\text{-variable).} \tag{50}$$

We want to construct an operator that enforces pure Neumann BC on the square. We make the choice that gives the 1D Neumann operator both in x- and y-variables. Hence, combining (43) and (44), we define the 2D pure Neumann operator as

$$\left(\mathcal{M}_{\mathrm{N}} - c\right) := -\mathcal{X}_{\mathrm{N}}\mathcal{Y}_{\mathrm{N}} = -\left(\mathcal{X}_{\mathrm{p}} P_{e,x'} + \mathcal{X}_{\mathrm{a}} P_{o,x'}\right)\left(\mathcal{Y}_{\mathrm{p}} P_{e,y'} + \mathcal{Y}_{\mathrm{a}} P_{o,y'}\right). \tag{51}$$

Similarly, combining (45) and (46), we define the 2D pure Dirichlet operator as

$$\left(\mathcal{M}_{\mathrm{D}} - c\right) := -\mathcal{X}_{\mathrm{D}}\mathcal{Y}_{\mathrm{D}} = -\left(\mathcal{X}_{\mathrm{a}} P_{e,x'} + \mathcal{X}_{\mathrm{p}} P_{o,x'}\right)\left(\mathcal{Y}_{\mathrm{a}} P_{e,y'} + \mathcal{Y}_{\mathrm{p}} P_{o,y'}\right). \tag{52}$$

Similarly, combining (47), (48), (49), and (50), we define the 2D mixed operators as follows:

$$\left(\mathcal{M}_{\mathrm{ND,ND}} - c\right) := -\mathcal{X}_{\mathrm{ND}}\mathcal{Y}_{\mathrm{ND}} = -\left(\mathcal{X}_{\mathrm{ap}} P_{e,x'} + \mathcal{X}_{\mathrm{pa}} P_{o,x'}\right)\left(\mathcal{Y}_{\mathrm{ap}} P_{e,y'} + \mathcal{Y}_{\mathrm{pa}} P_{o,y'}\right) \tag{53}$$

$$\left(\mathcal{M}_{\mathrm{N,DN}} - c\right) := -\mathcal{X}_{\mathrm{N}}\ \mathcal{Y}_{\mathrm{DN}} = -\left(\mathcal{X}_{\mathrm{p}} P_{e,x'} + \mathcal{X}_{\mathrm{a}} P_{o,x'}\right)\ \left(\mathcal{Y}_{\mathrm{pa}} P_{e,y'} + \mathcal{Y}_{\mathrm{ap}} P_{o,y'}\right). \tag{54}$$

Recalling (2), we immediately see that the operator $\mathcal{L}$ agrees in the bulk with the given operators above. Namely,

$$
\begin{aligned}
(\mathcal{L} - c) &= -\mathcal{C} = (\mathcal{M}_{\mathrm{N}} - c) \\
(\mathcal{L} - c) &= -\mathcal{C} = (\mathcal{M}_{\mathrm{D}} - c) \\
(\mathcal{L} - c) &= -\mathcal{C} = (\mathcal{M}_{\mathrm{ND,\,ND}} - c) \\
(\mathcal{L} - c) &= -\mathcal{C} = (\mathcal{M}_{\mathrm{N,\,DN}} - c).
\end{aligned}
$$

Remark 6. The operator $\mathcal{C}$ in (2) utilizes a 2D computational domain which is indicated by the integration variable $dx'dy' = d(x', y')$. We can show the construction of each operator by paying attention to the computational domain of each operator and rearranging (41) using the agreement of operators in the bulk in the following way:

$$
\begin{aligned}
\mathcal{C} = \mathcal{C} I_{x',y'} &= \mathcal{X} I_{x'}\, \mathcal{Y} I_{y'} \\
&= \mathcal{X}(P_{e,x'} + P_{o,x'})\, \mathcal{Y}(P_{e,y'} + P_{o,y'}) \\
&= (\mathcal{X} P_{e,x'} + \mathcal{X} P_{o,x'})\, (\mathcal{Y} P_{e,y'} + \mathcal{Y} P_{o,y'}) \\
&= (\mathcal{X}_{\mathrm{p}} P_{e,x'} + \mathcal{X}_{\mathrm{a}} P_{o,x'})\, (\mathcal{Y}_{\mathrm{p}} P_{e,y'} + \mathcal{Y}_{\mathrm{a}} P_{o,y'}) \quad =: -(\mathcal{M}_{\mathrm{N}} - c) \\
&= (\mathcal{X}_{\mathrm{a}} P_{e,x'} + \mathcal{X}_{\mathrm{p}} P_{o,x'})\, (\mathcal{Y}_{\mathrm{a}} P_{e,y'} + \mathcal{Y}_{\mathrm{p}} P_{o,y'}) \quad =: -(\mathcal{M}_{\mathrm{D}} - c) \\
&= (\mathcal{X}_{\mathrm{ap}} P_{e,x'} + \mathcal{X}_{\mathrm{ap}} P_{o,x'})\, (\mathcal{Y}_{\mathrm{ap}} P_{e,y'} + \mathcal{Y}_{\mathrm{ap}} P_{o,y'}) =: -(\mathcal{M}_{\mathrm{ND,\,ND}} - c) \qquad (55) \\
&= (\mathcal{X}_{\mathrm{p}} P_{e,x'} + \mathcal{X}_{\mathrm{a}} P_{o,x'})\, (\mathcal{Y}_{\mathrm{ap}} P_{e,y'} + \mathcal{Y}_{\mathrm{pa}} P_{o,y'}) \quad =: -(\mathcal{M}_{\mathrm{N,\,ND}} - c). \qquad (56)
\end{aligned}
$$

We construct the operators in higher dimensions by using the corresponding rearrangement; see section "Operators in Higher Dimensions" for the 3D construction. In addition, the 2D decomposition operator $I_{x',y'}$ given in (9) is indeed the product of the 1D decomposition operators $I_{x'}$ and $I_{y'}$ given in (8). More precisely,

$$
\begin{aligned}
I_{x',y'} &= I_{x'}\, I_{y'} \\
&= (P_{e,x'} + P_{o,x'})(P_{e,y'} + P_{o,y'}) \\
&= P_{e,x'} P_{e,y'} + P_{e,x'} P_{o,y'} + P_{o,x'} P_{e,y'} + P_{o,x'} P_{o,y'}.
\end{aligned}
$$

Verifying the Boundary Conditions

The operators $(\mathcal{M}_{\mathrm{N}} - c)$, $(\mathcal{M}_{\mathrm{D}} - c)$, $(\mathcal{M}_{\mathrm{ND,\,ND}} - c)$, and $(\mathcal{M}_{\mathrm{N,\,DN}} - c)$ given in (51), (52), (53), and (54), respectively, are the product of two 1D operators. As we mentioned, the limit in the definition of the BC can be interchanged with the integral sign due to the Lebesgue dominated convergence theorem and the Leibniz rule. Then, using the change in the order of integration as in (42) and (38), we can prove that the pure and mixed Neumann and Dirichlet BC are enforced.

First, we prove that the operators $\mathcal{M}_{\mathrm{N}}$ and $\mathcal{M}_{\mathrm{D}}$ enforce pure Neumann and pure Dirichlet BC in 2D:

$$\frac{\partial}{\partial n}\left[(\mathcal{M}_{\mathrm{N}}-c)\,u\right](x,+1) = -\left(\frac{\partial}{\partial y}\mathcal{X}_{\mathrm{N}}\mathcal{Y}_{\mathrm{N}}\right)u(x,+1)$$
$$= -\mathcal{X}_{\mathrm{N}}\left(\frac{d}{dy}\mathcal{Y}_{\mathrm{N}}\right)u(x,+1) = 0$$
$$\frac{\partial}{\partial n}\left[(\mathcal{M}_{\mathrm{N}}-c)\,u\right](x,-1) = \left(\frac{\partial}{\partial y}\mathcal{X}_{\mathrm{N}}\mathcal{Y}_{\mathrm{N}}\right)u(x,-1)$$
$$= \mathcal{X}_{\mathrm{N}}\left(\frac{d}{dy}\mathcal{Y}_{\mathrm{N}}\right)u(x,-1) = 0$$
$$\frac{\partial}{\partial n}\left[(\mathcal{M}_{\mathrm{N}}-c)\,u\right](+1,y) = -\left(\frac{\partial}{\partial x}\mathcal{Y}_{\mathrm{N}}\mathcal{X}_{\mathrm{N}}\right)u(+1,y)$$
$$= -\mathcal{Y}_{\mathrm{N}}\left(\frac{d}{dx}\mathcal{X}_{\mathrm{N}}\right)u(+1,y) = 0$$
$$\frac{\partial}{\partial n}\left[(\mathcal{M}_{\mathrm{N}}-c)\,u\right](-1,y) = \left(\frac{\partial}{\partial x}\mathcal{Y}_{\mathrm{N}}\mathcal{X}_{\mathrm{N}}\right)u(-1,y)$$
$$= \mathcal{Y}_{\mathrm{N}}\left(\frac{d}{dx}\mathcal{X}_{\mathrm{N}}\right)u(-1,y) = 0.$$
$$\big(\mathcal{M}_{\mathrm{D}}-c\big)u(x,\pm 1) = -\mathcal{X}_{\mathrm{D}}\mathcal{Y}_{\mathrm{D}}u(x,\pm 1) = 0$$
$$\big(\mathcal{M}_{\mathrm{D}}-c\big)u(\pm 1,y) = -\mathcal{Y}_{\mathrm{D}}\mathcal{X}_{\mathrm{D}}u(\pm 1,y) = 0.$$

Then, we prove that the operator $\mathcal{M}_{\mathrm{ND,ND}}$ enforces mixed (2+2) Neumann-Dirichlet, i.e., the West and South edges have Neumann and the East and North edges have Dirichlet BC:

$$\frac{\partial}{\partial n}\left[(\mathcal{M}_{\mathrm{ND,ND}}-c)\,u\right](-1,y) = \left(\frac{\partial}{\partial x}\mathcal{Y}_{\mathrm{ND}}\mathcal{X}_{\mathrm{ND}}\right)u(-1,y)$$
$$= \mathcal{Y}_{\mathrm{ND}}\left(\frac{d}{dx}\mathcal{X}_{\mathrm{ND}}\right)u(-1,y) = 0$$
$$\frac{\partial}{\partial n}\left[(\mathcal{M}_{\mathrm{ND,ND}}-c)\,u\right](x,-1) = \left(\frac{\partial}{\partial y}\mathcal{X}_{\mathrm{ND}}\mathcal{Y}_{\mathrm{ND}}\right)u(x,-1)$$
$$= \mathcal{X}_{\mathrm{ND}}\left(\frac{d}{dy}\mathcal{Y}_{\mathrm{ND}}\right)u(x,-1) = 0.$$
$$\big(\mathcal{M}_{\mathrm{ND,ND}}-c\big)u(+1,y) = -\mathcal{X}_{\mathrm{ND}}\mathcal{Y}_{\mathrm{ND}}u(+1,y) = 0$$
$$\big(\mathcal{M}_{\mathrm{ND,ND}}-c\big)u(x,+1) = -\mathcal{X}_{\mathrm{ND}}\mathcal{Y}_{\mathrm{ND}}u(x,+1) = 0.$$

Finally, we prove that the operator $\mathcal{M}_{\mathrm{N,DN}}$ enforces mixed Neumann-Dirichlet (3+1), i.e., the East, West, and North edges have Neumann and the South edge have Dirichlet BC:

$$\frac{\partial}{\partial n}\big[\,(\mathcal{M}_{\mathrm{N,DN}} - c)\,u\big](+1,y) = -\left(\frac{\partial}{\partial x}\mathcal{Y}_{\mathrm{DN}}\mathcal{X}_{\mathrm{N}}\right)u(+1,y)$$

$$= \mathcal{Y}_{\mathrm{DN}}\left(\frac{d}{dx}\mathcal{X}_{\mathrm{N}}\right)u(+1,y) = 0$$

$$\frac{\partial}{\partial n}\big[\,(\mathcal{M}_{\mathrm{N,DN}} - c)\,u\big](-1,y) = \left(\frac{\partial}{\partial x}\mathcal{Y}_{\mathrm{DN}}\mathcal{X}_{\mathrm{N}}\right)u(-1,y)$$

$$= \mathcal{Y}_{\mathrm{DN}}(\frac{d}{dx}\mathcal{X}_{\mathrm{N}})u(-1,y) = 0$$

$$\frac{\partial}{\partial n}\,[(\mathcal{M}_{\mathrm{N,DN}} - c)\,u]\,(x,+1) = -\left(\frac{\partial}{\partial y}\mathcal{X}_{\mathrm{N}}\mathcal{Y}_{\mathrm{DN}}\right)u(x,+1)$$

$$= -\mathcal{X}_{\mathrm{N}}\left(\frac{d}{dy}\mathcal{Y}_{\mathrm{DN}}\right)u(x,+1) = 0.$$

$$\big(\mathcal{M}_{\mathrm{N,DN}} - c\big)u(x,-1) = -\mathcal{X}_{\mathrm{N}}\mathcal{Y}_{\mathrm{DN}}u(x,-1) = 0.$$

Operators in Higher Dimensions

Let us consider the convolution in 3D and the domain be $\Omega := [-1,1]\times[-1,1]\times[-1,1]$. We define the convolution in 3D similarly using notation in (39).

$$\mathcal{C}u(x,y) = \iiint\limits_{\Omega} \widehat{C}(x'-x, y'-y, z'-z)u(x',y',z')dx'dy'dz'.$$

Note that $\mathcal{C} = -\big(\mathcal{L} - c\big)$. Hence we concentrate on finding suitable operators that agree with $\mathcal{C}$ in the bulk. Assuming a separable restricted kernel function similar to (11),

$$C(x,y,z) = X(x)Y(y)Z(z),$$

the operators $\mathcal{M}_{\mathrm{N}}$ and $\mathcal{M}_{\mathrm{D}}$ in 3D defined below enforce pure Neumann and Dirichlet BC and simultaneously agree with the operator $\mathcal{L}$ in the bulk. The construction process is an extension of the 2D case:

$$\begin{aligned}
\mathcal{C} &= \mathcal{C}I_{x',y',z'} = \mathcal{X}I_{x'}\,\mathcal{Y}I_{y'}\,\mathcal{Z}I_{z'} \\
&= \mathcal{X}(P_{e,x'} + P_{o,x'})\,\mathcal{Y}(P_{e,y'} + P_{o,y'})\mathcal{Z}(P_{e,z'} + P_{o,z'}) \\
&= \left(\mathcal{X}P_{e,x'} + \mathcal{X}P_{o,x'}\right)\left(\mathcal{Y}P_{e,y'} + \mathcal{Y}P_{o,y'}\right)\left(\mathcal{Z}P_{e,z'} + \mathcal{Z}P_{o,z'}\right) \\
&= \left(\mathcal{X}_{\mathrm{p}}P_{e,x'} + \mathcal{X}_{\mathrm{a}}P_{o,x'}\right)\left(\mathcal{Y}_{\mathrm{p}}P_{e,y'} + \mathcal{Y}_{\mathrm{a}}P_{o,y'}\right)\left(\mathcal{Z}_{\mathrm{p}}P_{e,z'} + \mathcal{Z}_{\mathrm{a}}P_{o,z'}\right) =: -(\mathcal{M}_{\mathrm{N}}-c) \\
&= \left(\mathcal{X}_{\mathrm{a}}P_{e,x'} + \mathcal{X}_{\mathrm{p}}P_{o,x'}\right)\left(\mathcal{Y}_{\mathrm{a}}P_{e,y'} + \mathcal{Y}_{\mathrm{p}}P_{o,y'}\right)\left(\mathcal{Z}_{\mathrm{a}}P_{e,y'} + \mathcal{Z}_{\mathrm{p}}P_{o,z'}\right) =: -(\mathcal{M}_{\mathrm{D}}-c).
\end{aligned}$$

The operators that enforce mixed Neumann and Dirichlet BC can be constructed in a similar fashion to the operators given in (55) and (56). The extension to arbitrary dimension can be performed by the same line of argument.

Numerical Experiments

We numerically solve the following nonlocal wave equation:

$$\begin{aligned}
u_{tt}(\mathbf{x},t) + \mathcal{M}_{\mathrm{BC}}u(\mathbf{x},t) &= b(\mathbf{x},t), \quad (\mathbf{x},t) \in \Omega \times [0,T], \\
u(\mathbf{x},0) &= u_0(\mathbf{x}), \\
u_t(\mathbf{x},0) &= 0
\end{aligned} \tag{57}$$

by employing the governing operators $\mathcal{M}_{\mathrm{N}}$ and $\mathcal{M}_{\mathrm{D}}$ in 1D, i.e., $\mathrm{BC} \in \{\mathrm{N},\mathrm{D}\}$, and the operators $\mathcal{M}_{\mathrm{N}}$, $\mathcal{M}_{\mathrm{D}}$, $\mathcal{M}_{\mathrm{ND},\mathrm{ND}}$, and $\mathcal{M}_{\mathrm{N},\mathrm{DN}}$ in 2D, i.e., $\mathrm{BC} \in \{\mathrm{N},\mathrm{D},(\mathrm{ND},\mathrm{ND}),(\mathrm{N},\mathrm{DN})\}$, with discontinuous and continuous initial displacement $u_0(\mathbf{x})$; see the definition of the governing operators in Theorem 1. For the discretization of the 1D problem, we use the Galerkin projection method with piecewise polynomials. For implementation details and theoretical construction, see Aksoylu et al. (2017a). Note that, for all time, BC are satisfied; see Fig. 3.

In 1D, as far as the boundary behavior goes, in nonlocal problems, we observe a similar wave reflection pattern from the boundary as in classical problems. In the classical case, we see that the Neumann and the Dirichlet BC create reflections of same and opposite signs, respectively; for the Neumann BC, see Fig. 4. A parallel behavior is observed for the nonlocal Neumann and Dirichlet cases; see Fig. 3.

For the discretization of the 2D problem, we use the Nyström method with the quadrature chosen as the trapezoidal rule (Fig. 5). For implementation details, see Aksoylu et al. (Submitted). We depict the solutions to the nonlocal wave equation domain with homogeneous pure Neumann, pure Dirichlet, and mixed Neumann-Dirichlet with vanishing initial velocity and discontinuous initial displacement; see Figs. 6, 7, 8, and 9. Also, for continuous initial displacement, see Fig. 10. The initial solutions are depicted in Fig. 5. Notice that, for all time, local BC are clearly satisfied. Furthermore, for pure Neumann problem, we have numerically verified that $\iint_{\Omega} u(\mathbf{x},t)d\mathbf{x}$ remains constant for all t. This is in agreement with the physical implication that homogeneous Neumann BC model insulated boundaries.

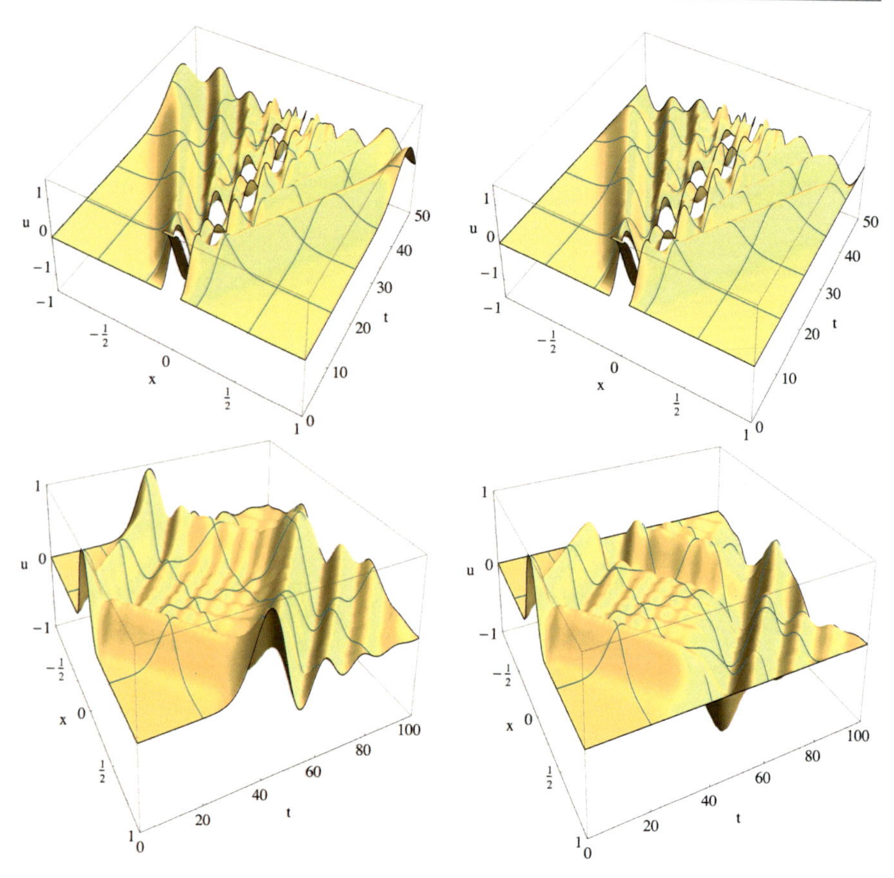

Fig. 3 Solution to the nonlocal wave equation on a 1D domain with discontinuous (top) and continuous (bottom) initial solution and vanishing initial velocity with Neumann (left) boundary condition using the governing operator $\mathcal{M}_{\mathrm{N}}$ and Dirichlet (right) boundary condition using the governing operator $\mathcal{M}_{\mathrm{D}}$. Note that, for all time, BC are satisfied and discontinuities remain stationary

The Treatment of General Nonlocal Problems Using Functional Calculus

Our main tool that allows us to incorporate local BC into nonlocal operators is *functional calculus*. More precisely, the novel governing operators are obtained by employing the functional calculus of self-adjoint operators, i.e., by replacing the classical governing operator A by a suitable function of A, $f(A)$. We call f the *regulating function*. Since classical BC is an integral part of the classical operator, these BC are automatically inherited by $f(A)$. One advantage of our approach is that every symmetry that commutes with A also commutes with $f(A)$. As a result,

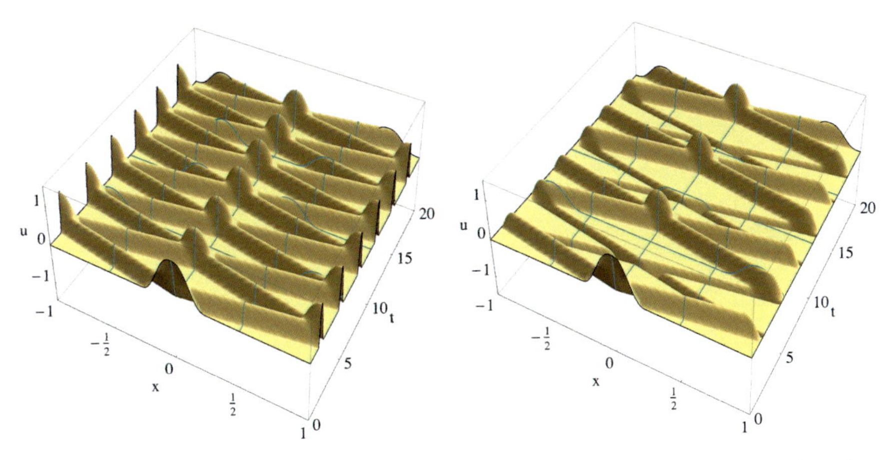

Fig. 4 Solution to the classical wave equation with Neumann (left) and Dirichlet (right) boundary conditions with the same continuous initial displacement as in Fig. 3 and vanishing initial velocity

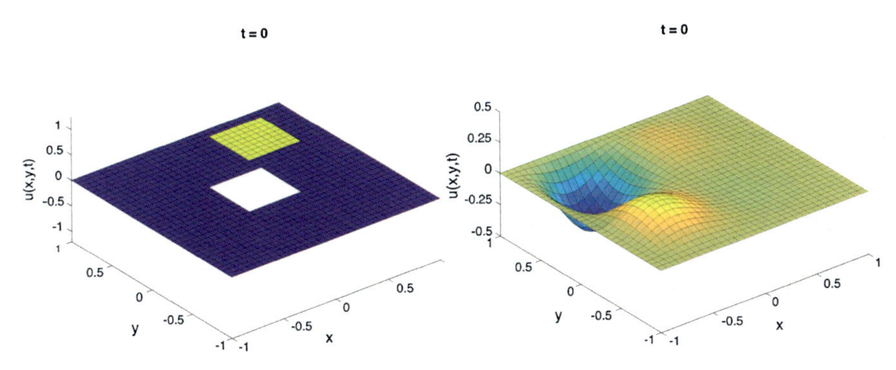

Fig. 5 Initial solutions to the nonlocal wave equation on a 2D domain with discontinuous (left) and continuous (right) initial solutions

required invariance with respect to classical symmetries such as translation, rotation, and so forth is preserved.

We illustrate the benefit of functional calculus, for instance, by comparing the Laplace operator Δ to the biharmonic operator Δ^2. By simply inspecting how the biharmonic operator is connected to the Laplace operator, one can guess that the regulating function would be $f(\lambda) = \lambda^2$. Before making a rigorous connection, one has to prescribe the BC for each operator. We choose the Laplace operator with homogeneous Dirichlet and Neumann BC and compare it to the biharmonic operator with simply supported (SS) and Cahn-Hilliard (CH) type BC for plate vibration utilizing the weak formulation of the following eigenvalue problems where the BC used are precisely the following:

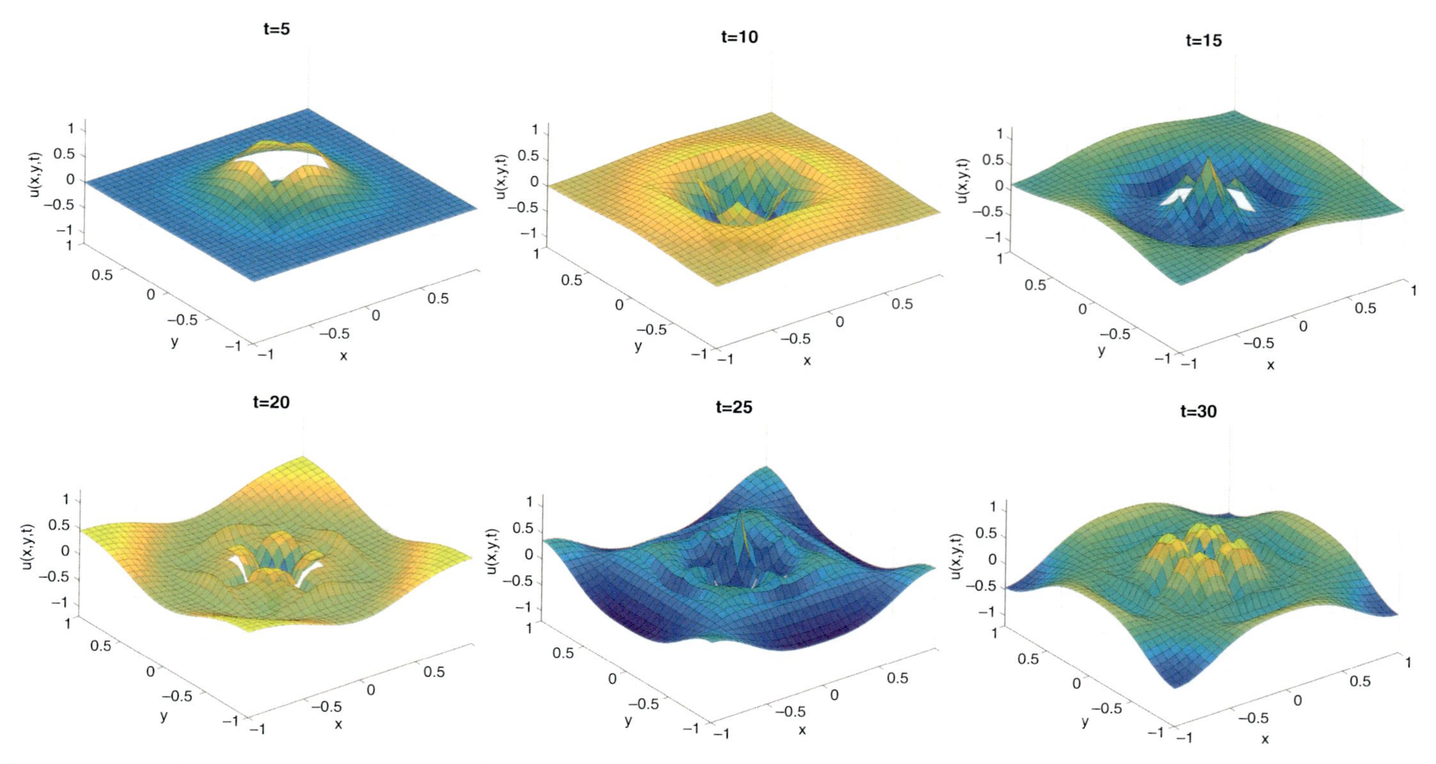

Fig. 6 Solution to the nonlocal wave equation on a 2D domain with discontinuous initial solution using the governing operator $\mathcal{M}_N$

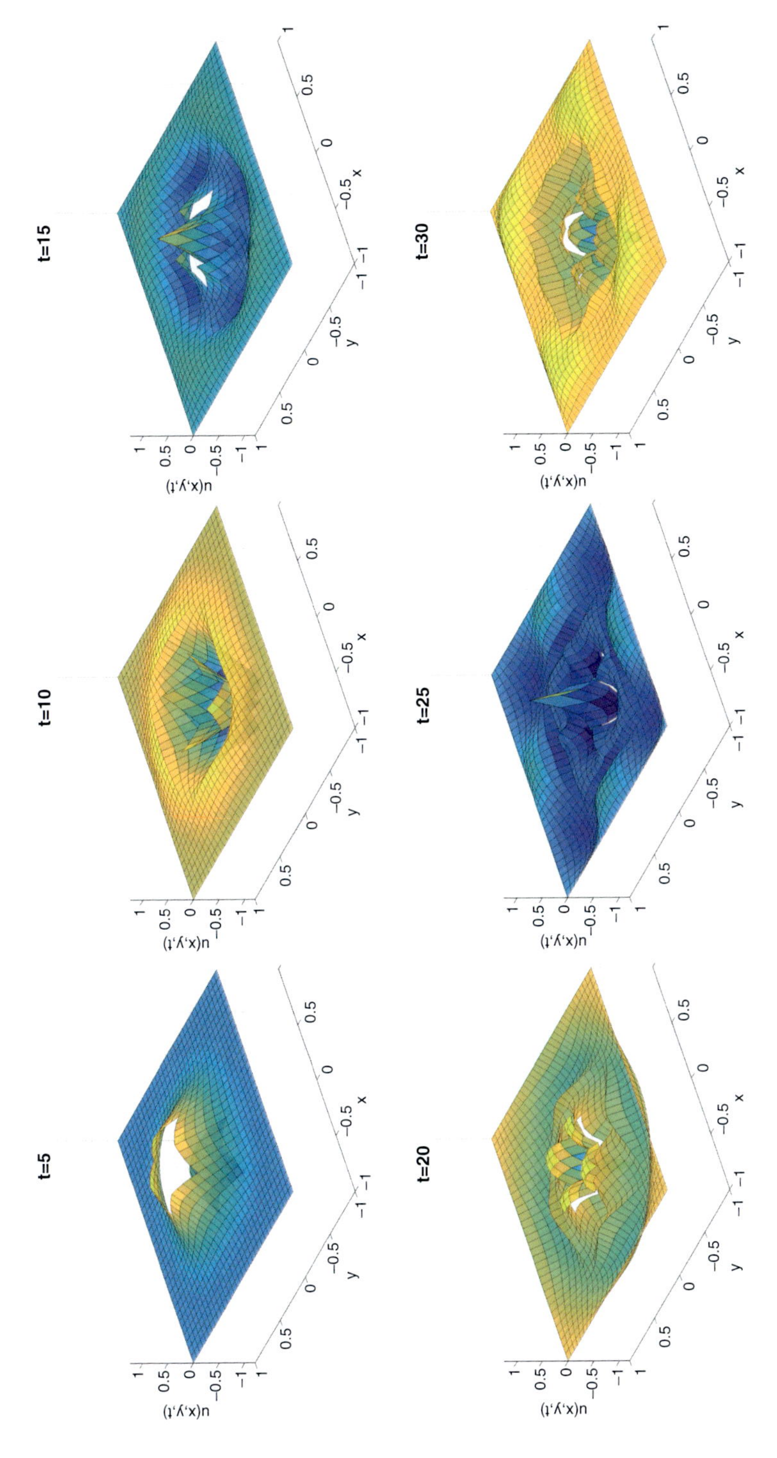

Fig. 7 Solution to the nonlocal wave equation on a 2D domain with discontinuous initial solution using the governing operator $\mathcal{M}_D$

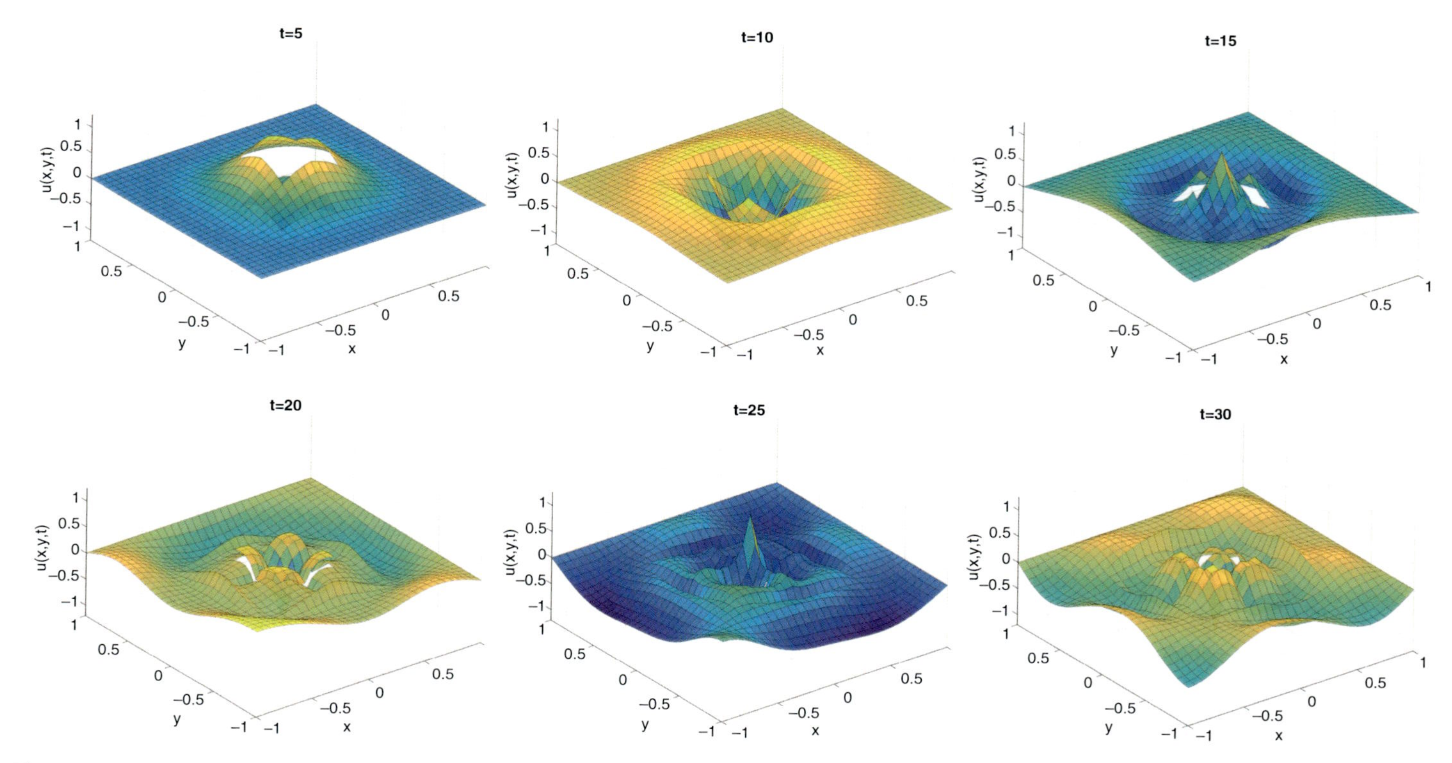

Fig. 8 Solution to the nonlocal wave equation on a 2D domain with discontinuous initial solution using the governing operator $\mathcal{M}_{\mathrm{ND,ND}}$

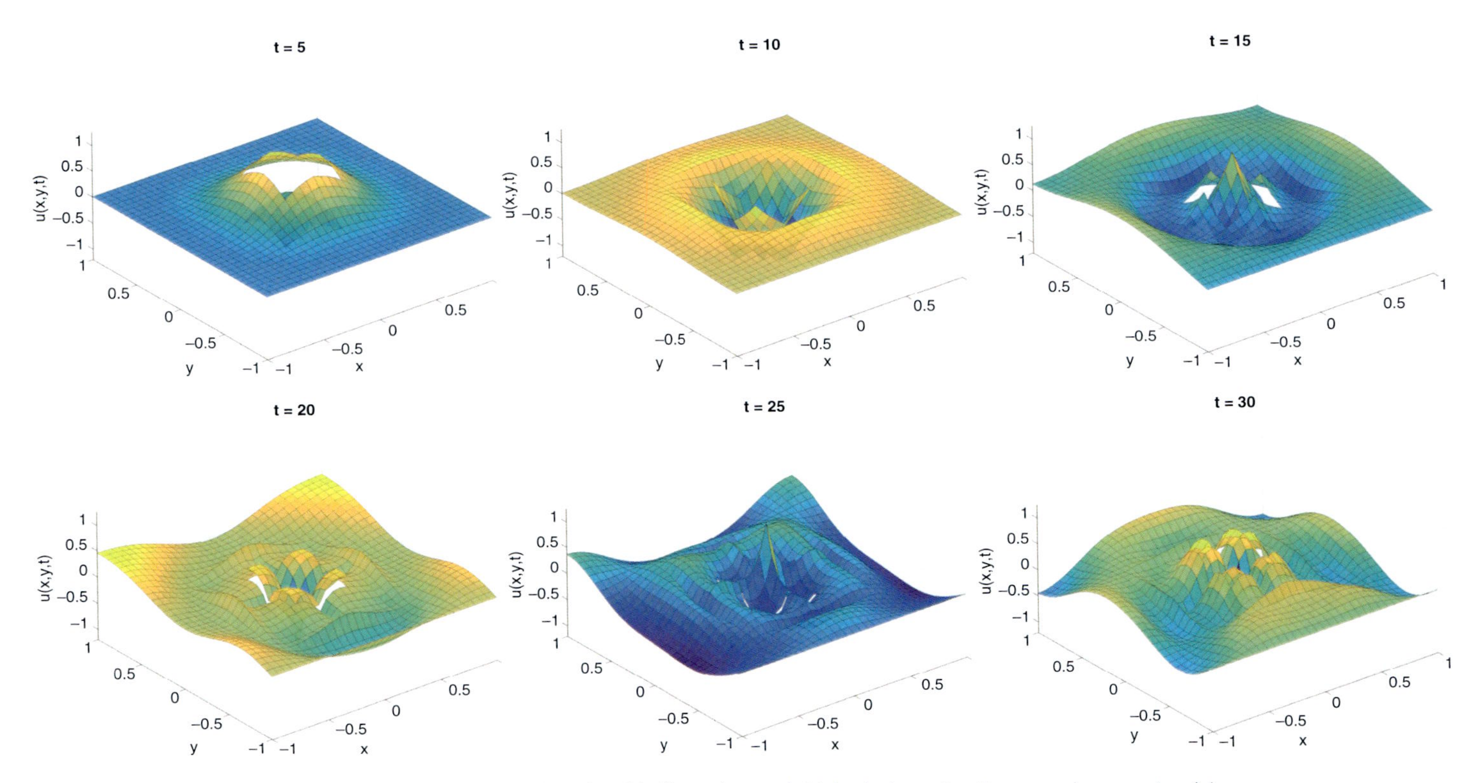

Fig. 9 Solution to the nonlocal wave equation on a 2D domain with discontinuous initial solution using the governing operator $\mathcal{M}_{\mathrm{N,DN}}$

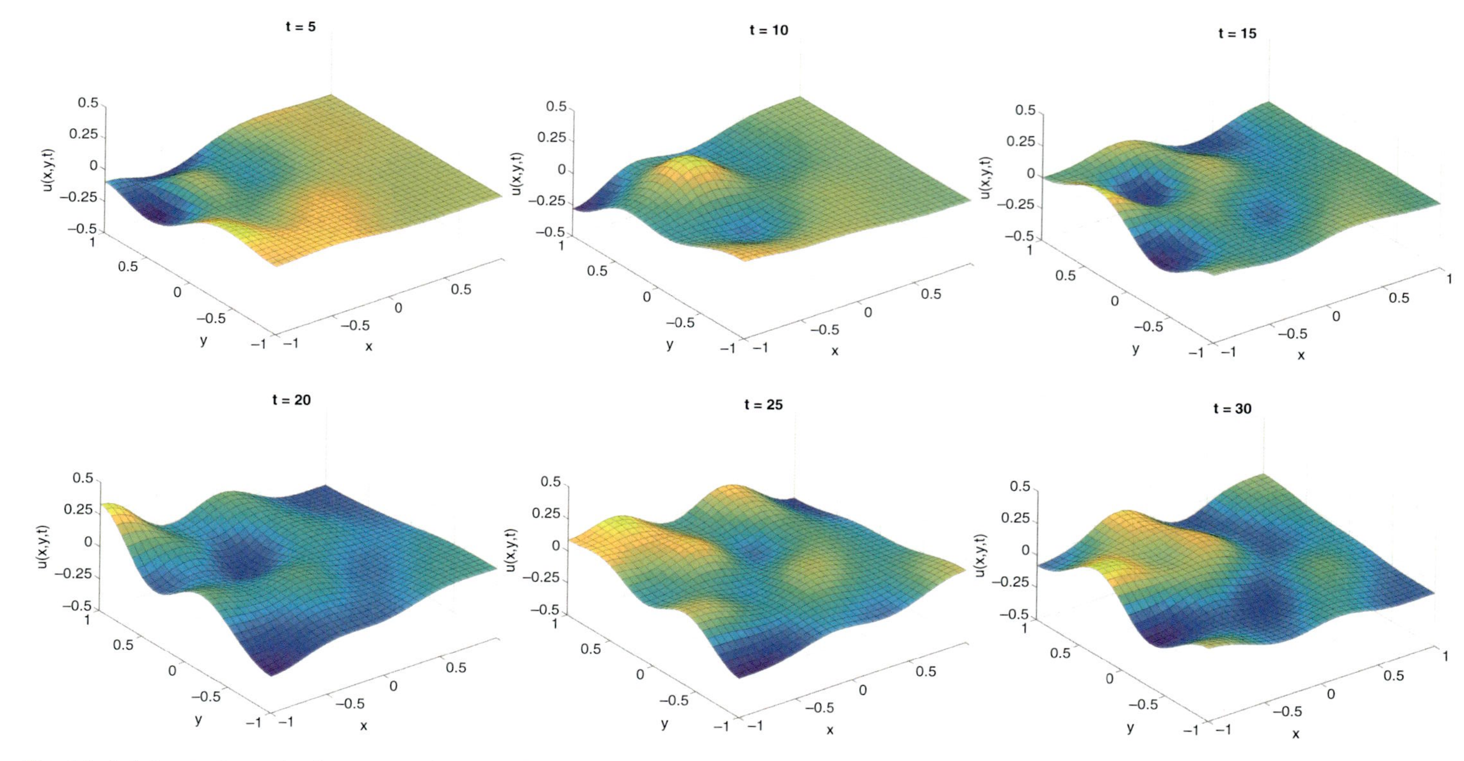

Fig. 10 Solution to the nonlocal wave equation on a 2D domain with continuous initial solution using the governing operator $\mathcal{M}_N$

$$\text{SS-BC:} \quad u = \Delta u = 0 \text{ on } \partial\Omega, \quad \text{CH-BC:} \quad \frac{\partial u}{\partial n} = \frac{\partial \Delta u}{\partial n} = 0 \text{ on } \partial\Omega.$$

$$L(u,v) := \int_\Omega \nabla u \cdot \nabla v \, dx \quad = \lambda R(u,v) \; := \lambda \int_\Omega uv \, dx, \quad v \in V_{L,D} \text{ or } V_{L,N}$$

$$B(u,v) := \int_\Omega \nabla^2 u : \nabla^2 v \, dx = \lambda R(u,v), \qquad v \in V_{B,SS} \text{ or } V_{B,CH},$$

$$V_{L,D} := H_0^1(\Omega), \qquad V_{L,N} := \left\{ v \in H^1(\Omega) : \frac{\partial v}{\partial n} = 0 \right\},$$

$$V_{B,SS} := H_0^1(\Omega) \cap H^2(\Omega), \quad V_{B,CH} := \left\{ v \in H^2(\Omega) : \frac{\partial v}{\partial n} = 0 \right\}.$$

Indeed, the eigenvalues of the biharmonic operator with SS-BC and CH-BC are the squares of those of the Laplace operator with Dirichlet and Neumann BC, respectively. Furthermore, the eigenfunctions are identical for Dirichlet and Neumann BC with SS and CH, respectively. We have provided this example as a proof of concept and, hence, chosen the BC carefully to establish the connection. One may not obtain such connection with arbitrary BC.

The convolution operators in (12) in the form of integrals are derived from their (original) series representation. We defined generalized convolution operators in Aksoylu et al. (2017a,b) in the following series form:

$$\mathcal{C}_{\mathrm{BC}} u(x) := \sum_k \langle e_k^{\mathrm{BC}} | C \rangle \, \langle e_k^{\mathrm{BC}} | u \rangle \, e_k^{\mathrm{BC}}, \tag{58}$$

where $\mathrm{BC} = \mathrm{p}, \mathrm{a}$ and $\langle \cdot | \cdot \rangle$ denotes the inner product in $L^2_{\mathbb{C}}(\Omega)$ and is defined by

$$\langle e_k^{\mathrm{BC}} | u \rangle := \int_\Omega \left(e_k^{\mathrm{BC}} \right)^* (x') u(x') dx'.$$

In addition, $\left(e_k^{\mathrm{BC}}\right)_k$ is chosen to be a basis associated with a multiple of the Laplace operator with appropriate BC, which we call as the classical operator and denote by Δ_{BC}. The spectrum of Δ_{BC} with classical BC such as periodic, antiperiodic, Neumann, and Dirichlet is purely discrete. Furthermore, we can explicitly calculate the eigenfunctions e_k^{BC} corresponding to each BC. These eigenfunctions form a Hilbert (complete and orthonormal) basis for $L^2_{\mathbb{C}}(\Omega)$ through which the generalized convolution operator is defined. The main reason why we discuss Δ_{BC} is the fact that the governing operator (1) turns out to be a function of Δ_{BC} (Aksoylu et al. 2017a,b; Beyer et al. 2016). Since the classical operator A_{BC} is defined through *local* BC, the eigenfunctions inherit this information. This observation opened a gateway to incorporate local BC to nonlocal theories on bounded domains (Aksoylu et al. 2017b).

The normalized eigenfunctions of the classical operators are as follows:

$$e_k^{\mathrm{p}}(x) := \frac{1}{\sqrt{2}} e^{i\pi k x}, \quad k \in \mathbb{N}, \qquad e_k^{\mathrm{a}}(x) := \frac{1}{\sqrt{2}} e^{i\pi(k+\frac{1}{2})x}, \quad k \in \mathbb{N},$$

$$e_k^{\mathrm{N}}(x) := \begin{cases} \frac{1}{\sqrt{2}}, & k = 0, \\ \cos\left(\frac{k\pi}{2}(x+1)\right), & k \in \mathbb{N}^*, \end{cases} \qquad e_k^{\mathrm{D}}(x) := \sin\left(\frac{k\pi}{2}(x+1)\right), \quad k \in \mathbb{N}^*. \tag{59}$$

Plugging the eigenfunctions in (59) into (58) and after hefty calculation, we proved that the operators $\mathcal{C}_{\mathrm{p}}$ and $\mathcal{C}_{\mathrm{a}}$ have integral representations given (12). For more details, see Aksoylu et al. (2017a).

Next, we present the steps how to apply functional calculus (FC). We denote a nonlocal operator by NL and its local counterpart by A. Note that both nonlocal diffusion and PD operators are defined initially on $\mathbb{R}^d$ and contain convolution. The size of nonlocality is determined by the parameter δ which is encoded in the kernel function.

FC-1. Apply limit to the horizon parameter, i.e., $\delta \to 0$, to identify a local counterpart A of NL.

FC-2. Apply the Fourier transform to "diagonalize" NL and A to obtain the corresponding spectra.

FC-3. Read off the regulating function f by comparing the spectra of NL and A. Spectra on $\mathbb{R}^d$ are continuous.

We apply the above steps to the concrete example of nonlocal diffusion on $\mathbb{R}^d$ where the classical operator A is the Laplace operator $-\Delta : W^2(\mathbb{R}^d) \to L^2(\mathbb{R}^d)$.

$$NLu(\mathbf{x},t) = f(A)u(\mathbf{x},t) := \left(\int_{\mathbb{R}^d} C(\mathbf{x}')d\mathbf{x}'\right)u(\mathbf{x},t) - \int_{\mathbb{R}^d} C(\mathbf{x}'-\mathbf{x})u(\mathbf{x}',t)d\mathbf{x}'. \tag{60}$$

We connect the nonlocal operator to A through Fourier transforms. Let $F_1 : L^1(\mathbb{R}^d) \to C_\infty(\mathbb{R}^d)$, $F_2 : L^2(\mathbb{R}^d) \to L^2(\mathbb{R}^d)$ be the Fourier transforms and the kernel function $C \in L^1(\mathbb{R}^d)$ be even:

$$A = F_2^{-1} \circ T_\lambda \circ F_2,$$

$$f(A) = F_2^{-1} \circ T_{f(\lambda)} \circ F_2, \tag{61}$$

where $T_{h(\lambda)}$ denotes the maximal multiplication operator by $h(\lambda)$. Then, we directly diagonalize $f(A)$ by using the expression given in (60):

$$f(A) = F_2^{-1} \circ T_{F_1C(0)-F_1C(\lambda)} \circ F_2. \tag{62}$$

Therefore, a comparison of (61) and (62) yields

$$f(\lambda) = F_1 C(0) - F_1 C(\lambda).$$

We explicitly identify to which function of the classical operator the nonlocal operator corresponds:

$$f(\lambda) = F_1 C(0) - F_1 C(\lambda) = \langle 1|C\rangle - \langle e_\lambda|C\rangle, \quad \langle e_\lambda|C\rangle := \int_{\mathbb{R}^d} e_\lambda(\mathbf{x}')^* C(\mathbf{x}')d\mathbf{x}',$$

where the spectral value of the classical operator $\lambda \in [0, \infty)$. Now, we extend the construction on $\mathbb{R}^d$ to a bounded domain Ω.

FC-4. Restrict A to Ω with a prescribed BC. Denote the new operator by A_{BC}. Spectrum of A_{BC}, $\sigma(A_{\mathrm{BC}})$ is now discrete. Find the eigenfunctions of A_{BC}.
FC-5. Define a generalized convolution as in (58) by using eigenfunctions of A_{BC}.
FC-6. Rewrite (recycle) the regulating function with discrete spectrum.

$$f_{\mathrm{BC}} : \sigma(A_{\mathrm{BC}}) \to \mathbb{R}, \quad f_{\mathrm{BC}}(\lambda_k^{\mathrm{BC}}) = \langle 1|C\rangle - \langle e_k^{\mathrm{BC}}|C\rangle, \quad \mathrm{BC} \in \{\mathrm{p}, \mathrm{a}\}. \tag{63}$$

FC-7. Construct $f_{\mathrm{BC}}(A_{\mathrm{BC}})$ using the spectral theorem. Namely, for $u = \sum_k \langle e_k^{\mathrm{BC}}|u\rangle\, e_k^{\mathrm{BC}}$, we have

$$f_{\mathrm{BC}}(A_{\mathrm{BC}})u = \sum_k f_{\mathrm{BC}}(\lambda_k^{\mathrm{BC}})\, \langle e_k^{\mathrm{BC}}|u\rangle\, e_k^{\mathrm{BC}}. \tag{64}$$

FC-8. Find a computationally feasible expression of $f_{\mathrm{BC}}(A_{\mathrm{BC}})$ such as an integral representation.

Now, we show how we use the **FC** steps to construct the governing operators $\mathcal{M}_{\mathrm{BC}}$, $\mathrm{BC} \in \{\mathrm{p}, \mathrm{a}\}$ in 1D. Namely, we want to verify $f_{\mathrm{BC}}(A_{\mathrm{BC}})u = \mathcal{M}_{\mathrm{BC}}u$.
Using (63) and (64), we have the following:

$$\begin{aligned} f_{\mathrm{BC}}(A_{\mathrm{BC}})u &= \sum_k [\langle 1|C\rangle - \langle e_k^{\mathrm{BC}}|C\rangle]\, \langle e_k^{\mathrm{BC}}|u\rangle\, e_k^{\mathrm{BC}} \\ &= \langle 1|C\rangle \sum_k \langle e_k^{\mathrm{BC}}|u\rangle\, e_k^{\mathrm{BC}} - \sum_k \langle e_k^{\mathrm{BC}}|C\rangle\, \langle e_k^{\mathrm{BC}}|u\rangle\, e_k^{\mathrm{BC}} \\ &= cu - \mathcal{C}_{\mathrm{BC}}u \\ &= \mathcal{M}_{\mathrm{BC}}, \quad \mathrm{BC} \in \{\mathrm{p}, \mathrm{a}\}. \end{aligned}$$

Expressing the regulating function for the case of $\mathrm{BC} \in \{\mathrm{N}, \mathrm{D}\}$ requires nontrivial manipulation of series and is more involved than the case of $\mathrm{BC} \in \{\mathrm{N}, \mathrm{D}\}$. We simply report them here:

$$f_{\mathrm{N}} : \sigma(A_{\mathrm{N}}) \to \mathbb{R}, \quad f_{\mathrm{N}}(\lambda_k^{\mathrm{N}}) = \langle 1|C\rangle - \begin{cases} \langle e_{k/2}^{\mathrm{p}}|C\rangle & \text{if } k \in \mathbb{N} \text{ is even,} \\ \langle e_{(k-1)/2}^{\mathrm{a}}|C\rangle & \text{if } k \in \mathbb{N} \text{ is odd.} \end{cases}$$

$$f_{\mathrm{D}} : \sigma(A_{\mathrm{D}}) \to \mathbb{R}, \quad f_{\mathrm{D}}(\lambda_k^{\mathrm{D}}) = \langle 1|C\rangle - \begin{cases} \langle e_{k/2}^{\mathrm{p}}|C\rangle & \text{if } k \in \mathbb{N}^* \text{ is even,} \\ \langle e_{(k-1)/2}^{\mathrm{a}}|C\rangle & \text{if } k \in \mathbb{N} \text{ is odd.} \end{cases}$$

In Aksoylu et al. (2017a), we showed that

$$f_{\mathrm{N}}(A_{\mathrm{N}})u = \big(c - \mathcal{C}_{\mathrm{p}} P_e - \mathcal{C}_{\mathrm{a}} P_o\big)u = \mathcal{M}_{\mathrm{N}} u$$
$$f_{\mathrm{D}}(A_{\mathrm{D}})u = \big(c - \mathcal{C}_{\mathrm{a}} P_e - \mathcal{C}_{\mathrm{p}} P_o\big)u = \mathcal{M}_{\mathrm{D}} u.$$

The operators $\mathcal{M}_{\mathrm{N}}$ and $\mathcal{M}_{\mathrm{D}}$ were used as governing operator in (57) to perform the numerical experiments in 1D.

Remark 7. Fractional diffusion and fractional PDEs also fall into the class of nonlocal problems; see some of the recent developments (Andreu-Vaillo et al. 2010; Caffarelli et al. 2007; Di Nezza et al. 2012; Nochetto et al. 2015). There is a fundamental difference between these operators and ours: our governing operators are bounded. Note that the regulating function in (63) is bounded and that is why the application of the spectral theorem in (64) is valid. Since our ultimate goal is to capture discontinuities or cracks, we are mainly interested in bounded governing operators. Fractional operators become unbounded for such discontinuities, and hence, we exclude them from our discussion.

Conclusion

We presented novel governing operators in arbitrary dimension for nonlocal diffusion. The operators agree with the original PD operator in the bulk of the domain and simultaneously enforce local BC. We presented methodically how to verify the BC by using a change in the order of integration. We presented different types of BC in 2D which include pure and mixed combinations of Neumann and Dirichlet BC. We presented numerical experiments for the nonlocal wave equation. We verified that the novel operators enforce local BC for all time. We also observed that the property we proved for 1D, namely, discontinuities remain stationary, also holds for 2D.

Our ongoing work aims to extend the novel operators to vector-valued problems which will allow the extension of PD to applications that require local BC. Furthermore, we hope that our novel approach potentially will avoid altogether the surface effects seen in PD.

Acknowledgements Burak Aksoylu was supported in part by the European Commission Marie Curie Career Integration 293978 grant, and Scientific and Technological Research Council of Turkey (TÜBİTAK) MFAG 115F473 grant.

References

B. Aksoylu, H.R. Beyer, F. Celiker, Application and implementation of incorporating local boundary conditions into nonlocal problems. Numer. Funct. Anal. Optim. **38**(9), 1077–1114 (2017a). https://doi.org/10.1080/01630563.2017.1320674

B. Aksoylu, H.R. Beyer, F. Celiker, Theoretical foundations of incorporating local boundary conditions into nonlocal problems. Rep. Math. Phys. **40**(1), 39–71 (2017b). https://doi.org/10.1016/S0034-4877(17)30061-7

B. Aksoylu, F. Celiker, Comparison of nonlocal operators utilizing perturbation analysis, in *Numerical Mathematics and Advanced Applications ENUMATH 2015*, vol. 112, ed. by B.K. et al. Lecture Notes in Computational Science and Engineering (Springer, 2016), pp. 589–606. https://doi.org/10.1007/978-3-319-39929-4_57

B. Aksoylu, F. Celiker, Nonlocal problems with local Dirichlet and Neumann boundary conditions. J. Mech. Mater. Struct. **12**(4), 425–437 (2017). https://doi.org/10.2140/jomms.2017.12.425

B. Aksoylu, F. Celiker, O. Kilicer, Nonlocal problems with local boundary conditions in higher dimensions (Submitted)

B. Aksoylu, A. Kaya, Conditioning and error analysis of nonlocal problems with local boundary conditions. J. Comput. Appl. Math. **335**, 1–19 (2018). https://doi.org/10.1016/j.cam.2017.11.023

B. Aksoylu, T. Mengesha, Results on nonlocal boundary value problems. Numer. Funct. Anal. Optim. **31**(12), 1301–1317 (2010)

B. Aksoylu, M.L. Parks, Variational theory and domain decomposition for nonlocal problems. Appl. Math. Comput. **217**, 6498–6515 (2011). https://doi.org/10.1016/j.amc.2011.01.027

B. Aksoylu, Z. Unlu, Conditioning analysis of nonlocal integral operators in fractional Sobolev spaces. SIAM J. Numer. Anal. **52**(2), 653–677 (2014)

F. Andreu-Vaillo, J.M. Mazon, J.D. Rossi, J. Toledo-Melero, *Nonlocal Diffusion Problems*. Mathematical Surveys and Monographs, vol. 165. (American Mathematical Society and Real Socied Matematica Espanola, 2010)

H.R. Beyer, B. Aksoylu, F. Celiker, On a class of nonlocal wave equations from applications. J. Math. Phys. **57**(6), 062902 (2016). https://doi.org/10.1063/1.4953252. Eid:062902

L. Caffarelli, L. Silvestre, An extension problem related to the fractional Laplacian. Commun. Part. Diff. Eqs. **32**, 1245–1260 (2007)

E. Di Nezza, G. Palatucci, E. Valdinoci, Hitchhiker's guide to fractional Sobolev spaces. Bull. Sci. Math. **136**(5), 521–573 (2012)

Q. Du, M. Gunzburger, R.B. Lehoucq, K. Zhou, Analysis and approximation of nonlocal diffusion problems with volume constraints. SIAM Rev. **54**, 667–696 (2012)

Q. Du, R. Lipton, Peridynamics, fracture, and nonlocal continuum models. SIAM News **47**(3) (2014)

E. Emmrich, O. Weckner, The peridynamic equation and its spatial discretization. Math. Model. Anal. **12**(1), 17–27 (2007)

G. Gilboa, S. Osher, Nonlocal operators with applications to image processing. Multiscale Model. Simul. **7**(3), 1005–1028 (2008)

B. Kilic, Peridynamic theory for progressive failure prediction in homogeneous and heterogeneous materials. Ph.D. thesis, Department of Aerospace and Mechanical Engineering, University of Arizona, Tucson (2008)

E. Madenci, E. Oterkus, *Peridynamic Theory and Its Applications* (Springer, New York/Heidelberg/Dordrecht/London, 2014). https://doi.org/10.1007/978-1-4614-8465-3

T. Mengesha, Q. Du, Analysis of a scalar peridynamic model for sign changing kernel. Disc. Cont. Dyn. Sys. B **18**, 1415–1437 (2013)

J.A. Mitchell, S.A. Silling, D.J. Littlewood, A position-aware linear solid constitutive model for peridynamics. J. Mech. Mater. Struct. **10**(5), 539–557 (2015)

R.H. Nochetto, E. Otarola, A.J. Salgado, A PDE approach to fractional diffusion in general domains: a priori error analysis. Found. Comput. Math. **15**, 733–791 (2015)

P. Seleson, M. Gunzburger, M.L. Parks, Interface problems in nonlocal diffusion and sharp transitions between local and nonlocal domains. Comput. Methods Appl. Mech. Eng. **266**, 185–204 (2013)

P. Seleson, M.L. Parks, On the role of the influence function in the peridynamic theory. Internat. J. Multiscale Comput. Eng. **9**(6), 689–706 (2011)

S. Silling, Reformulation of elasticity theory for discontinuities and long-range forces. J. Mech. Phys. Solids **48**, 175–209 (2000)

S. Silling, R.B. Lehoucq, Peridynamic theory of solid mechanics. Adv. Appl. Mech. **44**, 73–168 (2010)

S.A. Silling, M. Zimmermann, R. Abeyaratne, Deformation of a peridynamic bar. J. Elast. **73**, 173–190 (2003)

X. Tian, Q. Du, Analysis and comparison of different approximations to nonlocal diffusion and linear peridynamic equations. SIAM J. Numer. Anal. **51**(6), 3458–3482 (2013)

O. Weckner, R. Abeyaratne, The effect of long-range forces on the dynamics of a bar. J. Mech. Phys. Solids **53**, 705–728 (2005)

Peridynamics and Nonlocal Diffusion Models: Fast Numerical Methods

40

Hong Wang

Contents

Abstract

We outline the recent developments of fast numerical methods for linear nonlocal diffusion and peridynamic models in one and two space dimensions. We show how the analysis was carried out to take full advantage of the structure of the stiffness matrices of the numerical methods in its storage, evaluation, and assembly and in the efficient solution of the corresponding numerical schemes. This significantly reduces the computational complexity and storage of the numerical methods over conventional ones, without using any lossy compression. For instance, we would use the same numerical quadratures for conventional methods to evaluate the singular integrals in the stiffness matrices, except that we only need to evaluate $O(N)$ of them instead of $O(N^2)$ of them. Numerical results are presented to show the utility of these fast methods.

H. Wang (✉)
Department of Mathematics, University of South Carolina, Columbia, SC, USA
e-mail: hwang@math.sc.edu

G. Z. Voyiadjis (ed.), *Handbook of Nonlocal Continuum Mechanics for Materials and Structures*, https://doi.org/10.1007/978-3-319-58729-5_35

Keywords
Peridynamics · Nonlocal diffusion model · Fast numerical method · Toplitz matrix

Introduction

In the last couple of decades, many nonlocal models have been developed, which are emerging as powerful tools for modeling challenging phenomena in many disciplines and applications, including problems involving anomalous transport and long-range time memory or spatial interactions. For instance, in the scenario of diffusion processes, Fick first set up the diffusion equation at a macroscopic scale in a deterministic manner during his study on how nutrients travel through membranes in living organisms. But Fick's approach was phenomenological. It was Einstein who derived the diffusion equation from first principle as part of his work on Brownian motion. Alternatively, Pearson modeled diffusion process via a stochastic formulation at microscopic scale in terms of a random walk, leading to the stochastic differential equation driven by a Brownian motion (Meerschaert and Sikorskii 2011; Metzler and Klafter 2000; Øksendal 2010; Podlubny 1999).

The common assumptions between Einstein's explanation of diffusion and Pearson's random walk are (i) the existence of a mean free path and (ii) the existence of a mean waiting time. Under these assumptions, the central limit theorem concludes that the underlying stochastic process that describes the particle jumps is a Brownian motion in the Pearson's approach. Equivalently, the corresponding probability density function describing the stochastic process satisfies the Fickian diffusion equation as the Fokker-Planck equation of the Brownian motion in the Einstein's approach (Meerschaert and Sikorskii 2011). Note that the assumptions (i) and (ii), which lead to a stochastic process of Brownian motion, virtually hold for transport processes in homogeneous media. In highly heterogeneous media with faults and fractures, the underlying particle movements in the transport processes may experience long jumps and so will have large deviation from Brownian motion but converge to a Lévy process. Consequently, the corresponding probability density function satisfies a fractional partial differential equation (PDE) as its Fokker-Planck equation (Meerschaert and Sikorskii 2011). This justifies why a fractional PDE or a nonlocal diffusion model better describes anomalous diffusive processes in heterogeneous media than traditional integer-order PDEs (del-Castillo-Negrete et al. 2004; D'Elia et al. 2014; Du et al. 2012; Meerschaert and Sikorskii 2011; Metzler and Klafter 2000; Podlubny 1999).

Similar phenomena occur in the context of continuum solid mechanics. In this case, the classical theory assumes that all internal forces act locally and yields mathematical models that are expressed in terms of PDEs. These models have difficulties to describe problems with evolving discontinuities, due to its differentiability assumption on displacement fields. The peridynamic theory (Silling 2000; Silling et al. 2007) was proposed as a reformation of the classical theory of continuum solid mechanics and yields nonlocal mathematical formulations that are based on long-range interactions. Constitutive models in the peridynamic theory depend on finite deformation vectors, instead of deformation gradients in classical constitutive models. Consequently, peridynamic models are particularly suitable for the representation of discontinuities in displacement fields and the description of cracks and their evolution in materials. To date, peridynamic models have been successfully applied in many important applications, including failure and damage in composite laminates (Oterkus et al. 2012), crack propagation and branching (Ha and Bobaru 2011), crack nucleation (Silling et al. 2010; Parks et al. 2008; Seleson et al. 2009), phase transformations in solids (Dayal and Bhattacharya 2006), impact damage (Seleson and Parks 2011; Silling and Askari 2005), polycrystal fracture (Ghajari et al. 2014), crystal plasticity (Sun and Sundararaghavan 2014), damage in concrete (Gerstle et al. 2007), and geomaterial fragmentation (Lai et al. 2015).

Because these nonlocal models provide greatly improved modeling capability than traditional integer-order PDE models do, different numerical methods have been developed for these models with different computational expense, memory requirements, and implementational effort as well as accuracy, convergence, and stability (Emmrich and Weckner 2007). For example, collocation methods and mesh-free methods apply directly to the strong form of the nonlocal models and are relatively simple to implement (Seleson et al. 2016; Seleson and Littlewood 2016; Silling and Askari 2005). On the other hand, finite element discretizations of the nonlocal models are based on Galerkin weak forms and enjoy high convergence rates (Chen and Gunzburger 2011; Du et al. 2013; Oterkus et al. 2012; Tian and Du 2013; Zhou and Du 2010).

However, nonlocal models involve singular integral operators and present numerical difficulties that were not encountered in classical integer-order PDE models: (i) the numerical methods for integer-order PDE models generate sparse stiffness matrices which have a linear storage requirement of $O(N)$ where N is the number of spatial unknowns. The matrix-vector multiplication has computational work of $O(N)$. Hence, the convergence of any Krylov subspace iterative solver depends only on the number of iterations. In contrast, the numerical methods for nonlocal models yield dense stiffness matrices in which their bandwidths increase to infinity as the mesh size decreases to zero. Consequently, the numerical methods for nonlocal models have $O(N^2)$ memory requirement. Direct solvers have been widely used in the numerical simulation of nonlocal models, which have an $O(N^3)$ computational complexity in order to obtain the numerical solutions. On the other hand, the matrix-vector multiplication by the stiffness matrices has $O(N^2)$ computational complexity. (ii) Due to the impact of the singular kernels in the integral operators in nonlocal models, the evaluation and assembly of the stiffness matrices require the evaluation of $O(N^2)$ entries, which can be very expensive. Numerical experiments show that the evaluation and assembly of the coefficient matrix often constitute a very large portion of simulation times! (iii) While they have high-order convergence rates, Galerkin finite element methods double the number of spatial dimensions that need to be discretized, leading to a significantly increased computational work and memory requirement as compared to collocation methods. In summary, the significantly increased computational work and memory requirement of the nonlocal models over those for the classical integer-order PDE models severely limit the applications of these models, especially for problems in multiple space dimensions.

In this chapter we go over the recent developments of fast and accurate numerical methods for nonlocal models. We do not intend to develop a more accurate numerical discretization or numerical quadrature for these models. Rather, we explore the structure of the stiffness matrices of existing numerical methods in order to significantly reduce the computational work to evaluate and assemble the stiffness matrices and to compute the numerical solutions, as well as the memory requirement to store the stiffness matrices. The idea applies to different numerical methods and numerical quadratures in the literature. In this chapter we take some representative numerical methods as examples to illustrate the developments.

The rest of this chapter is organized as follows. In section "Nonlocal Diffusion and Peridynamic Models" we present a nonlocal diffusion model and a bond-based peridynamic model and discuss related numerical issues. In section "Fast Numerical Methods for a One-Dimensional Peridynamic Model" we go over the development of fast Galerkin finite element method and a fast collocation method for a one-dimensional peridynamic model. In section "A Fast Collocation Method for the Nonlocal Diffusion Model (1)" we outline the development of a fast collocation method for a nonlocal diffusion model in two space dimensions. In section "A Fast Collocation Method for the Peridynamic Model (3)" we go over the development of a fast collocation method for a bond-based peridynamic model in two space dimensions. In section "Concluding Remarks" we draw concluding remarks and discuss the future directions.

Nonlocal Diffusion and Peridynamic Models

In this section we use the presentative nonlocal diffusion model and peridynamic model in two space dimensions as an example to illustrate the corresponding numerical issues.

A Nonlocal Diffusion Model

A linear nonlocal diffusion model can be expressed as the following integral-differential equation (D'Elia et al. 2014; Du et al. 2012):

$$\begin{aligned} u_t(x,y,t) + \int_{B_\delta(x,y)} & K(x-x',y-y')\big(u(x,y,t)-u(x',y',t)\big)dx'dy' \\ &= f(x,y,t), \quad (x,y)\in\Omega,\ t\in(0,T], \\ u(x,y,0) &= u_o(x,y), \quad (x,y)\in\Omega, \\ u(x,y,t) &= g(x,y,t), \quad (x,y)\in\Omega_c,\ t\in[0,T]. \end{aligned} \tag{1}$$

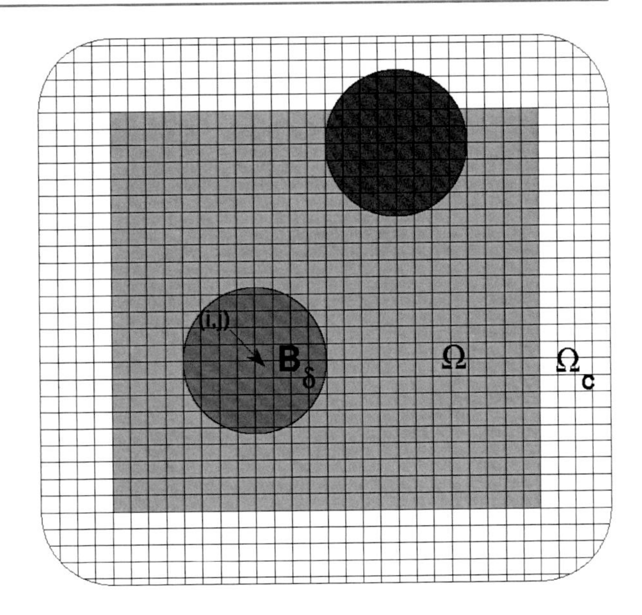

Fig. 1 Illustration of a spatial partition on $\Omega \cup \Omega_c$ with a circular horizon

Here Ω is the physical domain in the plane, Ω_c represents a boundary zone of width δ surrounding Ω with $\delta > 0$ being the horizon parameter of the material, u represents the concentration of the solute, u_o is the prescribed initial condition, and f is the source and sink term, respectively; $K(x, y)$ is the kernel function of the integral operator which is often of the form

$$K(x, y) := \frac{C}{\left(x^2 + y^2\right)^{1+\gamma}} \tag{2}$$

with C being a normalization constant; $B_\delta(x, y)$ represents the material horizon which is often chosen to be a disk centered at (x, y) with radius δ (cf. Fig. 1).

We note that in the nonlocal diffusion model (1), the boundary condition is "volume constrained" that is imposed on a two-dimensional boundary zone Ω_c. This is in contrast to the boundary conditions for classical integer-order PDE models that are imposed on the boundary $\partial\Omega$ of the physical domain Ω. It is clear that the integral operator in (1) can be viewed as a truncated version of the fractional Laplacian operator $(-\Delta)^\gamma$, which is the infinitesimal generator of a Feller semigroup that is generated by a Lévy process (Applebaum 2009; Meerschaert and Sikorskii 2011).

A Bond-Based Linear Peridynamic Model

In a bond-based linear peridynamic model for modeling a proportional, microelastic material in a reference domain in the plane, the equation of motion at any point

(x, y) in the reference configuration Ω at any time $t \in (0, T]$ can be expressed as (Silling 2000)

$$
\begin{aligned}
\rho \mathbf{u}_{tt}(x, y, t) + \int_{B_\delta(x,y)} & \mathbf{K}(x - x', y - y')(\mathbf{u}(x, y, t) - \mathbf{u}(x', y', t))dx'dy' \\
&= \mathbf{f}(x, y, t), \quad (x, y) \in \Omega, \ t \in (0, T], \\
\mathbf{u}(x, y, 0) &= \mathbf{u}_o(x, y), \quad (x, y) \in \Omega, \\
\mathbf{u}(x, y, t) &= \mathbf{g}(x, y, t), \quad (x, y) \in \Omega_c, \ \ t \in [0, T].
\end{aligned} \tag{3}
$$

Here $\mathbf{u}(x, y, t) = [v(x, y, t), w(x, y, t)]^T$ represents the displacement vector field with $v(x, y, t)$ and $w(x, y, t)$ being its components in the x and y directions, respectively; similarly, $\mathbf{f}(x, y, t) = \left[f^v(x, y, t), f^w(x, y, t)\right]^T$ represents an external force density field, and $\mathbf{g}(x, y, t) = \left[g^v(x, y, t), g^w(x, y, t)\right]^T$ is the prescribed nonlocal boundary data imposed on the domain Ω_c. ρ is the mass density in the reference configuration. The micromodulus function for the material or the kernel of the integral operator $\mathbf{K}$ is of the form

$$
\mathbf{K}(x, y) = \frac{c}{\left(x^2 + y^2\right)^{3/2}} \begin{bmatrix} x^2 & xy \\ xy & y^2 \end{bmatrix} \tag{4}
$$

where c is a constant that depends on the material properties and space dimensions.

Numerical Issues and Computational Complexity

The mathematical analysis for the nonlocal and peridynamic models and their corresponding numerical analysis can be found, e.g., in Chen and Gunzburger (2011), D'Elia et al. (2014), Du et al. (2012, 2013), Tian and Du (2013), Zhang et al. (2016), and Zhou and Du (2010). In particular, the following error estimate which was proved in Zhou and Du (2010) for the linear finite element method for a steady-state nonlocal diffusion or peridynamic model states that for any $0 < \varepsilon \ll 1$, there exists a positive constant $C = C(\varepsilon)$ which is independent of h such that

$$
\|u - u^h\|_{L^2} \leq Ch^{2-\varepsilon}\|u\|_{H^2}. \tag{5}
$$

The estimate looks similar to that for the finite element method for classical integer-order PDEs in which the stiffness matrix is sparse. In this case, the stiffness matrix can be stored in $O(N)$ memory and each matrix-vector multiplication (so each Krylov subspace iteration) can be carried out in $O(N)$ computations.

However, for the nonlocal diffusion model (1) and peridynamic model (3), the singular integral operator is defined on the material horizon $B_\delta(x, y)$, which yields a dense stiffness matrix that has $O(N)$ nonzero entries at each row, as the mesh size

h tends to zero. Asymptotically, the stiffness matrix requires $O(N^2)$ memory to store and each matrix-vector multiplication (and so each Krylov subspace iteration) has $O(N^2)$ computational complexity. Alternatively, a direct solver has $O(N^3)$ computational complexity to invert the linear system at each time step.

This has profound impact computationally. For example, for a three-dimensional analogue of the nonlocal diffusion model (1) or the peridynamic model (3), each time we refine the mesh size and time step by half, the memory requirement would increase 64 times, and the computational complexity by a direct solver increases by $8^3 \times 2 = 1024$ times!

If we use a finite difference or finite element method with 1000 by 1000 by 1000 nodes to solve an integer-order diffusion PDE or elasticity model in three space dimensions, a multigrid solver can solve the problem on the order of 10^9 operations. However, if we use a similar method with the same number of nodes and time steps to solve a three-dimensional analogue of the nonlocal diffusion model (1) or the peridynamic model (3), a direct solver would have a computational complexity on the order 10^{27} per time step. This will take state-of-the-art supercomputer at least years of CPU time to finish the simulation, to say the least. Here we have not taken into account for the extra computational complexity due to the nonlinearity and the evaluation and assembly of the stiffness matrices in the numerical simulation. The significantly increased computational complexity and memory requirements of the nonlocal diffusion model and the peridynamic model significantly limit their applications in reality, especially for problems in three space dimensions.

A simplified peridynamic model was proposed to reduce its computational complexity by assuming that the horizon parameter $\delta = O(h)$ (Chen and Gunzburger 2011). Under this assumption, the stiffness matrix of a numerical scheme becomes local like a high-order finite difference or finite element method for an integer-order PDE. However, it is not clear why physically relevant horizon parameter δ is proportional to the computationally chosen mesh size h. This inconsistency is also reflected in the error estimate of the corresponding finite element method, which now reduces to the following first-order estimate in this case:

$$\|u - u^h\|_{L^2} \leq Ch\|u\|_{H^2}. \tag{6}$$

In this chapter we go over the recent developments of fast and accurate, matrix-free numerical methods with an efficient matrix evaluation and assembly for the nonlocal diffusion model and the peridynamic model without the assumption of $\delta = O(h)$.

Fast Numerical Methods for a One-Dimensional Peridynamic Model

In this section we go over the development of fast and accurate numerical methods with efficient matrix assembly for a one-dimensional steady-state peridynamic model of a finite bar that consists of microelastic materials, which is given by the following pseudo-differential equation (Chen and Gunzburger 2011; Silling 2000):

$$\int_{\alpha}^{\beta} \frac{u(x) - u(y)}{|x - y|^{1+\gamma}} dy = f(x), \quad x \in (\alpha, \beta), \tag{7}$$

which is closed by the homogeneous Dirichlet boundary condition that is imposed on the complement of the interval (α, β) or simply in the neighborhood of the interval (α, β).

A Fast Galerkin Finite Element Method on a Uniform Partition

Define a uniform spatial partition on (α, β) by

$$x_i := \alpha + ih, \ 0 \le i \le N, \qquad h := \frac{\beta - \alpha}{N} \tag{8}$$

for a given positive integer N. Let S^h be the linear finite element space defined on $[\alpha, \beta]$ with respect to the given partition.

Let $\psi(\xi) := 1 - |\xi|$ for $\xi \in [-1, 1]$ or zero otherwise. Let $\{\phi_i\}_{i=1}^{N-1}$ be the standard nodal basis functions for S^h, i.e.,

$$\phi_i(x) := \psi\Big(\frac{x - x_i}{h}\Big) \tag{9}$$

Let $u^h \in S^h$ be the finite element approximation to the true solution u of problem (7). Then u^h can be expressed as follows:

$$u^h(x) = \sum_{j=1}^{N-1} u_j \phi_j(x). \tag{10}$$

Then the Galerkin finite element method can be expressed as follows:

$$\sum_{j=1}^{N-1} a(\phi_i, \phi_j) u_j = l(\phi_i), \quad 1 \le i \le N - 1, \tag{11}$$

where the bilinear form $a(\cdot, \cdot)$ and the linear functional l are defined by

$$\begin{aligned} a(\phi_j, \phi_i) &= \frac{1}{h} \int_{\alpha}^{\beta} \phi_i(x) \int_{\alpha}^{\beta} \frac{\phi_j(x) - \phi_j(y)}{|x - y|^{1+\gamma}} dy dx, \\ l(\phi_i) &= \frac{1}{h} \int_{\alpha}^{\beta} f(x) \phi_i(x) dx. \end{aligned} \tag{12}$$

Let $u := [u_1, u_2, \dots, u_{N-1}]^T$, $f := [f_1, f_2, \dots, f_{N-1}]^T$, and $A := [a_{i,j}]_{i,j=1}^{N-1}$ with $a_{i,j} = a(\phi_j, \phi_i)$ and $f_i = l(\phi_i)$. The finite element method (11) can be expressed in a matrix form

$$Au = f. \tag{13}$$

It can be proved that the stiffness matrix A is a symmetric and positive definite matrix. Moreover, it is clear from (12) that the stiffness matrix A is full. Consequently, the Galerkin finite element method (11) has a significantly increased memory requirement and computational complexity compared to that for an integer-order PDE model.

Theorem 1. *The linear system* (14) *can be solved via the conjugate gradient method in a matrix-free fashion, which requires $O(N)$ memory and has an $O(N \log N)$ computational complexity per conjugate gradient iteration.*

Outline of proof. We carefully studied all the entries of the stiffness matrix A in Wang and Wang (2010) and proved that the stiffness matrix A can be decomposed as

$$A = A_{tr} + T. \tag{14}$$

Here A_{tr} is a tridiagonal matrix and $T = [t_{i,j}]_{i,j=1}^{N-1}$ contains all remaining nonzero entries of A and is a symmetric Toeplitz matrix. We let q_{j-i} denote the common entry in the $(j-i)$th descending diagonal of T from left to right. Namely,

$$t_{i,j} = q_{j-i}, \qquad j \geq i. \tag{15}$$

Then the symmetric Toeplitz matrix A_o can be embedded into a $2(N-1)\times 2(N-1)$ circulant matrix C as follows (Chan and Ng 1996):

$$C := \begin{bmatrix} T & S \\ S & T \end{bmatrix} \qquad S := \begin{bmatrix} 0 & q_{N-2} & \cdots & q_2 & q_1 \\ q_{N-2} & 0 & q_{N-2} & \cdots & q_2 \\ \vdots & q_{N-2} & 0 & \ddots & \vdots \\ q_2 & \vdots & \ddots & \ddots & q_{N-2} \\ q_1 & q_2 & \cdots & q_{N-2} & 0 \end{bmatrix}. \tag{16}$$

The circulant matrix C can be diagonalized by the discrete Fourier transform matrix (Chan and Ng 1996)

$$C = F^{-1}\,\mathrm{diag}(Fc)\,F \tag{17}$$

where c is the first column vector of C and F is the $2(N-1) \times 2(N-1)$ discrete Fourier transform matrix.

Table 1 The efficiency of Gauss, CG and FCG

h	Gauss	CG		FCG	
	CPU	Iter.	CPU	Iter.	CPU
2^{-10}	47 s	36	1.19 s	36	0.18 s
2^{-14}	6 days 11 h	46	9 m 50 s	46	3.30 s
2^{-16}	Stopped test	53	3 h 9 m	53	18 s
2^{-17}	Out of memory	N/A	Out of memory	56	1 m 29 s
2^{-18}	N/A	N/A	N/A	58	2 m 51 s
2^{-28}	N/A	N/A	N/A	92	3 days 11 h

For any $v \in \mathbb{R}^{N-1}$, define v_{2N-2} by

$$v_{2N-2} = \begin{bmatrix} v \\ 0 \end{bmatrix}, \; C v_{2N-2} = \begin{bmatrix} T & S \\ S & T \end{bmatrix} \begin{bmatrix} v \\ 0 \end{bmatrix} = \begin{bmatrix} Tv \\ Sv \end{bmatrix}. \tag{18}$$

By (18) to evaluate Tv for any vector $v \in \mathbb{R}^{N-1}$, we need only to evaluate $C v_{2N}$ for v_{2N-2} defined in (7). Note that the latter can be evaluated using the decomposition (17) via the fast Fourier transform (FFT) in $O((2N)\log(2N)) = O(N \log N)$ operations. Hence, Tv can be evaluated in $O(N \log N)$ operations. Note that the matrix-vector multiplication is carried out using (17) and (18), which does not require the entire matrix T, but the first column vector c of the matrix C. Hence, it is matrix-free and requires only $O(N)$ memory.

The matrix A_{tr} is a tridiagonal matrix which can be stored in $O(N)$ memory, and $A_{tr} v$ can be evaluated in $O(N)$ operations. □.

The following table shows the numerical performance of Gaussian elimination (Gauss), the conventional conjugate gradient method (FG), and the fast conjugate gradient method (FCG) for a one-dimensional model problem (Wang and Wang 2010). The numerical results show that all the three solvers generate solutions with the same accuracy. The table below shows the efficiency of different solvers (Table 1).

A Fast Collocation Method on a Graded Mesh

Note that peridynamic model intends to describe problems with singularity. Hence, a fast numerical method on a uniform mesh is very efficient but not as effective. In Tian et al. (2013) we developed a fast collocation method for the peridynamic model (7) with $r = 0$ and an assumed singularity point at the right-end point $x = \beta$. In this case, a geometrically graded mesh is defined below

$$x_0 := \alpha, \qquad x_i := \alpha + \sum_{k=1}^{i} \frac{\beta - \alpha}{2^k}, \qquad 1 \le i \le N-1, \qquad x_N := \beta. \tag{19}$$

Let $\{\phi_i\}_{i=1}^{N-1}$ be the nodal basis functions defined in (9) except that the partition is graded now. In this case, the trial function u_h is still defined in (18). If we choose x_i $(i = 1 \ldots, N-1)$ to be the collocation points, then a collocation method can be formulated as follows:

$$\int_\alpha^\beta \frac{u_i - u(y)}{|x_i - y|} dy = f(x_i), \quad 1 \le i \le N-1. \tag{20}$$

Then the collocation method (20) can be expressed in the matrix form (14) except that now $f_i = f(x_i)$ and

$$a_{i,j} = \int_\alpha^\beta \frac{-\phi_j}{|x_i - y|} dy, \ i \ne j, \quad a_{i,i} = \int_\alpha^\beta \frac{1-\phi_i}{|x_i - y|} dy, \ 1 \le i \le N-1. \tag{21}$$

Since the partition (19) is geometrically graded, the basis functions $\{\phi_i\}_{i=1}^{N-1}$ are not translation invariant. We analyzed the structure of the stiffness matrix A and proved the following theorem in

Theorem 2. *The stiffness matrix A defined in* (21) *can be decomposed as*

$$A = T + L. \tag{22}$$

Here T is a Toeplitz matrix of order $(N-1)$ which is obtained by extending the matrix $A_{l,l}$ to an $(N-1)$-by-$(N-1)$ matrix, and L is a low-rank matrix of rank at most 2 which is nonzero only in the last row or last column.

Consequently, A can be stored in $O(N)$ memory. The linear system (14) *can be solved via any Krylov subspace iterative method with $O(N \log N)$ operations per iteration.*

Outline of proof. Note that the rightmost cell has the same size as its left neighbor, so the gridding violates the geometrically decreasing pattern which affects the structure of the matrix. To analyze the matrix structure, we express the stiffness matrix A as a two-by-two block matrix of the form

$$A = \begin{bmatrix} A_{l,l} & A_{l,r} \\ A_{r,l} & A_{r,r} \end{bmatrix}. \tag{23}$$

Here $A_{l,l}$ is an $(N-2)$-by-$(N-2)$ submatrix that consists of the first $N-2$ rows and columns in A, $A_{l,r}$ consists of the first $(N-2)$ entries in the last column of A, $A_{r,l}$ consists of the first $(N-2)$ entries in the last row of A, and $A_{r,r}$ consists of the entry in the last row and column of A.

We proved in Tian et al. (2013) that the submatrix $A_{l,l}$ is a Toeplitz matrix of order $(N-1)$. By extending $A_{l,l}$ to a Toeplitz matrix T of order $(N-1)$, we can get the decomposition (22). The rest of the theorem can be proved as in Theorem 1. □

For the graded mesh (19), the stiffness matrix A is nonsymmetric. Hence, a nonsymmetric Krylov subspace iterative method should be used. Numerical experiments were presented in Tian et al. (2013) which shows the utility of the fast method on graded mesh compared that on the uniform mesh. This coincides with the observations in the context of the numerical methods for fractional PDEs (Jia et al. 2014; Jia and Wang 2015).

A Fast Collocation Method for the Nonlocal Diffusion Model (1)

In this section we outline the development of a fast bilinear collocation method for the nonlocal diffusion model (1) on a rectangular domain $\Omega = (0, x_r) \times (0, y_r)$. As we focus on the study of the matrix structure, we focus on the steady-state analogue of problem (1). Without loss of generality, we assume that $g \equiv 0$ as it does not affect the structure of the stiffness matrix A.

Let the positive integers N_x and N_y denote the numbers of intervals in the x and y directions, respectively. We define a spatial partition on $\bar{\Omega}$ by

$$\begin{aligned} x_i &:= i h_x \ \text{ for } i = 0, 1, \cdots, N_x \ \text{ with } h_x := \frac{x_r}{N_x}, \\ y_j &:= j h_y \ \text{ for } j = 0, 1, \cdots, N_y \ \text{ with } h_y := \frac{y_r}{N_y}. \end{aligned} \tag{24}$$

To handle the discretization on the boundary zone Ω_c, we extend the partition to the nodes (x_i, y_j) for $i = -i_x, -i_x+1, \cdots, -1, 0, 1, \cdots, N_x, N_x+1, \cdots, N_x+i_x$ and $j = -j_y, -j_y+1, \cdots, -1, 0, 1, \cdots, N_y, N_y+1, \cdots, N_y+j_y$. Here

$$i_x := \left\lceil \frac{\delta}{h_x} \right\rceil, \qquad j_y := \left\lceil \frac{\delta}{h_y} \right\rceil \tag{25}$$

are the ceilings of δ/h_x and δ/h_y, respectively.

The two-dimensional pyramid functions $\phi_{i,j}(x, y)$ centered at (x_i, y_j) can be expressed as

$$\begin{aligned} \phi_{i,j}(x, y) := \psi\Big(\frac{x - x_i}{h_x}\Big)\psi\Big(\frac{y - y_j}{h_y}\Big), \quad (x, y) \in \big(\Omega \cup \Omega_c\big) \\ 0 \leqslant i \leqslant N_x, 0 \leqslant j \leqslant N_y. \end{aligned} \tag{26}$$

The trial function u is of the form

$$u(x,y) = \sum_{i'=1}^{N_x-1} \sum_{j'=1}^{N_y-1} u_{i',j'}\phi_{i',j'}(x,y), \quad (x,y) \in (\Omega \cup \Omega_c). \tag{27}$$

With the prescribed nonlocal homogeneous Dirichlet boundary data, the unknowns are $u_{i,j}$ for $i = 1,2,\ldots,N_x-1$ and $j = 1,2,\ldots,N_y-1$. In the numerical scheme we need $N := (N_x-1)(N_y-1)$ collocation points. We thus choose the spatial nodes (x_i, y_j) for $i = 1,2,\ldots,N_x-1$ and $j = 1,2,\ldots,N_y-1$ as collocation points. This yields the following collocation formulation:

$$\begin{aligned} &\int_{B_\delta(x_i,y_j)} K(x_i - x', y_j - y')\big(u(x_i,y_j) - u(x',y')\big)dx'dy' = f(x_i,y_j), \\ &\qquad 1 \le i \le N_x - 1, \ \ 1 \le j \le N_y - 1. \end{aligned} \tag{28}$$

If we incorporate (27) into (28), we obtain the following collocation scheme:

$$\begin{aligned} &\sum_{i'=1}^{N_x-1} \sum_{j'=1}^{N_y-1} u_{i',j'} \int_{B_\delta(x_i,y_j)} K(x_i - x', y_j - y') \\ &\qquad \big(u(x_i,y_j) - \phi_{i',j'}(x',y')\big)dx'dy' = f(x_i,y_j), \\ &\qquad 1 \le i \le N_x - 1, \ \ 1 \le j \le N_y - 1. \end{aligned} \tag{29}$$

Let u and f be the N-dimensional vectors defined by

$$\begin{aligned} u &= [u_{1,1},\ldots,u_{N_x-1,1},u_{1,2},\ldots,u_{N)x-1,2},\ldots,u_{1,N_y-1},\ldots,u_{N_x-1,N_y-1}]^T, \\ f &= [f_{1,1},\ldots,f_{N_x-1,1},f_{1,2},\ldots,f_{N_x-1,2},\ldots,f_{1,N_y-1},\ldots,f_{N_x-1,N_y-1}]^T, \end{aligned} \tag{30}$$

and $A = [a_{m,n}]_{m,n=1}^N$ be the N-by-N stiffness matrix. Then the collocation method (29) can be expressed in the matrix form (14) with f_m and $A_{m,n}$ being defined by

$$\begin{aligned} a_{m,n} &= \int_{B_\delta(x_i,y_j)} K(x_i - x', y_j - y')(\delta_{m,n} - \phi_{i',j'}(x',y'))dx'dy', \\ f_m &= f(x_i,y_j). \end{aligned} \tag{31}$$

Here $\delta_{m,n} = 1$ for $m = n$ or 0 otherwise. The global indices m and n are related to the indices (i,j) and (i',j') by

$$m = (j-1)(N_x-1)+i,\ 1 \le i \le N_x-1,\ 1 \le j \le N_y-1,$$
$$n = (j'-1)(N_x-1)+i'\ 1 \le i' \le N_x-1,\ 1 \le j' \le N_y-1. \tag{32}$$

It is clear that the stiffness matrix of the collocation method (28) has a block structure. Hence, assembly of the matrix requires the evaluation of $(2i_x+1)(N_x-1)-i_x(i_x+1)$ entries per matrix block and totally $(2j_y+1)(N_y-1)-j_y(j_y+1)$ blocks, which is totally $(2j_x+1)(2j_y+1)N-(2j_x+1)(N_y-1)j_x(j_x+1)-(2j_x+1)(N_x-1)j_y(j_y+1)+i_x(i_x+1)j_y(j_y+1)$ of entries in the assembly of the stiffness matrix. This is asymptotically $O(N^2)$ of entries as $h_x \to 0$ and $h_y \to 0$. Similarly, the collocation method requires asymptotically $O(N^2)$ of memories to store the matrix. In the numerical simulation of nonlocal diffusion models, the assembly of the stiffness matrix often consumes a large portion of CPU time, as the evaluation of each entry can be very costly.

A direct solver has $O(N^3)$ computational complexity asymptotically, while a Krylov subspace iterative method has $O(N^2)$ computational complexity per Krylov subspace iteration. As we have observed from section "Fast Numerical Methods for a One-Dimensional Peridynamic Model," the key to developing a fast numerical method is to explore the structure of its stiffness matrix.

In Wang and Basu (2012) we analyzed the structure of the stiffness matrix of a finite difference method for a space-fractional PDE in two space dimensions and developed a fast method accordingly. In that analysis we fully utilize the fact that the fractional derivatives in the space-fractional PDE are in the coordinate directions so that the stiffness matrix of the numerical scheme has a tensor product structure.

However, the singular integration operator in the context of the nonlocal diffusion model (1) couples the unknowns in all the directions and so the stiffness matrix of the numerical scheme (29) is not expected to have a tensor product structure. Hence, the analysis is more complicated. We proved the following theorem in Wang et al. (2014).

Theorem 3. *The stiffness matrix A of the numerical scheme* (29) *has a block-Toeplitz-Toeplitz-block structure with* (N_y-1) *by-*(N_y-1) *blocks*

$$A = \begin{bmatrix} T_0 & T_1 & \dots & T_{j_y} & 0 & 0 & \dots & 0 \\ T_{-1} & T_0 & T_1 & \dots & T_{j_y} & 0 & \dots & 0 \\ \vdots & T_{-1} & T_0 & \ddots & \ddots & T_{j_y} & \ddots & \vdots \\ T_{-j_y} & \ddots & \ddots & \ddots & T_1 & \ddots & \ddots & 0 \\ 0 & T_{-j_y} & \ddots & \ddots & T_0 & T_1 & \ddots & T_{j_y} \\ 0 & \ddots & \ddots & \ddots & T_{-1} & T_0 & \ddots & \vdots \\ \vdots & \ddots & \ddots & \ddots & \ddots & \ddots & \ddots & T_1 \\ 0 & 0 & \dots & 0 & T_{-j_y} & \dots & T_{-1} & T_0 \end{bmatrix}. \tag{33}$$

Each block T_j, $-j_y \le j \le j_y$, is a Toeplitz matrix of order $N_x - 1$

$$T_j = \begin{bmatrix} t_{0,j} & t_{1,j} & \dots & t_{i_x,j} & 0 & 0 & \dots & 0 \\ t_{-1,j} & t_{0,j} & t_{1,j} & \dots & t_{i_x,j} & 0 & \dots & 0 \\ \vdots & t_{-1,j} & t_{0,j} & \ddots & \ddots & t_{i_x,j} & \ddots & \vdots \\ t_{-i_x,j} & \ddots & \ddots & \ddots & t_{1,j} & \ddots & \ddots & 0 \\ 0 & t_{-i_x,j} & \ddots & \ddots & t_{0,j} & t_{1,j} & \ddots & t_{i_x,j} \\ 0 & \ddots & \ddots & \ddots & t_{-1,j} & t_{0,j} & \ddots & \vdots \\ \vdots & \ddots & \ddots & \ddots & \ddots & \ddots & \ddots & t_{1,j} \\ 0 & 0 & \dots & 0 & t_{-i_x,j} & \dots & t_{-1,j} & t_{0,j} \end{bmatrix}. \tag{34}$$

Outline of proof. It is easy to show that if we express the stiffness matrix A as a block matrix of order $(N_y - 1)$, i.e., $A = [B^{j,j'}]_{j,j'=1}^{N_y-1}$, then the matrix A is a block-banded matrix with bandwidth $2j_y + 1$. Similarly, each nonzero matrix block $B^{j,j'} = [b_{i,i'}^{j,j'}]_{i,i'=1}^{N_x-1}$ of order $(N_x - 1)$ is banded with bandwidth $2i_x + 1$.

Using the substitution

$$x' = x_i + x, \quad y' = y_j + y, \tag{35}$$

we can show the relation

$$\phi_{i',j'}(x', y') = \phi_{i'-i,j'-j}(x, y). \tag{36}$$

Accordingly, we can prove that if

$$j_1' - j_1 = j_2' - j_2, \tag{37}$$

then we have

$$b_{i,i'}^{j_1,j_1'} = b_{i,i'}^{j_2,j_2'}, \quad 1 \le i, i' \le N_x - 1. \tag{38}$$

Namely, $B^{j_1,j_1'} = B^{j_2,j_2'}$. That is, the matrix $A = [B^{j,j'}]_{j,j'=1}^{N_y-1}$ is block-Toeplitz. Similarly, we prove that if

$$i_3' - i_3 = i_4' - i_4, \tag{39}$$

then we have

$$b_{i_3,i_3'}^{j,j'} = b_{i_4,i_4'}^{j,j'}. \tag{40}$$

We have thus proved that each block $B^{j,j'}$ is a Toeplitz matrix of bandwidth $2i_x + 1$. □

From Theorem 3, we see that in the fast numerical method, we need only to evaluate and store $(2i_x + 1)(2j_y + 1)$ entries of the matrix. A fast matrix-vector multiplication algorithm can be developed as in section "Fast Numerical Methods for a One-Dimensional Peridynamic Model," in which each Krylov subspace iteration has computational complexity of $O(N \log N)$. Numerical experiments presented in Wang et al. (2014) showed the utility of the fast method.

A Fast Collocation Method for the Peridynamic Model (3)

As in section "A Fast Collocation Method for the Nonlocal Diffusion Model (1)," we consider the steady-state analogue of problem (3) with the nonlocal homogeneous Dirichlet data. We use the same partition (24) and the basis functions (26). Without loss of generality, we can assume $c = 1$ in (4), as we can always divide both sides of the model by c in this case. Then the x-component $v(x, y)$ and the y-component $w(x, y)$ of the displacement field $\mathbf{u}(x, y)$ can be expressed as

$$
\begin{aligned}
v(x, y) &= \sum_{i'=1}^{N_x-1} \sum_{j'=1}^{N_y-1} v_{i',j'} \phi_{i',j'}(x, y), \\
w(x, y) &= \sum_{i'=1}^{N_x-1} \sum_{j'=1}^{N_y-1} w_{i',j'} \phi_{i',j'}(x, y).
\end{aligned}
\tag{41}
$$

We insert (41) into (3) and enforce the resulting equations at the collocation points (x_i, y_j) for $i = 1, 2, \cdots, N_x - 1$ and $j = 1, 2, \cdots, N_y - 1$ to obtain the following scheme:

$$
\begin{aligned}
&\int_{B_\delta(x_i,y_j)} \frac{(x' - x_i)^2 \Big[v_{i,j} - \sum_{i'=1}^{N_x-1} \sum_{j'=1}^{N_y-1} v_{i',j'} \phi_{i',j'}(x', y') \Big]}{\big((x' - x_i)^2 + (y' - y_j)^2\big)^{3/2}} dx'dy' \\
&+ \int_{B_\delta(x_i,y_j)} \frac{(x' - x_i)(y' - y_j) \Big[w_{i,j} - \sum_{i'=1}^{N_x-1} \sum_{j'=1}^{N_y-1} w_{i',j'} \phi_{i',j'}(x', y') \Big]}{\big((x' - x_i)^2 + (y' - y_j)^2\big)^{3/2}} dx'dy' \\
&= f^v(x_i, y_j),
\end{aligned}
$$

$$\int_{B_\delta(x_i,y_j)} \frac{(x'-x_i)(y'-y_j)\Big[v_{i,j} - \sum_{i'=1}^{N_x-1}\sum_{j'=1}^{N_y-1} v_{i',j'}\phi_{i',j'}(x',y')\Big]}{\big((x'-x_i)^2+(y'-y_j)^2\big)^{3/2}} dx'dy'$$
$$+ \int_{B_\delta(x_i,y_j)} \frac{(y'-y_j)^2\Big[w_{i,j} - \sum_{i'=1}^{N_x-1}\sum_{j'=1}^{N_y-1} w_{i',j'}\phi_{i',j'}(x',y')\Big]}{\big((x'-x_i)^2+(y'-y_j)^2\big)^{3/2}} dx'dy'$$
$$= f^w(x_i,y_j), \quad 1 \le i \le N_x-1, \;\; 1 \le j \le N_y-1. \tag{42}$$

The collocation scheme (42) can be written in the matrix form

$$\mathbf{A}_{2N}\mathbf{u}_{2N} = \mathbf{f}_{2N}. \tag{43}$$

The $2N$-dimensional unknown displacement vector $\mathbf{u}_{2N}$ and the right-hand side loading vector $\mathbf{f}_{2N}$ of (42) are labeled in their natural order. At each node (x_i, y_j), the x component $v_{i,j}$ or $f^v_{i,j}$ is put first and the y component $w_{i,j}$ or $f^w_{i,j}$ second:

$$\begin{aligned}
\mathbf{u}_{2N} &:= \big[v_{1,1}, w_{1,1}, \cdots, v_{N_x-1,1}, w_{N_x-1,1}, v_{1,2}, w_{1,2}, \cdots, v_{N_x-1,2}, w_{N_x-1,2},\\
&\quad \cdots, v_{1,N_y-1}, w_{1,N_y-1}, \cdots, v_{N_x-1,N_y-1}, w_{N_x-1,N_y-1}\big]^T,\\
\mathbf{f}_{2N} &:= [f^v_{1,1}, f^w_{1,1}, \cdots, f^v_{N_x-1,1}, f^w_{N_x-1,1}, f^v_{1,2}, f^w_{1,2}, \cdots, f^v_{N_x-1,2}, f^w_{N_x-1,2},\\
&\quad \cdots, f^v_{1,N_y-1}, f^w_{1,N_y-1}, \cdots, f^v_{N_x-1,N_y-1}, f^w_{N_x-1,N_y-1}]^T.
\end{aligned} \tag{44}$$

The $2N$-by-$2N$ stiffness matrix $\mathbf{A}_{2N}$ can be written as the following N-by-N block matrix:

$$\mathbf{A}_{2N} = \begin{bmatrix} \mathbf{A}_{1,1} & \mathbf{A}_{1,2} & \ldots & \mathbf{A}_{1,N-1} & \mathbf{A}_{1,N} \\ \mathbf{A}_{2,1} & \mathbf{A}_{2,2} & \ldots & \mathbf{A}_{2,N-1} & \mathbf{A}_{2,N} \\ \vdots & \vdots & \ddots & \vdots & \vdots \\ \mathbf{A}_{N-1,1} & \mathbf{A}_{N-1,2} & \ldots & \mathbf{A}_{N-1,N-1} & \mathbf{A}_{N-1,N} \\ \mathbf{A}_{N,1} & \mathbf{A}_{N,2} & \ldots & \mathbf{A}_{N,N-1} & \mathbf{A}_{N,N} \end{bmatrix}, \tag{45}$$

where each entry of $\mathbf{A}_{2N}$ is a two-by-two matrix

$$\mathbf{A}_{m,n} = \begin{bmatrix} a^{v,v}_{m,n} & a^{v,w}_{m,n} \\ a^{w,v}_{m,n} & a^{w,w}_{m,n} \end{bmatrix}, \tag{46}$$

with the entries in $\mathbf{A}_{m,n}$ being defined by

$$
\begin{aligned}
a_{m,n}^{v,v} &:= \int_{B_\delta(x_i,y_j)} \frac{(x'-x_i)^2(\delta_{m,n}-\phi_{i',j'}(x',y'))}{\left((x'-x_i)^2+(y'-y_j)^2\right)^{3/2}} dx'dy', \\
a_{m,n}^{v,w} &:= \int_{B_\delta(x_i,y_j)} \frac{(x'-x_i)(y'-y_j)(\delta_{m,n}-\phi_{i',j'}(x',y'))}{\left((x'-x_i)^2+(y'-y_j)^2\right)^{3/2}} dx'dy', \\
a_{m,n}^{w,w} &:= \int_{B_\delta(x_i,y_j)} \frac{(y'-y_j)^2(\delta_{m,n}-\phi_{i',j'}(x',y'))}{\left((x'-x_i)^2+(y'-y_j)^2\right)^{3/2}} dx'dy'.
\end{aligned} \tag{47}
$$

$A_{m,n}^{w,v} = A_{m,n}^{v,w}$. The global indices m and n are related to the indices (i,j) and (i',j') by (32). The entries of the right-hand side vector $\mathbf{f}_{2N}$ are given by

$$
f_{i,j}^{v} = f^{v}(x_i,y_j), \quad f_{i,j}^{w} = f^{w}(x_i,y_j), \quad 1 \leqslant i \leqslant N_x-1, 1 \leqslant j \leqslant N_y-1. \tag{48}
$$

We use the similar idea as in section "A Fast Collocation Method for the Nonlocal Diffusion Model (1)" to conduct the analysis. However, due to the coupling between the x component v and the y component w of the displacement $\mathbf{u}_{2N}$ introduced by the kernel in the model, the stiffness matrix $\mathbf{A}_{2N}$ is not block-Toeplitz-Toeplitz-block. To develop a fast method, we analyze the displacement vector $\mathbf{u}_{2N}$ and the stiffness matrix $\mathbf{A}_{2N}$.

We first decompose the displacement vector $\mathbf{u}_{2N}$ as the following two auxiliary vectors, which replace the even or odd entries of $\mathbf{u}_{2N}$ by 0, respectively:

$$
\begin{aligned}
\mathbf{u}_{2N}^{o} &:= \big[v_{1,1}, 0, \cdots, v_{N_x-1,1}, 0, v_{1,2}, 0, \cdots, v_{N_x-1,2}, 0, \\
&\qquad \cdots, v_{1,N_y-1}, 0, \cdots, v_{N_x-1,N_y-1}, 0\big]^T, \\
\mathbf{u}_{2N}^{e} &:= \big[0, w_{1,1}, \cdots, 0, w_{N_x-1,1}, 0, w_{1,2}, \cdots, 0, w_{N_x-1,2}, \\
&\qquad \cdots, 0, w_{1,N_y-1}, \cdots, 0, w_{N_x-1,N_y-1}\big]^T.
\end{aligned} \tag{49}
$$

We remove the zero entries in the above vectors to obtain two N-dimensional vectors $\mathbf{v}_N$ and $\mathbf{w}_N$. It is clear that

$$
\mathbf{u}_{2N} = \mathbf{u}_{2N}^{o} + \mathbf{u}_{2N}^{e}. \tag{50}
$$

Next, we introduce the N-by-N matrices $\mathbf{A}_N^{v,v}$, $\mathbf{A}_N^{v,w}$, $\mathbf{A}_N^{w,v}$, and $\mathbf{A}_N^{w,w}$ as follows:

$$
\mathbf{A}_N^{v,v} := \begin{bmatrix}
a_{1,1}^{v,v} & a_{1,2}^{v,v} & \cdots & a_{1,N-1}^{v,v} & a_{1,N}^{v,v} \\
a_{2,1}^{v,v} & a_{2,2}^{v,v} & \cdots & a_{2,N-1}^{v,v} & a_{2,N}^{v,v} \\
\vdots & \vdots & \ddots & \vdots & \vdots \\
a_{N-1,1}^{v,v} & a_{N-1,2}^{v,v} & \cdots & a_{N-1,N-1}^{v,v} & a_{N-1,N}^{v,v} \\
a_{N,1}^{v,v} & a_{N,2}^{v,v} & \cdots & a_{N,N-1}^{v,v} & a_{N,N}^{v,v}
\end{bmatrix}, \tag{51}
$$

$$\mathbf{A}_N^{v,w} := \begin{bmatrix} a_{1,1}^{v,w} & a_{1,2}^{v,w} & \cdots & a_{1,N-1}^{v,w} & a_{1,N}^{v,w} \\ a_{2,1}^{v,w} & a_{2,2}^{v,w} & \cdots & a_{2,N-1}^{v,w} & a_{2,N}^{v,w} \\ \vdots & \vdots & \ddots & \vdots & \vdots \\ a_{N-1,1}^{v,w} & a_{N-1,2}^{v,w} & \cdots & a_{N-1,N-1}^{v,w} & a_{N-1,N}^{v,w} \\ a_{N,1}^{v,w} & a_{N,2}^{v,w} & \cdots & a_{N,N-1}^{v,w} & a_{N,N}^{v,w} \end{bmatrix}, \tag{52}$$

$$\mathbf{A}_N^{w,w} := \begin{bmatrix} a_{1,1}^{w,w} & a_{1,2}^{w,w} & \cdots & a_{1,N-1}^{w,w} & a_{1,N}^{w,w} \\ a_{2,1}^{w,w} & a_{2,2}^{w,w} & \cdots & a_{2,N-1}^{w,w} & a_{2,N}^{w,w} \\ \vdots & \vdots & \ddots & \vdots & \vdots \\ a_{N-1,1}^{w,w} & a_{N-1,2}^{w,w} & \cdots & a_{N-1,N-1}^{w,w} & a_{N-1,N}^{w,w} \\ a_{N,1}^{w,w} & a_{N,2}^{w,w} & \cdots & a_{N,N-1}^{w,w} & a_{N,N}^{w,w} \end{bmatrix}, \tag{53}$$

and $\mathbf{A}_N^{w,v} = \mathbf{A}_N^{v,w}$.

We decompose the matrix-vector multiplication $\mathbf{A}_{2N}\mathbf{u}_{2N}$ as follows:

$$\mathbf{A}_{2N}\mathbf{u}_{2N} = \mathbf{A}_{2N}\mathbf{u}_{2N}^o + \mathbf{A}_{2N}\mathbf{u}_{2N}^e. \tag{54}$$

In addition, if we collect all the odd numbered rows and all the even numbered rows of $\mathbf{A}_{2N}\mathbf{u}_{2N}^o$, we obtain the following matrix-vector multiplications:

$$\mathbf{A}_N^{v,v}\mathbf{v}_N, \quad \mathbf{A}_N^{w,v}\mathbf{v}_N, \tag{55}$$

respectively. Similarly, if we collect all the odd numbered rows and all the even numbered rows of $\mathbf{A}_{2N}\mathbf{u}_{2N}^e$, we obtain the following matrix-vector multiplications:

$$\mathbf{A}_N^{v,w}\mathbf{w}_N, \quad \mathbf{A}_N^{w,w}\mathbf{w}_N, \tag{56}$$

respectively.

In summary, to evaluate the matrix-vector multiplication $\mathbf{A}_{2N}\mathbf{u}_{2N}$ efficiently, we need only to efficiently evaluate the matrix-vector multiplications $\mathbf{A}_N^{v,v}\mathbf{v}_N$, $\mathbf{A}_N^{w,v}\mathbf{v}_N$, $\mathbf{A}_N^{v,w}\mathbf{w}_N$, and $\mathbf{A}_N^{w,w}\mathbf{w}_N$. Similarly, to efficiently store and assemble the stiffness matrix $\mathbf{A}_{2N}$, we need only to efficiently store and assemble the submatrices $\mathbf{A}_N^{v,v}$, $\mathbf{A}_N^{w,v}$, $\mathbf{A}_N^{v,w}$, and $\mathbf{A}_N^{w,w}$. We use the similar analysis as in section "A Fast Collocation Method for the Nonlocal Diffusion Model (1)" to prove that the matrices $\mathbf{A}_N^{v,v}$, $\mathbf{A}_N^{w,v}$, $\mathbf{A}_N^{v,w}$, and $\mathbf{A}_N^{w,w}$ are block-Toeplitz-Toeplitz-block matrices. So fast matrix-free numerical methods with efficient storage and evaluation of matrix entries can be developed (Zhang and Wang 2016). Numerical experiments show the utility of the method.

Concluding Remarks

In this chapter we outlined the recent developments of fast numerical methods for linear nonlocal diffusion models and bond-based peridynamic models in one and two space dimensions. We show how the analysis was carried out to take the full advantage of the structure of the stiffness matrices of the numerical methods in its storage, evaluation, and assembly and in the efficient solution of the corresponding numerical schemes. This would significantly reduce the computational complexity and storage of the numerical methods over conventional ones, without using any lossy compression. For instance, we would use the same numerical quadratures for conventional methods to evaluate the singular integrals in the stiffness matrices, except that we only need to evaluate $O(N)$ of them instead of $O(N^2)$ of them. Numerical results are presented to show the utility of these fast methods.

Technically, the contributions of this chapter are summarized as follows: (i) the development of the fast numerical methods for the nonlocal diffusion model (1) and peridynamic model (3) is motivated by that of the fast finite difference methods for FPDEs (Chen et al. 2015; Wang and Basu 2012; Wang and Wang 2010; Yang et al. 2011, 2014). However, the nonlocal diffusion model (1) and peridynamic model (3) give rise to the coupling in all the directions, in contrast to FPDEs in which the fractional derivatives are expressed only in the coordinate directions, and so represent a much stronger coupling than FPDEs do. In this chapter we outline the analysis of the stiffness matrices in the development of fast numerical methods with efficient matrix assembly and storage. (ii) The fast numerical methods developed for FPDEs (Wang and Basu 2012; Wang and Wang 2010; Wang and Du 2013) and nonlocal diffusion models (Wang and Tian 2012; Wang et al. 2014; Wang and Wang 2017) are primarily for scalar problems. However, the peridynamic model (3) is a vector equation. Due to the coupling between the x component v and the y component w of the displacement vector $\mathbf{u}$ introduced by the tensor product kernel, the stiffness matrix in the numerical method is not block-Toeplitz-Toeplitz-block as that for nonlocal diffusion models (Wang et al. 2014). Thus, the fast numerical methods developed for FPDEs and nonlocal diffusion models do not apply to the peridynamic model (3). We outlined the analysis of the coupling effect of the peridynamic model (3) on the stiffness matrix of the numerical method and utilize the structure in the development of a fast collocation method.

We end this section by discussing potential extensions: (i) a time-stepping algorithm needs to be used for time-dependent problems, which actually reduces the condition number of the stiffness matrix due to the impact of the time step size. (ii) The extension to three-dimensional problem is conceptually straightforward as the stiffness matrix should retain a similar structure. However, fractional implementation of a three-dimensional peridynamic model requires tremendous effort. (iii) Additional problems include peridynamic model with variable coefficients, state-based peridynamic model, and numerical methods discretized on locally refined composite meshes.

Acknowledgements This work was supported in part by the OSD/ARO MURI under grant W911NF-15-1-0562 and by the National Science Foundation under grants DMS-1620194 and DMS-1216923.

References

D. Applebaum, *Lévy Processes and Stochastic Calculus* (Cambridge University Press, Cambridge/New York , 2009)

H.G. Chan, M.K. Ng, Conjugate gradient methods for Toeplitz systems. SIAM Rev. **38**, 427–482 (1996)

S. Chen, F. Liu, X. Jiang, I. Turner, V. Anh, A fast semi-implicit difference method for a nonlinear two-sided space-fractional diffusion equation with variable diffusivity coefficients. Appl. Math. Comp. **257**, 591–601 (2015)

X. Chen, M. Gunzburger, Continuous and discontinuous finite element methods for a peridynamics model of mechanics. Comput. Methods Appl. Mech. Eng. **200**, 1237–1250 (2011)

K. Dayal, K. Bhattacharya, Kinetics of phase transformations in the peridynamic formulation of continuum mechanics. J. Mech. Phys. Solids **54**, 1811–1842 (2006)

D. del-Castillo-Negrete, B.A. Carreras, V.E. Lynch, Fractional diffusion in plasma turbulence. Phys. Plasmas **11**, 3854 (2004)

M. D'Elia, R. Lehoucq, M. Gunzburger, Q. Du, Finite range jump processes and volume-constrained diffusion problems, Sandia National Labs SAND, 2014–2584 (Sandia National Laboratories, Albuquerque/Livermore, 2014)

Q. Du, M. Gunzburger, R. Lehoucq, K. Zhou, Analysis and approximation of nonlocal diffusion problems with volume constraints. SIAM Rev. **54**, 667–696 (2012)

Q. Du, L. Ju, L. Tian, K. Zhou, A posteriori error analysis of finite element method for linear nonlocal diffusion and peridynamic models. Math. Comp. **82**, 1889–1922 (2013)

E. Emmrich, O. Weckner, The peridynamic equation and its spatial discretisation. Math. Model. Anal. **12**, 17–27 (2007)

W. Gerstle, N. Sau, S. Silling, Peridynamic modeling of concrete structures. Nucl. Eng. Des. **237**, 1250–1258 (2007)

M. Ghajari, L. Iannucci, P. Curtis, A peridynamic material model for the analysis of dynamic crack propagation in orthotropic media. Comput. Methods Appl. Mech. Eng. **276**, 431–452 (2014)

Y.D. Ha, F. Bobaru, Characteristics of dynamic brittle fracture captured with peridynamics. Eng. Fract. Mech. **78**, 1156–1168 (2011)

J. Jia, C. Wang, H. Wang, A fast locally refined method for a space-fractional diffusion equation. # IE0147, ICFDA'14 Catania, 23–25 June 2014. Copyright 2014 IEEE ISBN:978-1-4799-2590-2

J. Jia, H. Wang, A preconditioned fast finite volume scheme for a fractional differential equation discretized on a locally refined composite mesh. J. Comput. Phys. **299**, 842–862 (2015)

X. Lai, B. Ren, H. Fan, S. Li, C.T. Wu, R.A. Regueiro, L. Liu, Peridynamics simulations of geomaterial fragmentation by impulse loads. Int. J. Numer. Anal. Meth. Geomech **39**, 1304–1330 (2015)

M.M. Meerschaert, A. Sikorskii, *Stochastic Models for Fractional Calculus* (De Gruyter, Berlin, 2011)

R. Metzler, J. Klafter, The random walk's guide to anomalous diffusion: a fractional dynamics approach. Phys. Rep. **339**, 1–77 (2000)

B. Øksendal, *Stochastic Differential Equations: An Introduction with Applications* (Springer, Heidelberg, 2010)

E. Oterkus, E. Madenci, O. Weckner, S.A. Silling, P. Bogert, Combined finite element and peridynamic analyses for predicting failure in a stiffened composite curved panel with a central slot. Compos. Struct. **94**, 839–850 (2012)

M.L. Parks, R.B. Lehoucq, S.J. Plimpton, S.A. Silling, Implementing peridynamics within a molecular dynamics code. Comp. Phys. Comm. **179**, 777–783 (2008)

M.L. Parks, P. Seleson, S.J. Plimpton, S.A. Silling, R.B. Lehoucq, Peridynamics with LAMMPS: a user guide v0.3 beta, SAND report 2011–8523 (Sandia National Laboratories, Albuquerque/Livermore, 2011)

I. Podlubny, *Fractional Differential Equations* (Academic Press, San Diego, 1999)

P. Seleson, Improved one-point quadrature algorithms for two-dimensional peridynamic models based on analytical calculations. Comput. Methods Appl. Mech. Eng. **282**, 184–217 (2014)

P. Seleson, Q. Du, M.L. Parks, On the consistency between nearest-neighbor peridynamic discretizations and discretized classical elasticity models. Comput. Methods Appl. Mech. Eng. **311**, 698–722 (2016)

P. Seleson, D. Littlewood, Convergence studies in meshfree peridynamic simulations. Comput. Math. Appl. **71**, 2432–2448 (2016)

P. Seleson, M.L. Parks, On the role of the infuence function in the peridynamic theory. Int. J. Multiscale Comput. Eng. **9**, 689–706 (2011)

P. Seleson, M.L. Parks, M. Gunzburger, R.B. Lehoucq, Peridynamics as an upscaling of molecular dynamics. Multiscale Model. Simul. **8**, 204–227 (2009)

S.A. Silling, Reformulation of elasticity theory for discontinuous and long-range forces. J. Mech. Phys. Solids **48**, 175–209 (2000)

S.A. Silling, E. Askari, A meshfree method based on the peridynamic model of solid mechanics. Comput. Struct. **83**, 1526–1535 (2005)

S.A. Silling, M. Epton, O. Wecker, J. Xu, E. Askari, Peridynamic states and constitutive modeling. J. Elast. **88**, 151–184 (2007)

S.A. Silling, O. Weckner, E. Askari, F. Bobaru, Crack nucleation in a peridynamic solid. Int. J. Fract. **162**, 219–227 (2010)

S. Sun, V. Sundararaghavan, A peridynamic implementation of crystal plasticity. Int. J. Solids Struct. **51**, 3350–3360 (2014)

H. Tian, H. Wang, W. Wang, An efficient collocation method for a non-local diffusion model. Int. J. Numer. Anal. Model. **10**, 815–825 (2013)

X. Tian, Q. Du, Analysis and comparison of different approximations to nonlocal diffusion and linear peridynamic equations. SIAM J. Numer. Anal. **51**, 3458–3482 (2013)

C. Wang, H. Wang, A fast collocation method for a variable-coefficient nonlocal diffusion model. J. Comput. Phys. **330**, 114–126 (2017)

H. Wang, T.S. Basu, A fast finite difference method for two-dimensional space-fractional diffusion equations. SIAM J. Sci. Comput. **34**, A2444–A2458 (2012)

H.Wang, N. Du, Fast alternating-direction finite difference methods for three-dimensional space-fractional diffusion equations. J. Comput. Phys. **258**, 305–318 (2013)

H. Wang, H. Tian, A fast Galerkin method with efficient matrix assembly and storage for a peridynamic model. J. Comput. Phys. **231**, 7730–7738 (2012)

H. Wang, H. Tian, A fast and faithful collocation method with efficient matrix assembly for a two-dimensional nonlocal diffusion model. Comput. Methods Appl. Mech. Eng. **273**, 19–36 (2014)

H. Wang, K. Wang, T. Sircar, A direct $O(N \log^2 N)$ finite difference method for fractional diffusion equations. J. Comput. Phys. **229**, 8095–8104 (2010)

Q. Yang, I. Turner, F. Liu, M. Ilis, Novel numerical methods for solving the time-space fractional diffusion equation in 2D. SIAM Sci. Comput. **33**, 1159–1180 (2011)

Q. Yang, I. Turner, T. Moroney F. Liu, A finite volume scheme with preconditioned Lanczos method for two-dimensional space-fractional reaction-diffusion equations. Appl. Math. Model. **38**, 3755–3762 (2014)

X. Zhang, H. Wang, A fast method for a steady-state bond-based peridynamic model. Comput. Methods Appl. Mech. Eng. **311**, 280–303 (2016)

X. Zhang, M. Gunzburger, L. Ju, Nodal-type collocation methods for hypersingular integral equations and nonlocal diffusion problems. Comput. Methods Appl. Mech. Eng. **299**, 401–420 (2016)

K. Zhou, Q. Du, Mathematical and numerical analysis of linear peridynamic models with nonlocal boundary condition. SIAM J. Numer. Anal. **48**, 1759–1780 (2010)

Peridynamic Functionally Graded and Porous Materials: Modeling Fracture and Damage

41

Ziguang Chen, Sina Niazi, Guanfeng Zhang, and Florin Bobaru

Contents

Abstract

In this chapter, we present two peridynamic models for composite materials: a locally homogenized model (FH-PD model, based on results reported in Cheng

Z. Chen (✉)
Department of Mechanics, Huazhong University of Science and Technology, Wuhan, Hubei Sheng, China

Hubei Key Laboratory of Engineering, Structural Analysis and Safety Assessment, Wuhan, China
e-mail: chenziguang@huskers.unl.edu

S. Niazi · G. Zhang · F. Bobaru
Mechanical and Materials Engineering, University of Nebraska–Lincoln, Lincoln, NE, USA
e-mail: sinaniazi85@yahoo.com; zh.guanfeng@gmail.com; fbobaru2@unl.edu

G. Z. Voyiadjis (ed.), *Handbook of Nonlocal Continuum Mechanics for Materials and Structures*, https://doi.org/10.1007/978-3-319-58729-5_36

et al. (Compos Struct 133: 529–546, 2015)) and an intermediately homogenized model (IH-PD model). We use these models to simulate fracture in functionally graded materials (FGMs) and in porous elastic materials. We analyze dynamic fracture, by eccentric impact, of a functionally graded plate with monotonically varying volume fraction of reinforcements. We study the influence of material gradients, elastic waves, and of contact time and magnitude of impact loading on the crack growth from a pre-notch in terms of crack path geometry and crack propagation speed. The results from FH-PD and IH-PD models show the same cracking behavior and final crack patterns. The simulations agree very well, through full failure, with experiments. We discuss advantages offered by the peridynamic models in dynamic fracture of FGMs compared with, for example, FEM-based models. The models lead to a better understanding of how cracks propagate in FGMs and of the factors that control crack path and its velocity in these materials. The IH-PD model has important advantages when compared with the FH-PD model when applied to composite materials with phases of disparate mechanical properties. An application to fracture of porous and elastic materials (following Chen et al. (Peridynamic model for damage and fracture in porous materials, 2017)) shows the major effect local heterogeneities have on fracture behavior and the importance of intermediate homogenization as a modeling approach of crack initiation and growth.

Keywords

Peridynamics · Functionally graded materials · Composite materials · Dynamic fracture · Crack propagation · Impact · Brittle fracture · Porous elastic materials

Introduction

Functionally graded materials (FGMs) are a special class of composite materials characterized by spatially varying material properties with the goals of eliminating/reducing stress concentrations, relaxing residual stresses, and enhancing the bonding strength of the composite constituents. Generally, FGMs are multiphase materials having *continuously varying* volume fractions of constituent phases most often along a desired spatial direction. Benefits of using FGMs or FGM coatings in thermomechanical applications are shown in, for example, Kashtalyan and Menshykova (2009), Bobaru (2007), and Thai and Kim (2015). In designing components involving FGMs, it is important to consider imperfections, such as cracks, often preexisting as manufacturing defects or generated by external loads during service. Fracture mechanics of FGMs, especially their dynamic failure, plays a key role in the design of FGMs structures. The fracture behavior of FGMs has been studied especially in the past two decades (see, e.g., Eischen 1987; Jin and Batra 1996; Delale and Erdogan 1983; Wang et al. 2014; Marur and Tippur 2000; Anlas et al. 2000; Itou 2010). Theoretical studies conclude that the leading term in the crack-tip stress field is of the inverse-square-root-singularity for any form of elastic modulus variation and the elastic modulus has a significant effect on the crack-tip stress field, while the Poisson's ratio has little effect on it (see Delale and Erdogan 1983). Only recently have studies focused on dynamic fracture problems in FGMs. Dynamic analysis of cracks in FGMs under impact loading conditions has appeared in, e.g., Itou (2010), Guo et al. (2004), and Ma et al. (2005), and analysis of crack propagation in FGMs is also presented in Xia and Ma (2007), Kidane et al. (2010), Cheng and Zhong (2007), Cheng et al. (2010), Lee (2009), and Matbuly (2009), for example.

Due to the lack of symmetry in material properties, fracture is inherently mixed-mode when a crack in a FGM is not parallel to the direction of material properties variation, or when loading is asymmetric relative to the crack plane. Jin et al. (2009) experimentally investigated the quasi-static mixed-mode crack propagation in a FGM beam under offset loading. Abanto-Bueno and Lambros (2006) experimentally studied quasi-static mixed-mode crack initiation and growth in FGM through fracture experiments. Shukla and his coauthors (2006) presented a detailed experimental study to understand the dynamic fracture behavior of FGMs. Kirugulige and Tippur (2006) have conducted mixed-mode dynamic fracture experiments on FGM samples made of compositionally graded glass-filled epoxy plates with initial edge cracks along the material gradient. Rousseau and Tippur (2001) presented an experimental study of the crack-tip deformation and fracture parameter histories in compositionally graded glass-filled epoxy under low velocity impact loading.

Most efforts dedicated to simulating static or dynamic fracture behavior of FGM have used the finite element method (FEM). Kim and Paulino (2004) used local remeshing technique to predict the crack path for the mixed-mode quasi-static fracture test in Rousseau and Tippur (2001). Kirugulige and Tippur (2008) have reexamined the mixed-mode dynamic fracture experiment reported in Kirugulige and Tippur (2006) by using the cohesive-zone finite element method. Zhang and Paulino (2005) have conducted mode-I and mixed-mode dynamic fracture simulations in FGMs by cohesive finite elements. Bayesteh and Mohammadi (2013) have used the extended finite element method (XFEM) to analyze the fracture behavior of orthotropic FGMs. The approaches mentioned above used the classical continuum mechanics models to treat fracture. Such approaches are based on partial differential equations and, for describing discontinuities such as cracks, one has to introduce special modeling techniques (such as adaptive remeshing) that define crack surfaces as new boundaries of the updated domain. Continuum-type methods using the cohesive FEM and XFEM require special criteria to decide, for example, when to branch a dynamic crack (see Ha and Bobaru 2011; Song et al. 2008), as well as explicit tracking of the crack path. Brittle damage processes involved in branching of a crack depend on the flow of the strain energy density. A general, predictive theory for how energy is dissipated in the creation of experimentally observed damage around the region where a crack branches (the mirror–mist–hackle transition) is not available. Because of these difficulties, classical models that use a flat surface to represent the actual rough fracture surface are not be able to release the correct amount of energy for crack growth in a dynamic setting and, therefore, are not able to predict experimentally observed crack propagating speeds in dynamic brittle fracture (see Song et al. 2008). Cohesive-zone models need to modify the experimental values of the fracture energy by several factors in order to get propagation velocities in the range of measured ones (see Song et al. 2008).

An alternative approach to modeling dynamic fracture is the new nonlocal continuum model (proposed by Silling (2000) and Silling et al. (2007)), peridynamics, which eliminates spatial derivatives from the formulation with the goal of having a consistent mathematical model for problems with discontinuities in the displacement field. Peridynamics is a reformulation of continuum mechanics in which each material point is connected (via peridynamic bonds) with points in a certain region around it, and not only with its nearest neighbors. In order to overcome mathematical inconsistencies in the classical continuum mechanics models of problems in which cracks initiate and evolve in time, peridynamics uses an integral of forces over a nonlocal region (called horizon) around a point to replace the divergence of the stress tensor in the equations of motion. The model is particularly well suited for dealing with cracks and damage in solid mechanics especially in situations where the crack path is not known in advance.

Peridynamics has been applied to damage and failure analysis of homogeneous and nonhomogeneous materials. Silling and Askari (2005) employed the bond-based peridynamics to study crack growth and impact of a sphere on a brittle target. Bobaru and his coauthors (Ha and Bobaru 2010, 2011; Hu et al. 2012, 2013; Zhang and Bobaru 2015) conducted several peridynamic studies of dynamic fracture in brittle materials and composite materials, including crack branching, and impact loading and fragmentation.

In this chapter, we employ the bond-based peridynamic (PD) model to investigate the dynamic fracture behavior of FGMs under offset (asymmetric) impact loading and understand the factors that influence the crack propagation direction and crack growth in FGMs. We formulate a fully homogenized PD (FH-PD) model for an already homogenized FGM by defining the bond properties between PD nodes based on the elastic and fracture properties at the nodes' locations in the FGM given by experiments. In Cheng et al. (2015), we verified the model in terms of its elastic dynamic behavior (wave propagation induced by sudden loading) by comparing results, in the limit of the nonlocal region going to zero, with analytical ones of the classical mechanics (local) model.

In addition to the FH-PD model, we present a new model in which intermediate homogenization (IH-PD) is used to capture some information about the material microstructure. Different from the FH-PD model, in which the bond properties are computed locally based on an equivalent homogenized composite material, in the IH-PD model we introduce two types of bonds, intraphase bond and interphase bond, that represent the properties of the distinct composite phases and the interfaces between them. For example, for a two-phase composite with phases A and B, we will define A–A bonds (intraphase), B–B bonds (intraphase), and A–B bonds (interphase). The distribution of the bonds depends on the volume fraction of the phases at the nodes forming that particular bond. While the specific geometry of the microstructure is not preserved (unless the scale of the nonlocal region and the corresponding discretization is at the scale of the smallest geometrical features of the microstructure, rendering the model as an explicit model of the microstructure), the specific volume fraction is, and we shall see how this is critical in modeling failure. The model with this additional information that is preserved from the microscale level will be tested against crack growth in a particular two-phase composite in which one of the phases is void (a poroelastic material).

As examples used to test the models, we analyze dynamic crack growth in FGMs, specifically the mixed-mode dynamic fracture experiment from Kirugulige and Tippur (2006). Both the FH and the IH-PD models are employed to simulate this experiment and we point to similarities and differences between the results. A FGM plate (modeled under 2D plane stress conditions) with monotonically varying volume fraction of reinforcement is simulated under mixed-mode loading by eccentric impact relative to the pre-crack location. The role played by the boundary conditions on the crack path is studied by using two configurations for the impact loading: one with free boundaries and one with loads applied, at the same time as the impact loading, at the support locations, closer, but not identical, to conditions used in experiments. In particular, we notice an interesting phenomenon: duration of impact loading influences the crack path by "attracting" the propagating crack toward the moving location of strain energy concentration, generated by a release surface wave initiated when the load is removed. Both FH-PD and IH-PD models lead to the same cracking patterns for these dynamic loading conditions. One difference between the fully homogenized solution and the intermediately homogenized one is that in the IH-PD model, stress waves become less coherent, being locally dispersed ("noisier" strain energy density maps) by the more detailed representation of the composite microstructure.

While the differences between the two PD models for problems in which extreme dynamics controls the crack behavior (high-intensity impact stress waves) is small, in problems in which the microstructure dictates critical phenomena (like crack initiation), the IH-PD model is able to capture an evolution of the failure not seen with the FH-PD model. Indeed, when a porous material (two-phase composite in which one of the phases is void) is considered under quasi-static loading, the IH-PD model reproduces the experimentally observed dependence of crack initiation and growth on the size of a pre-notch, while the FH-PD model, due to ignoring of any microstructural details, fails to do so. We find that the IH model is as computationally efficient compared with the FH model, which lead us to recommend it against the FH-PD for modeling failure in composite and/or porous materials.

Peridynamic Formulation for Functionally Graded Materials (FGMs)

In this section, we first briefly review the peridynamic theory for elastic brittle materials. Then, we introduce the fully (FH-PD) and intermediate (IH-PH) homogenized peridynamic models for composite materials.

Brief Review of Peridynamic Theory for Elastic Brittle Materials

The peridynamic model is a framework for continuum mechanics based on the idea that pairs of material points exert forces on each other across a finite distance. This concept can be viewed as an effective treatment of material length-scale induced by, for example, the material microstructure. The peridynamic equations of motion for bond-based model are given as (Silling 2000):

$$\rho(\mathbf{x})\,\ddot{\mathbf{u}}(\mathbf{x},t) = \int_{H(\mathbf{x})} f\left(\mathbf{u}(\mathbf{x},t) - \mathbf{u}(\hat{\mathbf{x}},t)\,,\hat{\mathbf{x}} - \mathbf{x}\right) dV_{\hat{\mathbf{x}}} + \mathbf{b}(\mathbf{x},t) \tag{1}$$

where f is the pairwise force function in the peridynamic bond that connects point $\hat{\mathbf{x}}$ to $\mathbf{x}$, $\mathbf{u}$ is the displacement vector field, $\rho(\mathbf{x})$ is the spatially dependent density, and $\mathbf{b}(\mathbf{x},t)$ is the body force. The integral is defined over a region $H(\mathbf{x})$ called the "horizon region," or simply the "horizon." The horizon is the compact supported domain of the pairwise force function around a point $\mathbf{x}$ (see Fig. 1). The horizon region is taken here to be a circle of radius δ. We refer to δ also as the "horizon," and from the context there should be no confusion whether we refer to the region or its radius.

A microelastic material is defined as one for which the pairwise force derives from a potential:

$$f\,(\boldsymbol{\eta},\boldsymbol{\xi}) = \frac{\partial\omega\,(\boldsymbol{\eta},\boldsymbol{\xi})}{\partial\boldsymbol{\eta}} \tag{2}$$

where $\boldsymbol{\xi} = \hat{\mathbf{x}} - \mathbf{x}$ is the relative position in the reference configuration and $\boldsymbol{\eta} = \hat{\mathbf{u}} - \mathbf{u}$ is the relative displacement of $\mathbf{x}$ and $\hat{\mathbf{x}}$. A micropotential that leads to a linear microelastic material is given by

$$\omega = \frac{1}{2}c\,(\xi)\,s^2\xi \tag{3}$$

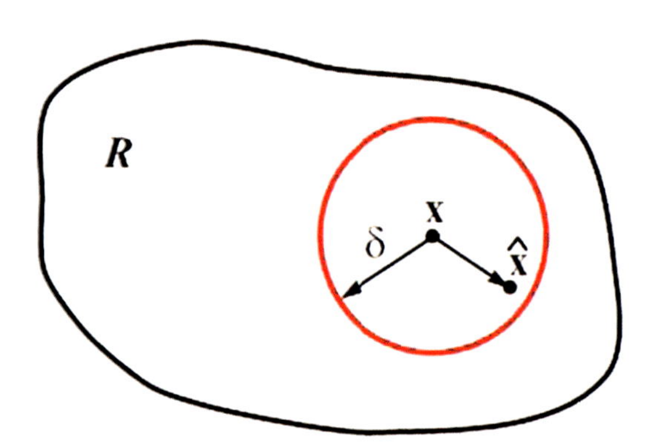

Fig. 1 Each point x interacts directly with the point $\hat{x}$ in the horizon region delimitated by the red circle

where $\xi = \|\boldsymbol{\xi}\|$, $s = \frac{\zeta - \xi}{\xi}$ is the relative elongation of a bond, and $\zeta = \|\boldsymbol{\xi} + \boldsymbol{\eta}\|$. The function $c(\xi)$ is called the micro-modulus and has the meaning of bond's elastic stiffness. The integrand in Eq. 1 may have different forms. For examples in diffusion models please see Chen and Bobaru (2015), and in elasticity models please see Chen et al. (2016). It has been observed that the crack propagation speed in brittle fracture is not influenced by the particular shape of the micro-modulus, once the horizon is reasonably small compared to the dimensions of the structure analyzed (see Ha and Bobaru 2010). The pairwise force corresponding to the micropotential given above has the following form:

$$f(\boldsymbol{\eta}, \boldsymbol{\xi}) = \begin{cases} \frac{\boldsymbol{\xi}+\boldsymbol{\eta}}{\|\boldsymbol{\xi}+\boldsymbol{\eta}\|} c(\xi)\, s & \xi \leq \delta \\ \mathbf{0} & \xi > \delta \end{cases} \tag{4}$$

In this paper, we use the conical 2D plane stress micro-modulus functions (see Ha and Bobaru 2010). Following the same procedure performed to calculate the micro-modulus functions in 1D (see Bobaru et al. 2009), one obtains the conical micro-modulus function in 2D, plane stress conditions (see Ha and Bobaru 2010):

$$c(\xi) = c_1 \left(1 - \frac{\xi}{\delta}\right) = \frac{24E}{\pi \delta^3 (1 - \upsilon)} \left(1 - \frac{\xi}{\delta}\right) \tag{5}$$

For a FGM, the elastic modulus depends on the location, in our case on the position of the ends of the peridynamic bond connecting $\mathbf{x}$ and $\hat{\mathbf{x}}$. The model used in this paper uses the location-dependent elastic modulus and this is discussed in the next section.

Failure is introduced in peridynamics by considering that the peridynamic bonds break when they are deformed beyond a critical value, called the critical relative elongation s_0, computed based on the material's fracture energy. In 2D, the energy per unit fracture length for complete separation of the two halves of the body is the fracture energy G_0. Equating it to the work done in a PD material to accomplish the separation of the body into two halves gives:

$$G_0 = 2 \int_0^{\delta} \int_z^{\delta} \int_0^{\cos^{-1}\left(\frac{z}{\xi}\right)} \left[\frac{c(\xi)\, s_0^2 \xi}{2} \right] \xi \, d\theta \, d\xi \, dz \tag{6}$$

Substituting the micro-modulus function from Eq. 5 into Eq. 6, s_0 is obtained as:

$$s_0 = \sqrt{\frac{5\pi G_0}{9E\delta}} \tag{7}$$

For the PD-FGM model, G_0 is a function of location of the ends of the peridynamic bond, and this is given in the next section.

A Fully Homogenized Peridynamic Model for FGMs

In this fully homogenized peridynamic (FH-PD) model, a FGM is understood as a material model *locally homogenized* in terms of its elastic properties, density, and fracture energy, corresponding to an actual composition of two phases. The FGM cannot be considered as a globally homogeneous material since its material properties change over large distances, but it can be viewed as a locally homogeneous material because we are not considering the explicit composition of the FGM and we assume that in a sufficiently small neighborhood of any point, material properties are, effectively, constant. We assume that only the volume fraction of each phase influences the elastic and fracture properties of the composite. This assumption is, obviously, a simplification since the particular shape and geometric distribution of the inclusions can affect the elastic and, even more strongly, the fracture properties in the FGM. For problems in which these dependencies are strong, one could first obtain a homogenized peridynamic model from the specific material microstructure. Such a model will lead to a specific horizon size that would capture one or several length-scales generated by the specific microstructure of the composite material (see Silling 2014).

We formulate the peridynamic model for the homogenized FGM concept, and, therefore, the nonlocal region size is not connected to material length-scales generated by a specific microstructure. In this way, we are able to consider the convergence of the nonlocal model solution to the local model, in the limit of the horizon size going to zero (see Silling and Lehoucq 2008; Ha and Bobaru 2010; Bobaru et al. 2009). Such a modeling approach may not be appropriate for modeling failure processes in certain FGMs under certain loading conditions. Part of our goal here is to see to what extent the proposed peridynamic model of a locally homogenized FGM is able to reproduce experimentally observed dynamic fracture behavior for an FGM prepared by varying the volume fraction of small inclusions (glass spheres) in an epoxy matrix as used in Kirugulige and Tippur (2006, 2008). For the type of FGM used in Kirugulige and Tippur (2006, 2008), the fracture behavior under impact conditions reveals a very high repeatability in terms of the crack path (see Fig. 4 in Jain and Shukla (2006)).

Consider two arbitrary material points $\mathbf{x}$ and $\hat{\mathbf{x}}$ connected by a peridynamic bond in an FGM with E_x, ρ_x, G_x and $E_{\hat{x}}$, $\rho_{\hat{x}}$, $G_{\hat{x}}$ being their effective Young's moduli, densities, and fracture energies, respectively. For a peridynamic model of the FGM we need to specify the micro-modulus of the bond connecting these two points. We propose a model in which the material properties used to compute the micro-modulus of the bond connecting points x and x̂ are the average values of the material properties at the two points. Therefore, in Eqs. 5 and 7 we will use:

$$E\left(x,\hat{x}\right)=\frac{\left(E_x+E_{\hat{x}}\right)}{2},\quad G_0\left(x,\hat{x}\right)=\frac{\left(G_x+G_{\hat{x}}\right)}{2} \tag{8}$$

The material density is a pointwise quantity and is introduced at each point in the peridynamic equation of motion (see Eq. 1). For the two-dimensional bond-based peridynamic model of an isotropic and homogeneous material in plane-stress conditions, the Poisson's ratio is 1/3. For the particular FGM used in the experiments in Kirugulige and Tippur (2006, 2008), Poisson's ratio varies from 0.33 to 0.37, and the authors of Kirugulige and Tippur (2006, 2008) select 0.34 in their modeling and calculations. While Poisson's ratio may influence the path of a growing crack, its variation is not expected to play a significant role in the fracture behavior of FGMs (see Delale and Erdogan (1983) for a discussion on this subject). Peridynamic modeling of other FGMs for which bond-based model's fixed Poisson ratio is not a good fit can be performed using the state-based peridynamic formulation (see Silling and Lehoucq 2008; Silling et al. 2007). The state-based PD formulation eliminates the Poisson ratio restriction, but it comes with a computational penalty, normally increasing the cost by at least a factor of two compared with the bond-based. Because the Poisson ratio for the material considered here is close to 1/3 and the fact that our primary interest here is to observe the capabilities of a PD model in capturing the evolution of fracture and failure in FGMs and to understand what are the factors that control crack growth in FGMs, in this book chapter, we use the bond-based peridynamic formulation.

An Intermediate-Homogenization Peridynamic Model for Two-Phase Composites

The FH-PD model presented in the previous section assumes a homogenized local material point. In reality, FGMs usually are composites of two or more material phases. The mechanical connections between material points could be interphase or intraphase bonding. An example of a two-phase (A and B) composite is shown in Fig. 2. We consider two arbitrary material points $\boldsymbol{x}$ and $\hat{\boldsymbol{x}}$ connected by a peridynamic bond with R and $\hat{R}$ being their Phase A volume fractions, respectively. The peridynamic bond between these two material points has a certain probability

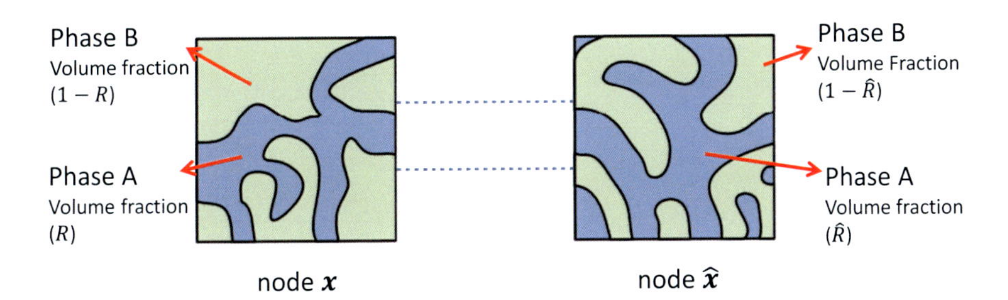

Fig. 2 Description of the intermediate-homogenization PD (IH-PD) model for a two-phase composite material. The bond between the two nodes depends on the local volume fractions of the two composite phases. When the nonlocal size and the discretization size reach the size of the smallest geometrical features of the microstructure, the model matches one in which the microstructural description is explicit

to be considered as a bonding between the same phase or between different phases, depending on the volume fraction of the phases in the two material volumes occupied by the points. We will refer to Phase A – Phase A bonding as A–A bond, Phase B – Phase B bonding as B–B bond, and Phase A – Phase B bonding as A–B bond.

In the new intermediately homogenized peridynamic model (IH-PD, see Chen et al. (2017) for details), we assume a linear relationship between the chance of a bond to be of a certain type and the *local* phase volume fractions at the two nodes. Thus, the bond between material points has chance $R\hat{R}$ to be an A–A bond, chance $(1-R)\left(1-\hat{R}\right)$ to be a B–B bond, and chance $1-(1-R)\left(1-\hat{R}\right)-R\hat{R}=R+\hat{R}-2R\hat{R}$ to be a A–B bond. Heterogeneity is thus introduced by the combination of different mechanical bonds (with different micro-moduli and critical bond strains) connecting at a particular material point.

The detailed steps for the algorithm that implements the IH-PD model, in the model generation stage (preprocessing step), at a node $\boldsymbol{x}$, are:

(i) For each of the undetermined bonds in the family of this node, generate a random number r from a uniform distribution in (0, 1).
(ii) If $\mathrm{r} < \mathrm{R}\hat{\mathrm{R}}$, then label the bond as A–A bond by assigning the micro-modulus and critical bond stretch of material A for this node. If $\mathrm{R}\hat{\mathrm{R}} < r < (1-\mathrm{R})\left(1-\hat{\mathrm{R}}\right)+\mathrm{R}\hat{\mathrm{R}}$, then label the bond as B–B bond. Otherwise, label the bond as an A–B bond.

Once the bonds are assigned their type, the mechanical properties (micro-modulus value, critical bond strain) are determined. For A–A bonds, we assign the properties corresponding to phase A, B–B bonds are from phase B, and for A–B bonds we have several options. Two possible ones are based on couplings in series (arithmetic average) or parallel (harmonic average):

$$M_{AB}=\frac{M_A+M_B}{2}\quad(\text{for the a}-\text{IH}-\text{PD model}),\qquad \text{and}\ \frac{2M_AM_B}{M_A+M_B}\quad(\text{for the h}-\text{IH}-\text{PD model}) \tag{9}$$

where M_A, M_B, and M_{AB} are the mechanical properties corresponding to A, B, and the interphase, phases, respectively. Another possibility for the A–B bonds is to find interface properties from experiments and assign them to A–B bonds. This option is especially useful when the interface is weaker than both of the phases.

Because the algorithm above contains random numbers, every different run that uses a newly seeded random number generator will lead to slightly different results for the same input data. For the elastic model, and a fixed horizon size, however, increasing the m-value (the ratio between the horizon size and the discretization size. See Appendix A for more details of the numerical discretization) results in attenuating/minimizing these differences. Obviously, with a fixed horizon size, once the node size is smaller than the smallest dominant geometrical feature of the microstructure (pore size in a porous material, fiber diameter in a fiber-reinforced composite, size of debonds or microcracks), the elastic response would no longer change. When the horizon size itself (and the corresponding discretization) is on the scale of the microstructural features, the model essentially becomes an "explicit microstructure" model. Since such models are expensive to compute, the goal for the IH-PD model is to serve as an efficient alternative to the explicit model while incorporating some of the microscale information that has an effect on fracture and failure behavior with the goal of providing more realistic failure response than the FH-PD model could.

To compare the effective properties of the IH-PD model with those from the FH-PD model, we "homogenize" the bonds in the IH-PD model (see Fig. 2) so that each bond has a mechanical property $M(\boldsymbol{x}, \hat{\boldsymbol{x}})$ that is computed by considering a combination of parallel bonds with properties M_A, M_B, and M_{AB} between $\boldsymbol{x}$ and $\hat{\boldsymbol{x}}$, as follows:

$$M = M_A R\hat{R} + M_B (1 - R)\left(1 - \hat{R}\right) + M_{AB}\left(R + \hat{R} - 2R\hat{R}\right) \tag{10}$$

By applying Eq. 9, we have:

$$M = \begin{cases} M_B + \frac{1}{2}(M_A - M_B)\left(R + \hat{R}\right), a - \mathrm{IH} - \mathrm{PD} \\ M_B + \frac{(M_A - M_B)\left[(M_A - M_B)R\hat{R} + M_B\left(R + \hat{R}\right)\right]}{(M_A + M_B)}, h - \mathrm{IH} - \mathrm{PD} \end{cases} \tag{11}$$

Note that for elasticity (and problems in which the microstructure is uniform), the IH-PD and FH-PD converge, in the limit of the horizon going to zero, to the classical composite model solution. However, when the nonlinear model is considered (in which damage is allowed via bond breakage), the results from IH-PD and FH-PD models do not have to be the same. The results from the two models will be different especially for cases with high contrast between the fracture toughness of the composite phases.

When applying the IH-PD model to FGMs with continuously varying volume fraction of reinforcement, taking the horizon size small enough (relative to the overall gradient in composition) leads to $R \cong \hat{R}$. Then, Eq. 11 reduces to:

$$M = \begin{cases} M_B + (M_A - M_B)\, R(\boldsymbol{x}), a - \mathrm{IH} - \mathrm{PD} \\ M_B + \frac{(M_A - M_B)\left[(M_A - M_B)R^2(\boldsymbol{x}) + 2M_B R(\boldsymbol{x})\right]}{(M_A + M_B)}, h - \mathrm{IH} - \mathrm{PD} \end{cases} \tag{12}$$

Here, $R(\boldsymbol{x})$ is the volume fraction of phase A at $\boldsymbol{x}$. When $R(\boldsymbol{x}) = 0$ (100% phase B), $M = M_B$; when $R(\boldsymbol{x}) = 1$ (100% phase A), $M = M_A$. This is the same as the effective properties in the FH-PD model for an FGM (see Fig. 4 below).

Note that although the effective mechanical properties from the "homogenization" in the IH-PD model could be the same as the ones in FH-PD model, the results from two models are different and in sections "Numerical Studies for Dynamic Crack Propagation in FGMs" and "Results and Discussion" we will test these differences in both dynamic fracture and quasi-static fracture.

The IH-PD model can also be used for modeling porous elastic materials, by treating the porous material as a composite with one of the phases being a void phase. We apply h-IH-PD model for a porous material. By assuming that phase B is void phase in Fig. 2, material porosity P is the volume fraction of phase B. Substituting $R(\boldsymbol{x}) = 1 - P$ and $M_B = 0$ into the h-IH-PD model in Eq. 12, we obtain the effective mechanical properties as follows:

$$M = M_A(1 - P)^2 \tag{13}$$

Applications of h-IH-PD model for quasi-static fracture in a porous elastic and brittle material are presented in section "Results and Discussion."

Numerical Studies for Dynamic Crack Propagation in FGMs

Problem Setting

We consider dynamic crack growth in an edge-notch FGM plate specimen under offset impact loading. This setup has been experimentally investigated in Kirugulige and Tippur (2006). The sample geometry and boundary conditions are shown in Fig. 3. We use two types of boundary conditions in our simulations, for two reasons:

(a) Different dynamic loading induce different wave propagation scenarios that are likely to drive crack growth to different results. These scenarios will allow us to understand the role wave propagation plays in dynamic brittle fracture of FGMs.
(b) The description given in Kirugulige and Tippur (2006) is not sufficiently detailed to allow us to specify the boundary conditions precisely as those used

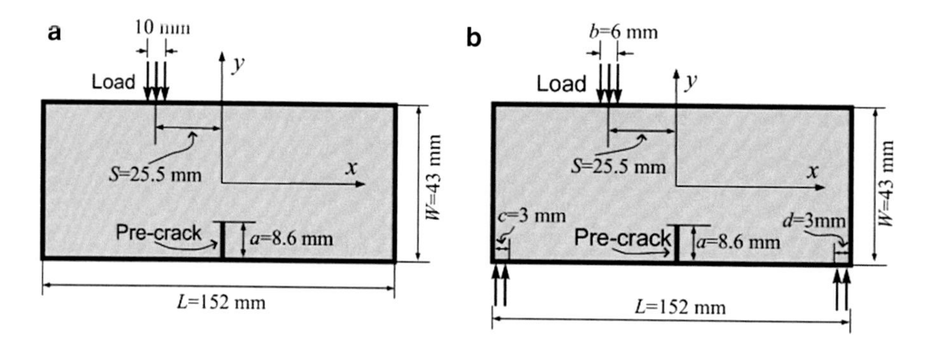

Fig. 3 Specimen geometry and boundary conditions: (**a**) sudden loading on the top surface, otherwise free boundaries, and (**b**) sudden loading applied at three points (satisfying static equilibrium)

in the actual experiment. Only the launch speed of the impactor is given. The authors of Kirugulige and Tippur (2006) also mentioned that "the specimen rests on two blocks of soft putty to preclude any interaction from the supports while the specimen undergoes stress wave loading." The size of the hammer used as the impactor in the experiments is not specified, the rebound time is not given, and the size and properties of the supports are also not reported. The three-point impact loading conditions used in our work, while not identical, are meant to be closer to those in the experiments than the impact loading on the free-free plate.

The cohesive-zone finite element-based model for this problem used in Kirugulige and Tippur (2008) employed only the "free–free" boundary conditions. The computational study in Kirugulige and Tippur (2008) only presents the initial stages of crack propagation, by following the crack growth only for its first 1.5–2.0 cm. The authors of Kirugulige and Tippur (2008) do not explain why the study was limited to this stage of crack growth and why they did not compare the crack paths through final failure of the sample with the experimental results. Perhaps, the FEM solution breaks down after a while. Our peridynamic simulations will track the crack growth through full failure.

The FGM sample used in experiments in Kirugulige and Tippur (2006) is made of epoxy with continuously varying volume fraction of glass filler particles (35 μm mean diameter) from 0 to 40%. The mechanical properties of the constituents of the FGM specimen studied are listed in Table 1. The variation of elastic modulus and mass density along the width of a sample can be found in Fig. 3 in Kirugulige and Tippur (2006). The elastic modulus varies from 10 GPa to 4 GPa over the width and the mass density varies from 1,750 kg/m^3 to 1,175 kg/m^3 over the same width. The FGM specimen is a rectangular plate (see Fig. 3) with width $W = 43$ mm, length $L = 152$ mm, and an edge crack of length $a = 8.6$ mm. The specimen is subjected to an impact loading at an offset distance of $S = 25.5$ mm with respect to the initial crack location.

In their FEM analysis of this problem, the authors of Kirugulige and Tippur (2008) use linear curve fitting to approximate the variation of fracture energy K_{ICR}, elastic modulus E, and the mass density ρ over the sample's width. We will do the same here, to have a fair comparison between the PD and FEM results. However, in section "Comparison with Experimental and FEM-Based Results: Nonlinear Material Variation" we will also test the crack propagation paths and speed by using the actual, nonlinear variation of elastic modulus and density measured from

Table 1 Nominal bulk properties of the constituent materials used in the FGM samples for the dynamic fracture examples (from Kirugulige and Tippur 2006)

	E (GPa)	V	ρ (kg/m^3)
Epoxy	3.2	0.34	1,175
Soda-lime glass	70	0.23	2,500

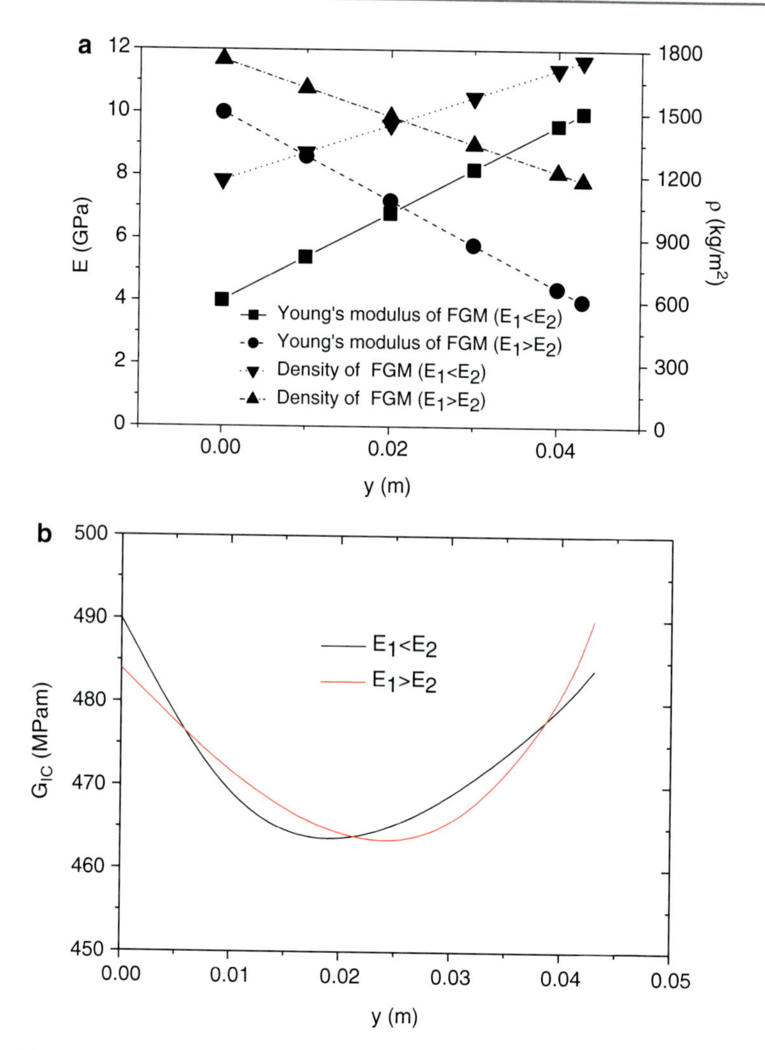

Fig. 4 Linear curve fits for elastic moduli and density across the width of the sample (**a**). Fracture energy variation across the sample's width based on linear curve fits for the elastic modulus and stress intensity factor (**b**) (same as those used in the FEM analysis in Kirugulige and Tippur (2008))

experiments and shown in Fig. 3 in Kirugulige and Tippur (2006) and repeated in Fig. 1 in Kirugulige and Tippur (2008).

Let the elastic modulus at the bottom and top surfaces of the specimen be E_1 and E_2, respectively. The maps with the linear curve fit for density and elastic modulus, as well as the variation of fracture energy along the width of the sample, are shown in Fig. 4.

For the case with the pre-crack on the stiffer side (i.e. $E_1 > E_2$), we have the following linear curve-fits (see Kirugulige and Tippur 2008):

$$K_{ICR}(y) = 2.2 - \frac{2.2 - 1.4}{43} y, \quad 0 \le y \le 43 \text{ mm} \tag{14}$$

$$E(y) = 10 - \frac{10 - 4}{43} y, \quad 0 \le y \le 43 \text{ mm} \tag{15}$$

$$\rho(y) = 1,750 - \frac{1,750 - 1,175}{43} y, \quad 0 \le y \le 43 \text{ mm} \tag{16}$$

$$G_{IC}(y) = \frac{K^2_{ICR}(Y)}{E(y)}, \quad G_{IC}(y) = G_{IIC}(y) \tag{17}$$

Similar linear approximations are used for the other case, when $E_1 < E_2$, when the pre-crack is on the compliant side of the specimen.

The profiles of parameters shown in Eqs. 14, 15, 16, and 17 can be directly applied in Eq. 8 to obtaine the bonds information in the peridynamic FH-PD model.

For the IH-PD model, we need the mechanical properties of each phase in the FGMs and their corresponding volume fraction variation over the volume of a peridynamic node, which is decided by the discretization scale used.

In the example here, we consider discretizations that are much larger than the size of the glass beads inclusions. Therefore, we can use the properties as shown above and assume a linear variation of volume fractions. We can thus take the material at the bottom and top surfaces of the sample as being of a single phase. For example, when the pre-crack is on the stiffer side ($E_1 > E_2$), the material at the bottom surface is phase A material, of material parameters shown in Eqs. 14, 15, 16, and 17 with $y = 0$, and the material at the top surface is phase B material, of material parameters shown in Eqs. 14, 15, 16, and 17 with $y = 43$ mm. The volume fraction of each phase is assumed to be linearly distributed in the y-direction. When $E_1 < E_2$, phase A switches with phase B.

Boundary and Loading Conditions

We now present the details about the two types of boundary/loading conditions used in this study. The first type conditions, shown in Fig. 3a, use impact loading of the free beam/plate since this was also used in the finite-element-based computational study in Kirugulige and Tippur (2008). In our case, we apply the normal stress as a uniform distribution of the equivalent body force over a length of 10 mm (from $S - 5$ mm to $S + 5$ mm) on the top surface (see Fig. 3a).

The second type of boundary/loading conditions mimics, to a certain extent, the three-point loading from the experiments. In this case, we apply the dynamic loading, uniformly distributed in space at three locations, corresponding to the two supports and the impact location (see Fig. 3b). The lengths over which the loads are applied are shown in Fig. 3b.

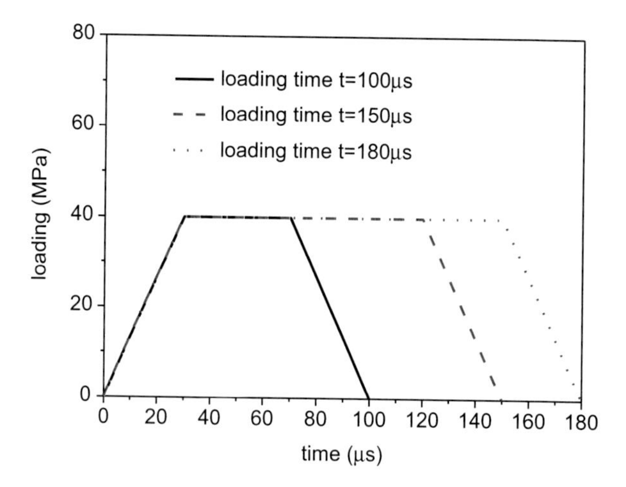

Fig. 5 Variations of the duration of impact loading stress tested in the peridynamic computations

In absence of data from experiments regarding the rebound time of the hammer used to impact the sample, we use a time variation of the impact load described by the following trapezoidal profile (see Fig. 5):

$$\sigma = \begin{cases} \sigma_0 \frac{t}{t_1} & 0 \le t \le t_1 \\ \sigma_0 & t_1 \le t \le t_2 \\ \sigma_0 \frac{t_3 - t}{t_3 - t_2} & t_2 \le t \le t_3 \end{cases} \tag{18}$$

where σ_0 is the maximum impact stress, the ramp-up time $t_1 = 30$ μs, t_3 is the total loading time, and the ramp-down time is $t_3 - t_2 = 30$ μs. We keep the ramp-up and ramp-down times fixed, but we vary the time spent at maximum stress level: $t_2 - t_1 = 40$ μs, 90 μs, and 120 μs resulting in total loading times $t_3 = 100$ μs, 150 μs, and 180 μs, respectively (see Fig. 5).

To decide about the actual stress magnitude and corresponding loading force in each of the boundary conditions types, we perform several tests at different loadings and choose the ones that are about the smallest that lead to full failure of the sample (complete separation into two pieces due to crack growth from the pre-notch). For the three-position loading configuration, we select $\sigma_0 = 26$ MPa, while for the impact with free boundary conditions, we choose $\sigma_0 = 40$ MPa. We also note that some no-fail zones are assigned to the regions around the loading location to avoid immediate failure generation at those locations. In these regions, the critical relative elongation for the peridynamic bonds is set to a large value that is never reached in the present loading conditions. We made sure that these areas do not interfere with the crack growth in any of the computational tests we performed.

For the remaining computational simulations, we use a uniform time step size of 50 ns, which is sufficiently small for all of the tests, based on the stability conditions from Silling and Askari (2005).

Convergence Studies for Crack Propagation Path in Dynamic Brittle Fracture of FGMs

In this subsection, we perform convergence studies in terms of the crack path shape that grows from the pre-cracked FGM specimen under the impact loading for the free–free boundary condition type. Convergence for peridynamic results in *dynamic brittle fracture* has been discussed in Ha and Bobaru (2010, 2011) for isotropic and homogeneous materials and in Hu et al. (2013) for homogeneous but anisotropic materials. Here we study convergence for FGMs with the FH-PD model, and try to answer whether the size of the nonlocal region is sufficiently small, relative to the gradient of material properties function in the FGM, to obtain a crack path that no longer changes when the horizon size decreases further.

Details of the discretized PD model were already given in section "Numerical Studies for Dynamic Crack Propagation in FGMs" of Cheng et al. (2015) and Appendix A. To those details, we append the combination of Eqs. 7 and 8 for the FH-PD model.

For m-convergence (see Appendix A), we perform tests for a fixed horizon size $\delta = 2$ mm and vary the grid density: $\Delta x = 1$ mm ($m = 2$), 0.5 mm ($m = 4$), and 0.25 mm ($m = 8$). For the case with $E_1 < E_2$ (pre-crack is on the compliant side), we select a total loading time of 150 μs and the resulting crack paths after 210 μs from impact are shown in Fig. 6. The results indicate that, for this horizon size and the particular material variation in the FGM chosen, when a sufficient number of nodes are inside the horizon of the node (when m is larger than 4), no further increases in the grid density changes the crack path. We also note that the lengths of the crack paths in Fig. 6b, c are very similar, meaning that the crack speed is similar between these two cases. The result in Fig. 6a shows that when there is a small number of bonds connecting a node with its nonlocal neighbors ($m = 2$ in 2D produces only 12–20 possible bonds for each node, depending whether nodes covered only partially by the horizon are included or not), the crack path suffers. This is because there are not enough bonds in enough possible directions that can break when stretched beyond their critical value to result in a crack growth that is consistent with the dynamics of the problem. We emphasize here that in peridynamics cracks grow autonomously, because of a cascading of bond rupture events.

For the case when $E_1 > E_2$ (pre-crack is on the stiffer side of the FGM), the m-convergence study is performed using the same numerical grids, the same loading magnitude, and the same total loading time as above. The crack paths for this case at 210 μs after impact are compared in Fig. 7. As in the previous case, it can be observed that the paths with $m = 4$ and 8 are very similar to each other, in both shape and length, meaning that the crack propagation speed between the two grids

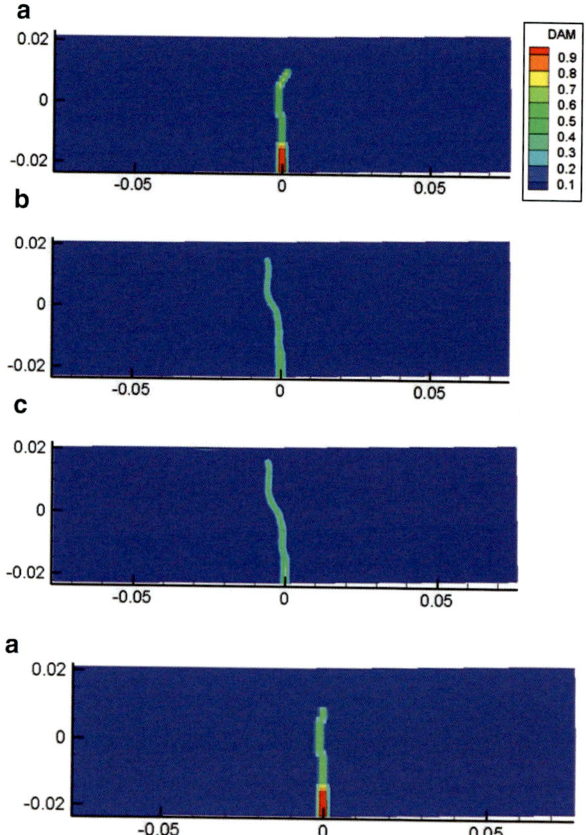

Fig. 6 m-convergence in terms of the crack path for an FGM specimen with $E_1 < E_2$ computed with different grids for $\delta = 2$ mm: (**a**) $m = 2$ ($\Delta x = 1$ mm), (**b**) $m = 4$ ($\Delta x = 0.5$ mm), and (**c**) $m = 8$ ($\Delta x = 0.25$ mm). The damage maps shown are at 210 μs after impact for the loading case with the total loading time of 150 μs. Dimensions on the axes are in meters

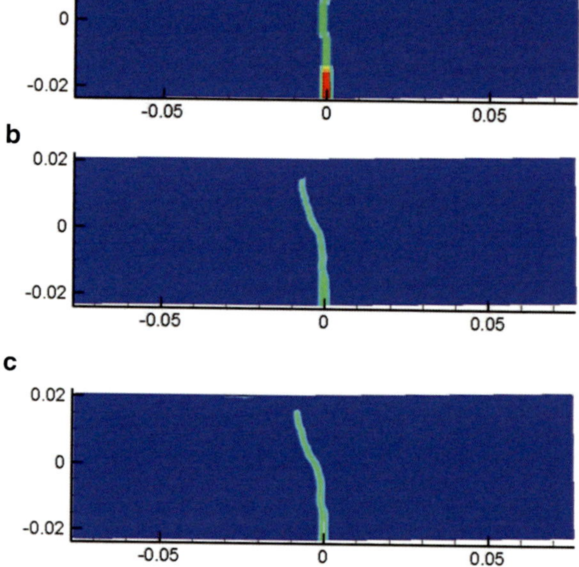

Fig. 7 m-convergence in terms of crack path for an FGM specimen with $E_1 > E_2$ computed with different grids for $\delta = 2$ mm: (**a**) $m = 2$ ($\Delta x = 1$ mm), (**b**) $m = 4$ ($\Delta x = 0.5$ mm), and (**c**) $m = 8$ ($\Delta x = 0.25$ mm). The damage maps shown are at 210 μs after impact for the loading case with the total loading time of 150 μs

are not much different from each other. From these two tests we conclude that, for a given horizon size, a sufficiently high grid density to produce results that no longer change with grid refinement is one produced by $m = 4$. Using a larger value of m

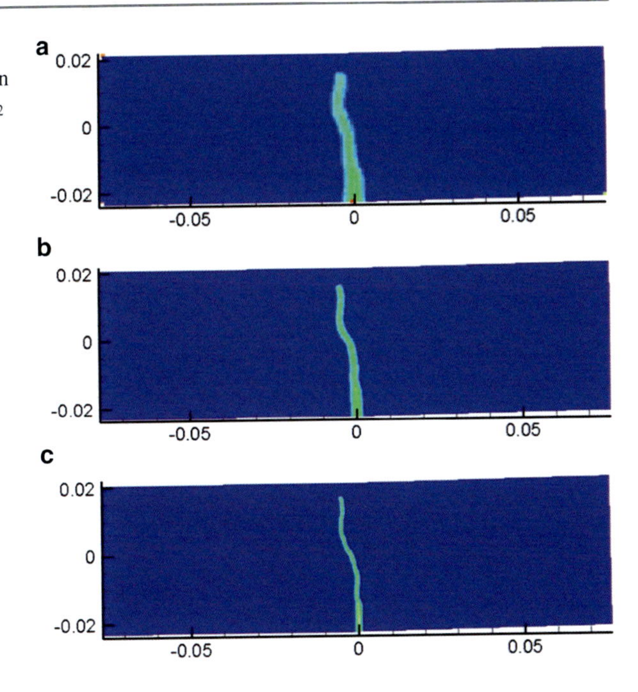

Fig. 8 δ–convergence in terms of the crack path for an FGM specimen with $E_1 < E_2$ computed with different horizon sizes (and grids corresponding to $m = \delta/\Delta x = 4$): (**a**) $\delta = 4$ mm, (**b**) $\delta = 2$ mm, and (**c**) $\delta = 1$ mm. The damage maps shown are at 210 μs after impact for the loading case with a total loading time of 150 μs

requires higher computational cost, while the results are not affected. Therefore, in the remaining computations we choose $m = 4$. Note also that under m-convergence, the "thickness" of the damage zone does not change (see remarks in Ha and Bobaru (2010)).

To assess whether the crack path is resolved in terms of the horizon size we now turn to δ-convergence (see Appendix A). Since the FH-PD model here is for a locally homogenized FGM model, the horizon size is not connected to an eventual material length-scale generated by specific microstructure architecture of the FGM. Therefore, the role of a δ-convergence study is to determine the largest horizon size that produces a crack path (crack propagation speed) that does not change when a smaller horizon size is used. Since in the examples considered here the cracks are generated from a pre-crack, and because the "size" of the initial damage caused by the pre-crack depends on the size of the horizon (see also Ha and Bobaru (2010) and Bobaru and Hu (2012) for a discussion on this topic), a large horizon size will not represent well the physical dimensions of the pre-crack, leading to stress profiles near the crack tip different from the actual ones. This will likely influence the dynamic crack growth and the crack path in the FGM. Moreover, dynamic fracture is strongly influenced by wave propagation and using a relatively large horizon size leads to a larger nonlocal dispersion for propagating stress waves that influences, in turn, crack growth evolution (Fig. 8).

We perform the δ-convergence study for $m = 4$ and use the following values for the horizon size: 4 mm, 2 mm (as the one used in the m-convergence study above), and 1 mm. Recall that the pre-crack length is close to 9 mm. Since m

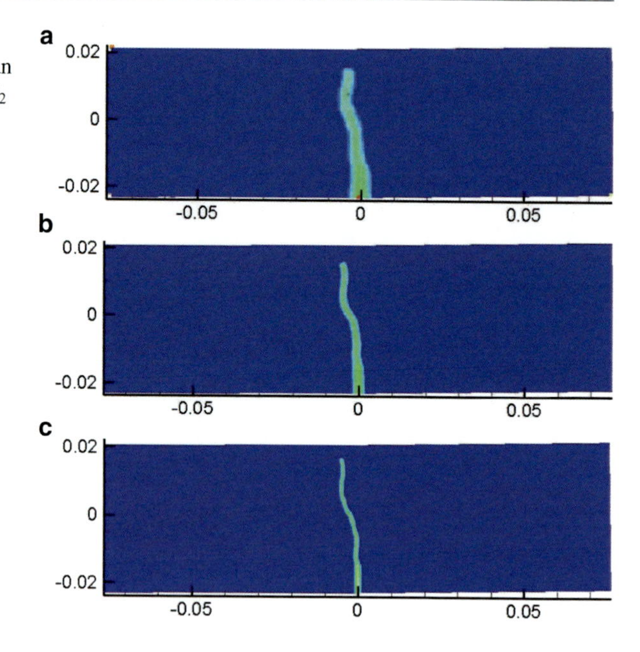

Fig. 9 δ–convergence in terms of the crack path for an FGM specimen with $E_1 > E_2$ computed with different horizon sizes (and grids corresponding to $m = \delta/\Delta x = 4$): (**a**) $\delta = 4$ mm, (**b**) $\delta = 2$ mm, and (**c**) $\delta = 1$ mm. The damage maps shown are at 210 μs after impact for the loading case with a total loading time of 150 μs

is fixed, each horizon size determines the grid spacing used with it, therefore the corresponding Δx values are: 1 mm, 0.5 mm, and 0.25 mm, respectively. The same impact conditions as those in the m-convergence test are applied. The crack path of both $E_1 < E_2$ and $E_1 > E_2$ at 210 μs for peridynamic simulations using the three cases specified above are shown in Figs. 9 and 13, respectively. It can be noticed that, as the horizon decreases under δ-convergence, the crack path does not change much, but it becomes more defined, consistent with the assertion that the "spread" or "thickness" of damage is related to the horizon size. Because the crack length at the same time instant is very similar between the last two horizon sizes used, we can also conclude that the crack propagation speed does not change as the horizon changes. Note that the width of the pre-crack mentioned in the experiments in Kirugulige and Tippur (2008, p. 272), is about 0.3 mm, while Fig. 3 in Kirugulige and Tippur (2008) shows the pre-notch width to be in the 1 mm range. Because of this, and because using an even smaller horizon size (while keeping a value of m of at least 4, as required by the results from the m-convergence tests above) would lead to higher computational cost with little difference in the results, in the computations that follow we will use the horizon size $\delta = 1$ mm and a discretization corresponding to $m = 4$, leading to a total of 105,184 nodes for the given geometry.

In the next section, we compare the results from the FH-PD model with those obtained with the IH-PD model for the smallest horizon size used with the FH-PD model.

Results and Discussion

In this section, we first apply the FH-PD model and discuss results in which we vary the time impact loading is applied at its maximum, for the FGM with $E_1 < E_2$ (the results for the case with $E_1 > E_2$ are available in Cheng et al. (2015)). We compare the results with the ones from the IH-PD model and find them to be similar. We also investigate differences in failure response between the free–free conditions and the three-point impact loading conditions. We then discuss the effect of using the actual material variation (nonlinear) versus using the linear curve-fit to those properties when modeling dynamic fracture in FGMs with the FH-PD model. Finally, we compare results with experimental ones on FGM samples and on homogeneous samples.

Influence of Loading Duration and Material Variation: The Crack Path "Attractor"

When the pre-cracked FGM sample with $E_1 < E_2$ is impacted with the 40 MPa equivalent stress in the free–free boundary conditions, the duration over which the sample is under that level of stress loading has a strong influence on the crack path direction growth. A close examination of the strain energy maps during the crack propagation process clearly shows regions where the strain elastic energy density is concentrated and the interaction of stress waves propagating through the structure (see results from FH-PD model, including Movies 1, 2, and 3, for the three loading times of 100 μs, 150 μs, and 180 μs from Fig. 5, respectively). The crack does not start when the first waves reach its tip, but much later when waves reflected from the various boundaries and influenced by the presence of the pre-crack, lead to mixed-mode conditions favorable for crack growth. In Fig. 10 we show some snapshots taken at 144 μs and 210 μs from the initial impact, and after the final failure of the sample, for the three cases of total loading times of 100 μs, 150 μs, and 180 μs (see Fig. 10).

The crack initiates around 120 μs after the impact in all three cases. The initial growth of the crack is close to straight in all cases, but soon after, it starts to deviate from the straight line and it curves to the right when loading time is the shortest (100 μs), while for the other two loading times cases the crack deviates to the left. We explain this phenomenon as follows: for the loading time of 100 μs, the load is removed *before* the crack starts to propagate. The Rayleigh wave generated during the unloading phase, then acts as a "*crack attractor*," and the crack, "sensing" the strain energy density concentration created by the moving Rayleigh wave on the top surface, starts to follow it and tilts to the right. This explanation is confirmed by the results shown in Fig. 10 and in Movies 1, 2, and 3: for the loading time of 150 μs, the crack tilts left, but when the released Rayleigh wave starts moving to the right, the crack path changes direction and is following the motion of the location with the largest strain energy density on the top surface of the sample, where bending stresses start to become significant due to the opening of the crack. In the case when the loading lasts 180 μs, the crack moves toward the location of the applied load,

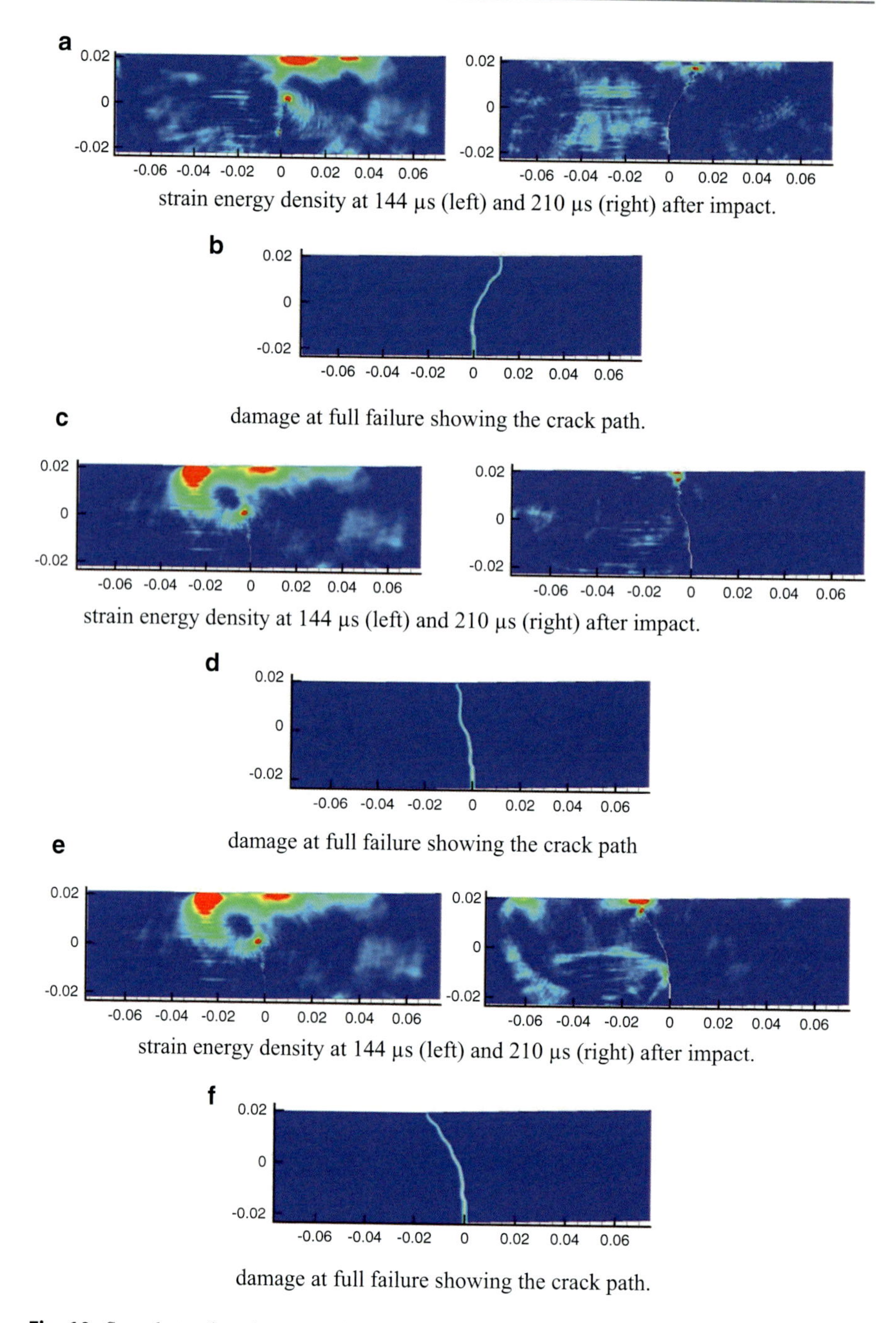

Fig. 10 Snapshots of strain energy density and crack propagation path in the FGM sample with $E_1 < E_2$ at various moments for different total loading times: $t_3 = 100\ \mu s$ for (**a**) and (**b**); $t_3 = 150\ \mu s$ for (**c**) and (**d**); $t_3 = 180\ \mu s$ for (**e**) and (**f**)

and by the time the load is released, the crack path has almost split the sample in two and the Rayleigh wave traveling to the right can change very little of this final stage of failure in the FGM.

The above observations suggest that it might be possible to control the crack path in dynamic fracture from remote locations by using Rayleigh waves.

Using the same loading conditions and solution parameters, the FGM sample with the pre-crack crack on the stiffer side ($E_1 > E_2$) was also tested, and the results are presented in Cheng et al. (2015). The crack initiates now at about 100 μs independent of the time at which the loading stays at the 40 MPa stress value. The explanation presented above for the $E_1 < E_2$ case remains true.

The peridynamic results shown above correctly predict a trend observed in experiments (see Fig. 8 in Kirugulige and Tippur (2006)), namely that the initiation of crack propagation is earlier in the $E_1 > E_2$ case than in the $E_1 < E_2$ case. Notably, this trend *is not* reproduced by the cohesive FEM model in Kirugulige and Tippur (2008). The FEM computations there show either the exact opposite to what is observed in experiments (conversion of stored strain energy into fracture energy starts earlier for the $E_1 < E_2$ case than the $E_1 > E_2$ case, see Sect. 5.2 in Kirugulige and Tippur (2008)), or that the crack initiation time is insensitive to these two different cases of material variation in the FGM (see Sect. 5.4 in Kirugulige and Tippur (2008)).

We also apply IH-PD model, including both a-IH-PD model and h-IH-PD model, for the cases shown in Fig. 10. The comparison of the strain energy density profile between the FH-PD model and the a-IH-PD model at 144 μs is shown in Fig. 11 (see Movies 4, 5, and 6, for strain energy evolution from a-IH-PD model, with the three loading times of 100 μs, 150 μs, and 180 μs, respectively). Only the cases with $t_3 = 100$ μs and 150 μs are included. The strain energy density profile (at 144 μs) for the case with $t_3 = 180$ μs is the same as the one with $t_3 = 150$ μs. In both cases, FH-PD model and a-IH-PD model generate the similar strain energy density profiles. One difference between the fully homogenized solution and the intermediately homogenized one, is that in the IH-PD model, stress waves become less coherent, being locally dispersed ("noisier" strain energy density maps) by the more detailed representation of the composite microstructure. Since in this particular impact loading case the damage profile, or the crack propagation, is controlled by the high-intensity impact stress waves, and the two models show similar wave profiles, the final crack patterns from the FH-PD and a-IH-PD models are almost identical, as shown in Fig. 12.

The final damage profiles from a-IH-PD model (see Fig. 12) are the same as the ones from FH-PD model. Only minor differences are seen between the results with the h-IH-PD model and those from the FH-PD model. The reason is that the effective mechanical properties from the a-IH-PD model (see Eq. 11) are the same as the ones in the FH-PD model, while the ones from the h-IH-PD model (see Eq. 11) are slightly smaller than the ones in the FH-PD model. Thus, h-IH-PD model leads to slightly earlier crack initiation and the final crack tip is closer to the right side than the one from FH-PD or a-IH-PD models.

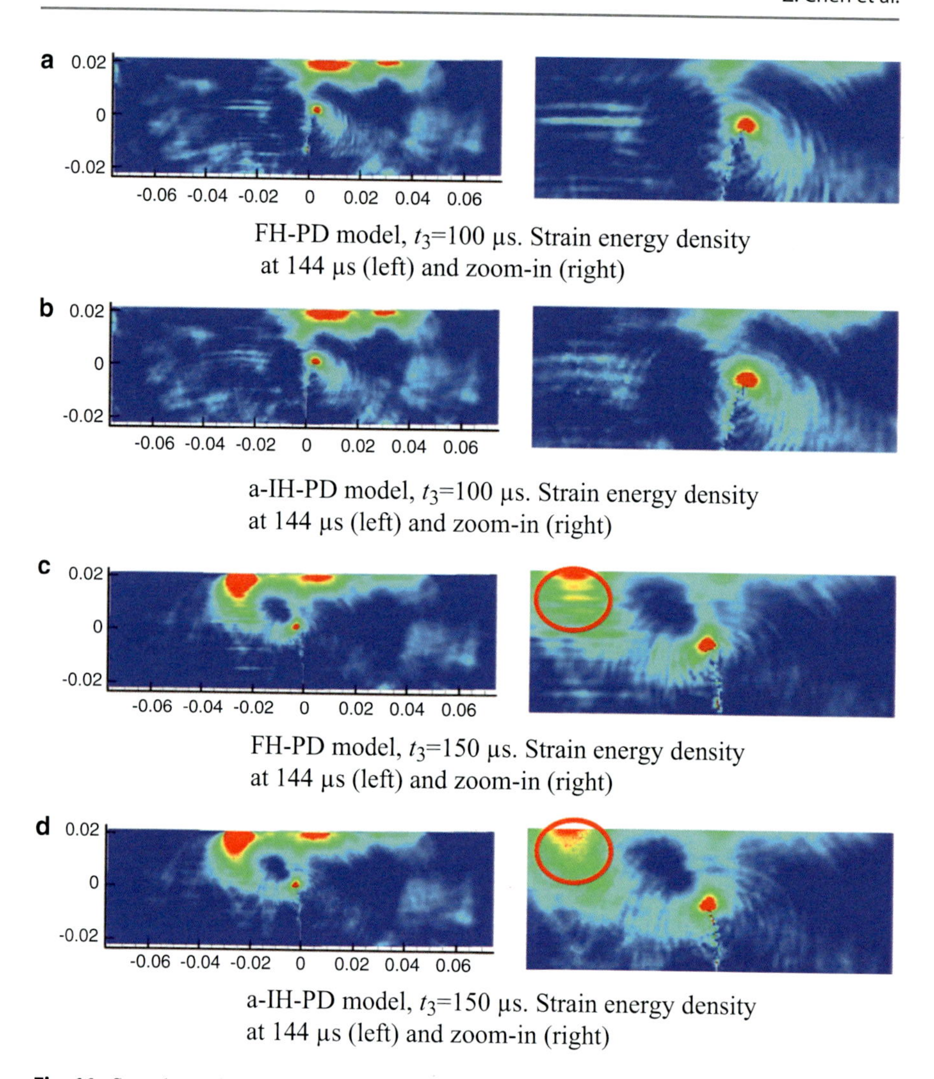

Fig. 11 Snapshots of strain energy density based on the FH-PD model and the a-IH-PD models

Evolution of Crack Propagation Path: Influence of Boundary Conditions

The impact loading used so far is different from the one used in the experiments, where soft supports are holding the impacted structure. In this section, we employ the second type of boundary conditions/loading mentioned in sections "Problem Setting" and "Boundary and Loading Conditions" and shown in Fig. 3b. The maximum applied stress (delivered on the peridynamic model as an equivalent body force) in this case needs to be lower than in the previous case in order to get a crack that splits the sample in two pieces because of the bending induced by the three-point impact loading used. We find that using $\sigma_0 = 26$ MPa (see Eq. 17 and Fig. 5) is sufficient to lead to full crack propagation through the width of the sample. The

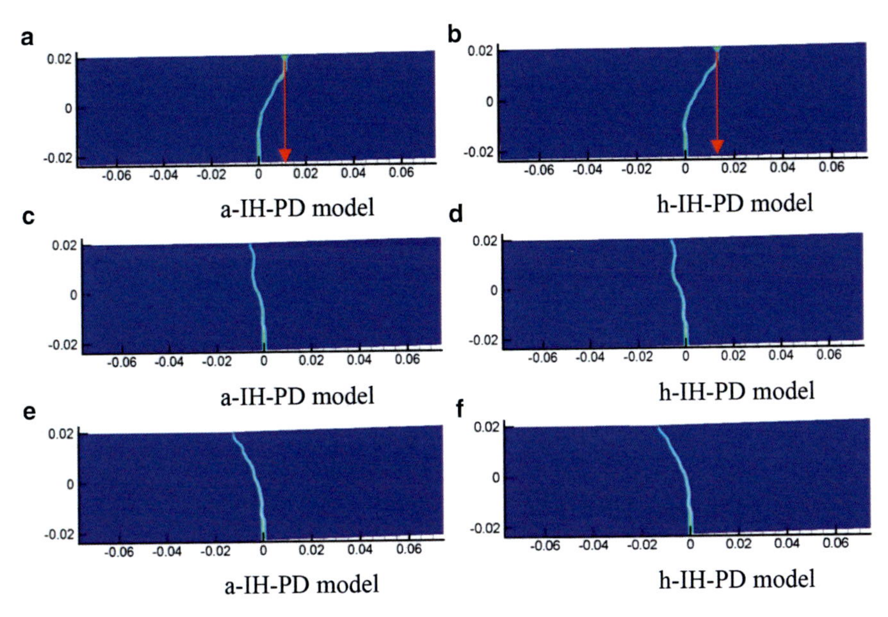

Fig. 12 Damage maps at full failure in the FGM sample with $E_1 < E_2$ for different total loading times: $t_3 = 100$ μs for (**a**) and (**b**); $t_3 = 150$ μs for (**c**) and (**d**); $t_3 = 180$ μs for (**e**) and (**f**)

rest of the computational parameters are identical with those used for the results obtained in the previous section. Note that the impact loads applied on the sides of the sample act simultaneously as the impact load on the top surface. In the actual experiment, the reaction from the sides will be felt only after the waves generated by the striking hammer reach the respective sides. The analysis of the results below will decide whether these differences are significant in terms of dynamic failure in the FGM samples used in our study.

Note that since in the previous section, we have shown that both FH-PD model and IH-PD model lead to similar results, for the remaining of section "Numerical Studies for Dynamic Crack Propagation in FGMs," we only present results with the FH-PD model.

The crack path for the FGM sample with the pre-crack on the compliant side ($E_1 < E_2$) for total loading times of 100 μs, 150 μs, and 180 μs is shown in Fig. 13. The initiation time for this case is about 125 μs, from the moment the loads are applied. The main qualitative difference when comparing simulation results for the three-point loading with free-free impact loading can be seen from the strain energy density movies: once the crack starts moving, the three-point loading leads to the creation of high strain energy density near the center of the top surface induced by the *bending* of the cracked beam. This is the main reason for which the crack path now does not stray much from a straight crack. The reasons for which the crack does not grow completely straight are two: (a) the asymmetric impact loading and (b) the release Rayleigh wave, induced by the unloading phase on the top surface, modifies the location of the "crack attractor" on the top surface.

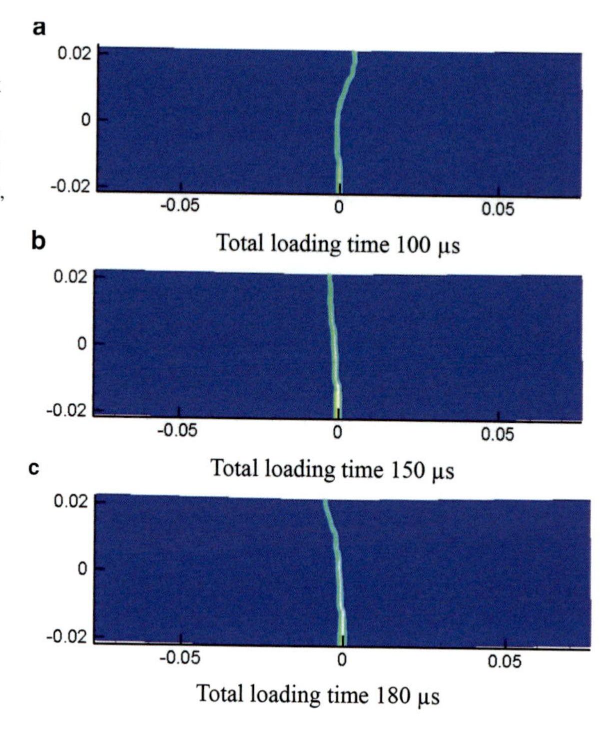

Fig. 13 Three-point impact loading conditions: damage maps showing the final crack path for the FGM sample with $E_1 < E_2$, under different total loading times. Compare results with those in Fig. 10b, d, and f, respectively

The differences between the impact loading on the free–free beam and the impact at three locations seen above are also seen for the sample with the pre-crack on the stiffer side ($E_1 > E_2$), as shown in Cheng et al. (2015). Interestingly, the differences between the two FGM samples for the total loading times of 150 μs and 180 μs show two trends: when the crack starts in the stiffer region, the crack initiates earlier and the deflection angle (or kink angle) is larger than when the crack starts in the softer region. These results *are confirmed* by the experimental observations in Kirugulige and Tippur (2006, 2008). A kink angle larger when $E_1 > E_2$ than when $E_1 < E_2$ is also predicted by the maximum tangential stress (MTS) criterion introduced in Erdogan and Sih (1963) for the kink angle in FGMs.

In summary, the three-point impact loading conditions is closer to the actual experimental conditions than the impact on the free–free beam, and the crack paths obtained support this assertion. In addition, our study of varying the total loading time tells us that, most likely, the time that the hammer used in the experiments to impact the structure spends between first touch to rebound is larger than 100 μs. The crack paths obtained for the total loading times of 150 μs and 180 μs are close to those seen from the experiments.

Comparison with Experimental and FEM-Based Results: Nonlinear Material Variation

In this section, we compare the results from the peridynamic simulations with those from the experiments in Kirugulige and Tippur (2006) and from the FEM-based simulations in Kirugulige and Tippur (2008). The analytical results in Kirugulige and Tippur (2006) and the FEM-based results in Kirugulige and Tippur (2008) used linear curve-fitting of the material properties as shown in Fig. 4. The peridynamic results presented so far used the same curve-fitting. In this section, we directly employ the estimated sigmoidal (described by with a piecewise-linear interpolation) elastic modulus and density measured in Kirugulige and Tippur (2006) and shown in Fig. 3 in that reference. This will determine how strongly the small variations in the elastic modulus and density influence the dynamic crack propagation behavior of FGMs. Finally, we select the peridynamic computational results that most resemble the experimental crack paths to compare among them. We also measure the crack speed growth from our peridynamic solution and compare with it the experimental and FEM-based numerical results from Kirugulige and Tippur (2008).

We extract the data from Fig. 3 in Kirugulige and Tippur (2006) and use it, with the piecewise-linear interpolation shown there, to compute the micro-moduli in our FG-PD model. We apply the three-point impact loading conditions used in the previous section and use the 180 μs total loading time. We obtain the final crack path shapes shown in Fig. 14a for the case with the pre-crack on the compliant side ($E_1 > E_2$), and in Fig. 14b for the case with the pre-crack on the stiffer

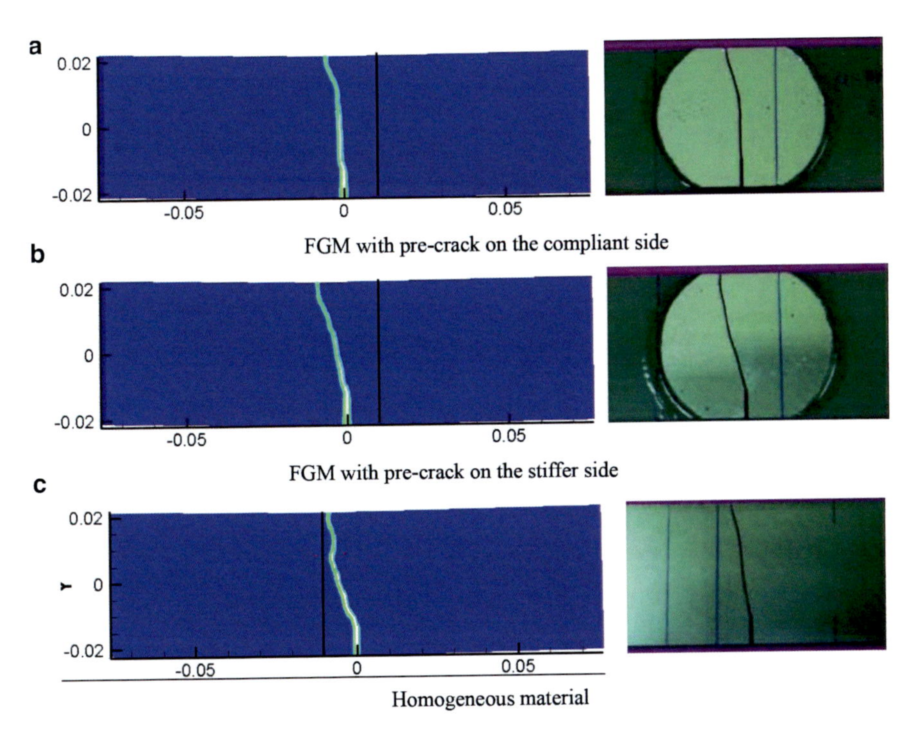

Fig. 14 Damage maps (left column) showing final crack path and experimental results (right column, reproduced from Kirugulige and Tippur (2006), showing only part of the beam). The three-point impact loading was used in the computations with 180 μs total loading time for (**a**) and (**b**), and 150 μs for (**c**). The PD computations for the FGM samples used the nonlinear (experimentally measured in Kirugulige and Tippur (2006)) variation of elastic modulus and density

side of the FGM ($E_1 < E_2$). The crack paths for the corresponding variation of material properties gradation obtained from experiments (taken from Fig. 10 in Kirugulige and Tippur (2006)) are placed next to the computed damage maps. To make it easier to compare the crack shapes, we superpose a black line at 1 cm to the right of the pre-crack tip, similar with what the authors of Kirugulige and Tippur (2006) did. While the dynamic loading used in our simulations is different from that used in the experiments (for the reasons mentioned in the beginning of section "Quasi-Static Fracture in Brittle Porous Elastic Materials"), we observe a close resemblance between the computed crack paths and the experimental ones. The authors of Kirugulige and Tippur (2006) concluded that: "the differences in the crack paths are attributable directly to the combined effects of elastic gradients as well as fracture toughness gradients." Our peridynamic model for FGM confirms this statement. Moreover, the peridynamic results shown in section "Results and Discussion" allow us to add to this statement the following:

(a) Crack path is influenced by stress wave, including surface waves that can modify the location of the largest bending stresses/strain that "attracts" the crack toward it.
(b) Stress waves modify the crack path especially after initial crack growth.
(c) Reflections of elastic waves from the boundaries and their subsequent self-reinforcement determine the time at which the crack starts propagating.

In Fig. 14c, we show the peridynamic result for the homogeneous material with three-point impact loading conditions and a total loading time of 150 μs. Next to this figure, we show the experimental results for the same material type. We attribute the slight differences in the crack path here to the dynamic loading conditions used in our computations. To make it easier to compare the shape of the crack, we superpose a black line at 1 cm to the left of the pre-crack tip, similar with what the authors of Kirugulige and Tippur (2006) did.

By comparing the results using the nonlinear (piecewise linear) variation in Fig. 14a, b with those using the linear curve-fit for the elastic modulus and density shown in Fig. 13c in this chapter and Fig. 19c in Cheng et al. (2015), we conclude that differences in terms of crack paths exist, but they are small. The main source of these differences are differences in stress waves propagation, caused by the slight differences in local stiffness and density between the two models.

Quasi-Static Fracture in Brittle Porous Elastic Materials

In the previous sections, we have shown that both FH-PD model and IH-PD model can be used to reproduce the dynamic fracture of an FGM plate. While only small differences are observed between the results produced by the FH-PD and IH-PD models when applied to heterogeneous materials in which mechanical properties of the phases are not orders of magnitude different, significant differences are expected when phases differ by orders of magnitude. Examples of such cases are fiber-reinforced composites, porous materials (in which the voids can be considered as one of the phases), etc. In Chen et al. (2017), we apply both FH-PD model and h-IH-PD model for modeling quasi-static fracture of a porous rock sample. We find that only the IH-PD model delivers similar damage patterns and crack profiles as seen in experiments. In this section, we briefly discuss this test.

The experiments shown in Lin et al. (2009) measure the damage evolution of Berea sandstone notched specimens under quasi-static three-point bending tests. The location of a pre-notch, as a stress concentrator, can control the development of fracture, either through mode I failure for a center-notch beam or mixed-mode failure for an off-center notch beam (Lin et al. 2009). Another critical factor for the development of fracture is the pre-notch length. For the case shown in Fig. 15a (from Lin et al. 2009), under the same loading conditions, different notch lengths lead to different crack patterns. For instance, when the notch is short (5.65 mm), the crack initiates from near the beam center and propagates vertically upward (see section "Results and Discussion" and Fig. 6 in Lin et al. (2009)). When the notch is long (12.23 mm), the beam fails by cracking from the tip of the pre-notch toward the point of loading on the top boundary (see section "Results and Discussion" and Fig. 8 in Lin et al. (2009)).

We apply both the FH-PD and the h-IH-PD models with quasi-static loading conditions to simulate the cases shown in Fig. 15 and to examine the effect the pre-notch length has on the crack pattern. In the FH-PD model of porous material, we use the effective material properties calculated from Eq. 13 for the bond information (micro-modulus and critical bond strain). In the h-IH-PD model, the porous material is treated as a multiphase material with one empty phase (zero modulus). In this model, the bonds involving the empty phase (A–B bond and B–B bond, if phase B is the void phase) bear zero force. These bonds are equivalent with broken bonds. Therefore, the effect of porosity can be represented by inserting "predamage" in the material. The predamage maps corresponding to different porosities, as well as detailed material parameters and model formulation, can be found in Chen et al. (2017).

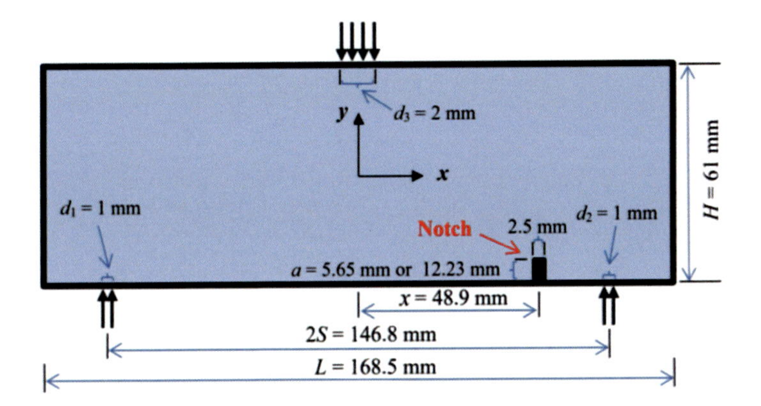

Fig. 15 Geometry and loading configuration used in experiments of Hu et al. (2012) and in the PD simulations below

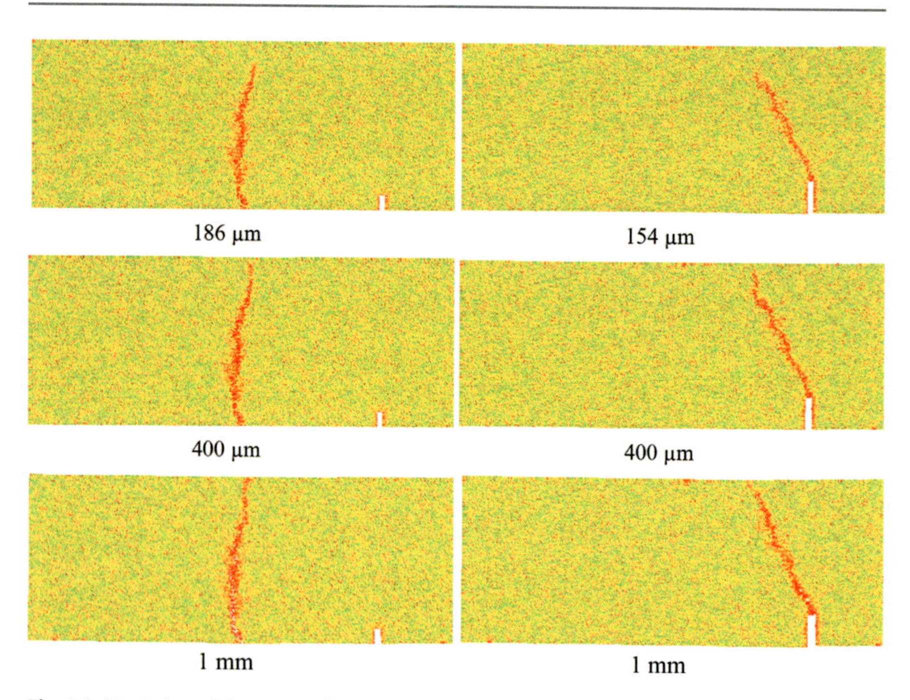

Fig. 16 Evolution of fracture paths (at different imposed displacements) for the short (left) and long (right) pre-notched sample using the h-IH-PD model (see Movies 7 and 8 for the damage evolution for the short and long pre-notched samples). The *color* represents the damage index with the *same color* legend in Fig. 6 (predamage is included in this plot)

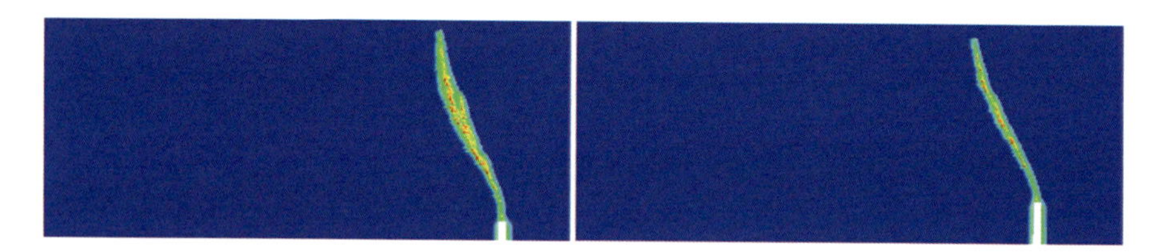

Fig. 17 Fracture paths for the short (left) and long (right) pre-notched samples using FH-PD model (see Movies 9 and 10 for the damage evolution for the short and long pre-notched samples, respectively)

Figure 16 shows the simulation results generated by h-IH-PD model for the developing fracture in the short and long pre-notched Berea sandstone samples. The damage maps show results consistent with the experimental findings on the effect of the notch length: the long pre-notch leads to crack initiating from the pre-notch tip, while in the short pre-notch case, the crack initiates from the bottom-center of the beam.

The FH-PD model does not capture this effect of the pre-notch length on the crack initiation and propagation in the porous sample, as shown in Fig. 17. With the FH-PD model, the crack always initiates from the tip of the pre-notch.

Conclusions

In this chapter, we answered the question of how much homogenization is too much when modeling fracture processes. We presented two peridynamic models, one that homogenizes the material to a greater extent (the fully homogenized peridynamic model, FH-PD) than the other (the intermediately homogenized peridynamic model (IH-PD). To assess the differences between these models, we studied two types of fracture problems: a dynamic impact problem in a functionally graded material (FGM), for which the crack growth is driven by stress waves, and a quasi-static crack growth problem in a brittle porous elastic material.

The solutions for the problem of mixed-mode dynamic crack propagation in functionally graded glass-filled epoxy matrix obtained with these peridynamic models were similar, and matched closely the experimental results reported in the literature. We provided a detailed computational analysis that showed the influence of loading time and loading conditions on the crack path through full failure of the sample.

In contrast, for a quasi-static crack growth in a porous material, the fully homogenized model fails to capture the experimentally observed fracture behavior in Berean sandstone that is controlled, in this case, by the size of a notch. The intermediately homogenized peridynamic model, however, does reproduce the observed crack growth behavior.

We concluded that for problems in which the microstructure plays an important role on the failure behavior, a fully homogenized strategy will not work. Some information about the microstructure is needed for a predictive model of crack growth in these cases. The new IH-PD model showed that one does not need an explicit description of the small-scale geometry of the material microstructure to predict crack growth in a porous material. Even for the stress-wave driven crack growth in the case of the asymmetric impact loading of the FGM plate, this model produces a landscape of strain energy density that is more realistic compared with the FH-PD model. The crack paths are the same in this case simply because the microstructure effects are rendered secondary by the high flow of energy in the system (stress waves generated by impact and reflecting from the boundaries).

Acknowledgments This work was supported by a grant from ONR (program manager: William Nickerson) and by the AFOSR MURI Center for Material Failure Prediction through Peridynamics (program managers: Dr. David Stargel, Dr. Ali Sayir, Dr. Fariba Fahroo, and James Fillerup). This work was completed utilizing the Holland Computing Center of the University of Nebraska.

Appendix A

Numerical Discretization

The peridynamic equations can be discretized using the finite element method, or any other method appropriate to compute solutions to integro-differential equations (or integration equations in static model). This approach, however, soon hits well-known obstacles and difficulties for problems with evolving topologies, like those in dynamic fracture and fragmentation. Instead, meshfree-type discretizations are preferred in peridynamics simulations of dynamic failure of materials. The discretization proposed in Silling and Askari (2005) uses the midpoint integration scheme (equivalent to a one-point Gaussian integration) for the domain integral. Numerical simulations are performed using the following discretized equation:

$$\int_{H_x} \boldsymbol{f}\left(\hat{\boldsymbol{x}} - \boldsymbol{x}, \boldsymbol{u}\left(\hat{\boldsymbol{x}}, t\right) - \boldsymbol{u}\left(\boldsymbol{x}, t\right)\right) \mathrm{d}V_{\hat{\boldsymbol{x}}} \simeq \sum_{j \in Fam(i)} c\left(\xi_{ij}\right) s_{ij} V_{ij}$$

where Fam(i) is the family of nodes j with their area (volume in 3D) covered, fully or partially, by the horizon region of nodes i, ξ_{ij} is the bond length between nodes i and j, s_{ij} is the relative elongation for the same bond, and V_{ij} is the area of node j estimated to be covered by the horizon of node i.

Note that node j may not be fully contained within the horizon of node I, so a partial volume integration, first introduced in Hu et al. (2010) and also shown in Zhang and Bobaru (2015), is used here to improve the accuracy of midpoint quadrature scheme. The main advantage of this algorithm compared with one that simply checks whether a node is inside or outside the horizon region is that as the grid density increases (for a fixed horizon value), the numerical convergence (in terms of strain energy density, for example) is monotonic (see Hu et al. 2010).

Both dynamic (see section "Numerical Studies for Dynamic Crack Propagation in FGMs") and static (see section "Quasi-Static Fracture in Brittle Porous Elastic Materials") simulations are performed in this work. In the dynamic fracture simulations of the FGM plate (section "Numerical Studies for Dynamic Crack Propagation in FGMs"), we apply Velocity-Verlet method with a time interval of 0.05 μs. For the quasi-static fracture tests in section "Quasi-Static Fracture in Brittle Porous Elastic Materials," the energy minimization method (see Shewchuk 1994; Zhang et al. 2016; Le and Bobaru 2017) is used, and the conjugate gradient (CG) method with secant line search is adopted to minimize the strain energy of the system. For all static simulations in this chapter, the CG method is used with a convergence tolerance defined by: $\frac{|W_i - W_{i-1}|}{W_{i-1}} < 10^{-6}$, in which W_i and W_{i-1} are the total strain energy at current (i) and previous ($i-1$) CG iterations.

m-convergence and δ-convergence: For a fixed horizon, the ratio $m = \delta/\Delta x$ describes how accurate the numerical quadrature for the integral in Eq. 1 will be. We call this ratio "the horizon factor." We recall that in the m-convergence we consider the horizon δ fixed and take $m \rightarrow \infty$. The numerical PD approximation will converge to the exact nonlocal PD solution for the given δ. In the case of δ-convergence, the horizon $\delta \rightarrow 0$ while m is fixed or increases with decreasing δ. For δ-convergence and in problems with no singularities, the numerical PD solutions are expected to converge to the classical local solution.

References

J. Abanto-Bueno, J. Lambros, An experimental study of mixed model crack initiation and growth in functionally graded materials. Exp. Mech. **46**, 179–186 (2006)

G. Anlas, M.H. Santare, J. Lambros, Numerical calculation of stress intensity factors in functionally graded materials. Int. J. Fract. **104**, 131–143 (2000)

H. Bayesteh, S. Mohammadi, XFEM fracture analysis of orthotropic functionally graded materials. Compos. Part B Eng. **44**, 8–25 (2013)

F. Bobaru, Designing optimal volume fractions for functionally graded materials with temperature-dependent material properties. J. Appl. Mech. **74**(5), 861–874 (2007)

F. Bobaru, W. Hu, The meaning, selection, and use of the peridynamic horizon and its relation to crack branching in brittle materials. Int. J. Fract. **176**, 215–222 (2012)

F. Bobaru, M. Yang, L.F. Alves, S.A. Silling, E. Askari, J.F. Xu, Convergence, adaptive refinement, and scaling in 1D peridynamics. Int. J. Numer. Methods Eng. **77**, 852–877 (2009)

Z. Chen, F. Bobaru, Selecting the kernel in a peridynamic formulation: a study for transient heat diffusion. Comput. Phys. Commun. **197**, 51–60 (2015)

Z. Chen, D. Bakenhus, F. Bobaru, A constructive peridynamic kernel for elasticity. Comput. Methods Appl. Mech. Eng. **311**, 356–373 (2016)

Z. Chen, S. Niazi, F. Bobaru, Peridynamic model for damage and fracture in porous materials (2017, in preparation)

Z.Q. Cheng, Z. Zhong, Analysis of a moving crack in a functionally graded strip between two homogeneous layers. Int. J. Mech. Sci. **49**, 1038–1046 (2007)

Z.Q. Cheng, D.Y. Gao, Z. Zhong, Crack propagating in functionally graded coating with arbitrarily distributed material properties bonded homogeneous substrate. Acta Mech. Solida Sin. **23**, 437–446 (2010)

Z. Cheng, G. Zhang, Y. Wang, F. Bobaru, A peridynamic model for dynamic fracture in functionally graded materials. Compos. Struct. **133**, 529–546 (2015)

F. Delale, F. Erdogan, The crack problem for a nonhomogeneous plane. J. Appl. Mech. **50**, 609–614 (1983)

J.W. Eischen, Fracture of nonhomogeneous material. Int. J. Fract. **34**, 3–22 (1987)

F. Erdogan, G.C. Sih, On the crack extension in plates under plane loading and transverse shear. Trans. ASME J. Basic Eng. **84D**(4), 519–525 (1963)

L.C. Guo, W. LZ, T. Zeng, L. Ma, Fracture analysis of a functionally graded coating-substrate structure with a crack perpendicular to the interface – part II: transient problem. Int. J. Fract. **127**, 39–59 (2004)

Y.D. Ha, F. Bobaru, Studies of dynamic crack propagation and crack branching with peridynamics. Int. J. Fract. **162**(1–2), 229–244 (2010)

Y.D. Ha, F. Bobaru, Characteristics of dynamic brittle fracture captured with peridynamics. Eng. Fract. Mech. **78**, 1156–1168 (2011)

W. Hu, Y.D. Ha, F. Bobaru, Numerical integration in peridynamics, in Technical report, University of Nebraska–Lincoln, Lincoln, 2010

W. Hu, Y.D. Ha, F. Bobaru, Peridynamic model for dynamic fracture in fiber-reinforced composites. Comput. Methods Appl. Mech. Eng. **217–220**, 247–261 (2012)

W. Hu, Y. Wang, J. Yu, C.F. Yen, F. Bobaru, Impact damage on a thin glass with a thin polycarbonate backing. Int. J. Impact Eng. **62**, 152–165 (2013)

S. Itou, Dynamic stress intensity factors for two parallel interface cracks between a nonhomogeneous bonding layer and two dissimilar elastic half-planes subject to an impact load. Int. J. Solids Struct. **47**, 2155–2163 (2010)

N. Jain, A. Shukla, Mixed mode dynamic fracture in particulate reinforced functionally graded materials. Exp. Mech. **46**(2), 137–154 (2006)

Z.H. Jin, R.C. Batra, Some basic fracture mechanics concepts in functionally graded materials. J. Mech. Phys. Solids **44**, 1221–1235 (1996)

X. Jin, W. LZ, L.C. Guo, Y. HJ, Y.G. Sun, Experimental investigation of the mixed-mode crack propagation in ZrO_2/NiCr functionally graded materials. Eng. Fract. Mech. **76**, 1800–1810 (2009)

M. Kashtalyan, M. Menshykova, Effect of a functionally graded interlayer on three-dimensional elastic deformation of coated plates subjected to transverse loading. Compos. Struct. **89**(2), 167–176 (2009)

A. Kidane, V.B. Chalivendra, A. Shulka, R. Chona, Mixed-mode dynamic crack propagation in graded materials under thermo-mechanical loading. Eng. Fract. Mech. **77**, 2864–2880 (2010)

J.H. Kim, G.H. Paulino, Simulation of crack propagation in functionally graded materials under mixed-mode and non-proportional loading. Int. J. Mech. Mater. Des. **1**, 63–94 (2004)

M.S. Kirugulige, H.V. Tippur, Mixed-mode dynamic crack growth in functionally graded glass-filled epoxy. Exp. Mech. **46**(2), 269–281 (2006)

M.S. Kirugulige, H.V. Tippur, Mixed-mode dynamic crack growth in a functionally graded particulate composite: experimental measurement and finite element simulations. J. Appl. Mech. **75**(5), 051102 (2008)

Q. Le, F. Bobaru, Surface corrections for peridynamic models in elasticity and fracture. Comput. Mech. (2017). https://doi.org/10.1007/s00466-017-1469-1

K.H. Lee, Analysis of a transiently propagating crack in functionally graded materials under mode I and II. Int. J. Eng. Sci. **47**, 852–865 (2009)

Q. Lin, A. Fakhimi, M. Haggerty, J.F. Labuz, Initiation of tensile and mixed-mode fracture in sandstone. Int. J. Rock Mech. Min. Sci. **46**, 489–497 (2009)

L. Ma, L.Z. Wu, L.C. Guo, Z.G. Zhou, On the moving Griffith crack in a non-homogeneous orthotropic medium. Eur. J. Mech. A. Solids **24**, 393–405 (2005)

P.R. Marur, H.V. Tippur, Numerical analysis of crack-tip fields in functionally graded materials with a crack normal to the elastic gradient. Int. J. Solids Struct. **37**, 5353–5370 (2000)

M.S. Matbuly, Multiple crack propagation along the interface of a nonhomogeneous composite subjected to anti-plane shear loading. Meccanica **44**, 547–554 (2009)

C.E. Rousseau, H.V. Tippur, Dynamic fracture of compositionally graded materials with cracks along the elastic gradient: experiments and analysis. Mech. Mater. **37**, 403–421 (2001)

J.R. Shewchuk, An introduction to the conjugate gradient method without the agonizing pain, in Technical report, School of Computer Science, Carnegie Mellon University, Pittsburgh, 1994

S.A. Silling, Reformulation of elasticity theory for discontinuities and long-range forces. J. Mech. Phys. Solids **48**, 175–209 (2000)

S.A. Silling, Origin and effects of nonlocality in a composite. J. Mech. Mater. Struct. **9**, 245–258 (2014)

S.A. Silling, E. Askari, A meshfree method based on the peridynamic model of solid mechanics. Comput. Struct. **83**, 1526–1535 (2005)

S.A. Silling, R.B. Lehoucq, Convergence of peridynamics to classical elasticity theory. J. Elast. **93**, 13–37 (2008)

S.A. Silling, M. Epton, O. Weckner, J. Xu, E. Askari, Peridynamic states and constitutive modeling. J. Elast. **88**(2), 151–184 (2007)

J. Song, H. Wang, T. Belytschko, A comparative study on finite element methods for dynamic fracture. Comput. Mech. **42**, 239–250 (2008)

H.T. Thai, S.E. Kim, A review of theories for the modeling and analysis of functionally graded plates and shells. Compos. Struct. **128**, 70–86 (2015)

Z.H. Wang, L.C. Guo, L. Zhang, A general modeling method for functionally graded materials with an arbitrarily oriented crack. Philos. Mag. **94**, 764–791 (2014)

C.H. Xia, L. Ma, Dynamic behavior of a finite crack in functionally graded materials subjected to plane incident time-harmonic stress wave. Compos. Struct. **77**(1), 10–17 (2007)

G. Zhang, F. Bobaru, Why do cracks branch? A peridynamic investigation of dynamic brittle fracture. Int. J. Fract. **196**, 59–98 (2015)
Z. Zhang, G.H. Paulino, Cohesive zone modeling of dynamic failure in homogeneous and functionally graded materials. Int. J. Plast. **21**, 1195–1254 (2005)
G. Zhang, Q. Le, A. Loghin, A. Subramaniyan, F. Bobaru, Validation of a peridynamic model for fatigue cracking. Eng. Fract. Mech. **162**, 76–94 (2016)

Numerical Tools for Improved Convergence of Meshfree Peridynamic Discretizations

42

Pablo Seleson and David J. Littlewood

Contents

Abstract

Peridynamic models have been employed to simulate a broad range of engineering applications concerning material failure and damage, with the majority of these simulations using a meshfree discretization. This chapter reviews

P. Seleson (✉)
Computer Science and Mathematics Division, Oak Ridge National Laboratory, Oak Ridge, TN, USA
e-mail: selesonpd@ornl.gov

D. J. Littlewood
Center for Computing Research, Sandia National Laboratories, Albuquerque, NM, USA
e-mail: djlittl@sandia.gov

G. Z. Voyiadjis (ed.), *Handbook of Nonlocal Continuum Mechanics for Materials and Structures*, https://doi.org/10.1007/978-3-319-58729-5_39

that meshfree discretization, related issues present in peridynamic convergence studies, and possible remedies proposed in the literature. In particular, we discuss two numerical tools, partial-volume algorithms and influence functions, to improve the convergence behavior of numerical solutions in peridynamics. Numerical studies in this chapter involve static and dynamic simulations for linear elastic state-based peridynamic problems.

Keywords
Peridynamics · Meshfree discretization · Partial volumes · Influence functions · Convergence · Statics · Dynamics

Introduction

Peridynamic models have been employed over the past years to simulate a broad range of engineering applications. These applications include failure and damage in composite laminates (Hu et al. 2012; Kilic et al. 2009; Oterkus and Madenci 2012; Oterkus et al. 2012; Xu et al. 2008), crack propagation and branching in glass (Ha and Bobaru 2010, 2011; Kilic and Madenci 2009), crack nucleation (Littlewood 2011; Silling et al. 2010), impact damage (Bobaru et al. 2012; Seleson and Parks 2011; Silling and Askari 2005; Tupek et al. 2013), fracture in polycrystalline materials (Askari et al. 2008; De Meo et al. 2016; Ghajari et al. 2014), structural health monitoring (Littlewood et al. 2012), and damage in concrete (Gerstle et al. 2007), among many others. There are two main distinctive discretization methods employed by the peridynamic community. These are the finite element method (see, e.g., Chen and Gunzburger 2011) and the meshfree method of Silling and Askari (2005). The former discretization method, in spite of robust supportive mathematical analysis, is not amenable in terms of computational cost and software implementation, particularly for simulations involving complex geometry in three dimensions. Two specific challenges of finite element discretizations of peridynamic models are proper quadratures for peridynamic weak formulations, which double the number of spatial dimensions to be discretized relative to strong formulations, and the necessity to adapt finite element meshes to conform to evolving crack surfaces. In contrast, the latter discretization method, i.e., the meshfree method, has been rapidly adopted by the engineering community, due to its implementation simplicity and its ability to handle material separation in a relatively simple manner. A downside of the meshfree approach is its poor convergence behavior. Specifically, it has been shown in Seleson and Littlewood (2016) that standard meshfree discretizations of peridynamic models exhibit irregular (non-smooth) convergence behavior and a low (average) first-order convergence. Nevertheless, numerical tools have been proposed and studied in Seleson (2014) and Seleson and Littlewood (2016) to correct such irregular convergence behavior, resulting in smoother convergence curves. The (average) first-order convergence is however tied to an underlying piecewise-constant representation of the displacement field in meshfree discretizations and could possibly be improved with higher-order approximations (as in finite element methods), although these are beyond the scope of this chapter.

Generally speaking, setting up computational studies of convergence of numerical solutions in peridynamics may or may not be a simple task. First of all, as discussed in Bobaru et al. (2009), the concept of convergence in peridynamics has more than one meaning. One may study, for instance, the convergence of a numerical solution of a peridynamic problem to the analytical solution of that problem; in this case, the horizon δ is kept fixed and the grid spacing h is taken to zero. On the other hand, one could study the convergence of a numerical solution of a peridynamic problem to the analytical solution of a corresponding classical (local) problem; in this case, both δ and h are simultaneously taken to zero, most probably at different rates. The various convergence avenues in peridynamics have been illustrated in Du and Tian (2014).

This chapter is dedicated to reviewing the meshfree discretization of peridynamic models, related issues present in convergence studies, and possible remedies proposed in the literature. This chapter also discusses important considerations for setting up convergence studies in peridynamics, such as the choice of horizon, the error norm, the convergence avenue, and the refinement path, as well as related computational costs. We focus on numerical convergence studies for linear elastic state-based peridynamic problems. Following Seleson and Littlewood (2016), we first study the convergence of numerical solutions of static peridynamic problems. Then, we present studies concerning wave propagation in a peridynamic medium to demonstrate how such convergence issues manifest in dynamic problems. To keep this chapter succinct while providing the most impactful results toward practical applications, we restrict the discussion to a select number of three-dimensional problems.

Peridynamic Models and Their Meshfree Discretization

This chapter is concerned with convergence studies in three-dimensional problems for ordinary state-based peridynamic (PD) models, given by the *linear peridynamic solid* (LPS) model (Silling et al. 2007). The LPS strain energy density is

$$W = \frac{1}{2}\left(K - \frac{5G}{3}\right)\vartheta^2 + \frac{1}{2}\frac{15G}{m}\left(\underline{\omega}\,\underline{e}\right) \bullet \underline{e} \tag{1}$$

and the corresponding force vector state field is given by

$$\underline{\mathbf{T}} = \left(\frac{3K - 5G}{m}\vartheta\,\underline{\omega}\,\underline{x} + \frac{15G}{m}\underline{\omega}\,\underline{e}\right)\underline{\mathbf{M}}, \tag{2}$$

where K is the bulk modulus and G is the shear modulus. In Eqs. 1 and 2, ϑ is the dilatation, $\underline{x}$ is the reference position state, $\underline{e}$ is the extension state, m is the weighted volume, $\underline{\mathbf{M}}$ is the deformed direction vector state, and $\underline{\omega} = \omega\left(|\boldsymbol{\xi}|\right)$ is a spherical influence function depending on the length of a bond $\boldsymbol{\xi}$ (Seleson and Parks 2011; Silling et al. 2007). For computational purposes, it is useful to express Eqs. 1 and 2, respectively, in a more explicit form as

$$W(\mathbf{x},t) = \frac{1}{2}\left(K - \frac{5G}{3}\right)(\vartheta[\mathbf{x},t])^2 + \frac{1}{2}\frac{15G}{m}\int_{\mathcal{H}_{\mathbf{x}}} \omega(|\boldsymbol{\xi}|)\left(|\mathbf{y}(\mathbf{x}+\boldsymbol{\xi},t) - \mathbf{y}(\mathbf{x},t)| - |\boldsymbol{\xi}|\right)^2 dV_{\boldsymbol{\xi}}$$

and

$$\underline{\mathbf{T}}[\mathbf{x},t]\langle\boldsymbol{\xi}\rangle = \frac{\omega(|\boldsymbol{\xi}|)\,|\boldsymbol{\xi}|}{m}\left[(3K-5G)\,\vartheta[\mathbf{x},t] + 15G\frac{|\mathbf{y}(\mathbf{x}+\boldsymbol{\xi},t) - \mathbf{y}(\mathbf{x},t)| - |\boldsymbol{\xi}|}{|\boldsymbol{\xi}|}\right] \times \frac{\mathbf{y}(\mathbf{x}+\boldsymbol{\xi},t) - \mathbf{y}(\mathbf{x},t)}{|\mathbf{y}(\mathbf{x}+\boldsymbol{\xi},t) - \mathbf{y}(\mathbf{x},t)|},$$

where

$$\vartheta[\mathbf{x},t] = \frac{3}{m}\int_{\mathcal{H}_{\mathbf{x}}} \omega(|\boldsymbol{\zeta}|)\,|\boldsymbol{\zeta}|\left(|\mathbf{y}(\mathbf{x}+\boldsymbol{\zeta},t) - \mathbf{y}(\mathbf{x},t)| - |\boldsymbol{\zeta}|\right) dV_{\boldsymbol{\zeta}} \tag{3}$$

and

$$m = \int_{\mathcal{H}_{\mathbf{x}}} \omega(|\boldsymbol{\xi}|)\,|\boldsymbol{\xi}|^2 dV_{\boldsymbol{\xi}} \tag{4}$$

with $\mathcal{H}_{\mathbf{x}}$ the family of $\mathbf{x}$, and where $\mathbf{y}(\mathbf{x},t)$ is the position of the material point $\mathbf{x}$ in the deformed configuration at time $t > 0$. For simplicity and for consistency with Seleson and Littlewood (2016), in this chapter the value of m is computed for a point in the bulk of a body and equally assigned to all points in that body, including those near the boundary of the body. This does not affect the static simulations in section "Convergence Studies of Static Peridynamic Problems," because in those studies m is only required for points within the bulk of the body, but may slightly influence the dynamic results in section "Convergence Studies of Dynamic Peridynamic Problems." Given a PD body $\mathcal{B}$, the LPS equation of motion for $\mathbf{x} \in \mathcal{B}$ at time $t \geq 0$ is

$$\rho(\mathbf{x})\frac{\partial^2\mathbf{u}}{\partial t^2}(\mathbf{x},t) = \int_{\mathcal{H}_{\mathbf{x}}} \frac{\omega(|\boldsymbol{\xi}|)\,|\boldsymbol{\xi}|}{m}\left\{(3K-5G)\left(\vartheta[\mathbf{x},t] + \vartheta[\mathbf{x}+\boldsymbol{\xi},t]\right) + 30G\frac{|\mathbf{y}(\mathbf{x}+\boldsymbol{\xi},t) - \mathbf{y}(\mathbf{x},t)| - |\boldsymbol{\xi}|}{|\boldsymbol{\xi}|}\right\} \times \frac{\mathbf{y}(\mathbf{x}+\boldsymbol{\xi},t) - \mathbf{y}(\mathbf{x},t)}{|\mathbf{y}(\mathbf{x}+\boldsymbol{\xi},t) - \mathbf{y}(\mathbf{x},t)|} dV_{\boldsymbol{\xi}} + \mathbf{b}(\mathbf{x},t), \tag{5}$$

where $\boldsymbol{\xi} = \mathbf{x}' - \mathbf{x}$, ρ is the mass density field, $\mathbf{u}$ is the displacement field, and $\mathbf{b}$ is the body force density field.

Linearized LPS Model

For a small deformation, the LPS model can be linearized giving the following strain energy density (Silling 2010)

$$\begin{aligned} W(\mathbf{x}, t) =& \frac{1}{2}\left(K - \frac{5G}{3}\right)\left(\vartheta^{\mathrm{lin}}[\mathbf{x}, t]\right)^2 \\ &+ \frac{1}{2}\frac{15G}{m}\int_{\mathcal{H}_{\mathbf{x}}} \frac{\omega(|\boldsymbol{\xi}|)}{|\boldsymbol{\xi}|^2}(\boldsymbol{\xi}\cdot(\mathbf{u}(\mathbf{x}+\boldsymbol{\xi}, t) - \mathbf{u}(\mathbf{x}, t)))^2 \mathrm{d}V_{\boldsymbol{\xi}} \end{aligned} \tag{6}$$

and force vector state

$$\begin{aligned} \underline{\mathbf{T}}[\mathbf{x}, t]\langle\boldsymbol{\xi}\rangle =& \frac{\omega(|\boldsymbol{\xi}|)}{m}\Bigg[(3K - 5G)\,\vartheta^{\mathrm{lin}}[\mathbf{x}, t]\,\boldsymbol{\xi} \\ &+ 15G\frac{\boldsymbol{\xi}\otimes\boldsymbol{\xi}}{|\boldsymbol{\xi}|^2}(\mathbf{u}(\mathbf{x}+\boldsymbol{\xi}, t) - \mathbf{u}(\mathbf{x}, t))\Bigg], \end{aligned} \tag{7}$$

where the linearized dilatation is given by

$$\vartheta^{\mathrm{lin}}[\mathbf{x}, t] = \frac{3}{m}\int_{\mathcal{H}_{\mathbf{x}}} \omega(|\boldsymbol{\zeta}|)\,\boldsymbol{\zeta}\cdot(\mathbf{u}(\mathbf{x}+\boldsymbol{\zeta}, t) - \mathbf{u}(\mathbf{x}, t))\,\mathrm{d}V_{\boldsymbol{\zeta}}. \tag{8}$$

The linearized LPS equation of motion is

$$\begin{aligned} \rho(\mathbf{x})\frac{\partial^2\mathbf{u}}{\partial t^2}(\mathbf{x}, t) =& \int_{\mathcal{H}_{\mathbf{x}}} \frac{\omega(|\boldsymbol{\xi}|)}{m}\Bigg\{(3K - 5G)\left(\vartheta^{\mathrm{lin}}[\mathbf{x}, t] + \vartheta^{\mathrm{lin}}[\mathbf{x}+\boldsymbol{\xi}, t]\right)\boldsymbol{\xi} \\ &+ 30G\frac{\boldsymbol{\xi}\otimes\boldsymbol{\xi}}{|\boldsymbol{\xi}|^2}(\mathbf{u}(\mathbf{x}+\boldsymbol{\xi}, t) - \mathbf{u}(\mathbf{x}, t))\Bigg\}\mathrm{d}V_{\boldsymbol{\xi}} + \mathbf{b}(\mathbf{x}, t), \end{aligned} \tag{9}$$

and the corresponding equilibrium equation can be expressed as

$$\begin{aligned} -\int_{\mathcal{H}_{\mathbf{x}}} \frac{\omega(|\boldsymbol{\xi}|)}{m}\Bigg\{&(3K - 5G)\left(\vartheta^{\mathrm{lin}}[\mathbf{x}] + \vartheta^{\mathrm{lin}}[\mathbf{x}+\boldsymbol{\xi}]\right)\boldsymbol{\xi} \\ &+ 30G\frac{\boldsymbol{\xi}\otimes\boldsymbol{\xi}}{|\boldsymbol{\xi}|^2}(\mathbf{u}(\mathbf{x}+\boldsymbol{\xi}) - \mathbf{u}(\mathbf{x}))\Bigg\}\mathrm{d}V_{\boldsymbol{\xi}} = \mathbf{b}(\mathbf{x}). \end{aligned} \tag{10}$$

Discretized Peridynamic Models

In this chapter, we employ a discretization similar to the meshfree method introduced in Silling and Askari (2005). In that method, no elements or other geometrical connections between computational nodes are used. In our case, we sometimes employ a reference mesh to approximate quadrature weights. However, once these weights are computed, the mesh is discarded and there is no need to computationally track mesh elements in a simulation, in contrast to traditional (mesh-based) finite element methods. For this reason, the methods below are also referred to as meshfree.

To discretize the PD equation of motion, we begin by discretizing the body $\mathcal{B}$ with a set of N^3 nodes $\mathcal{L} = \{\mathbf{x}_i\}_{i=1}^{N^3}$, where each node i is assigned a material volume V_i corresponding to its Voronoi cell, τ_i, such that $\tau_i \cap \tau_j = \emptyset$ for $i \neq j$. The resulting semi-discrete equation is given by

$$\rho_i \frac{d^2\mathbf{u}_i}{dt^2} = \sum_{j \in \mathcal{F}_i} \left\{ \underline{\mathbf{T}}[\mathbf{x}_i, t]\left\langle \mathbf{x}_j - \mathbf{x}_i \right\rangle - \underline{\mathbf{T}}[\mathbf{x}_j, t]\left\langle \mathbf{x}_i - \mathbf{x}_j \right\rangle \right\} V_j^{(i)} + \mathbf{b}_i, \tag{11}$$

where $\rho_i := \rho(\mathbf{x}_i)$, $\mathbf{u}_i := \mathbf{u}(\mathbf{x}_i,t)$, $\mathbf{b}_i := \mathbf{b}(\mathbf{x}_i,t)$, and $\mathcal{F}_i$ is the family of node i, representing the set of all nodes j interacting with node i. Here, the family of a computational node i, $\mathcal{F}_i$, is a discrete set and should not be confused with the family of a material point $\mathbf{x}$, $\mathcal{H}_{\mathbf{x}}$, which represents a finite continuum region. In Eq. 11, the quadrature weights $V_j^{(i)}$ represent the volume of the intersection between the cell j and the neighborhood of node i (Seleson 2014). An illustration of the meshfree discretization based on a uniform grid is given in Fig. 1. The quadrature weights $V_j^{(i)}$ may be computed by different numerical algorithms, as described in section "Numerical Tools for Improved Convergence." When a cell partially overlaps the neighborhood of a given node, e.g., $\mathcal{H}_{\mathbf{x}_i}$ in Fig. 1, the volume of that overlapping region is referred to as a partial volume; these overlaps are illustrated in light orange color in Fig. 1. Algorithms that attempt to approximate

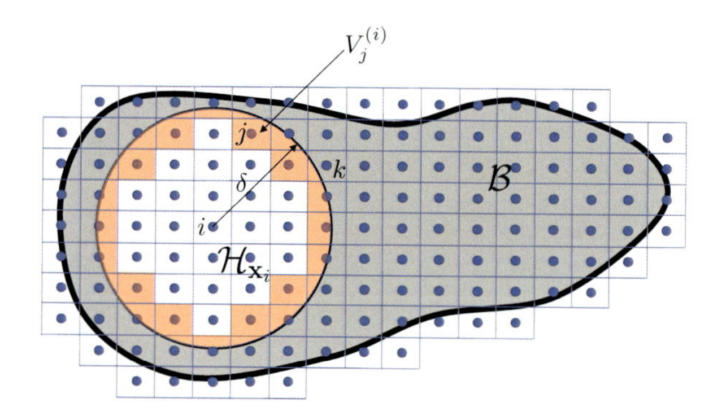

Fig. 1 Meshfree discretization of a peridynamic body $\mathcal{B}$

such volumes for use as quadrature weights are referred to as *partial-volume algorithms*.

Algorithm 1 : FV (from Silling and Askari (2005))
1: {Compute bond length}
2: $|\boldsymbol{\xi}| = |\mathbf{x}_j - \mathbf{x}_i|$
3: {Check if node j is in the family of node i}
4: **if** $|\boldsymbol{\xi}| \leq \delta$ **then**
5: $\quad V_j^{(i)} = V_j$
6: **else**
7: $\quad V_j^{(i)} = 0$
8: **end if**
9: Return $V_j^{(i)}$

Numerical Tools for Improved Convergence

In the meshfree method proposed in Silling and Askari (2005), the discretization of the PD equation of motion is given by Eq. 11 with a choice of uniform quadrature weights, i.e., $V_j^{(i)} = V_j$ for all $j \in \mathcal{F}_i$, where the family of node i is

$$\mathcal{F}_i = \left\{j \in \mathcal{L} : \left|\mathbf{x}_j - \mathbf{x}_i\right| \leq \delta\right\}. \tag{12}$$

To easily compare this discretization method with other discretization methods based on partial-volume algorithms, we first present the computation of quadrature weights in this method in Algorithm 1; we refer to it as the FV (Full Volume) algorithm. This discretization method has two drawbacks, as follows:

(1) **Low spatial integration accuracy**: The interactions between node i and nodes j for which $|\mathbf{x}_j - \mathbf{x}_i| < \delta$ and $\tau_j \not\subset \mathcal{H}_{\mathbf{x}_i}$ are not accurately estimated. Moreover, the interactions between node i and nodes k for which $|\mathbf{x}_k - \mathbf{x}_i| > \delta$ are completely omitted, even when $\tau_k \cap \mathcal{H}_{\mathbf{x}_i} \neq \emptyset$. See Fig. 1 for an illustration. This suggests a low accuracy in the numerical computation of the integral in the PD governing equation, due to a staircase approximation of the boundary of the neighborhood of a given point.
(2) **Irregular convergence behavior**: A node k for which $|\mathbf{x}_k - \mathbf{x}_i| > \delta$ does not interact with node i. However, under a small grid refinement, such that $\mathbf{x}_k \to \mathbf{x}_k + \epsilon\widehat{\boldsymbol{e}}_n$ with ϵ a small number and $\widehat{\boldsymbol{e}}_n$ a unit vector so that $|\mathbf{x}_k + \epsilon\widehat{\boldsymbol{e}}_n - \mathbf{x}_i| < \delta$, that node interacts with node i. This suggests that a small grid refinement may cause a significant change in the numerical integration, possibly resulting in abrupt variations in the numerical error. This may result in an irregular (nonsmooth) convergence behavior, as demonstrated in Seleson and Littlewood (2016).

Algorithm 2 : PV-PDLAMMPS (from Parks et al. (2008))
1: {Compute bond length}
2: $|\boldsymbol{\xi}| = |\mathbf{x}_j - \mathbf{x}_i|$
3: {Check if cell j is contained within the neighborhood of node i (exact in 1D only)}
4: **if** $|\boldsymbol{\xi}| + \frac{h}{2} \leq \delta$ **then**
5: $V_j^{(i)} = V_j$
6: {Check if node j is in the family of node i}
7: **else if** $|\boldsymbol{\xi}| \leq \delta$ **then**
8: $V_j^{(i)} = \frac{1}{h}\left[\delta - \left(|\boldsymbol{\xi}| - \frac{h}{2}\right)\right] V_j$
9: **else**
10: $V_j^{(i)} = 0$
11: **end if**
12: Return $V_j^{(i)}$

A way to mitigate the low integration accuracy and the irregular convergence behavior described above is either to account for all the nodal contributions to the spatial integration in the governing equation, by properly computing all cell-neighborhood intersection volumes, or to weaken the contribution to the spatial integration of nodes near the boundary of the neighborhood of a given node. The former approach has been used by implementing partial-volume algorithms, whereas the latter approach has been used by employing influence functions that smoothly approach zero as the bond length approaches the horizon value (Seleson 2014; Seleson and Littlewood 2016). We describe these approaches in the next sections.

Partial-Volume Algorithms

Partial-volume algorithms are aimed at numerically approximating (or analytically calculating when feasible) volumes of intersections between neighbor cells and the neighborhood of a given node. Those volumes are used as quadrature weights in meshfree discretizations of PD equations. We begin by describing an algorithm proposed in Parks et al. (2008) and implemented in PDLAMMPS (Parks et al. 2011); we refer to it as the PV-PDLAMMPS algorithm (PV is used as an acronym for Partial Volume). This algorithm is based on the family of a given node in Eq. 12 and is described in Algorithm 2. We observe that, as explained in Seleson (2014), the condition to determine whether a cell j is contained within the neighborhood of node i (line 4 of Algorithm 2) is not accurate. Furthermore, the partial-volume correction (line 8 of Algorithm 2) is only a rough approximation to a partial volume based on a simple linear correction.

To improve Algorithm 2, a method was proposed in Hu et al. (2010) and used in Bobaru and Ha (2011). That method is based on the idea of expanding the family of a given node in Eq. 12 to include neighboring nodes located at a distance larger than the horizon from that node with cells intersecting the neighborhood of that node.

The corresponding algorithm is described in Algorithm 3; we refer to it as the PV-HHB algorithm (HHB refers to the initials of the authors of Hu et al. (2010)). It only differs from the PV-PDLAMMPS algorithm in the inequality in line 7.

Although the PV-HHB algorithm represents an improvement to both the FV and PV-PDLAMMPS algorithms, it has the following limitations. First, it does not always account for all neighboring nodes with cells intersecting the neighborhood of a given node, as demonstrated in Seleson (2014). Second, the partial-volume correction is based on the same rough approximation used in the PV-PDLAMMPS algorithm. To correct these issues, a method was proposed in Seleson (2014) for two-dimensional models and expanded to three-dimensional models in Seleson and Littlewood (2016). The idea of that method is to take the family of a given node as all neighboring nodes with cells intersecting the neighborhood of that node, and compute the corresponding intersections as accurately as possible. In Seleson (2014), those intersections were calculated analytically for two-dimensional models discretized over uniform grids, whereas in Seleson and Littlewood (2016) a generalized computational method based on recursive subdivision and sampling was employed to numerically approximate intersection volumes. We refer to the corresponding algorithm as the PV-NC algorithm as in Seleson and Littlewood (2016) (NC is used as an acronym for Numerical Calculation). The algorithm name PV-NC was chosen in Seleson and Littlewood (2016) as a 3D numerically based analog of the 2D algorithm PA-AC (Partial Area-Analytical Calculation), which is based on analytical calculations of cell-neighborhood intersection areas.

Algorithm 3 : PV-HHB (from Hu et al. (2010))

1: {Compute bond length}
2: $|\boldsymbol{\xi}| = |\mathbf{x}_j - \mathbf{x}_i|$
3: {Check if cell j is contained within the neighborhood of node i (exact in 1D only)}
4: **if** $|\boldsymbol{\xi}| + \frac{h}{2} \leq \delta$ **then**
5: $\quad V_j^{(i)} = V_j$
6: {Check if node j is in the family of node i}
7: **else if** $|\boldsymbol{\xi}| - \frac{h}{2} \leq \delta$ **then**
8: $\quad V_j^{(i)} = \frac{1}{h}\left[\delta - \left(|\boldsymbol{\xi}| - \frac{h}{2}\right)\right] V_j$
9: **else**
10: $\quad V_j^{(i)} = 0$
11: **end if**
12: Return $V_j^{(i)}$

Influence Functions

The computational cost and geometrical challenge posed by the calculation of partial volumes, particularly for nonuniform grids in higher dimensions, motivate the study of a different approach. An alternative method to reduce errors in the numerical approximation of spatial integrations is to employ kernels that vanish for bonds of a horizon or longer length, particularly smoothly decaying kernels. Such kernels can result from the incorporation of highly regular influence functions. Influence functions in PD models have been originally described in Silling et al. (2007), and their role in peridynamics has been further studied in Seleson and Parks (2011). These functions can be used to select which bonds contribute to the material response of a given point, exclude damaged bonds, and modulate the strength of nonlocal interactions, as well as represent interfaces, free surfaces, and mixtures (Seleson 2010; Silling et al. 2007).

Following Seleson (2014) and Seleson and Littlewood (2016), we employ spherical influence functions of the form

$$\omega\left(|\boldsymbol{\xi}|\right) = \begin{cases} \frac{P_n(|\boldsymbol{\xi}|)}{|\boldsymbol{\xi}|^{\alpha}} & |\boldsymbol{\xi}| \leq \delta, \\ 0 & \text{otherwise}, \end{cases} \quad ; \; \alpha = 0, 1, \tag{13}$$

where $P_n(r)$ is a polynomial of order $n \in \mathbb{N}_0$, which satisfies $P_n(0) = 1$ and $P_n(\delta) = 0$ for $n > 0$, and $P_n'(r) = P_n''(r) = \ldots = P_n^{(k)}(r) = 0$ at $r = 0, \delta$ with $k = (n-1)/2$ for $n > 1$ odd. Specifically, we have

$$P_0(r) = 1, \tag{14a}$$

$$P_1(r) = 1 - \frac{r}{\delta}, \tag{14b}$$

$$P_3(r) = 1 - 3\left(\frac{r}{\delta}\right)^2 + 2\left(\frac{r}{\delta}\right)^3, \tag{14c}$$

$$P_5(r) = 1 - 10\left(\frac{r}{\delta}\right)^3 + 15\left(\frac{r}{\delta}\right)^4 - 6\left(\frac{r}{\delta}\right)^5, \tag{14d}$$

$$P_7(r) = 1 - 35\left(\frac{r}{\delta}\right)^4 + 84\left(\frac{r}{\delta}\right)^5 - 70\left(\frac{r}{\delta}\right)^6 + 20\left(\frac{r}{\delta}\right)^7. \tag{14e}$$

An illustration of these polynomials is given in Fig. 2.

Convergence Studies of Static Peridynamic Problems

In this section, we present convergence studies of static PD problems following Seleson and Littlewood (2016). Let a body $\mathcal{B} = (0,1)\times(0,1)\times(0,1)$ and let us define a subregion $\Omega = (2\delta,1-2\delta)\times(2\delta,1-2\delta)\times(2\delta,1-2\delta) \subset \mathcal{B}$. All points in Ω are at a distance of at least 2δ from the boundary $\partial\mathcal{B}$, which is equivalent to saying that Ω represents the bulk of the body. The choice of Ω allows us to properly impose displacement boundary conditions, given a boundary function $\mathbf{g}(\mathbf{x})$, in the boundary layer $\overline{\mathcal{B}}\backslash\Omega$ with $\overline{\mathcal{B}} := \mathcal{B} \cup \partial\mathcal{B}$, which contains a 2δ-width volumetric layer required in state-based PD problems. This is in contrast to bond-based PD

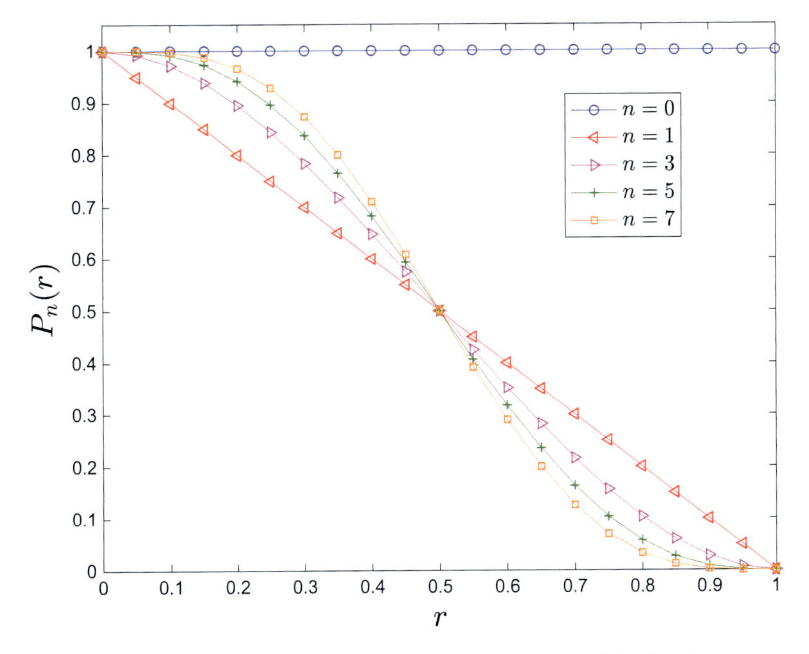

Fig. 2 Illustration of the polynomials $P_n(r)$, $n = 0, 1, 3, 5, 7$, in Eq. 14 for $\delta = 1$

problems, for which a layer of width δ is suffcient. The computational domain and the boundary layer are illustrated in Fig. 3. For simplicity, our static problem employs the linearized LPS model in Eq. 10 and is stated as

$$-\int_{\mathcal{H}_{\mathbf{x}}} \frac{\omega(|\boldsymbol{\xi}|)}{m} \Bigg\{ (3K - 5G)\left(\vartheta^{\mathrm{lin}}[\mathbf{x}] + \vartheta^{\mathrm{lin}}[\mathbf{x}+\boldsymbol{\xi}]\right)\boldsymbol{\xi} + 30G \frac{\boldsymbol{\xi} \otimes \boldsymbol{\xi}}{|\boldsymbol{\xi}|^2} \left(\mathbf{u}(\mathbf{x}+\boldsymbol{\xi}) - \mathbf{u}(\mathbf{x})\right) \Bigg\} \mathrm{d}V_{\boldsymbol{\xi}} = \mathbf{b}(\mathbf{x}) \quad \mathbf{x} \in \Omega, \tag{15a}$$

$$\mathbf{u}(\mathbf{x}) = \mathbf{g}(\mathbf{x}) \quad \mathbf{x} \in \overline{\mathcal{B}} \backslash \Omega. \tag{15b}$$

Preliminary Considerations

(1) *Manufactured solutions*: In this study, we investigate the convergence of numerical solutions of Problem (15) to analytical solutions. For this purpose, we employ the method of manufactured solutions to find a corresponding body force density, given an analytical solution. As proved in Seleson and Littlewood (2016), given a quadratic displacement field, the PD equilibrium Eq. 10 reduces to the classical Navier-Cauchy equation of static elasticity,

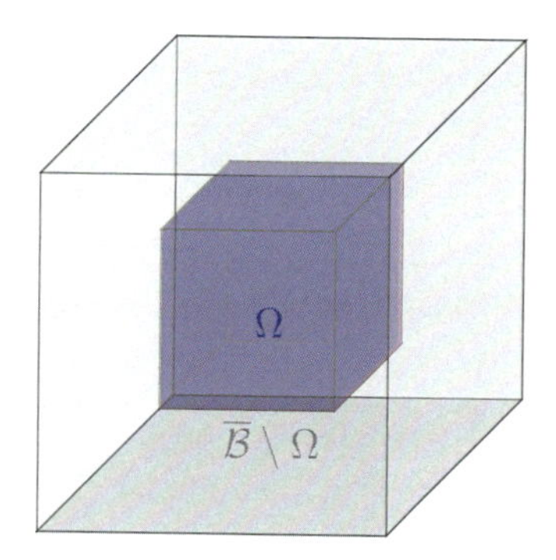

Fig. 3 Illustration of the computation domain Ω and boundary layer $\overline{\mathcal{B}}\backslash\Omega$ for Problem (15)

$$-\left[G\nabla^2\mathbf{u}(\mathbf{x})+\left(K+\frac{1}{3}G\right)\nabla(\nabla\cdot\mathbf{u})(\mathbf{x})\right]=\mathbf{b}(\mathbf{x}). \tag{16}$$

Consequently, we can employ Eq. 16 to find a corresponding body force density for our PD problem as follows. Let a displacement field $\mathbf{u} = (u, v, w)$, where

$$u(\mathbf{x}) = U_{11}x^2 + U_{22}y^2 + U_{33}z^2 + U_{12}xy + U_{13}xz + U_{23}yz, \tag{17}$$

$$v(\mathbf{x}) = V_{11}x^2 + V_{22}y^2 + V_{33}z^2 + V_{12}xy + V_{13}xz + V_{23}yz, \tag{18}$$

$$w(\mathbf{x}) = W_{11}x^2 + W_{22}y^2 + W_{33}z^2 + W_{12}xy + W_{13}xz + W_{23}yz, \tag{19}$$

where $\mathbf{x} = (x, y, z)$, and U_{11}, U_{22}, U_{33}, U_{12}, U_{13}, U_{23}, V_{11}, V_{22}, V_{33}, V_{12}, V_{13}, V_{23}, W_{11}, W_{22}, W_{33}, W_{12}, W_{13}, and W_{23} are constant coeffcients. Then, the components of the body force density $\mathbf{b} = (b_1, b_2, b_3)$ are:

$$b_1 = -\left[2G(U_{11}+U_{22}+U_{33})+\left(K+\frac{1}{3}G\right)(2U_{11}+V_{12}+W_{13})\right],$$

$$b_2 = -\left[2G(V_{11}+V_{22}+V_{33})+\left(K+\frac{1}{3}G\right)(U_{12}+2V_{22}+W_{23})\right],$$

$$b_3 = -\left[2G(W_{11}+W_{22}+W_{33})+\left(K+\frac{1}{3}G\right)(U_{13}+V_{23}+2W_{33})\right].$$

(2) *Error norm for numerical solutions*: Let the body $\mathcal{B}$ be discretized with a set of N^3 nodes $\mathcal{L} = \{\mathbf{x}_i\}_{i=1}^{N^3}$ over a uniform grid with grid spacing h, so that the Voronoi cell of node i is the cubic cell $\tau_i = \left(x_i - \frac{h}{2}, x_i + \frac{h}{2}\right) \times \left(y_i - \frac{h}{2}, y_i + \frac{h}{2}\right) \times \left(z_i - \frac{h}{2}, z_i + \frac{h}{2}\right)$, where $\mathbf{x}_i = (x_i, y_i, z_i)$. In particular, in the numerical simulations we employ a discretization with N nodes per direction, where each computational node possesses a full cubic cell inside the (cubic) body, i.e., all nodes are at a distance of a least $h/2$ from the boundary of

the body. To compute an error norm, we assume the numerical solution to be given by a piecewise-constant approximation $\mathbf{u}^h = (u^h, v^h, w^h)$, so that $\mathbf{u}^h(\mathbf{x}) = \left(u_i^h, v_i^h, w_i^h\right)$ for all $\mathbf{x} \in \tau_i$ with u_i^h, v_i^h, and w_i^h constants. The L^2-norm of the error is then

$$\begin{aligned}
\left\|\mathbf{u}^h - \mathbf{u}\right\|_2 &= \left[\int_{\mathcal{B}} \left(u^h(\mathbf{x}) - u(\mathbf{x})\right)^2 + \left(v^h(\mathbf{x}) - v(\mathbf{x})\right)^2 + \left(w^h(\mathbf{x}) - w(\mathbf{x})\right)^2 \mathrm{d}V_{\mathbf{x}}\right]^{\frac{1}{2}} \\
&= \left[\sum_{i=1}^{N^3} \int_{\tau_i} \left(u_i^h - u(\mathbf{x})\right)^2 + \left(v_i^h - v(\mathbf{x})\right)^2 + \left(w_i^h - w(\mathbf{x})\right)^2 \mathrm{d}V_{\mathbf{x}}\right]^{\frac{1}{2}} \\
&= \left[\sum_{i=1}^{N^3} \sum_{g=1}^{N_G} \sum_{\ell=1}^{N_L} \sum_{s=1}^{N_S} \left\{ \left(u_i^h - u\left(x_i + \frac{h}{2}\alpha_g, y_i + \frac{h}{2}\beta_\ell, z_i + \frac{h}{2}\gamma_s\right)\right)^2 \right.\right. \\
&\quad + \left(v_i^h - v\left(x_i + \frac{h}{2}\alpha_g, y_i + \frac{h}{2}\beta_\ell, z_i + \frac{h}{2}\gamma_s\right)\right)^2 \\
&\quad \left.\left. + \left(w_i^h - w\left(x_i + \frac{h}{2}\alpha_g, y_i + \frac{h}{2}\beta_\ell, z_i + \frac{h}{2}\gamma_s\right)\right)^2 \right\} \frac{h^3}{8} w_g \widehat{w}_\ell \tilde{w}_s \right]^{\frac{1}{2}},
\end{aligned} \tag{20}$$

where $\{\alpha_g\}_{g=1,\ldots,N_G}$ and $\{w_g\}_{g=1,\ldots,N_G}$;$\{\beta_\ell\}_{\ell=1,\ldots,N_L}$ and $\{\widehat{w}_\ell\}_{\ell=1,\ldots,N_L}$; and $\{\gamma_s\}_{s=1,\ldots,N_S}$ and $\{\tilde{w}_s\}_{s=1,\ldots,N_S}$ are Gauss quadrature points and weights in the standard element $[-1,1]$ along the x-, y-, and z-directions, respectively. A quadrature rule with N_G, N_L, and N_S Gauss quadrature points along the x-, y-, and z-directions, respectively, integrates exactly monomials of the form $x^g y^\ell z^s$ with $0 \le g \le 2N_G - 1$, $0 \le \ell \le 2N_L - 1$, and $0 \le s \le 2N_S - 1$. Since the analytical solution is assumed quadratic and we integrate a function of the solution squared, we would like to integrate exactly monomials up to $g = 4$, $\ell = 4$, and $s = 4$. We thus choose $N_G = N_L = N_S = 3$.

(3) *Computational cost and horizon choice*: In this type of convergence study, the horizon needs to be carefully chosen such that the size of the boundary layer is small relative to the size of the computational domain, yet the computational cost remains tractable. To clarify this, let a body $\mathcal{B}$ be a cube of edge length L and the computational domain Ω be a smaller cube of edge length $L - 4\delta$ (as illustrated in Fig. 3), and define $N_{\text{neig}} = \delta/h$ with h the grid spacing. Then, the total number of computational nodes required to discretize the body $\mathcal{B}$ is

$$N^3 = \left(\frac{L}{h}\right)^3 = \left(N_{\text{neig}}\frac{L}{\delta}\right)^3. \tag{21}$$

Table 1 Total number of computational nodes

N_{neig} (δ/h)	$\delta = 0.10L$	$\delta = 0.04L$	$\delta = 0.02L$	$\delta = 0.01L$
1	1,000	15,625	125,000	1,000,000
2	8,000	125,000	1,000,000	8,000,000
3	27,000	421,875	3,375,000	27,000,000
4	64,000	1,000,000	8,000,000	64,000,000
5	125,000	1,953,125	15,625,000	125,000,000
6	216,000	3,375,000	27,000,000	216,000,000

In Table 1, we show the total number of computational nodes required to discretize $\mathcal{B}$, for various choices of horizon δ. We observe that the horizon cannot be too small, because otherwise the total number of computational nodes may exceed the available computational resources. For instance, for $\delta = 0.01L$, a computation with $N_{\text{neig}} = 6$ would require 216 million nodes.

We now demonstrate that the horizon cannot be too large either. For this purpose, we look at the ratio of the boundary layer to the computational domain:

$$\frac{|\mathcal{B}\backslash\Omega|}{|\Omega|} = \frac{L^3 - (L - 4\delta)^3}{(L - 4\delta)^3} = \frac{1}{(1 - 4\delta/L)^3} - 1, \tag{22}$$

where $|\Omega|$ denotes the volume of the domain Ω. This ratio rapidly grows with increasing ratio δ/L, as shown in Fig. 4. In particular, for $\delta > 0.0516L$, the volume of the boundary layer $\mathcal{B} \setminus \Omega$ exceeds that of the computational domain Ω. To exemplify this, specific values of Eq. 22 are shown in Table 2.

(4) *Convergence avenue and refinement path*: As described in the Introduction, there exist different types of convergence in peridynamics. Here, we would like to study the convergence of numerical solutions of static PD problems to the corresponding analytical PD solutions. For this purpose, we fix the horizon δ and employ a grid refinement. This, in fact, is analogous to standard convergence studies in numerical partial differential equations (PDEs), where the grid (or mesh) is refined. Since the horizon is fixed, the ratio δ/h increases under grid refinement.

As demonstrated in Table 1, we cannot afford to refine the mesh beyond a certain ratio $N_{\text{neig}} = \delta/h$ because the number of required computational nodes would become intractable. On the other hand, to produce meaningful comparisons in our study, we need enough convergence data points. For the purpose of our studies, we took $L = 1$ and $\delta = 0.04$, and we performed a refinement given by $N_{\text{neig}} = 3, 4, 5, 6$. The data used here is a subset of the results presented in Seleson and Littlewood (2016). The reader should refer to that reference for a refinement study with additional data points.

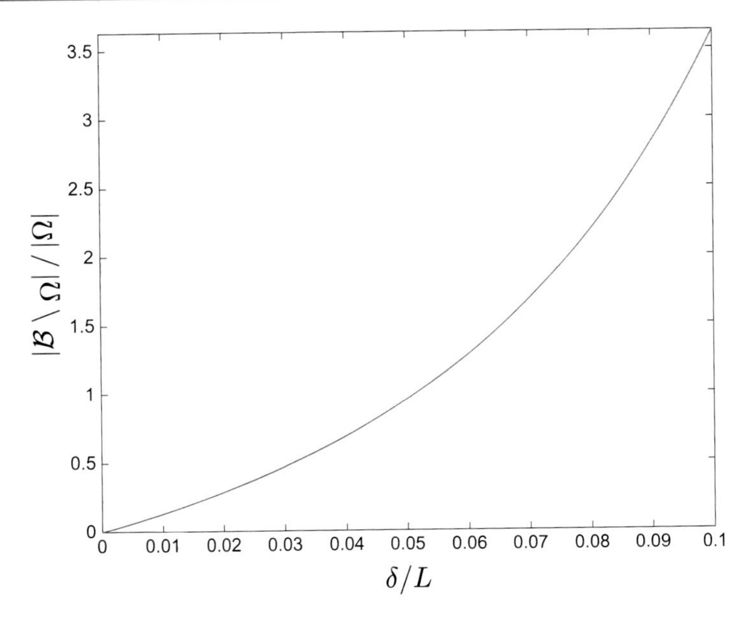

Fig. 4 Ratio $|\mathcal{B}\setminus \Omega|/|\Omega|$ as a function of the horizon δ in a cubic body $\mathcal{B}$ of edge length L with a cubic subdomain Ω of edge length $L - 4\delta$

Table 2 Volume ratio $|\mathcal{B}\setminus \Omega|/|\Omega|$

δ/L	0.000	0.001	0.005	0.010	0.050	0.100
$\lvert\mathcal{B}\setminus \Omega\rvert/\lvert\Omega\rvert$	0.000	0.012	0.062	0.130	0.953	3.630

Numerical Results

We study the convergence of numerical solutions of Problem (15) with the following choices: $\delta = 0.04$, $K = 1$, $G = 0.5$, $\mathbf{b} = \left(-\frac{10}{3}, 0, 0\right)$, and $\mathbf{g}(\mathbf{x}) = (x^2, 0, 0)$, and where the weighted volume m is computed analytically. The chosen elastic moduli correspond to a material with Poisson's ratio $\nu = 0.29$. The goal of this study is, on one hand, to compare the performance of the FV algorithm with the partial-volume algorithms PV-PDLAMMPS, PV-HHB, and PV-NC, and, on the other hand, to study the effect of incorporating specific influence functions within the PD force vector state field. The total number of computational nodes employed in the simulations appears in the third column of Table 1. The simulations were run in *Peridigm*, a PD code developed at Sandia National Laboratories (Parks et al. 2012); see Seleson and Littlewood (2016) for details regarding the solution procedure.

In Fig. 5, we present the convergence results for the FV algorithm as well as for the different partial-volume algorithms and influence functions for (a) $\alpha = 0$ and (b) $\alpha = 1$, where the error is computed with the L^2-norm given in Eq. 20. Specifically, we examine influence functions of the following form: piecewise linear (PWL) with $P_1(r)$ in Eq. 14b; piecewise cubic (PWC) with $P_3(r)$ in Eq. 14c;

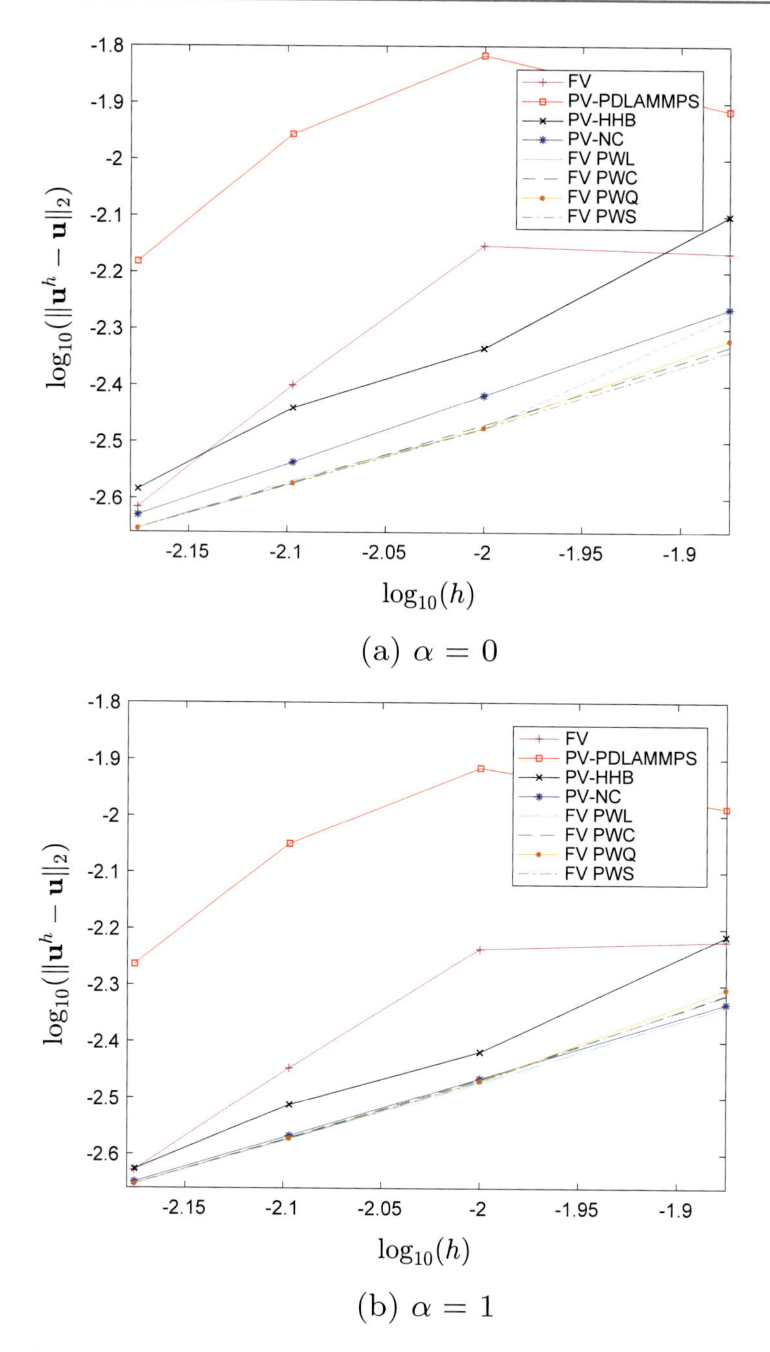

Fig. 5 Convergence of the numerical solution of Problem (15) using different partial-volume algorithms and different influence functions

Table 3 Convergence rates for the (static) convergence results in Fig. 5

Algorithm	$\alpha = 0$		$\alpha = 1$	
	$\overline{r}$	R	$\overline{r}$	R
FV	1.53	0.165	1.38	0.128
PV-PDLAMMPS	0.86	0.186	0.89	0.167
PV-HHB	1.56	0.035	1.34	0.030
PV-NC	1.22	0.003	1.05	0.001
FV PWL	1.24	0.036	1.05	0.004
FV PWC	1.07	0.005	1.11	0.009
FV PWQ	1.10	0.014	1.15	0.016
FV PWS	1.04	0.006	1.12	0.012

piecewise quintic (PWQ) with $P_5(r)$ in Eq. 14d; and piecewise septic (PWS) with $P_7(r)$ in Eq. 14e. In order to estimate a convergence rate, to each curve we fit a linear function (in a least-squares sense) of the form

$$f(h) = \overline{r}\,\log_{10}(h) + q, \tag{23}$$

where $\overline{r}$ and q are constant coeffcients. The value of $\overline{r}$ represents an "average" convergence rate. To estimate the quality of the linear fit, we also compute the norm of residuals,

$$R = \sqrt{\sum_{n=1}^{N_{\text{iter}}} \left(f\,(h_n) - \log_{10}(E_n)\right)^2}, \tag{24}$$

where E_n and h_n are the L^2-norm of the error (*cf.* Eq. 20) and the grid spacing, respectively, at the refinement step n, and N_{iter} is the total number of refinement steps. These results are presented in Table 3. We observe that both methods, accurate estimation of partial volumes (see PV-NC) and employment of smooth influence functions (see, e.g., FV PWS), provide a significant improvement relative to the FV algorithm, as clearly observed in the reduction in the value of R. Based on the convergence rates obtained, we conclude that those methods approach a first-order convergence.

Convergence Studies of Dynamic Peridynamic Problems

In this section, we present convergence studies of dynamic PD problems, based on unconstrained wave propagation. As in section "Convergence Studies of Static Peridynamic Problems," we investigate the effect of different partial-volume algorithms and influence functions on the convergence behavior, as the grid is refined under a fixed horizon.

Let a body $\mathcal{B} = \left(-\frac{1}{2}, \frac{1}{2}\right) \times \left(-\frac{1}{2}, \frac{1}{2}\right) \times \left(-\frac{1}{2}, \frac{1}{2}\right)$. Our dynamic problem employs the linearized LPS model in Eq. 9 and is stated as

$$\rho(\mathbf{x}) \frac{\partial^2 \mathbf{u}}{\partial t^2}(\mathbf{x}, t) = \int_{\mathcal{H}_\mathbf{x}} \frac{\omega(|\boldsymbol{\xi}|)}{m} \Big\{ (3K - 5G) \left(\vartheta^{\text{lin}}[\mathbf{x}, t] + \vartheta^{\text{lin}}[\mathbf{x} + \boldsymbol{\xi}, t] \right) \boldsymbol{\xi} + 30G \frac{\boldsymbol{\xi} \otimes \boldsymbol{\xi}}{|\boldsymbol{\xi}|^2} \left(\mathbf{u}(\mathbf{x} + \boldsymbol{\xi}, t) - \mathbf{u}(\mathbf{x}, t) \right) \Big\} dV_{\boldsymbol{\xi}} + \mathbf{b}(\mathbf{x}, t) \quad \mathbf{x} \in \overline{\mathcal{B}}, t \geq 0, \tag{25a}$$

$$\mathbf{u}(\mathbf{x}, 0) = \mathbf{u}_0(\mathbf{x}) \quad \mathbf{x} \in \overline{\mathcal{B}}, \tag{25b}$$

$$\frac{\partial \mathbf{u}}{\partial t}(\mathbf{x}, 0) = \mathbf{v}_0(\mathbf{x}) \quad \mathbf{x} \in \overline{\mathcal{B}}, \tag{25c}$$

where $\mathbf{u}_0$ and $\mathbf{v}_0$ are displacement and velocity initial conditions. As opposed to the static Problem (15), no displacement boundary conditions are imposed in these dynamic studies, eliminating the need to define a boundary layer.

Preliminary Considerations

(1) *Error norm for numerical solutions*: The error for the dynamic PD problems was computed by comparing a series of increasingly refined discretizations against results obtained using a highly refined reference discretization. This was necessitated by the lack of analytical solutions of the state-based PD Problem (25). While analytical solutions for wave propagation concerning linear elastic bond-based PD models have been presented in one dimension in Weckner and Abeyaratne (2005) and in higher dimensions in Weckner et al. (2009), analytical solutions are not currently available for the state-based PD Problem (25). Unfortunately, our approach for computing the error is extremely computationally demanding, because the number of PD bonds increases dramatically as the mesh is refined under a fixed horizon, as demonstrated in Table 4. To achieve numerical accuracy while remaining computationally tractable, a value of $N_{\text{neig}} = 10$ was employed for the reference numerical solution.

The calculation of an error norm in the dynamic studies followed assumptions analogous to those in the static simulations. We assumed the numerical

Table 4 Total number of peridynamic (PD) bonds in a simulation for $\delta = 0.05$

N_{neig} (δ/h)	Number of PD bonds	
	FV	PV-NC
3	12,433,244	25,077,672
4	62,022,592	110,046,364
5	242,986,412	384,681,876
6	753,964,092	1,040,684,328
7	1,838,660,296	2,552,461,732
8	4,080,378,204	5,479,353,788
9	8,456,684,628	10,782,968,496
10	15,752,838,172	19,683,573,672

solution for both the trial simulations and the reference simulation to be given by a piecewise-constant approximation $\mathbf{u}^h = (u^h, v^h, w^h)$, so that $\mathbf{u}^h(\mathbf{x},t) = \left(u_i^h, v_i^h, w_i^h\right)$ for all $\mathbf{x} \in \tau_i$ at time t with u_i^h, v_i^h, and w_i^h constants. We denote by τ_i^{ref} a Voronoi cell in the reference discretization. To avoid having the reference discretization being an exact subdivision of any of the trial discretizations, which could result in biased error calculations for those specific discretizations, the grid in the reference discretization was perturbed so that the internal nodes (i.e., nodes not directly adjacent to the boundary of the body) were shifted a distance of 0.3 times the reference grid spacing, h_{ref}, in each direction. This perturbation does not affect the error calculation in our studies because in the wave propagation problems considered here nodes adjacent to the boundary of the body have a zero displacement over the course of the simulations. Let the domain $\mathcal{B}$ be decomposed into N^{int} subdomains $\{\tilde{\tau}_n\}_{n=1,\ldots,N^{\text{int}}}$, so that $\tilde{\tau}_n = \tau_i \cap \tau_j^{\text{ref}}$ for given i and j; in other words, these subdomains are intersections of pairs of Voronoi cells, one from the trial discretization and one from the reference discretization. Then, the L^2-norm of the error for a given level of grid refinement, given by a grid spacing h, at the final time T can be computed as

$$\begin{aligned}\left\|\mathbf{u}^h - \mathbf{u}^{h_{\text{ref}}}\right\|_2 &= \left[\int_{\mathcal{B}} \left(u^h(\mathbf{x},T) - u^{h_{\text{ref}}}(\mathbf{x},T)\right)^2 + \left(v^h(\mathbf{x},T) - v^{h_{\text{ref}}}(\mathbf{x},T)\right)^2\right.\\ &\qquad\left. + \left(w^h(\mathbf{x},T) - w^{h_{\text{ref}}}(\mathbf{x},T)\right)^2 dV_{\mathbf{x}}\right]^{\frac{1}{2}}\\ &= \left[\sum_{n=1}^{N^{\text{int}}} \left(\left(u_n^h - u_n^{h_{\text{ref}}}\right)^2 + \left(v_n^h - v_n^{h_{\text{ref}}}\right)^2 + \left(w_n^h - w_n^{h_{\text{ref}}}\right)^2\right) V_n^{\text{int}}\right]^{\frac{1}{2}},\end{aligned} \tag{26}$$

where $\left(u_n^h, v_n^h, w_n^h\right)$ and $\left(u_n^{h_{\text{ref}}}, v_n^{h_{\text{ref}}}, w_n^{h_{\text{ref}}}\right)$ are the trial and reference numerical solutions, respectively, over the subdomain $\tilde{\tau}_n$ at the final time T, and V_n^{int} is the volume of that subdomain.

(2) *Reference solution*: The solution of the PD Problem (25) varies for different choices of influence function. This is true in general for PD problems, as demonstrated in Seleson and Parks (2011), but was not the case for the static simulations in section "Convergence Studies of Static Peridynamic Problems" due to the particular choice of PD model and applied body force density. On the other hand, even when the influence function is unchanged, the use of different partial-volume algorithms may produce different numerical solutions. As the reference solution is based on a (highly refined) numerical solution, and not on an analytical one, separate reference solutions were computed for the dynamic PD problems, one for each of the partial-volume algorithms and influence functions.

(3) *Simulation running time*: In these dynamic simulations, we would like to investigate the accuracy of the numerical results for various choices of partial-volume algorithms and influence functions. This suggests the need to let the solution evolve sufficiently such that the effect of the discretization on the system dynamics is clearly observable. On the other hand, we do not want to concern ourselves with wave reflections from domain boundaries or with interactions of a propagating wave with itself, as this may introduce another level of complexity in the dynamics of the system. As a result, we run simulations based on an initial pulse concentrated away from the center of the domain and away from the domain boundaries, and we allow the pulse to propagate as much as possible without interacting with the domain boundaries or with itself at the center of the domain.

(4) *Signal resolution and discretization*: As discussed in section "Preliminary Considerations" of the static PD problems, extremely refined discretizations must be avoided due to computational limitations. However, in the dynamic simulations, the discretization cannot be too coarse either because we need to resolve the shape of the propagating wave in the domain.

Numerical Results

We study the convergence of numerical solutions of Problem (25) with the following choices: $\delta = 0.05$, $K = 1$, $G = 0.5$, $\mathbf{b} = \mathbf{0}$, and $\rho = 100.0$; as in the static Problem (15), the weighted volume m is computed analytically. The initial conditions are given as follows:

$$\mathbf{u}_0(\mathbf{x}) = \begin{cases} a e^{-\frac{(|\mathbf{x}|-r_0)^2}{\ell^2}} \frac{\mathbf{x}}{|\mathbf{x}|} & \text{if } (r_0 - 3\ell) \leq |\mathbf{x}| \leq (r_0 + 3\ell), \\ \mathbf{0} & \text{otherwise}, \end{cases} \tag{27}$$

$$\mathbf{v}_0(\mathbf{x}) = \mathbf{0}, \tag{28}$$

where the constants were assigned values of $a = 0.001$, $r_0 = 0.25$, and $\ell = 0.07$. The initial displacement field is a spherical layer, with a Gaussian radial distribution. The parameter r_0 represents the center of the Gaussian and the parameter ℓ determines its width. Points in that layer are assigned an initial outward radial displacement as determined by the Gaussian function, whereas points outside that layer are given a zero initial displacement. The width parameter ℓ was chosen so that the coarsest discretization can resolve the shape of the Gaussian distribution. In addition, we observe that at a distance $r = 3\ell$ from the center of the Gaussian, the amplitude of the initial displacement decays to about $1.2E\text{-}7$, suggesting a reasonable transition to the zero displacement field. No initial velocity is given. Plots of the initial displacement field are given in Figs. 6a and 7a. The wave is allowed to propagate freely from the initial time $t_0 = 0$ to the final time $T = 0.6$. Time integration was achieved using the standard velocity-Verlet scheme with a fixed time step $\Delta t = 0.01$,

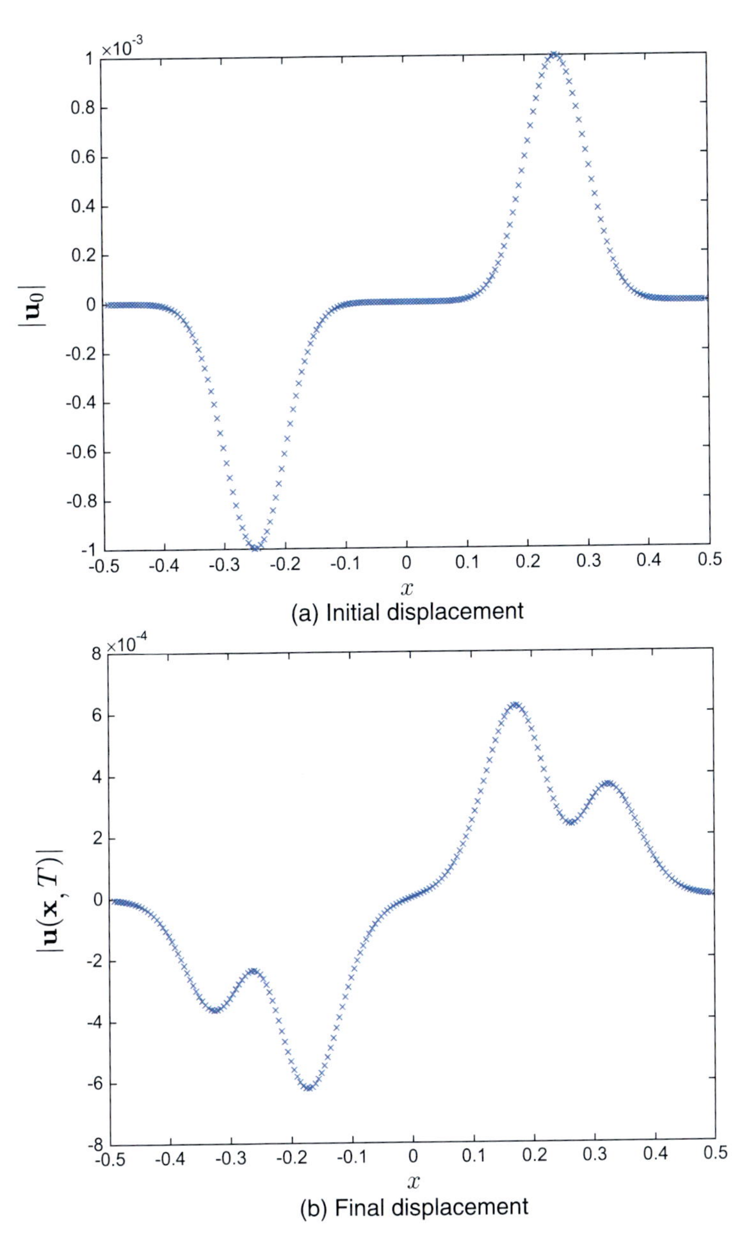

Fig. 6 Initial displacement (a) and final displacement (b) for nodes located along the x-axis for the three-dimensional wave propagation, based on the reference finest grid with grid spacing $h_{\text{ref}} = 0.005$ ($\delta/h_{\text{ref}} = 10$). The final displacement was computed with the FV algorithm for $\alpha = 0$

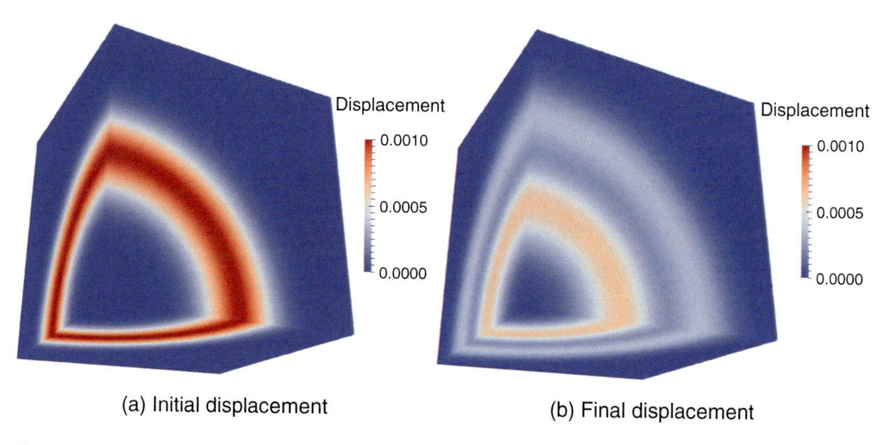

Fig. 7 Initial displacement (a) and final displacement (b) on an octant of the cubic domain for the three-dimensional wave propagation, based on the reference finest grid with grid spacing $h_{\text{ref}} = 0.005$ ($\delta/h_{\text{ref}} = 10$). The final displacement was computed with the FV algorithm for $\alpha = 0$. Coloration denotes displacement magnitude

which was stable for all the simulations. The final displacement field numerically computed with the FV algorithm for $\alpha = 0$ is given in Figs. 6b and 7b.

In Fig. 8, we present the convergence results for the different partial-volume algorithms and influence functions for (a) $\alpha = 0$ and (b) $\alpha = 1$, where the error is computed with the L^2-norm given by Eq. 26. As in the static Problem (15), to each curve we fit a linear function of the form of Eq. 23 to estimate a convergence rate; to also estimate the quality of the linear fit, we compute the norm of residuals with Eq. 24. These results are presented in Table 5. As in the static problems, both methods, accurate estimation of partial volumes (see PV-NC) and employment of smooth influence functions (see, e.g., FV PWS), provide a significant improvement relative to the FV algorithm, as clearly observed in the reduction in the value of R. Based on the convergence rates obtained, we conclude that those methods approach a first-order convergence.

To demonstrate the effect of the influence function on the solution of the dynamic problem, we compare in Fig. 9 the final displacement field for points along the x-axis, for the different partial-volume algorithms and influence functions. The full displacement field is presented in the top plots for (a) $\alpha = 0$ and (b) $\alpha = 1$. To emphasize the differences between the two numerical tools discussed in this chapter, partial-volume algorithms and influence functions, in the bottom plots, we zoom in to the positive x-axis and simply compare the FV algorithm to the PV-NC and FV PWS algorithms for (c) $\alpha = 0$ and (d) $\alpha = 1$. The results suggest that the solution for the smooth influence function is different than the one obtained with the FV algorithm; on the other hand, using a partial-volume correction seems to more closely preserve the dynamics of the original model. This is natural, since partial-volume algorithms are only aimed at improving the discretization of the PD governing equation, without modifying the continuum model. In contrast, the

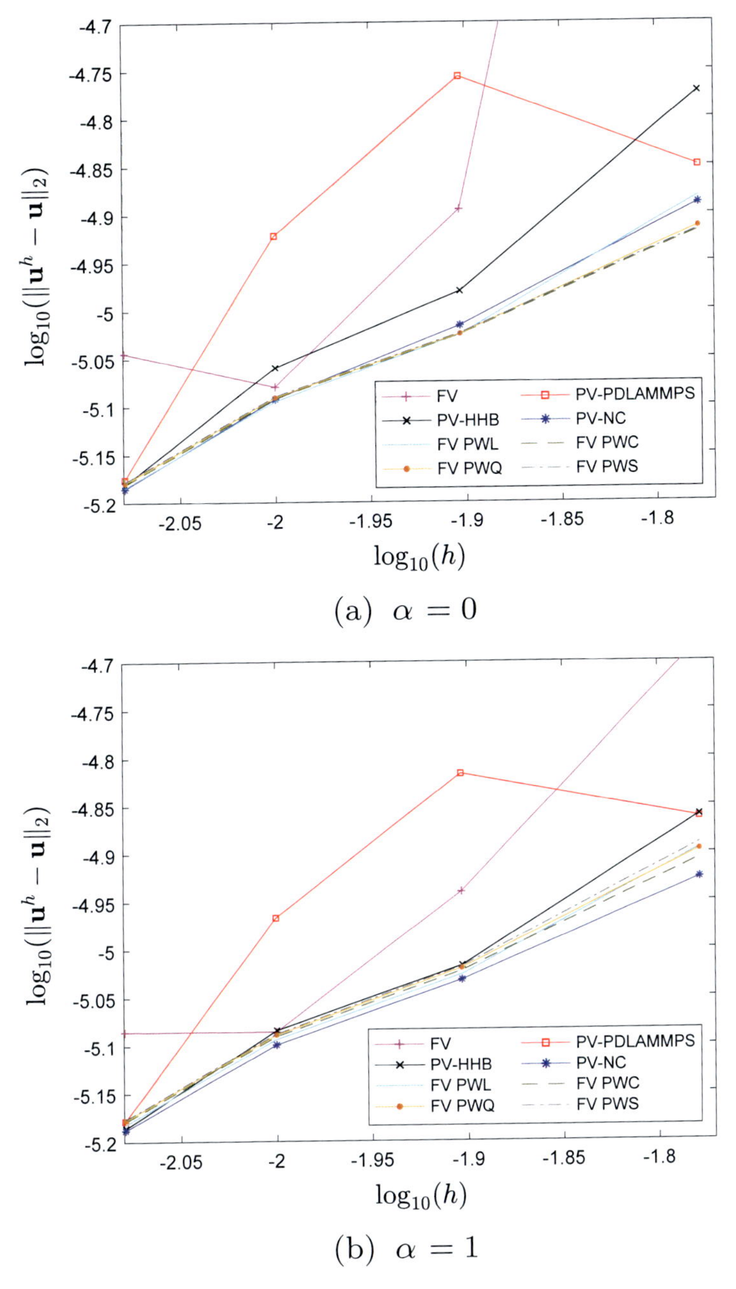

Fig. 8 Convergence of the numerical solution of Problem (25) at the final time T using different partial-volume algorithms and different influence functions

Table 5 Convergence rates for the (dynamic) convergence results in Fig. 8

Algorithm	$\alpha = 0$		$\alpha = 1$	
	$\overline{r}$	R	$\overline{r}$	R
FV	4.27	0.514	1.41	0.099
PV-PDLAMMPS	1.05	0.202	1.02	0.157
PV-HHB	1.31	0.038	1.04	0.026
PV-NC	0.96	0.013	0.85	0.016
FV PWL	0.98	0.019	0.93	0.017
FV PWC	0.85	0.016	0.88	0.015
FV PWQ	0.86	0.015	0.91	0.015
FV PWS	0.85	0.016	0.93	0.015

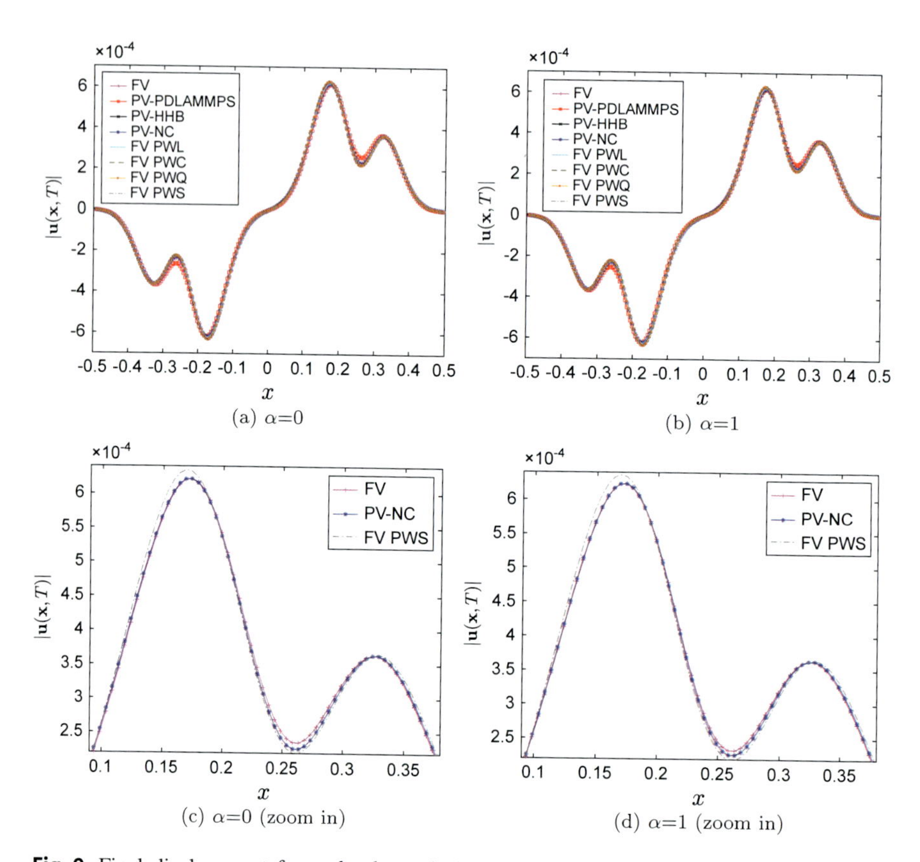

Fig. 9 Final displacement for nodes located along the x-axis for the three-dimensional wave propagation, for different partial-volume algorithms and influence functions, based on the reference finest grid with grid spacing $h_{\text{ref}} = 0.005$ ($\delta/h_{\text{ref}} = 10$). In the top plots, the entire displacement field is compared for (a) $\alpha = 0$ and (b) $\alpha = 1$. In the bottom plots, a zoom in to a portion of the curve along the positive x-axis is presented comparing the FV, PV-NC, and FV PWS algorithms for (c) $\alpha = 0$ and (d) $\alpha = 1$

incorporation of different influence functions alters the continuum model itself. This is consistent with the results presented in Seleson and Parks (2011). Further dynamic studies comparing the use of partial-volume algorithms and smooth influence functions are currently underway and will be reported in the future.

Concluding Remarks

This chapter considered convergence studies for linear elastic state-based peridynamic models, based on meshfree discretizations, and presented two numerical tools, partial-volume algorithms and influence functions, to improve the convergence behavior of numerical solutions of static and dynamic peridynamic problems. The work in this chapter drew from, and extended, the convergence studies presented for static peridynamic problems in Seleson and Littlewood (2016). We observed that both numerical tools can assist in improving the convergence behavior of static and dynamic peridynamic simulations, resulting in approximately first-order convergence. Depending on the application and problem under consideration, influence functions can, however, significantly affect the dynamical behavior of peridynamic systems.

An important contribution of this chapter is a discussion regarding various challenges present in convergence studies in peridynamic problems, many of which are not found in analogous convergence studies for PDEs. We hope that such discussion will serve as guidance for future related research studies.

Acknowledgments This study was supported in part by the Householder Fellowship which is jointly funded by: the U.S. Department of Energy, Office of Science, Office of Advanced Scientific Computing Research, Applied Mathematics program, under award number ERKJE45, and the Laboratory Directed Research and Development program at the Oak Ridge National Laboratory (ORNL), which is operated by UT-Battelle, LLC., for the U.S. Department of Energy under Contract DE-AC05-00OR22725; and by the Laboratory Directed Research and Development program at Sandia National Laboratories, which is a multimission laboratory managed and operated by National Technology and Engineering Solutions of Sandia, LLC., a wholly owned subsidiary of Honeywell International, Inc., for the U.S. Department of Energy's National Nuclear Security Administration under contract DE-NA0003525.

References

E. Askari, F. Bobaru, R.B. Lehoucq, M.L. Parks, S.A. Silling, O. Weckner, Peridynamics for multiscale materials modeling, in *SciDAC 2008, 13–17 July, Seattle*, Journal of Physics: Conference Series, vol. 125, IOP Publishing, 2008. 012078

F. Bobaru, Y.D. Ha, Adaptive refinement and multiscale modeling in 2D peridynamics. Int. J. Multiscale Comput. Eng. **9**, 635–659 (2011)

F. Bobaru, Y.D. Ha, W. Hu, Damage progression from impact in layered glass modeled with peridynamics. Centr. Eur. J. Eng. **2**, 551–561 (2012)

F. Bobaru, M. Yang, L.F. Alves, S.A. Silling, E. Askari, J. Xu, Convergence, adaptive refinement, and scaling in 1D peridynamics. Int. J. Numer. Meth. Eng. **77**, 852–877 (2009)

X. Chen, M. Gunzburger, Continuous and discontinuous finite element methods for a peridynamics model of mechanics. Comput. Methods Appl. Mech. Eng. **200**, 1237–1250 (2011)
D. De Meo, N. Zhu, E. Oterkus, Peridynamic modeling of granular fracture in polycrystalline materials. J. Eng. Mater. Technol. **138**, 041008–041008–16 (2016)
Q. Du, X. Tian, Asymptotically compatible schemes for peridynamics based on numerical quadratures, in *Proceedings of the ASME 2014 International Mechanical Engineering Congress and Exposition*, Montreal, 2014. Paper No. IMECE2014-39620
W. Gerstle, N. Sau, S. Silling, Peridynamic modeling of concrete structures. Nucl. Eng. Des. **237**, 1250–1258 (2007)
M. Ghajari, L. Iannucci, P. Curtis, A peridynamic material model for the analysis of dynamic crack propagation in orthotropic media. Comput. Methods Appl. Mech. Eng. **276**, 431–452 (2014)
Y.D. Ha, F. Bobaru, Studies of dynamic crack propagation and crack branching with peridynamics. Int. J. Fract. **162**, 229–244 (2010)
Y.D. Ha, F. Bobaru, Characteristics of dynamic brittle fracture captured with peridynamics. Eng. Fract. Mech. **78**, 1156–1168 (2011)
W. Hu, Y. D. Ha, F. Bobaru, Numerical integration in peridynamics, Technical report, University of Nebraska-Lincoln, Sept. 2010
W. Hu, Y.D. Ha, F. Bobaru, Peridynamic model for dynamic fracture in unidirectional fiber-reinforced composites. Comput. Methods Appl. Mech. Eng. **217–220**, 247–261 (2012)
B. Kilic, A. Agwai, E. Madenci, Peridynamic theory for progressive damage prediction in center-cracked composite laminates. Compos. Struct. **90**, 141–151 (2009)
B. Kilic, E. Madenci, Prediction of crack paths in a quenched glass plate by using peridynamic theory. Int. J. Fract. **156**, 165–177 (2009)
D.J. Littlewood, A nonlocal approach to modeling crack nucleation in AA 7075-T651, in *Proceedings of the ASME 2011 International Mechanical Engineering Congress and Exposition*, Denver, 2011. Paper No. IMECE2011-64236
D.J. Littlewood, K. Mish, K. Pierson, Peridynamic simulation of damage evolution for structural health monitoring, in *Proceedings of the ASME 2012 International Mechanical Engineering Congress and Exposition*, Houston, 2012. Paper No. IMECE2012-86400
E. Oterkus, E. Madenci, Peridynamics for failure prediction in composites, in *53rd AIAA/ASME/ASCE/AHS/ASC Structures, Structural Dynamics and Materials Conference*, Honolulu, 2012
E. Oterkus, E. Madenci, O.Weckner, S. Silling, P. Bogert, A. Tessler, Combined finite element and peridynamic analyses for predicting failure in a stiffened composite curved panel with a central slot. Compos. Struct. **94**, 839–850 (2012)
M.L. Parks, R.B. Lehoucq, S.J. Plimpton, S.A. Silling, Implementing peridynamics within a molecular dynamics code. Comput. Phys. Commun. **179**, 777–783 (2008)
M.L. Parks, P. Seleson, S.J. Plimpton, S.A. Silling, R.B. Lehoucq, Peridynamics with LAMMPS: a user guide v0.3 beta, Report SAND2011-8523, Sandia National Laboratories, Albuquerque and Livermore, 2011
M.L. Parks, D.J. Littlewood, J.A. Mitchell, S.A. Silling, Peridigm users' guide v1.0.0, Report SAND2012-7800, Sandia National Laboratories, Albuquerque and Livermore, 2012
P.D. Seleson, *Peridynamic Multiscale Models for the Mechanics of Materials: Constitutive Relations, Upscaling from Atomistic Systems, and Interface Problems*, PhD thesis, Florida State University, 2010. Electronic Theses, Treatises and Dissertations. Paper 273
P. Seleson, Improved one-point quadrature algorithms for two-dimensional peridynamic models based on analytical calculations. Comput. Methods Appl. Mech. Eng. **282**, 184–217 (2014)
P. Seleson, D.J. Littlewood, Convergence studies in meshfree peridynamic simulations. Comput. Math. Appl. **71**, 2432–2448 (2016)
P. Seleson, M.L. Parks, On the role of the influence function in the peridynamic theory. Int. J. Multiscale Comput. Eng. **9**, 689–706 (2011)
S.A. Silling, Linearized theory of peridynamic states. J. Elast. **99**, 85–111 (2010)
S.A. Silling, E. Askari, A meshfree method based on the peridynamic model of solid mechanics. Comput. Struct. **83**, 1526–1535 (2005)

S.A. Silling, M. Epton, O. Weckner, J. Xu, E. Askari, Peridynamic states and constitutive modeling. J. Elast. **88**, 151–184 (2007)

S.A. Silling, O. Weckner, E. Askari, F. Bobaru, Crack nucleation in a peridynamic solid. Int. J. Fract. **162**, 219–227 (2010)

M.R. Tupek, J.J. Rimoli, R. Radovitzky, An approach for incorporating classical continuum damage models in state-based peridynamics. Comput. Methods Appl. Mech. Eng. **263**, 20–26 (2013)

O. Weckner, R. Abeyaratne, The effect of long-range forces on the dynamics of a bar. J. Mech. Phys. Solids **53**, 705–728 (2005)

O. Weckner, G. Brunk, M.A. Epton, S.A. Silling, E. Askari, Green's functions in non-local three-dimensional linear elasticity. Proc. R. Soc. A **465**, 3463–3487 (2009)

J. Xu, A. Askari, O. Weckner, S. Silling, Peridynamic analysis of impact damage in composite laminates. J. Aerosp. Eng. Spec. Issue Impact Mech. Compos. Mater. Aerosp. Appl. **21**, 187–194 (2008)

Well-Posed Nonlinear Nonlocal Fracture Models Associated with Double-Well Potentials

43

Prashant K. Jha and Robert Lipton

Contents

Abstract

In this chapter, we consider a generic class of bond-based nonlocal nonlinear potentials and formulate the evolution over suitable function spaces. The peridynamic potential considered in this work is a differentiable version of the original bond-based model introduced in Silling (J Mech Phys Solids 48(1):175–

P. K. Jha
Department of Mathematics, Louisiana State University, Baton Rouge, LA, USA
e-mail: prashant.j16o@gmail.com; jha@math.lsu.edu

R. Lipton (✉)
Department of Mathematics and Center for Computation and Technology, Louisiana State University, Baton Rouge, LA, USA
e-mail: lipton@lsu.edu

G. Z. Voyiadjis (ed.), *Handbook of Nonlocal Continuum Mechanics for Materials and Structures*, https://doi.org/10.1007/978-3-319-58729-5_40

209, 2000). The potential associated with the model has two wells where one well corresponds to linear elastic behavior and the other corresponds to brittle fracture (see Lipton (J Elast 117(1):21–50, 2014; 124(2):143–191, 2016)). The parameters in the potential can be directly related to the elastic tensor and fracture toughness. In this chapter we show that well-posed formulations of the model can be developed over different function spaces. Here we will consider formulations posed over Hölder spaces and Sobolev spaces. The motivation for the Hölder space formulation is to show a priori convergence for the discrete finite difference method. The motivation for the Sobolev formulation is to show a priori convergence for the finite element method. In the following chapter we will show that the discrete approximations converge to well-posed evolutions. The associated convergence rates are given explicitly in terms of time step and the size of the spatial mesh.

Keywords
Peridynamic modeling · Numerical analysis · Finite difference approximation · Finite element approximation · Stability · Convergence

Introduction

The peridynamic formulation, introduced in Silling (2000), is a nonlocal model for crack propagation in solids. The basic idea is to redefine the strain in terms of the difference quotients of the displacement field and allow for nonlocal forces acting over some finite horizon. This generalized notion of strain allows for the participation of larger class of deformations in the dynamics. The modeling introduces a natural length scale given by the size of the horizon. The force at any given material point is computed by considering the deformation of all neighboring material points contained within the horizon. For linear peridynamic formulations (Silling and Lehoucq 2008; Emmrich et al. 2013; Mengesha and Du 2014; Jha and Lipton 2017c), it is shown that as the nonlocal length scale goes to zero, the peridynamic model collapses to the elastic equilibrium and elastodynamics models. For the nonlinear model introduced in Silling (2000), one may consider a smooth version to find that the energy of the evolution recovers the energy of Linear Elastic Fracture Mechanics as the nonlocal length scale goes to zero (Lipton 2014, 2016). One of the important points of this model is the fact that as the size of the horizon goes to zero, i.e., when we tend to the local limit, the model behaves as if it is an elastic model away from the crack set. Therefore, in the limit, the model not only converges to linear elasticity in regions with small deformation but also has finite Griffith fracture energy associated with a sharp fracture set. The nonlinear potential can be calibrated so that it gives the same fracture toughness as the Linear Elastic Fracture Mechanics model. Further, the slope of the nonlinear force for small strain is specified precisely by the elastic constant of the material. These results are summarized in Lipton (2016) and Jha and Lipton (2017a). On the other hand to use this model for numerical simulation, we take advantage the regularization given by the nonlocal formulation of the problem. With this in mind we show existence of solutions in more regular spaces for fixed but small horizon and develop a theory for the numerical simulation of fracture problems. In this chapter we present the foundations for the theory and exhibit initial data and boundary conditions so that solutions exist in the Hölder space $C^{0,\gamma}$, $\gamma \in (0, 1]$ and the Sobolev space H^2 (see Jha and Lipton 2017a,b; Diehl et al. 2016). A numerical implementation scheme using the finite difference model is proposed and demonstrated in Lipton et al. (2016). In the following chapters, we show a priori convergence for the finite difference method and finite element method. These results are reported in Jha and Lipton (2017a,b). We show that these discrete approximations converge to the well-posed evolutions described in this chapter. The associated convergence rates are given explicitly in terms of time step and the size of the spatial mesh.

In this chapter we begin by describing bond-based peridynamics and the double-well potential model. Here the nonlocal forces acting between points are given by the derivatives of the potential with respect to the strain (see section "Problem Formulation with Bond-Based Nonlinear Potentials"). The existence of a peridynamic evolution taking values in the space of Hölder continuous functions is presented in section "Existence of Solutions in Hölder Space." The proof uses the Hölder continuity of the nonlocal force with respect to the Hölder norm (see section "Lipschitz Continuity in the Hölder Norm and Existence of a Hölder Continuous Solution"). We then show the existence of a peridynamic evolution in the set of essentially bounded functions taking values in the Sobolev space H^2, the space of functions with function values, and derivatives of order one and two that are square integrable (section "Existence of Solutions in the Sobolev Space H^2"). As before the proof uses the Hölder continuity of the nonlocal force, but now with respect to a norm that is the sum of the H^2 norm and the L^∞ norm, see section "Lipschitz Continuity in the H^2 Norm and Existence of an H^2 Solution." We conclude the chapter observing that the well-posed evolutions over these regular spaces converge to sharp fracture evolutions posed over spaces of functions with jumps (section "Conclusions: Convergence of Regular Solutions in the Limit of Vanishing Horizon").

Problem Formulation with Bond-Based Nonlinear Potentials

Let $D \subset \mathbb{R}^d$, for $d = 2, 3$ be the material domain with characteristic length scale of unity. Every material point $x \in D$ interacts nonlocally with all other material points inside a horizon of length $\epsilon \in (0, 1)$. Let $H_\epsilon(x)$ be the ball of radius ϵ centered at x containing all points y that interact with x. After deformation the material point x assumes position $z = x + u(x)$. In this treatment we assume infinitesimal displacements $u(x)$ so the deformed configuration is the same as the reference configuration and the linearized strain is given by

$$S = S(y, x; u) = \frac{u(y) - u(x)}{|y - x|} \cdot \frac{y - x}{|y - x|}.$$

We let t denote time and the displacement field $u(t, x)$ evolves according to Newton's second law

$$\rho \partial_{tt}^2 u(t, x) = -\nabla PD^\epsilon(u(t))(x) + b(t, x) \tag{1}$$

for all $x \in D$. Here the body force applied to the domain D can evolve with time and is denoted by $b(t, x)$. Without loss of generality, we will assume $\rho = 1$. The peridynamic force denoted by $-\nabla PD^\epsilon(u)(x)$ is given by summing up all forces acting on x

$$-\nabla PD^\epsilon(u)(x) = \frac{2}{\epsilon^d \omega_d} \int_{H_\epsilon(x)} \partial_S W^\epsilon(S, y - x) \frac{y - x}{|y - x|} dy,$$

where $\partial_S W^\epsilon$ is the force exerted on x by y and is given by the derivative of the nonlocal two-point potential $W^\epsilon(S, y - x)$ with respect to the strain and ω_d is volume of unit ball in dimension d.

Let ∂D be the boundary of material domain D. The Dirichlet boundary condition on u is

$$u(t, x) = 0 \qquad \forall x \in \partial D, \forall t \in [0, T] \tag{2}$$

and initial condition is

$$u(0, x) = u_0(x) \qquad \text{and} \qquad \partial_t u(0, x) = v_0(x). \tag{3}$$

The initial data and solution $u(t, x)$ are extended by 0 outside D.

The total energy $\mathcal{E}^\epsilon(u)(t)$ is given by the sum of kinetic and potential energy given by

$$\mathcal{E}^\epsilon(u)(t) = \frac{1}{2}||\dot{u}(t)||_{L^2} + PD^\epsilon(u(t)), \tag{4}$$

where potential energy PD^ϵ is given by

$$PD^\epsilon(u) = \frac{1}{2} \int_D \left[\frac{1}{\epsilon^d \omega_d} \int_{H_\epsilon(x)} W^\epsilon(S(u), y - x) dy \right] dx.$$

Nonlocal Potential

We consider potentials W^ϵ of the form

$$W^\epsilon(S, y - x) = \omega(x)\omega(y) \frac{J^\epsilon(|y - x|)}{\epsilon} f(|y - x| S^2), \tag{5}$$

where $f : \mathbb{R}^+ \to \mathbb{R}$ is assumed to be positive, smooth, and concave with the following properties

$$\lim_{r\to 0+} \frac{f(r)}{r} = f'(0), \qquad \lim_{r\to\infty} f(r) = f_\infty < \infty. \tag{6}$$

The peridynamic force $-\nabla PD^\epsilon$ is written as

$$\begin{aligned} &-\nabla PD^\epsilon(u)(x) \\ &= \frac{4}{\epsilon^{d+1}\omega_d}\int_{H_\epsilon(x)} \omega(x)\omega(y)J^\epsilon(|y-x|)f'(|y-x|S(u)^2)S(u)e_{y-x}dy, \end{aligned} \tag{7}$$

where we write $S(u) = S(y, x; u)$ and $e_{y-x} = \frac{y-x}{|y-x|}$.

The function $J^\epsilon(|y-x|)$ models the influence of separation between points y and x. Here $J^\epsilon(|y-x|) = J(|y-x|/\epsilon)$, and we define J to be zero outside the ball $\{\xi : |\xi| < 1\} = H_1(0)$ and $0 \le J(|\xi|) \le M$ for all $\xi \in H_1(0)$.

The boundary function $\omega(x)$ is nonnegative and takes the value 1 for points x inside D of distance ϵ away from the boundary ∂D. Inside the boundary layer of width ϵ, the function $\omega(x)$ smoothly decreases from 1 to 0 taking the value 0 on ∂D.

The potential described in Eq. 5 gives the convex-concave dependence (see Fig. 1) of $W(S, y-x)$ on the strain S for fixed $y-x$. Here the potential has a well at zero strain and has a second well at infinite strain given by the horizontal asymptote. Initially the deformation is elastic for small strains and then softens as the strain becomes large; this is illustrated in Fig. 2. The critical strain where the force between x and y begins to soften is given by $S_c(y, x) := \bar{r}/\sqrt{|y-x|}$, and the force decreases monotonically for

$$|S(y, x; u)| > S_c. \tag{8}$$

Here $\bar{r}$ is the inflection point of $r \mapsto f(r^2)$ and is the root of the following equation:

$$f'(r^2) + 2r^2 f''(r^2) = 0. \tag{9}$$

In (Theorem 5.2, Lipton 2016), it is shown that in the limit $\epsilon \to 0$, the peridynamic solution has bounded linear elastic fracture energy, provided the initial data (u_0, v_0) has bounded linear elastic fracture energy and u_0 is bounded. The elastic constants (Lamé constant λ and μ) and energy release rate of the limiting energy are given by

$$\lambda = \mu = C_d f'(0) M_d, \quad \mathcal{G}_c = \frac{2\omega_{d-1}}{\omega_d} f_\infty M_d$$

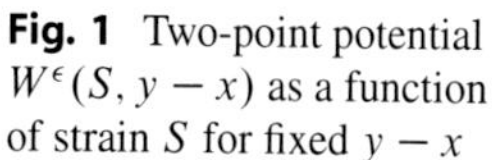

Fig. 1 Two-point potential $W^{\epsilon}(S, y - x)$ as a function of strain S for fixed $y - x$

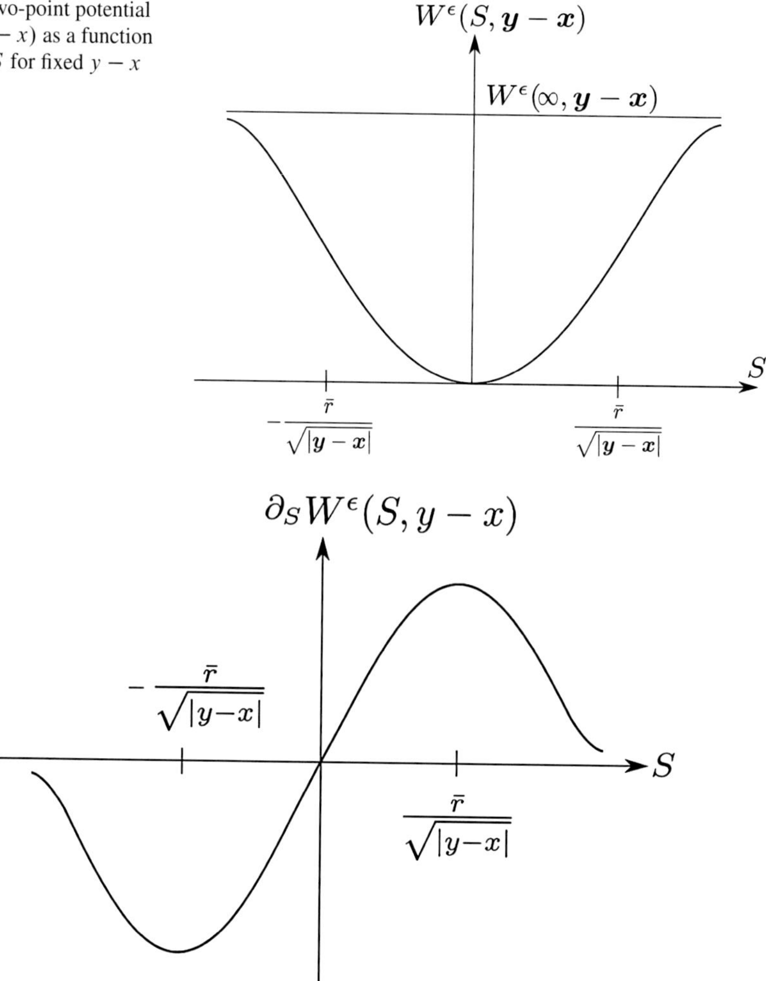

Fig. 2 Nonlocal force $\partial_S W^{\epsilon}(S, y - x)$ as a function of strain S for fixed $y - x$. Second derivative of $W^{\epsilon}(S, y - x)$ is zero at $\pm \bar{r}/\sqrt{|y - x|}$

where $M_d = \int_0^1 J(r) r^d dr$ and $f_\infty = \lim_{r \to \infty} f(r)$. $C_d = 2/3, 1/4, 1/5$ for $d = 1, 2, 3$, respectively, and $\omega_n = 1, 2, \pi, 4\pi/3$ for $n = 0, 1, 2, 3$. Therefore, $f'(0)$ and f_∞ are determined by the Lamé constant λ and fracture toughness $\mathcal{G}_c$.

Weak Formulation

We now give the weak formulation of the evolution. Multiplying Eq. 1 by a smooth test function $\tilde{u}$ with $\tilde{u} = 0$ on ∂D, we get

$$(\ddot{u}(t), \tilde{u}) = (-\nabla PD^\epsilon(u(t)), \tilde{u}) + (b(t), \tilde{u}).$$

We denote L^2 dot product of u, v as (u, v). An integration by parts easily shows for all smooth u, v taking zero boundary values that

$$(-\nabla PD^\epsilon(u), v) = -a^\epsilon(u, v),$$

where

$$a^\epsilon(u, v) = \frac{2}{\epsilon^{d+1}\omega_d} \int_D \int_{H_\epsilon(x)} \omega(x)\omega(y) J^\epsilon(|y - x|) f'(|y - x| S(u)^2) |y - x| S(u) S(v) dy dx. \tag{10}$$

Finally, the weak form of the evolution in terms of operator a^ϵ becomes

$$(\ddot{u}(t), \tilde{u}) + a^\epsilon(u(t), \tilde{u}) = (b(t), \tilde{u}). \tag{11}$$

Using definition of a^ϵ in Eq. 10, one easily sees that

$$\frac{d}{dt}\mathcal{E}^\epsilon(u)(t) = (\ddot{u}(t), \dot{u}(t)) + a^\epsilon(u(t), \dot{u}(t)). \tag{12}$$

In the sequel the notation $||\cdot||$ denotes the L^2 norm on D, and $||\cdot||_\infty$ is used for the L^∞ norm on D and $||\cdot||_2$ for Sobolev H^2 norm on D.

Existence of Solutions in Hölder Space

In this section, we establish the existence of solutions in Hölder space. Here we follow the approach developed in Jha and Lipton (2017a). Let $C^{0,\gamma}(D; \mathbb{R}^d)$ be the Hölder space with exponent $\gamma \in (0, 1]$. The closure of continuous functions with compact support on D in the supremum norm is denoted by $C_0(D)$. We identify functions in $C_0(D)$ with their unique continuous extensions to $\overline{D}$. It is easily seen that functions belonging to this space take the value zero on the boundary of D (see, e.g., Driver 2003). We introduce $C_0^{0,\gamma}(D) = C^{0,\gamma}(D) \cap C_0(D)$. Here we extend all functions in $C_0^{0,\gamma}(D)$ by zero outside D. The norm of $\boldsymbol{u} \in C_0^{0,\gamma}(D; \mathbb{R}^d)$ is taken to be

$$\|\boldsymbol{u}\|_{C^{0,\gamma}(D;\mathbb{R}^d)} := \sup_{\boldsymbol{x}\in D} |\boldsymbol{u}(\boldsymbol{x})| + [\boldsymbol{u}]_{C^{0,\gamma}(D;\mathbb{R}^d)},$$

where $[\boldsymbol{u}]_{C^{0,\gamma}(D;\mathbb{R}^d)}$ is the Hölder semi norm and given by

$$[\boldsymbol{u}]_{C^{0,\gamma}(D;\mathbb{R}^d)} := \sup_{\substack{x \neq y, \\ x,y \in D}} \frac{|\boldsymbol{u}(\boldsymbol{x}) - \boldsymbol{u}(\boldsymbol{y})|}{|\boldsymbol{x} - \boldsymbol{y}|^\gamma},$$

and $C_0^{0,\gamma}(D;\mathbb{R}^d)$ is a Banach space with this norm. Here we make the hypothesis that the domain function ω belongs to $C_0^{0,\gamma}(D;\mathbb{R}^d)$.

We write the evolution Eq. 1 as an equivalent first-order system with $y_1(t) = \boldsymbol{u}(t)$ and $y_2(t) = \boldsymbol{v}(t)$ with $\boldsymbol{v}(t) = \partial_t \boldsymbol{u}(t)$. Let $y = (y_1, y_2)^T$ where $y_1, y_2 \in C_0^{0,\gamma}(D;\mathbb{R}^d)$ and let $F^\epsilon(y,t) = (F_1^\epsilon(y,t), F_2^\epsilon(y,t))^T$ such that

$$F_1^\epsilon(y,t) := y_2 \tag{13}$$

$$F_2^\epsilon(y,t) := -\nabla PD^\epsilon(y_1) + \boldsymbol{b}(t). \tag{14}$$

The initial boundary value associated with the evolution Eq. 1 is equivalent to the initial boundary value problem for the first-order system given by

$$\frac{d}{dt} y = F^\epsilon(y,t), \tag{15}$$

with initial condition given by $y(0) = (\boldsymbol{u}_0, \boldsymbol{v}_0)^T \in C_0^{0,\gamma}(D;\mathbb{R}^d) \times C_0^{0,\gamma}(D;\mathbb{R}^d)$.

The function $F^\epsilon(y,t)$ satisfies the Lipschitz continuity given by the following theorem.

Proposition 1 (Lipschitz continuity and bound). *Let $X = C_0^{0,\gamma}(D;\mathbb{R}^d) \times C_0^{0,\gamma}(D;\mathbb{R}^d)$. The function $F^\epsilon(y,t) = (F_1^\epsilon, F_2^\epsilon)^T$, as defined in Eqs. 13 and 14, is Lipschitz continuous in any bounded subset of X. We have, for any $y, z \in X$ and $t > 0$,*

$$\|F^\epsilon(y,t) - F^\epsilon(z,t)\|_X \leq \frac{\left(L_1 + L_2\left(\|\omega\|_{C^{0,\gamma}(D)} + \|y\|_X + \|z\|_X\right)\right)}{\epsilon^{2+\alpha(\gamma)}} \|y - z\|_X \tag{16}$$

where L_1, L_2 are independent of $\boldsymbol{u}, \boldsymbol{v}$ and depend on peridynamic potential function f and influence function J and the exponent $\alpha(\gamma)$ is given by

$$\alpha(\gamma) = \begin{cases} 0 & \text{if } \gamma \geq 1/2 \\ 1/2 - \gamma & \text{if } \gamma < 1/2. \end{cases}$$

Furthermore for any $y \in X$ and any $t \in [0,T]$, we have the bound

$$\|F^\epsilon(y,t)\|_X \leq \frac{L_3}{\epsilon^{2+\alpha(\gamma)}}(1 + \|\omega\|_{C^{0,\gamma}(D)} + \|y\|_X) + b \tag{17}$$

where $b = \sup_t \|\boldsymbol{b}(t)\|_{C^{0,\gamma}(D;\mathbb{R}^d)}$ *and* L_3 *is independent of* y.

We easily see that on choosing $z = 0$ in Eq. 16 that $-\nabla PD^\epsilon(\boldsymbol{u})(\boldsymbol{x})$ is in $C^{0,\gamma}(D;\mathbb{R}^3)$ provided that $\boldsymbol{u}$ belongs to $C^{0,\gamma}(D;\mathbb{R}^3)$. Since $-\nabla PD^\epsilon(\boldsymbol{u})(\boldsymbol{x})$ takes the value 0 on ∂D, we conclude that $-\nabla PD^\epsilon(\boldsymbol{u})(\boldsymbol{x})$ belongs to $C_0^{0,\gamma}(D;\mathbb{R}^3)$.

The following theorem gives the existence and uniqueness of solution in any given time domain $I_0 = (-T, T)$.

Theorem 1 (Existence and uniqueness of Hölder solutions of cohesive dynamics over finite time intervals). *For any initial condition* $x_0 \in X = C_0^{0,\gamma}(D;\mathbb{R}^d) \times C_0^{0,\gamma}(D;\mathbb{R}^d)$, *time interval* $I_0 = (-T, T)$, *and right-hand side* $\boldsymbol{b}(t)$ *continuous in time for* $t \in I_0$ *such that* $\boldsymbol{b}(t)$ *satisfies* $\sup_{t\in I_0} ||\boldsymbol{b}(t)||_{C^{0,\gamma}} < \infty$, *there is a unique solution* $y(t) \in C^1(I_0; X)$ *of*

$$y(t) = x_0 + \int_0^t F^\epsilon(y(\tau), \tau)\, d\tau,$$

or equivalently

$$y'(t) = F^\epsilon(y(t), t), \textit{with } y(0) = x_0,$$

where $y(t)$ *and* $y'(t)$ *are Lipschitz continuous in time for* $t \in I_0$.

The proof of this theorem is given in the following section.

Lipschitz Continuity in the Hölder Norm and Existence of a Hölder Continuous Solution

In this section, we prove Proposition 1.

Proof of Proposition 1

Let $I = [0, T]$ be the time domain and $X = C_0^{0,\gamma}(D;\mathbb{R}^d) \times C_0^{0,\gamma}(D;\mathbb{R}^d)$. Recall that $F^\epsilon(y,t) = (F_1^\epsilon(y,t), F_2^\epsilon(y,t))$, where $F_1^\epsilon(y,t) = y^2$ and $F_2^\epsilon(y,t) = -\nabla PD^\epsilon(y^1) + \boldsymbol{b}(t)$. Given $t \in I$ and $y = (y^1, y^2), z = (z^1, z^2) \in X$, we have

$$\begin{aligned}
&\|F^\epsilon(y,t) - F^\epsilon(z,t)\|_X \\
&\leq \left\|y^2 - z^2\right\|_{C^{0,\gamma}(D;\mathbb{R}^d)} + \left\|-\nabla PD^\epsilon(y^1) + \nabla PD^\epsilon(z^1)\right\|_{C^{0,\gamma}(D;\mathbb{R}^d)}. \qquad (18)
\end{aligned}$$

Therefore, to prove the Eq. 16, we only need to analyze the second term in above inequality. Let $\boldsymbol{u}, \boldsymbol{v} \in C_0^{0,\gamma}(D;\mathbb{R}^d)$, then we have

$$
\begin{aligned}
&\|-\nabla PD^{\epsilon}(\boldsymbol{u}) - (-\nabla PD^{\epsilon}(\boldsymbol{v}))\|_{C^{0,\gamma}(D;\mathbb{R}^d)} \\
&= \sup_{\boldsymbol{x}\in D} |-\nabla PD^{\epsilon}(\boldsymbol{u})(\boldsymbol{x}) - (-\nabla PD^{\epsilon}(\boldsymbol{v})(\boldsymbol{x}))| \\
&\quad + \sup_{\substack{\boldsymbol{x}\neq \boldsymbol{y},\\ \boldsymbol{x},\boldsymbol{y}\in D}} \frac{|(-\nabla PD^{\epsilon}(\boldsymbol{u}) + \nabla PD^{\epsilon}(\boldsymbol{v}))(\boldsymbol{x}) - (-\nabla PD^{\epsilon}(\boldsymbol{u}) + \nabla PD^{\epsilon}(\boldsymbol{v}))(\boldsymbol{y})|}{|\boldsymbol{x}-\boldsymbol{y}|^{\gamma}}.
\end{aligned}
\tag{19}
$$

Note that the force $-\nabla PD^{\epsilon}(\boldsymbol{u})(\boldsymbol{x})$ can be written as follows:

$$
\begin{aligned}
&-\nabla PD^{\epsilon}(\boldsymbol{u})(\boldsymbol{x}) \\
&= \frac{4}{\epsilon^{d+1}\omega_d} \int_{H_{\epsilon}(\boldsymbol{x})} \omega(\boldsymbol{x})\omega(\boldsymbol{y}) J\left(\frac{|\boldsymbol{y}-\boldsymbol{x}|}{\epsilon}\right) f'(|\boldsymbol{y}-\boldsymbol{x}|\, S(\boldsymbol{y},\boldsymbol{x};\boldsymbol{u})^2) S(\boldsymbol{y},\boldsymbol{x};\boldsymbol{u}) \frac{\boldsymbol{y}-\boldsymbol{x}}{|\boldsymbol{y}-\boldsymbol{x}|} d\boldsymbol{y} \\
&= \frac{4}{\epsilon\omega_d} \int_{H_1(\boldsymbol{0})} \omega(\boldsymbol{x})\omega(\boldsymbol{x}+\epsilon\boldsymbol{\xi}) J(|\boldsymbol{\xi}|) f'(\epsilon\,|\boldsymbol{\xi}|\, S(\boldsymbol{x}+\epsilon\boldsymbol{\xi},\boldsymbol{x};\boldsymbol{u})^2) S(\boldsymbol{x}+\epsilon\boldsymbol{\xi},\boldsymbol{x};\boldsymbol{u}) \frac{\boldsymbol{\xi}}{|\boldsymbol{\xi}|} d\boldsymbol{\xi}.
\end{aligned}
$$

where we substituted $\partial_S W^{\epsilon}$ using Eq. 5. In the second step, we introduced the change in variable $\boldsymbol{y} = \boldsymbol{x} + \epsilon\boldsymbol{\xi}$.

Let $F_1 : \mathbb{R} \to \mathbb{R}$ be defined as $F_1(S) = f(S^2)$. Then $F_1'(S) = f'(S^2)2S$. Using the definition of F_1, we have

$$
2Sf'(\epsilon\,|\boldsymbol{\xi}|\, S^2) = \frac{F_1'(\sqrt{\epsilon\,|\boldsymbol{\xi}|}S)}{\sqrt{\epsilon\,|\boldsymbol{\xi}|}}.
$$

Because f is assumed to be positive, smooth, and concave and is bounded far away, we have the following bound on derivatives of F_1

$$
\sup_r |F_1'(r)| = F_1'(\bar{r}) =: C_1 \tag{20}
$$

$$
\sup_r |F_1''(r)| = \max\{F_1''(0), F_1''(\hat{u})\} =: C_2 \tag{21}
$$

$$
\sup_r |F_1'''(r)| = \max\{F_1'''(\bar{u}_2), F_1'''(\tilde{u}_2)\} =: C_3. \tag{22}
$$

where $\bar{r}$ is the inflection point of $f(r^2)$, i.e., $F_1''(\bar{r}) = 0$. $\{0, \hat{u}\}$ are the maxima of $F_1''(r)$. $\{\bar{u}, \tilde{u}\}$ are the maxima of $F_1'''(r)$. By chain rule and by considering the assumption on f, we can show that $\bar{r}, \hat{u}, \bar{u}_2, \tilde{u}_2$ exists and the C_1, C_2, C_3 are bounded. Figures 3, 4, and 5 show the generic graphs of $F_1'(r)$, $F_1''(r)$, and $F_1'''(r)$, respectively.

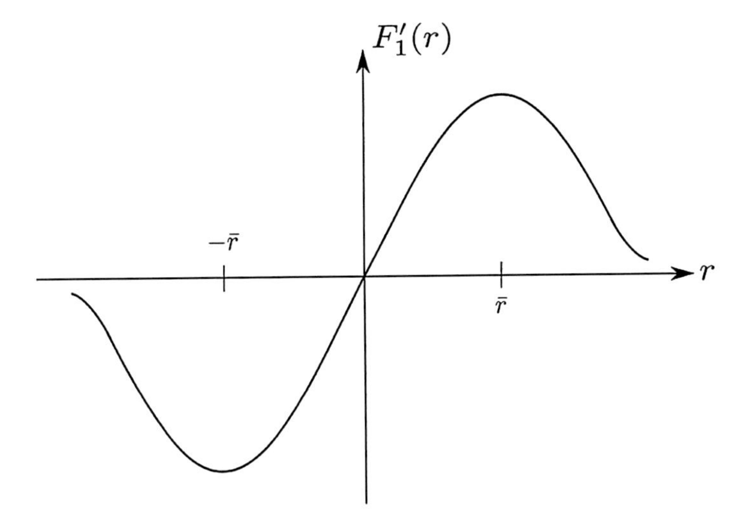

Fig. 3 Generic plot of $F_1'(r)$. $|F_1'(r)|$ is bounded by $\left|F_1'(\bar{r})\right|$

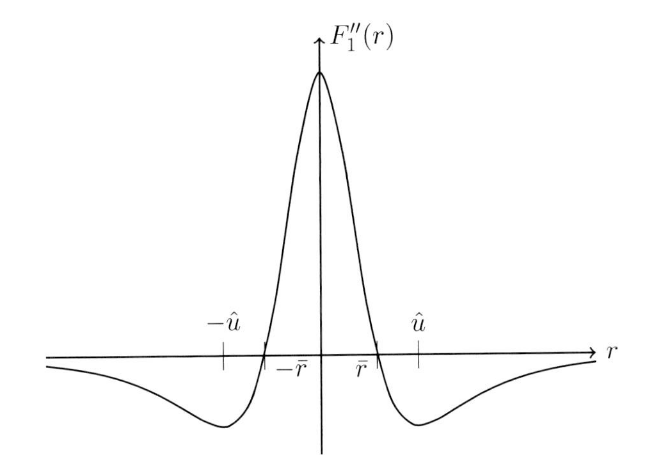

Fig. 4 Generic plot of $F_1''(r)$. At $\pm\bar{r}$, $F_1''(r) = 0$. At $\pm\hat{u}$, $F_1'''(r) = 0$

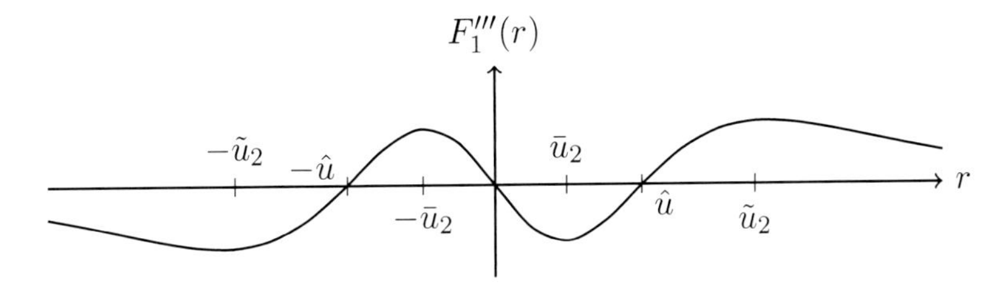

Fig. 5 Generic plot of $F_1'''(r)$. At $\pm\bar{u}_2$ and $\pm\tilde{u}_2$, $F_1'''' = 0$

The nonlocal force $-\nabla PD^\epsilon$ can be written as

$$-\nabla PD^\epsilon(\boldsymbol{u})(\boldsymbol{x}) = \frac{2}{\epsilon\omega_d}\int_{H_1(\boldsymbol{0})} \omega(\boldsymbol{x})\omega(\boldsymbol{x}+\epsilon\boldsymbol{\xi})J(|\boldsymbol{\xi}|)F_1'(\sqrt{\epsilon|\boldsymbol{\xi}|}S(\boldsymbol{x}+\epsilon\boldsymbol{\xi},\boldsymbol{x};\boldsymbol{u}))\frac{1}{\sqrt{\epsilon|\boldsymbol{\xi}|}}\frac{\boldsymbol{\xi}}{|\boldsymbol{\xi}|}d\boldsymbol{\xi}. \tag{23}$$

To simplify the calculations, we use following notation:

$$\bar{\boldsymbol{u}}(\boldsymbol{x}) := \boldsymbol{u}(\boldsymbol{x}+\epsilon\boldsymbol{\xi}) - \boldsymbol{u}(\boldsymbol{x}),$$
$$\bar{\boldsymbol{u}}(\boldsymbol{y}) := \boldsymbol{u}(\boldsymbol{y}+\epsilon\boldsymbol{\xi}) - \boldsymbol{u}(\boldsymbol{y}),$$
$$(\boldsymbol{u}-\boldsymbol{v})(\boldsymbol{x}) := \boldsymbol{u}(\boldsymbol{x}) - \boldsymbol{v}(\boldsymbol{x}),$$

and $\overline{(\boldsymbol{u}-\boldsymbol{v})}(\boldsymbol{x})$ is defined similar to $\bar{\boldsymbol{u}}(\boldsymbol{x})$. Also, let

$$s = \epsilon|\boldsymbol{\xi}|, \quad \boldsymbol{e} = \frac{\boldsymbol{\xi}}{|\boldsymbol{\xi}|}.$$

In what follows, we will come across the integral of type $\int_{H_1(\boldsymbol{0})} J(|\boldsymbol{\xi}|)|\boldsymbol{\xi}|^{-\alpha}d\boldsymbol{\xi}$. Recall that $0 \leq J(|\boldsymbol{\xi}|) \leq M$ for all $\boldsymbol{\xi} \in H_1(\boldsymbol{0})$ and $J(|\boldsymbol{\xi}|) = 0$ for $\boldsymbol{\xi} \notin H_1(\boldsymbol{0})$. Therefore, let

$$\bar{J}_\alpha := \frac{1}{\omega_d}\int_{H_1(\boldsymbol{0})} J(|\boldsymbol{\xi}|)|\boldsymbol{\xi}|^{-\alpha}d\boldsymbol{\xi}. \tag{24}$$

With notations above, we note that $S(\boldsymbol{x}+\epsilon\boldsymbol{\xi},\boldsymbol{x};\boldsymbol{u}) = \bar{\boldsymbol{u}}(\boldsymbol{x})\cdot\boldsymbol{e}/s$. $-\nabla PD^\epsilon$ can be written as

$$-\nabla PD^\epsilon(\boldsymbol{u})(\boldsymbol{x}) = \frac{2}{\epsilon\omega_d}\int_{H_1(\boldsymbol{0})} \omega(\boldsymbol{x})\omega(\boldsymbol{x}+\epsilon\boldsymbol{\xi})J(|\boldsymbol{\xi}|)F_1'(\bar{\boldsymbol{u}}(\boldsymbol{x})\cdot\boldsymbol{e}/\sqrt{s})\frac{1}{\sqrt{s}}\boldsymbol{e}\,d\boldsymbol{\xi}. \tag{25}$$

We first estimate the term $|-\nabla PD^\epsilon(\boldsymbol{u})(\boldsymbol{x}) - (-\nabla PD^\epsilon(\boldsymbol{v})(\boldsymbol{x}))|$ in Eq. 19.

$$\begin{aligned}
&|-\nabla PD^\epsilon(\boldsymbol{u})(\boldsymbol{x}) - (-\nabla PD^\epsilon(\boldsymbol{v})(\boldsymbol{x}))| \\
&\leq \left|\frac{2}{\epsilon\omega_d}\int_{H_1(\boldsymbol{0})} \omega(\boldsymbol{x})\omega(\boldsymbol{x}+\epsilon\boldsymbol{\xi})J(|\boldsymbol{\xi}|)\frac{\left(F_1'(\bar{\boldsymbol{u}}(\boldsymbol{x})\cdot\boldsymbol{e}/\sqrt{s}) - F_1'(\bar{\boldsymbol{v}}(\boldsymbol{x})\cdot\boldsymbol{e}/\sqrt{s})\right)}{\sqrt{s}}\boldsymbol{e}\,d\boldsymbol{\xi}\right| \\
&\leq \left|\frac{2}{\epsilon\omega_d}\int_{H_1(\boldsymbol{0})} J(|\boldsymbol{\xi}|)\frac{1}{\sqrt{s}}\left|F_1'(\bar{\boldsymbol{u}}(\boldsymbol{x})\cdot\boldsymbol{e}/\sqrt{s}) - F_1'(\bar{\boldsymbol{v}}(\boldsymbol{x})\cdot\boldsymbol{e}/\sqrt{s})\right|d\boldsymbol{\xi}\right|
\end{aligned}$$

$$\leq \sup_r \left|F_1''(r)\right| \left|\frac{2}{\epsilon\omega_d}\int_{H_1(\mathbf{0})} J(|\boldsymbol{\xi}|)\frac{1}{\sqrt{s}}\left|\bar{\boldsymbol{u}}(\boldsymbol{x})\cdot\boldsymbol{e}/\sqrt{s} - \bar{\boldsymbol{v}}(\boldsymbol{x})\cdot\boldsymbol{e}/\sqrt{s}\right| d\boldsymbol{\xi}\right|$$
$$\leq \frac{2C_2}{\epsilon\omega_d}\left|\int_{H_1(\mathbf{0})} J(|\boldsymbol{\xi}|)\frac{|\bar{\boldsymbol{u}}(\boldsymbol{x}) - \bar{\boldsymbol{v}}(\boldsymbol{x})|}{\epsilon|\boldsymbol{\xi}|}d\boldsymbol{\xi}\right|. \tag{26}$$

Here we have used the fact that $|\omega(\boldsymbol{x})| \leq 1$ and for a vector $\boldsymbol{e}$ such that $|\boldsymbol{e}| = 1$, $|\boldsymbol{a}\cdot\boldsymbol{e}| \leq |\boldsymbol{a}|$ holds and $|\alpha\boldsymbol{e}| \leq |\alpha|$ holds for all $\boldsymbol{a} \in \mathbb{R}^d, \alpha \in \mathbb{R}$. Using the fact that $\boldsymbol{u}, \boldsymbol{v} \in C_0^{0,\gamma}(D;\mathbb{R}^d)$, we have

$$\frac{|\bar{\boldsymbol{u}}(\boldsymbol{x}) - \bar{\boldsymbol{v}}(\boldsymbol{x})|}{s} = \frac{|(\boldsymbol{u}-\boldsymbol{v})(\boldsymbol{x}+\epsilon\boldsymbol{\xi}) - (\boldsymbol{u}-\boldsymbol{v})(\boldsymbol{x})|}{(\epsilon|\boldsymbol{\xi}|)^\gamma}\frac{1}{(\epsilon|\boldsymbol{\xi}|)^{1-\gamma}}$$
$$\leq \|\boldsymbol{u}-\boldsymbol{v}\|_{C^{0,\gamma}(D;\mathbb{R}^d)}\frac{1}{(\epsilon|\boldsymbol{\xi}|)^{1-\gamma}}.$$

Substituting the estimate given above, we get

$$|-\nabla PD^\epsilon(\boldsymbol{u})(\boldsymbol{x}) - (-\nabla PD^\epsilon(\boldsymbol{v})(\boldsymbol{x}))| \leq \frac{2C_2\bar{J}_{1-\gamma}}{\epsilon^{2-\gamma}}\|\boldsymbol{u}-\boldsymbol{v}\|_{C^{0,\gamma}(D;\mathbb{R}^d)}, \tag{27}$$

where C_2 is given by Eq. 21 and $\bar{J}_{1-\gamma}$ is given by Eq. 24.

We now estimate the second term in Eq. 19. To simplify notation, we write $\tilde{\omega}(\boldsymbol{x},\boldsymbol{\xi}) = \omega(\boldsymbol{x})\omega(\boldsymbol{x}+\epsilon\boldsymbol{\xi})$ and with the help of Eq. 25, we get

$$\frac{1}{|\boldsymbol{x}-\boldsymbol{y}|^\gamma}|(-\nabla PD^\epsilon(\boldsymbol{u}) + \nabla PD^\epsilon(\boldsymbol{v}))(\boldsymbol{x}) - (-\nabla PD^\epsilon(\boldsymbol{u}) + \nabla PD^\epsilon(\boldsymbol{v}))(\boldsymbol{y})|$$
$$= \frac{1}{|\boldsymbol{x}-\boldsymbol{y}|^\gamma}|\frac{2}{\epsilon\omega_d}\int_{H_1(0)} J(|\boldsymbol{\xi}|)\frac{1}{\sqrt{s}}\times\Big(\tilde{\omega}(\boldsymbol{x},\boldsymbol{\xi})(F_1'(\frac{\bar{\boldsymbol{u}}(\boldsymbol{x})\cdot\boldsymbol{e}}{\sqrt{s}}) - F_1'(\frac{\bar{\boldsymbol{v}}(\boldsymbol{x}))\cdot\boldsymbol{e}}{\sqrt{s}}))$$
$$-\tilde{\omega}(\boldsymbol{y},\boldsymbol{\xi})(F_1'(\frac{\bar{\boldsymbol{u}}(\boldsymbol{y})\cdot\boldsymbol{e}}{\sqrt{s}}) - F_1'(\frac{\bar{\boldsymbol{v}}(\boldsymbol{y}))\cdot\boldsymbol{e}}{\sqrt{s}})\Big)\boldsymbol{e}\,d\boldsymbol{\xi}|$$
$$\leq \frac{1}{|\boldsymbol{x}-\boldsymbol{y}|^\gamma}|\frac{2}{\epsilon\omega_d}\int_{H_1(0)} J(|\boldsymbol{\xi}|)\frac{1}{\sqrt{s}}\times$$
$$|\tilde{\omega}(\boldsymbol{x},\boldsymbol{\xi})(F_1'(\frac{\bar{\boldsymbol{u}}(\boldsymbol{x})\cdot\boldsymbol{e}}{\sqrt{s}}) - F_1'(\frac{\bar{\boldsymbol{v}}(\boldsymbol{x})\cdot\boldsymbol{e}}{\sqrt{s}})) - \tilde{\omega}(\boldsymbol{y},\boldsymbol{\xi})(F_1'(\frac{\bar{\boldsymbol{u}}(\boldsymbol{y})\cdot\boldsymbol{e}}{\sqrt{s}})$$
$$- F_1'(\frac{\bar{\boldsymbol{v}}(\boldsymbol{y})\cdot\boldsymbol{e}}{\sqrt{s}}))|d\boldsymbol{\xi}. \tag{28}$$

We analyze the integrand in above equation. We let H be defined by

$$H := \frac{|\tilde{\omega}(\boldsymbol{x},\boldsymbol{\xi})(F_1'(\frac{\bar{\boldsymbol{u}}(\boldsymbol{x})\cdot\boldsymbol{e}}{\sqrt{s}}) - F_1'(\frac{\bar{\boldsymbol{v}}(\boldsymbol{x})\cdot\boldsymbol{e}}{\sqrt{s}})) - \tilde{\omega}(\boldsymbol{y},\boldsymbol{\xi})(F_1'(\frac{\bar{\boldsymbol{u}}(\boldsymbol{y})\cdot\boldsymbol{e}}{\sqrt{s}}) - F_1'(\frac{\bar{\boldsymbol{v}}(\boldsymbol{y})\cdot\boldsymbol{e}}{\sqrt{s}}))|}{|\boldsymbol{x}-\boldsymbol{y}|^\gamma}.$$

Let $\boldsymbol{r} : [0,1] \times D \to \mathbb{R}^d$ be defined as

$$\boldsymbol{r}(l,\boldsymbol{x}) = \bar{\boldsymbol{v}}(\boldsymbol{x}) + l(\bar{\boldsymbol{u}}(\boldsymbol{x}) - \bar{\boldsymbol{v}}(\boldsymbol{x})).$$

Note $\partial \boldsymbol{r}(l,\boldsymbol{x})/\partial l = \bar{\boldsymbol{u}}(\boldsymbol{x}) - \bar{\boldsymbol{v}}(\boldsymbol{x})$. Using $\boldsymbol{r}(l,\boldsymbol{x})$, we have

$$F_1'(\bar{\boldsymbol{u}}(\boldsymbol{x})\cdot \boldsymbol{e}/\sqrt{s}) - F_1'(\bar{\boldsymbol{v}}(\boldsymbol{x})\cdot \boldsymbol{e}/\sqrt{s}) = \int_0^1 \frac{\partial F_1'(\boldsymbol{r}(l,\boldsymbol{x})\cdot \boldsymbol{e}/\sqrt{s})}{\partial l} dl \tag{29}$$

$$= \int_0^1 \frac{\partial F_1'(\boldsymbol{r}\cdot \boldsymbol{e}/\sqrt{s})}{\partial \boldsymbol{r}}|_{\boldsymbol{r}=\boldsymbol{r}(l,\boldsymbol{x})} \cdot \frac{\partial \boldsymbol{r}(l,\boldsymbol{x})}{\partial l} dl. \tag{30}$$

Similarly, we have

$$F_1'(\bar{\boldsymbol{u}}(\boldsymbol{y})\cdot \boldsymbol{e}/\sqrt{s}) - F_1'(\bar{\boldsymbol{v}}(\boldsymbol{y})\cdot \boldsymbol{e}/\sqrt{s}) = \int_0^1 \frac{\partial F_1'(\boldsymbol{r}\cdot \boldsymbol{e}/\sqrt{s})}{\partial \boldsymbol{r}}|_{\boldsymbol{r}=\boldsymbol{r}(l,\boldsymbol{y})} \cdot \frac{\partial \boldsymbol{r}(l,\boldsymbol{y})}{\partial l} dl. \tag{31}$$

Note that

$$\frac{\partial F_1'(\boldsymbol{r}\cdot \boldsymbol{e}/\sqrt{s})}{\partial \boldsymbol{r}}|_{\boldsymbol{r}=\boldsymbol{r}(l,\boldsymbol{y})} = F_1''(\boldsymbol{r}(l,\boldsymbol{x})\cdot \boldsymbol{e}/\sqrt{s})\frac{\boldsymbol{e}}{\sqrt{s}}. \tag{32}$$

Combining Eqs. 30, 31, and 32 gives

$$\begin{aligned} H &= \frac{1}{|\boldsymbol{x}-\boldsymbol{y}|^\gamma}\Big| \int_0^1 \big(\tilde{\omega}(\boldsymbol{x},\boldsymbol{\xi})F_1''(\boldsymbol{r}(l,\boldsymbol{x})\cdot \boldsymbol{e}/\sqrt{s})(\bar{\boldsymbol{u}}(\boldsymbol{x}) - \bar{\boldsymbol{v}}(\boldsymbol{x})) \\ &\quad -\tilde{\omega}(\boldsymbol{y},\boldsymbol{\xi})F_1''(\boldsymbol{r}(l,\boldsymbol{y})\cdot \boldsymbol{e}/\sqrt{s})(\bar{\boldsymbol{u}}(\boldsymbol{y}) - \bar{\boldsymbol{v}}(\boldsymbol{y}))\big)\cdot \frac{\boldsymbol{e}}{\sqrt{s}} dl\Big| \\ &\leq \frac{1}{|\boldsymbol{x}-\boldsymbol{y}|^\gamma}\frac{1}{\sqrt{s}}\Big| \int_0^1 |\tilde{\omega}(\boldsymbol{x},\boldsymbol{\xi})F_1''(\boldsymbol{r}(l,\boldsymbol{x})\cdot \boldsymbol{e}/\sqrt{s})(\bar{\boldsymbol{u}}(\boldsymbol{x}) - \bar{\boldsymbol{v}}(\boldsymbol{x})) \\ &\quad -\tilde{\omega}(\boldsymbol{y},\boldsymbol{\xi})F_1''(\boldsymbol{r}(l,\boldsymbol{y})\cdot \boldsymbol{e}/\sqrt{s})(\bar{\boldsymbol{u}}(\boldsymbol{y}) - \bar{\boldsymbol{v}}(\boldsymbol{y}))| dl\Big|. \end{aligned}$$

Adding and substracting $\tilde{\omega}(\boldsymbol{x},\boldsymbol{\xi})F_1''(\boldsymbol{r}(l,\boldsymbol{x})\cdot \boldsymbol{e}/\sqrt{s})(\bar{\boldsymbol{u}}(\boldsymbol{y}) - \bar{\boldsymbol{v}}(\boldsymbol{y}))$ and noting $0 \leq \tilde{\omega}(\boldsymbol{x},\boldsymbol{\xi}) \leq 1$ give

$$\begin{aligned} H &\leq \frac{1}{|\boldsymbol{x}-\boldsymbol{y}|^\gamma}\frac{1}{\sqrt{s}}\Big| \int_0^1 |F_1''(\boldsymbol{r}(l,\boldsymbol{x})\cdot \boldsymbol{e}/\sqrt{s})|\, |\bar{\boldsymbol{u}}(\boldsymbol{x}) - \bar{\boldsymbol{v}}(\boldsymbol{x}) - \bar{\boldsymbol{u}}(\boldsymbol{y}) + \bar{\boldsymbol{v}}(\boldsymbol{y})|\, dl\Big| \\ &\quad + \frac{1}{|\boldsymbol{x}-\boldsymbol{y}|^\gamma}\frac{1}{\sqrt{s}}\int_0^1 |(\tilde{\omega}(\boldsymbol{x},\boldsymbol{\xi})F_1''(\boldsymbol{r}(l,\boldsymbol{x})\cdot \boldsymbol{e}/\sqrt{s}) - \tilde{\omega}(\boldsymbol{y},\boldsymbol{\xi})F_1''(\boldsymbol{r}(l,\boldsymbol{y})\cdot \boldsymbol{e}/\sqrt{s}))| \\ &\qquad \times |\bar{\boldsymbol{u}}(\boldsymbol{y}) - \bar{\boldsymbol{v}}(\boldsymbol{y})|\, dl. \\ &=: H_1 + H_2. \end{aligned}$$

The H_1 term is estimated first. Note that $|F_1''(r)| \leq C_2$. Since $\boldsymbol{u}, \boldsymbol{v} \in C_0^{0,\gamma}(D;\mathbb{R}^d)$, it is easily seen that

$$\frac{|\bar{\boldsymbol{u}}(\boldsymbol{x}) - \bar{\boldsymbol{v}}(\boldsymbol{x}) - \bar{\boldsymbol{u}}(\boldsymbol{y}) + \bar{\boldsymbol{v}}(\boldsymbol{y})|}{|\boldsymbol{x} - \boldsymbol{y}|^\gamma} \leq 2\|\boldsymbol{u} - \boldsymbol{v}\|_{C^{0,\gamma}(D;\mathbb{R}^d)}.$$

Therefore, we have

$$H_1 \leq \frac{2C_2}{\sqrt{s}} \|\boldsymbol{u} - \boldsymbol{v}\|_{C^{0,\gamma}(D;\mathbb{R}^d)}. \tag{33}$$

We now estimate H_2. We add and subtract $\tilde{\omega}(\boldsymbol{x}, \boldsymbol{\xi}) F_1''(\boldsymbol{r}(l, \boldsymbol{y}) \cdot \boldsymbol{e}/\sqrt{s}))$ in H_2 to get

$$H_2 \leq H_3 + H_4,$$

where

$$H_3 = \frac{1}{|\boldsymbol{x}-\boldsymbol{y}|^\gamma} \frac{1}{\sqrt{s}} \int_0^1 |(F_1''(\boldsymbol{r}(l, \boldsymbol{x}) \cdot \boldsymbol{e}/\sqrt{s}) - F_1''(\boldsymbol{r}(l, \boldsymbol{y}) \cdot \boldsymbol{e}/\sqrt{s}))| \, |\bar{\boldsymbol{u}}(\boldsymbol{y}) - \bar{\boldsymbol{v}}(\boldsymbol{y})| \, dl,$$

and

$$H_4 = \frac{1}{|\boldsymbol{x} - \boldsymbol{y}|^\gamma} \frac{1}{\sqrt{s}} \int_0^1 |(\tilde{\omega}(\boldsymbol{x}, \boldsymbol{\xi}) - \tilde{\omega}(\boldsymbol{y}, \boldsymbol{\xi})| F_1''(\boldsymbol{r}(l, \boldsymbol{y}) \cdot \boldsymbol{e}/\sqrt{s}))| \, |\bar{\boldsymbol{u}}(\boldsymbol{y}) - \bar{\boldsymbol{v}}(\boldsymbol{y})| \, dl.$$

Now we estimate H_3. Since $|F_1'''(r)| \leq C_3$ (see Eq. 22), we have

$$\begin{aligned}
&\frac{1}{|\boldsymbol{x} - \boldsymbol{y}|^\gamma} |F_1''(\boldsymbol{r}(l, \boldsymbol{x}) \cdot \boldsymbol{e}/\sqrt{s}) - F_1''(\boldsymbol{r}(l, \boldsymbol{y}) \cdot \boldsymbol{e}/\sqrt{s})| \\
&\leq \frac{1}{|\boldsymbol{x} - \boldsymbol{y}|^\gamma} \sup_r \left| F'''(r) \right| \frac{|\boldsymbol{r}(l, \boldsymbol{x}) \cdot \boldsymbol{e} - \boldsymbol{r}(l, \boldsymbol{y}) \cdot \boldsymbol{e}|}{\sqrt{s}} \\
&\leq \frac{C_3}{\sqrt{s}} \frac{|\boldsymbol{r}(l, \boldsymbol{x}) - \boldsymbol{r}(l, \boldsymbol{y})|}{|\boldsymbol{x} - \boldsymbol{y}|^\gamma} \\
&= \frac{C_3}{\sqrt{s}} \left(\frac{|1 - l| \, |\bar{\boldsymbol{v}}(\boldsymbol{x}) - \bar{\boldsymbol{v}}(\boldsymbol{y})|}{|\boldsymbol{x} - \boldsymbol{y}|^\gamma} + \frac{|l| \, |\bar{\boldsymbol{u}}(\boldsymbol{x}) - \bar{\boldsymbol{u}}(\boldsymbol{y})|}{|\boldsymbol{x} - \boldsymbol{y}|^\gamma} \right) \\
&\leq \frac{C_3}{\sqrt{s}} \left(\frac{|\bar{\boldsymbol{v}}(\boldsymbol{x}) - \bar{\boldsymbol{v}}(\boldsymbol{y})|}{|\boldsymbol{x} - \boldsymbol{y}|^\gamma} + \frac{|\bar{\boldsymbol{u}}(\boldsymbol{x}) - \bar{\boldsymbol{u}}(\boldsymbol{y})|}{|\boldsymbol{x} - \boldsymbol{y}|^\gamma} \right).
\end{aligned} \tag{34}$$

Where we have used the fact that $|1-l| \leq 1, |l| \leq 1$, as $l \in [0,1]$. Also, note that

$$\frac{|\bar{\boldsymbol{u}}(\boldsymbol{x}) - \bar{\boldsymbol{u}}(\boldsymbol{y})|}{|\boldsymbol{x}-\boldsymbol{y}|^{\gamma}} \leq 2\|\boldsymbol{u}\|_{C^{0,\gamma}(D;\mathbb{R}^d)}$$
$$\frac{|\bar{\boldsymbol{v}}(\boldsymbol{x}) - \bar{\boldsymbol{v}}(\boldsymbol{y})|}{|\boldsymbol{x}-\boldsymbol{y}|^{\gamma}} \leq 2\|\boldsymbol{v}\|_{C^{0,\gamma}(D;\mathbb{R}^d)}$$
$$|\bar{\boldsymbol{u}}(\boldsymbol{y}) - \bar{\boldsymbol{v}}(\boldsymbol{y})| \leq s^{\gamma}\|\boldsymbol{u}-\boldsymbol{v}\|_{C^{0,\gamma}(D;\mathbb{R}^d)}.$$

We combine above estimates with Eq. 34 to get

$$\begin{aligned} H_3 &\leq \frac{1}{\sqrt{s}}\frac{C_3}{\sqrt{s}}\left(\|\boldsymbol{u}\|_{C^{0,\gamma}(D;\mathbb{R}^d)} + \|\boldsymbol{v}\|_{C^{0,\gamma}(D;\mathbb{R}^d)}\right) s^{\gamma}\|\boldsymbol{u}-\boldsymbol{v}\|_{C^{0,\gamma}(D;\mathbb{R}^d)} \\ &= \frac{C_3}{s^{1-\gamma}}\left(\|\boldsymbol{u}\|_{C^{0,\gamma}(D;\mathbb{R}^d)} + \|\boldsymbol{v}\|_{C^{0,\gamma}(D;\mathbb{R}^d)}\right)\|\boldsymbol{u}-\boldsymbol{v}\|_{C^{0,\gamma}(D;\mathbb{R}^d)}. \end{aligned} \tag{35}$$

Next we estimate H_4. Here we add and subtract $\omega(\boldsymbol{y})\omega(\boldsymbol{x}+\epsilon\boldsymbol{\xi})$ to get

$$\begin{aligned} H_4 = &\frac{1}{|\boldsymbol{x}-\boldsymbol{y}|^{\gamma}}\frac{1}{\sqrt{s}}\int_0^1 |(\omega(\boldsymbol{x},\boldsymbol{x}+\epsilon\boldsymbol{\xi})(\omega(\boldsymbol{x})-\omega(\boldsymbol{y}))+\omega(\boldsymbol{y})(\omega(\boldsymbol{x}+\epsilon\boldsymbol{\xi})-\omega(\boldsymbol{y}+\epsilon\boldsymbol{\xi})) \\ &\times |F_1''(\boldsymbol{r}(l,\boldsymbol{y})\cdot\boldsymbol{e}/\sqrt{s}))|\,|\bar{\boldsymbol{u}}(\boldsymbol{y})-\bar{\boldsymbol{v}}(\boldsymbol{y})|\,dl. \end{aligned}$$

Recalling that ω belongs to $C_0^{0,\gamma}(D;\mathbb{R}^d)$ and in view of the previous estimates, a straightforward calculation gives

$$H_4 \leq \frac{4C_2}{s^{1/2-\gamma}}\|\omega\|_{C^{0,\gamma}(D;\mathbb{R}^d)}\|\boldsymbol{u}-\boldsymbol{v}\|_{C^{0,\gamma}(D;\mathbb{R}^d)}. \tag{36}$$

Combining Eqs. 33, 35, and 36 gives

$$\begin{aligned} H \leq &\left(\frac{2C_2}{\sqrt{s}} + \frac{4C_2}{s^{1/2-\gamma}}\|\omega\|_{C^{0,\gamma}(D;\mathbb{R}^d)} + \right. \\ &\left. + \frac{C_3}{s^{1-\gamma}}\left(\|\boldsymbol{u}\|_{C^{0,\gamma}(D;\mathbb{R}^d)} + \|\boldsymbol{v}\|_{C^{0,\gamma}(D;\mathbb{R}^d)}\right)\right)\|\boldsymbol{u}-\boldsymbol{v}\|_{C^{0,\gamma}(D;\mathbb{R}^d)}. \end{aligned}$$

Substituting H in Eq. 28 gives

$$\begin{aligned} &\frac{1}{|\boldsymbol{x}-\boldsymbol{y}|^{\gamma}}|(-\nabla PD^{\epsilon}(\boldsymbol{u}) + \nabla PD^{\epsilon}(\boldsymbol{v}))(\boldsymbol{x}) - (-\nabla PD^{\epsilon}(\boldsymbol{u}) + \nabla PD^{\epsilon}(\boldsymbol{v}))(\boldsymbol{y})| \\ &\leq |\frac{2}{\epsilon\omega_d}\int_{H_1(\boldsymbol{0})} J(|\boldsymbol{\xi}|)\frac{1}{\sqrt{s}}H d\boldsymbol{\xi}| \end{aligned}$$

$$\leq \left(\frac{4C_2 \bar{J}_1}{\epsilon^2} + \frac{4C_2 \bar{J}_{1-\gamma}}{\epsilon^{2-\gamma}} \|\omega\|_{C^{0,\gamma}(D;\mathbb{R}^d)} \right.$$

$$\left. + \frac{2C_3 \bar{J}_{3/2-\gamma}}{\epsilon^{2+1/2-\gamma}} \left(\|\boldsymbol{u}\|_{C^{0,\gamma}(D;\mathbb{R}^d)} + \|\boldsymbol{v}\|_{C^{0,\gamma}(D;\mathbb{R}^d)} \right) \right) \|\boldsymbol{u} - \boldsymbol{v}\|_{C^{0,\gamma}(D;\mathbb{R}^d)}. \tag{37}$$

We combine Eqs. 19, 27, and 37 and get

$$\|-\nabla PD^\epsilon(\boldsymbol{u}) - (-\nabla PD^\epsilon(\boldsymbol{v}))\|_{C^{0,\gamma}}$$

$$\leq \left(\frac{4C_2 \bar{J}_1}{\epsilon^2} + \frac{2C_2 \bar{J}_{1-\gamma}}{\epsilon^{2-\gamma}} (1 + \|\omega\|_{C^{0,\gamma}}) + \frac{2C_3 \bar{J}_{3/2-\gamma}}{\epsilon^{2+1/2-\gamma}} (\|\boldsymbol{u}\|_{C^{0,\gamma}} + \|\boldsymbol{v}\|_{C^{0,\gamma}}) \right) \|\boldsymbol{u} - \boldsymbol{v}\|_{C^{0,\gamma}}$$

$$\leq \frac{\bar{C}_1 + \bar{C}_2 \|\omega\|_{C^{0,\gamma}} + \bar{C}_3 (\|\boldsymbol{u}\|_{C^{0,\gamma}} + \|\boldsymbol{v}\|_{C^{0,\gamma}})}{\epsilon^{2+\alpha(\gamma)}} \|\boldsymbol{u} - \boldsymbol{v}\|_{C^{0,\gamma}} \tag{38}$$

where we introduce new constants $\bar{C}_1, \bar{C}_2, \bar{C}_3$. We let $\alpha(\gamma) = 0$, if $\gamma \geq 1/2$, and $\alpha(\gamma) = 1/2 - \gamma$, if $\gamma \leq 1/2$. One can easily verify that, for all $\gamma \in (0, 1]$ and $0 < \epsilon \leq 1$,

$$\max \left\{ \frac{1}{\epsilon^2}, \frac{1}{\epsilon^{2+1/2-\gamma}}, \frac{1}{\epsilon^{2-\gamma}} \right\} \leq \frac{1}{\epsilon^{2+\alpha(\gamma)}}$$

To complete the proof, we combine Eqs. 38 and 18 and get

$$\|F^\epsilon(y,t) - F^\epsilon(z,t)\|_X \leq \frac{L_1 + L_2(\|\omega\|_{C^{0,\gamma}} + \|y\|_X + \|z\|_X)}{\epsilon^{2+\alpha(\gamma)}} \|y - z\|_X.$$

This proves the Lipschitz continuity of $F^\epsilon(y,t)$ on any bounded subset of X. The bound on $F^\epsilon(y,t)$ (see Eq. 17) follows easily from Eq. 25. This completes the proof of Proposition 1.

Existence of Solution in Hölder Space

In this section, we prove Theorem 1. We begin by proving a local existence theorem. We then show that the local solution can be continued uniquely in time to recover Theorem 1.

The existence and uniqueness of local solutions is stated in the following theorem.

Theorem 2 (Local existence and uniqueness). *Given* $X = C_0^{0,\gamma}(D;\mathbb{R}^d) \times C_0^{0,\gamma}(D;\mathbb{R}^d)$, $\boldsymbol{b}(t) \in C_0^{0,\gamma}(D;\mathbb{R}^d)$, *and initial data* $x_0 = (\boldsymbol{u}_0, \boldsymbol{v}_0) \in X$. *We suppose that* $\boldsymbol{b}(t)$ *is continuous in time over some time interval* $I_0 = (-T, T)$

and satisfies $\sup_{t \in I_0} \|\boldsymbol{b}(t)\|_{C^{0,\gamma}(D;\mathbb{R}^d)} < \infty$. *Then, there exists a time interval* $I' = (-T', T') \subset I_0$ *and unique solution* $y = (y^1, y^2)$ *such that* $y \in C^1(I'; X)$ *and*

$$y(t) = x_0 + \int_0^t F^\epsilon(y(\tau), \tau)\, d\tau, \; for\; t \in I' \tag{39}$$

or equivalently

$$y'(t) = F^\epsilon(y(t), t), with\; y(0) = x_0,\; for\; t \in I'$$

where $y(t)$ *and* $y'(t)$ *are Lipschitz continuous in time for* $t \in I' \subset I_0$.

To prove Theorem 2, we proceed as follows. We write $y(t) = (y^1(t), y^2(t))$ and $||y||_X = ||y^1(t)||_{C^{0,\gamma}} + ||y^2(t)||_{C^{0,\gamma}}$. Define the ball $B(0, R) = \{y \in X : ||y||_X < R\}$ and choose $R > ||x_0||_X$. Let $r = R - \|x_0\|_X$ and we consider the ball $B(x_0, r)$ defined by

$$B(x_0, r) = \{y \in X : ||y - x_0||_X < r\} \subset B(0, R), \tag{40}$$

(see Fig. 6).

To recover the existence and uniqueness, we introduce the transformation

$$S_{x_0}(y)(t) = x_0 + \int_0^t F^\epsilon(y(\tau), \tau)\, d\tau.$$

Introduce $0 < T' < T$ and the associated set $Y(T')$ of Hölder continuous functions taking values in $B(x_0, r)$ for $I' = (-T', T') \subset I_0 = (-T, T)$. The goal is to find appropriate interval $I' = (-T', T')$ for which S_{x_0} maps into the corresponding set $Y(T')$. Writing out the transformation with $y(t) \in Y(T')$ gives

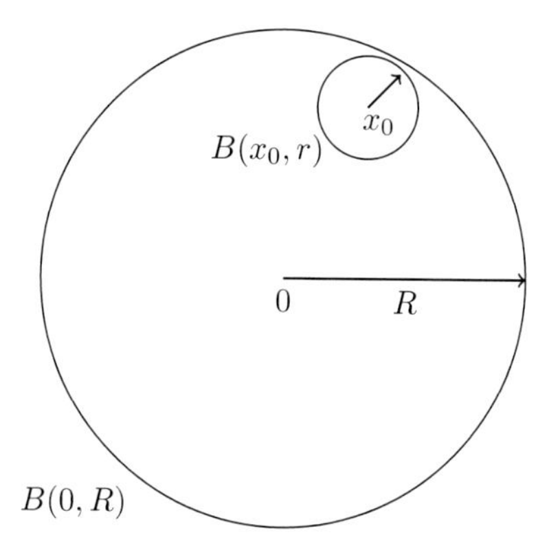

Fig. 6 Geometry

$$S^1_{x_0}(y)(t) = x^1_0 + \int_0^t y^2(\tau)\,d\tau \tag{41}$$

$$S^2_{x_0}(y)(t) = x^2_0 + \int_0^t (-\nabla PD^\epsilon(y^1(\tau)) + \boldsymbol{b}(\tau))\,d\tau, \tag{42}$$

and there is a positive constant $K = C/\epsilon^{2+\alpha(\gamma)}$ (see Eq. 17) independent of $y^1(t)$, for $-T' < t < T'$, such that estimation in Eq. 42 gives

$$||S^2_{x_0}(y)(t) - x^2_0||_{C^{0,\gamma}} \leq (K(1 + \frac{1}{\epsilon^\gamma} + \sup_{t\in(-T',T')} ||y^1(t)||_{C^{0,\gamma}}) + \sup_{t\in(-T,T)} ||\boldsymbol{b}(t)||_{C^{0,\gamma}})T' \tag{43}$$

and from Eq. 41

$$||S^1_{x_0}(y)(t) - x^1_0||_{C^{0,\gamma}} \leq \sup_{t\in(-T',T')} ||y^2(t)||_{C^{0,\gamma}} T'. \tag{44}$$

We write $b = \sup_{t\in I_0} ||\boldsymbol{b}(t)||_{C^{0,\gamma}}$ and adding Eqs. 43 and 44 gives the upper bound

$$||S_{x_0}(y)(t) - x_0||_X \leq (K(1 + \frac{1}{\epsilon^\gamma} + \sup_{t\in(-T',T')} ||y(t)||_X) + b)T'. \tag{45}$$

Since $B(x_0, r) \subset B(0, R)$ (see Eq. 40), we make the choice T' so that

$$||S_{x_0}(y)(t) - x_0||_X \leq ((K(1 + \frac{1}{\epsilon^\gamma} + R) + b)T' < r = R - ||x_0||_X. \tag{46}$$

For this choice we see that

$$T' < \theta(R) = \frac{R - ||x_0||_X}{K(R + 1 + \frac{1}{\epsilon^\gamma}) + b}. \tag{47}$$

Now it is easily seen that $\theta(R)$ is increasing with $R > 0$ and

$$\lim_{R\to\infty} \theta(R) = \frac{1}{K}. \tag{48}$$

So given R and $||x_0||_X$, we choose T' according to

$$\frac{\theta(R)}{2} < T' < \theta(R), \tag{49}$$

and set $I' = (-T', T')$. We have found the appropriate time domain I' such that the transformation $S_{x_0}(y)(t)$ as defined in Eqs. 41 and 42 maps $Y(T')$ into itself. We now proceed using standard arguments (see, e.g., Driver 2003, Theorem 6.10) to complete the proof of existence and uniqueness of solution for given initial data x_0 over the interval $I' = (-T', T')$.

We now prove Theorem 1. From the proof of Theorem 2 above, we see that a unique local solution exists over a time domain $(-T', T')$ with $\frac{\theta(R)}{2} < T'$. Since $\theta(R) \nearrow 1/K$ as $R \nearrow \infty$, we can fix a tolerance $\eta > 0$ so that $[(1/2K) - \eta] > 0$. Then given any initial condition with bounded Hölder norm and $b = \sup_{t \in [-T,T)} ||\boldsymbol{b}(t)||_{C^{0,\gamma}}$, we can choose R sufficiently large so that $||x_0||_X < R$ and $0 < (1/2K)) - \eta < T'$. Thus we can always find local solutions for time intervals $(-T', T')$ for T' larger than $[(1/2K) - \eta] > 0$. Therefore we apply the local existence and uniqueness result to uniquely continue local solutions up to an arbitrary time interval $(-T, T)$.

Existence of Solutions in the Sobolev Space H^2

We start by recalling that the space $H_0^2(D; \mathbb{R}^d)$ is the closure in the H^2 norm of twice differentiable functions with compact support in D. We denote the norm in H^m by $|| \cdot ||_m$, $m = 1, 2$, and the L^∞ norm by $|| \cdot ||_\infty$. In this section, we find that solutions of peridynamic evolutions exist for almost all times in $H_0^2(D; \mathbb{R}^d) \cap L^\infty(D; \mathbb{R}^d)$. For the sake of convenience, we let W denote the $H_0^2(D; \mathbb{R}^d) \cap L^\infty(D; \mathbb{R}^d)$ space. The norm on W is defined as

$$||u||_W := ||u||_2 + ||u||_\infty. \tag{50}$$

We will assume that $u \in H_0^2(D; \mathbb{R}^d)$ is extended by zero outside D; therefore, $u = 0, \nabla u = 0, \nabla^2 u = 0$ for $x \notin D$ and $||u||_{H^2(D;\mathbb{R}^d)} = ||u||_{H^2(\mathbb{R}^d;\mathbb{R}^d)}$.

Noting the Sobolev embedding property of $u \in H_0^2(D; \mathbb{R}^d)$ (see Theorem 2.31, Demengel and Demengel 2012) given by

$$||\nabla u||_{L^q(D;\mathbb{R}^{d\times d})} \leq C_e ||u||_{H_0^2(D;\mathbb{R}^d)} \tag{51}$$

for any q such that $2 \leq q \leq 6$ in case of $d = 3$ and $2 \leq q < \infty$ in case of $d = 2$. Constant C_e is independent of u.

In what follows, we will first prove the Lipschitz bound on $-\nabla PD^\epsilon(u)$, and then using Lipschitz bound, we will show the local existence of solution u in W. We write the peridynamic equation as an equivalent first-order system with $y_1(t) = u(t)$ and $y_2(t) = v(t)$ with $v(t) = \dot{u}(t)$. Let $y = (y_1, y_2)^T$ where $y_1, y_2 \in W$ and let $F^\epsilon(y, t) = (F_1^\epsilon(y, t), F_2^\epsilon(y, t))^T$ such that

$$F_1^\epsilon(y, t) := y_2, \tag{52}$$

$$F_2^\epsilon(y, t) := -\nabla PD^\epsilon(y_1) + b(t). \tag{53}$$

The initial boundary value is equivalent to the initial boundary value problem for the first-order system given by

$$\dot{y}(t) = F^{\epsilon}(y, t), \tag{54}$$

with initial condition given by $y(0) = (u_0, v_0)^T \in W \times W$.

Theorem 3 (Lipschitz bound on peridynamic force). *For any $u, v \in W$, we have*

$$\begin{aligned} & \| - \nabla PD^{\epsilon}(u) - (-\nabla PD^{\epsilon}(v))\|_W \\ & \leq \frac{\bar{L}_1 + \bar{L}_2(\|u\|_W + \|v\|_W) + \bar{L}_3(\|u\|_W + \|v\|_W)^2}{\epsilon^3} \|u - v\|_W \end{aligned} \tag{55}$$

where constants $\bar{L}_1, \bar{L}_2, \bar{L}_3$ are independent of ϵ, u, and v and are defined in (96). Also, for $u \in W$, we have

$$\| - \nabla PD^{\epsilon}(u)\|_W \leq \frac{\bar{L}_4 \|u\|_W + \bar{L}_5 \|u\|_W^2}{\epsilon^{5/2}}, \tag{56}$$

where constants are independent of ϵ and u and are defined in (105).

We state the theorem which shows the existence and uniqueness of solution in any given finite time interval $I_0 = (-T, T)$.

Theorem 4 (Existence and uniqueness of solutions over finite time intervals). *For any initial condition $x_0 \in X = W \times W$, time interval $I_0 = (-T, T)$, and right-hand side $b(t)$ continuous in time for $t \in I_0$ such that $b(t)$ satisfies $\sup_{t \in I_0} \|b(t)\|_W < \infty$, there is a unique solution $y(t) \in C^1(I_0; X)$ of*

$$y(t) = x_0 + \int_0^t F^{\epsilon}(y(\tau), \tau)\, d\tau,$$

or equivalently

$$y'(t) = F^{\epsilon}(y(t), t), \text{with } y(0) = x_0,$$

where $y(t)$ and $y'(t)$ are Lipschitz continuous in time for $t \in I_0$.

The proof of the Lipschitz continuity and existence is established in the following section.

Lipschitz Continuity in the H^2 Norm and Existence of an H^2 Solution

In this section we prove Theorems 3 and 4. To simplify the presentation, we denote the peridynamic force $-\nabla PD^{\epsilon}(u)$ by simply $P(u)$. Recall that we denote $H_0^2(D; \mathbb{R}^d) \cap L^{\infty}(D; \mathbb{R}^d)$ by W and

$$||u||_W = ||u|| + ||\nabla u|| + ||\nabla^2 u|| + ||u||_\infty.$$

We need to analyze $||P(u) - P(v)||_W$.

We use the following short notations:

$$s_\xi = \epsilon|\xi|,\ e_\xi = \frac{\xi}{|\xi|},\ \bar{J}_\alpha = \frac{1}{\omega_d}\int_{H_1(0)} J(|\xi|)\frac{1}{|\xi|^\alpha}d\xi,$$

$$S_\xi(u) = \frac{u(x+\epsilon\xi) - u(x)}{s_\xi}\cdot e_\xi,$$

$$S_\xi(\nabla u) = \nabla S_\xi(u) = \frac{\nabla u^T(x+\epsilon\xi) - \nabla u^T(x)}{s_\xi} e_\xi,$$

$$S_\xi(\nabla^2 u) = \nabla S_\xi(\nabla u) = \nabla\left[\frac{\nabla u^T(x+\epsilon\xi) - \nabla u^T(x)}{s_\xi} e_\xi\right].$$

In indicial notation, we have

$$\begin{aligned} S_\xi(\nabla u)_i &= \frac{u_{k,i}(x+\epsilon\xi) - u_{k,i}(x)}{s_\xi}(e_\xi)_k, \\ S_\xi(\nabla^2 u)_{ij} &= \left[\frac{u_{k,i}(x+\epsilon\xi) - u_{k,i}(x)}{s_\xi}(e_\xi)_k\right]_{,j} = \frac{u_{k,ij}(x+\epsilon\xi) - u_{k,ij}(x)}{s_\xi}(e_\xi)_k \end{aligned} \tag{57}$$

and

$$[e_\xi \otimes S_\xi(\nabla^2 u)]_{ijk} = (e_\xi)_i S_\xi(\nabla^2 u)_{jk}, \tag{58}$$

where $u_{i,j} = (\nabla u)_{ij}$, $u_{k,ij} = (\nabla^2 u)_{kij}$, and $(e_\xi)_k = \xi_k/|\xi|$.

Peridynamic Force

Let $F_1(r) := f(r^2)$ where f is described in section "Problem Formulation with Bond-Based Nonlinear Potentials." We have $F_1'(r) = f'(r^2)2r$. Thus, $2Sf'(\epsilon|\xi|S^2) = F_1'(\sqrt{\epsilon|\xi|}S)/\sqrt{\epsilon|\xi|}$. We define the following constants related to nonlinear potential

$$C_1 := \sup_r |F_1'(r)|,\ C_2 := \sup_r |F_1''(r)|,\ C_3 := \sup_r |F_1'''(r)|,\ C_4 := \sup_r |F_1''''(r)|.$$

The potential function f as chosen here satisfies $C_1, C_2, C_3, C_4 < \infty$. Let

$$\bar{\omega}_\xi(x) = \omega(x)\omega(x+\epsilon\xi), \tag{59}$$

and we choose ω such that

$$|\nabla \bar{\omega}_\xi| \le C_{\omega_1} < \infty \quad \text{and} \quad |\nabla^2 \bar{\omega}_\xi| \le C_{\omega_2} < \infty. \tag{60}$$

With notations described so far, we write peridynamic force $P(u)$ as

$$P(u)(x) = \frac{2}{\epsilon \omega_d} \int_{H_1(0)} \bar{\omega}_\xi(x) J(|\xi|) \frac{F_1'(\sqrt{s_\xi} S_\xi(u))}{\sqrt{s_\xi}} e_\xi d\xi. \tag{61}$$

The gradient of $P(u)(x)$ is given by

$$\begin{aligned} \nabla P(u)(x) &= \frac{2}{\epsilon \omega_d} \int_{H_1(0)} \bar{\omega}_\xi(x) J(|\xi|) F_1''(\sqrt{s_\xi} S_\xi(u)) e_\xi \otimes \nabla S_\xi(u) d\xi \\ &+ \frac{2}{\epsilon \omega_d} \int_{H_1(0)} J(|\xi|) \frac{F_1'(\sqrt{s_\xi} S_\xi(u))}{\sqrt{s_\xi}} e_\xi \otimes \nabla \bar{\omega}_\xi(x) d\xi \\ &= g_1(u)(x) + g_2(u)(x), \end{aligned} \tag{62}$$

where we denote first and second term as $g_1(u)(x)$ and $g_2(u)(x)$, respectively. We also have

$$\begin{aligned} \nabla^2 P(u)(x) &= \frac{2}{\epsilon \omega_d} \int_{H_1(0)} \bar{\omega}_\xi(x) J(|\xi|) F_1''(\sqrt{s_\xi} S_\xi(u)) e_\xi \otimes S_\xi(\nabla^2 u) d\xi \\ &+ \frac{2}{\epsilon \omega_d} \int_{H_1(0)} \bar{\omega}_\xi(x) J(|\xi|) \sqrt{s_\xi} F_1'''(\sqrt{s_\xi} S_\xi(u)) e_\xi \otimes S_\xi(\nabla u) \otimes S_\xi(\nabla u) d\xi \\ &+ \frac{2}{\epsilon \omega_d} \int_{H_1(0)} J(|\xi|) F_1''(\sqrt{s_\xi} S_\xi(u)) e_\xi \otimes S_\xi(\nabla u) \otimes \nabla \bar{\omega}_\xi(x) d\xi \\ &+ \frac{2}{\epsilon \omega_d} \int_{H_1(0)} J(|\xi|) \frac{F_1'(\sqrt{s_\xi} S_\xi(u))}{\sqrt{s_\xi}} e_\xi \otimes \nabla^2 \bar{\omega}_\xi(x) d\xi \\ &+ \frac{2}{\epsilon \omega_d} \int_{H_1(0)} J(|\xi|) F_1''(\sqrt{s_\xi} S_\xi(u)) e_\xi \otimes \nabla \bar{\omega}_\xi(x) \otimes S_\xi(\nabla u) d\xi \\ &= h_1(u)(x) + h_2(u)(x) + h_3(u)(x) + h_4(u)(x) + h_5(u)(x) \end{aligned} \tag{63}$$

where we denote first, second, third, fourth, and fifth terms as h_1, h_2, h_2, h_4, and h_5, respectively. Estimating $||P(u) - P(v)||$ and $||P(u) - P(v)||_\infty$. From (61), we have

$$
\begin{aligned}
&|P(u)(x) - P(v)(x)| \\
&\leq \frac{2}{\epsilon \omega_d} \int_{H_1(0)} J(|\xi|) \frac{1}{\sqrt{s_\xi}} |F_1'(\sqrt{s_\xi} S_\xi(u)) - F_1'(\sqrt{s_\xi} S_\xi(v))| d\xi \\
&\leq \frac{2}{\epsilon \omega_d} \left(\sup_r |F_1'(r)| \right) \int_{H_1(0)} J(|\xi|) \frac{1}{\sqrt{s_\xi}} |\sqrt{s_\xi} S_\xi(u) - \sqrt{s_\xi} S_\xi(v)| d\xi \\
&= \frac{2C_2}{\epsilon \omega_d} \int_{H_1(0)} J(|\xi|) |S_\xi(u) - S_\xi(v)| d\xi, \qquad (64)
\end{aligned}
$$

where we used the fact that $|\bar{\omega}_\xi(x)| \leq 1$ and $|F_1'(r_1) - F_1'(r_2)| \leq C_2 |r_1 - r_2|$. Since

$$
|S_\xi(u) - S_\xi(v)| \leq \frac{|u(x + \epsilon\xi) - v(x + \epsilon\xi)| + |u(x) - v(x)|}{\epsilon |\xi|}
$$

we have

$$
||P(u) - P(v)||_\infty \leq \frac{2C_2}{\epsilon \omega_d} \int_{H_1(0)} J(|\xi|) \frac{2||u - v||_\infty}{\epsilon |\xi|} d\xi = \frac{L_1}{\epsilon^2} ||u - v||_W \qquad (65)
$$

where we let $L_1 := 4C_2 \bar{J}_1$.

From (64), we have

$$
\begin{aligned}
&||P(u) - P(v)||^2 \\
&\leq \int_D \left(\frac{2C_2}{\epsilon \omega_d} \right)^2 \int_{H_1(0)} \int_{H_1(0)} \frac{J(|\xi|)}{|\xi|} \frac{J(|\eta|)}{|\eta|} |\xi| |S_\xi(u) - S_\xi(v)| |\eta| |S_\eta(u) - S_\eta(v)| \\
&\quad d\xi d\eta dx.
\end{aligned}
$$

Using the identities $|a||b| \leq |a|^2/2 + |b|^2/2$ and $(a + b)^2 \leq 2a^2 + 2b^2$, we get

$$
\begin{aligned}
&||P(u) - P(v)||^2 \\
&\leq \int_D \left(\frac{2C_2}{\epsilon \omega_d} \right)^2 \int_{H_1(0)} \int_{H_1(0)} \frac{J(|\xi|)}{|\xi|} \frac{J(|\eta|)}{|\eta|} \frac{|\xi|^2 |S_\xi(u) - S_\xi(v)|^2 + |\eta|^2 |S_\eta(u) - S_\eta(v)|^2}{2} d\xi d\eta dx \\
&= 2 \int_D \left(\frac{2C_2}{\epsilon \omega_d} \right)^2 \int_{H_1(0)} \int_{H_1(0)} \frac{J(|\xi|)}{|\xi|} \frac{J(|\eta|)}{|\eta|} \frac{|\xi|^2 |S_\xi(u) - S_\xi(v)|^2}{2} d\xi d\eta dx \\
&= \int_D \left(\frac{2C_2}{\epsilon \omega_d} \right)^2 \omega_d \bar{J}_1 \int_{H_1(0)} \frac{J(|\xi|)}{|\xi|} |\xi|^2 \frac{2|u(x + \epsilon\xi) - v(x + \epsilon\xi)|^2 + 2|u(x) - v(x)|^2}{\epsilon^2 |\xi|^2} d\xi dx \\
&= \left(\frac{2C_2}{\epsilon \omega_d} \right)^2 \omega_d \bar{J}_1 \int_{H_1(0)} \frac{J(|\xi|)}{|\xi|} \frac{1}{\epsilon^2} \left[2 \int_D \left(|u(x + \epsilon\xi) - v(x + \epsilon\xi)|^2 + |u(x) - v(x)|^2 \right) dx \right] d\xi \\
&\leq \left(\frac{2C_2}{\epsilon \omega_d} \right)^2 \omega_d \bar{J}_1 \int_{H_1(0)} \frac{J(|\xi|)}{|\xi|} \frac{1}{\epsilon^2} \left[4||u - v||^2 \right] d\xi, \qquad (66)
\end{aligned}
$$

where we used symmetry wrt ξ and η in second equation. This gives

$$||P(u) - P(v)|| \le \frac{L_1}{\epsilon^2}||u - v|| \le \frac{L_1}{\epsilon^2}||u - v||_W. \tag{67}$$

Estimating $||\nabla P(u) - \nabla P(v)||$. From (62), we have

$$||\nabla P(u) - \nabla P(v)|| \le ||g_1(u) - g_1(v)|| + ||g_2(u) - g_2(v)||.$$

Using $|\bar{\omega}_\xi(x)| \le 1$, we get

$$\begin{aligned}
&|g_1(u)(x) - g_1(v)(x)| \\
&\le \frac{2}{\epsilon\omega_d}\int_{H_1(0)} J(|\xi|)|F_1''(\sqrt{s_\xi}S_\xi(u))\nabla S_\xi(u) - F_1''(\sqrt{s_\xi}S_\xi(v))\nabla S_\xi(v)|d\xi \\
&\le \frac{2}{\epsilon\omega_d}\int_{H_1(0)} J(|\xi|)|F_1''(\sqrt{s_\xi}S_\xi(u)) - F_1''(\sqrt{s_\xi}S_\xi(v))||\nabla S_\xi(u)|d\xi \\
&+ \frac{2}{\epsilon\omega_d}\int_{H_1(0)} J(|\xi|)|F_1''(\sqrt{s_\xi}S_\xi(v))||\nabla S_\xi(u) - \nabla S_\xi(v)|d\xi \\
&\le \frac{2C_3}{\epsilon\omega_d}\int_{H_1(0)} J(|\xi|)\sqrt{s_\xi}|S_\xi(u) - S_\xi(v)||\nabla S_\xi(u)|d\xi \\
&+ \frac{2C_2}{\epsilon\omega_d}\int_{H_1(0)} J(|\xi|)|\nabla S_\xi(u) - \nabla S_\xi(v)|d\xi \\
&= I_1(x) + I_2(x)
\end{aligned} \tag{68}$$

where we denote first and second term as $I_1(x)$ and $I_2(x)$. Proceeding similar to ((66)), we can show

$$\begin{aligned}
||I_1||^2 &= \int_D \left(\frac{2C_3}{\epsilon\omega_d}\right)^2 \int_{H_1(0)}\int_{H_1(0)} \frac{J(|\xi|)}{|\xi|^{3/2}}\frac{J(|\eta|)}{|\eta|^{3/2}}|\xi|^{3/2}|\eta|^{3/2}\sqrt{s_\xi}\sqrt{s_\eta} \\
&\times |S_\xi(u) - S_\xi(v)||\nabla S_\xi(u)||S_\eta(u) - S_\eta(v)||\nabla S_\eta(u)|d\xi d\eta dx \\
&\le \int_D \left(\frac{2C_3}{\epsilon\omega_d}\right)^2 \omega_d \bar{J}_{3/2}\int_{H_1(0)} \frac{J(|\xi|)}{|\xi|^{3/2}}|\xi|^3 s_\xi|S_\xi(u) - S_\xi(v)|^2|\nabla S_\xi(u)|^2 d\xi dx.
\end{aligned} \tag{69}$$

Now

$$\int_D |S_\xi(u) - S_\xi(v)|^2|\nabla S_\xi(u)|^2 dx$$

$$\leq \frac{4||u-v||_{\infty}^{2}}{\epsilon^2|\xi|^2}\frac{1}{\epsilon^2|\xi|^2}\int_D 2(|\nabla u(x+\epsilon\xi)|^2+|\nabla u(x)|^2)dx$$

$$\leq \frac{16||\nabla u||^2||u-v||_{\infty}^{2}}{\epsilon^4|\xi|^4} \leq \frac{16||u||_W^2}{\epsilon^4|\xi|^4}||u-v||_W^2.$$

Substituting above in (69) to get

$$||I_1||^2 \leq \left(\frac{2C_3}{\epsilon\omega_d}\right)^2 \omega_d \bar{J}_{3/2}\int_{H_1(0)} \frac{J(|\xi|)}{|\xi|^{3/2}}|\xi|^3\epsilon|\xi|\frac{16||u||_W^2}{\epsilon^4|\xi|^4}||u-v||_W^2 d\xi$$

$$= \left(\frac{8C_3\bar{J}_{3/2}||u||_W}{\epsilon^{5/2}}\right)^2 ||u-v||_W^2.$$

Let $L_2 = 8C_3\bar{J}_{3/2}$ to write

$$||I_1|| \leq \frac{L_2(||u||_W + ||v||_W)}{\epsilon^{5/2}}||u-v||_W. \tag{70}$$

Similarly

$$||I_2||^2 = \int_D \left(\frac{2C_2}{\epsilon\omega_d}\right)^2 \int_{H_1(0)}\int_{H_1(0)} \frac{J(|\xi|)}{|\xi|}\frac{J(|\eta|)}{|\eta|}|\xi||\eta|$$

$$\times|\nabla S_\xi(u)-\nabla S_\xi(v)||\nabla S_\eta(u)-\nabla S_\eta(v)|d\xi d\eta dx$$

$$\leq \left(\frac{2C_2}{\epsilon\omega_d}\right)^2 \omega_d\bar{J}_1\int_{H_1(0)}\frac{J(|\xi|)}{|\xi|}|\xi|^2\left[\int_D|\nabla S_\xi(u)-\nabla S_\xi(v)|^2dx\right]d\xi.$$

This gives

$$||I_2|| \leq \frac{4C_2\bar{J}_1}{\epsilon^2}||u-v||_W = \frac{L_1}{\epsilon^2}||u-v||_W. \tag{71}$$

Thus

$$||g_1(u)-g_1(v)|| \leq \frac{\sqrt{\epsilon}L_1 + L_2(||u||_W+||v||_W)}{\epsilon^{5/2}}||u-v||_W. \tag{72}$$

We now work on $|g_2(u)(x)-g_2(v)(x)|$ (see (62)). Noting the bound on $\nabla\bar{\omega}_\xi$, we get

$$|g_2(u)(x)-g_2(v)(x)|$$

$$
\begin{aligned}
&= \left| \frac{2}{\epsilon \omega_d} \int_{H_1(0)} J(|\xi|) \left[\frac{F_1'(\sqrt{s_\xi} S_\xi(u))}{\sqrt{s_\xi}} - \frac{F_1'(\sqrt{s_\xi} S_\xi(v))}{\sqrt{s_\xi}} \right] e_\xi \otimes \nabla \bar{\omega}_\xi(s) d\xi \right| \\
&\leq \frac{2C_{\omega_1}}{\epsilon \omega_d} \int_{H_1(0)} J(|\xi|) \left| \frac{F_1' \sqrt{s_\xi} S_\xi(u))}{\sqrt{s_\xi}} - \frac{F_1'(\sqrt{s_\xi} S_\xi(v))}{\sqrt{s_\xi}} \right| d\xi \\
&\leq \frac{2C_{\omega_1} C_2}{\epsilon \omega_d} \int_{H_1(0)} J(|\xi|) |S_\xi(u) - S_\xi(v)| d\xi. \qquad (73)
\end{aligned}
$$

Above is similar to (64) and therefore we get

$$
||g_2(u) - g_2(v)|| \leq \frac{4C_{\omega_1} C_2 \bar{J}_1}{\epsilon^2} ||u - v||_W = \frac{C_{\omega_1} L_1}{\epsilon^2} ||u - v||_W. \qquad (74)
$$

Combining (72) and (74) to write

$$
||\nabla P(u) - \nabla P(v)|| \leq \frac{\sqrt{\epsilon}(1 + C_{\omega_1}) L_1 + L_2(||u||_W + ||v||_W)}{\epsilon^{5/2}} ||u - v||_W. \quad (75)
$$

Estimating $||\nabla^2 P(u) - \nabla^2 P(v)||$. From (63), we have

$$
\begin{aligned}
&||\nabla^2 P(u) - \nabla^2 P(v)|| \\
&\leq ||h_1(u) - h_1(v)|| + ||h_2(u) - h_2(v)|| + ||h_3(u) - h_3(v)|| \\
&+ ||h_4(u) - h_4(v)|| + ||h_5(u) - h_5(v)||. \qquad (76)
\end{aligned}
$$

We can show, using the fact $|\bar{\omega}_\xi(x)| \leq 1$ and $|F_1''(r_1) - F_1''(r_2)| \leq C_3 |r_1 - r_2|$, that

$$
\begin{aligned}
|h_1(u)(x) - h_1(v)(x)| &\leq \frac{2C_3}{\epsilon \omega_d} \int_{H_1(0)} J(|\xi|) \sqrt{s_\xi} |S_\xi(u) - S_\xi(v)| |S_\xi(\nabla^2 u)| d\xi \\
&+ \frac{2C_2}{\epsilon \omega_d} \int_{H_1(0)} J(|\xi|) |S_\xi(\nabla^2 u) - S_\xi(\nabla^2 v)| d\xi \\
&= I_3(x) + I_4(x). \qquad (77)
\end{aligned}
$$

Following similar steps used above, we can show

$$
||I_3|| \leq \frac{8C_3 \bar{J}_{3/2} ||u||_W}{\epsilon^{5/2}} ||u - v||_W \leq \frac{L_2(||u||_W + ||v||_W)}{\epsilon^{5/2}} ||u - v||_W \qquad (78)
$$

and

$$
||I_4|| \leq \frac{4C_2 \bar{J}_1}{\epsilon^2} ||u - v||_W = \frac{L_1}{\epsilon^2} ||u - v||_W, \qquad (79)
$$

where $L_1 = 4C_2\bar{J}_1$, $L_2 = 8C_3\bar{J}_{3/2}$.

Next we focus on $|h_2(u)(x) - h_2(v)(x)|$ and get

$$\begin{aligned}
&|h_2(u)(x) - h_2(v)(x)| \\
&\leq \frac{2}{\epsilon\omega_d}\int_{H_1(0)} J(|\xi|)\sqrt{s_\xi}|F_1'''(\sqrt{s_\xi}S_\xi(u)) - F_1'''(\sqrt{s_\xi}S_\xi(v))||S_\xi(\nabla u)|^2 d\xi \\
&+ \frac{2}{\epsilon\omega_d}\int_{H_1(0)} J(|\xi|)\sqrt{s_\xi}|F_1'''(\sqrt{s_\xi}S_\xi(v))||S_\xi(\nabla u)\otimes S_\xi(\nabla u) \\
&- S_\xi(\nabla v)\otimes S_\xi(\nabla v)|d\xi \leq \frac{2C_4}{\epsilon\omega_d}\int_{H_1(0)} J(|\xi|)s_\xi|S_\xi(u) - S_\xi(v)||S_\xi(\nabla u)|^2 d\xi \\
&+ \frac{2C_3}{\epsilon\omega_d}\int_{H_1(0)} J(|\xi|)\sqrt{s_\xi}|S_\xi(\nabla u)\otimes S_\xi(\nabla u) - S_\xi(\nabla v)\otimes S_\xi(\nabla v)|d\xi \\
&= I_5(x) + I_6(x).
\end{aligned} \tag{80}$$

Proceeding as below for $||I_5||^2$

$$\begin{aligned}
&||I_5||^2 \\
&\leq \int_D \left(\frac{2C_4}{\epsilon\omega_d}\right)^2 \int_{H_1(0)}\int_{H_1(0)} \frac{J(|\xi|)}{|\xi|^2}\frac{J(|\eta|)}{|\eta|^2}|\xi|^2 s_\xi |\eta|^2 s_\eta \\
&\times |S_\xi(u) - S_\xi(v)||S_\xi(\nabla u)|^2 |S_\eta(u) - S_\eta(v)||S_\eta(\nabla u)|^2 d\xi d\eta dx \\
&< \int_D \left(\frac{2C_4}{\epsilon\omega_d}\right)^2 \omega_d \bar{J}_2 \int_{H_1(0)} \frac{J(|\xi|)}{|\xi|^2}|\xi|^4 s_\xi^2 |S_\xi(u) - S_\xi(v)|^2 |S_\xi(\nabla u)|^4 d\xi dx \\
&\leq \left(\frac{2C_4}{\epsilon\omega_d}\right)^2 \omega_d \bar{J}_2 \int_{H_1(0)} \frac{J(|\xi|)}{|\xi|^2}|\xi|^4 s_\xi^2 \frac{4||u-v||_\infty^2}{\epsilon^2|\xi|^2}\left[\int_D |S_\xi(\nabla u)|^4 dx\right] d\xi.
\end{aligned} \tag{81}$$

We estimate the term in square bracket. Using the identity $(|a|+|b|)^4 \leq (2|a|^2 + 2|b|^2)^2 \leq 8|a|^4 + 8|b|^4$, we have

$$\begin{aligned}
\int_D |S_\xi(\nabla u)|^4 dx &\leq \frac{8}{\epsilon^4|\xi|^4}\int_D (|\nabla u(x+\epsilon\xi)|^4 + |\nabla u(x)|^4)dx \\
&\leq \frac{16}{\epsilon^4|\xi|^4}||\nabla u||^4_{L^4(D;\mathbb{R}^{d\times d})}.
\end{aligned} \tag{82}$$

where $||u||_{L^4(D,\mathbb{R}^d)} = \left[\int_D |u|^4 dx\right]^{1/4}$. Using Sobolev embedding property of $u \in H_0^2(D;\mathbb{R}^d)$ as mentioned in (51), we get

$$\int_D |S_\xi(\nabla u)|^4 dx \le \frac{16}{\epsilon^4 |\xi|^4} C_e^4 ||\nabla u||^4_{H^1(D;\mathbb{R}^{d\times d})} \le \frac{16 C_e^4}{\epsilon^4|\xi|^4}||u||^4_W. \tag{83}$$

Substituting above to get

$$||I_5||^2 \le \left(\frac{2C_4}{\epsilon\omega_d}\right)^2 \omega_d \bar{J}_2 \int_{H_1(0)} \frac{J(|\xi|)}{|\xi|^2}|\xi|^4 s_\xi^2 \frac{4||u-v||^2_\infty}{\epsilon^2|\xi|^2}\frac{16C_e^4}{\epsilon^4|\xi|^4}||u||^4_W d\xi$$

Above gives

$$||I_5|| \le \frac{16C_4C_e^2\bar{J}_2||u||^2_W}{\epsilon^3}||u-v||_W \le \frac{L_3(||u||_W + ||v||_W)^2}{\epsilon^3}||u-v||_W \tag{84}$$

where we let $L_3 = 16C_4C_e^2\bar{J}_2$.

Next, using

$$|S_\xi(\nabla u)\otimes S_\xi(\nabla u) - S_\xi(\nabla v)\otimes S_\xi(\nabla v)| \le (|S_\xi(\nabla u)| + |S_\xi(\nabla v)|)|S_\xi(\nabla u) - S_\xi(\nabla v)|$$

to estimate $||I_6||$ as follows:

$$\begin{aligned}
&||I_6||^2 \\
&\le \int_D \left(\frac{2C_3}{\epsilon\omega_d}\right)^2 \int_{H_1(0)}\int_{H_1(0)} \frac{J(|\xi|)}{|\xi|^{3/2}}\frac{J(|\eta|)}{|\eta|^{3/2}}|\xi|^{3/2}|\eta|^{3/2}\sqrt{s_\xi s_\eta} \\
&\times (|S_\xi(\nabla u)| + |S_\xi(\nabla v)|)|S_\xi(\nabla u) - S_\xi(\nabla v)| \times (|S_\eta(\nabla u)| + |S_\eta(\nabla v)|)|S_\eta(\nabla u) \\
&- S_\eta(\nabla v)| d\xi d\eta dx \le \int_D \left(\frac{2C_3}{\epsilon\omega_d}\right)^2 \omega_d \bar{J}_{3/2}\int_{H_1(0)} \frac{J(|\xi|)}{|\xi|^{3/2}}|\xi|^3 \epsilon|\xi|(|S_\xi(\nabla u)| \\
&+ |S_\xi(\nabla v)|)^2 |S_\xi(\nabla u) - S_\xi(\nabla v)|^2 d\xi dx = \left(\frac{2C_3}{\epsilon\omega_d}\right)^2 \omega_d \bar{J}_{3/2}\int_{H_1(0)} \frac{J(|\xi|)}{|\xi|^{3/2}}|\xi|^3\epsilon|\xi| \\
&\left[\int_D (|S_\xi(\nabla u)| + |S_\xi(\nabla v)|)^2 |S_\xi(\nabla u) - S_\xi(\nabla v)|^2 dx\right] d\xi.
\end{aligned} \tag{85}$$

We focus on the term in square bracket. Using Holder inequality, we have

$$\begin{aligned}
&\int_D (|S_\xi(\nabla u)| + |S_\xi(\nabla v)|)^2 |S_\xi(\nabla u) - S_\xi(\nabla v)|^2 dx \\
&\le \left(\int_D (|S_\xi(\nabla u)| + |S_\xi(\nabla v)|)^4 dx\right)^{1/2}\left(\int_D |S_\xi(\nabla u) - S_\xi(\nabla v)|^4 dx\right)^{1/2}.
\end{aligned} \tag{86}$$

Using $(|a| + |b|)^4 \leq 8|a|^4 + 8|b|^4$, we get

$$\int_D (|S_\xi(\nabla u)| + |S_\xi(\nabla v)|)^4 dx \leq 8\left[\int_D |S_\xi(\nabla u)|^4 dx + \int_D |S_\xi(\nabla v)|^4 dx\right]$$
$$\leq 8\left[\frac{8}{\epsilon^4|\xi|^4}\int_D (|\nabla u(x+\epsilon\xi)|^4 + |\nabla u(x)|^4)dx + \frac{8}{\epsilon^4|\xi|^4}\int_D (|\nabla v(x+\epsilon\xi)|^4\right.$$
$$\left.+|\nabla v(x)|^4)dx\right] \leq \frac{128}{\epsilon^4|\xi|^4}(||\nabla u||^4_{L^4(D;\mathbb{R}^{d\times d})} + ||\nabla v||^4_{L^4(D;\mathbb{R}^{d\times d})})$$
$$\leq \frac{128C_e^4}{\epsilon^4|\xi|^4}(||\nabla u||^4_{H^1(D;\mathbb{R}^{d\times d})} + ||\nabla v||^4_{H^1(D;\mathbb{R}^{d\times d})}) \leq \frac{128C_e^4}{\epsilon^4|\xi|^4}(||u||^4_W + ||v||^4_W)$$
$$\leq \frac{128C_e^4}{\epsilon^4|\xi|^4}(||u||_W + ||v||_W)^4. \tag{87}$$

where we used Sobolev embedding property (51) in third last step. Proceeding similarly to get

$$\int_D |S_\xi(\nabla u) - S_\xi(\nabla v)|^4 dx$$
$$\leq \frac{8}{\epsilon^4|\xi|^4}\left[\int_D |\nabla(u-v)(x+\epsilon\xi)|^4 dx + \int_D |\nabla(u-v)(x)|^4 dx\right]$$
$$\leq \frac{16}{\epsilon^4|\xi|^4}||\nabla(u-v)||^4_{L^4(D,\mathbb{R}^{d\times d})}$$
$$\leq \frac{16C_e^4}{\epsilon^4|\xi|^4}||u-v||^4_W. \tag{88}$$

Substituting (87) and (88) into (86) to get

$$\int_D (|S_\xi(\nabla u)| + |S_\xi(\nabla v)|)^2 |S_\xi(\nabla u) - S_\xi(\nabla v)|^2 dx$$
$$\leq \left(\frac{128C_e^4}{\epsilon^4|\xi|^4}(||u||_W + ||v||_W)^4\right)^{1/2}\left(\frac{16C_e^4}{\epsilon^4|\xi|^4}||u-v||^4_W\right)^{1/2}$$
$$= \frac{32\sqrt{2}C_e^4}{\epsilon^4|\xi|^4}(||u||_W + ||v||_W)^2||u-v||^2_W$$
$$\leq \frac{64C_e^4}{\epsilon^4|\xi|^4}(||u||_W + ||v||_W)^2||u-v||^2_W.$$

Substituting above in (85) to get

$$||I_6||^2 \leq \left(\frac{2C_3}{\epsilon\omega_d}\right)^2 \omega_d \bar{J}_{3/2} \int_{H_1(0)} \frac{J(|\xi|)}{|\xi|^{3/2}} |\xi|^3 \epsilon|\xi| \left[\frac{64C_e^4}{\epsilon^4|\xi|^4}(||u||_W + ||v||_W)^2 ||u-v||_W^2\right] d\xi.$$

From above we have

$$||I_6|| \leq \frac{16C_3C_e^2\bar{J}_{3/2}(||u||_W+||v||_W)}{\epsilon^{5/2}}||u-v||_W = \frac{L_4(||u||_W+||v||_W)}{\epsilon^{5/2}}||u-v||_W, \tag{89}$$

where we let $L_4 = 16C_3C_e^2\bar{J}_{3/2}$.

From the expression of $h_3(u)(x)$ and $h_5(u)(x)$, we find that it is similar to term $g_1(u)(x)$ from the point of view of L^2 norm. Also, $h_4(u)(x)$ is similar to $P(u)(x)$. We easily have

$$|h_4(u)(x) - h_4(v)(x)| \leq \frac{2C_2C_{\omega_2}}{\epsilon\omega_d} \int_{H_1(0)} J(|\xi|)|S_\xi(u) - S_\xi(v)|d\xi,$$

where we used the fact that $|\nabla^2\bar{\omega}_\xi(x)| \leq C_{\omega_2}$. Above is similar to the bound on $|P(u)(x) - P(v)(x)|$ (see (64)); therefore we have

$$||h_4(u) - h_4(v)|| \leq \frac{L_1C_{\omega_2}}{\epsilon^2}||u-v||_W. \tag{90}$$

Similarly, we have

$$\begin{aligned}|h_3(u)(x) - h_3(v)(x)| &\leq \frac{2}{\epsilon\omega_d}\int_{H_1(0)} J(|\xi|)|F_1''(\sqrt{s_\xi}S_\xi(u)) \\ &- F_1''(\sqrt{s_\xi}S_\xi(v))||\nabla\bar{\omega}_\xi(x)||S_\xi(\nabla u)|d\xi + \frac{2}{\epsilon\omega_d}\int_{H_1(0)} J(|\xi|)|F_1''(\sqrt{s_\xi}S_\xi(v))||e_\xi \\ &\otimes \nabla\bar{\omega}_\xi(x)\otimes S_\xi(\nabla u) - e_\xi\otimes\nabla\bar{\omega}_\xi(x)\otimes S_\xi(\nabla v)|d\xi \leq \frac{2C_3C_{\omega_1}}{\epsilon\omega_d}\int_{H_1(0)} \\ &J(|\xi|)\sqrt{s_\xi}|S_\xi(u) - S_\xi(v)||S_\xi(\nabla u)|d\xi + \frac{2C_2C_{\omega_1}}{\epsilon\omega_d}\int_{H_1(0)} J(|\xi|)|S_\xi(\nabla u) \\ &- S_\xi(\nabla v)|d\xi = C_{\omega_1}(I_1(x) + I_2(x)),\end{aligned} \tag{91}$$

where $I_1(x)$ and $I_2(x)$ are given by (68). From (70) to (71), we have

$$\begin{aligned}||h_3(u) - h_3(v)|| &\leq C_{\omega_1}(||I_1|| + ||I_2||) \\ &\leq \frac{\sqrt{\epsilon}C_{\omega_1}L_1 + C_{\omega_1}L_2(||u||_W + ||v||_W)}{\epsilon^{5/2}}||u - v||_W. \end{aligned} \tag{92}$$

Expression of $h_3(u)$ and $h_5(u)$ is similar and hence we have

$$\begin{aligned}||h_5(u) - h_5(v)|| &\leq C_{\omega_1}(||I_1|| + ||I_2||) \\ &\leq \frac{\sqrt{\epsilon}C_{\omega_1}L_1 + C_{\omega_1}L_2(||u||_W + ||v||_W)}{\epsilon^{5/2}}||u - v||_W. \end{aligned} \tag{93}$$

Collecting our results delivers the bound

$$\begin{aligned}&||\nabla^2 P(u) - \nabla^2 P(v)|| \\ &\leq \Bigg[\frac{\epsilon L_1 + \sqrt{\epsilon}L_2(||u||_W + ||v||_W) + L_3(||u||_W + ||v||_W)^2 + \sqrt{\epsilon}L_4(||u||_W + ||v||_W)}{\epsilon^3} \\ &+ \frac{\epsilon C_{\omega_2}L_1 + 2\epsilon C_{\omega_1}L_1 + 2\sqrt{\epsilon}C_{\omega_1}L_2(||u||_W + ||v||_W)}{\epsilon^3}\Bigg]||u - v||_W \\ &\leq \Bigg[\frac{\epsilon(1 + 2C_{\omega_1} + C_{\omega_2})L_1 + \sqrt{\epsilon}(L_2 + 2C_{\omega_1}L_2 + L_4)(||u||_W + ||v||_W)}{\epsilon^3} \\ &+ \frac{L_3(||u||_W + ||v||_W)^2}{\epsilon^3}\Bigg]||u - v||_W. \end{aligned} \tag{94}$$

We now combine (65), (67), (75), and (94) to get

$$\begin{aligned}&||P(u) - P(v)||_W \\ &\leq \Bigg[\frac{2\epsilon L_1 + \epsilon(1 + C_{\omega_1})L_1 + \sqrt{\epsilon}(||u||_W + ||v||_W)}{\epsilon^3} \\ &+ \frac{\epsilon(1 + 2C_{\omega_1} + C_{\omega_2})L_1 + \sqrt{\epsilon}(L_2 + 2C_{\omega_1}L_2 + L_4)(||u||_W + ||v||_W)}{\epsilon^3} \\ &+ \frac{L_3(||u||_W + ||v||_W)^2}{\epsilon^3}\Bigg]||u - v||_W. \end{aligned} \tag{95}$$

Finally we let

$$\bar{L}_1 := (4 + 3C_{\omega_1} + C_{\omega_2})L_1, \ \bar{L}_2 := (1 + 2C_{\omega_1})L_2 + L_4, \ \bar{L}_3 := L_3 \tag{96}$$

and write

$$\begin{aligned} &||P(u) - P(v)||_W \\ &\leq \frac{\bar{L}_1 + \bar{L}_2(||u||_W + ||v||_W) + \bar{L}_3(||u||_W + ||v||_W)^2}{\epsilon^3} ||u - v||_W. \end{aligned} \tag{97}$$

This completes the proof of (55).

We now obtain an upper bound on the peridynamic force. Note that $F_1'(0) = 0$ and $S_\xi(v) = 0$ if $v = 0$. Substituting $v = 0$ in (65) and (67) to get

$$||P(u)|| + ||P(u)||_\infty \leq \frac{2L_1}{\epsilon^2} ||u||_W. \tag{98}$$

For $||g_1(u)||$ and $||g_2(u)||$ we proceed differently. For $||g_2(u)||$, we substitute $v = 0$ in (74) to get

$$||g_2(u)|| \leq \frac{C_{\omega_1} L_1}{\epsilon^2} ||u||_W. \tag{99}$$

To estimate $||g_1(u)||$, we first estimate

$$\begin{aligned} |g_1(u)(x)| &\leq \frac{2C_2}{\epsilon\omega_d} \int_{H_1(0)} J(|\xi|) |\nabla S_\xi(u)| d\xi \\ &\leq \frac{2C_2}{\epsilon^2 \omega_d} \int_{H_1(0)} \frac{J(|\xi|)}{|\xi|} (|\nabla u(x + \epsilon\xi)| + |\nabla u(x)|) d\xi, \end{aligned} \tag{100}$$

and we obtain

$$\begin{aligned} ||g_1(u)||^2 &\leq \left(\frac{2C_2}{\epsilon^2 \omega_d} \right)^2 \omega_d \bar{J}_1 \int_{H_1(0)} \frac{J(|\xi|)}{|\xi|} \left[\int_D (|\nabla u(x + \epsilon\xi)| + |\nabla u(x)|)^2 dx \right] d\xi \\ &\leq \left(\frac{4C_2 \bar{J}_1}{\epsilon^2} \right)^2 ||\nabla u||^2 \end{aligned} \tag{101}$$

i.e.

$$||g_1(u)|| \leq \frac{L_1}{\epsilon^2} ||u||_W. \tag{102}$$

Combining (99) and (102) gives

$$||\nabla P(u)|| \leq \frac{(1 + C_{\omega_1}) L_1}{\epsilon^2} ||u||_W. \tag{103}$$

We need to estimate $||\nabla^2 P(u)||$. We have from (63)

$$||\nabla^2 P(u)|| \leq ||h_1(u)|| + ||h_2(u)|| + ||h_3(u)|| + ||h_4(u)|| + ||h_5(u)||.$$

From the expression of $h_1(u)$ and $h_2(u)$, we find that

$$||h_1(u)|| \leq \frac{4C_2\bar{J}_1}{\epsilon^2}||u||_W = \frac{L_1}{\epsilon^2}||u||_W \text{ and } ||h_2(u)|| \leq \frac{8C_3C_e^2\bar{J}_{3/2}}{\epsilon^{5/2}}||u||_W^2 \leq \frac{L_4}{\epsilon^{5/2}}||u||_W^2,$$

where $L_4 = 16C_3C_e^2\bar{J}_{3/2}$. Case of $||h_3(u)||$ and $||h_5(u)||$ is similar to $||g_1(u)||$, and case of $||h_4(u)||$ is similar to $||P(u)||$.

$$||h_4(u)|| \leq \frac{C_{\omega_2}L_1}{\epsilon^2}||u||_W$$

and

$$||h_3(u)|| \leq \frac{C_{\omega_1}L_1}{\epsilon^2}||u||_W \quad \text{and} \quad ||h_5(u)|| \leq \frac{C_{\omega_1}L_1}{\epsilon^2}||u||_W.$$

We combine the inequalities listed above to get

$$||\nabla^2 P(u)|| \leq \frac{\sqrt{\epsilon}(1 + C_{\omega_2} + 2C_{\omega_1})L_1 + L_4||u||_W}{\epsilon^{5/2}}||u||_W. \tag{104}$$

Finally, after combining (98), (103), and (104), we get

$$||P(u)||_W < \frac{\sqrt{\epsilon}(4 + 3C_{\omega_1} + C_{\omega_2})L_1 + L_4||u||_W}{\epsilon^{5/2}}||u||_W.$$

We let

$$\bar{L}_4 := \bar{L}_1 \qquad \text{and} \qquad \bar{L}_5 := L_4 \tag{105}$$

to write

$$||P(u)||_W \leq \frac{\bar{L}_4||u||_W + \bar{L}_5||u||_W^2}{\epsilon^{5/2}}. \tag{106}$$

This completes the proof of (56) and this completes the proof of Theorem 3.

Local and Global Existence of Solution in $H^2 \cap L^\infty$ Space

We now prove Theorem 4. We first prove local existence for a finite time interval. We find that we can choose this time interval independent of the initial data. We repeat the local existence theorem to uniquely continue the local solution over any finite time interval. The existence and uniqueness of local solutions is stated in the following theorem.

Theorem 5 (Local existence and uniqueness). *Given $X = W \times W$, $b(t) \in W$, and initial data $x_0 = (u_0, v_0) \in X$. We suppose that $b(t)$ is continuous in time over some time interval $I_0 = (-T, T)$ and satisfies $\sup_{t \in I_0} ||b(t)||_W < \infty$. Then, there exists a time interval $I' = (-T', T') \subset I_0$ and unique solution $y = (y^1, y^2)$ such that $y \in C^1(I'; X)$ and*

$$y(t) = x_0 + \int_0^t F^\epsilon(y(\tau), \tau)\, d\tau, \text{ for } t \in I' \tag{107}$$

or equivalently

$$y'(t) = F^\epsilon(y(t), t), \text{with } y(0) = x_0, \text{ for } t \in I'$$

where $y(t)$ and $y'(t)$ are Lipschitz continuous in time for $t \in I' \subset I_0$.

Proof. To prove Theorem 5, we proceed as follows. Write $y(t) = (y^1(t), y^2(t))$ with $||y||_X = ||y^1(t)||_W + ||y^2(t)||_W$. Let us consider $R > ||x_0||_X$ and define the ball $B(0, R) = \{y \in X : ||y||_X < R\}$. Let $r < \min\{1, R - ||x_0||_X\}$. We clearly have $r^2 < (R - ||x_0||_X)^2$ as well as $r^2 < r < R - ||x_0||_X$. Consider the ball $B(x_0, r^2)$ defined by

$$B(x_0, r^2) = \{y \in X : ||y - x_0||_X < r^2\}. \tag{108}$$

Then we have $B(x_0, r^2) \subset B(x_0, r) \subset B(0, R)$ (see Fig. 7).

To recover the existence and uniqueness, we introduce the transformation

$$S_{x_0}(y)(t) = x_0 + \int_0^t F^\epsilon(y(\tau), \tau)\, d\tau.$$

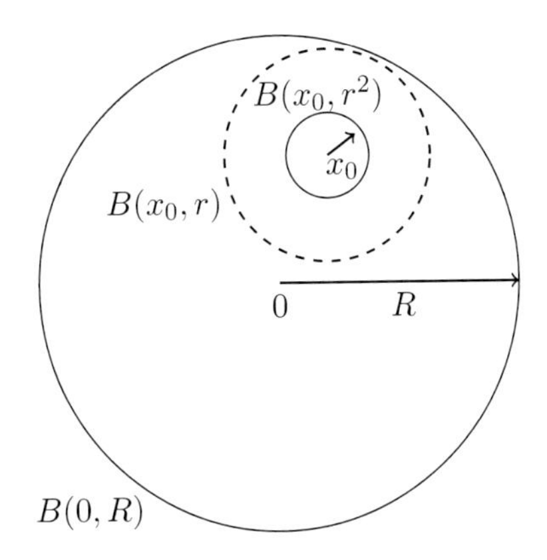

Fig. 7 Geometry

Introduce $0 < T' < T$ and the associated set $Y(T')$ of functions in W taking values in $B(x_0, r^2)$ for $I' = (-T', T') \subset I_0 = (-T, T)$. The goal is to find appropriate interval $I' = (-T', T')$ for which S_{x_0} maps into the corresponding set $Y(T')$. Writing out the transformation with $y(t) \in Y(T')$ gives

$$S^1_{x_0}(y)(t) = x^1_0 + \int_0^t y^2(\tau)\, d\tau \tag{109}$$

$$S^2_{x_0}(y)(t) = x^2_0 + \int_0^t (-\nabla PD^\epsilon(y^1(\tau)) + b(\tau))\, d\tau. \tag{110}$$

We have from (109)

$$||S^1_{x_0}(y)(t) - x^1_0||_W \leq \sup_{t \in (-T', T')} ||y^2(t)||_W T'. \tag{111}$$

Using bound on $-\nabla PD^\epsilon$ in Theorem 3, we have from (110)

$$||S^2_{x_0}(y)(t) - x^2_0||_W \leq \int_0^t \left[\frac{\bar{L}_4}{\epsilon^{5/2}} ||y^1(\tau)||_W + \frac{\bar{L}_5}{\epsilon^{5/2}} ||y^1(\tau)||^2_W + ||b(\tau)||_W \right] d\tau. \tag{112}$$

Let $\bar{b} = \sup_{t \in I_0} ||b(t)||_W$. Noting that transformation S_{x_0} is defined for $t \in I' = (-T', T')$ and $y(\tau) = (y^1(\tau), y^2(\tau)) \in B(x_0, r^2) \subset B(0, R)$ as $y \in Y(T')$, we have from (112) and (111)

$$||S^1_{x_0}(y)(t) - x^1_0||_W \leq RT',$$

$$||S^2_{x_0}(y)(t) - x^2_0||_W \leq \left[\frac{\bar{L}_4 R + \bar{L}_5 R^2}{\epsilon^{5/2}} + \bar{b} \right] T'.$$

Adding gives

$$||S_{x_0}(y)(t) - x_0||_X \leq \left[\frac{\bar{L}_4 R + \bar{L}_5 R^2}{\epsilon^{5/2}} + R + \bar{b} \right] T'. \tag{113}$$

Choosing T' as below

$$T' < \frac{r^2}{\left[\frac{\bar{L}_4 R + \bar{L}_5 R^2}{\epsilon^{5/2}} + R + \bar{b} \right]} \tag{114}$$

will result in $S_{x_0}(y) \in Y(T')$ for all $y \in Y(T')$ as

$$||S_{x_0}(y)(t) - x_0||_X < r^2. \tag{115}$$

Since $r^2 < (R - ||x_0||_X)^2$, we have

$$T' < \frac{r^2}{\left[\frac{\bar{L}_4 R + \bar{L}_5 R^2}{\epsilon^{5/2}} + R + \bar{b}\right]} < \frac{(R - ||x_0||_X)^2}{\left[\frac{\bar{L}_4 R + \bar{L}_5 R^2}{\epsilon^{5/2}} + R + \bar{b}\right]}.$$

Let $\theta(R)$ be given by

$$\theta(R) := \frac{(R - ||x_0||_X)^2}{\left[\frac{\bar{L}_4 R + \bar{L}_5 R^2}{\epsilon^{5/2}} + R + \bar{b}\right]}. \tag{116}$$

$\theta(R)$ is increasing with $R > 0$ and satisfies

$$\theta_\infty := \lim_{R\to\infty} \theta(R) = \frac{\epsilon^{5/2}}{\bar{L}_5}. \tag{117}$$

So given R and $||x_0||_X$, we choose T' according to

$$\frac{\theta(R)}{2} < T' < \theta(R), \tag{118}$$

and set $I' = (-T', T')$. This way we have shown that for time domain I' the transformation $S_{x_0}(y)(t)$ as defined in Eqs. 109 and 110 maps $Y(T')$ into itself. Existence and uniqueness of solution can be established using (Theorem 6.10, Driver 2003). □

We now prove Theorem 1. From the proof of Theorem 2 above, we have a unique local solution over a time domain $(-T', T')$ with $\frac{\theta(R)}{2} < T'$. Since $\theta(R) \nearrow \epsilon^{5/2}/\bar{L}_5$ as $R \nearrow \infty$, we can fix a tolerance $\eta > 0$ so that $[(\epsilon^{5/2}/2\bar{L}_5) - \eta] > 0$. Then for any initial condition in W and $b = \sup_{t\in[-T,T)} ||b(t)||_W$, we can choose R sufficiently large so that $||x_0||_X < R$ and $0 < (\epsilon^{5/2}/2\bar{L}_5) - \eta < T'$. Since choice of T' is independent of initial condition and R, we can always find local solutions for time intervals $(-T', T')$ for T' larger than $[(\epsilon^{5/2}/2\bar{L}_5) - \eta] > 0$. Therefore we apply the local existence and uniqueness result to uniquely continue local solutions up to an arbitrary time interval $(-T, T)$.

Conclusions: Convergence of Regular Solutions in the Limit of Vanishing Horizon

In this final section, we examine the behavior of bounded Hölder continuous solutions as the peridynamic horizon tends to zero. We find that the solutions converge to a limiting sharp fracture evolution with bounded Griffiths fracture energy and satisfy the linear elastic wave equation away from the fracture set. We look at a subset of Hölder solutions that are differentiable in the spatial variables to show that sharp fracture evolutions can be approached by spatially smooth evolutions in the limit of vanishing nonlocality. As ϵ approaches zero, derivatives can become large but must localize to surfaces across which the limiting evolution jumps. These conclusions are reported in Jha and Lipton (2017b). The same behavior can be recovered for bounded H^2 solutions in the limit of vanishing horizon. These results support the numerical simulation for more regular nonlocal evolutions that approximate sharp fracture in the limit of vanishing nonlocality. In the next chapter we provide a priori estimates of the finite difference and finite element approximations to fracture evolution for nonlocal models with horizon $\epsilon > 0$.

To fix ideas consider a sequence of peridynamic horizons $\epsilon_k = 1/k, k = 1, \dots$ and the associated Hölder continuous solutions $\boldsymbol{u}^{\epsilon_k}(t, \boldsymbol{x})$ of the peridynamic initial value problem Eqs. 1, 2, and 3. We assume that the initial conditions $\boldsymbol{u}_0^{\epsilon_k}, \boldsymbol{v}_0^{\epsilon_k}$ have uniformly bounded peridynamic energy and mean square initial velocity given by

$$\sup_{\epsilon_k} PD^{\epsilon_k}(\boldsymbol{u}_0^{\epsilon_k}) < \infty \text{ and } \sup_{\epsilon_k} ||\boldsymbol{v}_0^{\epsilon_k}||_{L^2(D;\mathbb{R}^d)} < \infty.$$

Moreover we suppose that $\boldsymbol{u}_0^{\epsilon_k}, \boldsymbol{v}_0^{\epsilon_k}$ are differentiable on D and that they converge in $L^2(D;\mathbb{R})$ to $\boldsymbol{u}_0^0, \boldsymbol{v}_0^0$ with bounded Griffith free energy given by

$$\int_D 2\mu|\mathcal{E}\boldsymbol{u}_0^0|^2 + \lambda|\operatorname{div}\boldsymbol{u}_0^0|^2\,dx + \mathcal{G}_c\mathcal{H}^{d-1}(J_{\boldsymbol{u}_0^0}) \leq C < \infty,$$

where $J_{\boldsymbol{u}_0^0}$ denotes an initial fracture surface given by the jumps in the initial deformation $\boldsymbol{u}_0^0$ and $\mathcal{H}^2(J_{\boldsymbol{u}^0(t)})$ is its *two*-dimensional Hausdorff measure of the jump set. Here $\mathcal{E}\boldsymbol{u}_0^0$ is the elastic strain and $\operatorname{div}\boldsymbol{u}_0^0 = Tr(\mathcal{E}\boldsymbol{u}_0^0)$. The constants μ, λ are given by the explicit formulas

$$\text{and } \mu = \lambda = \tfrac{1}{5}f'(0)\int_0^1 r^d J(r)dr,\ d = 2, 3$$

and

$$\mathcal{G}_c = \frac{3}{2}f_\infty\int_0^1 r^d J(r)dr,\ d = 2, 3,$$

where $f'(0)$ and f_∞ are defined by Eq. 6. Here $\mu = \lambda$ and is a consequence of the central force model used in cohesive dynamics. Last we suppose as in Lipton (2016) that the solutions are uniformly bounded, i.e.,

$$\sup_{\epsilon_k}\sup_{[0,T]} ||\boldsymbol{u}^{\epsilon_k}(t)||_{L^\infty(D;\mathbb{R}^d)} < \infty,$$

The Hölder solutions $\boldsymbol{u}^{\epsilon_k}(t, \boldsymbol{x})$ naturally belong to $L^2(D;\mathbb{R}^d)$ for all $t \in [0, T]$, and we can directly apply the Gronwall inequality (Equation (6.9) of Lipton

2016) together with Theorems 6.2 and 6.4 of Lipton (2016) to conclude similar to Theorems 5.1 and 5.2 of Lipton (2016) that there is at least one "cluster point" $\boldsymbol{u}^0(t,\boldsymbol{x})$ belonging to $C([0,T];L^2(D;\mathbb{R}^d))$ and subsequence, also denoted by $\boldsymbol{u}^{\epsilon_k}(t,\boldsymbol{x})$ for which

$$\lim_{\epsilon_k\rightarrow 0}\max_{0\leq t\leq T}\left\{\|\boldsymbol{u}^{\epsilon_k}(t)-\boldsymbol{u}^0(t)\|_{L^2(D;\mathbb{R}^d)}\right\}=0.$$

Moreover it follows from Lipton (2016) that the limit evolution $\boldsymbol{u}^0(t,\boldsymbol{x})$ has a weak derivative $\boldsymbol{u}_t^0(t,\boldsymbol{x})$ belonging to $L^2([0,T]\times D;\mathbb{R}^d)$. For each time $t\in[0,T]$, we can apply methods outlined in Lipton (2016) to find that the cluster point $\boldsymbol{u}^0(t,\boldsymbol{x})$ is a special function of bounded deformation (see Ambrosio et al. 1997; Bellettini et al. 1998) and has bounded linear elastic fracture energy given by

$$\int_D 2\mu|\mathcal{E}\boldsymbol{u}^0(t)|^2+\lambda|\mathrm{div}\,\boldsymbol{u}^0(t)|^2\,dx+\mathcal{G}_c\mathcal{H}^2(J_{\boldsymbol{u}^0(t)})\leq C,$$

for $0\leq t\leq T$ where $J_{\boldsymbol{u}^0(t)}$ denotes the evolving fracture surface The deformation-crack set pair $(\boldsymbol{u}^0(t),J_{\boldsymbol{u}^0(t)})$ records the brittle fracture evolution of the limit dynamics.

Arguments identical to Lipton (2016) show that away from sets where $|S(\boldsymbol{y},\boldsymbol{x};\boldsymbol{u}^{\epsilon_k})|>S_c$, the limit $\boldsymbol{u}^0$ satisfies the linear elastic wave equation. This is stated as follows: Fix $\delta>0$ and for $\epsilon_k<\delta$ and $0\leq t\leq T$ consider the open set $D'\subset D$ for which points $\boldsymbol{x}$ in D' and $\boldsymbol{y}$ for which $|\boldsymbol{y}-\boldsymbol{x}|<\epsilon_k$ satisfy,

$$|S(\boldsymbol{y},\boldsymbol{x};\boldsymbol{u}^{\epsilon_k}(t))|<S_c(\boldsymbol{y},\boldsymbol{x}).$$

Then the limit evolution $\boldsymbol{u}^0(t,\boldsymbol{x})$ evolves elastodynamically on D' and is governed by the balance of linear momentum expressed by the Navier-Lamé equations on the domain $[0,T]\times D'$ given by

$$\boldsymbol{u}_{tt}^0(t)=\mathrm{div}\boldsymbol{\sigma}(t)+\boldsymbol{b}(t),\text{on }[0,T]\times D',$$

where the stress tensor $\boldsymbol{\sigma}$ is given by

$$\boldsymbol{\sigma}=\lambda I_d Tr(\mathcal{E}\,\boldsymbol{u}^0)+2\mu\mathcal{E}\boldsymbol{u}^0,$$

where I_d is the identity on $\mathbb{R}^d$ and $Tr(\mathcal{E}\,\boldsymbol{u}^0)$ is the trace of the strain. Here the second derivative $\boldsymbol{u}_{tt}^0$ is the time derivative in the sense of distributions of $\boldsymbol{u}_t^0$, and $\mathrm{div}\boldsymbol{\sigma}$ is the divergence of the stress tensor $\boldsymbol{\sigma}$ in the distributional sense. This shows that sharp fracture evolutions can be approached by spatially smooth evolutions in the limit of vanishing nonlocality.

Acknowledgements This material is based upon work supported by the US Army Research Laboratory and the US Army Research Office under contract/grant number W911NF1610456.

References

L. Ambrosio, A. Coscia, G. Dal Maso, Fine properties of functions with bounded deformation. Arch. Ration. Mech. Anal. **139**, 201–238 (1997)

G. Bellettini, A. Coscia, G. Dal Maso, Compactness and lower semicontinuity properties. Mathematische Zeitschrift **228**, 337–351 (1998)

P. Diehl, R. Lipton, M. Schweitzer, Numerical verification of a bond-based softening peridynamic model for small displacements: deducing material parameters from classical linear theory. University of Bonn Technical report, Institut für Numerische Simulation (2016)

F. Demengel and G. Demengel, Functional spaces for the theory of elliptic partial differential equations. Springer Verlag, London (2012)

B.K. Driver, Analysis tools with applications. Lecture notes (2003), www.math.ucsd.edu/~bdriver/240-01-02/Lecture_Notes/anal.pdf

E. Emmrich, R.B. Lehoucq, D. Puhst, Peridynamics: a nonlocal continuum theory, in *Meshfree Methods for Partial Differential Equations VI*, ed. by M. Griebel, M.A. Schweitzer. Lecture notes in computational science and engineering, vol. 89 (Springer, Berlin/Heidelberg, 2013), pp. 45–65

P.K. Jha, R. Lipton, Finite element approximation of nonlinear nonlocal models. arXiv preprint arXiv:1710.07661 (2017a)

P.K. Jha, R. Lipton, Numerical analysis of peridynamic models in Hölder space. arXiv preprint arXiv:1701.02818 (2017b)

P.K. Jha, R. Lipton, Numerical convergence of nonlinear nonlocal continuum models to local elastodynamics. arXiv preprint arXiv:1707.00398 (2017c)

R. Lipton, Dynamic brittle fracture as a small horizon limit of peridynamics. J. Elast. **117**(1), 21–50 (2014)

R. Lipton, Cohesive dynamics and brittle fracture. J. Elast. **124**(2), 143–191 (2016)

R. Lipton, S. Silling, R. Lehoucq, Complex fracture nucleation and evolution with nonlocal elastodynamics. arXiv preprint arXiv:1602.00247 (2016)

T. Mengesha, Q. Du, Nonlocal constrained value problems for a linear peridynamic navier equation. J. Elast. **116**(1), 27–51 (2014)

S.A. Silling, Reformulation of elasticity theory for discontinuities and long-range forces. J. Mech. Phys. Solids **48**(1), 175–209 (2000)

S.A. Silling, R.B. Lehoucq, Convergence of peridynamics to classical elasticity theory. J. Elast. **93**(1), 13–37 (2008)

Finite Differences and Finite Elements in Nonlocal Fracture Modeling: A Priori Convergence Rates

44

Prashant K. Jha and Robert Lipton

Contents

Abstract

In this chapter we present a rigorous convergence analysis of finite difference and finite element approximation of nonlinear nonlocal models. In the previous chapter, we considered a differentiable version of the original bond-based model introduced in Silling (J Mech Phys Solids 48(1):175–209, 2000). There we showed, for a fixed horizon of nonlocal interaction ϵ, that well-posed

P. K. Jha
Department of Mathematics, Louisiana State University, Baton Rouge, LA, USA
e-mail: prashant.j16o@gmail.com; jha@math.lsu.edu

R. Lipton (✉)
Department of Mathematics, and Center for Computation and Technology, Louisiana State University, Baton Rouge, LA, USA
e-mail: lipton@math.lsu.edu; lipton@lsu.edu

G. Z. Voyiadjis (ed.), *Handbook of Nonlocal Continuum Mechanics for Materials and Structures*, https://doi.org/10.1007/978-3-319-58729-5_44

formulations of the model can be developed over Hölder spaces and Sobolev spaces. In this chapter we apply these formulations to show a priori convergence for the discrete finite difference and finite element methods. We show that the error made using the forward Euler in time and a finite difference (i.e., piecewise constant) discretization in space with time step Δt and spatial discretization h is of the order of $O(\Delta t + h/\epsilon^2)$. For a central difference approximation in time and piecewise linear finite element approximation in space, the approximation error is of the order of $O(\Delta t + h^2/\epsilon^2)$. We point out these are the first such error estimates for nonlinear nonlocal fracture formulations and are reported in Jha and Lipton (2017b Numerical analysis of nonlocal fracture models models in holder space. arXiv preprint arXiv:1701.02818. To appear in SIAM Journal on Numerical Analysis 2018) and Jha and Lipton (2017a, Finite element approximation of nonlocal fracture models. arXiv preprint arXiv:1710.07661). We then go on to prove the stability of the semi-discrete approximation and show that the energy of the discrete approximation is bounded in terms of work done by the body force and initial energy put into the system. We look forward to improvements and development of a posteriori error estimation in the coming years.

Keywords
Peridynamic modeling · Finite differences · Finite elements · Stability · Convergence

Introduction

In this chapter we present a rigorous convergence analysis of finite difference and finite element approximation of nonlinear nonlocal fracture models. The model considered in this work is a differentiable version of the original peridynamic bond-based model introduced in Silling (2000) and analyzed in Lipton (2014, 2016). It is a bond-based model characterized by a nonlinear double-well potential. As discussed in the previous chapter, the nonlocal evolution converges to a sharp fracture evolution with bounded Griffith fracture energy as the length scale of nonlocality ϵ tends to zero. In this limit the displacement field satisfies the linear elastic wave equation off the fracture set.

We first consider the forward Euler time discretization and finite difference approximations in space with a uniform square mesh in 2-d and cubic mesh in 3-d. The mesh size is taken to be h and the time step is Δt. An a priori bound on the error is obtained for solutions in the Hölder space $C_0^{0,\gamma}(D;\mathbb{R}^d)$, where $\gamma \in (0,1]$ is the Hölder exponent, D is the material domain, and $d = 2, 3$ is the dimension. The rate of convergence is shown to be no larger than $O(\Delta t + h^\gamma/\epsilon^2)$. We also show stability of the semi-discrete approximation. The semi-discrete evolution is shown to be uniformly bounded in time in terms of initial energy and the work done by body force. In this chapter we prove all results for the forward Euler in time discretization, and we refer to Jha and Lipton (2017b) for the general single-step time discretization.

Next we consider central differences in time and a finite element discretization in space using triangular or tetrahedral meshes and conforming linear elements. Assuming $H_0^2(D;\mathbb{R}^d)$ solutions, we estimate the error and obtain a convergence rate of $O(\Delta t + h^2/\epsilon^2)$. We show that the semi-discrete evolution for the finite element scheme is also stable in time. We provide a stability analysis of the fully discrete problem, for the linearized peridynamic force. For this case we exhibit a CFL-like stability condition for the time step Δt.

The results presented here show that convergence requires $h^\gamma < \epsilon^2$ for the finite difference case while $h^2 < \epsilon^2$ (or $h < \epsilon$) for the finite element case. The technical reason for the appearance of the factor $1/\epsilon^2$ in these convergence rates is that we are numerically approximating a nonlinear but Lipschitz continuous vector valued ODE. Here the vector space is the space of square integrable displacement fields, and the $1/\epsilon^2$ factor is proportional to the Lipschitz constant of the nonlocal nonlinear force acting on mean square integrable displacement fields. Our results requiring $h < \epsilon$ are consistent with the earlier work of Tian and Du (2014) for linear nonlocal forces and finite element approximations applied to equilibrium problems. We point out that the nonlocal nonlinear models treated here are identified with sharp fracture evolutions as $\epsilon \to 0$ (see Lipton 2014, 2016). However a convergence rate with respect to ϵ remains to be established. We discuss this aspect in the conclusions section.

There is a rapidly growing literature in peridynamic modeling and analysis (see, e.g., Emmrich et al. 2007; Du and Zhou 2011; Foster et al. 2011; Aksoylu and Parks 2011; Du et al. 2013a; Dayal 2017; Emmrich et al. 2013; Mengesha and Du 2013; Lipton 2014, 2016; Lipton et al. 2016; Emmrich and Puhst 2016; Du, Tao, and Tian 2017; Lipton et al. 2018; Aksoylu and Mengesha 2010; Mengesha and Du 2013, 2014; Aksoylu and Unlu 2014). In Macek and Silling (2007), Gerstle et al. (2007), Littlewood (2010), the finite element method is applied to the peridynamics formulation for the simulation of cracks. In Du et al. (2013b), the finite element approximation of linear peridynamic models for general quasistatic evolutions is analyzed. For linear elastic local models, the stability of the general Newmark time discretization is shown in Baker (1976), Grote and Schötzau (2009), and Karaa (2012). This behavior is shown to persist for elastic nonlocal models in Guan and Gunzburger (2015). In Chen and Gunzburger (2011), the finite element approximation with continuous and discontinuous elements is developed for nonlocal problems in one dimension. A numerical analysis of linear peridynamics models for a 1-d bar has been carried out in Weckner and Emmrich (2005) and Bobaru et al. (2009). In Tian and Du (2014) and Tian et al. (2016a,b), an asymptotically compatible approximation scheme is identified. In Askari et al. (2008), Silling et al. (2010), Ha and Bobaru (2011), Agwai et al. (2011), Bobaru and Hu (2012), and Zhang et al. (2016), crack prediction and crack branching phenomenon are analyzed through peridynamics. The list of references is by no means complete; additional references to the literature can be found in this handbook.

Review of the Nonlocal Model

We define the strain associated with the displacement field $\boldsymbol{u}(x)$ as

$$S(\boldsymbol{u}) = S(y, x; \boldsymbol{u}) = \frac{\boldsymbol{u}(y) - \boldsymbol{u}(x)}{|y - x|} \cdot e_{y-x} \text{ and } e_{y-x} = \frac{y - x}{|y - x|}. \tag{1}$$

We consider the following type of potential (see Figs. 1 and 2)

$$W^{\epsilon}(S, y - x) = \omega(x)\omega(y)\frac{J^{\epsilon}(|y - x|)}{\epsilon} f(|y - x|S^2), \tag{2}$$

where the function $f : \mathbb{R}^+ \to \mathbb{R}$ is positive, smooth, and concave and satisfies the following properties

$$\lim_{r \to 0^+} \frac{f(r)}{r} = f'(0), \qquad \lim_{r \to \infty} f(r) = f_\infty < \infty. \tag{3}$$

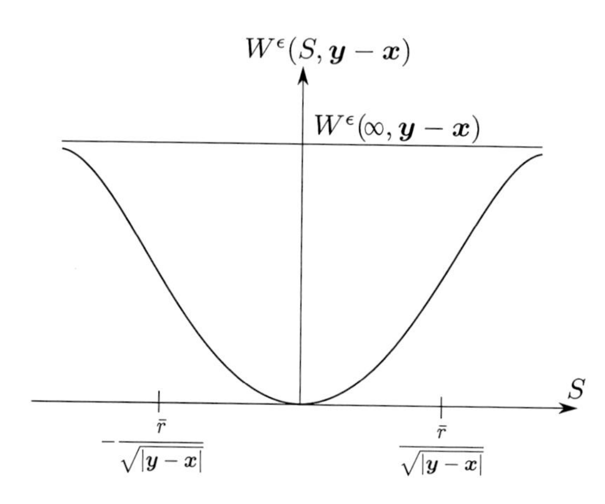

Fig. 1 Two-point potential $W^{\epsilon}(S, y - x)$ as a function of strain S for fixed $y - x$

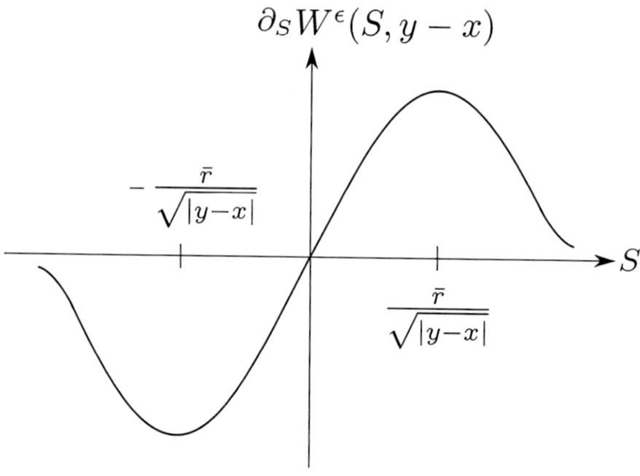

Fig. 2 Nonlocal force $\partial_S W^{\epsilon}(S, y - x)$ as a function of strain S for fixed $y - x$. Second derivative of $W^{\epsilon}(S, y - x)$ is zero at $\pm\bar{r}/\sqrt{|y - x|}$

The function $J^\epsilon(r) = J(r/\epsilon)$ is the influence function where $0 \leq J(|x|) \leq M$ for $x \in H_1(0)$ and $J = 0$ outside. The boundary function $0 \leq \omega(x) \leq 1$ takes the value 0 on the boundary ∂D of the material domain D. For $x \in D$, a distance ϵ away from boundary, $\omega(x)$ is 1 and smoothly decreases from 1 to zero as x approaches ∂D.

In the sequel we will set

$$\bar{\omega}(x) = \omega(x)\omega(x + \epsilon\xi) \tag{4}$$

and we assume

$$|\nabla\bar{\omega}| \leq C_{\omega_1} < \infty \quad \text{and} \quad |\nabla^2\bar{\omega}| \leq C_{\omega_2} < \infty.$$

The peridynamic force is written $-\nabla PD^\epsilon$ and given by

$$\begin{aligned}
&-\nabla PD^\epsilon(\boldsymbol{u})(x)\\
&= \frac{2}{\epsilon^d \omega_d} \int_{H_\epsilon(x)} \partial_S W^\epsilon(S, y - x) \frac{y - x}{|y - x|} dy\\
&= \frac{4}{\epsilon^{d+1}\omega_d} \int_{H_\epsilon(x)} \omega(x)\omega(y) J^\epsilon(|y - x|) f'(|y - x| S(\boldsymbol{u})^2) S(\boldsymbol{u}) e_{y-x} dy,
\end{aligned} \tag{5}$$

Peridynamics Equation

Let $u : [0, T] \times D \to \mathbb{R}^d$ be the displacement field such that it satisfies the following evolution equation

$$\rho\partial^2_{tt}\boldsymbol{u}(t, x) = -\nabla PD^\epsilon(\boldsymbol{u}(t))(x) + b(t, x), \tag{6}$$

where $b(t, x)$ is the body force and ρ is the density. We will assume $\rho = 1$ throughout the chapter. The initial condition is given by

$$\boldsymbol{u}(0, x) = \boldsymbol{u}_0(x) \qquad \text{and} \qquad \partial_t\boldsymbol{u}(0, x) = \boldsymbol{v}_0(x) \tag{7}$$

and the boundary condition is given by

$$\boldsymbol{u}(t, x) = 0 \qquad \forall x \in \partial D, \forall t \in [0, T]. \tag{8}$$

Throughout this chapter, we will assume $\boldsymbol{u} = 0$ on the boundary ∂D and is extended outside D by zero.

Additionally we can also write the evolution in weak form by multiplying Eq. 6 by a smooth test function $\tilde{\boldsymbol{u}}$ with $\tilde{\boldsymbol{u}} = 0$ on ∂D to get

$$(\ddot{\boldsymbol{u}}(t), \tilde{\boldsymbol{u}}) = (-\nabla PD^\epsilon(\boldsymbol{u}(t)), \tilde{\boldsymbol{u}}) + (b(t), \tilde{\boldsymbol{u}}).$$

We denote L^2 dot product of u, v as (u, v). An integration by parts easily shows for all smooth u, v taking zero boundary values that

$$(-\nabla PD^\epsilon(\boldsymbol{u}), \boldsymbol{v}) = -a^\epsilon(\boldsymbol{u}, \boldsymbol{v}),$$

where

$$\begin{aligned} &a^\epsilon(\boldsymbol{u}, \boldsymbol{v}) \\ &= \frac{2}{\epsilon^{d+1}\omega_d} \int_D \int_{H_\epsilon(x)} \omega(x)\omega(y) J^\epsilon(|y-x|) \\ &f'(|y-x|S(\boldsymbol{u})^2)|y-x|S(\boldsymbol{u})S(\boldsymbol{v}) dy dx. \end{aligned} \tag{9}$$

Weak form of the evolution in terms of operator a^ϵ becomes

$$(\ddot{\boldsymbol{u}}(t), \tilde{\boldsymbol{u}}) + a^\epsilon(\boldsymbol{u}(t), \tilde{\boldsymbol{u}}) = (b(t), \tilde{\boldsymbol{u}}). \tag{10}$$

Last we introduce the peridynamic energy. The total energy $\mathcal{E}^\epsilon(\boldsymbol{u})(t)$ is given by the sum of kinetic and potential energy given by

$$\mathcal{E}^\epsilon(\boldsymbol{u})(t) = \frac{1}{2}||\dot{\boldsymbol{u}}(t)||_{L^2(D;\mathbb{R}^d)} + PD^\epsilon(\boldsymbol{u}(t)), \tag{11}$$

where potential energy PD^ϵ is given by

$$PD^\epsilon(\boldsymbol{u}) = \int_D \left[\frac{1}{\epsilon^d \omega_d} \int_{H_\epsilon(x)} W^\epsilon(S(\boldsymbol{u}), y-x) dy \right] dx.$$

We state the following equation which will be used later in the chapter

$$\frac{d}{dt}\mathcal{E}^\epsilon(\boldsymbol{u})(t) = (\ddot{\boldsymbol{u}}(t), \dot{\boldsymbol{u}}(t)) - (-\nabla PD^\epsilon(\boldsymbol{u}(t)), \dot{\boldsymbol{u}}(t)). \tag{12}$$

In order to develop the approximation theory in the following sections, we find it convenient to write the evolution Eq. 6 as an equivalent first order system with $y_1(t) = \boldsymbol{u}(t)$ and $y_2(t) = \boldsymbol{v}(t)$ with $\boldsymbol{v}(t) = \partial_t \boldsymbol{u}(t)$. Let $y = (y_1, y_2)^T$ where at each time y_1, y_2 belong to the same function space V, and let $F^\epsilon(y,t) = (F_1^\epsilon(y,t), F_2^\epsilon(y,t))^T$ such that

$$F_1^\epsilon(y,t) := y_2 \tag{13}$$

$$F_2^\epsilon(y,t) := -\nabla PD^\epsilon(y_1) + \boldsymbol{b}(t). \tag{14}$$

The initial boundary value associated with the evolution Eq. 6 is equivalent to the initial boundary value problem for the first order system given by

$$\frac{d}{dt}y = F^\epsilon(y,t), \tag{15}$$

with initial condition given by $y(0) = (\boldsymbol{u}_0, \boldsymbol{v}_0)^T \in X = V \times V$.

To establish the error estimates, we will use the Lipschitz property of the peridynamic force in $X = L^2(D;\mathbb{R}^d) \times L^2(D;\mathbb{R}^d)$. It is given by Theorem 6.1 of Lipton 2016.

Theorem 1.

$$||F^\epsilon(y,t) - F^\epsilon(z,t)||_X \leq \frac{L}{\epsilon^2}||y - z||_X \qquad \forall y, z \in X, \forall t \in [0,T] \tag{16}$$

for all $y, z \in L^2(D;\mathbb{R}^d)^2$.
Here L does not depend on $\boldsymbol{u}, \boldsymbol{v}$.

Finite Difference Approximation

In this section, we present the finite difference scheme and compute the rate of convergence. We also consider the semi-discrete approximation and prove the bound on energy of semi-discrete evolution in terms of initial energy and the work done by body forces.

Let h be the size of a mesh and Δt be the size of time step. We will keep ϵ fixed and assume that $h < \epsilon < 1$. Let $D_h = D \cap (h\mathbb{Z})^d$ be the discretization of material domain. Let $i \in \mathbb{Z}^d$ be the index such that $\boldsymbol{x}_i = hi \in D$. Let U_i be the unit cell of volume h^d corresponding to the grid point $\boldsymbol{x}_i$, see Fig. 3. The exact solution evaluated at grid points is denoted by $(\boldsymbol{u}_i(t), \boldsymbol{v}_i(t))$.

Time Discretization

Let $[0,T] \cap (\Delta t \mathbb{Z})$ be the discretization of time domain where Δt is the size of time step. Denote fully discrete solution at $(t^k = k\Delta t, \boldsymbol{x}_i = ih)$ as $(\hat{\boldsymbol{u}}_i^k, \hat{\boldsymbol{v}}_i^k)$. Similarly, the exact solution evaluated at space-time grid points is denoted by $(\boldsymbol{u}_i^k, \boldsymbol{v}_i^k)$. We enforce the boundary condition $\hat{\boldsymbol{u}}_i^k = \boldsymbol{0}$ for all $\boldsymbol{x}_i \notin D$ and for all k.

We begin with the forward Euler time discretization, with respect to velocity, and the finite difference scheme for $(\hat{\boldsymbol{u}}_i^k, \hat{\boldsymbol{v}}_i^k)$ is written

$$\frac{\hat{\boldsymbol{u}}_i^{k+1} - \hat{\boldsymbol{u}}_i^k}{\Delta t} = \hat{\boldsymbol{v}}_i^{k+1} \tag{17}$$

$$\frac{\hat{\boldsymbol{v}}_i^{k+1} - \hat{\boldsymbol{v}}_i^k}{\Delta t} = -\nabla PD^\epsilon(\hat{\boldsymbol{u}}^k)(\boldsymbol{x}_i) + \boldsymbol{b}_i^k \tag{18}$$

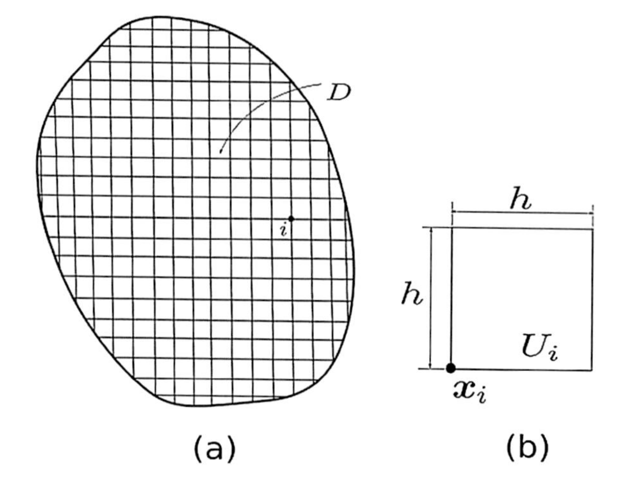

Fig. 3 (**a**) Typical mesh of size h. (**b**) Unit cell U_i corresponding to material point $\boldsymbol{x}_i$

The scheme is complemented with the discretized initial conditions $\hat{\boldsymbol{u}}_i^0 = (\hat{\boldsymbol{u}}_0)_i$ and $\hat{\boldsymbol{v}}_i^0 = (\hat{\boldsymbol{v}}_0)_i$. If we substitute Eq. 17 into Eq. 18, we get the standard central difference scheme in time for second order in time differential equation. Here we have assumed, without loss of generality, $\rho = 1$.

The piecewise constant extensions of the discrete sets $\{\hat{\boldsymbol{u}}_i^k\}_{i\in\mathbb{Z}^d}$ and $\{\hat{\boldsymbol{v}}_i^k\}_{i\in\mathbb{Z}^d}$ are given by

$$\hat{\boldsymbol{u}}^k(\boldsymbol{x}) := \sum_{i,\boldsymbol{x}_i\in D} \hat{\boldsymbol{u}}_i^k \chi_{U_i}(\boldsymbol{x})$$

$$\hat{\boldsymbol{v}}^k(\boldsymbol{x}) := \sum_{i,\boldsymbol{x}_i\in D} \hat{\boldsymbol{v}}_i^k \chi_{U_i}(\boldsymbol{x})$$

In this way we represent the finite difference solution as a piecewise constant function. We will show that this function provides an L^2 approximation of the exact solution.

Convergence Results

In this section we provide upper bounds on the rate of convergence of the discrete approximation to the solution of the peridynamic evolution. The L^2 approximation error E^k at time t^k, for $0 < t^k \leq T$, is defined as

$$E^k := \left\|\hat{\boldsymbol{u}}^k - \boldsymbol{u}^k\right\|_{L^2(D;\mathbb{R}^d)} + \left\|\hat{\boldsymbol{v}}^k - \boldsymbol{v}^k\right\|_{L^2(D;\mathbb{R}^d)}$$

The upper bound on the convergence rate of the approximation error is given by the following theorem.

Theorem 2 (Convergence of finite difference approximation (forward Euler time discretization)). *Let $\epsilon > 0$ be fixed. Let $(\boldsymbol{u}, \boldsymbol{v})$ be the solution of peridynamic equation Eq. 15. We assume $\boldsymbol{u}, \boldsymbol{v} \in C^2([0,T]; C_0^{0,\gamma}(D;\mathbb{R}^d))$. Then the forward Euler time discretization, and finite difference spatial discretization scheme given by Eqs. 17 and 18, is consistent in both time and spatial discretization and converges to the exact solution uniformly in time with respect to the $L^2(D;\mathbb{R}^d)$ norm. If we assume the error at the initial step is zero, then the error E^k at time t^k is bounded and to leading order in the time step Δt satisfies*

$$\sup_{0\le k\le T/\Delta t} E^k \le O\left(C_t\Delta t + C_s\frac{h^\gamma}{\epsilon^2}\right), \tag{19}$$

where constants C_s and C_t are independent of h and Δt and C_s depends on the Hölder norm of the solution and C_t depends on the L^2 norms of time derivatives of the solution.

Here we have assumed the initial error to be zero for ease of exposition only.

We remark that the explicit constants leading to Eq. 19 can be large. The inequality that delivers Eq. 19 is given to leading order by

$$\sup_{0\le k\le T/\Delta t} E^k \le \exp\left[T(1+6\bar{C}/\epsilon^2)\right] T\left[C_t\Delta t + (C_s/\epsilon^2)h^\gamma\right], \tag{20}$$

where the constants $\bar{C}$, C_t, and C_s are given by Eqs. 43, 45, and 46. The explicit constant C_t depends on the spatial L^2 norm of the time derivatives of the solution, and C_s depends on the spatial Hölder continuity of the solution and the constant $\bar{C}$. This constant is bounded independently of horizon ϵ. Although the constants are necessarily pessimistic, they deliver a priori error estimates. These constants are discussed in Jha and Lipton (2017a) in the context of fracture experiments. Fracture experiments are on the order of hundreds of μ-sec, and the size of the constants in the estimate for finite element simulations remains small for tens of μ-sec. For finite element schemes, we have a priori estimates with constants that stay small for same order of magnitude in time as fracture experiments. These features are discussed in Jha and Lipton (2017a,b).

Error Analysis

We define the L^2 projections of the actual solutions onto the space of piecewise constant functions defined over the cells U_i. These are given as follows. Let $(\tilde{\boldsymbol{u}}_i^k, \tilde{\boldsymbol{v}}_i^k)$ be the average of the exact solution $(\boldsymbol{u}^k, \boldsymbol{v}^k)$ in the unit cell U_i given by

$$\tilde{\boldsymbol{u}}_i^k := \frac{1}{h^d}\int_{U_i} \boldsymbol{u}^k(\boldsymbol{x})d\boldsymbol{x}$$

$$\tilde{\boldsymbol{v}}_i^k := \frac{1}{h^d}\int_{U_i} \boldsymbol{v}^k(\boldsymbol{x})d\boldsymbol{x}$$

and the L^2 projection of the solution onto piecewise constant functions is $(\tilde{\boldsymbol{u}}^k, \tilde{\boldsymbol{v}}^k)$ given by

$$\tilde{\boldsymbol{u}}^k(\boldsymbol{x}) := \sum_{i,\boldsymbol{x}_i \in D} \tilde{\boldsymbol{u}}_i^k \chi_{U_i}(\boldsymbol{x}) \tag{21}$$

$$\tilde{\boldsymbol{v}}^k(\boldsymbol{x}) := \sum_{i,\boldsymbol{x}_i \in D} \tilde{\boldsymbol{v}}_i^k \chi_{U_i}(\boldsymbol{x}) \tag{22}$$

The error between $(\hat{\boldsymbol{u}}^k, \hat{\boldsymbol{v}}^k)^T$ and $(\boldsymbol{u}(t^k), \boldsymbol{v}(t^k))^T$ is now split into two parts. From the triangle inequality, we have

$$\left\| \hat{\boldsymbol{u}}^k - \boldsymbol{u}(t^k) \right\|_{L^2(D;\mathbb{R}^d)} \le \left\| \hat{\boldsymbol{u}}^k - \tilde{\boldsymbol{u}}^k \right\|_{L^2(D;\mathbb{R}^d)} + \left\| \tilde{\boldsymbol{u}}^k - \boldsymbol{u}^k \right\|_{L^2(D;\mathbb{R}^d)}$$

$$\left\| \hat{\boldsymbol{v}}^k - \boldsymbol{v}(t^k) \right\|_{L^2(D;\mathbb{R}^d)} \le \left\| \hat{\boldsymbol{v}}^k - \tilde{\boldsymbol{v}}^k \right\|_{L^2(D;\mathbb{R}^d)} + \left\| \tilde{\boldsymbol{v}}^k - \boldsymbol{v}^k \right\|_{L^2(D;\mathbb{R}^d)}$$

In section "Error Analysis for Approximation of L^2 Projection of the Exact Solution" we will show that the error between the L^2 projections of the actual solution and the discrete approximation decays according to

$$\sup_{0 \le k \le T/\Delta t} \left(\left\| \hat{\boldsymbol{u}}^k - \tilde{\boldsymbol{u}}^k \right\|_{L^2(D;\mathbb{R}^d)} + \left\| \hat{\boldsymbol{v}}^k - \tilde{\boldsymbol{v}}^k \right\|_{L^2(D;\mathbb{R}^d)} \right) = O\left(\Delta t + \frac{h^\gamma}{\epsilon^2} \right). \tag{23}$$

In what follows we can estimate the terms

$$\left\| \tilde{\boldsymbol{u}}^k - \boldsymbol{u}(t^k) \right\|_{L^2()} \text{ and } \left\| \tilde{\boldsymbol{v}}^k - \boldsymbol{v}(t^k) \right\|_{L^2()} \tag{24}$$

and show they go to zero at a rate of h^γ uniformly in time. The estimates given by Eq. 23 together with the $O(h^\gamma)$ estimates for Eq. 24 establish Theorem 2. We now establish the L^2 estimates for the differences $\tilde{\boldsymbol{u}}^k - \boldsymbol{u}(t^k)$ and $\tilde{\boldsymbol{v}}^k - \boldsymbol{v}(t^k)$.

We write

$$\begin{aligned}
&\left\| \tilde{\boldsymbol{u}}^k - \boldsymbol{u}^k \right\|^2_{L^2(D;\mathbb{R}^d)} \\
&= \sum_{i,\boldsymbol{x}_i \in D} \int_{U_i} \left| \tilde{\boldsymbol{u}}^k(\boldsymbol{x}) - \boldsymbol{u}^k(\boldsymbol{x}) \right|^2 d\boldsymbol{x} \\
&= \sum_{i,\boldsymbol{x}_i \in D} \int_{U_i} \left| \frac{1}{h^d} \int_{U_i} (\boldsymbol{u}^k(\boldsymbol{y}) - \boldsymbol{u}^k(\boldsymbol{x})) d\boldsymbol{y} \right|^2 d\boldsymbol{x} \\
&= \sum_{i,\boldsymbol{x}_i \in D} \int_{U_i} \left[\frac{1}{h^{2d}} \int_{U_i} \int_{U_i} (\boldsymbol{u}^k(\boldsymbol{y}) - \boldsymbol{u}^k(\boldsymbol{x})) \cdot (\boldsymbol{u}^k(\boldsymbol{z}) - \boldsymbol{u}^k(\boldsymbol{x})) d\boldsymbol{y} d\boldsymbol{z} \right] d\boldsymbol{x}
\end{aligned}$$

$$\leq \sum_{i, \boldsymbol{x}_i \in D} \int_{U_i} \left[\frac{1}{h^d} \int_{U_i} \left| \boldsymbol{u}^k(\boldsymbol{y}) - \boldsymbol{u}^k(\boldsymbol{x}) \right|^2 d\boldsymbol{y} \right] d\boldsymbol{x} \tag{25}$$

where we used Cauchy's inequality and Jensen's inequality. For $\boldsymbol{x}, \boldsymbol{y} \in U_i$, $|\boldsymbol{x} - \boldsymbol{y}| \leq ch$, where $c = \sqrt{2}$ for $d = 2$ and $c = \sqrt{3}$ for $d = 3$. Since $\boldsymbol{u} \in C_0^{0,\gamma}$ we have

$$\begin{aligned} \left| \boldsymbol{u}^k(\boldsymbol{x}) - \boldsymbol{u}^k(\boldsymbol{y}) \right| &= |\boldsymbol{x} - \boldsymbol{y}|^\gamma \frac{\left| \boldsymbol{u}^k(\boldsymbol{y}) - \boldsymbol{u}^k(\boldsymbol{x}) \right|}{|\boldsymbol{x} - \boldsymbol{y}|^\gamma} \\ &\leq c^\gamma h^\gamma \left\| \boldsymbol{u}^k \right\|_{C^{0,\gamma}(D;\mathbb{R}^d)} \leq c^\gamma h^\gamma \sup_t \|\boldsymbol{u}(t)\|_{C^{0,\gamma}(D;\mathbb{R}^d)} \end{aligned} \tag{26}$$

and substitution in Eq. 25 gives

$$\begin{aligned} \left\| \tilde{\boldsymbol{u}}^k - \boldsymbol{u}^k \right\|^2_{L^2(D;\mathbb{R}^d)} &\leq c^{2\gamma} h^{2\gamma} \sum_{i, \boldsymbol{x}_i \in D} \int_{U_i} d\boldsymbol{x} \left(\sup_t \|\boldsymbol{u}(t)\|_{C^{0,\gamma}(D;\mathbb{R}^d)} \right)^2 \\ &\leq c^{2\gamma} |D| h^{2\gamma} \left(\sup_t \|\boldsymbol{u}(t)\|_{C^{0,\gamma}(D;\mathbb{R}^d)} \right)^2 . \end{aligned}$$

A similar estimate can be derived for $||\tilde{\boldsymbol{v}}^k - \boldsymbol{v}^k||_{L^2}$, and substitution of the estimates into Eq. 24 gives

$$\sup_k \left(\left\| \tilde{\boldsymbol{u}}^k - \boldsymbol{u}(t^k) \right\|_{L^2(D;\mathbb{R}^d)} + \left\| \tilde{\boldsymbol{v}}^k - \boldsymbol{v}(t^k) \right\|_{L^2(D;\mathbb{R}^d)} \right) = O(h^\gamma).$$

In the next section, we establish the error estimate (Eq. 23) for forward Euler in section "Error Analysis for Approximation of L^2 Projection of the Exact Solution."

Error Analysis for Approximation of L^2 Projection of the Exact Solution

In this subsection, we estimate the difference between approximate solution $(\hat{\boldsymbol{u}}^k, \hat{\boldsymbol{v}}^k)$ and the L^2 projection of the exact solution onto piecewise constant functions given by $(\tilde{\boldsymbol{u}}^k, \tilde{\boldsymbol{v}}^k)$ (see Eqs. 21 and 22). Let the differences be denoted by $\boldsymbol{e}^k(u) := \hat{\boldsymbol{u}}^k - \tilde{\boldsymbol{u}}^k$ and $\boldsymbol{e}^k(v) := \hat{\boldsymbol{v}}^k - \tilde{\boldsymbol{v}}^k$, and their evaluations at grid points are $\boldsymbol{e}_i^k(u) := \hat{\boldsymbol{u}}_i^k - \tilde{\boldsymbol{u}}_i^k$ and $\boldsymbol{e}_i^k(v) := \hat{\boldsymbol{v}}_i^k - \tilde{\boldsymbol{v}}_i^k$. Subtracting $(\tilde{\boldsymbol{u}}_i^{k+1} - \tilde{\boldsymbol{u}}_i^k)/\Delta t$ from Eq. 17 gives

$$\begin{aligned} &\frac{\hat{\boldsymbol{u}}_i^{k+1} - \hat{\boldsymbol{u}}_i^k}{\Delta t} - \frac{\tilde{\boldsymbol{u}}_i^{k+1} - \tilde{\boldsymbol{u}}_i^k}{\Delta t} \\ &= \hat{\boldsymbol{v}}_i^{k+1} - \frac{\tilde{\boldsymbol{u}}_i^{k+1} - \tilde{\boldsymbol{u}}_i^k}{\Delta t} \\ &= \hat{\boldsymbol{v}}_i^{k+1} - \tilde{\boldsymbol{v}}_i^{k+1} + \left(\tilde{\boldsymbol{v}}_i^{k+1} - \frac{\partial \tilde{\boldsymbol{u}}_i^{k+1}}{\partial t} \right) + \left(\frac{\partial \tilde{\boldsymbol{u}}_i^{k+1}}{\partial t} - \frac{\tilde{\boldsymbol{u}}_i^{k+1} - \tilde{\boldsymbol{u}}_i^k}{\Delta t} \right). \end{aligned}$$

Taking the average over unit cell U_i of the exact peridynamic equation Eq. 15 at time t^k, we will get $\tilde{\boldsymbol{v}}_i^{k+1} - \frac{\partial \tilde{\boldsymbol{u}}_i^{k+1}}{\partial t} = 0$. Therefore, the equation for $\boldsymbol{e}_i^k(u)$ is given by

$$\boldsymbol{e}_i^{k+1}(u) = \boldsymbol{e}_i^k(u) + \Delta t \boldsymbol{e}_i^{k+1}(v) + \Delta t \tau_i^k(u), \tag{27}$$

where we identify the discretization error as

$$\tau_i^k(u) := \frac{\partial \tilde{\boldsymbol{u}}_i^{k+1}}{\partial t} - \frac{\tilde{\boldsymbol{u}}_i^{k+1} - \tilde{\boldsymbol{u}}_i^k}{\Delta t}. \tag{28}$$

Similarly, we subtract $(\tilde{\boldsymbol{v}}_i^{k+1} - \tilde{\boldsymbol{v}}_i^k)/\Delta t$ from Eq. 18 and add and subtract terms to get

$$\begin{aligned}\frac{\hat{\boldsymbol{v}}_i^{k+1} - \hat{\boldsymbol{v}}_i^k}{\Delta t} - \frac{\tilde{\boldsymbol{v}}_i^{k+1} - \tilde{\boldsymbol{v}}_i^k}{\Delta t} &= -\nabla PD^\epsilon(\hat{\boldsymbol{u}}^k)(\boldsymbol{x}_i) + \boldsymbol{b}_i^k - \frac{\partial \boldsymbol{v}_i^k}{\partial t} + \left(\frac{\partial \boldsymbol{v}_i^k}{\partial t} - \frac{\tilde{\boldsymbol{v}}_i^{k+1} - \tilde{\boldsymbol{v}}_i^k}{\Delta t}\right) \\ &= -\nabla PD^\epsilon(\hat{\boldsymbol{u}}^k)(\boldsymbol{x}_i) + \boldsymbol{b}_i^k - \frac{\partial \boldsymbol{v}_i^k}{\partial t} \\ &\quad + \left(\frac{\partial \tilde{\boldsymbol{v}}_i^k}{\partial t} - \frac{\tilde{\boldsymbol{v}}_i^{k+1} - \tilde{\boldsymbol{v}}_i^k}{\Delta t}\right) + \left(\frac{\partial \boldsymbol{v}_i^k}{\partial t} - \frac{\partial \tilde{\boldsymbol{v}}_i^k}{\partial t}\right),\end{aligned} \tag{29}$$

where we identify $\tau_i^k(v)$ as follows

$$\tau_i^k(v) := \frac{\partial \tilde{\boldsymbol{v}}_i^k}{\partial t} - \frac{\tilde{\boldsymbol{v}}_i^{k+1} - \tilde{\boldsymbol{v}}_i^k}{\Delta t}. \tag{30}$$

Note that in $\tau^k(u)$ we have $\frac{\partial \tilde{\boldsymbol{u}}_i^{k+1}}{\partial t}$, and from the exact peridynamic equation, we have

$$\boldsymbol{b}_i^k - \frac{\partial \boldsymbol{v}_i^k}{\partial t} = \nabla PD^\epsilon(\boldsymbol{u}^k)(\boldsymbol{x}_i). \tag{31}$$

Combining Eqs. 29, 30, and 31 gives

$$\begin{aligned}\boldsymbol{e}_i^{k+1}(v) &= \boldsymbol{e}_i^k(v) + \Delta t \tau_i^k(v) + \Delta t \left(\frac{\partial \boldsymbol{v}_i^k}{\partial t} - \frac{\partial \tilde{\boldsymbol{v}}_i^k}{\partial t}\right) \\ &\quad + \Delta t \left(-\nabla PD^\epsilon(\hat{\boldsymbol{u}}^k)(\boldsymbol{x}_i) + \nabla PD^\epsilon(\boldsymbol{u}^k)(\boldsymbol{x}_i)\right)\end{aligned}$$

$$= \boldsymbol{e}_i^k(v) + \Delta t \tau_i^k(v) + \Delta t \left(\frac{\partial \boldsymbol{v}_i^k}{\partial t} - \frac{\partial \tilde{\boldsymbol{v}}_i^k}{\partial t} \right)$$
$$+ \Delta t \left(-\nabla PD^\epsilon(\hat{\boldsymbol{u}}^k)(\boldsymbol{x}_i) + \nabla PD^\epsilon(\tilde{\boldsymbol{u}}^k)(\boldsymbol{x}_i) \right)$$
$$+ \Delta t \left(-\nabla PD^\epsilon(\tilde{\boldsymbol{u}}^k)(\boldsymbol{x}_i) + \nabla PD^\epsilon(\boldsymbol{u}^k)(\boldsymbol{x}_i) \right).$$

The spatial discretization error $\sigma_i^k(u)$ and $\sigma_i^k(v)$ is given by

$$\sigma_i^k(u) := \left(-\nabla PD^\epsilon(\tilde{\boldsymbol{u}}^k)(\boldsymbol{x}_i) + \nabla PD^\epsilon(\boldsymbol{u}^k)(\boldsymbol{x}_i) \right) \tag{32}$$

$$\sigma_i^k(v) := \frac{\partial \boldsymbol{v}_i^k}{\partial t} - \frac{\partial \tilde{\boldsymbol{v}}_i^k}{\partial t}. \tag{33}$$

We finally have

$$\boldsymbol{e}_i^{k+1}(v) = \boldsymbol{e}_i^k(v) + \Delta t \left(\tau_i^k(v) + \sigma_i^k(u) + \sigma_i^k(v) \right)$$
$$+ \Delta t \left(-\nabla PD^\epsilon(\hat{\boldsymbol{u}}^k)(\boldsymbol{x}_i) + \nabla PD^\epsilon(\tilde{\boldsymbol{u}}^k)(\boldsymbol{x}_i) \right). \tag{34}$$

We now show the consistency and stability properties of the numerical scheme.

Consistency

We deal with the error in time discretization and the error in spatial discretization error separately. The time discretization error follows easily using the Taylor's series, while the spatial discretization error uses properties of the nonlinear peridynamic force.

Time discretization: We first estimate the time discretization error. A Taylor series expansion is used to estimate $\tau_i^k(u)$ as follows

$$\tau_i^k(u) = \frac{1}{h^d} \int_{U_i} \left(\frac{\partial \boldsymbol{u}^k(\boldsymbol{x})}{\partial t} - \frac{\boldsymbol{u}^{k+1}(\boldsymbol{x}) - \boldsymbol{u}^k(\boldsymbol{x})}{\Delta t} \right) d\boldsymbol{x}$$
$$= \frac{1}{h^d} \int_{U_i} \left(-\frac{1}{2} \frac{\partial^2 \boldsymbol{u}^k(\boldsymbol{x})}{\partial t^2} \Delta t + O((\Delta t)^2) \right) d\boldsymbol{x}.$$

Computing the L^2 norm of $\tau_i^k(u)$ and using Jensen's inequality gives

$$\left\| \tau^k(u) \right\|_{L^2(D;\mathbb{R}^d)} \leq \frac{\Delta t}{2} \left\| \frac{\partial^2 \boldsymbol{u}^k}{\partial t^2} \right\|_{L^2(D;\mathbb{R}^d)} + O((\Delta t)^2)$$
$$\leq \frac{\Delta t}{2} \sup_t \left\| \frac{\partial^2 \boldsymbol{u}(t)}{\partial t^2} \right\|_{L^2(D;\mathbb{R}^d)} + O((\Delta t)^2).$$

Similarly, we have

$$\left\|\tau^k(v)\right\|_{L^2(D;\mathbb{R}^d)} = \frac{\Delta t}{2}\sup_t \left\|\frac{\partial^2 \boldsymbol{v}(t)}{\partial t^2}\right\|_{L^2(D;\mathbb{R}^d)} + O((\Delta t)^2).$$

Spatial discretization: We now estimate the spatial discretization error. Substituting the definition of $\tilde{\boldsymbol{v}}^k$ and following the similar steps employed in Eq. 26 gives

$$\left|\sigma_i^k(v)\right| = \left|\frac{\partial \boldsymbol{v}_i^k}{\partial t} - \frac{1}{h^d}\int_{U_i}\frac{\partial \boldsymbol{v}^k(\boldsymbol{x})}{\partial t}d\boldsymbol{x}\right| \leq c^\gamma h^\gamma \int_{U_i}\frac{1}{|\boldsymbol{x}_i - \boldsymbol{x}|^\gamma}\left|\frac{\partial \boldsymbol{v}^k(\boldsymbol{x}_i)}{\partial t} - \frac{\partial \boldsymbol{v}^k(\boldsymbol{x})}{\partial t}\right| d\boldsymbol{x}$$
$$\leq c^\gamma h^\gamma \left\|\frac{\partial \boldsymbol{v}^k}{\partial t}\right\|_{C^{0,\gamma}(D;\mathbb{R}^d)} \leq c^\gamma h^\gamma \sup_t \left\|\frac{\partial \boldsymbol{v}(t)}{\partial t}\right\|_{C^{0,\gamma}(D;\mathbb{R}^d)}.$$

Taking the L^2 norm of error $\sigma_i^k(v)$ and substituting the estimate above delivers

$$\left\|\sigma^k(v)\right\|_{L^2(D;\mathbb{R}^d)} \leq h^\gamma c^\gamma \sqrt{|D|}\sup_t \left\|\frac{\partial \boldsymbol{v}(t)}{\partial t}\right\|_{C^{0,\gamma}(D;\mathbb{R}^d)}.$$

Now we estimate $\left|\sigma_i^k(u)\right|$. Note that the force $-\boldsymbol{\nabla} PD^\epsilon(\boldsymbol{u})(\boldsymbol{x})$ can be written as follows

$$-\boldsymbol{\nabla} PD^\epsilon(\boldsymbol{u})(\boldsymbol{x}) = \frac{4}{\epsilon^{d+1}\omega_d}\int_{H_\epsilon(\boldsymbol{x})}\omega(\boldsymbol{x})\omega(\boldsymbol{y})J\left(\frac{|\boldsymbol{y}-\boldsymbol{x}|}{\epsilon}\right)f'(|\boldsymbol{y}-\boldsymbol{x}|\,S(\boldsymbol{y},\boldsymbol{x};\boldsymbol{u})^2)$$
$$S(\boldsymbol{y},\boldsymbol{x};\boldsymbol{u})\frac{\boldsymbol{y}-\boldsymbol{x}}{|\boldsymbol{y}-\boldsymbol{x}|}d\boldsymbol{y} = \frac{4}{\epsilon\omega_d}\int_{H_1(\boldsymbol{0})}\omega(\boldsymbol{x})\omega(\boldsymbol{x}+\epsilon\boldsymbol{\xi})J(|\boldsymbol{\xi}|)f'(\epsilon\,|\boldsymbol{\xi}|$$
$$S(\boldsymbol{x}+\epsilon\boldsymbol{\xi},\boldsymbol{x};\boldsymbol{u})^2)S(\boldsymbol{x}+\epsilon\boldsymbol{\xi},\boldsymbol{x};\boldsymbol{u})\frac{\boldsymbol{\xi}}{|\boldsymbol{\xi}|}d\boldsymbol{\xi}.$$

where we substituted $\partial_S W^\epsilon$ using Eq. 2. In the second step, we introduced the change in variable $\boldsymbol{y} = \boldsymbol{x} + \epsilon\boldsymbol{\xi}$.

Let $F_1 : \mathbb{R} \to \mathbb{R}$ be defined as $F_1(S) = f(S^2)$. Then $F_1'(S) = f'(S^2)2S$. Using the definition of F_1, we have

$$2Sf'(\epsilon\,|\boldsymbol{\xi}|\,S^2) = \frac{F_1'(\sqrt{\epsilon\,|\boldsymbol{\xi}|}S)}{\sqrt{\epsilon\,|\boldsymbol{\xi}|}}.$$

Because f is assumed to be positive, smooth, and concave and is bounded far away, we have the following bound on derivatives of F_1

$$\sup_r \left|F_1'(r)\right| = F_1'(\bar{r}) =: C_1 \tag{35}$$

$$\sup_r \left| F_1''(r) \right| = \max\{F_1''(0), F_1''(\hat{u})\} =: C_2 \tag{36}$$

$$\sup_r \left| F_1'''(r) \right| = \max\{F_1'''(\bar{u}_2), F_1'''(\tilde{u}_2)\} =: C_3. \tag{37}$$

where $\bar{r}$ is the inflection point of $f(r^2)$, i.e., $F_1''(\bar{r}) = 0$. $\{0, \hat{u}\}$ are the maxima of $F_1''(r)$. $\{\bar{u}, \tilde{u}\}$ are the maxima of $F_1'''(r)$. By chain rule and by considering the assumption on f, we can show that $\bar{r}, \hat{u}, \bar{u}_2, \tilde{u}_2$ exist and the C_1, C_2, C_3 are bounded. Figures 4, 5, and 6 show the generic graphs of $F_1'(r)$, $F_1''(r)$, and $F_1'''(r)$, respectively.

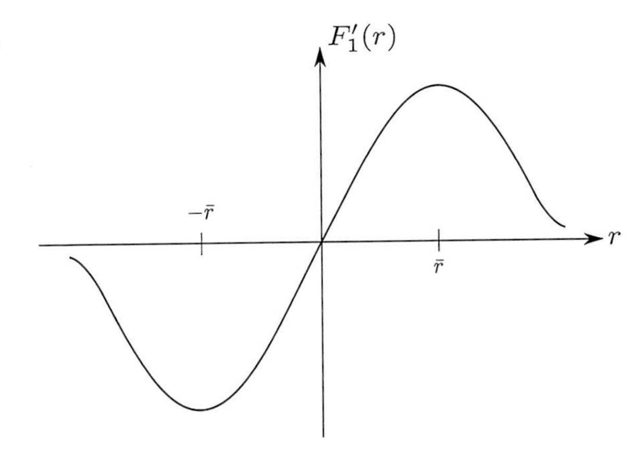

Fig. 4 Generic plot of $F_1'(r)$. $|F_1'(r)|$ is bounded by $|F_1'(\bar{r})|$

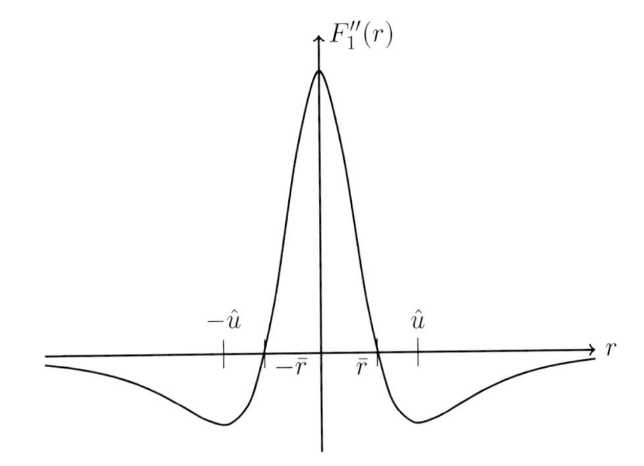

Fig. 5 Generic plot of $F_1''(r)$. At $\pm\bar{r}$, $F_1''(r) = 0$. At $\pm\hat{u}$, $F_1'''(r) = 0$

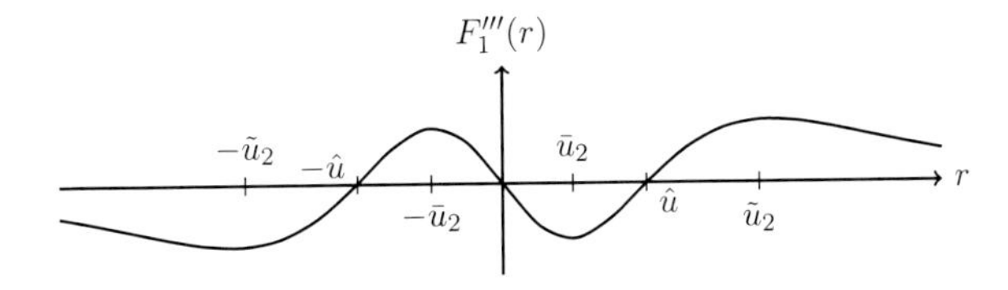

Fig. 6 Generic plot of $F_1'''(r)$. At $\pm\bar{u}_2$ and $\pm\tilde{u}_2$, $F_1'''' = 0$

The nonlocal force $-\boldsymbol{\nabla} PD^{\epsilon}$ can be written as

$$
\begin{aligned}
&-\boldsymbol{\nabla} PD^{\epsilon}(\boldsymbol{u})(\boldsymbol{x}) \\
&= \frac{2}{\epsilon\omega_d}\int_{H_1(\mathbf{0})} \omega(\boldsymbol{x})\omega(\boldsymbol{x}+\epsilon\boldsymbol{\xi})J(|\boldsymbol{\xi}|)F_1'(\sqrt{\epsilon\,|\boldsymbol{\xi}|}S(\boldsymbol{x}+\epsilon\boldsymbol{\xi},\boldsymbol{x};\boldsymbol{u}))\frac{1}{\sqrt{\epsilon\,|\boldsymbol{\xi}|}}\frac{\boldsymbol{\xi}}{|\boldsymbol{\xi}|}d\boldsymbol{\xi}.
\end{aligned} \tag{38}
$$

To simplify the calculations, we use the following notation

$$
\begin{aligned}
\bar{\boldsymbol{u}}(\boldsymbol{x}) &:= \boldsymbol{u}(\boldsymbol{x}+\epsilon\boldsymbol{\xi}) - \boldsymbol{u}(\boldsymbol{x}), \\
\bar{\boldsymbol{u}}(\boldsymbol{y}) &:= \boldsymbol{u}(\boldsymbol{y}+\epsilon\boldsymbol{\xi}) - \boldsymbol{u}(\boldsymbol{y}), \\
(\boldsymbol{u}-\boldsymbol{v})(\boldsymbol{x}) &:= \boldsymbol{u}(\boldsymbol{x}) - \boldsymbol{v}(\boldsymbol{x}),
\end{aligned}
$$

and $\overline{(\boldsymbol{u}-\boldsymbol{v})}(\boldsymbol{x})$ is defined similar to $\bar{\boldsymbol{u}}(\boldsymbol{x})$. Also, let

$$
s = \epsilon\,|\boldsymbol{\xi}|, \quad \boldsymbol{e} = \frac{\boldsymbol{\xi}}{|\boldsymbol{\xi}|}.
$$

In what follows, we will come across the integral of type $\int_{H_1(\mathbf{0})} J(|\boldsymbol{\xi}|)\,|\boldsymbol{\xi}|^{-\alpha}\,d\boldsymbol{\xi}$. Recall that $0 \le J(|\boldsymbol{\xi}|) \le M$ for all $\boldsymbol{\xi} \in H_1(\mathbf{0})$ and $J(|\boldsymbol{\xi}|) = 0$ for $\boldsymbol{\xi} \notin H_1(\mathbf{0})$. Therefore, let

$$
\bar{J}_\alpha := \frac{1}{\omega_d}\int_{H_1(\mathbf{0})} J(|\boldsymbol{\xi}|)\,|\boldsymbol{\xi}|^{-\alpha}\,d\boldsymbol{\xi}. \tag{39}
$$

With notations above, we note that $S(\boldsymbol{x}+\epsilon\boldsymbol{\xi},\boldsymbol{x};\boldsymbol{u}) = \bar{\boldsymbol{u}}(\boldsymbol{x})\cdot\boldsymbol{e}/s$. $-\boldsymbol{\nabla} PD^{\epsilon}$ can be written as

$$
-\boldsymbol{\nabla} PD^{\epsilon}(\boldsymbol{u})(\boldsymbol{x}) = \frac{2}{\epsilon\omega_d}\int_{H_1(\mathbf{0})} \omega(\boldsymbol{x})\omega(\boldsymbol{x}+\epsilon\boldsymbol{\xi})J(|\boldsymbol{\xi}|)F_1'(\bar{\boldsymbol{u}}(\boldsymbol{x})\cdot\boldsymbol{e}/\sqrt{s})\frac{1}{\sqrt{s}}\boldsymbol{e}\,d\boldsymbol{\xi}. \tag{40}
$$

We estimate $|-\boldsymbol{\nabla} PD^{\epsilon}(\boldsymbol{u})(\boldsymbol{x}) - (-\boldsymbol{\nabla} PD^{\epsilon}(\boldsymbol{v})(\boldsymbol{x}))|$.

$$
\begin{aligned}
&|-\boldsymbol{\nabla} PD^{\epsilon}(\boldsymbol{u})(\boldsymbol{x}) - (-\boldsymbol{\nabla} PD^{\epsilon}(\boldsymbol{v})(\boldsymbol{x}))| \\
&\le \left|\frac{2}{\epsilon\omega_d}\int_{H_1(\mathbf{0})} \omega(\boldsymbol{x})\omega(\boldsymbol{x}+\epsilon\boldsymbol{\xi})J(|\boldsymbol{\xi}|)\frac{\left(F_1'(\bar{\boldsymbol{u}}(\boldsymbol{x})\cdot\boldsymbol{e}/\sqrt{s}) - F_1'(\bar{\boldsymbol{v}}(\boldsymbol{x})\cdot\boldsymbol{e}/\sqrt{s})\right)}{\sqrt{s}}\boldsymbol{e}\,d\boldsymbol{\xi}\right| \\
&\le \left|\frac{2}{\epsilon\omega_d}\int_{H_1(\mathbf{0})} J(|\boldsymbol{\xi}|)\frac{1}{\sqrt{s}}\left|F_1'(\bar{\boldsymbol{u}}(\boldsymbol{x})\cdot\boldsymbol{e}/\sqrt{s}) - F_1'(\bar{\boldsymbol{v}}(\boldsymbol{x})\cdot\boldsymbol{e}/\sqrt{s})\right|d\boldsymbol{\xi}\right| \\
&\le \sup_r |F_1''(r)| \left|\frac{2}{\epsilon\omega_d}\int_{H_1(\mathbf{0})} J(|\boldsymbol{\xi}|)\frac{1}{\sqrt{s}}\left|\bar{\boldsymbol{u}}(\boldsymbol{x})\cdot\boldsymbol{e}/\sqrt{s} - \bar{\boldsymbol{v}}(\boldsymbol{x})\cdot\boldsymbol{e}/\sqrt{s}\right|d\boldsymbol{\xi}\right|
\end{aligned}
$$

$$\leq \frac{2C_2}{\epsilon\omega_d}\left|\int_{H_1(\mathbf{0})} J(|\boldsymbol{\xi}|)\frac{|\bar{\boldsymbol{u}}(\boldsymbol{x})-\bar{\boldsymbol{v}}(\boldsymbol{x})|}{\epsilon|\boldsymbol{\xi}|}d\boldsymbol{\xi}\right|. \tag{41}$$

Here we have used the fact that $|\omega(\boldsymbol{x})| \leq 1$ and for a vector $\boldsymbol{e}$ such that $|\boldsymbol{e}| = 1$, $|\mathbf{a}\cdot\boldsymbol{e}| \leq |\mathbf{a}|$ holds and $|\alpha\boldsymbol{e}| \leq |\alpha|$ holds for all $\mathrm{a} \in \mathbb{R}^d, \alpha \in \mathbb{R}$.

We use the notation $\bar{\boldsymbol{u}}^k(\boldsymbol{x}) := \boldsymbol{u}^k(\boldsymbol{x}+\epsilon\boldsymbol{\xi}) - \boldsymbol{u}^k(\boldsymbol{x})$ and $\bar{\tilde{\boldsymbol{u}}}^k(\boldsymbol{x}) := \tilde{\boldsymbol{u}}(\boldsymbol{x}+\epsilon\boldsymbol{\xi}) - \tilde{\boldsymbol{u}}^k(\boldsymbol{x})$ and choose $\boldsymbol{u} = \boldsymbol{u}^k$ and $\boldsymbol{v} = \tilde{\boldsymbol{u}}^k$ in Eq. 41 to find that

$$\begin{aligned}|\sigma_i^k(u)| &= \left|-\nabla PD^\epsilon(\tilde{\boldsymbol{u}}^k)(\boldsymbol{x}_i) + \nabla PD^\epsilon(\boldsymbol{u}^k)(\boldsymbol{x}_i)\right| \\ &\leq \frac{2C_2}{\epsilon\omega_d}\left|\int_{H_1(\mathbf{0})} J(|\boldsymbol{\xi}|)\frac{\left|\boldsymbol{u}^k(\boldsymbol{x}_i+\epsilon\boldsymbol{\xi})-\tilde{\boldsymbol{u}}^k(\boldsymbol{x}_i+\epsilon\boldsymbol{\xi})-(\boldsymbol{u}^k(\boldsymbol{x}_i)-\tilde{\boldsymbol{u}}^k(\boldsymbol{x}_i))\right|}{\epsilon|\boldsymbol{\xi}|}d\boldsymbol{\xi}\right|.\end{aligned} \tag{42}$$

Here C_2 is the maximum of the second derivative of the profile describing the potential given by Eq. 36. Following the earlier analysis (see Eq. 26), we find that

$$\left|\boldsymbol{u}^k(\boldsymbol{x}_i+\epsilon\boldsymbol{\xi}) - \tilde{\boldsymbol{u}}^k(\boldsymbol{x}_i+\epsilon\boldsymbol{\xi})\right| \leq c^\gamma h^\gamma \sup_t \|\boldsymbol{u}(t)\|_{C^{0,\gamma}(D;\mathbb{R}^d)}$$

$$\left|\boldsymbol{u}^k(\boldsymbol{x}_i) - \tilde{\boldsymbol{u}}^k(\boldsymbol{x}_i)\right| \leq c^\gamma h^\gamma \sup_t \|\boldsymbol{u}(t)\|_{C^{0,\gamma}(D;\mathbb{R}^d)}.$$

For reference, we define the constant

$$\bar{C} = \frac{C_2}{\omega_d}\int_{H_1(\mathbf{0})} J(|\boldsymbol{\xi}|)\frac{1}{|\boldsymbol{\xi}|}d\boldsymbol{\xi}. \tag{43}$$

We now focus on Eq. 42. We substitute the above two inequalities to get

$$\begin{aligned}|\sigma_i^k(u)| &\leq \frac{2C_2}{\epsilon^2\omega_d}\left|\int_{H_1(\mathbf{0})} J(|\boldsymbol{\xi}|)\frac{1}{|\boldsymbol{\xi}|}\right. \\ &\qquad \left.\left(\left|\boldsymbol{u}^k(\boldsymbol{x}_i+\epsilon\boldsymbol{\xi}) - \tilde{\boldsymbol{u}}^k(\boldsymbol{x}_i+\epsilon\boldsymbol{\xi})\right| + \left|\boldsymbol{u}^k(\boldsymbol{x}_i) - \tilde{\boldsymbol{u}}^k(\boldsymbol{x}_i)\right|\right)d\boldsymbol{\xi}\right| \\ &\leq 4h^\gamma c^\gamma\frac{\bar{C}}{\epsilon^2}\sup_t \|\boldsymbol{u}(t)\|_{C^{0,\gamma}(D;\mathbb{R}^d)}.\end{aligned}$$

Therefore, we have

$$\left\|\sigma^k(u)\right\|_{L^2(D;\mathbb{R}^d)} \leq h^\gamma\left(4c^\gamma\sqrt{|D|}\frac{\bar{C}}{\epsilon^2}\sup_t \|\boldsymbol{u}(t)\|_{C^{0,\gamma}(D;\mathbb{R}^d)}\right).$$

This completes the proof of consistency of numerical approximation.

Stability

Let e^k be the total error at the kth time step. It is defined as

$$e^k := \left\| \boldsymbol{e}^k(u) \right\|_{L^2(D;\mathbb{R}^d)} + \left\| \boldsymbol{e}^k(v) \right\|_{L^2(D;\mathbb{R}^d)}.$$

To simplify the calculations, we define new term τ as

$$\tau := \sup_t \Big(\left\| \tau^k(u) \right\|_{L^2(D;\mathbb{R}^d)} + \left\| \tau^k(v) \right\|_{L^2(D;\mathbb{R}^d)} + \left\| \sigma^k(u) \right\|_{L^2(D;\mathbb{R}^d)} + \left\| \sigma^k(v) \right\|_{L^2(D;\mathbb{R}^d)} \Big).$$

From our consistency analysis, we know that to leading order

$$\tau \leq C_t \Delta t + \frac{C_s}{\epsilon^2} h^\gamma \tag{44}$$

where,

$$C_t := \frac{1}{2} \sup_t \left\| \frac{\partial^2 \boldsymbol{u}(t)}{\partial t^2} \right\|_{L^2(D;\mathbb{R}^d)} + \frac{1}{2} \sup_t \left\| \frac{\partial^3 \boldsymbol{u}(t)}{\partial t^3} \right\|_{L^2(D;\mathbb{R}^d)}, \tag{45}$$

$$C_s := c^\gamma \sqrt{|D|} \left[\epsilon^2 \sup_t \left\| \frac{\partial^2 \boldsymbol{u}(t)}{\partial t^2} \right\|_{C^{0,\gamma}(D;\mathbb{R}^d)} + 4\bar{C} \sup_t \|\boldsymbol{u}(t)\|_{C^{0,\gamma}(D;\mathbb{R}^d)} \right]. \tag{46}$$

We take L^2 norm of Eqs. 27 and 34 and add them. Noting the definition of τ as above, we get

$$e^{k+1} \leq e^k + \Delta t \left\| \boldsymbol{e}^{k+1}(v) \right\|_{L^2(D;\mathbb{R}^d)} + \Delta t \tau + \Delta t \left(\sum_i h^d \left| -\nabla PD^\epsilon(\hat{\boldsymbol{u}}^k)(\boldsymbol{x}_i) + \nabla PD^\epsilon(\tilde{\boldsymbol{u}}^k)(\boldsymbol{x}_i) \right|^2 \right)^{1/2}. \tag{47}$$

We only need to estimate the last term in the above equation. Similar to the Eq. 42, we have

$$\begin{aligned}
&\left| -\nabla PD^\epsilon(\hat{\boldsymbol{u}}^k)(\boldsymbol{x}_i) + \nabla PD^\epsilon(\tilde{\boldsymbol{u}}^k)(\boldsymbol{x}_i) \right| \\
&\leq \frac{2C_2}{\epsilon^2 \omega_d} \left| \int_{H_1(\boldsymbol{0})} J(|\boldsymbol{\xi}|) \frac{1}{|\boldsymbol{\xi}|} \left| \hat{\boldsymbol{u}}^k(\boldsymbol{x}_i + \epsilon \boldsymbol{\xi}) - \tilde{\boldsymbol{u}}^k(\boldsymbol{x}_i + \epsilon \boldsymbol{\xi}) - (\hat{\boldsymbol{u}}^k(\boldsymbol{x}_i) - \tilde{\boldsymbol{u}}^k(\boldsymbol{x}_i)) \right| d\boldsymbol{\xi} \right| \\
&= \frac{2C_2}{\epsilon^2 \omega_d} \left| \int_{H_1(\boldsymbol{0})} J(|\boldsymbol{\xi}|) \frac{1}{|\boldsymbol{\xi}|} \left| \boldsymbol{e}^k(u)(\boldsymbol{x}_i + \epsilon \boldsymbol{\xi}) - \boldsymbol{e}^k(u)(\boldsymbol{x}_i) \right| d\boldsymbol{\xi} \right| \\
&\leq \frac{2C_2}{\epsilon^2 \omega_d} \left| \int_{H_1(\boldsymbol{0})} J(|\boldsymbol{\xi}|) \frac{1}{|\boldsymbol{\xi}|} \left(\left| \boldsymbol{e}^k(u)(\boldsymbol{x}_i + \epsilon \boldsymbol{\xi}) \right| + \left| \boldsymbol{e}^k(u)(\boldsymbol{x}_i) \right| \right) d\boldsymbol{\xi} \right|.
\end{aligned}$$

By $\boldsymbol{e}^k(u)(\boldsymbol{x})$ we mean evaluation of piecewise extension of set $\{\boldsymbol{e}_i^k(u)\}_i$ at $\boldsymbol{x}$. We proceed further as follows

$$\begin{aligned}&\left|-\nabla PD^\epsilon(\hat{\boldsymbol{u}}^k)(\boldsymbol{x}_i) + \nabla PD^\epsilon(\tilde{\boldsymbol{u}}^k)(\boldsymbol{x}_i)\right|^2 \\ &\leq \left(\frac{2C_2}{\epsilon^2 \omega_d}\right)^2 \int_{H_1(\boldsymbol{0})} \int_{H_1(\boldsymbol{0})} J(|\boldsymbol{\xi}|) J(|\boldsymbol{\eta}|) \frac{1}{|\boldsymbol{\xi}|} \frac{1}{|\boldsymbol{\eta}|} \\ &\left(\left|\boldsymbol{e}^k(u)(\boldsymbol{x}_i + \epsilon\boldsymbol{\xi})\right| + \left|\boldsymbol{e}^k(u)(\boldsymbol{x}_i)\right|\right)\left(\left|\boldsymbol{e}^k(u)(\boldsymbol{x}_i + \epsilon\boldsymbol{\eta})\right| + \left|\boldsymbol{e}^k(u)(\boldsymbol{x}_i)\right|\right) d\boldsymbol{\xi} d\boldsymbol{\eta}.\end{aligned}$$

Using inequality $|ab| \leq (|a|^2 + |b|^2)/2$, we get

$$\begin{aligned}&\left(\left|\boldsymbol{e}^k(u)(\boldsymbol{x}_i + \epsilon\boldsymbol{\xi})\right| + \left|\boldsymbol{e}^k(u)(\boldsymbol{x}_i)\right|\right)\left(\left|\boldsymbol{e}^k(u)(\boldsymbol{x}_i + \epsilon\boldsymbol{\eta})\right| + \left|\boldsymbol{e}^k(u)(\boldsymbol{x}_i)\right|\right) \\ &\leq 3\left(\left|\boldsymbol{e}^k(u)(\boldsymbol{x}_i + \epsilon\boldsymbol{\xi})\right|^2 + \left|\boldsymbol{e}^k(u)(\boldsymbol{x}_i + \epsilon\boldsymbol{\eta})\right|^2 + \left|\boldsymbol{e}^k(u)(\boldsymbol{x}_i)\right|^2\right),\end{aligned}$$

and

$$\begin{aligned}&\sum_{\boldsymbol{x}_i \in D} h^d \left|-\nabla PD^\epsilon(\hat{\boldsymbol{u}}^k)(\boldsymbol{x}_i) + \nabla PD^\epsilon(\tilde{\boldsymbol{u}}^k)(\boldsymbol{x}_i)\right|^2 \\ &\leq \left(\frac{2C_2}{\epsilon^2 \omega_d}\right)^2 \int_{H_1(\boldsymbol{0})} \int_{H_1(\boldsymbol{0})} J(|\boldsymbol{\xi}|) J(|\boldsymbol{\eta}|) \frac{1}{|\boldsymbol{\xi}|} \frac{1}{|\boldsymbol{\eta}|} \\ &\sum_{\boldsymbol{x}_i \in D} h^d 3\left(\left|\boldsymbol{e}^k(u)(\boldsymbol{x}_i + \epsilon\boldsymbol{\xi})\right|^2 + \left|\boldsymbol{e}^k(u)(\boldsymbol{x}_i + \epsilon\boldsymbol{\eta})\right|^2 + \left|\boldsymbol{e}^k(u)(\boldsymbol{x}_i)\right|^2\right) d\boldsymbol{\xi} d\boldsymbol{\eta}.\end{aligned}$$

Since $\boldsymbol{e}^k(u)(\boldsymbol{x}) = \sum_{\boldsymbol{x}_i \in D} \boldsymbol{e}_i^k(u)\chi_{U_i}(\boldsymbol{x})$, we have

$$\sum_{\boldsymbol{x}_i \in D} h^d \left|-\nabla PD^\epsilon(\hat{\boldsymbol{u}}^k)(\boldsymbol{x}_i) + \nabla PD^\epsilon(\tilde{\boldsymbol{u}}^k)(\boldsymbol{x}_i)\right|^2 \leq \frac{(6\bar{C})^2}{\epsilon^4} \left\|\boldsymbol{e}^k(u)\right\|^2_{L^2(D;\mathbb{R}^d)}. \tag{48}$$

where $\bar{C}$ is given by Eq. 43. In summary Eq. 48 shows the Lipschitz continuity of the peridynamic force with respect to the L^2 norm (see Eq. 16) expressed in this context as

$$\|\nabla PD^\epsilon(\hat{\boldsymbol{u}}^k)(\boldsymbol{x}) - \nabla PD^\epsilon(\tilde{\boldsymbol{u}}^k)\|_{L^2(D;\mathbb{R}^d)} \leq \frac{(6\bar{C})}{\epsilon^2} \|\boldsymbol{e}^k(u)\|_{L^2(D;\mathbb{R}^d)}. \tag{49}$$

Finally, we substitute above inequality in Eq. 47 to get

$$e^{k+1} \leq e^k + \Delta t \left\|\boldsymbol{e}^{k+1}(v)\right\|_{L^2(D;\mathbb{R}^d)} + \Delta t \tau + \Delta t \frac{6\bar{C}}{\epsilon^2} \left\|\boldsymbol{e}^k(u)\right\|_{L^2(D;\mathbb{R}^d)}$$

We add positive quantity $\Delta t||e^{k+1}(u)||_{L^2(D;\mathbb{R}^d)} + \Delta t 6\bar{C}/\epsilon^2||e^k(v)||_{L^2(D;\mathbb{R}^d)}$ to the right side of above equation to get

$$e^{k+1} \leq (1 + \Delta t 6\bar{C}/\epsilon^2)e^k + \Delta t e^{k+1} + \Delta t \tau$$
$$\Rightarrow e^{k+1} \leq \frac{(1 + \Delta t 6\bar{C}/\epsilon^2)}{1 - \Delta t}e^k + \frac{\Delta t}{1 - \Delta t}\tau.$$

We recursively substitute e^j on above as follows

$$\begin{aligned}
e^{k+1} &\leq \frac{(1 + \Delta t 6\bar{C}/\epsilon^2)}{1 - \Delta t}e^k + \frac{\Delta t}{1 - \Delta t}\tau \\
&\leq \left(\frac{(1 + \Delta t 6\bar{C}/\epsilon^2)}{1 - \Delta t}\right)^2 e^{k-1} + \frac{\Delta t}{1 - \Delta t}\tau\left(1 + \frac{(1 + \Delta t 6\bar{C}/\epsilon^2)}{1 - \Delta t}\right) \\
&\leq \dots \\
&\leq \left(\frac{(1 + \Delta t 6\bar{C}/\epsilon^2)}{1 - \Delta t}\right)^{k+1} e^0 + \frac{\Delta t}{1 - \Delta t}\tau\sum_{j=0}^{k}\left(\frac{(1 + \Delta t 6\bar{C}/\epsilon^2)}{1 - \Delta t}\right)^{k-j}.
\end{aligned} \quad (50)$$

Since $1/(1 - \Delta t) = 1 + \Delta t + \Delta t^2 + O(\Delta t^3)$, we have

$$\frac{(1 + \Delta t 6\bar{C}/\epsilon^2)}{1 - \Delta t} \leq 1 + (1 + 6\bar{C}/\epsilon^2)\Delta t + (1 + 6\bar{C}/\epsilon^2)\Delta t^2 + O(\bar{C}/\epsilon^2)O(\Delta t^3).$$

Now, for any $k \leq T/\Delta t$, using identity $(1 + a)^k \leq \exp[ka]$ for $a \leq 0$, we have

$$\begin{aligned}
&\left(\frac{1 + \Delta t 6\bar{C}/\epsilon^2}{1 - \Delta t}\right)^k \\
&\leq \exp\left[k(1 + 6\bar{C}/\epsilon^2)\Delta t + k(1 + 6\bar{C}/\epsilon^2)\Delta t^2 + kO(\bar{C}/\epsilon^2)O(\Delta t^3)\right] \\
&\leq \exp\left[T(1 + 6\bar{C}/\epsilon^2) + T(1 + 6\bar{C}/\epsilon^2)\Delta t + O(T\bar{C}/\epsilon^2)O(\Delta t^2)\right].
\end{aligned}$$

We write above equation in more compact form as follows

$$\begin{aligned}
&\left(\frac{1 + \Delta t 6\bar{C}/\epsilon^2}{1 - \Delta t}\right)^k \\
&\leq \exp\left[T(1 + 6\bar{C}/\epsilon^2)(1 + \Delta t + O(\Delta t^2))\right].
\end{aligned}$$

We use above estimate in Eq. 50 and get the following inequality for e^k

$$
\begin{aligned}
e^{k+1} &\leq \exp\left[T(1+6\bar{C}/\epsilon^2)(1+\Delta t+O(\Delta t^2))\right]\left(e^0+(k+1)\tau\Delta t/(1-\Delta t)\right) \\
&\leq \exp\left[T(1+6\bar{C}/\epsilon^2)(1+\Delta t+O(\Delta t^2))\right]\left(e^0+T\tau(1+\Delta t+O(\Delta t^2)\right).
\end{aligned}
$$

where we used the fact that $1/(1-\Delta t) = 1+\Delta t+O(\Delta t^2)$.

Assuming the error in initial data is zero, i.e., $e^0 = 0$, and noting the estimate of τ in Eq. 44, we have

$$
\sup_k e^k \leq \exp\left[T(1+6\bar{C}/\epsilon^2)\right]T\tau
$$

and we conclude to leading order that

$$
\sup_k e^k \leq \exp\left[T(1+6\bar{C}/\epsilon^2)\right]T\left[C_t\Delta t+(C_s/\epsilon^2)h^\gamma\right], \tag{51}
$$

Here the constants C_t and C_s are given by Eqs. 45 and 46.This shows the stability of the numerical scheme and Theorem 2 is proved.

Stability of the Energy for the Semi-discrete Approximation

We first spatially discretize the peridynamics equation (Eq. 6). Let $\{\hat{\boldsymbol{u}}_i(t)\}_{i,\boldsymbol{x}_i\in D}$ denote the semi-discrete approximate solution which satisfies the following, for all $t \in [0,T]$ and i such that $\boldsymbol{x}_i \in D$,

$$
\ddot{\hat{\boldsymbol{u}}}_i(t) = -\nabla PD^\epsilon(\hat{\boldsymbol{u}}(t))(\boldsymbol{x}_i)+\boldsymbol{b}_i(t) \tag{52}
$$

where $\hat{\boldsymbol{u}}(t)$ is the piecewise constant extension of discrete set $\{\hat{\boldsymbol{u}}_i(t)\}_i$ and is defined as

$$
\hat{\boldsymbol{u}}(t,\boldsymbol{x}) := \sum_{i,\boldsymbol{x}_i\in D}\hat{\boldsymbol{u}}_i(t)\chi_{U_i}(\boldsymbol{x}). \tag{53}
$$

The scheme is complemented with the discretized initial conditions $\hat{\boldsymbol{u}}_i(0) = \boldsymbol{u}_0(\boldsymbol{x}_i)$ and $\hat{\boldsymbol{v}}_i(0) = \boldsymbol{v}_0(\boldsymbol{x}_i)$. We apply boundary condition by setting $\hat{\boldsymbol{u}}_i(t) = \boldsymbol{0}$ for all t and for all $\boldsymbol{x}_i \notin D$.

We have the stability of semi-discrete evolution.

Theorem 3 (Energy stability of the semi-discrete approximation). *Let $\{\hat{\boldsymbol{u}}_i(t)\}_i$ satisfy Eq. 52 and $\hat{\boldsymbol{u}}(t)$ is its piecewise constant extension. Similarily let $\hat{\boldsymbol{b}}(t,\boldsymbol{x})$ denote the piecewise constant extension of $\{\boldsymbol{b}(t,\boldsymbol{x}_i)\}_{i,\boldsymbol{x}_i\in D}$. Then the peridynamic energy $\mathcal{E}^\epsilon$ as defined in Eq. 11 satisfies, $\forall t\in[0,T]$,*

$$
\mathcal{E}^\epsilon(\hat{\boldsymbol{u}})(t) \leq \left(\sqrt{\mathcal{E}^\epsilon(\hat{\boldsymbol{u}})(0)}+\frac{TC}{\epsilon^{3/2}}+\int_0^T ||\hat{\boldsymbol{b}}(s)||_{L^2(D;\mathbb{R}^d)}ds\right)^2. \tag{54}
$$

The constant C, defined in Eq. 59, is independent of ϵ and h.

Proof. We multiply Eq. 52 by $\chi_{U_i}(\boldsymbol{x})$ and sum over i and use definition of piecewise constant extension in Eq. 53 to get

$$\begin{aligned}\ddot{\hat{\boldsymbol{u}}}(t,\boldsymbol{x}) &= -\nabla \hat{PD^\epsilon}(\hat{\boldsymbol{u}}(t))(\boldsymbol{x}) + \hat{\boldsymbol{b}}(t,\boldsymbol{x}) \\ &= -\nabla PD^\epsilon(\hat{\boldsymbol{u}}(t))(\boldsymbol{x}) + \hat{\boldsymbol{b}}(t,\boldsymbol{x}) \\ &\quad + (-\nabla \hat{PD^\epsilon}(\hat{\boldsymbol{u}}(t))(\boldsymbol{x}) + \nabla PD^\epsilon(\hat{\boldsymbol{u}}(t))(\boldsymbol{x}))\end{aligned}$$

where $-\nabla \hat{PD^\epsilon}(\hat{\boldsymbol{u}}(t))(\boldsymbol{x})$ and $\hat{\boldsymbol{b}}(t,\boldsymbol{x})$ are given by

$$\begin{aligned}-\nabla \hat{PD^\epsilon}(\hat{\boldsymbol{u}}(t))(\boldsymbol{x}) &= \sum_{i,\boldsymbol{x}_i\in D} (-\nabla PD^\epsilon(\hat{\boldsymbol{u}}(t))(\boldsymbol{x}_i))\chi_{U_i}(\boldsymbol{x}) \\ \hat{\boldsymbol{b}}(t,\boldsymbol{x}) &= \sum_{i,\boldsymbol{x}_i\in D} \boldsymbol{b}(t,\boldsymbol{x}_i)\chi_{U_i}(\boldsymbol{x}).\end{aligned}$$

We define set as follows

$$\sigma(t,\boldsymbol{x}) := -\nabla \hat{PD^\epsilon}(\hat{\boldsymbol{u}}(t))(\boldsymbol{x}) + \nabla PD^\epsilon(\hat{\boldsymbol{u}}(t))(\boldsymbol{x}). \tag{55}$$

We use the following result which we will show after few steps

$$||\sigma(t)||_{L^2(D;\mathbb{R}^d)} \leq \frac{C}{\epsilon^{3/2}}. \tag{56}$$

We then have

$$\ddot{\hat{\boldsymbol{u}}}(t,\boldsymbol{x}) = -\nabla PD^\epsilon(\hat{\boldsymbol{u}}(t))(\boldsymbol{x}) + \hat{\boldsymbol{b}}(t,\boldsymbol{x}) + \sigma(t,\boldsymbol{x}). \tag{57}$$

Multiply above with $\dot{\hat{\boldsymbol{u}}}(t)$ and integrate over D to get

$$\begin{aligned}(\ddot{\hat{\boldsymbol{u}}}(t),\dot{\hat{\boldsymbol{u}}}(t)) &= (-\nabla PD^\epsilon(\hat{\boldsymbol{u}}(t)),\dot{\hat{\boldsymbol{u}}}(t)) \\ &\quad + (\hat{\boldsymbol{b}}(t),\dot{\hat{\boldsymbol{u}}}(t)) + (\sigma(t),\dot{\hat{\boldsymbol{u}}}(t)).\end{aligned}$$

Consider energy $\mathcal{E}^\epsilon(\hat{\boldsymbol{u}})(t)$ given by Eq. 11 and note the identity Eq. 12, to have

$$\begin{aligned}\frac{d}{dt}\mathcal{E}^\epsilon(\hat{\boldsymbol{u}})(t) &= (\hat{\boldsymbol{b}}(t),\dot{\hat{\boldsymbol{u}}}(t)) + (\sigma(t),\dot{\hat{\boldsymbol{u}}}(t)) \\ &\leq \left(||\hat{\boldsymbol{b}}(t)||_{L^2(D;\mathbb{R}^d)} + ||\sigma(t)||_{L^2(D;\mathbb{R}^d)}\right) ||\dot{\hat{\boldsymbol{u}}}(t)||_{L^2(D;\mathbb{R}^d)},\end{aligned}$$

where we used Hölder inequality in last step. Since $PD^\epsilon(\boldsymbol{u})$ is positive for any $\boldsymbol{u}$, we have

$$||\dot{\hat{\boldsymbol{u}}}(t)|| \leq 2\sqrt{\frac{1}{2}||\dot{\hat{\boldsymbol{u}}}(t)||^2_{L^2(D;\mathbb{R}^d)} + PD^\epsilon(\hat{\boldsymbol{u}}(t))} = 2\sqrt{\mathcal{E}^\epsilon(\hat{\boldsymbol{u}})(t)}.$$

Using above, we get

$$\frac{1}{2}\frac{d}{dt}\mathcal{E}^\epsilon(\hat{\boldsymbol{u}})(t) \leq \left(||\hat{\boldsymbol{b}}(t)||_{L^2(D;\mathbb{R}^d)} + ||\sigma(t)||_{L^2(D;\mathbb{R}^d)}\right)\sqrt{\mathcal{E}^\epsilon(\hat{\boldsymbol{u}})(t)}.$$

Let $\delta > 0$ be some arbitrary but fixed real number and let $A(t) = \delta + \mathcal{E}^\epsilon(\hat{\boldsymbol{u}})(t)$. Then

$$\frac{1}{2}\frac{d}{dt}A(t) \leq \left(||\hat{\boldsymbol{b}}(t)||_{L^2(D;\mathbb{R}^d)} + ||\sigma(t)||_{L^2(D;\mathbb{R}^d)}\right)\sqrt{A(t)}.$$

Using the fact that $\frac{1}{\sqrt{A(t)}}\frac{d}{dt}A(t) = 2\frac{d}{dt}\sqrt{A(t)}$, we have

$$\begin{aligned}\sqrt{A(t)} &\leq \sqrt{A(0)} + \int_0^t \left(||\hat{\boldsymbol{b}}(s)||_{L^2(D;\mathbb{R}^d)} + ||\sigma(s)||_{L^2(D;\mathbb{R}^d)}\right) ds \\ &\leq \sqrt{A(0)} + \frac{TC}{\epsilon^{3/2}} + \int_0^T ||\hat{\boldsymbol{b}}(s)||_{L^2(D;\mathbb{R}^d)} ds.\end{aligned}$$

where we used bound on $||\sigma(s)||_{L^2(D;\mathbb{R}^d)}$ from Eq. 56. Noting that $\delta > 0$ is arbitrary, we send it to zero to get

$$\sqrt{\mathcal{E}^\epsilon(\hat{\boldsymbol{u}})(t)} \leq \sqrt{\mathcal{E}^\epsilon(\hat{\boldsymbol{u}})(0)} + \frac{TC}{\epsilon^{3/2}} + \int_0^T ||\hat{\boldsymbol{b}}(s)|| ds,$$

and Eq. 54 follows by taking square of above equation.

It remains to show in Eq. 56. To simplify the calculations, we use the following notations: let $\boldsymbol{\xi} \in H_1(\boldsymbol{0})$ and let

$$\begin{aligned}s_{\boldsymbol{\xi}} &= \epsilon|\boldsymbol{\xi}|, e_{\boldsymbol{\xi}} = \frac{\boldsymbol{\xi}}{|\boldsymbol{\xi}|}, \bar{\omega}(\boldsymbol{x}) = \omega(\boldsymbol{x})\omega(\boldsymbol{x} + \epsilon\boldsymbol{\xi}), \\ S_{\boldsymbol{\xi}}(\boldsymbol{x}) &= \frac{\hat{\boldsymbol{u}}(t, \boldsymbol{x} + \epsilon\boldsymbol{\xi}) - \hat{\boldsymbol{u}}(t, \boldsymbol{x})}{s_{\boldsymbol{\xi}}} \cdot e_{\boldsymbol{\xi}}.\end{aligned}$$

With above notations and using expression of $-\nabla PD^\epsilon$ from Eq. 38, we have for $\boldsymbol{x} \in U_i$

$$\begin{aligned}|\sigma(t, \boldsymbol{x})| &= |-\nabla PD^\epsilon(\hat{\boldsymbol{u}}(t))(\boldsymbol{x}_i) + \nabla PD^\epsilon(\hat{\boldsymbol{u}}(t))(\boldsymbol{x})| \\ &= \left|\frac{2}{\epsilon\omega_d}\int_{H_1(\boldsymbol{0})} \frac{J(|\boldsymbol{\xi}|)}{\sqrt{s_{\boldsymbol{\xi}}}} \left(\bar{\omega}(\boldsymbol{x}_i)F_1'(\sqrt{s_{\boldsymbol{\xi}}}S_{\boldsymbol{\xi}}(\boldsymbol{x}_i)) - \bar{\omega}(\boldsymbol{x})F_1'(\sqrt{s_{\boldsymbol{\xi}}}S_{\boldsymbol{\xi}}(\boldsymbol{x}))\right) e_{\boldsymbol{\xi}} d\boldsymbol{\xi}\right|\end{aligned}$$

$$\leq \frac{2}{\epsilon \omega_d} \int_{H_1(\mathbf{0})} \frac{J(|\boldsymbol{\xi}|)}{\sqrt{s_{\boldsymbol{\xi}}}} \left| \bar{\omega}(\boldsymbol{x}_i) F_1'(\sqrt{s_{\boldsymbol{\xi}}} S_{\boldsymbol{\xi}}(\boldsymbol{x}_i)) - \bar{\omega}(\boldsymbol{x}) F_1'(\sqrt{s_{\boldsymbol{\xi}}} S_{\boldsymbol{\xi}}(\boldsymbol{x})) \right| d\boldsymbol{\xi}$$

$$\leq \frac{2}{\epsilon \omega_d} \int_{H_1(\mathbf{0})} \frac{J(|\boldsymbol{\xi}|)}{\sqrt{s_{\boldsymbol{\xi}}}} \left(\left| \bar{\omega}(\boldsymbol{x}_i) F_1'(\sqrt{s_{\boldsymbol{\xi}}} S_{\boldsymbol{\xi}}(\boldsymbol{x}_i)) \right| + \left| \bar{\omega}(\boldsymbol{x}) F_1'(\sqrt{s_{\boldsymbol{\xi}}} S_{\boldsymbol{\xi}}(\boldsymbol{x})) \right| \right) d\boldsymbol{\xi}. \tag{58}$$

Using the fact that $0 \leq \omega(\boldsymbol{x}) \leq 1$ and $|F_1'(r)| \leq C_1$, where C_1 is $\sup_r |F_1'(r)|$, we get

$$|\sigma(t, \boldsymbol{x})| \leq \frac{4 C_1 \bar{J}_{1/2}}{\epsilon^{3/2}}.$$

where $\bar{J}_{1/2} = (1/\omega_d) \int_{H_1(\mathbf{0})} J(|\boldsymbol{\xi}|) |\boldsymbol{\xi}|^{-1/2} d\boldsymbol{\xi}$.

Taking the L^2 norm of $\sigma(t, \boldsymbol{x})$, we get

$$||\sigma(t)||^2_{L^2(D;\mathbb{R}^d)} = \sum_{i, \boldsymbol{x}_i \in D} \int_{U_i} |\sigma(t, \boldsymbol{x})|^2 d\boldsymbol{x} \leq \left(\frac{4 C_1 \bar{J}_{1/2}}{\epsilon^{3/2}} \right)^2 \sum_{i, \boldsymbol{x}_i \in D} \int_{U_i} d\boldsymbol{x}$$

thus

$$||\sigma(t)||_{L^2(D;\mathbb{R}^d)} \leq \frac{4 C_1 \bar{J}_{1/2} \sqrt{|D|}}{\epsilon^{3/2}} = \frac{C}{\epsilon^{3/2}}$$

where

$$C := 4 C_1 \bar{J}_{1/2} \sqrt{|D|}. \tag{59}$$

This completes the proof.

Finite Element Approximation

Let V_h be the approximation of $H_0^2(D, \mathbb{R}^d)$ associated with linear continuous interpolation associated with the mesh $\mathcal{T}_h$ (triangular or tetrahedral) where h denotes the size of finite element mesh. Let $\mathcal{I}_h(u)$ be defined as below

$$\mathcal{I}_h(u)(x) = \sum_{T \in \mathcal{T}_h} \left[\sum_{i \in N_T} u(x_i) \phi_i(x) \right]$$

where N_T is the set of global indices of nodes associated to finite element T, ϕ_i is the linear interpolation function associated to node i, and x_i is the material coordinate of node i.

Assuming that the size of each element in the mesh $\mathcal{T}_h$ is bounded by h, we have (see, e.g., [Theorem 4.6, Arnold 2011])

$$||u - \mathcal{I}_h(u)|| \leq ch^2||u||_2, \qquad \forall u \in H_0^2(D; \mathbb{R}^d). \tag{60}$$

Projection of Function in FE Space

Let $r_h(u) \in V_h$ be the projection of $u \in H_0^2(D; \mathbb{R}^d)$. It is defined as

$$||u - r_h(u)|| = \inf_{\tilde{u} \in V_h} ||u - \tilde{u}||. \tag{61}$$

It also satisfies the following

$$(r_h(u), \tilde{u}) = (u, \tilde{u}), \qquad \forall \tilde{u} \in V_h. \tag{62}$$

Since $\mathcal{I}_h(u) \in V_h$, we get an upper bound on right-hand side term and we have

$$||u - r_h(u)|| \leq ch^2||u||_2 \qquad \forall u \in H_0^2(D; \mathbb{R}^d). \tag{63}$$

Semi-discrete Approximation

Let $u_h(t) \in V_h$ be the approximation of $u(t)$ which satisfies the following

$$(\ddot{u}_h, \tilde{u}) + a^\epsilon(u_h(t), \tilde{u}) = (b(t), \tilde{u}), \qquad \forall \tilde{u} \in V_h. \tag{64}$$

We now show that the semi-discrete approximation is stable, i.e., energy at time t is bounded by initial energy and work done by the body force.

Theorem 4 (Stability of the semi-discrete approximation). *The semi-discrete scheme is energetically stable and the energy $\mathcal{E}^\epsilon(u_h)(t)$, defined in (11), satisfies the following bound*

$$\mathcal{E}^\epsilon(u_h)(t) \leq \left[\sqrt{\mathcal{E}^\epsilon(u_h)(0)} + \int_0^t ||b(\tau)||d\tau\right]^2.$$

We note that, while proving the stability of semi-discrete scheme corresponding to nonlinear peridynamics, we do not require any assumption on strain $S(y, x; u_h)$.

Proof. Letting $\tilde{u} = \dot{u}_h(t)$ in (10) and noting the identity (12), we get

$$\frac{d}{dt}\mathcal{E}^\epsilon(u_h)(t) = (b(t), \dot{u}_h(t)) \leq ||b(t)||||\dot{u}_h(t)||$$

We also have

$$||\dot{u}_h(t)|| \leq 2\sqrt{\frac{1}{2}||\dot{u}_h||^2 + PD^\epsilon(u_h(t))} = 2\sqrt{\mathcal{E}^\epsilon(u_h)(t)}$$

where we used the fact that $PD^\epsilon(u)(t)$ is nonnegative. We have

$$\frac{d}{dt}\mathcal{E}^\epsilon(u_h)(t) \leq 2\sqrt{\mathcal{E}^\epsilon(u_h)(t)}||b(t)||.$$

Fix $\delta > 0$ and let $A(t) = \mathcal{E}^\epsilon(u_h(t)) + \delta$. Then, from above equation, we easily have

$$\frac{d}{dt}A(t) \leq 2\sqrt{A(t)}||b(t)|| \quad \Rightarrow \frac{1}{2}\frac{\frac{d}{dt}A(t)}{\sqrt{A(t)}} \leq ||b(t)||.$$

Noting that $\frac{1}{\sqrt{a(t)}}\frac{da(t)}{dt} = 2\frac{d}{dt}\sqrt{a(t)}$, integrating from $t = 0$ to τ, and relabeling τ as t, we get

$$\sqrt{A(t)} \leq \sqrt{A(0)} + \int_0^t ||b(s)||ds.$$

Letting $\delta \to 0$ and taking the square of both sides proves the claim.

Central Difference Time Discretization

For illustration, we consider the central difference scheme and present the convergence rate for the central difference scheme for the fully nonlinear problem. We remark that the extension of these results to the general Newmark scheme is straightforward. We then consider a linearized peridynamics and demonstrate CFL-like conditions for stability of the fully discrete scheme.

Let Δt be the time step. The exact solution at $t^k = k\Delta t$ (or time step k) is denoted as (u^k, v^k), with $v^k = \partial u^k/\partial t$, and the projection onto V_h at t^k is given by $(r_h(u^k), r_h(v^k))$. The solution of the discrete problem at time step k is denoted as (u_h^k, v_h^k).

We approximate the initial data on displacement u_0 and velocity v_0 by their projection $r_h(u_0)$ and $r_h(v_0)$. Let $u_h^0 = r_h(u_0)$ and $v_h^0 = r_h(v_0)$. For $k \geq 1$, (u_h^k, v_h^k) satisfies, for all $\tilde{u} \in V_h$,

$$\left(\frac{u_h^{k+1} - u_h^k}{\Delta t}, \tilde{u}\right) = (v_h^{k+1}, \tilde{u}),$$

$$\left(\frac{v_h^{k+1} - v_h^k}{\Delta t}, \tilde{u}\right) = (-\nabla PD^\epsilon(u_h^k), \tilde{u}) + (b_h^k, \tilde{u}), \tag{65}$$

where we denote projection of $b(t^k)$, $r_h(b(t^k))$, as b_h^k. Combining the two equations delivers central difference equation for u_h^k. We have

$$\left(\frac{u_h^{k+1} - 2u_h^k + u_h^{k-1}}{\Delta t^2}, \tilde{u}\right) = (-\nabla PD^\epsilon(u_h^k), \tilde{u}) + (b_h^k, \tilde{u}), \qquad \forall \tilde{u} \in V_h. \quad (66)$$

For $k = 0$, we have $\forall \tilde{u} \in V_h$

$$\left(\frac{u_h^1 - u_h^0}{\Delta t^2}, \tilde{u}\right) = \frac{1}{2}(-\nabla PD^\epsilon(u_h^0), \tilde{u}) + \frac{1}{\Delta t}(v_h^0, \tilde{u}) + \frac{1}{2}(b_h^0, \tilde{u}). \quad (67)$$

We now show that finite element discretization converges to the exact solution.

Convergence of Approximation

In this section, we establish uniform bound on the discretization error and prove that approximate solution converges to the exact solution at the rate $C_t \Delta t + C_s h^2/\epsilon^2$ for fixed $\epsilon > 0$. We first compare the exact solution with its projection in V_h and then compare the projection with approximate solution. We further divide the calculation of error between projection and approximate solution in two parts, namely, consistency analysis and error analysis.

Error E^k is given by

$$E^k := ||u_h^k - u(t^k)|| + ||v_h^k - v(t^k)||.$$

We split the error as follows

$$E^k \leq \left(||u^k - r_h(u^k)|| + ||v^k - r_h(v^k)||\right) + \left(||r_h(u^k) - u_h^k|| + ||r_h(v^k) - v_h^k||\right),$$

where first term is error between exact solution and projections and second term is error between projections and approximate solution. Let

$$e_h^k(u) := r_h(u^k) - u_h^k \quad \text{and} \quad e_h^k(v) := r_h(v^k) - v_h^k \quad (68)$$

and

$$e^k := ||e_h^k(u)|| + ||e_h^k(v)||. \quad (69)$$

Using (63), we have

$$E^k \leq C_p h^2 + e^k, \quad (70)$$

where

$$C_p := c\left[\sup_t ||u(t)||_2 + \sup_t ||\frac{\partial u(t)}{\partial t}||_2\right]. \tag{71}$$

We have the following main result.

Theorem 5 (Convergence of the central difference approximation). *Let (u, v) be the exact solution of peridynamics equation in (6). Let (u_h^k, v_h^k) be the central difference approximation in time and piecewise linear finite element approximation in space solution of (65). If $u, v \in C^2([0,T], H_0^2(D;\mathbb{R}^d))$, then the scheme is consistent, and the error E^k satisfies the following bound*

$$\begin{aligned}
&\sup_{k\leq T/\Delta t} E^k \\
&= C_p h^2 + \exp[T(1 + L/\epsilon^2)(1 + \Delta t + O(\Delta t^2))] \\
&\left[e^0 + T(1 + \Delta t + O(\Delta t^2))\left(C_t \Delta t + C_s \frac{h^2}{\epsilon^2}\right)\right]
\end{aligned} \tag{72}$$

where constants C_p, C_t, and C_s are given by (71) and (79). The constant L/ϵ^2 is the Lipschitz constant of $-\nabla PD^\epsilon(u)$ in L^2 (see Theorem 1).

If the error in initial data is zero, then E^k is of the order of $C_t\Delta t + C_s h^2/\epsilon^2$.

Error Analysis

We derive the equation for evolution of $e_h^k(u)$ as follows

$$\begin{aligned}
&\left(\frac{u_h^{k+1} - u_h^k}{\Delta t} - \frac{r_h(u^{k+1}) - r_h(u^k)}{\Delta t}, \tilde{u}\right) \\
&= (v_h^{k+1}, \tilde{u}) - \left(\frac{r_h(u^{k+1}) - r_h(u^k)}{\Delta t}, \tilde{u}\right) \\
&= (v_h^{k+1}, \tilde{u}) - (r_h(v^{k+1}), \tilde{u}) + (r_h(v^{k+1}), \tilde{u}) - (v^{k+1}, \tilde{u}) \\
&\quad + (v^{k+1}, \tilde{u}) - \left(\frac{\partial u^{k+1}}{\partial t}, \tilde{u}\right) \\
&\quad + \left(\frac{\partial u^{k+1}}{\partial t}, \tilde{u}\right) - \left(\frac{u^{k+1} - u^k}{\Delta t}, \tilde{u}\right) \\
&\quad + \left(\frac{u^{k+1} - u^k}{\Delta t}, \tilde{u}\right) - \left(\frac{r_h(u^{k+1}) - r_h(u^k)}{\Delta t}, \tilde{u}\right).
\end{aligned}$$

Using property $(r_h(u), \tilde{u}) = (u, \tilde{u})$ for $\tilde{u} \in V_h$ and the fact that $\frac{\partial u(t^{k+1})}{\partial t} = v^{k+1}$ where u is the exact solution, we get

$$\left(\frac{e_h^{k+1}(u) - e_h^k(u)}{\Delta t}, \tilde{u} \right) = (e_h^{k+1}(v), \tilde{u}) + \left(\frac{\partial u^{k+1}}{\partial t}, \tilde{u} \right) - \left(\frac{u^{k+1} - u^k}{\Delta t}, \tilde{u} \right). \quad (73)$$

Let $(\tau_h^k(u), \tau_h^k(v))$ be the consistency error in time discretization given by

$$\tau_h^k(u) := \frac{\partial u^{k+1}}{\partial t} - \frac{u^{k+1} - u^k}{\Delta t},$$

$$\tau_h^k(v) := \frac{\partial v^k}{\partial t} - \frac{v^{k+1} - v^k}{\Delta t}.$$

With above notation, we have

$$(e_h^{k+1}(u), \tilde{u}) = (e_h^k(u), \tilde{u}) + \Delta t (e_h^{k+1}(v), \tilde{u}) + \Delta t (\tau_h^k(u), \tilde{u}). \quad (74)$$

We now derive the equation for $e_h^k(v)$ as follows

$$\begin{aligned}
&\left(\frac{v_h^{k+1} - v_h^k}{\Delta t} - \frac{r_h(v^{k+1}) - r_h(v^k)}{\Delta t}, \tilde{u} \right) \\
&= (-\nabla PD^\epsilon(u_h^k), \tilde{u}) + (b_h^k, \tilde{u}) - \left(\frac{r_h(v^{k+1}) - r_h(v^k)}{\Delta t}, \tilde{u} \right) \\
&= (-\nabla PD^\epsilon(u_h^k), \tilde{u}) + (b^k, \tilde{u}) - \left(\frac{\partial v^k}{\partial t}, \tilde{u} \right) \\
&\quad + \left(\frac{\partial v^k}{\partial t}, \tilde{u} \right) - \left(\frac{v^{k+1} - v^k}{\Delta t}, \tilde{u} \right) \\
&\quad + \left(\frac{v^{k+1} - v^k}{\Delta t}, \tilde{u} \right) - \left(\frac{r_h(v^{k+1}) - r_h(v^k)}{\Delta t}, \tilde{u} \right) \\
&= (-\nabla PD^\epsilon(u_h^k) + \nabla PD^\epsilon(u^k), \tilde{u}) + (b_h^k - b(t^k), \tilde{u}) \\
&\quad + \left(\frac{\partial v^k}{\partial t}, \tilde{u} \right) - \left(\frac{v^{k+1} - v^k}{\Delta t}, \tilde{u} \right) + \left(\frac{v^{k+1} - v^k}{\Delta t}, \tilde{u} \right) - \left(\frac{r_h(v^{k+1}) - r_h(v^k)}{\Delta t}, \tilde{u} \right) \\
&= (-\nabla PD^\epsilon(u_h^k) + \nabla PD^\epsilon(u^k), \tilde{u}) + \left(\frac{\partial v^k}{\partial t} - \frac{v^{k+1} - v^k}{\Delta t}, \tilde{u} \right)
\end{aligned}$$

where we used the property of $r_h(u)$ and the fact that

$$(-\nabla PD^\epsilon(u^k), \tilde{u}) + (b^k, \tilde{u}) - \left(\frac{\partial v^k}{\partial t}, \tilde{u}\right) = 0, \quad \forall \tilde{u} \in H_0^2(D; \mathbb{R}^d).$$

We further divide the error in peridynamics force as follows

$$\begin{aligned}&\left(-\nabla PD^\epsilon(u_h^k) + \nabla PD^\epsilon(u^k), \tilde{u}\right)\\&= \left(-\nabla PD^\epsilon(u_h^k) + \nabla PD^\epsilon(r_h(u^k)), \tilde{u}\right) + \left(-\nabla PD^\epsilon(r_h(u^k)) + \nabla PD^\epsilon(u^k), \tilde{u}\right).\end{aligned}$$

We will see in the next section that the second term is related to consistency error in spatial discretization. Therefore, we define another consistency error term $\sigma^k_{per,h}(u)$ as follows

$$\sigma^k_{per,h}(u) := -\nabla PD^\epsilon(r_h(u^k)) + \nabla PD^\epsilon(u^k). \tag{75}$$

After substituting the notations related to consistency errors, we get

$$\begin{aligned}(e_h^{k+1}(v), \tilde{u}) = (e_h^k(v), \tilde{u}) + \Delta t(-\nabla PD^\epsilon(u_h^k) + \nabla PD^\epsilon(r_h(u^k)), \tilde{u})\\ + \Delta t(\tau_h^k(v), \tilde{u}) + \Delta t(\sigma^k_{per,h}(u), \tilde{u}).\end{aligned} \tag{76}$$

Since u, v are C^2 in time, we can easily show

$$||\tau_h^k(u)|| \leq \Delta t \sup_t ||\frac{\partial^2 u}{\partial t^2}|| \qquad \text{and} \qquad ||\tau_h^k(v)|| \leq \Delta t \sup_t ||\frac{\partial^2 v}{\partial t^2}||.$$

To estimate $\sigma^k_{per,h}(u)$, we note the Lipschitz property of peridynamics force in L^2 norm (see Theorem 1). This leads us to

$$||\sigma^k_{per,h}(u)|| \leq \frac{L}{\epsilon^2}||u^k - r_h(u^k)|| \leq \frac{Lc}{\epsilon^2} h^2 \sup_t ||u(t)||_2, \tag{77}$$

where we have relabeled the L^2 Lipschitz constant L_1 as L.

Let τ be given by

$$\begin{aligned}\tau &:= \sup_k \left(||\tau_h^k(u)|| + ||\tau_h^k(v)|| + ||\sigma^k_{per,h}(u)||\right)\\ &\leq C_t \Delta t + C_s \frac{h^2}{\epsilon^2}.\end{aligned} \tag{78}$$

where

$$C_t := ||\frac{\partial^2 u}{\partial t^2}|| + ||\frac{\partial^2 v}{\partial t^2}|| \quad \text{and} \quad C_s := Lc \sup_t ||u(t)||_2. \tag{79}$$

In equation for $e_h^k(u)$ (see (74)), we take $\tilde{u} = e_h^{k+1}(u)$. Note that $e_h^{k+1}(u) = u_h^k - r_h(u^k) \in V_h$. We have

$$||e_h^{k+1}(u)||^2 = (e_h^k(u), e_h^{k+1}(u)) + \Delta t(e_h^{k+1}(v), e_h^{k+1}(u)) + \Delta t(\tau_h^k(u), e_h^{k+1}(u)).$$

Using the fact that $(u, v) \leq ||u||||v||$, we get

$$\begin{aligned}||e_h^{k+1}(u)||^2 \leq ||e_h^k(u)||||e_h^{k+1}(u)|| + \Delta t||e_h^{k+1}(v)||||e_h^{k+1}(u)|| \\ + \Delta t||\tau_h^k(u)||||e_h^{k+1}(u)||.\end{aligned}$$

Canceling $||e_h^{k+1}(u)||$ from both sides gives

$$||e_h^{k+1}(u)|| \leq ||e_h^k(u)|| + \Delta t||e_h^{k+1}(v)|| + \Delta t||\tau_h^k(u)||. \tag{80}$$

Similarly, if we choose $\tilde{u} = e_h^{k+1}(v)$ in (76) and use the steps similar to above, we get

$$\begin{aligned}||e_h^{k+1}(v)|| \leq ||e_h^k(v)|| + \Delta t|| - \nabla PD^\epsilon(u_h^k) + \nabla PD^\epsilon(r_h(u^k))|| \\ + \Delta t\left(||\tau_h^k(v)|| + ||\sigma_{per,h}^k(u)||\right).\end{aligned} \tag{81}$$

Using the Lipschitz property of the peridynamics force in L^2, we have

$$|| - \nabla PD^\epsilon(u_h^k) + \nabla PD^\epsilon(r_h(u^k))|| \leq \frac{L}{\epsilon^2}||u_h^k - r_h(u^k)|| = \frac{L}{\epsilon^2}||e_h^k(u)||. \tag{82}$$

After adding (80) and (81) and substituting (82), we get

$$\begin{aligned}||e_h^{k+1}(u)|| + ||e_h^{k+1}(v)|| \leq ||e_h^k(u)|| + ||e_h^k(v)|| + \Delta t||e_h^{k+1}(v)|| \\ + \frac{L}{\epsilon^2}\Delta t||e_h^k(u)|| + \Delta t\tau\end{aligned}$$

where τ is defined in (78).

Let $e^k := ||e_h^k(u)|| + ||e_h^k(v)||$. Assuming $L/\epsilon^2 \geq 1$, we get

$$\begin{aligned}e^{k+1} &\leq e^k + \Delta t e^{k+1} + \Delta t\frac{L}{\epsilon^2}e^k + \Delta t\tau \\ \Rightarrow e^{k+1} &\leq \frac{1 + \Delta t L/\epsilon^2}{1 - \Delta t}e^k + \frac{\Delta t}{1 - \Delta t}\tau.\end{aligned}$$

Substituting e^k recursively in above equation, we get

$$e^{k+1} \le \left(\frac{1+\Delta t L/\epsilon^2}{1-\Delta t}\right)^{k+1} e^0 + \frac{\Delta t}{1-\Delta t}\tau \sum_{j=0}^{k} \left(\frac{1+\Delta t L/\epsilon^2}{1-\Delta t}\right)^{k-j}.$$

Noting $1/(1-\Delta t) = 1 + \Delta t + \Delta t^2 + O(\Delta t^3)$,

$$\frac{1+\Delta t L/\epsilon^2}{1-\Delta t} \le 1 + (1+L/\epsilon^2)\Delta t + (1+L/\epsilon^2)\Delta t^2 + O(L/\epsilon^2)O(\Delta t^3),$$

and $(1+a\Delta t)^k \le \exp[ka\Delta t] \le \exp[Ta]$ for $a > 0$, we get

$$\begin{aligned}\left(\frac{1+\Delta t L/\epsilon^2}{1-\Delta t}\right)^{k} &\le \exp[k\Delta t(1+L/\epsilon^2) + k\Delta t^2(1+L/\epsilon^2) + kO(L/\epsilon^2)O(\Delta t^3)] \\ &\le \exp[T(1+L/\epsilon^2) + T\Delta t(1+L/\epsilon^2) + O(TL/\epsilon^2)O(\Delta t^2)] \\ &= \exp[T(1+L/\epsilon^2)(1+\Delta t + O(\Delta t^2))].\end{aligned}$$

Substituting above estimates, we can easily show that

$$\begin{aligned} e^{k+1} &\le \exp[T(1+L/\epsilon^2)(1+\Delta t + O(\Delta t^2))] \\ &\quad \left[e^0 + \Delta t(1+\Delta t + O(\Delta t^2))\tau \sum_{j=0}^{k} 1\right] \\ &\le \exp[T(1+L/\epsilon^2)(1+\Delta t + O(\Delta t^2))]\left[e^0 + k\Delta t(1+\Delta t + O(\Delta t^2))\tau\right].\end{aligned}$$

Finally, we substitute above into (70) to have

$$\begin{aligned} E^k \le C_p h^2 &+ \exp[T(1+L/\epsilon^2)(1+\Delta t + O(\Delta t^2))] \\ &\left[e^0 + k\Delta t(1+\Delta t + O(\Delta t^2))\tau\right].\end{aligned}$$

After taking sup over $k \le T/\Delta t$, we get the desired result and proof of Theorem 2 is complete.

We now consider the stability of linearized peridynamics model.

Stability Condition for Linearized Peridynamics

In this section, we linearize the peridynamics model and obtain a CFL-like stability condition. For problems where strains are small, the stability condition for the linearized model is expected to work for nonlinear model. The slope of peridynamics potential f is constant for sufficiently small strain, and therefore for small strain, the nonlinear model behaves like a linear model. When displacement field is smooth, the difference between the linearized peridynamics force and nonlinear peridynamics force is of the order of ϵ. See [Proposition 4, Jha and Lipton 2017c].

In (5), linearization gives

$$-\nabla PD_l^\epsilon(u)(x) = \frac{4}{\epsilon^{d+1}\omega_d} \int_{H_\epsilon(x)} \omega(x)\omega(y)J^\epsilon(|y-x|)f'(0)S(u)e_{y-x}dy. \quad (83)$$

The corresponding bilinear form is denoted as a_l^ϵ and is given by

$$a_l^\epsilon(u,v) = \frac{2}{\epsilon^{d+1}\omega_d} \int_D \int_{H_\epsilon(x)} \omega(x)\omega(y)J^\epsilon(|y-x|)f'(0)|y-x|S(u)S(v)dydx. \quad (84)$$

We have

$$(-\nabla PD_l^\epsilon(u), v) = -a_l^\epsilon(u,v).$$

We now discuss the stability of the FEM approximation to the linearized problem. We replace $-\nabla PD^\epsilon$ by its linearization denoted by $-\nabla PD_l^\epsilon$ in (66) and (67). The corresponding approximate solution in V_h is denoted by $u_{l,h}^k$ where

$$\left(\frac{u_{l,h}^{k+1} - 2u_{l,h}^k + u_{l,h}^{k-1}}{\Delta t^2}, \tilde{u}\right) = (-\nabla PD_l^\epsilon(u_{l,h}^k), \tilde{u}) + (b_h^k, \tilde{u}), \qquad \forall \tilde{u} \in V_h \quad (85)$$

and

$$\left(\frac{u_{l,h}^1 - u_{l,h}^0}{\Delta t^2}, \tilde{u}\right) = \frac{1}{2}(-\nabla PD^\epsilon(u_{l,h}^0), \tilde{u}) + \frac{1}{\Delta t}(v_{l,h}^0, \tilde{u}) + \frac{1}{2}(b_h^0, \tilde{u}), \qquad \forall \tilde{u} \in V_h. \quad (86)$$

We will adopt the following notations

$$\begin{aligned} \overline{u}_h^{k+1} &:= \frac{u_h^{k+1} + u_h^k}{2}, \ \overline{u}_h^k := \frac{u_h^k + u_h^{k-1}}{2}, \\ \bar{\partial}_t u_h^k &:= \frac{u_h^{k+1} - u_h^{k-1}}{2\Delta t}, \ \bar{\partial}_t^+ u_h^k := \frac{u_h^{k+1} - u_h^k}{\Delta t}, \ \bar{\partial}_t^- u_h^k := \frac{u_h^k - u_h^{k-1}}{\Delta t}. \end{aligned} \quad (87)$$

With above notations, we have

$$\bar{\partial}_t u_h^k = \frac{\bar{\partial}_t^+ u_h^k + \bar{\partial}_h^- u_h^k}{2} = \frac{\overline{u}_h^{k+1} - \overline{u}_h^k}{\Delta t}.$$

We also define

$$\bar{\partial}_{tt} u_h^k := \frac{u_h^{k+1} - 2u_h^k + u_h^{k-1}}{\Delta t^2} = \frac{\bar{\partial}_t^+ u_h^k - \bar{\partial}_t^- u_h^k}{\Delta t}.$$

We introduce the discrete energy associated with $u_{l,h}^k$ at time step k as defined by

$$\mathcal{E}(u_{l,h}^k) := \frac{1}{2}\left[||\bar{\partial}_t^+ u_{l,h}^k||^2 - \frac{\Delta t^2}{4} a_l^\epsilon(\bar{\partial}_t^+ u_{l,h}^k, \bar{\partial}_t^+ u_{l,h}^k) + a_l^\epsilon(\overline{u}_{l,h}^{k+1}, \overline{u}_{l,h}^{k+1}) \right]$$

Following [Theorem 4.1, Karaa 2012], we have

Theorem 6 (Stability of the central difference approximation of linearized peridynamics).

Let $u_{l,h}^k$ be the approximate solution of (85) *and* (86) *with respect to linearized peridynamics. In the absence of the body force $b(t) = 0$ and for all t, if Δt satisfies the CFL-like condition*

$$\frac{\Delta t^2}{4} \sup_{u \in V_h \setminus \{0\}} \frac{a_l^\epsilon(u,u)}{(u,u)} \leq 1, \tag{88}$$

then the discrete energy is positive and satisfies

$$\mathcal{E}(u_{l,h}^k) = \mathcal{E}(u_{l,h}^{k-1}), \tag{89}$$

and we have the stability

$$\mathcal{E}(u_{l,h}^k) = \mathcal{E}(u_{l,h}^0). \tag{90}$$

Proof. Set $b(t) = 0$. Noting that a_l^ϵ is bilinear, after adding and subtracting term $(\Delta t^2/4) a_l^\epsilon(\bar{\partial}_{tt} u_{l,h}^k, \tilde{u})$ to (85), and noting the following

$$u_{l,h}^k + \frac{\Delta t^2}{4} \bar{\partial}_{tt} u_{l,h}^k = \frac{\overline{u}_{l,h}^{k+1}}{2} + \frac{\overline{u}_{l,h}^k}{2}$$

we get

$$(\bar{\partial}_{tt} u_{l,h}^k, \tilde{u}) - \frac{\Delta t^2}{4} a_l^\epsilon(\bar{\partial}_{tt} u_{l,h}^k, \tilde{u}) + \frac{1}{2} a_l^\epsilon(\overline{u}_{l,h}^{k+1} + \overline{u}_{l,h}^k, \tilde{u}) = 0.$$

We let $\tilde{u} = \bar{\partial}_t u_{l,h}^k$, to write

$$(\bar{\partial}_{tt} u_{l,h}^k, \bar{\partial}_t u_{l,h}^k) - \frac{\Delta t^2}{4} a_l^\epsilon(\bar{\partial}_{tt} u_{l,h}^k, \bar{\partial}_t u_{l,h}^k) + \frac{1}{2} a_l^\epsilon(\overline{u}_{l,h}^{k+1} + \overline{u}_{l,h}^k, \bar{\partial}_t u_{l,h}^k) = 0.$$

It is easily shown that

$$(\bar{\partial}_{tt}u^k_{l,h}, \bar{\partial}_t u^k_{l,h}) = \left(\frac{\bar{\partial}^+_t u^k_{l,h} - \bar{\partial}^-_t u^k_{l,h}}{\Delta t}, \frac{\bar{\partial}^+_t u^k_{l,h} + \bar{\partial}^-_t u^k_{l,h}}{2} \right)$$
$$= \frac{1}{2\Delta t}\left[||\bar{\partial}^+_t u^k_{l,h}||^2 - ||\bar{\partial}^-_t u^k_{l,h}||^2 \right]$$

and

$$a^\epsilon_l(\bar{\partial}_{tt}u^k_{l,h}, \bar{\partial}_t u^k_{l,h}) = \frac{1}{2\Delta t}\left[a^\epsilon_l(\bar{\partial}^+_t u^k_{l,h}, \bar{\partial}^+_t u^k_{l,h}) - a^\epsilon_l(\bar{\partial}^-_t u^k_{l,h}, \bar{\partial}^-_t u^k_{l,h}) \right].$$

Noting that $\bar{\partial}_t u^k_{l,h} = (\overline{u}^{k+1}_{l,h} - \overline{u}^k_{l,h})/\Delta t$, we get

$$\frac{1}{2\Delta t} a^\epsilon_l(\overline{u}^{k+1}_{l,h} + \overline{u}^k_{l,h}, \overline{u}^{k+1}_{l,h} - \overline{u}^k_{l,h})$$
$$= \frac{1}{2\Delta t}\left[a^\epsilon_l(\overline{u}^{k+1}_{l,h}, \overline{u}^{k+1}_{l,h}) - a^\epsilon_l(\overline{u}^k_{l,h}, \overline{u}^k_{l,h}) \right].$$

After combining the above equations, we get

$$\frac{1}{\Delta t}\left[\left(\frac{1}{2}||\bar{\partial}^+_t u^k_{l,h}||^2 - \frac{\Delta t^2}{8} a^\epsilon_l(\bar{\partial}^+_t u^k_{l,h}, \bar{\partial}^+_t u^k_{l,h}) + \frac{1}{2} a^\epsilon_l(\overline{u}^{k+1}_{l,h}, \overline{u}^{k+1}_{l,h}) \right) \right.$$
$$\left. - \left(\frac{1}{2}||\bar{\partial}^-_t u^k_{l,h}||^2 - \frac{\Delta t^2}{8} a^\epsilon_l(\bar{\partial}^-_t u^k_{l,h}, \bar{\partial}^-_t u^k_{l,h}) + \frac{1}{2} a^\epsilon_l(\overline{u}^k_{l,h}, \overline{u}^k_{l,h}) \right) \right] = 0. \quad (91)$$

We recognize the first term in bracket as $\mathcal{E}(u^k_{l,h})$. We next prove that the second term is $\mathcal{E}(u^{k-1}_{l,h})$. We substitute $k = k-1$ in the definition of $\mathcal{E}(u^k_{l,h})$ to get

$$\mathcal{E}(u^{k-1}_{l,h}) = \frac{1}{2}\left[||\bar{\partial}^+_t u^{k-1}_{l,h}||^2 - \frac{\Delta t^2}{4} a^\epsilon_l(\bar{\partial}^+_t u^{k-1}_{l,h}, \bar{\partial}^+_t u^{k-1}_{l,h}) + a^\epsilon_l(\overline{u}^k_{l,h}, \overline{u}^k_{l,h}) \right].$$

We clearly have $\bar{\partial}^+_t u^{k-1}_{l,h} = \frac{u^{k-1+1}_{l,h} - u^{k-1}_{l,h}}{\Delta t} = \bar{\partial}^-_t u^k_{l,h}$, and this implies that the second term in (91) is $\mathcal{E}(u^{k-1}_{l,h})$. It now follows from (91) that $\mathcal{E}(u^k_{l,h}) = \mathcal{E}(u^{k-1}_{l,h})$.

The stability condition is such that discrete energy is positive. In the definition of $\mathcal{E}(u^k_{l,h})$, we see that the second term is negative. We now obtain a condition on the time step that insures the sum of the first two terms is positive, and this will establish the positivity of $\mathcal{E}(u^k_{l,h})$. Let $v = \bar{\partial}^+_t u^k_{l,h} \in V_h$, and then we require

$$||v||^2 - \frac{\Delta t^2}{4} a^\epsilon_l(v, v) \geq 0 \quad \Rightarrow \frac{\Delta t^2}{4} \frac{a^\epsilon_l(v, v)}{||v||^2} \leq 1 \quad (92)$$

Clearly if Δt satisfies

$$\frac{\Delta t^2}{4} \sup_{v \in V_h \setminus \{0\}} \frac{a_l^\epsilon(v, v)}{||v||^2} \leq 1 \tag{93}$$

then (92) is also satisfied and the discrete energy is positive. Iteration gives $\mathcal{E}(u_{l,h}^k) = \mathcal{E}(u_{l,h}^0)$ and the theorem is proved.

Conclusion

In this chapter we computed the a priori error incurred in finite element and finite difference discretizations of peridynamics. We show that for finite element approximation with linear elements, the rate of convergence is better as compared to rate of convergence of finite difference approximation. A CFL-like condition for the stability of linearized peridynamics is obtained. For the fully nonlinear problem, we find that for the semi-discrete approximation the energy at any instant is bounded by initial energy and work done by the body force.

This model has been analyzed using a quadrature-based finite element approximation in detail in Jha and Lipton (2017c) for nonlinear nonlocal models and their linearization assuming an a priori higher regularity of solutions. If one assumes more regular solutions with three continuous spatial derivatives (no cracks), then solutions of the nonlinear nonlocal model converge to those of the classical local elastodynamic model at the rate ϵ uniformly in time in the H^1 norm (see (Theorem 5, Jha and Lipton 2017c)). The numerical simulation of problems using finite differences for this model is carried out in Lipton et al. (2016) and Diehl et al. (2016). In earlier work (Tian and Du 2014) develop a framework for asymptotically compatible finite element schemes for linear problems where the solutions of the nonlocal problem are known to converge to a unique solution of the local problem. For the problems treated there, the discrete approximations associated with asymptotically compatible schemes converge if $h \to 0$ and $\epsilon \to 0$.

For the bond-based prototypical microelastic brittle material model analyzed here, the uniqueness property for the $\epsilon = 0$ problem is much less clear. The nonlinear nonlocal model treated in this chapter is an evolution in taking values in the vector space L^2 and can be identified with a sharp fracture evolution as $\epsilon \to 0$ (see Lipton 2014, 2016). The limit evolution is shown to be an element of the vector space, the space of special functions of bounded deformation referred to as SBD. The description and properties of this vector space can be found in Ambrosio et al. (1997). Unlike the linear nonlocal models, we do not necessarily have a unique sharp fracture evolution in the $\epsilon = 0$ limit. The uniqueness of the limit evolution for the nonlocal nonlinear model is an open question and remains to be established. The issue of nonuniqueness arises as the limit evolution is not completely characterized. What is currently missing is a limiting kinetic relation relating crack growth to crack driving force. Future work will seek to account for the missing information and address the issue of uniqueness for the limit problem.

Acknowledgements This material is based upon work supported by the US Army Research Laboratory and the US Army Research Office under contract/grant number W911NF1610456.

References

A. Agwai, I. Guven, E. Madenci, Predicting crack propagation with peridynamics: a comparative study. Int. J. Fract. **171**(1), 65–78 (2011)

B. Aksoylu, T. Mengesha, Results on nonlocal boundary value problems. Numer. Funct. Anal. Optim. **31**(12), 1301–1317 (2010)

B. Aksoylu, ML Parks, Variational theory and domain decomposition for nonlocal problems. Appl. Math. Comput. **217**(14), 6498–6515 (2011)

B. Aksoylu, Z. Unlu, Conditioning analysis of nonlocal integral operators in fractional sobolev spaces. SIAM J. Numer. Anal. **52**, 653–677 (2014)

L. Ambrosio, A. Coscia, G. Dal Maso, Fine properties of functions with bounded deformation. Arch. Ration. Mech. Anal. **139**, 201–238 (1997)

D.N. Arnold, Lecture notes on numerical analysis of partial differential equations (2011), http://www.math.umn.edu/~arnold/8445/notes.pdf

E. Askari, F. Bobaru, R. Lehoucq, M. Parks, S. Silling, O. Weckner, Peridynamics for multiscale materials modeling. J Phys Conf Ser **125**, 012078 (2008). IOP Publishing

G.A. Baker, Error estimates for finite element methods for second order hyperbolic equations. SIAM J. Numer. Anal. **13**(4), 564–576 (1976)

F. Bobaru, W. Hu, The meaning, selection, and use of the peridynamic horizon and its relation to crack branching in brittle materials. Int. J. Fract. **176**(2), 215–222 (2012)

F. Bobaru, M. Yang, L.F. Alves, S.A. Silling, E. Askari, J. Xu, Convergence, adaptive refinement, and scaling in 1d peridynamics. Int. J. Numer. Meth. Eng. **77**(6), 852–877 (2009)

X. Chen, M. Gunzburger, Continuous and discontinuous finite element methods for a peridynamics model of mechanics. Comput. Meth. Appl. Mech. Eng. **200**(9), 1237–1250 (2011)

K. Dayal, Leading-order nonlocal kinetic energy in peridynamics for consistent energetics and wave dispersion. J. Mech. Phys. Solids **105**, 235–253 (2017)

P. Diehl, R. Lipton, M. Schweitzer, Numerical verification of a bond-based softening peridynamic model for small displacements: deducing material parameters from classical linear theory. Institut für Numerische Simulation Preprint, (2016)

Q. Du, M. Gunzburger, R. Lehoucq, K. Zhou, Analysis of the volume-constrained peridynamic navier equation of linear elasticity. J. Elast. **113**(2), 193–217 (2013a)

Q. Du, L. Tian, X. Zhao, A convergent adaptive finite element algorithm for nonlocal diffusion and peridynamic models. SIAM J. Numer. Anal. **51**(2), 1211–1234 (2013b)

Q. Du, K. Zhou, Mathematical analysis for the peridynamic nonlocal continuum theory. ESAIM Math. Model. Numer. Anal. **45**(2), 217–234 (2011)

Q. Du, Y. Tao, X. Tian, A peridynamic model of fracture mechanics with bond-breaking. J. Elast. (2017). https://doi.org/10.1007/s10659-017-9661-2

E. Emmrich, R.B. Lehoucq, D. Puhst, Peridynamics: a nonlocal continuum theory, in *Meshfree Methods for Partial Differential Equations VI* (Springer, Berlin/Heidelberg, 2013), pp. 45–65

E. Emmrich, O. Weckner, et al. On the well-posedness of the linear peridynamic model and its convergence towards the navier equation of linear elasticity. Commun. Math. Sci. **5**(4), 851–864 (2007)

E. Emmrich, D. Puhst, A short note on modeling damage in peridynamics. J. Elast. **123**, 245–252 (2016)

J.T. Foster, S.A. Silling, W. Chen, An energy based failure criterion for use with peridynamic states. Int. J. Multiscale Comput. Eng. 9(6), 675–688 (2011)

W. Gerstle, N. Sau, S. Silling, Peridynamic modeling of concrete structures. Nucl. Eng. Des. **237**(12), 1250–1258 (2007)

M.J. Grote, D. Schötzau, Optimal error estimates for the fully discrete interior penalty dg method for the wave equation. J. Sci. Comput. **40**(1), 257–272 (2009)

Q. Guan, M. Gunzburger, Stability and accuracy of time-stepping schemes and dispersion relations for a nonlocal wave equation. Numer. Meth. Partial Differ. Equ. **31**(2), 500–516 (2015)

Y.D. Ha, F. Bobaru, Characteristics of dynamic brittle fracture captured with peridynamics. Eng. Fract. Mech. **78**(6), 1156–1168 (2011)

P.K. Jha, R. Lipton, Finite element approximation of nonlocal fracture models (2017a). arXiv preprint arXiv:1710.07661

P.K. Jha, R. Lipton, Numerical analysis of nonlocal fracture models models in holder space (2017b). arXiv preprint arXiv:1701.02818. To appear in SIAM Journal on Numerical Analysis 2018

P.K. Jha, R. Lipton, Numerical convergence of nonlinear nonlocal continuum models to local elastodynamics (2017c). arXiv preprint arXiv:1707.00398. To appear in International Journal for Numerical Methods in Engineering 2018

S. Karaa, Stability and convergence of fully discrete finite element schemes for the acoustic wave equation. J. Appl. Math. Comput. **40**(1–2), 659–682 (2012)

R. Lipton, Dynamic brittle fracture as a small horizon limit of peridynamics. J. Elast. **117**(1), 21–50 (2014)

R. Lipton, Cohesive dynamics and brittle fracture. J. Elast. **124**(2), 143–191 (2016)

R. Lipton, S. Silling, R. Lehoucq, Complex fracture nucleation and evolution with nonlocal elastodynamics (2016). arXiv preprint arXiv:1602.00247

R. Lipton, E. Said, P. Jha, Free damage propagation with memory. Journal of Elasticity, 1–25 (2018). https://doi.org/10.1007/s10659-018-9672

D.J. Littlewood, Simulation of dynamic fracture using peridynamics, finite element modeling, and contact, in *Proceedings of the ASME 2010 International Mechanical Engineering Congress and Exposition (IMECE)* (2010)

R.W. Macek, S.A. Silling, Peridynamics via finite element analysis. Finite Elem. Anal. Des. **43**(15), 1169–1178 (2007)

T. Mengesha, Q. Du, Analysis of a scalar peridynamic model with a sign changing kernel. Discrete Contin. Dynam. Syst. B **18**, 1415–1437 (2013)

T. Mengesha, Q. Du, Nonlocal constrained value problems for a linear peridynamic navier equation. J. Elast. **116**(1), 27–51 (2014)

S. Silling, O. Weckner, E. Askari, F. Bobaru, Crack nucleation in a peridynamic solid. Int. J. Fract. **162**(1–2), 219–227 (2010)

S.A. Silling, Reformulation of elasticity theory for discontinuities and long-range forces. J. Mech. Phys. Solids **48**(1), 175–209 (2000)

X. Tian, Q. Du, Asymptotically compatible schemes and applications to robust discretization of nonlocal models. SIAM J. Numer. Anal. **52**(4), 1641–1665 (2014)

X. Tian, Q. Du, M. Gunzburger, Asymptotically compatible schemes for the approximation of fractional laplacian and related nonlocal diffusion problems on bounded domains. Adv. Comput. Math. **42**(6), 1363–1380 (2016a)

X. Tian, Q. Du, M. Gunzburger, Asymptotically compatible schemes for the approximation of fractional laplacian and related nonlocal diffusion problems on bounded domains. Adv. Comput. Math. **42**(6), 1363–1380 (2016b)

O. Weckner, E. Emmrich, Numerical simulation of the dynamics of a nonlocal, inhomogeneous, infinite bar. J. Comput. Appl. Mech. **6**(2), 311–319 (2005)

G. Zhang, Q. Le, A. Loghin, A. Subramaniyan, F. Bobaru, Validation of a peridynamic model for fatigue cracking. Eng. Fract. Mech. **162**, 76–94 (2016)

Dynamic Damage Propagation with Memory: A State-Based Model

45

Robert Lipton, Eyad Said, and Prashant K. Jha

Contents

Abstract

A model for dynamic damage propagation is developed using nonlocal potentials. The model is posed using a state-based peridynamic formulation. The resulting evolution is seen to be well posed. At each instant of the evolution, we identify a damage set. On this set, the local strain has exceeded critical values either for tensile or hydrostatic strain, and damage has occurred. The damage set is nondecreasing with time and is associated with damage state variables defined at each point in the body. We show that a rate form of energy balance holds at

R. Lipton (✉)
Department of Mathematics and Center for Computation and Technology, Louisiana State University, Baton Rouge, LA, USA
e-mail: lipton@lsu.edu

E. Said · P. K. Jha
Department of Mathematics, Louisiana State University, Baton Rouge, LA, USA
e-mail: esaid1@lsu.edu; prashant.j16o@gmail.com

G. Z. Voyiadjis (ed.), *Handbook of Nonlocal Continuum Mechanics for Materials and Structures*, https://doi.org/10.1007/978-3-319-58729-5_45

each time during the evolution. Away from the damage set, we show that the nonlocal model converges to the linear elastic model in the limit of vanishing nonlocal interaction.

Keywords
Damage model · Nonlocal interactions · Energy dissipation · State-based peridynamics

Introduction

In this chapter, we address the problem of damage propagation in materials. Here the damage evolution is not known a priori and is found as part of the problem solution. Our approach is to use a nonlocal formulation with the purpose of using the least number of parameters to describe the model. We will work within the small deformation setting, and the model is developed within a state-based peridynamic formulation. Here strains are expressed in terms of displacement differences as opposed to spatial derivatives. For the problem at hand, the nonlocality provides the flexibility to simultaneously model non-differentiable displacements and damage evolution. The net force acting on a point x is due to the strain between x and neighboring points y. The neighborhood of nonlocal interaction between x and its neighbors y is confined to ball of radius δ centered at x denoted by $B_\delta(x)$. The radius of the ball is called the called the horizon. Numerical implementations based on nonlocal peridynamic models exhibit formation and localization of features associated with phase transformation and fracture (see, e.g., Dayal and Bhattacharya 2006; Silling and Lehoucq 2008; Silling et al. 2010; Foster et al. 2011; Agwai et al. 2011; Lipton et al. 2016; Bobaru and Hu 2012; Ha and Bobaru 2010; Silling and Bobaru 2005; Weckner and Abeyaratne 2005; Gerstle et al. 2007; Weckner and Emmrich 2005). A recent review can be found in Bobaru et al. (2016).

The recent model studied in Lipton (2014, 2016), Lipton et al. (2016), and Jha and Lipton (2017) is defined by double-well two-point strain potentials. Here one potential well is centered at the origin and associated with elastic response, while the other well is at infinity and associated with surface energy. The rational for studying these models is that they are shown to be well posed, and in the limit of vanishing nonlocality, the dynamics recovers features associated with sharp fracture propagation (see Lipton 2014, 2016). While memory is not incorporated in this model, it is seen that the inertia of the evolution keeps the forces in a softened state over time as evidenced in simulations (Lipton et al. 2016). This modeling approach is promising for fast cracks, but for cyclic loading and slowly propagating fractures, an explicit damage-fracture modeling with memory is needed. In this work, we develop this approach for more general models that allow for three-point nonlocal interactions and irreversible damage. The use of three-point potentials allows one to model a larger variety of elastic properties. In the lexicon of peridynamics, we adopt an ordinary state-based formulation (Silling 2000; Silling et al. 2007). We introduce nonlocal forces that soften irreversibly as the shear strain or dilatational strain increases beyond critical values. This model is shown to deliver a mathematically well-posed evolution. Our proof of this is motivated by recent work Emrich and Phulst (2016) where existence of solution for bond-based peridynamic models with damage is established. Recently another well-posed bond-based model with damage has been proposed in Du et al. (2017) where fracture simulations are carried out.

In addition to being state based, our modeling approach differs from Emrich and Phulst (2016) and earlier bond-based work Silling and Askari (2005) and uses differentiable damage variables. This feature allows us to establish an energy balance equation relating kinetic energy, potential energy, and energy dissipation at each instant during the evolution. At each instant, we identify the set undergoing damage where the local energy dissipation rate is positive. On this set, the local strain has exceeded a critical value, and damage has occurred. Damage is irreversible, and the damage set is monotonically increasing with time. Explicit damage models are illustrated, and stress strain curves for both cyclic loading and strain to failure are provided. These models are illustrated in two numerical examples. In the first example, we consider a square domain and apply a time periodic y-directed displacement along the top edge while fixing the bottom, left and right edges. We track the strain and force over three loading periods. The simulations show that bonds suffer damage and the strain vs force plot is similar to the one predicted by the damage law (see Fig. 14). In the second example, we apply a shear load to the top edge while fixing the bottom edge and leaving left and right edges free. As expected, we find that damage appears along the diagonal of square (see Fig. 15).

We conclude by noting that for this model the forces scale inversely with the length of the horizon. With this in mind, we consider undamaged regions, and we are able to show that the nonlocal operator converges to a linear local operator associated with the elastic wave equation. In this limit, the elastic tensor can have any combination of Poisson's ratio and Young's modulus. The Poisson's ratio and Young modulus are determined uniquely by explicit formulas in terms of the nonlocal potentials used to define the model. This result is consistent with small horizon convergence results for convex energies (see Emmrich and Weckner 2007; Mengesha and Du 2014; Silling and Lehoucq 2008). Further reading and complete derivations can be found in the recent monograph Lipton et al. (2018).

Formulation

In this work, we assume the displacements u are small (infinitesimal) relative to the size of the three-dimensional body D. The tensile strain is denoted $S = S(y, x, t; u)$ and given by

$$S(y, x, t; u) = \frac{u(t, y) - u(t, x)}{|y - x|} \cdot e_{y-x}, \qquad e_{y-x} = \frac{y - x}{|y - x|}, \tag{1}$$

where e_{y-x} is a unit direction vector and $\cdot$ is the dot product. It is evident that $S(y,x,t;u)$ is the tensile strain along the direction e_{y-x}. We introduce the nonnegative weight $\omega^\delta(|y-x|)$ such that $\omega^\delta = 0$ for $|y-x| > \delta$ and the hydrostatic strain at x is defined by

$$\theta(x,t;u) = \frac{1}{V_\delta} \int_{D \cap B_\delta(x)} \omega^\delta(|y-x|)|y-x| S(y,x,t;u)\, dy, \tag{2}$$

where V_δ is the volume of the ball $B_\delta(x)$ of radius δ centered at x. The weight is chosen such that $\omega^\delta(|y-x|) = \omega(|y-x|/\delta)$ and

$$\ell_1 = \frac{1}{V_\delta} \int_{B_\delta(x)} \omega^\delta(|y-x|)\, dy < \infty. \tag{3}$$

We follow Silling (2000) and Emrich and Phulst (2016) and introduce a nonnegative damage factor taking the value one in the undamaged region and zero in the fully damaged region. The damage factor for the force associated with tensile strains is written $H^T(u)(y,x,t)$; the corresponding factor for hydrostatic strains is written $H^D(u)(x,t)$. Here we assume no damage and $H^T(u)(y,x,t) = 1$ until a critical tensile strain S_c is reached. For tensile strains greater than S_c, damage is initiated and $H^T(y,x,t)$ drops below 1. The fully damaged state is $H^T(y,x,t) = 0$. For hydrostatic strains, we assume no damage until a critical positive dilatational strain θ_c^+ or a negative compressive strain (θ_c^-) is reached. Again $H^D(x,t) = 1$ until a critical hydrostatic strain is reached and then drops below 1 with the fully damaged state being $H^D(x,t) = 0$. We postpone description of the specific form of the history-dependent damage factors until after we have defined the nonlocal forces.

The force at a point x due to tensile strain is given by

$$\mathcal{L}^T(u)(x,t) = \frac{2}{V_\delta} \int_{D \cap B_\delta(x)} \frac{J^\delta(|y-x|)}{\delta |y-x|} H^T(u)(y,x,t) \partial_S$$
$$f(\sqrt{|y-x|} S(y,x,t;u)) e_{y-x}\, dy, \tag{4}$$

Here $J^\delta(|y-x|)$ is a nonnegative bounded function such that $J^\delta = 0$ for $|y-x| > \delta$ and $M = \sup\{y \in B_\delta(x); J^\delta(|y-x|)\}$ and

$$\ell_2 = \frac{1}{V_\delta} \int_{B_\delta(x)} \frac{J^\delta(|y-x|)}{|y-x|^2}\, dy < \infty \text{ and } \ell_3 = \frac{1}{V_\delta} \int_{B_\delta(x)} \frac{J^\delta(|y-x|)}{|y-x|^{3/2}}\, dy < \infty. \tag{5}$$

Both J^δ and ω^δ are prescribed and characterize the influence of nonlocal forces on x by neighboring points y. Here ∂_S is the partial derivative with respect to strain. The function $f = f(r)$ is twice differentiable for all arguments r on the real line, and f' and f'' are bounded. Here we take $f(r) = \alpha r^2/2$ for $r < r_1$ and $f = r$ for

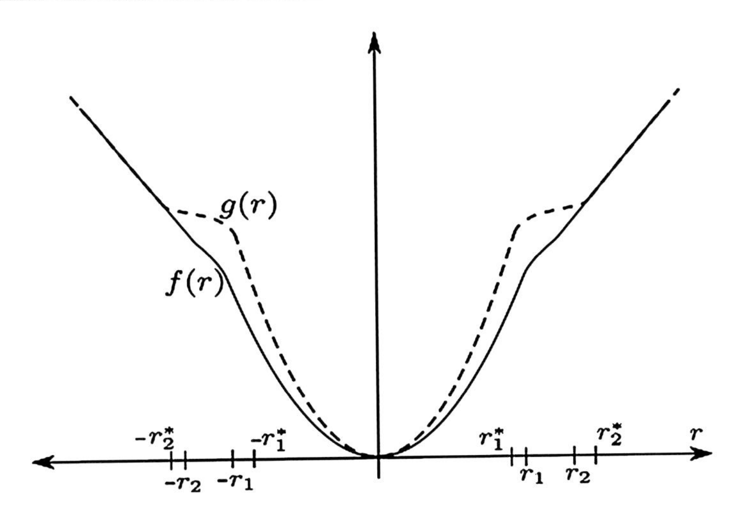

Fig. 1 Generic plot of $f(r)$ (Solid line) and $g(r)$ (Dashed line)

$r_2 < r$, with $r_1 < r_2$ (see Fig. 1). The factor $\sqrt{|y-x|}$ appearing in the argument of $\partial_S f$ ensures that the nonlocal operator $\mathcal{L}^T$ converges to the divergence of a stress tensor in the small horizon limit when it's known a priori that displacements are smooth (see section "Linear Elastic Operators in the Small Horizon Limit").

The force at a point x due to the hydrostatic strain is given by

$$\mathcal{L}^D(u)(x,t) = \frac{1}{V_\delta}\int_{D\cap B_\delta(x)} \frac{\omega^\delta(|y-x|)}{\delta^2}\left[H^D(u)(y,t)\partial_\theta g(\theta(y,t;u))+ \right. \tag{6}$$

$$\left. H^D(u)(x,t)\partial_\theta g(\theta(x,t;u))\right] e_{y-x}\, dy, \tag{7}$$

where the function $g(r) = \beta r^2/2$ for $r < r_1^*$, $g = r$ for $r_2^* < r$, with $r_1^* < r_2^*$ and g is twice differentiable and g' and g'' are bounded (see Fig. 1). It is readily verified that the force $\mathcal{L}^T(u)(x,t) + \mathcal{L}^D(u)(x,t)$ satisfies balance of linear and angular momentum.

The damage factor for tensile strain $H^T(u)(y,x,t)$ is given in terms of the functions $h(x)$ and $j_S(x)$. Here h is nonnegative, has bounded derivatives (hence Lipschitz continuous), takes the value one for negative x and for $x \geq 0$ decreases, and is zero for $x > x_c$ (see Fig. 2). Here we are free to choose x_c to be any small and positive number. The function $j_S(x)$ is nonnegative, has bounded derivatives (hence Lipschitz continuous), takes the value zero up to a positive critical strain S_C, and then takes on positive values. We will suppose $j_S(x) \leq \gamma|x|$ for some $\gamma > 0$ (see Fig. 3). The damage factor is now defined to be

$$H^T(u)(y,x,t) = h\left(\int_0^t j_S(S(y,x,\tau;u))\,d\tau\right). \tag{8}$$

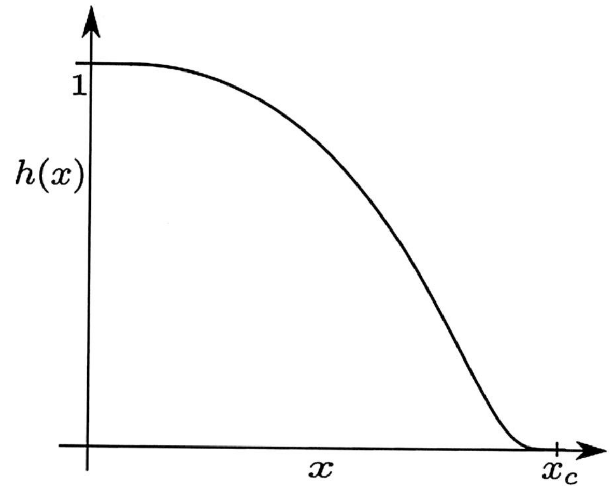

Fig. 2 Generic plot of $h(x)$

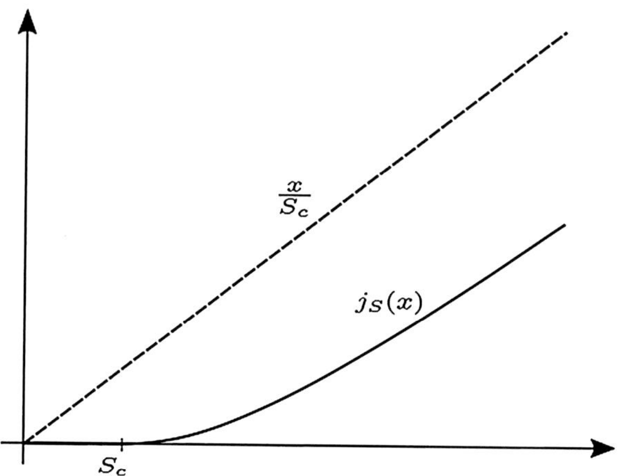

Fig. 3 Generic plot of $j_S(x)$ with S_c

It is clear from this definition that damage occurs when the stress exceeds S_c for some period of time and the bond force decreases irrevocably from its undamaged value. The damage function defined here is symmetric, i.e., $H^T(u)(y, x, t) = H^T(u)(x, y, t)$. For hydrostatic strain, we introduce the nonnegative function j_θ with bounded derivatives (hence Lipschitz continuous). We suppose $j_\theta = 0$ for an interval containing the origin given by (θ_c^-, θ_c^+) and take positive values outside this interval (see Fig. 4). As before we will suppose $j_\theta(x) \leq \gamma|x|$ for some $\gamma > 0$. The damage factor for hydrostatic strain is given by

$$H^D(u)(x, t) = h\left(\int_0^t j_\theta(\theta(x, \tau; u))\, d\tau\right). \tag{9}$$

For this model, it is clear that damage can occur irreversibly for compressive or dilatational strain when the possibly different critical values θ_c^- or θ_c^+ are exceeded.

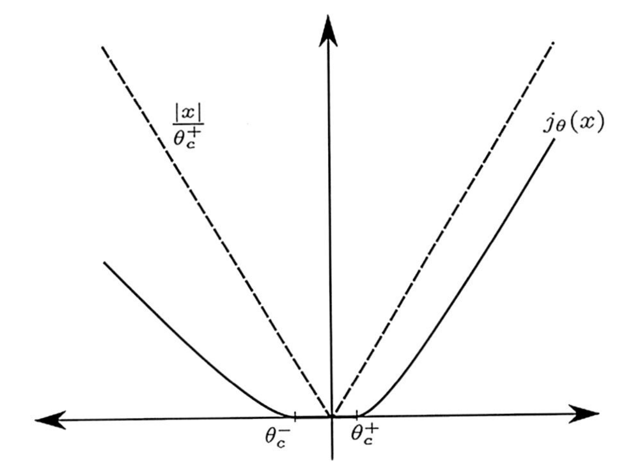

Fig. 4 Generic plot of $j_\theta(x)$ with θ_c^+, and θ_c^-

The *damage set* at time t is defined to be the collection of all points x for which $H^T(y,x,t)$ or $H^D(u)(x,t)$ is less than one. This set is monotonically increasing in time. The *process zone* at time t is the collection of points x undergoing damage such that $\partial_t H^T(y,x,t) < 0$ or $\partial_t H^D(x,t) < 0$. Explicit examples of $H^T(u)(y,x,t)$ and $H^D(u)(x,t)$ are given in section "Explicit Damage Models, Cyclic Loading, and Strain to Failure."

We define the body force $b(x,t)$, and the displacement $u(x,t)$ is the solution of the initial value problem given by

$$\rho\partial_t^2 u(x,t) = \mathcal{L}^T(u)(x,t) + \mathcal{L}^D(u)(x,t) + b(x,t) \text{ for } x \in D \text{ and } t \in (0,T), \tag{10}$$

with initial data

$$u(x,0) = u_0(x), \qquad \partial_t u(x,0) = v_0(x). \tag{11}$$

It is easily verified that this is an ordinary state-based peridynamic model. We show in the next section that this initial value problem is well posed.

Existence of Solutions

The regularity and existence of the solution depend on the regularity of the initial data and body force. In this work, we choose a general class of body forces and initial conditions. The initial displacement u_0 and velocity v_0 are chosen to be integrable and bounded and belonging to $L^\infty(D;\mathbb{R}^3)$. The space of such functions is denoted by $L^\infty(D;\mathbb{R}^3)$. The body force $b(x,t)$ is chosen such that for every $t \in [0,T_0]$, b takes values in $L^\infty(D,\mathbb{R}^3)$ and is continuous in time. The associated norm is defined to be $\|b\|_{C([0,T_0];L^\infty(D,\mathbb{R}^3))} = max_{t\in[0,T_0]}\|b(x,t)\|_{L^\infty(D,\mathbb{R}^3)}$. The associated space of continuous functions in time taking values in $L^\infty(D;\mathbb{R}^3)$ for which this norm is finite is denoted by $C([0,T_0];L^\infty(D,\mathbb{R}^3))$. The space of functions twice differentiable in time taking values in $L^\infty(D,\mathbb{R}^3)$ such that both derivatives belong to $C([0,T_0];L^\infty(D,\mathbb{R}^3))$ is written as $C^2([0,T_0];L^\infty(D,\mathbb{R}^3))$. We now assert the existence and uniqueness for the solution of the initial value problem.

Theorem 1 (Existence and uniqueness of the damage evolution). *The initial value problem given by* (10) *and* (11) *has a solution $u(x,t)$ such that for every $t \in [0,T_0]$, u takes values in $L^\infty(D,\mathbb{R}^3)$ and is the unique solution belonging to the space $C^2([0,T_0];L^\infty(D,\mathbb{R}^3))$.*

To prove the theorem, we will show

(1) The operator $\mathcal{L}^T(u)(x,t) + \mathcal{L}^D(u)(x,t)$ is a map from $C([0,T_0];L^\infty(D,\mathbb{R}^3))$ into itself.
(2) The operator $\mathcal{L}^T(u)(x,t) + \mathcal{L}^D(u)(x,t)$ is Lipschitz continuous with respect to the norm of $C([0,T_0];L^\infty(D,\mathbb{R}^3))$.

The theorem then follows from an application of the Banach fixed point theorem.

To establish properties (1) and (2), we state and prove the following lemmas for the damage factors.

Lemma 1. *Let $H^T(u)(y,x,t)$ and $H^D(u)(x,t)$ be defined as in* (8) *and* (9). *Then for $u \in C([0,T_0];L^\infty(D,\mathbb{R}^3))$, the mappings*

$$(y,x) \mapsto H^T(u)(y,x,t) : D \times D \to \mathbb{R}, \qquad x \mapsto H^D(u)(x,t) : D \to \mathbb{R} \quad (12)$$

are measurable for every $t \in [0,T_0]$, and the mappings

$$t \mapsto H^T(u)(y,x,t) : [0,T_0] \to \mathbb{R}, \qquad t \mapsto H^D(u)(x,t) : [0,T_0] \to \mathbb{R} \quad (13)$$

are continuous for almost all (y,x) and x, respectively. Moreover for almost all $(y,x) \in D \times D$ and all $t \in [0,T_0]$, the map

$$u \mapsto H^T(u)(y,x,t) : C([0,T_0];L^\infty(D,\mathbb{R}^3)) \to \mathbb{R} \quad (14)$$

is Lipschitz continuous, and for almost all $x \in D$ and all $t \in [0,T_0]$, the map

$$u \mapsto H^D(u)(x,t) : C([0,T_0];L^\infty(D,\mathbb{R}^3)) \to \mathbb{R} \quad (15)$$

is Lipschitz continuous.

Proof. The measurability properties are immediate. In what follows, constants are generic and apply to the context in which they are used. We establish continuity in time for $H^D(u)$. For $\hat{t}$ and t in $[0,T_0]$, we have

$$
\begin{aligned}
&|H^D(u)(x,\hat{t}) - H^D(u)(x,t)| \\
&= |h\left(\int_0^{\hat{t}} j_\theta(\theta(x,\tau;u))\,d\tau\right) - h\left(\int_0^{t} j_\theta(\theta(x,\tau;u))\,d\tau\right)| \\
&\leq C_1 \int_{\min\{\hat{t},t\}}^{\max\{\hat{t},t\}} j_\theta(\theta(x,\tau;u))d\tau \\
&\leq \gamma\, C_1 \int_{\min\{\hat{t},t\}}^{\max\{\hat{t},t\}} |\theta(x,\tau;u)|d\tau \\
&\leq \gamma\,\ell_1 C_1 C_2 |\hat{t} - t| 2\|u\|_{C([0,T_0];L^\infty(D,\mathbb{R}^3))}.
\end{aligned} \tag{16}
$$

The first inequality follows from the Lipschitz continuity of h, the second follows from the growth condition on j_θ, and the third follows from (3).

We establish continuity in time for $H^T(u)$. For $\hat{t}$ and t in $[0, T_0]$, we have

$$
\begin{aligned}
&|H^T(u)(x,\hat{t}) - H^T(u)(x,t)| \\
&= |h\left(\int_0^{\hat{t}} j_S(S(y,x,\tau;u))\,d\tau\right) - h\left(\int_0^{t} j_S(S(y,x,\tau;u))\,d\tau\right)| \\
&\leq C_1 \int_{\min\{\hat{t},t\}}^{\max\{\hat{t},t\}} j_S(S(y,x,\tau;u))d\tau \\
&\leq \gamma\, C_1 \int_{\min\{\hat{t},t\}}^{\max\{\hat{t},t\}} |S(y,x,\tau;u)|d\tau \\
&\leq \gamma\, C_1 C_2 \frac{|\hat{t} - t|}{|y - x|} 2\|u\|_{C([0,T_0];L^\infty(D,\mathbb{R}^3))}.
\end{aligned} \tag{17}
$$

The first inequality follows from the Lipschitz continuity of h, the second follows from the growth condition on j_S, and the third follows from the definition of strain (1).

To demonstrate Lipschitz continuity for $H^D(u)(x,t)$, we write

$$
\begin{aligned}
&|H^D(u)(x,t)) - H^D(v)(x,t)| \\
&= |h\left(\int_0^{t} j_\theta(\theta(x,\tau;u))\,d\tau\right) - h\left(\int_0^{t} j_\theta(\theta(x,\tau;v))\,d\tau\right)| \\
&\leq C_1 |\int_0^t (j_\theta(\theta(x,\tau;u) - j_\theta(\theta(x,\tau;v))\,d\tau| \\
&\leq C_1 C_2 \int_0^t |\theta(x,\tau;u) - \theta(x,\tau;v)|\,d\tau \\
&\leq 2t\ell_1 C_1 C_2 \|u - v\|_{C([0,t];L^\infty(D,\mathbb{R}^3))}.
\end{aligned} \tag{18}
$$

The first inequality follows from the Lipschitz continuity of h, the second follows from the Lipschitz continuity of j_θ, and the third follows from (3). The Lipschitz continuity for $H^S(u)(y,x,t)$ follows from similar arguments using the Lipschitz continuity of h, j_S, and (1), and we get

$$|H^T(u)(y,x,t)) - H^T(v)(y,x,t)| \leq \frac{2tC_1C_2C_3}{|y-x|}\|u-v\|_{C([0,t];L^\infty(D,\mathbb{R}^3))}. \tag{19}$$

□

Proof of Theorem 1. We establish (1) by first noting that

$$|\mathcal{L}^T(u)(x,t) + \mathcal{L}^D(u)(x,t)| \leq \frac{C}{\delta^2}, \tag{20}$$

where C is a constant. This estimate follows from the boundedness of f', g', $H^T(u)$, and $H^D(u)$ and the integrability of the ratios $J^\delta(|y-x|)/|y-x|^2$, $J^\delta(|y-x|)/|y-x|^{3/2}$, and $\omega^\delta(|y-x|)$. Thus $\|\mathcal{L}^T(u)(x,t)+\mathcal{L}^D(u)(x,t)\|_{L^\infty(D,\mathbb{R}^3)}$ is uniformly bounded for all $t \in [0,T_0]$.

To complete the demonstration of (1), we point out that the force functions $\partial_S f$ and $\partial_\theta g$ are Lipschitz continuous in their arguments. The key features are given in the following lemma.

Lemma 2. *Given two functions v and w in $L^\infty(D,\mathbb{R}^3)$, then*

$$|\partial_S f(\sqrt{|y-x|}S(y,x;v)) - \partial_S f(\sqrt{|y-x|}S(y,x;w)) | \leq \frac{2C}{\sqrt{|y-x|}}\|v-w\|_{L^\infty(D,\mathbb{R}^3)}. \tag{21}$$

and

$$|\partial_\theta g(\theta(x;v)) - \partial_\theta g(\theta(x;w))| \leq 2\ell_1 C\|v-w\|_{L^\infty(D,\mathbb{R}^3)}. \tag{22}$$

Proof.

$$\begin{aligned}&|\partial_S f(\sqrt{|y-x|}S(y,x;v)) - \partial_S f(\sqrt{|y-x|}S(y,x;w))|\\ &\leq C\sqrt{|y-x|}|S(y,x;v)-S(y,x;w)| \leq \frac{2C}{\sqrt{|y-x|}}\|v-w\|_{L^\infty(D,\mathbb{R}^3))},\end{aligned} \tag{23}$$

where the first inequality follows from the Lipschitz continuity of $\partial_S f$ and the second follows from the definition of S.

For $\partial_\theta g$, we have

$$|\partial_\theta g(\theta(x;v)) - \partial_\theta g(\theta(x;w))| \leq C|\theta(x;v) - \theta(x;w)| \leq 2\ell_1 C_1 \|v - w\|_{L^\infty(D,\mathbb{R}^3))}, \tag{24}$$

where the first inequality follows from the Lipschitz continuity of $\partial_\theta g$ and the second follows from the definitions of θ and S. □

We have

$$\begin{aligned} &|\mathcal{L}^T(u)(x,\hat{t}) - \mathcal{L}^T(u)(x,t)| \\ &\leq \frac{2}{V_\delta}\int_{D\cap B_\delta(x)} \frac{J^\delta(|y-x|)}{\delta|y-x|}|\partial_S f(\sqrt{y-x}S(y,x,\hat{t};u)) \\ &- \partial_S f(\sqrt{y-x}S(y,x,t;u))|\,dy \\ &+ \frac{2}{V_\delta}\int_{D\cap B_\delta(x)} \frac{J^\delta(|y-x|)}{\delta|y-x|}|H^T(u)(y,x,\hat{t}) - H^T(u)(y,x,t)|\,dy. \end{aligned} \tag{25}$$

From the above, (19), and Lemma 2, we see that

$$\begin{aligned} &\|\mathcal{L}^T(u)(x,\hat{t}) - \mathcal{L}^T(u)(x,t)\|_{L^\infty(D,\mathbb{R}^3)} \\ &\leq \frac{\ell_3 C_3}{\delta}\|u(x,\hat{t}) - u(x,t)\|_{L^\infty(D,\mathbb{R}^3)} + \frac{\ell_2\gamma C_1 C_2}{\delta}|\hat{t} - t|2\|u\|_{C([0,T_0];L^\infty(D,\mathbb{R}^3))} \end{aligned} \tag{26}$$

and we see $\mathcal{L}^T$ is well defined and maps $C([0,T_0];L^\infty(D,\mathbb{R}^3))$ into itself.

We show the continuity in time for $\mathcal{L}^D(u)(x,t)$. Now we have

$$\begin{aligned} &|\mathcal{L}^D(u)(x,\hat{t}) - \mathcal{L}^D(u)(x,t)| \\ &\leq \frac{1}{V_\delta}\int_{D\cap B_\delta(x)} \frac{\omega^\delta(|y-x|)}{\delta^2}|\partial_\theta g(\theta(y,\hat{t};u)) - \partial_\theta g(\theta(y,t;u))|\,dy \\ &+ \frac{1}{V_\delta}\int_{D\cap B_\delta(x)} \frac{\omega^\delta(|y-x|)}{\delta^2}|H^D(u)(y,\hat{t}) - H^D(u)(y,t)|\,dy \\ &+ \frac{1}{V_\delta}\int_{D\cap B_\delta(x)} \frac{\omega^\delta(|y-x|)}{\delta^2}|\partial_\theta g(\theta(x,\hat{t};u)) - \partial_\theta g(\theta(x,t;u))|\,dy \\ &+ \frac{1}{V_\delta}\int_{D\cap B_\delta(x)} \frac{\omega^\delta(|y-x|)}{\delta^2}|H^D(u)(x,\hat{t}) - H^D(u)(x,t)|\,dy \end{aligned} \tag{27}$$

and applying Lemma 2 and (18), (19), (20), (21), (22), (23), (24), (25), (26), and (27), we get the continuity

$$\begin{aligned} &|\mathcal{L}^D(u)(x,\hat{t}) - \mathcal{L}^D(u)(x,t)| \leq \frac{4\ell_1^2 C_1}{\delta^2}\|u(\hat{t},x) - u(t,x)\|_{L^\infty(D,\mathbb{R}^3)} \\ &+ \frac{\gamma\, 4\ell_1^2 C_1 C_2}{\delta^2}|\hat{t} - t|\|u\|_{C([0,T_0];L^\infty(D,\mathbb{R}^3)}. \end{aligned} \tag{28}$$

We conclude that $\mathcal{L}^D$ is well defined and maps $C([0,T_0];L^\infty(D,\mathbb{R}^3))$ into itself and item (1) is proved.

To show Lipschitz continuity, consider any two functions u and w belonging to $C([0,T_0];L^\infty(D,\mathbb{R}^3))$, $t \in [0,T_0]$ to write

$$\begin{aligned}
&|\mathcal{L}^T(u)(x,t) + \mathcal{L}^D(u)(x,t) - [\mathcal{L}^T(w)(x,t) + \mathcal{L}^D(w)(x,t)]| \\
&\le \frac{2}{V_\delta}\int_{D\cap B_\delta(x)} \frac{J^\delta(|y-x|)}{\delta|y-x|}|\partial_S f(\sqrt{|y-x|}S(y,x,t;u)) \\
&- \partial_S f(\sqrt{|y-x|}S(y,x,t;w))|\,dy \\
&+ \frac{2}{V_\delta}\int_{D\cap B_\delta(x)} \frac{J^\delta(|y-x|)}{\delta|y-x|}|H^T(u)(y,x,t) - H^T(w)(y,x,t)|\,dy \\
&+ \frac{1}{V_\delta}\int_{D\cap B_\delta(x)} \frac{\omega^\delta(|y-x|)}{\delta^2}|\partial_\theta g(\theta(y,t;u)) - \partial_\theta g(\theta(y,t;w))|\,dy \\
&+ \frac{1}{V_\delta}\int_{D\cap B_\delta(x)} \frac{\omega^\delta(|y-x|)}{\delta^2}|H^D(u)(y,t) - H^D(w)(y,t)|\,dy \\
&+ \frac{1}{V_\delta}\int_{D\cap B_\delta(x)} \frac{\omega^\delta(|y-x|)}{\delta^2}|\partial_\theta g(\theta(x,t;u)) - \partial_\theta g(\theta(x,t;w))|\,dy \\
&+ \frac{1}{V_\delta}\int_{D\cap B_\delta(x)} \frac{\omega^\delta(|y-x|)}{\delta^2}|H^D(u)(x,t) - H^D(w)(x,t)|\,dy.
\end{aligned} \tag{29}$$

Applying (18) and (19), (20), (21), (22), (23), (24), (25), (26), (27), (28), and (29) delivers the estimate

$$\begin{aligned}
&\|\mathcal{L}^T(u)(x,t) + \mathcal{L}^D(u)(x,t) - [\mathcal{L}^T(w)(x,t) + \mathcal{L}^D(w)(x,t)]\|_{C([0,t];L^\infty(D,\mathbb{R}^3))} \\
&\le \frac{C_1 + tC_2}{\delta^2}\|u-w\|_{C([0,t];L^\infty(D,\mathbb{R}^3))},
\end{aligned} \tag{30}$$

where C_1 and C_2 are constants not depending on time u or w. For $T_0 > t$, we can choose a constant $L > (C_1 + T_0C_2)/\delta^2$ and

$$\begin{aligned}
&\|\mathcal{L}^T(u)(x,t) + \mathcal{L}^D(u)(x,t) - [\mathcal{L}^T(w)(x,t) + \mathcal{L}^D(w)(x,t)]\|_{C([0,t];L^\infty(D,\mathbb{R}^3))} \\
&\le L\|u-w\|_{C([0,t];L^\infty(D,\mathbb{R}^3))}, \text{ for all } t \in [0,T_0].
\end{aligned} \tag{31}$$

This proves the Lipschitz continuity, and item (2) of the theorem is proved. Note that $u(\tau) = w(\tau)$ for all $\tau \in [0,t]$ implies $\mathcal{L}^T(u)(x,t) + \mathcal{L}^D(u)(x,t) = [\mathcal{L}^T(w)(x,t) + \mathcal{L}^D(w)(x,t)]$ and $\mathcal{L}^T(u)(x,t) + \mathcal{L}^D(u)(x,t)$ is a Volterra operator.

We write evolutions $u(x,t)$ belonging to $C([0,t];L^\infty(D,\mathbb{R}^3))$ as $u(t)$ and $(Vu)(t)$ is the sum

$$(Vu)(t) = \mathcal{L}^T(u)(t) + \mathcal{L}^D(u)(t). \tag{32}$$

We seek the unique fixed point of $u(t)=(Iu)(t)$ where I maps $C([0,t];L^\infty(D,\mathbb{R}^3))$ into itself and is defined by

$$(Iu)(t) = u_0 + t v_0 + \int_0^t (t-\tau)(Vu)(\tau) + b(\tau)\,d\tau. \tag{33}$$

This problem is equivalent to finding the unique solution of the initial value problem given by (10) and (11). We now show that I is a contraction map, and by virtue of the Banach fixed point theorem, we can assert the existence of a fixed point in $C([0,t];L^\infty(D,\mathbb{R}^3))$. To see that I is a contraction map on $C([0,t];L^\infty(D,\mathbb{R}^3))$, we introduce the equivalent norm

$$|||u|||_{C([0,t];L^\infty(D,\mathbb{R}^3))} = \max_{t\in[0,T_0]}\{e^{-2LT_0t}\|u\|_{L^\infty(D,\mathbb{R}^3)}\}, \tag{34}$$

and show I is a contraction map with respect to this norm. We apply (30) to find for $t \in [0,T_0]$ that

$$\begin{aligned}
\|(Iu)(t)-(Iw)(t)\|_{L^\infty(D,\mathbb{R}^3)} &\leq \int_0^t (t-\tau)\|(Vu)(\tau)-(Vw)(\tau)\|_{L^\infty(D,\mathbb{R}^3)}\,d\tau \\
&\leq LT_0\int_0^t \|u-w\|_{C([0,\tau];L^\infty(D,\mathbb{R}^3))}\,d\tau \\
&\leq LT_0\int_0^t \max_{s\in[0,\tau]}\{\|u(s)-w(s)\|_{L^\infty(D,\mathbb{R}^3)}e^{-2LT_0s}\}e^{2LT_0\tau}\,d\tau \\
&\leq \frac{e^{2LT_0t}-1}{2}|||u-w|||_{C([0,T_0];L^\infty(D,\mathbb{R}^3))},
\end{aligned} \tag{35}$$

and we conclude

$$|||(Iu)(t)-(Iw)(t)|||_{C([0,T_0];L^\infty(D,\mathbb{R}^3))} \leq \frac{1}{2}|||u-w|||_{C([0,T_0];L^\infty(D,\mathbb{R}^3))}, \tag{36}$$

so I is a contraction map. From the Banach fixed point theorem, there is a unique fixed point $u(t)$ belonging to $C([0,T_0];L^\infty(D,\mathbb{R}^3))$, and it is evident from (33) that $u(t)$ also belongs to $C^2([0,T_0];L^\infty(D,\mathbb{R}^3))$. This concludes the proof of Theorem 1.

Energy Balance

The evolution is shown to exhibit a balance of energy at all times. In this section, we describe the potential and the energy dissipation rate and show energy balance in rate form. The potential energy at time t for the evolution is denoted by $U(t)$ and is given by

$$\begin{aligned} U(t) =& \frac{2}{V_\delta}\int_D \int_{D\cap B_\delta(x)} \frac{J^\delta(|y-x|)}{\delta} H^T(u)(y,x,t) f(\sqrt{|y-x|}S(y,x,t;u))\, dy dx \\ &+ \int_D \frac{1}{\delta^2} H^D(u)(x,t) g(\theta(x,t;u))\, dx. \end{aligned} \tag{37}$$

The energy dissipation rate $\partial_t R(t)$ is

$$\begin{aligned} \partial_t R(t) =& -\frac{2}{V_\delta}\int_D \int_{D\cap B_\delta(x)} \frac{J^\delta(|y-x|)}{\delta} \partial_t H^T(u)(y,x,t) f(\sqrt{|y-x|}S(y,x,t;u))\, dy dx \\ &- \int_D \frac{1}{\delta^2} \partial_t H^D(u)(x,t) g(\theta(x,t;u))\, dx. \end{aligned} \tag{38}$$

The derivatives $\partial_t H^T(u)(y,x,t)$ and $\partial_t H^D(u)(x,t)$ are easily seen to be nonpositive, and the dissipation rate satisfies $\partial_t R(t) \geq 0$. The kinetic energy is

$$K(t) = \rho \int_D \frac{|\partial_t u(x,t)|^2}{2} dx. \tag{39}$$

The energy balance in rate form is given in the following theorem.

Theorem 2. *The rate form of energy balance for the damage-fracture evolution is given by*

$$\partial_t K(t) + \partial_t U(t) + \partial_t R(t) = \int_D b(x,t)\cdot \partial_t u(x,t)\, dx. \tag{40}$$

Proof of Theorem 2. We multiply both sides of the evolution Eq. (10) by $\partial_t u(x,t)$ and integrate over D to get

$$\rho \int_D \partial_t^2 u(x,t)\cdot \partial_t u(x,t)\, dx = \int_D \mathcal{L}^T(u)(x,t)\cdot \partial_t u(x,t)\, dx \tag{41}$$

$$\begin{aligned} &+ \int_D \mathcal{L}^D(u)(x,t)\cdot \partial_t u(x,t)\, dx \\ &+ \int_D b(x,t)\cdot \partial_t u(x,t)\, dx. \end{aligned} \tag{42}$$

The term on the left side of the equation is immediately recognized as $\partial_t K(t)$. The first and second terms on the right-hand side of the equation are given in the following lemma.

Lemma 3. *One has the following integration by parts formulas given by*

$$\int_D \mathcal{L}^T(u)(x,t)\cdot\partial_t u(x,t)\,dx$$
$$= -\frac{2}{V_\delta}\int_D\int_{D\cap B_\delta(x)}\frac{J^\delta(|y-x|)}{\delta}H^T(u)(y,x,t)\partial_t f(\sqrt{|y-x|}S(y,x,t;u))\,dydx. \tag{43}$$

and

$$\int_D \mathcal{L}^D(u)(x,t)\cdot\partial_t u(x,t)\,dx = -\int_D\frac{1}{\delta^2}H^D(u)(x,t)\partial_t g(\theta(x,t;u))\,dx. \tag{44}$$

Now note that

$$\partial_t U(t)+\partial_t R(t)$$
$$= \frac{2}{V_\delta}\int_D\int_{D\cap B_\delta(x)}\frac{J^\delta(|y-x|)}{\delta}H^T(u)(y,x,t)\partial_t f(\sqrt{|y-x|}S(y,x,t;u))\,dydx$$
$$+\int_D\frac{1}{\delta^2}H^D(u)(x,t)\partial_t g(\theta(x,t;u))\,dx, \tag{45}$$

and the energy balance theorem follows from (41) and (45).

We conclude by proving the integration by parts Lemma 3. We start by proving (44). We expand $\partial_t g(\theta(x,t))$

$$\partial_t g(\theta(x,t;u))$$
$$= \partial_\theta g(\theta(x,t;u))\frac{1}{V_\delta}\int_{D\cap B_\delta(x)}\omega^\delta(|y-x|)|y-x|\frac{\partial_t u(y)-\partial_t u(x)}{|y-x|}\cdot e_{y-x}\,dy \tag{46}$$

and write

$$-\int_D\frac{1}{\delta^2}H^D(u)(x,t)\partial_t g(\theta(x,t;u))\,dx = A(t)+B(t), \tag{47}$$

where

$$A(t)=-\int_D\frac{1}{\delta^2}H^D(u)(x,t)\partial_\theta g(\theta(x,t;u))\frac{1}{V_\delta}\int_{D\cap B_\delta(x)}\omega^\delta(|y-x|)\partial_t u(y)\cdot e_{y-x}\,dydx \tag{48}$$

and

$$B(t)=\int_D \frac{1}{\delta^2} H^D(u)(x,t)\partial_\theta g(\theta(x,t;u))\frac{1}{V_\delta}\int_{D\cap B_\delta(x)} \omega^\delta(|y-x|)\partial_t u(x)\cdot e_{y-x}\,dydx. \tag{49}$$

Next introduce the characteristic function of D denoted by $\chi_D(x)$ taking the value one inside D and zero outside, and together with the properties of $\omega^\delta(|y-x|)$, we rewrite $A(t)$ as

$$A(t) = -\int_{\mathbb{R}^3\times\mathbb{R}^3} \chi_D(x)\chi_D(y)\omega^\delta(|y-x|)\frac{1}{\delta^2} H^D(u)(x,t)\partial_\theta g(\theta(x,t;u)) \\ \frac{1}{V_\delta}\partial_t u(y)\cdot e_{y-x}\,dydx; \tag{50}$$

we switch the order of integration and note $-e_{y-x} = e_{x-y}$ to obtain

$$A(t)=\int_D \frac{1}{V_\delta}\int_{D(x)\cap B_\delta(y)} \frac{\omega^\delta(|y-x|)}{\delta^2} H^D(u)(x,t)\partial_\theta g(\theta(x,t;u))e_{x-y}\,dx\cdot\partial_t u(y)dy. \tag{51}$$

We can move $\partial_t u(x)$ outside the inner integral, regroup factors, and write $B(t)$ as

$$B(t)=\int_D \frac{1}{V_\delta}\int_{D\cap B_\delta(x)} \frac{\omega^\delta(|y-x|)}{\delta^2} H^D(u)(x,t)\partial_\theta g(\theta(x,t;u))e_{y-x}\,dy\cdot\partial_t u(x)dx. \tag{52}$$

We rename the inner variable of integration y and the outer variable x in (51) and add equations (51) and (52) to get

$$A(t) + B(t) = \int_D \mathcal{L}^D(u)(x,t)\cdot\partial_t u(x,t)\,dx \tag{53}$$

and (44) is proved.

The steps used to prove (43) are similar to the proof of (44), so we provide only the key points of its derivation below. We expand $\partial_t f(\sqrt{|y-x|}S)$ to get

$$\begin{aligned}&\partial_t f(\sqrt{|y-x|}S(y,x,t;u)) \\ &\quad= \partial_S f(\sqrt{|y-x|}S(y,x,t;u))\frac{\partial_t u(y)-\partial_t u(x)}{|y-x|}\cdot e_{y-x},\end{aligned} \tag{54}$$

and write

$$-\frac{2}{V_\delta}\int_D\int_{D\cap B_\delta(x)} \frac{J^\delta(|y-x|)}{\delta} H^T(u)(y,x,t)\partial_t f(\sqrt{|y-x|}S(y,x,t;u))\,dydx \\ = A(t) + B(t), \tag{55}$$

where

$$A(t) = \\ = -\int_D \frac{1}{V_\delta} \int_{D\cap B_\delta(x)} \frac{J^\delta(|y-x|)}{\delta|y-x|} H^T(u)(y,x,t)\partial_S \\ f(\sqrt{|y-x|}S(y,x,t;u))\partial_t u(y) \cdot e_{y-x}\, dydx \tag{56}$$

and

$$B(t) = \\ = \int_D \frac{1}{V_\delta} \int_{D\cap B_\delta(x)} \frac{J^\delta(|y-x|)}{\delta|y-x|} H^T(u)(y,x,t)\partial_S \\ f(\sqrt{|y-x|}S(y,x,t;u))\partial_t u(x) \cdot e_{y-x}\, dydx. \tag{57}$$

We note that $S(y,x,t;u) = S(x,y,t;u)$ and $H^T(u)(y,x,t) = H^T(u)(x,y,t)$, and proceeding as in the proof of (44), we change the order of integration in (56) noting that $-e_{y-x} = e_{x-y}$ to get

$$A(t) = \\ = \int_D \frac{1}{V_\delta} \int_{D\cap B_\delta(y)} \frac{J^\delta(|y-x|)}{\delta|y-x|} H^T(u)(x,y,t)\partial_S \\ f(\sqrt{|y-x|}S(x,y,t;u)))e_{x-y}\, dx \cdot \partial_t u(y)\, dy. \tag{58}$$

Taking $\partial_t u(x)$ outside the inner integral in (57) gives

$$B(t) = \\ = \int_D \frac{1}{V_\delta} \int_{D\cap B_\delta(x)} \frac{J^\delta(|y-x|)}{\delta|y-x|} H^T(u)(y,x,t)\partial_S \\ f(\sqrt{|y-x|}S(y,x,t;u))e_{y-x}\, dy \cdot \partial_t u(x)\, dx. \tag{59}$$

We conclude noting that now

$$A(t) + B(t) = \int_D \mathcal{L}^T(u)(x,t) \cdot \partial_t u(x,t)\, dx, \tag{60}$$

and (43) is proved.

Explicit Damage Models, Cyclic Loading, and Strain to Failure

In this section, we provide concrete examples of the damage functions $H^T(u)(y,x,t)$ and $H^D(u)(x,t)$. We provide an example of cyclic loading and the associated degradation in the nonlocal force-strain law as well as the strain to failure curve for monotonically increasing strains. In this work, both damage functions H^T and H^D are given in terms of the function h. Here we give an example of $h(x) : \mathbb{R} \to \mathbb{R}^+$ as follows

$$h(x) = \begin{cases} \bar{h}(x/x_c), & \text{for } x \in (0, x_c), \\ 1, & \text{for } x \leq 0, \\ 0, & \text{for } x \geq x_c. \end{cases} \tag{61}$$

with $\bar{h} : [0,1] \to \mathbb{R}^+$ is defined as

$$\bar{h}(x) = \exp[1 - \frac{1}{1 - (x/x_c)^a}] \tag{62}$$

where $a > 1$ is fixed. Clearly, $\bar{h}(0) = 1$, $\bar{h}(x_c) = 0$ (see Fig. 5).

For a given critical strain $S_c > 0$, we define the threshold function for tensile strain $j_S(x)$ as follows

$$j_S(x) := \begin{cases} \bar{j}(x/S_c), & \forall x \in [S_c, \infty), \\ 0, & \text{otherwise.} \end{cases} \tag{63}$$

where $\bar{j} : [1, \infty) \to \mathbb{R}^+$ is given by

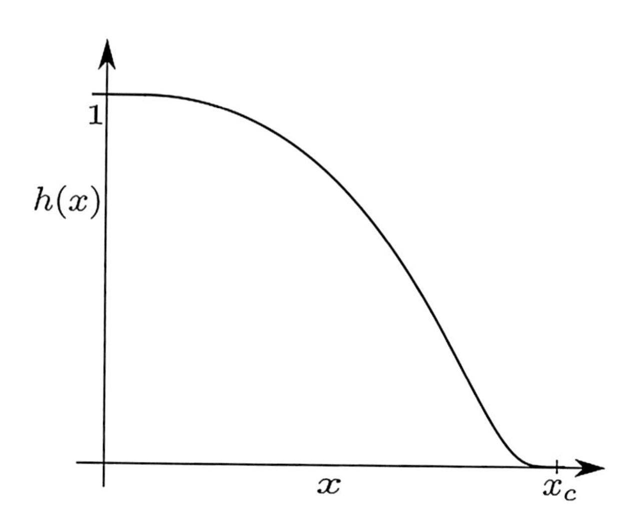

Fig. 5 Plot of $h(x)$ with $a = 2$

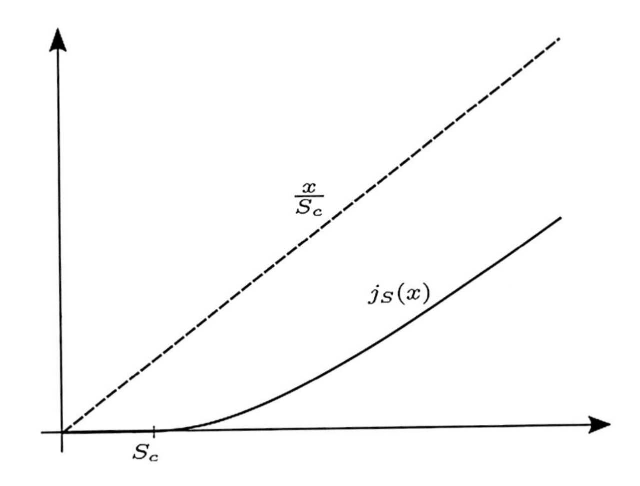

Fig. 6 Plot of $j_S(x)$ with $a = 4, b = 5$ and $S_c = 2$

$$\bar{j}(x) = \frac{(x-1)^a}{1+x^b} \tag{64}$$

with $a > 1$ and $b \geq a - 1$ fixed. Note that $j_S(1) = 0$. Here the condition $b \geq a - 1$ insures the existence of a constant $\gamma > 0$ for which

$$j_S(x) \leq \gamma |x|, \qquad \forall x \in \mathbb{R} \tag{65}$$

(see Fig. 6).

For a given critical hydrostatic strains $\theta_c^- < 0 < \theta_c^+$, we define the threshold function $j_\theta(x)$ as

$$j_\theta(x) := \begin{cases} \bar{j}(x/\theta_c^+), & \forall x \in [\theta_c^+, \infty), \\ \bar{j}(-x/\theta_c^-), & \forall x \in (-\infty, -\theta_c^-], \\ 0, & \text{otherwise}, \end{cases} \tag{66}$$

where $\bar{j}(x)$ is defined by (64), and we plot j_θ in Fig. 7. We summarize noting that an explicit form for $H^T(u)(y, x, t)$ is obtained by using (61) and (63) in (8) and an explicit form for $H^D(u)(x, t)$ is obtained by using (61) and (66) in (9).

We first provide an example of cyclic damage incurred by a periodically varying tensile strain. Let x, y be two fixed material points with $|y - x| < \delta$, and let $S(y, x, t; u) = S(t)$ correspond to a temporally periodic strain (see Fig. 8a). Here $S(t)$ periodically takes excursions above the critical strain S_c. During the first period, we have

$$S(t) = \begin{cases} t, & \forall t \in [0, S_C + \epsilon], \\ 2(S_C + \epsilon) - t & \forall t \in (S_C + \epsilon, 2(S_c + \epsilon)] \end{cases}$$

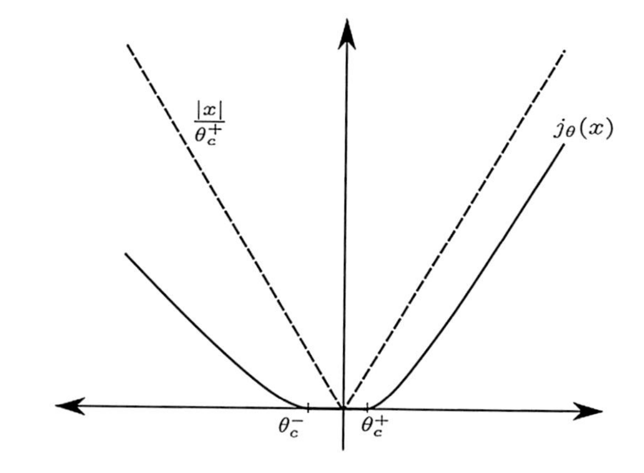

Fig. 7 Plot of $j_\theta(x)$ with $a = 4, b = 5, \theta_c^+ = 2$, and $\theta_C^- = 3$

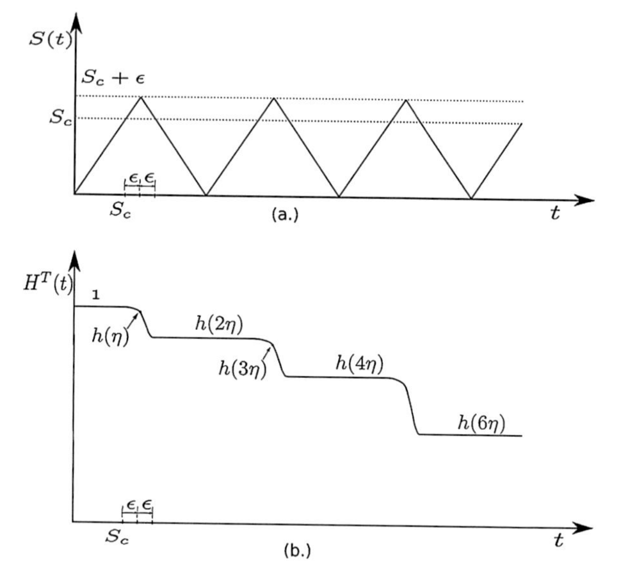

Fig. 8 (**a**) Strain profile. (**b**) Damage function plot corresponding to strain profile

and $S(t)$ is extended to $\mathbb{R}^+$ by periodicity (see Fig. 8a). For this damage model, we let η be the area under the curve $j_S(x)$ from $x = S_c$ to $x = S_c + \epsilon$. It is given by

$$\eta = \int_{S_c}^{S_c+\epsilon} j_S(x)dx = \int_{S_c}^{S_c+\epsilon} j_S(S(t))dt.$$

From symmetry, the area under the curve $j_S(x)$ under unloading from $S_c + \epsilon$ to S_c is also η. The corresponding damage function $H^T(u)(y, x, t)$ is plotted in Fig. 8b.

In Fig. 9, we plot the strain-force relation where S is the abscissa and the tensile force given by $H^T((u)(y, x, t))\partial_S f(\sqrt{|y-x|}S(y, x, t; u)))$ is the ordinate. Here the damage factor $H^T(u)(y, x, t)$ drops in value with each cycle of strain loading. After each cycle, the slope (elasticity) in the linear and recoverable part of the force-

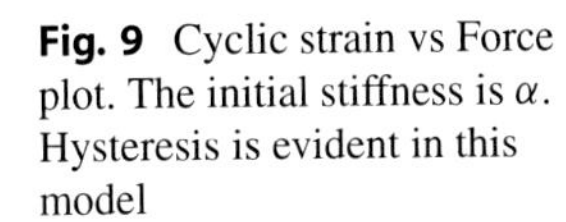

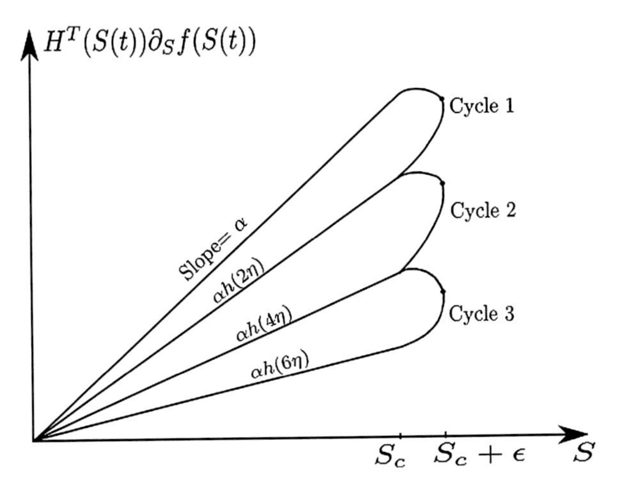

Fig. 9 Cyclic strain vs Force plot. The initial stiffness is α. Hysteresis is evident in this model

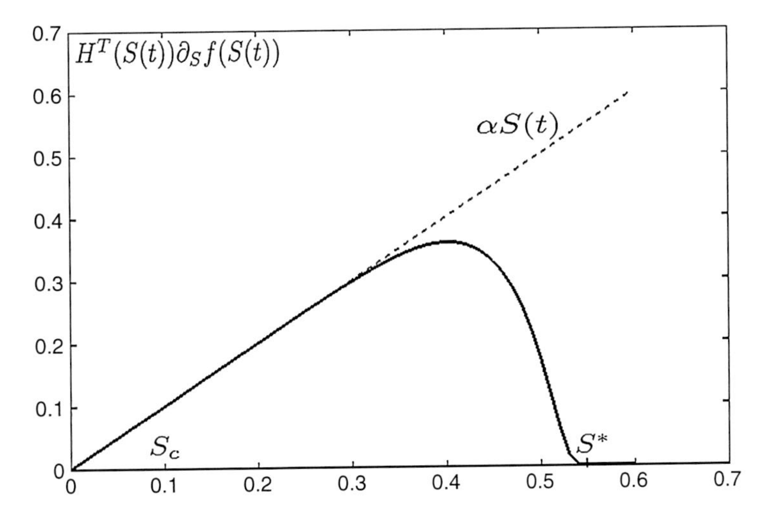

Fig. 10 Strain vs force plot where $S(t) = t$. $H^T(S(t))$ begins to drop at $S_c = 0.1$ and $S^* \approx 0.55025$

strain curve decreases due to damage. The force needed to soften the material is the strength, and it is clear from the model that the strength decreases after each cycle due to damage.

Application of this rigorously established model to fatigue is a topic of future research but beyond the scope of this article. We note that fatigue models based on peridynamic bond softening are introduced in Oterkus (2010) and with fatigue crack nucleation in the context of the Paris law in Silling and Askari (2014).

The next example is strain to failure for a monotonically increasing strain. Here we let

$S(y, x, t; u) = S(t) = t$ and plot the corresponding force-strain curve in Fig. 10.

We see that the force-strain relation is initially linear until the strain exceeds S_c; the force then reaches its maximum and subsequently softens to failure. At $S^* \approx 0.55025$, we have $\int_0^{S^*} j_S(t)dt = x_c$ and $H^T = 0$. Here we take $\alpha = 1$.

Numerical Results

In this section, we present numerical results. Explicit expressions of the functions described in e the previous section are used in simulating the problem. The damage function h is defined similar to Eq. 61 with exponents $a = 1.01$ and $x_c = 0.2$. The function j_S is given by Eq. 63 with $a = 5, b = 5, S_c = 0.01$. The function j_θ is given by Eq. 66 with $a = 4, b = 5, \theta_c^+ = 0.3, \theta_c^- = 0.4$. Nonlinear potential function f is given by $f(r) = \alpha r^2$ for $r < r_1$ and $f(r) = r$ for $r > r_2$. We let $\alpha = 10$ and let $r_1 = r_2 = 0.05$. Similarly, the nonlinear potential function g is given by $g(r) = \beta r^2$ for $r < r_1^*$ and $g(r) = r$ for $r > r_2^*$. We let $\beta = 1$ and let $r_1^* = r_2^* = 0.05$. The influence function is given by $J^\delta(|y-x|) = \omega^\delta(|y-x|) = 1 - \frac{|y-x|}{\delta}$ for $0 \leq |y-x| \leq \delta$ and $J^\delta(|y-x|) = \omega^\delta(|y-x|) = 0$ otherwise. We consider θ_c^+ and θ_c^- sufficiently high so that we only see damage due to tensile forces and not hydrostatic forces.

In both numerical problems, we consider the material domain $D = [0,1]^2$. We also keep the initial condition fixed to $u_0 = 0$ and $v_0 = 0$. Further, we apply no body force, i.e., $b = 0$. However we will consider boundary loading that is periodic in time. Let $x = (x_1, x_2)$ where x_1 corresponds to the component along horizontal axis and x_2 corresponds to the component along vertical axis.

Periodic Loading

We apply boundary condition $u = 0$ on edge $x_1 = 0$, $x_1 = 1$, and $x_2 = 0$. We consider function $\bar{u}$ of form

$$\bar{u}(t) = \begin{cases} \alpha_{bc} t, & \forall t \in [0, T_{bc}], \\ \alpha_{bc} T_{bc} - t & \forall t \in (T_{bc}, 2T_{bc}] \end{cases} \tag{67}$$

and periodically extend the function for any time t. For point x on edge $x_2 = 1$, we apply $u(t,x) = (u_1(t,x), u_2(t,x)) = (0, \bar{u}(t))$. We consider $\alpha_{bc} = 0.01$ and $T_{bc} = 0.216$.

To numerically approximate the evolution equation, we discretize the domain D uniformly with mesh size $h = \delta/5$, where $\delta = 0.15$ in this problem. For time discretization, we consider the velocity Verlet scheme for second order in time differential equation and a midpoint quadrature for the spatial discretization. Final time is $T = 1.2$ and size of time step is $\Delta t = 10^{-5}$.

To obtain the hysteresis plot, we chose bonds as shown in Fig. 11. We track the bond strain $S(y, x, t; u)$ and other relevant quantities. While we track all the bonds

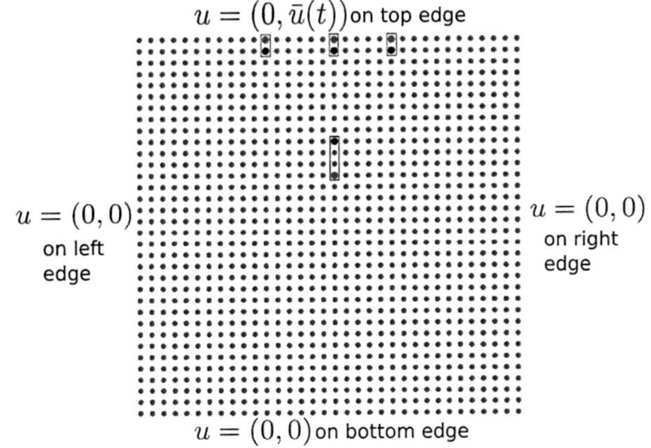

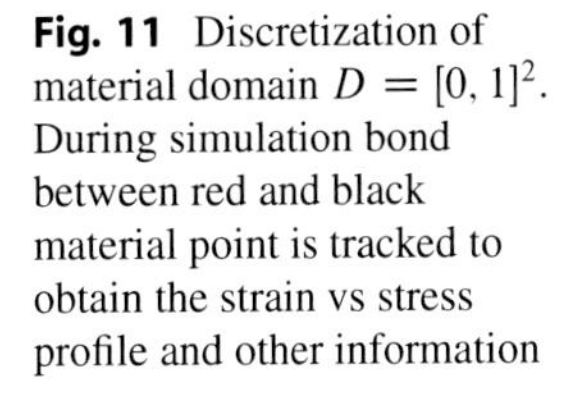
Fig. 11 Discretization of material domain $D = [0, 1]^2$. During simulation bond between red and black material point is tracked to obtain the strain vs stress profile and other information

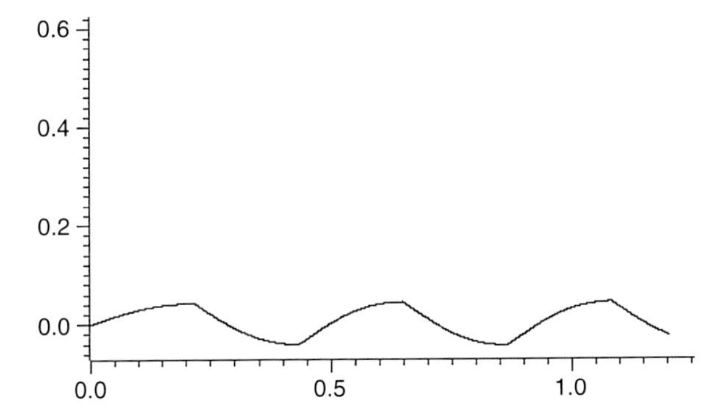

Fig. 12 Time vs Strain $S(y, x, t; u)$ plot

shown in Fig. 11, we only provide plots for the bond which is near to middle top edge. For the bonds in either left and right of the bond at middle top edge, the response is the same. For the bond inside the material, the strains are never greater than S_c, and therefore it experiences no damage.

Figures 12 and 13 show the strain of the bond and damage H^T of the bond as function of time. It is quite similar to the plots shown in Figs. 8 and 9. In Fig. 14, we show the strain vs force plot. Red line shows response of bond when damage function is taken to be unity. We further note that the damage is defined for positive strains above critical strain.

Shear Loading

We apply $u = 0$ on bottom edge and keep left and right edge free. On top, we apply $u(t, x) = (u_1(t, x), u_2(t, x)) = (\gamma t x_2, 0)$. We chose $\gamma = 0.0001$ and simulate the problem up to time $T = 750$. Time step is $\Delta t = 10^{-5}$.

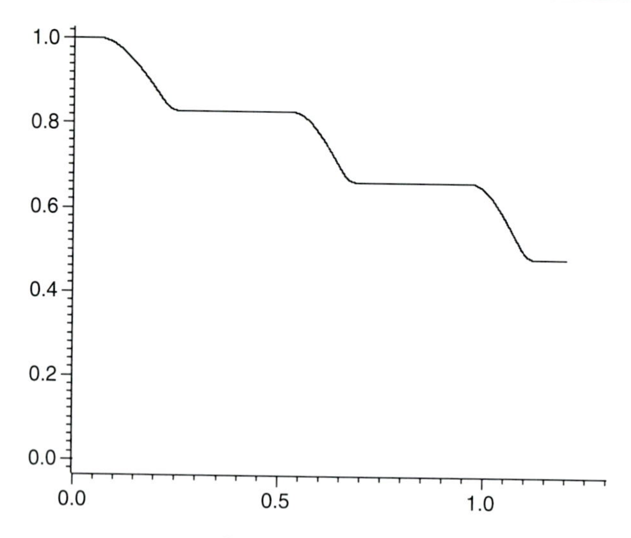

Fig. 13 Time vs Damage function $H^T((u)(y,x,t))$ plot

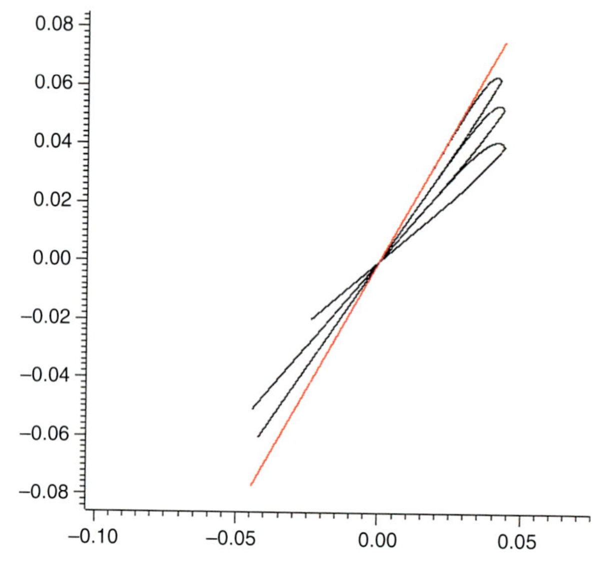

Fig. 14 Strain $S(y,x,t;u)$ vs Stress $H^T((u)(y,x,t))\partial_S f(\sqrt{|y-x|}S(y,x,t;u)))$ plot for the bond near middle top edge. Red color corresponds to $\partial_S f(\sqrt{|y-x|}S(y,x,t;u)))$

We choose the size of horizon to be $\delta = 0.05$ and mesh size $h = \delta/5$. As noted in the beginning of the section that we choose hydrostatic parameters large enough such that the damage is only due to the tensile interaction between material points. For tensile interaction, the extent of damage experienced by a material is defined as

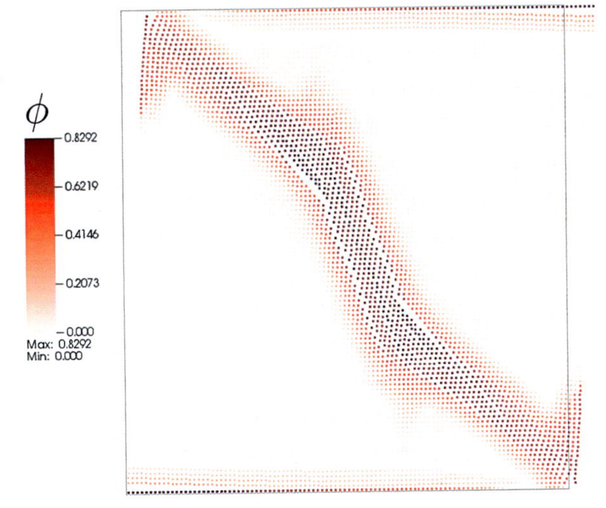

Fig. 15 Each point in figure shows the discretized mesh node. Strength of color shows the damage ϕ experienced by the mesh node. Box shows reference material domain $[0, 1]^2$

$$\phi(t, x; u) = 1 - \frac{\int_{D \cap B_\delta(x)} H^T(u)(y, x, t) dy}{\int_{D \cap B_\delta(x)} dy}. \tag{68}$$

Clearly, if all bonds in a horizon of material point x suffer no damage, then ϕ will be 0. As the damage of bonds increases, ϕ also increases. In Fig. 15, we show ϕ at final time $t = 750$. As we can see, the damage is along the diagonal of square.

Linear Elastic Operators in the Small Horizon Limit

In this section, we consider smooth evolutions u in space and show that away from damage set, the operators $\mathcal{L}^T + \mathcal{L}^D$ acting on u converge to the operator of linear elasticity in the limit of vanishing nonlocality. We denote the damage set by $\tilde{D}$ and consider any open undamaged set D' interior to D with its boundary a finite distance away from the boundary of D and the damage set $\tilde{D}$. In what follows, we suppose that the nonlocal horizon δ is smaller than the distance separating the boundary of D' from the boundaries of D and $\tilde{D}$.

Theorem 3. *Convergence to linear elastic operators. Suppose that* $u(x, t) \in C^2([0, T_0], C^3(D, \mathbb{R}^3))$ *and no damage, i.e.,* $H^T(y, x, t) = 1$ *and* $H^D(x, t) = 1$, *for every* $x \in D' \subset D \setminus \tilde{D}$, *then there is a constant* $C > 0$ *independent of nonlocal horizon* δ *such that for every* (x, t) *in* $D' \times [0, T_0]$, *one has*

$$|\mathcal{L}^T(u(t)) + \mathcal{L}^D(u(t)) - \nabla \cdot \mathbb{C}\,\mathcal{E}(u(t))| < C\delta, \tag{69}$$

where the the elastic strain is $\mathcal{E}(u) = (\nabla u + (\nabla u)^T)/2$ *and the elastic tensor is isotropic and given by*

$$\mathbb{C}_{ijkl} = 2\mu\left(\frac{\delta_{ik}\delta_{jl} + \delta_{il}\delta_{jk}}{2}\right) + \lambda\delta_{ij}\delta_{kl}, \tag{70}$$

with shear modulus μ *and Lamé coefficient* λ *given by*

$$\mu = \frac{f''(0)}{10}\int_0^1 r^3 J(r)\,dr \text{ and } \lambda = g''(0)\left(\int_0^1 r^3 J(r)\,dr\right)^2 + \frac{f''(0)}{10}\int_0^1 r^3 J(r)\,dr. \tag{71}$$

The numbers $f''(0) = \alpha$ *and* $g''(0) = \beta$ *can be chosen independently and can be any pair of real numbers such that* $\mathbb{C}$ *is positive definite.*

Proof. We start by showing

$$|\mathcal{L}^T(u(t)) - \frac{f''(0)}{2\omega_3}\int_{B_1(0)} e|\xi|J(|\xi|)e_i e_j e_k\,d\xi\partial^2_{jk}u_i(x)| < C\delta, \tag{72}$$

where $\omega_3 = 4\pi/3$ and $e = e_{y-x}$ are unit vectors on the sphere; here repeated indices indicate summation. To see this, recall the formula for $\mathcal{L}^T(u)$ and write $\partial_S f(\sqrt{|y-x|}S) = f'(\sqrt{|y-x|}S)\sqrt{|y-x|}$. Now Taylor expand $f'((\sqrt{|y-x|}S)$ in $\sqrt{|y-x|}S$, and Taylor expand $u(y)$ about x, denoting e_{y-x} by e to find that all odd terms in e integrate to zero and

$$\begin{aligned}|\mathcal{L}^T(u(t))_l - \frac{2}{V_\delta}\int_{B_\delta(x)} \frac{J^\delta(|y-x|)}{\delta|y-x|}\frac{f''(0)}{4}|y-x|^2\partial^2_{jk}u_i(x)e_i e_j e_k e_l,dy| \\ < C\delta,\ l = 1,2,3.\end{aligned} \tag{73}$$

On changing variables $\xi = (y-x)/\delta$, we recover (72). Now we show

$$\begin{aligned}|\mathcal{L}^D(u(t))_k - \frac{1}{\omega_3}\int_{B_1(0)} |\xi|\omega(|\xi|)e_i e_j\,d\xi\frac{g''(0)}{\omega_3}\int_{B_1(0)} |\xi|\omega(|\xi|)e_k e_l\,d\xi\partial^2_{lj}u_i(x)| \\ < C\delta,\ k = 1,2,3.\end{aligned} \tag{74}$$

We note for $x \in D'$ that $D \cap B_\delta(x) = B_\delta(x)$ and the integrand in the second term of (6) is odd and the integral vanishes. For the first term in (6), we Taylor expand $\partial_\theta g(\theta)$ about $\theta = 0$ and Taylor expand $u(z)$ about y inside $\theta(y,t)$ noting that terms odd in $e = e_{z-y}$ integrate to zero to get

$$|\partial_\theta g(\theta(y,t)) - g''(0)\frac{1}{V_\delta}\int_{B_\delta(y)} \omega^\delta(|z-y|)|z-y|\partial_j u_i(y)e_i e_j\,dz| < C\delta^3. \tag{75}$$

Now substitution for the approximation to $\partial_\theta g(\theta(y,t)$ in the definition of $\mathcal{L}^D$ gives

$$\Big|\mathcal{L}^D(u) - \frac{1}{V_\delta}\int_{B_\delta(x)} \frac{\omega^\delta(|y-x|)}{\delta^2} e_{y-x} \frac{1}{V_\delta}\int_{B_\delta(y)} \omega^\delta(|z-y|)|z-y| g''(0)\partial_j u_i(y) e_i e_j \, dz\, dy\Big| < C\delta. \tag{76}$$

We Taylor expand $\partial_j u_i(y)$ about x, note that odd terms involving tensor products of e_{y-x} vanish when integrated with respect to y in $B_\delta(x)$, and we obtain (74).

We now calculate as in (Lipton 2016, equation (6.64)) to find that

$$\begin{aligned}&\frac{f''(0)}{2\omega_3}\int_{B_1(0)} |\xi| J(|\xi|) e_i e_j e_k e_l \, d\xi \partial_{jk}^2 u_i(x)\\ &\quad = \left(2\mu_1\left(\frac{\delta_{ik}\delta_{jl}+\delta_{il}\delta_{jk}}{2}\right) + \lambda_1 \delta_{ij}\delta_{kl}\right)\partial_{jk}^2 u_i(x),\end{aligned} \tag{77}$$

where

$$\mu_1 = \lambda_1 = \frac{f''(0)}{10}\int_0^1 r^3 J(r)\, dr. \tag{78}$$

Next observe that a straight forward calculation gives

$$\frac{1}{\omega_3}\int_{B_1(0)} |\xi|\omega(|\xi|) e_i e_j \, d\xi = \delta_{ij}\int_0^1 r^3 \omega(r)\, dr, \tag{79}$$

and we deduce that

$$\begin{aligned}&\frac{1}{\omega_3}\int_{B_1(0)} |\xi|\omega(|\xi|) e_i e_j \, d\xi \frac{g''(0)}{\omega_3}\int_{B_1(0)} |\xi|\omega(|\xi|) e_k e_l \, d\xi \partial_{lj}^2 u_i(x)\\ &= g''(0)\left(\int_0^1 r^3\omega(r)\, dr\right)^2 \delta_{ij}\delta_{kl}\partial_{lj}^2 u_i(x).\end{aligned} \tag{80}$$

Theorem 3 follows on adding (77) and (80) □

Conclusions

We have introduced a simple nonlocal model for free damage propagation in solids. In this model, there is only one equation, and it describes the dynamics of the displacement using Newton's law $F = ma$. The damage is a consequence of displacement history and diminishes the force-strain law as damage accumulates. The modeling allows for both cyclic damage or damage due to abrupt loading. The damage is irreversible, and the damage set grows with time. The dissipation energy due to damage together with the kinetic and potential energy satisfies energy balance at every instant of the evolution. Future work will address the question of localization of damage using this model. We believe that if the loading is such that large monotonically increasing strains are generated, then damage localization based on material softening and inertia could be anticipated.

In this treatment, we have considered dynamic problems only. For this case, we have shown uniqueness for the model. The analysis of this model in the absence of inertial forces leads to the quasi-static case where the effects of inertia are absent but memory of the load history is still present. Future work aims to explore this model for this case and understand regimes of body force specimen geometry and boundary loads for which there is loss of uniqueness and associated instability. Such nonuniqueness is well known for quasi-static gradient damage models (Pham and Marigo 2013).

Acknowledgements This material is based upon the work supported by the U S Army Research Laboratory and the U S Army Research Office under contract/grant number W911NF1610456.

References

A. Agwai, I. Guven, E. Madenci, Predicting crack propagation with peridynamics: a comparative study. Int. J. Fract. **171**, 65–78 (2011)

F. Bobaru, W. Hu, The meaning, selection, and use of the peridynamic horizon and its relation to crack branching in brittle materials. Int. J. Fract. **176**, 215–222 (2012)

F. Bobaru, J.T. Foster, P.H. Geubelle, S.A. Silling, *Handbook of Peridynamic Modeling* (CRC Press, BOCA Ratone, 2016)

K. Dayal, K. Bhattacharya, Kinetics of phase transformations in the peridynamic formulation of continuum mechanics. J. Mech. Phys. Solids **54**, 1811–1842 (2006)

Q. Du, Y. Tao, X. Tian, A peridynamic model of fracture mechanics with bond-breaking. J Elast. (2017). https://doi.org/10.1007/s10659-017-9661-2

E. Emmrich, D. Phust, A short note on modeling damage in peridynamics. J. Elast. **123**, 245–252 (2016)

E. Emmrich, O. Weckner, On the well-posedness of the linear peridynamic model and its convergencee towards the Navier equation of linear elasticity. Commun. Math. Sci. **5**, 851–864 (2007)

J.T. Foster, S.A. Silling, W. Chen, An energy based failure criterion for use with peridynamic states. Int. J. Multiscale Comput. Eng. **9**, 675–688 (2011)

W. Gerstle, N. Sau, S. Silling, Peridynamic modeling of concrete structures. Nuclear Eng. Des. **237**, 1250–1258 (2007)

Y.D. Ha, F. Bobaru, Studies of dynamic crack propagation and crack branching with peridynamics. Int. J. Fract. **162**, 229–244 (2010)

P. K. Jha, R. Lipton, Numerical analysis of peridynamic models in Hölder space, arXiv preprint arXiv:1701.02818 (2017)

R. Lipton, Dynamic brittle fracture as a small horizon limit of peridynamics. J. Elast. **117**, 21–50 (2014)

R. Lipton, Cohesive dynamics and brittle fracture. J. Elast. **124**(2), 143–191 (2016)

R. Lipton, S. Silling, R. Lehoucq, Complex fracture nucleation and evolution with nonlocal elastodynamics. arXiv preprint arXiv:1602.00247 (2016)

R. Lipton, E. Said, P.K. Jha, Free damage propagation with memory. J. Elast. To appear in J. Elasticity (2018)

T. Mengesha, Q. Du, Nonlocal constrained value problems for a linear peridynamic Navier equation. J. Elast. **116**, 27–51 (2014)

E. Oterkus, I. Guven, E. Madenci, Fatigue failure model with peridynamic theory,in *IEEE Intersociety Conference on Thermal and Thermomechanical Phenomena in Electronic Systems (ITherm)*, Las Vegas, June 2010, pp. 1–6

K. Pham, J.J. Marigo, From the onset of damage to rupture: construction of responses with damage localization for a general class of gradient damage models. Contin. Mech. Thermodyn. **25**, 147–171 (2013)

S.A. Silling, Reformulation of elasticity theory for discontinuities and long-range forces. J. Mech. Phys. Solids **48**, 175–209 (2000)

S.A. Silling, E. Askari, A meshfree method based on the peridynamic model of solid mechanics. Comput. Struct. **83**, 1526–1535 (2005)

S.A. Silling, E. Askari, Peridynamic model for fatigue cracking. Sandia Report, SAND2014-18590, 2014

S.A. Silling, F. Bobaru, Peridynamic modeling of membranes and fibers. Int. J. Nonlinear Mech. **40**, 395–409 (2005)

S.A. Silling, M. Epton, O. Weckner, J. Xu, E. Askari, Peridynamic states and constitutive modeling. J. Elast. **88**, 151–184 (2007)

S.A. Silling, R.B. Lehoucq, Convergence of peridynamics to classical elasticity theory. J. Elast. **93**, 13–37 (2008)

S. Silling, O. Weckner, E. Askari, F. Bobaru, Crack nucleation in a peridynamic solid. Int. J. Fract. **162**, 219–227 (2010)

O. Weckner, R. Abeyaratne, The effect of long-range forces on the dynamics of a bar. J. Mech. Phys. Solids **53**, 705–728 (2005)

O. Weckner, E. Emmrich, Numerical simulation of the dynamics of a nonlocal, inhomogeneous, infinite bar. J. Comput. Appl. Mech, **6**, 311–319 (2005)

Index

G. Z. Voyiadjis (ed.), *Handbook of Nonlocal Continuum Mechanics for Materials and Structures*, https://doi.org/10.1007/978-3-319-58729-5

G

Q

R

V

W

X

Y

PGSTL